WORLD CLIMATE

Make the most of your study time—with

WORLD REGIONAL Geography ⊕ Now™

World Regional GeographyNow™ is an online learning companion that helps you gauge your individual learning needs and identify the concepts on which you most need to focus your study time. Access to this program is FREE with your purchase of a new text. Here's how it works:

Personalized learning plans and cool animations

After reading a chapter in the text, take the diagnostic pre-test ("What Do I Know?") for a quick assessment of how well you already understand the material. (Your instructor may also assign these pre-tests as homework.)

Your responses will automatically generate a personalized learning plan ("What Do I Need to Learn?") that outlines the interactive animations, tutorials, and exercises you need to review. The animations—many based on maps and figures in the text—help you understand geographic concepts in ways that a printed page simply cannot.

A chapter quiz/post-test ("What Have I Learned?") helps you measure how well you "get" the core concepts from the chapter. If you still need more help to improve your score, the system will encourage you to work through the chapter's tutorial again.

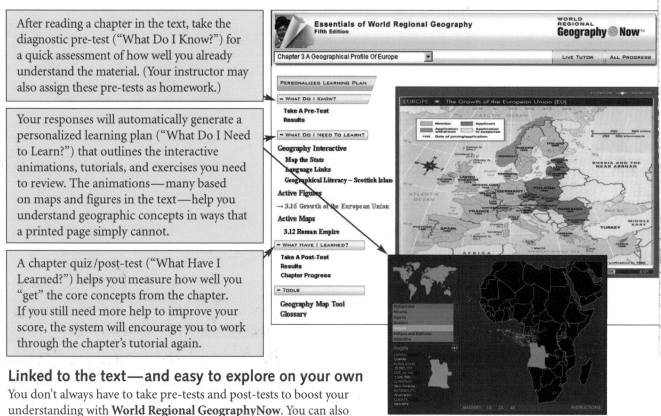

Linked to the text—and easy to explore on your own

You don't always have to take pre-tests and post-tests to boost your understanding with **World Regional GeographyNow**. You can also explore the system at any point—it's as easy as surfing the web. And, as you work through the text, you'll see **World Regional GeographyNow** icons that direct you to the program's media-enhanced activities and animations.

Study smarter—and make every minute count!

Log on today at **http://earthscience.brookscole.com/wrg5e** with the access code packaged with this text. If you didn't buy a new book, you can visit the website above to purchase electronic access.

THOMSON
BROOKS/COLE

Essentials of World Regional Geography

FIFTH EDITION

Hobbs • Salter

Essentials of World Regional Geography

Fifth Edition

Joseph J. Hobbs
Christopher L. Salter

University of Missouri—Columbia

Cartography by Andrew Dolan
University of Missouri—Columbia

BROOKS/COLE

™

THOMSON LEARNING

Australia • Canada • Mexico • Singapore • Spain • United Kingdom • United States

Essentials of World Regional Geography, Fifth Edition
Joseph J. Hobbs
Christopher L. Salter

Earth Science Editor: *Keith Dodson*
Development Editor: *Marie Carigma-Sambilay*
Assistant Editor: *Carol Ann Benedict*
Editorial Assistant: *Megan Asmus, Melissa Newt*
Technology Project Manager: *Ericka Yeoman-Saler*
Marketing Manager: *Jennifer Somerville*
Marketing Assistant: *Michele Colella*
Marketing Communications Manager: *Kelley McAllister*
Project Manager, Editorial Production: *Andy Marinkovich*
Art Director: *Vernon T. Boes*
Print Buyer: *Karen Hunt*
Permissions Editor: *Stephanie Lee*
Production Service: *Nancy Shammas, New Leaf Publishing Services*
Text Designer: *Patrick Devine*
Photo Researcher: *Kathleen Olson*

Copy Editor: *Frank Hubert*
Indexer: *James Minkin*
Illustrator: *Andrew Dolan (cartography) and Rolin Graphics*
Chapter Opener Satellite Images: CO 2, CO 24: NASA; CO 3, CO 19: *Tom Van Sant/Corbis;* CO 15: *Tom Van Sant/SPL/Photo Researchers, Inc.;* all other COs: *Digital Wisdom, Inc.*
Cover Designer: *Patrick Devine*
Cover Image: *Maps © 2005 Map Resources; Aborigine portrait (Australia), © Goodshoot/Wonderfile (top left); Aymara Indian Woman (Bolivia), © Angelo Cavall/Getty Images (top right); Tanzanian Man Wearing Headdress on Savanna, © Royality-Free/CORBIS (bottom left); Indian Woman on Stone Steps, © David Buffington/Getty Images. (bottom right)*
Compositor: *G & S Typesetters*
Printer: *Transcontinental Printing/Interglobe*

Printed in Canada
1 2 3 4 5 6 7 09 08 07 06 05

For more information about our products, contact us at:

Thomson Learning Academic Resource Center
1-800-423-0563
For permission to use material from this text or product, submit a request online at http://www.thomsonrights.com. Any additional questions about permissions can be submitted by email to thomsonrights@thomson.com.

Library of Congress Control Number: 2004117714
Student Edition: ISBN 0-534-46600-1
Instructor's Edition: ISBN 0-534-46608-7

Thomson Higher Education
10 Davis Drive
Belmont, CA 94002-3098
USA

Asia (including India)
Thomson Learning
5 Shenton Way
#01–01 UIC Building
Singapore 068808

Australia/New Zealand
Thomson Learning Australia
102 Dodds Street
Southbank, Victoria 3006
Australia

Canada
Thomson Nelson
1120 Birchmount Road
Toronto, Ontario M1K 5G4
Canada

UK/Europe/Middle East/Africa
Thomson Learning
High Holborn House
50/51 Bedford Row
London WC1R 4LR
United Kingdom

Latin America
Thomson Learning
Seneca, 53
Colonia Polanco
11560 Mexico
D.F. Mexico

Spain (including Portugal)
Thomson Paraninfo
Calle Magallanes, 25
28015 Madrid, Spain

Brief Contents

This book is dedicated to the Students who use it,

with hopes that it will open your eyes

further to the wonders of the

Earth, and stimulate your

appetite to see for

yourself as much

of the world

as you

can.

Contents

Maps

Credits

Figure 2.1: Adapted from UNEP/GRID, "Mean Annual Precipitation," Geneva Data Sets—Atmosphere, GNV174 (http://www.grid.unep.ch/data).   **Figure 2.4:** Adapted from Tom L. McKnight, *Physical Geography: A Landscape Appreciation,* 5th ed. (Prentice Hall, 1996), p. 122.   **Figure 2.6:** Adapted from Rand McNally's *Classroom Atlas,* 2003. **Figure 2.7:** Adapted from World Wildlife Fund ecoregions data, 1999.   **Figure 2.9:** Adapted from NASA, "Global Tectonic Activity Map of the Earth," DTAM-1, 2002.   **Figure 2.10:** Adapted from Conservation International, "Hotspots Explorer," 2002 (http://www.biodiversityhotspots.org/xp/Hotspots).   **Figure 2.18:** Data from *World Factbook,* CIA, 2004. **Figure 2.23:** Data from Population Reference Bureau, 2004.   **Figure 3.2 (a):** *The Oxford School Atlas* (Oxford University Press, 1997).   **Figure 3.7 (a):** Adapted from Rand McNally's *Classroom Atlas,* 2003.   **Figure 3.7 (b):** Adapted from World Wildlife Fund ecoregions data, 1999.   **Figure 3.8:** *The Oxford School Atlas* (Oxford University Press, 1997). **Figure 3.12:** Adapted from John Haywood, ed., *Atlas of World History* (Barnes and Noble Books, 1997).   **Figure 3.C:** Adapted from a map in *The Economist,* Sept. 20, 1997, p. 53. © 1998 The Economist Newspaper Ltd. All rights reserved. Reprinted with permission. Further reproduction prohibited.   **Figure 3.16:** EU map adapted from "European Union Survey," *The Economist,* May 31, 1997, p. 5. © 1997 The Economist Newspaper Ltd. All rights reserved. Reprinted with per-

mission. Further reproduction prohibited. NATO map data from EUROSTAT (Statistical Office of the European Communities), 1996; *World Factbook,* CIA, 1995.   **Figure 4.2:** Data from Global Forest Watch, 2004 (http://www.globalforestwatch.org/english/datawarehouse/index.asp).   **Figure 4.4:** After Jordan-Bychkov and Bychkova Jordan, 2002, p. 300. **Figure 4.9:** After Jordan-Bychkov and Bychkova Jordan, 2002, p. 300.   **Figure 5.22:** Adapted from Ian Barnes and Robert Hudson, *The History Atlas of Europe* (Macmillan Inc., 1998), p. 151.   **Figure 6.2(a):** *The Oxford School Atlas* (Oxford University Press, 1997).   **Figure 6.4(a):** Adapted from Rand McNally's *Classroom Atlas,* 2003.   **Figure 6.4(b):** Adapted from World Wildlife Fund ecoregions data, 1999.   **Figure 6.6:** *The Oxford School Atlas* (Oxford University Press, 1997).   **Figure 6.11:** Bernard Comrie, et al., *The Atlas of Languages,* Revised Edition (New York: Facts on File, 2003). **Figure 8.3(a):** *The Oxford School Atlas* (Oxford University Press, 1997).   **Figure 8.5(a):** Adapted from Rand McNally's *Classroom Atlas,* 2003.   **Figure 8.5(b):** Adapted from World Wildlife Fund ecoregions data, 1999.   **Figure 8.7:** *The Oxford School Atlas* (Oxford University Press, 1997).   **Figure 8.16:** Bernard Comrie, et al., *The Atlas of Languages,* Revised Edition (New York: Facts on File, 2003).   **Figure 8.C:** Adapted from National Geographic's *Atlas of the Middle East.* **Figure 9.1:** Adapted from John Haywood, ed., *Atlas of World History* (Barnes and Noble Books, 1997).   **Figure 9.2:** From *The Middle East and North Africa: A Political Geography,* by A. Drysdale and G. Blake. Copyright © 1985 Oxford University Press, Inc. Used by permission of Oxford University Press, Inc.   **Figure 9.3:** From *The Middle East and North Africa: A Political Geography,* by A. Drysdale and G. Blake. Copyright © 1985 Oxford University Press, Inc. Used by permission of Oxford University Press, Inc.   **Figure 9.A:** Zone control data from Foundation for Middle East Peace, 2002 (www.fmep.org); security wall route from IDF maps.   **Figure 9.9:** Held, Colbert. 1994. *Middle East Patterns.*   **Figure 9.25:** Data from CIA maps.   **Figure 9.35:** Ethnicity and poppy data from CIA maps.   **Figure 10.3(a):** *The Oxford School Atlas* (Oxford University Press, 1997).   **Figure 10.5(a):** Adapted from Rand McNally's *Classroom Atlas,* 2003. **Figure 10.5(b):** Adapted from World Wildlife Fund ecoregions data, 1999.   **Figure 10.7:** *The Oxford School Atlas* (Oxford University Press, 1997).   **Figure 10.10:** Bernard Comrie, et al., *The Atlas of Languages,* Revised Edition (New York: Facts on File, 2003).   **Figure 11.7:** A. Smith, *Atlas of Mesozoic and Cenozoic Coastlines* (New York: Cambridge University Press, 1994).   **Figure 12.4:** Data from Global Forest Watch, 2004 (http://www.globalforestwatch.org/english/datawarehouse/index.asp).   **Figure 12.6:** From Wayne Arnold, "A Gas Pipeline to the World Outside," *The New York Times,* 10-26-01. Copyright © 2001 The New York Times Co.   **Figure 12.F:** From "Indonesia Cracks Down in Separatist Irian Jaya," by Calvin Sims, *The New York Times,* 12-4-00, A3. Copyright © 2000 The New York Times Co. Reprinted by permission.   **Figure 13.D:** Sources: Erik Eckholm, "China Will Move Waters to Quench Thirst of Cities," *The New York Times,* 8-27-02, A1; "Water in China: In Deep," *The Economist,* 08-18-01, 31.   **Figure 13.10:** *China Atlas of Population and Environment;* http://141.211.136.211/eng/macro/atlas2000/ybtableview.asp?ID=4).   **Figure 15.3(a):** *The Oxford School Atlas* (Oxford University Press, 1997).   **Figure 15.5(a):** Adapted from Rand McNally's *Classroom Atlas,* 2003. **Figure 15.5(b):** Adapted from World Wildlife Fund ecoregions data, 1999.   **Figure 15.7:** *The Oxford School Atlas* (Oxford University Press, 1997).   **Figure 17.3(a):** *The Oxford School Atlas* (Oxford University Press, 1997).   **Figure 17.A (left):** Joint United Nations Program on H.I.V./AIDS.   **Figure 17.A (right):** Data from UNAIDS.   **Figure 17.8(a):** Adapted from Rand McNally's *Classroom Atlas,* 2003.   **Figure 17.8(b):** Adapted from World Wildlife Fund ecoregions data, 1999.   **Figure 17.9:** *The Oxford School Atlas* (Oxford University Press, 1997).   **Figure 17.13:** Bernard Comrie, et al., *The Atlas of Languages,* Revised Edition (New York: Facts on File, 2003).   **Figure 17.16:** Adapted from John Haywood, ed., *Atlas of World History* (Barnes and Noble Books, 1997).   **Figure 17.19:** USGS country mineral maps. **Figure 18.1:** Robert Stock, *Africa South of the Sahara.*   **Figure 18.B:** *World Factbook,* CIA, 2004.   **Figure 19.3A:** *The Oxford School Atlas* (Oxford University Press, 1997).   **Figure 19.5(a):** Adapted from Rand McNally's *Classroom Atlas,* 2003.   **Figure 19.5(b):** Adapted from World Wildlife Fund ecoregions data, 1999.   **Figure 19.7:** *The Oxford School Atlas* (Oxford University Press, 1997).   **Figures 19.A and 19.B:** From *Living in the Environment,* 13th edition, by Miller. © 2004. Reprinted with permission of Brooks/Cole, a division of Thomson Learning, Inc.   **Figure 19.10:** *The History Atlas of South America,* Edwin Early, ed., (Macmillan, 1998).   **Figure 19.13:** Bernard Comrie, et al., *The Atlas of Languages,* Revised Edition (New York: Facts on File, 2003).   **Figure 19.22:** *The Economist,* April 11, 1998, p. 125.   **Figure 19.E:** Data from the U.N. Drug Control Program (UNDCP), 2003.   **Figure 21.12:** Adapted from *Atlas of World History* (Barnes and Noble Books, 1997).   **Figure 21.E:** Andrew Downie, "Amazon Destruction Rising Fast," *Christian Science Monitor,* 4/22/04, p. 5.   **Figure 22.2(a):** *The Oxford School Atlas* (Oxford University Press, 1997).   **Figure 22.3:** Data from the U.S. Census Bureau and Statistics Canada.   **Figure 22.5(a):** Adapted from Rand McNally's *Classroom Atlas,* 2003. **Figure 22.5(b):** Adapted from World Wildlife Fund ecoregions data, 1999.   **Figure 22.6:** *The Oxford School Atlas* (Oxford University Press, 1997).   **Figure 22.8:** Carl Waldman, *Atlas of the North American Indian* (Facts on File, 1985). **Figure 22.12:** Adapted from *Atlas of World History* (Barnes and Noble Books, 1997).   **Figure 22.14:** Bernard Comrie, et al., *The Atlas of Languages,* Revised Edition (New York: Facts on File, 2003).   **Figure 22.15:** *New Historical Atlas of Religion in America* (Oxford University Press, 2001).   **Figure 23.7:** Getis, Arthur and Judith Getis. 1995. *The United States and Canada: The Land and Its People.* Dubuque, Iowa: Wm. C. Brown Publishers, p. 231.   **Figure 24.A:** Timothy Egan, "Vanishing Point: Amid Dying Towns of Rural Plains, One Makes a Stand," *New York Times,* 12/01/03, p. A18.

About the Authors

Joseph J. Hobbs received his B.A. at the University of California—Santa Cruz in 1978 and his M.A. and Ph.D. at the University of Texas—Austin in 1980 and 1986 respectively. He is a professor at the University of Missouri—Columbia, and a geographer of the Middle East with many years of field research on biogeography and Bedouin peoples in the deserts of Egypt. Joe's interest in the region grew from a boyhood lived in Saudi Arabia and India. His research in Egypt has been supported by Fulbright fellowships, the American Council of Learned Societies, the American Research Center in Egypt, and the National Geographic Society Committee for Research and Exploration. He served as the team leader of the Bedouin Support Program, a component of the St. Katherine National Park project in Egypt's Sinai Peninsula. His current research interests are indigenous peoples' participation in protected areas in the Middle East, Southeast Asia and Central America, human uses of caves worldwide, the global narcotics trade, and human trafficking. He is the author of *Bedouin Life in the Egyptian Wilderness* and *Mount Sinai* (both University of Texas Press), co-author of *The Birds of Egypt* (Oxford University Press), and co-editor of *Dangerous Harvest: Drug Plants and the Transformation of Indigenous Landscapes* (Oxford). He teaches graduate and undergraduate courses in world regional geography, environmental geography, the geography of the Middle East, the geography of caves, and the geography of global current events, the geographies of drugs and terrorism, and a field course on the ancient Maya geography of Belize. He has received the University of Missouri's highest teaching award, the Kemper Fellowship. During many summers, he leads "adventure travel" tours to remote areas in Latin America, Africa, the Indian Ocean, Asia, Europe, and the High Arctic. Joe lives in Missouri with his wife, Cindy, daughters Katherine and Lily, and an animal menagerie.

Christopher L. "Kit" Salter did his undergraduate work at Oberlin College, with a major in Geography and Geology. He spent 3 years teaching English at a Chinese university in Taiwan immediately following Oberlin. Graduate work for both the M.A. and the Ph.D. was done at the University of California, Berkeley. He taught at UCLA from 1968 to 1987, when he took on full-time employment with the National Geographic Society in Washington with his wife, Cathy. They were both involved in the Society's campaign to bring geography back into the American school system. Kit has been a professor and chair of the Department of Geography at the University of Missouri Columbia since moving to the heartland in 1988.

Themes that have made Salter glad he chose geography as a life field include landscape study and interpretation in both domestic and foreign settings; landscape and literature to show students that geography occurs in all writing, not just in textbooks; and geography education to help learners at all levels see the critical nature of geographic issues. He has written more than 125 articles in various aspects of geography; has traveled to a lot of the places that he writes about in this text; and has been wise enough to have a son who is an architect in Los Angeles, a daughter who is a teacher in the San Francisco Bay Area, and a neat wife who is a writer in the heart of the heartland in Breakfast Creek, Missouri.

Preface

Approach and Scope

It was not so long ago that many scholars were writing of a "unipolar world," a post-Cold War global environment in which the central organizing theme was the supremacy of the United States. Much has happened, though, to bring more humble perspectives. The events of 9/11, the 2004 expansion of the European Union, the abstention of the United States from climate change and other international agreements, the widespread resentment of U.S. action in Iraq, the surge in China's wealth and influence, and the diffusion of dangerous weapons systems have given Americans more reason than ever before to look on the world with curiosity and deference rather than with arrogance or indifference.

This is not a time for shortcuts or simplifications. This is not a small book, but it is called *Essentials of World Regional Geography* because, in order to come to grips with how the world works today, it is essential for the reader to have a thorough grounding in the environmental, cultural, economic, and geopolitical contexts of the world's regions and nations. The book examines the issues that underlie the major issues of our time, and those issues most likely to be of concern in the near future. If, for example, a crisis breaks out on the Korean Peninsula, this book will provide an excellent tool for appreciating the geographic context of those unfolding events. Between the book's covers are the answers to many of the follow-up questions that would arise in a less comprehensive world regional geography text. Of course, there is much more to follow up on, and Brooks/Cole has designed a rich and easy-to-use set of ancillary resources. Not a "world light" book, then, this book's comprehensive scope allows for maximum flexibility in use. May it afford you a satisfying and interesting journey through the world!

Features of This Edition

This is the most thorough revision in the history of *Essentials of World Regional Geography*. New content includes completely revised chapters on Europe, the profile of Monsoon Asia, China, the United States and Canada, Middle America, and South America; and extensive revision of the remaining chapters. Greater emphasis is given to geopolitical issues throughout. There are updated perspectives on economic geography, particularly the rise of China and India and the emergence of postindustrial economies in Europe and North America. There is more attention than in earlier editions to issues of ethnicity, language, and indigenous peoples. As much as globalization is bringing new commonalities to peoples around the world, ancient traditions and worldviews still thrive and deserve attention. Migration and energy issues are more prominent than in any previous edition. This edition puts a premium on relevancy and timely issues, to

help readers appreciate current events around the world. Principal sources were current journals and newspapers, interpreted with an eye on the issues underlying tomorrow's news. The completely revised art program includes more than 100 new maps and more than 100 revised maps, created by an unprecedented collaborative effort between author and cartographer. There is a new consistency in the presence and format of the maps throughout the text. In addition, numerous maps and figures in the text have been made "Active" as part of the new, innovative World Regional GeographyNow website that supports the text. New author photos bring the geographer's eye to the text. This edition has been updated so that all of the maps, statistical tables, and text reflect the most current information available in late 2004 and early 2005. All city population data are for metropolitan areas.

Users of the fifth edition will find many familiar features, including the Problem Landscape, Regional Perspective, Definitions + Insights, Perspectives from the Field, and the end-of-chapter summary, review, and discussion elements. The end of chapter pedagogical elements have been improved, and chapter objectives at the beginning of each chapter are new. There are several new "Geography of . . ." box features bringing consistent attention to cultural, political, physical, and other issues. These include "Geography of Energy," "Ethnic Geography," "Medical Geography," "Geography of the Sacred," "Geography of Terrorism," "Natural Hazards," and "Biogeography." There is more attention to hydropolitics and to freshwater and marine issues, including the box feature "The World's Great Rivers." There is also more attention to thematic and conceptual issues in the more detailed country chapters following each profile chapter.

Organization and Use of the Text

A major goal in this edition has been to make it easier and more satisfying for use in a single semester course, while still meeting all the needs of a two-semester sequence. Each of the eight regions has a regional "Geographic Profile" chapter that follows the consistent format of Area and Population, Physical Geography and Human Adaptations, Cultural and Historical Geographies, Economic Geography, and Geopolitical Issues. Each profile chapter is followed by one or more subregional chapters that explore the region in more detail. These days it is more important than ever in a world regional text to provide access to the particular environmental, ethnic, economic, and other features of the world's countries. In a single-semester course and in some other cases, the instructor may wish to assign only the introductory chapters (1 and 2) and the regional profile chapters (3, 6, 8, 10, 15, 17, 19, 22) and then encourage students to explore the subsequent chapters on their own or in supplementary assignments. In the Middle East, for example, following a thorough reading of the Chapter 8 profile, readers may want to focus on just the Arab-Israeli conflict in Chapter 9. In Africa, following Chapter 17, readers may want to select a particular problem area that is timely, such as Nigeria (for its oil and ethnic issues), Sudan (for the crisis in Darfur), or Zimbabwe (for land control problems).

While the regional coverage begins with Europe and concludes with North America, it is not necessary to pursue the regions in any particular order. Tying together many significant themes, issues, and characteristics throughout the text, and making the book easy to use with whatever regional sequence is chosen, are cross-references noted in the margins. These are the blue page and figure numbers in the margins that generally mean "turn to this page to get more information on this issue or to see how this problem is paralleled in another country or culture."

Chapters 1 and 2 provide the basic concepts, tools, and vocabulary for world regional geography. In the first chapter, geogaphy's uniquely spatial approach to the world is introduced, along with some of the discipline's milestone concepts and its considerable appeal for students interested in career possibilities. The second chapter briefs students in the essential characteristics of the world's physical processes, traces the modification of landscapes as a result of human activities, describes trends and projections of population growth, and considers the Kyoto Protocol, Sustainable Development, and other efforts to slow destructive global trends.

Chapters 3, 4, and 5 explore Europe. A major concern in the profile chapter (3) is the expansion of the European Union in the "big bang" of 2004, and its implications for the continent's economic and political influence in the rest of the world. The subsequent chapters are organized around the concepts of core and periphery, acknowledging the economic supremacy of the western portion of the continent but also examining events and characteristics in the wide periphery of north, east, and south. In this edition, the Baltic countries have joined Europe—where by almost all measures they belong today, rather than in the Russian orbit. Ethnic and political dilemmas in the Balkans are spelled out with a new map and updated text.

Chapters 6 and 7 describe and discuss the geography of Russia and the Near Abroad. This is still a region in transition. The legacy of Soviet rule lingers with Russia's primacy in many affairs, especially in some of the 12 countries that make up this region. The disturbing downward trend in quality of life in Russia is finally tempered by hope that surging oil revenues will boost well-being. There is close attention to the dynamic and in some cases dangerous situations in the Caucasus and Central Asia.

Containing approximately 60 percent of the world's oil, the region of the Middle East and North Africa has immense relevance to everyone. Chapters 8 and 9 take readers deep into a region that is likely to make headlines throughout their lives. These chapters establish a strong foundation for

understanding the circumstances underlying strife in the Middle East, including the underpinnings for the 9/11 attacks and the subsequent regrouping of al-Qa'ida and its allies. Many critical questions are asked about the region's future in the wake of the U.S. occupation of Iraq. Understanding of the Arab–Israeli conflict begins here with comprehension of the importance of sacred places in Judaism and Islam, the exile of the Jews, the Zionist movement, and the establishment of Israel. The subsequent wars are described thoroughly, because understanding of each helps the reader grapple with the difficult issues confronting would-be peacemakers in the region. The Palestinian–Israeli peace process, which is above all a set of geographic problems, is updated through the death of Yasser Arafat. These chapters also contain a wealth of material unrelated to the questions of conflict and oil, and students should come away with a strong appreciation of the cultural and environmental features of the nations that comprise the region.

Monsoon Asia is covered in Chapters 10 through 14. The world's two demographic giants—China and India—are also the region's most dynamic economies, and their clout is a major theme throughout this section. The profile chapter (10) gives new insight into the spiritual and other rich cultural traditions of this region. In Chapter 12, Southeast Asia's demographic giant, Indonesia, receives the attention it deserves including, tragically, its place at the epicenter of the December 2004 tsunami. One of the strong features of Chapter 14 is its thorough background briefing on tensions between the two Koreas and between North Korea and the West.

Chapters 14 and 15 offer what few regional texts do, a thorough treatment of the Pacific World. Chapter 14's discussion of the environmental setting of this vast region includes new maps on island formation, along with several case studies of historical and contemporary problems in environmental management. Chapter 15 considers the perceived dilemma that the majority of white Australians face in deciding whether to forge a new, multiethnic Australia oriented to the Pacific Rim and Asia or to retain ties with the Commonwealth and the British monarchy. Ancient Easter Island and modern Nauru tell much about the liabilities of island living, however romantic that life may seem to far-off continent-dwellers.

Africa has long troubled students of geography, who have had difficulty coming to terms with such a vast and complex region. This region can be studied and comprehended as well as any other, and Chapters 17 and 18 make that possible. Chapter 17 lays out the major themes that unite and differentiate the subregions and countries that make up Africa south of the Sahara. There are case studies of compelling issues such as HIV/AIDS and "dirty diamonds." Africa's importance in the global oil picture is often overlooked, but is prominent here. Chapter 18 is a comprehensive and thoroughly updated journey through each of the seven subregions.

Chapters 19 to 22 cover Latin America. Both the pre-Columbian and post-conquest geographies of the region are considered. The outcomes and prospects for NAFTA and other free trade agreements are discussed, as are "fair trade" coffees and other commodities. Immigration and other hemispheric issues like remittances, the drug trade, and manifestations of the Monroe Doctrine figure prominently in these chapters. U.S. interests in the oil of Venezuela, Colombia, and Ecuador are explained, as are the reasons why indigenous people in the region often oppose oil development.

The text concludes where most of its readers live, in the United States and Canada. Profile Chapter 22 considers the United States in the post-9/11 world, introduces the United States and Canada as the postindustrial peers of Europe, and considers the ever-important role of immigration in shaping the two countries' cultures. In Chapter 23, Canada is depicted in the shadow of the United States in some affairs, but in all of its distinctiveness—including its unique French qualities and aspirations—in others. The book's concluding chapter is a detailed exploration of the five vernacular regions of the United States, with an approach that emphasizes the geographic rationale behind the development of cities and economic cores. The latest census data are interpreted cartographically—and dramatically—to depict the depopulation of the Great Plains and the booms elsewhere.

Ancillaries

This text is accompanied by a number of ancillary publications to assist instructors and enhance student learning:

Online Instructor's Manual with Test Bank
by Andrew Dolan
Contains chapter outlines, key terms, and themes for general discussion as well as perspectives on selected discussion questions from the text. The Test Bank offers nearly 1,000 multiple-choice, classroom-tested questions and answers keyed to the test.

Study Guide and Student Resources
Includes chapter objectives, key terms, critical thinking problems, and a variety of self-test questions and answers.

Places of the World: Achieving Fundamental Geographical Literacy, Fifth Edition
by Dr. James W. Lett, Indian River Community College
Places of the World is the only place-name geography book on the market that systematically identifies, maps, and describes each and every one of the world's 193 independent countries. With an organization that parallels Hobbs and Salter's text, this unique workbook provides students with organized practice for matching the names of places with their physical location on a map. The new Fifth Edition is thoroughly updated and includes more than 300 maps—200 new. A consistent framework summarized by the acronym PLACES (Population, Language, Area, Capital, Economy, and Sovereignty) is used throughout.

Transparency Acetates
A set of full-color transparencies of selected images from the text.

Exam View® Computerized Testing
Create, deliver, and customize tests and study guides (both print and online) in minutes with this easy-to-use assessment and tutorial system.

JoinIn™ on TurningPoint®
Book-specific JoinIn content for Response Systems tailored to Hobbs and Salter's text, allows you to assess your students' progress with instant in-class quizzes and polls. You can pose questions and display students' answers seamlessly within the Microsoft PowerPoint slides of your own lecture, in conjunction with the "clicker" hardware of your choice. (For college and university adopters only. Contact your local Thomson representative to learn more.)

Multimedia Manager. A Microsoft® PowerPoint® Link Tool
This one-stop presentation tool makes it easy to assemble, edit, and present custom lectures using Microsoft PowerPoint. The Multimedia Manager includes animated versions of many figures and maps from the text, allowing you to integrate dynamic visuals into your course in a snap. You'll also find chapter outlines, summaries of key chapter concepts, and all of the art and photographs from this text.

Rand McNally® Atlas of World Geography
This comprehensive world atlas can be packaged with this text.

World Regional GeographyNow™
Access to the first assessment-centered student tutorial system for this course, World Regional GeographyNow, is automatically packaged free with each new copy of the text. This interactive online resource helps students gauge their unique study needs, then gives them a personalized learning plan for each text chapter that focuses their study time on the concepts they most need to master.

Acknowledgments
Many reviewers and advisors helped to bring this fifth edition to completion. Thanks to each and every one of these distinguished geographers and outstanding teachers and students! They include Elizabeth Leppman, Saint Cloud State University; Kefa Otiso, Bowling Green State University; Mark Guizlo, Lakeland Community College; Tarek Joseph, Henry Ford Community College; Joy Adams, Texas State University; Kyong (Keith) Chu, Bergen Community College; Keith Bell, Volunteer State Community College; William Strong, University of North Alabama; Mary Kimsey, James Madison University; Rebecca A. Howze, South Florida

Community College; Robert Maguire, Trinity College; Jeffrey Bury, San Francisco State University; J. Duncan Shaeffer, Arizona State University; Kelly Ann Victor, Eastern Michigan University; David Wall, Saint Cloud State University; Norman Carter, California State University—Long Beach; Gail Hobbs, Pierce College; Jim Nemeth, University of Toledo; Fritz Gritzner, South Dakota State University; Wayland Bauer, St. Ambrose University; James Lindberg, University of Iowa; and at the University of Missouri—Columbia, using and advising on improvements from the Fourth Edition, Larry Brown, Jennifer Presberry, Victor Lozano, Jason Jindrich, Nathaniel Albers, Rob Long, Jeff Pickles, Sunny Stevens, Johnny Finn, Trent Holmes, and Paul Hammond.

We would especially like to thank Elizabeth Leppman, who has also been very actively involved in the development of the World Regional GeographyNow materials for the book.

Andy Dolan, also of the University of Missouri—Columbia Department of Geography, did exceptional work on the maps in this edition. With very tight deadlines, Andy met every request, and then went way beyond what was asked of him—just because he was dedicated to making the best product possible. Andy also updated and added new statistics, freeing up valuable time for the writing of the text.

Deep thanks and appreciation go to the Brooks/Cole editorial staff who labored so diligently to help us all meet the tight deadlines and the technical challenges of the Fifth Edition. Keith Dodson never flagged in his support of the project, which forged ahead also with the confidence of David Harris, Hal Humphrey, and Andy Marinkovich. Melissa Newt kept up with the flurry of arrangements as the book got off the ground, and subsequently Megan Asmus very ably managed the constant barrage of pressing requests from the authors and editorial production. Frank Hubert was the kind and very exacting copy editor who helped improve the prose from start to finish. Nancy Shammas of New Leaf Publishing Services had the patience of a saint and extrordinary fortitude in seeing the manuscript through its various stages of production. Shutterbug and photo researcher Kathleen Olsen did a wonderful job tracking down a wide range of photo requests, some quite difficult to fill, while Marie Carigma-Sambilay brought the photo program together. Azadeh Poursepanj kept the pages coming fast and in great shape. Carol Benedict's insistence on top quality in the ancillaries package is much appreciated. The development of the new World Regional GeographyNow website for the book was carried out under the very talented supervision of Ericka Yeoman-Saler. Kelley McAllister, who has been involved with this text in many critical roles over the last decade, did an extraordinary job of pulling together the numerous marketing materials. Thanks also go to Jennifer Somerville, the marketing manager for the book, who came on board late but hit the ground running and has done a great job.

On the personal side, Joe Hobbs is ever grateful to his mother, Mary Ann Hobbs-Frakes, for raising him all over the place and always agreeing to join him on his next hairbrained travel idea ("Meet me in Ethiopia, Mom?"); to all of his great, extended family; and to his loving, very tolerant wife and best friend Cindy and daughters Katie and Lily. Kit Salter's ultimate thanks are always to Cathy, for bringing such grace to his life and labors.

Joseph J. Hobbs
Christopher L. Salter

Objectives and Tools of World Regional Geography

Joe Hobbs

Karst landscape in northern Vietnam. Geographers use diverse perspectives to understand and depict a landscape such as this one. They can map it at various scales, characterize the rural pattern it represents, and explain how it inspires artistic and poetic traditions.

chapter objectives

This chapter should enable you to:

- Identify the six essential elements of geography
- Appreciate the book's overall objectives
- Learn the basic "language" of maps
- Become familiar with some of the fundamental concepts of geography
- Know what geographers do and what the job prospects in geography are

WORLD
REGIONAL
Geography Now™

Look for this logo in the text and go to GeographyNow at http://earthscience.brookscole.com / wrg5e to explore interactive maps, view animations, sharpen your factual knowledge and geographic literacy, and test your critical thinking and analytical skills with unique interactive resources.

1.1 Welcome to World Regional Geography!

In a 2002 study commissioned by the National Geographic Society, a population of United States citizens aged 18–24 was asked to locate on maps many of the most important places in their lives and in the news. Eleven percent of those surveyed could not locate the United States on a blank map of the world. Forty-nine percent could not find New York City, ground zero for the most spectacular of the 9/11 attacks. Eighty-three percent did not know where Afghanistan is, despite that country's omnipresence in news of the war on terrorism. Eighty-seven percent did not know where to situate Iraq, which at the time was also prominent in the news as U.S. forces prepared to invade the country. The National Geographic survey also tested the geographic awareness of 18–24 year olds from Canada, France, Germany, Italy, Japan, Mexico, Sweden, and Great Britain. The Americans came in next to last, above only the youth of Mexico. (Sweden, incidentally, was number one, followed by Germany and Italy.)

Most of you reading this book are in that group of 18–24-year-old Americans. The dismal findings related above are not meant to embarrass you or confront you with how little you may know about the world. Instead, they pose a challenge. They compel us to ask questions about our role in the world and how important it is to know about that world. So what if a lot of us don't know where we are on a map? So what if we don't know where other people and countries are? Geography long ago earned a reputation for inflicting rote memorization of places on students. By itself, knowledge of these facts probably means very little. In context, though, they have great power to transform a person's life and con-

tribute to the welfare of an entire community or nation. What is geography? What good is geography, really, and what will you get out of learning world regional geography? These questions may begin to be answered by identifying the essential elements of geography and explaining the organization and objectives of this book.

In the 1990s, the National Geographic Society commissioned a team of expert geographers to identify the core features of the discipline of geography. They came up with the **six essential elements of geography.** These are widely regarded as accurately reflecting what geographic research and education encompass today and what distinguishes geography from other disciplines. All six elements are represented strongly in this book. They are:

1. *The World in Spatial Terms.* Geography studies the relationships between people, places, and environments by mapping information about them into a spatial context.

2. *Places and Regions.* The identities and lives of individuals and peoples are rooted in particular places and in those human constructs called regions.

3. *Physical Systems.* Physical processes shape Earth's surface and interact with plant and animal life to create, sustain, and modify ecosystems.

4. *Human Systems.* People are central to geography; human activities, settlements, and structures help shape Earth's surface, and humans compete for control of Earth's surface.

5. *Environment and Society.* The physical environment is influenced by the ways in which human societies value and use Earth's physical features and processes.

6. *Uses of Geography.* Knowledge of geography enables people to develop an understanding of the relationships between people, places, and environments over time—that is, of Earth as it was, is, and might be.

Geography thus has a broad, encompassing scope, and because the world is a very large subject of study, there is a vast amount of material of potential interest to you as a student of geography. The text is designed to simplify the challenge of grappling with the world, first by dividing it into eight regions, then by presenting a thematic profile of each region, and finally by examining in greater detail the geographic qualities of countries within each region.

Each regional profile chapter (Chapters 3, 6, 8, 10, 15, 17, 19, and 22) has five thematic sections. The first is area and population, identifying the region, its major cultures, and the numbers and distributions of people living in it. Then come physical geography and human adaptations, integrating elements of standards 3, 4, and 5 listed above by describing the region's environment and how people have accommodated themselves within it. The section on cultural and historical geographies explores in more detail the identities and experiences of the peoples inhabiting the region. Economic geography is vital to explaining the movements of materials and ideas around the world, and the broad patterns of rich and poor nations, and so it is treated separately. Finally, in the post-September 11 world in which conflicts over resources and between ideologies are so prominent, each profile chapter examines the region's geopolitical issues.

This book should answer your questions about the world. It has been written with four objectives in mind. First, in using this book, you will come *to understand important geographical problems and their potential solutions.* Geography literally means "description of the earth," but it has been more appropriately described as "the study of the earth as the home of humankind." Earth is our home, and to learn what there is to know about it, there is arguably no better discipline than geography, and no better geographical approach than world regional geography. As a home, Earth requires maintenance; it has some problems that need attention and can be remedied.

Many of these problems are environmental, and like geography in general, world regional geography is quite concerned with environmental problems and their solutions. Some of these, such as climate change, are global issues that may be addressed with global agreements, like the Kyoto Protocol described in Chapter 2. Others are national and international. Many regional and international security problems have roots that are environmental. Why can't some countries feed themselves? In some cases, their populations have exceeded their resources bases. What are their prospects then? The least appealing outcome is that famine will lower the population. Another unsavory option is for the country to aggressively seek the resources of others. We will see plenty of examples of how countries that have felt pushed into corners by agricultural failure, debt, or geographic con-

strictions have lashed out against their neighbors (and your attention will be brought to North Korea because of such concerns).

Another option is for rich countries to assist the country in need. As the world's only superpower, what can the United States do to help a country that is in trouble (and is therefore perhaps more likely to cause trouble)? We know we can help, by investing in education, employment, and infrastructure, for example. But should we help? Some authorities (like Garrett Hardin, discussed in the next chapter) have said we should turn our backs. But if we choose to be engaged, there are achievable solutions available. The next chapter will introduce some of these.

Second, in using this book, you should develop the skill *to make connections between different kinds of information as a means of understanding the world.* To understand the problems of North Korea, for example, you must consider many variables, including resources, population, economic development, ethnicity, history, and geopolitical interests. Is that too much information for you to take in? Not if you have the right tool with which to filter and synthesize the information. Geography is that tool. Geography is all about making connections. In geography, we look at relationships between places and people. We bring together information, techniques, and perspectives from both the natural and social sciences. As a student of world regional geography, you will be exposed to issues in political science, history, economics, anthropology, sociology, geology, atmospheric science, and more. You will pull these issues and perspectives together and find the links between them. In doing that, you will be doing geography, which is above all a holistic and integrative science.

This approach will almost certainly be rewarding for you. Your overall university experience should become richer as you become better able to tie together the various courses you are taking. In the future, you will be more likely to interact with the world with new wisdom that can be rewarding both professionally and personally. For example, more complete knowledge of the world—good geography—is good business. American businesses working around the world are doing much better now than in the past, in large part because of greater understanding of local cultures and environments.

Third, this book should help you *to understand current events.* You should become able to pick up a newspaper or turn on the TV news with a much-enhanced understanding of the issues underlying the world's news. Incidents of violence in the Middle East, earthquakes in the western Pacific, and the HIV/AIDS epidemic in Africa are not random, unpredictable events but instead are rooted in consistent, recognizable problems that have geographic dimensions. You should find it very satisfying to feel like something of an expert on a problem that appears in the news.

Finally, this book should help you develop the ability *to interpret places and read landscapes.* In geography, we are concerned both with **space**—the precise placement of loca-

Figure 1.1 Can you identify the location of this photograph?

tions on the face of the earth—and with **place**—the physical and cultural setting of a location. As you use this book, you should be increasingly able to look at a photograph of a place, or travel to a place, with new insight into many of the features that make up its identity: its climate and vegetation, for example, and the language, religion, history, and livelihoods of the people who live there. As an example of how you can come to appreciate place, please study the photograph above (Figure 1.1) and, before reading the footnote below, try to identify what country or region this photograph was taken in.[1]

The text uses photographs that will teach you a lot about place. There are aerial and satellite images, too, that will help you identify place on a broader scale and appreciate the world as if you were looking at it from space.. These images, plus something you cannot be given in a book—a globe—are important tools in your gaining the "whole earth" perspective that is geography's.

1.2 The Language of Maps

Geography has many concerns overlapping with those of other sciences, but its preoccupations with space, place, location, landscapes, and maps distinguish this field. Above all, geographers recognize spatial distributions and patterns in Earth's physical and human characteristics. The term *spatial* comes from the noun "space," and it relates to the distribution of various phenomena on the earth's surface. Geographers portray spatial data cartographically—that is, with maps. There are a few concepts related to these distributions and patterns that you need to know so you can better study

world regional geography, and you need to know the "language" of the maps that are geography's most basic tools.

The art and science of making maps is called cartography. Cartographers are the geographers and other skilled technicians who create maps through the research and presentation of geographic information. Using maps, the cartographer shows where things and places are located in relation to each other. Some maps show the distribution and interrelation of things that are fixed in place, such as structures and terrain features, whereas other maps portray the flow of people, goods, and ideas from place to place. Cartographers today almost always use computers to interpret and display spatial data, but it was not so long ago that they drafted maps laboriously by hand.

Because maps are so basic to geographic understanding, it is important to learn their language. Major elements of this language (beyond the title, date, and north arrow) are scale, coordinate systems, projections, and symbolization.

Scale

A map is a reducer; it shrinks an area to the manageable size of a chart, piece of paper, or computer monitor. The amount of reduction appears on the map's scale, which shows the actual distance on Earth as represented by a given linear unit on the map. A common way of denoting scale is to use a representative fraction, such as 1:10,000 or 1:10,000,000. In other words, one linear unit on the map (for example, an inch or centimeter) represents 10,000 or 10,000,000 such real-world units on the ground. A large-scale map is one with a relatively large representative fraction (for example, 1:10,000 or even 1:100) that portrays a relatively small area in more detail. A small-scale map has a relatively small representative fraction (such as 1:1,000,000 or 1:10,000,000) that, in contrast, portrays a relatively large area in more generalized terms. Figure 1.2 is a small-scale map depicting almost the entire world on a single page.

Coordinate Systems

A principal concern in maps is location. There are two general types of locational information: absolute location and relative location. **Relative location** defines a place in relationship to other places. It is one of the most basic reference tools of everyday life; you might say you live south of the city, just west of the shopping mall, or next door to a good friend. Relative location will become part of your basic geographic literacy as you learn, for example, that Sumatra is northwest of Java and that Ethiopia is tantalizingly close to, but is cut off from, the Red Sea. Understanding the implications of relative location can be quite useful in following world affairs; be watchful for Ethiopia to go to war with Eritrea because Ethiopia wants access to the sea.

Absolute location, also known as mathematical location, uses different ways to label places on the earth so that every place has its own unique location, or "address." Coordinate systems are used to determine absolute location. These coordinate systems use grids consisting of horizontal and vertical

---

[1] This is the Nile Valley in Egypt. If you did not get it right, chances are that you would after reading Chapters 8 and 9, which will immerse you in the landscapes of the Middle East. There are some geographic clues in the photograph that say "Egypt." It is among the few places in the world where there is such a stark boundary between lush cultivation and barren desert. In the limestone cliffs, there is a line of ancient rock-cut tombs, another characteristic landscape signature of Egypt.

Definitions + Insights

Core Location and Peripheral Location

Among the concepts useful to you in studying world regional geography are **core location** and **peripheral location.** Some locales have greater importance in local, regional, or world affairs because they have a central, or core, location relative to others. Other locales are less important because they are situated far from "where the action is."

A comparison of two countries, Great Britain and New Zealand, provides an example (Figure 1.a). Both are island countries. Their climates are remarkably similar, although they are in opposite hemispheres and are about as far apart as two places on Earth can be. Westerly winds blow off the surrounding seas, bringing abundant rain and moderate temperatures throughout the year to both.

But there are important differences. Great Britain is located in the Northern Hemisphere, which has the bulk of the world's land (and may thus be described as the **land hemisphere**) and most of its principal centers of population and industry; New Zealand is on the other side of the equator, in the Southern Hemisphere, where there is less land and less economic activity. Great Britain is located near the center of the world's land masses and is separated by only a narrow channel from the densely populated industrial areas of Western continental Europe; New Zealand is surrounded by vast expanses of ocean in the **water hemisphere.** Great Britain is located in the western seaboard area of Europe, where many major ocean routes of the world converge; New Zealand is far away from the centers of world commerce. For more than four centuries, Great Britain has shared in the development of northwestern Europe as an organizing center for the world's economic and political life; New Zealand, meanwhile, has languished in comparative isolation.

Great Britain, in other words, has a core or central location in the modern framework of human activity on Earth, whereas New Zealand has a peripheral location. Centrality of location is a very important consideration when it comes to assessing the economic and political geographies of places, countries, and regions.

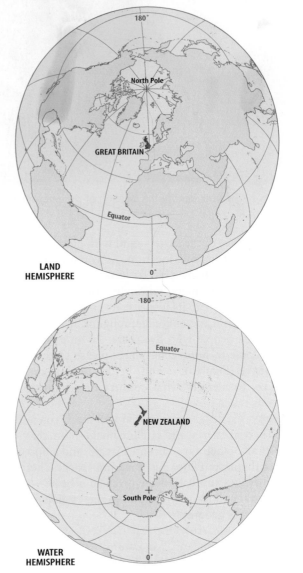

Figure 1.A In the top map, note how the major land masses are grouped around the margins of the Atlantic and Arctic Oceans. The British Isles and the northwestern coast of Europe lie in the center of the "land hemisphere," which constitutes 80 percent of the world's total land area and has approximately 91 percent of the world's population. New Zealand lies near the center of the opposite hemisphere, or "water hemisphere," which has only 20 percent of the land and 9 percent of the population.

lines covering the entire globe. The intersections of these lines create the addresses in the global coordinate system, giving each location a specific, unique, and mathematical situation.

The most common coordinate system on maps uses *parallels of latitude and meridians of longitude* (see Figure 1.2). The term *latitude* denotes position with respect to the equator and the poles. Latitude is measured in **degrees** (°), min-utes ('), and **seconds** ("). Each degree of latitude, which is made up of 60 minutes of latitude, is about 69 miles (111 km) apart; these distances vary a little because of Earth's slightly ellipsoid shape. Each minute of latitude, which is made up of 60 seconds of latitude, is thus roughly a mile apart. The **equator,** which circles the globe east and west midway between the poles, has a latitude of 0°. All other latitudinal lines are parallel to the equator and to each other and are

Major World Regions

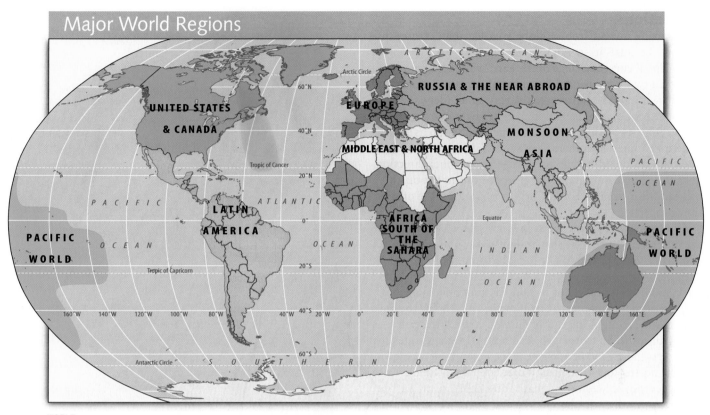

Active Figure 1.2 A small-scale map of the world regions of this text. *See an animation based on this figure, and take a short quiz on the concept.*

therefore called parallels. Every point on a given parallel has the same latitude. Places north of the equator are in **north latitude;** places south of the equator are in **south latitude.** The highest latitude a place can have is 90°N (the **North Pole**) or 90°S (the **South Pole**). Places near the equator are said to be in **low latitudes;** places near the poles are in **high latitudes.** The **Tropic of Cancer** and the **Tropic of Capricorn,** at 23.5°N and 23.5°S, respectively, and the **Arctic Circle** and **Antarctic Circle,** at 66.5°N and 66.5°S, respectively, form the most commonly recognized boundaries of the low and high latitudes. Places occupying an intermediate position with respect to the poles and the equator are said to be in the **middle latitudes.** Incidentally, there are no universally accepted definitions for the boundaries of the high, middle, and low latitudes. The northern half of Earth between the equator and the North Pole is called the **Northern Hemisphere,** and the southern half between the equator and the South Pole is the **Southern Hemisphere.**

Meridians of longitude are straight lines connecting the poles. Every meridian is drawn due north and south. They converge at the poles and are farthest apart at the equator. Longitude, like latitude, is measured in degrees, minutes, and seconds, but because the longitude lines are not equidistant across the globe, their distance values vary. At the equator, the distance between lines of longitude is about 69.15 statute miles (111.29 km), while at the Arctic Circle it is only about

27.65 miles (44.50 km). The meridian most used as a base or starting point is the one running through the Royal Astronomical Observatory in Greenwich, England. It is known as the meridian of Greenwich, or the prime meridian, and has a longitude of 0°. Places east of the prime meridian are in **east longitude;** places west of it are in **west longitude.**

The meridian of 180°, exactly halfway around the world from the prime meridian, is the other dividing line between places east and west of Greenwich and is called the **International Date Line.** This line, which has a few zigzags in it (see, for example, Figure 15.1, page 405), separates two consecutive calendar days. The date west of the line is one day ahead of the date east of the line. All of the earth's surface eastward from the prime meridian to the International Date Line is in the **Eastern Hemisphere,** and all of the earth's surface westward from the prime meridian to the International Date Line is in the **Western Hemisphere.**

With these simple tools, it is possible for you to interpret the absolute location of any given place. The author's **GPS (Global Positioning System)** device, an increasingly common and popular tool that uses satellite technology to pinpoint absolute locations around the earth, was manufactured by Garmin Industries in suburban Kansas City, Kansas, at latitude N 38°51′20″ (north latitude 38 degrees, 51 minutes, 20 seconds), longitude W 94°47′55″ (west longitude 94 degrees, 47 minutes, 55 seconds).

Projections

Transferring images from a globe to a flat map creates distortions because it is impossible to represent the three-dimensional curved surface of the earth with complete accuracy on a two-dimensional flat sheet of paper or a computer monitor. If the area represented is very small (depicted on a large-scale map), the distortion may be slight enough to disregard, but maps that represent larger spatial areas (small-scale maps) may introduce very serious distortions. Distortions affect four different properties of a globe: direction, distance, shape, and area. The purpose of the various map projections is to minimize distortion in one or more of these four properties.

Conformal map projections preserve shape over small areas and retain directions for straight lines. The best-known conformal map projection is the **Mercator projection** (Figure 1.3a). To serve the projection's original purpose of aiding navigation, the parallels of latitude and meridians of longitude are straight and meet at right angles. In keeping the lines of longitude straight instead of curving toward the poles, as on a globe, the Mercator projection greatly exaggerates the east–west dimension of areas near the poles. It also exaggerates the north–south dimension because the parallels are not spaced evenly between the equator and poles, as on a globe, but gradually draw farther and farther apart with increasing distance from the equator. The angular array of parallels and meridians makes the absolute latitudinal and longitudinal location of every place accurate, but it distorts the size and shape of polar regions. This explains why on this projection Greenland appears larger than all of South America, although it is actually only slightly larger than Mexico. To appreciate Greenland's true size relative to the rest of the earth, your best tool is a globe, and it is a good idea to have and use one as a supplement to this textbook.

Equal-area projections, such as the Lambert Equal-area projection (Figure 1.3b), Goode's Interrupted Homolosine projection (Figure 1.3c), and the Peters projection, minimize distortions relating to how big an area looks and are second best to a globe. Equal-area projections must distort shape to retain equal-area characteristics. Goode's Interrupted Homolosine projection tries to minimize some of the shape distortion by segmenting the map into lobes (the cartographic equivalent of sections if Earth were an orange), thereby minimizing shape distortion in midlatitudes while preserving area over all.

Compromise projections, such as the Robinson projection (Figure 1.3d), do not eliminate distortion in any of the four properties. Rather, the Robinson projection allows different degrees of shape, area, scale, and distance distortion to

WORLD
REGIONAL
Geography ⊕ Now™   **Active Figure 1.3** Each of these map projections attempts to represent the three-dimensional surface of the earth on a two-dimensional sheet of paper. Each has advantages and drawbacks. *Source: Goode's 20th edition, published by Rand McNally. See an animation based on this figure, and take a short quiz on the concept.*

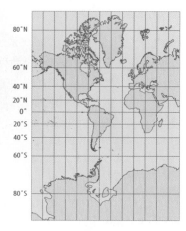

(a) Mercator Projection

(b) Lambert Equal-area Projection

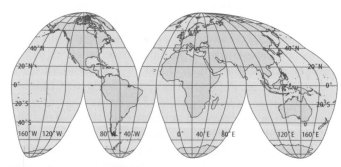

(c) Goode's Interrupted Homolosine Projection

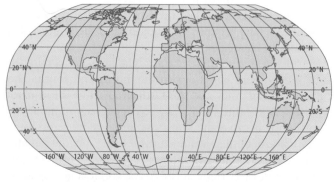

(d) Robinson Projection

minimize the severity of the other distorted properties, leading to an aesthetically pleasing map.

Symbolization

Maps enable us to extract certain information from the totality of things, to see patterns of distribution, and to compare these patterns with one another. No map is a complete record of an area; it represents instead a selection of certain details, shown by symbols, that a cartographer uses for a particular purpose. Unprocessed data must be classified to provide categories that the symbols will represent. Symbolization of details may be categories of physical or cultural forms like rivers or roads, aggregates such as 100,000 people or 1 million barrels of crude oil, or averages such as population density (the number of persons per square mile or kilometer within a defined area). Aggregates or averages are frequently ordered into ranked categories in a graded series (for example, population densities of 0–49, 50–99, or 100 or more people per square mile) for portrayal on the map. The categories are not self-evident but are selected by the cartographer, sometimes through elaborate statistical procedures. To represent varied phenomena, the cartographer may choose from a wide range of symbols, including lines, dots, circles, squares, shadings, colors, and more.

The **dot map** is one of the most common types of maps used to show distributions of people or things on Earth. Dots usually portray quantities; on a dot map of population, for example, each dot represents a stated number of people and is placed as near as possible to the center of the area that these people occupy. In interpreting such a map, we are not greatly concerned with the individual dots but with the way the dots are arranged. We look for the pattern and distribution of a given phenomenon.

One complexity of cartography is that different scales and ways of showing distributions can give different impressions. For example, methods of symbolization other than dots could be used to show the distribution of world population and might convey rather different ideas from those presented in the dot map. One alternative would be to include the use of a single symbol for each country, with the size of the symbol proportional to the country's population. This sort of a proportional map is called a cartogram (see Figure 1.4a). One of the most common map types is the choropleth map, in which each political unit on a map is filled with a distinguishing pattern or color representing some derived value (for example, population density within a political unit or the number of university students per 100,000 people in that political unit). A good example of a choropleth map is Figure 2.18 on page 33, which depicts relative wealth of the world's countries.

Both Figures 1.4a and 1.4b illustrate clearly that China and India have the most populous regions on Earth; they simply do it in different ways, respectively, as a cartogram and as an **isarithmic map** (one in which lines join points of equal value, such as elevations on a contour map). The cartogram emphasizes that the countries of China and India are the world's demographic giants, but it does not indicate

where populations are concentrated. The isarithmic map underscores where the people are actually clustered but, in this instance, does not provide the political names that would tell you in which countries those people are located. Different kinds of maps, in other words, serve different purposes, and there is no single best way to represent spatial information.

Maps need to be used critically, too, and you will gradually acquire the skill of evaluating map quality. There is a rather well-known book in the field of geography entitled *How to Lie with Maps,* suggesting, like the motto "let the buyer beware," that just because something is mapped does not guarantee it is mapped accurately.

WORLD REGIONAL
Geography⊕Now™

Click Geography Literacy to make your own map to show how maps can give different impressions.

World Regions and Other Mental Maps

Our understanding of location is not completely objective, for each of us has a personal sense of space and place and their associations. A mental map is the collection of personal geographic information that each of us uses to order the images and facts we have about places, both local and distant. Sometimes we use that information to create actual maps, as when a friend asks you to draw a map showing how to get to her house or when a nomad in the Middle East draws lines in the sand diagraming the territories of his and others' tribes (Figure 1.5). Such maps are not accurate, precise, or scientific, but they portray useful information and tell us much about the individuals and cultures that create them. They are subjective, just as are the eight regions identified and used in this book.

We have divided the world into eight regions (see Figure 1.2 and Table 1.1) for this text because it is impossible to introduce order and logic to something as massive, complex, and diverse as the earth's surface without an organized framework built around smaller spatial units. Regions are human constructs, not "facts on the ground." People create and draw boundaries around regions in an effort to define distributions of relatively homogeneous characteristics. The Middle East and North Africa comprise a region where, very generally, there is a majority Muslim population and an arid environment and where Europe, Asia, and Africa meet. But different sources draw different boundaries for this region, and within those boundaries—however they are drawn—there are many non-Arab peoples and nonarid environments. There are often recognizable transition zones between regions, too; the country of Sudan, for example, is both Middle Eastern and African in culture. A region is simply a convenience and a generalization, helping us become acquainted with the world and preparing us for more detailed insight.

Within the eight world regions, this text also employs other regional concepts and tools, and a brief introduction to

World Population Cartogram

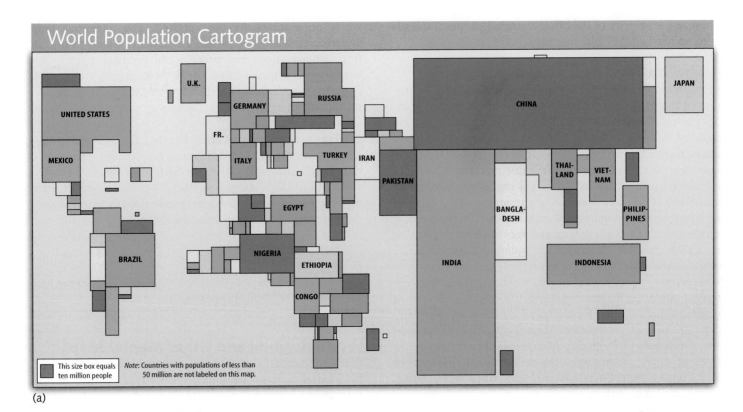

This size box equals ten million people

Note: Countries with populations of less than 50 million are not labeled on this map.

(a)

World Population Density

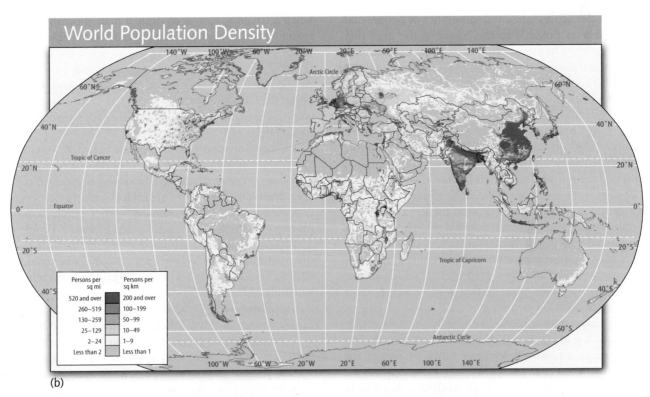

Persons per sq mi	Persons per sq km
520 and over	200 and over
260–519	100–199
130–259	50–99
25–129	10–49
2–24	1–9
Less than 2	Less than 1

(b)

Figure 1.4 (a) This cartogram illustrates one of the ways cartographers represent different patterns on Earth. In a cartogram such as this one of world population by country, the shape and size of each country named differ in direct relationship to the size of its population. The cartogram provides the viewer with a graphic and geometric sense of a world ranking of national population or of any other item that the cartogram is designed to represent. (b) This is an isarithmic map designed to show the approximate distribution of global population. By comparing 1.4a and 1.4b, you can see the different visual messages sent by different maps.

Table 1.1   The Major World Regions: Basic Data

Region	Area (thousand/sq mi)	Area (thousand/sq km)	Estimated Population (millions)	Annual Rate of Increase (%)	Estimated Population Density (sq mi)	Estimated Population Density (sq km)	Human Development Index[a]	Urban Population (%)	Arable Land (%)	GDP PPP Per Capita ($US)
Europe	1959.2	5072.2	523.5	0.1	258	100	0.901	74	24	22611
Russia and the Near Abroad	8533.2	22100.8	280.3	-0.1	32	12	0.771	64	8	6577
Middle East and North Africa	5916.7	15324.2	479.0	1.9	160	62	0.690	54	6	5345
Monsoon Asia	8013.5	20755.0	3504.2	1.2	437	169	0.677	37	20	5019
Pacific World	3305.7	8561.7	32.7	1.0	10	4	0.846	71	5	21553
Africa South of the Sahara	8406.5	21772.5	693.7	2.4	85	33	0.439	31	6	1767
Latin America	7946.2	20580.7	549.4	1.6	69	27	0.773	74	7	7262
North America	8403.8	21765.8	325.6	0.5	38	15	0.939	79	10	37013
World Regional Totals	**52484.8**	**135935.6**	**6396.0**	**1.3**	**123**	**47**	**0.696**	**48**	**10**	**7590**
USA (for comparison)	3717.8	9629.1	293.6	0.6	79	30	0.939	79	10	37800

[a] HDI is a numeric index developed by the United Nations that reflects gross domestic product per capita, literacy rate, level of education on average, and life expectancy. The maximum HDI possible is 1.0.

Source: *World Population Data Sheet,* Population Reference Bureau, 2004; *U.N. Human Development Report,* United Nations, 2004; *World Factbook,* CIA, 2004.

Joe Hobbs

Figure 1.5 As we grow, our minds fill with mental maps, which we use consciously or otherwise to orient and navigate ourselves through daily life. In drawing these maps for others, we often discover the importance of scale, direction, and other map essentials.

the different types of regions is useful. A formal region (also called a **uniform** or **homogeneous region**) is one in which all the population shares a defining trait. A good example is an administrative political unit such as a county or a state, where the regional boundaries are made explicit on a map. A functional region (also called a **nodal region**) is a spatial unit characterized by a central focus on some activity (often an economic one). At the center of a functional region—a good example is the marketing area for a metropolitan newspaper—the activity is most intense, while toward the edges of the region, the defining activity diminishes in importance.

A vernacular region (or **perceptual region**) is a region that exists in the mind of a large number of people, and it may play an important role in cultural identity but does not necessarily possess explicit or objective borders. Good examples are concepts of "the South" or "the West" or "the Rust Belt" of the United States. These regional terms have economic and cultural connotations, but 10 people might have 10 different definitions of the states and boundaries of these regions. Vernacular or perceptual regions, constructed by individuals and cultures, serve as shorthand for place and regional identity, helping us organize, simplify, and make sense of the world around us. In this text, regions will help you make generalizations about different parts of the world and also encourage you to explore their diverse and often unique components.

1.3   More Key Concepts of Geography

There are other key concepts of geography that underlie the world regional approach. As mentioned earlier, geography is concerned with the delicate and often troubled relationship between people and the natural environment, and it seeks to explain and help solve some of Earth's problems. It is useful to have some understanding of how geographers deal with these issues today and how these perspectives have developed from roots among geographers of the past.

Humankind's role in changing the face of the earth has emerged time and again as one of geography's central themes. The great German geographer Alexander von Humboldt (1769–1859) initiated geography's modern era in a series of classic studies on this theme. From field observations in Venezuela, he concluded, "Felling the trees which cover the sides of the mountains provokes in every climate two disasters for future generations: a want of fuel and a scarcity of water."[1] Today, as we observe, for example, the direct connection between deforestation in Nepal's Himalaya Mountains and the devastating floods in downstream Bangladesh, we can appreciate the foresight in von Humboldt's warnings.

In his wake, other geographers in Europe and the United States wrote about climatic changes resulting from human destruction of forests and from the expansion of farmlands. Civilization leads to aridity, they concluded, and they chastised people for violating the harmony and balance that seemed inherent in the natural world. George Perkins Marsh (1801–1882), who served as President Abraham Lincoln's ambassador in Italy, was troubled by the legacy of ancient Greece and Rome on the landscapes of the eastern Mediterranean region. He used the past as a cautionary tale for the future, writing in his book *Man and Nature,* "man has too long forgotten that the earth was given to him for usufruct alone, not for consumption, still less for profligate waste."[2]

A century later, geographer Carl Sauer (1889–1975), another American, wrote:

> We have accustomed ourselves to think of ever expanding productive capacity, of ever fresh spaces of the world to be filled with people, of ever new discoveries of kinds and sources of raw materials, of continuous technical progress operating indefinitely to solve problems of supply. We have lived so long in what we have regarded as an expanding world, that we reject in our contemporary theories of economics and of population the realities that contradict such views. Yet our modern expansion has been effected in large measure at the cost of an actual and permanent impoverishment of the world.[3]

These words might have been written today, but the University of California–Berkeley geographer penned them in 1938.

Sauer is credited with founding the **landscape perspective** in American geography, based on the method of studying the transformation through time of a **natural landscape** to a **cultural landscape**. The agent of that metamorphosis is humankind. Sauer collaborated in the production of one of the classics in the field of geography, a 1956 tome entitled *Man's Role in Changing the Face of the Earth*. He defined the geographer as the scientist determined to "grasp in all of its meaning and color the varied terrestrial scene."[4] Geographers continue to be interested in identifying the distinct forces of nature and culture that have been at work in the creation of the landscape—the collection of physical and human geographic features on the earth's surface—and in

Figure 1.6 Environmental determinists argued that climate, soils, and other physical factors had decisive impacts on personalities and cultures. Such associations, however, are more caricature and stereotype than science. The Maya people who relax in a hammock on a warm afternoon are also energetic and highly capable.

evaluating the role of landscape in economic and social development.

Ironically, since its earliest days, this discipline so concerned with humanity's impact on the environment has periodically embraced the notion that nature irreversibly shapes people's ideals and actions. This doctrine of environmental determinism holds that individuals and entire societies follow predictable patterns of thought and behavior molded inevitably by climate, elevation, and other environmental factors. In the fifth century B.C., the Greek physician and scholar Hippocrates, for example, asserted that people living in hot climates were always light and agile in physique and passionate and violent in temperament.

Modern environmental determinists took the nature–culture connections to great extremes. The writings of the American geographer Ellsworth Huntington (1876–1947) seem painfully racist today, yet in their time, they were accepted by many as rational interpretations of interactions between people and places (Figure 1.6). Huntington wrote of the people of southern Mexico:

> As the amount of Spanish blood in the people of Yucatan increases, their energy and resourcefulness also increase. They also become more light-hearted and gay than the silent, sober Indians, but at the same time the degree of honesty and morality is said to decrease markedly. All classes of people are slow compared to Americans . . . Something in the new environment seems to make people slow. In part, this may be due to contact with an inferior race, but more probably it is a matter of climate. Possibly the heat has something to do with it.[5]

If fascinating to read, the determinists' arguments were often flawed; one need only think of how the "slow" Maya built one of the world's great civilizations in the hot forests of the Yucatán.

[1] Geoffrey J. Martin and Preston E. James, *All Possible Worlds: A History of Geographical Ideas* (New York: Wiley, 1993), p. 150.

[2] Ajavit Gupta, *Ecology and Development in the Third World* (New York: Routledge, Chapman and Hall, 1988), p. 2.

[3] Andrew Goudie, *The Human Impact on the Natural Environment* (Oxford: Basil Blackwell, 1986), p. 6.

[4] Larry Grossman, "Man–Environment Relations in Anthropology and Geography." *Annals of the Association of American Geographers, 67,* 1977, p. 129.

[5] Ellsworth Huntington, "The Peninsula of Yucatan." *Bulletin of the American Geographical Society, XLIV* (11), 1912, pp. 812–813.

While the natural environment may have some effects on personality and culture, these are difficult to identify and defend scientifically, and most geographers avoid doing so. It has thus been said that environmental determinism was never disproved, just disapproved of. More acceptable, more valid scientifically, and generally embraced today is the concept of environmental possibilism, or **possibilism.** Developed by Paul Vidal de la Blache (1845–1918) and other French geographers and historians in the late 19th century, possibilism is a direct refutation of environmental determinism. It argues that environment does not have a sovereign influence on personalities and cultures but rather offers a range of possible adaptations, which different individuals and cultures embrace differently. "The Pennsylvania forest can be transformed into Amish farms or into golf courses," the possibilist René Dubos wrote. "The American desert can be turned into a Mormon settlement or Las Vegas. The Tuaregs continue to live as nomads in the Sahara, while the Israelis make the sand bloom in the Negev Desert."[6]

In the 1970s, the related concept of environmental perception emerged, and it also prevails today. It holds that the manner in which people perceive the environment influences their use of environment and ultimately the condition of the environment. Geographer David Lowenthal (b. 1923) argued, "we are all artists and landscape architects, creating order and organizing space, time and causality in accordance with our perceptions and predilections."[7] Factors like gender, age, ethnicity, socioeconomic status, religion, and culture influence decision making in the environment and thus shape landscapes. For example, Muslims in Morocco regard the environs of saints' tombs as inviolate precincts, where no woodchopping or hunting is allowed. These islands of relatively lush vegetation and wildlife, preserved because they are seen as sacred space, now stand surrounded by denuded mountains and serve as windows on what mature or climax ecosystems would look like throughout the region if the human agency were removed. Unlike environmental determinism, then, environmental possibilism and environmental perception shift the focus back to the role of humankind in changing the earth.

This people-oriented perspective on the environment, so favored in modern geography, has important practical applications today. In the subfield of **natural hazards,** for example, geographers identify what factors compel people to move to, stay in, or return to areas prone to recurrent drought, storm, or earthquake. The varied explanations, from a lack of choice imposed by poverty (property rents and prices are cheaper in floodplains) to male tendencies to take risks ("don't worry, honey, there hasn't been a hurricane on this coastline in 100 years"), provide vital information to planners and policymakers in hazard-prone areas.

---

[6]René Dubos, *A God Within* (New York: Charles Scribner's Sons, 1972), p. 151.

[7]Larry Grossman, "Man–Environment Relations in Anthropology and Geography." *Annals of the Association of American Geographers,* 67, 1977, p. 141.

1.4 Geography, Geographers, and You

The subfield of natural hazards is just one of many within geography's renaissance. Geography is making a strong comeback as an academic discipline and, even better, is delivering more jobs to its graduates. Geography students typically specialize in either regional, systematic (the topical study of various physical or cultural geographic elements), or technical fields. The **Association of American Geographers (AAG)** lists 53 specialty groups among its approximately 7,500 members (Table 1.2). **Regional specialties** of the members show a preference for Western (especially North American and European) rather than non-Western areas. The relative shortage of U.S. geographers with professional expertise in non-Western and developing countries is a disadvantage, but it does open opportunities for younger geographers willing to

Table 1.2 Specialty Groups of the Association of American Geographers

Africa	Geomorphology
Applied Geography	Hazards
Asian Geography	Historical Geography
Bible	History of Geography
Biogeography	Human Dimensions of Global Change
Canadian Studies	Indigenous Peoples
Cartography	Latin American
China	Medical Geography
Climate	Microcomputers
Coastal and Marine	Middle East
Communication Geography	Military Geography
Cryosphere	Mountain Geography
Cultural and Political Ecology	Political Geography
Cultural Geography	Population
Developing Areas	Qualitative Research
Disability	Recreation, Tourism and Sport
Economic Geography	Regional Development and Planning
Energy and Environment	Remote Sensing
Environmental Perception and Behavioral Geography	Rural Geography
Ethics, Justice and Human Rights	Russian, Central Asian and East European
Ethnic Geography	Sexuality and Space
European	Socialist and Critical Geography
Geographic Information Systems	Spatial Analysis and Modeling
Geographic Perspectives on Women	Transportation Geography
Geography Education	Urban Geography
Geography of Religion and Belief Systems	Water Resources
	Worldwide Web

Source: www.aag.org

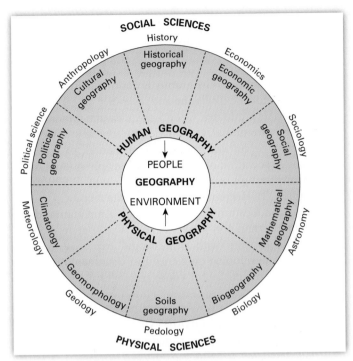

Figure 1.7 Selected subfields of geography. This diagram identifies the main subject-matter areas within human geography and physical geography and links them with the most closely related disciplines in the social sciences and the physical sciences.

undertake the challenges of foreign language study and fieldwork. As might be expected, there is particularly high demand among intelligence agencies for students acquainted with regions deemed to be strategic in the national interest, such as the Middle East. However, most geographers with foreign-area expertise have university research and teaching positions.

Most U.S. geographers emphasize **systematic specialties** over regional specialties (Figure 1.7). Specialists in physical geography study spatial patterns and associations of natural features. Prominent subfields include geomorphology (the study of landforms and a field in which geography intersects with geology), climatology (climatic processes and patterns), **biogeography** (the geography of plants and animals), and **soils geography.** Closely related to physical geography is the large interdisciplinary field of **environmental studies,** which is concerned with reciprocal relationships between society and the environment. Another allied specialty is medical geography, which focuses on spatial associations between the environment and human health and on locational aspects of disease and healthcare delivery.

Specialists in **economic geography** study spatial aspects of human livelihood. Major occupations and products are their main concerns, and they have an active theoretical component that intersects with the field of economics. **Marketing geography** is a relatively small but active applied offshoot of economic geography and is of particular importance in the continuing expansion of residential areas on the edges of cities of all sizes. **Agricultural geography, manufacturing ge-**

ography, and **transportation geography** are important subfields that deal with these major aspects of networks of interaction and the transformation of land and resources. **Urban geography** studies the locational associations, internal spatial organization, and functions of cities. It offers an alternative pathway to the applied field of urban planning.

Geographers specializing in cultural geography concern themselves with places of origin, diffusion, interactions, landscape evidence and analysis, and regionalization of culture. Closely allied to cultural geography is the field of **cultural ecology,** which focuses on the relationship between culture and environment and uses some of the tools of anthropology. **Social geography** deals with spatial aspects of human social relationships, generally in urban settings. **Population geography** assesses population composition, distribution, migration, and demographic shifts. **Political geography** studies topics such as spatial organization of geopolitical units, international power relationships, nationalism, boundary issues, military conflicts, and regional separatism within states. Historical geography studies the geography of past periods and the evolution of geographic phenomena such as cities, industries, agricultural systems, and rural settlement patterns.

Technical specialties in geography include cartography, computer cartography, remote sensing, quantitative methods (mathematical model building and analysis), air photo interpretation, and Geographic Information Systems (GIS). GIS, which involves the computer manipulation and analysis of spatial data, represents the "cutting edge" of geography today (Figure 1.8). GIS is essentially cartography in its most sophisticated and dynamic form. With GIS, a cartographer can create different "layers" of landscape data and with a few clicks of a computer mouse can depict, for example, any

WORLD REGIONAL **Geography ⊕ Now™**   **Active Figure 1.8** The most in-demand tool of geography today is Geographic Information Systems (GIS). Here the principal cartographer for this book, Andrew Dolan, is using GIS to create a population map of the United States. *See an animation based on this figure, and take a short quiz on the concept.*

A Geographic Information System

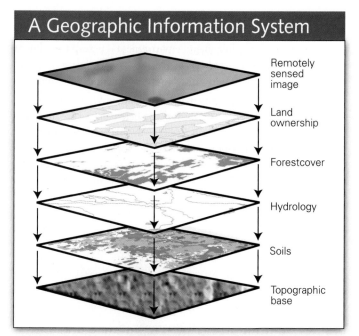

Remotely sensed image

Land ownership

Forestcover

Hydrology

Soils

Topographic base

Practical GIS

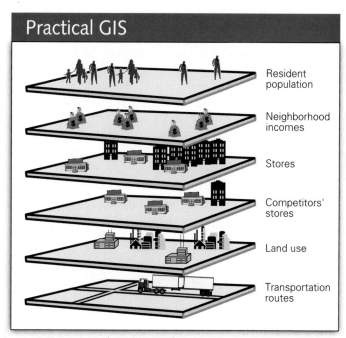

Resident population

Neighborhood incomes

Stores

Competitors' stores

Land use

Transportation routes

WORLD REGIONAL Geography ⊛ Now™

Active Figure 1.9 Geographic Information Systems (GIS) create and use layers of different types of spatial data. The computer-generated images and models that result have a wide range of academic and practical applications. *Source: (left) After Bergman, 1995; (right) After Rubenstein, 1995. See an animation based on this figure, and take a short quiz on the concept.*

combination of information such as a neighborhood's or city's street plan, elevation above sea level, distribution of trees, sewage and electricity systems, residents' income distribution, and more (Figure 1.9). With GIS, geographic information can be created, manipulated, and compared with great efficiency, especially in comparison with the capabilities of old-fashioned manual cartography. The sky is the limit for how GIS can be used, from concerns ranging from archeology to zoology and for employers across the spectrum of private and public interests. This is by far geography's leading area of growth and employment, and geographers increasingly use GIS as tools in their research on both human and environmental issues. Remote sensing and air photo interpretation use various kinds of satellite imagery and photo coverage to assess land use or other geographic patterns. These are very valuable tools in a wide variety of applications, from keeping track of deforestation in the Amazon to evaluating the movement of soldiers and matériel on the battlefield.

Increasingly, departments of geography are developing courses and laboratories to train students in the study of GIS, computer cartography, remote sensing, digital image processing, and spatial statistics. Many graduates with these technical skills find employment in government agencies at the local, state, regional, and federal levels. Firms working with land-use decisions at all scales use geography and geographers. Retail firms find the problem-solving skills of geographic analysis useful in management, survey design and implementation, estimation of public response to innovations or expansion to new locations, and policy shifts. There are other kinds of employment, too, notably in the travel and tourism industry. Although you might not be a geography major, and might not see classified ads that proclaim "Geographer Wanted," chances are that many opportunities will arise if you work at cultivating knowledge of the world in which you live. This book will help you do just that.

SUMMARY

- A recent study showed that U.S. citizens aged 18–24 generally have poor knowledge of world geography.

- There are six essential elements of the national geography standards: the world in spatial terms, places and regions, physical systems, human systems, environment and society, and the uses of geography. These may be used as a guide to what students should know about and be able to do with geography.

- This book divides the world into eight major regions. It is organized so that regional profile chapters introduce the distinguish-

ing characteristics of each region, and subsequent chapters present more detail about the countries, landscapes, and peoples within each region.

- Geography means "description of the earth" and is also known as "the study of the earth as the home of humankind."

- Four main objectives of the text are for readers (1) to understand important geographic problems and their potential solutions, (2) to become better able to make connections between different kinds of information as a means of understanding the world,

(3) to understand current events, and (4) to develop skills in interpreting places and reading landscapes.

- Maps are the geographers' most basic tools. To be able to use maps effectively, one must understand their basic language, especially the concepts and terms of scale, coordinate systems, projection, and symbolization. There are a variety of ways in which maps can depict spatial data.

- Individuals and cultures generate their own unique "mental maps," which are as various as the people who use them. Regions are in effect mental maps that help us make sense of a complex world; they are not objective and static phenomena.

- Modern geographic thought derives from a long legacy of intellectual interest in how people interact with the environment. The dominant approach has been to understand how people have changed the landscape or face of the earth (especially with the perspectives of possibilism and environmental perception), but at times, geographers have been distracted by the apparent impact of nature on people and culture (environmental determinism).

- The discipline of geography may be divided into regional and systematic specialties, with the systematic fields having the most followers by far. Their concerns are very diverse and overlap with a variety of disciplines in the natural and social sciences. Geographers are employed in many private and public capacities. The strongest growth area, with the most jobs, is in Geographic Information Systems (GIS).

KEY TERMS + CONCEPTS

Terms in blue are also defined in the glossary.

absolute location (p. 3)
agricultural geography (p. 12)
Antarctic Circle (p. 5)
Arctic Circle (p. 5)
Association of American Geographers (AAG) (p. 11)
biogeography (p. 12)
cartogram (p. 7)
cartography (p. 3)
choropleth map (p. 7)
climatology (p. 12)
computer cartography (p. 12)
coordinate systems (p. 3)
core location (p. 4)
cultural ecology (p. 12)
cultural geography (p. 12)
cultural landscape (p. 10)
degrees, minutes, and seconds (p. 4)
dot map (p. 7)
east longitude (p. 5)
Eastern Hemisphere (p. 5)
economic geography (p. 12)
environmental determinism (p. 10)
environmental perception (p. 11)
environmental possibilism (p. 11)
environmental studies (p. 12)
equator (p. 4)
formal (uniform or homogeneous) region (p. 9)
functional (nodal) region (p. 9)

Geographic Information Systems (GIS) (p. 12)
geomorphology (p. 12)
Global Positioning System (GPS) (p. 5)
high, low, and middle latitudes (p. 5)
historical geography (p. 12)
humankind's role in changing the face of the earth (p. 9)
International Date Line (p. 5)
isarithmic map (p. 7)
land hemisphere (p. 4)
landscape (p. 10)
landscape perspective (p. 10)
location (p. 3)
manufacturing geography (p. 12)
marketing geography (p. 12)
medical geography (p. 12)
mental map (p. 7)
meridian of Greenwich (prime meridian) (p. 5)
meridian of longitude (p. 4)
natural hazards (p. 11)
natural landscape (p. 10)
north latitude (p. 5)
North Pole (p. 5)
Northern Hemisphere (p. 5)
parallel of latitude (p. 4)
parallels (p. 5)
peripheral location (p. 4)
physical geography (p. 12)
place (p. 3)

political geography (p. 12)
population geography (p. 12)
prime meridian (meridian of Greenwich) (p. 5)
projections (p. 6)
  compromise (p. 6)
  conformal (p. 6)
  equal area (p. 6)
  Mercator (p. 6)
regional specialties (p. 11)
relative location (p. 3)
remote sensing (p. 12)
scale (p. 3)
six essential elements of geography (p. 1)
social geography (p. 12)
soils geography (p. 12)
south latitude (p. 5)
South Pole (p. 5)
Southern Hemisphere (p. 5)
space (p. 2)
symbolization (p. 7)
systematic specialties (p. 11)
transportation geography (p. 12)
Tropic of Cancer (p. 5)
Tropic of Capricorn (p. 5)
urban geography (p. 12)
vernacular (perceptual) region (p. 9)
water hemisphere (p. 4)
west longitude (p. 5)
Western Hemisphere (p. 5)

REVIEW QUESTIONS

WORLD
REGIONAL
Geography ⊛ Now™

Assess your understanding of this chapter's topics with additional quizzing and concept-based problems at http://earthscience.brookscole.com/wrg5e.

1. What are the six essential elements of geography, as defined by the National Geographic Society? What does each element indicate about geography's concern with space, place, or the environment?

2. What does *spatial* mean and how does geography's interest in space differentiate it from other disciplines?

3. What geographic features make Great Britain and New Zealand different?

4. What are the major terms and concepts associated with scale, coordinate systems, projections, and symbolization?

5. What is the latitude and longitude of the community in which you live?

6. Why is a map made with the Mercator projection more suitable for navigation than a map made with the Robinson projection?

7. What is the difference between a dot map and a choropleth map?

8. What is a mental map?

9. What geographic concepts are associated with the following geographers: Alexander von Humboldt, Carl Sauer, Ellsworth Huntington, Paul Vidal de la Blache, David Lowenthal?

10. What is GIS and what typically makes it different from old-fashioned manual cartography?

DISCUSSION QUESTIONS

1. The chapter opened with a discussion of how unfamiliar many young Americans are with world geography. Is that unfamiliarity a problem? Why or why not?

2. These questions were posed: What is geography? What good is geography, really, and what will you get out of learning world regional geography? Discuss your answers using, if you like, the book's four objectives.

3. In addition to Great Britain and New Zealand, what are other examples of core and peripheral locations?

4. Examine any published map and discuss its essential elements. Are there any properties of a map that you can now recognize that you did not know about before?

5. Identify a category of spatial data you would like to represent and discuss whether it would be best to render those data with a dot, choropleth, or other type of map.

6. What are the advantages and drawbacks of the Mercator and the Robinson projections?

7. Discuss the differences between large-scale and small-scale maps.

8. What are these hemispheres: water, land, Northern, Southern, Western, Eastern?

9. Draw and compare your "mental maps" of your campus, community, or some other landscape you know in common. Who drew the best map and what made it best? Did your maps pay attention to scale and direction? What other elements were present or missing?

10. Discuss the conceptual differences between environmental determinism and environmental possibilism.

11. Discuss some of the subfields of geography and what geographers do for a living.

12. Discuss any uses or applications of Geographic Information Systems (GIS) that you may have seen.

CHAPTER

2

Physical and Human Processes That Shape World Regions

The trail to Mt. Everest, Nepal. The human imprint is almost everywhere on the landscape, even here, close to Earth's highest mountain. (Above) Earth at night. The bright lights of prosperous North America and Europe are in marked contrast with the darkness of Africa, the least developed continent.

chapter objectives

This chapter should enable you to:

- Recognize consistent global patterns in the distribution of temperatures, precipitation, and vegetation types

- Identify the natural areas most threatened by human activity and explain how natural habitat loss may endanger human welfare

- Describe the potential impacts of global climate change and international efforts to prevent them

- Gain a historical perspective on the capacity of human societies to transform environments and landscapes

- Understand why some countries are rich and others poor and recognize the geographic distribution of wealth and poverty

- Explain the simultaneous trends of falling population growth in the richer countries and rapid population growth worldwide

- Learn how "sustainable development" proposes alternatives to destructive patterns of resource use

WORLD
REGIONAL
Geography ⊕ Now™

Look for this logo in the text and go to GeographyNow at http://earthscience.brookscole.com/wrg5e to explore interactive maps, view animations, sharpen your factual knowledge and geographic literacy, and test your critical thinking and analytical skills with unique interactive resources.

Many issues in world regional geography relate to the interaction of people and the natural environment. This chapter is an introduction to geography's basic vocabulary about how the natural world works and how people have interacted with the environment to change the face of the earth. It begins with the identification of climate and vegetation that sets the global environmental stage for human activities. Then it considers how that stage may be changing as a result of human impacts on the atmosphere. Modern trends in people–land associations are examined as the products of revolutionary changes in the past: the arrival of agriculture and industrialization. The chapter explains where rich and poor countries are located on Earth's surface and accounts for some of these patterns of prosperity and poverty. It considers where and why populations are increasing and what the implications of that growth are. Finally, there is a look at ideas about how to solve some of the most important global problems of our time

2.1 Patterns of Climate and Vegetation

"Everybody talks about the weather, but nobody does anything about it," Mark Twain reportedly said. The day's weather certainly does influence our conversations and decisions. More important for us as students of geography, the cumulative impacts of weather shape landscapes and influence human activities and settlements in patterns recognizable around the world.

As you experience a warm, dry, cloudless summer day or a cold, wet, overcast winter day, you are encountering the weather—the atmospheric conditions occurring at a given time and place. Climate is the average weather of a place over a long time period. Along with surface conditions such as elevation and soil type, climatic patterns have a strong correlation with patterns of natural vegetation and in turn with human opportunities and activities on the landscape. Precipitation and temperature are the key variables in weather and climate.

Precipitation

Water is essential for life on Earth, and it is very useful to understand the processes of weather and climate that distribute this precious resource (Figure 2.1). Warm air holds more moisture than cool air, and **precipitation**—rain, snow, sleet, and hail—is best understood as the result of processes that cool the air to release moisture. Precipitation results when water vapor in the atmosphere cools to the point of condensation, changing from a gaseous to a liquid or solid form. The amount of cooling needed for this change depends on

World Precipitation

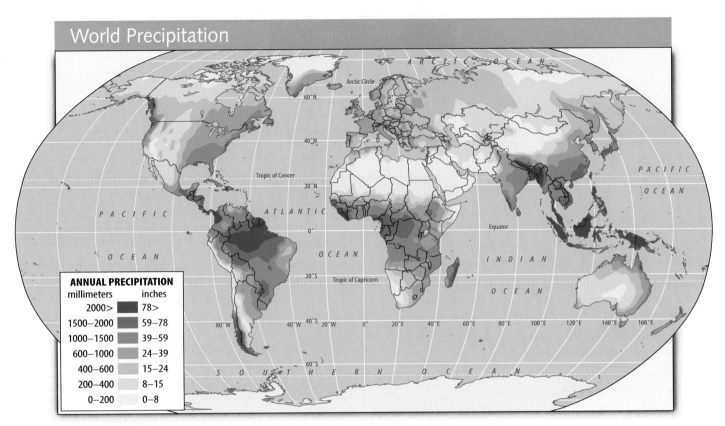

ANNUAL PRECIPITATION

millimeters	inches
2000>	78>
1500–2000	59–78
1000–1500	39–59
600–1000	24–39
400–600	15–24
200–400	8–15
0–200	0–8

Figure 2.1 World precipitation map

the original temperature and the amount of water vapor in the air.

For this cooling and precipitation to occur, generally air must rise in one of several ways. In equatorial latitudes or in the high-sun season (summer, when the sun's rays strike Earth's surface more vertically) elsewhere, air heated by intense surface radiation can rise rapidly, cool, and produce a heavy downpour of rain, an event that occurs often in summer over much of the United States. Precipitation that originates in this way and is released from tall cumuloform clouds is called convectional precipitation (Figure 2.2).

Orographic (mountain-associated) precipitation results when moving air strikes a topographic barrier and is forced upward. Most of the precipitation falls on the windward side of the barrier, and the lee (sheltered) side is likely to be very dry. Such dry areas—for example, Nevada, which is on the lee side of the Sierra Nevada in the western United States—are in a rain shadow. Rain shadows are the primary cause of arid and semiarid lands in some regions.

Cyclonic, or frontal, precipitation is generated in traveling low-pressure cells, called **cyclones,** that bring air masses with different characteristics of temperature and moisture into contact (Figures 2.2 and 2.3). In the atmosphere, air moves from areas of high pressure to areas of low pressure. Air from masses of different temperature and moisture content is drawn into a cyclone, which has low pressure. One air

mass is normally cooler, drier, and more stable than the other. These masses do not mix readily but instead come in contact within a boundary zone 3 to 50 miles wide (c. 5 to 80 km) called a front. A front is named after which air mass is advancing to overtake the other. In a cold front, the colder air wedges under the warmer air, forcing it upward and back. In a warm front, the warmer air rides up over the colder air, gradually pushing it back. Whether from a warm or cold front, precipitation is likely to result because the warmer air mass rises and condensation takes place.

Large areas of the earth receive little precipitation because of subsiding air masses having high atmospheric pressure, called high-pressure cells, or anticyclones (Figure 2.4). In a high-pressure cell, the air is descending and becoming warmer under the increased pressure (weight) of the air above it. As it warms, its capacity to hold water vapor increases, its relative humidity decreases, and its potential to make precipitation falls. Streams of dry, stable air moving outward from the anticyclones often bring prolonged drought to the areas below their path. Most famous of these air streams are the trade winds that originate in semipermanent anticyclones on the margins of the tropics and are attracted equatorward (becoming more moisture-laden as they do so) by a semipermanent low-pressure cell, the **equatorial low.** As Figure 2.4 illustrates, high-pressure cells rotate clockwise in the Northern Hemisphere and counterclock-

Types of Precipitation

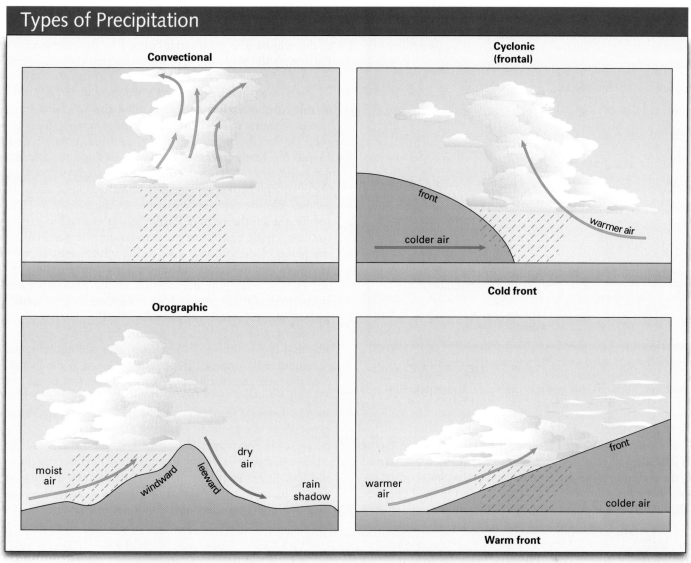

Active Figure 2.2 Diagrams showing origins of convectional, orographic, and cyclonic precipitation. Note that cyclonic precipitation results from both cold fronts and warm fronts. *See an animation based on this figure, and take a short quiz on the concept.*

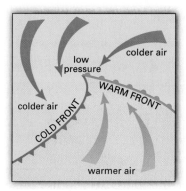

Figure 2.3 Diagram of a cyclone in the northern hemisphere

wise in the Southern Hemisphere. Conversely, low-pressure systems rotate counterclockwise in the Northern Hemisphere and clockwise in the Southern Hemisphere.

Cold ocean waters are responsible for coastal deserts in some parts of the world, such as the Atacama Desert in Chile and the Namib Desert in southwestern Africa. Here, air moving from sea to land is warmed. Instead of yielding precipitation, the air's capacity to hold water vapor is increased and its relative humidity is decreased; the result is little precipitation. Many areas of extremely low precipitation result from a combination of influences. The Sahara of northern Africa, for example, is mainly a product of high atmospheric pressure, but the rain-shadow effect of the Atlas Mountains and the presence of cold Atlantic Ocean waters along its western coast also contribute to the Sahara's dryness.

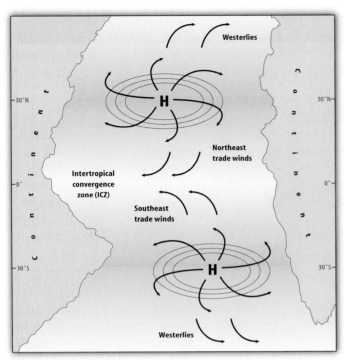

Figure 2.4 Idealized wind and pressure systems. Irregular shading indicates wetter areas. *Source: Adapted from Tom L. McKnight,* Physical Geography: A Landscape Appreciation, *5th ed. (Prentice Hall, 1996), p. 122.*

Temperature

In the middle and high latitudes, the most significant factor in determining temperatures is seasonality, which is related to the inclination of Earth's polar axis as the planet orbits the sun over a period of 365 days (Figure 2.5). On or about June 22, the first day of summer in the Northern Hemisphere, the northern tip of Earth's axis is inclined toward the sun at an angle of 23.5 degrees from a line perpendicular to the plane of the ecliptic (the great circle formed by the intersection of Earth with the plane of Earth's orbit around the sun). This is the summer solstice in the Northern Hemisphere and the winter solstice in the Southern Hemisphere. A larger portion of the Northern Hemisphere than of the Southern Hemisphere remains in daylight, and warmer temperatures prevail. On or about September 23, and again on or about March 20, Earth reaches the equinox position. Its axis does not point toward or away from the sun, so days and nights are of equal length at all latitudes on Earth. On or about December 22, the first day of winter in the Northern Hemisphere, the southern tip of Earth's axis is inclined toward the sun at an angle of 23.5 degrees from a line perpendicular to the plane of the ecliptic. This is the winter solstice in the Northern Hemisphere and the summer solstice in the Southern Hemisphere. A larger portion of the Southern Hemisphere than of the Northern Hemisphere remains in daylight, and warmer temperatures prevail.

Great regional differences exist in the world's annual and seasonal temperatures. In lowlands near the equator, temperatures remain high throughout the year, while in areas near the poles, temperatures remain low for most of the year. The intermediate (middle) latitudes have well-marked seasonal changes of temperature, with warmer temperatures generally in the summer season of high sun and cooler temperatures in the winter season of low sun (when sunlight's angle of impact is more oblique and daylight hours are shorter). Intermittent incursions of polar and tropical air masses increase the variability of temperature in these latitudes, bringing unseasonably cold or warm weather.

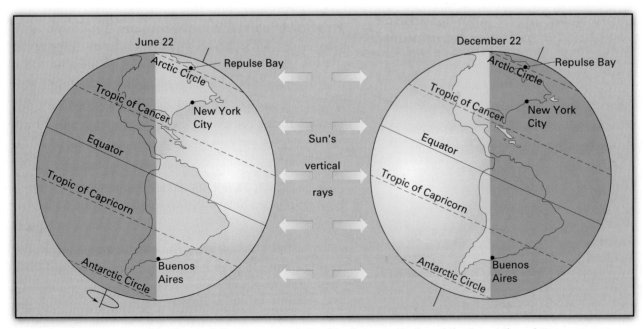

Active Figure 2.5 Geometric relationships between Earth and the Sun at the solstices. *See an animation based on this figure, and take a short quiz on the concept.*

Climate and Vegetation Types

The varied combinations of precipitation, temperature, latitude, and elevation produce a great variety of local climates. Geographers group these local climates into a limited number of major climate types, each of which occurs in more than one part of the world (Figure. 2.6) and is associated closely with other types of natural features, especially vegetation. Geographers recognize 10 to 20 major types of terrestrial ecosystems, called biomes, which are categorized by dominant types of natural vegetation (Figures 2.7 and 2.8). Climate plays the main role in determining the distribution of biomes, but differing soils and landforms may promote different types of vegetation where the climatic pattern is essentially the same. Vegetation and climate types are sufficiently related that many climate types take their names from vegetation types—for example, the tropical rain forest climate and the tundra climate. You may easily see the geographic links between climate and vegetation by comparing the maps in Figures 2.6 and 2.7. The spatial distributions of climate and vegetation types do not overlap perfectly, but there is a high degree of correlation.

In the icecap, tundra, and subarctic climates, the dominant feature is a long, severely cold winter, making agriculture difficult or impossible. The summer is short and cool. Perpetual ice in the form of glaciers may be found at very high elevations in the lower latitudes (even in equatorial regions) and in the polar realms; the icecap biome (Figure 2.8a) is devoid of vegetation, except in those very few spots where enough ice or snow melts in the summer to allow tundra vegetation to grow. Tundra vegetation (Figure 2.8b) is composed of mosses, lichens, shrubs, dwarfed trees, and some grasses. Needleleaf evergreen coniferous trees can stand long periods when the ground is frozen, depriving them of moisture. Thus, coniferous forests, often called **boreal forests** or *taiga,* their Russian name (Figure 2.8c), occupy large areas where there is subarctic climate.

In desert and **semiarid/steppe** climates, the dominant feature is aridity or semiaridity. Deserts and steppes occur in both low and middle latitudes. Agriculture in these areas usually requires irrigation. Earth's largest dry region extends in a broad band across northern Africa and southwestern and central Asia. The deserts of the middle and low latitudes are generally too dry for either trees or grasslands. They have desert shrub vegetation (Figure 2.8d), and some areas have practically no vegetation at all. The bushy desert shrubs are xerophytic (literally, "dry plant"), having small leaves, thick bark, large root systems, and other adaptations to absorb and retain moisture. Grasslands dominate in the moister steppe climate, a transitional zone between very arid deserts

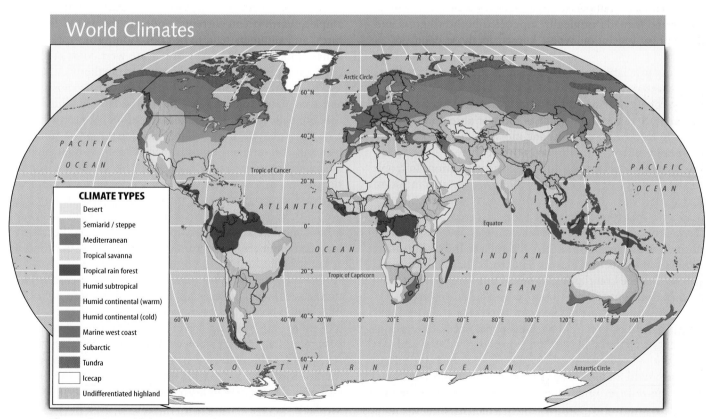

Figure 2.6 World distribution of the types of climate discussed in this text

World Biomes

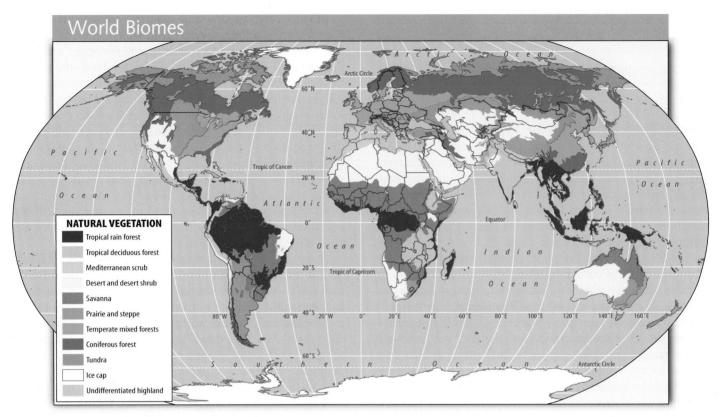

Figure 2.7 World biomes (natural vegetation) map

and humid areas. The biome composed mainly of short grasses is also called the steppe, or temperate grassland (Figure 2.8e). The temperate grassland region of the United States and Canada originally supported both tall grass and short grass vegetation types, known in those countries as prairies.

Rainy low-latitude climates include the tropical rain forest climate and the tropical savanna climate. The critical difference between them is that the tropical savanna type has a pronounced dry season, which is short or absent in the tropical rain forest climate. Heat and moisture are almost always present in the tropical rain forest biome (Figure 2.8f), where broadleaf evergreen trees dominate the vegetation. In tropical areas with a dry season but still having enough moisture for tree growth, tropical deciduous forest (Figure 2.8g) replaces the rain forest. Here, the broadleaf trees are not green throughout the year; they lose their leaves and are dormant during the dry season and then add foliage and resume their growth during the wet season. The tropical deciduous forest approaches the luxuriance of tropical rain forest in wetter areas but thins out to low, sparse scrub and thorn forest (Figure 2.8h) in drier areas. Savanna vegetation (Figure 2.8i), which has taller grasses than the steppe, occurs in areas of greater overall rainfall and more pronounced wet and dry seasons.

The humid middle-latitude regions have mild to hot summers and winters ranging from mild to cold, with several

types of climate. In the marine west coast climate, occupying the western sides of continents in the higher middle latitudes (the Pacific Northwest region of the United States, for example), warm ocean currents moderate the winter temperatures, and summers tend to be cool. Coniferous forest dominates some cool, wet areas of marine west coast climate; a good example is the redwood forests of northern California.

The Mediterranean climate (named after its most prevalent area of distribution, the lands around the Mediterranean Sea) typically has an intermediate location between a marine west coast climate and the lower latitude steppe or desert climate. In the summer high-sun period, it lies under high atmospheric pressure and is rainless. In the winter low-sun period, it lies in a westerly wind belt and receives cyclonic or orographic precipitation. Mediterranean scrub forest (Figure 2.8j), known locally by such names as *maquis* and *chaparral,* characterizes Mediterranean climate areas. Because of hot, dry summers, the natural vegetation consists primarily of xerophytic shrubs.

The humid subtropical climate occupies the eastern portion of continents between approximately 20° and 40° of latitude and is characterized by hot summers, mild to cool winters, and ample precipitation for agriculture. The humid continental climate lies poleward of the humid subtropical type; it has cold winters, warm to hot summers, and enough rainfall for agriculture, with the greater part of the precipitation in the summer. This climate type is often subdivided into warm and cold subtypes, indicating the greater severity of winter in the zones closer to the poles. In middle-latitude areas with these humid subtropical and humid continental climate types, a **temperate mixed forest** with mostly broadleaf but also coniferous trees (as in the U.S. midwest and northeast; Figure 2.8k) is found. As cold winter temperatures freeze the water within reach of plant roots, broadleaf trees shed their then-colorful leaves and cease to grow, thus reducing water loss. They then produce new foliage and grow vigorously during the hot, wet summer. Coniferous forests can thrive in some hot and moist locations where porous sandy soil allows water to escape downward, giving conifers (which can withstand drier soil conditions) an advantage over broadleaf trees. Pine forests on the coastal plains of the southern United States are an example.

Undifferentiated highland climates have a range of conditions according to elevation and exposure to wind and sun. Undifferentiated highland vegetation (Figure 2.8l) types likewise differ greatly depending on elevation, degree and direction of slope, and other factors. They are "undifferentiated" in the context of world regional geography because a small mountainous area may contain numerous biomes, and it would be impossible to map them on a small scale. The world's mountain regions have a complex array of natural conditions and opportunities for human use. Increasing elevation lowers temperatures by a predictable **lapse rate,** on average about 3.6°F (2.0°C) for each increase of 1,000 feet (305 m) in elevation. In climbing from sea level to the summit of a high mountain peak near the equator (for example, in western Ecuador), a person would experience many of the major climate and biome types found in a sea level walk from the equator to the North Pole!

Landforms—and such factors as their elevation, latitude, soil type, and the processes that change them—have enormous relevance to the distributions of plants and animals and to human uses of resources. This chapter does not introduce the world's landforms; these are dealt with mainly in the profile chapter introducing each region. For a global view of landforms, you may wish to look at the physical map of Earth inside the book's front cover. A world map of plate tectonics is a very useful reference for understanding Earth's major landform patterns and is included here (Figure 2.9). Plate tectonics refers to the process in which there are slow but steady movements of massive plates of the earth's crust and underlying mantle. Where these plates collide (in Turkey, Japan, and the northwest United States, for example), the common results are mountain building, volcanoes, the **seismic activity** of earthquakes, and deep ocean trenches. Where they separate, there is **rifting** that produces very low land elevations (well below sea level at the Dead Sea of Israel and Jordan, for example) or the emergence of new crust on ocean floors (in the middle Atlantic Ocean, for example). These processes of plate tectonics account for the past, present, and future locations of the continents.

2.2 Biodiversity

Anyone who has admired the nocturnal wonders of the "barren" desert or appreciated the variety of the "monotonous" Arctic can vouch for an astonishing diversity of life even in the biomes least hospitable to life. But geographers and ecologists recognize the exceptional importance of some biomes because of their biological diversity (or biodiversity)—the number of plant and animal species present and the variety of genetic materials these organisms contain.

The most diverse biome is the tropical rain forest. From a single tree in the Peruvian Amazon region, entomologist Terry Erwin recovered about 10,000 insect species. From another tree several yards away, he counted another 10,000, many of which differed from those of the first tree. Alwyn Gentry of the Missouri Botanical Garden recorded 300 tree species in a single-hectare (2.47-acre) plot of the Peruvian rain forest. Less than 50 years ago, scientists calculated that there were 4 to 5 million species of plants and animals on Earth. Now, however, their estimates are much higher, in the range of 40 to 80 million. This startling revision is based on research, still in its infancy, on species inhabiting the rain forest.

Such diversity is important in its own right, but it also has vital implications for nature's ongoing evolution and for people's lives on Earth. Humankind now relies on a handful of crops as staple foods. In our agricultural systems, the trend in recent decades has been to develop high-yield varieties of grains and to plant them as vast monocultures (single-crop plantings). This trend, which is the cornerstone of the so-called Green Revolution, is controversial. On the one hand, it puts more food on the global table. But on the other, it may render agriculture more vulnerable to pests and diseases and thus pose long-term risks of famine. In evolutionary terms, the Green Revolution has reduced the natural diversity of crop varieties that allows nature and farmer to turn to alternatives when adversity strikes. At the same time, while we remove tropical rain forests and other natural ecosystems to provide ourselves with timber, agriculture, or living space, we may be eliminating the foods, medicines, and raw materials of tomorrow, even before we have collected them and assigned them scientific names. "We are causing the death of birth," laments biologist Norman Myers.[1]

Regions where human activities are rapidly depleting a rich variety of plant and animal life are known as biodiversity hot spots, which scientists believe deserve immediate attention for study and conservation (Figure 2.10). These 25 priority regions as identified by Conservation International, are in South, Central, and North America: the tropical Andes, the Chocó-Darién-Western Ecuador region, central Chile, the Atlantic coastal forest of Brazil, the Brazilian Cerrado,

---

[1] In Edward O. Wilson, "Threats to Biodiversity." *Scientific American,* 261(3), 1989, pp. 60–66.

(a) Icecap, glacier in British Columbia, Canada

(b) Tundra, northern Norway

(c) Coniferous forest, British Columbia, Canada

(d) Desert shrub, southern Sinai Peninsula, Egypt

(e) Steppe, eastern Turkey

(f) Tropical rain forest, Dominica, West Indies

Figure 2.8 An album of Earth's biomes

(g) Tropical deciduous forest, Gir Forest, western India

(h) Scrub and thorn forest, northern Zimbabwe

(i) Savanna, southern Kenya

(j) Mediterranean scrub forest, southern California, U.S.

(k) Temperate mixed forest, southern Missouri, U.S.

Figure 2.8 An album of Earth's biomes (*continued*)

(l) Undifferentiated highland vegetation, San Juan Mountains, Colorado, U.S.

Tectonic Plates

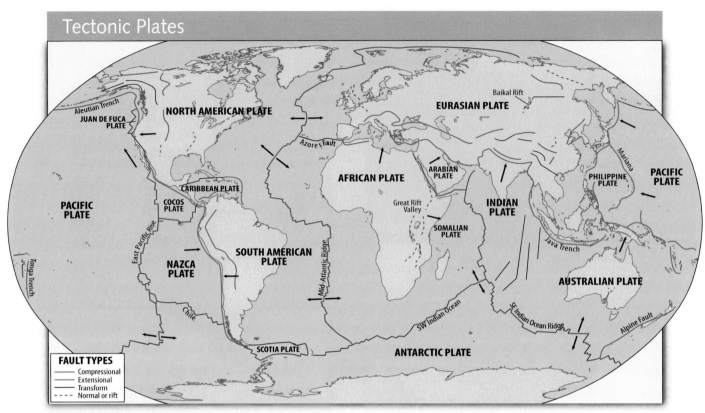

Figure 2.9 Major tectonic plates and their general direction of movement. Most tectonic activity occurs along the plate boundaries where plates separate, collide, or slide past one another.

World Biodiversity Hot Spots

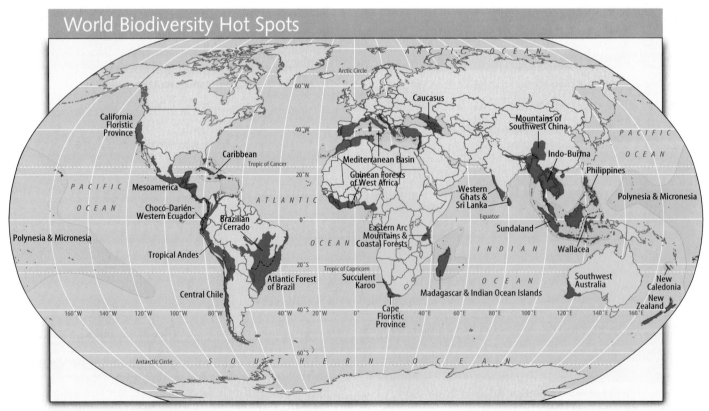

Figure 2.10 World biodiversity hot spots

the Caribbean, Mesoamerica, and the California floristic province; in Asia: Sri Lanka and the Western Ghats of India, the mountains of southwest China, Indo-Burma, the Philippines, Sundaland, and Wallacea; in the Pacific: southwest Australia, New Zealand, New Caledonia, and Polynesia/Micronesia; in Africa and adjacent regions: the Guinean forests of West Africa, the Cape floristic province of South Africa, the succulent Karoo of South Africa and Namibia, the Eastern Arc Mountains and coastal forests of Tanzania and Kenya, and Madagascar and the Indian Ocean islands; and in Europe, North Africa, and Southwest Asia: the Mediterranean Basin and the Caucasus.

By referring to the map of Earth's biomes in Figure 2.7, you will see that most of these hot spots are within tropical rain forest areas. Many are islands that tend to have high biodiversity because species on them have evolved in isolation to fulfill special roles in these ecosystems and because human pressures on island ecosystems are particularly intense. Efforts are underway in most of these hot spots to establish national parks and other protected areas.

2.3   Global Environmental Change

Efforts to preserve biodiversity, especially with the development of protected areas, are often hard-fought. Ironically, once established, these islands of wilderness in a sea of humanized landscapes might end up losing the species they were established to defend, not as a direct result of human activities within them but because of what is happening overhead. The global climate is, or may be, changing, and if it is, the spatial distributions of plants and animals will be changing, too.

There is a debate between those who insist that human activities are responsible for a documented warming of Earth's surface and those who either insist the warming is not occurring or, if it is, that a natural climatic cycle is responsible. However, an emerging consensus, supported by improvements in computer modeling, is that Earth's atmosphere is warming because of human production of "greenhouse gases" such as carbon dioxide (Figure 2.11). In 2000, an international team of atmospheric scientists known as the **Intergovernmental Panel on Climate Change (IPCC)** abandoned its previously neutral stance on the question of whether or not people were to blame for global warming. "There is now stronger evidence for a human influence on the climate," the IPCC concluded. "There is increasing evidence from many sources that the signal of human influence on the climate has emerged from natural variability, sometime around 1980 . . . Man-made greenhouse gases have contributed substantially to the observed warming over the last 50 years."[2]

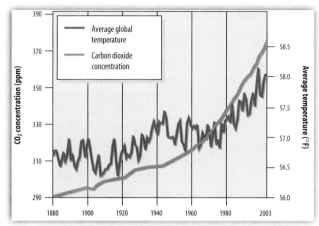

Figure 2.11 Industrialization and the burning of tropical forests have produced a steady increase in carbon dioxide emissions. Many scientists believe these increased emissions explain the corresponding steady increase in the global mean temperature.

In 1827, French mathematician Baron Jean Baptiste Fourier established the concept of the greenhouse effect, noting that Earth's atmosphere acts like the transparent glass cover of a greenhouse (Figure 2.12) (for modern purposes, think of a car's windows). Visible sunlight passes through the glass to strike Earth's surface. Ocean and land (the floor of the greenhouse or the car upholstery) reradiate the incoming solar energy as invisible infrared radiation (heat). Acting like the greenhouse glass or car window, Earth's atmosphere traps some of that heat.

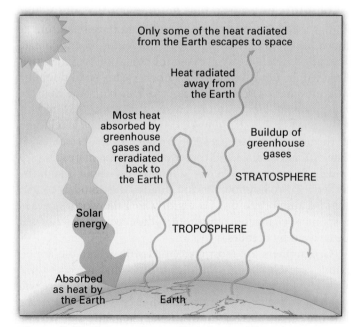

Figure 2.12 The greenhouse effect. Some of the solar energy radiated as heat (infrared radiation) from Earth's surface escapes into space, while greenhouse gases trap the rest. Naturally occurring greenhouse gases make Earth habitable, but carbon dioxide and greenhouse gases emitted by human activities accentuate the greenhouse effect, making Earth unnaturally warmer.

---

[2] Quoted in "Panel: Earth to Become Hotter." *Columbia Daily Tribune*, October 26, 2000, p. 4A, and in Andrew C. Revkin, "A Shift in Stance on Global Warming Theory." *The New York Times*, October 26, 2000, p. A18.

In the atmosphere, naturally occurring **greenhouse gases** such as carbon dioxide (CO_2) and water vapor make Earth habitable by trapping heat from sunlight. Concern over global warming focuses on human-made sources of greenhouse gases, which trap abnormal amounts of heat. **Carbon dioxide** released into the atmosphere from the burning of coal, oil, and natural gas is the greatest source of concern, but **methane** (from rice paddies and the guts of ruminating animals like cattle), **nitrous oxide** (from the breakdown of nitrogen fertilizers), and **chlorofluorocarbons,** or **CFCs** (from cooling and blowing agents), are also human-made greenhouse gases (CFCs also destroy stratospheric ozone, a gas that has the important effect of preventing much of the sun's harmful ultraviolet radiation from reaching Earth's surface). In the car-as-Earth metaphor, continued production of these greenhouse gases has the effect of rolling up the car windows on a sunny day, with the result of increased temperatures and physical problems for the occupants.

Although formal records of meteorological observations began just over a century ago, past climates left evidence in the form of marine fossils, corals, glacial ice, fossilized pollen, and annual growth rings in trees. These indirect sources, along with formal records, indicate that the 20th century was the warmest century in the last 600 years, and the warmest years in that 600-year period were 1990, 1995, 1997, 1998, 1999, 2002 and 2003. Direct sources indicate that the 1990s was the hottest decade, and 1998 the hottest year, since instrument recording began in 1861.

In 1896, Swedish scientist Svante Arrhenius feared that Europe's growing industrial pollution would eventually double the amount of carbon dioxide in Earth's atmosphere and as a consequence raise the global mean temperature as much as 9° Fahrenheit (5°C). Computer-based climate change modelers today also use the scenario (assuming that little will be done to reduce greenhouse gases) of a doubling of atmospheric carbon dioxide, and they conclude that the mean global temperatures might warm from 3° to 11° Fahrenheit (1.7° to 6.1°C) by 2100; this would be in addition to the increase of just over 1° Fahrenheit (0.5°C) since 1975. The estimated range and consequences of the increases vary widely because of different assumptions about the little-understood roles of oceans and clouds. Most models concur that because of the slowness with which the world's oceans respond to changes in temperature, the effects of this rise will be delayed by decades.

Geographers are most concerned with where the anticipated climatic changes will occur and what impacts these changes will have. Computer models conclude that there will not be a uniform temperature increase across the entire globe but rather that the increases will vary spatially and seasonally. Differing models produce contradictory results about the timing and impacts of warming. Two teams of scientists publishing their results in the journal *Nature,* for example, first assumed a doubling of CO_2. Both predicted rises in temperature, but because they differed on what season of year the increases would be greatest, they reached different con-

clusions about the effects on agriculture in the United States. One predicted warmer summers and forecast that crop productivity would decline by 20 percent. The other predicted warmer winters and forecast a 10 percent increase in crop yields.

Geographically, the impacts of global warming appear to be greatest at the higher latitudes. The volume of the north polar ice declined by more than 40 percent between 1958 and 2004, and the ice shelves surrounding Antarctica have also retreated substantially. Sailing on a Russian nuclear-powered icebreaker, the author saw open water at the North Pole in August 1996 (Figure 2.13), and similar observations from the same vessel in subsequent summers produced worldwide headlines about polar ice melting. Some quarters are cheering the trend. A further retreat of polar ice would bode well for maritime shipping through the long-icebound "Northwest Passage" across Arctic Canada and through the northern sea route across the top of Russia, which has been navigable only in summer and only with the aid of icebreakers.

There is general agreement that with global warming the distribution of climatic conditions typical of biomes will shift poleward worldwide and upward in mountainous regions. Many animal species will be able to migrate to keep pace with changing temperatures, but plants, being stationary, will not. Conservationists who have struggled to maintain islands of habitat as protected areas are particularly concerned about rapidly changing climatic conditions. For example, ecologists have documented that two-thirds of Europe's butterfly species have already shifted their habitat ranges northward by 22 to 150 miles (35 to 240 km), coinciding with the continent's warming trend. The World Wildlife Fund recently issued a report warning that, due to global warming,

F7.2
176

Joe Hobbs

Figure 2.13 The passengers and crew of the Russian icebreaker *Yamal* found open water at the North Pole on August 16, 1996. Much of the voyage from northeast Siberia was ice-free. Similar observations since then have contributed to widespread concern about global warming. The ship's GPS instrument indicated that the vessel's bow was on 90°N latitude when this photo was taken.

Geography of Energy

The Kyoto Protocol

In 1997, 84 countries (of 160 countries represented at the conference) signed the Kyoto Protocol, a landmark international treaty on climate change. The agreement requires 38 more developed countries (known as the Annex I countries) to reduce carbon dioxide emissions to at least 5 percent below their 1990 levels by the year 2012. The United States pledged to cut its emissions to 7 percent below 1990 levels by that date. This would be a huge cut; in 2003, the United States produced 16 percent more carbon dioxide than it did in 1990. The European Union made a promise of 8 percent below 1990 levels, and Japan promised 6 percent.

Meeting these pledges would require substantial legislative, economic, and behavioral changes in these more affluent countries. Transportation and other technologies that use fossil fuels would have to become more energy efficient, making these technologies at least temporarily more expensive. Gasoline prices would rise, so consumers would feel the pinch. Advocates of the protocol argue that the initial sacrifices would soon be rewarded by a more efficient and competitive economy powered by cleaner and cheaper sources of energy derived from the sun. Opponents, however, feel that higher fuel prices will be too costly for the United States and other industrialized economies to bear. Their position made it difficult for some signatories of the Kyoto Pro-

tocol to ratify the treaty. For the protocols to take effect, at least 55 countries must ratify them, and the industrialized Annex I countries that ratify must have collectively produced at least 55 percent of the world's total greenhouse gas emissions in 1990. By mid 2004, 120 countries—including all 25 European Union members and Japan—had ratified, and the critical threshold of 55 percent of the 1990 emissions by Annex I countries was near. Ratification by either the United States or Russia would cross that threshold, putting the treaty into effect.

Much to the dismay of the European Union countries, U.S. President George W. Bush rejected the Kyoto Protocol soon after taking office in 2001. The Bush administration had two objections: the potentially high economic cost of implementing the treaty and the fact that China, along with all the world's less developed countries, is not required by the Kyoto Protocol to take any steps to reduce greenhouse gas emissions. The European Union countries and Japan dismissed these concerns and expressed outrage that the United States broke ranks with them on global warming. All eyes turned to Russia, whose ratification of the Kyoto Protocol would fulfill the 55 percent of emissions requirement and therefore finally put the treaty into effect in all ratifying countries. Late in 2004, Russia ratified the treaty. Now the world waits to see if Kyoto's promises can be fulfilled.

---

as much as 70 percent of the natural habitat could be lost, and 20 percent of the species could become extinct, in the arctic and in subarctic regions of Canada, Scandinavia, and Russia. The report predicted that more than a third of existing habitats in the American states of Maine, New Hampshire, Oregon, Colorado, Wyoming, Idaho, Utah, Arizona, Kansas, Oklahoma, and Texas would be irrevocably altered by global warming.

There is also general agreement that if global temperatures rise, so will sea levels as glacial ice melts and as seawater warms and thus occupies more volume. Sea levels have already risen 8 inches (20 cm) in the past century, possibly due to global warming, and are continuing to rise at a rate of 0.1 inch (2 mm) each year. Decision makers in coastal cities throughout the world, in island countries, and in nations with important lowland areas adjacent to the sea are worried about the implications of rising sea levels. The coastal lowland countries of the Netherlands and Bangladesh have for many years been outspoken advocates for reductions of greenhouse gases. The president of the Maldives, an Indian Ocean country comprised entirely of islands barely above sea level, pleaded, "We are an endangered nation!"

There is a growing international political will to take the strong measures that may be necessary to prevent or reverse

global warming. Present efforts to reduce the production of greenhouse gases focus on the world's wealthiest nations, where about 85 percent of the emissions originate. The United States alone now accounts for 24 percent of all emissions, China for 13 percent, and in the less developed countries (excluding China) together for 21 percent. However, China's share is likely to grow substantially as this coal-rich country continues to industrialize, with generally less stringent emissions standards; China is projected to overtake the United States as the world's largest carbon dioxide producer by 2018.

There are precedents proving that countries can unite in effective action against global environmental change. Thanks to the Montreal Protocol and its amendments signed by 37 countries in the late 1980s, the production of CFCs worldwide has all but ceased; the wealthier countries no longer produce them, and the less developed countries are scheduled to cease production by 2010. The anticipated result is that there will be a marked reduction in the size of stratospheric ozone "holes" that have been observed seasonally over the southern and northern polar regions since about 1985. However, CFC molecules have very long life spans (up to 110 years), and many years may pass before the lasting impact of CFC elimination will be felt. The current forecast is that the effects of the Montreal Protocol will be noticeable by 2010, with the ozone layer recovering to pre-1980 levels by 2050.

Compared with richer nations, the less developed countries, which tend to use fuelwood to complement scarce fossil fuel supplies, release relatively little CO_2 into the atmosphere, but they do contribute to the greenhouse effect. When burned, trees not only release the CO_2 they stored in the process of photosynthesis, but trees are also eliminated as organisms that in the process of photosynthesis remove CO_2 from the atmosphere. However, because these countries produce relatively little CO_2, policymakers are considering innovative ways to reward them and encourage them to develop clean industrial technologies. One system being considered is that of tradable permits, in which each country would be assigned the right to emit a certain quantity of CO_2, according to its population size, setting the total at an acceptable global standard. The United States, already an overproducer by this scale, could purchase emission rights from a populous country like India, which currently "underproduces" CO_2. India would be obliged to use the income to invest in energy-efficient and nonpolluting technologies.

Most notable about this system of emissions trading is that it views global warming as a global problem. However, international talks on the Kyoto Protocol have often broken down over this concept. At those negotiations, when the United States argued that it should be able to buy emissions rights from poorer countries, the European Union countries responded that in doing so the United States would be avoiding the important work of reducing carbon dioxide emissions in its own country. The Europeans were also angry over U.S. insistence that the U.S. should not have to reduce carbon dioxide emissions as much as demanded in the Kyoto Protocol, because the United States has huge areas of forest and farmland that absorb this gas and act as **carbon dioxide "sinks."** Here again, the United States was viewed as trying to find the easy way out. Early on, Russia balked at ratification in large part because it lost the prospect of earning big profits by selling its emission credits to the United States. In the end, Russia agreed to ratify because the Europeans finally decided to allow emissions trading.

2.4 Two Revolutions That Have Changed the Earth

Climate change is a very recent expression of human impact on the environment. The geographer's approach to understanding a current landscape—in almost all cases, a cultural landscape that has been fashioned by human activity—is sometimes deeply historical, involving study of its development from the prehuman or early human **natural landscape.** With a perspective of great historic depth, the current spatial patterns of our relationship with Earth may be seen as products of two "revolutions": the Agricultural Revolution— also known as the Neolithic (New Stone Age) Revolution—

that began in the Middle East about 10,000 years ago and the Industrial Revolution that began in 18th-century Europe. Each of these revolutions transformed humanity's relationship with the natural environment. Each increased substantially our capacity to consume resources, modify landscapes, grow in number, and spread in distribution.

Hunting and Gathering

Until about 10,000 years ago, our ancestors practiced a hunting and gathering livelihood (also known as **foraging**). Joined in small bands consisting of extended family members, they were nomads practicing extensive land use in which they covered large areas to locate foods such as seeds, tubers, foliage, fish, and game animals (Figure 2.14). Moving from place to place in small numbers, they had a relatively limited impact on natural environments, at least compared with agricultural and industrial societies. Many scholars have praised these preagricultural people for the apparent harmony they maintained with the natural world in both their economies and their spiritual systems. Hunters and gatherers have even been described as the **"original affluent society"** because, after short periods of work to collect the foods they needed, they enjoyed long stretches of leisure time. Studies of those few hunter-gatherer cultures that lingered into modern times, such as the San (Bushmen) of southern Africa and several Amerindian groups of South America, suggest that although their life expectancy was low, they suffered little from the mental illnesses and broken family structures that characterize industrial societies.

Hunters and gatherers were not always at peace with one another or with the natural world, however. With upright posture, stereoscopic vision, opposable thumbs, an especially large brain, and no mating season, *Homo sapiens* became after its emergence in southern Africa about 125,000

Figure 2.14 Until the relatively recent past—just a few thousand years ago—people were exclusively hunters and gatherers. This rock art in Egypt's Sinai Peninsula was created over a long time span, as it depicts Neolithic period hunting of ibex, later uses of camels and horses, and writing from the Nabatean period (first century A.D.).

Joe Hobbs

years ago an ecologically dominant species—one that competes more successfully than other organisms for nutrition and other essentials of life or that exerts a greater influence than other species on the environment. Using fire to flush out or create new pastures for the game animals they hunted, preagricultural people shaped the face of the land on a vast scale relative to their small numbers. Many of the world's prairies, savannas, and steppes where grasses now prevail developed as hunters and gatherers repeatedly set fires. These people also overhunted and in some cases eliminated animal species. The controversial Pleistocene overkill hypothesis states that rather than being at harmony with nature, hunters and gatherers of the Pleistocene Era (2 million to 10,000 years ago) hunted many species to extinction, including the elephantlike mastodon of North America.

The Agricultural Revolution

Despite these excesses, the environmental changes that hunters and gatherers could cause were limited. Humankind's power to modify landscapes took a giant step with domestication, the controlled breeding and cultivation of plants and animals. Domestication brought about the Agricultural Revolution, also known as the Neolithic Revolution and the Food-Producing Revolution. Why people began to produce rather than continue to hunt and gather plant and animal foods—first in the Middle East and later in Asia, Europe, Africa, and the Americas—is uncertain. Two theories prevail. Climatic change in the form of increasing drought and reduced plant cover may have forced people and wild plants and animals into smaller areas, where people began to tame wild herbivores and sow wild seeds to produce a more dependable food supply. A more widely accepted theory is that their own growing populations in areas originally rich in wild foods compelled people to find new food sources, so they began sowing cereal grains and breeding animals. The latter process may have begun about 8000 B.C. in the Zagros Mountains of what is now Iran. The culture of domestication spread outward from there but also developed independently in several world regions.

In choosing to breed plants and animals, people slowly abandoned the nomadic, extensive land use of hunting and gathering for the sedentary, intensive land use of agriculture and animal husbandry. The new system enabled them to create large and reliable surpluses of food. Through dry farming, or planting and harvesting according to the seasonal rainfall cycle, population densities could be 10 to 20 times higher than they were in the hunting and gathering mode. By about 4000 B.C., people along the Tigris, Euphrates, and Nile Rivers began crop irrigation—bringing water to the land artificially—an innovation that allowed them to grow crops year-round, independently of seasonal rainfall or river flooding (Figure 2.15). Irrigation technology allowed even more people to make a living off the land; irrigated farming yields about five to six times more food per unit area than dry farming. In ecological terms, the expanding food surpluses of the Agricultural Revolution raised Earth's carrying

Figure 2.15 The Tigris River in southeastern Turkey. The brown areas are rainless in the long, hot summers and are capable of producing only one crop a year through dry (unirrigated) farming. The green areas along the river are irrigated and can produce two or three crops a year. Irrigation was thus a revolutionary technology that greatly increased the number of people the land could support.

capacity—that is, the size of a species' population (in this case, humans) that an ecosystem can support.

This steep increase in food surpluses freed more people from the actual work of producing food, and they undertook a wide range of activities unrelated to subsistence needs. Irrigation and the dependable food supplies it provided thus set the stage for the development of civilization, the complex culture of urban life characterized by the appearance of writing, economic specialization, social stratification, and high population concentrations. By 3500 B.C., for example, 50,000 people lived in the southern Mesopotamian city of Uruk, in what is now Iraq. Other culture hearths—regions where civilization followed the domestication of plants and animals—emerged between 8000 and 2500 B.C. in China, Southeast Asia, the Indus River Valley, Egypt, West Africa, Mesoamerica, and the Andes.

The agriculture-based urban way of life that spread from these culture hearths had larger and more lasting impacts on the natural environment than either hunting and gathering or early agriculture. Acting as agents of humankind, domesticated plants and animals proliferated at the expense of the wild species that people came to regard as pests and competitors. Agriculture's permanent and site-specific nature magnified the human imprint on the land, while the pace and distribution of that impact increased with growing numbers of people.

The Industrial Revolution

The human capacity to transform natural landscapes took another giant leap with the Industrial Revolution, which began in Europe around A.D. 1700. This new pattern of human–land relations was based on breakthroughs in tech-

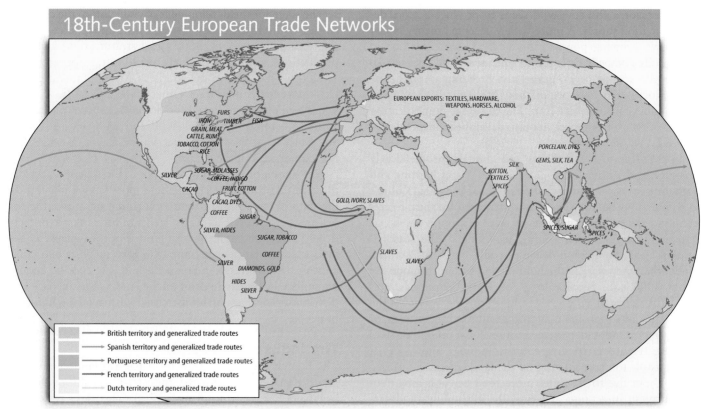

Figure 2.16 A Tweed Mill in Stornoway on Scotland's Lewis Island. The process of industrialization that began in 18th-century Europe has rapidly transformed the face of the earth.

nology that several factors made possible (Figure 2.16). First, western Europe had the economic capital necessary for experimentation, innovation, and risk. Much of this money was derived from the lucrative trade in gold and slaves undertaken initially in the Spanish and Portuguese empires after 1400. Second, in Europe prior to 1500, there had been significant improvements in agricultural productivity, particularly with new tools such as the heavy plow and with more intensive and sustainable use of farmland. Crop yields increased, and human populations grew correspondingly. A

third factor was population growth itself. More people freed from work in the fields represented a greater pool of talent and labor in which experimentation and innovation could flourish. As agricultural innovations and industrial productivity improved, an increasingly large proportion of the growing European population was freed from farming, and for the first time in history, a region had more city dwellers than rural folk. The process of industrialization continues to promote urbanization today.

Most geographers see population growth today as a drain on resources, but the Industrial Revolution illustrates that, given the right conditions, more people do create more resources. Innovations such as the steam engine tapped the vast energy of fossil fuels—initially, coal and, later, oil and natural gas. This energy, the photosynthetic product of ancient ecosystems, allowed Earth's carrying capacity for humankind to be raised again—this time into the billions.

As they began to deplete their local supplies of resources needed for industrial production, Europeans started to look for these materials abroad. As early as their Age of Discovery, which began in the 15th century, Europeans probed ecosystems across the globe to feed a growing appetite for innovation, economic growth, and political power. The process of European colonization was thus linked directly to the Industrial Revolution. Mines and plantations from such faraway places as central Africa and India supplied the copper and cotton that fueled economic growth in Belgium and England (Figure 2.17).

Figure 2.17 In the 18th century, European merchant fleets carried goods, slaves, and information all around the world, profoundly transforming cultures and natural environments.

No longer dependent on the foods and raw materials they could procure within their own political and ecosystem boundaries, European vanguards of the Industrial Revolution had an impact on the natural environment that was far more extensive and permanent than that of any other people in history. There are many measures of the unprecedented changes that the Industrial Revolution and its wake have wrought on Earth's landscapes. Between 1700 and the present, the total forested area on Earth declined by more than 20 percent. During the same period, total cropland grew nearly 500 percent, with more expansion in the period from 1950 to today than in the 150 years from 1700 to 1850. Human use of energy increased more than a hundredfold from 1700 to now. Today, fully 40 percent of Earth's land-based photosynthetic output is dedicated to human uses, especially in agriculture and forestry. Of particular interest to geographers is how the costs and benefits of such expansion are distributed in unequal patterns across the earth.

2.5 The Geography of Development

One of the most striking characteristics of human life on Earth is the large disparity between wealthy and poor people, both within and between countries. At a high level of generalization, the world's countries can be divided into "haves" and "have-nots" (Figure 2.18 and Table 2.1). Writers refer to these distinctions variously as "developed" and "underdeveloped," "developed" and "developing," "more developed" and "less developed," "industrialized" and "nonindustrialized," and "north" and "south" based on the concentration of wealthier countries in the middle latitudes of the Northern Hemisphere and the abundance of poorer nations in the Southern Hemisphere. This text uses the terms more developed countries (MDCs) and less developed countries (LDCs). It must be emphasized that this framework is an introductory tool and cannot account for the tremendous variations and ongoing changes in economic and social welfare that characterize the world today. Some countries, including those known as the "Asian Tigers," are best described as newly industrializing countries (NICs) because they do not fit either the MDC or LDC idealized type. The relevant regional chapters describe these cases.

290

Measures of Development

MDCs may be distinguished from LDCs by comparing countries' annual **per capita gross domestic product (GDP)**—the total output of goods and services a country produces for home use in a year divided by the country's population. Even more useful for comparative purposes is that figure adjusted for **purchasing power parity (PPP)**, which applies standardized international dollar price weights to the quantities of final goods and services produced in a given economy. The resulting measure, per capita GDP PPP, which is the one used in this text, provides the best available starting point for comparisons of economic strength and well-being among countries. Definitions vary, but generally the MDC is a coun-

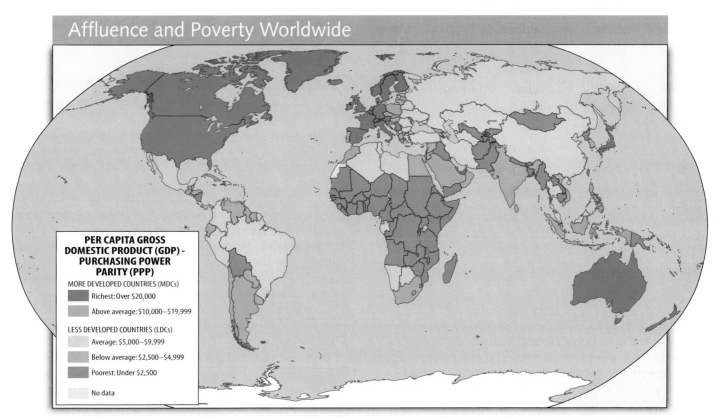

Affluence and Poverty Worldwide

PER CAPITA GROSS DOMESTIC PRODUCT (GDP) - PURCHASING POWER PARITY (PPP)

MORE DEVELOPED COUNTRIES (MDCs)

Richest: Over $20,000

Above average: $10,000–$19,999

LESS DEVELOPED COUNTRIES (LDCs)

Average: $5,000–$9,999

Below average: $2,500–$4,999

Poorest: Under $2,500

No data

Figure 2.18 Wealth and poverty by country. Note the concentration of wealth in the middle latitudes of the Northern Hemisphere.

Table 2.1   Characteristics of More Developed Countries (MDCs) and Less Developed Countries (LDCs)

Characteristic	MDC	LDC
Per capita GNP and income	High	Low
Percent in middle class	High	Low
Percent in manufacturing	High	Low
Energy use	High	Low
Percent urban	High	Low
Percent rural	Low	High
Birth rate	Low	High
Death rate	Low	Low[a]
Population growth rate	Low	High
Percent under age 15	Low	High
Percent literate	High	Low
Leisure time available	High	Low
Life expectancy	High	Low

[a] Although death rates are high in the LDCs relative to the MDCs, they are quite low compared to what they were previously in the LDCs.

TABLE 2.2   Top and Bottom Countries in Per Capita GDP PPP (U.S. Dollars, 2003)[a]

Top Countries	GDP PPP, per capita	Bottom Countries	GDP PPP, per capita
Luxembourg	$55,100	East Timor	$500
Norway	$37,800	Somalia	$500
United States	$37,800	Sierra Leone	$500
Switzerland	$32,700	Malawi	$600
Denmark	$31,100	Tanzania	$600
Iceland	$30,900	Burundi	$600
Austria	$30,000	Congo Republic	$700
Canada	$29,800	Congo, Democratic Republic	$700
Ireland	$29,600	Comoros	$700
Belgium	$29,100	Ethiopia	$700
		Eritrea	$700
		Afghanistan	$700

[a] This table excludes city-states, territories, colonies, and dependencies
Source: *CIA World Factbook* 2004

poorest with only $500. By another measure, more than 1.5 billion people, or more than 25 percent of the world's population, were "abjectly poor," living on less than $1 per day. These raw numbers suggest that economic productivity and income alone characterize development, which, according to a common definition, is a process of improvement in the material conditions of people through diffusion of knowledge and technology.

Such economic definitions reveal little about measures of well-being such as income distribution, gender equality, literacy, and life expectancy. Recognizing the shortcomings of strictly economic definitions, the United Nations Development Programme created the Human Development Index (HDI), a scale that considers attributes of quality of life (and which this textbook uses in the Basic Data Tables of each regional profile chapter). According to this index, Norway is "the world's best place to live," although it is third in per capita GDP PPP ranking. Following Norway, in descending order, are Iceland, Sweden, Australia, Netherlands, Belgium, United States, Canada, and Japan. In HDI terms, Sierra Leone is the world's worst place to live.

WORLD
REGIONAL
Geography ⊛ Now™

Click Geography Literacy for an activity to make a map of the Human Development Index.

On the basis of per capita GDP PPP, 1.2 billion, or 19 percent, of the world's 6.4 billion people inhabit the MDCs. Most citizens of these countries, such as the United States, Canada, Japan, Australia, New Zealand, and the nations of Western Europe, enjoy an affluent lifestyle with freedom from hunger. Employed in industries or services, most of the people live in cities rather than in rural areas. Disposable income, or money that people can spend on goods beyond their subsistence needs, is generally high. There is a large middle class. Population growth is low as a result of low birth rates and low death rates (these demographic terms are explained in the section on population that follows). Life expectancy is long, and the literacy rate is high.

Life for the planet's other 81 percent, or about 5.2 billion people, is very different. In the LDCs, including most countries in Latin America, Africa, and Asia, poverty and often hunger prevail. The leading occupation is subsistence agriculture, and the industrial base is small. The middle class tends to be small but growing, with an enormous gulf between the vast majority of poor and a very small wealthy elite, which owns most of the private landholdings. With high birth rates and falling death rates, population growth is high. Life expectancy is short, and the literacy rate is low.

With four-fifths of the world's people living in the poorer, less developed countries, it is important to understand the root causes of underdevelopment and to appreciate how wealth and poverty affect the global environment in very different but equally profound ways.

try with a per capita GDP PPP of over $10,000, while the LDC is below that figure. The gulf between the world's richest and poorest countries is startling (Table 2.2). The average per capita GDP PPP in the MDCs is nearly *six times* greater than in the LDCs. In 2003, with a per capita GDP PPP of $55,100, Luxembourg was the world's richest country, whereas East Timor, Sierra Leone, and Somalia were the

Causes of Disparities

Many theories attempt to explain the disparities between MDCs and LDCs. The most widely debated, and also the most embraced in the LDCs, is dependency theory, which argues that the worldwide economic pattern established by the Industrial Revolution and the attendant process of colonialism persists today. In his book *Ecological Imperialism*, historian and geographer Alfred Crosby explains how dependency led to the rich–poor divide by depicting the two very different ways in which European powers used foreign lands during the Industrial Revolution. In the pattern of settler colonization, Europeans sought to create new Europes, or **neo-Europes,** in lands much like their own: temperate midlatitude zones with moderate rainfall and rich soils where they could raise wheat and cattle. Thus, between 1630 and 1930, more than 50 million Europeans emigrated from their homelands to create European-style settlements in what are now Canada, the United States, Argentina, Uruguay, Brazil, South Africa, Australia, and New Zealand. These lands were destined to become some of the world's wealthier regions and countries.

In contrast to their preference to settle familiar midlatitude environments, Europeans viewed the world's tropical lands mainly as sources of raw materials and markets for their manufactured goods. The environment was too different from home to make settlement attractive. In establishing a pattern of mercantile colonialism, Crosby explains, Europeans were less inhabitants than conquering occupiers of the colonies, overseeing indigenous peoples and resettled slaves in the production of primary or unfinished products: sugar in the Caribbean; rubber in Latin America, West Africa, and Southeast Asia; and gold and copper in southern Africa, for example. Colonialism required huge migrations of people to extract the earth's resources, including 30 million slaves and contracted workers from Africa, India, and China to work mines and plantations around the globe.

In the mercantile system, the colony provided raw materials to the ruling country in return for finished goods; thus, people in India would purchase clothing made in England from the raw cotton they themselves had harvested. The relationship was most advantageous to the colonizer. England, for example, would not allow its colony India to purchase finished goods from any country but England. It prohibited India from producing any raw materials the empire already had in abundance, such as salt (India's Mohandas Gandhi defiantly violated this prohibition in his famous "March to the Sea"). Finished products are **value-added products,** meaning they are worth much more than the raw materials they are made from, so manufacturing in the ruling country concentrated wealth there while limiting industrial and economic development in the colony.

The colony was obliged to contribute to, but was prohibited from competing with, the economy of the ruling country—a relationship that dependency theorists insist continues today. Dependency theory asserts that to participate in the world economy, the former colonies but now-independent countries continue to depend on exports of raw materials to, and purchases of finished goods from, their former colonizers and other MDCs, and this disadvantageous position keeps them poor. Dependency theorists call this relationship neocolonialism. With independence, the former colonies needed revenue. To earn that money, they continued to produce the goods for which markets already existed—generally the same unprocessed primary products they supplied in colonial times. Dependency theorists argue that when the former colonies try to break their dependency by becoming exporters of manufactured goods, the MDCs impose **trade barriers** and quotas to block that development (see Definitions and Insights, page 36).

Many geographers, however, view dependency theory as too simplistic and politicized. Rather, they consider a wider and more complex set of factors, including culture, location, and natural environment, to explain why some countries are wealthy and others are poor. For example, because it is situated close to a great mainland with which to trade, the island of Great Britain enjoys a core location favorable for economic development. Japan has a similar location relative to the Asian landmass. In contrast, landlocked nations such as Bolivia in South America and numerous nations in Africa have locations unfavorable for trade and economic development, and they have not overcome this disadvantage. But it is important to recognize that geographic location is never the sole decisive factor in development. Like Japan and Great Britain, Madagascar and Sri Lanka are island nations situated close to large mainlands, but neither has experienced prosperity.

A superabundance of one particular resource (for example, oil in the Persian/Arabian Gulf countries) or a diversity of natural resources has helped some countries to become more developed than others. The former Soviet Union and the United States achieved superpower status in the 20th century in large part by using the enormous natural resources of both countries. In other cases, human industriousness has helped to compensate for resource limitations and to promote development. For example, Japan has a rather small territory with few natural resources (including almost no petroleum). Yet, in the second half of the 20th century, it became an industrial powerhouse largely because the Japanese people united in common purpose to rebuild from wartime devastation, placing priorities on education, technical training, and seaborne trade from their advantageous island location. Conversely, cultural or political problems like corruption and ethnic factionalism can hinder development in a resource-rich nation, as in the mineral-wealthy Democratic Republic of Congo.

Whether because of neocolonialism or a more complex array of variables, many developing countries continue to rely heavily on income from the export of a handful of raw materials. This makes them vulnerable to the whims of nature and the world economy. The economy of a country heavily dependent on rubber exports, for example, may suffer if an insect pest wipes out the crop or if a foreign

Definitions + Insights

Globalization: The Process and the Backlash

One of the most remarkable international trends of recent decades has been globalization—the spread of free trade, free markets, investments, and ideas across borders and the political and cultural adjustments that accompany this diffusion. There has been much debate and sometimes violent conflict over the pros and cons of globalization.

Advocates of globalization argue that the newly emerging global economy will bring increased prosperity to the entire world. Innovations in one country will be transferred instantly to another country, productivity will increase, and standards of living will improve. They propose that one obvious solution to the problem of the LDCs' inability to compete in the world economy, and therefore escape their dependency, is to reduce the trade barriers that MDCs have erected against them (some of these barriers are discussed on page 622 in Chapter 22). With free trade, free enterprise will prosper, pumping additional capital into national economies and raising incomes for all. Much of the support for globalization comes from multinational companies (also called transnational companies, meaning companies with operations outside their home countries), as they increase their investments abroad. Most of the companies are based in the MDCs, but multinational corporations also have grown in LDCs such as Mexico.

Opponents of globalization argue that the process will actually increase the gap between rich and poor countries; a selected few developing nations will prosper from increased foreign investment and resulting industrialization, but the hoped-for "global" wealth will bypass other countries altogether. And the increasing interdependence of the world economies will make all the players more vulnerable to economic and political instability. The multinational companies will recognize huge profits at the expense of poor wage laborers. In addition, environments will be harmed if environmental regulations are reduced to a lowest common denominator (for example, the high standards of air quality demanded by the U.S. Clean Air Act would be deemed noncompliant with World Trade Organization rules because they make it harder for countries with "dirtier" technologies to compete in the marketplace). Such concerns have already led to massive protests against "corporate-led globalization," especially at meetings of the World Trade Organization, the International Monetary Fund, and the World Bank. Many more confrontations like these can be anticipated. Typical protestors' demands are that working conditions be improved in the foreign "sweatshops," where textiles and other goods are produced at low cost for U.S. corporations, and that Starbucks Coffee should sell only "fair trade" coffee beans bought at a price giving peasant coffee growers a living wage rather than at the "exploitive" price typically paid.

538

laboratory develops a synthetic substitute. When demand for rubber rises, that country may actually harm itself trying to increase its market share (by producing more rubber) because in the process it drives down the price (or consider oil: OPEC countries drove oil prices to all-time lows in the mid-1980s when they overproduced oil in a bid to earn more revenue). If the country withholds production to shore up rubber's price, it provides consuming countries with an incentive to look for substitutes and alternative sources (again look at oil: After OPEC embargoed shipments of oil to the United States in the 1970s, the United States began developing domestic oil supplies and becoming more energy efficient, thus reducing oil prices and OPEC's revenues). The developing country is in a dependent and disadvantaged position.

Environmental Impacts of Underdevelopment

Geographers are very interested in the environmental impacts of relations between MDCs and LDCs, especially on the poorer countries. LDCs generally lack the financial resources needed to build roads, dams, energy grids, and other

infrastructure critical to development. They turn to the World Bank, International Monetary Fund (IMF), and other institutions of the MDCs to borrow funds for these projects. Many borrowers are unable to pay even the interest on these loans, which is sometimes huge; Tanzania, for example, spends 40 percent of its annual revenue on interest payments to the IMF, World Bank, and other lenders, more than it spends on education and health combined.

When lender institutions threaten to cut off assistance, borrowing countries often try to raise cash quickly to avoid this prospect. One method is to dedicate more quality land to the production of cash crops (also known as commercial crops), luxuries such as coffee, tea, sugar, coconuts, and bananas exported to the MDCs. Governments or foreign corporations often displace or "marginalize" subsistence farmers in the search for new lands on which to grow these commercial crops. In the process of marginalization, poor subsistence farmers are pushed onto fragile, inferior, or marginal lands that cannot support crops for long and that are degraded by cultivation. In Brazil's Amazon Basin, for example, peasant migrants arrive from Atlantic coastal regions

The problems of hot money and the digital divide are also cited as drawbacks to globalization. Hot money refers to short-term (and often volatile) flows of investment that can cause serious damage to the "emerging market" economies of less developed countries. Individual and corporate investors in the MDCs can invest such money heavily in stock market securities of the LDCs, reasoning that these developing nations have much greater economic growth potential than the mature economies of the MDCs. This investment can bring rapid wealth to at least some sectors of the economies of LDCs. The problem, however, is that the capital can be withdrawn as quickly as it is pumped in, resulting in huge economic consequences.

It is often assumed that the rapid growth and spread of computer and wireless technologies are benefiting all of humankind, but the notion of the digital divide challenges this view. The divide is between the handful of countries that are the technology innovators and users and the majority of nations that have little ability to create, purchase, or use new technologies. The United States, most European countries, and Japan are leaders in information technology (IT), and the LDCs are well behind. For example, more than 70 percent of U.S. citizens use the World Wide Web, compared with 4 percent of Russians, 2 percent of Indians, and 7 percent of Middle Easterners (Figure 2.A). Statistically, an American must save a month's salary to buy a computer, but a Bangladeshi must save 8 years' wages to buy one.

U.S. companies now earn about 85 percent of the revenues from the global Internet business and represent 95 percent of the stock market value of Internet corporations. The fear is that the growth of e-commerce will concentrate the wealth generated by that commerce in the technology leadership countries, enabling them to make further advances in technology and realize even bigger economic gains, while the technology laggards fail to catch up and simply become poorer. Will globalization prove to be good or bad? It may depend on where and whom you are.

Figure 2.A An Internet café in Mérida, Yucatán, Mexico. Information technology—especially computers, the Internet, and cellular telephones—is spreading rapidly around the world. Some people believe there is now an unfolding "Information Revolution," similar to the Agricultural and Industrial Revolutions in its capacity to change humanity's relationship with Earth.

where government and wealthy private landowners cultivate the best soils for sugarcane and other cash crops. The newcomers to Amazonia slash and burn the rain forest to grow rice and other crops that exhaust the soil's limited fertility in a few years. They move on to cultivate new lands and in their wake come cattle ranchers whose land use further degrades the soil.

National decision makers often face a difficult choice between using the environment to produce more immediate or more long-term economic rewards. In most cases, they feel compelled to take short-term profits and, by cash cropping and other strategies, initiate a sequence having sometimes tragic environmental consequences. Most LDCs have resource-based economies that rely not on industrial productivity but on stocks of productive soil, forests, and fisheries. The long-term economic health of these countries could be assured by the perpetuation of these natural assets, but to pay off international debts and meet other needs, the LDCs generally draw on their ecological capital faster than nature can replace it. In ecosystem terms, they exceed sustainable yield or the natural replacement rate, the highest rate at which a renewable resource can be used without decreasing its potential for renewal. In tropical biomes, for example, people cut down 10 trees for every tree they plant. In Africa, that ratio is 29 to 1. In 1950, about 30 percent of Ethiopia's land surface was forested. Today, less than 1 percent is in forest.

Such countries have experienced **ecological bankruptcy;** they have exhausted their environmental capital. Many political and social crises result from this bankruptcy: Revolutions, wars, and refugee migrations in developing nations often have underlying environmental causes. Such problems are becoming more central to the national interests of the United States and other powerful MDCs. The U.S. Central Intelligence Agency, for example, is increasingly using Geographic Information Systems (GIS) and other geographic techniques to analyze the natural phenomena and human misuses of the environment that are root causes of war and threats to global security.

People in the LDCs feel the impacts of environmental degradation more directly than do people in the MDCs. The world's poor tend to drink directly from untreated water supplies and to cook their meals with fuelwood rather than

Definitions + Insights

The Fuelwood Crisis

The removal of tree cover in excess of sustainable yield (a characteristic problem in Nepal, for example) illustrates the many detrimental effects that a single human activity can have in the LDCs (Figure 2.b). These impacts comprise the complex problem, widespread in the world's LDCs, known as the fuelwood crisis. As people remove trees to use as fuel, for construction, or to make room for crops, their existing crop fields lose protection against the erosive force of wind. Less water is available for crops because, in the absence of tree roots to funnel water downward into the soil, it runs off quickly. Increased salinity (salt content) generally accompanies increased runoff so that the quality of irrigation and drinking water downstream declines. Eroded topsoil resulting from reduced plant cover can choke irrigation channels, reduce water deliv-

ery to crops, raise floodplain levels, and increase the chance of floods destroying fields and settlements. As reservoirs fill with silt, hydroelectric generation, and therefore industrial production, is diminished. Upstream, where the problem began, fewer trees are available to use as fuel.

Owing to depletions of wood, people living in rural areas must now change their behavior (the fuelwood crisis often refers only to this part of the problem). Women, most often the fuelwood collectors, must walk farther to gather fuel. The use of animal dung and crop residues as sources of fuel deprives the soil of the fertilizers these poor people most often use when farming. Reduced food output is the result. A family may eventually give up one cooked meal a day or tolerate colder temperatures in their homes because fuel is lacking—steps that negatively impact the family's health.

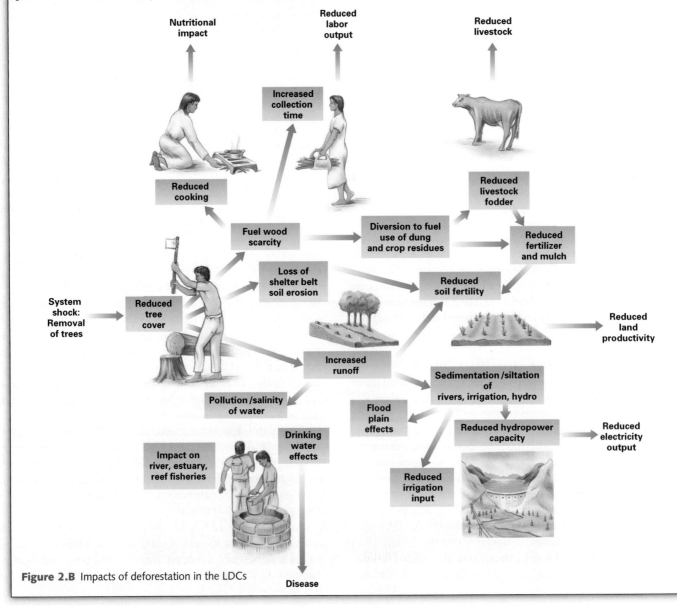

Figure 2.B Impacts of deforestation in the LDCs

Definitions + Insights

The Second Law of Thermodynamics

It is useful to understand how energy flows through ecosystems because this process is central to the question of how many people Earth and its regions can support and at what levels of resource consumption. Food chains (for example, in which an antelope eats grass and a leopard eats an antelope) are short, seldom consisting of more than four feeding (trophic) levels. The collective weight (known as biomass) and absolute number of organisms decline substantially at each successive trophic level (there are fewer leopards than antelopes). The reason is a fundamental rule of nature known as the second law of thermodynamics, which states that the amount of high-quality usable energy is lost as energy passes through an ecosystem. In living and dying, organisms use and lose the high-quality, concentrated energy that green plants produce and that is passed up the food chain. As an organism is consumed in any given link in the food chain, about 90 percent of that organism's energy is lost, in the form of heat and feces, to the environment. The amount of animal biomass that can be supported at each successive level thus declines geometrically, and little energy remains to support the top carnivores. So, for example, there are many more deer than there are mountain lions in a natural system.

The second law of thermodynamics has important consequences and implications for human use of resources. The higher we feed on the food chain (the more meat we eat), the more energy we use. The question of how many people Earth's resources can support thus depends very much on what we eat. The affluent consumer of meat demands a huge expenditure of food energy because that energy flows from grain (producer) to livestock animal (primary consumer, with a 90 percent energy loss) and then from livestock animal to human consumer (secondary consumer, with another 90 percent energy loss). Each year, the average U.S. citizen eats 100 pounds of beef, 50 pounds of pork, and 45 pounds of poultry. By the time it is slaughtered, a cow has eaten about 10 pounds of grain per pound of its body weight; a pig, 5 pounds; and a chicken, 3 pounds. Thus, in a year, a single American consumes the energy captured by two-thirds of a ton of grain (1,330 lb) in meat products alone. By skipping the meat and eating just the corn that fattens these animals, many people could live on the energy required to sustain just one meat-eater. The point is not to feel guilty about eating meat but to realize there is no answer to the question: How many people can the world support? The question needs to be rephrased: How many people can the world support based on people consuming _____? (with the blank filled in with a specific number of calories per day or with a general lifestyle).

Development means a more affluent lifestyle, including eating more meat (feeding higher on the food chain). Some analysts fear the burden that the changing dietary habits accompanying development will pose to the planet's food energy supply. Lester Brown of the WorldWatch Institute wrote a provocative essay entitled "Who Will Feed China?" Keeping in mind the second law of thermodynamics, he argued that if China continues on its present course of growing affluence, with a move away from reliance on rice as a dietary staple to a diet rich in meats and beer, the entire planet will shudder. How could more than a billion people be sustained so high on the food chain?

---

fossil fuels. They are more dependent on nature's abilities to replenish and cleanse itself, and thus, they suffer more when those abilities are diminished (see Definitions and Insights, page 38).

Two Types of Overpopulation

High rates of human population growth intensify the environmental problems characteristic of LDCs. More people cut more trees, a phenomenon that suggests there is a problem of people overpopulation in the poorer countries. Many persons, each using a small quantity of natural resources daily to sustain life, have a great collective impact on the environment and may add up to too many people for the local environment to support (Figure 2.19). The common consequences are malnutrition and even the famine emergencies in which richer countries are called on to provide relief.

There is another type of overpopulation, consumption overpopulation, which is characteristic of the MDCs. In the wealthier countries, there are fewer persons, but each uses a large quantity of natural resources from ecosystems across the world. Their collective impacts also degrade the environment, and even their smaller numbers may be too many at such unsustainably high levels of consumption. With less than 5 percent of the world's population, the United States may be seen as the world's leading overconsumer, accounting, for example, for about one-quarter of the world's annual energy consumption. One analyst calculated that the average U.S. citizen, in his or her lifetime, will consume more than 250 times as many goods as the average person in Bangladesh. Such disparities suggest that, just as Bangladesh is "underdeveloped," the United States is "overdeveloped."

If the vast majority of the world's population were to consume resources at the rate that U.S. citizens do, the environmental results might be ruinous (see Definitions and Insights, above). Even if consumption levels in the LDCs do not rise substantially, the sheer increase in numbers of people in those countries suggests that degradation of the environment will accelerate in the coming decades. The basic attributes of the world's population geography provide an insight into this dilemma.

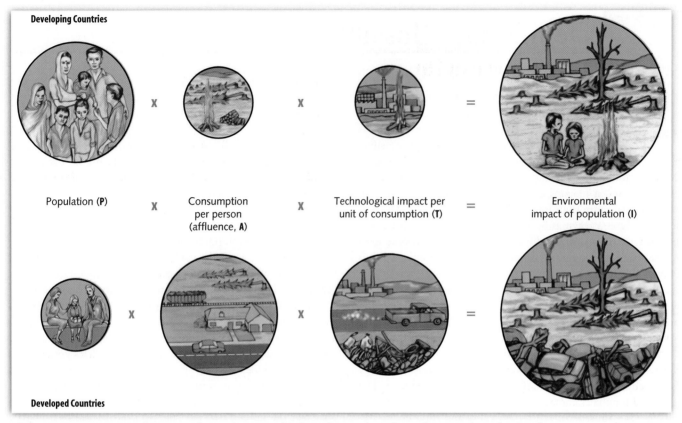

Figure 2.19 Two types of overpopulation depicting this formula: number of people × number of units of resources used per person × environmental degradation and pollution per unit of resource used = environmental impact. Circle size shows relative importance of each factor. People overpopulation is caused mostly by growing numbers of people and is typical of LDCs. Consumption overpopulation is caused mostly by growing affluence and is typical of MDCs. *Source: From* Environmental Science, *4th edition, by G. T. Miller. © 1993. Reprinted with permission of Wadsworth, a division of Thomson Learning, Inc.*

2.6 Population Geography

Earth's human population 10,000 years ago, before the Agricultural Revolution, was probably about 5.3 million. By A.D. 1, it was probably between 250 and 300 million, or about the population of the United States today. The first billion was reached about 1800. Then, a staggering population explosion occurred in the wake of the Industrial Revolution. The second billion came in 1930, the fourth in 1975, and the sixth in 1999. As one measure of ecological dominance, humankind is now by far the most populous large mammal on Earth and has succeeded where no other animal has in extending its range to the world's farthest corners. Some world regions and parts of some countries have especially dense populations (see Figures 1.4a and b). Notable are China and India, where a long history of productive agriculture is the main reason for high density, and Western Europe and the northeastern United States, where industrial productivity is the main factor.

The population explosion is clearly manifest as the classic J-shaped curve of exponential growth (Figure 2.20). Most

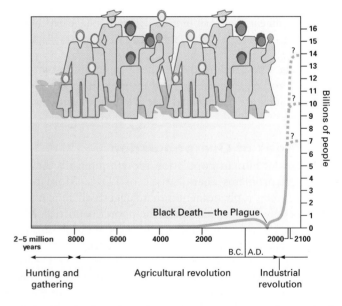

Figure 2.20 The j-shaped curve tracing exponential growth of the human population

Definitions + Insights

The Language of Human Migration

Migration refers to the movement of people from one location to another in any setting, whether within a community, within a country, or between countries (Figure 2.c). Migration is one of the most dynamic and problematic human processes on Earth. A migrant is always both an **emigrant** (one who moves from a place) and an **immigrant** (one who moves to a place). Emigration may be caused by so-called **push factors,** as when hunger or lack of land pushes peasants out of rural areas into cities, or by **pull factors,** as when an educated villager responds to a job opportunity in the city. People responding to push factors are often referred to as **nonselective migrants,** whereas those reacting to pull factors are called **selective migrants.** Both push and pull forces are behind the **rural to urban migration** that is characteristic of most countries.

Migration is almost never clearly "good" or "bad" but is almost always a controversial mixed blessing. Migration is bad for the country whose most talented people emigrate from, but it is good for the country they immigrate to; the Indian doctor who treats you benefits your community in the United States but may be sorely missed in India, for example. The doctor is illustrative of the brain drain, the problematic emigration of ed-ucated and talented people from one place to another. The Mexican immigrant who does the low wage labor in the United States may be seen as either the **illegal alien** threatening to overwhelm social services and take jobs away from local people or the guest worker who performs important services that no one else wants to do. Some host country peoples are more welcoming of new cultures that enrich their ethnic mosaic, but others fear losing their ethnic majorities and privileges.

Within and between countries today, there are also considerable movements of **refugees,** the victims of such severe push factors as persecution, political repression, and war. They may be on the move either as illegal immigrants or may have been granted **asylum,** meaning permission to immigrate on the grounds that they deserve protection in the host country. Immigration and asylum laws and quotas vary widely around the world depending on host countries' political, economic, and social traits. Among the most disadvantaged of the world's peoples are the **internally displaced persons** (IDPs) who are dislodged and impoverished by strife in their home country and have little prospect of emigration. The African country of Sudan has the world's highest number of internally displaced persons. This book will make frequent reference to other places characterized by the migration terms introduced here.

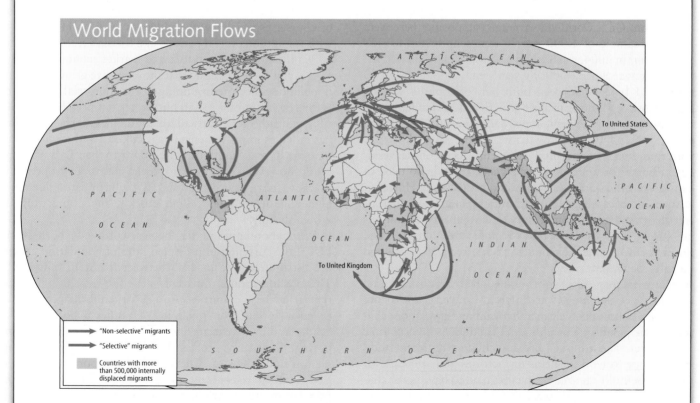

Figure 2.C Major global migration flows

striking is the upward curve of growth after the dawn of the Industrial Revolution, with the greatest increase in the population growth rate taking place since 1950. At the 2004 rate of 1.3 percent annual growth, our population would double in 54 years.

Already large and growing numbers pose a fundamental question: Will we exceed Earth's carrying capacity for our species, and what will happen if we do? Early in the Industrial Revolution, English clergyman Thomas Malthus (1766–1834) postulated that human populations, which can grow geometrically or exponentially, would exceed food supplies, which usually grow only arithmetically or linearly. He predicted a catastrophic human die-off as a result of this irreconcilable equation. He could not have foreseen that the exploitation of new lands and resources, including tapping into the energy of fossil fuels, would permit food production to keep pace with or even outpace population growth for at least the next two centuries. However, this Malthusian scenario of the lost race between food supplies and mouths to feed remains a source of constant and important debate today.

Population Concepts and Characteristics

Two principal variables determine population change in a given village, city, country, or over the entire Earth: birth rate and death rate. The birth rate is the annual number of live births per 1,000 people in a population (in a city, state, country, or the entire world). The death rate is the annual number of deaths per 1,000 people in that population. The population change rate is the birth rate minus the death rate in that population. On a worldwide average (supposing there were a perfect sample of 1,000 people representing the world's population) in 2004, the birth rate was 21 per 1,000 and the death rate was 9 per 1,000. By year's end, among the 1,000 people, 21 babies had been born but 9 people of varying ages had died, resulting in a net growth of 12. That figure, 12 per 1,000, or 1.2 percent, represents the 2004 population change rate (population growth rate) for the world (however, with rounding it is officially 1.3 percent). As no one leaves or enters Earth for or from other planets, there is no need to consider a third variable, migration, at this scale. However, at the scale of a village, city, or country, migration does affect the population. Compulsory and voluntary migrations are strong forces in the world's regions and countries, and this text cites frequent examples of both (see Definitions and Insights, page 41).

Many factors affect birth rates. Better educated and wealthier people have fewer children. Conversely, less educated and poorer people generally want and have more children. Poor parents view additional children as an economic asset rather than a burden because children represent additional labor to work in fields or factories and will care for them in their old age. People in cities tend to have fewer children than those in rural areas. Those who marry earlier generally have more children. Couples with access to and understanding of contraception may have fewer children.

Figure 2.21 These graffiti on the back of a road sign in Agra, India, sum up the attitude many of the world's people, particularly the rural poor, have about children. For many, family planning means trying to go against divine will.

Cultural norms are important because even where contraception is available, for religious or social reasons, a couple may decide not to interfere with what they perceive as God's will or may seek the social status associated with a larger family (Figure 2.21).

Death rates are affected mainly by health factors. Improvements in food production and distribution help reduce death rates. Better sanitation, hygiene, and drinking water eliminate infant diarrhea, a common cause of infant mortality in LDCs. The availability of antibiotics, immunizations, insecticides, and other improvements in medical and public health information and technologies have a marked correlation with declining death rates. Death rates sometimes rise, of course, especially with epidemics such as HIV/AIDS or with tumultuous social periods like Russia's transition in the 1990s from communism to a free-market economy.

It is vital to understand that the explosive growth in world population since the beginning of the Industrial Revolution is the result not of a rise in birth rates but of a dramatic decline in death rates, particularly in the LDCs. The death rate has fallen as improvements in agricultural and medical technologies have diffused from the MDCs to the LDCs. Until recently, however, there were no strong incentives for people in LDCs to have fewer children. With birth rates remaining high and death rates falling quickly, the population has grown sharply.

With about nine of ten babies worldwide born in the LDCs, the current and projected rates of population growth are distributed quite unevenly between the MDCs and LDCs. This phenomenon is apparent in the age-structure diagrams typical of these countries. An age-structure diagram (often called a "population pyramid") classifies a population by gender and by 5-year age increments (Figure 2.22). One important index these profiles show is the percentage of a population under age 15. The very high proportion typical of LDCs, about 33 percent, is remarkable for it means that populations will continue to grow in these poor countries as these children enter their reproductive years. The bottom-heavy, pyramid-shaped age-structure diagram of LDCs con-

144
448

Population by Age and Sex

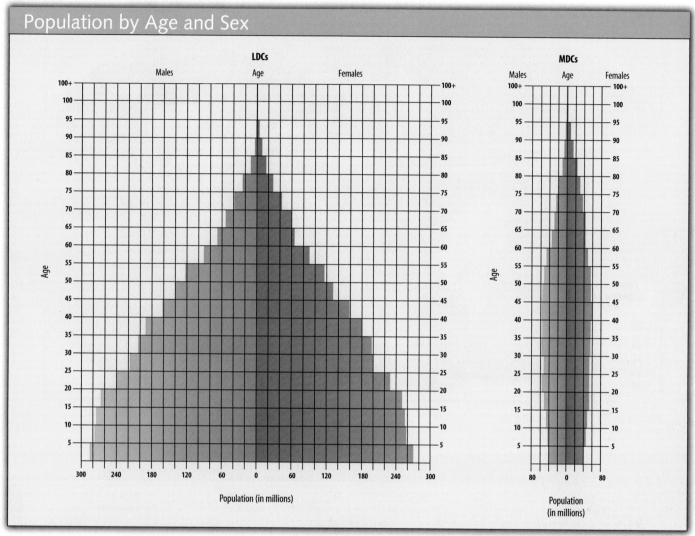

Active Figure 2.22 Age-structure profiles of the LDCs and MDCs in 2004
See an animation based on this figure, and take a short quiz on the concepts.

trasts markedly with the more chimney-shaped structure of the MDCs. The wealthier countries have a much more even distribution of population through age groups, with a modest share of 17 percent under age 15. Such profiles suggest that, not considering migration, their population growth will be low in the near future.

The Demographic Transition: Will the LDCs Complete It?

The history of population change in the MDCs provides a model whose usefulness in predicting the population future for LDCs is still uncertain. Known as the demographic transition, this model depicts the change from high birth rates and high death rates to low birth rates and low death rates that accompanied economic growth in the MDCs. Four stages comprise the model (Figure 2.24). In the **first (preindustrial) stage** (from the earliest humans to about A.D. 1800), birth rates and death rates were high, and population

growth was negligible. In the **second (transitional) stage,** birth rates remained high, but death rates dropped sharply after about 1800 due to medical and other innovations of the Industrial Revolution. In the **third (industrial) stage,** beginning around 1875, birth rates began to fall as affluence spread. Finally, after about 1975, some of the industrialized countries entered the **fourth (postindustrial) stage,** with both low birth rates and low death rates and, therefore (once again, as in the first stage), low population growth.

Some MDCs, including in 2004 Portugal, Slovakia, and Greece, officially registered zero population growth (ZPG), a rate of equal birth rates and death rates, and many in Europe (notably, most of those east of an imaginary line drawn from Germany to Italy, including most of the newest European Union members) were experiencing a negative growth rate with more deaths than births. Some of these, like Russia and Ukraine, were unfortunately experiencing population losses due more to rising death rates than to

Population Change Rates

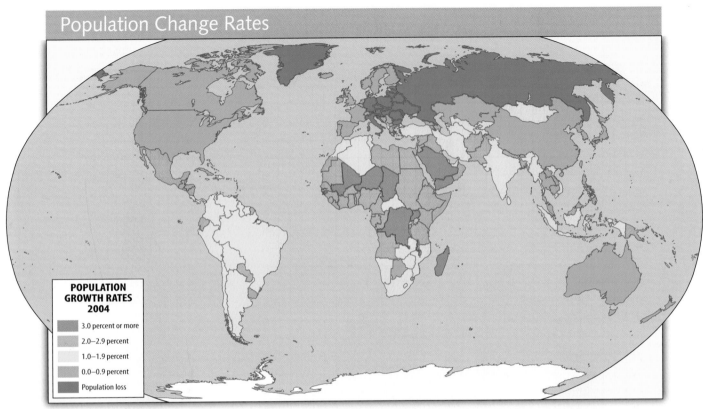

Figure 2.23 Population change rates worldwide. *Source: Data from Population Reference Bureau, 2004.*

Demographic Transition

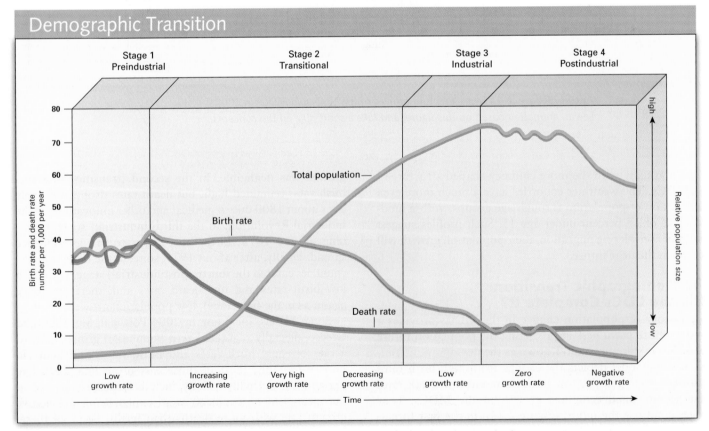

Figure 2.24 The demographic transition model. In the MDCs, economic development has resulted in a transition from high birth rates and high death rates to low birth rates and low death rates. Population growth rates at stages 1 and 4 are low because birth rates and death rates are nearly equal.

Definitions + Insights

Lifeboat Ethics

Neo-Malthusian ecologist Garrett Hardin introduced lifeboat ethics—the question of whether or not the wealthy should rescue the "drowning" poor—in a distressing but challenging essay. "People turn to me," wrote Hardin, "and say 'my children are starving. It's up to you to keep them alive.' And I say, 'The hell it is. I didn't have those children.'"[a] He described the world not as a single "spaceship earth" or "global village" with a single carrying capacity, as many environmentalists do, but as a number of distinct "lifeboats," each occupied by the citizens of single countries and each having its own carrying capacity. Each rich nation is a lifeboat comfortably seating a few people. The world's poor are in lifeboats so overcrowded that many fall overboard. They swim to the rich lifeboats and beg to be brought aboard. What should the passengers of the rich lifeboats do? The choices pose an ethical dilemma.

Hardin proposed the following: There are 50 rich passengers in a boat with a capacity of 60. Around them are 100 poor swimmers who want to come aboard. The rich boaters have three choices. First, they could take in all the swimmers, capsizing the boat with "complete justice, complete catastrophe." Second, as they enjoy an unused excess capacity of 10, they could admit just 10 from the water. But which 10? And what about the margin of comfort that excess capacity allows them? Finally, the rich could prevent any of the doomed from coming aboard, ensuring their own safety, comfort, and survival.

Translating this metaphor into reality, as an occupant of the rich lifeboat United States, for example, what would you do for the drowning refugees from lifeboat Haiti, Liberia, or Afghanistan?

Hardin's choice was the third: drowning. To preserve their own standard of living and ensure the planet's safety, the wealthy countries must cease to extend food and other aid to the poor and must close their doors to immigrants from poor countries. "Every life saved this year in a poor country diminishes the quality of life for subsequent generations," Hardin

concluded. "For the foreseeable future, survival demands that we govern our actions by the ethics of a lifeboat."[b] Garrett Hardin, incidentally, committed suicide in 2003.

In the United States and other MDCs, there is often a sense of donor fatigue, a trend that Hardin probably would have applauded as beneficial for the global environment. Wearied by constant images of people in need, feeling their contributions might be only marginally useful, and perhaps experiencing some economic hardship themselves, people who can afford to give are simply tired of thinking about giving. Governmental support has flagged, too; even with a post 9/11 boost, development aid fell by 20 percent between 1990 and 2004. That decline is widely attributed to the end of the Cold War, when winning and maintaining influence in foreign lands began to be regarded as less critical to the security of Western governments. Critics of foreign aid point out that corrupt governments often steal or squander the assistance money, and that too much of the aid is "tied," so that the recipient country is obliged to buy goods or services from the donor country.

So why shouldn't we be fed up with giving? For one reason, neglect may come back to harm us. Some analysts think that American neglect of Afghanistan after the Soviets withdrew in 1989 made that country the breeding ground for the 9/11 attacks. Recently, U.S. foreign policy analysts have begun to identify certain strategically "pivotal" countries for enhanced aid, while assistance to other nations would be drawn down. Pivotal countries are those whose collapse would cause international refugee migration, war, pollution, disease epidemics, or other international security problems. According to one model, nine countries are clearly pivotal: Mexico, Brazil, Algeria, Egypt, South Africa, Turkey, India, Pakistan, and Indonesia.

---

[a]Garrett Hardin, "The Tragedy of the Commons." *Science 162*, 1968, pp. 1243–1248. His original lifeboat essay was entitled "Living on a Lifeboat" and was published in *Psychology Today* in September 1974.
[b]Ibid.

---

falling birth rates; these problems are explored in subsequent chapters.

Viewed as a group, the LDCs are in the latter part of stage two of the demographic transition. Birth rates in the LDCs are now falling and, consequently, so is the world's rate of population growth (the actual numbers are dropping, too, from an all-time annual growth of 88 million per year in 1994 to 77 million in 2004. In view of declining birth rates worldwide, the United Nations in 1997 revised its projection for future population growth downward. The agency predicted 9.4 billion by 2050 (half a billion fewer than that pro-

jected only 2 years earlier) and a stabilization at 10.7 billion in 2120 (the earlier projection was 12.6 billion). This revision prompted a rash of popular and academic articles proclaiming "the population explosion is over," and some essays even argued there would soon be too few people on Earth, particularly in the MDCs. Other sources cautioned that it was too soon to declare the population bomb defused. "World population growth turned a little slower," a Population Institute report argued. "The difference, however, is comparable to a tidal wave surging toward one of our coastal cities. Whether the tidal wave is 80 feet or 100 feet high, the impact will be similar." [3] Not content with its own projected figure for stabilization, the United Nations has pledged to do its best to help stabilize the global population at no more than 9.8 billion after the year 2050. The organization's plan is to focus on enhancing the education and employment of women in the LDCs as a means of bringing down birth rates.

There are two very distinct points of view on where these LDCs will proceed from stage two of the demographic transition, with their low death rates, high but falling birth rates, and rapid population growth. Some observers believe that the poorer countries will follow the wealthier ones through the transition, achieving prosperity and a stabilized population. Most recognizable among these optimists are the so-called technocentrists, or cornucopians, who argue that human history provides insight into the future: Through their technological ingenuity, people always have been able to conquer food shortages and other problems, and therefore, they always will. The late Julian Simon, a University of Maryland economist, argued that far from being a drain on resources, additional people create additional resources. Technocentrists thus insist that people can raise Earth's carrying capacity indefinitely and that the die-off that Malthus predicted will always be averted. Our more numerous descendants will instead enjoy more prosperity than we do.

In contrast, the neo-Malthusians (heirs of the reasoning of Thomas Malthus) argue that although successful so far, we cannot indefinitely increase Earth's carrying capacity. There is an upper limit beyond which growth cannot occur, with calculations ranging from 8 to 40 billion. Neo-Malthusians insist that LDCs cannot remain indefinitely in the transitional stage. Either they must intentionally bring birth rates down further and make it successfully through the demographic transition, or they must unwillingly suffer nature's solution, a catastrophic increase in death rates. Thus, by either the **birth rate solution** or **death rate solution** to the problem, the neo-Malthusians argue, the less developed world must confront its population crisis (see Definitions and Insights, page 45).

Whereas technocentrists view the equation between people and resources passively, insisting that no corrective action is needed, neo-Malthusians tend to be activists who describe terrible scenarios of a death rate solution to motivate people to adopt the birth rate solution. Biologist Paul Ehrlich thinks that we are increasingly vulnerable to a Malthusian catastrophe, particularly as HIV and other viruses diffuse around the globe with unprecedented speed. "The only big question that remains," wrote Ehrlich, "is whether civilization will end with the bang of an all-out nuclear war, or the whimper of famine, pestilence and ecological collapse." [4] Such dire warnings have earned many neo-

Malthusians the reputation of being "gloom and doom pessimists."

2.7 Where to from Here?

Within the last two decades, new concepts and tools for managing Earth in an effective, long-term way have emerged. Known collectively as sustainable development, or ecodevelopment, these ideas and techniques consider what both MDCs and LDCs can do to avert the possible Malthusian dilemma and improve life on the planet. Sustainable development offers an activists' agenda without the peril of doom depicted by the neo-Malthusians.

The World Conservation Union defines sustainable development as "improving the quality of human life while living within the carrying capacity of supporting ecosystems." [5] Sustainable development refutes what its proponents perceive as the current pattern of unsustainable development, whereby economic growth is based in large part on excessive resource use. Advocates of sustainable development point out that a country that depletes its resource base for short-term profits gained through deforestation increases its gross national product (GNP) and appears to be more "developed" than a country that protects its forests for a long-term harvest of sustainable yield. Deforestation appears to be beneficial to a country because it raises GNP through the production of pulp, paper, furniture, and charcoal. However, GNP does not measure the negative impacts of deforestation, such as erosion, flooding, siltation, and malnutrition. These consequences are known as external costs, or externalities, and are not priced into goods and services. Advocates of sustainable development argue that these externalities should be added or "internalized" before a good or service is marketed because the true high costs of these goods and services would then be recognized. The lower true costs of less destructive practices would then be evident, providing stronger incentives for individuals, companies, and nations to invest in sustainable practices and technologies.

Sustainable development is a complex assortment of theories and activities, but its proponents call for eight essential changes in the way people perceive and use their environments:

1. People must *change their worldviews and value systems*, recognizing the finiteness of resources and reducing their expectations to a level more in keeping with Earth's environmental capabilities. Proponents of sustainable development argue that this change in perspective is needed especially in the MDCs, where, instead of trying to "keep up with the Joneses," people should try to enjoy life through more social rather than material pursuits.

2. People should *recognize that development and environmental protection are compatible*. Rather than viewing envi-

---

[3] Quoted in Steven A. Holmes, "Global Crisis in Population Still Serious, Group Warns." *The New York Times*, December 31, 1997, p. A7.

[4] Paul R. Ehrlich, "Populations of People and Other Living Things." In *Earth '88: Changing Geographic Perspectives*, Harm De Blij, ed. (Washington, D.C.: National Geographic Society, 1988), pp. 302–315.

[5] World Wildlife Fund, *Sustainable Use of Natural Resources: Concepts, Issues and Criteria* (Gland, Switzerland: World Wildlife Fund, 1995), p. 5.

ronmental conservation as a drain on economies, we should see it as the best guarantor of future economic well-being. This is especially important in the LDCs, with their resource-based economies.

3. People all over the world should *consider the needs of future generations* more than we do now. Much of the wealth we generate is in effect borrowed or stolen from our descendants. Our economic system values current environmental benefits and costs far more than future benefits and costs, and thus, we try to improve our standard of living today without regard to tomorrow.

4. Communities and countries should *strive for self-reliance,* particularly through the use of appropriate technologies. For example, remote villages could rely increasingly on solar power for electricity rather than be linked into national grids of coal-burning plants.

5. LDCs need to *limit population growth* as a means of avoiding the destructive impacts of "people overpopulation." Advances in the status of women, improvements in education and social services, and effective family planning technologies can help limit population growth.

6. Governments need to *practice land reform,* particularly in the LDCs. Poverty is often not the result of too many people on too little total land area but of a small, wealthy minority holding a disproportionably high share of quality land. To avoid the environmental and economic consequences of marginalization, a more equitable distribution of land is needed.

7. *Economic growth in the MDCs should be slowed* to reduce the effects of "consumption overpopulation." If economic growth, understood as the result of consumption of natural resources, continues at its present rate in excess of sustainable yield, Earth's environmental "capital" will continue to diminish rapidly.

8. *Wealth should be redistributed between the MDCs and LDCs.* Because poverty is such a fundamental cause of environmental degradation, the spread of a reasonable level of prosperity and security to the LDCs is essential. Proponents argue that this does not mean rich countries should give cash outright to poor countries. Instead, the lending institutions of MDCs could forgive some existing debts owed by LDCs or use such innovations as debt-for-nature swaps, in which a certain portion of debt is forgiven in return for the borrower's pledge to invest that amount in national parks or other conservation programs. Reducing or eliminating trade barriers that MDCs impose against products from LDCs would also help redistribute global wealth.

Some geographers and other scientists believe that sustainable development (rather than information technology) will bring about the "Third Revolution," a shift in human ways of interacting with Earth so dramatic that it will be compared with the origins of agriculture and industry. The formidable changes called for in sustainable development are attracting increasing attention, perhaps because at present there are no comprehensive alternative strategies for dealing with some of the most critical issues of our time.

SUMMARY

⚭ Weather refers to atmospheric conditions prevailing at one time and place. Climate is a typical pattern recognizable in the weather of a region over a long period of time. Climatic patterns have a strong correlation with patterns of vegetation and in turn with human opportunities and activities in the environment.

⚭ Warm air holds more moisture than cool air, and precipitation is best understood as the result of processes that cool the air to release moisture. Convectional precipitation is the result of air being heated by surface radiation and rising rapidly to cool and produce rain. Orographic precipitation results when moving air confronts a topographic barrier and is forced upward to cool and produce rain. Cyclonic (frontal) precipitation is generated in cyclones or traveling low-pressure cells that bring air masses with different characteristics of temperature and moisture into contact.

⚭ Most of the sun's visible short-wave energy that reaches Earth is absorbed, but some of it returns to the atmosphere in the form of infrared long-wave radiation, which generates heat and helps to warm the atmosphere. This is Earth's natural greenhouse effect.

⚭ Geographers group local climates into major climate types, each of which occurs in more than one part of the world and is associated with other natural features, particularly vegetation. Geog-

raphers recognize 10 to 20 major types of ecosystems or biomes, which are categorized by the type of natural vegetation. Vegetation and climate types are so sufficiently related that many climate types are named from the vegetation types—for example, the tropical rain forest climate and the tundra climate.

⚭ Geographers and ecologists recognize the importance of some biomes because of their biological diversity—the number of plant and animal species and the variety of genetic materials these organisms contain. Regions where human activities are rapidly depleting a rich variety of plant and animal life are known as biodiversity hot spots, a list of places scientists believe deserve immediate attention for study and conservation.

⚭ Computer-based climate change models use the scenario of a doubling of atmospheric carbon dioxide and indicate that the mean global temperature might warm up to 11° Fahrenheit by the year 2100 due to human contributions to the greenhouse effect. The general agreement is that with global warming the distribution of climatic conditions typical of biomes will shift poleward and upward in elevation.

⚭ The Kyoto Protocol is an international agreement requiring the industrialized countries that ratified it to make substantial cuts in their carbon dioxide emissions to reduce global warming. The

United States dropped its support for the treaty. It was set to go into effect for the ratifying countries after the critical nation of Russia ratified it in 2004.

- A useful perspective on the current spatial patterns of our relationship with Earth is to view these patterns as products of two "revolutions" in the relatively recent past: the Agricultural, or Neolithic (New Stone Age), Revolution and the Industrial Revolution. Each transformed humanity's relationship with the natural environment.

- High rates of human population growth intensify environmental problems characteristic of the LDCs. More people use more resources, a phenomenon that suggests there is a problem of people overpopulation in the poorer countries. Consumption overpopulation is more characteristic of the MDCs, where fewer people use large quantities of natural resources from across the world.

- Considering population projections, we may ask how and where global resources can support 10 billion or more people. Our demands for more food and other photosynthetic products are growing, and it is questionable whether supplies can keep pace. The Malthusian scenario, which has not been realized so far, insists that a catastrophic die-off of people will occur when their numbers exceed food supplies.

- The explosive growth in world population since the beginning of the Industrial Revolution is not the result of a rise in birth rates but of a dramatic decline in death rates, particularly in the LDCs.

- Within the last two decades, new concepts and tools for managing Earth in an effective, long-term way have emerged. Known collectively as sustainable development, or ecodevelopment, these ideas and techniques consider what humanity can do to avert the Malthusian dilemma and improve life on the planet.

KEY TERMS + CONCEPTS

Terms in blue are also defined in the glossary.

Age of Discovery (p. 32)
age-structure diagram (p. 42)
Agricultural Revolution (Neolithic or Food-Producing Revolution) (p. 31)
Anticyclone (high-pressure cells) (p. 18)
asylum (p. 41)
biodiversity hot spots (p. 23)
biological diversity (biodiversity) (p. 23)
biomes (p. 21)
    coniferous forest (boreal forest or *taiga*) (p. 21)
    desert shrub (p. 21)
    Mediterranean scrub forest (p. 22)
    savanna (p. 22)
    steppe (temperate grassland, prairie) (p. 22)
    temperate mixed forest (p. 23)
    tropical deciduous forest (p. 22)
birth rate (p. 42)
birth rate solution (p. 46)
brain drain (p. 41)
carbon dioxide "sinks" (p. 30)
carrying capacity (p. 31)
cash crops (commercial crops) (p. 36)
civilization (p. 31)
climate (p. 17)
climates (p. 21)
    desert (p. 21)
    humid continental (p. 22)
    humid subtropical (p. 22)
    icecap (p. 21)
    marine west coast (p. 22)
    Mediterranean (p. 22)
    semiarid/steppe (p. 21)
    subarctic (p. 21)
    tropical rain forest (p. 22)
    tropical savanna (p. 22)
    tundra (p. 21)
    undifferentiated highland (p. 23)
consumption overpopulation (p. 39)
cultural landscape (p. 30)
culture hearth (p. 31)

cyclone (p. 18)
death rate (p. 42)
death rate solution (p. 46)
debt-for-nature swap (p. 47)
demographic transition (p. 43)
    first (preindustrial) stage (p. 43)
    second (transitional) stage (p. 43)
    third (industrial) stage (p. 43)
    fourth (postindustrial) stage (p. 43)
dependency theory (p. 35)
development (p. 34)
digital divide (p. 37)
domestication (p. 31)
donor fatigue (p. 45)
dry farming (p. 31)
ecological bankruptcy (p. 37)
ecologically dominant species (p. 31)
emigrant (p. 41)
equatorial low (p. 18)
equinox (p. 20)
extensive land use (p. 31)
external costs (externalities) (p. 46)
front (p. 18)
fuelwood crisis (p. 38)
globalization (p. 36)
Green Revolution (p. 23)
greenhouse effect (p. 27)
greenhouse gases (p. 28)
    carbon dioxide (CO_2) (p. 28)
    chlorofluorocarbons (CFCs) (p. 28)
    methane (p. 28)
    nitrous oxide (p. 28)
guest worker (p. 41)
high pressure cell (anticyclone) (p. 18)
hot money (p. 37)
Human Development Index (HDI) (p. 34)
hunting and gathering (foraging) (p. 30)
illegal alien (p. 41)
immigrant (p. 41)
Industrial Revolution (p. 31)
information technology (IT) (p. 37)

intensive land use (p. 31)
Intergovernmental Panel on Climate Change (IPCC) (p. 27)
internally displaced persons (p. 41)
irrigation (p. 31)
Kyoto Protocol (p. 29)
lapse rate (p. 23)
less developed countries (LDCs) (p. 33)
lifeboat ethics (p. 45)
Malthusian scenario (p. 42)
marginalization (p. 36)
mercantile colonialism (p. 35)
migration (p. 41)
monoculture (p. 23)
Montreal Protocol (p. 29)
more developed countries (MDCs) (p. 33)
multinational companies (p. 36)
natural landscape (p. 30)
natural replacement rate (p. 37)
neocolonialism (p. 35)
neo-Europes (p. 35)
neo-Malthusians (p. 46)
newly industrializing countries (NICs) (p. 33)
nonselective migrants (p. 41)
"original affluent society" (p. 30)
people overpopulation (p. 39)
per capita GDP PPP (p. 33)
per capita gross domestic product (GDP) (p. 33)
pivotal countries (p. 45)
plate tectonics (p. 23)
Pleistocene overkill hypothesis (p. 31)
population change rate (p. 42)
population explosion (p. 40)
precipitation (p. 17)
    convectional (p. 18)
    cyclonic (frontal) (p. 18)
    orographic (p. 18)
pull factors (p. 41)
push factors (p. 41)

purchasing power parity (PPP) (p. 33)
rain shadow (p. 18)
refugees (p. 41)
rifting (p. 23)
rural to urban migration (p. 41)
second law of thermodynamics (p. 39)
seismic activity (p. 23)

selective migrants (p. 41)
settler colonization (p. 35)
summer solstice (p. 20)
sustainable development (p. 46)
sustainable yield (p. 37)
technocentrists (cornucopians) (p. 46)
tradable permits (p. 30)

trade barriers (p. 35)
trade winds (p. 18)
value-added products (p. 35)
weather (p. 17)
winter solstice (p. 20)
xerophytic (p. 21)
zero population growth (ZPG) (p. 43)

REVIEW QUESTIONS

WORLD REGIONAL
Geography Now™

Assess your understanding of this chapter's topics with additional quizzing and concept-based problems at http://earthscience.brookscole.com/wrg5e.

1. What is the difference between weather and climate? What are the main forces that produce precipitation and aridity?

2. What are the major climate types and their associated biomes? Where do they tend to occur on Earth?

3. Where are the biodiversity hot spots? What kinds of locations and biomes do many of them occur in?

4. What apparent long-term effects on Earth's atmosphere can be attributed to modern technology? What predictions are there for future changes and what steps are being taken or considered to avert these changes?

5. What are the Agriculture and Industrial Revolutions? In what ways did they represent important changes in human–Earth relationships?

6. What are the typical differences between MDCs and LDCs?

7. According to dependency theory, what are the causes of disparities between MDCs and LDCs? What other factors may explain global wealth and poverty?

8. What is the fuelwood crisis? What are some of the effects of deforestation on human lives and economies in the LDCs?

9. What are the two types of overpopulation and how do they differ?

10. What has been the main cause of the world's explosive population growth since the beginning of the Industrial Revolution?

11. What variables distinguish the four stages of the demographic transition and what explains them?

12. What do Hardin's lifeboats represent—what is his metaphor?

DISCUSSION QUESTIONS

1. Using the album of Earth's biomes in Figure 2.8, discuss the variables of precipitation, latitude, and elevation that likely give rise to each type.

2. What are the indications that Earth is getting warmer? What might these changes mean for agriculture, plant and animal life, and human adaptations?

3. Discuss the Kyoto Protocol. What would be the advantages and disadvantages of its implementation? Do you agree or disagree with the position taken against the treaty by the administration of George W. Bush?

4. What does the Green Revolution mean for agriculture and biodiversity? Why does Norman Myers lament "we are causing the death of birth"?

5. Explain how *Homo sapiens* is an ecologically dominant species.

6. Explain how *Homo sapiens* is an ecologically dominant species.

7. What is development and what are some of the main standards and indexes used to measure it?

8. Why are some countries rich and some poor? Discuss and debate the arguments presented in this chapter and use your own perspectives.

9. Discuss the graph of the fuelwood crisis on page 38. Can you explain the causes and effects of the initial system shock?

10. What is overdevelopment? Is it a useful concept for describing the way of life in the country you live in? What would you regard as indicators of overdevelopment?

11. What would happen to the global environment if China became a prosperous country with personal habits to match?

12. Compare and contrast the birth rate solution and the death rate solution. Is one or the other really inevitable? Which is Earth's population heading for?

13. Is immigration good or bad for your country? How do you, your neighbors, and your politicians view immigration?

14. Is globalization good or bad for the world's peoples and economies?

15. Debate Garrett Hardin's lifeboat ethics. Do you agree or disagree with his view and why?

16. Discuss sustainable development. Are these notions pie in the sky or are they achievable? Should achieving them be attempted? Why or why not?

A Geographic Profile of Europe

The Standing Stones of Callanish on the island of Lewis in Scotland's Hebrides Island group, date to about 2000 B.C.

chapter objectives

This chapter should enable you to:

- Recognize Europe as a postindustrial region with a wealthy, declining population

- Become familiar with Europe's immigration issues

- Learn the region's distinguishing geographic characteristics, landforms, and climates

- Get acquainted with Europe's major ethnic groups, languages, and religions

- Understand how Europe rose to global political and economic dominance and then declined

- Trace Europe's emergence from wartime divisions to supranational unity within the European Union and know why the EU is important

WORLD REGIONAL
Geography ⊛ Now™

Look for this logo in the text and go to GeographyNow at http://earthscience.brookscole.com/wrg4e to explore interactive maps, view animations, sharpen your factual knowledge and geographic literacy, and test your critical thinking and analytical skills with unique interactive resources.

Many culture hearths have exerted influences on peoples and landscapes around the world, but none more impressively than Europe. For more than 500 years, the experiences of European nation building, scientific and technological developments, and colonization of far-flung lands have reshaped cultures and ecosystems around the world. Mainly as a result of self-inflicted wounds in the form of two world wars, Europe has taken a backseat to the United States in global influence but still today wields enormous economic and political clout. Beginning with this profile, these three chapters explore the human and physical geographies of this influential world region.

3.1 Area and Population

Europe has traditionally been classified as one of the world's seven continents, but one look at the globe reveals what an arbitrary designation that is. Europe is not a distinct landmass such as Australia and Africa. It is rather an appendage or a subcontinent of the world's greatest landmass, Eurasia (a conflation of "Europe" and "Asia"). Nevertheless, the popular and scholarly designation of Europe as a region distinct from the rest of Eurasia is a time-honored and useful organizing device. In this text, Europe is the culture region made up of the countries of Eurasia lying west of Russia and the Near Abroad and Turkey. (Incidentally, the traditional *physical* dividing line between Europe and Asia is drawn from the Ural Mountains down to the Caucasus, technically placing a good part of Russia and the Near Abroad within Europe.)

Defined this way, Europe is a great peninsula, fringed itself by lesser peninsulas and islands and bounded by the Arctic and Atlantic Oceans, the Mediterranean Sea, Turkey, the Black Sea, and Russia and the Near Abroad (Figure 3.1).

This text divides Europe into four subregions (see Table 3.1). The Coreland is Northwest and North Central Europe. It includes the United Kingdom, Ireland, France, the Benelux Countries (Belgium, the Netherlands, and Luxembourg), Switzerland, Austria, and Germany. These countries have the largest populations and play major economic and political roles in contemporary Europe. The North is made up of Denmark, Iceland, Norway, Sweden, and Finland. The South includes Portugal, Spain, Italy, Greece, Malta, and Cyprus. The East is made up of Estonia, Latvia, Lithuania, Poland, the Czech Republic, Slovakia, Hungary, Romania, Bulgaria, Albania, Yugoslavia, Bosnia-Herzegovina, Croatia, Macedonia, and Slovenia. There are also the microstates of Andorra, Monaco, and Liechtenstein.

All of Europe has an area only about two-thirds as large as that of the 48 contiguous United States. The average European country is some 50,000 square miles (130,000 sq km), or about the size of Arkansas. Four countries—Germany, the United Kingdom, France, and Italy—far outsize the others in Europe in population. Their respective populations range from about 83 million for Germany to approximately 60 million each for France, Italy, and the United Kingdom. Together, the four countries represent about one-half of Europe's population, or about 90 percent of the population of the entire United States.

TABLE 3.1 Europe: Basic Data

Political Unit	Area (thousand/ sq mi)	Area (thousand/ sq km)	Estimated Population (millions)	Annual Rate of Increase (%)	Estimated Population Density (sq mi)	Estimated Population Density (sq km)	Human Development Index	Urban Population (%)	Arable Land (%)	GDP PPP Per Capita ($US)
European Core										
Andorra	0.2	0.5	0.1	0.8	500	193	N/A	92	2	19,000
Austria	32.4	83.8	8.1	0.0	250	97	0.934	54	17	30,000
Belgium	11.8	30.5	10.4	0.1	881	340	0.942	97	23	29,100
France	212.9	551.2	60	0.4	282	109	0.932	74	33	27,600
Germany	137.8	356.7	82.6	−0.2	599	231	0.925	88	34	27,600
Ireland	27.1	70.1	4.1	0.8	151	58	0.936	60	15	29,600
Liechtenstein	1	2.6	0.03	0.5	30	12	N/A	21	25	25,000
Luxembourg	0.06	0.1	0.5	0.3	8333	3218	0.933	91	23	55,100
Monaco	0.001	0.002	0.03	0.6	30,000	11,583	N/A	100	0	27,000
Netherlands	15.8	40.9	16.3	0.4	1032	398	0.942	62	26	28,600
Switzerland	15.9	41.1	7.4	0.1	465	180	0.936	68	10	32,700
United Kingdom	94.5	244.6	59.7	0.1	632	244	0.936	89	23	27,700
Total	**549.5**	**1422.6**	**249.3**	**0.2**	**453**	**175**	**0.931**	**81**	**28**	**28,059**
Northern Europe										
Denmark	16.6	43	5.4	0.1	325	126	0.932	72	54	31,100
Finland	130.6	338.1	5.2	0.2	40	15	0.935	62	7	27,400
Iceland	39.8	103	0.3	0.8	8	3	0.941	94	0	30,900
Norway	125.1	323.8	4.6	0.3	37	14	0.956	78	3	37,800
Sweden	173.7	449.7	9	0.1	52	20	0.946	84	6	26,800
Total	**485.8**	**1257.7**	**24.5**	**0.2**	**50**	**19**	**0.944**	**77**	**6**	**30,276**
Eastern Europe										
Albania	11.1	28.7	3.2	1.2	288	111	0.781	42	21	4500
Bosnia and Herzegovina	19.7	51	3.9	0.1	198	76	0.781	43	13	6100
Bulgaria	42.8	110.8	7.8	−0.6	182	70	0.796	70	40	7600

Europe contains one of the world's great clusters of human population (see Figures 1.4a, 1.4b, 3.2, and Table 3.1). Europe's approximately 524 million people in 2004 were nearly twice those of the United States. In Europe, 1 of every 12 people in the world live in a space half the size of the United States. Except for some northern, rugged, and infertile areas, European population density is everywhere greater than the world average (258/sq mi; 100/sq km). Europe's population density varies widely, from 1032 persons per square mile (398/sq km) in the intensively developed Netherlands to about 8 per square mile (3/sq km) in Iceland. Comparisons with the United States reveal about the same range, but in general, the European countries are more densely populated.

The greatest densities are in two belts of industrialization and urbanization near coal and hydroelectric power sources; as will be seen in the following chapters, these energy resources were critical in the location and rapid development of many European cities. One belt extends north–south from Great Britain to Italy. In addition to large parts of Great Britain, it includes extreme northeastern France, most of Belgium and the Netherlands, Germany's Rhineland, northern Switzerland, and northern Italy (across the Alps). The second belt of dense population extends west–east from Great Britain to southern Poland and continues out of the region into the Ukraine. It corresponds to the first belt as far east as the Rhineland, where it forms a narrow strip eastward across Germany, the western Czech Republic, and Poland.

Although the two belts represent a minor part of Europe's total extent, they contain more large cities and a greater value of industrial output than the rest of Europe combined. Only in eastern North America and Japan are there urban-

TABLE 3.1   (continued)

Political Unit	Area (thousand/ sq mi)	Area (thousand/ sq km)	Estimated Population (millions)	Annual Rate of Increase (%)	Estimated Population Density (sq mi)	Estimated Population Density (sq km)	Human Develop- ment Index	Urban Popula- tion (%)	Arable Land (%)	GDP PPP Per Capita ($US)
Croatia	21.8	56.4	4.4	−0.2	202	78	0.830	56	26	10,600
Czech Republic	30.4	78.7	10.2	−0.2	336	130	0.868	77	40	15,700
Estonia	17.4	45	1.3	−0.4	75	29	0.853	69	16	12,300
Hungary	35.9	92.9	10.1	−0.4	281	109	0.848	65	50	13,900
Latvia	24.9	64.4	2.3	−0.5	92	36	0.823	68	29	10,200
Lithuania	25.2	65.2	3.4	−0.3	135	52	0.842	67	45	11,400
Macedonia	9.9	25.6	2	0.5	202	78	0.793	59	22	6700
Poland	124.8	323.1	38.2	0.0	306	118	0.850	62	46	11,100
Romania	92	238.1	21.7	−0.3	236	91	0.778	53	40	7000
Serbia and Montenegro	39.4	102	10.7	0.2	272	105	N/A	52	33	2200
Slovakia	18.9	48.9	5.4	0.0	286	110	0.842	56	30	13,300
Slovenia	7.8	20.2	2	−0.1	256	99	0.895	51	8	19,000
Total	**522**	**1351.4**	**126.6**	**−0.1**	**242**	**93**	**0.820**	**60**	**37**	**9655**
Southern Europe										
Cyprus	3.6	9.3	0.9	0.5	250	97	0.883	65	7	19,200
Greece	51	132	11	0.0	216	83	0.902	60	21	20,000
Italy	116.3	301.1	57.8	−0.1	497	192	0.920	90	27	26,700
Malta	0.1	0.2	0.4	0.2	4000	1544	0.875	91	28	17,700
Portugal	35.5	91.9	10.5	0.0	296	114	0.897	53	21	18,000
San Marino	0.02	0.05	0.03	0.3	1500	579	N/A	84	16	34,600
Spain	195.4	505.9	42.5	0.1	218	84	0.922	76	26	22,000
Total	**401.9**	**1040.5**	**123.1**	**0.0**	**306**	**118**	**0.916**	**79**	**25**	**22,728**
Summary Total	**1959.2**	**5072.2**	**523.5**	**0.1**	**258**	**100**	**0.901**	**74**	**24**	**22,611**

Sources: *World Population Data Sheet*, Population Reference Bureau, 2004; *U.N. Human Development Report*, United Nations, 2004; *World Factbook*, CIA, 2004.

industrial belts of the same order of complexity and importance. All of these regions have populations that are overwhelmingly urban. Europe overall has a population that is 74 percent urban. In Iceland and Belgium, more than 90 percent of the population is urban, with Belgium reaching 97 percent. More than 85 percent of the population of Germany, Italy, the United Kingdom, and Luxembourg resides in cities. The least urban are the countries of Albania, Bosnia, and Slovenia, with urban percentages of 42, 43, and 51, respectively. Europe essentially defined the demographic transition, with a trajectory from preindustrial high birth rates and high death rates to postindustrial low birth rates and low death rates (see Figure 2.24). Over a period of decades, Europe recovered from the demographic setbacks of two world wars (France lost more than 3 percent of its prewar population in World War I, and Poland lost 19 percent in World

War II, for example). The population peaked in 1997 and then began a slow, steady decline. Today, birth rates are low in this region of relative affluence and high urbanization. In fact, the region's fertility rate is below **population replacement level** (the number of new births required to keep the population steady), so the population will age and shrink in coming decades. Typical of Europe are the United Kingdom and Belgium, both with annual population change rates of 0.1 percent, with the highest growth rate in Albania (1.2 percent) and the lowest in Bulgaria (−0.6 percent) and Latvia (−0.5 percent). Europe's population is aging faster than that of any other world region. The median age in Italy is projected to rise from 41 in 2004 to 53 by 2050. Spain's population is expected to contract by 22 percent by 2050, and the Netherlands' to shrink by 10 percent. The Italian government is trying to slow a similar trend by offering parents a

Political Geography of Europe

Figure 3.1 Modern Europe. The continent's demographic heavyweights are Germany, the United Kingdom, France, and Italy. In nearly all European countries, the political capital is also the largest and most important city.

"**baby bounty**" equivalent to $1,200 for each child after their first—but only if the mother is Italian or carries a European Union (EU) passport.

Europe's shrinking, aging population could use a youthful boost from immigration, which is already the main source of population growth in most European countries. Studies have shown what additional levels of immigration would be needed to prevent major population declines in Europe and

to maintain the existing ratio of workers aged 15–64 to those over 65 needing support. The European Union countries would need an annual inflow of about 3 million migrants a year (about twice the current rate) to prevent the future support ratio from dropping sharply. But that prospect raises a number of concerns in the individual countries and for the European Union (see the Regional Perspective on page 56).

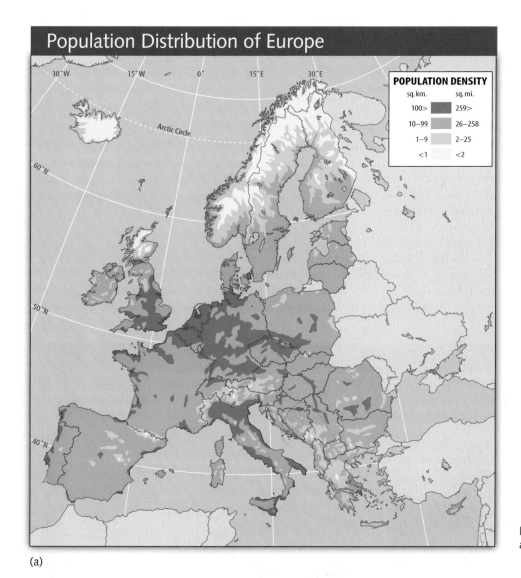

Population Distribution of Europe

POPULATION DENSITY

sq. km.	sq. mi.
100>	259>
10–99	26–258
1–9	2–25
<1	<2

(a)

Figure 3.2 (a) Population distribution and (b) population cartogram of Europe.

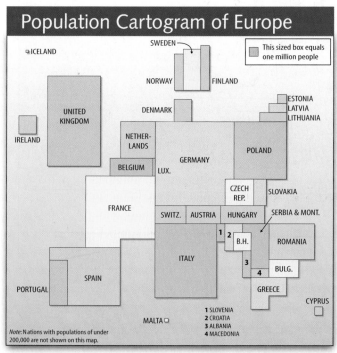

Population Cartogram of Europe

This sized box equals one million people

ICELAND, SWEDEN, NORWAY, FINLAND, UNITED KINGDOM, DENMARK, ESTONIA, LATVIA, LITHUANIA, IRELAND, NETHERLANDS, GERMANY, POLAND, BELGIUM, LUX., FRANCE, CZECH REP., SLOVAKIA, SWITZ., AUSTRIA, HUNGARY, SERBIA & MONT., 1, 2, B.H., ROMANIA, 3, ITALY, 4, BULG., SPAIN, GREECE, PORTUGAL, CYPRUS, MALTA

1 SLOVENIA
2 CROATIA
3 ALBANIA
4 MACEDONIA

Note: Nations with populations of under 200,000 are not shown on this map.

(b)

3.2 Physical Geography and Human Adaptations

Distinguishing Physical Characteristics

There is a broad range of environmental settings in Europe. Among the most distinctive physical characteristics are Europe's irregular outline, its high latitude, and its temperate climate. Europe has a remarkably irregular coastal outline. The main peninsula of Europe is fringed by numerous smaller ones, most notably the Scandinavian, Jutland, Iberian, Italian, and Balkan Peninsulas (Figure 3.3). Offshore are numerous islands, including Great Britain, Ireland, Iceland, Sicily, Sardinia, Corsica, and Crete. The island of Cyprus lies far in the eastern Mediterranean. Around the indented shores of Europe, arms of the sea penetrate the land in the form of significant estuaries (the tidal mouths of rivers), and harbors offer protection for shipping. This complex mingling of land and water provides many opportunities for maritime activity, and much of Europe's history has focused on seaborne trade, sea fisheries, and sea power.

Regional Perspective

Europe's Immigration Issues

There are an estimated 12 to 14 million people of non-European origin living in Europe (Figure 3.A). The major groups are Turks, Kurds, Arabs, Indians, Pakistanis, Sri Lankans, and a variety of peoples from sub-Saharan Africa, the Caribbean, and Latin America. Some of the immigrant communities have been in Europe for a generation or more, including the West Indians who came into Britain to help with worker shortages in the 1950s and the Turks that Germany encouraged to come and work during the boom years beginning around 1960. There are new streams of migrants, too, including the Bangladeshis coming to Britain. Some host countries specialize in importing people from certain countries. In Portugal, for example, Brazilians make up 11 percent of resident foreigners, and Moroccans and Algerians constitute 30 percent of the immigrants in France. There are also internal migrations in Europe, such as Albanians moving to Italy and Bulgarians to Great Britain.

During periods of more rapid economic growth, prior to a downturn in the late 1990s and early 2000s, the immigrants often provided needed labor resources and were generally welcomed. They were the guest workers encouraged to come to industrial and agricultural centers that were not able to meet their labor needs with domestic populations. But recent events, including growing domestic unemployment, security concerns, and tax burdens, have changed native attitudes toward immigrants considerably. Europeans overall are not as fa-

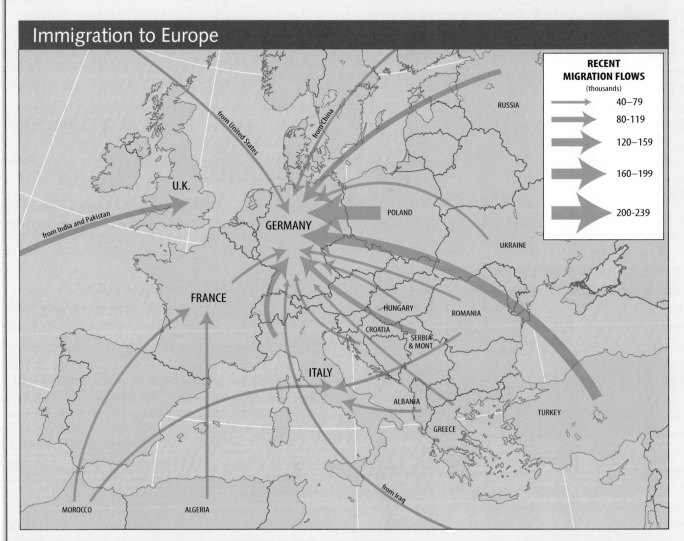

Figure 3.A Immigration, both legal and illegal, is a major issue in Europe. Additional, smaller population flows, such as of Moroccans to Spain, are not indicated on this map.

miliar with immigration as Americans are, and they are trying to figure out what to do with their increasingly multiethnic mosaic (Figure 3.B).

Some governments believe that immigration is good for the economy and continue to encourage and accept more foreigners; Germany is one. Other countries stress that immigration does not always help the receiving country and may harm it. Some countries prohibit students and asylum seekers from working. Not all employable immigrants are fully employed. Many are either in very low-paying and part-time jobs or are out of work. The unemployment rate for migrants is twice that of citizens in Denmark, three times that of citizens in Finland, and four times that of citizens in the Netherlands. Immigrants often have large families whose members end up on welfare,

thus costing the state and its taxpayers rather than generating new tax revenue.

In some countries, notably Denmark and the Netherlands, welfare benefits are extremely generous and in themselves have provided a strong magnet for further immigration. There is now a backlash. Denmark has reduced welfare benefits for new arrivals. Increasing numbers of Dutch nationals are expressing weariness with paying the tax burden to support literacy, technical training, and healthcare costs for the immigrants. Many point to the fact that the country's second largest city, Rotterdam, has almost as many nonnative (mainly Moroccan and Turkish) as native Dutch inhabitants. They fear they may become a minority in their own country, pointing out that Moroccan-born women who reside in the Netherlands have a birth rate four times that of Dutch women. The country is now exploring ways to slow down immigration.

Powerful economic forces tend to trump governmental restrictions on migration. With the rich countries of Europe creating jobs, the underemployed young in the poor countries want to fill them; simple laws of supply and demand keep the flow going. Governments are often reluctant to impose draconian measures that would restrict migration. Borders are porous. Today, about 1.2 million people enter the European Union legally each year, and about 500,000 come illegally. Germany has an estimated 1 million illegals (also known as **undocumented workers**), and France has 500,000. Most of the illegals cross land or sea borders illegally, but perhaps a third are visitors who overstay tourist or student visas. Concerned about a new flood of undocumented workers that might inundate the older, richer EU countries, the European Union will not allow citizens of the 10 new EU countries the freedom of movement between member states now enjoyed by the others.

In Europe, especially Northern Europe, a large source of legal immigration is seeking **asylum.** A 1967 protocol signed by the European countries gives everyone the right to seek asylum in Europe and gives refugees the right not to be sent to a place of torture or persecution. At first, few people took advantage of this opportunity, but a tide of asylum seekers rose during the conflicts in the former Yugoslavia in the 1990s. Asylum claims in EU countries have fallen 50 percent since the early 1990s but still number about half a million people each year. Iranians, Afghans, and others still want to come, claiming asylum from persecution but perhaps more often motivated by poverty. This accounts for why about 80 percent of the applicants are rejected.

Figure 3.B Muslim women in Aberdeen, Scotland. Europeans are not of one voice in answering questions about their ethnic and religious diversity.

Physical Geography of Europe

Figure 3.3 Reference map of the main islands, peninsulas, seas, straits, topographic features, and rivers of Europe

A striking environmental characteristic of Europe is its northerly location: Much of Europe lies north of the 48 conterminous United States (Figure 3.4). Despite their moderate climate, the British Isles are at the same latitude as Hudson Bay in Canada. Athens, Greece, is only slightly farther south than St. Louis, Missouri. The high latitudes have predictable implications for people. Northern Europeans revel in the long summer days but pay a price with long, dark nights of winter. Scandinavians in particular celebrate the summer solstice as "midsummer" with great fervor.

Europe has a much milder climate than other regions lying on the same parallels of latitude. Winter temperatures are

Figure 3.4 Europe in terms of latitude and area compared with the United States and Canada. Most islands have been omitted.

especially mild for such high latitudes. London, for example, has about the same average temperature in January as Richmond, Virginia, which is 950 miles (c. 1,500 km) farther south. These anomalies of temperature are caused by relatively warm currents of water. The Gulf Stream and its appendage the North Atlantic Drift, which originate in tropical western parts of the Atlantic Ocean, flow to the north and east and cause the waters around Europe to be much warmer in winter than the latitude would warrant. With winds blowing mainly from the west (from sea to land) along the Atlantic coast of Europe, the moving air in winter absorbs heat from the ocean and transports it to the land. This makes winter temperatures abnormally mild for this latitude, especially along the coast. In the summer, the climatic roles of water and land are reversed: Instead of being warmer than the land, the ocean is now cooler, so the air brought to the land by westerly winds (also known as *westerlies*) in the summer has a cooling effect. These influences are carried as far north as the Scandinavian Peninsula and into the Barents Sea, giving the Russians a warm-water port at Murmansk, even though it lies within the Arctic Circle.

The same winds that bring warmth in winter and coolness in summer also bring abundant moisture. Most of this falls as rain, although the higher mountains and more northerly areas have considerable snow. Ample, well-distributed, and relatively dependable moisture has always been one of Europe's major assets (see the world precipitation map, Figure 2.1 on page 18). The average precipitation in most places in the European lowlands is 20 to 40 inches (c. 50 to 100 cm) per year. This is sufficient for a wide range of crops because mild temperatures and high atmospheric humidity reduce the rate of evapotranspiration.

Europe's Varied Landscapes

Europe's topographic features are highly diversified, with plains, plateaus, hill lands, mountains, and water bodies all well represented. Overspread with varied plant life and enriched by the human associations and constructions of an eventful history, Europe presents a series of distinctive and often highly scenic landscapes.

One of the most prominent surface features of Europe is a plain that extends without a break from near the French–Spanish border, across western and northern France, central and northern Belgium, the Netherlands, Denmark, northern Germany, Poland, and far into Russia. Known as the North European Plain (see Figure. 3.3), it has outliers in Great Britain, the southern part of the Scandinavian Peninsula, and southern Finland. For the most part, the plain is undulating or rolling. The North European Plain contains the greater part of Europe's cultivated land, and it is underlain in some places by deposits of coal, iron ore, potash (used in the production of soaps, glass, and the manufacture of other potassium compounds), and other minerals that were important in the region's industrial development. Many of the largest European cities, including London, Paris, and Berlin, developed on the plain. From northeast France eastward, a band of especially dense population extends along the southern edge of the plain. It coincides with fertile soils formed from deposits of loess (a windblown soil of generally high fertility), an important natural transportation route skirting the highlands to the south, and a large share of the region's mineral deposits.

South of this northern plain, Europe is mainly mountainous or hilly, with scattered plains, valleys, and plateaus. The hills are geologically old, and erosion has lowered their elevations over long periods. But many mountains in southern Europe are geologically young, and often high and ruggedly spectacular, with jagged peaks and snowcapped summits. They reach their peak of height and grandeur in the Alps, with the highest summit of Mt. Blanc (15,771 feet, 4,807 m)

Figure 3.5 The Alps near Mt. Blanc in January

Maximum Glaciation Across Europe

Figure 3.6 The maximum extent of glaciation in Europe, about 18,000 years ago   *Source: After Hoffman, 1990*

on the French–Italian border (Figure 3.5). Another formidable mountain chain, the Pyrenees, forms a wall up to 10,695 feet (3,404 m) along the border of Spain and France. Lower but legendary chains of mountains, the Transylvanian Alps and the Carpathian Mountains, join in Romania.

Glaciation has played a major role in shaping the landscapes of Europe. During periods of ice advance in the **Ice Age** of the Pleistocene (c. 2,000,000 to 10,000 years ago), continental ice sheets formed over Europe, beginning on the Scandinavian Peninsula and Scotland (Figure 3.6). Evidence suggests that there were four major periods of widespread glacial coverage. Today, landscapes of glacial scouring—the erosive action of ice masses in motion—characterize most of Norway and Finland, much of Sweden, parts of the British Isles, and Iceland. These changes have often created favorable sites for hydroelectric installations. Glacial deposition—the process of offloading rock and soil in glacial retreat or lateral movement—had a major effect on the present landscape. Glacial deposits of varying thickness were left behind on most of the North European Plain and are productively farmed today.

The North European Plain is bordered by glaciated lowlands and hill lands in Finland and eastern Sweden and by rugged, ice-scoured mountains and fjords in western Sweden and most of Norway. The British Isles also have extensive areas of glacially scoured hill country and low mountains, along with lowlands where glacial deposition occurred.

Climates of Europe

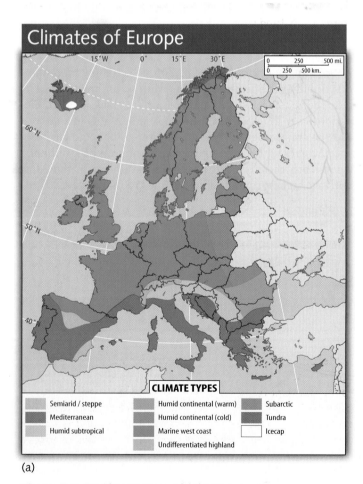

CLIMATE TYPES

Semiarid / steppe	Humid continental (warm)	Subarctic
Mediterranean	Humid continental (cold)	Tundra
Humid subtropical	Marine west coast	Icecap
	Undifferentiated highland	

(a)

Biomes of Europe

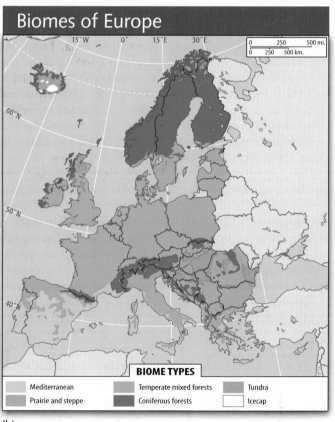

BIOME TYPES

Mediterranean	Temperate mixed forests	Tundra
Prairie and steppe	Coniferous forests	Icecap

(b)

Figure 3.7 (a) Climate types and (b) Biomes

Climate and Vegetation Patterns

Europe, despite its modest spatial dimensions, has remarkable climatic and biotic diversity (Figures 3.7a and b; see also Figures 2.6 and 2.7, pages 21 and 22). A marine west coast climate extends from the coast of Norway to northern Spain and inland to West Central Europe, or the Coreland. Comparable to the climate found in the U.S. Pacific Northwest, the main characteristics are mild winters, cool summers, and ample rainfall, with many drizzly, cloudy, and foggy days. Throughout the year, changes of weather follow each other in rapid succession as different air masses temporarily dominate or collide with each other along weather fronts. Most precipitation is frontal in origin or results from a combination of frontal and orographic (highland) influences (see Chapter 2, pages 17–19). In lowlands, winter snowfall is light, and the ground is seldom covered for more than a few days at a time. Summer days are longer, brighter, and more pleasant than the short, cloudy days of winter, but even in summer, there are many chilly and overcast days. The frost-free season of 175 to 250 days is long enough for most crops grown in the middle latitudes to mature, although most areas have summers that are too cool for heat-loving crops such as corn (maize) to ripen (Figure 3.8).

Inland from the coast, in Western and Central Europe, the marine climate gradually changes. Winters become colder and summers are hotter; cloudiness and annual precipitation decrease. Influences of maritime air masses from the Atlantic diminish and are modified by continental air masses from inner Asia. Farther inland, conditions are different, and two new climate types are apparent: the humid continental short-summer (cold) climate in the north (principally in southern parts of Sweden and Finland, Estonia, Latvia, Lithuania, Poland, Slovakia, eastern Hungary, and northern Romania) and the humid continental long-summer (warm) climate in the warmer south (mainly in southern Romania, Serbia, and northern Bulgaria). The natural vegetation is mostly temperate mixed forest, and soils vary in quality. Among the best soils are those formed from alluvium and loess along the 1,776-mile (2,842-km) valley of the Danube River.

Southernmost Europe has a distinctive climate: the Mediterranean climate, with a pattern of dry summers and wet winters. This pattern results from seasonal shifts in atmospheric belts. In winter, the belt of westerly winds shifts southward, bringing precipitation at a time of relatively low evaporation and helping to replenish subsurface waters. In summer, the belt of subsiding high atmospheric pressure over the Sahara shifts northward, bringing desert conditions. Mediterranean summers are warm to hot, and little precipitation occurs during the summer months when temperatures are most advantageous for crop growth. Winters are mild and frosts are few. Drought-resistant trees originally covered Mediterranean lands, but little of that forest remains, having been depleted by centuries of human use. It has been replaced by the wild scrub that the French call *maquis* (called *chaparral* in the United States and Spain) or by cultivated

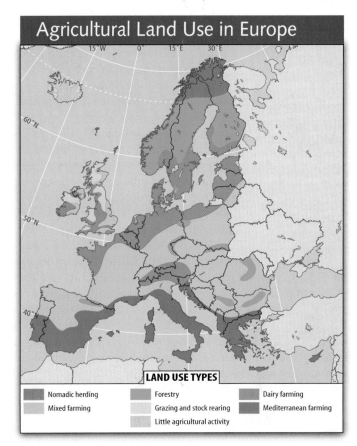

Agricultural Land Use in Europe

LAND USE TYPES

- Nomadic herding
- Mixed farming
- Forestry
- Grazing and stock rearing
- Little agricultural activity
- Dairy farming
- Mediterranean farming

Figure 3.8 Land use in Europe

Joe Hobbs

Figure 3.9 Mediterranean landscape, Greece

fields, orchards, or vineyards (Figure 3.9). Much of the land consists of rugged, rocky, and eroded slopes where thousands of years of deforestation, overgrazing, and excessive cultivation have taken their toll. The subtropical temperatures allow a great variety of crops, but irrigation is a must in the dry summers. With irrigation, an enormous variety of crops can be raised productively. This climate is also a draw for summer travelers, and tourism has immense economic importance in Mediterranean Europe.

Some northerly sections of Europe experience the harsh conditions associated with subarctic and tundra climates. The subarctic climate, characterized by long, severe winters and short, rather cool summers, covers most of Finland, much of Sweden, and some of Norway. A short frost-free season, coupled with thin, highly leached, acidic soils, makes agriculture difficult. Human settlement is sparse, and forest of needle-leaf conifers, such as spruce and fir, covers most of the land. In the tundra climate of northernmost Norway and much of Iceland, cold winters combine with brief, cool summers and strong winds to create conditions hostile to tree growth. An open, windswept landscape results, covered with lichens, mosses, grass, low bushes, dwarf trees, and wildflowers (a typical Norwegian tundra landscape may be seen in Figure 2.8b on page 23). Human inhabitants are few. Agriculture in a normal sense is not feasible. In the taiga region, south of the tundra, a slight increase in moisture and temperature promotes growth of coniferous forest.

The higher mountains of Europe, like high mountains in other parts of the world, have an undifferentiated highland climate varying with elevation and differential exposure to sun, wind, and precipitation. The variety can be startling. The Italian slope of the Alps, for instance, ascends from subtropical conditions at the base of the mountains to tundra and icecap climates at the highest elevations. The icecap climate experiences temperatures that historically have averaged below freezing every month of the year. At the higher latitudes, these extreme conditions are found even at sea level. Europe's glaciers are following the almost worldwide trend of retreat in the past half century due mainly, climatologists believe, to increasing levels of greenhouse gases in the Earth's atmosphere and the accompanying global warming.

Rivers and Waterways

Europe has many river systems that are important economically for transport, water supply, electricity generation, and recreation. Rivers were an important part of Europe's transport system at least as far back as Roman times. Today, they still facilitate movement of cargo at low cost, mostly in motorized barges. The more important rivers for transportation are those of the highly industrialized areas in the Coreland area of West North Central Europe. An extensive system of canals connects and supplements the rivers. Some are quite old; the Romans built canals throughout northern Europe and Britain, mainly for military transport.

The Dutch developed the pound (pond) lock for canals in the late 14th century. This led to the expansion of regional connections of waterways through canals and to the use of canals to link farmlands with the coasts. Starting in the late 1700s, Britain built canals to move raw materials toward factory locations (the longest of these was the 140-mile [230-km] Leeds–Liverpool Canal). By the mid-19th century, the ex-

panding influence of railroads and the steam engine competed with canals. Although canals continue to play a significant transport role, since the 1950s railroads and highways have been the main means to transport goods throughout Europe.

Important seaports developed along the lower courses of many rivers, and some became major cities. London on the Thames, Antwerp on the Scheldt, Rotterdam in the delta of the Rhine, and Hamburg on the Elbe are outstanding examples. Often, the river mouths are wide and deep, allowing ocean ships to travel a considerable distance upstream and inland. This is true even of short rivers such as the Thames and Scheldt.

The Rhine and the Danube are particularly important European rivers, touching or crossing the territory of many countries (see Figure 3.3). The Rhine countries include Switzerland, Liechtenstein, Austria, France, Germany, and the Netherlands. Highly scenic for much of its course, the Rhine River is Europe's most important inland waterway (Figure 3.10). Along and near it, an axis of intense urban-industrial development and high population density developed. At its North Sea end, the Rhine axis connects to world commerce by the world's most active seaport, Rotterdam, in the Netherlands. The Danube (German: Donau) River, on its journey from the Black Forest of southwestern Germany to the Romanian shore of the Black Sea, touches or crosses more countries than any other river in the world. Within Europe, the Danube is unusual in its southeasterly course. The river is an important artery for the flow of goods and is also held in deep esteem by the cultures that share its waters. The Danube delta region of eastern Romania is one of the world's most important wetlands for waterfowl, and it draws growing numbers of ecotourists.

Figure 3.10 The Rhine River

3.3 Cultural and Historical Geographies

The apparent crowding of so many countries into the relatively small land area of Europe is an indication of the region's extraordinary cultural diversity. Travelers in Europe experience this richness. A brief train ride, for example, takes the visitor from French-speaking western Switzerland eastward into the country's mainly Germanic region, and southward through mountain tunnels into the Italian world. Today, almost all of Europe's different peoples coexist harmoniously, but this harmony is something rather recent for Europe.

Language and Ethnicity

Europe emerged from prehistory as the homeland of many different peoples. In ancient and medieval times, some of them expanded vigorously, and their languages and cultures became widely diffused. The first millennium B.C. witnessed a great expansion of the **Greek** and Celtic (pronounced KEL-tik)

peoples. In peninsulas and islands bordering the Aegean and Ionian Seas, the early Greeks developed a civilization that reached unsurpassed heights of philosophical inquiry and literary and artistic expression. Greek adventurers, traders, and colonists used the Mediterranean Sea as their highway to spread classical Greek civilization and its language along much of the Mediterranean shoreline. Evidence of the geographic range and influence of the Greek language and culture is apparent in the many Greek elements in modern European languages. But over time, the use of Greek in most areas disappeared as new peoples and languages were introduced and expanded. Several language families are represented in Europe today, and language is one of they key components of European ethnicity (Figure 3.11).

WORLD
REGIONAL
Geography Now™

Click Geography Literacy to hear words in the different languages and language families.

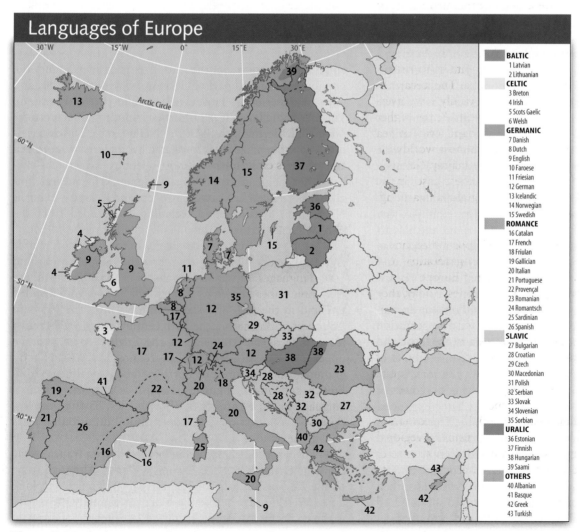

Figure 3.11 The languages of Europe   *Source: After Comrie et al., 2003*

Europe's **Celtic languages** expanded at roughly the same time as Greek and, like Greek, are represented today only by remnants. Preliterate Celtic-speaking tribes radiated from a culture hearth in what is now southern Germany and Austria, eventually occupying much of continental Europe and the British Isles. Conquest and cultural influence by later arrivals, mainly Romans and Germans, eventually eliminated the Celtic languages except for a few traces that survive today in isolated pockets in the British Isles (notably in Wales, Scotland, and Ireland) and in French Brittany.

WORLD
REGIONAL
Geography⊛Now™

Click Geography Literacy to take a virtual tour of the Scottish Islands.

In present-day Europe, the overwhelming majority of the people speak Romance, Germanic, and Slavic languages. Like Greek and Celtic, these are **Indo-European languages.** The **Romance languages** evolved from **Latin,** originally the language of ancient Rome and a small district around it. In the few centuries before and just after the birth of Christ, the Romans subdued territories extending through western Europe as far west as Great Britain, and the use of Latin spread to this large empire (Figure 3.12). Latin had the greatest impact in the less developed and less populous western parts of the empire. Over a long period, well beyond the collapse of this western part of the empire in the fifth century, regional dialects of Latin survived and evolved into **Italian, French, Spanish, Catalan** (still spoken in the northeast corner of Spain), **Portuguese, Romanian,** and other Romance languages of today.

It is worth noting here that language frontiers do not always coincide with political frontiers. For example, the French language extends to western Switzerland and also to southern Belgium, where it is known as **Walloon.** French was Europe's dominant language as recently as the 18th century, when France was the continent's intellectual capital. French is still very widely spoken outside the handful of countries in which it is the leading language and is second only to English as a leading tongue in European affairs.

In the middle centuries of the first millennium A.D., the power of Rome declined and a prolonged expansion by Germanic and Slavic peoples began. Germanic (Teutonic) peoples first appear in history as groups inhabiting the coasts of Germany and much of Scandinavia. They later expanded southward into Celtic lands east of the Rhine. They repelled Roman attempts at conquest, and the Latin language had little impact in Germany. In the fifth and sixth centuries A.D., Germanic incursions overran the western Roman Empire, but in many areas, the conquerors eventually were absorbed into the culture and language of their Latinized subjects.

However, the **German** language expanded into, and remains, the language of present-day Germany, Austria, Lux-

Roman Empire at Greatest Extent

WORLD
REGIONAL
Geography⊛Now™

Active Figure 3.12 The Mediterranean became a "Roman lake" in the first century A.D. *See an animation based on this figure, and take a short quiz on the concepts.*

embourg, Liechtenstein, the greater part of Switzerland, the previously Latinized part of Germany west of the Rhine, and parts of easternmost France (Alsace and part of Lorraine). The **Germanic languages** of Europe include many languages other than German itself. In the Netherlands, **Dutch** developed as a language closely related to dialects of northern Germany, and **Flemish**—almost identical to Dutch—became the language of northern Belgium. The present languages of Denmark, Norway, Sweden, and Iceland—**Danish, Norwegian, Swedish,** and **Icelandic**—descended from the same ancient Germanic tongue, although Finnish did not.

English is a Germanic language, but it has many words and expressions derived from French, Latin, Greek, and other languages. Originally, English was the language of the Germanic tribes known as Angles and Saxons who invaded England in the fifth and sixth centuries A.D. The Norman Conquest of England in the 11th century established French for a time as the language of the English court and the upper classes. Modern English retains the Anglo-Saxon grammatical structure but borrows much vocabulary from French and other languages. English is now the principal language in most parts of the British Isles, having been imposed by conquest or spread by cultural diffusion to areas outside England.

English is also becoming the de facto lingua franca of the 25 member nations of the European Union. Although some of the key institutions of the EU are in French-speaking cities, French usage across the bloc is declining. One reason is the recent accession of 10 new EU member countries in Eastern Europe, where about 60 percent of the population

Definitions + Insights

Devolution

A notable political process in modern Europe is devolution, the dispersal of political powers to ethnic minorities and other subnational groups. Like that of many other world regions, Europe's historical geography is rife with dominant groups occupying the traditional lands of smaller minority groups. Minorities are sometimes peacefully absorbed into the larger culture and political system, but more often than not, they retain some degree of distinctiveness and a desire to control their own affairs.

This text relates numerous examples around the world of strife and war taking place because the dominant power refuses to yield any measure of authority to minority groups. But in Europe, the process of devolution is taking place mainly peacefully and more energetically now than ever. This may be because the European countries, after centuries of strife, are finally self-assured of their sovereignty and well-being, and are so focused on international cooperation, that demands for autonomy among their minorities are no longer perceived as threats. However, the European Union—already preoccupied with meeting the diverse needs of 25 member states—sees devolution as a vexing problem. The following two chapters describe devolution in Wales, Scotland, Northern Ireland, Spain, and elsewhere; Figure 3.C is a map of the major locales where devolution has occurred or is being appealed for in modern Europe.

Figure 3.C Power is being devolved from central governments to regional authorities in many parts of Europe. This map depicts places like Scotland and Wales, where the transfer has taken place, and others like Corsica, where devolution is anticipated or sought.

speaks English as a second language, compared with only 20 percent speaking French as a second language. English is by far the most widely taught second language in non-English speaking European countries, where it is studied by over 90 percent of secondary school students (compared with 33 percent studying French and 13 percent German).

Slavic languages—also in the Indo-European language group—are dominant in most of Eastern Europe. The main ones today include the **Russian** and **Ukrainian** of the adjoining region of Russia and the Near Abroad, **Polish, Czech,** and **Slovak,** and **Serbian, Croatian,** and **Bulgarian** in the Balkan Peninsula. They apparently originated in eastern Europe and Russia and were spread and differentiated from each other during the Middle Ages as people traded and migrated. Closely related to the Slavic languages are the **Baltic languages** of **Latvian** and **Lithuanian.** Slavic ethnicity is described in more detail in Chapter 5.

A few languages in present-day Europe are not related to any of the groups just discussed. Some are ancient languages that have persisted from prehistoric times in isolated, and usually mountainous, locales. Two outstanding examples are **Albanian** (of the Indo-European language family) in the Balkan Peninsula and **Basque** (an indigenous European language without Indo-European roots), spoken in and near the western Pyrenees Mountains of Spain and France. Some languages unrelated to others in Europe have relatives in Russia and reached their present locales through migrations of peoples westward. The prime examples are **Finnish, Saami, Estonian,** and **Hungarian** (also called Magyar). These belong to the **Uralic language family,** which is thought to have originated in the Urals region of Russia. **Turkish,** within the **Altaic language family** that originated in central Asia, is spoken by the ethnic Turkish inhabitants of northern Cyprus. The Roma (Gypsies) speak **Romany,** in the **Indic** branch of the Indo-European languages.

There are seven ethnic patterns in Europe, with language serving as important components of ethnicity.[1] Several nations are populated overwhelmingly by a single ethnic group—for example, the Germans of Germany, the **Poles** of Poland, and the Dutch of the Netherlands. There are also substantial populations of ethnic groups that originated in other nations living in "foreign" countries (for example, the Poles of Lithuania), where they are known as **national minorities.** Third, there are groups distinguished more by culture than by ethnicity; they share a common culture but have many different ethnic origins (for example, the French and the Italians). Fourth, there are a few ethnolinguistic minorities, like the **Saami** (**Lapps**) of Scandinavia and the **Ladin** of Italy, who have resided in an area for a very long time but have retained cultural distinctiveness.

There are also regional ethnic minorities that have a major presence in a particular nation, where they are culturally very different from the country's other ethnic groups; examples are the Basques and Catalans in Spain and the **Bretons** in France. Such groups often call for the devolution of power from the host country to their ethnic subregion (see Definitions and Insights, page 65). Sixth, there are recent immigrants from outside Europe, including "guest workers" mainly from former European colonies in Asia, Africa, and the Americas; examples are the Indians of Great Britain and the North Africans of France. Finally, there are ethnic groups with members all across Europe but without a homeland there; the **Jews** (who originated in modern Israel and adjacent areas; see Chapter 8, page 212) and the **Roma,** or **Gypsies** (who probably originated in India), are the most prominent of these. In subsequent chapters, the characteristics, the contributions, and the conflicts of Europe's ethnic groups are discussed in more detail.

Belief Systems

One of the principal culture traits of Europe is the dominance of **Christianity** (see the map of Europe's faiths in Figure 3.13 and the discussion of Christianity in Chapter 8, page 215). Initially outlawed by the Romans, Christianity was embraced as the faith of the realm by the Emperor Constantine in the fourth century, and it became a dominant institution in the late stages of the empire. It survived the empire's fall and continued to spread, first within Europe and then to other parts of the world. In total number of adherents, the **Roman Catholic Church,** headed by the pope in the Vatican, is Europe's largest religious group (with 280 million followers, or 45 percent of Europe's Christians), as it has been since the Christian Church was first established. Today, it remains the principal faith of a highly secularized Europe. The main areas that are predominantly Roman Catholic are Italy, Spain, Portugal, France, Belgium, the Republic of Ireland, large parts of the Netherlands, Germany, Switzerland, Austria, Poland, Hungary, the Czech Republic, Slovakia, Croatia, and Slovenia.

During the Middle Ages, a center of Christianity evolved in Constantinople (now Istanbul, Turkey) as a rival to Rome. From that seat of power, the **Eastern Orthodox Church** spread to become, and remains, dominant in Greece, Bulgaria, Romania, Serbia and Montenegro, and Macedonia, as well as in the Ukraine and Russia to the east.

In the 16th century, the **Protestant Reformation** took root in various parts of Europe, and a subsequent series of religious wars, persecutions, and counterpersecutions left **Protestantism** dominant in Great Britain, northern Germany, the Netherlands, Denmark, Norway, Sweden, Iceland, and Finland. Except for the Netherlands, where a higher Catholic birth rate has since reversed the balance, these areas are still mainly Protestant, with the **Church of England** and **Calvinism** the strongest sects in Great Britain and **Lutheran Protestantism** dominant in Germany and Scandinavia.

F5.15
118

---

[1]This characterization is by David Levinson in *Ethnic Groups Worldwide* (Oryx Press, 1998), p. 1.

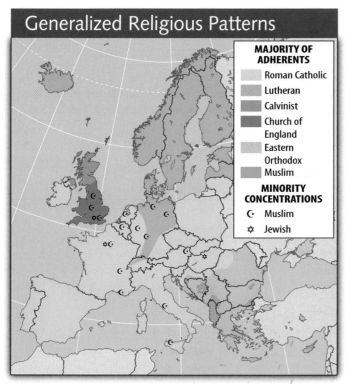

Generalized Religious Patterns

MAJORITY OF ADHERENTS
- Roman Catholic
- Lutheran
- Calvinist
- Church of England
- Eastern Orthodox
- Muslim

MINORITY CONCENTRATIONS
- ☾ Muslim
- ✡ Jewish

Figure 3.13 The religions of Europe. Sects of Christianity dominate, but Islam has a strong presence in the Balkans.

Table 3.2 European Union Countries with the Biggest Muslim Populations

Country	Muslim Population	% of Population
France	4–5 million	7.1–8.9%
Germany	3 million	3.7%
Britain	1.4 million	2.6%
Italy	700,000	1.2%
Netherlands	696,000	4.4%
Belgium	370,000	3.6%
Greece	370,000	3.7%
Spain	300,000–400,000	0.7–1%
Austria	300,000	3.7%
Sweden	250,000–300,000	2.8–3.4%

Source: *The Christian Science Monitor* (Ehrlich and Vandyck, 2003).

Among the three Baltic countries, Lutheran Protestantism is the dominant faith in Estonia and Latvia, while most Lithuanians are Catholic.

That Christians form a vast majority is a somewhat misleading characterization of religion in Europe. Mainstream Christianity is in obvious decline as Europeans have become increasingly secularized in recent decades. Recent polls indicate that large majorities of Europeans believe in God, but even larger majorities shun traditions like attending church. In predominantly Catholic France, about 1 in 20 people attends a religious service every week (compared with one in three in the United States). Various polls suggest that somewhere between 15 and 33 percent of Italians attend church in mainly Catholic Italy. Polls indicate that, overall, 21 percent of all Europeans regard religion as "very important" to them (compared with 58 percent of Americans). Those Christian churches that are growing—even surging—in attendance tend to be independent **Pentecostal Churches** attended by converts from Nigeria, Sierra Leone, and other African countries.

The faith of **Islam** (practiced by **Muslims**) has the fastest growing number of adherents of any world religion and is the fastest growing religion in Europe (for the description of Islam, see Chapter 8, pages 216–218). Muslims are a majority in just two European countries: Albania and Bosnia and Herzegovina on the Balkan Peninsula. Islam was once widespread in the Balkans (as this region is often called), where

the Ottoman Turks established it during their period of rule from the 14th to 19th centuries. It was also the religion of the **Moors** who ruled the Iberian Peninsula from the 8th century until the end of the 15th century, when Spain's Catholic monarchy ousted them. Today, Islam is growing throughout the European realm, particularly among immigrant communities in Western Europe (Table 3.2). Estimates vary, but France has as many as 5 million Muslims, followed by Germany with about 3 million, Britain approaching 2 million, and Italy and the Netherlands with almost 1 million each. Muslim immigrants from the Balkans and from North Africa and the Middle East continue to arrive, and resident Muslims in Europe have higher birth rates than their native counterparts, so the overall numbers of Muslims are poised to rise substantially.

One of humankind's greatest tragedies has been the story of the Jews of Europe (see also Chapter 8, page 215). The Romans forcibly dispersed these followers of **Judaism** to Europe from Palestine in the first century A.D., and over time, they grew to have more numbers in Europe than anywhere else. In 1880, before many emigrated to the United States, 90 percent of the world's Jews lived in Europe. In 1939, as World War II erupted, there were 9.7 million Jews, fully 60 percent of the world total, in Europe. By the end of the war, systematic murder by the Nazis in the Holocaust reduced their numbers to 3.9 million. About half of those subsequently emigrated to Israel and elsewhere, and there are about 1.6 million Jews in Europe today.

European Impacts Abroad

Far more than any other region, Europe has shaped the human geography of the modern world. Yet, before the late 15th century, Europe played a minor role in world trade patterns. Western Europe saw goods moving from southern France and northern Italy northwest to Britain. A more active European trade connected Italy with the margins of the Mediterranean Sea and beyond to Southwest Asia. But the most important trade routes had centers much farther east and south. The longest link was the 5,000-mile (9,000-km) **Silk Road** that moved goods overland (and on the Mediterranean Sea) from Chang-an (Xi'an) in China to Venice.

The balance of world affairs started shifting to Europe with the beginning of the Age of Discovery in the 15th century. European sailors, missionaries, traders, soldiers, and colonists burst upon the world scene. Although these agents had different motives, they shared a conviction that European ways were superior to all others, and by various means, they sought to bring as much of the world as possible under their sway. Economic benefits for the European homelands were among their main motivations. By the end of the 19th century, Europeans had created a world in which they were economically dominant and exercised great influence on indigenous cultures.

The process of exploration and discovery by which Europeans filled in the world map began with 15th century Portuguese expeditions down the west coast of Africa. In 1488, a Portuguese expedition headed by Bartholomeu Dias rounded the Cape of Good Hope at the southern tip of Africa and opened the way for European voyages eastward into the Indian Ocean. Then, in 1492, North America was brought into contact with Europe when a Spanish expedition commanded by a Genoese (Italian) man named Christopher Columbus (Cristóbal Colón) crossed the Atlantic to the Caribbean Sea. Less than half a century later, these early feats of exploration culminated in the first circumnavigation of the globe by a Spanish expedition commanded initially by a Portuguese man, Ferdinand Magellan. Worldwide exploration continued apace for centuries, with the ascending Dutch, French, and English wresting leadership from the Spanish and Portuguese.

The explorers were the vanguards of a global European invasion that in turn would bring missionaries, soldiers, traders, settlers, and administrators. They gradually built the frameworks of European colonization of the Americas, Africa, Asia, and the Pacific. The colonial patterns are described on pages 31–35 in Chapter 2 and will not be repeated in detail here. Briefly, their main manifestations were the transfer of wealth from the colonies to Europe, the movement of great numbers of indigenous laborers to serve European interests in other lands, the settlement of Europeans in agriculturally productive colonies (where their descendants are great majorities or significant minorities today), and the attempted subjugation of indigenous cultures to European ones. These are among the most prominent issues in world geography; every chapter of this text deals with the impacts of European colonization on cultures and landscapes beyond Europe.

Of great importance in reshaping the world's biogeography was the transfer of plants and animals from one place to another following Europe's conquest of the Americas. A few

TABLE 3.3 Diffusion of Foodstuffs Between New World and Old World: The Columbian Exchange

This table shows some of the most significant foodstuffs diffused from the New World to the Old World after 1492, as well as the major New World foodstuffs introduced into world trade after 1492 by the ships and crews that came to the newly found lands of North, Middle, and South America.

Diffused from Old World to New World	Diffused from New World to Old World
Wheat	Maize
Grapes	Potatoes
Olives	Cassava (manioc)
Onions	Tomatoes
Melons	Beans (lima, string, navy, kidney, etc.)
Lettuce	Pumpkins
Rice	Squash
Soybeans	Sweet potatoes
Coffee	Peanuts
Bananas	Cacao
Yams	

major examples include the introduction of hogs and cattle and the reintroduction of horses to the Americas from Europe; of tobacco and corn (maize) from the Americas to Europe and other parts of the world; of rubber from South America to Asia; of the potato from the Americas to Europe, where it became a major food; and of coffee from Africa to Latin America, where it became the principal basis of a number of national economies (Table 3.3).

3.4 Economic Geography

Supremacy and Decline

The colonial enterprise made Europe the world's wealthiest region for several centuries. Europe had significant material and cultural wealth in its own right, however, and the colonial system was built on that wealth. The European capitalist system and huge strides in science and technology helped launch the region into economic supremacy. Parts of Europe had already developed capitalist institutions by the end of the Middle Ages, and the Colonial Age saw further rapid development of economic enterprise in the region. Profit became an acceptable motive, and the relative freedom of action afforded by the capitalistic system provided opportunities for taking risks and gaining wealth. Energetic entrepre-

360

Edmund O. Rhodes

Figure 3.14 In the 20th century, two devastating world wars originated in and centered on Europe, causing enormous suffering and eclipsing the region's global economic dominance. These are the ruins of the Reichstag in Berlin in 1945.

neurs mobilized capital into large companies and utilized both European and overseas labor.

Europeans reached a level of technology generally superior to that of non-Europeans with whom they came in contact during the early Age of Discovery. Achievements in shipbuilding, navigation, and the manufacture and handling of weapons gave them decided advantages. Nations such as China, once technologically superior, quickly fell behind. Europe gained almost free rein in control and expansion of technological innovation for several centuries. The foundations of modern science were also constructed almost entirely in Europe, and the great names of science between about 1500 and 1900 were overwhelmingly the names of Europeans.

These material, cultural, and intellectual forces combined to launch the Industrial Revolution in Europe, which was the first world region to evolve from an agricultural into an industrial society. In the first half of the 18th century, and mainly in Britain, industrial innovations made water power and then coal-fueled steam power increasingly available to turn machines and drive gears in new factories. James Watt's invention of a steam engine in 1769 made coal a major resource and greatly increased the amount of power available. New processes and equipment, including the development of coke (coal with most of its volatile constituents burned off), made iron smelting cheaper and more abundant. The invention of new industrial machinery made of iron and driven by steam engines multiplied the output of manufactured products, initially textiles. Then, in the 19th century, British inventors developed processes that allowed steel, a metal superior to iron in strength and versatility, to be made inex-

pensively and on a large scale for the first time. These developments soon diffused to new hearths of massive, mechanized industrial production and brought unprecedented changes to landscapes and ecosystems around the world (for more on the revolutionary nature of the Industrial Revolution, see Chapter 2, pages 31–33). Among the most immediate and significant processes accompanying industrialization were rapid population growth and a dramatic shift of people from rural areas into cities.

By 1900, the people and machines of western European cities created about 90 percent of the world's manufacturing output. Europe led the world in production of textiles, steel, ships, chemicals, and other industries. But in the 20th century (and especially after 1960), Europe's preeminence in world trade and industry diminished to about 25 percent of the world's manufacturing output for several reasons. One was warfare (Figure 3.14). Europe suffered enormous casualties and damage in World Wars I and II, which were initiated and fought mainly in Europe. Recovery was eventually achieved with U.S. aid, but the region did not regain its former supremacy.

Second, during the 20th century, rising nationalism—the quest by colonies and ethnic groups to possess their own homelands—brought an end of the European colonial empires. Taking advantage of a weakened Europe and mounting disapproval of colonialism in the world at large, one European colony after another gained independence quickly in the decades following the end of World War II. Opposition to continued European control was often spearheaded by colonial leaders who had been educated in Europe and had absorbed nationalistic ideas there.

Third, Europe's predominance was seriously eroded by the rising economic and political stature of the United States and the Soviet Union. Each far larger than any European nation, these countries outpaced Europe in military power, economic resources, and world influence, particularly in the postwar years 1950–1970.

Fourth, a major shift took place in global manufacturing patterns. Europe once enjoyed a near monopoly in exports of manufactured goods, but in recent decades, manufacturing has accelerated in many countries outside Europe. The United States reached industrial maturity in the 20th century and became a vigorous competitor of Europe. More recently, the Asian countries Japan, South Korea, China, and Taiwan have became industrial forces capable of competing with Europe in markets all over the world.

Finally, Europe's leadership was weakened by new dependence on outside sources of energy. The region's traditionally significant coal fields became outmoded by technological developments and environmental concerns, and the energy balance shifted to petroleum and natural gas. Despite the development of North Sea oil and gas resources (which benefit primarily Great Britain and Norway), Europe became, and remains, very dependent on imported Middle Eastern oil.

A striking pattern in Europe's economic geography is that Western Europe is wealthier than Eastern Europe. This trend dates to at least the 1870s, when per capita incomes in the West were twice those in the East. The gap grew even wider through the 20th century. After World War II, Eastern European countries were in effect colonized by the Soviet Union, serving as vassal states that gave up human and material resources to service the motherland. Only now, with the dissolution of the Soviet Union and their new membership within the European Union, can the Eastern European countries hope to catch up with their Western counterparts. They stand to benefit enormously from joining the European Union, as it all but guarantees income redistribution from West to East and, through agricultural subsidies and other instruments, promises to breathe new life into the faltering farms and factories of the East.

Postindustrial Europe

Europe's economy today, like its population trend, is best characterized as postindustrial. In the European context, this means that since about 1960, the region has experienced **deindustrialization**, shifting by choice and by force away from energy-hungry, labor-costly, and polluting industries toward an economy based on production of high-tech goods and on services. The high-technology products include electronic devices, data processing hardware, robotics, and telecommunications equipment. Other industries like drug and pesticide manufacturing make use of these technological products to create their own, so there is an important interconnectedness centered on high technology. Europe is also renowned for the production and export of high-quality, expensive luxury goods, including automobiles such as the

Ferrari and watches such as the Rolex. In the service sector, growing proportions of Europeans are employed in transportation, energy production, healthcare, banking, retailing, wholesaling, advertising, legal services, consulting, information processing, research and development, education, and tourism.

Neither high-tech nor service industries employ as many people as the old manufacturing sector did, and like the United States, the European countries are grappling with problems resulting from industrial unemployment. There are concerns over where new jobs will come from, how to make the transition from manufacturing to service roles without too much social dislocation, and how to allocate income between a few highly productive industrial workers who tend robots and the mass of service employees whose output per worker is less and whose basic employment is therefore at risk.

Land and Sea in the European Economies

Despite the dominance of a postindustrial, urban economy, some Europeans continue to live off the land and the sea. Agriculture was the original foundation of Europe's economy, and it is still very important. Food provided by the region's agriculture allowed Europe to become a relatively well-populated area at an early time. After about 1500, a period of steady agricultural improvement began. Introduction of important new crops such as the potato played a part (see Table 3.3 on crop diffusion), but so did such practical improvements as new systems of crop rotation and scientific advances that produced better knowledge of the chemistry of fertilizers. The continuing expansion of industrial cities provided growing markets for European farmers, who received some protection through **tariffs** (tax penalties imposed on imports) or direct **subsidies** (payments made to domestic producers) to encourage production and support rural incomes (these are also discussed on page 622 in Chapter 22, in the context of similar protective measures by the United States). As a result of European Union policies, surpluses of many products are common, including grains, butter, cheese, olive oil, and table wines. However, Europe is not agriculturally self-sufficient and depends on a vital trade of foodstuffs (see Definitions and Insights, page 71).

Throughout history, fishing has been an important part of the European food economy, and the coasts still have many busy fishing ports (Figure 3.15). At times, control of fishing grounds has been a major commercial and political objective of nations, even resulting in warfare. Fisheries are particularly important in shallow seas that are rich in plankton, the organisms that are the principal food for schools of herring, cod, and other fish of commercial value. The Dogger Bank (to the east of Great Britain) in the North Sea and the waters off Norway and Iceland are major fishing areas, and Norway, Iceland, and Denmark are Europe's leading nations in total catch. Fishing has long been especially crucial to countries such as Norway and Iceland, which have little agriculturally productive land.

Definitions + Insights

Genetically Modified (GM) Foods and "Food Fights"

The United States is Europe's leading trade partner, but the exchange is not always open or friendly, especially when it comes to food. Countries often pay lip service to the concept of free trade while actually limiting trade in some products to protect their own industries and interests. Nations on both sides of the Atlantic use these trade barriers, sometimes precipitating **food fights** and other kinds of **trade wars.** In the late 1990s, for example, the United States wanted

Figure 3.D Biotechnological processes can manipulate the genetic makeup of rice and other grains to make them more productive, increasing yields—and also, say many Europeans, endangering local plant varieties and evolutionary processes.

to sell bananas its corporations grew in Latin America to the countries of the European Union. The Europeans, who wanted to protect their investments in banana farming in former European colonies in Africa, the Caribbean, and the Pacific, refused, and the so-called "**banana war**" ensued.

In the new century, an ongoing trade dispute involves **genetically modified (GM) foods** produced in the United States. These are products such as rice, soybeans, corn, and tomatoes that have been genetically manipulated to be more productive and resilient (Figure 3.D). The EU countries refuse to import most of them, citing uncertain long-term health and environmental risks (for example, a GM plant cross-pollinating with neighboring plants, causing unpredictable ecological consequences). If they were to be imported, the Europeans say, they should be product-labeled as GM foods (many Europeans disdainfully label them in conversation as "**Frankenfoods**"). The United States replies that labeling would make them almost impossible to sell to GM-conscious Europeans and argues that the ban on GM foods defies the free trade principles of the World Trade Organization (WTO). For their part, Europeans are insistent on labeling because of their experience with mad cow disease and foot and mouth disease; they are loathe to buy products that are not clearly traceable.

Meanwhile, the United States and European Union have had skirmishes in other fronts of the trade war. The EU threatened punitive tariffs on billions of dollars worth of American goods if the U.S. did not end tax breaks for American corporations' overseas operations. To placate Europe, and to yield to accusations of unfair trade practices by the WTO, the United States abandoned protective tariffs on steel imported from Europe.

Figure 3.15 Fishing boats sail from the harbor of Skagen, Denmark.

3.5 **Geopolitical Issues**

In the past 100 years, Europe's geopolitical situation changed more profoundly and violently than that of any other world region. Europe experienced two world wars that wrought unprecedented devastation. World War I, until 1939 labeled as the "Great War," saw Europe fracture on fault lines of military alliances. Germany and the Ottoman Empire (based in what is now Turkey) were solidly defeated by an alliance led by Britain, France, Russia (early on), and the United States (later on). The war accomplished little and destroyed much, but it did shoulder a costly burden of reparations on the defeated Germany.

The economic and social humiliation of Germany helped plant the seeds for the rise of Adolph Hitler's fascist Nazi (National Socialist) Party Movement. Hitler's Germany surged in strength and allied with Italy, the Soviet Union (early), and Japan (later) in a military quest to establish European and Far Eastern empires. Early in the war, Germany advanced swiftly on the European mainland and Scandinavia and set its sights on an occupation of Great Britain. When, in 1941, Japan bombed the United States at Pearl Harbor, the U.S. was brought firmly into the war on Britain's side. Soon joined by the Soviet Union, these forces, assisted by many smaller ones, began to turn the tide against Germany. In 1945, with German cities in ruin, and two Japanese cities nearly obliterated, the war ended. In the European theater, about 50 million people had died. "What is Europe now?" the British Prime Minister Winston Churchill asked. "It is a rubble heap, a charnel house, a breeding ground of pestilence and hate." [2]

World War II veterans are now aging and dying, and Europe is on a course to establish, in the form of the European Union (EU), what might be called the "United States of Europe," a federation of states akin to that of the United States. The European Union is one of several postwar European **supranational organizations,** meaning that their member countries are united beyond the authority of any single national government, and they are planned and controlled by a group of nations. There are many reasons, largely economic ones, for this push toward greater unity. One of the most important but understated reasons is simply that Europeans want their continent to be war-proof, with its member countries so intertwined in economics, foreign policy, and other ways that war among them would be all but impossible.

East versus West

Europe is recovering and reorganizing from a second momentous geopolitical shift, the end of the Cold War that came with the collapse of the Soviet Union. When World War II ended, Soviet forces and political institutions moved in on Eastern Europe. As Churchill described it, an **Iron Cur-**

tain descended from Poland southward to Yugoslavia. West of it, the capitalist democracies led by Britain, France, West Germany, Italy, and the United States prevailed; to the east, the Soviet Union controlled authoritarian, state-run regimes among its European "satellites," including East Germany, Poland, Czechoslovakia, Hungary, Romania, and Bulgaria. The United States nurtured and gave a huge financial boost to Western Europe's interests. From 1947–1952, it pumped billions of dollars into the reconstruction of the devastated European infrastructure in a program known as the Marshall Plan. The United States was also the linchpin of the military alliance known as the North Atlantic Treaty Organization (NATO), founded in 1949 between the U.S., Canada, most of the European countries west of the Iron Curtain, and Turkey. NATO faced off against the Warsaw Pact comprised of the Soviet Union and its Eastern European satellites. On more than one occasion, notably in the Cuban Missile Crisis of 1962 and in the Arab–Israeli War of 1967, these military alliances came all too close to apocalyptic nuclear war.

In 1989, the Berlin Wall, which separated the German city that was a microcosm of the Cold War, was dismantled in a popular and peaceful uprising. The Soviet grip on East Germany and its satellites had been slipping for years, and now it crumbled. The Soviet Union itself disbanded in 1991 (see Chapter 6, page 139). An exhilarated public in "liberated" Eastern Europe embraced the prospects of democracy, capitalism, and material improvement. The Warsaw Pact was dissolved. The nuclear arsenals of the respective alliances were reduced substantially.

After the end of the Cold War, NATO's attention focused mainly on strife in the Balkans, where it used force to reverse the advance of the Serbs against their neighbors in the former Yugoslavia (see page 123). NATO remains, and in 2004 its 19 members grew to 26 with the addition of seven former Soviet allies (Estonia, Latvia, Lithuania, Slovakia, Slovenia, Romania, and Bulgaria; Figure 3.16). Russia grudgingly accepted this Western expansion into its former sphere of influence, in part because of its own growing alignment with Western defense and security concerns. NATO's future role is uncertain. The Europeans are, however, united in their certainty that they do not want to spend too much money on defense.

The European Union

The most important of Europe's supranational organizations is the European Union (EU) (see Figure 3.16). It began in 1957 as the European Economic Community (EEC), also called the Common Market, made up of France, West Germany, Italy, Belgium, Luxembourg, and the Netherlands. By 1996, three years after it acquired the name European Union, the organization had an additional nine members: Great Britain, Denmark, Ireland, Greece, Portugal, Spain, Austria, Finland, and Sweden.

The EEC was initially designed, as the EU is today, to secure the benefits of large-scale production by pooling the re-

---

[2] Quoted in A. A. Byatt, "What Is a European?" *The New York Times Magazine,* October 13, 2002, p. 58.

European Union and NATO Membership

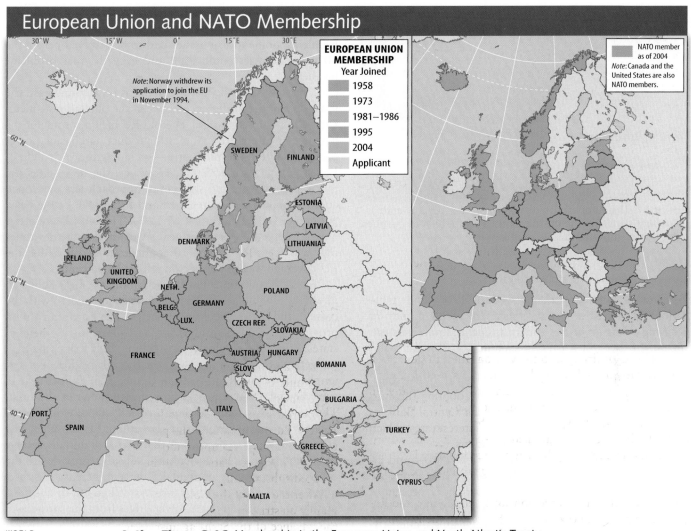

EUROPEAN UNION MEMBERSHIP
Year Joined
- 1958
- 1973
- 1981–1986
- 1995
- 2004
- Applicant

Note: Norway withdrew its application to join the EU in November 1994.

NATO member as of 2004
Note: Canada and the United States are also NATO members.

Active Figure 3.16 Membership in the European Union and North Atlantic Treaty Organization. *See an animation based on this figure, and take a short quiz on the concepts.*

sources—natural, human, and financial—and markets of its members. Tariffs were eliminated on goods moving from one member state to another, and restrictions on the movement of labor and capital between member states were eased (in theory, a citizen of any EU country may work in any other EU country, but the process is not simple; a work permit must be obtained from the host country). Monopolistic trusts and cartels that formerly restricted competition were discouraged.

Meanwhile, a common set of external tariffs was established for the entire area to regulate imports from the outside world, and a common system of price supports for agriculture replaced the individual systems of member states. The founders anticipated that free trade within such a populous and highly developed bloc of countries would stimulate investment in mass-production enterprises, which could, wherever they were located, sell freely into all member countries. This trade would encourage productive geographic specialization, with each part of the union expanding lines of production for which it was best suited. Each country might thereby achieve greater production, larger exports, lower costs to consumers, higher wages, and a higher standard of living than it could achieve on its own.

The European Union, under provisions of the Maastricht Treaty of European Union that went into force in 1993, is now pursuing major new steps in unification. These include the removal of nontariff trade barriers, such as "quality standards," that still exist between member countries. Major energy focused on the implementation of a single EU currency, the **euro,** in 1999, as the centerpiece of the **European Economic and Monetary Union (EMU)** (Figure 3.17). The common currency was promoted on the basis of economic theories suggesting that sharing a currency across borders had many advantages, such as lower transaction costs, more certainty for investors, enhanced competition, and more consistent pricing. A common currency could theoretically

Figure 3.17 The European Union's common currency, the euro, is an important symbol of the organization's unity, but not all member countries have embraced it.

also restrain public spending, reduce debt, and tame inflation. In practical terms, consumers spend money in euro-using countries just as Americans do at home; they can compare the price of a car, for example, in Germany and Italy, and decide to buy the cheaper car in Italy, just as an American might buy a car in Kentucky instead of Missouri. Besides sharing a currency, all countries adopting the euro have agreed to allow the strong European Central Bank to make critical decisions on issues such as interest rates, which would be identical for all euro-using nations.

Citizens of various EU countries view the euro—and indeed the European Union itself—differently. The euro is not all that popular; recent polls showed that only 52 percent of its users felt the currency was "advantageous overall." There is strong euro-phobia among the Swedes, Danish, and British, and they rejected the currency. Sweden feared a loss of fiscal and monetary independence. Danes expressed fears that the euro would threaten their generous welfare benefits or even their national unity. Britain's economy has recently been stronger than the economies of Germany and France, suggesting to the British that they may as well shun the euro for now. The new Eastern European members of the European Union will have to make wrenching economic reforms to meet the requirements of adopting the euro, as they would like to do.

Ten Eastern European nations joined the European Union in what was called the **"big bang"** of May 2004. The newcomers are Poland, the Czech Republic, Hungary, Slovakia, Estonia, Latvia, Lithuania, Slovenia, Malta, and officially all of Cyprus, although de facto EU rule applies only to the Greek portion (see Chapter 5, page 120). Turkey's application to join was rejected. Their inclusion created a mega-Europe of 450 million people, stretching from the Atlantic to Russia, with an economy valued at almost $10 trillion, nearly as strong as the U.S. economy. The EU's statistics grew enormously, with the addition of 75 million people (about 20 percent, almost half of them in Poland) and a geographic expansion equal to almost one-quarter of its present terri-

134

263

tory. This growth is accompanied by some clear advantages for the organization and its new members but also by some problems. Some thorny issues had to be resolved before the 10 new countries could join, and even thornier problems lie ahead.

Consistent with the historic patterns of west and east, the most outstanding difference between the old and new members is in their economies. The old EU includes some of the highest per capita gross domestic product purchasing power parity (GDP PPP) in the world—including the very highest, Luxembourg, with $55,000. However, the new states are much less affluent, and some even approach less developed country status (see Figures 2.18 and Table 3.1). Latvia has a per capita GDP PPP of just $10,200. The old EU countries have 95 percent of the continent's wealth, and the new ones only 5 percent. When the 10 countries joined in 2004, the EU's average wealth per person fell by 13 percent.

Naturally, questions have arisen about the EU's ability to afford its new members. There is particular concern with the costs of the EU's agricultural policies. The European Union provides generous subsidies for agriculture in its member states. Prior to the big bang, $50 billion, fully half of the EU's budget, was spent every year on these subsidies. For example, the EU pays an Austrian farmer $28,000 per year in subsidies for 75-acre production of seed potatoes and milk. That farmer, like many in the European Union, would probably not be able to farm profitably without EU subsidies. The 10 new EU members are upset that they will initially receive only one-fourth of the agricultural subsidies provided to the older members (with an annual 5 percent increase in subsequent years). They fear their farmers will go out of business, unable to compete with the lower prices from the subsidized, heavily mechanized farms in the West.

For its part, the European Union argues that the new countries do not have the high production costs and quality controls of the old countries. Mainly, though, the old European Union fears that providing full subsidies to the new countries would bankrupt the organization. Germany is especially fearful. It is the wealthiest member of the European Union and carries the largest cost burden. Germany had its own economic pain absorbing the costs of integration with the poor former East Germany and is concerned about a far more expensive integration on a much grander scale. Despite hopes among many in the newly joining states that membership will bring them prosperity, there will also be at least short-term pain. Food prices are expected to rise in those countries almost immediately by at least 20 percent, while salaries will only slowly increase.

The new countries will be denied, at least for now, another benefit generally enjoyed by the old: the freedom of movement across borders of member countries. That is supposed to come in 2011. For now, the fears of a tidal wave of cheap eastern labor and of opening Western Europe's doors to terrorism and trafficking in drugs or humans have demanded special rules for the eastern countries, long regarded as abodes for these shady activities. In the meantime, labor-

179

ers will continue to move illegally—for example, shuttling from Poland to Belgium where they will work at the low-paying jobs that no one else wants.

In theory, the European Union would like to move toward a situation in which there were no passport, visa, or other control issues at any internal land, sea, and airport frontiers. To date, only a framework outside the European Union has attempted this kind of integration. Known as the **Schengen Agreement,** it allows free circulation of people between 15 signatories to the agreement: Austria, Belgium, Denmark, Finland, France, Germany, Greece, Iceland, Italy, Luxembourg, the Netherlands, Norway, Portugal, Spain, and Sweden. To maintain internal security, within this greater "Schengenland" the member states are supposed to exercise common visa, asylum, and other policies at their external borders (for example, between Germany and Poland). Citing security concerns, some countries have used a safeguard clause of the treaty to impose stronger controls; France, for example, has at least temporarily reinstituted passport controls on its borders with Belgium and Luxembourg. Some countries, notably Great Britain, are skeptical of the effectiveness of control of external borders and have stayed out of Schengen. Truly open borders will probably not exist among all the EU member countries for a very long time, if ever.

Houses Divided?

Beyond its unifying economic principles, many questions remain about what the European Union is or should be. Remarkably, most citizens in the EU are not sure what to make of the organization. In a recent poll, only 30 percent said they understood how the European Union works. A recent EU Commission poll found that only 48 percent of EU citizens agreed that their country's membership was "a good thing." With the EU's expansion came a struggle within to achieve a new constitution, in which all members would agree on matters such as defense and foreign affairs. The debate was analogous to the effort in the United States in the 1780s to form a confederation of states agreeing to give a national government more power. There were differences about how power should be wielded: Some envisioned a federal system patterned after that of the United States, but others wanted individual national governments to make most of the decisions. A basic conflict had to do with distribution of power between big and small states. Initially, Poland and Spain had almost as many votes each in the European Council as Germany, which has about twice as many people as either country. That system was dispensed with to give greater weight to the bigger countries—Italy, France, Germany, and Great Britain—in an arrangement called the **double majority.**

There were other struggles, too. There was a hard-argued debate on whether or not the preamble to the EU Constitution should mention God. He was dropped. There was acrimony over whether France and Germany should be allowed to run budget deficits that exceed statutory limits governing the euro, leading to suspicions that one set of rules existed for big countries and another for small ones. There were concerns with how to articulate a defense policy that did not seem to conflict with that of NATO. And although a stated goal of the European Union is to develop a common voice in foreign affairs, the U.S.-led invasion of Iraq in 2003 underscored how far away that goal was. Britain's Prime Minister Tony Blair was a staunch advocate of the war (even if not backed by a majority of Britons), and Italy and Spain expressed support. So did most of the then Eastern European members-to-be countries, led by Poland. Germany and France led the very strong dissenting voices.

The rift that developed between the United States and most of its European peers over the Iraq war raised a lot of questions on both sides of the Atlantic on how much we do, or do not, have in common. These are fitting questions for the end of this chapter. Many of us instinctively equate the United States with Europe when we speak about global patterns; after all, we share many common cultural roots, share a great deal of the world's wealth, and have many of the same measures of quality of life. But there are differences, and these seem to have grown in recent years. "Americans are from Mars and Europeans are from Venus," wrote policy analyst Robert Kagan.

Here are a few exemplary differences. European governments and people have a stronger sense of the social compact between the state and its citizens. Governments are expected to provide "free" public services like education and healthcare, while the people understand they must foot the bill for these "free" services by paying very high taxes. (Incidentally, undergraduate public education in most European nations is free. However, per capita spending on public education is higher in the U.S., and American college education is widely regarded as superior to that of Europeans.) Europeans complain about how U.S. "cultural industries" such as Hollywood films humiliate other cultures and languages that cannot defend themselves against English-speaking cultural industries. Europeans despise the death penalty, which is outlawed in all EU countries. Europeans tend to take a "precautionary" approach to things they think may be harmful: There may be risks from GM foods, so ban them; there probably is global warming, so take action to reduce greenhouse gases. The U.S. has more of a "show me" attitude, with at least a commonly expressed official stand that there is not enough scientific proof, so better to reject the Kyoto Protocol (see page 29).

For Americans, September 11, 2001, was the watershed time when it reevaluated its place in the world and began to feel less secure. For Europeans, the fall of the Berlin Wall in 1989, leading to the collapse of the Soviet bloc, was the watershed that made them start to feel more secure. It is a sad reflection of relations that many Europeans feel insecure about the United States. A Gallup survey in the wake of the 2003 Iraq war revealed that more than two-thirds of the Europeans polled regarded the United States as the second largest threat to world peace, behind only Israel, and tied with North Korea and Iran.

SUMMARY

- Europe is physically part of the great continent of Eurasia. However, because of historic distinctiveness, it is generally labeled a continent and is treated as a separate region. Europe's population is about twice that of the United States, and Europe is more densely settled. Europe is demographically postindustrial, with a slowly declining and aging population.

- Among Europe's most distinctive physical geographic traits are its northerly location, temperate climate, and varied topography. The North European Plain is a major belt of settlement and agricultural productivity. The marine west coast climate, continental climates, and Mediterranean climate are characteristic. The Rhine and the Danube are the two most important rivers.

- European languages derive primarily from Indo-European roots and include Romance, Germanic, and Slavic languages. English is a Germanic language, and like most European languages, it is enriched by many other tongues. The dominant religion is Christianity, with major followings of Protestant, Roman Catholic, and Eastern Orthodox churches. There are significant populations of Muslims (most of them recent immigrants) and of Jews (whose population is a small remnant of the pre-Holocaust community). Immigration is enriching Europe's ethnic mosaic but also presenting economic and security dilemmas for the European countries.

- From the beginning of the 16th century until late in the 19th century, Europe was at the center of global patterns of colonization and foreign settlement, long-distance trade, and agricultural and industrial innovation. During this time, Europeans diffused crops and animals between the Old and New Worlds. The Industrial Revolution originated in Europe, with energy derived from coal and factory technology focused on textiles and iron. Industrialization and colonization launched Europe to global economic and political supremacy.

- Recent decades have seen a global shift in power away from Europe. Factors such as war dislocation, rising nationalism, the ascendancy of the United States, a shift in world manufacturing patterns, and new energy sources have combined to diminish Europe's global centrality. Europe is nevertheless a very strong force in world economic, political, and social affairs, and its peoples are among the most prosperous in the world.

- Europe's economy is postindustrial, making the transition from energy-hungry, labor-costly, and polluting industries to leaner high-tech industries and to services. This shift has caused unemployment. European nations often try to protect their domestic industries and have been involved in "trade wars" with the United States, which likewise wants to protect its industries. Eastern Europe has long been much poorer than Western Europe.

- In recent decades, Europe has been reorganizing itself to ensure that nothing like the two world wars will happen again and to strengthen its economies. Its principal military alliance, NATO, is growing in membership and redefining its focus toward peacekeeping. The most important development is the growth of the 25-member European Union (EU), a supranational organization that pools the economic and human resources of its member countries. There have been obstacles to its achievement of common EU policies in money matters, defense, and foreign affairs. The U.S. war in Iraq highlighted major differences among EU members, and overall, there are several marked differences in the ways Europeans and Americans view the world.

KEY TERMS + CONCEPTS

Terms in blue are also defined in the glossary.

Age of Discovery (p. 68)
Albanian (p. 66)
Altaic languages (p. 66)
  Turkish (p. 66)
asylum (p. 57)
"baby bounty" (p. 54)
"banana war" (p. 71)
Basque (p. 66)
"big bang" (p. 74)
Bretons (p. 66)
Christianity (p. 66)
coke (p. 69)
Cold War (p. 72)
culture hearth (p. 64)
deindustrialization (p. 70)
demographic transition (p. 53)
devolution (p. 65)
double majority (p. 75)
Eastern Orthodox Church (p. 66)
estuary (p. 55)
European Economic and Monetary Union (EMU) (p. 73)
European Economic Community (EEC; or Common Market) (p. 72)

European Union (EU) (p. 72)
food fights (p. 71)
"Frankenfoods" (p. 71)
genetically modified (GM) foods (p. 71)
glacial deposition (p. 60)
glacial scouring (p. 60)
guest workers (p. 56)
Gulf Stream (p. 59)
Holocaust (p. 67)
Ice Age (p. 60)
Indo-European languages (p. 64)
  Baltic (p. 66)
    Latvian (p. 66)
    Lithuanian (p. 66)
  Celtic (p. 64)
  Germanic (p. 64)
    Danish (p. 64)
    Dutch (p. 64)
    English (p. 64)
    Flemish (p. 64)
    German (p. 64)
    Icelandic (p. 64)
    Norwegian (p. 64)
    Swedish (p. 64)

Greek (p. 63)
Indic (p. 66)
  Romany (p. 66)
Romance (p. 64)
  Catalan (p. 64)
  French (p. 64)
  Italian (p. 64)
  Latin (p. 64)
  Portuguese (p. 64)
  Romanian (p. 64)
  Spanish (p. 64)
  Walloon (p. 64)
Slavic (p. 66)
  Bulgarian (p. 66)
  Croatian (p. 66)
  Czech (p. 66)
  Polish (p. 66)
  Russian (p. 66)
  Serbian (p. 66)
  Slovak (p. 66)
  Ukrainian (p. 66)
Industrial Revolution (p. 69)
Iron Curtain (p. 72)
Islam, Muslims (p. 67)

Jews (p. 66)
Judaism (p. 67)
Ladin (p. 66)
loess (p. 59)
Maastrict Treaty of European Union
   (p. 73)
Marshall Plan (p. 72)
microstates (p. 51)
Moors (p. 67)
national minorities (p. 66)
nationalism (p. 69)
North Atlantic Drift (p. 59)
North Atlantic Treaty Organization
   (NATO) (p. 72)

Poles (p. 66)
population replacement level (p. 53)
postindustrial (p. 70)
Protestant Reformation (p. 66)
Protestantism (p. 66)
   Calvinist (p. 66)
   Church of England (p. 66)
   Lutheran (p. 66)
   Pentecostal Churches (p. 67)
Roma (Gypsies) (p. 66)
Roman Catholic Church (p. 66)
Saami (Lapps) (p. 66)
Schengen Agreement (p. 75)
Silk Road (p. 68)

subsidies (p. 70)
supranational organization (p. 72)
tariffs (p. 70)
trade wars (p. 71)
undocumented workers (p. 57)
Uralic languages (p. 66)
   Estonian (p. 66)
   Finnish (p. 66)
   Hungarian (p. 66)
   Saami (p. 66)
Warsaw Pact (p. 72)
westerly winds (p. 59)

REVIEW QUESTIONS

WORLD
REGIONAL
Geography ⊛ Now™

Assess your understanding of this chapter's topics with additional quizzing and concept-based problems at http://earthscience.brookscole.com/wrg4e.

1. Why is Europe usually treated as a separate region?

2. What terms and trends best describe Europe's population today? Why is immigration so critical to Europe's demographic future and what problems does immigration pose?

3. What are Europe's major physical and environmental characteristics?

4. What roles have rivers played in Europe's development and which are the most important ones? Where is industrial and other economic activity concentrated in Europe and why?

5. What are the dominant languages, religions, and other ethnic traits of Europe?

6. What factors led to Europe's global dominance in economic and political affairs? What impacts did that dominance have on other peoples and environments?

7. Explain the origins and significance of the Industrial Revolution, including the role of resources in that revolution. How does Europe's present urban pattern reflect its industrial past?

8. What are Europe's main economic traits? How have these changed in recent decades?

9. What happened to Europe in the 20th century? How are Europe's political and economic institutions of today trying to create a different Europe?

10. What are the main goals and principles of the European Union? What successes and difficulties has the organization had?

DISCUSSION QUESTIONS

1. Why are Europe's climates so mild, considering the region's latitudes?

2. Europe's population is declining. Is immigration the answer?

3. Why, of all the world's regions, did Europe become a dominant one after about 1500?

4. What makes Europe postindustrial in economic terms? What economic characteristics do Europe and the United States seem to have in common?

5. What do Europeans think about genetically modified foods? How do these perceptions affect trade between Europe and the United States?

6. What political and military events shaped Europe in the 20th century? What were the major features of Western and Eastern Europe after World War II?

7. What can European countries accomplish within the European Union that they could not on their own?

8. The European Union is not popular with all Europeans. Which nations stayed out? Which did not adopt the euro and why? Why were there problems in drafting an EU Constitution?

9. What is likely to be the fallout of the "big bang" that brought 10 new nations into the European Union in 2004?

10. How do Europeans and Americans differ on such issues as national security, global warming, and social welfare? In what other ways are they different? How are they similar?

The European Core

The European Coreland has some of the most modified landscapes on Earth. Some have been carefully crafted to be pleasing getaways from the hustle and bustle of urban life. These are the Bodnant Gardens in Wales, with the peaks of Snowdonia visible in the distance.

chapter outline

4.1 The British Isles	4.4 Benelux: Tolerance and Trade in the Low Countries
4.2 France: *Vive la Difference!*	4.5 Switzerland and Austria: Prosperous Mountain Fastness
4.3 Great Germany	

chapter objectives

This chapter should enable you to:

- Identify the concurrence of economic and other forces that have made this a core region

- Recognize the spatial relations between heavy industrial resources and cities in Europe and the factors that have caused the deindustrialization of Europe

- Relate ways in which this portion of Europe represents one of the most heavily modified landscapes in the world—for example, through deforestation and land reclamation

- Contrast the primacy of Paris with the decentralized urban pattern of Germany and other European countries

- Become acquainted with the roots and issues of some of Europe's persistent ethnic and political struggles, including The Troubles of Northern Ireland and French policies affecting Muslims

- Appreciate the significance of a reunified Germany, especially the economic costs of reintegration for the more prosperous west, an experience that is important globally because of its relevance to the two Koreas and other pairs of nations with prospects for reunification

WORLD
REGIONAL
Geography ⊕ Now™

Look for this logo in the text and go to GeographyNow at http://earthscience.brookscole.com/wrg5e to explore interactive maps, view animations, sharpen your factual knowledge and geographic literacy, and test your critical thinking and analytical skills with unique interactive resources.

Europe has a recognizable **core,** a subregion that has long played a dominant role in the continent's political, economic, and cultural development. Today, the European core is recognizable as the subregion of Europe with these traits: densest, most urbanized population; most prosperous economy; lowest unemployment; most productive agriculture; most conservative politics; greatest concentration of highways and railroads; and the highest levels of crowding, congestion, and pollution.[1]

Defined in this way, the Coreland consists primarily of three of Europe's major countries and of smaller nations in the British Isles and the west central portions of the European mainland (Figure 4.1). The United Kingdom and the Republic of Ireland share the British Isles. The countries of West Central Europe are Germany and France and their smaller neighbors of the Netherlands, Belgium, Luxembourg, Switzerland, and Austria. Also fitting the Coreland traits cited above are roughly the northern half of Italy, Denmark, southern Sweden, and parts of Poland, but other characteristics of these countries place them more appropriately in the "European Periphery" chapter that follows. Because they are nestled among the core countries described in this chapter, the microstates of Liechtenstein, Monaco, and Andorra are included here.

The European core includes some of the world's most geopolitically and economically significant countries (see Table 3.1, page 52). This subregion also has one of the world's most transformed landscapes, shaped by millennia of productive agriculture and hundreds of years of industrial development and affluence (Figure 4.2). The European core is one of just four regions in the world classified as a **major cluster of continuous settlement,** meaning that all habitations lie no more than 3 miles (5 km) from other habitations in at least six different directions. In addition, roads or railroads lie no farther than 10 to 20 miles (16 to 32 km) away in at least three directions.[2] Europe's cultural landscape is unusually intricate. Different layers of history appear at every turn, and there are sharp differences in human geographies and national perspectives of its peoples. There is also a wide range of environments and resources. Local resources (especially coal and iron ore) and transportation opportunities were the basis for industrial and urban development in past

---

[1] From Terry G. Jordan and Bella Bychova Jordan, *The European Culture Area: A Systematic Geography* (Oxford: Rowman and Littlefield, 2002), p. 402.

[2] From Terry G. Jordan and Bella Bychova Jordan, *The European Culture Area: A Systematic Geography* (Oxford: Rowman and Littlefield, 2002), p. 163.

Political Units of the European Core

Figure 4.1 The political geography of Europe's core

Deforestation of Europe

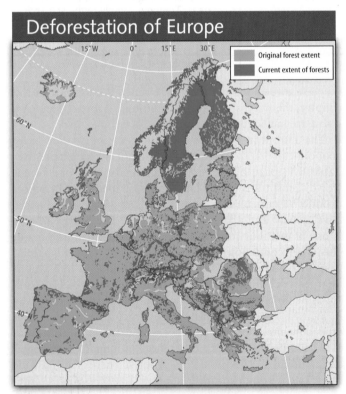

Original forest extent

Current extent of forests

WORLD
REGIONAL
Geography⊛Now™ **Active Figure 4.2** Human activities over long periods of time have extensively transformed the natural landscapes of Europe, particularly in the core region. This maps depicts areas that once sustained temperate mixed forests and coniferous forests but have lost that tree cover. *See an animation based on this figure, and take a short quiz on the facts or concepts.*

Historical Industrial Concentrations

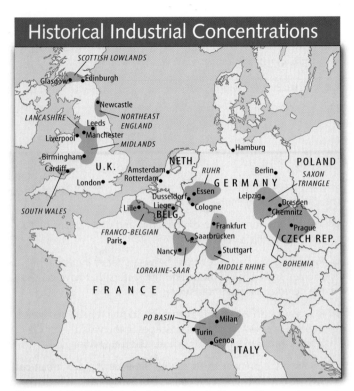

Figure 4.3 Historic industrial concentrations, cities, and seaports of the European core. Older industries such as coal mining, heavy metallurgy, heavy chemicals, and textiles clustered in these congested districts. Local coal deposits provided fuel for the Industrial Revolution in most of these areas, which have shifted increasingly to newer forms of industry as coal has lost its economic significance and older industries have declined.

centuries (Figure 4.3), but the modern Coreland has a distinctly postindustrial geography (Figure 4.4).

4.1 The British Isles

The islands of Great Britain and Ireland, off the northwest coast of Europe, are known as the British Isles (Figure 4.5). They are home to two countries: the Republic of Ireland, with its capital at Dublin, and the larger United Kingdom of Great Britain and Northern Ireland (U.K.), with its capital at London. The United Kingdom is made up of the entire island of Great Britain, with its political units of England, Scotland, and Wales, plus the northeastern corner of Ireland—known as Northern Ireland—and most of the smaller outlying islands. Northern Ireland separated from Ireland in 1922 and maintained its link with the United Kingdom, but Ireland achieved full independence. Altogether, the U.K. encompasses about four-fifths of the area and over 90 percent of the population of the British Isles.

The United Kingdom: Once an Empire

Lying only 21 miles (34 km) from France at the closest point (see Figure 4.5), England is the largest of the four main subdivisions of the United Kingdom. Like Scotland, Wales, and

Decline and Growth in Industries of the European Core

DECLINES IN MANUFACTURING AND MINING

- Most severe decline, 1965–1990
- Major decline, 1965–1990
- Most severe decline, 1990–1995
- Major decline, 1990–1995
- High-tech manufacturing
- High-quality and luxury goods manufacturing
- Borders of the new, prosperous industrial core of Europe

Map labels: Arctic Circle; North Sea oil and gas development; "Silicon Glen"; "Silicon Bog"; Randstad; "Silicon Saxony"; "Silicon Fen" M4 Corridor; "Scientific City"; "Techno City"; "Third Italy"

WORLD REGIONAL
Geography ⊛ Now ™

Active Figure 4.4
Deindustrialization and new manufacturing in the European core, 1960–2000. Declining and growing industries tend to be occurring in geographically distinct areas. The new, mainly high-tech, industries occupy smaller areas than the older industries. *Source: After Jordan and Bychova Jordan, 2002, p. 300.*
See an animation based on this figure, and take a short quiz on the facts or concepts.

Northern Ireland, England was originally an independent political unit. England twice conquered mainly Catholic Ireland, including Northern Ireland.

Scotland was first joined to England when a Scottish king inherited the English throne in 1603, and the two became one country under the Act of Union in 1707. Nearly three centuries later, in 1997, Great Britain yielded to increasing Scottish demands and allowed for the devolution (the dispersal of political power and autonomy to smaller spatial units) of many administrative powers (including agriculture, education, health, housing, taxation, and lawmaking, but excluding defense and social security) to a new Scottish parliament.

England conquered Wales in the Middle Ages, but Wales still has considerable cultural distinctiveness, especially in the Welsh language. A bare majority in a 1997 referendum voted for the creation of a Welsh Assembly. While this body has some powers independent of the United Kingdom, growing numbers of Welsh are calling for a stronger devolution of power that would transfer more authority from the national government to a full-fledged parliament, like Scotland has now.

The United Kingdom became the world's strongest country in the 19th century, maintaining that status between the defeat of Napoleonic France in 1815 and the outbreak of World War I in 1914. Its overseas empire covered a quarter of the earth at the time of its maximum extent; "the sun never sets on the British Empire" was a proud boast. The influence of English law, education, and culture spread still farther. Until the late 19th century, the United Kingdom was the world's leading manufacturing and trading nation. Britain's Royal Navy dominated the seas, and the British merchant marine moved half or more of the world's ocean trade. London became the center of a free trade and financial system that invested its profits from industry and commerce around the world.

The U.K.'s economic decline began late in the 19th century. World War I damaged the free trade and financial system based in London, and the downward trend accelerated after World War II. A drop in political stature accompanied economic decline. All the large colonies of the former British Empire have gained independence, and Britain's remaining overseas possessions are scattered and small. The United Kingdom, however, is still a country of consequence. It plays

65

United Kingdom and Ireland

Figure 4.5 Principal features of the British Isles. The "Tees-Exe" line connecting the English towns of those names is the general dividing line between Highland Britain to the northwest and Lowland Britain to the southeast. *Source: After Hoffman, 1990, p. 245*

a major role in the European Union and is associated with many of its former colonies in the worldwide **Commonwealth of Nations.** Its strong allegiance to the United States, even in times of great adversity such as during the recent war in Iraq, is another of Britain's characteristic traits on the international stage.

Highland and Lowland Great Britain

The island of Great Britain is about 600 miles (c. 1,000 km) long by 50 to 300 miles (80 to nearly 500 km) wide. For its small size, this island has a wide variety of landscapes and modes of life. Some of this variety can be attributed to a broad division of the island into two contrasting physical areas: Highland Britain and Lowland Britain.

The high regions of Highland Britain are the Scottish Highlands, the Southern Uplands of Scotland, the Pennine Chain and the Lake District of northern England, the mountains that occupy most of Wales, and the uplands of Cornwall and Devon in southwestern England (see Figure 4.5). These mountains are low by most standards—Ben Nevis in the Scottish Highlands is the highest at 4,406 feet (1,343 m)—but many slopes are steep, and the mountains often look high because they rise precipitously from a base at or near sea level. Highland Britain's most important low-lying area is the Scottish Lowlands, a densely populated industrialized valley separating the rugged Scottish Highlands from the gentler and more fertile Southern Uplands of Scotland. The Scottish Lowlands compose only about one-fifth of

Definitions + Insights

Hedgerows

One of the most distinct landscape features of England and Wales is the **hedgerow,** a long, trailing land-use boundary consisting of shrubs and trees or of cleared field stones overgrown with heather (Figure 4.A). Farmers and ranchers have used them to mark perimeters, to contain sheep and cattle, and simply to dispose of rocks since very ancient times; the Romans found them when they arrived in Britain. Many hedgerows were created in the **Enclosure Movement** of the 18th and 19th centuries, when large landowners appropriated local "commons," or unfenced land tilled by villagers, and enclosed them in hedges for sheep breeding. This closure of lands that had historically been worked by peasant farmers helped promote the migration of peasant youth to the newly industrializing English cities.

As recently as 1985, there were an estimated 400,000 miles (640,000 km) of hedgerows in England and Wales. But housing and commercial developments, suburban sprawl, and the rearrangement of land for larger farming operations have taken their toll, destroying more than 100,000 miles (160,000 km) of hedgerows. Although hedgerow defenders decry the loss of these historic boundaries and wildlife habitats, provincial and national governments have done little to stem the losses.

Figure 4.A Hedgerow in Cornwall, southwestern England

Scotland's area but contain more than four-fifths of its population and its two main cities: Glasgow (1.4 million) to the west and Edinburgh, Scotland's political capital (670,000), to the east (Figure 4.6).

The highland landscapes have long been popular tourist destinations because of their stark grandeur and quaint settlements, and recent immigration by wealthier English has reversed a long trend of depopulation. The Scots who live there have a history of proud and often defiant distinction from the more powerful English to the south. That independent streak is the stuff of romantic and tragic tales, recounted in history books, novels, and films such as Mel Gibson's *Braveheart.*

Practically all of Lowland Britain (the English Lowland) lies in England. It has better soils than the highlands and a gentler topography. It is a mosaic of well-kept pastures, meadows, and crop fields fenced by hedgerows (see Definitions and Insights, above) and punctuated by closely spaced villages, market towns, and industrial cities. The largest industrial and urban districts, aside from London, lie in the midlands and in the north, around the margins of the Pennine Chain. Here, clusters of manufacturing cities such as Birmingham, Manchester, Leeds, and Sheffield arose near coal fields, which were a major locational factor in the region's development.

Urbanization and Industrialization in Britain

The Industrial Revolution gave a powerful impetus to the growth of cities at the same time that Britain's total popula-

Figure 4.6 Historic Edinburgh, capital of Scotland

tion was increasing rapidly. In 1851, the United Kingdom became the world's first predominantly urban nation. Except for London, most of the large cities are on or near the coal fields that supplied the power for early industrial growth (see Geographic Spotlight, page 84). Five industries were of major significance in Britain's urban/industrial rise: coal mining, iron and steel manufacture, cotton textiles, woolen textiles, and shipbuilding. The advantage of an early British start in these enterprises gave the United Kingdom industrial preeminence through most of the 19th century. Exported

Geographic Spotlight

London Town

Britain's premiere city, London (Figures 4.B and 4.C), is the country's single significant metropolis that was not located near coal fields. Instead, other important locational factors account for the establishment and rise of this great city. It is located in the corner of Great Britain closest to major trading partners on the European continent. It is in a productive agricultural area; even in the preindustrial period, the fertile English Lowland supplied most of Britain's wool for exports and domestic consumption. London's location facilitated transport connections with the surrounding region. The Thames River and its tributaries provided major routes to the interior, where canals connected the Thames drainages with other rivers. Before the 19th-century railroad and the 20th-century truck, water transport had greater advantages over land transport than it has today, and these natural inland waterways were used extensively for transporting goods to and from London.

Figure 4.B Big Ben, the world's most renowned clock, stands aside the British Houses of Parliament in London.

Its river location allowed London to act as an inland seaport, serving more trading territory in the surrounding countryside than would a coastal port.

London's network of highway connections to the rest of Britain now facilitates port activities linking the island with global trade. However, getting around the city itself can be difficult. London's traffic jams finally prompted authorities to begin a program of "congesting charging" in 2003. Except for residents and those with exemptions, motorists using the 8 square mile (21 sq km) area of central London must pay a daily fee of £5 (about $8). The system relies on camera surveillance and a hefty fine.

London (population: 11.3 million) is no longer one of Europe's or the world's major port cities or manufacturing centers. The city's main industries are government, insurance, business administration and service, communications media, tourism, and finance. London is Europe's leading financial center, a status it also held at the height of Britain's colonial empire prior to World War I. London's ethnic geography is a vibrant, living legacy of that colonial past, reflecting the city's function as the empire's core.

More than one-third of Londoners belong to an ethnic minority, and collectively, they speak an estimated 300 languages. Most arrived with British decolonization following World War II, drawn by the economic opportunities perceived in their former "mother country." These include more than 500,000 Asians (referring to people of Indian, Pakistani, and Bangladeshi origins), about 300,000 West Indians (mainly Jamaicans), and more than 160,000 West Africans. Tourists in London not only take in the traditional offerings of palaces, museums, and theater shows but feel themselves at a cultural crossroads of the world. Entire neighborhoods are ethnic Indian or Arab, for example, and many members of the minorities within them wear their homeland's traditional clothing, making the visitor feel "abroad" within London.

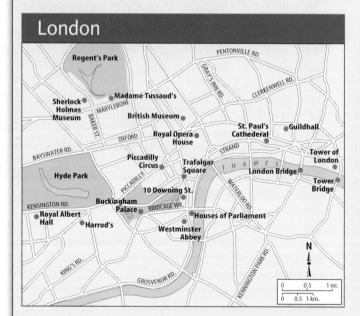

Figure 4.C Downtown London and some of its world-famous attractions

surpluses paid for food and raw material imports and made Britain a wealthy hub of world trade. Britain has recently had difficulties in these same industries due to growing international competition, changing patterns of demand, and shifting urban populations.

These industrial innovations began with 17th-century coal mining and associated steam technology. Coal became a major industrial resource between 1700 and 1800, when it fueled the blast furnace and the steam engine. Coal was coked (cleansed of its volatile constituents) for use in blast furnaces, and coke replaced the charcoal previously used for smelting iron ore. With Britain's development of an economically practical steam engine, coal replaced human muscle and running water as the principal source of industrial energy.

Coal production rose in England and Wales until World War I. Then, between 1914 and the 1990s, output fell by about two-thirds, the number of jobs in coal mining was cut by more than four-fifths, and practically all of the coal export trade disappeared. Major factors contributing to this decline included increasing competition from foreign coal in world markets, depletion of Britain's own coal, which was the easiest and cheapest to mine, the environmental unattractiveness of coal (the dirtiest of the fossil fuels), and increasing substitution of other energy sources for coal both overseas and in Britain itself (see Geography of Energy, page 86). By the 1980s, once-dominant coal's principal remaining role in Britain had been reduced to supplying fuel for generating domestic electricity. Beginning in the late 1960s and the 1970s, large-scale development of newly discovered North Sea oil and gas fields revolutionized Britain's energy situation (see Figure 4.4 and Figure 5.7, page 110). Those offshore fields continue to be very productive, providing welcome revenue to the United Kingdom and especially to the historically marginal Shetland Isles of Scotland. These fields may be described as "middle aged"; by 2010, their output is projected to decline slightly to about 5 million barrels of oil per day for Britain.

Steel was a scarce and expensive metal before the Industrial Revolution. Iron, not too cheap or plentiful itself, was more commonly used. It was made in small blast furnaces by heating iron ore over a charcoal fire, with temperatures raised by an air blast from a primitive bellows. Supplies of iron ore in Great Britain met the needs of the time, but the island was largely deforested by the 18th century, and charcoal was in short supply. By 1740, Britain was importing nearly two-thirds of its iron from countries with better charcoal supplies, notably the American colonies, Sweden, and Russia.

This heavy import dependence changed when 18th-century British ironmasters made revolutionary changes in iron production, replacing charcoal with coal and coke, improving the blast mechanism, inventing a new refining process called "puddling," and adopting the rolling mill in place of the hammer for final processing. By 1855, Britain was producing about one-half of the world's iron, much of which was exported. With British development of the Bessemer converter and the open-hearth furnace in the 1850s, steel, a metal superior in strength and versatility to iron, began to be made inexpensively and on a large scale.

The early iron industry in the United Kingdom used domestic ores (mainly from the Midlands), but these eventually became inadequate, and the U.K. turned to imported ores. British inventors and entrepreneurs led the world into the age of cheap, mass-produced, and extensively used iron and steel, but because of growing resource shortages and rising labor costs, the country lost its leadership in global totals of steel production. The United States and Germany surpassed the British output before 1900. Britain now produces only about 12 million tons of steel annually, well behind the European leader of Germany with its 45 million tons, the United States with its 90 million tons, and the global leader, China, with 180 million tons.

During Britain's period of industrial and commercial ascendancy, cotton textiles were its leading export, amounting to almost one-quarter of all exports by value in 1913. The enormous cotton industry eventually became concentrated in the district of Lancashire (especially in its leading cities of Manchester and Liverpool) in northwestern England. Imported raw cotton was the basis for this industry. Lancashire inventors first mechanized the spinning and weaving processes. Then, power was applied to drive improved machines, first using falling water from Pennine streams and then steam fueled by coal from underlying Lancashire coal fields. In 1793, the invention of the cotton gin by American Eli Whitney provided a machine that separated seeds from raw cotton fibers economically, making cotton a relatively cheap material from which inexpensive cloth could be manufactured. Industrial growth acted like a magnet to bring farm youth to the cities in quest of factory labor. Manufactured textiles were exported through the Mersey River port of Liverpool. Manchester itself became a supplementary port after completion of a ship canal to the Mersey Estuary in 1894.

For more than a century, until World War I, Lancashire dominated world trade in cotton and cotton textiles. But 1913 saw the peak of British production; after that, the decline of the industry was nearly as spectacular as its earlier rise. The root of the trouble was increased foreign competition. Throughout the 20th century, numerous countries surpassed the United Kingdom in cotton textile production and exports, with a resulting slide in employment in Lancashire. Other countries such as China and South Korea had lower labor costs and were increasingly able to put their own or imported textile machinery to use. Its supremacy in cotton eclipsed, Lancashire's economy still languishes and has not yet attracted the high-tech industries that city fathers hope will resurrect it. Manchester and Liverpool now comprise a metropolitan area with a population of about 3.6 million. Liverpool's claim to global fame is as the birthplace of The Beatles, four sons of working-class families who helped transform Western music and society in the 1960s.

Great Britain is well endowed with grazing land and, as early as the Middle Ages, exported raw wool and woolen

Geography of Energy

Europe's Energy Alternatives

One way in which the European nations are quite distinct from the United States is in their embrace of alternative energy. The EU countries have set a target of deriving 22 percent of their electricity, and 12 percent of all their energy, from renewable sources by 2010. Some EU members are especially aggressive with alternatives. Denmark uses wind power to meet more than 20 percent of its electricity needs—the highest rate in the world from this source. Even with its bountiful North Sea oil reserves, Great Britain aims to meet 20 percent of its energy needs from renewable sources by 2020 (as of 2004, only 3 percent came from renewables). A major component of this effort is a $12 billion project to construct more than 1,000 wind turbines off England's coast, mainly in the Thames estuary, the East Coast area, and the northwest coast. These will supply an estimated 7 percent of the country's energy needs. Great Britain is seeking to establish itself as the EU leader in carbon dioxide emissions reductions. Going well beyond its pledge in the Kyoto Protocol, the country's stated aim is to cut carbon dioxide emissions to 60 percent below their 1990 levels by 2050.

Although an alternative to fossil fuels, nuclear power is not a benign or renewable power resource, and it has fallen out of favor in Europe since the 1986 Chernobyl disaster in nearby Ukraine. Sweden, already reliant on renewable energy for 32 percent of its power, is in the process of phasing out its nuclear energy program. Germany vows to cease nuclear production completely by 2023. Germany has established itself as the world's leading manufacturer of wind turbines, and more than 12,000 of these mills dot Germany itself, providing 8 percent of the country's electricity needs. That share is projected to rise to 20 percent by 2020.

Figure 4.D

Figure 4.E

Figure 4.F

Figure 4.G

Traditional and modern sources of energy in Europe: (D) peat from a Scottish bog, (E) coal in Poland, and (F and G) Danish windmills, both old and new

Medical Geography

Mad Cow Disease

British agriculture was hit hard in the late 1990s by an outbreak of **mad cow disease** (more properly, bovine spongiform encephalopathy, or **BSE**). Probably originating from sheep body parts in the feed offered to animals, this debilitating and ultimately fatal disease spread to more than 200,000 British cattle. BSE also manifest itself as a variant of a degenerative brain disorder in humans called Creutzfeldt-Jakob disease (CJD), leading to the deaths of some 140 Britons.

Public anxiety grew rapidly, and the British beef industry all but collapsed. The European Union banned all exports of cattle, beef, or beef products from the United Kingdom from 1996 until 1999. More than two million cattle were killed, and their bodies incinerated in great pyres. The spread of mad cow disease was finally halted, but not before it had spread to France, Germany, Switzerland, Portugal, Spain, Poland, Greece, and other nations, causing economic disruption and significant changes in diet.

cloth to continental Europe. As mechanization of the wool industry improved, Britain's domestic raw wool supplies were supplemented with imports. Exports of manufactured woolen textiles peaked before World War I, but Britain is still a major world exporter. The wool manufacturing industry is centered on the east slope of the Pennines in western Yorkshire (known as the West Riding area), especially in Leeds–Bradford (population: 1.5 million).

When wood was the principal material used in building ships, Great Britain, as a major seafaring and shipbuilding force of global significance, created difficulties for itself by deforesting the land. Large quantities of timber were imported from the Baltic Sea area and from North America, and the importation of completed ships from the American colonies, principally New England, was so great that one-third of the British merchant marine was American-built by the time of the American Revolution. During the first half of the 19th century, American wooden ships posed a serious threat to Britain's commercial dominance on the seas. In the later 19th century, iron and steel ships propelled by steam transformed ocean transportation. By the end of the century, Britain led the way and built four-fifths of the world's seagoing tonnage.

Two districts developed as the world's greatest centers of shipbuilding: the Northeast England district, with shipyards concentrated along the lower Tyne River between Newcastle and the North Sea, and the western Scottish Lowlands, with shipyards lining the River Clyde for miles downstream from Glasgow. Both districts had a seafaring tradition, both were adjacent to centers of the iron and steel industry, and both had suitable waterways for location of the shipyards. Their ships were built mainly for the United Kingdom's own merchant marine, as they still are, but those built for foreign shipowners were important. Like the other three industries of early importance and current decline in the United Kingdom, shipbuilding lost major significance in the last decades of the 20th century. But shipbuilding rebounded sharply in the early 2000s, especially as a result of new defense contracts.

As a whole, then, resources and industries that put the major British cities (aside from London) on the map are

played out or are in serious decline. This process of **deindustrialization** has hit Britain especially hard, but other regions in the Coreland, notably the Ruhr in Germany, have also suffered (see Figure 4.4). Relatively high unemployment is a major consequence of this trend, but a positive result is that the cities are cleaner and more livable. Many British cities, notably Birmingham—second in size only to London—are replacing old industrial areas with greenbelts and attractive architecture. Ideally, the improved environment will lure in high-tech and other clean industries, services, and tourists that will drive a new economy. The process is underway in select areas, such as Scotland's **Silicon Glen**, London's **M-4 Corridor**, and southeast England's **Silicon Fen**, as Figure 4.4 illustrates.

Food and Agriculture in Britain

Like other more developed countries, the United Kingdom has few people employed in farming and other rural pursuits. Only about 2 percent of the working population of Great Britain is in agriculture, forestry, and fishing. About one-quarter of the total land area is given over to crops, with nearly half used for pasture. Grass farming for milk and meat production is a major rural landscape use (see Medical Geography above). Even with its broad arable base, more than 60 percent of Britain's food supply is imported. Britain is more self-reliant in fish and seafood, although its fishing industry has been in steady decline since the 1960s. EU policies have led to fishing restrictions and a decline in catch, but even with this, Britain possesses the third-largest fishing fleet in the EU. Major fishing ports include Hull, Grimsby, Fleetwood, and Plymouth in England and Peterhead in Scotland.

Total food production increased sharply after World War II, mainly through government policies aimed at reducing imports. Expensive subsidies promoted large increases in wheat and barley production. The United Kingdom is like most other EU countries following a policy of subsidized agriculture, with high import tariffs on farm products from non-EU countries. As long as this policy holds, British farmers will continue to receive incentives in the form of relatively high and protected prices, and the British people, like the other peoples in the European Union, will eat expensive food produced within the home country or within the European Union.

Ireland: The Emerald Isle

Ireland (known as Eire in the indigenous Celtic language) is a land of hills and lakes, marshes and peat bogs, cool dampness, and the verdant grassland that earned its epithet "the emerald isle"(Figure 4.7). The island consists of a central plain surrounded on the north, south, and west by hills and low rounded mountains. The Republic of Ireland, which occupies a little over four-fifths of the island, is in the midst of a transition from an agricultural to an industrial economy.

Ireland ranked near the bottom of Western Europe's countries for many decades, but recent economic growth changed this pattern. High-tech industries such as electronic products and software in particular boosted Ireland's economy. The size of the economy doubled in the 1990s, earning Ireland the nickname **Celtic Tiger** and placing it on par with the United Kingdom in per capita GDP PPP (see Table 3.3, page 52). Unemployment dropped sharply. This surge of de-

Figure 4.8 Dublin's Grafton street

velopment was accomplished under a governmental program designed to attract foreign-owned industries through a combination of cheap labor, tax concessions, and help in financing plant construction. Hundreds of industrial plants, owned mainly by U.S., British, and German companies, produced computers, electronics, foods, textiles, office machinery, organic chemicals, and clothing. Most of these plants sprang up around the two main cities, Dublin (population: 1 million; Figure 4.8) and Cork, where they comprised an economic region dubbed **Silicon Bog** (see Figure 4.4), and in the vicinity of Shannon International Airport in western Ireland. They thrived until Ireland fell victim to the global technology bust of 2000–2002. The Gateway computer and some other plants closed, and Ireland's economic growth returned to more modest levels.

Ireland's agriculture, now employing only about 10 percent of the labor force, has had a troubled career. England conquered Ireland in the 17th century and made the island a formal part of the United Kingdom. Ireland was effectively an English colony. Most of the land was expropriated and divided among English landlords into large estates, where Irish peasants worked as tenants. The income from this enterprise generally found its way to Britain, and Ireland became a land of poverty, sullen hostility, and periodic violence. A crop disease known as blight caused the notorious **potato famine** of 1845–1851. Nearly 10 percent of Ireland's prefamine population of 8 million died of starvation or disease during these years. An even larger number emigrated to England, Wales, Scotland, Australia, and North America. The large Irish American community in the United States traces much of its ancestry to this migration. Today, its members are just a few of the eager consumers abroad of Irish culture, especially dance and music.

Northern Ireland, now a majority Protestant part of the United Kingdom, has had a particularly difficult struggle that has colonial roots. One outcome of the second English conquest of Ireland, in the 17th century, was the settlement of

Figure 4.7 A typical irish landscape, beheld from the Blarney Castle

English and Scottish Presbyterians and other Protestants in the north, where they became numerically dominant. They were intended to form a nucleus of loyal population in a hostile, conquered, largely Roman Catholic country. In 1921, when the Irish Free State (later to become the Republic of Ireland) was established in the Catholic part of the island, the predominantly Protestant north, for economic and religious reasons, elected to remain with the United Kingdom. Northern Ireland (also known as Ulster) was given much autonomy, including its own parliament in Belfast. Northern Ireland's Catholic minority felt increasingly marginalized by the pro-British Protestant majority and, in the late 1960s, began agitating, often violently, for change. The British imposed direct rule, and Catholic discontent soon focused on the perceived British occupation. The **Irish Republican Army (IRA)** began a campaign of bombings, shootings, and arson, sometimes into the heart of England, ostensibly designed to drive the British army from Northern Ireland. However, "The Troubles," as the locals call them, mainly involved an often violent struggle between **Catholic Republicans** and **Protestant Unionists** in Northern Ireland.

In 1998, by which time more than 3,400 had died in the conflict, the U.S.-brokered **Good Friday Agreement** was signed. Approved by the IRA (and by its political counterpart, **Sinn Fein**) and the Unionists, the agreement called for the devolution of British power. Direct rule by Britain would be replaced by the **Northern Ireland Assembly,** in which Protestants and Catholics would share power. Violent challenges to the peace agreement persisted from both sides, however, leading the British to halt devolution and reimpose direct rule in 2002. If achieved, peace could help bring prosperity to Northern Ireland, which has had periods of strong economic growth based on shipbuilding and linen textile industries. The main industrial and residential city is Belfast (population: 482,000).

4.2 France: *Vive la Difference!*

Thanks to the Channel Tunnel, or Chunnel (also called the Eurotunnel), running 16 miles (26 km) along the floor of the English Channel, since 1994 it has been possible to travel by train from London to Paris in about 3 hours (Figures 4.5 and 4.9). Britain and France were often enemies in the past, but the Chunnel is an appropriate symbol for how they are linked today.

France has long been one of the world's most celebrated countries. Europe's largest country in area, it is also the region's leading agricultural producer. It is one of the world's major industrial and trading nations, is a nuclear power, and is one of the five permanent members of the United Nations Security Council. Although France has lost its former significance as a major colonial power, it still maintains special relations with a large group of former French colonies. More than a dozen are now Departments of France and send representatives to the Assembly in Paris. Still another important foundation for France's international importance is the high

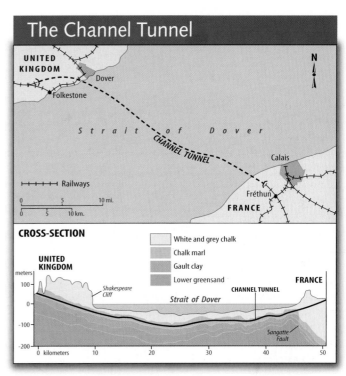

Figure 4.9 Location and geological and topographical cross section of the Chunnel *Source: After Jordan and Bychova Jordan, 2002, p. 300*

prestige of French culture in many parts of the world. It is a country of great self-esteem and recently has distinguished itself in world affairs by breaking some patterns of alignment with the West, especially the United States.

Framed and Fortified

About four-fifths the size of Texas, France lies mostly within an irregular hexagon framed on five sides by seas and mountains (Figure 4.10). In the south, the Mediterranean Sea and Pyrenees Mountains form two sides of the hexagon; in the west and southwest, the Bay of Biscay and the English Channel form two sides; in the southeast and east, the Alps and the Jura Mountains, and farther north, the Vosges Mountains, form a fifth side, with the Rhine River a short distance to the east. Part of the sixth or northeastern side of the hexagon is formed by the low Ardennes Upland, lying mostly in Belgium and Luxembourg, with lowland passageways leading into France from Germany both north and south of the Ardennes.

Aside from the large Massif Central in south central France, all the country's major highlands are peripheral. Even the Massif Central, a hilly upland with low mountains in the east, is skirted by lowland corridors that connect France's extensive plains of the north and west with the Mediterranean coast. These corridors include the Rhône–Saône Valley between the Massif to the west and the Alps and Juras to the east and the Carcassonne Gap between the Massif and the Pyrenees. France is thus a country with no serious internal barriers to movement but with barriers at its bor-

Figure 4.10 Principal features of France

ders. These have often failed to prevent the incursions of foreign armies into France; Germany, for example, attacked France three times and occupied the country throughout World War II.

Although they have been breached at times, France has natural fortifications, with two of its three land frontiers lying in rugged mountain areas. The Pyrenees are a formidable barrier, lowest in the west, where the independent-minded Basques, who are neither French nor Spanish in language, occupy an area extending into both France and Spain (see discussion of Basque issues in Chapter 5, page 129). Also straddling the mountainous Pyrenees border is the microstate of Andorra, whose 67,000 inhabitants enjoy a high standard of living from tourist revenues. In southeastern France, the Alps and Jura Mountains follow France's boundaries with Italy and Switzerland. The sparsely populated Alpine frontier between France and Italy is even higher than that of the Pyrenees: Mt. Blanc, the highest summit, reaches 15,771 feet (4,807 m). The number of important routes through the mountains is limited, and the Alps have been an important defensive rampart throughout France's history. The Jura Mountains, on the French–Swiss frontier, are lower but also difficult to cross, as they are arranged in long ridges separated by deep valleys.

Flourishing Agriculture

Although France's population is 74 percent urban, agriculture is a very important part of the French economy, far more than is typical of MDCs. Some 35 percent of the French landscape is agricultural, and another 28 percent is wooded. In 2003, France was the world's largest exporter of wine, barley, and sugar, and a major exporter of wheat, milk, butter, eggs, and meat. Since World War II, rural to urban migration has accompanied economic change, and increasing mechanization has replaced lost farm workers. The French continue, nevertheless, to see theirs as an agricultural nation, and the customs of the family farm still play a strong role in French identity and sense of well-being. One of the most celebrated efforts to diminish the force of globalization, seen as particularly ruinous to the family farm in France, was the destruction of a McDonald's restaurant under construction outside Paris in 1999. French sheep farmer Jose Bove, long active in efforts to slow down French accommodation and support of American cultural and economic expansion, led a militant farming union in this civil disobedience. Bove has become a national folk hero.

France enjoys some advantages in the agricultural sector. Its membership in the European Union gives French farmers free access to a huge and affluent consuming market and pro-

Definitions + Insights

The Primate City

By a common geographic definition, a primate city is a country's city that is larger than the country's second and third largest cities combined. Paris, for example, has a metropolitan area population of 11.4 million, far exceeding the combined populations of the second and third largest cities of Lille (with 1.7 million people) and Lyon (with 1.7 million people). In practical terms, a primate city dwarfs all others in its demographic, economic, political, and cultural importance. The primate city holds a disproportionately high share of the country's goods and services, and many national resources—such as bureaucratic offices and educational opportunities—are available only in the primate city. The primate city there-fore acts as a great magnet for **rural to urban migration** and tends to absorb more investment than other communities. Urban primacy is often bad for a country's development because funds that could be spent to improve services and quality of life in smaller cities and rural areas are often diverted instead to the primate city. With greater opportunities available in the primate than in the smaller cities of the hinterland, the most educated and talented people abandon the hinterland for the primate city. This internal brain drain contributes further to underdevelopment outside the primate city. Primate cities are rather rare in the more developed countries but are quite typical of the LDCs where regional development is a critical issue. Cairo, Karachi, and Mexico City are exemplary primate cities of Africa, Asia, and the Americas.

41

tection by tariffs from non-EU producers. The EU budget has generally provided generous price supports at rates above world prices for many agricultural products, with French farmers being the leading beneficiaries.

France is blessed on its own, independent of the European Union. Compared with the rest of continental Europe, the country has a relatively low population density. France's physical assets include superior topographic, climatic, and soil conditions. The main topographic advantage is the large area, especially in the north and west, level enough for cultivation. Climatically, most of the country has the moderate temperatures and year-round moisture provided by a marine west coast climate. The wettest lowland areas are in the northern part of the country, where crop and dairy production is especially intense. Southern France often experiences warmer temperatures, allowing production of corn in the southwestern plains. Productive vineyards are also concentrated in the south and southwest, although viticulture is practiced as far north as the Champagne region, east of Paris (Figure 4.11). Finally, northeastern France from the frontier to beyond Paris has fine soils developed from loess. As in much of Europe, these very fertile soils are used principally to produce wheat and sugar beets. The wheat belt in northeastern France is the country's chief breadbasket. Beet production from the same area provides both sugar and livestock feed (beet residue from processing) for meat and milk production.

French Cities and Industries

From the dawn of the Industrial Revolution until recent decades, France tended to fall behind its European neighbors, except for Spain and Italy, in industrialization and modernization. But the country shifted gears after World War II. Government subsidies supporting children spurred popula-tion growth. New energy sources replaced traditionally inadequate coal resources. France's membership in the European Union ended the protection of many inefficient producers. The French government promoted the combination of companies into larger units capable of higher efficiency and international competitiveness. France was well placed geographically and structurally to move into newly developing industries, technologies, and markets. It had a large pool of labor available for transfer from an oversupplied agriculture sector, a relatively modest share of its labor force and capital tied up in older industries, and an excellent educational system. As the economic pace picked up, the government took steps to allocate investment and development more equitably around the country, deliberately avoiding the traditional tendency to funnel resources into Paris. Planners use restrictions

Figure 4.11 Wine country of Southern France. The country produces about 20 percent of the world's wine.

and penalties on various types of new development in the Paris region, financial rewards for firms locating facilities in regions of greater need, and direct location or relocation of government-owned facilities in the less concentrated regional centers.

In terms of industrial productivity, France ranks only behind the United States, Japan, and Germany. Automobile manufacturing (of Peugeot and Renault cars, for example) is the most important single industry. Based in Toulouse, France, the Airbus corporation produces 60 percent of the world's medium and large jet aircraft, eclipsing the American company Boeing. In addition to automobiles and aircraft, major products include armaments, rubber, chemicals, textiles, and processed foods.

Paris is the greatest urban and industrial center of France, a primate city completely overshadowing all other cities in both population and economic activity (see Definitions and Insights, page 91). With its 11.4 million people, it is by far the largest city on the mainland of Europe. Paris is located at a strategic point on the Seine River relative to natural lines of transportation, but it is more the product of the growth and centralization of the French government and of the transportation system it created. Like London, it has no major natural resources for the industry in its immediate vicinity, yet it is the greatest industrial center of the country. Despite its economic clout, Paris maintains its international reputation as the romantic and monument-studded **"City of Light"** (Figures 4.12 and 4.13). It is the world's leading urban tourist destination, and the main attraction in what makes France the world's number one tourist destination.

In the Middle Ages, Paris became the capital of the kings who gradually extended their effective control over all of France. As the French monarchy became more absolute and

Figure 4.13 The entrance to the Louvre, one of the world's great museums and a landmark of Paris

centralized, its seat of power grew in size and came to dominate the cultural and political life of France. As national road and rail systems were built, their trunk lines were laid out to connect Paris with outlying regions. The result was a radial pattern with Paris at the hub, and this is essentially the pattern that exists today (Figure 4.14). As the city grew in

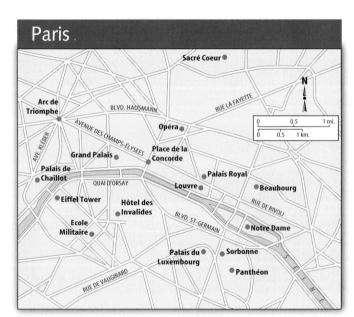

Figure 4.12 Downtown Paris and some of its world-famous attractions

Figure 4.14 Paris is the primate city at the hub of France's transportation network.

population and wealth, it became an increasingly large and rich market for goods.

The local market, plus transportation advantages and proximity to the government, provided the foundations for a huge industrial complex. This development has involved two major classes of industries. Paris is the principal producer of the high-quality luxury items (fashions, perfumes, cosmetics, jewelry, for example) for which that city and France have long been famous. Paris is also the country's leading center of engineering industries, secondary metal manufacturing, and diversified light industries. They are concentrated in a ring of industrial suburbs that sprang up in the 19th and 20th centuries. Paris inverts the usual pattern of the industrial city by maintaining a core of low-profile historic monuments, while the skyscrapers of the modern economy tower away from the city center.

With the more recent growth of Marseilles, Lyon, and Lille, the dominance of Paris has diminished somewhat. Two ports near the mouth of the Seine River handle much of the overseas trade of Paris: Rouen (population: 530,000), the medieval capital of Normandy, is located at the **head of navigation** (the farthest inland a ship can go on a river) for smaller ocean vessels using the Seine. With the increasing size of ships in recent centuries, more of Rouen's port functions have been taken over by Le Havre (population: 190,000), located at the entrance to the wide estuary of the Seine.

France's second-ranking urban and industrial area is the Nord (North), near the Belgian border, stretching along the western edge of the Ruhr District coal and industrial region. The largest city, Lille (population: 1.7 million), was established alongside France's once-leading (now closed) coal field. The Nord was an important contributor to, and is now a major problem in, the economy of France. In the 19th and early 20th centuries, it developed a typical concentration of the coal-based industries: coal mining, steel production, textile plants, coal-based chemical production, and heavy engineering. As in Britain, these operations have declined in recent decades, and the region suffers from high unemployment.

France's third largest urban-industrial cluster centers on the metropolis of Lyon (population: 1.7 million) in southeastern France. The valleys of the Rhône and Saône Rivers join here, providing routes through the mountains to the Mediterranean, northern France, and Switzerland. Since prehistoric times, the valley has been a major connection between the North European Plain and the Mediterranean. Today, it forms the leading transportation artery in France, providing links between Paris and the coast of the Mediterranean via superhighways, rails that carry the TGV—one of the fastest trains in the world—and a barge waterway formed by the Rhône, Saône, and connecting rivers and canals to the north.

France has promoted hydroelectric power and nuclear energy to compensate for inadequate fossil fuel supplies, and the Rhône Valley plays a major role in both. A series of massive dams on the Rhône and its tributaries produce about 10 percent of France's electricity, and further power comes from several large nuclear plants in the Rhône Valley using ura-

nium from the nearby Massif Central and foreign sources. France produces a greater proportion of its electricity from nuclear power—about 75 percent—than any other country in the world.

France's most important Mediterranean cities and metropolitan areas—Marseille, Toulon, and Nice (pronounced NIECE)—are strung along the indented coast east of the Rhône delta. Marseille (population: 1.5 million is France's leading seaport. Recent growth of port traffic and manufacturing in the Marseille area has been associated with increasingly large oil imports and the related petrochemical industry. Toulon, southeast of Marseille, is France's main Mediterranean naval base, and Nice, near the Italian border, is the principal city of the French Riviera, the most famous resort district in Europe. Along this easternmost section of the French coast, the Alps come down to the sea to provide a spectacular shoreline dotted with the beaches of the Côte d'Azur. Near Nice, a string of resort towns and cities along the coast includes the tiny principality of Monaco, a microstate that is nominally independent but closely related to France economically, culturally, and administratively. About three times the size of the National Mall in Washington, D.C., Monaco makes a living from gambling and other forms of recreation.

The south central wedge of western France and the Massif Central is characterized by a rich mix of manufacturing, agriculture, and tourism. This region is less populous, urban, and industrial than the parts of France to the northeast and east. Each of the largest urban centers, Toulouse, Bordeaux, and Nantes, has a population of less than 1 million. Bordeaux is a seaport on the Garonne River, which discharges into the Gironde Estuary on the Bay of Biscay. The city passed its name to the wines from the surrounding region, exported through Bordeaux for centuries. Nantes is the corresponding seaport where the other main river of western France, the Loire River, reaches the Atlantic. Toulouse, a historic city with modern machinery, chemical, and aircraft plants, developed on the Garonne well upstream from Bordeaux at the point where the river comes closest to the Carcassonne Gap. Lying between the Pyrenees and the Massif Central, the Gap provides a lowland route from France's Southwestern Lowland to the Mediterranean.

The Social Landscape

However proud and confident the French may seem in the international arena, at home they have their share of troubles. The economy has been sluggish. French law prescribes a 35-hour workweek, and many in France wonder if that has contributed to the troubled economy. French regulations make firing a substandard employee almost impossible, so trimming the fat is not as easy as in most other industrialized countries. The French enjoy their national rights to a 6-week vacation and comprehensive pensions and healthcare. Desperate to reduce the economic costs of pensions, the government has been trying to legislate lower retirement payments and more working years for public employees. The workers have responded with national transportation strikes. Caught between their privileges and the social costs these impose, the French are frustrated; a 2004 poll showed that about two-thirds of the French think their country is in decline.

One of the most serious issues France faces is how to deal with its large immigrant population, especially of Muslim Arabs. France has about 5 million Muslims, or about 9 percent of the population, which is the highest percentage of Muslims in the European core. Most are either immigrants or their offspring, mainly from the former French colonies of North Africa. France has a vigorously secular constitution, and laws interpreting it have provoked discord with the largely religious Muslim minority. In 2004, France's National Assembly issued a ban against wearing head scarves and other items with religious connotations in public schools, a motion affecting Sikhs, Jews, and others but targeted mainly at Muslim women. Except for support by feminist Muslim women, the Islamic community in France responded indignantly, and many Muslim schoolchildren have turned to private schools or home schooling. The Islamist militant group al-Qa'ida, typically focused against the United States and its allies in the Middle East, condemned the head-scarf ban and promised its own unique form of retaliation against France.

4.3   Great Germany

Germany (formally, the Federal Republic of Germany) reappeared on the map of Europe as a unified country in 1990. Between 1949 and 1990, there had been two Germanys: West Germany (German Federal Republic), formed from the occupation zones of Britain, France, and the United States after Germany's defeat in World War II in 1945, and Communist East Germany (German Democratic Republic), formed from territory occupied postwar by the Soviet Union. The urban landscape and political climate of West Berlin were a part of West Germany for those decades, although the city of Berlin was entirely surrounded by East Germany. The withdrawal of Soviet support for East Germany in 1989, and the destruction of the Berlin Wall, led to the rapid collapse of its Communist government in 1990 and to the jubilant reunification of Germany's 16 states (*lander*).

The new Germany is Europe's dominant country. Its population of 83 million is much greater than that of any other nation in the region. Politically, it is seen, along with France, as the cornerstone of the European Union. Economic considerations are even more important. The former West Germany alone was Europe's leading industrial and trading country during the latter half of the 20th century. The former East Germany was an advanced country by Soviet bloc standards but an economic disaster by West German standards. Many billions of dollars and much time have been required to rehabilitate Germany's East, and the task is far from done. This area, however, can be expected to contribute increasingly to overall German dominance in the European economy.

Figure 4.15 Principal features of Germany

A Central Location

Germany arrived at its favorable economic position by skillfully exploiting the advantages of its centrality within Europe together with the advantages of a diverse environment (Figure 4.15). Broadly, the terrain of Germany can be divided into a low-lying, undulating plain in the north and higher country to the south. The lowland North German Plain is a part of the much larger North European Plain. The central and southern countryside is much less uniform. To the extreme south are the moderately high German (Bavarian) Alps, fringed by a flat to rolling piedmont or foreland that slopes gradually down to the east-flowing upper Danube River. Between the Danube and the northern plain is a complex series of uplands, highlands, and depressions. The higher lands are mainly composed of rounded, forested hills and low mountains. Interspersed with these are agricultural lowlands draining to the Rhine, Danube, Weser, and Elbe Rivers, which drain into three different seas.

Germany has a maritime climate in the northwest and increasingly continental climate toward the east and south. The annual precipitation is moderate, with most places in the lowlands receiving an average of 20 to 30 inches (c. 50 to 75 cm) per year. Farmers generally must contend with infertile soils. German agriculture has long been characterized by crops that do well in infertile ground, such as rye, cabbages, and potatoes, and by hogs, which can consume farm waste. Two areas are more fertile, supporting the more demanding wheat, sugar beets, and cattle: the southern margin of the North German Plain, forming a belt of fertile loess soils, and the fertile alluvial Upper Rhine Plain, extending from Mainz to the Swiss border and mostly enclosed by highlands on both sides.

German Cities and Industries

Within this environmental framework, Germany's major cities tend to be located in the most economically advantageous areas and those most strategic with respect to transportation. A string of major metropolitan areas is located in the western part of Germany, along and near the Rhine River. This waterway has been a major route since prehistoric times. Four of the eight largest German metropolises— Düsseldorf, Cologne, Wiesbaden–Mainz, and Mannheim–Ludwigshafen–Heidelberg—are on the Rhine itself. So is Bonn, which was made the capital of West Germany in 1949 just in case the traditional capital city of Berlin fell into the Soviet grip.

The other four metropolises—Essen, Wuppertal, Frankfurt, and Stuttgart—are on important Rhine tributaries. This north–south belt of cities also includes several other large metropolitan centers, including Dortmund, Duisburg, and Aachen, and numerous smaller places. The concentration of seven of the forenamed cities at or just north of the boundary between the Rhine Uplands and the North German Plain reflects other geographic factors. Historically, some of the seven cities profited from their location on the agriculturally excellent loess soil just north of the Uplands. This concentration of large and medium-sized urban centers is a remarkable contrast with France's urban pattern, dominated by Paris.

Other German cities developed as early industrial centers before the Industrial Revolution, drawing on the Uplands for ores, waterpower, wood for charcoal, and surplus labor. Then, in the 19th and early 20th centuries, all seven cities grew in connection with the development of both the Ruhr coal field (Europe's largest), located at the south edge of the plain and east of the Rhine, and the smaller Aachen field. Except for Aachen and Cologne, the cities became part of the Ruhr, Europe's greatest concentration of coal mines and heavy industries, linked by a network of active river and canal transport (see Figure 4.3). Still other important German cities lie apart from the Rhine and the Ruhr. These include Hanover, Leipzig, and Dresden, spread along the loess belt across Germany; the North Sea ports of Hamburg, located well inland on the Elbe estuary, and Bremen, up the Weser estuary; two cities in the southern uplands—Munich, north of a pass route across the Alps and Nuremberg; and Berlin (population: 3.9 million; Figure 4.16), whose large size

Figure 4.16 Berlin

belies its unlikely location in a sandy and infertile area of the northern plain. Berlin is mainly an artificial product of the central governments, first of Prussia and then of Germany, which developed it as their capital. Since reunification, Berlin has once again become Germany's capital.

Much of Germany's present industry is not within the Ruhr or within Saxony, a southern region of former East Germany that had an industrial past. Instead, industrial and urban centers are widely scattered. Many cities are large, but none is as nationally dominant as London or Paris. This dispersed pattern came about because Germany was divided for a long time into petty states, many of which eventually became internal states within German federal structures. These often had their own capitals, which would become Germany's large cities.

Many of these industrial cities were bombed heavily in World War II, but industrial capacity was not completely devastated. Several factors promoted rapid economic reconstruction of West Germany after the war. The massive U.S.-funded aid package known as the Marshall Plan, supplying $13 billion in capital goods between 1948 and 1952, helped rebuild West Germany as a means of staving off the Communist threat from the east. The skills of the population were still there, and labor was inexpensive for a while, as workers were abundant and eager to be working again. Markets were hungry for industrial products. By the 1960s, the pace of growth slowed, but it was still fast enough to provide jobs through subsequent decades for Europe's largest numbers of foreign workers, especially Turks and Kurds from Turkey, who replaced upwardly mobile German workers. The export of German industrial products to world markets became more difficult as Japanese and other foreign competition stiffened. Nevertheless, Germany still ranked as the world's second largest exporter of goods (after the United States) in 2004.

East Germany (1949–1990) was Communist ruled and Soviet dominated, but it was still German. Even Soviet exploitation and bureaucratic inefficiency were unable to destroy the traditional German ethics of hard work, efficiency, attention to detail, and high standards. East Germany was the most productive of the Soviet "satellite" countries that lay between Russia and Western Europe, but its economy lagged far behind West Germany's. When East Germany collapsed in 1989, a major migration stream began flowing from East to West Germany. By 2004, more than 2 million people had streamed westward from what Germans call the "new states" of the former East Germany, and the flow continued. Most of those who remained were educated but unemployed or low-wage earners, a situation that may turn in the region's favor. There is a growing semiconductor industry in what local promoters like to call **"Silicon Saxony,"** especially in Dresden and Leipzig (see Figure 4.4). Taking advantage of the skilled and inexpensive labor, more than 70 American chipmaking companies had set up operations in the region by 2004.

Billions of dollars of investment are helping to raise economic standards in the east and to clean up the severe envi-

ronmental damage incurred under Communist rule there. To complete that transition, the former West Germans will have to bear much of the tax and social burden of bringing the former East Germans up to the standards long enjoyed in the west. Economically, Germany is a **welfare state** that subsidizes its citizens generously, so the costs have mounted quickly. These costs are partly to blame for Germany's economic downturn in the early 2000s, when economic growth was the slowest of all EU countries and unemployment grew to a national average of about 10 percent. The rate of unemployment in the former East Germany was more than twice that in the west. Germany is now making a painful transition from a welfare state to one in which its citizens must pay more out of pocket for healthcare and other vital services.

WORLD
REGIONAL
Geography ⊛ Now™

Click Geography Literacy to make a map of changes in economic well-being of eastern and western Germany before and after reunification.

4.4   Benelux: Tolerance and Trade in the Low Countries

Belgium, the Netherlands, and Luxembourg have been closely associated throughout their long histories. They are known collectively as the Benelux Countries or the Low Countries. The northern two-thirds of Belgium and practically all of the Netherlands consist of a low plain facing the North Sea (Figure 4.17). This plain is the narrowest section of the North European Plain. Luxembourg is also included in the Low Countries designation because of the long historic connection among all three countries. Large sections near the coast, especially in the Netherlands, are indeed below sea level and have been made into major farming and urban landscapes through the engineering of polders—land below sea level that has been protected from the sea by dikes, canal systems, and pumps (see Figure 4.h and Definitions and Insights, page 98). It is not surprising that the Netherlands has played a leading role in negotiations to reduce global emissions of greenhouse gases, which might lead to rising sea levels and increase the threat of catastrophic flooding from the North Sea.

Small Countries, Great Trade

The economic life of the Benelux countries is characterized by an intensive and interrelated development of industry, agriculture, and trade. An especially distinctive feature of their economies for more than four centuries has been a dependence on international trade. Successful pursuit of international trade is reflected by the presence of three of the world's major port cities within a distance of about 80 miles (c. 130 km): Rotterdam and Amsterdam in the Netherlands

Netherlands, Belgium, and Luxembourg

Figure 4.17 Principal features of the Benelux countries

and Antwerp in Belgium. Amsterdam (population: 740,000) is the largest metropolis and the constitutional capital of the Netherlands. The government is actually located at The Hague, which is important in international circles as the seat of the International Court of Justice, the judiciary branch of the United Nations.

Amsterdam was the main port of the Netherlands (also known as Holland, and its people as the Dutch) from the 17th through 19th centuries, when the Dutch colonial empire was expanding. The most important component of this empire was the Netherlands East Indies (now Indonesia). Until Indonesia's independence after World War II, Amsterdam built a growing trade in imported tropical crops and oil from its Spice Islands. Many of these imports were reexported, often after processing, to other European countries. Profits that accumulated during these centuries supplied funds for the large overseas investments of the Netherlands and made Amsterdam an important financial center, which it continues to be. Amsterdam also remains an important port. Its original approach by sea via the Zuider Zee, a landlocked inlet of the North Sea, proved too shallow for the larger ships of the 19th century. The opening in 1876 of the

North Sea Canal solved this problem for a time, but efforts to widen the canal could not give Amsterdam access to the sea comparable to that of Rotterdam.

Continuing ties with its former colony have given Amsterdam a large Indonesian community that imparts a distinctive ethnic flavor to the city. The city has earned an international reputation as a cosmopolitan and tolerant city with a very high quality of life (Figure 4.18). Prostitution is legal and carefully regulated, and cannabis may be consumed in licensed "cafés." Pedestrian and bicycle traffic replace cars to a large degree in the downtown district, making for a clean, quiet, and enjoyable urban environment. Such amenities help compensate for the fact that the Netherlands has a higher population density than any other European country, aside from the small islands and the microstates (see Table 3.1). All Dutch cities are trying to limit urban sprawl, provide more outdoor and waterfront recreational areas, and maintain an environment that is pleasant for their residents and attractive to tourists. Planners are especially concerned with maintaining the open space semi-encircled by Amsterdam, Utrecht, Rotterdam, The Hague, and other cities. These urban areas are tending to coalesce into what is referred to as Randstad Holland, the "Ring City" of Holland.

Rotterdam (population: 600,000; Figure 4.19), the world's largest port in tonnage handled, is on one of the navigable distributaries of the Rhine River and is better situated for Rhine shipping than either Amsterdam or Antwerp. Rotterdam controls and profits from the major portion of the river's transit trade. It receives goods by sea and dispatches them upstream by barge and also receives goods downstream by barge—especially from Germany's Ruhr district—and sends them on to global destinations. Part of the Rotterdam facilities is the harbor and industrial development of Europoort, designed to handle supertankers and international ferry traffic. Antwerp (population: 964,000), located about 50 miles (80 km) up the Scheldt River, is mainly a port for

Figure 4.18 Cannibis café in Amsterdam

Owen Franken /Corbis

345

Definitions + Insights

The Polder

One of the most characteristic Low Countries landscapes is land reclaimed from the North Sea. Polder land is low-lying land, often below sea level, that has been surrounded by dikes and artificially pumped dry. The process of turning former swamps, lakes, and shallow seas into agricultural land has been going on for more than seven centuries in Holland. About 50 percent of the Netherlands now consists of an intricate patchwork of polders and canals (Figure 4.H), and Belgium has a narrow strip of such lands behind its coastal sand dunes. Even though the polders are the best agricultural lands in the Netherlands and production from them is the heart of Dutch farming, this land has become increasingly important in accommodating the continuing growth of cities.

Figure 4.H An aerial view of the Netherlands illustrating its amphibious and low-lying, vulnerable situation

Belgium itself but also has a share of the Rhine transit trade. Belgium's coast is straight and its rivers are shallow, so the deep estuary of the Scheldt gives Antwerp the best harbor in the country, even though it must be reached through the Netherlands.

Industrial Patterns

The Benelux countries are highly industrialized. Before major discoveries of natural gas were made in the Netherlands in the 1950s, Belgium and Luxembourg had more mineral

Figure 4.19 The port of Rotterdam

resources. This helped them initially to become more industrialized than the Netherlands. Now, however, the Netherlands, with its advantageous trade position and natural gas, is the leading industrial Benelux country.

The Netherlands has long depended on imported fuels other than natural gas. More oil imports now enter Europe via Rotterdam–Europoort than any other port, with some arriving as crude oil and some refined further in the Rotterdam area. This activity has generated one of the world's largest concentrations of refineries. Petrochemical plants near the refineries represent the main branch of the Dutch chemical industry. Other leading Dutch industries are food processing, electrical machinery, textiles and apparel, and electrical and electronics industries (of which Philips is the most renowned).

One of Europe's major coal fields crosses Belgium in a narrow east–west belt about 100 miles (161 km) long, roughly following the valleys of the Sambre and Meuse Rivers and extending into France on the west and Germany on the east. Liège (population: 180,000) and smaller industrial cities along this Sambre–Meuse field accounted for most of Belgium's metallurgical, chemical, and other heavy industrial production until those industries followed the familiar decline prompted by depleted reserves and increased international competition. An iron and steel industry developed here even before the Industrial Revolution, using local iron ore and charcoal. Local coal and the early adoption of British techniques soon made Liège the first city in continental Europe to develop modern large-scale iron and steel manufacturing. Newer industrial plants are mainly in northern Belgium, where Belgium's capital and largest city, Brussels (population: 1.8 million), is the leading center of a breed of new, less energy-dependent cities. Brussels plays a vital regional role as the location of the European Union's headquarters. Belgium is internally secure, despite the divide between the majority Flemish speakers and the large French-speaking Walloon minority. The Flanders region of Belgium, where the Flemish are concentrated, periodically calls for greater devolution of power from Brussels.

Luxembourg's most important exports previously came from the iron and steel industry based near the southern border where the Lorraine iron ore deposits of France overlap into Luxembourg. Its steel days numbered, Luxembourg continues to follow the European pattern in diversifying its manufacturing industries and exports, especially with chemicals and rubber products. It has modified its economy successfully; its annual per capita GDP PPP of more than $55,000 makes it the richest country in the world by that standard.

Benelux Farming

Despite their small sizes, the Netherlands and Belgium are very productive agriculturally, with high yields per unit of land. Parts of each country have very fertile soils, especially in the polders of the Netherlands and the loess lands of Belgium. There is great investment in water control, agricultural machines, knowledge, technical expertise, and fertilizer. Agriculture in the Low Countries has profited from European Union tariff protection and subsidies for agricultural producers, stimulating spectacular increases in output to the extent that overproduction of certain products such as milk and butter has become a more worrisome problem than food supply. Belgium and the Netherlands specialize in labor-intensive farm commodities such as commercial flowers (notably Holland's tulips) and dairy products. The dense and affluent populations in and near the Low Countries provide a large market for such specialties.

4.5 Switzerland and Austria: Prosperous Mountain Fastness

On the high southern margin of Europe's core lie Switzerland and Austria (Figure 4.20). More than half of each country is occupied by the lofty, rugged Alps. North of the Alps, both countries include strips of rolling, once glaciated plains and foothills of the mountain foreland. The Swiss section of the Alpine foreland, called the Swiss Plateau, extends between Lake Geneva on the French border and Lake Constance (German: Bodensee) on the German border. The Austrian section of the foreland lies between the Alps and the Danube River from Salzburg on the west to Vienna on the east. A third portion of the foreland in southern Germany separates the Swiss and Austrian sections. On their northern edges, Switzerland and Austria have mountains that are much lower and smaller in extent than the Alps. These are the Jura Mountains, on the border between Switzerland and France, and the Bohemian Hills of Austria, on the border with the Czech Republic.

Resources and Development

Economic success is often linked to a generous resource base, but in this case, less well-endowed Switzerland has become somewhat more prosperous than Austria. Based on natural resources alone, Austria should be the more successful country economically. With fewer mountains, it has more arable land than Switzerland and more land per person. It also has more forested land and a better mineral resource base. Both countries depend heavily on abundant hydroelectric potential, but Austria's resource base is greater.

In economic success, however, the advantage lies with Switzerland, whose GDP PPP is about 9 percent higher than that of Austria. The historical developments that favored Switzerland outweighed Austria's natural resource advantage. Switzerland enjoyed peace within stable boundaries, while Austria was rocked and chipped away at by two world

Figure 4.20 Principal features of Switzerland and Austria

wars. Switzerland's peace is the product of its position as a buffer state between greater powers, the comparative defensibility of much of the country's terrain, and the relatively small value of Swiss economic production to a would-be aggressor. In addition, Switzerland has historic and political value as an intermediary between belligerents in wartime and as a refuge for people and money. It also has a policy of strict and heavily armed **neutrality.** Surrounded by countries in conflict, Switzerland managed to stay out of both world wars.

Switzerland

With peace prevailing, the Swiss have had the opportunity to develop an economy finely tuned to the country's potentials and opportunities. They have exploited with special skill the fields of banking, tourism, manufacturing, and agriculture. Favored by Swiss law, the country's banks enjoy a worldwide reputation for security, discretion, and service. The result has been a massive inflow of capital—less often of dubious origins these days—to Swiss banks. Switzerland's largest city, Zürich (population: 980,000), is one of the major centers of international finance, and other Swiss cities are also heavily involved in banking.

Switzerland's scenic resources make it a major tourist destination. The country is a mecca for winter sports, and Alpine resorts such as Zermatt (Figure 4.21), Davos, and St. Moritz are world famous. Numerous tunnels and costly highway engineering marvels link highland Switzerland with surrounding countries. Because most of its raw materials and fuels must be imported, industrial specialties in Switzerland minimize the importance of bulky materials and derive value from skilled design and workmanship. Swiss watches are the most renowned, but other high-tech instrumentation—for example, in biotechnology and medical devices—is also typical of Switzerland. Swiss-based multinational firms produce foods (Nestlé) and pharmaceuticals (Roche). The country has many small industrial centers, mostly on the Swiss Plateau, and these obtain their electricity from hydropower. Five of the country's six largest cities, including the capital of Bern, are located on the plateau. The remaining city, Basel, lies beyond the Juras at the point where the Rhine River turns north between Germany and France. Basel is the head of navigation for Rhine barges that connect Switzerland with ocean ports and other destinations via river and canal.

Switzerland's internal political organization reflects its linguistic and ethnic diversity. German is the primary language of 64 percent of the population, French of 19 percent, and Italian of 8 percent. The city of Geneva in the west has a distinctively French feel, and Zurich in the east is strikingly German. The country was originally a loose alliance of small sovereign units known as *cantons.* With the development of a stronger central authority in the 19th century, the cantons retained many of their powers, except for limited functions assigned to the central government. Central government responsibilities have since grown, but each of the 26 local units has preserved a large measure of authority. Switzerland's government often submits important legislation directly to the

Figure 4.21 With the iconic Matterhorn towering to 14,691 feet (4,478 m) above it, the village of Zermatt is carefully managed as a pedestrian-only jewel of international tourism.

people. Guarantees of fundamental rights, local autonomy, and close governmental responsiveness to the will of the people are used to foster national unity, despite the potential difficulties of cultural diversity. Partly because of these strong political traditions, Switzerland has so far resisted joining the European Union. The country also fears it would be a "small fish" in the big EU pond and that its special interests such as banking and farming might lose their privileges within the European Union.

The central government has pushed economic development vigorously. Two outstanding accomplishments are the railroad and hydroelectric power networks. Despite the difficult terrain, the dense railway network carries enough international transit traffic to earn important revenues (Figure 4.22). The hydroelectric system, taking advantage of the country's mountainous terrain and abundant precipitation, yields about 60 percent of the country's electric output, making Switzerland one of the highest per capita nations in the world in hydropower produced. A country with much

Figure 4.22 Swiss rail

unrealized hydropower potential is Liechtenstein, a microstate sandwiched between Switzerland and Austria. With just over 30,000 people, Liechtenstein is a kingdom that has prospered. Its decades-long reliance on dubious profits from money laundering and tax evasion has given way to a cleaner financial industry and the unique niche of false-teeth production.

Austria

Austria emerged in its present form as a remnant of the Austro-Hungarian Empire when that empire disintegrated in defeat in 1918, at the end of World War I. Austria's population is essentially German in language and culture—and the countries have been closely linked historically—but unification with Germany was forbidden by the victors. The economy was wracked by the loss of its empire, and the country limped through the interwar period until it was absorbed by Nazi Germany in 1938. In 1945, after defeat in World War II, Austria was reconstituted as a separate state but was divided into occupation zones administered, respectively, by the United States, the United Kingdom, France, and the Soviet Union. During 10 years of occupation, the Soviet Union removed a vast amount of industrial equipment as war reparations from its Austrian Soviet Zone in the east (which included Vienna). It was only in 1955 that this occupation was ended by agreement among the four powers, and Austria became an independent, neutral state.

As the core area of the earlier Austro-Hungarian (Hapsburg) Empire in southeastern Europe, Austria developed a diversified industrial economy prior to 1914. Its industries depended on resources and markets provided by the broader empire that covered much of the North European Plain. Austrian iron ore was smelted mainly with coal drawn from Bohemia and Moravia, now in the Czech Republic. Austria's textile industry specialized in spinning, leaving much of the weaving to be done in Bohemia. Austrian industry had the benefit of a large protected market in which one currency was in use. In turn, the producers of foodstuffs and raw ma-

terials also had a market within an empire protected by external tariffs.

When the empire disintegrated at the end of World War I, the areas that had formed Austria's protected markets were incorporated into independent countries. Motivated by a desire to develop industries of their own, these countries erected tariff barriers, and other industrialized nations competed with Austria in their markets. The resulting decrease in Austria's ability to export made it more difficult for the country to secure the imports of food and raw materials that its unbalanced economy required. Such difficulties were further increased after World War II by the absorption of East Central Europe into the Soviet Communist sphere. The necessary 1990s reorientation of Austrian industries toward new markets, principally in Western Europe, was made more difficult because Austria did not join the European Union until 1995. As a consequence, Austrian exports of goods were consistently inadequate to pay for imports. Despite these obstacles, Austria has built a successful economy suited to its status as a small independent state. Major elements in the economy are forest industries, tourism, motor vehicle production, hydroelectric development, manufacturing of traditional clothing and crafts, and larger scale engineering and textile industries.

Since the end of the empire, Austrian agriculture has become more specialized in dairy and livestock farming. Enlarged acreage devoted to feed crops such as barley and corn accompanies the increased emphasis on animal products. But Austria is still far behind Western Europe's leading countries in agricultural efficiency. Many small and poor farms still exist, and 5 percent of the labor force is still in agriculture—a high proportion by Western European standards.

Austria's capital and largest city, Vienna (population: 2 million; Figure 4.23) is not as prominent in national and regional affairs as it once was. People and industry have shifted away from the capital to the smaller cities and mountain districts. Vienna's importance, however, has persisted since Roman times and is based on more than purely Austrian

Figure 4.23 The Graben, one of Vienna's main shopping areas

circumstances. The city is located at the crossing of two of the European continent's major natural routes: the Danube Valley route through the highlands separating Germany from the Hungarian plains and southeastern Europe and the route from Silesia and the North European Plain to the head of the Adriatic Sea. The latter follows the lowland of Moravia to the north and makes use of the passes of the eastern Alps, especially the Semmering Pass, to the south. Thus, Vienna has long been a major focus of transportation and trade and also a major strategic objective in time of war. In the geography of the early 21st century, Vienna finds itself particularly well-positioned to capitalize on the growth potential of the formerly Communist East European nations, which are considered in the following chapter.

CHAPTER SUMMARY

- The nations that are considered to comprise the European core include the United Kingdom, Ireland, France, Germany, the Low Countries (the Netherlands, Belgium, and Luxembourg), Austria, and Switzerland. The giant nations in this region are Germany, the United Kingdom, and France. They are the primary engines for the economic momentum in this part of the world. Three microstates—Andorra, Monaco, and Liechtenstein—are also included here.

- The United Kingdom is made up of England, Scotland, Wales, and Northern Ireland. It forms the greater part of two major islands: Great Britain and Ireland, which are called the British Isles. The island of Great Britain has distinctive highlands and a lowlands regions.

- Most of the 18th and 19th century industrial and urban development of Britain related to the location of coal and iron ore, the two leading resources of the Industrial Revolution. London arose on the Thames estuary with neither coal nor iron as a local resource and is today a major global financial center.

- Britain's five leading industries during and after the Industrial Revolution were coal, iron and steel, cotton textiles, woolens, and shipbuilding.

- While the country of Ireland has recently experienced strong economic growth based on high-tech industries, Northern Ireland has not yet been able to resolve a longstanding conflict between historically indigenous Roman Catholics and more recent residents of Protestant, English background.

- France is territorially the largest of the European core countries. The nation has an internal coherence and reasonably clear national borders made up of rivers, mountains, and seacoasts. Agriculture is more important in French economy and culture than in any other country of the European core. Paris is an unusual example of a primate city in a more developed country. The French are at odds with themselves over quality of life, immigration, and other issues.

- The EU countries have established an ambitious schedule to replace fossil fuel energy with wind and other renewable energy alternatives. Britain, Germany, and Sweden are notable leaders in alternative energy technology.

- Germany, re-formed when West Germany reunified with East Germany in 1990, has the largest economy and population of all Europe. German urban centers have a much broader distribution than do the industrial cities of France and even Britain. As in Britain, these cities generally were located by the availability of coal and other resources and of rivers for transport, and as in Britain, they have been deindustrializing in recent decades. Germany's economy has been performing poorly because of the burdens of being a welfare state and because of the costs of reunification.

- The Netherlands, Belgium, and Luxembourg are called the Low Countries, or Benelux. They have waged a struggle to hold back the North Sea and still derive significant income from agriculture. They have developed global trading networks. Three major port cities in Benelux are Rotterdam–Europoort and Amsterdam in the Netherlands and Antwerp in Belgium.

- Switzerland and Austria have similar geographic features and economic potential, but Switzerland is the slightly more prosperous. Switzerland has gained more stability with its traditional neutrality, whereas Austria has had its fate tied more closely to Germany and its battles on the North European Plain. Both countries are partly in the Alps and are significant tourist destinations. Neither has particularly abundant natural resources. Hydroelectric power is a major source of energy for both countries. Switzerland is world famous for its banking and high-tech instrumentation industries. Austria is a member of the European Union, but Switzerland is not.

- The microstates of Andorra, Monaco, and Liechtenstein have become prosperous from a combination of financial services and recreation, including tourism and gambling.

KEY TERMS + CONCEPTS

Terms in blue are also defined in the glossary.

brain drain (p. 91)
buffer state (p. 100)
Catholic Republicans (p. 89)
Celtic Tiger (p. 88)

Channel Tunnel (Chunnel, Eurotunnel) (p. 89)
"City of Light" (p. 92)
Commonwealth of Nations (p. 82)

core (p. 79)
deindustrialization (p. 87)
devolution (p. 81)
Enclosure Movement (p. 83)

Good Friday Agreement (p. 89)
head of navigation (p. 93)
hedgerow (p. 83)
Industrial Revolution (p. 83)
Irish Republican Army (IRA) (p. 89)
M-4 Corridor (p. 87)
mad cow disease (BSE) (p. 87)
major cluster of continuous settlement
  (p. 79)

Marshall Plan (p. 96)
microstate (p. 79)
neutrality (p. 100)
Northern Ireland Assembly (p. 89)
polder (p. 98)
potato famine (p. 88)
primate city (p. 91)
Protestant Unionists (p. 89)
rural to urban migration (p. 91)

Silicon Bog (p. 88)
Silicon Fen (p. 87)
Silicon Glen (p. 87)
Silicon Saxony (p. 96)
Sinn Fein (p. 89)
The Troubles (p. 89)
welfare state (p. 96)

REVIEW QUESTIONS

WORLD
REGIONAL
Geography Now™

Assess your understanding of this chapter's topics with additional quizzing and concept-based problems at http://earthscience.brookscole.com/wrg5e.

1. What are the countries of the European core and on what basis have they been designated as belonging to the Coreland?

2. What was the spatial relationship between coal fields and cities in industrial Britain and mainland Europe?

3. What five industries were central to the Industrial Revolution in Britain? To what extent do they exist in modern times?

4. What are the political affiliations of Ireland, Northern Ireland, England, Scotland, Wales?

5. What are "The Troubles" of Northern Ireland?

6. What impact did mad cow disease have on British agriculture?

7. Why is French agriculture particularly strong when compared with neighboring countries?

8. What is a primate city? What examples exist in Europe and elsewhere?

9. What was the impact of German reunification on the country's economy?

10. What has made Rotterdam the most active and prosperous port in Europe?

11. What role has geography had in the distinct histories of Switzerland and Austria?

12. Why hasn't Switzerland joined the European Union?

13. How do the microstates of Europe make a living?

14. What are the principal physical geographic features of the United Kingdom, France, and Germany, the big three of the European core?

DISCUSSION QUESTIONS

1. "Europe has some of the world's most transformed landscapes." What are some examples of that transformation?

2. What powers has the Scottish Parliament had since Britain allowed devolution of power to Scotland? What do the Welsh want?

3. Could "congestion charging" work in an American city?

4. What is the basis of strife between Catholics and Protestants in Northern Ireland? Is progress being made toward a resolution of this conflict?

5. Why has the Irish economy apparently done well, while the German and others struggle?

6. What are the apparent drawbacks and benefits of a primate city for a country and its people?

7. Many of you have been to London, Paris, or both. What would you add to the depictions of these cities in the text?

8. During the Iraq war of 2003, there was a lot of anti-French sentiment in the United States. But America also has a love affair with many aspects of French culture (and agriculture). Discuss the French–American relationship.

9. Should the United States follow Europe's lead in embracing renewable sources of energy? Does the United States have any obstacles to this shift that Europe does not face?

10. There was great euphoria among Germans when the two Germanys reunified. Has that optimism been rewarded? What are some of the likely benefits and growing pains of a reunified country? Could the two Koreas follow the path taken by the Germanys?

11. What do the Dutch apparently appreciate as measures of quality of life? Could American cities be like Dutch cities in any or all of these ways?

12. How have the people of the Netherlands managed to hold back the North Sea?

13. In what ways is Switzerland unique compared to its European neighbors?

14. Discuss ways in which coal, hydropower potential, rivers, mountain passes, and other natural features have affected the location and development of some European cities.

The European Periphery

A woman of Norway. From the northern tip of Scandinavia to the boot of Italy, there is an extraordinarily rich range of ethnicities.

Joe Hobbs

chapter objectives

This chapter should enable you to:

- Recognize the geographic, economic, and political factors that have kept these subregions secondary to the European core in power and influence

- Appreciate the benefits and drawbacks that people realize from different political and economic systems, including the welfare state and collectivization

- Recognize the impacts of the wars and their aftermath on the political geographies of the region

- Observe the trend toward greater unity under the European Union and NATO and the concurrent trend of devolution of power from central governments to provinces

- Relate the often-troubled histories of ethnic minorities in Europe and how aims of ethnic peoples have led to war, terrorism, and the redrawing of national boundaries

- See that some of the environmental obstacles to development (mountains, dry lands, and ice, for example) are also potential tourist assets that may promote development

WORLD
REGIONAL
Geography ⊛ Now™

Look for this logo in the text and go to GeographyNow at http://earthscience.brookscole.com/wrg5e to explore interactive maps, view animations, sharpen your factual knowledge and geographic literacy, and test your critical thinking and analytical skills with unique interactive resources.

Just as there is a recognizable European core, where the region's demographic, economic, and political weight is centered, there is also a European **periphery.** This rimland consists of countries whose interests are tied closely to those of the core but which are strongly influenced by, rather than influential upon, the core countries. Even integration within the European Union does not make the smaller countries equal partners with the larger.

Although they are grouped here within the single heading of periphery, these countries are diverse. To introduce them, it is useful to subdivide these countries into three broad categories: those of Northern, Eastern, and Southern Europe (Figure 5.1). Even within those subgroups, the countries have distinct national identities and experiences. In this chapter, we explore the geographic personalities and economic and political dynamics of these countries outside the European center. The major traits of each nation are listed in Table 3.1 on page 53.

5.1   Northern Europe: Prosperous, Wild, and Wired

Denmark, Norway, Sweden, Finland, and Iceland are the countries of Northern Europe (Figure 5.2). The countries are often grouped geographically under the regional term Norden, or "the North," and their traits are qualified by the adjective Nordic. These countries are also known collectively as Scandinavia, or the Scandinavian countries. But this regional name is ambiguous. It sometimes refers only to the two countries of the Scandinavian Peninsula, Norway and Sweden. More often, it includes these countries plus Denmark, and sometimes it includes these three plus Iceland. When Finland is included in the group, the term Fennoscandia, or the Fennoscandian countries, is used to acknowledge the cultural and historical differences between the Finns and the other northern nations.

Traits of Northern Europe

No other highly developed countries in the world have their populated areas at such high latitudes (see Figure 3.4, page 59). Located in the general latitude of Alaska, the countries of Northern Europe represent the northernmost concentration of advanced industrial nations in the world.

Westerly winds from the Atlantic, warmed in winter by the North Atlantic Drift, moderate the climatic effects of this northern location. Temperatures average above freezing in winter over most of Denmark and along the coast of Norway. But away from the direct influence of the west winds, winter temperatures average below freezing and are more severe at elevated, interior, and northern locations. In summer, the ocean has a cooling rather than a warming influence, and

59

Political Units of the European Periphery

Figure 5.1 The subregions and countries of the European Periphery

most of Northern Europe's Fahrenheit temperatures in July average no higher than the 50s or low 60s (10° to 18°C). The populations of the countries tend to cluster in the southern sections. All except Denmark have large areas of sparsely populated terrain where the rigors of a high-latitude environment pose problems for development.

Historical interconnections and cultural similarities are important factors in the regional unity of Northern Europe. Historically, each country has been more closely related to others of the group than to any outside power. In the past, the countries often waged war among themselves. But since the early 19th century, relations among the Northern European peoples have been close and peaceful. Recently, they have differed on how to deal with integration into the European Union. Only Norway and Iceland have opted out, due in large part to their interests in marine resources (see the Problem Landscape—in this case, Seascape—issue on page 108). The other three joined the European Union, with Finland the most enthusiastic to do so.

The languages of Denmark, Norway, and Sweden descend from the same ancient Germanic tongue and are mutually intelligible. Icelandic, although a branch of the same root, has changed less from the original Germanic language and has borrowed less from other languages. It derives from Old Norse, the language of the Vikings, and Icelanders have been proud of their linguistic conservatism. Only Finnish, which belongs to the entirely distinct Uralic language family, is fully

different from the others. Even in Finland, however, about 6 percent of the population speaks Swedish as a native tongue, and Swedish is a second official language. Remarkably, English is the lingua franca of the Nordic countries, and many businesses insist on conducting all affairs in English.

The five countries have other traits in common. About 90 percent of the population belongs nominally to the Evangelical Lutheran Church, but only a very small percentage attends services; people tend to be quite secular. The Lutheran Church is a state church, supported by taxes, and is the most all-embracing organization aside from the state itself. The countries share many similarities in law and political institutions. They all have a long tradition of individual rights, broad political participation, limited governmental powers, and democratic control. They are such peaceful democracies that rare events like the assassinations of Sweden's prime minister in 1986 and foreign minister in 2003 shock Scandinavians to the core.

High standards of health, education, individual security, and creative achievement are common to the Nordic nations. They have increasingly high-tech and service oriented economies, and per capita they are the most Internet-connected people in the world. They are **welfare states,** with the governments supplying services such as healthcare and pensions. These services, generally of very high quality, are supported by some of the highest tax rates in the world. Taxes on alcoholic beverages are especially high in an effort to discourage the kind of epidemic alcoholism that plagued Sweden in the 19th century.

Globally, the Nordic countries always rate near the top by quality of life measures like the Human Development Index. For example, Finland has the world's lowest corruption rate; its economy is the most competitive after that of the United States; Norwegians have the highest overall standard of living; the Nordic peoples are happiest with their jobs; and Nordic women have the greatest equality with men. High quality of life may seem improbable given the region's challenging physical environments, but the countries have specialized economies that compensate for their resource limitations. Following is a summary by country of these specialties and other unique features of the Nordic countries.

Denmark

Denmark has the largest city in Northern Europe and is the most densely populated of the five countries, yet it is the most dependent on agriculture. The Danish capital of Copenhagen (population: 1.1 million), with one-fifth of Denmark's population in its metropolitan area, is clearly recognizable as a **primate city.** Copenhagen lies on the island of Sjaelland (Zealand) at the extreme eastern edge of Denmark (see Figure 5.2). Sweden lies only 12 miles (c. 20 km) away across the Sound (Oresund), the main passage between the Baltic and the North Seas. The city grew beside a natural harbor well placed to control traffic through the Sound. Before the 17th century, Denmark controlled adjacent southern Sweden and levied tolls on all shipping passing to and from the

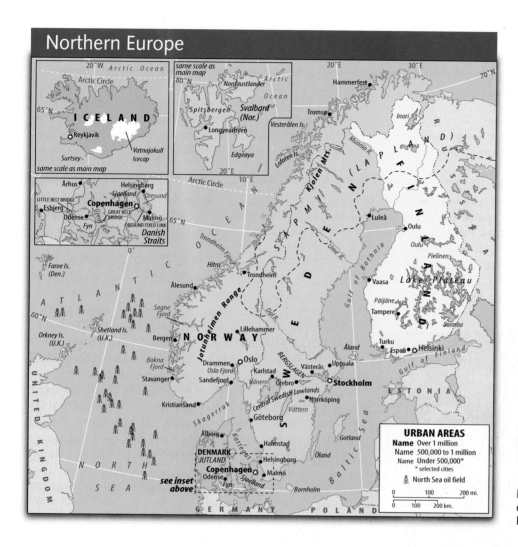

Figure 5.2 Principal features of Northern Europe. Nearly one-third of Northern Europe lies within the Arctic Circle.

Baltic. Copenhagen still does a large transit and **entrepôt** (commercial center) business in North Sea–Baltic Sea trade.

The first land link to connect Denmark and Sweden—and thus the Scandinavian Peninsula with mainland Europe—was opened in 2000. The 10-mile (16-km) Oresund Fixed Link bridge and tunnel for rail and auto traffic connects Malmö, Sweden, with Copenhagen's airport (Figure 5.3). This link is like the Channel Tunnel in breaching a historic gap between the European mainland and a vital outlying region. Danes use the bridge to buy Sweden's low-taxed children's shoes and clothing, while Swedes come mainly to buy cheap Danish alcoholic beverages. Denmark's food, metal, and chemical exports reach markets across this bridge and around the world from its Copenhagen port and airport.

Membership in the European Union and a location adjacent to the largest country in Western Europe (Germany) have provided a ready market for Danish agricultural products. Quality farmland is Denmark's greatest natural resource. About three-fifths of Denmark's land is arable, representing the largest proportion of any European country. Although not productive for crops, the Faeroe Islands (between Norway and Iceland) and Greenland are considered integral, although self-governing, parts of Denmark. Their

Figure 5.3 The Oresund Fixed Link during construction. This strategic bridge links the Scandinavian Peninsula with mainland Europe.

Problem Landscape

Save the Whales . . . For Dinner

One reason that both Norway and Iceland have refused to join the European Union is that both have economically vital fishing industries, and they fear that common EU policies on fishing might diminish those profits. The two countries, along with Japan, also are distinguished by their unique perspective on another marine resource: the world's whales. People in all three countries maintain that populations of some whale species, especially the minke (pronounced MING–kie), have rebounded to levels that should allow regular, limited harvesting for human consumption. Whale meat has long been prized by their palettes; "save the whale for the dinner table" is the Norwegian response to a popular Western bumper sticker (Figure 5.A).

In a landmark agreement honored by most of the world's marine countries, the International Whaling Commission (IWC) banned commercial whaling worldwide in 1986. The Norwegians never accepted the ban and defiantly continue to take about 700 minkes each year. Norwegians eat whale meat without remorse, arguing that the rest of the world is foolish in its insistence that whales are a "sacred cow." Even Norway's famously staunch environmentalists see no problem with the whale harvest.

The other whaling countries are more circumspect. The 1986 ban forbids commercial whaling but allows whales to be killed for scientific purposes and for subsistence (the latter clause allows Native Americans in Alaska, for example, to hunt whales). Iceland and Japan have exploited the loophole of the scientific purpose and in the process put whale meat on the table. They argue that there is a legitimate need for the science: Iceland estimates that its territorial waters are inhabited by 43,000 whales that eat perhaps 1 to 2 million tons of fish each year. Only by studying the whales' stomach contents can reliable estimates of the whales' impact on Iceland's vital fishing industry be evaluated. The studies accomplish two other goals: They reduce the pressure on the fish resource, and they allow whale meat to find its way to markets for ordinary consumers.

Killing whales does carry some costs. When it resumed whaling in 2002, Iceland said it intended to take about 100 minkes each year. Most Icelanders support the hunt, but others argue that whale watching—which previously had drawn 60,000 tourists yearly—was the more economically sensible way to manage the resource. Many whale watching groups have canceled tours to Iceland to protest whaling. And with "save the whales" an international mantra, Iceland, Norway, and Japan are subjected to intense criticism from around the world.

Joe Hobbs

Figure 5.A Norwegians are not shy about buying and selling whale meat.

peoples have the rights of Danish citizens. The Faeroes are a group of treeless islands where some 46,000 people make a living by fishing and grazing sheep. As a geological extension of North America, Greenland is discussed in the final section of this book.

Norway

Norway stretches for well over 1,000 miles (c. 1,600 km) along its west side and around the north end of the Scandinavian Peninsula. The peninsula, as well as Finland, occupies part of the **Fennoscandian Shield**, a block of ancient, stable metamorphic rocks. The western margins of the block in Norway are extremely rugged. Most areas have little or no soil to cover the rock surface, which was scraped bare by the continental glaciation that centered in Scandinavia. Most of Norway is rugged wilderness of coniferous forest and tundra, and only 3 percent of the land is arable.

The terrain hinders both agriculture and transportation. Glaciers deepened the valleys of streams flowing from Norway's mountains into the sea, and when the ice melted, the sea invaded the valleys. This process created the famous fjords—long, narrow, and deep extensions of the sea into the land, edged by steep cliff faces and valley walls (Figure 5.4). It is difficult to build highways and railroads parallel to such a coast. The fjords provide some of the finest harbors in the world, but most of them have practically no productive

F3.6
60

hinterland that would justify port construction. Much of Norway's population is scattered along these waterways in small, isolated clusters, interconnected by boat traffic or long roundabout roads (Figure 5.5). One of the largest and most scenic of the settlements is Bergen, established as a trading post for the medieval merchant network known as the **Hanseatic League.**

In Norway's far north, there are about 40,000 **Saami,** known by outsiders as **Lapps.** Formerly nomadic reindeer herders, the Saami of Norway, Sweden, Finland, and Denmark—now totaling around 80,000 in the arc of land they call **Sapmi**—have generally settled down to take up manufacturing and service jobs in their home countries. Norway's Saami, recognizing that significant gold and diamond reserves underlie their traditional territories, are calling for self-determination, especially greater rights to minerals and other resources. The Norwegian Saami already have their own parliament, but they want even more devolution of power from Oslo.

Norway also controls the Arctic island group of Svalbard, whose largest island is Spitsbergen. Although largely covered by ice, the main island has the only substantial deposits of high-grade coal in Northern Europe. Norwegian and Russian companies now exploit these resources (see Figure 6.5, page 145).

More than half of all Norwegians live in the southeastern core region, centered on the capital and primate city of Oslo (population: 800,000). Located at the head of Oslo Fjord, where several valleys converge, Oslo is Norway's principal seaport and industrial, commercial, and cultural center. Valleys are wider and the land is less rugged in this region,

65

Figure 5.4 Norway's scenic fjords were carved by glaciers and then filled by rising sea levels. This is Geiranger Fjord.

Figure 5.5 Alesund, on Norway's south central coast, is the country's busiest fishing port and one of its most beautiful cities.

where the most extensive agricultural lands and the largest forests in the country are located.

Apart from its sparse agricultural base, Norway has rich natural resources relative to its small population of less than 5 million. Given Norway's beauty, it is difficult to imagine anyone wanting to leave. But the benefits of industrialization came late to this challenging land, and emigration was the ticket to prosperity imagined by many hardscrabble farmers and fishermen. Key resources that keep Norwegians home today include spectacular scenery (a major draw for international tourists), hydropower, forests, and a variety of minerals—notably oil. Hydroelectricity powers sawmilling, pulp and paper production, metallurgy, electrochemical industries, and industries that produce various goods for Norwegian consumption and some export trade (Figure 5.6). Large oil and natural-gas deposits in the Norwegian sector of the North Sea floor prompted an economic boom for Norway, beginning in the 1970s (Figure 5.7). In just a few years, Norway's oil and gas exports grew from nothing to the country's largest by value (representing about 35 percent of exports), and Norway advanced from being reasonably well-off to being one of Europe's richest countries as measured by per capita GNP PPP (see Table 3.1, page 52). Meanwhile, older economic emphases remain important. A seafaring tradition going back to the Vikings is reflected today in Norway's large merchant, cruising, and fishing fleets.

Norway is one of the few European countries that does not belong to the European Union; in 1994, Norwegians voted by a narrow margin to stay out of the EU. With the economic stability provided by Norway's energy resources, most Norwegians felt they could thrive without EU protection and intervention. Coalitions of farmers, fishermen, environmentalists, and nationalists insisted on the relative independence of Norway from bureaucrats in Brussels, the EU's administrative home. Farmers and fishermen also wanted to retain their Norwegian state subsidies, which are

Figure 5.7 One of Norway's offshore oil platforms in the North Sea. Built to withstand the notoriously violent weather of the region, these are also supplied with many amenities to keep the workers comfortable.

even more generous than those offered by the European Union. Supporters of EU membership warned that Norway's petroleum income would decline in the future, and the country would do well to seek the long-term economic security envisioned by EU policy.

Sweden

In the Middle Ages and early modern times, Sweden was a powerful and imperialistic country. By the 17th century, the Baltic Sea was essentially a Swedish lake. During the 18th century, however, the rising power of Russia and Prussia put an end to Swedish imperialism, aside from a brief campaign in 1814 in which Sweden won control of Norway from Denmark. Although it keeps a strong army as a deterrent, Sweden has not engaged in a war since 1814, and it has become known as one of Europe's most successful neutrals. The long period of peace has helped promote economic success and social well-being in what is today the largest and most populous (9 million) country of Northern Europe. Sweden is also renowned as the most multicultural of the five countries. While Denmark and Norway have imposed strict controls on immigration, Sweden has few restrictions; already 25 percent of the population is of non-Swedish ethnicity, and that number is growing steadily.

In the northwest, Sweden shares the mountains of the Scandinavian Peninsula with Norway. To the north, between the mountains and the shores of the Baltic Sea and Gulf of Bothnia, there are ice-scoured forested uplands, and a similar small area of this type is around Lake Vättern, south of Stockholm. In its central area of great lakes lies a block of good farmland, and in the south, there are rolling, fertile farmlands like those of Denmark (Figure 5.8). Unlike Denmark, Sweden has not developed a specialized export-oriented agriculture, and unlike Norway, Sweden commands enough good land to supply all but a minor share of its small population's food requirements. The introduction of new

Figure 5.6 Norway is blessed with mountains and rivers in the right combination for producing hydroelectricity. Hydropower provides more than 99 percent of the country's electricity. The barrier just right of center is part of a small hydropower station.

labor-saving technologies in the forestry industries has reduced job opportunities in this most traditional occupation in the country's remote lands.

Engineering and metallurgical industries distinguish the Swedish economy. These industries have evolved out of a centuries-old tradition of mining and smelting. The iron ores of the Bergslagen region, just north of the Central Swedish Lowland, have been made into iron and steel for more than a millennium. Originally, the fuel used to smelt the ore was charcoal made from the surrounding forest. When coal and coke began to replace charcoal as fuel in Europe's iron industries in the late 17th century, Sweden was disadvantaged by its scarcity of coal. But in the later age of electricity, the Swedes responded to this problem by a heavy reliance on electric furnaces using current from well-sited hydropower plants. The modern Swedish steel industry continues to focus on its famous high-quality steel, especially from large plants on the Baltic and Bothnian coasts. Swedish exports employing this steel include automobiles (Volvo and Saab), industrial machinery (including robots), office machinery, telephone equipment, medical instruments and machinery, agricultural machinery, and electric transmission equipment. The country's reputation for skill in design and quality of product also extends to items such as ball bearings, cutlery, tools, glassware, and furniture.

Overall, Swedish manufacturing is centered in the Central Lowland, an area favorably situated with respect to minerals, forests, water power, labor, food supplies, and trading potential. The Central Lowland is the historic core of Sweden and has long maintained important agricultural development and a relatively dense population. Sweden is now trying to promote its sparsely settled northlands as centers for growth in the telecommunications industry, beginning with telephone call center services.

The location of Sweden's capital, Stockholm (population: 1.6 million; Figure 5.9), reflects the core role of the Central

Figure 5.9 Stockholm's harbor is the heart of the city.

Lowland and Sweden's early orientation toward the Baltic and trans-Baltic lands. Stockholm is the principal administrative, financial, and cultural center of the country and shares in many of the manufacturing activities typical of the Central Lowland. But in the past century, as Sweden has traded more via the North Sea, Stockholm has been displaced by Göteborg, located at the western edge of the Central Lowland, as the leading port of the country and in fact of all Northern Europe. Göteborg has a harbor that is ice-free year round, whereas icebreakers are needed to keep Stockholm's harbor open in winter.

Finland

Conquered and Christianized by the Swedes in the 12th and 13th centuries, Finland was ceded by Sweden to Russia in 1809 and was controlled by Russia until 1917. Under both the Swedes and the Russians, the country's people developed their own culture and feelings of nationality. Finns seized the opportunity for independence provided by the collapse of Tsarist Russia in 1917. During World War II, Finns fought the Soviets, losing much territory, life, and infrastructure. Finland made a rapid postwar recovery, in part because of pressure applied by the USSR. The Soviets required that a large part of the reparations paid by Finland to the USSR were in metal goods. Several small steel plants and other metalworking establishments were built to meet this demand, and they continue to operate today.

One-third of Finland lies within the Arctic Circle. Most of Finland is a sparsely populated, glacially scoured, subarctic wilderness of coniferous forest, ancient igneous and metamorphic rocks, lakes, and marshes (Figure 5.10). About two-thirds of Finland is forested, mainly in pine or spruce, and exports of wood products including paper, pulp, and timber are important to the economy. Forest production is concentrated in the south central part of the country, known as the Lake Plateau. The majority of the population lives in the rel-

Figure 5.8 Long daylight allows crops to do well where good soil is present in Scandinavia. This small farm is near Karlstad in central Sweden.

Figure 5.10 Coniferous forests, lakes, and granite are characteristic of Finland's challenging environment.

atively fertile and warmer lowland districts scattered through the southern half of the country. Agriculture focuses on hay, oats, barley, and livestock production, especially dairying. Wherever they live and whatever their occupations, Finns are nature lovers who enjoy retreating to lakeside and seaside saunas where they take the customary hot steam treatment followed by an invigorating plunge.

Helsinki (population: 1.2 million; Figure 5.11), located on the coast of the Gulf of Finland, is Finland's capital, largest city, main seaport, and principal commercial and cultural center. It is also the country's most important and diversified industrial center. Hydroelectricity originally powered Fin-

land's industries, but this source has now been surpassed by both coal and nuclear power. Manufacturing is the most important economic sector, with exports including paper and forestry products, metals, electronics, and telecommunications equipment, notably the Nokia products that make up the world's largest market share of cell phones.

Iceland

Iceland is a fairly large (39,699 sq mi/102,819 sq km), mountainous island in the Atlantic Ocean just south of the Arctic Circle. Reflecting the island's position astride the **Mid-Atlantic Ridge,** birthplace of two major plates of the earth's crust, Iceland's landscapes have been shaped by vulcanism (Figure 5.12). A volcanic eruption in 1783 wiped out most of what crops Icelanders were growing at the time, causing a famine that killed 20 percent of the population. An underwater eruption in 1963 gave birth to the brand-new island of Surtsey, off Iceland's southeast coast. The country derives about 10 percent of its electricity from its volcanic resources in the form of geothermal energy; hydropower supplies another 80 percent. One enterprising community in northern Iceland wants to harness its hot waters to develop alligator farming, which the locals say will attract tourists and produce exports of alligator meat and skin.

Because of the proximity of the relatively warm North Atlantic Drift, the coasts of Iceland have winter temperatures that are mild for the latitude. The capital, Reykjavik (64°N latitude), has an average January temperature that is practically the same as New York City's (41°N latitude). Iceland's vegetation is primarily tundra. Summer temperatures average about 52°F (11°C) or below, limiting agriculture to po-

Figure 5.11 Helsinki's city center is dominated by its Lutheran cathedral.

Figure 5.12 Vulcanism is a serious natural hazard in Iceland.

Emory Kristof/National Geographic/Getty Images

WORLD
REGIONAL
Geography ® Now™

Click Geography Literacy to see an animation of a divergent plate
boundary, and take a short quiz on the facts and concepts.

5.2 Eastern Europe: Out from Behind the Curtain

As defined in this text, the Eastern European countries are
those in an arc extending from the Gulf of Finland in the
north to the Balkan Peninsula in the south (Figure 5.13).
They are Estonia, Latvia, Lithuania, Poland, the Czech Republic, Slovakia, Hungary, Romania, Bulgaria, Slovenia,
Croatia, Bosnia and Herzegovina, Serbia and Montenegro,
Albania, and Macedonia. Other sources, especially those
that follow the strict definition of including western Russia
in Europe, refer to these countries as "Central Europe" or
"East Central Europe." The French Geological Institute has
calculated that a spot in eastern Lithuania is the geographic
center of Europe.

These countries are in the process of reinventing themselves after more than four decades of direct or indirect control by the Soviet Union. Prior to German reunification in
1990, the former East Germany was also part of Eastern
Europe, and it is brought into the present discussion when
appropriate. Three of the countries—Estonia, Latvia, and
Lithuania—became parts of the Soviet Union in World
War II, while many others became **satellites** of the Soviet empire, with local Communist governments effectively governed
from Moscow. Exceptions were Albania and the former Yugoslavia, where national Communist resistance forces took
power on their own as German and Italian power collapsed.
Majority Slavic ethnicity, former Communist status, and/or
subjugation to Soviet interests were among the few unifying
themes of the region prior to the end of the Cold War. Now
the true complexity of the region is more apparent, and this
section of the chapter explores various themes of unification
and distinction within the Eastern European mosaic.

Physical Geography

Eastern Europe is a checkerboard of mountains and plains
(Figure 5.14). Located mainly in the center and south, the
mountains vary widely in elevations and degrees of ruggedness. The Carpathian Mountains and lower ranges farther
west along the Czech frontiers form a central mountain zone.
The Carpathians extend in a giant arc for about 1,000 miles
(c. 1,600 km) from Slovakia and southern Poland to south
central Romania. These mountains still contain much forested wilderness and are home to Europe's largest concentrations of carnivores, including wolves, bears, and lynx. Under communism, much of this wilderness was untouched
state-owned land, and there are fears now that privatization

tatoes and some other hardy vegetables. There is some cattle
and sheep ranching. Mineral resources are almost nonexistent. The economy depends mainly on exports of fish and fish
products, which represents about 70 percent of export earnings. Aluminum is the second most valuable export product,
but next to fish, tourists bring in the most money. While international visitors soak in hot springs and enjoy the night-
life, Icelanders prefer Spain for their holidays.

Despite its environmental limitations, Iceland has been
continuously inhabited since at least the ninth century and is
now the home of a progressive and democratic republic with
a population of about 292,000. Practically the entire population lives in coastal settlements, with the largest concentration in and near Reykjavik (population: 187,000). Its ice-free
ports and strategic location in the North Atlantic have given
Iceland geopolitical significance, and it passed the Cold War
years as a Western ally. It continues to be a member of the
North Atlantic Treaty Organization (NATO).

URBAN AREAS
Name   Over 1 million
Name   500,000 to 1 million
Name   Under 500,000*
       * selected cities
       Major industrial area

0   100   200 mi.
0   100   200 km.

WORLD
REGIONAL
Geography Now™

Active Figure 5.13
Principal features of Eastern Europe.
Positioned between stronger powers to
the east and west, the region is a classic
"shatter belt." *See an animation based
on this figure, and take a short quiz on
the facts and concepts.*

The World's Great Rivers

The Danube

The Danube River, which supplies a vital navigable water connection between these lowlands and the outside world, rises in the Black Forest of southwestern Germany and follows a winding course of some 1,750 miles (2,800 km) to the Black Sea, making it the longest river in Europe. It is also one of the most storied, immortalized best by Johann Strauss' waltz *The Blue Danube* (Figure 5.B).

Figure 5.B The Danube in Austria. Some of Eastern Europe's greatest cities developed on its banks, and its delta is one of the world's premiere bird habitats.

Traffic on the Danube is modest compared with the tonnage carried by barges on the Rhine or by road and rail in Eastern Europe. The fact that the Danube flows into the Black Sea while the Rhine pours into the North Sea has made a great difference in the economic roles of the two waterways. The industrial world served by the Rhine, and the very active trading world into which it flows, is very distinct from the more agricultural world served by the Danube. The Danube flows mostly through agrarian areas, and heavy industries—the most important generators of barge traffic—are found in only a few places along its banks.

Upstream from Vienna, the river is swift and hard to navigate, although large barges go upstream as far as Regensburg, Germany. Below Vienna, the Danube flows slowly across the Hungarian and Serbian plains past Budapest and Belgrade. In the border zone between Serbia and Romania, the river follows a series of gorges through a belt of mountains about 80 miles (c. 130 km) wide. The easternmost gorge is the Iron Gate gorge (also called Portile de Fier), a 2-mile-long gorge with rapids that serves as the border between Romania and Serbia and Montenegro. Beyond it, the Danube forms the boundary between the plains of southern Romania and northern Bulgaria.

The river then turns northward into Romania and enters the Black Sea through the Danube Delta, Europe's largest wetland. Resided in or transited by more than 300 bird species, this was one of Europe's greatest wilderness areas until, in the 1980s, Romania's Communist government began diking and draining the land for new farms and shooting off bird populations as "pests." Efforts are still underway to remove these engineering works and restore life to the marshes.

Figure 5.14 Romania's Transylvanian Alps

of the land and the expansion of regional economies will lead to rapid habitat and species loss.

West of the Carpathians, lower mountains enclose the hilly, industrialized Bohemian basin in the Czech Republic. On the north, the Sudeten Mountains and Ore Mountains separate the Czech Republic from Poland and the former East Germany. Eastern Europe's southern mountain zone occupies the greater part of the Balkan Peninsula. Bulgaria, the Yugoslav successor states, and Albania, which share this zone, are very mountainous but have some productive lowlands. In Bulgaria, the principal mountains are the Balkans, extending east–west across the center of the country, and the Rhodope Mountains in the southwest. These are rugged mountains reaching over 9,000 feet (c. 2,800 m) in a few places.

In southern Serbia and Montenegro and neighboring areas to the west and south, hills and mountains form a major barrier to travel. Through this difficult region, a historic lowland passage connecting the Danube Valley with the Aegean

Sea follows the trough of the Morava and Vardar Rivers. At the Aegean end of the passage is the Greek seaport of Salonika (Thessaloniki). The Morava–Vardar corridor is linked with the Maritsa Valley by an east–west route that leads through the high basin in which Sofia, the capital of Bulgaria, is located. Along the rugged, island-fringed Dalmatian Coast, the Dinaric Alps rise steeply from the Adriatic Sea. The principal ranges run parallel to the coast and are crossed by only a few significant passes.

The plains, concentrated in the north and in the center along the Danube, have varying degrees of flatness, fertility, warmth, and humidity. Agriculturally, they range from some of Europe's better farming areas to some of the poorest. Between mountain-rimmed Bohemia and the Carpathians of Slovakia, the lowland corridor of Moravia in the Czech Republic provides a busy passageway. Through this corridor run major routes connecting Vienna and the Danube Valley with the plains of Poland. Near the Polish frontier, the corridor narrows at the Moravian Gate between the Sudeten Mountains and the Carpathians. Just beyond this gateway, East Central Europe's most important concentration of coal mines and iron and steel plants developed in the Upper Silesian–Moravian coal field of Poland and the Czech Republic (see Figure 4.3, page 80).

Two major lowlands, bordered by mountains and drained by the Danube River and its tributaries, compose the Danubian plains. One of these, the Great Hungarian Plain, occupies two-thirds of Hungary and smaller adjoining portions of Romania, Serbia, and Croatia. The second major lowland is composed of the plains of Walachia and Moldavia in Romania, together with the northern fringe of Bulgaria. The Danubian plains represent the most fertile large agricultural regions in East Central Europe.

A handful of Europe's largest rivers carries the region's waters to the Black Sea (via the Danube), the Baltic Sea (via the Vistula), and the North Sea (via the Elbe). Between the Sudeten and Ore Mountains, at the Saxon Gate, the valley of the Elbe River provides a lowland connection and a navigable waterway from Bohemia to the industrial region of Saxony in Germany and, farther north, to the German seaport of Hamburg and the North Sea. To the southwest, the Bohemian forest occupies the border zone between Bohemia and the former West Germany.

In most of Eastern Europe, winters are colder and summers are warmer than in Western Europe. On the Danubian plains, the humid continental long-summer climate is comparable to that of the corn and soybean region (Corn Belt) in the U.S. Midwest, but the plains of Poland have a less favorable humid continental short-summer climate, comparable to that of the American Great Lakes region. The most exceptional climatic area is the Dalmatian Coast, on the Balkan Peninsula side of the Adriatic Sea. Here, temperatures are subtropical and precipitation, concentrated in the winter, reaches 180 inches (c. 260 cm) per year on some slopes facing the Adriatic. This is Europe's greatest precipitation. The climate of the Dalmatian Coast is classed as Mediterranean (dry-summer subtropical). Together with spectacular scenery, this environment made the area one of Europe's prime tourist destinations before the conflicts of the 1990s.

Political and Ethnic Geography of the "Shatter Belt"

The extension of Soviet power into Eastern Europe at the end of World War II was a replay of the region's history. In the Middle Ages, several peoples in the region—the Poles, Czechs, Magyars, Bulgarians, and Serbs—enjoyed political independence for long periods and at times controlled extensive territories outside their homelands. Their situation deteriorated as stronger powers—Germans, Austrians, Ottoman Turks, and Russians—pushed into East Central Europe and carved out empires. These empires frequently collided, and the local peoples were caught in wars that devastated great areas, often resulted in a change of authority, and sometimes brought about large transfers of populations from one area to another. In geopolitical terms, Eastern Europe is a classic **shatter belt**—a large, strategically located region composed of conflicting states caught between the conflicting interests of great powers.

There have been large population transfers in the region since the beginning of World War II. Jews, Germans, Poles, Hungarians, Italians, and others were uprooted, often without notice, losing all their possessions, and were dumped as refugees in so-called "homelands" that many had never seen. During the war, Nazi Germany systematically killed as many as 6 million **Jews** and 500,000 to 1.5 million Gypsies in what is known as the Holocaust. The prewar populations of Germans in Poland and Czechoslovakia were expelled at the end of World War II and forced into East and West Germany. Many died in this process of forced migration. Ethnic minorities now constitute only 2 percent of the population in Poland, which transferred most of its German population to Germany and whose territories containing Lithuanians, Russians, and Ukrainians were absorbed by neighboring countries. A number of countries still have large minorities, including the Gypsies (see Ethnic Geography, page 117).

These population transfers helped to create an ethnic map of Eastern Europe that was overwhelmingly Slavic. The Slavic peoples, or Slavs, are typically grouped into three large divisions: **East Slavs** of the former Soviet region; **West Slavs,** including **Poles, Czechs,** and **Slovaks;** and **South Slavs,** including **Serbs, Croats, Slovenes, Bulgarians,** and **Macedonians.** Although the various Slavic peoples speak related languages, such languages are often mutually unintelligible; for example, a Czech will not easily understand a Pole, although both are grouped as Slavs. West Serbian and Croatian are essentially one language in spoken form, but the Serbs, like the Bulgarians and most of the East Slavs, use the Cyrillic alphabet (based on the Greek language), whereas the Croats use the Latin alphabet, as do the Slovenes and the West Slavs. The language of the **Romanians** is classed with the Romance languages derived from Latin; however, the contemporary

223

215

149

Ethnic Geography

The Roma

One of Europe's largest ethnic minorities is the **Gypsies**—properly known as **Roma**—who number 8 to 10 million (Figure 5.C). Romania has the highest number—as many as 2.5 million—but estimates of Roma population vary widely, because the Roma live on the road and at the margins.

The Roma began their great odyssey in what is now India, and their Romany language is very close to languages still spoken on the Indian subcontinent. Having in early times traveled thousands of miles from their homeland, and still often moving in caravans, they have come to be the archetypal people on the move. Throughout the Roma realm, host governments and majority populations have for centuries regarded the Gypsies with disdain. They are typically depicted as a rootless, lawless, and violent people apart, not deserving of the educational and economic opportunities offered to majority populations. Wherever they are, they have more children and are poorer than the majority population; in Hungary, for example, the Roma unemployment rate is five times that of non-Roma, and in some Roma communities, the unemployment rate is as high as 90 percent. They were the first to lose their jobs when the Eastern European countries gained economic freedom from the Communists.

The Roma typically live in shantytowns without water, sewage, or other

Figure 5.C This Roma woman and her children pass their days begging on a street corner in Szczecin, Poland.

services and make their living on scant child benefit payments, minor mechanical work, begging, foraging for food, and petty crime. In 1999, one Czech city built a wall between Gypsy and Czech neighborhoods, insisting it was needed to protect other townspeople from Gypsy criminals, noise, and visual squalor. Most Roma children do not attend schools because their parents do not believe that education will improve their job prospects, given the discrimination against them. Many that want to attend school are funneled into institutions for the mentally disabled. There is also violence; Romanies are often the targets of attacks by skinheads. These assaults, especially in the Czech Republic and Slovakia, have resulted in recent waves of Gypsy emigration to Canada, France, Britain, and Finland.

Romany activists have responded to discrimination by demanding more rights to welfare, pensions, and other benefits offered to regular citizens. Already feeling overwhelmed by the economic burden of their welfare services, the wealthier countries of the European core are worried that the new EU membership of Eastern European countries will unleash a tide of Roma immigration. A 2004 editorial in the British *Daily Express* declared, "The Roma gypsies of Eastern Europe are heading to Britain to leach on us. We do not want them here!"[a] The new EU members have about 1.5 million Roma, and another 3 million are in Romania and Bulgaria, which will join the European Union in 2007.

[a] *The Economist*, February 7, 2004, p. 54.

Romanian language contains many Slavic words and expressions. There is also a religious division between the West Slavs, Croats, and Slovenes, most of whom are Roman Catholics, and the East Slavs, Romanians, Bulgarians, Serbs, and Macedonians, who are mainly Eastern Orthodox Christians (Figure 5.15).

The principal non-Slavic elements in Eastern Europe are the Estonians, the Latvians and Lithuanians, the Hungarians, the Albanians, and the Romanians. The **Estonians,** who form an ethnic majority in Estonia, belong to the same Uralic ethnolinguistic group as the Finns and the Hungarians. **Latvians** and **Lithuanians** speak Baltic languages descending from the same ancestry as the Slavic languages but do not consider themselves Slavs. Also known as **Magyars,** the **Hungarians** are the descendants of nomads from the east who settled in Hungary in the ninth century. Roman Catholicism is the dominant religious faith in Hungary, although substantial Calvinist and Lutheran groups have long existed there. The **Albanians** speak an ancient Indo-European language only distantly related to the Slavic languages. Albania is the only European country in which Muslims are a majority. Serbia and Montenegro, Macedonia, and Bulgaria have many Albanian and Turkish Muslims, whose presence is a legacy of former Turkish rule.

Joe Hobbs

Figure 5.15 An Orthodox church in Tallinn, Estonia. Eastern Europe and adjacent Russia and Turkey make up the cradle of Orthodox Christianity.

Wrenching Reforms, One: Communism

In becoming satellites of the Soviet Union, the nations of Eastern Europe (and East Germany) were reformed by **communism,** which has these principal traits: one-party dictatorial governments; national economies planned and directed by organs of the state; abolition of private ownership (with some exceptions) in the fields of manufacturing, mining, transportation, commerce, and services; abolition of independent trade unions; and varying degrees of **socialization** (state ownership) of agriculture. Soviet military force crushed attempts by East Germany in 1953, Hungary in 1956, and Czechoslovakia in 1968 to break away from Soviet control.

In agriculture, the new Communist governments liquidated the remaining large private holdings and in their place introduced programs of collectivized agriculture on the Soviet model. Some farmland was placed in large state-owned farms on which the workers were paid wages, but most was organized into collective farms owned and worked jointly by peasant families who shared the proceeds after operating expenses of the collective had been met. This collectivization—the bringing together of individual land holdings into a government-organized and government-controlled agricultural unit—met with strong resistance, and it was discontinued in the 1950s in Yugoslavia and Poland. Today, all but a minor share of the cultivated land in Poland is privately owned. The remaining countries are pursuing programs to reprivatize their farmlands. Conversion to private farming after four decades of collectivization presents painful obstacles, and progress is slow.

Despite recent attempts to diversify and intensify agriculture, farming in Eastern Europe remains primarily a crop-growing enterprise based on corn and wheat, the leading crops in the region from Hungary and Romania southward through the Balkan Peninsula. In the Czech and Slovak republics and Poland, with their cooler climates, corn is difficult to grow, but wheat is a major crop as far north as southern Poland's belt of loess soils. Most of Poland lies north of the loess belt and has relatively poor sandy soils; here, rye, beets, and potatoes are the main crops. Raising livestock is a prominent secondary part of agriculture throughout the region. The meat supply is much less abundant than in Western Europe and comes primarily from pigs. The more southerly areas contain poor uplands that pasture millions of sheep, the main source of meat in Bulgaria, Albania, and parts of the former Yugoslavia.

In industry, the ascension of communism in Eastern Europe inaugurated a new era, also along Soviet lines. Most industries were taken out of private hands, and national economic plans were developed. Communist planners did not aim at a balanced development of all types of industry and instead stressed those seen as essential to regional industrial development as a whole: mining, iron and steel, machinery, chemicals, construction materials, and electric power. These were favored at the expense of consumer-type industries and agriculture.

The central planning agencies of the Communist governments maintained rigid control over individual industries. Plant managers were directed to produce certain goods in quantities determined by state governmental planners, and the success of a plant was judged by its ability to meet production targets rather than its ability to sell its products competitively and at a profit. Political reliability and conformity to the central plan were qualities much desired in plant managers. This system frequently resulted in shoddy goods that were often in short supply or in mountainous oversupply because production was not being driven or keyed by buyer demand and satisfaction. The planned expansion of mining and industry in Eastern Europe under communism produced marked increases in the total output of minerals, manufactured goods, and power. But it also produced inefficient, overmanned industries ineffective in their later competition for the world markets of the post-Communist period—a problem shared by the economies of all the successor states of the Soviet Union.

Under communism, the economies of Eastern Europe were closely tied to that of the Soviet Union. Eastern European industries relied on imports of Soviet iron ore, coal, oil, natural gas, and other minerals. In return, the Soviet Union was a large importer of East Central Europe's industrial products. During the 1970s and 1980s, however, the region's trade relationship with the Soviet Union weakened. Trade and financial relations with Western countries, companies, and banks expanded rapidly. With relatively little to export, the Eastern European countries borrowed massively from Western governments and banks to pay for imports from the West. Poland and Romania pursued this course especially strongly. New industrial plants and equipment acquired in this way were supposed to be paid for by goods that would be produced for export to the West by industries that had learned efficient production and marketing techniques from the West.

But then the Western economies began to slump, lowering the demand for imports. This led to a situation by the 1980s in which large debts to Western governments and banks needed to be repaid if countries were to maintain any credit at all. However, with the help of mismanagement by Communist bureaucrats, exports with which to pay were not being produced or, if produced, could not be sold. The result was a severe impediment to economic expansion and falling standards of living as the governments squeezed out the needed money from their people. In Poland, social action through the formation of an independent trade union called **Solidarity** paralyzed the country and threatened to upset Communist dominance (Figure 5.16). These conditions, along with the USSR's own growing economic and political distress, set the stage for the withdrawal of Soviet control and de-Communization in 1989 and 1990.

Wrenching Reforms, Two: Revolution and Capitalism

As late as the early summer of 1989, East Germany and the Eastern European countries seemed firmly in the control of totalitarian Communist governments. Later that year, the Communist order began to crumble. Public demands for freedom, democracy, and a better life gathered momentum in one country after another. Communist dictators who had ruled for many years were forced out, and reformist governments took charge. By mid-1991, democratic multiparty elections had been held in all countries. This liberalizing process continued throughout the 1990s and into the 2000s.

Figure 5.16 The rise of the Solidarity trade union was the beginning of the end for communism in Eastern Europe. This is a shrine, in a Polish Catholic church, dedicated to Solidarity.

In 1989 and 1990, the Soviet Union withdrew its backing for the region's Communist order. That support was too costly for the USSR, which also recognized the inevitable victory of "people power" in Eastern Europe. Several Communist regimes (including East Germany's) quickly collapsed and were replaced by democratically elected non-Communist governments. However, some "reformed" and renamed Communist parties and ex-Communist officials still play a role in the region, notably in Albania.

Governments are now struggling to build democracies with capitalist economies out of the economic wreckage of unproductive, unprofitable, uncompetitive, state-owned enterprises. Many countries pursued reforms aggressively to meet qualifications for joining the European Union. Economic restructuring has required many difficult steps, including the **privatization** (shift to nongovernmental ownership) of state-owned enterprises, an increase in the efficiency of state-run enterprises by allowing noncompetitive enterprises to fail and removing support of poor performers, and encouragement of the development of new private enterprises, often with foreign capital and management playing a role. In addition, economic restructuring has seen the termination of price controls so that prices reflect competition in the market, the development of new institutions required by a market-oriented economy (such as banks, insurance companies, stock exchanges, and accounting firms), the fostering of joint enterprises between state-owned firms and foreign firms, and the elimination of bureaucratic restrictions on the private sector. Furthermore, internationally convertible currencies have been created to increase trade and thus enhance competition and access to international communications media of all sorts has been expanded to help local entrepreneurs learn from foreign examples.

One of the most remarkable economic trends has been Western European and Chinese **outsourcing** of investments that take advantage of relatively cheap labor in Eastern Europe. In the late 1990s, Austrian and German electronics firms established large factories in the East, especially in Hungary. Some of these closed and shifted their assets to China, where labor was even cheaper. But recently, China has opened television and other electronics factories of its own in Eastern Europe. The challenge for the Eastern European newcomers to the European Union will be to raise their standards of living while still offering competitive terms for international investment.

Countries of the "New Europe"

The U.S. Secretary of Defense Donald Rumsfeld raised more than a few eyebrows when he declared in 2003 that countries like France and Germany (which opposed the U.S. war in Iraq) were part of an **"Old Europe,"** while the countries of Eastern Europe (which generally supported the war) represented **"New Europe."** The implication was that Eastern Europe represented a dynamic, up-and-coming, progressive bloc of countries that the United States was very happy to do business with. These certainly are dynamic countries, and it is worth considering the qualities of each. With the Balkan countries, it is important to consider the troubles they must overcome to gain parity with their more prosperous and peaceful neighbors.

The Baltics

Estonia, Latvia, and Lithuania, known as the Baltic States, or the Baltics, lie on the eastern shores of the Baltic Sea. These countries share many environmental, economic, and historical traits with the Slavic neighbors, including the Russians. However, the indigenous inhabitants of the Baltics are not Slavs, and they have long struggled to safeguard their national identities, especially against Russian encroachment. They were part of the Russian empire before World War I but became independent following the Russian Revolution. The Soviet Union reabsorbed them in 1940, and in 1991, they regained independence with the collapse of the USSR.

As independent countries again, the Baltics are striving to become more European, to be recognized as part of the up-and-coming New Europe, and to join European economic and military organizations. But their geographic past has

been hard to shake, and they are still linked to parts of the former Soviet Union in various ways. Russia uses Baltic ports to transport goods abroad and exports fossil fuels and a variety of other goods and services to the Baltic nations. But Estonia, Latvia, and Lithuania want to minimize their ties with their giant neighbor to the east. They refused to become members of the Russian-oriented Commonwealth of Independent States (CIS) and instead joined the European Union in 2004. During that same year, they joined the North Atlantic Treaty Organization (NATO), despite Russia's objection that this would pose a serious threat to its security. Russian energy firms have bought majority ownership of some Lithuanian energy industries, sparking new fears that Russia still has ambitions in the region.

Dairy farms alternate with forests in the hilly Baltic landscape. Mineral resources are scarce; the most notable are deposits of oil shale in Estonia, mined for use as fuel in electric power plants. The Baltics have usually had to rely on natural gas and petroleum brought from Russian fields located hundreds and even thousands of miles away. Expansion of industry after World War II helped make the area one of the more prosperous and technically advanced parts of the former USSR. The largest city and manufacturing center is the port of Riga (population: 880,000), the capital of Latvia (population: 2.3 million). Riga is typical of all of the capitals of Eastern Europe in that it serves as the country's primate city.

About 40 percent of Latvia's inhabitants and 85 percent of Latvia's non-Latvian ethnic minorities speak Russian; they are mostly retired Soviet military officers and their offspring. While independent Lithuania and Estonia established procedures early on to grant citizenship to ethnic Russians, Latvia still makes this process very difficult. The Latvian government fears that enfranchised Russians might destabilize the country in future elections. Russia protested by cutting off shipments of its oil through Latvia in 2003 (the transit fees generate important revenue for Latvia) and by threatening to stop buying Latvian agricultural exports. Latvia has made a few concessions, allowing all children born on Latvian soil since 1991 to become Latvian citizens and permitting more non-Latvians to apply for citizenship, but ethnic Russians still feel discriminated against.

The presence of Russian troops in the Baltic countries was a serious foreign relations problem after the countries gained independence. With independence, only Lithuania (population: 3.4 million; capital: Vilnius, population 542,000) successfully negotiated the immediate withdrawal of Russian troops. A large number of these Russian forces resettled in neighboring Kaliningrad, an anomalous Russian enclave two countries away from Russia.

Since independence, Estonia (population: 1.4 million; capital: Tallinn, population 400,000; Figure 5.17) has developed close economic ties with Finland, which quickly replaced Russia as the country's dominant trading partner. Estonia's indigenous people are closely related to the ethnic Finns of Finland, a mere 2.5-hour ferry ride away across the

Figure 5.17 Estonia's capital, Tallinn, is a lively and colorful city.

Gulf of Finland. Through low taxes and tariff-free trade, mainly with Scandinavia, Estonia has becoming the most economically prosperous of the three Baltic countries.

Poland

Poland is Eastern Europe's leading country in size (about the same as the U.S. state of New Mexico) and population (38 million). Most of Poland lies on part of the North European Plain between the Carpathian and Sudeten Mountains on the south and the Baltic Sea on the north. The country's most fertile soils are in the south, where the plain rises to low uplands that touch Europe's loess belt. This is where most of Poland's rural population (about 40 percent of the total) is located (Figure 5.18). The north is wilder, with some wetlands and forests that were protected as hunting reserves by centuries of aristocrats and communists. Some environmentalists fear that commercial logging and development in the new economy will destroy these resources.

Poland's largest rivers, the Vistula (Polish: Wisla) and Oder (Polish: Odra), wind across the northern plain from Upper Silesia to the Baltic. The Oder is the main internal waterway, carrying some barge traffic between the important Upper Silesian industrial area (to which it is connected by a canal) and the Baltic. Poland's main seaports—Gdansk (formerly, Danzig), Gdynia, and Szczecin (formerly, Stettin)—are along or near the lower courses of these rivers. Gdansk had a catalyst role in the breakaway of Eastern Europe from the Soviet grip, beginning in the early 1980s with Solidarity's first uprising in this port city's shipyards. The **Gdansk Accords** gave Polish workers the right to strike and to form free trade unions.

WORLD
REGIONAL
Geography ⊛ Now™

Click Geography Literacy for a link to the Solidarnosc website.

Gdansk is eclipsed in economic importance by Warsaw (population: 1.6 million), Poland's capital and primate city. Warsaw was the setting for one of history's worst war crimes when, during World War II, hundreds of thousands of Jews were forced to resettle in the prisonlike **Warsaw ghetto** and then were systematically shipped out to work and be exterminated in concentration camps like Auschwitz (in southern Poland, near Krakow). Krakow developed as an industrial city on the eastern end of the the Upper Silesian–Moravian coal field, a kind of "Ruhr in miniature" comparable to that area in Germany (Figure 5.19).

With Germany as its major trading partner, Poland gains more than two-thirds of its export dollars from the sale of machinery, chemicals, and manufactured goods. The economy has been suffering, with unemployment close to 20 percent, and traditionally large employers like the Szczecin shipyard are now bankrupt. Poland became a NATO member in 1999 and was the heavyweight of the countries that joined the European Union in 2004's "big bang."

Figure 5.18 Almost 40 percent of Poles are rural people. The agrarian nature of Eastern Europe contrasts sharply with the more urbanized West.

Figure 5.19 Polish coal, having been shipped from the Silesian coal fields, awaits export from the port of Swinoujscie at the mouth of the Oder River.

The Czech Republic

The economically diversified Czech Republic, where communism had undercut a long tradition of economic accomplishment, is enjoying an economic revival. The country's transformation began within Czechoslovakia with a bloodless overthrow of the Communist government during 6 weeks late in 1989. This **"Velvet Revolution"** culminated in the presidential election of Vaclav Havel, a gifted playwright. The Czech Republic was peacefully cleaved in its **"Velvet Divorce"** from Slovakia in 1993, thus ending the political entity of Czechoslovakia.

Prague, like Warsaw in Poland, is both the capital and primate city of the Czech Republic. This architecturally graced and lively city of 1.4 million people is located in the country's industrialized Bohemian basin, where most of the country's important output of machinery and transport equipment is based (Figure 5.20). In the extreme east of the Czech Republic is another industrial area, Moravia, that grew up on the Silesian–Moravian coal field. The Czech economy, which is now more than 90 percent privatized, still depends heavily on industry and has struggled with recession sparked by strong competition abroad.

Figure 5.20 Prague

The borderlands north, west, and south of Prague have long been a sore spot in relations between the Czechs and stronger Germany to the west. On the eve of World War II, much to the dismay of (then) Czechoslovakia, Britain and other European powers allowed Germany to annex Sudetenland and other portions of this borderland, whose inhabitants were primarily ethnic Germans. When Nazi Germany was defeated, vengeful Czechs regained the lands and forced the expulsion of more than 2 million ethnic Germans from these lands into Germany. So far, the Czech Republic has refused German demands for apologies or return of properties. A similar stalemate exits between Poland and the ethnic Germans likewise expelled from Poland's borders after the war. These issues became very significant with the new membership of the Czech Republic and Poland in the European Union, as farming interests had to be reassured that prewar claims to the land would not honored.

Slovakia

With the 1993 creation of the Slovak Republic, or Slovakia, the Danube River city of Bratislava (population: 428,000) became its national capital. The country is one-third arable and two-fifths forested and is more ethnically diverse than either the Czech Republic or Poland. Slovakia has a potential problem of further political separatism within its minority community of about 600,000 ethnic Hungarians. Slovakia still suffers from the fact that many professionals, and much potential foreign investment, stayed with the Czech Republic after the 1993 split. In the early 2000s, Slovakia's unemployment rate hovered around 20 percent. Plans were being made to generate more revenue from tourism along the Danube, in the Tatra Mountains in the east, and in the country's many caves and thermal spas.

Hungary

Hungary has been a relatively bright spot economically in Eastern Europe. Its move toward free-market capitalism began early and has been aided by Western and Japanese investments larger than those received by any other nation in this subregion. Hungary has a manufacturing edge because of local bauxite and aluminum production capabilities. Industries are concentrated in the beautiful capital city of Budapest (population: 2.6 million; Figure 5.21), dubbed the "Pearl of the Danube." A relatively small but growing class of "newly rich" private entrepreneurs has emerged, often as participants in joint business ventures with foreign companies. There are still poor people, especially among Hungary's one-third rural population, but they officially make up less than 10 percent of the population.

Romania

Romania is showcasing itself as a strategic, reliable ally of the United States, well located to assist American interests in the Caucasus and Middle East. The government in Bucharest (population: 2.2 million) supported the 2003 U.S. war on Iraq and became a NATO member in 2004. Because of its

Digital Vision/Getty Images

Figure 5.21 Budapest

poor economic status, however, Romania's bid to join the European Union was turned down, and it probably will not become a member until 2012. It is well below the average in measures of economic well-being in Eastern Europe. More than one-third of its 22 million people live in poverty. The country has to import oil and gas to provide its own energy needs. The export of textile, minerals, chemicals, and machinery make up approximately 50 percent of Romania's modest export income, with Germany and Italy the primary export destinations. With a long eastern coastline on the Black Sea, hundreds of miles of Danube River landscape, and the storied mountains of Transylvania, Romania has hoped for an influx of foreign exchange from the Western tourist trade. But the tourist traffic has so far been largely from the less affluent east, mainly Russia.

Bulgaria

Between the Rhodope Mountains and the Balkan Mountains is the productive valley of the Maritsa River. Together with the adjoining Sofia Basin, this is the economic core of Bulgaria. Only about a third of the country is arable, and Bulgaria is trying to industrialize. Clothing, footwear, iron and steel, and machinery, mainly produced in the capital Sofia (population: 1 million), are the main exports. By 2003, Bulgaria had made such progress in restructuring its economy that the European Union declared it to have a "functioning market economy." One of the preconditions for being an EU member, this qualified Bulgaria for possible admission to the European Union in 2007. One of the most important economic resources the country possesses is the rail route that links Istanbul, Turkey, with points in Central and Western Europe.

Albania

After World War II, this small Communist country, populated by non-Slavic ethnic Albanians, linked itself economically with the People's Republic of China. Economic support from its large benefactor failed to move Albania up from its position at the bottom of Europe's economic ladder. Its

leader, Enver Hoxha, kept the country isolated from the rest of the world and, to defend it from any threat, had the landscape peppered with almost a million fortified "pillbox" bunkers. Those few outsiders who visited Albania described it as a bizarre throwback to a medieval agrarian Europe. Hoxha was finally ousted in 1984, and in the 1990s, following the dissolution of the Communist government and a period of internal political turmoil, Albania began to open up.

However, more than half the population is still rural, and Albania has few resources or other prospects for industrialization. Receipts from **remittances**—money earned by Albanians working abroad, mainly in Greece and Italy—represent most of the country's wealth. A "pyramid scheme" in the late 1990s swindled much of what little wealth there was from hundreds of thousands of Albanians. Waves of Albanian emigrants have swept into Europe, especially into Italy, where Italians routinely blame the newcomers for theft, robbery, rape, gun-running, prostitution, and a host of other ills. The Albanian government in the capital, Tirana (population: 350,000), estimates that since 1991, more than 6,000 Albanian children—mainly Roma—have been sold or traded by their parents, ostensibly for the welfare of the children. However, most are handled by **human traffickers** who sell them into prostitution rings or to Europeans eager to adopt a child.

179, 602

Successors of the Fractured Yugoslavia

While economic progress of varying degrees characterized most of Eastern Europe in the 1990s, Yugoslavia was plagued by ethnic warfare following the dissolution of the Yugoslav federal state in 1991. Previously, under the Communist regime instituted by Marshal Josip Broz Tito during World War II, Yugoslavia ("Land of the South Slavs") had been organized into six Socialist People's Republics (Serbia, Croatia, Slovenia, Bosnia, Montenegro, and Macedonia) and two "autonomous provinces" (Kosovo and Vojvodina). As long as Tito and his successors could retain a firm grip, these ethnic groups could coexist within the artificial boundaries of a single nation. However, this apparent unity belied many historic underlying tensions and conflicts based on linguistic, ethnic, and religious distinctions: Orthodox Serb versus Muslim Kosovar, Catholic Croat versus Orthodox Serb, and more.

As the Iron Curtain dissolved across Europe, Yugoslavia began to fracture along its ancient ethnic fault lines (Figure 5.22). In 1988, Serbia took direct control of Kosovo and Vojvodina as part of Serbia's push (under an elected Communist president, Slobodan Milosevic) for greater influence within federal Yugoslavia. The quasi-independent states within Yugoslavia initially demanded a looser federation with more autonomy for each republic, but in 1991 and 1992, Croatia, Slovenia, Macedonia, and Bosnia insisted on independence. With Croatia's declaration of independence, fighting occurred between ethnic Croats and Serbs within the country. Croatian Serbs took over one-third of the new country and began ethnic cleansing, the forced emigration or murder of one ethnic group by another within a certain

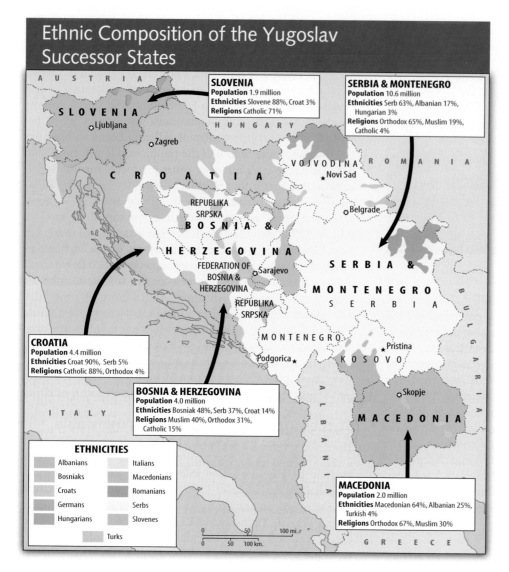

Ethnic Composition of the Yugoslav Successor States

SLOVENIA
Population 1.9 million
Ethnicities Slovene 88%, Croat 3%
Religions Catholic 71%

SERBIA & MONTENEGRO
Population 10.6 million
Ethnicities Serb 63%, Albanian 17%, Hungarian 3%
Religions Orthodox 65%, Muslim 19%, Catholic 4%

CROATIA
Population 4.4 million
Ethnicities Croat 90%, Serb 5%
Religions Catholic 88%, Orthodox 4%

BOSNIA & HERZEGOVINA
Population 4.0 million
Ethnicities Bosniak 48%, Serb 37%, Croat 14%
Religions Muslim 40%, Orthodox 31%, Catholic 15%

MACEDONIA
Population 2.0 million
Ethnicities Macedonian 64%, Albanian 25%, Turkish 4%
Religions Orthodox 67%, Muslim 30%

ETHNICITIES

- Albanians
- Bosniaks
- Croats
- Germans
- Hungarians
- Italians
- Macedonians
- Romanians
- Serbs
- Slovenes
- Turks

WORLD
REGIONAL
Geography ⊛ Now™

Active Figure 5.22
Ethnic composition of the Yugoslav successor states. *See an animation based on this figure, and take a short quiz on the facts or concepts.*

territory. While Bosnia's Muslims (known as **Bosniaks**) and Croats voted for independence from Yugoslavia, ethnic Serbs in Bosnia, and the Yugoslav government of Slobodan Milosevic, opposed this.

War erupted in 1992 between the Bosnian government and local Serbs. Supported by Serbia and Montenegro, the Bosnian Serbs fought to partition Bosnia along ethnic lines and join Serb-held areas to a "greater Serbia." Bosnian Serbs laid siege to the mainly Bosniak city of Sarajevo (Figure 5.23). The United Nations withdrew recognition of Yugoslavia because of the government's failure to halt Serbian atrocities against non-Serbs in Croatia and Bosnia. In 1994, the Bosniaks and Croats agreed to create the joint federation of Bosnia and Herzegovina.

In 1995, the presidents of Bosnia and Herzegovina, Croatia, and Serbia signed the **Dayton Accord**. This peace agreement retained Bosnia and Herzegovina's international boundaries and created a joint multiethnic, democratic government charged with conducting foreign and economic policies. The agreement also recognized a second tier of government com-

Figure 5.23 A woman of Sarajevo returns to the ruins of her home.

posed of two similarly sized entities charged with overseeing internal functions: the Bosniak/Croat Federation of Bosnia and Herzegovina and the Bosnian Serb-led Serb Republic (Republika Srpska). To help implement and monitor the agreement, NATO fielded a 60,000-strong force known as IFOR. It was later replaced by a smaller NATO force known as SFOR, which remains on the ground in Bosnia.

While the Bosnian situation quieted, in 1998 Serbia turned its sights on neighboring Kosovo, a majority ethnic Albanian and Muslim region with a minority of ethnic Serbians in the north. Here again, Serbs aspired to ethnically cleanse Kosovo of non-Serbs. Serbs asserted that Kosovo was the cradle of their culture and the location of the decisive battle of Kosovo Polje in 1389, which Serbs lost to the Ottoman Turks. NATO responded to the Serb offensive with 11 weeks of bombing. When the bombing stopped, the United Nations took control of Kosovo and oversaw a tense truce between its Serb and Muslim inhabitants. The fate of more than 200,000 ethnic Serbs who fled during the conflict from Kosovo to Serbia and Montenegro has not been resolved. Nominally still a part of Serbia and Montenegro, Kosovo may yet evolve into an independent country.

The ouster of Serbian leader Slobodan Milosevic in 2000, and his capture and subsequent transfer to The Hague for a war crimes trial, opened the way for a period of stable and peaceful borders in the early 2000s. The international community insisted on no further redrawing of Balkan boundaries. The region had undergone a profound and violent process of balkanization, or fragmentation into ethnically based, contentious units, a political-geographic term that takes its name from the characteristic disharmony of this region.

The Balkan conflicts unfolded on television and other media through the 1990s. They reminded Europeans and Americans of the earlier world wars on the continent and seemed a horrible anomaly in a world region then accustomed to peace and prosperity. Western military and political institutions eventually seemed to extinguish the fires of conflict, and the Balkan countries were set on the backburners of Europe. By all measures of quality of life and economic prosperity, they are clearly the poor stepchildren of the continent, and most cannot expect to join the European Union for many years. Only Slovenia—the sole Balkan newcomer to the EU in 2004—has done reasonably well, mainly through exports of metallurgical products and electronics. Croatia has failed to recapture its former status as major tourist destination, but it still has the attractions that include the historic walled city of Dubrovnik, nearby beaches of the Dalmatian Coast, and the inland Plitvice Lakes National Park (Figure 5.24).

Warfare devastated the still-moribund economy of Bosnia and Herzegovina, which, with Macedonia, is one of the two poorest successor states of Yugoslavia. However, since 1981 and the reported apparition of the Mother of Christ, its site of Medjugorje has grown to be a prospering pilgrimage destination for Catholics from all around the world. In Macedonia, there are continuing tensions between the majority

348

Figure 5.24 Croatia's Plitvice Lakes National Park, a World Heritage Site, is a majestic wilderness—and a difficult one to reach in winter.

(70 percent) ethnic Macedonians, who are Slavs and Orthodox Christians, and the mainly Muslim ethnic Albanians who make up 25 percent of its population. The ethnic Albanians, Roma, Jews, and other minorities insist they are being increasingly marginalized by strengthening Macedonian nationalism. The landlocked country relies on access to the Greek port of Salonika (Thessaloniki) and struggles to maintain good relations with its more powerful neighbor.

The country of Serbia and Montenegro is simply a loose federation of the two mainly Serbian states that succeeded what was left of Yugoslavia when the other countries broke away. Few observers think this union will last long. The federation was the result of deal making in the West, in which Serbia was called on to surrender more suspects for war crimes trials; in return, the West would not support the independence of Montenegro. The nominal union was also meant to blunt, for the time being, Kosovar desire for independence for Kosovo. Ethnic tensions between the ethnic Slavic Montenegrin majority and the ethnic Albanian minority threaten even this province's unity.

5.3 Southern Europe: The Mediterranean World

The four main countries of Southern Europe, as defined in this text, are Portugal, Spain, Italy, and Greece. They are on peninsulas extending southward from the main body of Europe. Portugal and much larger Spain share the Iberian Peninsula, Italy has the Italian Peninsula, and most of Greece is on a subpeninsula, between the Ionian and Aegean Seas, of the larger Balkan Peninsula. Southern Europe also includes the microstates of Vatican City (enclosed within Rome), San Marino (enclosed within Italy), and the British colony of Gibraltar. Three of the peninsular countries include islands in the Mediterranean, the largest of which are Italy's Sicily and Sardinia, the Greek island of Crete, and the Spanish

Balearic Islands. Some Mediterranean islands lie outside these countries, including Corsica, which is a part of France, and the independent countries of Malta and Cyprus. These are the lands of Mediterranean Europe (see Figure 5.1).

The Environment of Mediterranean Agriculture

In terrain, population distribution, climate, and agriculture, the countries of Mediterranean Europe have many similarities. Rugged terrain predominates. Plains tend to be small, facing the sea, and separated from each other by mountainous areas. Population distribution corresponds generally with topography, with the lowland plains being densely populated and the mountainous areas much less so, although some rough areas have surprisingly high densities. The Alps and the Pyrenees have long served as a permeable barrier between the European core and Southern Europe.

A distinctive natural characteristic of Southern Europe is its Mediterranean climate, combining mild, rainy winters with hot, dry summers. Characteristics associated with this climate become increasingly pronounced toward the south. The northern extremities of both Italy and Spain have different climatic characteristics. Except for a strip along the Mediterranean Sea, northern Spain has a marine west coast climate like that of northwestern Europe. It is cooler and wetter in summer than the typically Mediterranean climate areas. The basin of the Po River in northern Italy has a relatively wet summer and cold-month temperatures in the lowlands averaging just above freezing. Much of the high interior plateau of Spain, called the Meseta, is cut off from rain-bearing winds and has somewhat less precipitation and colder winters than is typical of the Mediterranean climate, although the seasonal regimen of precipitation with dry summers is a characteristic Mediterranean climate trait. The Meseta is enormously productive agriculturally.

Classic Mediterranean crops are olives, grapes, figs, citrus fruits (especially oranges and lemons), vegetables such as tomatoes and eggplants, and herbs such as thyme and oregano (Figure 5.25). All are components of the **Mediter-**ranean diet characteristic of peoples around the Mediterranean basin for centuries and touted by many nutritionists today as an excellent model for the health conscious. The olive tree and the grapevine have extensive root systems and certain other adaptations that allow them to survive the summer droughts. Olive oil is a major source of fat in the typical Mediterranean diet, and virtually all of the world supply is produced in countries that touch or lie near the Mediterranean Sea or are in a Mediterranean climate zone. The principal use of grapes is for wine, a standard household beverage in Southern Europe and a major export product.

Winter wheat has historical and modern importance in the region. This hardy grain crop is generally planted at the end of the harvest of a summer crop. Harvested in May or June, it tends to give a higher yield than does spring wheat, which is planted in late spring and harvested in early fall. Barley, a more adaptable grain, tends to replace wheat in some areas that are particularly dry or infertile. Where irrigation is lacking, vegetables are grown that will mature during the wetter winter or in the spring. The most important are several kinds of beans and peas. These are a source of protein in an area where, due to parched summer pastures and other limitations, livestock make only a limited contribution to the food supply. Extensive areas too rough for cultivation are used for grazing, but the carrying capacity of such lands is relatively low. Sheep and goats, which can survive on a sparser pasturage than cattle, are favored, and their meats are characteristic dishes. In many places, grazing depends on **transhumance**—utilizing lowland pastures during the wetter winter and mountain pastures during the summer. In some areas, nonfood crops supplement the basic Mediterranean products. An example is tobacco, an export of Greece.

Mediterranean agriculture is at its peak productivity where the land is irrigated. Since ancient times, people have made extraordinary efforts to deliver water from sources to fields during the dry, hot summer. The Roman aqueducts were constructed particularly well, and many remain today as unique elements of the cultural landscape (Figure 5.26).

Figure 5.25 Figs are among the characteristic Mediterranean tree crops. This orchard is in Greece.

Andrew Brown / Ecoscene / Corbis

Figure 5.26 A Roman aqueduct in Nerja, Spain

Irrigation on a small scale is found in many parts of Southern Europe, but there are a few large irrigated areas: northern Portugal, the Mediterranean coast of Spain, the northern coast of Sicily, and the Italian coastal areas near Naples. On the largest of all, the plain of the Po River in northeast Italy, irrigation supplements year-round rainfall.

In the Mediterranean coastal regions of Spain, irrigation supports extensive orchards. Oranges are the most important product, with growth concentrated around the city of Valencia, which has given its name to a type of orange. Lemons are particularly important on the island of Sicily. The district around Naples, known as Campania, is more intensively farmed and densely populated than any other agricultural area of comparable size in Southern Europe. Irrigation water applied to exceptionally fertile soils formed of volcanic debris from Mount Vesuvius supports a remarkable variety of products, including almost every crop grown in Southern Europe.

An Economic Revival

While more prosperous than Eastern Europe, Southern Europe has historically lagged behind Northern Europe and the European core in economic terms (see Table 3.1). The main reason for this poverty was a lagging development of industry related to unfavorable social and resource conditions during the 19th and early 20th centuries, when many European countries were industrializing rapidly. Lack of industrial employment left too many people in crowded small farm areas where output per person was low. The absence of large-scale trade and therefore of capital to invest in manufacturing fed this underdevelopment. Mainly poor and uneducated populations offered little skilled labor or dynamic market potential. Wealthy landowners often controlled both land and government and had little interest in overhauling the agrarian system that made them rich and powerful. Unstable, undemocratic, and ineffective governments were dis-

interested in economic development. Resource deficiencies included a lack of good coal and iron ore deposits, except in northern Spain; scarce wood supplies due to aridity and deforestation; and lack of hydropower in the areas of Mediterranean climate, where streams often dry up in summer. Finally, the rugged topography of much of the region made construction of adequate rail and road systems difficult and expensive.

Because of such impediments, only three sizable industrial areas developed in Southern Europe before World War II. The most important was in the Po Plain of northern Italy (see Figure 4.3, page 80). Here, imported coal and raw materials, hydroelectric power from the surrounding mountains, and cheap labor formed the basis for an area specializing in textiles but including some heavy industry, automobile production, an aircraft industry, and engineering. A second, smaller area similar in its foundations and specialties developed in the northeastern corner of Spain in and around Barcelona. A third area, focused on heavy industry, developed in the coal and iron mining section of northern coastal Spain. All three areas lie near the foot of mountains with humid conditions that provided the basis for hydroelectric power development.

In the later 20th century, the economies of Southern Europe finally began to turn around. Italy, Spain, Portugal, and Greece have evolved from agricultural to industrial and service-oriented countries. Although the South still lags behind Northern and Western Europe, recent growth has been shaped by both industrial modernization and an expanding role of the European Union, to which all the countries belong. Industrial and service activities led the way, and there was a sharp decline in the proportion of the population still employed in agriculture.

A number of factors played a part in this revitalization. Improved transportation, particularly ocean transport, made the importation of fuels and materials for manufacturing more economical. Sources of energy changed; coal was replaced by imported oil and, in the case of Italy, domestic natural gas and geothermal power. Market conditions for Southern European exports—fruits, vegetables, automobiles, and labor—improved after World War II, largely as a result of the rising prosperity of northwestern Europe and the expansion of European markets. Streams of temporary migrants from Southern Europe found jobs in booming, labor-short industries to the north. Absence of the migrants relieved unemployment, and their remittances added purchasing power in their home areas. There were other infusions of foreign money into the Southern European countries. Each country received U.S. economic and military aid, and favorable economic circumstances and government policies attracted foreign investment in industrial and other facilities.

The region also saw a tremendous boom in tourism. To increasingly prosperous Europeans and Americans, Southern Europe offers a combination of spectacular scenery, historic monuments, pleasant villages, sunny skies, the sea, and the

Figure 5.27 Mediterranean landscapes and settlements are irresistible attractions for tourists. This is the harbor on the Greek island of Aegina.

beaches (Figure 5.27). Following is more insight into each country's attractions, resources, and characteristic features.

Spain

Spanish power and influence reached a zenith in 1492, when a centuries-long struggle, the *Reconquista,* ended with Spain's defeat of the Kingdom of Granada, the last part of Iberia held by the Muslim **Moors** (Figure 5.28). Spanish Catholics expelled both Moors and a distinct population of **Sephardic Jews** from Iberia and ushered in **the Inquisition,** a repressive program designed to purge Spain of "false Christians." Also in 1492, Christopher Columbus, a Genoese Italian navigator in the service of Spain, landed in the Americas in search of a sea route to India and East Asia. For a century thereafter, Spain was the greatest power in Europe and probably in the world. It rapidly built an empire spreading to Italy, the Low Countries, the Americas, Africa, and the

Figure 5.28 The Alhambra in Granada, dating to the ninth century, is one of the greatest of all Muslim architectural accomplishments.

Figure 5.29 Principal features of Spain and Portugal.

Philippines. These possessions were subsequently lost at various times during a long, slow decline of Spanish power over the next three centuries. Of those lands of the empire, only the Canary Islands, off the Atlantic coast of Morocco, remain an integral part of Spain. The Balearic Islands of Ibiza, Majorca, and Minorca, off Spain's east coast, belong to the country and are its main destination for sun-loving tourists.

Most of the population of the Iberian Peninsula lives on discontinuous coastal lowlands (Figure 5.29). The greater part of the Iberian Peninsula consists of a more lightly inhabited plateau, the Meseta, with a surface lying between 2,000 and 3,000 feet (c. 600–900 m) in elevation. Mountains and deep river valleys interrupt the plateau surface, whose edges are steep and rugged. The Pyrenees and the Cantabrian Mountains border the Meseta on the north, and the Betic Mountains, culminating in the Sierra Nevada, border it on the south.

The three largest cities of Spain are the capital, Madrid (population: 6.1 million), and the two Mediterranean ports of Barcelona and Valencia. Madrid was chosen as the capital of Spain in the 16th century because of its location near the geographic center of the Iberian Peninsula, approximately equidistant from the various peripheral areas of dense population. Located in a poor region, Madrid has had little economic excuse for existence during most of its history. But its position as the capital has made it the center of the Spanish road and rail networks and thus has given it nodal business advantages. Its critical function as a railway hub put Madrid in the cross hairs of a ferocious terrorist bombing in 2004. Islamist militants claimed they carried out the bombings as revenge for Spain's support of the U.S.-led invasion of Iraq in

66

Ethnic Geography

The Basque Country

The 2.2 million **Basques** of Spain and 250,000 Basques of France have a unique ethnicity and culture, completely unrelated to those of their host country majorities. Like other minorities, they have often been the targets of discrimination and violence. This is especially true in Spain, where in 1937 a German bombing called for by Spanish fascists led to tragic loss of life among Basque civilians in the city of Guernica (Figure 5.D). By the 1960s, Basque desire for independence led to the emergence of the militant, often terrorist, group **ETA** (in Basque, an acronym for **Basque Homeland and Liberty**). Spanish authorities gradually allowed limited autonomy for the Basques so that they now control their own affairs in taxation, education, and policing. These concessions have not been enough for ETA, which has continued attacks on Spanish interests and triggered a vigorous crackdown by Spanish security forces. Basque efforts to retain their unique customs and language are mainly peaceful and have proved increasingly successful. Basque television now broadcasts on the Spanish side. Across the border, French Basques yearn for at least the kind of autonomy enjoyed by their kinfolk in Spain. For both populations, independence seems a distant and likely inaccessible goal.

Figure 5.D The Spanish Basque city of Guernica is known around the world as a place of civilian suffering thanks in large part to Pablo Picasso's depiction of its trauma in his painting *Guernica*.

Erich Lessing/Art Resource, NY

2003. In a national election just days later, bereaved and angered Spanish voters ousted the ruling party and ushered in a less pro-American government that quickly removed Spanish troops from Iraq.

Much of Spain's fishing fleet, the largest in the European Union, is based at Europe's largest fishing port, the northwestern city of Vigo. Barcelona, capital of the province of Catalonia, is Spain's major port and the center of a diversified industrial district. Barcelona's industrialization has been supported by hydroelectric power (mainly from the Pyrenees), cheap and skilled labor, and imported raw materials.

So prosperous is this region relative to others in Spain that Catalonian politicians are calling increasingly for the devolution of power from Madrid. Among other things, they want assurances that water from the Ebro River in Catalonia and Aragon will not be diverted south to more arid Valencia, Murcia, and Andalusia, as called for in Spain's national hydrological plan. They say they want an independent nation within Spain that would have its own voice in the European Union and be able to attract foreign investment independently of Madrid. Politicians in Galicia, Spain's northwesternmost province, are watching Catalonia to see if they want to demand a similar course of devolution. One of the outstanding political challenges for Spain today is the future of its independence-minded Basque population (see Ethnic Geography, page 129).

After a devastating civil war from 1936–1939 and then World War II, Spain gradually developed a stable democratic system. Like the United Kingdom, it is a constitutional monarchy. Relations between Spain and Britain are generally quite good, and they tend to see eye to eye on important international issues. However, there is a thorn in their relations: the tiny enclave of Gibraltar, on the northern side of the Strait of Gibraltar, which Britain has held for three centuries. Britain claims that its sovereignty over Gibraltar helps guarantee the security of one of the most economically critical waterways in the world, while Spain sees Britain's claim as an anachronistic, unjustified vestige of colonialism. To defuse this embarrassing row between two fellow EU and NATO members, both countries are looking for a sovereignty-sharing formula. This would require approval by Gibraltar's 30,000 inhabitants. Mainly of Genoese, Maltese, and Jewish ancestry, they are at least now overwhelmingly in favor of remaining British colonial subjects. They fear losing any of the defense, tourism, and tax-free shopping revenues they enjoy under British rule.

Portugal

In the 15th century, Portugal took the lead in seeking a sea route around Africa to southern and eastern Asia. Henry the Navigator of Portugal explored the coast of northwest Africa between 1418 and 1460. In terms of trade that would follow, the 1499 Vasco da Gama voyage to the major trading port of Calicut, near the southwest tip of the peninsula of India, was even more important than the 1492 Columbus voyage to the Americas. For the better part of a century thereafter, Portugal dominated European trade with the East and built an empire there and across the Atlantic in Brazil. Then there was a rapid decline in Portugal's fortunes, except for its control of Brazil until 1822. Portugal also held large African territories until the 1970s.

The two large cities of Portugal are seaports on the lower courses of rivers crossing the Meseta and reaching the Atlantic through Portugal. Lisbon (population: 2.6 million), on a magnificent natural harbor at the mouth of the Tagus River, is the country's capital and leading seaport. Little architecture survives from before 1755, when an earthquake destroyed the city. The smaller city of Oporto, at the mouth of the Douro River, is the regional capital of northern Portugal and the commercial center for trade in Portugal's famous port wine, which comes from terraced vineyards along the hills overlooking the Douro Valley.

Since joining the European Union in 1986, Portugal has developed a diversified economy based strongly on services, including finances and telecommunications. Key exports are clothing, footwear, machinery, chemicals, and cork products made from its distinct and traditional cork industry.

Italy

Italy has bestowed an enormous heritage on Western civilization. Its main period of preeminence occupied five centuries, during which the Roman Empire ruled over the whole Mediterranean basin and some lands beyond. Within the Empire, Christianity began and spread, and by the time the Empire collapsed in the fifth century A.D., Rome had become the seat of the popes and the Catholic Church. During the later Middle Ages, a number of Italian cities, notably Venice, Genoa, Milan, Bologna, and Florence, became powerful independent states. Venice and Genoa controlled maritime empires within the Mediterranean region, and all the cities profited as traders between the eastern Mediterranean and Europe north of the Alps. After unification in 1870, Italy attempted to emulate past glories, but the empire that it acquired in Libya and Ethiopia was lost when Italy was defeated in World War II.

Reflecting its position astride continental plates, Italy has a diverse and often rugged topography. The Alps and the Apennines are the principal mountain ranges. Northern Italy includes most of the southern slopes of the Alps. The Apennines, lower in elevation than the Alps, form the north–south trending backbone of the peninsula, extending from their junction with the southwestern end of the Alps to the toe of the Italian boot, appearing again across the Strait of Messina in Sicily, where Mount Etna, a volcanic cone, reaches 10,902 feet (3,323 m). Near Naples, volcanic Mount Vesuvius, at 4,190 feet (1,277 m), is renowned as the killer of the thriving Roman cities of Pompeii and Herculaneum, which were frozen in time under a massive ashfall in A.D. 79 (Figure 5.30). Still active, Vesuvius threatens more than 3 million people in the area of Naples. West of the Apennines, most of the land between Florence on the north and Naples on the south is occupied by lower hills and moun-

Figure 5.30 The eruption of Mount Vesuvius—part of the mountain complex seen in the distance—buried the Roman city of Pompeii under a vast deposit of ash, freezing the city in time until modern archeologists brought it to light.

Figure 5.31 Principal features of Italy.

Italy's area, this zone accounts for approximately one-half of its agricultural production and two-thirds of its industrial output. A rapid expansion and diversification of Po Basin industry, previously dominated by textiles, occurred after World War II. There was growth in engineering and in the chemical, iron, and steel industries, fueled by domestic natural gas and hydroelectric power and by imported oil.

Many cities in the basin share in the region's industry, but the leading ones are Milan (population: 4 million), Turin, and Genoa. Milan is Italy's largest industrial city and a dominant center of fashion, finance, business administration, and railway transportation. It is advantageously located between the port of Genoa and important passes, now shortened by railway tunnels through the Alps to Switzerland and the North Sea countries. Turin is the leading center of Italy's automobile industry (including Fiat) and the heart of its high-tech **Technocity** region (see Figure 4.4, page 81).

The port of Genoa, reached from the Milan–Turin area by passes across the narrow but rugged northwestern end of the Apennines, functions as a part of the Po Basin industrial complex. The city dominates the overseas trade of the Po area and is Italy's leading seaport. Genoa is now much larger and economically more important than Venice, its old medieval commercial and political rival at the eastern edge of the Po Plain. Venice is one of the world's most prized tourist destinations. Its canals, monuments, and people have exerted an irresistible and romantic charm on centuries of visitors. It was built on low-lying and reclaimed land, and today, it wages a war to hold back the Adriatic as the land sinks and the waters rise. In 2003, construction began on 79 giant sea gates that will rise from the ocean floor whenever high tides threaten Venice's lagoon (Figure 5.32).

The largest cities outside the northern industrial area are the capital, Rome (population: 3.6 million), and Naples (population: 3.6 million). Rome has a location that is roughly central within Italy and is a governmental, religious, and tourist center. Within Rome is the microstate of the Holy See (State of the Vatican City), the independent entity that is the center of the Catholic Church (Figure 5.33). Yet another microstate, San Marino, lies to the north of Rome. It prospers from tourism and banking. Naples, located south of Rome on the west coast of the peninsula, is the port for the populous and productive, but poor, Campanian agricultural region. It is also the main urban center of one of Italy's major tourist regions, with attractions such as Mount Vesuvius, the ruins of Pompeii, and the island of Capri in the vicinity.

In Italy, there is a longstanding vernacular regional distinction between north and south. Recently, yet another regional term has evolved, a **Third Italy** comprised of the relatively prosperous northeast and central portions of the country, with their textile, leather, ceramic, and furniture industries. But the north–south distinction is most deeply rooted and has dimensions beyond the economic. Northerners tend to see themselves as more sophisticated and cosmopolitan than the more earthy and provincial southerners.

tains, many of volcanic origin. Much of the Mediterranean's monumental architecture is built of Italian marble quarried in this region.

Italy has a population density of about 500 per square mile (c. 190 per sq km), nearly twice that of Portugal, the second most densely populated country of Southern Europe (Figure 5.31). The largest lowland, the Po Plain, has about one-third of the entire Italian population. Some other lowland areas with extremely high densities include the narrow Ligurian Coast centering on Genoa, the plain of the Arno River inland to Pisa and Florence, the lower valley of the Tiber River including Rome, and Campania around Naples.

A country evocative of romance and tradition, Italy is also a major component of the European industrial and postindustrial engines. Italy's per capita GDP PPP rivals that of the United Kingdom and stands clearly above that of any other country in Southern Europe. The Po Basin is the economic heart of modern Italy and the most highly developed and productive part of Southern Europe. With its competitive high-tech and high-fashion output, it ranks with other Western European centers of innovation. Functionally, this section of northern Italy belongs with the rest of the European core described in Chapter 4. Comprising just under a fifth of

F4.4, 81

Jonathan Blair/Corbis

Figure 5.32 Venice from the air. The amphibious and vulnerable setting of the city is most apparent from this perspective.

Southern Italians, whose region is known as the Mezzogiorno, often acknowledge their agrarian roots as the source of their superior kinship values and enjoyment of life. Statistics underscore the economic discrepancies between the two Italys; for example, northern Italy has labor shortages while the south has unemployment; northern labor and industries are more productive, and income levels in the north are much higher.

The feeling among many northerners that they would be better off if they did not have to subsidize the south has even led to a secessionist movement, championed by an organization called the **Northern League,** that calls for the creation of an autonomous or independent state of **Padania** in the north. Another organization, **Liga Veneta,** wants to create an autonomous region of Veneto, centered around the affluent city of Venice. Regardless of their differences, both northern and southern Italians take pride in the heritage bestowed by civilization from the Romans to the Renaissance and beyond, and both live on some of the most picturesque and storied landscapes on Earth.

Figure 5.33 A view from the Vatican Basilica dome, looking out over Rome, on the banks of the Tiber River.

Joe Hobbs

Greece

Greece, like Italy, has been a vital source area for Western life. Centuries before the Christian era, Greek artists, architects, authors, philosophers, and scholars produced many of the ideas and works that laid the foundation for Western civilization (Figure 5.34). Between about 600 and 300 B.C., the sailors, traders, warriors, and colonists of the Greek city-states spread their culture throughout the Mediterranean area. A second period of Greek power and influence occurred in the Middle Ages, when Constantinople (modern Istanbul, Turkey) was the capital of the large Byzantine Empire. The Eastern Orthodox branch of Christianity developed in and diffused from this empire.

Greece's rugged topography and seismic history reveal that it is even more a product of tectonic activity than is Italy (Figure 5.35). The collision of the African and European plates of Earth's crust in this region sometimes produces devastating earthquakes and volcanic eruptions. In about 1650 B.C., a massive volcanic explosion tore apart the Mediterranean island of Thera (Santorini), producing catastrophic **tsunami** (**tidal waves**) and ash clouds large enough to ruin crops and change climates for decades. There is growing evidence that this event and its aftermath brought an end to the Minoan civilization on the island of Crete.

The islands of the Ionian and Aegean Sea, which are some of the most prized holiday destinations in the world, are seaward extensions of the Pindus Mountains that occupy most of the peninsula north of the Gulf of Corinth. Greece south of the Gulf of Corinth, known as the Peloponnesus, is composed mainly of the Arcadian mountain knot. Along the coasts of Greece, the many small lowlands facing the sea between mountain spurs support the majority of the people. An especially famous lowland is the Attica Plain, still dominated as in ancient times by Athens and its seaport, Piraeus.

Figure 5.35 Principal features of Greece and Cyprus.

Athens (population: 3.3 million), the primate city of Greece, is home to nearly a third of the country's population. It is the capital, main port, and main industrial center. Its development is based on imported fuels, and manufactured products are sold almost entirely in a domestic market whose main center is Athens itself. Tourism is the mainstay of the Greek economy, with nearly $10 billion a year coming from this trade. Greece boasts the beauty of the Aegean Sea, the Parthenon and other monuments of early Greek culture, and vibrant modern arts. Greece prides itself as the birthplace of the Olympics and, in 2004, hosted the summer games.

The Mediterranean Islands

In addition to Crete, Rhodes, and the many smaller islands of Greece, there are other Mediterranean islands of considerable size and importance. North of Italy's island of Sardinia is the island of Corsica, which was an independent country before France absorbed it in 1768. Corsicans are calling for greater autonomy or even independence. France has answered by agreeing to allow for the devolution of some powers, giving the island's parliament some legislative autonomy, and letting the Corsican language to be taught in schools.

Malta is an independent country of less than half a million people, the least populous of the European Union countries. Representing their crossroads island location between Italy

Figure 5.34 Ancient Greek architecture is part of the "classic" tradition that is the model for major government and other institutional buildings in the West today. This is the temple of Aphea on the island of Aegina.

and North Africa, the Maltese are descendents of ancient Carthaginians, Phoenicians, Italians, and other Mediterranean peoples. Most speak both Maltese—an amalgam of languages spoken by its diverse ancestors—and English. Malta was a British colony until 1964. Its economy depends on tourism, financial services, and capitalizing on its strategic mid-Mediterranean location, shipping services.

The large Mediterranean island of Cyprus (see Figure 5.35), located near southeastern Turkey, came under British control in 1878 after centuries of Ottoman Turkish rule. In 1960, it gained independence as the Republic of Cyprus. The critical problem on the island is the division between the Greek Cypriots, who are Greek Orthodox Christians comprising about three-fourths of the estimated population of 900,000, and the Turkish Cypriots, who are Muslims comprising about one-fourth. Agitation by the Greek majority for union (*enosis*) with Greece was prominent after World War II and led in the 1950s to widespread terrorism and guerrilla warfare by Greek Cypriots against the occupying British. Violence also erupted between Greek advocates of enosis and the Turkish Cypriots, who greatly feared a transfer from British to Greek sovereignty.

In 1974, a major national crisis erupted when a short-lived coup by Greek Cypriots, led mainly by Greek military officers, temporarily overthrew Cyprus's President Makarios, who had followed a conciliatory policy toward the Turkish minority. Turkey then launched a military invasion that overran the northern part of the island. Cyprus was soon partitioned between the Turkish North and the Greek South. A buffer zone (the **Attila Line,** or **Green Line**) sealed off the two sectors from each other, and even the main city of Nicosia was divided. A separate government was established in the North, and in 1983, an independent Turkish Republic of Northern Cyprus was proclaimed. Only Turkey recognized this state. Meanwhile, the internationally recognized Republic of Cyprus functioned in the Greek-Cypriot sector, which comprises about three-fifths of the island. Both republics had their capitals in Nicosia (population: 200,000).

Prior to the partitioning of Cyprus, the North had dominated the economy, but after that, the North has had severe economic difficulties while Greek Cyprus prospered. Most outside nations refused to trade directly with Turkish Cyprus after the invasion, and the trade and economic assistance offered by an economically weak Turkey were insufficient to provide much momentum. The depressed North remained tied to the struggling economy of the Turkish mainland. Meanwhile the Greek sector was able to make effective use of economic aid from Greece, Britain, the United States, and the United Nations. A construction program provided new housing, commercial buildings, roads, and port facilities. Tourism based on both beach and mountain resorts flourished. New businesses were attracted.

Events leading to the EU's expansion in the "big bang" of 2004 provided a strong incentive for the two sides to resolve their differences and join the union. The United Nations devised a plan for the two halves to vote separately in an April 2004 referendum on reunification. Had both sides voted in favor of reunification, a united Cyprus would have taken up membership in the European Union the following month. However, the United Nations and European Union agreed that should either the north or south, or both of them, reject the referendum, only the Greek south would actually join the European Union. In the vote, the Turkish north voted overwhelmingly in favor of reunification, while the Greek south overwhelmingly rejected it. The somewhat perplexing result is that in EU terms, "Cyprus" nominally refers to both sides of the island, but only the Greek south is a de facto member of the European Union. This allows the UN, EU, and other international bodies to continue their policy of not officially recognizing the north as a legitimate, separate political entity. Turkey, long seen as the obstacle to reunification of Cyprus, is likely to benefit from the Turkish Cypriots' strong vote in favor of reunification. Turkey is less likely now to be rebuffed in its own efforts to join the European Union.

This concludes the three-chapter survey of Europe. From here, the text moves eastward to the world's largest country, Russia, and many of the countries within its realm of great influence.

CHAPTER SUMMARY

- The core region of Europe is ringed by a periphery of three subregions: Northern, Eastern, and Southern Europe. These have been less integrated into the core, and the Eastern and Southern countries are less prosperous.

- Most of these peripheral countries are members of or are seeking membership in the European Union and NATO. Norway has chosen to remain outside the European Union, and some of the poorer European countries do not qualify for membership.

- The Northern, or Nordic, countries include the Scandinavian countries of Norway, Sweden, and Denmark, plus Iceland. With Finland, this group is often called Fennoscandia, or the Fennoscandian countries.

- The Northern countries are the world's most northerly region that is highly settled and economically developed. Human activity is possible at such a high latitude because of the warming waters of the North Atlantic Drift.

- Except for Finland, the Nordic countries have close connections through language (Finnish has its origins in Asia). Other qualities that link these countries are settlements in the southern reaches of the countries (except Denmark) and sparse populations in the north and in the mountains of all the countries (except for flat Denmark). There are high levels of literacy, education, and Internet connections. Of major economic importance are forestry, fishing, skilled manufacturing, and in Denmark and part of Sweden, very productive agriculture. Norway also has very significant petroleum reserves in the North Sea. Hydroelectricity is also a major resource in all of the Northern countries except Denmark.

- Eastern Europe is made up of Estonia, Latvia, Lithuania, Poland, the Czech Republic, Slovakia, Hungary, Romania, Bulgaria, Slovenia, Croatia, Bosnia and Herzegovina, Serbia and Montenegro, Albania, and Macedonia. These countries have been shaped powerfully by struggles between stronger countries and make up a geopolitical "shatter belt." Most critical was their role as satellite states of the Communist Soviet Union between World War II and 1990.

- Slavs are the dominant ethnolinguistic groups in Eastern Europe and include Serbs, Croats, Slovenes, Bulgarians, Macedonians, Poles, Czechs, and Slovaks. Important non-Slavic groups include Jews, Roma (Gypsies), Estonians, Latvians and Lithuanians, Hungarians (Magyars), Albanians, and Romanians.

- The environment of the Eastern countries includes the northern plain, the central mountain zone, the Danubian plains, and the southern mountain zone, with a variety of climates.

- Collectivization under Communist rule during the Soviet era from 1947 until the 1980s was the agricultural model in Eastern Europe. Privatization of farmland has taken place since the fall of communism.

- Mining, iron and steel, machinery production, construction materials, and electrical power were the highlights of the industrial effort during the Soviet years, with the majority of raw materials coming from Russia. The shift to the market economy has left many firms uncompetitive, and many have been abandoned. There has been a major effort to make the industrial base of the East more productive, including increasing the efficiency of state-run enterprises (not everything has been privatized), encouraging new private enterprises, fostering joint enterprises, eliminating price controls and bureaucratic restrictions, expanding international connections, and increasing the quality of the region's manufactured products. Both Western European countries and China have taken advantage of cheap labor in East Europe to outsource some production in the region.

- Politically, the disintegration of the former Yugoslavia was the most disabling phenomenon of the post-Soviet era in Eastern Europe. Yugoslavia has been dismantled and replaced by several countries, some with precarious balances among ethnic groups. NATO and the UN have used various military and diplomatic means to prevent further balkanization of the countries. The future of Kosovo, now in the hands of Serbia and Macedonia, is still a concern.

- Southern Europe, or Mediterranean Europe, includes Portugal, Spain, Italy, and Greece, as well as a number of major islands, island countries, and microstates, including San Marino, Vatican City, the British colony of Gibraltar, Malta, Corsica, and Cyprus. The mainland countries are parts of three peninsulas: the Iberian, Italian, and Balkan.

- The Southern region has lagged in development because of trade deficiency, low educational levels, land tenure patterns, and resource deficiencies. These traditional characteristics have changed somewhat in recent decades due to transportation changes, energy shifts, market expansion, and capital infusion.

- Southern Europe generally has a Mediterranean climate, with summer drought and winter precipitation. Irrigation has been an important factor in this region's agricultural history. Important crops are winter wheat, olives, figs, citrus, grapes and associated winemaking, and vegetables.

- The Islamic Moors and the Sephardic Jews were important cultures of the Iberian Peninsula until the Spanish ousted them in 1492. Spain and Portugal undertook crucial exploration around the world and were important colonial powers. Italy had some colonial influence in Africa and is currently the most industrial and economically significant nation of the South. The Po Valley is central to industrial northern Italy and houses two-fifths of the nation's population. There is an imbalance between northern and southern Italy, with the north far more prosperous. Some northerners want to develop an autonomous or independent region of Padania.

- As the result of a 2004 referendum in which both sides of divided Cyprus voted on whether they should be reunited, the Greek south (which rejected reunification) was allowed to join the European Union, while the Turkish north (which favored reunification) was not allowed to join.

KEY TERMS + CONCEPTS

Terms in blue are also defined in the glossary.

Padania (p. 132)
periphery (p. 105)
primate city (p. 106)
privatization (p. 119)
Reconquista (p. 128)
remittances (p. 123)
Saami (Lapps) (p. 109)
Sapmi (p. 109)
Sephardic Jews (p. 128)
shatter belt (p. 116)
socialization; socialism (p. 118)

Solidarity (p. 119)
South Slavs (p. 116)
  Bulgarians (p. 116)
  Croats (p. 116)
  Macedonians (p. 116)
  Serbs (p. 116)
  Slovenes (p. 116)
Soviet satellites (p. 113)
Technocity (p. 131)
the Inquisition (p. 128)
Third Italy (p. 131)

transhumance (p. 126)
tsunami (tidal wave) (p. 133)
Velvet Divorce (p. 122)
Velvet Revolution (p. 122)
Warsaw ghetto (p. 121)
welfare state (p. 106)
West Slavs (p. 116)
  Czechs (p. 116)
  Poles (p. 116)
  Slovaks (p. 116)

REVIEW QUESTIONS

WORLD
REGIONAL
Geography ⊛ Now ™

Assess your understanding of this chapter's topics with additional quizzing and concept-based problems at http://earthscience.brookscole.com/wrg5e.

1. What traits do the countries of Northern Europe share and which are unique?

2. Describe the climate patterns of the European North, comparing this region with North America at the same latitudes. What geographic feature is most responsible for the regional warmth of the Northern region?

3. What is the significance of the Oresund Fixed Link? How do peoples on either end use it?

4. What are the major countries and ethnolinguistic groups of Eastern Europe?

5. Why is Eastern Europe considered a "shatter belt" in geopolitical terms?

6. How did communism shape agriculture and industry in Eastern Europe? What economic processes have occurred there since 1990?

7. What are the main forces behind the breakup of Yugoslavia and the current borders and ethnic components of its successor countries?

8. What are some of the main urban and natural historical points of interest on the Danube River?

9. What are the countries, microstates, and islands of Southern Europe?

10. What are the typical components of the Mediterranean climate, agriculture, and diet? What technology is critical for successful agriculture in parts of this realm?

11. What geographic, economic, or other factors account for the locations of major cities and belts of industrial activity in Southern Europe?

12. Why has Southern Europe generally lagged behind the European core in economic development? What countries or parts of countries have run counter to this trend? How is the region as a whole faring now?

13. Of all the countries in the European periphery, which belong to the European Union and which belong to NATO? Which want membership in the European Union and which do not?

DISCUSSION QUESTIONS

1. In what ways do the Nordic countries appear to have a high quality of life? How do they achieve these standards?

2. Should Iceland, Norway, and Japan hunt whales?

3. Why isn't Norway in the European Union?

4. How does nature both endow and threaten the people of Iceland?

5. What are some of the main challenges Eastern Europe has faced in moving from Communist to capitalist economies and societies?

6. How might cheap labor in Eastern Europe lead to more rapid economic development in the region?

7. In what ways have the countries of the "New Europe" satisfied U.S. geopolitical interests?

8. How have the Baltic countries handled relations with their powerful neighbor to the east?

9. What happened to ethnic Germans in the Czech Republic and Poland after World War II? Why is their fate still an issue?

10. What fractured Yugoslavia and what is in its place? How stable are the new units?

11. What are the main tourist attractions of the European periphery? Where would future tourist development be most beneficial to the economy?

12. What are some of the legacies of ancient Greece and Rome in modern cultures?

13. Why is Gibraltar still British?

14. What are the chief natural hazards of Southern Europe?

15. What were the results of the 2004 referendum on the reunification of Greek and Turkish portions of Cyprus?

16. There has long been discussion of how Communist industries helped despoil the environments of Eastern Europe. But this chapter refers twice to the possibility that post-Communist economies in the region will transform the environment. Discuss these situations.

17. Did the terrorists who bombed trains in Madrid in 2004 achieve their goals through violence?

18. Discuss these important ethnic groups in the European periphery: Jews, Roma, Basques. How have majority people interacted with them? What are milestone events in their history? What is their status today? What entities in the region are seeking devolution of power from national governments?

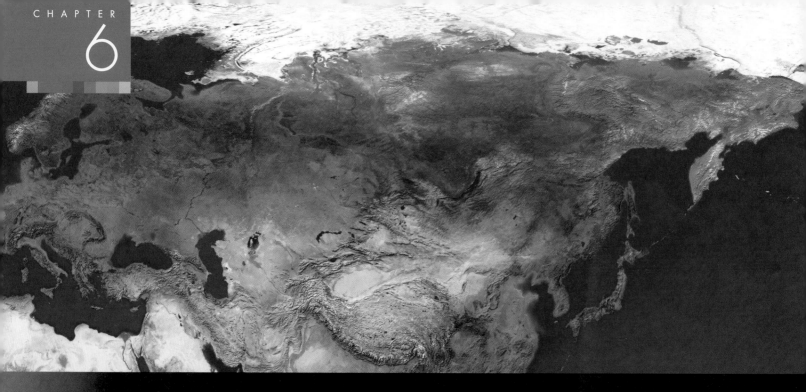

A Geographic Profile of Russia and the Near Abroad

The Gegard Monastery in Armenia was fashioned from the site's volcanic rock in the thirteenth century.

chapter objectives

This chapter should enable you to:

- Appreciate the environmental obstacles to development in vast areas of the world's largest country and nearby nations

- Become familiar with the ethnic complexity of a huge region until recently held together—against great odds—as a single country

- Learn the significant milestones in Russia's and the Soviet Union's historical-geographical development, some of them accompanied by unimaginable loss of life

- Recognize the differences between command and free market economies and the post-Soviet difficulties in shifting from one to the other

- Understand the reasons for the reversal of Russia's progress through the demographic transition

- Come to know the geopolitical and ethnic forces threatening the unity of Russia and pitting various groups and countries within and outside the region against one another

WORLD
REGIONAL
Geography ⊕ Now™

Look for this logo in the text and go to GeographyNow at http://earthscience.brookscole.com/wrg5e to explore interactive maps, view animations, sharpen your factual knowledge and geographic literacy, and test your critical thinking and analytical skills with unique interactive resources.

From 1917 to 1991, the huge region known in this text as Russia and the Near Abroad, plus the Baltic countries of Estonia, Latvia, and Lithuania, made up a single Communist-controlled country called (after 1922) the Union of Soviet Socialist Republics (USSR). The Russian-dominated government in Moscow controlled the affairs of many non-Russian peoples in the country and after World War II also effectively controlled its Communist **satellite countries** in Eastern Europe. This vast empire—dubbed in 1983 by U.S. President Ronald Reagan as the **"Evil Empire"**—engaged with the United States in a worldwide contest for political and economic supremacy. For four decades, that Cold War between the Soviet bloc of nations and the Western bloc led by the United States dominated world politics. The perspectives and actions of the Soviet Union had major impacts on world events.

Suddenly, in 1991, the country split into 15 independent nations (see Table 6.1 and Definitions and Insights, page 141). Russia—officially known as the Russian Federation—remained by far the largest in area, population, and political and economic influence. From the Russian perspective, the remaining 14 countries comprised the "Near Abroad," in which Russia's special interests and influence should be exerted and preserved. By 2004, Russian hopes to keep a dominant position in the Baltic region were dashed when Estonia, Latvia, and Lithuania joined the European Union.

What remained was a vast geographic complex consisting of Russia; the more Russia-like and mostly or partially Slavic countries of Ukraine, Belarus, and Moldova; the decidedly non-Russian but heavily Russian-influenced five "stan" countries of Central Asia: Kazakhstan, Uzbekistan, Turkmenistan, Kyrgyzstan, and Tajikistan; and the fractious countries of Georgia, Armenia, and Azerbaijan in the Caucasus. These 12 countries, designated in this book as Russia and the Near Abroad, are (as of 2005) also the 12 member states of the Commonwealth of Independent States (CIS), which is essentially an economic rather than political association.

Geographers are not having an easy time dealing with the successor countries of the Soviet Union. As discussed in Chapter 1, regions are organizing tools, not facts on the ground. There is no acknowledged best way to classify this region. Some geographers have chosen to cleave the Central Asian "stans" from Russia, establishing them as an entirely separate region, or include them as part of a greater Middle Eastern region. Some still consider the Baltics part of the greater Russian realm. The variety of geographers' names for the region suggests how unsettled the terminology is: "Post-Soviet Region," "Russia and Its Neighbors," "Russia and the Newly Independent States," "Russia and Neighboring Countries," "The Commonwealth of Independent States," and our own "Russia and the Near Abroad" are among them.

TABLE 6.1 Russia and the Near Abroad: Basic Data

Political Unit	Area (thousand/ sq mi)	Area (thousand/ sq km)	Estimated Population (millions)	Annual Rate of Natural Increase (%)	Estimated Population Density (sq mi)	Estimated Population Density (sq km)	Human Develop-ment Index	Urban Popula-tion (%)	Arable Land (%)	Per Capita GDP PPP ($US)
Slavic States and Moldova										
Belarus	80.2	207.7	9.8	−0.6	122	47	0.790	72	29	6100
Moldova	13	33.6	4.2	−0.1	323	125	0.681	45	55	1800
Russia	6592.8	17,075.3	144.1	−0.6	22	8	0.795	73	7	8900
Ukraine	233.1	603.7	47.4	−0.8	203	79	0.777	68	56	5400
Total	**6919.1**	**17,920.4**	**205.5**	**−0.6**	**29**	**11**	**0.788**	**72**	**9**	**7814**
Caucasus Region										
Armenia	11.5	29.8	3.2	0.2	278	107	0.754	64	17	3500
Azerbaijan	33.4	86.5	8.3	0.8	249	96	0.746	51	19	3400
Georgia	26.9	69.6	4.5	0.0	167	65	0.739	52	11	2500
Total	**71.8**	**185.9**	**16**	**0.4**	**222**	**86**	**0.745**	**53**	**15**	**3166**
Central Asia										
Kazakhstan	1049.2	2717.4	15	0.6	14	6	0.766	57	8	6300
Kyrgyzstan	76.6	198.4	5.1	1.4	67	26	0.701	35	7	1600
Tajikistan	55.3	143.2	6.6	1.9	119	46	0.671	27	6	1000
Turkmenistan	188.5	488.2	5.7	1.6	30	12	0.752	47	3	5800
Uzbekistan	172.7	447.3	26.4	1.6	153	59	0.709	37	11	1700
Total	**1542.3**	**3994.5**	**58.8**	**1.3**	**38**	**15**	**0.722**	**41**	**7**	**3183**
Summary Total	**8533.2**	**22,100.8**	**280.3**	**−0.1**	**32**	**12**	**0.771**	**64**	**8**	**6577**

Sources: *World Population Data Sheet*, Population Reference Bureau, 2004; *Human Development Report*, United Nations, 2004; *World Factbook*, CIA, 2004.

WORLD REGIONAL
Geography ⊛ Now™

Click Geography Literacy to make a map showing economic and cultural regions.

Russia and the Near Abroad is preferred here not to impose a Russian-centered view of the region, as the name implies, but because of the enormous power or hoped-for power that Russia wields there. The former Soviet countries have lasting and often uncomfortable strategic and economic associations with Russia. For example, the Soviet-era oil refinery and pipeline system still links Russia with Kazakhstan, Azerbaijan, Ukraine, Belarus, and other countries. Since the breakup of the Soviet Union, Russia has periodically shut off the pipelines supplying natural gas to obtain political concessions from some of these nations. The needs of many of the Near Abroad countries to buy and sell oil provide an incentive to hold themselves together in some kind of economic or political union with Russia.

Whatever forces remain favoring integration around Russia, further political fragmentation and decentralization are the dominant political themes in this region. Without the USSR's Red Army to impose order, long-simmering ethnic conflicts have boiled over. The Russian government in Moscow will try vigorously to maintain influence in the former republics. Some analysts even fear that Russia may attempt to reexert control over some of the recently independent nations, especially those seen as vital to its economic and political security. The future of this huge territory is uncertain. This chapter is an introduction to a major world region in which momentous changes are underway.

6.1 Area and Population

With an area (including inland waters) of 8.5 million square miles (22.1 million sq km), the region of Russia and the Near Abroad is the largest of the world regions recognized in this text. A good indication of its staggering size is the fact that this region (even Russia alone) spans eleven time zones; by

Definitions + Insights

Regional Names of Russia and the Near Abroad

Here are some useful geographic terms to know when studying this region. The region of Russia and the Near Abroad consists of what used to be the Union of Soviet Socialist Republics, also known as the Soviet Union or USSR, minus the Baltic countries of Estonia, Latvia, and Lithuania. The USSR came into existence in 1922 following the overthrow of the last Romanov tsar in the Russian Revolution of 1917 and the subsequent civil war. Prerevolutionary Russia is known as Old Russia, Tsarist Russia, Imperial Russia, or the Russian Empire.

The name Russia now refers to the independent country of Russia. It is known politically as the Russian Federation, which was the largest of the 15 Soviet Socialist Republics (Union Republics, or SSRs) that made up the Soviet Union. The full name of the Russian Federation during the Communist period was the Russian Soviet Federated Socialist Republic, or RSFSR; the name appears on many older maps.

The loosely aligned Commonwealth of Independent States (CIS) was formed by 12 of the 15 former Union Republics late in 1991. Estonia, Latvia, and Lithuania are not members of this organization.

The area west of the Ural Mountains and north of the Caucasus Mountains has been known historically as European Russia. The Caucasus and the area east of the Urals have been called Asiatic Russia, Soviet Asia, or the eastern regions. Here, Transcaucasia is the region of the mountains plus the area south of the Caucasus Mountains; Siberia is a general name for the area between the Urals and the Pacific; and Central Asia is the arid area occupied by the five countries with large Muslim populations immediately east and north of the Caspian Sea. They were known collectively in pre-Soviet times as "Turkestan" and are known informally today as "the stans" because all the country names end in *-stan*, Persian for "land of."

comparison, the United States from Maine to Hawaii spans seven. Russia is about 1.8 times the size of the United States, including Alaska.

The eventful geopolitical history of Russia and the Near Abroad has given this region land frontiers with 15 countries in Eurasia (Figure 6.1). Between the Black Sea and the Pacific, the region borders Turkey, Iran, Afghanistan, China, Mongolia, and North Korea. Pakistan and India also lie very close by. In the Pacific Ocean, narrow water passages separate the Russian-held islands of Sakhalin and the Kurils from Japan. In the west, the region has frontiers with Romania, Hungary, Slovakia, Poland, Lithuania, Latvia, Estonia, Finland, and Norway.

So much of this vast region is sparsely populated that its estimated population of 280 million in 2004 ranked it 7th among the eight world regions (Figure 6.2). The average population density of 32 per square mile (12/sq km) is less than half that of the United States. Great stretches of economically unproductive terrain separate many outlying populated areas from one another. The mountainous frontiers in Asia have especially few inhabitants. In contrast, on the frontier between the Black and Baltic Seas, international boundaries pass through populous lowlands that have long been disputed territory between Russia and other countries.

Russia has by far the largest population, about 144 million. The next largest country is Ukraine, with about 47 million people. Uzbekistan is the most populous Central Asian nation, with about 26 million people. All the other countries in the region trail far behind, each with populations gener-

ally under 15 million. Population growth rates are highest, around 1.5 percent annually, among the predominantly Muslim populations of the Central Asian countries. At the other end of the spectrum, Russia, Ukraine, and Belarus are losing populations at a rate of about 0.5 to 1 percent per year. This trend is due to the plummeting quality of life since the breakup in the Soviet Union in 1991 (see Definitions and Insights, page 144). In an atmosphere of economic and political uncertainty, couples have chosen to have fewer children, and there are more broken marriages. More disturbingly, death rates have soared due to declining healthcare, increasing alcoholism, violence, suicide, and more frequent botched abortions.

6.2 Physical Geography and Human Adaptations

Stretching nearly halfway around the globe in northern Eurasia, the region of Russia and the Near Abroad has formidable problems associated with climate, terrain, and distance. Most of the region is burdened economically by cold temperatures, infertile soils, marshy terrain, aridity, and ruggedness. Natural conditions are more similar to those of Canada than to those of the United States. Interaction with a complex and demanding environment was a major theme in Russian and Soviet expansion and development. Nature continues to provide large assets but also poses great problems for these 12 countries.

Political Geography of Russia and the Near Abroad

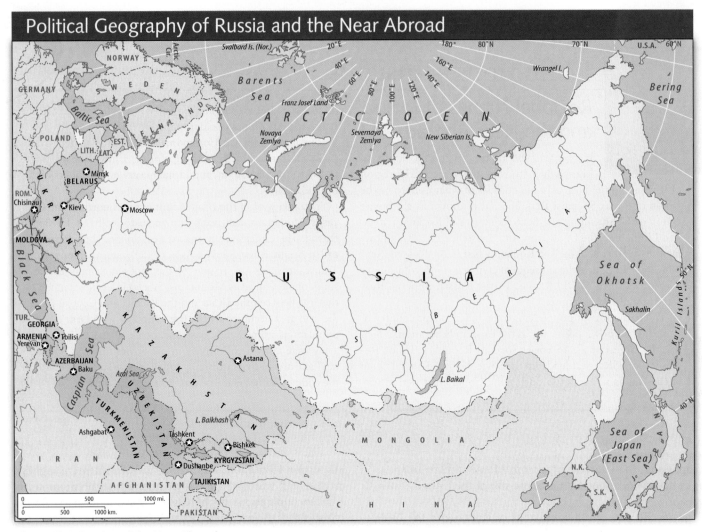

Figure 6.1 Russia and the Near Abroad

Population Distribution

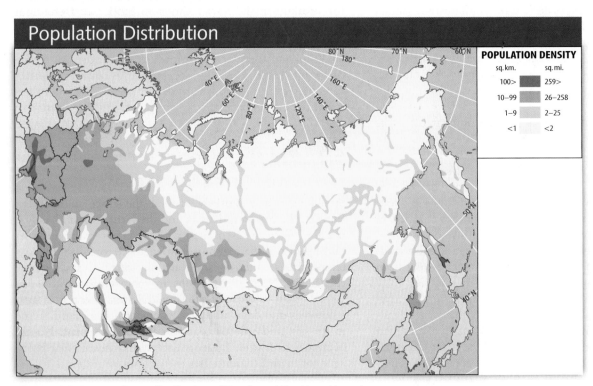

POPULATION DENSITY

sq. km.		sq. mi.
100>		259>
10–99		26–258
1–9		2–25
<1		<2

Figure 6.2 (a) Population distribution of Russia and the Near Abroad

The Roles of the Climates and Vegetation

The region of Russia and the Near Abroad has a harsh climatic setting. Severe winter cold, short growing seasons, drought, and hot, crop-shriveling winds are major disadvantages. But there are also advantages in the form of good soils, the world's largest forests, natural pastures for livestock, and diverse wild fauna. However, many of these resources are hard to access because of the logistical difficulties posed by an unfavorable climate and the vast distances involved.

Most parts of the region have continental climatic influences, with long cold winters, short warm summers, and low to moderate precipitation. Severe winters, for which Siberia (roughly the eastern two-thirds of Russia) is particularly infamous, prevail because the region is generally at a high latitude and has few of the moderating influences of oceans. Four-fifths of the total area is farther north than any point in the conterminous United States (Figure 6.3). The most extreme continental climate of the world is here; the lowest official temperature ever recorded in the Northern Hemisphere (−90°F; −68°C) occurred in the Siberian settlement of Verkhoyansk (67°N, 135°E). Westerly winds from the Atlantic moderate the winter temperatures somewhat in the west, but their effects become steadily weaker toward the east, where it is much colder.

The average frost-free season of 150 days or less in most areas (except the extreme south and west) is too short for many crops to mature. Most places are relatively warm during the brief summer, and the southern steppes and deserts are hot. Partially offsetting the summer's shortness are the long summer days that encourage plant growth. This is also a time of accelerated human activity outdoors. Russians are passionate campers and sun-worshippers, and at every opportunity, they get away from it all to enjoy the fleeting pleasures of summer. Urban residents never lose the traditional Russian passion for nature, and most retreat on weekends and longer holidays to rural cabins called *dachas*.

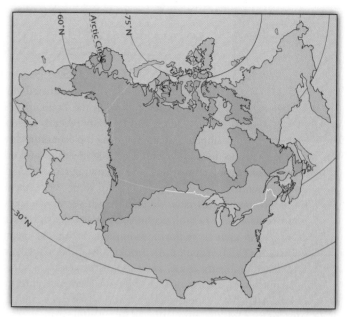

Figure 6.3 Russia and the Near Abroad compared in latitude and area with the continental United States and Canada

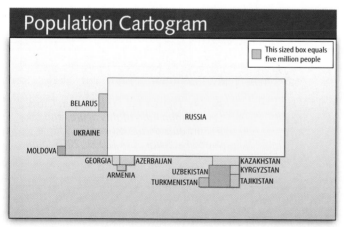

Population Cartogram

This sized box equals five million people

BELARUS

RUSSIA

UKRAINE

MOLDOVA

GEORGIA | AZERBAIJAN | KAZAKHSTAN

ARMENIA | UZBEKISTAN | KYRGYZSTAN

TURKMENISTAN | TAJIKISTAN

Figure 6.2 *(continued)* (b) Population cartogram of Russia and the Near Abroad

Aridity and drought are additional problems for agriculture. The annual average precipitation is less than 20 inches (50 cm) nearly everywhere except in the extreme west, along the eastern coast of the Black Sea and the Pacific coast north of Vladivostok, and in some of the higher mountains.

There are five main climatic belts in the region of Russia and the Near Abroad: tundra, subarctic, humid continental, steppe, and desert. In this order, these belts, each with its associated vegetation and soils, succeed the other from north to south (Figure 6.4). There are also smaller scattered areas of subtropical, Mediterranean, and undifferentiated highland climates.

Huge areas in the tundra and subarctic climatic zones are permanently frozen at a depth of a few feet. The frozen ground, called permafrost, makes construction difficult. Heat generated by buildings melts the upper layers of permafrost and causes foundations and walls to sink and tilt. Most buildings are elevated on pilings to reduce this risk (Figure 6.5). Pipelines carrying crude oil (which is hot when it comes from the ground) would likewise melt the permafrost and sink, and so they are heavily insulated or built on elevated supports.

The most productive human activities take place in the subarctic, humid continental, and steppe climatic zones (Figure 6.6). The subarctic climate zone corresponds largely with the Russian taiga, or northern coniferous forest, the largest continuous area of forest on Earth. The main trees are spruce, fir, larch, and pine, which are useful for pulpwood and firewood but often do not make good lumber. Large reserves of timber suitable for lumber do exist in parts of the taiga, and this is one of the major areas of lumbering in the world. Many observers fear that Russia, in its drive to

Definitions + Insights

The Russian Cross

Russia's economic health plummeted for at least a decade following the breakup of the Soviet Union and has only recently begun to improve. Accompanying the economic downturn were rising unemployment, a decline in health and other services, increasing crime rates, and a growing sense of despair at the individual level. These gave rise to some very unhealthful habits among Russians, which have not yet shown signs of improvement. The rate of alcoholism has soared; the national average for annual vodka consumption is an eye-blearing 4.4 gallons (17 l) (Figure 6.A). Smoking is a national pastime. Organized and petty crime, and simple drunken brawls, result in an alarmingly high incidence of physical violence. Infections of HIV/AIDS have increased more than 10-fold since 2000, first surging through the population of intravenous drug users and then into heterosexual partners.

These and other factors have led to a drop in Russians' life expectancy, from 68 in 1990 (average for men and women) to 65 in 2004, the same as in the developing nations of Guatemala and Mongolia. Between 1990 and 2001, the death rate rose almost one-third, to one of the world's highest levels (17 per 1,000), surpassed only by many countries in Africa, and out-side Africa only in Afghanistan. The leading causes of death, in descending order, are heart disease, accidents, violence, and cancer. The universal system of healthcare under the Soviets was not efficient, but a healthcare system close to collapse has replaced it, and hospitals are ill-equipped to help stem the rising tide of deaths.

This obviously is not the benign demographic transition, in which a country's population declines because of falling birth rates that reflect growing affluence. Instead, Russia is moving backward through the transition. As quality of life has deteriorated, death rates have soared and birth rates have plummeted. The birth rate in 2004 was 10 per 1,000, down 17 percent from 12 per 1,000 in 1990, a trend aided by routine abortion: In 2004, there were two abortions for every live birth. The disturbing trend lines of growing death rates and falling birth rates in Russia intersected in the mid-1990s, forming the grim graphic that demographers dubbed the "**Russian cross**" (Figure 6.B). It is an ominous intersection. Russia's rate of annual population loss of 700,000 (as of 2004) is the highest in the world and has seldom been seen on Earth except in times of warfare, famine, and epidemic. One United Nations estimate puts Russia's population at 104 million in 2050, down 28 percent from its 2004 figure of 144 million.

44

Figure 6.A Alcoholism is a major killer in Russia.

Joe Hobbs

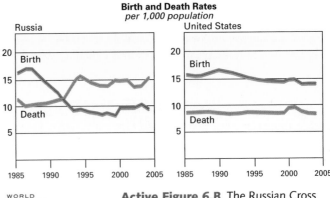

Active Figure 6.B The Russian Cross
See an animation of this figure, and take a short quiz on the concept. Source: World Factbook, *CIA, 2002, 2003, and 2004.*

WORLD REGIONAL
Geography⊛Now™

Climates

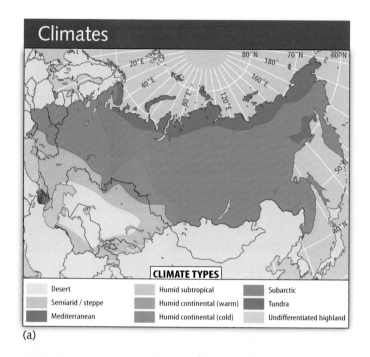

CLIMATE TYPES

Desert	Humid subtropical	Subarctic
Semiarid / steppe	Humid continental (warm)	Tundra
Mediterranean	Humid continental (cold)	Undifferentiated highland

(a)

Joe Hobbs

Figure 6.5 In the High Arctic, buildings must be erected on pilings so that they do not melt the permafrost below. This is the Russian coal-mining settlement of Barentsburg in Norway's Svalbard (Spitsbergen) Archipelago.

Biomes

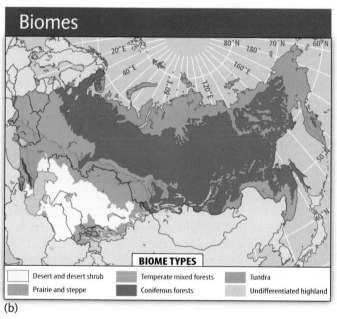

BIOME TYPES

Desert and desert shrub	Temperate mixed forests	Tundra
Prairie and steppe	Coniferous forests	Undifferentiated highland

(b)

Figure 6.4 (a) Climates and (b) biomes of Russia and the Near Abroad

Agricultural Land Use

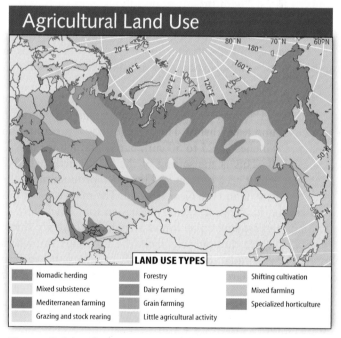

LAND USE TYPES

Nomadic herding	Forestry	Shifting cultivation
Mixed subsistence	Dairy farming	Mixed farming
Mediterranean farming	Grain farming	Specialized horticulture
Grazing and stock rearing	Little agricultural activity	

Figure 6.6 Land use in Russia and the Near Abroad

advance the economy, will deplete the taiga (Figure 6.7). Russia's conifers are believed to act as major "carbon sinks" that help absorb human-made greenhouse gases, so their removal could contribute to global warming. There is some marginal farming in the taiga, especially toward the south. The dominant soils are the acidic spodosols (from the Greek word for "wood ash" and known in Russian as *podzols*), which have a grayish, bleached appearance when plowed, lack well-decomposed organic matter, and are low in natural fertility.

A humid continental climate occupies a triangular area south of the subarctic climate, narrowing eastward from the region's western border to the vicinity of Novosibirsk. This area has the short-summer ("cold") subtype of humid continental climate, comparable to that of the Great Lakes region and the northern Great Plains of the United States and

Figure 6.7 The taiga in Russia is Earth's largest continuous forest biome, and Russia's economic growth is based in part on its exploitation. These milled conifers are awaiting shipment from St. Petersburg's docks.

Figure 6.8 A grazier is with his cattle on the steppe near the Don River in southern Russia.

adjacent parts of Canada. Here, evergreen trees occur in a "mixed forest" with broadleaf deciduous trees such as oak, ash, maple, and elm. Both climate and soil are more favorable for agriculture in the humid continental climate than in the subarctic. Soils developed under broadleaf deciduous or mixed forest are normally more fertile than spodosols, although less so than grassland soils.

The steppe climate characterizes the grassy plains south of the forest in Russia and is the main climate of Ukraine, Moldova, and Kazakhstan. The average annual precipitation is 10 to 20 inches (c. 25 to 50 cm), barely enough for unirrigated crops. This is the setting of the black-earth belt, the most important area of crop and livestock production in the region of Russia and the Near Abroad. The main soils of this belt are the chernozem, which is Russian for "black earth." Among the best soils to be found anywhere, these are thick, productive, and durable soils from the soil order known as mollisols. A similar belt of mollisols occupies the eastern portion of the Great Plains of North America. Their great fertility is due to an abundance of humus in the topsoil. The major natural threat to productive agriculture here is the occurrence of severe droughts and hot, desiccating winds (*sukhovey*) that damage or destroy crops. The steppe also includes extensive areas of chestnut or alfisol soils in the zones of lighter rainfall. These are lighter in color than chernozems and lack their superb fertility but are among the world's better soils.

The most characteristic natural vegetation of the steppe is short grass. Pastoralists, including the Scythians who in the fourth century B.C. produced astonishing artwork of gold in the area that is now Ukraine, grazed their herds from an early time on the treeless steppe grasslands, which stretched over a vast area between the forest and the southern mountains and deserts. Today, much of this area is culti-

vated, with wheat as the main crop. The steppe is also a major producer of sugar beets, sunflowers grown for vegetable oil, various other crops, and all the major types of livestock (Figure 6.8).

Growing crops is difficult in the desert climate areas east and just north of the Caspian Sea in the Central Asian countries. Cotton is the most important crop where irrigation water is available from the Syr Darya, Amu Darya, and other rivers flowing from high mountains to the south and east.

The Role of Rivers

In the early history of Russia, rivers formed natural passageways for trade, conquest, and colonization. They were especially crucial in the settlement of Siberia, which is drained by some of the greatest rivers on Earth: the Ob, Yenisey, Lena, and Kolyma, all flowing northward to the Arctic Ocean, and the Amur, flowing eastward to the Pacific. By following these rivers and their lateral tributaries, the Russians advanced from the Urals to the Pacific in less than a century.

The Moscow region lies on a low upland from which a number of large rivers radiate like the spokes of a wheel (Figure 6.9). The longest ones lead southward: the Volga to the landlocked Caspian Sea, the Dnieper (Dnepr) to the Black Sea, and the Don to the Sea of Azov, which connects with the Black Sea through a narrow strait (see The World's Great Rivers, page 148). Shorter rivers lead north and northwest to the Arctic Ocean and Baltic Sea. These river systems are accessible to each other by portages.

A major link in the inland waterway system is the Volga-Don Canal, opened in 1952 to tie together the two rivers where they approach each other in the vicinity of Volgograd (Figure 6.10). The completion of the canal meant that the White Sea and Baltic Sea in the north were linked to the Black Sea and Caspian Sea in the south in a single water transport

Physical Geography

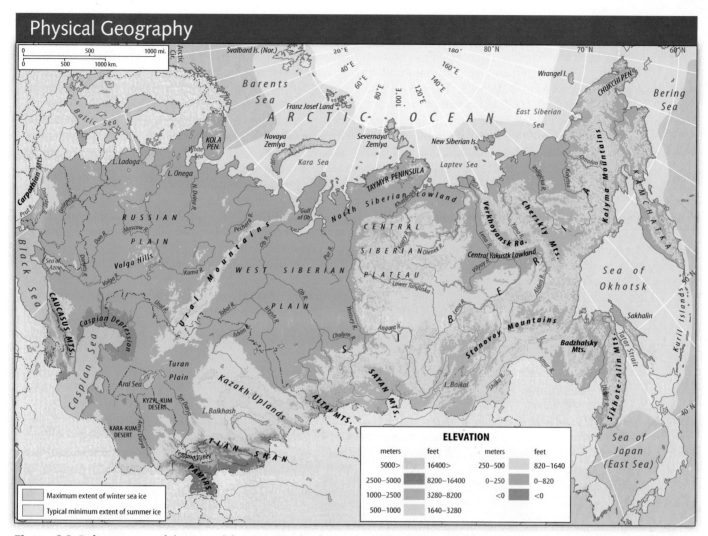

Figure 6.9 Reference map of the rivers, lakes, topographic features, and seas of Russia and the Near Abroad

Figure 6.10 A lock in the Volga-Don Canal, a vital link for trade between the Volga watershed and the Black Sea

The Volga

The Volga is arguably this region's most important river; Russians call it *Matushka*—"Mother." It rises about 200 miles (320 km) southwest of St. Petersburg and flows 2,300 miles (3,680 km) to the Caspian Sea. Before railroads supplemented river traffic, the Volga was Russia's premier commercial artery. Traditionally, wheat, coal, and pig iron from Ukraine, fish from the Caspian Sea, salt from the lower Volga, and oil from Baku, on the western shore of the Caspian, traveled upriver toward Moscow and the Urals. Timber and finished products moved downriver to the lower Volga and the Ukraine (Figure 6.C). Boatmen towed barges upstream on a 70-day journey from Astrakhan, near the Volga mouth, to Kazan, on the middle Volga. The steamboat arrived in the late 1800s, bringing an end to the way of life recalled in the famous Russian song "The Volga Boatmen."

The Volga was difficult to navigate until the latter half of the 20th century. There were shoals and shallows, especially in

very dry summers, and ice still closes the waterway each winter for 120 to 160 days. During the Soviet era, the **Great Volga Scheme** transformed the river. The goal was to control completely the flow of the river with a stairway of huge reservoirs, each of which reaches upstream to the dam forming the next reservoir, thus assuring complete navigability during the 6 months when the river is ice-free and supplying hydroelectric power and water for irrigation. Behind the dams that are the backbone of the Great Volga Scheme, the reservoirs are so vast that they are generally called "seas."

With these improvements in navigation, the Volga secured its place as Russia's most important internal waterway. Large numbers of barges and log rafts, as well as a fleet of passenger vessels that carry tourists, use it. Alteration of rivers for power, navigation, and irrigation became an important component of economic development under the Soviets. Large dams were also raised across the main courses or major tributaries of the Dnieper, Don, Kama, Irtysh, Ob, Yenisey, and Angara Rivers.

Joe Hobbs

Figure 6.C Timber moving along the Volga near Saratov. Russian literary depictions of life along the river are sometimes reminiscent of Mark Twain's Mississippi. Down along the Volga wharves at Kazan, Maxim Gorki found "a whirling world where men's instincts were coarse and their greed was naked and unashamed."

system. The accomplishment was no small task. The Don River, which empties into the Black Sea via the Sea of Azov, is about 150 feet (45 m) higher than the Volga. The ground that separates these two rivers rises to almost 300 feet (90 m). Engineers solved this discrepancy with 13 locks, each with a lift of about 30 feet (9 m), which carry water 145 feet (44 m) above the Don and drop it 290 feet (87 m) to the Volga.

The Role of Topography

Most of the important rivers of Russia and the Near Abroad wind slowly for hundreds or thousands of miles across large plains. Such plains, including low hills, compose nearly all the terrain from the Yenisey River to the western border of the region. The only mountains in this lowland are the

Urals, a low and narrow range (average elevation: less than 2,000 ft/610 m) that separates Europe from Asia and European Russia from Siberia. The Urals trend almost due north–south but do not occupy the full width of the lowland. A wide lowland gap between the southern end of the mountains and the Caspian Sea permits uninterrupted east–west movement by land. Cut by river valleys offering easy passageways, the Urals are not a serious barrier to transportation.

Between the Urals and the Yenisey River, the West Siberian Plain is one of the flattest areas on Earth. Immense wetlands, through which the Ob River and its tributaries slowly wind their way, cover much of this vast flatland. This waterlogged country, underlain by permafrost that blocks downward seepage of water, is a major barrier to land transport and is quite uninviting for settlement. Tremendous floods occur in the spring when the breakup of ice in the upper basin of the Ob releases great quantities of water, while the river channels farther north are still frozen and act as natural dams. Russian aircraft sometime bomb ice jams on the rivers to prevent worse flooding.

The area between the Yenisey and Lena Rivers is occupied by the hilly Central Siberian Uplands, which have a general elevation of 1,000 to 1,500 feet (c. 300 to 450 m). Mountains dominate the landscape east of the Lena River and Lake Baikal. Extreme northeastern Siberia is an especially bleak and difficult country for human settlement. High mountains rim the region of Russia and the Near Abroad on the south from the Black Sea to Lake Baikal, and lower mountains rim the region from Lake Baikal to the Pacific. Peaks rise to over 15,000 feet (c. 4,500 m) in the Caucasus Mountains between the Black and Caspian Seas and in the Pamir, Tien Shan, and Altai Mountains east of the Caspian Sea. From the feet of the lofty Pamir, Tien Shan, and Altai ranges, and the lower ranges between the Pamirs and the Caspian Sea, arid and semiarid plains and low uplands extend northward and gradually merge with the West Siberian Plain and the broad plains and low hills west of the Urals.

6.3 Cultural and Historical Geographies

A Babel of Languages

Russia and the Near Abroad comprise a complex cultural and linguistic mosaic. The region's peoples belong to about 30 major ethnic groups and speak more than 100 languages (Figure 6.11). In Russia, Belarus, and Ukraine, the majority are Slavs who originated in east central Europe as speakers of an ancestral **Slavic** language (a member of the Indo-European language family) and who now speak **Russian, Belarusian, and Ukrainian,** respectively. Along the upper Volga River, in the small regions of Tatarstan, Chuvashiya, and Mariy-El, there are majority speakers of **Finno-Ugric** (in the **Uralic** family) and **Turkic** languages. Yet another outpost of Turkic speakers is Yakutia (the Sakha Republic) in northeastern Russia. In Moldova, the majority ethnic Moldovans speak **Moldovian,** a dialect of the Romance language Romanian.

The Caucasus is extremely diverse in its ethnicity and languages and has long been a magnet for linguistic research. The **Armenians** have a unique language within the Indo-European group. Their neighbors in the Caucasus—the Georgians and Azerbaijanis—speak, respectively, a **Kartvelian (South Caucasian)** and a Turkic language in the **Altaic** family. Most smaller ethnic groups in the Caucasus speak languages unrelated to Georgian, belonging to the **Abkhaz-Adyghean** and **Nakh-Dagestanian** families.

The "stan" countries are less complex, linguistically. Turkic languages, within the Altaic language family, are spoken by most of the ethnic groups in the Central Asian countries of Turkmenistan, Kazakhstan, Uzbekistan, and Kyrgyzstan. In Tajikistan, however, the dominant group of **Tajiks** speaks an **Iranian** language within the **Indo-European** family. In extreme northeastern Russia, there are speakers of the **Chukotko-Kamchatkan** languages, including **Chukchi** and **Koryak,** belonging to the **Proto-Asiatic** language family and related to some of the languages that ancestors of Native Americans carried eastward into the Americas.

This is just a brief glimpse of the region's tongues; there are many smaller languages. As will be seen in this and the next chapter, this rich multiethnicity has at once bestowed great cultural wealth on the region and threatened to tear apart the nations within it.

The cultural histories of the minority ethnic peoples have long been influenced and dominated by the Russians, whose origins reach more than 1,000 years into the past. The central figures were Slavic peoples who colonized Russia from the west, interacted with many other peoples, stood off or outlasted invaders, and acquired the giant territory that would become Russia and then the Soviet Union.

Vikings, Byzantines, and Tatars

Slavic peoples have inhabited Russia since the early centuries of the Christian era. During the Middle Ages, Slavic tribes living in the forested regions of western Russia came under the influence of Viking adventurers from Scandinavia known as **Rus** or **Varangians.** The newcomers carved out trade routes, planted settlements, and organized principalities along rivers and portages connecting the Baltic and Black Seas. In the ninth century, the principality of Kiev, ruled by a mixed Scandinavian and Slavic nobility, achieved mastery over the others and became a powerful state. The culture it developed was the foundation on which the Russian, Ukrainian, and Belarusian cultures later arose.

Contacts with Constantinople (now Istanbul, Turkey) greatly affected Kievan Russia. Located on the straits connecting the Black Sea with the Mediterranean, Constantinople was the capital of the Eastern Roman or Byzantine Empire, which endured for nearly 1,000 years after the collapse of the Western Roman Empire in the fifth century A.D. Constantinople became an important magnet for Russian trade, and the Russians borrowed heavily from its culture. In 988, the ruler of Kiev, Grand Duke Vladimir I, formally accepted the Christian faith from the Byzantines and had his subjects baptized. Following its cleavage from the Roman Catholic faith in A.D. 1054, Orthodox Christianity became a permanent feature of Russian life and culture (see Figure 6.12 for the map of religions). Moscow eventually came to be known as the "**Third Rome**" (after Rome itself and Constantinople) for its importance in Christian affairs.

The Bolshevik Revolution in 1917 began a 75-year period of official repression and neglect of the Orthodox Church and other religions. Since the breakup of the Soviet Union, however, there has been a renaissance in religious observance—

Languages

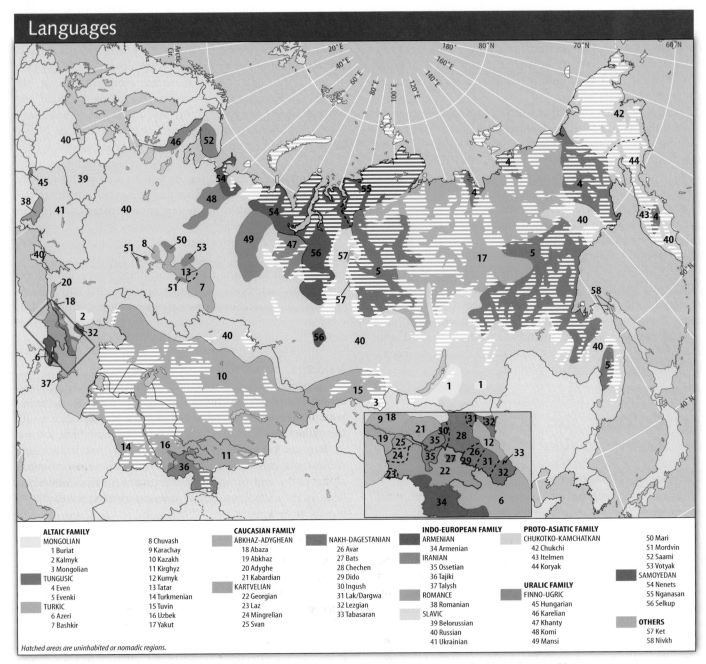

ALTAIC FAMILY
MONGOLIAN
  1 Buriat
  2 Kalmyk
  3 Mongolian
TUNGUSIC
  4 Even
  5 Evenki
TURKIC
  6 Azeri
  7 Bashkir

8 Chuvash
9 Karachay
10 Kazakh
11 Kirghyz
12 Kumyk
13 Tatar
14 Turkmenian
15 Tuvin
16 Uzbek
17 Yakut

CAUCASIAN FAMILY
ABKHAZ-ADYGHEAN
  18 Abaza
  19 Abkhaz
  20 Adyghe
  21 Kabardian
KARTVELIAN
  22 Georgian
  23 Laz
  24 Mingrelian
  25 Svan

NAKH-DAGESTANIAN
26 Avar
27 Bats
28 Chechen
29 Dido
30 Ingush
31 Lak/Dargwa
32 Lezgian
33 Tabasaran

INDO-EUROPEAN FAMILY
ARMENIAN
  34 Armenian
IRANIAN
  35 Ossetian
  36 Tajiki
  37 Talysh
ROMANCE
  38 Romanian
SLAVIC
  39 Belorussian
  40 Russian
  41 Ukrainian

PROTO-ASIATIC FAMILY
CHUKOTKO-KAMCHATKAN
  42 Chukchi
  43 Itelmen
  44 Koryak

URALIC FAMILY
FINNO-UGRIC
  45 Hungarian
  46 Karelian
  47 Khanty
  48 Komi
  49 Mansi

50 Mari
51 Mordvin
52 Saami
53 Votyak
SAMOYEDAN
54 Nenets
55 Nganasan
56 Selkup

OTHERS
57 Ket
58 Nivkh

Hatched areas are uninhabited or nomadic regions.

Figure 6.11 Ethnolinguistic distributions in Russia and the Near Abroad, where there is a close correlation of languages and the ethnic groups using them. The Soviet Union had difficulties trying to hold together such a vast collection of culture groups. The Russian Federation faces many of the same challenges.

not just among Christians but also Muslims, Buddhists, and others—across the vast region of Russia and the Near Abroad (Figure 6.13). There has even been a small reversal in the emigration of **Jews** to Israel and the West. Ten million Jews lived in the region before the Nazi Holocaust killed three million of them, and perhaps another million assimilated to escape official Soviet anti-Semitism. In the decade following the breakup of the USSR, half of Russia's Jews fled the country. Some are coming back to the somewhat more tolerant Russia of today.

Yet another cultural influence reached the Russians from the heart of Asia. The steppe grasslands of southern Russia had long been the habitat of nomadic horsemen of Asian origin. During the later days of the Roman Empire and in the Middle Ages, these grassy plains, stretching far into Asia, provided the Huns, Bulgars, and other nomads a passageway into Europe. In the 13th century, the **Tatars** (also known as Tartars) of Central Asia took this route. Many steppe and desert peoples of Turkic origins were in their ranks, but **Mongols** led them. In 1237, Batu Khan ("Batu the Splen-

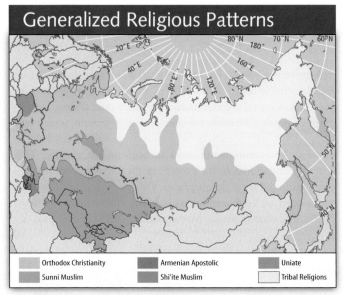

Figure 6.12 Religions of Russia and the Near Abroad. The Uniate Church of northwestern Ukraine is an Eastern Catholic Church that accepts the Catholic dogma and the primacy of the Pope in Rome but is not subject directly to the Pope's control. Sunni and Shi'ite Islam are described on pages 216–218 in Chapter 8, and the Armenian Church is discussed in Chapter 7, page 190.

Figure 6.13 Some of the faithful in a Russian Orthodox church in Vyborg, Russia. Since the breakup of the USSR, there has been a resurgence of organized worship throughout the region.

did"), grandson of Genghis Khan, launched a devastating invasion that brought all the Russian principalities except the northern one of Novgorod under Tatar rule.

The Tatars collected taxes and tributes from the Russian principalities but generally allowed the rulers of these units to be autonomous. Even the princes of Novgorod, who were not technically under Tatar control, paid tribute to avoid trouble. The Tatars established the Khanates (governmental units) of Kazan (on the upper Volga), Astrakhan (on the lower Volga), and Crimea (on the Black Sea). Russians called the Kazan Tatars **"the Golden Horde,"** after the brightly colored tents in which they lived. When Tatar power declined in the 15th century, the rulers of the Moscow principality were able to begin a process of territorial expansion that resulted in the formation of present-day Russia. Today, oil-rich Tatarstan, with its capital at Kazan, is one of Russia's most important "autonomous" political units. Its distinct ethnic Tatar population has proved remarkably resilient to centuries of Russian supremacy.

The Empire of the Russians

The Russian monarchy reached outward from Muscovy, its original domain in the Moscow region. From the 15th century until the 20th, the tsars created an immense Russian empire by building onto this core. This imperialism by land, in the era when the maritime powers of Western Europe were expanding by sea, brought a host of alien peoples under tsarist control (see Definitions and Insights, page 152). Russian motivations for expansion were diverse. Quelling raids

by troublesome neighbors, particularly nomadic Muslim peoples in the southern steppes and deserts, was one objective. Many Russian cities, such as Volgograd (originally named Tsaritsyn and then Stalingrad) on the Volga River, were founded as fortified outposts on the steppe frontier. In the wilderness of Siberia, the search for valuable furs and minerals, especially gold, stimulated early expansion, and the missionary impulse of Orthodox priests also played a role. Land hunger and a desire to escape serfdom and taxation led to the flight of many peasants into the fertile black-earth belt of southwestern Siberia, and many landlords also moved there with their serfs.

The initial outward thrust from Muscovy under Ivan the Great (reigned 1462–1505) was mainly northward. The rival principality of Novgorod, near present-day St. Petersburg, was annexed. This secured a domain that extended northward to the Arctic Ocean and eastward to the Ural Mountains. Later tsars pushed the frontiers of Russia westward toward Poland and southward toward the Black Sea. Peter the Great (reigned 1682–1725) defeated the Swedes under Charles XII to gain a foothold on the Baltic Sea, where he established St. Petersburg as Russia's capital and its **"Window on the West."** Catherine the Great (reigned 1762–1796) secured a frontage on the Black Sea at the expense of Turkey.

The eastward conquest was even more impressive. Ivan the Terrible (reigned 1533–1584) added large new territories in conquering the Tatar khanates of Kazan and Astrakhan, thus giving Russia control over the entire Volga River. At the end of Ivan the Terrible's reign, traders and Cossack military pioneers were already penetrating wild and lonely Siberia, where there were only scattered indigenous inhabitants. The **Cossacks** were peasant-soldiers in the steppes, formed originally of runaway serfs and others fleeing from tsardom. They

Definitions + Insights

Russia and Other Land Empires

Russia, and later the Soviet Union, developed as a **land empire.** Rather than establishing its colonies overseas, as did imperial powers such as Spain and Great Britain, Russia founded its colonies in its own vast continental hinterland (Figure 6.D). Many of the colonized peoples had little in common with the ethnic Russians ruling from faraway Moscow. Like former colonies of overseas empires such as Great Britain, non-Russian regions of the periphery of Russia and later the Soviet Union were drawn into a relationship of economic dependence on the imperial Russian core. The Central Asian republics of the Soviet Union, for example, followed Moscow's demands to grow cotton, which was shipped to Moscow to be manufactured into value-added clothing that was subsequently sold throughout the empire. This pattern contributed to the growth of the USSR's economy but often inhibited local development. Other entities characterized historically as land empires include the United States, China, Brazil, and the Ottoman, Mogul, Aztec, and Inca empires.

Growth of Russia from 1300

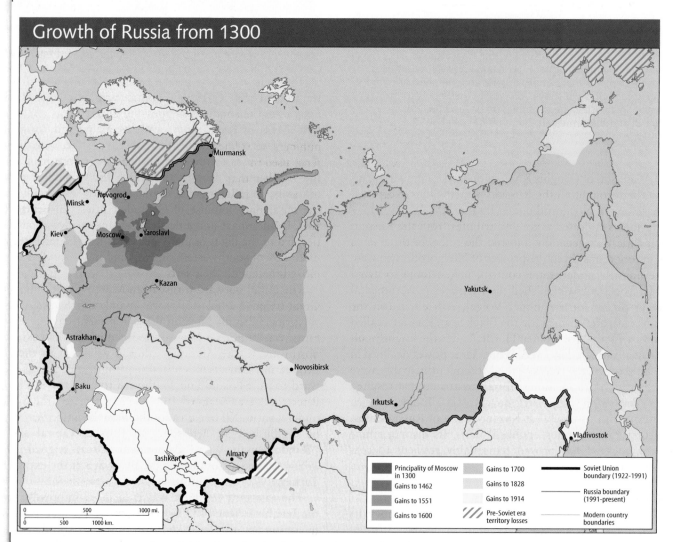

Legend:
- Principality of Moscow in 1300
- Gains to 1462
- Gains to 1551
- Gains to 1600
- Gains to 1700
- Gains to 1828
- Gains to 1914
- Pre-Soviet era territory losses
- Soviet Union boundary (1922–1991)
- Russia boundary (1991–present)
- Modern country boundaries

WORLD REGIONAL GeographyNow™   **Active Figure 6.D** The development of Russia's land empire, 1300–1914
Source: After Milner-Gulland, 1998. See an animation based on this figure, and take a short quiz on the facts and concepts.

eventually gained special privileges as military communities serving the tsars.

In 1639, a Cossack expedition reached the Pacific. Russian expansion toward the east did not stop at the Bering Strait but continued down the west coast of North America as far as northern California, where the Russian trading post of Fort Ross was active between 1812 and 1841. In 1867, however, Russia sold Alaska to the United States (for two cents per acre) and withdrew from North America. The relentless quest of Russian fur traders for sable, sea otter, and other valuable furs brought great cultural changes to aboriginal peoples in Siberia and Pacific North America. These effects became increasingly pronounced, especially in the last decades of the Russian empire when millions of Russians moved into Siberia.

682

There was one stumbling block in Russia's eastern expansion. In the Amur River region near the Pacific, the Russians were not able to consolidate their hold for nearly two centuries because of opposition by strong Manchu emperors who claimed this territory for China. They were thus barred from the east Siberian area best suited to grow food for the fur-trading enterprise and best endowed with good harbors for maritime expansion. This situation changed in 1858–1860 following the victory of European sea powers over China in the Opium Wars. Russia was able to add the Amur region to its earlier gains in the Ob, Yenisey, and Lena basins. Finally, in a series of military actions during the 19th and early 20th centuries, Russian tsars annexed most of the Caucasus region and Turkestan, the name for the Central Asia region east of the Caspian Sea.

The Soviet state that succeeded tsarist Russia had to grapple with difficult questions of how to govern and interact with a mosaic of over 100 languages and cultures. Soviet authorities permitted the different ethnolinguistic groups to retain their own languages and other elements of their traditional cultures and even created alphabets for those that previously had no written language. Politically, the Soviet Union officially recognized many of the groups as non-Russian nationalities. The USSR established 16 Autonomous Soviet Socialist Republics (ASSRs) as homelands for large ethnic minorities, in theory endowing them with limited self-governing powers. Smaller nationalities were organized into politically subordinate autonomous units, including five autonomous regions (autonomous *oblasts*) and ten autonomous areas (autonomous *okrugs*). Soviet labeling of these units as *autonomous* came back to haunt the nation of Russia in the 1990s.

160–161

In spite of such apparent concessions to non-Russians, the Russian-dominated regime implemented a deliberate policy of Russification, an effort to implant Russian culture in non-Russian regions and to make non-Russians more like Russians. Large numbers of Russians migrated to work in the factories and state farms in non-Russian republics. Russians were prominent in positions of responsibility even within the non-Russian republics; throughout the USSR, they held the majority of top posts in the Communist Party, the govern-

ment, and the military. But Russian ways had little impact on non-Slavic cultures such as the Georgians and Armenians. Even Slavic kinspeople of the Russians, such as the Ukrainians, insisted on retaining their cultural autonomy. In general, the policy of Russification was a failure because of strong nationalist sentiments throughout the Soviet Union.

Russia and the Soviet Union: Tempered by Revolution and War

Russia and the Soviet Union repeatedly triumphed over powerful invaders, notably the Swedish forces led by King Charles XII in 1709, the French and associated European forces under Napoleon I in 1812, and the German and other European forces sent by Hitler into the Soviet Union in World War II. In each case, the Russians lost early battles and much territory but eventually inflicted a crushing and decisive defeat on the invaders. The success of Old Russia and the Soviet Union in withstanding invasions by such formidable armies was due in part to the environmental rigors (particularly, the brutal Russian winter) that invaders faced, the overwhelming distances of a huge country with poor roads, and the defenders' love of their homeland. It was also due to talented Russian military leadership and to the willingness of good and poor generals alike to lose great numbers of soldiers in combat. Finally, the successful defense used the scorched earth strategy to protect the motherland; rather than leave Russian railways, crops, and other resources to fall into the invaders' hands, the defenders destroyed them.

The **Russian Revolution** of 1917, which set the stage for the formation of the Soviet Union, was really two revolutions that occurred against the backdrop of World War I. In 1914, when a Serb in Sarajevo (now in Bosnia and Herzegovina) assassinated the heir to the throne of Austria-Hungary, a complicated series of alliances required Tsar Nicholas II to commit Russian troops to fight with Serbia, France, and Great Britain against Austria-Hungary and Germany in World War I. The first revolution in Russia began early in 1917 as a general protest against the terrible sacrifices of Russian forces on the Eastern Front during this war. That revolt overthrew Nicholas II, the last of the Romanov tsars.

The second was the **Bolshevik Revolution** that came later that year. Led by Vladimir Ilyich Lenin (1870–1924), the Bolshevik faction of the Communist Party seized control of the government. The new regime made a separate peace with Germany and its allies and survived a difficult period of civil war and foreign intervention between 1917 and 1921. Lenin presided over the establishment of Russia's successor state, the Soviet Union, in 1922.

World War II found Russians again allied with France and Britain in a far more ferocious war against Germany. The Soviet Union's success in withstanding the German onslaught that began as **Operation Barbarossa** in June 1941 surprised many outside observers who had predicted that the Soviets would prove too weak and disunited to resist for more than a few weeks or months. The crucially important cities of

Geography of the Sacred

Stalingrad

Sacred space or sacred place may be defined as any locale that people hold in reverence. The most obvious sacred places in the realms of ordinary experience are places of worship such as synagogues, churches, and mosques. Cemeteries, too, evoke a special code of behavior in visitors and are managed as sacred sites. Places on Earth where large numbers of people have lost their lives are among the most significant sacred spaces—the Gettysburg Battlefield and Manhattan's Ground Zero are examples in the United States. Russia has the world's most extensive network of sacred places associated with the loss of life.

Perhaps in no other country is the memory of World War II etched so vividly in the national consciousness as it is in Russia. Even today, millions of pilgrims annually visit a large number of war monuments and cemeteries in an effort to heal deep emotional wounds and to keep alive the memory of Russia's costly wartime resistance. These rites of visitation and commemoration pass to each new generation. Immediately after their wedding ceremony, for example, it is common for a newlywed couple to place flowers at the local tomb of the unknown soldier. Contemporary Russian pride and nationalism have strong roots in wartime sacrifice, and the experiences of those who survived the war continue to influence Russian politics and international relations.

During the war, superiors urged Russian soldiers to fight to the death for the motherland: Russia was sacred ground to be defended at any cost. Today, the battlegrounds where those soldiers fell are sacred places. They are kept hallowed by the continuous ritual visitation of veterans, war widows, and two generations of descendants of war survivors.

Stalingrad (now Volgograd) is the greatest of all these sites. The visitor to Volgograd cannot help but feel the pain of war that has lingered more than 50 years. All over the city are monuments to remind the living of the dead. The devastated shell of a mill stands as the only physical artifact of the past, but the memorials built after the war rekindle the emotional losses most strongly. Volgograd's central memorial is the complex on Mamayev Hill, a mecca for Russians. Old soldiers, still wearing their medals, look war-weary even now as they shuffle through. Grandmothers lead small children to place flowers at the feet of statues. Over the mass grave of an estimated 300,000 Soviet and German soldiers, a huge hand raises an eternal torch. The honor guard changes in goosesteps once each hour. This sacred place is dominated by the world's largest statue, a

Joe Hobbs

Figure 6.E The Mamayev Hill monument to the memory of the defenders of Stalingrad. Within the hill lie the bodies of hundreds of thousands of the city's defenders and attackers. This sword-wielding figure called "the Russian Motherland" is the world's largest statue, 279 feet (85 m) high and weighing 8,000 tons.

female sword-wielding figure called "the Russian Motherland" (Figure 6.E).

WORLD
REGIONAL
Geography ⊛ Now™

Click Geography Literacy for a virtual tour of Russia's sacred places.

Hitler studied his geography before attacking the city. Located on the border of the steppe and semidesert where the Volga and Don Rivers are closest together, Volgograd is situated strategically. The city became an important grain and livestock center after a railway connected it to the Don Valley. Later, the rail connected the city to the Caucasus region, between the Caspian and Black Seas. Early in World War II, Hitler resolved to capture the oil fields of the Caucasus. Rostov-on-Don and Stalingrad stood in his way. Though strategically vital, Stalingrad was to Hitler as much a symbolic as a military prize: Because it bore the name of Russia's leader, its fall would be of great propaganda value to the German war effort. Equally, its salvation from the invader was a goal of nearly religious significance for the Stalin regime.

The German offensive on the city began in August 1942. On September 7, from a bunker in Moscow's Kremlin, Stalin ordered: "Not another step backward." He issued this directive:

> Comrades and citizens of Stalingrad: Each of us must apply ourselves to the task of defending our beloved town, our homes, and our families. Let us barricade every street, transform every district, every block, every house, into an impregnable fortress.[a]

Russian resistance at Stalingrad is an extraordinary chapter in the history of warfare. The invaders were unprepared for the resolve of Russian fighters, as these passages from the diary of a German soldier reveal:

> September 11 [1942]. Our battalion is fighting in the suburbs of Stalingrad. Firing is going on all the time. Wherever you look, fire and flame. Russian cannon and machine guns are firing out of the burning city. Fanatics!

> September 16. Our battalion, plus tanks, is attacking the grain elevator. The battalion is suffering heavy losses. The elevator is occupied not by men, but by devils, that no bullets or flames can destroy.

> September 18. Fighting is going on inside the elevator. If all the buildings of Stalingrad are defended like this, then none of our soldiers will get back to Germany!

> September 26. We don't see them at all. They've established themselves in houses, and cellars; they're firing from all sides, including from our rear. Barbarians! They use gangster methods!

As the fighting persisted, a Russian soldier wrote this scene:

> Stalingrad is no longer a town. By day, it is an enormous cloud of burning, blinding smoke; it is a vast furnace, lit by the reflection of the flames. And when night arrives, one of those very hot, noisy, bloody nights, the dogs plunge into the Volga and swim desperately to gain the other bank. The nights of Stalingrad are a terror for them. Animals flee from this hell. The hardest stones cannot bear it for long. Only men endure.

On November 8, when his soldiers held nine-tenths of the city, Hitler said in an after-dinner speech in Munich:

> I wanted to get to the Volga, at a particular point where stands a certain town—bears the name of Stalin himself. I wanted to take the place, and you know, we've done it. We've got it really, except for a few enemy positions still holding out. Now, people say, "Why don't they finish the job more quickly?" Well, I prefer to do the job with quite small assault troops. Time is of no consequence at all.

Hitler was mistaken, of course. The severe Russian winter had begun, fierce Russian resistance continued in the city, and time was very much against his forces. The Russians struck back on November 19 and soon confined the German Sixth Army within a pincer. The invaders were now on the defensive. Radio Moscow broadcast (in German) this demoralizing Christmas Day message to them:

> Every seven seconds, a German soldier dies in Russia. Stalingrad is a mass grave.

The final Russian assault began on January 11, 1943. Two weeks later, Hitler denied General Von Paulus's request to surrender in order to save German lives. On January 31, Hitler made Von Paulus a field marshal, reminding him that no field marshal had ever been taken alive by the enemy. On the day of his promotion, Field Marshal Von Paulus surrendered his forces to the Russians.

In Germany, news of the Russian victory at Stalingrad was announced on February 3, 1943. The communiqué noted that German forces had succumbed to superior enemy strength and to "unfavorable circumstances." It was a devastating defeat for the Germans: Two armies consisting of 24 generals, 2,000 officers, and 90,000 soldiers were taken prisoner. Enough matériel was lost to equip one-fourth of the German army. Two years earlier, the Germans could not have imagined such a defeat. But the victory for the Soviet Union exacted an unimaginable cost. After the 5-month battle, Stalingrad lay in ruins; Soviet authorities dubbed it a "city without an address" and, in honor of its defenders, a "hero-town." Fifty years after the battle, Russian military authorities finally released figures on the number of dead. In this sacred ground lay the bodies of 3.5 million soldiers and civilians, of whom 2.7 million were Soviets.

[a]Excerpts from the soldiers' diaries and Hitler's speeches, and an excellent analysis of this decisive battle, may be found in Antony Beevor's *Stalingrad* (London: Penguin, 1999).

Leningrad (now St. Petersburg) and Moscow held out through brutal sieges. Late in 1942, Soviet forces halted the German push eastward at the Volga River in the huge Battle of Stalingrad (see Geography of the Sacred, pages 154–155). This engagement was the turning point in the war, when Soviet and Allied forces began to reverse Nazi advances.

The failure of powerful Germany to conquer the Soviet Union was a clear indication that the strength of the Soviet Union had been underrated and that the country's power would henceforth be a major feature of world affairs. But the war's impact on the USSR would linger. The German invasion took an estimated 20 million or more Soviet lives and caused the relocation of millions of people. It did enormous damage to settlements, factories, and livestock. As a strategic precaution during the war, Lenin's successor, Josef Stalin, directed major Soviet industries to relocate eastward away from the front, a move that has had a lasting imprint on the region's economic geography.

6.4 Economic Geography

The Communist Economic System

The Soviet Union's Communist economic system was an attempt to put into practice the economic and social ideas of the 19th-century German philosopher Karl Marx. According to Marx, the central theme of modern history is a struggle between the capitalist class (**bourgeoisie**) and the industrial working class (**proletariat**). He forecast that exploitation of workers by greedy capitalists would lead the workers to revolt, overthrow the capitalists, and turn over ownership and management of the means of production to new workers' states. In the classless societies of these states, there would be social harmony and justice, with little need for formal government.

Marx's utopian vision did not materialize anywhere, but it did provide guidelines for antigovernment revolts and Communist political systems in many countries. The ideas of Marx and of Lenin (who died in 1924), as Lenin's successor Josef Stalin interpreted and implemented them, provided the philosophical basis for the Soviet Union's centrally planned command economy. Beginning in 1928, a series of 5-year economic plans demanded—thus, the term *command economy*—the fulfillment of quotas for the nation: types and quantities of minerals, manufactured goods, and agricultural commodities to be produced; factories, transportation links, and dams to be constructed or improved; and residential areas to be built for industrial workers. The goals were to abolish the old aristocratic and capitalist institutions of tsarist Russia and to develop a strong socialist state of equal stature with the major industrial nations of the West.

In the command economy, an agency in Moscow called **Gosplan** (**Committee for State Planning**) formulated the national plans, which were then transmitted downward through the bureaucracy until they reached individual factories, farms, and other enterprises. This was an unwieldy and inefficient process in many ways. First, the planners in Moscow were essentially required to act as CEOs of a giant corporation, effectively "USSR Inc.," that would manage the economy of an area larger than North America—a gargantuan, impossible task. Further, the Soviet planning bureaucracy had no free market to guide it, so the system produced goods that people would not buy or failed to produce goods that people would have liked to buy. Third, Gosplan stated production targets in quantitative rather than qualitative terms, churning out abundant but substandard products. There was often an obsession with fulfilling huge quotas or implementing grandiose schemes, a Soviet preoccupation sometimes known as **gigantomania.** Finally, fearing reprisals from people higher up the ladder, no one wanted to suggest ways to increase quality and efficiency.

Although cumbersome, this system succeeded in propelling the Soviet Union to superpower status, improved the overall standard of living, prompted rapid urbanization and industrialization, and altered the landscape profoundly. The principal goal of Soviet national planning after 1928 was a large increase in industrial output, with emphasis on heavy machinery and other capital goods, minerals, electric power, better transportation, and military hardware. Masses of peasants were converted into factory workers. New industrial centers were created, and old ones were enlarged. There were huge investments in defense, and the country grew strong enough to survive Germany's onslaught in World War II. After the war, the Soviets maintained large armed forces and accumulated a massive arsenal of conventional and nuclear weapons in the arms race against the world's only other superpower, the United States (Figure 6.14). By the height of the Cold War in the early 1980s, 15 to 20 percent of the country's GNP was dedicated to the military (in contrast to less than 10 percent in the United States), representing an

Figure 6.14 A nuclear-powered aircraft carrier in Russia's port of Murmansk. The arms buildup during the Cold War was unbearably costly for the Soviet Union.

enormous diversion of investment away from the country's overall economic development.

Farmers and farming had troubled careers under the Soviet system of **collectivized agriculture.** Between 1929 and 1933, about two-thirds of all peasant households in the Soviet Union were collectivized. Their landholdings were confiscated and reorganized into two types of large farm units: the collective farm (*kolkhoz*) and the factory-type state farm (*sovkhoz*). The consolidation of individual farmsteads and villages into fewer but larger communities on the collective farms was supposed to permit the government to more efficiently and cheaply administer, monitor, and indoctrinate the rural population and to provide them with services, including education, healthcare, and electricity.

But the rural people resisted collectivization fiercely. In their own version of scorched earth, peasants and nomads slaughtered millions of livestock and burned crops to avoid turning them over to the "socialized sector." Government reprisals followed, including wholesale imprisonments and executions, together with confiscation of food at gunpoint (often including the peasants' own food reserves and seed). The more prosperous private farmers, known as *kulaks,* were killed, exiled, sent to labor camps, or left to starve. Famines took millions of lives. Soviet leaders disregarded these costs, and collectivization was virtually complete by 1940.

The drive to increase the national supply of farm products also demanded an enlargement of cultivated area. In the 1950s, the Soviet Union began a program to increase the amount of grain (mainly spring wheat and spring barley) produced by bringing tens of millions of acres of what were called "virgin and idle lands" or "new lands" into production in the steppes of northern Kazakhstan and adjoining sections of western Siberia and the Volga region (Figure 6.15).

Soviet enterprises both in agriculture and industry harnessed the energies and resources of the whole country to achieve specific objectives. The government called on the people to sacrifice and to make the country strong, especially by working on so-called Hero Projects such as the construction of tractor plants, dams, railways, and land reclamation. The government in turn operated as a collectivized **welfare state,** providing guaranteed employment, low-cost housing, free education and medical care, and old-age pensions. Social services were often minimal but in some sectors were quite successful; the literacy rate, for example, rose from 40 percent in 1926 to 98.5 percent in 1959. However, military-industrial superpower status was generally achieved at the expense of Soviet consumers, whose needs were slighted in favor of heavy metallurgy and the manufacture of machinery, power-generating and transportation equipment, and industrial chemicals. Lines of consumers waited in stores to purchase scarce items of clothing and everyday conveniences. The exasperation of shoppers confronted by long lines and empty shelves was an important factor generating dissatisfaction with the economic system and demands that it be redesigned.

Figure 6.15 Soviet agricultural expansion, into the "Virgin and Idle Lands," 1954–1957.

Economic Roots of the Second Russian Revolution

Internal freedoms and prosperity did not accompany superpower status for Soviet citizens. Near paralysis gripped the flow of goods and services in the late 1980s, when large demonstrations and strikes underscored public anger at a political and economic system that was sliding downhill rapidly. The Communist system came under open challenge on the grounds that it stifled democracy, failed to provide a good living for most people, and thwarted the ambitions of the country's many ethnic groups for a greater voice in running their own affairs. The outpouring of dissent was unprecedented in Soviet history. Worsening economic conditions led to official calls for fundamental reform in the mid-1980s.

The revamping of the economic system became an urgent priority during the regime of Mikhail Gorbachev that began in 1985. Gorbachev initiated new policies of *glasnost* ("openness") and *perestroika* ("restructuring") to facilitate a more democratic political system, more freedom of expression, and a more productive economy with a market orientation (Figure 6.16). At the same time, there were increasing problems of disunity as the different republics and the ethnic groups organized by the Soviet system into nominally "autonomous" units took advantage of new freedoms to resurrect old quarrels and demand that their units be given more autonomy. Fighting among ethnic groups erupted in several republics. Meanwhile, in all of the republics, declarations of sovereignty and in some cases outright independence challenged the authority of the central government. In 1991, the USSR and most other nations recognized the independence of the three Baltic republics of Estonia, Latvia, and Lithuania. The process of the empire's disintegration had begun.

At the center of the crumbling empire, having failed to reverse the downward slide of the economy, Gorbachev faced

120

Joe Hobbs

Figure 6.16 Freedom of expression erupted all over the USSR during the Gorbachev era. This was an artist's kiosk in Leningrad (now St. Petersburg) in 1991.

growing sentiment to scrap the command system and move as rapidly as possible to a market-oriented economy. The reformers' cause was championed by Boris Yeltsin, a leading advocate of rapid movement toward a market-oriented economy and greater control by individual republics over their own resources, taxation, and affairs. Yeltsin's political stature grew when, in June 1991, the citizens of Russia chose him president of the giant Russian Federation in the Soviet Union's first open democratic election. Gorbachev, himself elected president of the USSR (but by a Congress of People's Deputies rather than by the people as a whole), opposed Yeltsin's proposals for radical reform. He advocated more gradual movement toward free-market orientation within the state-controlled planned economy and insisted on the need for unity and continued political centralization within the Soviet state.

Matters came to a head in August 1991, when an attempted coup by hardliners failed. Yeltsin gained enhanced stature from his defiant opposition to the coup. Gorbachev resigned his position as head of the Communist Party and began to work with Yeltsin to reconstruct the political and economic order. But his efforts to preserve the Soviet Union and a modified form of the Communist economic system could not stand against the growing tide of change. During the autumn of 1991, the Communist Party was disbanded, and the individual republics seized its property. On December 25, 1991, Gorbachev resigned the presidency, and the national parliament formally voted the Soviet Union out of existence on the following day. A powerful empire had quickly and quietly faded away, to be replaced by 15 independent countries, in what was dubbed the Second Russian Revolution.

Russia: A Misdeveloped Country?

The unprecedented experiment to transform a colossal Communist state into separate bastions of free-market democracy has produced strange and often unfortunate results, particularly in Russia. Some observers classify Russia today not as a less developed or more developed country but as a **"misdeveloped country."** The disturbing demographic trend of a plummeting population due to rising death rates—the Russian cross described earlier—is the strongest indicator of Russia's backward momentum in the years following independence. The roots of this trend were mainly economic.

Despite heightened public expectations for a better life in Russia, criticism of the triumphant President Yeltsin and his policy of economic reform toward freer markets increased throughout the course of his two terms in office (1991–1999). Early in his first term, Yeltsin introduced a program of rapid economic reform, known as economic shock therapy, designed to replace the Communist system with a free-market economy. It removed price controls and encouraged privatization of businesses. Products and services from the private sector had already been indispensable during the Communist era, but now they were to be the centerpiece of the economy (Figure 6.17).

The results of shock therapy were generally poor for most people. Former employees of the Communist Party resented the loss of their jobs and privileges. Printing of money to meet state obligations caused inflation. Poor people living on fixed incomes struggled to survive high prices for basic necessities. A new consumer-oriented society developed along class lines. There was growing unemployment and homelessness and a widening gap between rich and poor (Figure 6.18). Access to choice goods and services was no longer forbidden

Figure 6.17 Free enterprise thrives throughout what was once the Soviet Union. Customers are lined up to buy vegetables from a street vendor in Vyborg, Russia. Note the seller's abacus "calculator."

Figure 6.18 Having fallen through the cracks in Russia's transition from a command economy to free enterprise, these women are begging for help on a St. Petersburg sidewalk. The message of their placards may be summarized as: "Help us dear brothers and sisters. Please give us money for food, for God's sake. We are invalids with cancer. We have given of ourselves but are now forgotten by the state. We are poor and hungry, with no protection, just left to rot."

to ordinary citizens as it was during the Communist era, but prices were so high that most people could not afford them.

One of the major components that emerged in Russia's new economic geography was the underground economy, also known as the *countereconomy* or *second economy* (or economy *na levo*, meaning "on the left"). Rampant black-marketeering developed. Although much of this exchange was illegal, there was a general tendency to overlook such transactions because they were essential to the economy. Widespread barter—the exchange of goods and service in the absence of cash—resulted from the declining value of the ruble. Many people resorted to selling personal possessions to buy high-priced food and other necessities. Most Russians grew economically worse off. Many privately owned industries were too unproductive to pay their employees, or paid workers under the table to avoid taxation. Unpaid workers could not pay taxes to the government, and unofficially paid workers did not. Russia's new private entrepreneurs came to include a large criminal "Mafia" that preyed on government, business, and individuals. Organized crime and corruption became all pervasive and made "free enterprise" far from free. A company wanting to build a factory, for example, would have to pay off officials at every stage of construction; without the payouts, organized crime would terminate the project. Russia became a kleptocracy, an economic and political system based on crime.

The biggest beneficiaries of the new economic system have been the so-called oligarchs (from the Greek *oligarchy*, meaning "rule of the few"), Russia's leading businesspeople. Having in the 1990s acquired massive stakes in former state enterprises through a series of rigged auctions called "loans for shares," these tycoons still control an estimated 70 percent of Russia's economy. They wield enormous political power and helped usher Vladimir Putin, a former KGB (state

security) officer of the Soviet Union into Russia's highest office, the presidency, in 2000. Once in office, Putin vowed to crack down on the oligarchs. He did purge some of them, only to strengthen others that served his interests. Putin's increasingly authoritarian style proved to be popular with voters, who awarded him a second term in office due to end in 2008. His consolidation of power with little public opposition seemed to reaffirm what Stalin often said: "The Russians need a czar."

Russia's economic decline following independence raised public fears and prompted some people to call for a return to the relative stability and security of the Soviet state. Others emigrated, including 380,000 from the former Soviet Union to the United States in the decade 1990–2000. But most have stayed, struggling in creative and often brilliant ways with limited resources to make ends meet and, ideally, help build the middle class and the normal way of life that communism's collapse was supposed to bring.

Recently, there has been much good news that may forecast the resurrection of Russian well-being. Russia's economy and Russians' personal incomes grew steadily in the early 2000s. Between 2000 and 2004, according to official Russian statistics, the economic wealth of the average Russian household grew by 53 percent, and wages increased by 86 percent (to an average of $190 per month). One-fourth of Russia's population still lived in poverty, but poverty and unemployment declined by a third in those years. Ironically, the improvement resulted partly from the severe economic crisis Russia experienced in 1998. When Russia's currency was devalued that year, sales of imported goods plummeted, and Russian entrepreneurs boosted production of textiles, clothing, and automobiles. A surge in oil prices after 1999 also stimulated the economy. Russia has enormous energy reserves; its 50 billion barrels of proven oil reserves represent 5 percent of the world total, and its 50 trillion cubic meters of natural gas represent a third of proven global reserves. After years of belt tightening due to low world market prices, Russia finally began to realize energy revenues. Energy exports in the early 2000s, when Russia was the world's largest oil producer, accounted for about half of Russia's export earnings.

There was also considerable growth in exports of aluminum and other metals and minerals. The Russian companies selling these resources have used their growing profits to purchase co-ownership in automobile and other industries, and in agricultural companies, helping to spark a revival in some sectors of the Russian economy.

However, as long as raw material prices are high, there is little incentive to develop other parts of the economy. The biggest danger to Russia's economy lies in excessive dependence on oil and other natural resources, including natural gas, metals, and timber, to drive the economy overall. Should oil prices in particular slump, revenues would plummet and *35–36* drag the entire economy into a deep recession.

6.5   Geopolitical Issues

There are three concentric spheres of major geopolitical concern in the region of Russia and the Near Abroad: the unity of Russia itself, Russia's relationships with its Near Abroad, and the relations between this region and the rest of the world. There are perhaps more critical geopolitical issues in this world region than in any other, and getting to know these problems is worth the effort.

Within Russia

Internally, Russia is struggling to maintain a cohesive whole that is fashioned from diverse and sometimes volatile ethnic units. As mentioned earlier, the Soviet Union included a number of "autonomous" units based on ethnicity. When Russia *153* emerged from the Soviet Union, it kept those designations within its borders. The Russian Federation today includes 89 areal "subjects" divided into six categories: 49 *oblasts* (regions), 6 *krays* (territories), 21 autonomous republics (initially 20, before Chechnya and Ingushetia divided; these were the 16 Soviet-era ASSRs plus 4 autonomous regions upgraded to republic status), 10 autonomous *okrugs* (ethnic subdivisions of *oblasts* or *krays*), 2 federal cities, and 1 autonomous oblast (Figure 6.19). The 32 autonomies (nation-

Autonomous Regions of Russia

KEY TO NUMBERS
1 KARACHAYEVO-CHERKESIYA
2 KABARDINO-BALKARIYA
3 NORTH OSSETIA
4 INGUSHETIA
5 CHECHNYA

Autonomous republics
Autonomous okrugs
Autonomous oblasts

Note: names of other Russian administrative divisions are given on map 6.20.

Figure 6.19 Autonomous units of the Russian Federation. Russia fears that independence for any of them, particularly Chechnya, might fracture the country.

ality-based republics and lesser units) in total occupy over two-fifths of Russia's total area and contain about a fifth of the country's population.

Nearly half of the people in Russia's autonomies are ethnic Russians, who form a majority in many units. The titular nationalities of the autonomies are ethnically diverse, however (see Figure 6.11, page 150). Some, such as the Karelians, Mordvins, and Komi, are Uralic (also known as Finnic, related to the Finns and Magyars); others are Turkic (for example, Tatars, Bashkirs, and Yakuts), Mongolian (for example, Buriats near Lake Baikal), and members of many other ethnic groups. Their largest autonomous units form a nearly solid band stretching across northern Russia from Karelia, bordering Finland, to the Bering Strait. The Karelian, Komi, and the largest of all, Sakha (Yakut), Republics are in this group. This band also includes several large but thinly inhabited units of aboriginal peoples that the Soviets designated as "autonomous *okrugs.*"

What gives some of these units particular geopolitical significance is the presence of major mineral resources: oil and gas in the Volga-Urals fields in Tatarstan and Bashkhortostan, high-grade coal deposits at Vorkuta in the Komi Republic, and a quarter of all diamonds produced in the world in Sakha. After the breakup of the USSR, many of the former Soviet ASSRs (see page 153), recalling Moscow's longstanding promises of self-rule for them, issued declarations of sovereignty asserting their right to greater self-direction of their internal affairs and greater control over their own resources. Eighteen of Russia's twenty republics signed a 1992 Federation Treaty that granted them considerable autonomy. The treaty called for the devolution of power centralized in Moscow and for more cooperation between regional and federal governments. Each republic of the Russian Federation was legally entitled to have its own constitution, president, budget, tax laws and other legislation, and foreign and domestic economic partnerships. But many republics, and the *oblasts* and *krays* that have had their own elected governors since 1997, are complaining that Moscow has not honored the treaty, and they are looking for regional solutions to their economic and other problems.

Moscow has relented to some of these demands. The Sakha (Yakut) Republic, for example, won the right to keep 45 percent of hard currency earnings from foreign sales of Sakha diamonds, compared with only a small fraction during the Soviet era. In turn, Sakha must now pay for the government subsidies that Moscow previously paid to local industries. Without credits from Moscow, Sakha no longer pays taxes to Moscow. Salaries and other indexes of living standards in the Sakha Republic have risen since this agreement was implemented. The majority Muslim, oil-rich Tatarstan, which did not sign the Federation Treaty in 1992, now has "limited sovereignty" with its own constitution, parliament, flag, and official language and has since signed the treaty. The autonomous republic of Chechnya—of particular geopolitical importance because of its crossroads location for oil pipelines—never did sign the treaty, and its fu-

ture is critical to the Russian Federation (see Problem Landscape, page 164).

As Russia sees it, these regional demands for greater self-rule threaten the unity of the Russian Federation and threaten to deprive Russia of valuable resources. Other autonomous regions are keeping eyes on Tatarstan and Chechnya. Independence for either one might inspire the oil-rich Turkic-speaking Bashkirs and the Chuvash to push for their own independence. This could begin a process that would virtually cut Russia in half. Although some analysts argue that this process of devolution in Russia may actually bring more stability and prosperity to the country, the government in Moscow worries that what happened to the USSR might happen to Russia itself. Moscow has responded by creating seven of its own administrative regions (Northwestern, Central, Northern Caucasus, Volga, Urals, Siberian, and Far Eastern) and appointed a presidential envoy (known as a governor general) to manage national defense, security, and justice matters in each (Figure 6.20). The message is that Moscow rules. It remains to be seen whether yet another layer of bureaucracy will do anything to keep the country's vast periphery tied to its core.

Russia and the Near Abroad

A second sphere of geopolitical concerns involves Russia's relations with the other successor states of the USSR. With the collapse of the Soviet Union, the Russian Federation took over the property of the former government within the federation's borders, including Moscow's Kremlin (Figure 6.21). Russia also took custody of the international functions of the USSR, including its seat at the United Nations. In late 1991, the new countries of Russia, Ukraine, and Belarus (formerly Belorussia) formed a loose political and economic organization called the **Commonwealth of Independent States (CIS)**, headquartered in Minsk, Belarus. Except for Estonia, Latvia, and Lithuania, the other republics eventually joined the CIS, in which Russia took a strong leadership role. To date, the Commonwealth of Independent States has been concerned mainly with establishing common policies on economic concerns. Russia's President Putin has proposed that its members unite on other issues, particularly security, with the formation of an antiterrorism alliance.

Since the Soviet Union dissolved in 1991, many important links between Russia and the other 14 successor states have persisted. Economic, ethnic, and strategic ties are especially crucial. For example, one major economic link is the vital flow of Russian oil and gas to other states such as Ukraine (which still gets 70 percent of its energy from Russia), Belarus, the Baltic countries, and Kyrgyzstan, which are highly dependent on this supply of energy. There is some reverse flow of energy, particularly with Turkmenistan's exports of natural gas to Russia.

Ethnic issues are important, too. About 25 million ethnic Russians lived in the 14 smaller states at the time of independence. Although a considerable number have migrated to Russia since 1991, many have had difficulties finding hous-

Figure 6.20 Administrative units of the Russian Federation

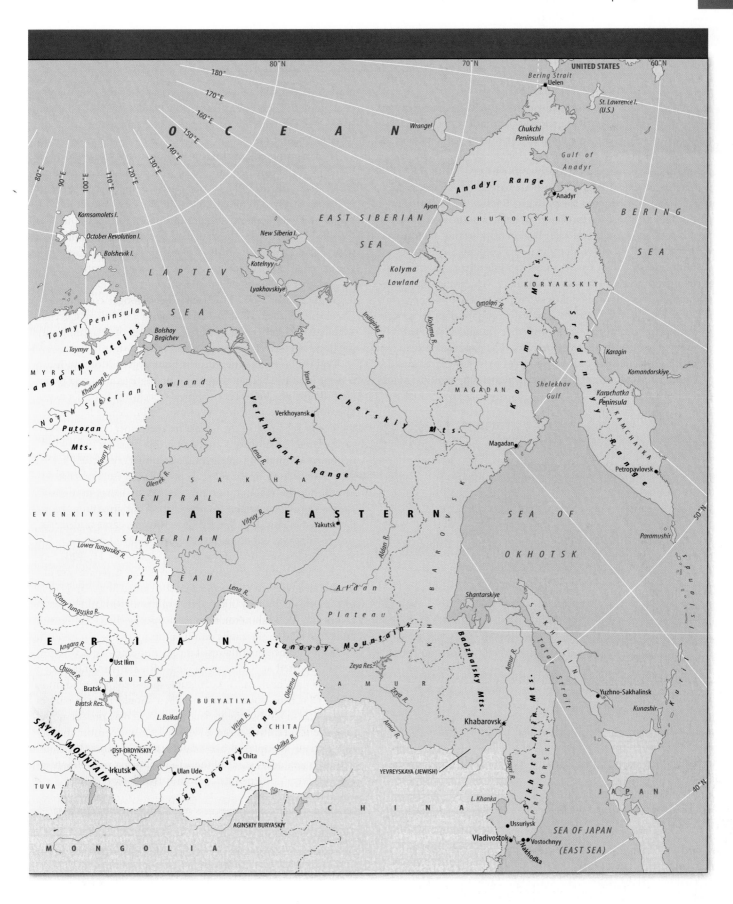

O C E A N

180°

170°E

160°E

150°E

140°E

130°E

120°E

110°E

100°E

90°E

80°E

80°N

70°N

60°N

UNITED STATES

Bering Strait
Uelen

St. Lawrence I.
(U.S.)

Wrangel

Chukchi
Peninsula

Gulf of
Anadyr

Anadyr Range

Ayon

Anadyr

B E R I N G

E A S T   S I B E R I A N

C H U K O T S K I Y

S E A

Komsomolets I.

October Revolution I.

Bolshevik I.

New Siberia I.

Kotelnyy

Lyakhovskiye

L A P T E V

Kolyma
Lowland

S E A

Karagin

K O R Y A K S K I Y

Omolon R.

Komandorskiye

Taymyr Peninsula

Bolshoy
Begichev

L. Taymyr

Khatanga R.

MYRSKIY

anga Mountains

North Siberian Lowland

Putoran
Mts.

Koyr R.

Olenek R.

S A K H A

C E N T R A L

EVENKIYSKIY

S I B E R I A N

P L A T E A U

Lower Tunguska R.

Stony Tunguska R.

E R I A N

Angara R.

Chuna R.

Ust Ilim

Bratsk

Bratsk Res.

IRKUTSK

SAYAN MOUNTAIN

TUVA

UST-ORDYNSKIY

Irkutsk

Ulan Ude

L. Baikal

BURYATIYA

Vitim R.

Yablonovyy Range

Shilka R.

CHITA

Chita

AGINSKIY BURYASKIY

M O N G O L I A

Indigirka R.

Kolyma R.

Yana R.

Verkhoyansk

Cherskiy Mts.

Kolyma Mts.

Shelekhov
Gulf

MAGADAN

Magadan

Sredinnyy Range

Karachatka
Peninsula

KAMCHATKA

Petropavlovsk

Paramushir

S E A   O F

O K H O T S K

Kuril Islands

50°N

Verkhoyansk Range

Lena R.

Yakutsk

F A R   E A S T E R N

Vilyuy R.

Aldan R.

A l d a n
P l a t e a u

KHABAROVSK

Stanovoy Mountains

Zeya Res.

A M U R

Zeya R.

Amur R.

Oliokma R.

Shantarskiye

Badzhalsky Mts.

Khabarovsk

YEVREYSKAYA (JEWISH)

C H I N A

L. Khanka

Amur R.

Ussuri R.

Sikhote Alin Mts.

PRIMORSKIY

S A K H A L I N

Tatar Strait

Yuzhno-Sakhalinsk

Kunashir

J A P A N

Ussuriysk

Vladivostok

Vostochnyy

Nakhodka

SEA OF JAPAN
(EAST SEA)

40°N

Problem Landscape

Chechnya

Nominally part of the Russian Federation, the Caucasus autonomy of Chechnya (population: 1 million) has insisted on independence ever since the USSR broke up, and is of particular concern to Russia (Figure 6.F). The Chechens, who are Sunni Muslims, have long resisted Russian rule, and Moscow has fought back. Russian troops attempted but failed to seize control over Chechnya soon after its 1991 declaration of independence. Beginning a second offensive in 1994, and at a cost of an estimated 30,000 to 80,000 lives on both sides, Russian troops succeeded in exerting physical control over most of Chechnya. But in 1996, Chechen forces recaptured the capital city of Groznyy (renaming it Jokhar-Gala). At that time, the Chechen war was deeply unpopular among Russians, and the prospect of further Russian losses led to a peace agreement with the Chechens. The pact required an immediate withdrawal of Russian forces from Chechnya and deferred the question of Chechnya's permanent political status

Figure 6.F Groznyy street scene. Will peace or victory come to the Chechens?

Sergei Uzakov/ITAR-TASS /Corbis

until 2001. Russia allowed Chechens to elect their own president (they did, choosing Aslan Maskhadov in 1997) but insisted that Chechnya was, for the time being, still part of the Russian Federation.

The cessation of hostilities was short-lived. Chechen rebels kidnapped and tortured hundreds of Russian citizens in southern Russia and in 1999 carried their campaign to the heart of Russia. They detonated bombs in Moscow and other vulnerable civilian centers, killing more than 300 Russians. The Chechen resistance by this time had become internationalized, with Islamic militant groups from abroad, including from Osama bin Laden's notorious al-Qa'ida organization based in Afghanistan, supplying men and war matériel to their Islamic guerilla brethren in Chechnya. No longer controlled by Chechen President Maskhadov, in August 1999 these Chechen Islamists invaded the neighboring province of Dagestan with the hope of creating an independent Islamic state.

That invasion, and the terrorist bombings in Moscow, led to a surge in anti-Chechen sentiment among Russians and popular calls for resolute action against Chechnya. Russia's newly appointed prime minister, Vladimir Putin, decided to pursue an all-out second Chechen War. In September 1999, Russian forces launched a furious attack that this time put them firmly in control of most of Chechnya. Putin's military success helped secure his victory in the Russian presidential election that followed, but the reaction abroad was mostly negative. Russian troops were documented to have carried out atrocities against Chechen civilians, and Western governments came under pressure to apply sanctions against Russia. None did, however, and were even less inclined to do so after September 11, 2001. After that, President Putin argued that the Chechens are just like the Islamist terrorists who attacked the United States and could only be dealt with by force. His argument was all the more compelling to Washington because many Chechen Islamist militants are among the ranks of al-Qa'ida.

With Chechen guerilla resistance continuing—still with terrorism directed at Russian civilians—Russia is maintaining a vise grip on Chechnya. In a Russian-sponsored referendum in 2003, Chechens voted that Chechnya should remain part of Russia and in a subsequent election chose a new president. Behind the scenes, Russia ran a rigged election to ensure the victory of their handpicked candidate, reviled by most Chechens, and assassinated by some of them the following year. Prospects for Chechnya's independence from Russia look remote.

ing and employment, and the great majority still live in the other countries. In certain areas, notably eastern Ukraine and Crimea, many ethnic Russians are causing political instability because they want to secede and form their own state or join Russia. Russians in the 14 non-Russian nations

complain that governments and peoples discriminate against them. In response, the Russian government has said that it has a right and a duty to protect Russian minorities in the other countries. Russia also encourages those Russian minorities to maintain and strengthen their ethnic identities, 120,192

Figure 6.21 Moscow's Red Square and Kremlin (far left), seat of power for the USSR and now the Russian Federation. The Lenin Mausoleum, just to the right of the line of conifers on the left side of this image, backs up to the Kremlin wall.

creating prospects for future political secession from host countries. In political geography terms, such a movement by an ethnic group in one country to revive or reinforce kindred ethnicity in another country—often in an effort to promote secession there—is known as irredentism.

Ukraine's postindependence struggles with Russia reflect these ethnogeographic and other concerns, including control of strategic territory. An early issue of contention was the Crimean Peninsula, a picturesque and verdant resort region. Russia's Catherine the Great annexed the Crimea in 1783,

but Soviet leader Nikita Khrushchev returned it to Ukraine in 1954. After the Soviet Union collapsed, Russian irredentists agitated to get back the Crimea, with its 70 percent Russian ethnic population. Ukraine yielded a bit by granting the Crimea special status as an autonomous republic. Crimeans then elected their own president, a pro-Russian politician who advocated the Crimea's reunification with Russia.

For now, Russia seems resigned to the Crimea's belonging to Ukraine but continues to be at odds with Ukraine over a strategic strait separating the Crimea from Russia's Taman Peninsula to the east. This Kerch Strait is the vital sea connection between the Azov Sea and the Black Sea—and therefore between southern Russia and the wider world (see Figure 7.3, page 177). Russia wants to share sovereignty of the strait with Ukraine, but Ukraine claims most of the waters as its own and charges Russia $200 million per year in transit fees. Thanks to a successfully resolved dispute between the two powers, Russia continues to operate its Black Sea naval fleet from the Ukrainian port of Sevastopol.

Russia has a number of geopolitical concerns with countries and ethnic groups of the Caucasus south of Chechnya (Figure 6.22). The Armenians embrace Russia as a strategic ally against their historic enemy, the Turks, who virtually surround them. Russia embraces a faction within Georgia that is opposed to Georgian rule: the primarily Muslim South Ossetians, who would like to free themselves from Georgian control and establish an Ossetian nation. The North Ossetians, whom they wish to join, live adjacent to them in Russia in a region the North Ossetians call Alania. South Ossetia now exists as a Russian puppet state within Georgia's borders, plagued recently by a violent anti-Georgian insurgency.

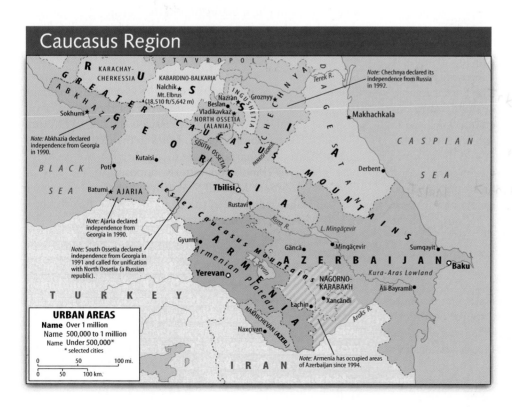

Figure 6.22 The Caucasus

Geography of Terrorism

Georgia's Pankisi Gorge

One of the post-9/11 hot spots that emerged in the "war on terrorism" was the Pankisi Gorge, some 90 miles (150 km) northwest of the Georgian capital, Tbilisi (see Figure 6.22). Adjacent to the war-torn, would-be breakaway Russian province of Chechnya, it is not really a gorge but a fertile valley some 18 miles (30 km) long. It is home to several ethnic groups, including Georgians, Kists (who are ethnically Chechen), and about 4,000 Chechens who fled from the war in Chechnya in 1999. There are fears in both Washington and Moscow that the sympathies of Chechen refugees in the Pankisi lie strongly with radical Islamic groups as far afield as Afghanistan. American authorities believed that al-Qa'ida fighters moved into the valley, perhaps to regroup and plan more attacks against Western targets. Some feared the Pankisi would become a "mini-Afghanistan," a new lawless refuge for al-Qa'ida and other militant organizations. The Bush administration gained Georgia's permission in 2002 to allow U.S. military advisers to train Georgian troops in antiterror tactics. The U.S. personnel became a sore point in Moscow because Russia considers Georgia to be in its sphere of influence, not that of its former Cold War adversary.

Russia sees the Pankisi in a different light than the United States does. To Moscow, it is not so much a mini-Afghanistan that would threaten the West as it is a haven for Chechen rebels who use it as a refuge and base in their continuing war against Russia. Russian aircraft sometimes cross into Georgian airspace in pursuit of Chechen rebels. The United States insists that the region's security is entirely a Georgian matter and that Russia should stay out. Georgia's parliament, already seething about the presence of Russian troops in the Georgian province of Abkhazia, called for Russia to get out and stay out of Georgia and for Georgia to withdraw from the Commonwealth of Independent States (which it joined and stayed in at Russia's insistence). Russian President Putin responded by threatening to send Russian troops into the Pankisi. What took shape was reminiscent of an old-fashioned Cold-War style proxy conflict between the United States and Russia. Russia appeared poised to invade, or at least raid, a country strongly allied with the United States, which already had a military presence there.

Its major source of income is from transit fees charged for a little-used tunnel running into Russia.

Also seeking independence from Georgia, sometimes with Russian help, is another Muslim people, the Abkhazians, who make up about 2 percent of Georgia's population and are concentrated in the province known as Abkhazia. In 1993, Abkhazian separatists captured the Georgian Black Sea port of Sokhumi in Abkhazia. Russian forces initially aided them, both to regain access to Black Sea resorts and to take revenge on Georgian President Eduard Shevardnadze, the former Soviet foreign minister whom many Russians held partly responsible for the breakup of the USSR. Russia was also putting pressure on Georgia to rejoin the CIS. Shevardnadze responded by having Georgia rejoin the CIS and agreeing to allow four Russian military bases to remain for a while on Georgian soil and to allow Russian troops to be stationed on Georgia's border with Turkey. Russian forces then put a stop to the Abkhazian offensive but did not drive the rebels from the territory they had captured.

In 2001, when Georgia insisted that it was time for Russia to scale back its military presence in the region, Russia responded by cutting off supplies of natural gas to Georgia and by insisting that Georgians working in Russia obtain proper visas. Those expatriate workers are a vital source of revenue for Georgia, where 60 percent of the people live below the poverty line, so Russia has strong leverage in hinting they might be sent home.

Georgians see Russia as using the Abkhazian problem, energy supplies, and the visa issue as means of retaining leverage in the strategically vital Caucasus region and, above all, securing access to Abkhazia's 110 miles (180 km) of Black Sea coastline. For its part, Georgia wants to loosen ties with Russia in favor of better relations with the West and its allies (for example, it wants to join the European Union and the North Atlantic Treaty Organization). Most road and rail links between the countries have been severed, and Turkey has already replaced Russia as Georgia's main trading partner. Georgia also wants to reestablish its historically important position in overland trade with the Central Asian countries, particularly by serving as an outlet for the export of oil from the Caspian Basin. A critical issue is the fate of Batumi, the Black Sea oil-shipping port. It lies in the rebellious province of Ajaria (Adzharia), home to another Russian military base and to a population bitterly opposed to the rule of Georgian President Mikhail Saakashvili, who ousted former President Shevardnadze in the peaceful "Rose Revolution" of 2003. Saakashvili vowed to reassert firm Georgian control over all of the separatist regions within Georgia, thereby risking a direct military confrontation with Russia.

In sum, Russia's perspective on the other former members of the USSR—even the Baltics, which are now firmly in the European orbit—is that they constitute a special foreign policy region. Some observers contend that Russia's diplomatic, economic, and limited military involvement in these countries is an effort to establish a buffer zone between Russia and the "Far Abroad." Along the outer frontiers of the cordon of successor states, Russia maintains a chain of lightly-staffed military bases that could be quickly ramped up for

Figure 6.23 Russian soldiers outside Russia are variously seen as peacekeepers or conquerors.

combat if conditions warrant. Within the Near Abroad, notably in Moldova, Tajikistan, Georgia, Armenia, and Azerbaijan, Russian troops have become engaged frequently as "peacekeepers" in local ethnic conflicts (Figure 6.23). In Armenia and Tajikistan, at least, the governments fear what would happen without Russian military support.

There are some in the West and among Russia's neighbors, however, who fear that peacekeeping is merely a euphemism for a plan to restore Russian imperial rule. Some observers contend that Russia wants access to important resources and economic assets in its former republics, such as uranium in Tajikistan, aviation plants in Georgia, military plants in Moldova, and the Black Sea coast and naval fleet in Ukraine's Crimea. Noting Moscow's reluctance to recall Russian troops from the Near Abroad, some analysts fear that Russia's reasoning may be that Moscow's empire will again be expanded, and it is not worth dismantling bases only to have to rebuild them later.

Within Russia, there is in fact a body of public opinion favoring reassertion of Russian control over its former empire, but the strength of this feeling is unknown. President Putin spoke of Russia's urgent priority to draw the countries that were republics of the former Soviet Union closer to Russia. Such reintegration could, of course, take place through peaceful political and economic means. In one case it has; in 2000, Russia and Belarus officially established a political union and elected a joint parliament. Years later, it is still uncertain what this union means. Some observers believe it represents the push of Russian influence even farther to the west.

Kazakhstan favors close coordination with Russia, and in 1996, it joined Russia and Kyrgyzstan in an agreement calling for the creation of a common market to ensure the free flow of goods, services, and economic capital among them and to promote coordinated industrial and agricultural policies. In an attempt to create a common market and integrate their economies more effectively, the five "stan" countries of Central Asia had already formed a **supranational organization** called the United States of Central Asia in 1993.

The Far Abroad

Finally, there are outstanding geopolitical issues in relations between the region of Russia and the Near Abroad and the rest of the world. In the years following the breakup of the USSR, most of these involved the development of a new model of relations between East and West and a peaceful succession to the Cold War. During those years, important plans were also made about how to ship Central Asia's critical oil supplies westward. Since 9/11, these issues have been complemented by others related to the "war on terrorism," as perceived both by the United States and Russia.

Russia and the other successors of the Soviet Union have generally enjoyed improving relations with the West since the USSR dissolved. Russia has ramped up oil production and advertised itself to the United States as a more stable source of energy than the volatile Persian Gulf countries. The Warsaw Pact, a military alliance formed around the Soviet Union to take on the West's North Atlantic Treaty Organization, has dissolved and Russia has grudgingly accepted the survival and expansion of NATO in its backyard. Both sides have reduced their nuclear arsenals, with Russia taking the lead in calling for even deeper cuts. Russia's economic problems have led to a dramatic decline in the size and quality of its armed forces and arsenal. Ukraine became a nuclear weapon-free nation by disposing of the 1,800 nuclear warheads that had briefly made it the world's third largest nuclear military power. The United States rewarded that move with a much-increased foreign aid package to Ukraine. Ukraine's growing orientation to the West may also be seen in its membership in the NATO's "Partnership for Peace" program and in its application to become a member of the European Union. Many European countries are dependent on natural gas shipped by pipeline across Ukrainian land from Russia.

The West is no longer worried by the prospect of a conventional nuclear strike by any of the USSR's successors. However, the United States and its European allies are terrified that low budgets and lax security at nuclear facilities throughout the region could help divert nuclear weapons or nuclear fuel to terrorists. There have already been hundreds of thefts of radioactive substances at nuclear and industrial institutions in the former USSR. The destination points of these materials remain largely unknown. The West also fears that underpaid nuclear scientists will sell their know-how to governments or organizations such as al-Qa'ida with nuclear strike intentions. Former Soviet nuclear experts are known to have worked in Iran, Iraq, Algeria, India, Libya, and Brazil, and it is possible that some have cooperated with stateless organizations like al-Qa'ida. Osama bin Laden repeatedly stated his intention to acquire and use nuclear weapons, so the issue of "**loose nukes**" of Russia and the Near Abroad has great importance.

Since 9/11, and particularly since the U.S. invasion of Iraq in 2003, the geopolitical significance of the Central Asian "stan" countries has increased enormously. This region borders unsettled Afghanistan, postrevolutionary Iran, and the great power of China and is itself rich in mineral resources.

Geography of Energy

Oil in the Caspian Basin

Estimated oil reserves beneath and adjacent to the Caspian Sea total as much as 200 billion barrels, or about one-tenth of the world's total (and there are an estimated 600 billion cubic meters of natural gas in the region). Reserves in Azerbaijan, Kazakhstan, and Turkmenistan represent the world's third most important oil region, behind the Persian/Arabian Gulf and Siberia. The potential wealth of this resource will be realized only through a difficult and delicate resolution of several geographic and political problems (Figure 6.G).

One of the first to be dealt with was the question of which countries owned the fossil fuels under Caspian waters. With relatively small reserves of oil lying adjacent to their Caspian shorelines, Iran and for a while Russia argued that the Caspian should be regarded as a lake, meaning its lakebed resources would be treated as common property, with shares divided equally among the five surrounding states. Eventually, Russia conceded that the Caspian should instead be regarded as a sea consisting of five separate sovereign territorial waters, with each country having exclusive access to the reserves in its domain. Iran unhappily accepted this outcome, which gave Azerbaijan the largest share of Caspian oil. But in 2001, Iran and Azerbaijan skirmished over rights to an oilfield below the Caspian. Joined by Turkmenistan, Iran and Russia again began insisting that the Caspian should be treated as a lake rather than a sea.

Azerbaijan's oil situation epitomizes the economic prospects and obstacles facing the southern countries of the region of

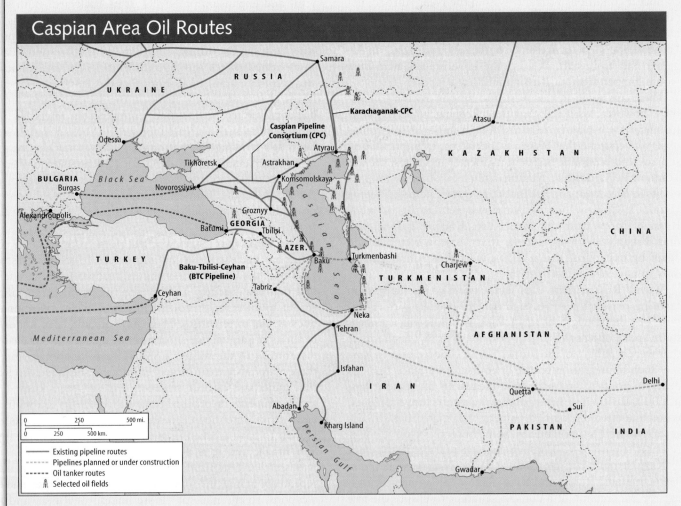

Figure 6.G Numerous physical and political obstacles stand in the way of oil exports from the Caspian region.

Russia and the Near Abroad. The Azerbaijani resource is large; there are huge reserves of oil and gas under the bed of the Caspian Sea. Planners studied Azerbaijan's oil with a difficult question in mind: What is the best way to get this oil to market? They considered a variety of export pipeline routes, each of which had a different set of geopolitical, technical, and ecological problems. Since the Caspian is a lake, Azerbaijan is landlocked, and Azerbaijani oil must cross the territories of other nations to reach market. These territories, however, are embroiled in the wars and political unrest left in the wake of the collapse of the USSR. Russia wants to retain strong influence over Azerbaijan in part by insisting that Azerbaijani oil reach export terminals on the Black Sea by passing through Russian territory, as it now does, through a pipeline to the Russian port of Novorossiysk.

Turkey opposes this strongly on the grounds that oil-laden ships navigating the narrow Bosporus and Dardanelles Straits imperil Turkey's environment (and Turkey would not receive transit fees for those shipments). Environmental concerns have not, however, kept energy-hungry Turkey from importing new supplies of natural gas from Russia via the Blue Stream pipeline (the world's deepest) across the Black Sea floor.

Iran offered a swap with Azerbaijan: Iran would import Azeri oil for Iran's domestic needs and would export equivalent amounts of its own oil from established ports on the Persian Gulf. Iran also argued that its territory offers the shortest and safest pipeline route for fossil fuel exports from Turkmenistan, Kazakhstan, and Azerbaijan. Iran is also in an excellent geographic position to facilitate transportation of other goods to and from the Central Asian nations, using its Gulf of Oman and Persian/Arabian Gulf ports. To stimulate this trade, Turkmenistan and Iran have begun linking their rail systems.

Azerbaijan's leaders, however, want more ties with the West (which is generally hostile to Iran and stands firmly against any pipelines routed through Iran) and independence from Moscow. The United States and its Western allies anticipate that Azerbaijan and Kazakhstan will soon be major counterweights to the volatile oil-producing states of the Middle East, and they want to be able to obtain Caspian Sea oil without relying on Russia's or Iran's goodwill. Therefore, in 1995, Azerbaijan signed an agreement with a U.S.-led Western oil consortium to export oil through a combination of new and old pipelines to the Georgian port of Batumi on the Black Sea. To avoid a direct confrontation with Russia over this sensitive issue, some of the initial production passed through the existing Russian pipeline to Novorossiysk.

This compromise applied only to the first stage of production, through 1997. A permanent route still had to be established, and both Turkey and Russia vied for the pipeline to pass through their territories. Russia insisted that the environmental hazard of navigating the Turkish straits could be averted by offloading oil at the Bulgarian Black Sea port of Burgas, shipping it overland through a new pipeline to the Greek Aegean Sea port of Alexandroupolis, and reloading it there. The alternative, a 1,080-mile (1,725-km) Turkish route, would be far from the straits, running across Azerbaijan from Baku, through Georgia (skirting hostile Armenia), and then across Turkey to the Mediterranean port of Dörtyol, near Ceyhan. The Western powers preferred this BTC (Baku-Tbilisi-Ceyhan) pipeline, mainly as a political means to block either Russian or Iranian reassertion of power in the region, and it was built. Georgia welcomed the projected $50 million in transit fees that the pipeline would generate but also worried about the security costs of protecting its 154 miles (c. 246 km) of the route. Turkey had security in mind when it routed the pipeline to detour around the volatile Kurdish-dominated region of southeastern Turkey; this explains the odd (and costly) bend in the Turkish portion.

Other Caspian Basin countries face similar questions about how best to export their resources. Turkmenistan, on the lake's southeast shore, has large oil deposits and natural gas reserves. At present, these are exported mainly in pipelines passing around the Caspian Sea through Russia, and Turkmenistan is seeking an alternative through Iran to Turkey (it already exports some natural gas to Iran via pipeline), across Afghanistan to Pakistan (a hazardous route, given conditions in Afghanistan), or under the Caspian to link up with Azerbaijani export routes. The country's leadership is in favor of that third alternative, the export route promoted by the United States, with oil and gas traveling westward through pipelines on the bed of the Caspian Sea to Baku and then through Georgian and Turkish pipelines to the Mediterranean Sea.

Of the five Central Asian countries, Kazakhstan has the largest fossil fuel endowment. With recent discoveries of huge reserves in the Kashagan oil field of the Caspian Sea, some have predicted that Kazakhstan will eclipse Russia and Saudi Arabia as the world's leading oil producers by 2015. In the late 1990s, Kazakhstan began exporting its oil from fields around Tengiz to join the Baku-to-Novorossiysk pipeline at Komsomolskaya. Unwilling for the time being to find an alternative to a Russian route, the Kazakhs worked with Omani, Russian, and American firms to build a pipeline bypass from Komsomolskaya to Novorossiysk. Kazakhstan is also studying the possibility of building a 2,000-mile (c. 3,000-km) pipeline to export oil eastward into China, but the construction costs are daunting.

257

Diverse kinship, spiritual, economic, and political relations exist between the Central Asian Muslims and their neighbors on the other side of the international frontiers. Along with internal ethnic conflicts and dissatisfaction over living conditions, these international affiliations may make the region a major political problem area in the future.

Both Iran and Turkey are vying for increased influence in the region. Except in Tajikistan, support for Iran is largely lacking, in part because the majority of Central Asians practice Sunni rather than Shi'ite Islam. Ethnicity is also important; the four Turkic states are inclined to orient with Turkey, with which they have already established cultural ties such as educational exchanges and shared media. Turkey also has extended economic assistance and established small business ventures and large construction contracts in Central Asia. Turkmenistan has good relations with both Turkey and Iran and has completed a rail link with Iran. Turkmenistan also tries to maintain good ties with Russia.

The region has become a focus for rivalry between the historical enemies Turkey and Russia. Russians are concerned that Turkey is winning too much influence in the young Muslim countries of Central Asia. For their part, many Turks believe Russia will try to take over the Transcaucasus region again and then turn to Central Asia—a fear shared to some extent by the United States, which is intent on maintaining secure access to the region's oil (see Geography of Energy, pages 168–169). Some Turks uphold a dream of Pan-Turkism, uniting all the Turkic peoples of Asia from Istanbul to the Sakha Republic in Russia's Siberia region—a prospect that Moscow does not like.

The twin scourges of terrorism and narcotics have put Central Asia on the geopolitical hot-spot map. Uzbekistan has erected concrete and barbed wire barriers and sown land mines on its borders with Tajikistan and Kyrgyzstan. The Uzbek government is trying to prevent the incursion of what it sees as two huge threats: an epidemic in heroin use and a militant Islamic insurgency. Both issues have preoccupied the region's governments. The major routes for exports of heroin from Pakistan and Afghanistan are north through Central Asia and thence into Russia and Western Europe. Along this pathway, large numbers of Central Asians have fallen prey to inexpensive, pure heroin. The associated health and social problems, including HIV, hepatitis, and prostitution, have also taken root (in the early 2000s, HIV infection rates grew faster here than in any other world region). So far, only Kyrgyzstan is fighting an aggressive public health education campaign, including a needle-exchange program for intravenous drug users and free condoms for prostitutes and their clients. Tajikistan's economy is now heavily dependent on drug trafficking, and most Western investors and aid agencies avoid the country.

The significance of an Islamist threat to Central Asian stability is uncertain. Russia has expressed great concern about an Islamist wave spreading into Central Asia and Russia itself. Tajikistan and Kyrgyzstan have allowed Russia to deploy thousands of Russian troops in their countries, in part

to stave off such a threat. Some observers believe Russia is using the Islamist threat as a pretense to maintain influence in the region or even to counter the growing U.S. military presence there since 9/11. Others believe the Islamist danger is real. Tajikistan is already home to a group called the Islamic Movement of Uzbekistan (IMU), which aims to overthrow the Uzbek government of President Islam Karimov and establish an Islamic state in the Fergana Valley. Following a 1999 attempted assassination blamed on the rebels, Karimov cracked down hard on suspected Islamists and even began to restrict symbols of Islamist sympathies, such as beards on men and head scarves on women. Uzbekistan erected an impressive cordon of barbed wire and line mines on its border with Kyrgyzstan to keep out any insurgents. Anxious to repress any external support for Islamist insurgency within its borders, and to court political and economic favor with Washington, Uzbekistan welcomed the U.S. military to use its land and air space in attacks against al-Qa'ida and the Taliban in 2001 and 2002. The United States subsequently maintained a military base there (known as Camp Stronghold Freedom) and in nearby Kyrgyzstan.

Most recently, China has sought to build security relationships with the Central Asian countries, especially to blunt any aspirations by Muslim insurgents in China's far western region. China has economic interest in the region, too; China and Kazakhstan are cooperating in the construction of a railway across Kazakhstan that will ultimately link producers and markets in China and Europe. Kazakhstan meanwhile continues to build strong ties with the West. Having given up its nuclear arsenal, Kazakhstan wants the West to help guarantee the security of its borders with Russia, China, the Caspian Sea, and three other Central Asian nations.

Maintaining security and other geopolitical interests in the region are costly enterprises for the West. Western governments and their financial institutions have extended large amounts of economic aid to many Soviet successor countries, particularly to Russia. It has been difficult to gauge how useful this assistance has been, and there is resentment on both sides about it. Many critics complain that such aid has failed to reach its intended targets and instead has lined the pockets of the oligarchs. One commented that while under communism the arrangement between the worker and the state was, "They pretend to pay us, we pretend to work." The new arrangement between the West and Russia is, "You Russians pretend to be creating a law-based, market-friendly liberal democracy. We westerners pretend to believe you—and what's more, we pay you for it."[1] For their part, many Russians perceive their leaders as having "sold out" to the West and resent their growing dependence on its institutions, exports, media, and culture. Russian poet Yevgeny Yevtushenko lamented what he called "the McDonaldization of Russia."

[1] Quoted in "Fuelling Russia's Economy." *The Economist*, August 28, 1999, p. 13.

CHAPTER SUMMARY

- From the Russian perspective, the other 14 countries of the former Soviet Union make up the region of the "Near Abroad," a special realm of policy and interaction. The three Baltic countries have struggled to divorce themselves from this orbit. The remaining countries, along with Russia, are the member states of the Commonwealth of Independent States. All continue to have significant economic and other ties with Russia, and Russia aspires to increase its influence on them.

- The area west of the Ural Mountains and north of the Caucasus Mountains is known as European Russia; the Caucasus and the area east of the Urals is Asiatic Russia. Siberia is the area between the Urals and the Pacific. Central Asia is the name for the arid area occupied by the five countries immediately east and north of the Caspian Sea.

- Russia and the Near Abroad span 11 time zones. Stretching nearly halfway around the globe, the region has formidable problems associated with climate, terrain, and distance.

- Birth rates have fallen and death rates have risen dramatically in Russia, leading to steep declines in population.

- Vast stretches of the region are at very high latitudes, making agriculture difficult. The most productive agricultural area is the steppe region south of the forest in Russia and in Ukraine, Moldova, and Kazakhstan. Here the low and variable rainfall is partially offset by the fertility of the mollisol soils.

- In early Russia, rivers formed natural passageways for trade, conquest, and colonization. In Siberia, the Russians followed tributaries to advance from the Urals to the Pacific. More recently, alteration of rivers for power, navigation, and irrigation became an important aspect of economic development under the Soviet regime.

- Large plains dominate the terrain from the Yenisey River to the western border of the country. The area between the Yenisey and Lena Rivers is occupied by the hilly Central Siberian Uplands. In the southern and eastern parts of the region, mountains, including the Caucasus, Pamir, Tien Shan, and Altai ranges, dominate the landscape.

- Slavic and Scandinavian peoples were prominent in the early cultural development that was the foundation for the Russian, Ukrainian, and Belarusian cultures. Although Slavs came to dominate economic and political life, the region has large and diverse populations of non-Slavic peoples.

- Early contacts with the Eastern Roman or Byzantine Empire led to the establishment of Orthodox Christianity in Russian life and culture. Although it was repressed after the Bolshevik Revolution, this and the region's other religions have been experiencing a rebirth since the breakup of the Soviet Union.

- From the 15th century until the 20th, the tsars built an immense Russian empire around the small nuclear core of Muscovy (Moscow). This imperialism brought huge areas under tsarist control. Russian expansion extended to the Pacific, across the Bering Strait, and down the west coast of North America as far as northern California.

- Russia has often triumphed over powerful foreign invaders. These achievements have been due in part to the environmental obstacles the invaders faced, the overwhelming distances involved, and the defenders' willingness to accept large numbers of casualties and implement a scorched earth strategy to protect the motherland.

- Policies of the Soviet Union permitted most ethnic groups to retain their own languages and other elements of traditional cultures. However, the Soviet regime implemented a deliberate policy of Russification in an effort to implant Russian culture in non-Russian regions. Millions of Russians settled in non-Russian areas, but local cultures were not converted to Russian ways.

- To meet the needs of its people, the USSR became a collective welfare state, with the government providing guaranteed employment, low-cost housing, free education and medical care, and old-age pensions. However, military-industrial superpower status was generally achieved at the expense of the ordinary consumer.

- Current efforts to privatize industry and agriculture are slowed by the difficulties of reforming the old, inefficient state-run systems of the Communists.

- The sudden transition from a command economy to capitalism in Russia and other countries in the region led to a widening gap between the rich and the poor, with growing ranks of poor people. Organized crime and an "underground economy" became common. Agricultural and industrial production fell dramatically. However, high oil prices in the early 2000s improved Russia's economic outlook.

- There are three important realms of geopolitical concern in the region. Within Russia, the aspirations of non-Russian people like the Chechens and Tatars appear to pose a danger to the unity of the Russian Federation. Between Russia and the Near Abroad countries, there are challenging issues including energy shortages and supplies, desires of Russians outside Russia to achieve their own rights and territories, stationing of Russian troops, and Islamist terrorism. Between this region and the rest of the world— the "Far Abroad"—there are concerns about the fate of Soviet-era nuclear materials, whether the oil-rich Central Asian countries are more sympathetic to Russia or Turkey, and what should be done to stem the tide of narcotics and terrorism.

- The Russian government has said that it has the right and duty to protect Russian minorities in the other countries. Some observers contend that Russia is attempting to reestablish the Near Abroad as buffers between Russia and the Far Abroad.

- The Caspian Sea and adjacent areas make up one of the world's greatest petroleum and natural gas regions. Russia, Iran, and the West, particularly the United States, are vying for influence in this increasingly important area. Of particular concern to the rivals is how the energy supplies should be routed via pipeline to reach ocean terminals.

⊙ Because of nuclear weapons and other critical issues, relations between Western countries and the countries of the former Soviet Union, particularly Russia, are vital to global security. The West gives much economic aid to Russia and some of the other countries but worries that it is misspent. For their part, some of the countries of Russia and the Near Abroad resent growing dependence on the West.

KEY TERMS + CONCEPTS

Terms in blue are also defined in the glossary.

Abkhaz-Adyghean language family (p. 149)
alfisol soils (p. 146)
Altaic language family (p. 149)
   Turkic subfamily (p. 149)
arms race (p. 156)
autonomies (p. 160)
barter (p. 159)
black-earth belt (p. 146)
Bolshevik Revolution (p. 153)
bourgeoisie (p. 156)
chernozem soils (p. 146)
Cold War (p. 139)
collectivized agriculture (p. 157)
command economy (p. 156)
Commonwealth of Independent States (CIS) (p. 161)
Cossacks (p. 151)
demographic transition (p. 144)
devolution (p. 161)
economic shock therapy (p. 158)
"Evil Empire" (p. 139)
gigantomania (p. 156)
glasnost ("openness") (p. 157)
Gosplan (Committee for State Planning) (p. 156)

Great Volga Scheme (p. 148)
Indo-European language family (p. 149)
   Armenian (p. 149)
   Iranian subfamily (p. 149)
     Tajik (p. 149)
   Romance language subfamily (p. 149)
     Moldovian (p. 149)
irredentism (p. 165)
Jews (p. 150)
Kartvelian (South Caucasian) language (p. 149)
kleptocracy (p. 159)
land empire (p. 152)
"loose nukes" (p. 167)
misdeveloped country (p. 158)
mollisol soils (p. 146)
Mongols (p. 150)
Nakh-Dagestanian language family (p. 149)
oligarchs (p. 159)
Operation Barbarossa (p. 153)
Proto-Asiatic language family (p. 149)
   Chukotko-Kamchatkan language subfamily (p. 149)
     Chukchi (p. 149)
     Koryak (p. 149)

perestroika ("restructuring") (p. 157)
permafrost (p. 143)
proletariat (p. 156)
Rus (Varangians) (p. 149)
Russian Revolution (p. 153)
Russification (p. 153)
sacred space (p. 154)
satellite countries (p. 139)
scorched earth (p. 153)
Slavic language family (p. 149)
   Belarusian (p. 149)
   Russian (p. 149)
   Ukrainian (p. 149)
spodosol soils (p. 145)
supranational organization (p. 167)
taiga (p. 143)
Tatars (p. 150)
"The Golden Horde" (p. 151)
"Third Rome" (p. 149)
underground economy (p. 159)
Uralic language family (p. 149)
   Finno-Ugric subfamily (p. 149)
"virgin and idle lands" ("new lands") (p. 157)
welfare state (p. 157)
"Window on the West" (p. 151)

REVIEW QUESTIONS

WORLD
REGIONAL
Geography ⊛ Now™

Assess your understanding of this chapter's topics with additional quizzing and concept-based problems at http://earthscience.brookscole.com/wrg5e.

1. What are the main climatic belts and corresponding vegetation types of Russia and the Near Abroad? Which are most and least productive for agriculture?

2. What are the region's major river systems?

3. Why is Russia's population declining?

4. Why is Russia known as a "land empire"? How did it acquire its empire?

5. What significant conflicts have occurred in this region? Why has Russia so often won them? What was the geographic significance of Stalingrad in World War II?

6. What were the perceived advantages of "collectivized agriculture"?

7. What major Soviet Communist projects changed the landscape of this region?

8. What were some of the successes and failures of efforts to industrialize the Soviet Union?

9. What were some of the results of Russia's "shock therapy" economic reform and the subsequent conditions of the country and its people? What accounts for the recent turnaround in Russia's economy?

10. Which of the non-Russian ethnic groups in Russia are particular security concerns to Russia today?

11. How does Caspian Sea oil reach markets abroad?

12. Where outside Russia are Russian and U.S. military troops stationed? What are the reasons for their presence?

DISCUSSION QUESTIONS

1. What roles have the natural environment played in this region's history and development?

2. What cultures were important in the region's development? What were the major contributions of these cultures?

3. What was Russification? Was it successful? Which countries of the former Soviet Union are more and less inclined to embrace Moscow today? How does Russia try to ensure that its interests in the region are served?

4. Why was Moscow known as the "Third Rome"? What are the significant religions of the region?

5. What were the foundations for the Communist economic system? What is this system's legacy for today's economy?

6. Why did the Second Russian Revolution occur? What are some of the difficulties that the present government is facing? What are the prospects for Russia's economy?

7. Why is Russia so intent on pacifying Chechnya?

8. Why are ethnic Russians troubling to the governments of Moldova and Ukraine?

9. What problems is Georgia having with ethnic minorities and with Russia?

10. What are the perceived terrorist threats in this region to Russia and the United States?

11. What are the interests of Russia, the United States, Turkey, and Iran in Central Asia? Which Caspian Sea oil export routes do Russia and the United States prefer and why?

12. How might events in this region impact you, your community, or your country?

Fragmentation and Redevelopment in Russia and the Near Abroad

Children at a day-care center in Ulyanovsk, Russia. They are growing up in the new and uncertain world that follows decades of Soviet rule.

chapter objectives

This chapter should enable you to:

- Recognize core and peripheral subregions of Russia and the Near Abroad

- Identify geographic obstacles (for example, vast distances) and opportunities (for example, transpolar routes) for Russia in east–west trade

- Appreciate the importance of oil in the development of Russia's peripheral regions

- See that free enterprise in former Communist countries has introduced a new set of environmental problems

- View cotton as a legacy of Soviet "colonialism" in Central Asia

WORLD
REGIONAL
Geography Now™

Look for this logo in the text and go to GeographyNow at http://earthscience.brookscole.com/wrg5e to explore interactive maps, view animations, sharpen your factual knowledge and geographic literacy, and test your critical thinking and analytical skills with unique interactive resources.

Few geographers foresaw the collapse of the Soviet Union. Like most other observers, we assumed that, when threatened with disintegration, the core of Russia would assert its military strength to retain the countries of the periphery at any cost. From the geographer's perspective, the survival of the Soviet Union and its status as a global superpower always depended on Russia's ability to control non-Russian resources such as the oil of the Caucasus, the soil of Ukraine, and the cotton of Central Asia. Asked prior to 1991 whether Russia would let go of these, most of us answered "no," not without a violent struggle.

But the union of so many disparate peoples was so unlikely that, in retrospect, it seems absurd. National unity was, in hindsight, an oxymoron in the Soviet Union. Marxism–Leninism had assumed that the cultural and linguistic differences among more than 200 separate groups would disappear as the ideal, universal proletarian society emerged. In effect, Moscow's rulers hoped they would all become Russian. Russification was supposed to level out the differences, but it could not and did not.

Now Russia has lost all but the core: the Russian Federation, itself an amalgam of many non-Russian entities cemented artificially to the core. The other former republics have gone their own ways as independent countries. How are these entities faring today and what is their future? What has happened to the Russian and non-Russian resources, agricultures, and industries that once supported a global power? This chapter will answer these questions and more.

An outstanding feature in the geography of this huge region is that resources are distributed unevenly, favoring development in certain subregions and countries. Agricultural and industrial resources are in fact clustered in a core region encompassing western Russia, northern Kazakhstan, Ukraine, Belarus, and Moldova. Nearly three-fourths of the region's people—the great majority of them Slavic—and an even larger share of the cities, industries, and cultivated land of this immense region are packed into this roughly triangular **Fertile Triangle** composing about one-fifth of the region's total area. Also known as the **Agricultural Triangle** and the **Slavic Coreland,** this is the distinctive functional **core** of the region (Figures 7.1 and 7.2).

WORLD
REGIONAL
Geography Now™

Click Geography Literacy to make a map of the core and periphery of this region.

A core suggests that there is a **periphery,** and one does exist—the world's largest. The remaining four-fifths of the region of Russia and the Near Abroad, lying mostly in Asia, consist of land supporting only spotty human settlements in conditions of sometimes huge environmental adversity. Most of the region's non-Slavic peoples live in these lands, but many immigrant Slavs also reside there, generally in cities. Although these areas are marginal when compared with the well-endowed Fertile Triangle, they do include particular resource and production centers that represent realized or potential wealth. Several countries in the Caspian Sea region

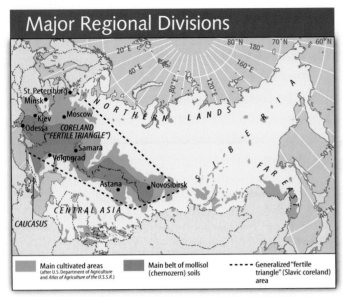

Figure 7.1 Major regional divisions of Russia and the Near Abroad as discussed in the text.

are rich in minerals. There are pockets of diverse and productive agriculture in the Caucasus. Russia's Siberia is a storehouse of minerals, timber, and waterpower. The geographic pattern that has emerged is a series of discrete production nodes widely separated by taiga, tundra, mountains, deserts, wetlands, and frozen seas.

7.1 Peoples and Nations of the Fertile Triangle

Slavic peoples are the dominant ethnic groups in most of the region of Russia and the Near Abroad in both numbers and political and economic power. The major groups are the Russians (discussed extensively in the previous chapter), Ukrainians, and Belarusians (Byelorussians). Most of these Slavs live in the Coreland extending from the Black and Baltic Seas to the neighborhood of Novosibirsk in Siberia. This area contains about one-half the area and more than four-fifths the population of the United States. Moscow

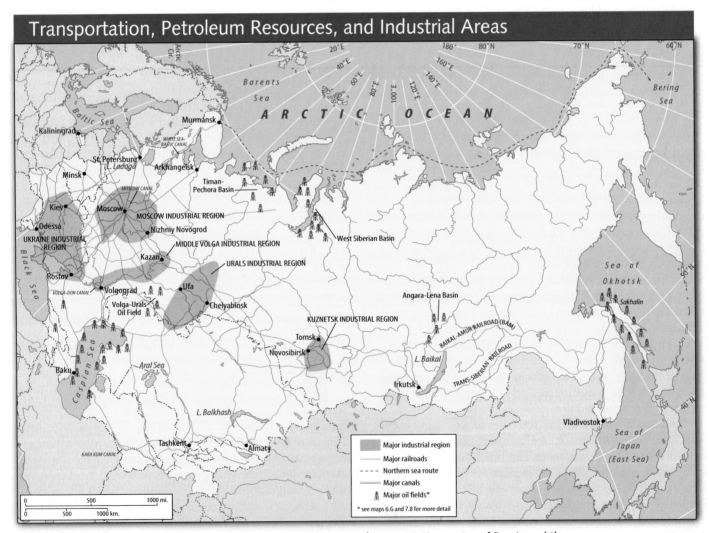

Active Figure 7.2 Major economic zones and transportation routes of Russia and the Near Abroad. The map reveals that the Fertile Triangle is also the industrial core of the region. Moscow's function as a transportation hub is evident. *See an animation based on this figure, and take a short quiz on the facts and concepts.*

(population: 11.8 million) and St. Petersburg (formerly Leningrad; population: 4.8 million) are the largest cities, followed by the Ukrainian capital, Kiev (population: 3 million). This is mainly a region of lowlands; the only mountains are the Urals, a small segment of the Carpathians, and a minor range in the Crimean Peninsula. The original vegetation was mixed coniferous and deciduous forest in the northern part and steppe in the south.

Ukraine

Although closely related to the Russians in language and culture, Ukrainians—the second largest ethnic group in the Coreland—are a distinct national group. The name Ukraine translates as "at the border" or "borderland," and this region has long served as a buffer between Russia and neighboring lands (Figure 7.3). Here, armies of the Russian tsars fought for centuries against nomadic steppe peoples, Poles, Lithuanians, and Turks, until the 18th century when the Russian empire finally absorbed Ukraine. For three centuries, Ukrainians were subordinate to Moscow, first as part of the Russian empire and then as a Soviet republic. Ukraine's industrial and agricultural assets were always vital to the Soviet Union; the Bolshevik leader Lenin once declared, "If we lose the Ukraine, we lose our head."

Slightly smaller than Texas, Ukraine (population: 47.4 million) lies partly in the forest zone and partly in the steppe. On the border between these biomes, and on the banks of the

Figure 7.4 Kiev, the historic capital of Ukraine

Dnieper River, is the historic city of Kiev, Ukraine's political capital and a major industrial and transportation center (Figure 7.4). Although its agricultural output has plummeted since the USSR's demise, Ukraine was traditionally a great "breadbasket" of wheat, barley, livestock, sunflower oil, beet sugar, and many other products, grown mainly in the black-earth soils of the steppe region south of Kiev.

North of Kiev is Chernobyl, where an explosion at a nuclear power station in 1986 rendered parts of northern Ukraine and adjoining Belarus incapable of safe agricultural production for years to come. Neurological, endocrinological, gastrointestinal, and urinary illnesses are 15 times greater in the parts of Ukraine contaminated by this accident than in other regions of the country. Having shut down the Chernobyl plant in 2000, Ukraine has promised to close all of its Chernobyl-type nuclear plants within a few years. This will be a costly proposition, considering the country's shortage of affordable alternative fuels.

Ukraine has paid a high economic cost for independence. Previously, the Soviet economy subsidized the provision of Russian oil, gasoline, natural gas, and uranium to Ukraine. But with independence, Ukraine was compelled to pay much higher world market prices for these commodities. Ukraine now maintains a costly debt for the natural gas it obtains from Russia. Ukraine is, however, well endowed with other mineral resources, including large deposits of coal, iron, manganese, and salt.

Since independence, Ukraine has remained far more reliant on heavy industrial production than has Russia, so these resources are actively exploited. Coal is concentrated in the Donets Basin (Donbas) coal field near the industrial center of Donetsk, where most of Ukraine's iron and steel plants are also located. Iron ore comes by rail from central Ukraine in the vicinity of Krivoy Rog, the most important iron-mining district in the region of Russia and the Near Abroad. Most of the ore is used in Ukrainian iron and steel plants. This area is also a center of chemical manufacturing. Surrounding the inner core of mining and heavy-metallurgical districts in Ukraine is an outer ring of large industrial cities, including Kiev, the machine-building and rail center of

Figure 7.3 Belarus, Ukraine, and Moldova were among the most productive republics of the Soviet Union and are now struggling to develop their economies independently or in association with Russia.

Kharkov (east of Kiev), and the seaport and diversified industrial center of Odessa on the Black Sea.

Along with Belarus, Ukraine is the former Soviet republic most like Russia—"Russia writ small," as one author described it.[1] Many Ukrainians would like to cultivate their Slavic distinctiveness and pursue closer ties with their Slavic Russian kin rather than seek a closer orbit around the West. Since achieving independence from Moscow in 1991, Ukraine has restored the Ukrainian language to its educational system. Still, a majority of Ukrainians wants Russian to be a second official language. Ukrainians share many of the Russians' postindependence problems. Most Ukrainians are poorer today than they were under communism. There are problems of corruption, and as in Russia, much business and accompanying wealth are in the hands of oligarchs and organized crime.

Belarus

The Kansas-sized country of Belarus (formerly called Belorussia, or Byelorussia, and "White Russia"; population: 9.8 million) adjoins Ukraine on the north. Minsk (population: 1.6 million) is the political capital and main industrial city. Belarus developed slowly in the past, mainly because of a lack of fuels other than peat. Recently, however, oil and gas pipelines from Siberia and other areas have provided raw material for oil-refining and petrochemical industries. There is some oil production in Belarus itself. Under Soviet rule, Belarus became an industrial powerhouse, shipping televisions, farming equipment, automobiles, textiles, and machinery to both Moscow and its Eastern European satellites. Belarus's foreign trade is now almost exclusively with Russia, much of it in the form of barter; for example, Belarus has sometimes paid off debts to Russia's natural gas monopoly by supplying it with refrigerators, tractors, and other hard goods. Unlike Russian industries, most manufacturing enterprises in Belarus are still state-run, under what the country's leaders call "market socialism." Industrial output is low, but Belarus has not suffered the extensive economic disruption typical of Russia and most of the other Soviet successor countries.

Belarusian farms produce potatoes (particularly for the production of vodka), small grains, hay, flax, and livestock products. Southern Belarus contains the greater part of the Pripyat (Pripet) Marshes, large sections of which were drained for agriculture in Soviet times. Agricultural productivity declined after 1986, when much ground was contaminated by the nuclear accident in nearby Chernobyl. To compensate for these losses, Belarus stepped up production of agricultural machinery for export.

Belarus's strategic geographic location between the Russian Coreland and Western Europe brought it enormous suffering in World War II. En route to Moscow and Leningrad, the Nazi army laid waste to Belarus. One-fourth of the population was killed, and three-quarters of the cities were destroyed. One and a half million Belarusians fled eastward to escape the fighting. The Jews of Belarus were decimated by Hitler's "final solution."

Of all the former Soviet republics, the Slavic nation of Belarus has the closest ties with Russia. The two countries unified their monetary systems in the mid-1990s. In this agreement, Belarus gave up sovereignty over its currency and banking system in exchange for preferential access to Russian energy and other resources. Soon after this, an overwhelming majority of Belarusians voted for even closer economic integration with Russia and for the adoption of Russian as their official language. This led to the official designation of Belarus and Russia as a **union state** with plans to integrate their economies, political systems, and cultures. During its subsequent financial crisis, Russia evaluated the costs of absorbing some of the financial liabilities of Belarus, and based on the evaluation, it decided not to implement the union fully. Militarily, however, the two nations are already intertwined; Belarus has reintegrated its air force, ground forces, intelligence, and arms production with those of Russia. And apparently to prepare for possible absorption into Russia, the government has removed Belarusian history and language from school curricula.

Moldova

Maryland-sized Moldova (known formerly as Moldavia and, in the 19th and early 20th centuries, as Bessarabia; population: 4.2 million), made up largely of territory that the Soviet Union took from Romania in 1940, adjoins Ukraine at the southwest. The indigenous Moldovan majority (about two-thirds of the country's population) is ethnically and linguistically Romanian, not Slavic. There are serious problems of ethnic antagonism and political separatism within Moldova's borders. In the eastern part of the country, in a sliver of territory along the Dniestr River known as Transdniestria, ethnic Russians and Ukrainians have been agitating, sometimes violently, to form a separate state (see Figure 7.3). Their self-proclaimed Trans-Dniester Republic is effectively beyond control of the government and has become a haven for smuggling and money laundering. Despite protests from the European Union and other international organizations, Russia maintains troops there, ostensibly to protect the ethnic Russians. Elsewhere in the country, Moldovans who claim that Moldovans are in fact Romanians want Moldova to join Romania. They engaged supporters of Russia in a civil war in 1992 and now are protesting mandatory Russian-language instruction in the country's schools.

Mainly a fertile, black-earth steppe upland, Moldova has no major mineral resources and is primarily agricultural. The Moldovan specialty is cultivation of vegetables and fruits, especially grapes for winemaking. Industrial production in Moldova jumped after World War II, especially around the capital of Chisinau (also known as Kishinev; population: 770,000). When the country gained independence, it embraced free-market enterprise. But Moldova has

---

[1] "The Ukrainian Question." *Economist*, November 29, 1999, p. 20.

Geographic Spotlight

The Sex Trade

One of the outcomes of the collapse of communism and the dissolution of the Soviet Union has been the unleashing of long-restrained expressions of sexuality. Entrepreneurs in Russia and other former Soviet countries quickly adopted an axiom of Western advertising: Sex sells. Russian television has surpassed even the spiciest fare on standard American television.

Many sexual awakenings in the region are applauded or accepted as good clean fun, but others have a decidedly dark side, and some send ripples far abroad. Women of the former Soviet Union have become commodities (Figure 7.A). Russian, Ukrainian, and Kazakh partners or brides can in effect be purchased by anyone in the world with access to the Internet. Dot com agencies act as brokers for these New Age "mail order brides." Some of the subsequent unions end happily, of course, but others collapse with physical abuse and heartache. Frequently, the woman ends up abandoned and destitute in a strange land.

The sex trade takes the greatest toll. Women are driven into it often unwittingly and almost always out of poverty. Women have paid the greatest price economically since communism's demise; two-thirds of Russia's unemployed, for example, are women. Criminal gangs use false promises of conventional employment to lure young women into the sex trade. The typical victim is an 18- to 30-year-old Russian, Ukrainian, Belarusian, or Moldovan—often with children to support—who is approached by a trafficker promising her a job as waitress, barmaid, babysitter, or maid in Eastern or Western Europe. She pays a fee of $800 to $1,000 for travel and visa costs. Arriving in Bosnia and Herzegovina, the Czech Republic, or Germany, she is met by another trafficker who confiscates her passport and identity papers and then sells her for several thousand dollars to a brothel owner. She is escorted to her guarded apartment. There she is compelled to reside with several other women who have taken a similar journey, and she is informed of her real job: prostitute.

Berlin papers advertise the merchandise: "The Best from Moscow!" "New! Ukrainian Pearls!" The working women meet their clients in clubs or brothels or are driven to their homes. The going rate for their services is $75 per half hour, but less than 10 percent goes to the women. Their pimps and other gang members take the lion's share, and the women use their meager earnings to buy food and pay rent. Most end up in debt. Unable to pay off her debts, lacking identification papers, fearful of going to the police, and terrified the gang might harm her family members back home if she tries to escape, the woman is trapped. There are common accounts of depression, isolation, venereal disease, beatings, and rape.

This is a big business with striking demographics. The sex industry in Europe generates revenues of about $9 billion per year. Ukrainian sources estimate that since the country's independence in 1991, more than half a million Ukrainian emigrant women have ended up in the European sex trade. Figures vary widely, but tens and perhaps even hundreds of thousands of women of the former Soviet Union immigrate to Europe each year, the great majority of them becoming prostitutes. There is an overflow into other regions. These so-called "Natashas" work in Turkey, Israel, Egypt, and other countries in the Middle East in similar deplorable circumstances.

Joe Hobbs

Figure 7.A Women of Russia and the Near Abroad are some of the main participants in and victims of the global sex trade.

had little to offer in the world market, and the economy has suffered. Economic dislocation has led to the involvement of many Moldovan women in the sex trade (see Geographic Spotlight, above). Voter dissatisfaction with the economy has brought Communists back to power.

Kaliningrad

Kaliningrad is the odd piece of Russia stranded between Lithuania and Poland (see Figure 5.13, page 114). As the northern half of the former German East Prussia, this territory—then known as Konigsberg—was transferred to the USSR at the end of World War II to become the Kaliningrad Oblast. Soviet authorities expelled nearly all of Kaliningrad's Germans to Germany, and Russians now make up more than three-fourths of the population of 970,000. Kaliningrad remains a massive military and naval establishment, with defense workers and their dependents and retired military families comprising most of the population. Moscow would like Kaliningrad to continue to be principally a military state, but many in Kaliningrad have other ideas about how to capitalize on their strategic geographic location. Seeing themselves as **Euro-Russians**, they want Kaliningrad to become a more vibrant Eastern European economy. Some dream of making Kaliningrad one of the world's great free-trade zones, the "Hong Kong on the Baltic."

Kaliningrad does have some geographic and other factors in its favor: Its port seldom freezes, labor is inexpensive, and it is a logical way station for east–west trade. Already much legitimate trade and also smuggling are carried on through Kaliningrad, helping to raise standards of living for its people. The fear, however, is now that Lithuania and Poland have joined the European Union, Kaliningrad will be left out to drift in a state of underdevelopment. Its people will also find it harder to transit EU territory en route to the Russian "mainland."

7.2 Agriculture and Industry in the Russian Coreland

Most of the Fertile Triangle is within Russia, which faces huge problems in transforming state-run into free-market farming. Throughout the second half of the 20th century, the Fertile Triangle was a global-scale producer of farm commodities such as wheat, barley, oats, rye, potatoes, sugar beets, flax, sunflower seeds, cotton, milk, butter, and mutton. The overall high output, however, did not reflect high agricultural productivity per unit of land and labor. There were many inefficiencies in the state-operated and collective farms that dominated the agricultural sector. On the state-operated farms, workers received cash wages in the same manner as industrial workers. Bonuses for extra performance were paid. Workers on collective farms received shares of the income after obligations of the collective had been met. As on the state farms, there were bonuses for superior output.

These methods, however, failed to provide enough incentives for highly productive agriculture. Farm machinery stood idle because of improper maintenance; since no individual owned the machine, the incentive to repair it was diminished. There were also shortages of spare parts. Poor storage, transportation, and distribution facilities, plus wholesale pilfering, caused alarming losses after the harvest. Younger people were deserting the farms for urban work, often living with the family in the countryside but commuting to their city jobs. With the shortage of younger male workers, the elderly, women, and children did a large share of the farm work.

In an effort to reverse such disturbing trends, all of the countries of the Fertile Triangle are promoting land reform by privatizing collective and state farms and developing more independent or "peasant" farming. More than 90 percent of Russia's farmland is now privately owned, at least on paper; much of "private" land is still under collective shareholding. Bureaucratic obstacles slow down the purchase and sale of land, and agricultural efficiency is hard to achieve.

However, there was unmistakable improvement in the agricultural sector in the early 2000s. For the first time since independence, Russia began to record surpluses in grain production. The country continued to be a net importer of meat and dairy products, but their production was also rising. Most analysts felt that the worst was over for Russia's agriculture.

Russia had the lion's share of Soviet industries and natural resources but lacked certain vital commodities and manufacturing capabilities. Access to those resources was not a problem in Soviet times, when Moscow commanded production of steel in Ukraine or cotton in Central Asia for "USSR Inc." Now, however, Russia must compete in an open market to buy and sell raw materials, components, and manufactured goods. The very abrupt transition from a command to a free-market economy has been a wrenching experience for Russia. That transition is still very much underway, and the outcome is uncertain. The industrial resources are there, the country has a large and highly skilled work force, and the markets both within and outside its borders are vast. But Russia has squandered much of its industrial promise in recent years. The process of privatizing formerly state-controlled industries has been corrupted by the Russian Mafia and by "cronyism," the deal making and favoritism among Russia's new tycoons that thwart the emergence of a genuine free market. The largest industries are extremely inefficient and continue the Soviet practice of churning out products that no one needs. A single steel factory typically employs more than 10,000 workers—far more than are needed—in great contrast to the lean and efficient plants of the West. Russia became a giant "rust belt," where aging industries suffered mechanical breakdowns that were seldom fixed.

Russia simply found it too expensive to buy new equipment or maintain Soviet-era equipment. Russia's gross domestic product (GDP) shrunk by almost half in the 1990s, the largest fall in production any industrialized country has ever experienced in peacetime. It is little wonder that Russia sought to profit by selling rights to emit carbon dioxide and other greenhouse gases to the richer countries of the West: Its carbon dioxide emissions in 2004 were a third lower than they were in 1990, a sure sign of a deeply troubled industrial sector.

The worst may be over for Russia's industries, however. If revenues from oil and other natural resource exports continue to be high, the benefits could spread to many areas of manufacturing and the country could enjoy a more stable, diversified economy. There is no way to know whether that will happen, but in the meantime, it is worthwhile to consider the geography of Russia's industrial assets. Industries and industrial resources are concentrated in four areas: the regions of Moscow and St. Petersburg, the Urals, the Volga, and the Kuznetsk region of southwestern Siberia (see Figure 7.2).

The industrialized area surrounding Moscow is known as the Central Industrial Region, Old Industrial Region, or Moscow-Tula-Nizhniy Novgorod Region. This region has a

157
158–159
30
160

central location physically within the populous western plains and is clearly the country's economic core, with an estimated 70 percent of Russia's financial wealth located there. It lies at the center of rail and air networks reaching Transcaucasia, Central Asia, and the Pacific and is connected by river and canal to the Baltic, White, Azov, Black, and Caspian Seas (Moscow is known as the **Port of Five Seas**). The Moscow region's favorable geographic connections help compensate for its lack of mineral resources. The area's well-developed railway connections, partly a product of political centralization, provide good facilities for an inflow of minerals, materials, and foods and for a return outflow of finished products to all parts of Russia and the Near Abroad. Moscow, capital of Russia before 1713, of the Soviet Union from 1918 to 1991, and of the Russian Federation today, is the most important manufacturing city, transportation hub, and cultural, educational, and scientific center in the region.

This region ranks with Ukraine as one of the two most important industrial areas in the region of Russia and the Near Abroad. South of Moscow lies the metallurgical and machine-building center of Tula, and to the east is the diversified industrial center of Nizhniy Novgorod (formerly Gorkiy) on the Volga. Textile milling, mainly using imported U.S. cotton and Russian flax, was the earliest form of large-scale manufacturing developed in the Moscow region. The area still has the main concentration of textile plants (cottons, woolens, linens, and synthetics) in the region of Russia and the Near Abroad, with Moscow and Ivanovo the leading centers. The breakup of the Soviet Union has done much damage to this textile industry, however. The price of Central Asian cotton quadrupled with independence, forcing mills to close and consumers to turn to cheap clothing imports.

The St. Petersburg area (Figure 7.5) was never as great an industrial center as the Moscow region, Ukraine, or the Urals, but it still developed a significant presence in industrial development and is a critical port. As in the case of Moscow, local minerals did not form the basis of this industrialization. Instead, metals from outside the area provide material for the metal-fabricating industries that are the leaders in St. Petersburg's diversified industrial structure. Supported by university and technological-institute research workers, the city's highly skilled labor force played an extremely significant role in early Soviet industrialization. They pioneered the development of many complex industrial products, such as power-generating equipment and synthetic rubber, and supplied groups of experienced workers and technicians to establish new industries in other areas. In the first decade of the 2000s, Russia has invested heavily in the development and expansion of port facilities in the St. Petersburg region. The main goal is to reduce the high transit fees charged by Latvia and Estonia for exports of Russian oil and other commodities through those ports. New port facilities at Primorsk and Ust-Luga, northwest and southwest of St. Petersburg, and the expansion of St. Petersburg's port, will allow Russian oil, coal, iron ore, timber, and other cargo to be exported at lower cost.

Figure 7.5 St. Petersburg was established on the Baltic Sea at the mouth of the Neva River to serve as Russia's "Window on the West." For its canals and distinctive architecture, it is also known as the "Venice of the North."

Other industrial cities are strung along the Volga River from Kazan southward, in the region Russians know as the Povolzhye. The largest are Samara (formerly Kuybyshev); Volgograd (formerly Stalingrad); Saratov, heart of the Volga German Autonomous Republic until World War II; and Kazan. These four cities in the middle Volga industrial region grew up around diversified machinery, chemical, and food-processing plants. Under the Soviets, Samara was one of the military-industrial "closed cities" (closed to foreigners because of security concerns) whose industries Stalin commanded to be hastily built as a defensive measure against German attack in World War II. Like Nizhniy Novgorod upriver on the Volga, the city is undergoing a process known officially as **"conversion"**—that is, the retooling of military enterprises for civilian-oriented manufacturing. Even while it converts, Russia hopes to reestablish dominance in weapons exporting. Although more than 90 percent of its 1,700 Soviet-era military factories shut down, Russia is still the world's leading arms exporter.

Industries in these cities were powered by the construction of dams and hydroelectric stations along the Volga River and its large tributary, the Kama. The Volga region took early leadership in the automobile industry. Russia's "Detroit" is the Volga city of Togliatti (Figure 7.6), by far the leading center for the production of passenger cars in the region of Russia and the Near Abroad. The cars come from the Volga Automobile Plant, which was built and equipped for the Soviet government by Italy's Fiat Company; the city itself takes its incongruous name from a former leader of Italy's Communist Party.

Large-scale exploitation of petroleum in the nearby Volga–Urals fields (which stretch from the Volga River to the western foothills of the Ural Mountains) contributed to the industrial rise of the Volga cities. Prior to the opening of fields in western Siberia during the 1970s, these were the Soviet Union's most important areas of oil production. Ufa and Samara are the region's principal oil-refining and petrochemical centers.

Joe Hobbs

Figure 7.6 Renowned for their drab functionality, most Russian apartment complexes were built in the 1960s and 1970s and are known, not always fondly, as "Khrushchevskies." These apartment blocks for autoworkers in the factories at Togliatti do have their amenities, including fountains enjoyed on hot summer days.

The Ural Mountains contain a diverse collection of useful minerals, including iron, copper, nickel, chromium, manganese, bauxite, asbestos, magnesium, potash, and industrial salt. Low-grade bituminous coal, lignite, and anthracite are mined. The former Communist regime fostered the development of the Urals as an industrial region well removed from the exposed western frontier of the Soviet Union. The major industrial activities are a legacy of Soviet investment and include heavy metallurgy, emphasizing iron and steel and the smelting of nonferrous ores; the manufacture of heavy chemicals based on some of the world's largest deposits of potassium and magnesium salt; and the manufacture of machinery and other metal-fabricating activities. Yekaterinburg (formerly Sverdlovsk), located at the eastern edge of the Ural Mountains, is the largest city of the Urals and the region's preeminent economic, cultural, and transportation center. The second most important center is Chelyabinsk, located 120 miles (193 km) to the south. This grimy steel town was infamous in Old Russia as the point of departure for exiles sent to Siberia.

In western Siberia, between the Urals and the large industrial and trading center of Omsk, rail lines from Yekaterinburg and Chelyabinsk join to form the Trans-Siberian Railroad, the main artery linking the Far East with the Coreland (Figure 7.7; see also Figure 7.2). Omsk is a major metropolitan base for Siberian oil and gas development. It is also Siberia's most important center of oil refining and petrochemical manufacturing.

From Omsk, the Trans-Siberian Railroad leads eastward to the Kuznetsk industrial region, the most important concentration of manufacturing east of the Urals. The principal localizing factor for industry here is an enormous reserve of coal. The manufacture of iron and steel is also a major industrial activity of the Kuznetsk region, whose main center is Novokuznetsk. The industry draws its iron ore from various sources among Russia's Central Asian neighbors.

WORLD
REGIONAL
Geography⊛Now™

Click Geography Literacy for a virtual ride on the Trans-Siberian Railroad.

The largest urban center of the Kuznetsk region is Novosibirsk, a diversified industrial, trading, and transportation center located on the Ob River at the junction of the Trans-Siberian and Turkestan-Siberian (Turk-Sib) Railroads. Some-

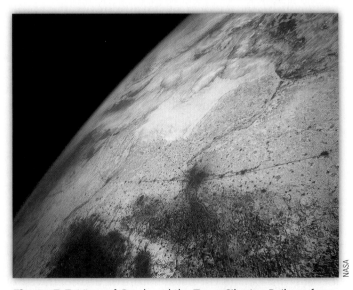

NASA

Figure 7.7 View of Omsk and the Trans-Siberian Railway from space. The railway draws a line through the April snow, and Osmk—the dark smudge just below the center of the photo—lies where it crosses the Irtysh River.

times called the "**Chicago of Siberia,**" Novosibirsk has developed from a town of a few thousand at the turn of the 20th century to more than a million today. In Soviet times, hundreds of factories produced mining, power-generating, and agricultural machinery, tractors, machine tools, and a wide range of other products.

The city of Akademgorodok (Academy Town or Science Town), Siberia's main center of scientific research, is an outlying satellite community of Novosibirsk. The city was established in 1957 as a parklike haven where some of the Soviet Union's greatest scientific minds could research and develop both civilian and military-industrial innovations. Reflecting a general crisis in the post-Soviet military-industrial establishment, Akademgorodok's funding has shrunk dramatically, forcing many prominent scientists to seek menial jobs.

7.3 The Russian Far East

The Far East is Russia's mountainous Pacific edge (see Figures 6.9 and 6.20). Most of it is a thinly populated wilderness in which the only settlements are fishing ports, lumber and mining camps, and the villages and camps of aboriginal peoples. Port functions, fisheries, and forest industries provide the main support for most Far Eastern communities, and until now, the output of coal, oil, and a few other minerals has been small. Most of the Russians and Ukrainians who make up the majority of its people live in a narrow strip of lowland behind the coastal mountains in the southern part of the region. This lowland, drained by the Amur River and its tributary, the Ussuri, is the region's main axis of industry, agriculture, transportation, and urban development.

Several small- to medium-sized cities form a north–south line along two important arteries of transportation: the Trans-Siberian Railroad and the lower Amur River. At the south on the Sea of Japan is the port of Vladivostok, which is kept open throughout the winter by icebreakers. About 50 miles (80 km) east of the city, the main commercial seaport area of the Far East has developed at Nakhodka and nearby Vostochnyy (East Port). Both ports are nearly ice-free and well positioned for trade in goods manufactured in Korea, China, and Japan. Seaborne shipments of cargo containers from northeast Asia to Scandinavia require 40 days, but the cargo can make the 6,000-mile (9,600-km) journey from Vostochnyy to Finland by rail in only 12 days. Business in Vostochnyy is picking up. A new oil terminal is being built at nearby Krylova Cape to support exports of oil, to be brought in initially on the Trans-Siberian Railroad, to Japan and elsewhere.

North of Vladivostok lies a small district that is the most important center of the Far East's meager agriculture, producing cereals, soybeans, sugar beets, and milk for Far Eastern consumption. The Far East is far from self-sufficient in food and consumer goods; large shipments from the Russian Coreland and from overseas supplement local production.

The diversified industrial and transportation center of Khabarovsk is located at the confluence of the Amur and Ussuri Rivers, where the main line of the Trans-Siberian Railroad turns south to Vladivostok and Nakhodka and the Amur River turns north toward the Sea of Okhotsk.

Before World War II, the Soviet Union and Japan held the northern and southern halves, respectively, of the large island of Sakhalin, which today has extremely important petroleum and natural gas reserves, along with some coal and forest and fishing industries. At the end of the war, the USSR annexed southern Sakhalin and the Kuril Islands and repatriated the Japanese population. Russian control of these former Japanese territories continues to be a problem in relations between the two countries and has kept them from reaching a formal post–World War II treaty.

Both countries are trying to put these differences aside in favor of economic cooperation, particularly in the energy realm. Japan has negotiated contracts to buy natural gas from Sakhalin and to build an oil pipeline from the eastern Siberian fields near Irkutsk, in which it would also like to invest, to the port of Vostochnyy or of nearby Nakhodka (Figure 7.8). Russia is happy to have the foreign investment, and Japan, the world's second largest oil consumer, is happy to have the energy from a source outside the volatile Middle East. With its booming economy, China is desperate for more energy, especially from closer and therefore cheaper sources, and is also negotiating with Russia to obtain eastern Siberian oil through a proposed pipeline. Russian sources initially estimated that there was not enough oil in the eastern Siberian fields to supply both markets, and the total reserves are still unknown.

The Kurils are small volcanic islands that screen the Sea of Okhotsk from the Pacific. Fishing is the main economic

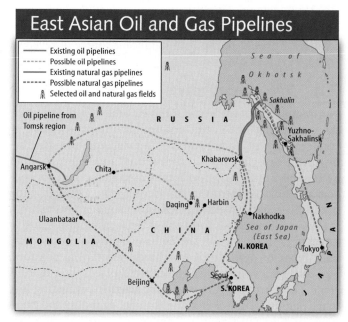

Figure 7.8 Existing and proposed pipelines for eastern Siberian and far eastern oil and gas exports

Kevin Schafer / Corbis. #AF005777

Figure 7.9 Koryaksky volcano towers over the port of Petropavlovsk on the Kamchatka Peninsula.

activity. At the north, the Kurils approach the mountainous peninsula of Kamchatka (total population: 400,000, with 195,000 living in Petropavlovsk). Like the islands, Kamchatka is located on the Pacific Ring of Fire—the geologically active perimeter of the Pacific Ocean—and has 23 active volcanoes (Figure 7.9). Soviet authorities protected Kamchatka for decades as a military area, and no significant development activities occurred there. A land of rushing rivers, extensive coniferous forests, and gurgling hot springs, Kamchatka is now one of the last great wilderness areas on Earth, resembling the U.S. Pacific Northwest landscape of a century ago. A struggle is underway between those who want to develop Kamchatka's gold and oil resources and those who wish to set the land aside in national parks and reserves where ecotourism would be the only significant source of revenue.

7.4 The Northern Lands of Russia

North and east of the Fertile Triangle and west of (and partially including) the Pacific coast lie enormous stretches of coniferous forest (taiga) and tundra extending from the Finnish and Norwegian borders to the Pacific (see Figures 6.9 and 6.20). These outlying wilderness areas in Russia may for convenience be designated the Northern Lands, although parts of the Siberian taiga extend to Russia's southern border.

These difficult lands form one of the world's most sparsely populated vast regions. The climate largely prevents ordinary types of agriculture, but hardy vegetables, potatoes, hay, and barley are grown in scattered places, and there is some dairy farming. A few primary activities, including logging, mining, reindeer herding, fishing, hunting, and trapping, support most people. Significant towns and cities are limited to a small number of sawmilling, mining, transporta-

tion, and industrial centers, generally along the Arctic coast to the west of the Urals, along the major rivers, and along the Trans-Siberian Railroad between the Fertile Triangle and Khabarovsk.

The oldest city on the Arctic coast is Arkhangelsk (Archangel), located on the Northern Dvina River inland from the White Sea. Tsar Ivan the Terrible established the city in 1584 for the purpose of opening seaborne trade with England. Today, the city is the most important sawmilling and lumbershipping center in the region of Russia and the Near Abroad. Despite its restricted navigation season from late spring to late autumn, it is one of the more important seaports.

Another Arctic port, Murmansk (population: 308,000, the world's largest city within the Arctic Circle; Figure 7.10), is located on a fjord along the north shore of the Kola Peninsula west of Arkhangelsk. It is the headquarters for important fishing trawler fleets that operate in the Barents Sea and

Joe Hobbs

Figure 7.10 The ice-free port of Murmansk is situated on a fjord at the northern end of the Kola Peninsula.

Joe Hobbs

Figure 7.11 The Russian nuclear-powered icebreaker *Yamal* at the North Pole. The former Soviet icebreaker fleet still escorts commercial ships on the Northern Sea Route but during the summer carries Western tourists such as these to the high Arctic.

North Atlantic. Murmansk also has a major naval base and cargo port and is home port to the icebreakers that escort cargo vessels and carry Western tourists to the North Pole and other high Arctic destinations (Figure 7.11). It is connected to the Coreland by rail. The harbor is open to shipping all year thanks to the warming influence of the North Atlantic Drift, an extension of the Gulf Stream. Murmansk and Arkhangelsk played a vital role in World War II, when they continued to receive supplies by sea from the Soviet Union's Western allies after Nazi forces captured and closed the other ports of the western Soviet Union.

Murmansk and Arkhangelsk are western terminals of the **Northern Sea Route (NSR)**, a waterway the Soviets developed to provide a connection with the Pacific via the Arctic Ocean (see Figure. 7.2). Navigation along the whole length of the route is possible for only up to 4 months per year, despite the use of powerful icebreakers (including nine nuclear-powered vessels; see Problem Landscape, page 186) to lead convoys of ships. Areas along the route provide cargoes such as the timber of Igarka and the metals and ores of Norilsk. Some supplies are shipped north on Siberia's rivers from cities along the Trans-Siberian Railroad and then loaded onto ships plying the Northern Sea Route for delivery to settlements along the Arctic coast.

The Northern Sea Route is now open to international transit shipping, and if not for the capricious sea ice, it would be an attractive route. In recent winters, typical sea ice coverage in the Arctic Ocean has retreated substantially—some authorities fear because of global warming—and the Russian shipping industry is hoping this trend will continue. The distance between Hong Kong and all European ports north of London is shorter via the Northern Sea Route than

through the Suez Canal, and Russia could earn valuable transit fees from this route. Russia is also anxious to develop the Northern Sea Route for its own exports because Russian exports by land to Western Europe must now pass over the territories of Ukraine, Belarus, and the Baltic countries, which collect their transit fees from Russia.

The railroad that connects Murmansk with the Coreland serves important mining districts in the interior of the Kola Peninsula and logging areas in Karelia, which adjoins Finland. The Kola Peninsula is a diversified mining area, holding nickel, copper, iron, aluminum, and other metals, as well as phosphate for fertilizer. The peninsula and Karelia are physically an extension of Fennoscandia, located on the same ancient, glacially scoured, granitic shield that underlies most of the Scandinavian Peninsula and Finland. Karelia, with its thousands of lakes, short and swift streams, extensive softwood forests, timber industries, and hydroelectric power stations, bears a close resemblance to nearby parts of Finland (which annexed some of this area during World War II).

Important energy resources, including good coking coal, petroleum, and natural gas, are present in the basin of the Pechora River, east of Arkhangelsk. The main coal-mining center is Vorkuta, infamous for the labor camp where great numbers of political prisoners died during the Stalin regime. Russian, U.S., and Norwegian oil companies are working together to explore for and develop oil for export from the Pechora region. The proven reserves of this region are enormous. The Timan Pechora field (see Figure 7.2) contains an estimated 126 billion barrels of oil, one of the world's largest reserves and enough oil to supply the world at current levels of oil use for 4 years. Natural gas reserves in this area and

Problem Landscape

Nuclear Waste Simmers in Russia's High Arctic

On the map, the Russian islands of Novaya Zemlya ("New Land"), separating the Barents and Kara Seas in the high Arctic, appear remote, wild, and untouched. These ice- and tundra-covered islands are indeed an isolated wilderness, which is precisely why Soviet authorities perceived them as prized grounds for nuclear weapons testing and nuclear waste dumping.

Soon after the breakup of the Soviet Union in 1991, a former Soviet radiation engineer stepped forward with disturbing information: For decades, Soviet authorities had ordered the dumping of highly radioactive materials off the coast of Novaya Zemlya (Figure 7.B). From the 1950s through 1991, the Soviet navy and icebreaker fleet secretly and illegally used the shallow waters off the eastern shore of Novaya Zemlya as a nuclear dumping ground. At least twelve nuclear reactors removed from submarines and warships, six of which still contained their highly radioactive nuclear fuel, were sent to the shallow seabed. Some of these individual reactors contain roughly seven times the nuclear material contained in the Chernobyl reactor that exploded in 1986. Between 1964 and 1990, 11,000 to 17,000 containers of solid radioactive waste were also dumped here at shallow depths between 200 and 1,000 feet (61 and 305 m). Soviet seamen reportedly cut holes in those "sealed" containers, which otherwise would not have sunk. This dumping occurred even though the Soviet Union joined other nations in a 1972 treaty that allowed only low-level

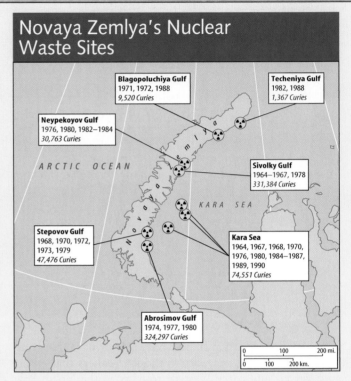

Figure 7.B Sites around Novaya Zemlya used as Soviet nuclear waste dumping grounds between 1964 and 1990. A curie is a measure of the intensity of radioactivity, approximately equal to the radiation in 1 gram of radium.

nearby in the Barents Sea are also huge. One of the big challenges is to find a way to transport these hydrocarbons to market. Plans are underway to construct a pipeline to send oil and gas south to the Gulf of Finland near St. Petersburg.

East of the Urals, the swampy plain of western Siberia north of the Trans-Siberian Railroad saw a surge of petroleum and natural gas production after the 1960s. The western Siberian fields (see Figure 7.2) are the largest producers in the region of Russia and the Near Abroad. However, an estimated half of the reserves of these fields have already been pumped, and so Russia looks to the Arctic for its energy future. Extraction and shipment of oil and gas take place in western Siberia under frightful difficulties caused by severe winters, permafrost, and swampy terrain. Huge amounts of steel pipe and pumping equipment have been required to connect the remote wells with markets in the Coreland and Europe (Figure 7.12).

Still farther east, the town of Igarka, located about 425 miles (c. 680 km) inland on the deep Yenisey River, is an important sawmilling center accessible to ocean shipping during the summer and autumn navigation season. North-

east of Igarka lies Norilsk, the most northerly mining center of its size on the globe and the site of a huge metal-refining complex. Rich ores yielding nickel, copper, platinum, and cobalt have justified the large investment in this Arctic city, including a railway connecting the mines with the port of Dudinka on the Yenisey downstream from Igarka. During the Soviet era, thousands of political prisoners labored and died in the nickel mines of Norilsk. Russian authorities have resisted appeals from surviving laborers to erect a memorial to those lost in this hub of the **gulag**, Soviet Russia's network of forced-labor camps where as many as 30 million prisoners toiled in the 1930s.

In central and eastern Siberia, the taiga extends southward beyond the Trans-Siberian Railroad. A vast but sparsely settled hinterland is served by a few cities spaced at wide intervals along the railroad, notably Krasnoyarsk on the Yenisey River and Irkutsk on the Angara tributary of the Yenisey near Lake Baikal. From east of Krasnoyarsk, a branch line of the Trans-Siberian leads eastward to the Lena River. Years of construction in the 1970s and 1980s under

radioactive waste to be dumped at sea, and then only at depths greater than 12,000 feet (3,657 m).

The danger posed by the nuclear material is not limited to the islands' shores but extends through the entire ecosystem of Russia's Arctic seas and possibly beyond. Contamination can begin at the base of the food chain (the phytoplankton that "bloom" in the area each spring) and pass upward to higher levels, where fish, seals, walrus, polar bears, and people eat. Alaskan authorities are worried, and Norwegians are particularly concerned. Norway's prime minister described the dumping as a "security risk to people and to the natural biology of northern waters." Western European and North American markets for Norwegian and Russian fish, which now represent significant exports for these countries, may refuse purchases if radiation levels rise.

Due to the long half-life of the nuclear materials, the consequences of 30 years of haphazard dumping near Novaya Zemlya may last thousands of years. The threat can be reduced greatly by retrieving the waste and disposing of it in terrestrial sites deep underground, which are considered safer. (The Russian military supports a proposal to convert some of the shafts used in 1950s atomic weapons tests to permanent radioactive waste repositories.) The major obstacle to this step is the extreme costliness of such an operation—estimated to be in the hundreds of billions of dollars—which is well beyond Russia's economic means. An arms-agreement obligation for Russia to decommission about 80 nuclear submarines based in Murmansk is also costly and, because of the radioactive materials involved, potentially dangerous. Confessing and confronting the nuclear legacy of the Soviet Union, Russia has asked the United States and other nations to help pay for an environment with fewer nuclear hazards. The United States in turn is asking Russia not to export its nuclear power technologies to Iran and other countries where the West fears they may be used to develop weapons.

The predicament of Novaya Zemlya and the future of the vast wilderness in Russia's Far East reflect important questions about environment and development all across Russia and the Near Abroad. Like the United States and other industrialized countries, the Soviet Union experienced problems of environmental pollution caused by rapid economic growth. Russia and the other countries must now deal with the Soviet legacy of decades of environmental neglect. Environmental cleanup is hindered by the massive cost of repairing past damage and a reluctance to bear the expenses of stringent enforcement of pollution-abatement measures. Some Russian sources estimate that 80 percent of all industrial enterprises in Russia would go bankrupt if forced to comply with environmental laws. A desperate search for hard currency and widespread corruption have led to other environmental problems, including the wide-scale selling of natural resources like timber and the poaching of animals. With the breakup of the USSR, many environmentalists had predicted more protection for nature. They have been disappointed. One of President Putin's first acts in office was to shut down the State Committee for Environment, and federal monitoring of the environment is negligible.

Figure 7.12 Laying an oil pipeline in Russia's northern lands. Immense quantities of steel have been required for thousands of miles of oil and gas pipelines that have been built in Russia and other parts of Russia and the Near Abroad since World War II.

very difficult conditions resulted in continuation of this line to the Pacific under the name Baikal-Amur Mainline, or BAM (see Figure 7.2). The railroad was built to open important mineralized areas east of Lake Baikal and to reduce the strategic vulnerability caused by the close proximity of the Trans-Siberian Railroad to the Chinese frontier. The Pacific terminus is located on the Tatar Strait, which connects the Sea of Okhotsk with the Sea of Japan.

There is growing support in Russia, the United States, and Canada to construct a rail link between Siberia and North America. This would require laying another 2,000 miles (c. 3,200 km) of track from the BAM to the Bering Strait. An underwater tunnel would carry the track across the Bering Strait, where another 750 miles (c. 1,200 km) of track would be laid to Fairbanks, Alaska. New lines would connect Alaska with western Canada and the lower 48 United States. Proponents of this route say that a Russia–North America railway link would reduce trans-Pacific shipping times by weeks and that it could carry 30 billion tons of cargo each year.

The BAM railway passes through Bratsk, site of a huge dam and hydroelectric power station on the Angara River.

Figure 7.13 Fishermen at work in Lake Baikal, Earth's deepest lake

Lake Baikal, which the Angara drains, lies in a mountain-rimmed rift valley and is the deepest body of fresh water in the world at more than a mile deep in places (Figure 7.13). Lake Baikal is at the center of a long-running environmental controversy over the estimated 8.8 million cubic feet of waste-water that pour into it daily from two large paper and pulp mills on its shores. Environmentalists insist that the waste jeopardizes Baikal's unique natural history, as the lake contains an estimated 1,800 endemic plant and animal species. "Nature as the sole creator of everything still has its favorites, those for which it expends special effort in construction, to which it adds finishing touches with a special zeal, and which it endows with special power," wrote Valentin Rasputin, a native of Siberia. "Baikal, without a doubt, is one of these. Not for nothing is it called the pearl of Siberia." [2]

Siberia's development is based in large part on hydroelectric plants located along the major rivers (Figure 7.14). The largest hydropower stations are on the Yenisey (at Krasnoyarsk and in the Sayan Mountains) and its tributary the Angara (at Bratsk and Ust Ilim). These power stations and dams are among the largest in the world. They supply inexpensive electricity, consumed mainly in mechanized industries that use power voraciously. These include cellulose plants that process Siberian timber and aluminum plants making use of aluminum-bearing material (alumina) derived from ores in Siberia, the Urals, and other areas and imported from abroad via the Black Sea.

Scattered throughout central and eastern Siberia are important gold-mining centers, mainly in the basins of the Kolyma and Aldan Rivers and in other areas in northeastern

[2] Valentin Rasputin, "Baikal." In Valentin Rasputin, *Siberia on Fire: Stories and Essays* (DeKalb: Northern Illinois University Press, 1989), pp. 188–189.

Figure 7.14 Siberia's hydroelectric power potential is vast. This river in extreme northeastern Siberia is one of the abundant northward-flowing drainages in the region.

Figure 7.15 Yakut reindeer herders in the northeastern Siberian region of Sakha

Siberia. In Stalin's time, the Kolyma mines, located in one of the world's harshest climates, developed an evil reputation as death camps for great numbers of political prisoners forced to work there. Large coal deposits are mined along the Trans-Siberian Railroad at various points between the Kuznetsk Basin and Irkutsk, Russia's leading center for the production of military aircraft. Major oil production will soon begin around Irkutsk in the eastern Siberian oil fields, which are expected to supply both Japan and China.

Aside from these scattered centers of logging, mining, transportation, and industry, where most people are Slavs, the Northern Lands are home mainly to non-Slavic peoples who make a living by herding reindeer, trapping, fishing, and in the more favored areas of the taiga, by raising cattle and precarious forms of cultivation. The domesticated reindeer is especially valuable to the tundra peoples, providing meat, milk, hides for clothing and tents, and draft power. The Yakuts, a Turkic people living in the basin of the Lena River, are among the most prominent of the non-Slavic ethnic groups in the region (Figure 7.15). Their political unit, Sakha (formerly Yakutia), is larger in area than any former Soviet Union republic except the Russian Federation, within which it lies. Sakha has only about 1 million people inhabiting an area of 1.2 million square miles (3.1 million sq km). The capital city of Yakutsk on the middle Lena River has road connections with the BAM and Trans-Siberian Railroads and with the Sea of Okhotsk. Riverboats on the broad and deep Lena provide a connection with the Northern Sea Route and the southern rail system during the warm season. Sakha will figure prominently in future mineral and timber development, with its large reserves of diamonds, natural gas, coking coal, other minerals, and timber.

One of the most interesting trends to watch in coming years will be Moscow's effort to "de-Russify" the Northern Lands. Soviet development of many parts of this vast wilderness was often based on political or demographic ambitions rather than on anticipated economic returns. Moscow must pay huge subsidies to support the heating, fuel, food, and social program costs for people to live in these cold and remote settlements. To reduce that burden, the Russian government wants the region's population to decline by 600,000 to 800,000 people, or roughly 10 percent. In 2003, the government began offering housing and other subsidies of up to $18,000 to each family willing to relocate southward and westward from cities like Norilsk and Vorkuta. So far, most residents have turned down the offer as insufficient to cover the cost of leaving the Northern Lands.

7.5 The Caucasus

The Caucasus is the name given to the isthmus between the Black and Caspian Seas (for the Caucasus map, see Figure 6.22, page 165). In the north of the isthmus is the Greater Caucasus Range, forming an almost solid wall from the Black Sea to the Caspian (Figure 7.16). It is similar in age and character to the Alps but much higher, up to 18,510 feet (5,642 m) on Mount Elbrus, on the Russia–Georgia border. The southern Caucasus is known as Transcaucasia. In southern Transcaucasia is the mountainous, volcanic Armenian Plateau. Between the Greater Caucasus Range and the Armenian Plateau are subtropical valleys and coastal plains where the majority of people in Transcaucasia live. A humid subtropical climate prevails in coastal lowlands and valleys of Georgia, south of the high Caucasus Mountains. Mild winters and warm to hot summers combine with the heaviest precipitation found anywhere in the region of Russia and the Near Abroad—50 to 100 inches (c. 125 to 250 cm) a year. The lowland area bordering the Black Sea in Georgia is the most densely populated part of Transcaucasia. Specialty crops such as tea, tung oil, tobacco, silk, and wine grapes, together with some citrus fruits, grow there.

Figure 7.16 Mountains and glaciers seen from Mount Elbrus in the Caucasus

Figure 7.17 Ethnic Armenian actors on a film set at the ancient site of Garni

The Caucasian isthmus has been an important north–south passageway for thousands of years, and the population includes dozens of ethnic groups that have migrated into this region (see Figure 6.11). Most of these are small ethnic populations confined to mountain areas that became their refuges in past times. The tiny republic of Dagestan, within Russia on its southeastern border with Georgia and Azerbaijan, is by itself home to 34 different ethnic groups. Some mountain villages, particularly in southeastern Azerbaijan, have achieved international fame as the abodes of the world's longest lived people—up to the much-challenged claim of 168 years! In addition to Russians and Ukrainians, most of whom live north of the Greater Caucasus Range, the most important nationalities are the Georgians, Azerbaijanis, and Armenians, each represented by an independent country (Figure 7.17). Throughout history, all have defiantly maintained their cultures in the face of pressure by stronger intruders.

These nationalities have ethnic characteristics and cultural traditions that are primarily Asian and Mediterranean in origin. Their religions differ. The Azerbaijanis (also known as Azeri Turks) are Muslims. Most Georgians belong to one of the Eastern Orthodox churches. The Armenian Apostolic Church is a very ancient, independent Christian body that is an offshoot of Eastern Orthodoxy; in A.D. 301, Armenia became the world's first Christian country when its government pronounced the young religion as its national faith (Figure 7.18). There has been a history of animosity between the Armenians and Azeri Turks (including a war between them in 1905) growing out of the Turks' persecution of Armenians in the Ottoman Empire prior to and during World War I. Turkey and Azerbaijan still have not accepted responsibility for their roles in the **Armenian genocide,** in which an estimated 1.5 million Armenians died between 1915 and 1918. The descendants of Armenians who fled this holocaust have established themselves in many places around the world

(for example, there is a large population in Hollywood's film industry) and today outnumber Armenians living within the country by almost two to one. One indicator of Armenia's declining economic fortunes following independence in 1991 is that 40 percent of the country's population has moved abroad since then.

There are some outstanding problems among the Caucasus countries in addition to the international geopolitical issues discussed in the previous chapter (pages 164–166). Early in the 1990s, historical animosity flared into massive violence between Armenians and Azeris over the question of Nagorno-Karabakh, a predominantly Armenian 1,700 square mile (4,420 sq km) enclave within Azerbaijan, then governed by Azerbaijan but claimed by Armenia (see Figure 6.22). From 1992 until 1994, when Armenian forces finally secured the region (which is still technically part of Azerbaijan), fighting over Nagorno-Karabakh took thousands of lives and created nearly a million refugees. Large

Figure 7.18 The Armenian Church of the Holy Cross on the island of Akhtamar, now in eastern Turkey. Armenia was the world's first Christian country, and its ancient borders included Mount Ararat and other land later lost to Turkey.

numbers of Armenians fled from Azerbaijan to Armenia as refugees. There was also a flight of Azeris from Armenia to Azerbaijan, where hundreds of thousands continue to live as refugees. Armenia and Azerbaijan have not yet made peace, but for several years, they have had a working framework for an agreement. The Armenians would return to Azerbaijan six of the seven regions they captured. Nagorno-Karabakh and the adjacent Lachin region would be granted self-governing status. Azerbaijan's compensation would be the construction of an internationally protected road linking Azerbaijan with its enclave of Nakhichevan, now cut off from Azerbaijan by Armenian territory.

Oil and gas are the most valuable minerals of the Caucasus region. The main oil fields are along and underneath the Caspian Sea around Azerbaijan's capital, Baku (population: 1.8 million) and north of the mountains near Groznyy in Chechnya. Dry and windswept Baku, the largest city of the Caucasus region, was the leading center of petroleum production in Old Russia and, at the dawn of the 20th century, was the world's leading center of oil production. Under the Soviets, its leadership was decisively surpassed in the 1950s by the Volga–Urals fields and again in the 1970s by the western Siberian fields. However, petroleum development is now reestablishing the Caucasus as a vital world region (see Geography of Energy in Chapter 6, pages 168–169). Under the Soviets, diverse manufacturing industries developed in Transcaucasia. Most of these factories, which now produce only a fraction of their Soviet-era output, are located in and near the political capitals: Azerbaijan's Baku, Georgia's Tbilisi (population: 1.1 million), and Armenia's Yerevan (population: 1.3 million).

Georgia's economy has fared the worst in post-Soviet years. By the time the "Rose Revolution" swept the old regime out of power in 2003, more than half of the population lived below the poverty line, and unemployment exceeded 25 percent. Up to 80 percent of the country's wealth was generated by a "shadow economy" fueled by smuggling and trafficking in drugs, guns, oil, and prostitutes. The new government faced daunting challenges in reversing Georgia's reputation as one of the world's most corrupt countries.

The Caspian Sea, which is the world's largest freshwater lake in area, has resources other than oil. It is an important fishery, especially for its three commercial species of sturgeon, which supply over 70 percent of the world's caviar (sturgeon eggs) (Figure 7.19). When only the USSR and Iran bordered the Caspian, this resource was managed sustainably. Now, however, the smaller, recently independent countries bordering the shallow Caspian are exerting tremendous pressure on this resource in an attempt to boost their market shares in competition with Russia and Iran. The result has been a rapid drop in the lake's sturgeon population and soaring prices for caviar on the world market (a pound from the prized beluga species retails for $1,000). Despite police patrols, sturgeon poaching is big business, netting five to ten

Figure 7.19 Caviar from the prehistoric-looking beluga sturgeon (*Huso huso*) of the Caspian Sea is one of the world's most prized and expensive delicacies. This is a captive beluga.

times the licensed harvest of these fish. There are growing fears about what an oil spill would do to the caviar industry.

7.6 The Central Asian Countries

Across the Caspian Sea from the Caucasus region lie large deserts and dry grasslands, bounded on the south and east by high mountains. Like Transcaucasia, this Central Asian region is inhabited by ethnic groups that outnumber Slavs. The largest populations are four Muslim peoples speaking closely related Turkic languages—the Uzbeks, Kazakhs, Kyrgyz, and Turkmen—and a Muslim people of Iranian origins, the Tajiks (Figure 7.20). Each has its own independent nation (see Table 6.1). Their total population of 59 million is growing faster than that of any other subregion of Russia and the Near Abroad.

Figure 7.20 The elders of this community in Tajikistan are about to partake in the traditional meal celebrating Novruz, the Muslim New Year.

Division and Conquest

The peoples of this region are heirs of ancient oasis civilizations and of the diverse cultures of nomadic peoples. Small-scale social and political units such as clans and tribes were traditionally associated with particular oases. Conquerors have repeatedly sought to possess this area because of its strategic location on the famed Silk Road, an ancient route across Asia from China to the Mediterranean. Among them were Alexander the Great, the Arabs who brought the Muslim faith in the eighth century, the Mongols, and the Turks. By the time of the Russian conquests in the 19th century, Turkestan (as this region was then known) was contested between feuding Muslim khanates with political ties to China, Persia, Ottoman Turkey, and British India. Cultural and political fragmentation continued up to the Russian Revolution and into the period of civil war that followed.

Nationalist identities existed historically in Central Asia, but nation-states did not. During World War I, Central Asian peoples revolted against Russia's tsarist government, which had begun to draft the region's Muslims for menial labor at the front. The Bolshevik government gained control in the early 1920s and organized the area into nationality-based units, effectively setting the boundaries for what are now independent countries.

During the 20th century, Russians and Ukrainians poured into Central Asia, mainly into the cities. Meant to Russify this non-Russian area, they included political dissidents banished by the Communists, administrative and managerial personnel, engineers, technicians, factory workers, and in the north of Kazakhstan, farmers in the "new lands" wheat region. Most of the Slavic newcomers lived apart from the local Muslims, as most of those who remain today still do.

Ethnic Russians have problems in post-Soviet Central Asia, especially in Kazakhstan. At the time of independence in 1991, Kazakhstan's population of 17 million was roughly 40 percent Kazakh and 39 percent Russian, with the rest a mix of nationalities. Ethnic Kazakhs, who generally believe they were treated as second-class citizens until independence, have been working to diminish the Russian cultural footprint. They are increasingly predominant in government and business. Kazakh is now the official language, and Russian is the language of "interethnic communication." Even ethnic Kazakhs have a lot of catching up to do; half of them cannot speak Kazakh. With rising ethnic tensions throughout the 1990s, about 1 million Russians and 700,000 ethnic Germans emigrated from Kazakhstan. Meanwhile, the government is calling for an estimated 4.5 million ethnic Kazakhs who live abroad to come home.

Environment and Agriculture

The five Central Asian "stans" are mainly flat, with plains and low uplands, except for Tajikistan and Kyrgyzstan, which are spectacularly mountainous and contain the highest summits in the region of Russia and the Near Abroad in the Pamir and Tien Shan Ranges (Figure 7.21; see also Figure 6.9). Central Asia is almost entirely a region of interior drainage. Only the waters of the Irtysh, a tributary of the Ob, reach the ocean; all the other streams either drain to enclosed lakes and seas or gradually lose water and disappear in the Central Asian deserts.

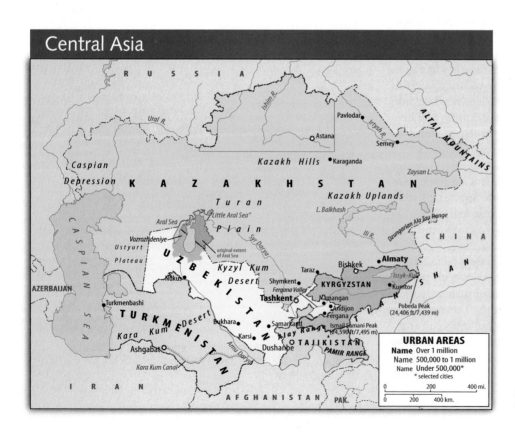

Figure 7.21 Principal features of Central Asia

Many of this region's people previously were pastoral nomads who grazed their herds on the natural forage of the steppes and in mountain pastures in the Altai and Tien Shan Ranges. Over the centuries, there was a slow drift away from nomadism. The Soviet government accelerated this process by forcibly collectivizing the remaining nomads and settling them in permanent villages. Raising livestock became a mainly sedentary form of ranching in the process. Sheep and cattle are the principal livestock animals, followed by goats, horses, donkeys, and camels. Horses remain central to the national personality, if not the livelihoods, of the indigenous people of Kyrgyzstan. In the post-Soviet period, they have resurrected the oral epics that tell the story of Manas, a great warrior horseman who, 1,000 years ago, fought off the enemies of the Kyrgyz and unified their disparate tribes for the first time. The Kyrgyz are passionate horsemen, and even townspeople keep horses in the countryside for weekend races and other games.

Most people today live in irrigated valleys at the base of the southern mountains. There, the wind-blown loess and other soils are fertile, the growing season is long, and rivers flowing from mountains provide irrigation water. The principal rivers in the heart of the region are the Amu Darya (Oxus) and Syr Darya, both of which empty into the enclosed Aral Sea.

Only northern Kazakhstan, which extends into the black-earth belt, receives enough moisture for dry farming (unirrigated agriculture). Irrigation is needed everywhere else, providing water to mulberry trees (grown to feed silkworms), rice, sugar beets, vegetables, vineyards, and fruit orchards. Commercial orchard farming is focused on the area around Almaty ("City of Apples"; population: 1.3 million), the former capital of Kazakhstan. But most of the irrigated land is in Uzbekistan. The most important farming area is the fertile Fergana Valley on the upper Syr Darya, nearly enclosed by high mountains, and the oases around Uzbekistan's cities of Tashkent (the nation's capital and Central Asia's largest city; population: 3.5 million), Samarkand, and Bukhara. These old cities became powerful political and economic centers on the Silk Road. Samarkand was the capital of Tamerlane's empire, which encompassed much of Asia in the 14th century. Samarkand and nearby Bukhara have superb examples of Islamic architecture that the government of Uzbekistan has recently restored at great cost (Figure 7.22).

Cotton began to be cultivated here during the American Civil War, when U.S. exports abroad virtually ceased. The Soviets essentially fostered a "plantation economy" of their own in Central Asia, exploiting the region's potential to produce large quantities of this valuable crop (known for a time as **"white gold"**) for the central Russian textile mills. Cotton remains the region's major crop. Production flows to textile mills in Tashkent, the Moscow region (at levels far below those of Soviet times), and locations outside the region.

A significant environmental price was paid to develop Central Asian agriculture. Driven by the need for foreign exchange from export sales, Soviet planners reshaped the

Figure 7.22 A 16th-century *madrasa* (Islamic school) on the Registan Square in the heart of the storied ancient city of Samarkand, Uzbekistan

Karen Sherlock /Stone /Getty Images

basins of the Syr Darya and Amu Darya with 20,000 miles (c. 32,000 km) of canals, 45 dams, and more than 80 reservoirs. As the area under irrigated cotton grew, the rivers and the Aral Sea became highly polluted with agricultural chemicals draining from the fields. Diversion of water from the rivers caused the Aral Sea to shrink rapidly in volume and area, virtually destroying a vast natural ecosystem and an important regional fishery (see Figure 7.21). Since the 1960s, the Aral Sea has lost 50 percent of its surface area and 75 percent of its volume, and has divided into two small lakes. Kazakhstan is trying to muster the financial resources to build a dam to contain the smaller lake, known as the Little Aral Sea, completely within its borders, in an effort to reduce its salinity and restore its fishery.

Both Kazakhstan and Uzbekistan use tremendous quantities of water to flood paddies to produce rice—an extraordinary land-use decision in such a thirsty region. Rice cultivation is one reason Uzbekistan is desperate for water and accuses Kyrgyzstan of hoarding it. Kyrgyzstan insists it needs to store water for hydroelectric power because the downstream countries will not provide it with the fossil fuels it needs. Meanwhile, Turkmenistan has declared it will harness Amu Darya waters to create a new "Lake of the Golden Turkmen" to irrigate an additional 1.2 million acres of land. Where all the water will come from is not clear, especially since the five "stans" signed an agreement in the 1990s promising to freeze in place their Soviet-era water distribution allotments.

Another environmental problem—this one with international security dimensions—exists on the former island of Vozrozhdeniye, on Uzbekistan's side of the Aral Sea. In 1988, as the Soviet Union hastened to build new bridges with the West, Soviet biological warfare specialists buried hundreds of tons of living anthrax bacteria, encased in steel canisters, on the island. As the Aral Sea continued to shrink, the island became a peninsula, raising fears that these more accessible biohazards could be exposed and blown into populated areas or be carried off by would-be terrorists. Following a number of incidents of anthrax-tainted letters reaching U.S. Senate offices and other facilities in the wake of the 9/11 attacks, the American government pledged the funds needed to clean up "Anthrax Island"; this effort is underway.

Economic and Political Prospects

In a familiar pattern, the economies of the five Central Asian countries worsened after an initial period of exuberance and growth following independence. The long-term prospects for some of the countries are bright, however. Central Asia has a generous endowment of minerals, particularly natural gas, coal, oil, iron ore, nonferrous metals, and ferroalloys. Kazakhstan (population: 15 million), the region's richest country, has the greatest natural resource wealth, including perhaps as much as 70 percent of the Caspian Sea's oil reserves. Political stability under the nominally democratic (but quite repressive) government of Nursultan Nazarbaev since independence has made it the biggest target for Western investment in Central Asia. Large reserves of bituminous coal are at Karaganda in central Kazakhstan. Despite Kazakhstan's coal wealth, however, equipment and management problems have resulted in declining production since the country's independence. In the late 1990s, the country's new capital was established at Astana (formerly Akmola and Tselinograd; population: 312,000), northwest of Karaganda, as part of a larger plan to redevelop Kazakhstan's industrial heartland. Prospecting has begun in Kazakhstan in what may be the world's third largest gold deposit. Kazakhstan also has impressive reserves of iron ore, chrome, and nickel. Industrial growth would help compensate for the plummeting productivity of Kazakhstan's agriculture; without the Soviet-era subsidies they once enjoyed, farmers have not been able to buy pesticides and fertilizers or maintain equipment. Since independence, almost two-thirds of the country's farmland has been abandoned.

After Kazakhstan, Turkmenistan (population: 5.7 million; capital, Ashkabat, population: 773,000) has the most

promising economic resources. The country borders the Caspian Sea with its rich natural gas and petroleum fields. Natural gas exports by pipeline to Western Europe are the country's main source of hard currency; the rest of the economy has been in shambles since independence. Unlike Kazakhstan, Turkmenistan has no pretensions of democracy. The country's Communist leader for life, Saparmurat Niyazov, has dubbed himself the Turkmenbashi ("Father of the Turkmen"), renaming cities and even calendar months after himself. His regime maintains an iron grip on the Turkmen majority and the ethnic Russian and Uzbek minorities.

Uzbekistan, Central Asia's most populous country (population: 26.4 million), and Kazakhstan's main rival for dominance in Central Asia, does not benefit from Caspian oil fields. The country is largely agricultural, and cotton exports are the most important source of hard currency. To diversify its important agricultural sector, the country is trying to reverse the ecological damage done to the Aral Sea and its drainage basin and to cut the area planted in cotton in favor of food crops.

Kyrgyzstan (population: 5.1 million; capital, Bishkek, population: 886,000), the second poorest country of the region after Tajikistan, is struggling to modernize. With independence, the country's first president, Askar Akaev, was hailed as a Thomas Jefferson-like figure who would bring freedom and prosperity to his country. Most of the Soviet-era state-owned businesses were quickly privatized, and a stock exchange opened. But Akaev lost the "Jefferson" label when he won a rigged election in 2000, and hopes for an economic boom have yet to be realized. Kyrgyzstan has few industrial resources on which to plan its future. Fully 40 percent of the country's export revenues come from the Kumtor gold mine. There are hopes for growth in the tourism sector. Central Asia's greatest scenic assets are here. The Tien Shan—the Celestial Mountains—dominate the land and hold much potential to draw Western ecotourists interested in rafting and trekking, particularly in the region around Lake Issyk Kul, one of the world's largest bodies of fresh water. At the same time, there is much concern about the environmental hazards posed by uranium waste at several locations in Kyrgyzstan where uranium had been processed for Soviet nuclear weapons.

Tajikistan (population: 6.6 million; capital, Dushanbe, population: 829,000) is still a Communist country, but its government has the backing of Washington and Moscow, which see the alternative as a radical Islamic regime. After the country gained independence, a coalition of prodemocracy and Islamic rebels, comprised largely of ethnic groups that had little economic and political power in the country, lost a bloody civil war against the Russian-backed government. More than 50,000 people died in the fighting, and an estimated 20 percent of the country's population fled, mostly into neighboring Afghanistan. After 1997, as part of a United Nations-brokered peace agreement between the rebels and the government, the former rebel fighters were absorbed into the national army and their political leaders brought into the government.

Despite Tajikistan's formal independence, it is now in fact a "client state" of Russia. Moscow currently funds 70 percent of the Tajik national budget, and there is little to fuel its economy other than Russian aid, smelted aluminum, and cotton. Uranium was once mined in Tajikistan for Soviet nuclear weapons, and cotton was produced there for Moscow's textile mills, but those markets have nearly evaporated. Tajikistan now sells its precious cotton to Switzerland and other distant markets. Tajikistan is the poorest of the Soviet Union's successor countries.

The Central Asian countries are like those of the Middle East in many respects, particularly in their overwhelmingly Muslim populations, their traditions of pastoral nomadism and irrigated agriculture, and their petroleum endowments. The text now turns to that critical world region of the Middle East and North Africa. Chapter 8's section on Islam provides useful background for further appreciation of Central Asia and the Caucasus.

CHAPTER SUMMARY

- Nearly three-fourths of the people, and an even larger share of the cities, industries, and cultivated land of the region of Russia and the Near Abroad, are packed into the Fertile Triangle composing one-fifth of the total area. This area is also called the Agricultural Triangle and the Slavic Coreland. The rest lies mostly in Asia and consists of land where settlement is spotty and limited by environmental obstacles.

- Situated partly in the forest zone and partly in the steppe, Ukraine translates as "at the border" or "borderland." Today, Ukraine is one of the most densely populated and productive areas of the region. Ukraine's industrial and agricultural assets were vital to the Soviet Union. Post-Soviet Ukraine remains more reliant on heavy industry than most of the other successor states of the Soviet Union.

- Under Soviet rule, Belarus became an industrial power, shipping commodities to both Moscow and its Eastern European satellites. Belarus's trade is now almost exclusively with Russia, much of it in the form of barter. Belarus and Russia have formed a "union state" to more closely link these predominantly Slavic countries.

- Moldova is made up of territory that the Soviet Union took from Romania in 1940. Moldova has serious ethnic and political problems, with the ethnic Moldovan population oriented toward their kin in Romania, while the ethnic Russian population has formed a self-proclaimed Trans-Dniester Republic. Agriculture is the economic base.

- Kaliningrad is an enclave of Russia separated from Russia proper by the Baltic countries. It is populated mainly by ethnic Russians who would like to become Euro-Russians, developing stronger ties with Western Europe while loosening dependence on Russia. Historically a military base, Kaliningrad has a favorable location for future east–west trade.

- Young women from Moldova, Ukraine, Russia, and other successor states of the Soviet Union are among the main participants in and victims of sex trafficking to Europe. Poverty and the illusory promise of legitimate employment lure them in prostitution and in situations from which they find it difficult to escape.

- In an effort to reverse the inefficiencies left by the state-run system, all of the countries—with varying degrees of speed and success—are promoting land reform by privatizing collective and state farms and developing more independent or "peasant" farming. Following the period of decline following independence in 1991, Russian agricultural production has improved.

- Russian industrial production plummeted after independence. Profits from the recent boom in oil exports have begun to revitalize industries, but long-term prospects are uncertain.

- The industrialized area around Moscow is known as the Central Industrial Region, Old Industrial Region, or Moscow-Tula-Nizhniy Novgorod Region. It has a central location within the populous western plains and is functionally the country's most important industrial region.

- Large-scale exploitation of petroleum in the Volga–Urals fields contributed to the industrial rise of the Volga cities. Prior to the opening of the fields in western and eastern Siberia, the Soviet Union's most important area of oil production was the Volga–Urals fields.

- Coal made the Kuznetsk area the most important industrial region east of the Urals. Commerce between this region and those to the west and east is facilitated by the Trans-Siberian Railroad and lines linked to it.

- Most of the Russian Far East is a thinly populated wilderness in which the only settlements are fishing ports, lumber and mining camps, and villages of aboriginal peoples. Several small- to medium-sized cities form a north–south line along the Trans-Siberian Railroad and the lower Amur River, the two main transportation arteries. The coastal ports of Vladivostok, Nakhodka, and Vostochnyy are coming to life as shipments of oil destined for Japan and other foreign markets have picked up. Oil from the eastern Siberian fields will flow through new pipelines to China and Vostochnyy. The island of Sakhalin is becoming an important oil and gas producer for the Japanese market.

- The Northern Lands comprise one of the world's most sparsely populated areas. Here lie enormous stretches of coniferous forest (taiga) and tundra extending from the Finnish and Norwegian borders to the Pacific. Murmansk and Archangelsk are the important ports and are situated on the Northern Sea Route, a transpolar route for ocean shipping that can be kept open seasonally with icebreakers.

- Soviet-era dumping of nuclear waste off the Arctic island of Novaya Zemlya has created a long-lasting environmental problem that will be very expensive to remedy.

- Siberia's development is based in large part on hydroelectric plants located on the major rivers. These power stations and dams are among the largest in the world. They supply inexpensive electricity, consumed mainly in mechanized industries that use power voraciously. The world's deepest lake, Lake Baikal, is a unique natural ecosystem apparently threatened by effluent from paper and pulp mills.

- Siberia's Northern Lands have some rich mineral resources that began to be exploited in Soviet times, often with forced labor. Now it is expensive for Russia to maintain populations in some of the northern settlements, and the government is trying, using housing subsidies, to attract people to more southerly latitudes.

- The far southern Caucasus region borders Russia between the Black and Caspian Seas. It includes the Caucasus Mountains, a fringe of foothills and level steppes to the north, and the area south known as Transcaucasia. Russians and Ukrainians are the majority north of the Greater Caucasus Range. To the south, the important nationalities are the Georgians, Armenians, and Azerbaijanis, each represented by an independent country and maintaining their unique cultures. Numerous smaller ethnic groups also inhabit the region, which has been a cauldron of conflict

since independence from the Soviet Union. One of the outstanding internal problems is the fate of Nagorno-Karabakh, a region claimed by Armenia but controlled by Azerbaijan.

- Caspian Sea sturgeon are the source of the world's most prized caviar. This resource is threatened by the overfishing that is in part a legacy of the USSR's breakup.

- The Central Asian states lie immediately east and north of the Caspian Sea and are populated mainly by a variety of non-Slavic and Muslim peoples. Those with the largest populations are the Uzbeks, Kazakhs, Kyrgyz, and Turkmen—all speaking closely related Turkic languages—and a people of Iranian origins, the Tajiks. Most of these people live in irrigated valleys at the base of the southern mountains that lie on the path of the famed Silk Road, an ancient route from China to the Mediterranean. These countries have varied economic fortunes, with the wealthiest, Kazakhstan, enjoying large natural resource wealth while the poorest, Tajikistan, has few resources and is unstable politically.

- Soviet-era colonization of Central Asia for a cotton "plantation economy" had pronounced and lingering effects on the environment. Most notable was the diversion of water for irrigated agriculture and the subsequent shrinkage of the Aral Sea. The post-Soviet countries of the region are locked in a critical competition over the region's water supplies from the Amu Darya and Syr Darya rivers.

KEY TERMS + CONCEPTS

Terms in blue are also defined in the glossary.

Armenian genocide (p. 190)	Fertile Triangle (Agricultural Triangle,	periphery (p. 175)
"Chicago of Siberia" (p. 183)	Slavic Coreland) (p. 175)	Port of Five Seas (p. 181)
conversion (p. 181)	gulag (p. 186)	Ring of Fire (p. 184)
core (p. 175)	"Hong Kong on the Baltic" (p. 180)	union state (p. 178)
Euro-Russians (p. 180)	Northern Sea Route (NSR) (p. 185)	"white gold" (p. 193)

REVIEW QUESTIONS

WORLD
REGIONAL
Geography ⊛ Now ™

Assess your understanding of this chapter's topics with additional quizzing and concept-based problems at http://earthscience.brookscole.com/wrg5e.

1. What resources and boundaries are associated with the Fertile Triangle?

2. To what extent are Ukraine, Belarus, and Moldova dependent on Russia or interested in developing further ties with Russia? Which are most and least like Russia?

3. What is Ukraine's main industrial region? Which Russian industrial regions are within the Fertile Triangle? What industries do these regions and their major cities specialize in?

4. Why is Moscow known as the Port of Five Seas?

5. Which city is dubbed Russia's "Detroit" and which is its "Chicago"? Why?

6. What are the physical aspects of Russia's Far East? What are the main transportation arteries there?

7. What are the resources and economic development prospects of Sakhalin and Kamchatka? Why are the Kurils important in Russian–Japanese relations?

8. What two distinct climate and vegetative zones dominate Russia's Northern Lands? What are the main settlements and resources of the Northern Lands?

9. What are the major countries, cultures, and products of the Caucasus?

10. What are the major physical aspects of the Central Asian countries? What are the main economic activities of the "stans"? Which countries are better and worse off economically?

DISCUSSION QUESTIONS

1. What country has been dubbed "Russia writ small"? Why? What country has the closest practical ties with Russia? How have those ties been expressed?

2. Using the text and outside sources, discuss the legacy of the nuclear accident at Chernobyl.

3. How do women of the former Soviet countries end up in the sex trade? What could be done to prevent this form of human trafficking?

4. What is the Trans-Dniester Republic? What are its prospects for independence?

5. Is Kaliningrad an "Eastern" or a "Western" entity? Discuss its unique position, problems, and opportunities.

6. How have Russia's farms and factories fared since Soviet times? What opportunities exist for future development?

7. What important roles do the Trans-Siberian Railroad and its trunk lines play?

8. The ports of Russia's Far East may be poised for an economic boom. Why?

9. Why are Japan and China interested in oil from eastern Siberia? How would they obtain this oil?

10. How might Russia benefit from global warming?

11. What should be done about the nuclear waste lying on the Russian Arctic seabed?

12. How could the railway systems of the lower 48 United States and those of Russia be linked?

13. What was the Armenian genocide?

14. What is the problem of Nagorno-Karabakh? How might it be settled?

15. Who introduced cotton into Central Asia and why? What impacts has cotton production had on the region's environment?

16. How is the shrinking of the Aral Sea related to international concerns about terrorism?

A Geographic Profile of the Middle East and North Africa

Egypt's St. Katherine Monastery, built in the sixth century, lies at the foot of Jebel Musa—a mountain believed by many to be the Biblical Mount Sinai.

chapter objectives

This chapter should enable you to:

↺ Understand and explain the mostly beneficial relationships between villagers, pastoral nomads, and city dwellers in an environmentally challenging region

↺ Know the basic beliefs and sacred places of Jews, Christians, and Muslims

↺ Recognize the importance of petroleum to this region and the world economy

↺ Identify the geographic chokepoints and oil pipelines that are among the world's most strategically important places and routes

↺ Appreciate the problems of control over fresh water in this arid region

↺ Know what al-Qa'ida and other Islamist terrorist groups are and what they want

WORLD
REGIONAL
Geography ⊛ Now™

Look for this logo in the text and go to GeographyNow at http://earthscience.brookscole.com/wrg5e to explore interactive maps, view animations, sharpen your factual knowledge and geographic literacy, and test your critical thinking and analytical skills with unique interactive resources.

T*he Middle East* is a fitting designation for the places and the cultures of this vital world region because they are literally in the middle. This is a physical crossroads, where the continents of Africa, Asia, and Europe meet and the waters of the Mediterranean Sea and the Indian Ocean mingle. Its peoples—Arab, Jew, Persian, Turk, Kurd, Berber, and others—express in their cultures and ethnicities the coming together of these diverse influences. Occupying as they do this strategic location, the nations of the Middle East and North Africa have through five millennia been unwilling hosts to occupiers and empires originating far beyond their borders. They have also bestowed upon humankind a rich legacy that includes the ancient civilizations of Egypt and Mesopotamia and the world's three great monotheistic faiths of Judaism, Christianity, and Islam.

People outside the region tend to forget about such contributions because they associate the Middle East and North Africa with war and terrorism. These negative connotations are often accompanied by muddled understanding. This is perhaps the most inaccurately perceived region in the world. Does sand cover most of the area? Are there camels everywhere? Does everyone speak Arabic? Are Turks and Persians Arabs? Are all Arabs Muslims? The answer to each of these questions is no, yet popular Western media suggest otherwise. This and the following chapter attempt to make the misunderstood, complex, sometimes bloody, but often hopeful events and circumstances in the Middle East and North

Africa more intelligible by illuminating the geographic context within which they occur.

8.1 Area and Population

What is the Middle East and where is it? The term itself is Eurocentric, created by the British who placed themselves in the figurative center of the world. They began to use the term prior to the outbreak of World War I, when the *Near East* referred to the territories of the Ottoman Empire in the Eastern Mediterranean region, the *East* to India, and the *Far East* to China, Japan, and the western Pacific Rim. With *Middle East*, they designated as a separate region the countries around the Persian Gulf (known to Arabs as the Arabian Gulf, and in this text as the Persian/Arabian Gulf and simply The Gulf). Gradually, the perceived boundaries of the region grew.

Sources today vary widely in their interpretation of which countries are in the Middle East. For some, the Middle East includes only the countries clustered around the Arabian Peninsula. For others, it spans a vast 6,000 miles (9,700 km) west to east from Morocco in northwest Africa to Afghanistan in central Asia, and a north–south distance of about 3,000 miles (4,800 km) from Turkey, on Europe's southeastern corner, ·to Sudan, which adjoins East Africa (Figure 8.1). This is the region covered in these chapters, where it is referred to as "The Middle East and North

Political Geography

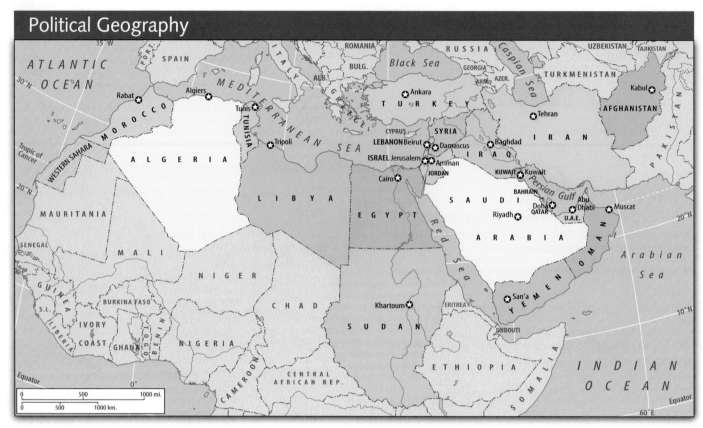

Figure 8.1 The Middle East and North Africa

Africa." The North African peoples of Morocco, Algeria, and Tunisia generally do not perceive themselves as Middle Easterners; they are, rather, from what they call *the Maghreb,* meaning "the western land." Many geographers would place Sudan in Africa South of the Sahara and Afghanistan in either Central Asia or South Asia. Both are clearly border or transitional countries in regional terms, and in this text, they are placed in the Middle East and North Africa with consideration given to their characteristics of the other regions.

Thus defined, the Middle East and North Africa include 21 countries, the Palestinian territories of the West Bank and Gaza Strip, and the disputed Western Sahara (Table 8.1), occupying 5.91 million square miles (15.32 million sq km) and inhabited by about 479 million people as of mid-2004. This area is about 1.8 times the size of the lower 48 United States and is generally situated at latitudes equivalent to those between Boston, Massachusetts, and Bogotá, Colombia (Figure 8.2).

These nearly half-billion people are not distributed evenly across the region but are concentrated in major clusters (Figures 8.3a and b). Three countries contain the lion's share of the region's population: Turkey, Iran, and Egypt, each with more than 65 million people. One look at a map of precipitation or vegetation explains why people are clustered this way (see Figure 2.1, page 18, and Figures 8.5a and b). Where water is abundant in this generally arid region, so are people.

Egypt has the Nile River, and parts of Iran and Turkey have bountiful rain and snow. Conversely, where rain seldom falls, as in the Sahara of North Africa and in the Arabian Peninsula, people are few.

The Middle East and North Africa as a whole have a high rate of population growth. The rapid growth is a general in-

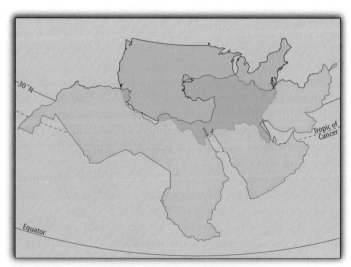

Figure 8.2 The Middle East and North Africa compared in latitude and area with the conterminous United States

Table 8.1 The Middle East and North Africa: Basic Data

Political Unit	Area (thousand/ sq mi)	Area (thousand/ sq km)	Estimated Population (millions)	Annual Rate of Natural Increase (%)	Estimated Population Density (sq mi)	Estimated Population Density (sq km)	Human Development Index	Urban Population (%)	Arable Land (%)	Per Capita GDP PPP ($U.S.)
Middle East										
Afghanistan	251.8	652.2	28.5	2.7	113	44	N/A	22	12	700
Bahrain	0.3	0.8	0.7	1.7	2333	901	0.843	87	3	16,900
Iran	630.6	1633.3	67.4	1.2	107	41	0.732	67	8	7000
Iraq	169.2	438.2	25.9	2.7	153	59	N/A	68	13	1500
Israel	8.1	21.0	6.8	1.6	840	324	0.908	92	16	19,800
Jordan	34.4	89.1	5.6	2.4	163	63	0.750	79	2	4300
Kuwait	6.9	17.9	2.5	1.7	362	140	0.838	100	1	19,000
Lebanon	4	10.4	4.5	1.7	1125	434	0.758	87	16	4800
Oman	82	212.4	2.7	2.2	33	13	0.770	76	0	13,100
Palestinian Territories	2.4	6.2	3.8	3.5	1583	611	0.726	57	17	800
Qatar	4.2	10.9	0.7	1.6	167	64	0.833	92	1	21,500
Saudi Arabia	830	2149.7	25.1	3.0	30	12	0.768	86	1	11,800
Syria	71.5	185.2	18	2.4	252	97	0.710	50	25	3300
Turkey	299.2	774.9	71.3	1.4	238	92	0.751	59	31	6700
United Arab Emirates	32.3	83.7	4.2	1.4	130	50	0.824	78	0	23,200
Yemen	203.8	527.8	20	3.3	98	38	0.482	26	2	800
Total	**2630.7**	**6813.5**	**287.7**	**2.0**	**207**	**80**	**0.728**	**60**	**8**	**6153**
North Africa										
Algeria	919.6	2381.8	32.3	1.5	35	14	0.704	49	3	6000
Egypt	386.7	1001.6	73.4	2.0	190	73	0.653	43	3	4000
Libya	679.4	1759.6	5.6	2.4	8	3	0.794	86	1	6400
Morocco	172.4	446.5	30.6	1.5	177	69	0.620	57	19	4000
Sudan	967.5	2505.8	39.1	2.8	40	16	0.505	31	7	1900
Tunisia	63.2	163.7	10	1.1	158	61	0.745	63	18	6900
Western Sahara	97.2	251.7	0.3	2.1	3	1	N/A	0	0	N/A
Total	**3286**	**8510.7**	**191.3**	**1.9**	**123**	**47**	**0.634**	**46**	**5**	**4130**
Summary Total	**5916.7**	**15,324.2**	**479**	**1.9**	**160**	**62**	**0.690**	**54**	**6**	**5345**

Source: *World Population Data Sheet,* Population Reference Bureau, *2004;* U.N. *Human Development Report,* United Nations, 2004; *World Factbook,* CIA, 2004.

dication that this is a developing rather than industrialized region and also reflects the majority Muslim culture that favors larger families. The average annual rate of population change for the 21 countries, the Palestinian territories, and Western Sahara was 1.9 percent in 2004. The lowest rate of population growth (1.1 percent) is in Tunisia. The highest is 3.5 percent in the Palestinian territories, followed closely by 3.3 percent in Yemen. These are some of the highest population growth rates in the world. In the Palestinian territories,

such rapid growth may be ascribed in large part to the Palestinians' poverty and perhaps to the wishes of many Palestinians to have more children to counterbalance the demographic weight of their perceived Israeli foe.

Between these extremes are countries with modest rates of population growth of 1.5 to 2 percent per year; these include Israel, Morocco, Algeria, Bahrain, Kuwait, Lebanon, and Qatar. Generally, their governments have regarded most of these countries as too populous for their resource and

Population Distribution of the Middle East and North Africa

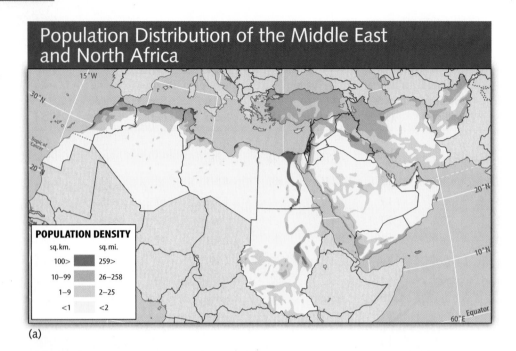

POPULATION DENSITY

sq. km.		sq. mi.
100>		259>
10–99		26–258
1–9		2–25
<1		<2

(a)

Population Cartogram of the Middle East and North Africa

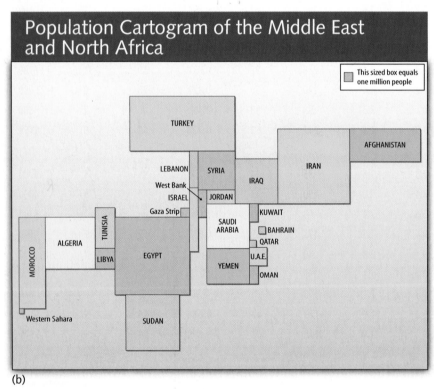

This sized box equals one million people

(b)

Figure 8.3 (a) Population distribution and (b) population cartogram of the Middle East and North Africa

industrial base and have encouraged family planning. They have been successful in lowering birth rates. On the other hand, oil-rich Saudi Arabia, with its 3 percent annual growth rate, is a good example of how rapid population growth is not always a sign of poverty, particularly in this region. In this case, an oil-rich nation has encouraged its citizens to create more citizens so that in the future they will not need to import foreign laborers and technicians and will be more self-sufficient in their development.

Many developing countries have economies largely dependent on subsistence agriculture and have low percentages of urban inhabitants. Perhaps surprisingly, however, the Middle East and North Africa have more urbanites than country folk. The average urban population among the 23 countries and territories is 54 percent. The most prosperous countries are also the most urban. Essentially a city-state, Kuwait is 100 percent urban. The other oil-wealthy Arab Gulf countries also have urban populations over 70 percent. Consistent with its profile as a Western-style industrialized country without oil resources, Israel is 92 percent urban. At the other end of the spectrum, desperately poor Afghanistan, Yemen, and Sudan have urban populations of less than 32 percent.

8.2 Physical Geography and Human Adaptations

The margins of the Middle East and North Africa are mainly oceans, seas, high mountains, and deserts (Figure 8.4). To the west lies the Atlantic Ocean; to the south, the Sahara and the highlands of East Africa; to the north, the Mediterranean, Black, and Caspian Seas, together with mountains and deserts lining the southern land frontiers of Russia and the Near Abroad; and to the east, the Hindu Kush mountains on the Afghanistan–Pakistan frontier and the Baluchistan Desert straddling Iran and Pakistan. The land is composed mainly of arid plains and plateaus, together with large areas of rugged mountains and isolated "seas" of sand. Despite the environmental challenges, this region has given rise to some of the world's oldest and most influential ways of living.

Hot, Dry, and Barren but, in Places, Damp and Forested

Aridity dominates the Middle East and North Africa (Figures 8.5a and b). At least three-fourths of the region has average yearly precipitation of less than 10 inches (25 cm), an amount too small for most types of dry farming (unirrigated agriculture). Sometimes, however, localized cloudbursts release moisture that allows plants, animals, and small populations of people—the Bedouin, Tuareg, and other pastoral nomads—to live in the desert. Even the vast Sahara, the world's largest desert, supports a surprising diversity and abundance of life. Plants, animals, and even people have developed strategies of drought avoidance and drought endurance to live in this harsh biome. Plants either avoid drought by completing their life cycle quickly wherever rain has fallen or endure drought by using their extensive root systems, small leaves, and other adaptations to take advantage of subsurface or atmospheric moisture. Animals endure drought by calling on extraordinary physical abilities (for example, a camel can sweat away a third of its body weight and still live) or by avoiding the worst conditions by being active only at night or by migrating from one moist place to another. Migration to avoid drought is also the strategy pastoral nomads use. Populations of people, plants, and animals are all but nonexistent in the region's vast sand seas, including the Great Sand Sea of western Egypt (Figure 8.6) and the Empty Quarter of the Arabian Peninsula.

The region's climates have the comparatively large daily and seasonal ranges of temperature characteristic of dry lands. Desert nights can be surprisingly cool. Most days and nights are cloudless, so the heat absorbed on the desert

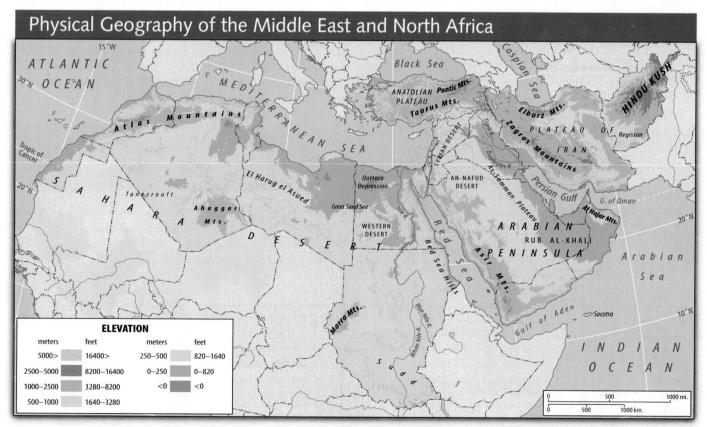

Figure 8.4 Reference map of the topographic features, deserts, and seas of the Middle East and North Africa

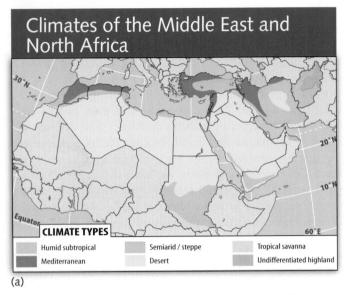

Climates of the Middle East and North Africa

CLIMATE TYPES
- Humid subtropical
- Mediterranean
- Semiarid / steppe
- Desert
- Tropical savanna
- Undifferentiated highland

(a)

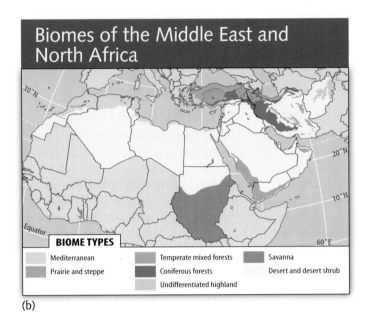

Biomes of the Middle East and North Africa

BIOME TYPES
- Mediterranean
- Prairie and steppe
- Temperate mixed forests
- Coniferous forests
- Undifferentiated highland
- Savanna
- Desert and desert shrub

(b)

Figure 8.5 (a) Climates and (b) biomes of the Middle East and North Africa

surface during the day is lost by radiational cooling to the heights of the atmosphere at night. Summers in the lowlands are very hot almost everywhere. The hottest shade temperature ever recorded on Earth, 136°F (58°C), occurred in Libya in September 1922. Many places regularly experience daily maximum temperatures over 100°F (38°C) for weeks at a time.

Human settlements located near the sand seas often experience the unpleasant combination of high temperatures and hot, sand-laden winds, creating the sandstorms known locally by such names as *simuum* ("poison") and *sirocco*. Only in mountainous sections and in some places near the sea do higher elevations or sea breezes temper the intense midsummer heat. The population of Alexandria, on the Mediterranean Sea coast, explodes in summer as Egyptians flee from

Cairo and other hot inland locations. In Saudi Arabia, the government relocates from Riyadh to the highland summer capital of Taif to escape the lowland furnace.

Lower winter temperatures bring relief from the summer heat, and the more favored places receive enough precipitation for dry farming of winter wheat, barley, and other cool-season crops (Figure 8.7). In general, winters are cool to mild. But very cold winters and snowfalls occur in the high interior basins and plateaus of Iran, Afghanistan, and Turkey. These locales generally have a steppe climate. Only in the southernmost reaches of the region, notably Sudan, do temperatures remain consistently high throughout the year. A savanna climate and biome prevail there.

Most areas bordering the Mediterranean Sea have 15 to 40 inches (38 to 102 cm) of precipitation a year, falling al-

Figure 8.6 Sand "seas" cover large areas in Saudi Arabia, Iran, and parts of the Sahara. This is the edge of the Great Sand Sea near Siwa Oasis in Egypt's Western Desert.

Joe Hobbs

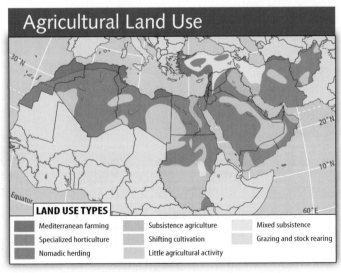

Agricultural Land Use

LAND USE TYPES
- Mediterranean farming
- Specialized horticulture
- Nomadic herding
- Subsistence agriculture
- Shifting cultivation
- Little agricultural activity
- Mixed subsistence
- Grazing and stock rearing

Figure 8.7 Land use in the Middle East and North Africa

most exclusively in winter, while the summer is dry and warm—a typical Mediterranean climate pattern. Throughout history, people without access to perennial streams have stored this moisture to make it available later for growing crops that require the higher temperatures of the summer months. The Nabateans, for example, who were contemporaries of the Romans in what is now Jordan, had a sophisticated network of limestone cisterns and irrigation channels (Figure 8.8). Rainfall sufficient for dry farming during the summer is concentrated in areas along the southern and northern margins of the region. The Black Sea side of Turkey's Pontic (also known as Kuzey) Mountains is lush and moist in the summer, and tea grows well there (Figure 8.9). In the southwestern Arabian Peninsula, a monsoonal climate brings summer rainfall and autumn harvests to Yemen and

Figure 8.9 Rainfall is heavy on the Black Sea side of Turkey's Pontic (Kuzey) Mountains. Note that roofs are pitched to shed precipitation; in the region's drier areas, most roofs are flat. The crop growing here is corn.

Figure 8.8 The Treasury, a temple carved in red sandstone, probably in the first century A.D., by the Nabateans at their capital of Petra in southern Jordan. English poet Dean Burgen called Petra "A Rose Red City, Half as Old as Time." The Nabateans built sophisticated networks for water storage and distribution. For scale, note the man at the lower right.

Oman, probably accounting for the Roman name for the area: *Arabia Felix,* or "Happy Arabia."

Mountainous areas in the region, like the river valleys and the margins of the Mediterranean, play a vital role in supporting human populations and national economies. Due to orographic or elevation-induced precipitation, the mountains tend to receive much more rainfall than surrounding lowland areas.

There are three principal mountainous regions of the Middle East and North Africa (see Figure 8.4). In northwestern Africa between the Mediterranean Sea and the Sahara, the Atlas Mountains of Morocco, Algeria, and Tunisia reach over 13,000 feet (3,965 m) in elevation. Mountains also rise on both sides of the Red Sea, with peaks up to 12,336 feet (3,760 m) in Yemen. These are the result of tectonic processes that are pulling the African and Arabian plates apart, creating the northern part of the Great Rift Valley. The hinge of this crustal movement is the Bekaa Valley of Lebanon, where the widening fault line follows the Jordan River valley southward to the Dead Sea (Figure 8.10). This

Figure 8.10 The surface of the Dead Sea is the lowest point on Earth, a consequence of the rifting of continental crust.

Figure 8.11 The Taurus Mountains of southeastern Turkey

valley is the deepest depression on Earth's land surface, lying about 600 feet (183 m) below sea level at Lake Kinneret (the Sea of Galilee or Lake Tiberius) and nearly 1,300 feet (400 m) below sea level at the shore of the Dead Sea, the lowest point on Earth. The rift then continues southward to the very deep Gulf of Aqaba and Red Sea before turning inland into Africa at Djibouti and Ethiopia. In tectonic processes, one consequence of rifting in one place is the subsequent collision of Earth's crustal plates elsewhere. There are several such collision zones in the Middle East and North Africa, particularly in Turkey and Iran, where mountain building has resulted. These are seismically active zones—meaning earthquakes occur there—and rarely does a year go by without a devastating quake rocking Turkey, Iran, or Afghanistan.

WORLD
REGIONAL
Geography ⊛ Now™

Click Geography Literacy to see an animation of divergent plate boundaries.

A large area of mountains, including the region's highest peaks, stretches across Turkey, Iran, and Afghanistan. These mountains are products of the collision of continental plates. On the eastern border with Pakistan, the Hindu Kush Range has peaks over 25,000 feet (7,600 m). The loftiest mountain ranges in Turkey are the Pontic, Taurus (Figure 8.11), and Anti-Taurus, and in Iran, the Elburz and Zagros Mountains. These chains radiate outward from the rugged Armenian Knot in the tangled border country where Turkey, Iran, and the countries of the Caucasus meet. Mt. Ararat is an extinct, glacier-covered volcano of 16,804 feet (5,122 m) towering over the border region between Turkey and Armenia. Many Biblical scholars and explorers think the ark of Noah lies high on the mountain (and some go in search of it), for in the book of Genesis, this boat was said to have come to rest "in the mountains of Ararat."

Extensive forests existed in early historical times in the Middle East and North Africa, particularly in these mountainous areas, but overcutting and overgrazing have almost eliminated them. Since the dawn of civilization in this area, around 3000 B.C., people have cut timber for construction and fuel faster than nature could replace it. Egyptian King Tutankhamen's funerary shrines and Solomon's Temple in Jerusalem were built of cedar of Lebanon. So prized has this wood been through the millennia that only a few isolated groves of cedar remain in Lebanon. Described in ancient times as "an oasis of green with running creeks" and "a vast forest whose branches hide the sky," Lebanon is now largely barren (Figure 8.12). Lumber is still harvested commercially in a few mountain areas such as the Atlas region of Morocco

Figure 8.12 It was believed that the solar boat of King Cheops of Egypt (c. 2500 B.C.), builder of the Great Pyramid, would carry the Pharoah's spirit through the firmament. It had to be made of the best wood—cedar of Lebanon. The solar boat was interred next to the pyramid, and now stands in a specially constructed museum.

and Algeria, the Taurus Mountains of Turkey, and the Elburz Mountains of Iran, but supply falls far short of demand.

The Middle Eastern Ecological Trilogy: Villager, Pastoral Nomad, and Urbanite

In the 1960s, American geographer Paul English developed a useful model for understanding relationships between the three ancient ways of life that still prevail in the Middle East and North Africa today: villager, pastoral nomad, and urbanite. Each of these modes of living is rooted in a particular physical environment. Villagers are the subsistence farmers of rural areas where dry farming or irrigation is possible; pastoral nomads are the desert peoples who migrate through arid lands with their livestock, following patterns of rainfall and vegetation; and urbanites are the inhabitants of the large towns and cities, generally located near bountiful water sources, but sometimes placed for particular trade, religious,

or other reasons. Describing these ways of life as components of the Middle Eastern ecological trilogy, English explained how each of them has a characteristic, usually mutually beneficial, pattern of interaction with the other two (Figure 8.13).

The peasant farmers of Middle Eastern and North African villages (the **villagers**) represent the cornerstone of the trilogy. They grow the staple food crops such as wheat and barley that feed both the city-dweller and the pastoral nomad of the desert. Neither urbanite nor nomad could live without them. The village also provides the city, often unwillingly, with tax revenue, soldiers, and workers. And before the mid-20th century, villages provided **pastoral nomads** with plunder as the desert-dwellers raided their settlements and caravan supply lines. Generally, however, the exchange is beneficial. The nomads provide villagers with livestock products, including live animals, meat, milk, cheese, hides, and wool, and with desert herbs and medicines. Educated and progressive

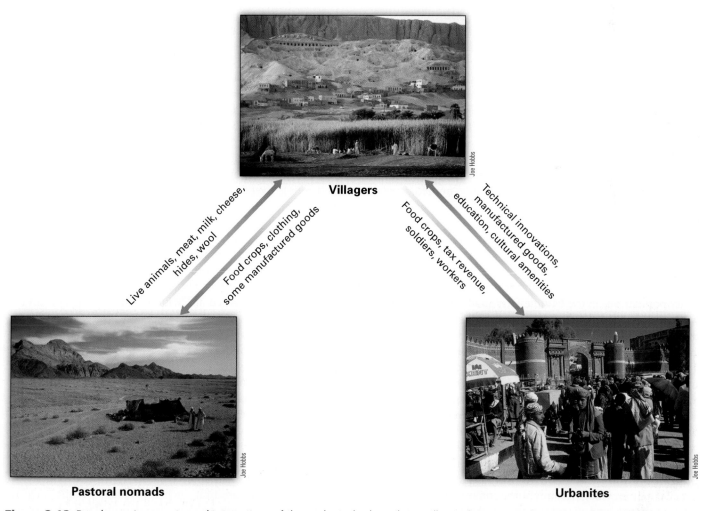

Villagers

Live animals, meat, milk, cheese, hides, wool

Food crops, clothing, some manufactured goods

Food crops, tax revenue, soldiers, workers

Technical innovations, manufactured goods, education, cultural amenities

Pastoral nomads

Urbanites

Figure 8.13 People, environments, and interactions of the ecological trilogy (here, villagers harvesting sugarcane in upper Egypt, Bedouin at camp in Egypt's Eastern Desert, and shoppers at the main gate to the historic city of San'a, Yemen). This relationship is generally symbiotic, although historically both urbanites and pastoral nomads preyed on the villagers, who are the trilogy's cornerstone.

urbanites provide technological innovations, manufactured goods, religious and secular education and training, and cultural amenities (today, including films and music).

There is little direct interaction between urbanites and pastoral nomads, although some manufactured goods such as clothing travel from city to desert, and some desert-grown folk medicines pass from desert to city. Historically, the exchange has been violent, as urban-based governments have sought to control the movements and military capabilities of the elusive and sometimes hostile nomads. Pastoral nomads once plundered rich caravans plying the major overland trade routes of the Middle East and North Africa. Governments did not tolerate such activities and often cracked down hard on the nomads they were able to catch.

Later, in the 1970s, Paul English wrote an article marking the "passing of the ecological trilogy." He noted that cities were encroaching on villages, villagers were migrating into cities and giving some neighborhoods a rural aspect, and pastoral nomads were settling down—thus, the trilogy no longer existed. In reality, although the makeup and interactions of its parts have changed somewhat, the trilogy model is still valid and useful as an introduction to the major ways of life in the Middle East and North Africa. It is especially significant that a given man or woman in the region strongly identifies himself or herself as either a villager, a pastoral nomad, or an urbanite. This perception of self has an important bearing on how these people of very different backgrounds interact, even when they live in close proximity. Urban officials may work in rural village areas, but they remain at heart and in their perspectives city people and usually live apart from farmers. Extended families of pastoral nomads may settle down and become farmers, but they continue to identify themselves by affiliation with the nomadic tribe. Many continue to harvest desert resources on a seasonal basis and retain marriage and other ties with desert-dwelling relatives.

The Village Way of Life

Agricultural villagers historically represented by far the majority populations in the Middle East and North Africa; only within recent decades have urbanites begun to outnumber them. In this generally dry environment, the villages are located near a reliable water source with cultivable land nearby. They are usually made up of closely related family groups, with the land of the village often owned by an absentee landlord. Most often, the villagers live in closely spaced flat-roofed houses made of mud brick or concrete blocks. Production and consumption focus on a staple grain such as wheat, barley, or rice. As land for growing fodder is often in short supply, villagers keep only a small number of sheep and goats and rely in part on nomads for pastoral produce. Residents of a given village usually share common ties of kinship, religion, ritual, and custom, and the changing demands of agricultural seasons regulate their patterns of activity.

Village life has been increasingly exposed to outside influences since the mid-18th century. Contacts with European colonialism brought significant economic changes, including the introduction of cash crops and modern facilities to ship them. Improved and expanded irrigation, financed initially with capital from the West, brought more land under cultivation. Recent agents of change have been the countries' own government doctors, government teachers, and land reform officers. Modern technologies such as sewing machines, motor vehicles, gasoline powered water pumps, radio, television, and even the Internet and cell phones have modified old patterns of living. The educated and ambitious, as well as the unskilled and desperately poor—motivated, respectively, by pull and push factors—have been drawn to urban areas. Improved roads and communications in turn have carried urban influences to villages, prompting villagers to become more integrated into national societies.

The Pastoral Nomadic Way of Life

Pastoral nomadism emerged as an offshoot of the village agricultural way of life not long after plants and animals were first domesticated in the Middle East (about 7000 B.C.). Rainfall and the wild fodder it produces, although scattered, are sufficient resources to support small, mobile groups of people who migrate with their sheep, goats, and camels (and in some locales, cattle) to take advantage of this changing resource base (see Perspectives from the Field, page 209). In mountainous areas, they follow a pattern of vertical migration (sometimes referred to as *transhumance*), moving with their flocks from lowland winter to highland summer pastures. In the flatter expanses that comprise most of the region, the nomads practice a pattern of horizontal migration over much larger areas where rainfall is typically far less reliable than in the mountains. In addition to selling or trading livestock to obtain foods, tea, sugar, clothing, and other essentials from settled communities, pastoral nomads hunt, gather, work for wages, and where possible, grow crops. Their multifaceted livelihood has been described as a strategy of **risk minimization** based on the exploitation of multiple resources so that some will support them if others fail.

Although renowned in Middle Eastern legends and in popular Western films like *Lawrence of Arabia*, pastoral nomads have been described as "more glamorous than numerous." It is still impossible to obtain adequate census figures on the number living in the deserts of the Middle East and North Africa, though estimates range from 5 to 13 million. Recent decades have witnessed the rapid and progressive settling down, or sedentarization, of the nomads—a process attributed to a variety of causes. In some cases, prolonged drought virtually eliminated the resource base on which the nomads depended. Traditionally, they were able to migrate far enough to find new pastures, but modern national boundaries now prohibit such movements. Some have returned with the rains to their desert homelands, but others have chosen to remain as farmers or wage laborers in villages and towns. On the Arabian Peninsula, the prosperity and technological changes prompted by oil revenues made rapid

Perspectives from the Field

Way-Finding in the Desert

The research for my doctoral dissertation in geography at the University of Texas–Austin was an 18-month journey with the Khushmaan Ma'aza, a clan of Bedouin nomads living in the northern Eastern Desert of Egypt. I studied their perceptions, knowledge, and uses of their desert resources and tried to understand how their worldviews and kinship patterns apparently helped them to devise ways of protecting these scarce resources. I have worked with these people from the early 1980s to the present and have always been astonished by how different they are from me and how much better they are in their ability to "read" the ground and recall landscape details. In the following passage from my book Bedouin Life in the Egyptian Wilderness, *I attempt to explain why.*

The process by which the Khushmaan nomads have developed roots in their landscape, fashioning subjective "place" from anonymous "space," and the means by which they orient themselves and use places on a daily basis deserve special attention: these are essential parts of the Bedouins' identity and profoundly influence how they use resources and affect the desert ecosystem.

The Bedouins have little need or knowledge of maps. The Mercator projection is meaningless. One evening as three of the men watched television news in Hurghada, they argued over whether the world map behind the anchorman was a map or a depiction of clouds. On the other hand, when shown a map or aerial image of countryside they know, the nomads accurately orient and interpret it, naming the mountains and drainages depicted. Saalih Ali (Figure 8.A) especially enjoyed my star chart. He would tell me what time the Pleiades would rise, for example, and ask me what time the chart indicated. The two were almost always in agreement.

Many desert travelers have been astonished by the nomads' navigational and tracking ability, calling it a "sixth sense." The Khushmaan are exceptional way-finders and topographical interpreters able, for instance, to tell from tracks whether a camel was carrying baggage or a man; whether gazelle tracks were made by a male or female; which way a car was traveling and what make it was; which man left a set of footprints, even if he wore sandals; and how old the tracks are. Bedouins are proud of their geographical skills which, they believe, distinguish them from settled people. Saalih told me, "Your people don't need to know the country but we do, to know exactly where things are in order to live." Pointing to his head, he said, "My map is here."

The difference in the way-finding abilities of nomads and settled persons may be due to the greater survival value of these skills for nomads and to the more complex meanings they attribute to locations. Bedouin places are rich in ideological and practical significance. The nomads interpret and interact with these meanings on a regular basis and create new places in their lifetimes. Theirs is an experience of belonging and becoming with the landscape, whereas settled people are more likely to inherit and accommodate themselves to a given set of places.

The nomads' homeland is so vast, and the margin of survivability in it so narrow, that topographic knowledge must be encyclopedic. Places either are the resources that allow human life in the desert, or are the signposts that lead to resources. A Khushmaan man pinpointed the role of places in desert survival: "Places have names so that people do not get lost. They can learn where water and other things are by using place names." Tragedy may result if the nomad has insufficient or incorrect information about places: many deaths by thirst are attributed to faulty directions for finding water.[a]

Figure 8.A Saalih Ali, a Bedouin of the Khushmaan Ma'aza

Joe Hobbs

[a] From *Bedouin Life in the Egyptian Wilderness*, by J. Hobbs, pp. 81–82. Copyright © 1989 University of Texas Press. Reprinted by permission.

inroads into the material culture—and then the livelihood preferences—of the desert people; many preferred the comforts of settled life. Some governments, notably those of Israel and prerevolutionary Iran, were frustrated by their inability to count, tax, conscript, and control a sizable migrant population and therefore compelled the nomads to settle.

Pastoral nomads of the Middle East and North Africa identify themselves primarily by their tribe, not by their na-

tionality. The major ethnic groups from which these tribes draw are the Arabic-speaking Bedouin of the Arabian Peninsula and adjacent lands, the Berber and Tuareg of North Africa, the Kababish and Bisharin of Sudan, the Yoruk and Kurds of Turkey, the Qashai and Bakhtiari of Iran, and the Pashtun of Afghanistan. Members of a tribe claim common descent from a single male ancestor who lived countless generations ago; their kinship organization is thus a patrilineal descent system. It is also a segmentary kinship system, so called because there are smaller subsections of the tribe, known as clans and lineages, that are functionally important in daily life. Members of the most closely related families comprising the lineage, for example, share livestock, wells, trees, and other resources. Both the larger clans, made up of numerous lineages, and the tribes possess territories. Members of a clan or tribe typically allow members of another clan or tribe to use the resources within its territory on the basis of **usufruct**, or nondestructive mutual use.

Although some detractors have depicted pastoral nomads as the "fathers" rather than "sons" of the desert, blaming them for wanton destruction of game animals and vegetation, there are numerous examples of pastoral nomadic groups who have developed indigenous and very effective systems of resource conservation. Most of these practices depend on the kinship groups of family, lineage, clan, and tribe to assume responsibility for protecting plants and animals.

The Urban Way of Life

The city was the final component to emerge in the ecological trilogy, beginning in about 4000 B.C. in Mesopotamia (modern Iraq) and 3000 B.C. in Egypt. Unlike the villages they resembled in many ways, the early cities were distinguished by their larger populations (more than 5,000 people), the use of written languages, and the presence of monumental temples and other ceremonial centers. The early Mesopotamian city and, after the seventh century, the classic Islamic city, called the medina, had several structural elements in common (Figure 8.14). The medina had a high surrounding wall built for defensive purposes. The congregational mosque and often an attached administrative and educational complex dominated the city center. Although Islam is often charac-

terized as a faith of the desert, religious life has always been focused in, and diffused from, the cities. The importance of the city's congregational mosque in religious and everyday life is often emphasized by its large size and outstanding artistic execution.

A large commercial zone, known as a *bazaar* in Persian and *suq* in Arabic, and recognizable as the ancestor of the modern shopping mall, typically adjoined the ceremonial and administrative heart of the city (Figure 8.15). Merchants and craftspeople of different commodities occupied separate spatial areas within this complex, and visitors to an old medina today can still expect to find sections where spices, carpets, gold, silver, traditional medicines, and other goods are sold exclusively. Smaller clusters of shops and workshops were located at the city gates.

Residential areas were differentiated as quarters not by income group but by ethnicity; the medina of Jerusalem, for example, still has distinct Jewish, Arab, Armenian, and non-Armenian Christian quarters (see Figure 8.b, page 214). Homes tended to face inward toward a quiet central courtyard, buffering the occupants from the noise and bustle of the street. The narrow, winding streets of the medina were intended for foot traffic and small animal-drawn carts, not for large motor vehicles, a fact that accounts for the traffic jams in some old Middle Eastern and North African cities today and for the wholesale destruction of the medina in others.

The medinas that survive today are gently decaying vestiges of a forgotten urban pattern. Periods of European colonialism and subsequent nationalism changed the face and orientation of the city. During the colonial age, resident Europeans preferred to live in more spacious settings at the outer edges of the city, and later, the national elite followed this pattern. In recent times, independent governments have adopted Western building styles, with broad traffic arteries cutting through the old quarters and large central squares near government buildings. This opening up of the cityscape has spread commercial activity along the wide avenues, thereby diluting the prime importance of the central bazaar as the focus of trade.

Rural–urban migration and the city's own internal growth contribute to a rapid rate of urbanization that puts enormous pressure on services in the region's poorer countries. Governments often build high-rise public housing to accommodate the growing population, contributing to a cycle in which the urban poor move into the new dwellings, only to leave their old quarters as a vacuum to draw in still more rural migrants. In Cairo, millions of former villagers now live in the "City of the Dead," an extraordinary urban landscape composed of multistory dwellings erected above graves—a last resort for the poor who have no other place to go. The overwhelmingly largest city, or primate city, so characteristic of Middle Eastern and North African capitals, thus grows at the expense of the smaller city.

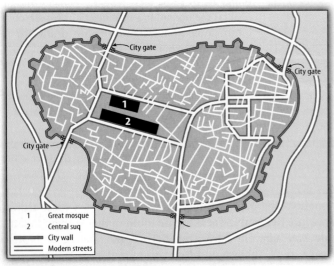

1 Great mosque
2 Central suq
   City wall
   Modern streets

Figure 8.14 An idealized model of the classic *medina,* or Muslim middle eastern city

91

Joe Hobbs

Figure 8.15 A characteristic Middle Eastern *suq,* or bazaar. This one is in Sanliurfa, Turkey.

Much of the rural–urban migration and subsequent urban gridlock and squalor could probably be avoided if governments invested more in the development of villages and smaller cities. The oil-rich countries with relatively small populations generally enjoy an urban standard of living equaling that of affluent Western countries. Modern industrial cities such as Saudi Arabia's Jubail and others founded on oil wealth were built virtually overnight, providing fascinating contrast to the region's colorful, complex ancient cities.

8.3 Cultural and Historical Geographies

Cultures of the Middle East and North Africa have made many fundamental contributions to humanity. Many of the plants and animals upon which the world's agriculture is based were first domesticated in the Middle East between 5,000 and 10,000 years ago in the course of the Agricultural Revolution. The list includes wheat, barley, sheep, goats, cattle, and pigs, whose wild ancestors were processed, manipulated, and bred until their physical makeup and behavior changed to suit human needs. The interaction between people and the wild plants and animals they eventually domesticated took place mainly in the well-watered Fertile Crescent, the arc of land stretching from Israel to western Iran.

By about 6,000 years ago, people sought higher yields by irrigating crops in the rich but often dry soils of the Tigris, Euphrates, and Nile River valleys. Their efforts produced the enormous crop surpluses that allowed civilization—a cultural complex based on an urban way of life—to emerge in Mesopotamia (literally, "the land between the rivers" Tigris and Euphrates) and Egypt. Accomplishments in science, technology, art, architecture, language, mathematics, and other areas diffused outward from these centers of civilization. Egypt and Mesopotamia are thus among the world's great culture hearths.

The Middle East and North Africa are sometimes mistakenly referred to as the *Arab World;* in fact, the region has huge populations of non-Arabs. It is true that most of the region's inhabitants are Arabs. An **Arab** is best defined as anyone who is of Semitic Arab ethnicity and whose ancestral language is **Arabic.** That language is spoken by about 280 million, or 58 percent, of the region's people. Originally, the Arabs were inhabitants of the Arabian Peninsula, but conquests after their majority conversion to Islam took them, their language, and their Islamic culture as far west as Morocco and Spain.

The region is also the homeland of the **Jews,** whose definition today is complex. Originally, Jews were both a distinct ethnic and linguistic group of the Middle East who practiced the religion of Judaism. It is possible (although difficult) for non-Jews to convert to Judaism (the convert is known as a proselyte, or "immigrant"), so a strictly ethnic definition does not apply today. Many Jews do not practice their religion but still consider themselves ethnically or culturally Jewish. Whatever debate exits about what defines a Jew, Jewish identity is extremely strong and resilient. It is important

31

Languages of the Middle East and North Africa

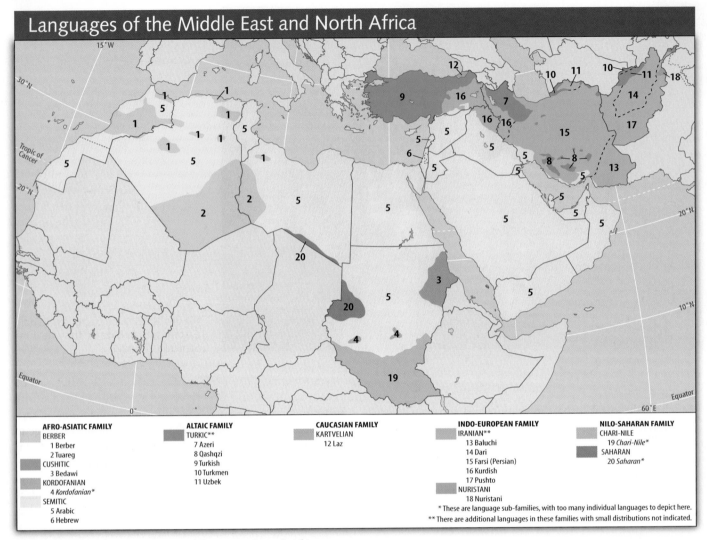

Figure 8.16 Languages of the Middle East and North Africa

to recognize that although political circumstances made them enemies in the 20th century, Jews and Arabs lived in peace for centuries, and they share many cultural traits. Both recognize Abraham as their patriarch. Arabic is a **Semitic** language, in the same Afro-Asiatic language family as **Hebrew,** which is spoken by most of the 5 million Jewish inhabitants of Israel (see the language map, Figure 8.16).

There are other very large populations of non-Semitic ethnic groups and languages in the region. The greatest are the 57 million **Turks** of Turkey, who speak **Turkish,** a member of the **Altaic language family,** and the 34 million **Persians** of Iran, who speak **Farsi,** or **Persian,** in the **Indo-European language family.** Both Persian and Arabic are written in Arabic script and so appear related, but they are not. Turkish also was written in Arabic script until early in the 20th century, but since then, it has been written in a Latin script. The ethnic **Pashtun** majority of Pakistan speaks **Pashto,** a language closely related to Persian, and the country's official language is **Dari,** or **Afghan Persian.** About 26 million **Kurds**—a

people living in Turkey, Iraq, Iran, and Syria—speak **Kurdish** (in the Indo-European language family). Many people in North Africa speak **Berber** and **Tuareg** (in the **Afro-Asiatic language family**). Sudan is ethnically and linguistically a transition zone between the Middle East and North Africa and Africa South of the Sahara. In that country, particularly in the south, there are many speakers of **Chari-Nile languages,** within the **Nilo-Saharan language family.**

The Middle East also gave the world the closely related monotheistic faiths of Judaism, Christianity, and Islam. It is impossible to consider the human and political geographies of this region without attention to these religions, the ways of life associated with them, and their holy places, so here is an introduction (Figure 8.17).

The Promised Land of the Jews

The world's first significant monotheistic faith, Judaism is today practiced by about 13 million people worldwide, mostly in Israel, Europe, and North America. The year 2005 corre-

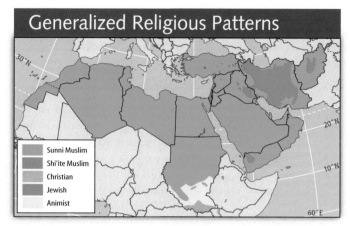

Figure 8.17 Religions of the Middle East and North Africa

sponded to the year 5765 in the Jewish calendar, but unlike its kindred faiths—Christianity and Islam—Judaism does not have an acknowledged starting point in time. Also unlike Christianity, Judaism does not have a fixed creed or doctrine. Jews are encouraged to behave in this life according to God's laws, which He gave to Moses on Mt. Sinai as a covenant with His people. Those laws are part of the Jewish Bible, which Christians know as the Old Testament. Jews do not accept the Christian New Testament because they do not recognize Jesus Christ as the Messiah ("Anointed One" in Hebrew) or savior prophesied in the Jewish scriptures.

WORLD
REGIONAL
Geography⊛Now™

Click Geography Literacy for a virtual tour of Mt. Sinai.

Another distinction between Christianity and Islam is that Judaism is not a proselytizing religion; it does not seek converts. Jewish identity is based strongly on a common historical experience shared over thousands of years. That historical experience has included deep-seated geographic associations with particular sacred places in the Middle East—particularly with places in Jerusalem, capital of ancient Judah (Judea), the province from which Jews take their name. Tragically, the Jewish history also has included unparalleled persecution.

Depending on one's perspective, the Jewish connection with the geographic region known as Palestine, essentially the area now composed of Israel, the West Bank, and the Gaza Strip, is most significant on a time scale of either about 4,000 years or about 100 years. According to the Bible, around 2000 B.C., God commanded Abraham and his kinspeople, known as **Hebrews** (later as Jews), to leave their home in what is now southern Iraq and settle in Canaan. God told Abraham that this land of Canaan—geographic Palestine—would belong to the Hebrews after a long period of persecution. The Bible says that the Hebrews did settle in

Canaan, until famine struck that land. At the command of Abraham's grandson Jacob, the Hebrews—known then as **Israelites**—relocated to Egypt, where grain was plentiful. That began the long sojourn of the Israelites in Egypt, which, according to the Bible, ended in about 1200 B.C. when Moses led them out in the journey known as the **Exodus.**

According to Jewish history, the prophecy of Abraham was first fulfilled when the Israelites settled once again in their **Promised Land** of Canaan. The Jewish King Saul unified the 12 tribes that descended from Jacob into the first united Kingdom of Israel in about 1020 B.C. In about 950 B.C. in Jerusalem—the capital of a kingdom enlarged by Saul's successor, David—King Solomon built Judaism's **First Temple.** He located it atop a great rock known to the Jews as *Even HaShetiyah,* the "Foundation Stone," plucked from beneath the throne of God to become the center of the world and the core from which the entire world was created (see Geography of the Sacred, page 214). The Ark of the Covenant, containing the commandments that God gave to Moses atop Mt. Sinai, was placed in the Temple's Holy of Holies.

The united Kingdom of Israel lasted only about 200 years before splitting into the states of Israel and Judah. Empires based in Mesopotamia destroyed these states: The Assyrians attacked Israel in 721 B.C., and the Babylonians sacked Judah in 586 B.C. The Babylonians destroyed the First Temple (at which point the Ark of the Covenant disappeared) and exiled the Jewish people to Mesopotamia, where they remained until conquering Persians allowed them to return to their homeland. In about 520 B.C., the Jews who returned to Judah rebuilt the temple (the **Second Temple**) on its original site. A succession of foreign empires came to rule the Jews and Arabs of Palestine: Persian, Macedonian, Ptolemaic, Seleucid, and around the time of Christ, Roman. Herod, the Jewish king who ruled under Roman authority and was a contemporary of Christ, greatly enlarged the temple complex.

The Jews of Palestine revolted against Roman rule three times between A.D. 64 and 135. The Romans quashed these rebellions in a series of famous sieges, including those of Masada and Jerusalem. The Romans destroyed the Second Temple, and a third has never been built. All that remains of the Second Temple complex is a portion of the surrounding wall built by Herod. Today, this **Western Wall,** known to non-Jews as the Wailing Wall, is the most sacred site in the world accessible to Jews (Figure 8.18). Some religious traditions prohibit Jews from ascending the **Temple Mount** above, the area where the Temple actually stood, because it is too sacred. After the temple's destruction, that site was occupied by a Roman temple and then in 691 by the Muslim shrine called the **Dome of the Rock,** which still stands today (also in Figure 8.18). The mostly Muslim Arabs know the Temple Mount as *al-Haraam ash-Shariif,* meaning "**The Noble Sanctuary.**" Supercharged with meaning, this place has in modern times often been the spark of conflagration between Jews and Palestinian Arabs.

Geography of the Sacred

Jerusalem

The old city of Jerusalem is filled with sacred places and is one of the world's premier pilgrimage destinations. Over thousands of years, Jerusalem has been coveted and conquered by people of many different cultures and faiths. In the process, a place held sacred by one group has often come to be held sacred by a second and even a third. In some cases, this is not a problem. In others, it is a recipe for long-term discord and violence.

In the previous page, it is important to note that the location of the First and Second Jewish Temples is identical to that of the Muslims' Dome of the Rock. Chapter 9 reveals the importance of this fact to the peace process between the Israelis and Palestinians. Figure 8.B depicts the major places sacred to Jews, Christians, and Muslims and also the ethnic quarters of old Jerusalem.

WORLD
REGIONAL
Geography ⊗ Now™

Active Figure 8.B Sacred sites and ethnic quarters of the old city of Jerusalem. *See an animation based on this figure, and take a short quiz on the facts and concepts.*

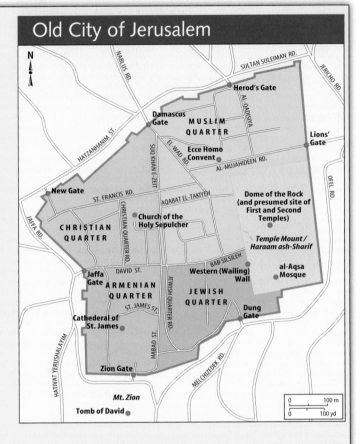

Old City of Jerusalem

[Map labels: N, NABLUS RD., SULTAN SULEIMAN RD., JERICHO RD., Herod's Gate, AL-QADISYA, Damascus Gate, MUSLIM QUARTER, Lions' Gate, HATZANHANIM ST., EL-WAD RD., Ecce Homo Convent, AL-MUJAHIDEEN RD., OFEL RD., New Gate, ST. FRANCIS RD., AQABAT EL-TAKIYEH, Dome of the Rock (and presumed site of First and Second Temples), SUQ KHAN-E-ZEIT, CHRISTIAN QUARTER RD., Church of the Holy Sepulcher, CHRISTIAN QUARTER, Temple Mount / Haraam ash-Sharif, JAFFA RD., BAB SILSILEH, Jaffa Gate, DAVID ST., Western (Wailing) Wall, al-Aqsa Mosque, ARMENIAN QUARTER, JEWISH QUARTER RD., JEWISH QUARTER, ST. JAMES ST., HABAD ST., Dung Gate, Cathederal of St. James, HATIVAT YERUSHALAYIM, Zion Gate, MELCHIZEDEK RD., Mt. Zion, Tomb of David, 0 100 m, 0 100 yd]

Figure 8.18 Jerusalem's Western Wall, Temple Mount, and Noble Sanctuary. In this view are some of Judaism's and Islam's holiest places. At left, below the golden dome, is the Western Wall, which is all that remains of the structure that surrounded the Jews' Second Temple. The dome is the Muslims' Dome of the Rock. At far right, with the black dome, is the al-Aqsa Mosque, another very holy place in Islam.

Joe Hobbs

The victorious Romans scattered the defeated Jews to the far corners of the Roman world. Thus began the Jewish exile, or Diaspora. In their exile, the Jews never forgot their attachment to the Promised Land. The Passover prayer ends with the words "Next year in Jerusalem!" In Europe, where their numbers were greatest, Jews were subjected to systematic discrimination and persecution and were forbidden to own land or engage in a number of professions. Known as anti-Semitism, the hatred of Jews developed deep roots in Europe. This sentiment in part grew out of the perception that Jews were responsible for the murder of Christ and the fact that Christian Europeans were generally prohibited from practicing money lending. Christians assigned the money-lending role to Jews but then resented having to pay interest to them.

In the 1930s, anti-Semitism became state policy in Germany under the Nazis, led by Adolph Hitler. Many German Jews, including Albert Einstein, fled to the United States, and others emigrated to Palestine in support of the Zionist movement, which aimed at establishing a Jewish homeland in Palestine with **Zion** (a synonym for Jerusalem) as its capital. Most Jews were not as fortunate as the emigrants. Within the boundaries of the Nazi empire that dominated most of continental Europe during World War II, Hitler's regime executed its "final solution" to the Jewish "problem." Nazi Germans and their allies killed an estimated 6 million Jews, along with other "inferior" minorities, including Roma (Gypsies) and homosexuals. It was this Holocaust that prompted the victorious allies of World War II, from their powerful position in the newly formed United Nations, to create a permanent homeland for the Jewish people in Palestine.

Christianity: Death and Resurrection in Jerusalem

Nearly 1,000 years after Solomon established the Jewish Temple, a new but closely related monotheistic faith emerged in Palestine. This was **Christianity,** named after Jesus Christ. Jesus, a Jew, was born near Jerusalem in Bethlehem, probably around 4 B.C. Tradition relates that when he was about 30 years old, Jesus began spreading the word that he was the Messiah, the deliverer of humankind long prophesied in Jewish doctrine. After his death, his surname became Christ, meaning "Anointed One" in Greek. A small group of disciples accepted that he was the Messiah and followed him for several years as he preached his message. He taught that love, sacrifice, and faith were the keys to salvation. He had come to redeem humanity's sins through his own death. To his followers, he was the Son of God, a living manifestation of God Himself, and the only path to eternal life was by accepting his divinity.

Jesus Christ's teachings denied the validity of many Jewish doctrines, and in about A.D. 29, a growing chorus of Jewish protesters called for his death. Palestine was then under Roman rule, and Roman administrators in Jerusalem placated the mob by ordering that he be put on trial. He was found guilty of being a claimant to Jewish kingship, and Roman soldiers put him to death by the particularly degrading and painful method of crucifixion. The cornerstone of Christian faith is that Jesus Christ was resurrected from the dead 2 days after his death and ascended into heaven. Christians believe that he continues to intercede with his Father on their behalf and that he will come again on Judgment Day, at the end of time.

After a period of relative tolerance, the Romans began actively persecuting Christians. Nevertheless, Christian ranks and influence grew. The turning point in Christianity's career came after A.D. 324, when the Roman Emperor Constantine embraced Christianity and quickly established it as the official religion of the empire. The Christian Byzantine civilization that developed in the **"New Rome"** that Constantine established—Constantinople, now Istanbul, Turkey—created fine monuments at places associated with the life and death of Jesus Christ. These include the place long acknowledged as the center of the Christian world, Jerusalem's **Church of the Holy Sepulcher** (Figure 8.19). This extraordinary, sprawling building, now administered by numerous separate Chris-

Figure 8.19 Jerusalem's Church of the Holy Sepulcher, containing the spots where many Christians believe Christ was crucified and buried.

6–117

tian sects, contains the locations where tradition says Jesus Christ was crucified and buried.

Christianity has seldom been the majority religion in the land where it was born; only until Islam arrived in Palestine in A.D. 638 was the region primarily Christian. From then until the 20th century, most of Palestine's inhabitants were Muslim, and since the mid-20th century, Muslims and Jews have been the major groups. Between the 11th and 14th centuries, European Christians dispatched military expeditions to recapture Jerusalem and the rest of the Holy Land from the Muslims. These bloody campaigns, known as the Crusades, resulted in a series of short-lived Christian administrations in the region.

There are significant minority populations of Christians throughout the Middle East, including members of distinct sects such as the **Copts** of Egypt and **Maronites** of Lebanon. Their population percentages have generally been declining, both because of emigration and lower birth rates than those of the majority Muslims. Nevertheless, the Middle East remains the cradle of their faith for Christians the world over, and Jerusalem and nearby Bethlehem are the world's premier Christian pilgrimage sites.

The Message of Islam

Islam is by far the dominant religion in the Middle East and North Africa; only Israel within its pre-1967 borders has a non-Muslim majority. Because of Islam's powerful influence not merely as a set of religious practices but as a way of life, an understanding of the religious tenets, culture, and diffusion of Islam is vital for appreciating the region's cultural geography.

Islam is a monotheistic faith built upon the foundations of the region's earliest monotheistic faith, Judaism, and its off-spring, Christianity. Indeed, **Muslims** (people who practice Islam) call Jews and Christians **"People of the Book,"** and their faith obliges them to be tolerant of these special peoples. Muslims believe that their prophet Muhammad was the very last in a series of prophets who brought the Word to humankind. Thus, they perceive the Bible as incomplete but not entirely wrong—Jews and Christians merely missed receiving the entire message. Muslims do not accept the Christian concept of the divine trinity and regard Jesus as a prophet rather than as God.

Muhammad was born in A.D. 570 to a poor family in the western Arabian (now Saudi Arabian) city of Mecca. Located on an important north–south caravan route linking the frankincense-producing area of southern Arabia (now Yemen and Oman) with markets in Palestine (now Israel) and Syria, Mecca was a prosperous city at the time. It was also a pilgrimage destination because more than 300 deities were venerated in a shrine there called the **Ka'aba** (the "Cube"; Figure 8.20). Muhammad married into a wealthy family and worked in the caravan trade. Muslim tradition holds that when he was about 40 years old, Muhammad was meditating in a cave outside Mecca when the Angel Gabriel appeared to him and ordered him to repeat the words of God that the angel would recite to him. Over the next 22 years, the prophet related these words of God (Allah) to scribes who wrote them down as the **Qur'an** (or **Koran**), the holy book of Islam.

During this time, Muhammad began preaching the new message, "There is no god but God," which the polytheistic people of Mecca viewed as heresy. As much of their income depended on pilgrimage traffic to the Ka'aba, they also viewed Muhammad and his small band of followers as an economic threat. They forced the Muslims to flee from Mecca and take

Figure 8.20 The black-shrouded cubical shrine known as the Ka'aba (just right of center) in Mecca's Great Mosque is the object toward which all Muslims face when they pray and is the center-piece of the pilgrimage to Mecca required of all able Muslims. Note the carpet of humanity spread across this image.

Nabeel Turner/Stone/Getty Images

Definitions + Insights

Sunni and Shi'ite Muslims

A schism occurred very early in the development of Islam, and it persists today. The split developed because the Prophet Muhammad had named no successor to take his place as the leader (caliph) of all Muslims. Some of his followers argued that the person with the strongest leadership skills and greatest piety was best qualified to assume this role. These followers became known as Sunni, or orthodox, Muslims. Others argued that only direct descendants of Muhammad, specifically through descent from his cousin and son-in-law Ali, could qualify as successors. They became known as Shi'a, or Shi'ite, Muslims.

The military forces of the two camps engaged in battle south of Baghdad at Karbala in A.D. 680, and in the encounter, Sunni troops caught and brutally murdered Hussein, a son of Ali (Karbala is thus sacred to Shi'ites, as is nearby al-Najaf, where Ali is buried). The rift thereafter was deep and permanent. The martyrdom of Hussein became an important symbol for Shi'ites, who still today regard themselves as oppressed peoples struggling against cruel tyrants, including some Sunni Muslims.

Today, only two Muslim countries, Iran and Iraq, have Shi'ite majority populations. Significant minority populations of Shi'ites are in Syria, Lebanon, Yemen, and the Arab states of the Persian/Arabian Gulf (see Figure 8.17).

---

refuge in Yathrib (modern Medina), where a largely Jewish population had invited them to settle. There were subsequent skirmishes between the Meccans and Muslims, but in 630, the Muslims prevailed and peacefully occupied Mecca. The Muslims destroyed the idols enshrined in the Ka'aba, which became a pilgrimage center for their one God.

The Ka'aba is Islam's holiest place, and Mecca and Medina are its holiest cities. Jerusalem is also sacred to Muslims. Muslim tradition relates that on his **Night Journey,** the Prophet Muhammad ascended briefly into heaven from the great rock now beneath the Dome of the Rock (the same rock Jews regard as the Foundation Stone). Nearby on the Temple Mount/*al-Haraam ash-Shariif* is the **al-Aqsa Mosque,** a sacred congregational site. The proximity, even the duplication, of holy places between Islam and Judaism came to be the most difficult issue in peace negotiations between Palestinians and Israelis, as explained in Chapter 9.

After Muhammad's death in 632, Arabian armies carried the new faith far and quickly. The two decaying empires that then prevailed in the Middle East and North Africa—the Byzantine or Eastern Roman Empire, based in Constantinople (now Istanbul), and the Sassanian Empire, based in Persia (now Iran) and adjacent Mesopotamia (now Iraq)—put up only limited military resistance to the Muslim armies before capitulating. Local inhabitants generally welcomed the new faith, in part because administrators of the previous empires had not treated them well, whereas the Muslims promised tolerance. Soon the Syrian city of Damascus became the center of a Muslim empire. Baghdad assumed this role in A.D. 750.

Arab science and civilization flourished in the Baghdad immortalized in the legends of *The Thousand and One Nights.* There were important accomplishments and discoveries in mathematics, astronomy, and geography. Scholars translated the Greek and Roman classics, and if not for their efforts,

many of these works would never have survived to become part of the modern European legacy. It was an age of exploration, when Arab merchants and voyagers visited China and the remote lands of southern Africa. Many important discoveries by the Arab geographers were recorded in Arabic, a language unfamiliar to contemporary Europeans, and had to be rediscovered centuries later by the Portuguese and Spaniards. Arab merchants carried their faith on the spice routes to the East. One result, surprising to many today, is that the world's most populous Muslim country is not in the Middle East and North Africa, and its people are not Arabs; it is Indonesia, 5,000 thousand miles (8,000 km) east of Arabia.

Whether in Arabia or Indonesia, whether they be Sunni Muslims or Shi'ite Muslims (see Definitions and Insights, above), all believers are united in support of the five fundamental precepts, or pillars of Islam. The first of these is the **profession of faith:** "There is no god but God, and Muhammad is His Messenger." This expression is often on the lips of the devout Muslim, both in prayer and as a prelude to everyday activities.

The second pillar is **prayer,** required five times daily at prescribed intervals. Two of these prayers mark dawn and sunset. Business comes to a halt as the faithful prostrate themselves before God. Muslims may pray anywhere, but wherever they are, they must turn toward Mecca. There also is a congregational prayer at noon on Friday, the Muslim Sabbath.

The third pillar is **almsgiving.** In earlier times, Muslims were required to give a fixed proportion of their income as charity, similar to the concept of the tithe in the Christian church. Today, the donations are voluntary. Even Muslims of very modest means give what they can to those in need.

The fourth pillar is **fasting** during Ramadan, the ninth month of the Muslim lunar calendar. Muslims are required to abstain from food, liquids, smoking, and sexual activity from dawn to sunset throughout Ramadan. The lunar month of Ramadan falls earlier each year in the solar calendar and thus periodically occurs in summer. In the torrid Middle East and North Africa, that timing imposes special hardships on the faithful, who, even if they are performing manual labor, must resist the urge to drink water during the long, hot days.

The final pillar is the **pilgrimage (*hajj*) to Mecca,** Islam's holiest city. Every Muslim who is physically and financially capable is required to make the journey once in his or her lifetime. A lesser pilgrimage may be performed at any time, but the prescribed season is in the 12th month of the Muslim calendar. Those days witness one of Earth's greatest annual migrations, as about 2 million Muslims from all over the world converge on Mecca. Hosting these throngs is an obligation the government of Saudi Arabia fulfills proudly and at considerable expense. However, there has been some trepidation in recent years because of the security threat foreign visitors may pose to the host country, and because accidents such as stampedes and tent city fires have cost numerous lives. Many pilgrims also visit the nearby city of Medina, where Muhammad is buried. Most Muslims regard the *hajj* as one of the most significant events of their lifetimes. All are required to wear simple seamless garments, and for a few days, the barriers separating groups by income, ethnicity, and nationality are broken. Pilgrims return home with the new stature and title of "*hajj*" but also with humility and renewed devotion.

All Muslims share the five pillars and other tenets, but they vary widely in other cultural practices related to their faith depending on what country they live in, whether they are from the desert, village, or city, and their education and income. The governments and associated clerical authorities in Saudi Arabia and Iran insist on strict application of **Islamic law (*shari'a*)** to civil life; in effect, there is no separation between church and state. The Qur'an does not state that women are required to wear veils, but it does urge them to be modest, and it portrays their roles as different from those of men. Clerics in Saudi Arabia insist that women wear floor-length, long-sleeved black robes and black veils in public, that they travel accompanied by a male member of their families, and that they not drive cars. In Egypt, by contrast, Muslim women are free to appear in public unveiled if they choose. However, in most Muslim countries, conservative ideas about the role of women are still very strong: They should be modest, retiring, good mothers, and keepers of the home. The Qur'an portrays women as equal to men in the sight of God, and in principle, Islamic teachings guarantee the right of women to hold and inherit property.

Most Muslim women argue that what others often see as "backward" cultural practices are in fact progressive. For example, their modest dress compels men to evaluate them on the basis of their character and performance, not their attractiveness. Segregation of the sexes in the classroom makes it easier for both women and men to develop their confidence and skills. Sexual assault is rare. A married woman retains her maiden name. These apparent advantages can be weighed against the drawbacks that women are generally subordinate to men in public affairs and have fewer opportunities for education and for work outside the home.

8.4 Economic Geography

Overall, this is a poor region; per capita GDP PPP for the 23 countries and territories averages only $5,345 (see Table 8.1). This may seem surprising in view of the "rich Arab" stereotype. Only the oil-endowed states of the Persian/Arabian Gulf deserve reputations for wealth, and only non-Arab Israel is truly a more developed country (MDC) by measures other than per capita wealth. Israel's prosperity comes from its innovation in computer and other high-technology industries, the processing and sale of diamonds, large amounts of foreign (mostly U.S.) aid, and investment and assistance by Jews and Jewish organizations around the world.

Vital to the industrialized countries as a source of fuels, lubricants, and chemical raw materials, petroleum is one of the world's most important natural resources. A crucial feature of world geography is the concentration of approximately two-thirds of the world's proven petroleum reserves in a few countries that ring the Persian/Arabian Gulf. By coincidence, the countries rich in oil tend to have relatively small populations, whereas the most populous nations have few oil reserves; Iran is an exception. All but about 1 percent of the Persian/Arabian Gulf oil region's proven reserves of crude oil are located in Saudi Arabia, Iraq, Kuwait, Iran, and the United Arab Emirates (U.A.E.), with smaller reserves in Oman and Qatar. Saudi Arabia, by far the world leader in reserves, has about 25 percent of the proven crude oil reserves on the globe. Iraq has an additional 11 percent, Kuwait 9 percent, Iran 9 percent, and the U.A.E. 6 percent. (These are the top five in the world's known oil reserves as of 2004.) By comparison, Venezuela has 5 percent of world reserves, Russia another 5 percent, and the world's largest oil consumer, the United States, only 2 percent. In the Gulf region, the great thickness of the region's oil-bearing strata and high reservoir pressures have made it possible to secure an immense amount of oil from a small number of wells. The productivity of each well makes each barrel inexpensive to extract and makes it simple to increase or reduce production quickly in response to world market conditions. Gulf oil is thus "cheap" to produce unless expenditures to maintain huge military forces to defend it are factored in. In 2004, the United States maintained about 170,000 military personnel in the region at a cost of more than $50 billion per year. A U.S. Navy secretary once remarked that the real price of oil, with military expenditures factored in, is about $100 per barrel, not the typical market price of less than $50.

Production, export, and profits of Middle Eastern and North African oil were once firmly in the hands of foreign companies. That situation changed after 1960 when most of the Gulf countries and other exporting nations formed the **Organization of Petroleum Exporting Countries (OPEC)** with the aim of joint action to demand higher profits from oil. It changed again after 1972 when the oil producing countries began to nationalize the foreign oil companies. OPEC was relatively obscure until the Arab–Israel war of 1973, after which the organization began a series of dramatic price increases. In 1980, the organization's price reached $37 (U.S.) per barrel, compared with $2 a barrel in early 1973.

These events had enormous repercussions for the world economy. Immense wealth was transferred from the more developed countries to the OPEC countries to pay for indispensable oil supplies. The skyrocketing cost of gasoline and other oil products helped cause serious inflation in the United States and many other countries and contributed to the 1973 energy crisis in the United States. Desperately poor, less developed countries (LDCs) found that high oil prices not only hindered the development of their industries and transportation but also reduced food production because of high prices for fertilizer made from oil and natural gas. In the Gulf countries, the oil bonanza produced a wave of spending for military hardware, showy buildings, luxuries for the elite, and ambitious development projects of many kinds. Per capita benefits to the general populace were greatest in the Arabian Peninsula, where small populations and immense inflows of oil money made possible the abolition of taxes, the development of comprehensive social programs, and heavily subsidized amenities such as low-cost housing and utilities, including water distilled from the Gulf by desalination plants.

Then, at the beginning of the 1980s, the era of continually expanding oil production, sales, and profits by the OPEC states came to an end. After 1973, the high price of oil stimulated oil development in countries outside OPEC. Oil conservation measures such as a shift to more fuel-efficient vehicles and furnaces were instituted. Substitution of cheaper fuels for oil increased. Coal replaced oil in many electricity-generating stations. Oil refineries were converted to make gasoline from cheaper "heavy" oils rather than the more expensive "light" oils previously used. Meanwhile, the world entered a period of economic recession due in part to high oil prices. Decreased business activity reduced the demand for oil. Profits of the world oil industry (and taxes paid to governments) were severely cut, large numbers of refineries had to close, and much of the world tanker fleet was idled.

Oil prices rose temporarily in 1990–1991 when the flow of Iraqi and Kuwaiti oil was cut off following Iraq's military takeover of Kuwait, but the prices soon fell again. In the 1990s, Saudi Arabia and other oil-rich Gulf states implemented internal economic austerity measures for the first time. However, the immense oil and gas reserves still in the ground guaranteed that the Gulf region would continue to have a major long-term impact on the world and would remain relatively prosperous as long as these finite resources are in demand in the MDCs. The lasting economic clout of the region was apparent early in the 21st century as the price of oil once again climbed to record levels (thanks mainly to OPEC decisions to reduce production), contributing to economic slowdowns in many nations.

There are other resources and industries in the region's economic geography—remittances (earned income) sent home by guest workers in the oil-rich countries, revenues from ship traffic through the Suez Canal, exports of cotton, rice, and other commercial crops, for example—but oil dominates the region's economy and is central to the global economy. The economies of the respective countries are described in more detail in the following chapter. Here the focus remains on Middle Eastern oil and its critical role in geopolitical affairs.

8.5 Geopolitical Issues

This has long been a vital region in world affairs and a target of outside interests. Its strategic crossroads location often has made it a cauldron of conflict. From very early times, overland caravan routes, including the famous Silk Road, crossed the Middle East and North Africa with highly prized commodities traded between Europe and Asia. The security of these routes was vital, and countries on either end could not tolerate threats to them. In more recent times, geopolitical concerns have focused on narrow waterways, access to oil, access to fresh water, and terrorism.

Chokepoints

One of the striking characteristics of the geography of the Middle East and North Africa is how many seas border and penetrate the region. In many cases, these seas are connected to one another though narrow straits and other passageways. In geopolitical terms, such constrictions may be viewed as chokepoints—strategic narrow passageways on land or sea that may be easily closed off by use, or even the threat, of force (Figure 8.21). Chokepoints must be unimpeded if

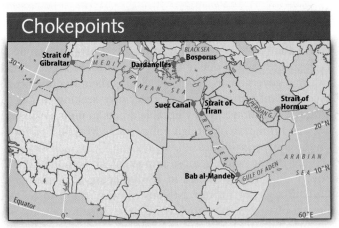

Active Figure 8.21 Chokepoints of the Middle East and North Africa. *See an animation based on this figure, and take a short quiz on the facts and concepts.*

Geography of Energy

Middle Eastern Oil Pipelines

The oil exporting countries of the Middle East, often with the financial support of the United States and other leading oil consuming countries, have invested enormous resources to ensure the safe passage of oil to world markets. The Middle East is not landlocked like the oil-rich "stans" of Central Asia are, but some of the same kinds of geopolitical issues are involved in oil-export planning. These concerns often point toward pipelines as the best possible means of routing oil shipments (Figure 8.C). Pipelines are attractive because they shorten the time and expense involved in seaborne transport, and they bypass chokepoints. However, they are very vulnerable to disruption. Weapons as small as grenades can disrupt supplies. A major challenge is how to route a pipeline so that it will not cross through a potential enemy's territory, and another is how to maintain friendships with the countries the oil crosses.

One of the first major pipelines in region was the 1,100-mile (1,760-km) Trans-Arabian Pipeline (Tapline), leading from the oil producing eastern province of Saudi Arabia to a terminal on the Mediterranean coast of Lebanon. It has the advantage of bypassing three chokepoints (Strait of Hormuz, Bab el Mandeb, and Suez Canal). However, at the time it was completed (1950), it could not have been anticipated that it would cross two of the worst conflict zones in the Middle East: the Golan Heights of Syria (which fell to Israel in the 1967 war) and southern Lebanon (a major theater of the Lebanese civil war of the 1970s and the Israeli–Lebanese Shi'ite struggles of the 1980s and 1990s). Tapline was knocked out early by its unfortunate geography.

Another Middle East conflict, the 1967 Six Day War between Israel and its neighbors, led to the closure of the Suez Canal (for about a decade) and thus to the construction of two new pipelines to bypass Suez: one across southern Israel from the Gulf of Aqaba to the Mediterranean Sea and another across Egypt from the Gulf of Suez to the Mediterranean (thus its acronym, SUMED).

No Middle Eastern country has been more dependent on pipelines than Iraq, and none has so systematically experienced the liabilities of pipelines. The Iran–Iraq war of 1980–1988 had major impacts on pipeline geography. Saudi Arabia supported Iraq in the war and so tried to ensure it could get oil to market without being threatened by Iran; this meant bypassing the Strait of Hormuz. That led to the construction of Petroline (opened in 1981), the east–west pipeline across the Arabian Peninsula. Iraq built a pipeline to ship some of its oil south to Petroline, thereby bypassing the Strait of Hormuz as well. Beginning in 1961, Iraq was also able to ship oil through pipelines almost due west though Syria to the Mediterranean Sea. But when it attacked Iran in 1980, Iraq lost that route because Syria was Iran's ally.

Iraq quickly responded to that setback by building a new pipeline leading almost due north to bypass Syrian territory and then making a sharp 90 degree turn westward through southern Turkey to the Mediterranean Sea. But in 1990, Iraq attacked Kuwait, threatened Saudi Arabia, and prompted a U.S.-led counterattack. Saudi Arabia responded by closing the southern link with Petroline. Turkey, a NATO ally of the United States, responded by closing the northern link. The United Nations responded by issuing strict controls on Iraqi oil exports. In sum, by its military actions, Iraq did almost everything imaginable to deprive itself of oil-export capabilities. Only in post-Saddam Hussein Iraq is the country's oil production recovering.

168–169

258

world commerce is to carry on normally. Keeping them open is therefore usually one of the top priorities of regional and external governments. Similarly, closing them is a priority to a combatant nation or a terrorist entity seeking to gain a strategic advantage. Many notable events in military history and the formation of foreign policy in the Middle East focus on these strategic places.

One of the world's most important chokepoints is the Suez Canal, which opened in 1869. Slicing 107 miles (172 km) through the narrow Isthmus of Suez, the British- and French-owned Suez Canal linked the Mediterranean Sea with the Indian Ocean, saving cargo, military, and passenger ships a journey of many thousands of miles around the southern tip of Africa (Figure 8.22). Keeping the Suez Canal in friendly hands was one of Britain's major military concerns during World War II. That objective led to a very hard-

Figure 8.22 The strategically vital Suez Canal zone saw bitter fighting in the Middle East wars of 1956, 1967, and 1973.

Joe Hobbs

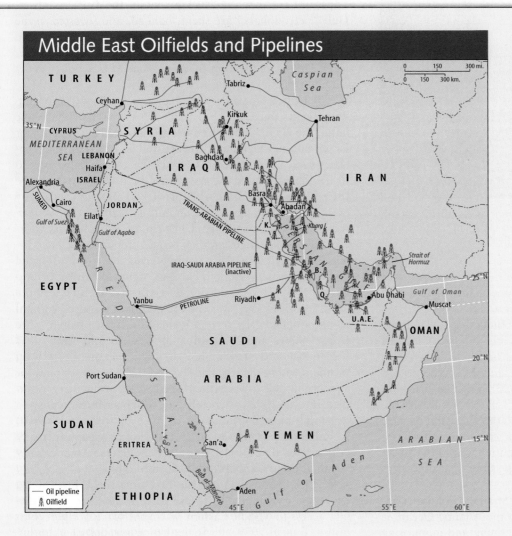

Middle East Oilfields and Pipelines

Figure 8.C Principal oil fields and pipelines in the heart of the Middle East. Vulnerable chokepoints and volatile political relations have led to the construction and often indirect routing of many pipelines.

fought and eventually successful British and Allied campaign against Nazi Germany in North Africa. Egyptian President Gamal Abdel Nasser's nationalization of the canal in 1956 led immediately to a British, French, and Israeli invasion of Egypt and a conflict known as the **Suez Crisis** and the **1956 Arab–Israeli War.** Egypt effectively won that war when international pressure caused the invading forces to withdraw, leaving the canal in Egypt's hands.

Nearby, another chokepoint played a critical role in the most important Arab–Israeli war, the 1967 Six Day War. One of the events that precipitated the war was Egyptian President Nasser's closure of the Strait of Tiran (at the southern end of the Gulf of Aqaba) to Israeli shipping. Israel had won the right of navigation through the strait after its war of independence and would not accept its closure; therefore, it attacked Egypt.

On either side of the Arabian Peninsula are two more critical chokepoints. One is extraordinarily vital to the world's economy: the Strait of Hormuz, connecting the Persian/Arabian Gulf with the Gulf of Oman and Arabian Sea. Much of the world's oil supply passes through here in the holds of giant supertanker ships. Closure of the Strait of Hormuz would have devastating impacts on the world's industrial and financial systems. Iran's plans to station Chinese-made silkworm missiles on the Strait of Hormuz and thus threaten international oil shipments led to a new level of U.S. involvement late in the Iran–Iraq War of 1980–1988. That war also prompted a flurry of new oil pipeline construction designed to bypass the Strait of Hormuz chokepoint (see the Geography of Energy, above). One of these pipelines, the Petroline route running east–west across the Arabian Peninsula, was built to also bypass the other important choke-

Joe Hobbs

Figure 8.23 The Dardanelles chokepoint is the southernmost of the two Turkish straits. This is a view from the Asian side, looking across to Europe near the Gallipoli Battlefield.

point: the Bab al-Mandeb, which connects the Red Sea with the Gulf of Aden and the Indian Ocean.

Turkey controls two more chokepoints that together are known as the Turkish Straits. The northernmost strait is the Bosporus, which cleaves the city of Istanbul into western (European) and eastern (Asian) sides, and the southern strait is called the Dardanelles (Figure 8.23). Their security has long been critical to the successful passage of goods between Europe and Asia and even more so to the successful passage of vessels between Russia (and the Soviet Union) and the rest of the world. Throughout the 20th century, one of the Soviet Union's constant strategic priorities was the right of navigation through the Turkish Straits.

Finally, it should be noted that the Strait of Gibraltar, connecting the Mediterranean Sea with the Atlantic Ocean, is also a chokepoint. Here, too, maintaining and monitoring the flow of maritime traffic have long been important concerns. Britain's insistence on maintaining control over its enclave of Gibraltar, decades after the decolonization of most of the world, is an excellent indicator of how critical this chokepoint is.

130

Access to Oil

The Suez Canal and other chokepoints, the cotton of the Egypt's Nile Delta, and the strategic location of the region were important during and since colonial times. But oil has been, and will remain (as long as fossil fuels drive the world's economies), what keeps the rest of the world interested in the Middle East and North Africa. The region's oil is sold to many countries, but most of it is marketed in Western Europe and Japan. The United States also imports large amounts of Gulf oil but has a much smaller relative dependence on this source than do Japan and Europe. However, the Gulf region is very important to the United States because of the heavy dependence of close American allies on Gulf oil

and because of the importance of the oil as a future reserve. American companies are also heavily involved in oil operations and oil-financed development in the Gulf countries. Gulf "petrodollars" are spent, banked, and invested in the United States, contributing significantly to the American economy. Maintaining a secure supply of Gulf oil has therefore been one of the longstanding pillars of U.S. policy in the Middle East.

The United States has long had a precarious relationship with the key players in the Middle Eastern arena. On the one hand, the United States has pledged unwavering support for Israel, but on the other, it has courted Israel's traditional enemies such as oil-rich Saudi Arabia. That Arab kingdom and its neighbors around the Persian/Arabian Gulf possess more than 60 percent of the world's proven oil reserves. Thus, they are vital to the long-term economic security of the Western industrial powers and Japan. The United States and its Western allies made it clear they would not tolerate any disruption of access to this supply when Iraqi troops directed by Iraqi President Saddam Hussein occupied Kuwait on August 2, 1990. U.S. President George Bush drew a "line in the sand," proclaiming "we cannot permit a resource so vital to be dominated by one so ruthless—and we won't." The United States and a coalition of Western and Arab allies fielded a massive array of military might that ousted the Iraqi invaders within months and secured the vital oil supplies for Western markets. When the United States invaded and occupied Iraq in 2003, ostensibly to extinguish Iraq's ability to develop and deploy **weapons of mass destruction** (**WMD**), including chemical, biological and nuclear weapons, many critics insisted this was just another example of America's determination to control Middle Eastern oil.

The Gulf War was not the first time the United States expressed its willingness to use force if necessary to maintain access to Middle Eastern oil. In the wake of the revolution in Iran in 1979, the Soviet Union invaded neighboring Afghanistan. U.S. military analysts feared the Soviets might use Afghanistan as a launch pad to invade oil-rich Iran. The United States deemed this prospect unacceptable, and President Jimmy Carter issued the policy statement that came to be known as the Carter Doctrine: The United States would use any means necessary to defend its vital interests in the region. Vital interests meant oil, and any means necessary meant the United States was willing to go to war with the Soviet Union, presumably nuclear war, to defend those interests.

Middle Eastern wars had already become proxy wars for the superpowers, with oil always looming as the prize: In the 1967 and 1973 Arab–Israeli wars, for example, Soviet-backed Syrian forces fought U.S.-backed Israeli troops. American support of Israel in this war prompted Arab members of OPEC to impose an embargo on sales of their oil to the United States, precipitating the nation's first energy crisis. During the 1973 war, the United States put its forces on an advanced state of readiness to take on the Soviets in a nuclear exchange if necessary. All of these events illustrate that the Middle East and North Africa comprise, in political geography terms, a shatter belt—a large, strategically located region composed of conflicting states caught between the conflicting interests of great powers.

Access to Fresh Water

Some of the most serious geopolitical issues in the Middle East and North Africa relate to **hydropolitics,** or political leverage and control over water. In this arid region, where most water is available either from rivers or from underground aquifers that cross national boundaries, control over water is an especially difficult and potentially explosive issue. An estimated 90 percent of the usable fresh water in the Middle East crosses one or more international borders.

Water is one of the most problematic issues in the Palestinian–Israeli conflict (Figure 8.24). Fresh-water aquifers underneath the West Bank supply about 40 percent of Israel's water. Palestinians point to Israel's control over West Bank water as one of the most problematic elements of its occupation. The average Jewish settler on the West Bank uses 74 gallons (278 liters) per day, whereas the average West Bank Palestinian uses 19 (72 liters). The World Health Organization calculates that 27 gallons (102 liters) per person per day is needed for minimal health and sanitation standards, but Israeli policies prohibit Palestinians from increasing their water usage. Many Israeli policymakers insist that water resources on the West Bank must remain under strict Israeli control and, on these grounds, oppose the creation of a Palestinian state on the West Bank. Critically, it is estimated that the West Bank aquifers will not contain enough water to support the region's population at current levels of consumption for more than a few more years (even taking into account anticipated replenishment from rainfall).

More promisingly, Jordan and Israel are working on agreements to share waters from the Jordan River (which forms a portion of their common border) and its tributary, the Yarmuk River. They are discussing a joint venture to build the "Red-Dead" Canal, which would connect the Gulf of Aqaba with the Dead Sea. The gravity flow of seawater to the Dead Sea would spin turbines and run generators to produce electricity the two nations could share.

Water is a critical issue blocking a peace treaty between Israel and Syria. If Syria were to recover all of the Golan Heights area (which it lost to Israel in the 1967 war), it would have shorefront on Lake Kinneret and therefore, presumably, rights to use its water. That prospect is unacceptable to Israel. This body of water, also known as Lake Tiberias and the Sea of Galilee, is Israel's principal supply of fresh water, feeding the National Water Carrier system that transports water south to the Negev Desert. In occupying the Golan Heights, Israel also controls some of the northern bank of the Yarmuk River on the border with Jordan. For many years, Israel has stated it would never allow Syria and Jordan to construct their proposed Unity Dam that would

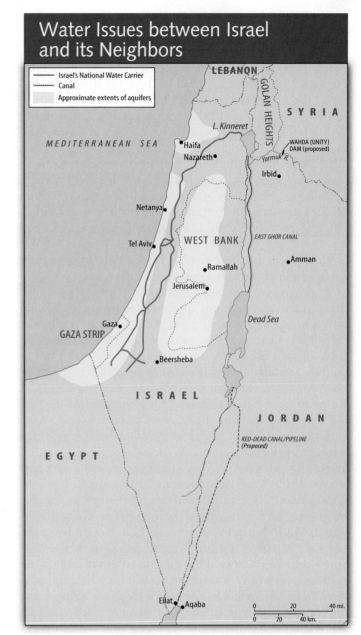

Figure 8.24 Locations of water issues between Israel and its neighbors

store waters of the Yarmuk River to be shared between those countries. Israel thus implied it would bomb the dam rather than allow it to deprive Israel of Jordan River water.

A useful way to think about the geography of hydropolitics is in terms of **"upstream" and "downstream" countries.** Simply because water flows downhill, an upstream country usually is able to maximize its water use at the expense of a downstream country. However, the situation between Israel and the countries upstream on the Yarmuk shows that this is not always true. Israel is far more powerful militarily and can use the threat of force to wrest more water out of the system.

Historically, the same has been true for Egypt. It is the ultimate downstream country, at the mouth of a great river than runs through 10 countries and sustains 160 million people (a population that is expected to double in 25 years; Figure 8.25). However, it has long been the strongest country in the Nile Basin and has threatened to use its greater force if it does not get the water it wants. In 1926, when the British ruled Egypt and many other colonies in Africa, 10 of Egypt's upstream countries were compelled to sign the **Nile Water Agreement.** This guaranteed Egyptian access to 56 billion cubic meters of the Nile's water, or fully two-thirds of its 84 billion cubic meters—even though barely a drop of the Nile's waters actually originates in Egypt. The treaty forbids any projects that might threaten the volume of water reaching Egypt, prohibits use of Lake Victoria's water without Egypt's permission, and gives Egypt the right to inspect the entire length of the Nile to ensure compliance.

In recent years, though, one country after another has defied the treaty, calling it an outmoded legacy of colonialism. Kenya and Tanzania have plans to build pipelines to carry Lake Victoria waters to thirsty towns and villages inland. Uganda is building its controversial Bujagali Dam on the Nile, mainly for hydroelectricity production. With Chinese assistance, Ethiopia is building the huge Takaze Dam, for hydropower and irrigation, on a tributary of the Blue Nile. Sudan is building the Merowe (Hamdab) and Kajbar Dams on its northern stretch of the Nile. Predictably, Egypt has had a bellicose response to these developments. For example, Egypt called Kenya's stated intention to withdraw from the Nile Water Agreement an "act of war."

Meanwhile, Egypt's demands on Nile waters are increasing. Egypt recently excavated the multibillion-dollar Toshka Canal, which transports water from Lake Nasser over a distance of 300 miles (500 km) to the Kharga Oasis of the Western Desert. Proponents of the canal insist it will result in the cultivation of nearly 1.5 million acres (600,000 hectares) of "new" land and provide a living for hundreds of thousands of people. Critics argue that it is a waste of money and that salinization and evaporation will take a huge toll on the cultivated land and the country's water supply.

The source of four-fifths of Syria's water and two-thirds of Iraq's, Turkey is an upstream country that exercises its upstream advantage on the Tigris and Euphrates Rivers (Figure 8.26). Turkey's position has long been that water in Turkey belongs to Turkey, just as Saudi Arabian oil belongs to Saudi Arabia. Not surprisingly, downstream Syria and Iraq reject this position and are distraught by the diminished flow and quality of water resulting from Turkey's comprehensive Southeast Anatolia Project. When completed, the project is expected to reduce Syria's share of the Euphrates waters by 40 percent and Iraq's by 60 percent. Also increasing the likelihood of serious future tension is a history of strained relations among Turkey, Syria, and Iraq, accompanied by the fact that no commonly accepted body-of-water law governs the allocation of water in such international situations.

262–263

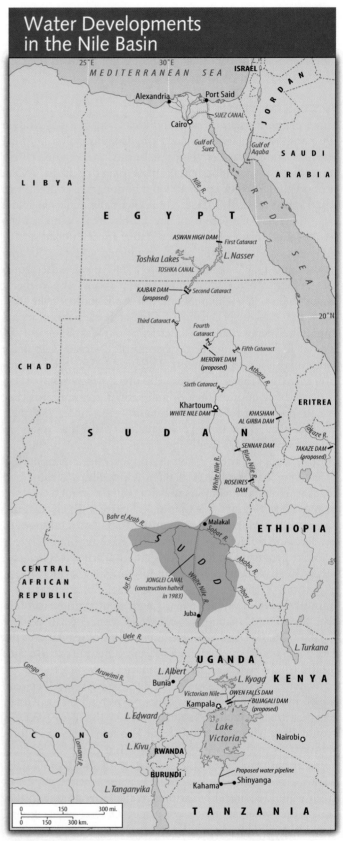

Water Developments in the Nile Basin

Figure 8.25 Recent and proposed water developments in the Nile basin

Figure 8.26 The Tigris and Euphrates Rivers rise in Turkey, giving this non-Arab country control over a resource vital to the lives of millions of Arabs in downstream Syria and Iraq. This waterfall is on a tributary of the Tigris in far eastern Turkey.

Turkish leaders have said they will never use water as a political weapon, but Turkey has wielded water to its advantage. In 1987, for example, Turkey increased the Euphrates flow into Syria in exchange for a Syrian pledge to stop support of Kurdish rebels inside Turkey. Turkey now says it wants to use water to promote peace in the Middle East by shipping it in converted supertankers for sale to such thirsty (and therefore potentially combative) countries as Israel, Jordan, Saudi Arabia, Tunisia, and Algeria.

Terrorism

Viewed from the perspective of the U.S. government, almost all of the geopolitical issues related to this region—oil, economic development, trade, aid, the Arab–Israeli conflict, and more—have recently been subsumed beneath the broader rubric of the "war on terrorism." Even the 2003 invasion of Iraq was explained in part as essential to the war on terror (see that discussion in Chapter 9, pages 257–258). How the United States pursues that war in the coming years will have enormous impacts on societies and economies in the Middle East and North Africa and perhaps in the United States as well.

The terrorists pursued by the United States are almost without exception Islamist militants (best known as **Islamists**), and so it is useful to understand the nature of Islamic "fundamentalism" and radicalism. Not all Islamists are militant or terrorist, but they all reject what they view as the materialism and moral corruption of Western countries and the political and military support these countries lend to Israel. Both Sunni and Shi'ite Muslims have advanced a wide range of Islamic movements, notably in Iran, Lebanon, Egypt, Afghanistan, Sudan, and Algeria.

Although nominally religious, the more radical of these movements have political and cultural aims, particularly the destabilization or removal of U.S. and Israeli interests in the region and abroad. In 1993, followers of the radical Egyptian cleric Sheikh Umar Abdel-Rahman bombed New York City's World Trade Center as a protest against American support of Israel and Egypt's pro-Western government. In an attempt to destabilize and replace Egypt's government, which they viewed as an illegitimate regime too supportive of the United States, another Egyptian Islamist group attacked and killed foreign tourists in Egypt in the 1990s. In the 1980s, members of the pro-Iranian **Hizbullah,** or Party of God, in Lebanon kidnapped foreigners as bargaining chips for the release of comrades jailed in other Middle Eastern countries.

Within Israel and the autonomous Palestinian territories of the West Bank and Gaza Strip, Palestinian members of **Hamas** (an Arabic acronym for the Islamic Resistance Movement) and another organized called the **al-Aqsa Martyrs Brigade** have carried out terrorist attacks on Israeli civilians and soldiers in an effort (apparently successful) to derail implementation of the peace agreements reached between the Israeli government and the Palestine Liberation Organization (PLO).

In Algeria, years of bloodshed followed the government's annulment of 1991 election results that would have given the **Islamic Salvation Front (FIS)** majority control in the parliament. Muslim sympathizers carried the battle to France, bombing civilian targets in protest against the French government's support for the Algerian regime.

In Western capitals, concern about "state-sponsored terrorism" has long focused on Iran. Iran has extended direct or clandestine assistance to a variety of Islamist terrorist groups, including Hizbullah and Hamas. There is great concern about Iran's nuclear weapons potential because such weapons might find their way to terrorist groups or be delivered by Iran itself on its own missiles against Israel or another target. Iran denies that it is developing the weapons, but U.S. and Israeli intelligence agencies believe Iran will possess them by 2008. Israeli officials have said publicly that Israel (which has its own nuclear weapons) regards Iran as an "existential threat" and will prevent that development, presumably by an air strike like the one Israel carried out against Iraq's nuclear facility in 1981.

In 1998, the world began to hear about Osama bin Laden, a former Saudi businessman living in exile in Afghanistan, whose **al-Qa'ida** organization bombed U.S. embassies in Kenya and Tanzania as part of a worldwide armed struggle, or *jihad,* against American imperialism and immorality (see Geography of Terrorism, pages 226–227). Bin Laden's organization was also responsible for the 2000 bombing of the American naval destroyer the U.S.S. *Cole* in Yemen's harbor of Aden. However shocking those assaults were, they pale in comparison to al-Qa'ida's attacks against targets in the United States on September 11, 2001. In the most ferocious terrorist actions ever undertaken to that date, members of al-Qa'ida cells in the United States hijacked four civilian jetliners and succeeded in piloting two of them into New York City's World Trade Center towers and one into Washington, D.C.'s Pentagon. More than 3,000 people, mainly civilians, perished. Bin Laden and his followers cheered the carnage as

Geography of Terrorism

What Does al-Qa'ida Want?

This is an important question. Perhaps if an answer could be found, there would be a way either to defeat this organization or to address the root causes of its existence in such a way that it would no longer have a reason to exist. Here, we will look at the question mainly from the inside, examining what al-Qa'ida says it wants. The answer may be fundamentally a geographic one.

To begin, it is useful to know what al-Qa'ida is. Al-Qa'ida ("the Base" in Arabic) is a transnational organization that seeks to unite Islamist militant groups worldwide in a common effort to achieve its goals. An Islamist or *jihadist* organization is one that employs Islamic faith, culture, and history to legitimize its philosophy and actions.[a] The principal sources Islamists cite are the Qur'an, the *Hadith* (sayings of the Prophet Muhammad), and the writings of earlier Islamic militants such as Ibn Taymiyya, Sayyid Qutb, and Muhammad al-Faraj. According to Rohan Gunaratna, the leading academic authority on the organization, al-Qa'ida is "above all else a secret, almost virtual, organization, one that denies its existence in order to remain in the shadows."[b] This desire for secrecy explains why al-Qa'ida seldom takes direct responsibility for terrorist acts. Instead, these are attributed to the other names and identities employed by al-Qa'ida, particularly the "World Islamic Front for the Jihad against the Jews and the Crusaders," a coalition of seven Islamist militant groups (three Egyptian, two Pakistani, one Bangladeshi, and one Afghan). Osama bin Laden, formerly a Saudi national, emerged as the figurehead of this organization.

Al-Qa'ida certainly is a terrorist organization if one employs the U.S. Department of State's definition of **terrorism**: "premeditated, politically motivated violence perpetrated against noncombatant targets by subnational groups or clandestine agents, usually intended to influence an audience."[c] Ayman al-Zawahiri, al-Qa'ida's second in command, called for an escalation of attacks with "the need to inflict the maximum casualties against the opponent, for this is the language understood by the West, no matter how much time and effort such operations take."[d] Osama bin Laden, as evident in his own words below, called for the killing of American and other Western civilians. Gunaratna regards al-Qa'ida as an extremely unusual terrorist group in the category he calls "apocalyptic," one that believes "it has been divinely ordained to commit violent acts, and likely to engage in mass casualty, catastrophic terrorism."[e] He warns that al-Qa'ida "will have no compunction about employing chemical, biological, radiological and nuclear weapons against population centers."[f]

But why? Why do they want to kill Western, and particularly American, civilians? Al-Qa'ida leaders have been very explicit in their rationale. Osama bin Laden explained that the main reason for the terrorist attacks is to convince the United States that it should withdraw its military forces and other interests from the Islamic Holy Land. In the following statement, the "Land of the Two Holy Places" means Saudi Arabia. The two holy places are the Saudi Arabian cities of Mecca (site of the Ka'aba, and Islam's most sacred city) and Medina (Islam's second holiest place, where the Prophet Muhammad is buried). The Dome of the Rock is the shrine in Jerusalem, described earlier, containing the sacred rock from where the Prophet Muhammad was said to have ascended into heaven on the Night Journey.

> The Arabian Peninsula has never—since God made it flat, created its desert, and encircled it with seas—been stormed by any forces like the crusader armies spreading in it like locusts, eating its riches and wiping out its plantations . . . The latest and greatest of these aggressions, incurred by the Muslims since the death of the Prophet . . . is the occupation of the Land of the Two Holy Places—the foundation of the house of Islam, the place of the revelation, the source of the message, and the place of the noble Ka'aba, the *qibla* of all Muslims—by the armies of the American crusaders and their allies. We bemoan this and can only say "No power and power acquiring except through Allah" . . . To push the enemy—the greatest *kufr* [infidel]—out of the country is a prime duty. No other duty after Belief is more important than this duty. Utmost effort should be made to prepare and instigate the *umma* [Islamic community] against the enemy, the American–Israeli alliance—occupying the country of the two Holy Places . . . to the al-Aqsa mosque in Jerusalem . . . The cru-

[a] John L. Esposito, *Unholy War: Terror in the Name of Islam* (Oxford: Oxford University Press, 2002), p. 28.

[b] Rohan Gunaratna, *Inside Al Qaeda: Global Network of Terror* (New York: Cambridge University Press, 2002), p. 3.

[c] Quoted in Rex A. Hudson, *Who Becomes a Terrorist and Why: The 1999 Government Report on Profiling Terrorists* (Guilford, Conn.: Lyons Press, 1999), p. 18.

[d] Ayman Al-Zawahiri, 2002b. "Why Attack America (January 2002)." In *Anti-American Terrorism and the Middle East,* Barry Rubin and Judith Colp Rubin, eds. (Oxford University Press, 2002), p. 133.

[e] Gunaratna, op. cit., p. 93.

[f] Ibid., p. 11.

saders and the Jews have joined together to invade the heart of Dar al-Islam—the Abode of Islam: our most sacred places in Saudi Arabia, Mecca and Medina, including the prophet's mosque and Dome of the Rock in Jerusalem, al-Quds.[g]

In what was arguably his most important policy statement, announcing the formation of the "Islamic World Front for the Jihad against the Jews and the Crusaders," bin Laden used the occupation of Islamic sacred space as the principal justification for war against the United States:

> In compliance with God's order, we issue the following *fatwa* [religious injunction] to all Muslims: The ruling to kill the Americans and their allies—civilians and military—is an individual duty for every Muslim who can do it in any country in which it is possible to do it, in order to liberate the al-Aqsa mosque and the holy mosque [in Mecca] from their grip, and in order for their armies to move out of all the lands of Islam, defeated and unable to threaten any Muslim.[h]

In an earlier *fatwa* (1996), bin Laden laid out these goals for al-Qa'ida: to drive U.S. forces out of the Arabian Peninsula, overthrow the Saudi government, and liberate the holy places of Mecca and Medina. Overthrowing the Saudi government and the other autocratic dynasties of the Persian Gulf region, along with other secular and pro-Western regimes of the Middle East—notably Mubarak's Egypt—is a theme that is very often articulated in al-Qa'ida ideology.

What is al-Qa'ida's ultimate goal? Is the organization satisfied now that U.S. troops actually have been withdrawn from Saudi Arabia? What would be achieved if all Western interests were driven from the Islamic Holy Land and if all of the governments sympathetic to the West were overthrown? According to Gunaratna, the ultimate aim is to reestablish the caliphate—the empire of Islam's early golden age—and thereby empower a formidable array of truly Islamic states to wage war on the United States and its allies.[i]

With these goals in mind, what should the United States do? Al-Qa'ida's strategists clearly hope that terrorism will inflict unacceptable losses of American lives, forcing the United States to withdraw its troops from Iraq and neighboring countries. That conviction may stem from the withdrawal of U.S. troops from Somalia after the loss of 18 American soldiers in Mogadishu in 1993 (the "Blackhawk Down" episode; see page 501) and the withdrawal of U.S. forces from Lebanon following the suicide bombing of the Marine barracks in Beirut in 1983 (see page 243). Al-Qa'ida thinks the United States has no stomach for sustained sacrifice.

Al-Qa'ida poses a dilemma for the United States. If the United States were to withdraw its troops from the region, it might only reaffirm al-Qa'ida's belief that its adversary is weak and vulnerable, thus encouraging more attacks. And if al-Qa'ida's ultimate goal is to take on the United States once it has established an Islamic empire, there is no reason to believe that a unilateral withdrawal would bring a cessation of hostilities. On the other hand, if the United States continues to conduct a war on al-Qa'ida and a broader war against terrorism in a host of Muslim countries, al-Qa'ida can use the American presence, and especially unintended civilian losses, to incite widespread hatred and violence directed against the United States. Al-Qa'ida has used the U.S. occupation of Iraq to bolster its case that "the Crusaders" *are* waging a war against Muslims and the Islamic world, and it is time to use terrorism because violence is the only language they understand. And by selecting Saudi citizens as 15 of the 19 hijackers of the 9/11 attacks, al-Qa'ida successfully introduced a major, lasting strain in relations between the United States and its vital ally Saudi Arabia.

Using Geographic Information Systems (GIS) and a variety of other tools, many geographers are interpreting the geographic dimensions of terrorism. One of the most profound questions for geographers is where future attacks will take place. Gunaratna estimates that al-Qa'ida maintains a reserve of at least 100 targets worldwide.[j] Where are most of these? Suleiman Abu Ghaith, one of bin Laden's top aides, offered a clue, along with a geographic answer to the question of what al-Qa'ida wants: "Let the United States know that with God's permission, the battle will continue to be waged on its territory until it leaves our lands."[k]

[g] Osama bin Ladin, "Declaration of War (August 1996)." In *Anti-American Terrorism and the Middle East,* Barry Rubin and Judith Colp Rubin, eds. (Oxford University Press, 2002a), pp. 137, 139; Osama bin Ladin, "Statement: Jihad against Jews and Crusaders" (February 23, 1998)." In *Anti-American Terrorism and the Middle East,* Barry Rubin and Judith Colp Rubin, eds. (Oxford University Press, 2002b), p. 149; Osama bin Ladin, "Al-Qa'ida Recruitment Video (2000)." In *Anti-American Terrorism and the Middle East,* Barry Rubin and Judith Colp Rubin, eds. (Oxford University Press, 2002c), p. 174.

[h] Bin Laden 2002b, op. cit., footnote vii, p. 150.

[i] Gunaratna 2002, op. cit., footnote ii, pp. 55, 89.

[j] Ibid., p. 188.

[k] Suleiman Abu Ghaith, "Al-Qa'ida Statement (October 10, 2001)." In *Anti-American Terrorism and the Middle East,* Barry Rubin and Judith Colp Rubin, eds. (Oxford University Press, 2002), p. 251.

Major al-Qa'ida-Related Terrorist Attacks, 1998–2004

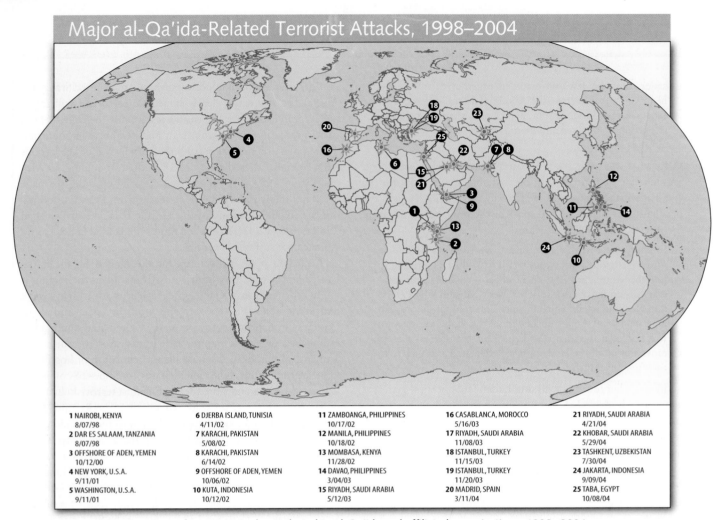

1 NAIROBI, KENYA 8/07/98	**6** DJERBA ISLAND, TUNISIA 4/11/02	**11** ZAMBOANGA, PHILIPPINES 10/17/02	**16** CASABLANCA, MOROCCO 5/16/03	**21** RIYADH, SAUDI ARABIA 4/21/04
2 DAR ES SALAAM, TANZANIA 8/07/98	**7** KARACHI, PAKISTAN 5/08/02	**12** MANILA, PHILIPPINES 10/18/02	**17** RIYADH, SAUDI ARABIA 11/08/03	**22** KHOBAR, SAUDI ARABIA 5/29/04
3 OFFSHORE OF ADEN, YEMEN 10/12/00	**8** KARACHI, PAKISTAN 6/14/02	**13** MOMBASA, KENYA 11/28/02	**18** ISTANBUL, TURKEY 11/15/03	**23** TASHKENT, UZBEKISTAN 7/30/04
4 NEW YORK, U.S.A. 9/11/01	**9** OFFSHORE OF ADEN, YEMEN 10/06/02	**14** DAVAO, PHILIPPINES 3/04/03	**19** ISTANBUL, TURKEY 11/20/03	**24** JAKARTA, INDONESIA 9/09/04
5 WASHINGTON, U.S.A. 9/11/01	**10** KUTA, INDONESIA 10/12/02	**15** RIYADH, SAUDI ARABIA 5/12/03	**20** MADRID, SPAIN 3/11/04	**25** TABA, EGYPT 10/08/04

Figure 8.27 Locations of terrorist attacks attributed to al-Qa'ida and affiliated organizations, 1998–2004

justifiable combat against an infidel nation whose military troops occupied the holy land of Arabia, where Mecca and Medina are located.

As the U.S. counterattacked, ousting al-Qa'ida from its bases in Afghanistan and cracking down hard on its leadership around the world, al-Qa'ida evolved into a far more geographically diffuse organization. It carried out or supported attacks in Indonesia, Morocco, Tunisia, Turkey, Spain, Uzbekistan, Egypt, and Saudi Arabia (Figure 8.27). A succession of al-Qa'ida videotapes and audiotapes promised an unrelenting and costly continuation of *jihad* against the United States and its allies. It will be many years before Americans in particular might be able to emerge from the shadow of this threat. Al-Qa'ida clearly will not hesitate to use the most devastating weapons, even against large numbers of civilians. In fact, this combination is its tactical priority.

In 9/11 and other atrocities, a tiny minority of Muslims carried out terrorist actions that the great majority of Muslims condemned. Islamic scholars and clerics pointed out in each case that the murder of civilians is prohibited in Islamic law and that the attacks had no legitimate religious grounds.

Mainstream Islamic movements are not military or terrorist organizations but have distinguished themselves through public service to the needy and through encouragement of strong moral and family values. For most Muslims, the growing Islamist trend means a reembrace of traditional values like piety, generosity, care for others, and Islamic legal systems, which have proven effective for centuries. These values pose a reasonable alternative to Western cultural influences and often repressive political and administrative systems. For many people outside the region, however, "Muslim" and "terrorist" have become synonymous—an erroneous association that can be overcome in part by careful study of the complex Middle East and North Africa.

This concludes an overview of land and life in the Middle East and North Africa. The following chapter offers more insight into the peoples and nations of this vital region. It begins with a continuation of the geopolitical theme, examining one of the world's most problematic, persistent, and influential conflicts, the one between Israel and its Arab neighbors.

SUMMARY

- Events in the Middle East and North Africa profoundly affect the daily lives of people around the world, yet this region is often misunderstood. Misleading stereotypes about its environment and people are common, and people outside the region often associate it solely with military conflict and terrorism.

- The region has bestowed upon humanity a rich legacy of ancient civilizations, including those of Egypt and Mesopotamia, and the three great monotheistic faiths of Judaism, Christianity, and Islam.

- Middle Easterners include Jews, Arabs, Turks, Persians, Pashtuns, Berbers, people of sub-Saharan African origin, and other ethnic groups who practice a wide variety of ancient and modern livelihoods.

- Arabs are the largest ethnic group in the Middle East and North Africa, and there are also large populations of ethnic Turks, Persians (Iranians), and Kurds. Islam is by far the largest religion. Jews live almost exclusively in Israel, and there are minority Christian populations in several countries.

- Population growth rates in the region are moderate to high. Oil wealth is concentrated in a handful of countries, and as a whole, this is a developing region.

- The Middle East has served as a pivotal global crossroads, linking Asia, Europe, Africa, and the Mediterranean Sea with the Indian Ocean. These countries have historically been unwilling hosts to occupiers and empires originating far beyond their borders.

- The margins of this region are occupied by oceans, high mountains, and deserts. The land is composed mainly of arid and semi-arid plains and plateaus, together with considerable areas of rugged mountains and isolated "seas" of sand.

- Aridity dominates the environment, with at least three-fourths of the region receiving less than 10 inches (25 cm) of yearly precipitation. Plants, animals, and people have developed strategies of drought avoidance and drought endurance to live here. In addition, great river systems and freshwater aquifers have sustained large human populations.

- Many of the plants and animals upon which the world's agriculture depends were first domesticated in the Middle East in the course of the Agricultural Revolution.

- The Middle Eastern "ecological trilogy" consists of peasant villagers, pastoral nomads, and city-dwellers. The relationships among them have been mainly symbiotic and peaceful, but city-dwellers have often dominated the relationship, and both pastoral nomads and urbanites have sometimes preyed upon the villagers, who are the trilogy's cornerstone.

- About two-thirds of the world's oil is here, making this one of the world's most vital economic and strategic regions.

- Since World War II, several international crises and wars have been precipitated by events in the Middle East. Strong outside powers depend heavily on this region for their current and future industrial needs. Unimpeded access to Gulf oil is one of the pillars of U.S. foreign policy.

- The Middle East and North Africa are characterized by a high number of chokepoints, strategic marine narrows that may be shut off by force, triggering conflict and economic disruption.

- Oil pipelines in the Middle East are routed both to shorten sea tanker voyages and to reduce the threat to sea tanker traffic through chokepoints but are themselves vulnerable to disruption.

- Access to fresh water is a major problem in relations between Turkey and its downstream neighbors, Egypt and its upstream neighbors, and Israel and its Palestinian, Jordanian, and Syrian neighbors.

- Al-Qa'ida and affiliated Islamist terrorist groups aim to drive the United States and its allied governments from the region and to replace them with an Islamic caliphate. Al-Qa'ida is an apocalyptic group that seeks to inflict mass casualties on its enemies, particularly on Americans in their home country.

KEY TERMS + CONCEPTS

Terms in blue are also defined in the glossary.

Afro-Asiatic language family (p. 212)
  Berber subfamily (p. 212)
    Berber (p. 212)
    Tuareg (p. 212)
al-Aqsa Martyrs Brigade (p. 225)
al-Aqsa Mosque (p. 217)
al-Qa'ida (p. 225)
Altaic language family (p. 212)
  Turkish (p. 212)
anti-Semitism (p. 215)
Arab (p. 211)
Carter Doctrine (p. 222)

chokepoints (p. 219)
Christianity (p. 215)
Church of the Holy Sepulcher (p. 215)
civilization (p. 211)
Copts (p. 216)
Crusades (p. 216)
culture hearth (p. 211)
Diaspora (p. 215)
Dome of the Rock (p. 213)
drought avoidance (p. 203)
drought endurance (p. 203)
dry farming (p. 203)

energy crisis (p. 219)
Exodus (p. 213)
Fertile Crescent (p. 211)
First and Second Temples (p. 213)
Hamas (p. 225)
Hebrew (p. 213)
Hizbullah (p. 225)
Holocaust (p. 215)
horizontal migration (p. 208)
hydropolitics (p. 223)
Indo-European language family (p. 212)
  Dari (Afghan Persian) (p. 212)

REVIEW QUESTIONS

WORLD
REGIONAL
Geography⊛Now™

Assess your understanding of this chapter's topics with additional quizzing and concept-based problems at http://earthscience.brookscole.com/wrg5e.

1. What countries constitute the Middle East and North Africa? What are the three most populous? Which encourage and discourage population growth?

2. What are the major climatic patterns of the Middle East? Where are the principal mountains, deserts, rivers, and areas of high rainfall?

3. Why can this region be described as a culture hearth? What major ideas, commodities, and cultures originated there?

4. What are the major ethnic groups and the countries in which they are found? What is an Arab? A Jew? A Turk? A Kurd? A Persian? A Muslim?

5. What are the principal beliefs and historical geographic milestones of Jews, Christians, and Muslims?

6. What is the difference between Shi'ite and Sunni Islam?

7. Where is oil concentrated in this region?

8. What is the Carter Doctrine?

9. What are "upstream" and "downstream" countries, and which are usually the more powerful? What are the exceptions to this rule?

10. What are Hizbullah, Hamas, and al-Qa'ida?

DISCUSSION QUESTIONS

1. What is the origin of the term *Middle East*?

2. What makes the Dead Sea the lowest place on Earth?

3. Why is Lebanon barren today? What other significant environmental changes have occurred in the region?

4. Three ancient ways of life—making up the ecological trilogy—prevail in the Middle East. What are these and what are some of the important characteristics and interrelationships of each? (This may be answered with an exercise involving three groups, with each group representing one livelihood.)

5. What are the main elements that make up the classic medina, or Muslim Middle Eastern city? List them and then draw a model city. What city mapped in this chapter conveys most of those essential elements?

6. Using a map, discuss the sacred places of Judaism, Christianity, and Islam in Jerusalem.

7. Why is Islam often described as a way of life? What are the five pillars of Islam? What do they require of a Muslim?

8. Discuss the importance of the region's oil to the United States

and other countries. How has that importance shaped U.S. foreign policy?

9. What are the major chokepoints in the Middle East? Who has closed them off and what has happened as a result? How has the threat of their closure affected the routing of oil pipelines? What was the shifting geography of Saddam Hussein's pipeline network?

10. Discuss hydropolitical problems between Israel and its neighbors, Turkey and its neighbors, or Egypt and its neighbors.

11. What does al-Qa'ida want and why? What should be done about this threat?

The Middle East and North Africa: Modern Struggles in an Ancient Land

Bedouin of the southern Sinai Peninsula, Egypt

Joe Hobbs

chapter objectives

This chapter should enable you to:

- Understand the major issues of the Arab–Israeli conflict and the obstacles to their resolution

- Appreciate how Lebanon's political system failed to represent the country's ethnic diversity, setting the stage for civil war

- Balance the pros and cons of Egypt's Aswan High Dam, Libya's Great Manmade River, and other large water engineering projects

- View the Sudanese civil war as an effort by one ethnic group to dominate others

- Recognize how the U.S. war on terrorism has transformed political systems and international relations

- Trace the trajectory of Gulf countries from underpopulated, wealthy welfare states to populous, vulnerable states overly dependent on a single commodity

- See the U.S. wars on Iraq as the result of tragic miscalculations by Saddam Hussein, combined with strategic American interests in the region

- Recognize Turkey as an "in-between" country, a less developed Asian nation aspiring to be a European power

- Consider how both superpower preoccupation and profound neglect sowed the seeds of conflict in Afghanistan

WORLD
REGIONAL
Geography ⊛ Now™

Look for this logo in the text and go to GeographyNow at http://earthscience.brookscole.com/wrg5e to explore interactive maps, view animations, sharpen your factual knowledge and geographic literacy, and test your critical thinking and analytical skills with unique interactive resources.

Conflict between Israel and its neighbors has been one of the most persistent and dangerous problems in global affairs since the end of World War II. This chapter begins with an introduction to the key players and issues in conflict. This depiction reveals why the Palestinian–Israeli dispute is so difficult to solve and how beneficial its resolution would be. The chapter continues with a survey of the major subregions, nations, and geographical problems of the diverse, fascinating Middle East.

9.1 The Arab–Israeli Conflict and Its Setting

The Arab–Israeli conflict persists as one of the world's most intractable disputes. It has not been resolved in part because the central issues are closely tied to such life-giving resources as land and water and to deeply held religious beliefs. The Arab–Israeli conflict is above all a conflict over who owns the land—sometimes very small pieces of land—and is therefore of extreme interest to geographers and geography. A geographic understanding of the Middle East to-

day requires familiarity with the events leading up to the creation of the state of Israel and with the wars that have followed, particularly as they have rearranged the boundaries of nations and territories.

The modern state of Israel was carved from lands whose fate had been undetermined since the end of World War I. The Ottoman Empire, based in what is now Turkey, had ruled Palestine (roughly the area now made up of Israel and the Occupied Territories) and surrounding lands in the eastern Mediterranean since the 16th century. After the British and French defeated the Turks in World War I and destroyed their empire, they divided the region between them (Figure 9.1). The British received the mandate for Palestine, Transjordan (modern Jordan), and Iraq (Mesopotamia), while the French received the mandate for Syria (now Syria and Lebanon). During World War I, British administrators of Palestine had made conflicting promises to Jews and Arabs. On the one hand, they implied they would create an independent Arab state in Palestine and, on the other, vowed to promote Jewish immigration to Palestine with an eye to the eventual establishment of a Jewish state there. The **Palestinians**—Arabs who historically formed the vast

Middle East and North Africa in 1920

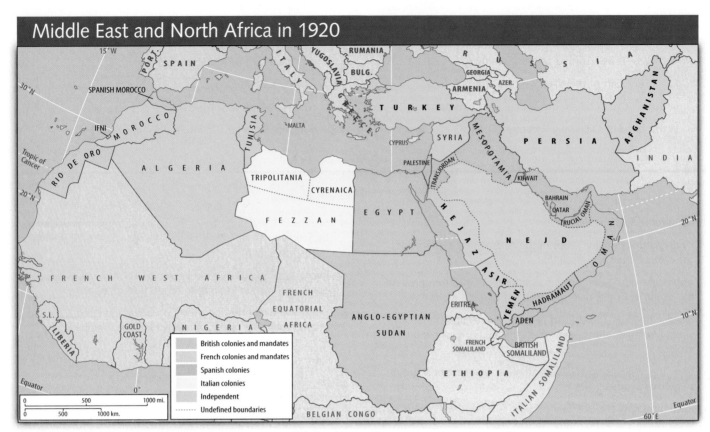

Legend:
- British colonies and mandates
- French colonies and mandates
- Spanish colonies
- Italian colonies
- Independent
- ----- Undefined boundaries

WORLD REGIONAL
Geography Now™

Active Figure 9.1 The victorious allies of World War I carved up the Middle East amongst themselves. Growing difficulties of administration would drive them from the region within a few decades. *See an animation based on this figure, and take a short quiz on the facts and concepts.*

majority of the region's inhabitants—did not welcome the ensuing Jewish immigration.

Placing themselves in a no-win position with these conflicting promises, and under increasing attack from both Jews and Arabs, in 1947 the British decided to withdraw from Palestine and leave the young United Nations with the task of determining the region's future. The United Nations responded in 1947 with a **two-state solution** to the problem of Palestine. It established an Arab state (which would have been called Palestine) and a Jewish state (Israel). The plan was deeply flawed. The states' territories were long, narrow, and almost fragmented, giving each side a sense of vulnerability and insecurity (Figure 9.2). War broke out in May 1948 between newborn Israel and the armies of the neighboring Arab countries of Transjordan, Egypt, Iraq, Syria, and Lebanon. The smaller but better organized and more highly motivated Israeli army defeated the Arab armies, and Israel acquired its **pre-1967 borders** (designated as the 1949 Armistice Agreement borders in Figure 9.2).

Prior to and during the fighting of the **1948–1949 war,** approximately 800,000 Palestinian Arabs were both forced and chose to flee from the new state of Israel to neighboring Arab countries (Figure 9.3). The United Nations established refugee camps for these displaced persons in Jordan, Egypt, Lebanon, and Syria. Little was done to resettle them in per-

manent homes, and both the Arab governments and the refugees themselves continued to insist on the return of the refugees to Israel and the restoration of their properties there. This Palestinian right of return is one of the central problems of the ongoing peace process, which will be discussed further.

Other countries assumed control of those parts of the proposed Palestine that Israel did not absorb. Egypt occupied the Gaza Strip, a piece of land on the Mediterranean shore adjacent to Egypt's Sinai Peninsula and inhabited mostly by Palestinian Arabs. Transjordan occupied the West Bank, the predominantly Arab hilly region of central Palestine on the west side of the Jordan River, and the entire old city of Jerusalem, including the Western Wall and Temple Mount, Judaism's holiest sites. The Palestinian Arab state envisioned in the UN Partition Plan was thus stillborn.

The **1967 war** fundamentally rearranged the region's political landscape in Israel's favor, setting the stage for subsequent struggles and the peace process. This conflict was precipitated in part when, by positioning arms at the Strait of Tiran chokepoint, Egypt closed the Gulf of Aqaba to Israeli shipping. Egypt's President Nasser and his Arab allies took several other belligerent but nonviolent steps toward a war they were ill prepared to fight. Israel elected to make a preemptive strike on its Arab neighbors, virtually destroying

F8.B, 214

219

Palestine and Israel, 1947–1949

U.N. PLAN FOR PALESTINE, 1947

- Jewish state
- Arab state
- International zone

1949 ARMISTICE AGREEMENT

- - - - Boundaries of the state of Israel according to the 1949 armistice agreement

LEBANON

SYRIA

•Haifa
Nazareth•

Netanya•

MEDITERRANEAN SEA

Tel Aviv•

•Ramallah
Jerusalem•
Bethlehem•

Gaza•

•Beersheba

Dead Sea

JORDAN

EGYPT

Note: The proposed Jewish state in 1947 follows modern Israel's borders south to the Gulf of Aqaba.

0   20   40 mi.
0   20   40 km.

WORLD REGIONAL
Geography ⊛ Now™

Active Figure 9.2 The 1947 UN partition plan for Palestine and Israel's pre-1967 borders. The 1948–1949 war, which began as soon as Britain withdrew from Palestine and Israel proclaimed its existence, aborted the UN Partition Plan and created a tense new political landscape in the region. *See an animation based on this figure, and take a short quiz on the facts and concepts.*

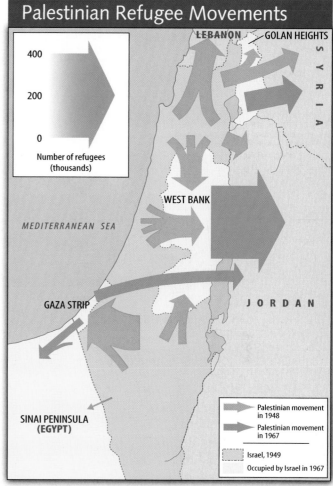

Palestinian Refugee Movements

400

200

0

Number of refugees (thousands)

LEBANON GOLAN HEIGHTS

SYRIA

WEST BANK

MEDITERRANEAN SEA

GAZA STRIP

JORDAN

SINAI PENINSULA (EGYPT)

→ Palestinian movement in 1948
➤ Palestinian movement in 1967
☐ Israel, 1949
☐ Occupied by Israel in 1967

WORLD REGIONAL
Geography ⊛ Now™

Active Figure 9.3 Palestinian refugee movements in 1948 and 1967. Many who fled in the first conflict relocated in the second. *See an animation based on this figure, and take a short quiz on the facts and concepts.*

the Egyptian and Syrian air forces on the ground. Israel gave Jordan's King Hussein an opportunity to stay out of the conflict. However, Jordan went to war and quickly lost the entire West Bank and the historic Old City of Jerusalem. The entire nation of Israel was transfixed by the news that Jewish soldiers were praying at the Western Wall. The Israeli army (Israeli Defense Forces, or IDF) also seized the Gaza Strip, Egypt's Sinai Peninsula (thus closing the Suez Canal), and the strategic Golan Heights section of Syria overlooking Israel's Galilee region. Israel had tripled its territory in six days of fighting; this conflict is also known as the **Six Day War** (Figure 9.4).

The Palestinian refugee situation became more complex as a result of the war. Israel took over the areas where most of the camps were located. Many persons in the camps again took flight, and Palestinians fleeing villages and towns in the newly Occupied Territories joined them as refugees (see Fig-

ure 9.3). Some later returned to their homes, but an estimated 116,000 either did not attempt to return or were denied permission by Israel authorities to do so.

The **1973 war** was a multifront Arab attempt to reverse the humiliating losses of 1967. On October 6, 1973, the Jewish holy day of Yom Kippur, Egypt and Syria launched a surprise attack on Israel with the hope of rearranging the stalemated political map of the Middle East. Egypt's army initially showed surprising strength, penetrating deep into the Sinai Peninsula and overturning Israel's image of invulnerability. Israeli troops soon reversed the tide and surrounded Egypt's army, but in the ensuing disengagement talks, Egypt won back the eastern side of the Suez Canal and by 1975 was able to reopen it to commercial traffic. In 1979, Israel agreed to return Sinai to Egypt under the U.S.-sponsored **Camp David Accords,** and Egypt recognized Israel's rights as a sovereign state. The Gaza Strip, West Bank,

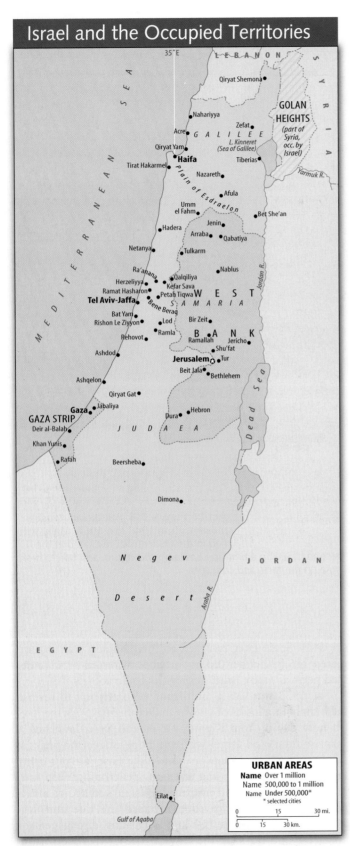

Israel and the Occupied Territories

URBAN AREAS
Name Over 1 million
Name 500,000 to 1 million
Name Under 500,000*
* selected cities

Figure 9.4 Israel and Arab territories occupied by Israel (West Bank, Gaza Strip, Golan Heights) as of 2005. Israel occupied the Sinai Peninsula in the 1967 war but returned it to Egypt in the early 1980s following the Camp David Accords.

and Golan Heights continued to be held by Israel and are known today as the Occupied Territories. Israel has variously held them as lands that rightfully belong to Jews and as cards to be played in the regional game of peacemaking.

Arabs and Jews: The Demographic Dimension

As well as issues of land, water, politics, and ideology, the Palestinian–Israeli conflict is about sheer numbers of people. Each side has wanted to maximize its numbers to the disadvantage of the other. To realize the Zionist dream of establishing Israel, Jews began immigrating to Palestine early in the 20th century. Jews were 11 percent of Palestine's population in 1922, 16 percent in 1931, and 31 percent in 1946, on the eve of Israel's creation. In keeping with national legislation known as the Law of Return, the Jewish state of Israel has always granted citizenship to any Jew who wishes to live there. Following the 1948–1949 war, waves of new immigrants from Europe (**Ashkenazi Jews**) joined the Middle Eastern **Sephardic Jews** who had inhabited Palestine and other parts of the Middle East (and until 1492, Spain) since early times.

In the 1990s, more waves of Jewish immigration followed the dissolution of the Soviet Union and a change of government in Ethiopia, where an ancient Jewish group called **Falashas** had lived in isolation for thousands of years. In 2004, Jews made up 80 percent of the population of 6.8 million people living within Israel's pre-1967 borders. About 1.3 million Palestinian and non-Palestinian Arabs, or 20 percent of the country's population, are now Israeli citizens residing within the pre-1967 boundaries of Israel; they are known as "Israeli Arabs." Palestinian Arabs without Israeli citizenship, and a variety of other ethnic groups, comprise the balance of the population within Israel's pre-1967 borders.

WORLD
REGIONAL
Geography ⊕ Now ™

Click Geography Literacy to watch a short video about honor killing in Israel.

Another 3.8 million people live in the West Bank, Gaza Strip, and Golan Heights—areas occupied by Israel in the wars of 1967 and 1973. About 200,000 of those are Jewish settlers, and most of the rest are Palestinian Arabs (mainly in the West Bank and Gaza Strip), **Druze** (members of a small offshoot of Shi'ite Islam, living mainly in the Golan Heights), and other Arabs. Another 180,000 Jewish settlers live in East Jerusalem, which is also part of the territories Israel conquered in 1967. There are 5 million more Palestinians living outside Israel and the Occupied Territories. Jordan has the largest number (2.8 million), followed by Syria (436,000) and Lebanon (415,000). The remaining 1.2 million are in other Arab countries and in numerous nations around the world, comprising a sizable diaspora of their own that Palestinians

Figure 9.5 This Jewish settlement on the West Bank is a well-built, seemingly permanent set of "facts on the ground." Many of the settlements, like this one, are situated strategically on hilltops. This settlement is a particularly important one because it is on the eastern edge of Jerusalem, serving as a declaration that the holy city belongs to Israel.

call the *ghurba*. In this respect, the Palestinians are much like the Jews prior to Israel's creation.

After 1967, and especially after the election of its conservative **Likud Party** to power in 1977, Israel moved to strengthen its grip on the Occupied Territories through security measures and the government-sponsored establishment of new **Jewish settlements.** Most of these settlements were, and are still, spread through the West Bank, where 200,000 Jewish settlers live (see Figure 9.A, page 239). Much smaller numbers of Jews settled in the troubled Gaza Strip and in the Golan Heights. These are generally not frontier farming settlements but communities inhabited by relatively prosperous middle-class Jews who commute to jobs in Israel. They were not built as temporary encampments but as permanent fixtures meant to create what have been described as **"facts on the ground"**—an Israeli presence so entrenched that its withdrawal would be almost inconceivable (Figure 9.5). Some of the motivation for this settlement has been ideological, as the West Bank is composed of Biblical **Judea and Samaria,** which many devout Jews regard as part of Israel's historic and inalienable homeland.

The proliferation of Jewish settlements worsened relations between Israel and its Arab neighbors and even drove a wedge into the Jewish population of Israel itself. Many Israelis strongly oppose further settlement in the Occupied Territories and annexation of these territories to the Jewish state because if the territories were annexed, Israel would become a country whose population was only about 60 percent Jewish. With an Arab minority that has a far higher birth rate than that of the Jews, Jews could eventually become the minority population in their own country. There is also international pressure against the settlements because they violate Geneva Conventions on activities allowed in territories

captured in war. **United Nations Resolutions 242 and 338** (of 1967 and 1973, respectively) called on Israel to withdraw completely from the Occupied Territories.

The settlements and other aspects of Israeli occupation, including the periodic closing off of the territories that prevents many Palestinians from reaching their jobs inside Israel proper, have fostered widespread and long-lasting Palestinian resistance against Israel. In 1987, Palestinians instigated a popular uprising in the Occupied Territories known as the **Intifada** (Arabic for "the shaking"). Initially, they relied on rocks and bottles to engage Israeli troops, who answered with rubber and metal bullets. International media coverage of a Palestinian "David" fighting an Israeli "Goliath" did much to damage Israel's image abroad, and many Israelis came to question the justice and importance to security of Israel's continued occupation. Popular opinion and, by 1992, a new Israeli government led by the liberal **Labor Party** began to consider the previously unthinkable: giving the Palestinians at least some control over land and internal affairs in the Occupied Territories (see Problem Landscape, pages 238–240).

A Geographic Sketch of Israel

Israel is much more than a country in a perpetual state of conflict. It is a physically and culturally diverse and vibrant nation with a promising future, especially if relations with its Arab neighbors improve.

Perhaps the most significant feature of Israel's geography is the country's tiny size. Within its legally recognized, pre-1967 borders, its area is only 8,019 square miles (20,770 sq km), about the sizes of New Jersey or Slovenia (Figure 9.6). Israelis often cite the small size of their country, nestled within the vastness of the Arab Middle East and North Africa,

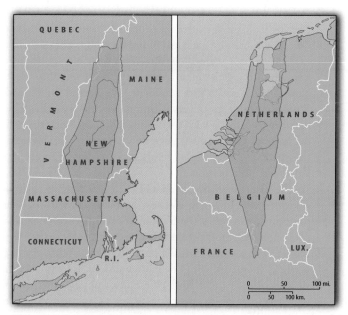

Figure 9.6 Israel and the occupied territories compared in size with New England and the Benelux countries

Problem Landscape

Land for Peace

President Anwar Sadat of Egypt set the precedent in 1979: Arabs could make peace with Israel on the formula of "land for peace," with Israel swapping Arab lands it occupied in 1967 in exchange for peace with its neighbors. The Arab countries scorned Egypt for a decade for its peacemaking, and Sadat was assassinated because of it. However, in the process, Israel restored the Sinai to Egyptian control, and the United States rewarded Egypt handsomely with about $2 billion in aid annually. Along with about $3 billion awarded Israel annually, for over two decades these countries have been the two largest recipients of U.S. foreign aid. Israel and its remaining Arab adversaries stand to gain much more through the implementation of a comprehensive framework for peace. The United States, Russia, and Norway initiated negotiations in the 1990s that made such a prospect appear possible. The process began in earnest in 1993, when Israel recognized the legitimacy of and began to negotiate with the Palestine Liberation Organization (PLO), long recognized by the Palestinians as their legitimate governing body. In return, the PLO, under the leadership of Yasser Arafat, recognized Israel's right to exist and renounced its longstanding use of military and terrorist force against Israel. U.S. President Clinton orchestrated a historic handshake between Arafat and the Israeli Prime Minister Yitzhak Rabin, and the leaders signed an agreement known as the **Gaza–Jericho Accord** (in its implementation, it came to be known as the **Oslo I Accord,** after the Norwegian capital where it was negotiated). It was designed to pave a pathway to peace by Israel's granting of limited autonomy, or self-rule, in the Gaza Strip and in the West Bank town of Jericho. Autonomy meant the Palestinians were responsible for their own affairs in matters of education, culture, health, taxation, and tourism. Israel also pulled its troops out of these areas. Palestinians were allowed to form their own government, known as the **Palestinian Authority (PA),** and they elected Yasser Arafat as its first president. The accord established a 5-year timetable for the resolution of much more difficult matters, the so-called final status issues. These included the political status of Jewish and Muslim holy places in Jerusalem and of the city itself, the possible return of Palestinian refugees, and Palestinian statehood.

The calendar proved impossible to abide by as each side struggled with lingering and new problems. Arafat's PLO had to respond to a series of violent attacks on Jewish civilians and soldiers by the radical Palestinian organization Hamas. While these terrorist attacks caused many Israelis to reject the peace process, subsequent PLO detention and prosecution of Palestinian extremists also hurt Palestinian support for the PLO, which many Palestinians came to see as an agent for Israeli interests. Many Palestinians turned against the peace process after 1994, when a right-wing Israeli extremist opposed to Israel's negotiations with the Palestinians shot and killed Muslim Palestinian worshipers in the Tomb of the Patriarchs, a holy place in the West Bank town of Hebron where the prophet Abraham is said to be buried.

Many of the Jewish settlers living in the West Bank are vehemently opposed to the peace process. Regarding Jewish presence in the area as a God-given right, they fear the peace process will result in an independent Palestinian state and the forced withdrawal of Israeli Jews. On November 4, 1995, after a peace rally in Tel Aviv, an Israeli Jew who held this view assassinated Israeli Prime Minister Yitzhak Rabin. The assassin proclaimed that his purpose was to destroy the peace process.

The peace process remained on track, however. In 1995, Israel and the PLO signed another agreement (known as **Oslo II Accord**) to implement the withdrawal of Israeli troops from several large Palestinian cities in the West Bank. Palestinians then acquired self-rule over about 30 percent of the West Bank, with the rest, including all Jewish settlements, remaining in Israeli hands.

In a subsequent deal known as the **Wye Agreement,** two Israeli prime ministers (Netanyahu and Barak) promised to transfer more West Bank lands from Israeli to Palestinian control, and Arafat promised increased Palestinian efforts to crack down on Palestinian terrorists and so guarantee the security of Israelis in the Palestinian territories and in Israel. Most critically, Israel pledged to transfer an additional 13 percent of the territory it controlled in the West Bank from the area designated in the peace process as Zone C (complete Israeli control) to Zone B (Palestinian civilian control, Israeli security control) and pledged to transfer 14 percent of the West Bank from Zone B to Zone A (complete Palestinian control). If finally implemented (it had not been as of 2005), this arrangement would have brought the total of West Bank lands under complete Palestinian control to 17 percent, leaving 57 percent completely in Israeli hands, and 26 percent under joint control (Figure 9.A). Israel also released some Palestinian prisoners from Israeli jails, allowed the opening of an international Palestinian airport in Gaza, and agreed to allow a corridor of "safe passage" for Palestinians to be constructed between the West Bank and the Gaza Strip.

During his final year in office in 2000, U.S. President Clinton sought to solidify his legacy as peacemaker by brokering a historic final settlement between Israelis and Palestinians. PLO Chairman Arafat and Israeli Prime Minister Barak huddled with Clinton and his advisers in the presidential retreat at Camp David, a site chosen because of its historic significance in Middle East peacemaking.

Over weeks of tough negotiations, the two sides came close to a deal. It would have included the creation of an independent Palestinian country on the Gaza Strip and approximately

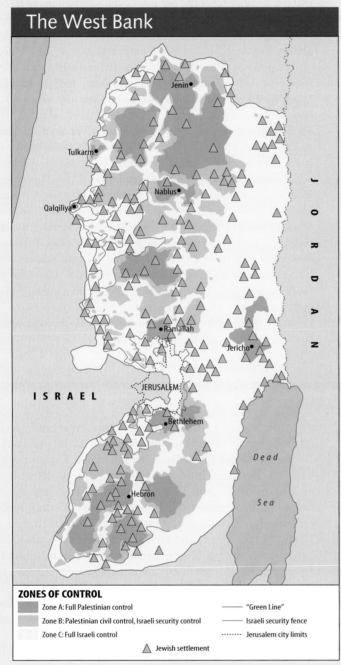

The West Bank

ZONES OF CONTROL

▨ Zone A: Full Palestinian control	—— "Green Line"
▨ Zone B: Palestinian civil control, Israeli security control	—— Israeli security fence
☐ Zone C: Full Israeli control	······· Jerusalem city limits
	△ Jewish settlement

WORLD
REGIONAL
Geography ⊛ Now™

Active Figure 9.A The West Bank. Israeli and Palestinian areas of control were delimited in the peace process of the 1990s, but because of recent violence, almost all areas are effectively controlled by Israel. Also depicted are Jewish settlements, the completed and planned portions of the Israeli-built security fence, and the Green Line delimiting the internationally recognized border between sovereign Israel and the Occupied West Bank. *See an animation based on this figure, and take a short quiz on the facts and concepts.*

95 percent of the West Bank, with the other 5 percent representing clusters of Jewish settlements that would remain under Israeli control while Palestine received 5 percent of Israel's land. However, the talks broke down over two thorny issues. One was the status of Palestinian refugees abroad; Israel was unwilling to permit the right of return of the large numbers demanded by the Palestinians. But the deal breaker was Jerusalem, which both Israelis and Palestinians proclaim as their capital. Israel agreed to have the Old City of Jerusalem divided, with Israel assuming sovereignty over the Jewish and Armenian quarters and Palestinians assuming sovereignty over the Arab and Christian quarters (see Figure 8.B). These issues were not as problematic as were those of the places held holy by each side. Israel agreed to allow the Palestinians "soft sovereignty" over the Temple Mount/*al-Haraam ash-Shariif,* meaning Palestinians could administer the area and fly a flag over it, but Israel would technically "own" it. Israel saw this point as essential because the First and Second Temples had stood on this spot. The Palestinians, however, insisted on full sovereignty over the Temple Mount/*al-Haraam ash-Shariif* and walked out on the negotiations.

Tragically, within weeks, this historic opportunity for peace evaporated and was replaced by a state of war. On September 28, 2000, the right-wing Israeli opposition leader (subsequently, the prime minister) Ariel Sharon walked up to Temple Mount/ *al-Haraam ash-Shariif* under heavily armed escort. He meant to make the point that this sacred precinct belonged to Israel. Many observers claim he also meant to disrupt the peace process. Whether or not it was his intention, a disruption in the peace process resulted. Beginning that day and stretching into a period of years, Palestinians took to the streets in the **al-Aqsa Intifada** to challenge Israeli forces with rocks, bottles, and guns. Hundreds of Palestinians and Israelis died.

In a ceaseless cycle of revenge killings, Hamas stepped up its shootings and suicide bombings against Israeli civilians and soldiers, while Israelis assassinated suspected Palestinian militants and leaders of Hamas, including its founder, Sheikh Ahmad Yassin. New militant Palestinian groups emerged including the al-Aqsa Martyrs Brigade, which was affiliated with Yasser Arafat's Fatah wing of the PLO. Along with Hamas, the al-Aqsa group carried out a series of ferocious suicide bombings of civilian Israeli targets, including a seaside hotel where Jews were beginning their Passover dinner in March 2002. Prime Minister Sharon reacted to the "Passover Massacre" with **Operation Defensive Shield.** This full-scale Israeli army invasion of several autonomous Palestinian towns in the West Bank, including Bethlehem, Ramallah, Nablus, and Jenin, destroyed much of the infrastructure of Arafat's Palestinian

(continued on page 240)

Authority leveled entire sections of some towns, and reestablished Israeli control over Palestinian autonomous areas.

Following the collapse of the Camp David talks and the escalation of suicide bombings of Israeli civilians, Israel chose to effectively suspend the peace process. Prime Minister Sharon branded Yasser Arafat as an untrustworthy partner for peace negotiations, isolating him physically in the West Bank town of Ramallah and systematically dismantling the Palestinian Authority he headed. Israel also persuaded the United States to stop negotiating with Arafat, who died a forlorn figure in 2004. Israel and the United States insisted on the creation of the new post of Palestinian prime minister, hoping it would create a new figure to negotiate with. The United States led an effort joined by the European Union, the United Nations, and Russia to establish a **"road map for peace"** between the Israelis and Palestinians. It was based largely on the 2000 Camp David formula and would have led to an independent Palestinian state on the West Bank and Gaza Strip. That effort faltered, however, as violence instigated by both sides continued.

Prime Minister Sharon appeared to give up on negotiations altogether, deciding instead to unilaterally withdraw Israeli settlers and troops from the Gaza Strip and to build an immense **security fence** separating the West Bank from Israel. The Gaza Strip had long been isolated from Israel by a barrier of this kind, and few suicide bombers were able to pass through it.

Israel's stated goal in constructing the new barrier was to prevent Palestinian suicide bombers from reaching Israel. Noting that the security fence did not follow the internationally recognized Green Line of Israel's pre-1967 border but instead penetrated significant portions of the West Bank to pick up Jewish settlements there, Palestinians condemned it as a "land grab."

The peace process has both progressed and faltered on other fronts. In 1994, following the PLO's lead, Jordan signed a treaty with Israel. It called for an end to hostility between the two nations; establishment of full diplomatic relations; the opening of exchanges in trade, tourism, science, and culture; and an end to economic boycotts. Meanwhile, Syria and Israel have failed to reach a peace agreement because of a deadlock on the issues of recognition of Israel and return of occupied land: Israel refuses to return the Golan Heights to Syria until Syria recognizes the sovereignty and legitimacy of Israel, and Syria refuses to recognize the sovereignty and legitimacy of Israel until Israel returns the Golan Heights. Syrian President Hafez al-Assad (who died in 2000) and his son, Bashar, who succeeded him, apparently wanted to see how successful Israel's existing peace treaties would be before concluding Syria's own with Israel. And because Syria is the dominant power in Lebanon, peace between Israel and Lebanon is improbable unless Israel and Syria first come to terms.

to highlight their vulnerability on the world stage. When and if Palestinians acquire their country, Palestine will be even smaller.

One of Israel's greatest achievements since independence has been the expansion and intensification of agriculture within its small territory; it is almost self-sufficient in agriculture, excluding grains. Production concentrates on citrus fruits (including the famous Jaffa oranges), which provide export revenue, and on dairy, beef, and poultry products, as well as flowers, vegetables, and animal feedstuffs. Israel's intensive, mechanized agriculture resembles the agriculture of densely populated areas in Western Europe. Collectivized settlements called *kibbutzim* (sing. *kibbutz*) are a distinctively Israeli feature on the agricultural landscape. Many of these lie near the frontiers and have defensive as well as agricultural and industrial functions. They have also been important in the development of Israeli identity, helping to instill a sense of work for the common good and reminding Israelis that their ancestors were often deprived of the right to own and till the land. Far more numerous and important in Israel's agricultural economy are other types of villages, including the small farmers' cooperatives called *moshavim* (sing. *moshav*) and villages of private farmers.

Agricultural development is concentrated in the northern half of the country, with its heavier rainfall (concentrated during the winter in this Mediterranean climate; see Figure 8.5a) and more ample supplies of surface and underground water for irrigation. Annual precipitation averages 20 inches (50 cm) or more in most sections north of the city of Tel Aviv and in the interior hills where Jerusalem lies. Due to the rain shadow effect, very little rain falls in the deep rift valley of the Jordan River. From Tel Aviv southward to Beersheba, annual rainfall decreases to about 8 inches (20 cm), and in the central and southern Negev Desert, it drops to 4 inches (10 cm) or less. To expand agriculture along the coastal plain and in the northern Negev, Israel transfers large quantities of water from the north by the aqueduct and pipeline network of the National Water Carrier (see Figure 8.24). The largest source for this network is Lake Kinneret (Lake Tiberias, or Sea of Galilee), which is fed by the upper Jordan River. Since occupying the West Bank in 1967, Israel has intensively developed irrigated agriculture along the lower Jordan River and in scattered areas of the drier West Bank hills. Israel would lose most of this production to any future Palestinian state.

Jerusalem (population: 500,000, excluding East Jerusalem in Occupied Territory) is the country's capital, but few countries recognize it as such because of the political sensibilities involved. Tel Aviv–Jaffa (population: 3 million) on the Mediterranean coast is Israel's greatest industrial center. The leading seaport is Haifa in northern Israel. The ancient town of Beersheba is the administrative and industrial center of the Negev Desert region. To the south lies Israel's small Red Sea port and resort town of Eilat. Thanks to the successful peace accords with Egypt and Jordan, open doors now link Eilat with the neighboring resort towns of Taba in the Egyptian Sinai and Aqaba in Jordan. Whenever conflict subsides, tourism booms, and developers are building and promoting a three-nation "Red Sea Riviera" to satisfy mainly European sun-worshippers. Almost incessant conflict since 2000 has taken a significant toll on Israel's economy, which relies heavily on the exports of high-technology equipment, cut diamonds, agricultural products, and on assistance from the United States.

Jordan: Between Iraq and Israel

Jordan's economic and political situation has always been precarious. Throughout his long reign, ending with his death in 1999, the country's King Hussein withstood various crises. Many of these related to the large numbers of Palestinian refugees in Jordan. Palestinians still make up about 50 percent of Jordan's population of 5.6 million, but they are effectively refugees without citizenship. After the 1967 war, Jordan was the chief base for military operations of the PLO against Israel, but in 1970 and 1971, the king expelled the armed Palestinian forces, which relocated to Lebanon.

Jordan (Figure 9.7) is a land of desert and semidesert. The main agricultural areas and settlement centers, including the largest city and capital, Amman (population: 2.7 million), are in the northwest. Prior to the 1967 war with Israel, northwestern Jordan included two upland areas—the West Bank on the west (captured during the war) and the East Bank, or Transjordanian Plateau, on the east—separated by the deep

rift valley occupied by the Jordan River, Lake Kinneret, and the Dead Sea (see Figure 9.4). Most of the valley bottom lies in deep rain shadow and receives less than 4 inches (10 cm) a year. The bordering uplands, rising to more than 2,000 feet (610 m) above sea level and lying in the path of moisture-bearing winds from the Mediterranean, receive precipitation averaging 20 inches (50 cm) or more annually. But overall, rainfall is so scarce that only small areas can be cultivated without irrigation, and the water available for irrigation is limited. Only about 4 percent of Jordan is cultivated, although larger areas are used to graze sheep and goats. Irrigated vegetables, together with olives, citrus fruits, and bananas, provide substantial exports, but some food has to be imported.

Natural resources are limited, and there is little manufacturing. Rock phosphate and potash extracted from the Dead Sea are shipped out of Jordan's small port of Aqaba on the Gulf of Aqaba. During the 1990s, when Iraq suffered from international economic sanctions and periodic disruptions in exports from Persian Gulf ports, Jordan earned revenues through Iraqi exports from Aqaba. Tourism is a growing industry, with the Nabatean rose-red city of Petra the chief attraction (see Figure 8.8).

Lebanon and Syria

Israel's immediate neighbors on the north are Lebanon and Syria (see Figure 9.7). Their physical and climatic features are similar to those of Israel and Jordan. As in Israel, narrow coastal plains backed by highlands front the sea. The highlands of Lebanon and Syria are loftier than those of Israel and Jordan. Mt. Hermon and the Lebanon and Anti-Lebanon Ranges rise above 10,000 feet (c. 3,050 m). Cyclonic precipitation brought by air masses off the Mediterranean is supplemented by orographic precipitation, the rain and snow generated along the mountain chains paralleling the coast.

Lebanon and Syria have agricultural patterns that conform to the Mediterranean climatic pattern of winter rain and summer drought. The coastal plains and seaward-facing mountain slopes are settled by villagers growing Mediterranean crops such as winter grains, vegetables, and drought-resistant trees and vines. Lebanon's Bekaa Valley, drained in the south by the Litani River, lies between the Lebanon and Anti-Lebanon Ranges and is a northward continuation of the Jordan rift valley and the African Great Rift Valley (Figure 9.8). Baalbek, its chief city and the site of an important ancient Roman temple, became a Syrian military stronghold in 1976 and a center of pro-Iranian Hizbullah activities against Israel after 1982. The political and military volatility of the Bekaa Valley was conducive to the emergence of a flourishing drug industry, producing both hashish (from marijuana plants) and opium (from poppies), from the mid-1970s until the 1990s. Marijuana cultivation made a comeback in the early 2000s as alternative crops proved to bring farmers far less income.

Syria is more industrialized than Lebanon, but neither country compares with Israel industrially. Textile milling,

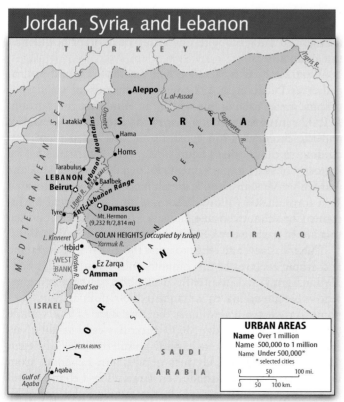

Figure 9.7 Principal features of Jordan, Syria, and Lebanon

Figure 9.8 Lebanon has strikingly beautiful mountain landscapes and a rich historical legacy, including this Roman Temple of the Sun in Baalbek in the northern Bekaa Valley.

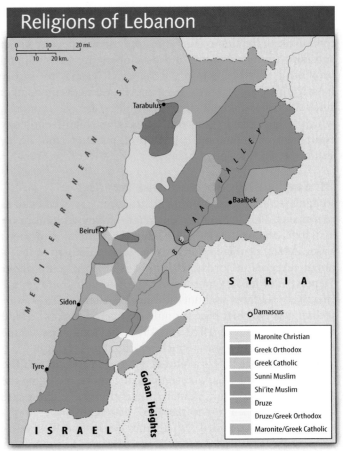

Figure 9.9 Generalized map of the distribution of religious groups in Lebanon

agricultural processing, and other light industries predominate. Both countries are deficient in minerals, although Syria has an oil field in the east. Its reserves are expected to be exhausted by 2010.

There are three large, historically important cities in Syria and Lebanon. Damascus (population: 2.4 million), Syria's capital, is in a large irrigated district at the foot of the Anti-Lebanon Range. Its Umayyad Mosque, originally a Byzantine cathedral, is one of the great monuments of the Middle East. The main city of northern Syria, Aleppo, is an ancient caravan center located in the Syrian Saddle between the Euphrates Valley and the Mediterranean. The old Phoenician city of Beirut (population: 1.9 million), on the Mediterranean, is Lebanon's capital, largest city, and main seaport. Equipped with modern harbor facilities and an important international airport, Beirut before 1975 was one of the busiest centers of transportation, commerce, finance, and tourism in the Middle East.

For its prosperity and its natural beauty, pre-1975 Lebanon had earned the name **"Switzerland of the Middle East."** Then a precarious political balance broke down, and the country was plunged into civil war in 1975 and 1976. The war's origins were rooted in Lebanon's mixture of strongly localized religious communities. Seventeen separate ethnic and religious groups, known as *confessions,* are recognized officially (Figure 9.9). Major confessions include the Maronite Christians, who predominate in the Lebanon Range north of Beirut; Sunni Muslims close to the northern border with Syria; Shi'ite Muslims in the Bekaa Valley and along Lebanon's southern border with Israel; and Druze Muslims in the Shouf Mountains, part of the Lebanon range south of Beirut. Eastern Orthodox, Greek Catholic, and nearly a dozen other Christian sects are also present in Lebanon. Beirut itself is divided between the predominantly Muslim west and Christian east. Conflict between the communities broke out from time to time, and foreign powers

that gained influence with one group or another heightened the antagonisms.

What the French carved out of greater Syria and put together as Lebanon in 1920 was a collection of small geographic areas, each dominated by an ethnic/religious group. A 1932 census revealed a majority of Christians in Lebanon, and the country's 1946 constitution distributed power according to those census figures. Maronite Christians held a majority of seats in parliament, followed by Sunni Muslims and Shi'ite Muslims. Top leadership positions were allocated by a formula still followed today, with the president a Maronite Christian, the prime minister a Sunni Muslim, and the speaker of parliament a Shi'ite Muslim.

There was never an effort to unify the disparate geographic and ethnic enclaves of this mountainous land. Muslim populations grew and eventually outnumbered the Christians, who were unwilling to relinquish their political and economic privileges. Palestinians became part of the volatile mix. Pushed from Jordan, PLO guerrillas established a virtual "state within a state" in the southern slums of Beirut and in Shi'ite-controlled areas of southern Lebanon, from which they launched attacks on Israel. The **Lebanese civil war** started in 1975 as the have-not Sunnis, soon joined by Shi'ites and Palestinians, took on the more prosperous Maronite Christians. Syria and Israel supported opposing sides. By the end of the war in 1976, as many as 100,000 people, most of them civilians, had been killed (Figure 9.10). The war accomplished little. Lebanon effectively split into a number of geographically distinct microstates or fiefdoms—a politically unstable situation that set the stage for the next war in 1982.

The **1982 war** was one facet of a larger war in Lebanon involving both civil strife and international intervention. Lebanon had become the main stronghold of the Palestine Liberation Organization, which used the country as a base for guerrilla operations against Israel. In June 1982, Israel invaded Lebanon with the avowed intention of smashing the PLO and establishing security for the northern Israeli frontier. Syria resisted the Israeli advance and gave support to the PLO. Israeli troops routed and inflicted heavy casualties on

Figure 9.10 Much of the fighting in the Lebanese civil war of 1975–1976 took place on streets like this, along the line separating Muslim west from Christian east Beirut. This portion of what had been a thriving, prosperous city became a desolate moonscape. A gunfight was underway here as the photographer took this picture in 1976; a taxi driver dumped him here and sped off.

both the Syrians and the PLO. The Israeli push continued northward to the suburbs and vicinity of Beirut, culminating in a ferocious aerial and artillery bombardment of the city—the **Battle of Beirut**—that ended only when U.S. President Ronald Reagan exerted pressure on Israel to cease fire. Under supervision by U.S. and other international peacekeeping troops, the PLO withdrew from Beirut to Tunisia. After the multinational troops withdrew, Muslim assailants killed Lebanon's Christian President Bashir Gemayal. Angry Christian forces then massacred hundreds of Palestinians and other civilians in the Sabra and Shatila refugee camps south of Beirut. U.S. and other multinational forces were again deployed to Beirut, and American ships used firepower to assist the new Christian president in widening his area of authority.

Lebanese Shi'ite Muslim guerrillas, seeking to avenge U.S. attacks on Lebanon and its support of Israel, soon bombed American and French targets in Beirut, killing over 200 U.S. Marines in a barracks near Beirut's airport in 1983. The multinational forces withdrew, but Syrian troops remained in Lebanon and allowed Iranian forces and Iran's Shi'ite Muslim allies in Lebanon to establish a stronghold in the Bekaa Valley town of Baalbek. Subjected to persistent guerrilla attacks, Israeli troops withdrew from all of Lebanon except a self-declared "security zone" in the south, inhabited by both hostile Shi'ite Muslims and Israel's de facto Christian Lebanese allies. Israel exerted control over this zone and engaged in regular skirmishes with Hizbullah, a Shi'ite Muslim organization sympathetic to Iran and hostile to the peace process between Israel and its Arab neighbors. With growing numbers of casualties among Israeli forces in the security zone, the occupation became increasing unpopular in Israel, and troops were finally withdrawn in 2000.

Lebanon is now on a steady course of reconciliation. Beirut is rising from the ashes in an ambitious, expensive reconstruction program. Political power has been redistributed to reflect the new demographic composition in which Muslims outnumber Christians. Once dominated by Christians, the Lebanese army is now comprised of soldiers from all of Lebanon's ethnic confessions. There are numerous social and education programs underway that mix children from different ethnic groups, with the hope they will learn tolerance. Lebanon is still not in control of its own destiny, however. The government in Damascus has never accepted the 1920 division of greater Syria, which gave a portion of Syria to Lebanon, and as of 2004, 20,000 Syrian soldiers remained in Lebanon. Surprisingly, perhaps, many Lebanese from all the major ethnic groups want the Syrians to remain and keep the peace while the country's former enemies continue the long, slow healing process.

Syria, like Lebanon, is a country beset by religious factionalism. In Syria, however, Muslims outnumber Christians nine to one. In 1970, a coup overthrew the ruling Sunni Muslim establishment and brought to power Hafez al-Assad, a member of the **Alawite** sect of Shi'ite Islam. Under his authoritarian rule as president, dominance of the government quickly passed to the Alawites, who make up only 12 percent of the country's population. The police and army crushed opposition within other communities, notably Sunni Islamists who rejected Alawites as heretics. For three decades, until his death in 2000, President Assad was a skilled politician who played Middle Eastern affairs cautiously. His regime survived the economic and political setbacks accompanying the downfall of its benefactor, the Soviet Union, and slowly adopted a more accommodating relationship with the United States and the West. His son and successor, Bashar al-Assad, has continued the cautious and generally unaccommodating approach to Israel established by Hafez al-Assad.

9.2 Egypt: The Gift of the Nile

At the southwest, Israel borders the Arab Republic of Egypt, the most populous Arab country (73.4 million) (Figure 9.11). This ancient land, strategically situated at Africa's northeastern corner between the Mediterranean and Red Seas, is utterly dependent on a single river. The Nile has created the floodplain and delta on which more than 95 percent of all Egyptians live (see Figure 8.25). The Greek historian and geographer Herodotus called Egypt "**the gift of the Nile.**" The river's bountiful waters and fertile silt helped Egypt become a culture hearth that produced remarkable achievements in technology and cultural expression over a period of almost 3,000 years until about 500 B.C., when a succession of foreign empires came to rule the land.

The Nile Valley may appropriately be described as a "river oasis," for stark, almost waterless desert borders this lush ribbon (see Figure 1.1). Only 3 percent of Egypt is cultivated, and nearly all this land lies along and is watered by the great river. The conversion of the original papyrus marshes and other wetlands along the Nile to the thickly settled, irrigated landscape of today is a process that has been unfolding for more than 50 centuries. It is difficult to imagine that, in the time of the Pharaohs, crocodiles and hippopotami swam in the Nile and game animals typical of East Africa roamed the nearby plateaus.

Ancient Egyptian civilization was based on a system of basin irrigation, in which the people captured and stored in built-up embankments the floodwaters that spread over the Nile floodplain each September. Egyptian farmers allowed the water to stand for several weeks in the basins, where it supersaturated the soil and left a beneficial deposit of fertile volcanic silt washed down by the Blue Nile from Ethiopia. They drained the excess water back into the Nile and planted their winter crops in the muddy fields. They were able to harvest one crop a year with this method. In limited areas, they could sustain perennial irrigation, producing a second or third crop by using water-lifting devices such as a weighted lever, or *shaduf* (a bucket on a counterweighted pole, manually operated to raise water from canal to field), and an Archimedes' screw (a cylinder containing a screw, also manually turned to raise the water).

Achieving perennial irrigation on a vast scale, however, required the construction of barrages and dams, which were

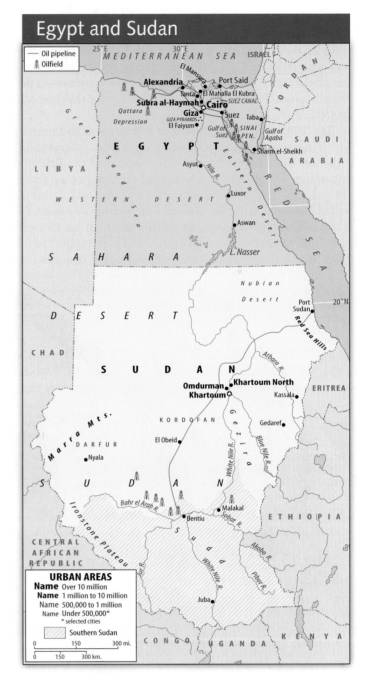

Figure 9.11 Principal features of Egypt and the Sudan

innovations of the 19th and 20th centuries in this part of the world. Late in the 19th century, French and British colonial occupiers of Egypt, anxious to boost Egypt's exports of cotton (a summer crop demanding perennial irrigation), began Egypt's conversion from basin to perennial irrigation by constructing a number of **barrages,** or low barriers designed to raise the level of the river high enough that the water flows by gravity into irrigation canals. Barrages are not designed to store large amounts of water, a function now performed by two dams in Upper Egypt (the old Aswan Dam, completed in 1902, and the huge Aswan High Dam, completed in 1970; see Figure 8.25) and several others along the Nile and its tributaries in Sudan and Uganda.

Figure 9.12 Egypt's controversial Aswan High Dam.

The enormous Aswan High Dam (Figure 9.12) is located on a stretch of the Nile known as the First Cataract (the Nile's many cataracts are areas where the valley narrows and rapids form in the river). Its reservoir, Lake Nasser, stretches for more than 300 miles (480 km) and reaches into northern Sudan. Like all giant hydrological projects, the Aswan High Dam has generated benefits and liabilities, and its construction was controversial. On the plus side, by storing water in years of high flood for use in years of low volume, the High Dam provides a perennial water supply to the Nile Valley's 6 million irrigated acres (2.4 million hectares), thus extending the cultivation period. The dam also made possible the "reclamation" of previously uncultivated land on the desert fringe adjoining the western Nile Delta. Navigation of the Nile downstream from (north of) Aswan is much easier, and the country's industries rely heavily on the hydroelectricity produced at the dam. Egypt weathered a 2-year drought in the late 1980s thanks to the excess waters stored in Lake Nasser. The dam also eliminates the prospect of catastrophic flooding, which sometimes occurred before it was built.

The Aswan High Dam has also had drawbacks. The dam has caused the water table to rise, making it harder for irrigated soil to drain properly. When farmers use too much water, standing water evaporates and leaves a deposit of mineral salts—a problem known as salinization—that causes the once-fertile soil to lose its productivity. In addition, perennially available canal waters are ideal breeding grounds for the snail that hosts the parasite that causes schistosomiasis, or bilharziasis, a debilitating disease affecting a large proportion of Egypt's rural population. Mediterranean sardine populations, now deprived of the rich silt that nurtured their feeding ground, have plummeted off the Nile Delta, and the sardine industry has faltered. Without the free fertilizer the silt offered, Egypt's bill for artificial fertilizers has increased. Generally, however, Egyptians are very proud of the High Dam, and it has increased food output dramatically for Egypt's mushrooming population.

Egypt's food crops include corn, wheat, barley, rice, millet, fruit, vegetables, and sugarcane. The country was once a great food exporter, but its population has grown so explosively that food now comprises over 30 percent of Egypt's import expenditures. Long-staple Egyptian cotton has long been the main commercial export crop for the hard currency Egypt desperately needs. Rural population densities already are among the highest in the world, and the crowding is growing worse despite a steady migration to Egypt's primate city, Cairo (population: 16 million; Figure 9.13)—the largest metropolis in Africa and the Middle East—and other cities. So far, the government's efforts to lure urban-dwellers and factories out of the Nile Valley to desert "satellite cities" have borne little fruit. Egypt is much more successful in exporting its brains and brawn abroad: Egyptians are prominent among the teachers, doctors, lawyers, engineers, and laborers working in the wealthy Gulf countries, and the **remittances** they send home pump much-needed cash into Egypt's economy.

Egypt's industries are concentrated in Cairo and in the Mediterranean port of Alexandria. They include cotton textiles, food processing, clothing manufacture, chemical fertilizers, and cement. Automobiles and other consumer durables are assembled primarily from imported compo-

Figure 9.13 View of modern downtown Cairo. In the distance, with a much lower profile, is the city's ancient medina.

nents. The country is self-sufficient in some vital minerals, including oil, and exports some oil. But Egypt lacks the mineral wealth to support extensive industrialization, and for its development, the country depends heavily on foreign loans and grants. Income from tourism, particularly to the outstanding temples and tombs of ancient Egypt and to the world-famous scuba diving sites on the Red Sea and Gulf of Aqaba, helps redress the unfavorable trade balance, as do transit fees from ships using the Suez Canal. However, tourism is a very vulnerable resource, as the number of visitors plummets each time violence rocks the Middle East.

Egypt is the giant of the Arab world, its largest country in terms of population, and its most influential in regional and international politics. The Western powers recognize its key role in mediation between Palestinians and Israelis. The West is anxious to see that Egypt does not become poorer than it already is and that the already huge gap between the rich and poor does not widen, perhaps throwing its population into the embrace of militant Islam. And Egypt's strategic location astride the Suez Canal is also a major concern for the West. The United States is Egypt's chief non-Arab ally, lavishing huge sums of civilian and military aid in support of the autocratic regime led by President Hosni Mubarak. To quell real and perceived threats by Islamists, Mubarak has maintained a state of emergency for more than two decades and has effectively stifled most opposition. A grass-roots Islamic movement has grown at the same time. Mostly peaceful, it has involved the widespread adoption of traditionally Islamic values in dress, education, relations between the sexes, and other social and cultural practices. A radical fringe has periodically attacked the interests of Egypt's 9 million strong Coptic Christian minority and foreign targets such as the tourism industry.

9.3 Sudan: Bridge between the Middle East and Africa

Egypt is bordered on the south by Africa's largest country, the vast, tropical, and sparsely populated republic of Sudan (population: 39.1 million) (see Figure 9.11). Formerly controlled by Great Britain and Egypt, Sudan gained its independence in 1956. This southern neighbor is vitally important to Egypt, for it is from Sudan that the Nile River brings the water that sustains the Egyptian people. The Nile receives its major tributaries in Sudan, and storage reservoirs exist there on the Blue Nile, the White Nile, and the Atbara River, all of which benefit Egypt as well as Sudan itself (see Figure 8.25).

The Nile's main branches—the White Nile and Blue Nile—originate respectively at the outlets of Lake Victoria in Uganda and Lake Tana in Ethiopia and flow eventually into Sudan. In southern Sudan, the river passes through a vast wetland called the Sudd, where much water is lost to evaporation. An artificial channel to straighten the river's course and increase the volume and velocity of its flow through the Sudd was under construction for a long time. Called the Jonglei Canal, its future is uncertain. Civil war in Sudan has halted its progress since 1983. Predicting the death of the Sudd wetlands if the canal is completed, many environmentalists would be pleased if work did not resume.

Sudan is overwhelmingly rural. The largest urban district, formed by the capital, Khartoum (population: 5.8 million, including the adjoining cities of Omdurman and Khartoum North), is located strategically at the junction of the Blue and White Niles. Transportation within Sudan as a whole is poorly developed. Several cataracts prohibit navigation on the Nile north of Khartoum, but the White Nile is continuously navigable at all seasons from Khartoum to Juba in the deep South. Most of the country lacks railways or good highways. The main route for rail freight connects Khartoum with Port Sudan, a modern, well-equipped port that handles most of Sudan's seaborne trade.

About 7 percent of Sudan is cultivated, but only 12 percent of the cultivated land is irrigated. The largest block of irrigated land is found in the Gezira (Arabic for "island") between the Blue and White Niles. The country's main cash crop is irrigated Egyptian-type cotton. Many Sudanese support themselves by raising cattle, camels, sheep, and goats. The country's industries are meager. The most important resource is oil, which Sudan began exporting in modest quantities in 1999.

Most of Sudan is an immense plain, broken frequently by hills and mountains. Climatically and culturally, the area is transitional between the Middle East and Africa south of the Sahara. A contrast exists between Sudan's arid Saharan North, peopled mainly by Arabic-speaking Muslims, and the seasonally rainy equatorial South, with its grassy savannas, papyrus wetlands, and non-Arabic-speaking peoples of many tribal and ethnic affiliations who practice both Christian and animistic faiths. About three-fourths of Sudan's total population is Muslim.

There has been much political friction between the two sections. A civil war raged between northern and southern Sudan between 1983 and 2003, with southern factions of the **Sudanese People's Liberation Army (SPLA)** engaging government army troops from the North. As many as 2 million people, mostly civilians, died, and another 4.5 million were driven from their homes. Among the reasons for this conflict were historical antagonisms and economic disparities between the more developed North and the less developed and poorer South, along with efforts by the Arab-dominated government to impose Islamic law (*shari'a*) and the Arabic language on the South and to exploit the South economically. The conflict helped keep Sudan mired in poverty. SPLA rebels repeatedly attacked the country's single export pipeline leading from the oil producing area around Bentui to the coast at Port Sudan. Bentui is perilously close to the southern SPLA stronghold, and the county's largest oil reserves are within SPLA territory.

Ongoing negotiations between the government and SPLA rebels may lead to a peace agreement and a brighter future for Sudan's economy. In a tentative agreement signed in 2004, the Sudanese government at the SPLA agreed to split oil revenues evenly to benefit both northerners and southerners. They also agreed to hold a referendum in 2010 in which southerners would decide if they wanted to separate from the North. Meanwhile, some of the 3 to 5 million refugees displaced from southern Sudan by two decades of civil conflict have been returning home. Most are **internally displaced persons (IDPs)**, meaning they sought refuge within their own borders. They face the challenge of rebuilding a homeland in which virtually all infrastructure has been destroyed.

Drought has regularly visited southern Sudan and adjacent Ethiopia during recent years, and Sudan has periodically hovered near catastrophic famine. Huge tides of refugees have flowed back and forth between Sudan and Ethiopia. International relief agencies have been allowed to provide food and medical aid in only a small portion of the stricken zone. Sudan's military regime has meanwhile been accused of gross human rights violations, including ethnic cleansing of the Nuba, Nuer, Dinka, and other tribespeople in the South. Human rights groups have accused the government of promoting the abduction and enslavement of non-Muslim, mostly Dinka, southerners, who usually end up serving Arab masters in the North. A new humanitarian crisis emerged in 2004 when government backed militias called **Janjawiid** and Sudanese army regulars carried out ethic cleansing in Sudan's westernmost province, Darfur. The victims (up to 70,000 dead by late 2004) were blacks of the Zaghawa, Massaliet, and Fur tribes. They were apparently targeted because of competition with Arabs over access to land and water and because of government fears that these groups might link up with antigovernment rebels in the South.

Sudan's foreign relations have changed dramatically twice since 1989, when an Islamic military regime overthrew an elected government. The new rulers allied themselves with militant Iran and also briefly provided refuge for Osama bin Laden. In 1993, after obtaining information linking Sudanese officials with the first bombing of New York City's World Trade Center, the United States branded Sudan a terrorist nation and halted economic aid. Following the analysis of intelligence information about bombings of U.S. embassies in Kenya and Tanzania, in August 1998 American cruise missiles obliterated a pharmaceutical factory, alleged to be a chemical weapons factory, in Sudan's capital. After that, and especially after 9/11, Sudanese relations with the West warmed considerably, and the United States now regards Sudan as an emerging ally in its war on terrorism.

9.4 Libya: Deserts, Oil, and a Reformed Survivor

Egypt's neighbor to the west, Libya—formerly an Italian colony—became an independent kingdom in 1951. In 1969, Libya became a republic after Colonel Muammar al-Qaddafi and other army officers led a coup against the monarchy. Some 97 percent of the country's people are Muslim Arabs and Berbers.

Libya is made up of three major areal divisions: Tripolitania in the northwest, Cyrenaica in the east, and the Fezzan in the southwest (Figure 9.14). The principal cities are the

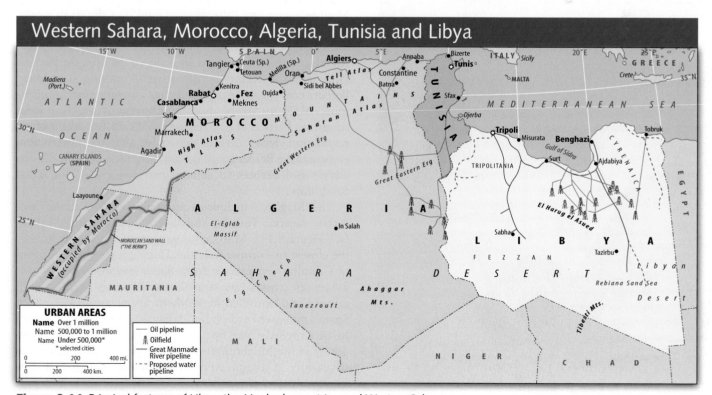

Figure 9.14 Principal features of Libya, the Maghreb countries, and Western Sahara

capital, Tripoli (population: 1.7 million) in Tripolitania, and Benghazi in Cyrenaica.

Most of Libya lies in the Sahara and is too dry to support cultivation except in scattered oases watered by wells and springs. The country is sparsely inhabited; its 5.6 million people are heavily concentrated in coastal lowlands and low highlands along the Mediterranean, where limited winter rains and irrigation with groundwater allow small-scale cultivation of typical Mediterranean crops. Center-pivot irrigation of wheat is carried on in southern Libya. Deep wells there tap nonreplenished fossil waters that accumulated in deep limestone aquifers in ancient times when rains fell abundantly in what is now the Sahara. In an ongoing project called the **Great Manmade River,** these same fossil waters are diverted from 1,155-feet (347-m) deep wells around Tazirbu in east central Libya and Hasquina in west central Libya by pipelines 13 feet (4 m) in diameter stretching hundreds of miles to the Mediterranean (Figure 9.15). The goal of this effort, which Colonel Qaddafi calls "the eighth wonder of the world" and critics abroad decry as "Libya's pipe dream," is agricultural self-sufficiency through the irrigation of fruits, vegetables, and wheat in northern Libya.

Crude oil, petroleum products, and natural gas represent nearly all of Libya's exports (primarily to Western Europe) and have brought prosperity to the country since the 1960s.

The government spends much of the oil revenue on public works, education, housing, aid to agriculture, and other projects for social and economic development. The country has also spent large sums on weapons.

Libya's revolutionary government was long involved in an exchange of hostile rhetoric and sometimes hostile action with the United States, its Western allies, and Israel. In 1986, American warplanes bombed Tripoli and Benghazi in reprisal for alleged terrorist acts against American citizens instigated by Libya, which was on the U.S. list of "terrorist states." After Western intelligence authorities linked two Libyans to the 1988 bombing of Pan American Airlines' Flight 103 over Lockerbie, Scotland, Libya was subjected to an international air embargo. Apparently conceding that his country's isolation was devastating the economy, Colonel Qaddafi decided to make amends with the West. Libya admitted to its role in the Flight 103 bombing and agreed to pay billions of dollars in restitution to the victims' families.

Libya also announced that it was giving up development of weapons of mass destruction, including nuclear weapons. That proclamation led to much praise from Western governments and to the progressive renewal of diplomatic and economic relations. U.S. oil companies are gradually getting back to work in this oil-blessed country. Growing numbers of international tourists are visiting Libya's outstanding Roman sites, pristine beaches, and desert wilderness.

9.5 Northwestern Africa: The Maghreb

Arabs know the northwestern fringes of Africa as the Maghreb (meaning "the western land") of the Arab world. Most of this area's people are Muslim Arabs. The ethnically distinct Berbers, who are most numerous in Morocco and Algeria, converted to Islam after the Arabs brought the new faith into the region in the seventh century A.D. Many Berbers now speak Arabic, and most have adopted Arab customs. Majority Arabs discriminate against them in some countries. Berbers in Algeria, who make up 10 percent of the population, have recently demanded that the government recognize the Berber language and help the economically disadvantaged Berbers to achieve better working and social conditions.

The Maghreb includes four main political units: Morocco, Algeria, Tunisia, and the disputed Western Sahara (see Figure 9.14). Morocco is geographically and culturally the closest to Europe; only the 11-mile-wide (18-km) Strait of Gibraltar separates it from Spain (Figure 9.16). Islamic influences crossed the strait beginning early in the eighth century and prevailed in southern Spain until 1492, when Spain's monarchs ordered both Jews and the Muslim Moors to be expelled from Spain. France became the dominant power in the Maghreb during the 19th and early 20th centuries. French investment developed mines, industries, irrigation works, power stations, railroads, highways, and port

Figure 9.15 Laying pipe for Libya's great manmade river

Akwa Berote/H₂0/Corbis

128

Figure 9.16 The Strait of Gibraltar is the gateway to the Mediterranean. Much of southern Spain is visible in the left of this image, with northern Morocco to the right. This is a photograph from the space shuttle *Endeavour*.

facilities, but French rule was very unpopular. In 1956, after long agitation by local nationalists, Tunisia and Morocco secured independence. In Algeria, independence in 1962 came only after a bitter civil war lasting 8 years.

Most people in the Maghreb live in the coastal belt of Mediterranean climate (called "The Tell" in Algeria), with its cool-season rains and hot, rainless summers. Here, in a landscape of hills, low mountains, small plains, and scrubby Mediterranean vegetation, people grow the familiar products associated with the Mediterranean climate (Figure 9.17). The region's major cities are on the coasts: on the Mediterranean, Tunisia's capital, Tunis (population: 1.7 million), and Algeria's capital, Algiers (population: 3.2 million), and on the Atlantic, Morocco's fabled Casablanca (population: 3.7 million), and the capital, Rabat (population: 1.6 million).

Inland, the Atlas Mountains extend in almost continuous chains from southern Morocco to northwestern Tunisia, with the highest peaks in the High Atlas Mountains of Morocco. They block the path of moisture-bearing winds from the Atlantic and the Mediterranean, and some mountain areas receive 40 to 50 inches (c. 100 to 130 cm) or more of precipitation annually. The precipitation nourishes forests of cork oak or cedar in some places. In Algeria, the mountains form two east–west chains: the Tell Atlas nearer the coast and the Saharan Atlas farther south.

All the countries reach the Sahara to the south. Some pastoral nomadic tribespeople, including the legendary Tuareg, still migrate through the Sahara with their camels, sheep, and goats. Clusters of oases exist in a few places, such as in the

dramatic sandstone Ahaggar Mountains of southern Algeria, which rise high enough to catch moisture from the passing winds. A line of oases fed by springs, wells, mountain streams, and *foggaras* (tunnels from mountain water sources, known as *qanats* in Iran) lies along or near the southern base of the Atlas Mountains. Several large areas of sandy desert (*ergs*) and a barren gravel plain, the Tanezrouft, occupy portions of the Algerian Sahara.

The Maghreb economies benefit from a valuable endowment of minerals. Algeria's oil and natural gas, located in the Sahara, are the region's greatest assets (see Figure 9.14). The oil flows through pipelines to shipping points on the Mediterranean in Algeria and Tunisia, and a pipeline is being built to carry Algerian natural gas to Europe. Petroleum products, crude oil, and natural gas make up almost all of Algeria's exports by value. Morocco's exports also include some minerals, notably phosphate, plus agricultural exports and clothing. Consumer industries, including plants that assemble foreign-made components, are the second leading industries in the Maghreb countries.

Algeria's oil boom went bust in the 1980s, but population growth continued, fueling an economic crisis and shortages of food and consumer goods that contributed to widespread unrest in the 1990s. Algeria built up a staggering foreign debt that contributed to the instability. Discord in Algeria reverberated through the entire Maghreb and across the Mediterranean to the region's former colonial power, France. During almost 8 years after 1992, when Algeria's military government canceled the results of parliamentary elections that would have given majority seats to the Islamic Salvation Front (FIS), civil strife claimed an estimated 150,000 lives within Algeria. Violent sympathizers of the FIS also carried their insurgency to France in the 1990s, bombing government and civilian targets to protest France's support of the

225

Figure 9.17 The western Maghreb has breathtaking physical landscapes, from the Mediterranean regions in the north to the Atlas Mountains and Sahara Desert in the south. Traditional village and urban architecture lends a unique character to the human environment. This is Vallée du Dades in Morocco, with the Atlas Mountains in the background.

military government in Algiers. The early 2000s saw Algeria's 32.3 million people reconciling with themselves and with the West and enjoying the strong economic growth prompted by high energy prices.

Morocco and Tunisia viewed events in neighboring Algeria with anxiety. Citing the violence in Algeria, Tunisia banned Islamist political parties in 1994. The government of Tunisia (population: 10 million) has promoted family planning and women's rights and is often cited as an "oasis" of openness and progressive change in the region. Tunisia also has the Maghreb's strongest economy, with active agricultural, mining, energy, tourism, and manufacturing sectors.

In Morocco, King Muhammad VI rules over a constitutional monarchy that promises political accommodation of Islamic groups, in contrast with Algeria's experience, but human rights monitors are increasingly alarmed by the young king's muzzling of opposition voices. Morocco's close alliance with the United States in the war on terrorism made it a target of an al-Qa'ida attack in 2003. Morocco is the poorest of the Maghreb countries. But there has been some progress in this country of 30.6 million in recent years. The country has shown exceptional success in bringing down birth rates; while the average woman had seven children in 1980, she now has three.

There are some significant international problems focused on this region. Morocco would like Spain to return to Moroccan control the two Spanish enclaves of Ceuta and Melilla on Morocco's Mediterranean coast, and the two countries have contesting claims to a number of small islands near the Moroccan coast. Both Morocco and Algeria have long been at odds over the future of Spain's former dependency known as Western Sahara. This sparsely populated coastal desert (estimated population: 300,000, a figure that includes 170,000 refugees living in Algeria), which contains rich phosphate deposits, adjoins Morocco on the south and is claimed by that country. When Spain relinquished control in 1975, the territory was partitioned between Morocco and Mauritania, but in 1979, Mauritania withdrew its presence. Algeria objected to Moroccan control and, together with other outside parties, gave assistance to an armed resistance movement in Western Sahara against Morocco called the **Polisario Front.** In 1989, Morocco and the Polisario Front agreed to a referendum on independence, and a peace treaty was signed in 1991. To influence the outcome of the referendum in its favor, Morocco relocated 40,000 Moroccans into the region in 1991; they remain there today under Moroccan guard. The Moroccan government has also used tax breaks and other subsidies to attract new Moroccan settlers there.

Such questionable actions led the United Nations to repeatedly suspend the referendum and finally call it off in 2000. Morocco continued to assert its claim to Western Sahara. It sold Western Sahara's offshore fishing rights to European countries. To stave off Algerian interests, Moroccan technicians also erected a 700-mile (1,120-km) sand barrier parallel to Western Sahara's border with Algeria, reinforcing

it with land mines and 150,000 troops (see Figure 9.14). West of this berm, in Morocco-held territory, is the world's largest reserve of phosphate.

The United Nations is studying the prospect of helping to establish limited Western Saharan autonomy under Moroccan rule, but 75 countries recognize the Polisario-proclaimed Saharawi Arab Democratic Republic as a country with a government in exile. While the status of Western Sahara remains unresolved, more than 170,000 **Saharawis** (as the indigenous people of Western Sahara are known) live in refugee camps in the Algerian desert, as they have for decades. Despite their bleak conditions, the Saharawis have achieved a 95 percent literacy rate and have a representative government with a constitution that insists on religious tolerance and gender equality.

9.6 The Gulf Oil Region

As discussed in Chapter 8, the Persian/Arabian Gulf region has by far the world's largest oil reserves. Foreign companies, mainly British and American, originally developed the oil industry in this region. These firms made huge profits from their oil concessions, but with a few exceptions, after 1972 the companies saw their properties nationalized by the Gulf governments. However, many outside companies still work closely with national oil companies in the Gulf countries, providing them with technical and managerial expertise, skilled workers, equipment, marketing channels, and varied services.

Gulf oil is shipped out by a combination of pipelines and ships (see Figure 8.C, page 221). Most oil produced in the Gulf region is transported to market by oceangoing supertankers. These ships load oil at terminals spaced along the Gulf. Some crude oil moves by pipelines from oil fields in the Gulf countries to tanker terminals on the Mediterranean and Red Sea coasts. A growing proportion of extracted crude oil is refined in the Gulf countries for local use and export. Petrochemical industries exist in most of the producing countries, with the largest development in Saudi Arabia. The industries use local natural gas as their primary source of energy. Huge gas reserves are associated with oil deposits, and other major reserves exist apart from the oil fields, in "nonassociated" deposits. Growing amounts of gas are exported in liquefied form; Saudi Arabia is the world's largest exporter.

Saudi Arabia: The Oil Colossus

With an area of roughly 830,000 square miles (2.1 million sq km), Saudi Arabia occupies the greater part of the Arabian Peninsula, the homeland of Arab civilization and the faith of Islam (Figure 9.18). The interior of the country is an arid plateau fronted on the west by mountains that rise steeply from a narrow coastal plain bordering the Red Sea. Most of the plateau lies at an elevation of 2,000 to 4,000 feet (c. 600 to 1,200 m), with the lowest elevations near the Gulf and in the eastern part of the gigantic sea of sand called the

226–228

218

Arabian Peninsula

URBAN AREAS
Name Over 1 million
Name 500,000 to 1 million
Name Under 500,000*
* selected cities
0 150 300 mi.
0 150 300 km.

Figure 9.18 Principal features of the Arabian Peninsula

Empty Quarter (*al-Rub' al-Khali*). Mountains (Arabic: *jebels*) rise locally from the plateau, and the plateau surface is cut by many valleys (Arabic: *wadis*) that carry water whenever it happens to rain; Saudi Arabia has no permanent rivers. Most interior sections average less than 4 inches (10 cm) of precipitation annually, but years may pass at a given locale with no precipitation at all. Rainfall is greatest along the mountainous western margin of the peninsula, especially in the southwestern Asir region, near the border with Yemen, which benefits from summer monsoon rains. About 95 percent of Saudi Arabia's crop production depends on irrigation. Cultivated land makes up only 2 percent of the country's area, yet surprisingly, Saudi Arabia is self-sufficient in many agricultural staples. It is an exporter of wheat, irrigated mostly in the central part of the country by gigantic center-pivot sprinkler systems tapping fossil waters.

Saudi Arabia's oil reserves make up more than one-fourth of the world's total. Most of the country's oil fields, including the world's largest, the Ghawar field, are located in the eastern margins of the country, including its adjacent seabed. Some Saudi Arabian oil is processed in refineries within the country and in nearby Bahrain, but most of the production leaves Saudi Arabia in crude form by tankers, using loading terminals that jut into both the Gulf and the Red Sea. To meet anticipated future demand, huge refining and petrochemical complexes have been developed at Jubail on the Gulf and Yanbu on the Red Sea. These cities sprang up almost overnight in the 1980s. Saudi Arabia's best-known city is ancient Mecca, Islam's holiest city. The interior city of Riyadh (population: 3.6 million) is Saudi Arabia's capital. The Red Sea port of Jidda, some 40 miles (64 km) from Mecca, is the main point of entry for foreigners who make the pilgrimage to Mecca.

216–218

Saudi Arabia's population of 25.1 million is young and growing; half of all Saudis are younger than 21, and the annual growth rate is 3 percent. About one-quarter of the population consists of foreign workers. The country was long concerned that its native population was too small to run the country's economy and stave off cultural domination by foreign workers. The government promoted population growth and passed "**Saudization**" laws to put Saudis in jobs previously held by foreigners. Now the fear is that the indigenous Saudi population is growing much faster than job opportunities will, perhaps creating a young, unemployed, and disaffected population that might agitate for change. Since the height of the oil boom in 1981, the population has

Joe Hobbs

Figure 9.19 Saudi Arabia's Shi'ite Muslim minority is significantly poorer than the Sunni majority. These are Shi'ites in the Eastern Province town of Qatif.

more than doubled and per capita income has fallen by three-fourths. The country's transition to a standard of living comparable to the more developed countries has been stalled. The infant mortality rate remains high.

There are striking discrepancies in income; although the sharing of oil wealth has brought once undreamed-of prosperity to most of the population, the Shi'ite Muslim minority in the Eastern Province is poor (Figure 9.19). Despite a recent surge in energy prices and corresponding revenue flow to the country, a long-term trend of declining oil revenues has compelled many Saudis to take menial jobs once reserved for foreigners. The oil will, of course, run out one day, so the biggest challenge for Saudi Arabia is to diversify away from its near complete dependence on that commodity.

Despite the Western-style luxuries that oil wealth has brought to Saudi Arabia, the conservative Saudi family that rules this absolutist monarchy insists on maintaining local Muslim traditions. The legal and religious systems are based on interpretations of Islam by the Wahhabist sect, followers of the 18th-century Islamic reformist Muhammad bin Abd al-Wahhab who joined forces with the founder of the Saudi dynasty. **Wahhabism** emphasizes fastidious piety, rejection of non-Muslim beliefs and cultures, corporal and capital punishment, and complete segregation of men and women in public life. There are restrictions on the freedoms of women, who are prohibited from driving automobiles. Although comprising almost 60 percent of university graduates, women make up only 5 percent of the workforce (Figure 9.20).

The vast Saudi royal family (with more than 7,000 princes) keeps a wary eye on many internal and external threats. Al-

Joe Hobbs

Figure 9.20 Education of women in the Gulf countries has increased dramatically in recent decades, but strong traditions keep most women out of the workforce. This is a GIS class in the Department of Geography, United Arab Emirates University.

Qa'ida has bombed Saudi civilian compounds and U.S. military facilities in an effort to purge the region of secular and "imperialist" Western interests and to destabilize the monarchy in order to replace it with an Islamic government. U.S. forces did withdraw from Saudi Arabia after the 2003 Iraq war. Countercurrents are now buffeting the desert kingdom. While struggling to prevent outside influences from changing Saudi ways, the government is finally allowing tourists to visit, encouraging foreign investment, and signing international trade and human rights treaties.

226–228

Bahrain, Qatar, and the United Arab Emirates

The sheikdom of Bahrain (ruled by a monarch called a *Sheikh* or *Emir*; population: 700,000) occupies a small archipelago in the Gulf between the Qatar Peninsula and Saudi Arabia (see Figure 9.18). A modern four-lane causeway connects the main island to nearby Saudi Arabia. In addition to being an oil producer, Bahrain is a center of oil refining, seaborne trade, ship repair, air transportation, business administration, international banking, and aluminum manufacturing using imported alumina and local natural gas. Like Saudi Arabia, Bahrain is an oil-financed welfare state. However, in contrast to its giant neighbor, Bahrain's petroleum reserves are expected to be exhausted within two decades. Bahrain is home to the U.S. Navy's Fifth Fleet, making this tiny country a vital strategic ally of the United States in a critical region. Within the region, Bahrain is notable for women's rights to vote and run for office; in Qatar, women can vote but not be elected, and in the other Gulf countries, they can do neither.

East of Bahrain along the Gulf lie the small emirates of Qatar (population: 700,000) and the United Arab Emirates (U.A.E.) (population: 4.2 million). Qatar is atop the world's third largest reserve of natural gas. Only 150,000 of the country's 750,000 inhabitants are indigenous Qataris; the rest are imported workers, mainly from South and Southeast Asia, here to turn that mineral wealth into tangible assets. Qatar established itself as a vital U.S. ally during the 2003 war against Iraq, permitting American warplanes and warships to be based there. Qatar is also the home of al-Jazeera, the satellite television network that upsets many Gulf autocracies with its frank criticisms and that serves as the media outlet for al-Qa'ida videotapes and audiotapes.

The United Arab Emirates is a loose confederation of seven units, three of which produce oil; only Abu Dhabi and Dubai pump it in major quantities. Qatar and the United Arab Emirates are extremely arid and, prior to the oil era, supported only a meager subsistence economy based on fishing, seafaring, herding, and a little agriculture, augmented by smuggling and piracy. Today, the United Arab Emirates is capitalizing on its central location between Asia, Africa, and Europe with the construction of sophisticated air and seaport facilities. The recently built seaports of Abu Dhabi and Dubai are colossal facilities that have dramatically changed the coastal landscapes and are easily visible from space (Figure 9.21). The United Arab Emirates is successfully diversifying its economies beyond the oil industry. Dubai is a well-established international trade and banking center (Figure 9.22). Abu Dhabi, the largest and richest unit of the United Arab Emirates, is hoping to establish regional supremacy as a financial center through the construction of a massive commercial development on Saadiyat Island. Unlike its more conservative neighbors, the United Arab Emi-

(a)

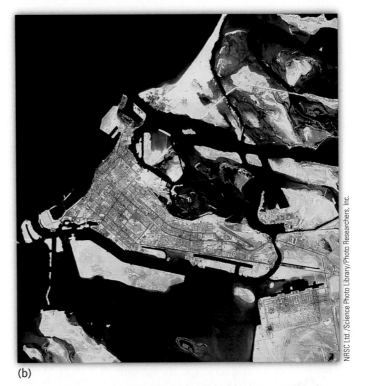

(b)

Figure 9.21 These extraordinary before-and-after satellite images reveal how oil wealth has transformed some Middle Eastern landscapes. (a) Abu Dhabi, capital of the United Arab Emirates, in 1972. The city occupies much of the natural island in the center. (b) This is the same area, recorded in 1984, and reveals the outcome of huge investments, including human-made islands and deepwater port facilities.

NRSC Ltd./Science Photo Library/Photo Researchers, Inc.

Figure 9.22 Like most cities in the oil producing countries of the Gulf, Dubai is a thoroughly modern metropolis with only hints of the past—here, the storied wooden sailing vessels known as *dhows*.

rates has promoted itself as a destination for sun-worshipping international tourists and shoppers. Dubai is the most liberal unit within the United Arab Emirates and, in marked contrast with its teetotaling neighbors, tolerates bars and brothels.

The Southern Margins of Arabia

Known to the ancient Romans as *Arabia Felix,* or "Happy Arabia," because of their verdant landscapes, Yemen (population: 20 million) and Oman (population: 2.7 million) occupy, respectively, the mountainous southwestern and southeastern corners of the Arabian Peninsula (see Figure 9.18). The present Republic of Yemen was created in 1990 by the union of the former North Yemen and South Yemen. They remain united despite a brief civil war in 1994. Its political capital and largest city is San'a (population: 1.5 million), located in the former North Yemen. Until the 1960s, North Yemen was an autocratic monarchy, extremely isolated and little known. In the mid-1960s, a republican government supported by Egypt and opposed by Saudi Arabia took control of large sections at a cost of much bloodshed.

The former South Yemen, known until 1967 as Aden, a British colony and protectorate, lies along the Indian Ocean between the former North Yemen and Oman. The main city, Aden, is located on a fine harbor near the straits. Aden is perfectly located for refueling vessels that transit the Suez Canal and is close to the strategic chokepoint of Bab el-Mandeb connecting the Red Sea with the Indian Ocean. Possession or use of this port has therefore long been prized by outside powers. After British colonial times, South Yemen became a Marxist state supported by the Soviet Union and other Eastern-bloc nations. Since the two Yemens were unified, Aden has been an important refueling station for the United States and its Western allies.

Security at the port of Aden has been stepped up since the bombing there of the U.S.S. *Cole* in 2000, and Western governments have pressured Yemen's government to crack down on Islamist groups that advocate struggle against Western interests in the region. Yemen gave the United States permission to use a remote-controlled predator drone to kill an important al-Qa'ida leader on a remote Yemeni road in 2002. Some of Yemen's tribal peoples live in areas well beyond the effective control of the government in San'a and are known to be strongly sympathetic to al-Qa'ida. The mountainous north-central region of Yemen, known as the Hadramaut, is the ancestral homeland of Osama bin Laden's family.

Yemen receives the heaviest rainfall in the Arabian Peninsula, enjoying a summer rainfall maximum derived from the same monsoonal wind system that brings summer rain to the Horn of Africa and the Indian subcontinent. Its main subsistence crops are millet and sorghums. Most of the country's best agricultural land consists of terraces and is dedicated to the cultivation of the valuable cash crop *qat (Cathya edulis),* a stimulant shrub chewed by virtually all adults in the country on a daily basis (Figure 9.23). Most of the country's people live in the rainier highlands, and the coastal plain is extremely arid and sparsely populated.

Yemen's meager agricultural output and lack of industrial production have been compensated for during recent decades by remittances, the earnings sent home by Yemeni workers in the oil fields and services of the wealthy Gulf countries. The support of Yemen's government for Saddam Hussein's regime in Iraq during the 1991 Gulf War prompted Saudi Arabia to expel millions of these expatriate Yemeni workers, leading to a precipitous decline in Yemen's economy. Oil production began in Yemen itself only in the 1980s. The fields are located in a contested border area between Yemen and Saudi Arabia and subjected periodically to raids

Figure 9.23 The mountainous landscape near San'a, Yemen's capital. The only crop under cultivation in this scene is the stimulant known as *qat (Cathya edulis).*

and kidnappings by groups opposed to Yemen's government, so production has been slow to increase.

Most of the Sultanate of Oman lies along the Indian Ocean and the southern shore of the Arabian Sea, with a small, detached part bordering the Strait of Hormuz and the Gulf. As in Yemen, the coastal region is arid, and most people live in interior highlands where mountains induce a fair amount of rain. For about 2,000 years, beginning around 1000 B.C., Oman was the source of most of the world's frankincense, a prized commodity that made the region a hub of important oceanic and continental trade routes. The capital, Muscat (population: 650,000), has long been a vital port in trade between the Middle East and South Asia. Crude oil now furnishes almost all of Oman's exports. The country's strategic position on the Strait of Hormuz, through which much of the oil produced by the Gulf countries is transported, has made it an object of concern in industrial nations, particularly Great Britain, with which it has historic ties, and the United States. These Western powers have rights of access to base facilities in Oman in the event of any military emergency that threatens the continued flow of oil from the Gulf region. Allied forces used the bases after Oman joined the United Nations coalition against Iraq in the Gulf War of 1991, and again in the U.S.-led invasion of Iraq in 2003.

Kuwait and the Gulf War of 1991

With 9 percent of the world's proven oil reserves within its borders, tiny Kuwait (6,900 sq mi/17,900 sq km—about the size of the U.S. state of New Jersey; Figure 9.24) has the world's fourth largest store of proven petroleum resources (after Saudi Arabia, Iraq, and Iran). Associated with the oil is much natural gas, which Kuwait both uses and exports. Oil revenues have changed Kuwait (population: 2.5 million) from an impoverished country of boatbuilders, sailors, small traders, and pearl fishermen to an extraordinarily prosperous, completely urbanized welfare state. Generous benefits have flowed to Kuwaiti citizens but not to the Asian and other non-Kuwaiti expatriates drawn by employment opportunities to this emirate. About 55 percent of the people residing in Kuwait are foreigners. To reduce possible security threats and become more self-reliant, the government is trying to reduce dependence on foreign workers. A huge challenge will be to provide jobs for its young population (60 percent of Kuwaitis are younger than 18) in non-oil-related industries and services. Kuwait produces only fossil fuel commodities. Therefore, it is vulnerable to developments such as alternative energy and, of course, to the eventual depletion of its own reserves.

Kuwait took center stage in world affairs in August 1990, when Iraqi military forces invaded the country. Claiming that it was historically part of Iraq and calling it Iraq's "19th province," the Baghdad regime attempted to annex Kuwait. Saudi Arabia and other Gulf oil states seemed in danger of invasion. Worldwide opposition to Iraq's aggression resulted in a sweeping embargo on Iraqi foreign trade imposed by the Security Council of the United Nations. Led by the United States, a coalition of 28 UN members (including several Arab states) staged a rapid buildup of armed forces in the Gulf region.

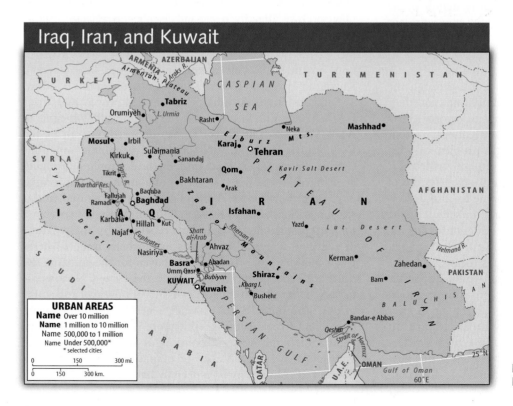

Figure 9.24 Principal features of Iraq, Iran, and Kuwait

Beginning in January 1991, the coalition's crushing air war drove the Iraqi air force from the skies, massively damaged Iraq's infrastructure, and hammered the Iraqi forces in Kuwait and adjacent southern Iraq with intensive bombardment. In the ensuing ground war, the Iraqis were driven from Kuwait within 100 hours. Overall, an estimated 110,000 Iraqi soldiers and tens of thousands of Iraqi civilians were killed, while the technologically superior coalition forces suffered 340 combat deaths, of which 148 were Americans. During the conflict, which came to be known as the Gulf War, the Iraqis also made war on the environment, setting fire to hundreds of oil wells in Kuwait and allowing damaged wells to discharge huge amounts of crude oil into the Gulf.

Iraq: Troubled Oil State of Mesopotamia

Iraq occupies a broad, irrigated plain drained by the Tigris and Euphrates Rivers, together with fringing highlands in the north and deserts in the west (Figure 9.24). This riverine land of Mesopotamia has a complex geopolitical history, having been a seat of empires, a target of conquerors, and in the 20th and early 21st centuries, a focus of oil development and political and military contention involving many other nations.

An estimated 70 percent of Iraq's 25.9 million people are urban-dwellers. Most of the rest are agricultural villagers. A critical feature of Iraq's geography is its ethnic and religious complexity, with three major groups present: the Shi'ite Arabs, mainly in the south, who make up about 52 percent of the country's population; the Sunni Arabs, about 30 percent of the total and living mainly in the center (especially in the so-called "Sunni Triangle"; and the Kurds, mainly in the north, most of whom are Sunni Muslims and who represent 16 percent of the total (Figure 9.25; see also Ethnic Geography, page 257). There are numerous smaller minorities, including Yazidi and Sunni Turkoman.

Iraq's government has expanded irrigation through the construction of dams on the Tigris and Euphrates and their tributaries for multiple uses of flood control, hydroelectric development, and irrigation storage. Agricultural output in Iraq is poor, however, due mainly to the problem of overwatering, which causes the salinization of once-fertile soils. Small bands of nomadic Bedouins still raise camels, sheep, and goats in the western deserts. Some of the Kurdish tribespeople in the mountain foothills of the north are seminomadic grazers, although most Iraqi Kurds are sedentary farmers.

Baghdad (population: 7.1 million), Iraq's political capital, largest city, and main industrial center, is located strategically on the Tigris near its closest approach to the Euphrates River. The country invested heavily in rapid reconstruction of its capital city after the 1980–1988 war and again after the 1991 and 2003 conflicts. Basra is Iraq's main seaport, although it is located upriver on the Shatt al-Arab. Iraq suffers from the fundamental geographic disadvantage of not having natural deep-water ports in its tiny 35-mile (55-km) win-

Figure 9.25 Ethnicity in Iraq and neighboring regions

dow on the Gulf. One of the main reasons Iraq attacked Kuwait in 1990 and Iran in 1980 was to widen this window.

Saddam Hussein's military adventures proved disastrous for the country. A formal cease-fire to the Gulf War came in April 1991, with harsh conditions imposed on Iraq by the UN Security Council. UN-sponsored economic sanctions created shortages of food and medical supplies, bringing great hardship to the country's population. Iraq agreed to pay reparations for the damage its forces had caused, and it renounced its claims to Kuwait. Saddam Hussein's surviving forces then mercilessly put down rebellions by Shi'ite Arabs in southern Iraq and by Kurds in northern Iraq. To deter Iraq's military and provide a safe haven for the Kurds in the north and the Shi'ites in the south, the United Nations established "no-fly zones" north of the 36th parallel and south of the 32nd (later, 33rd) parallel, where Iraqi aircraft were prohibited from operating. But there was widespread condemnation of the Western coalition for apparently encouraging the Kurds and Shi'ites to rebel and then failing to assist them. Iraq's Shi'ites fared especially badly.

Ethnic Geography

The Kurds

A mostly Sunni Muslim people of Indo-European origins, the Kurds are by far the largest of several non-Arab minorities in Iraq. They have revolted against Iraqi central authority on numerous occasions, most recently during the Gulf War of 1991. The Kurdish question has international implications because the area occupied by an estimated 26 million Kurds also extends into Iran, Turkey, Syria, and the southern Caucasus region (see Figure 9.25); an estimated 8 million more live outside the Middle East. European powers promised an independent state of Kurdistan at the end of World War I, but the governments concerned never took steps to create it. Kurds today remain the world's largest ethnic group without a country.

A Kurdish rebellion against Turkish rule has smoldered and sometimes flamed since 1990. The main Kurdish resistance comes from the **Kurdistan Workers Party,** known by its Kurdish acronym **PKK.** The organization suffered a major setback in 1999 when Turkey caught, jailed, and sentenced to death its leader, Abdullah Ocalan. Following the Gulf War, Iraq's Kurds enjoyed a decade of relative autonomy and prosperity in northern Iraq. Under Saddam Hussein, Iraq tried to "Arabize" its Kurds by forcing them to renounce their ethnicity and sign forms saying they were Arabs. Turkish officials have also downplayed the identity of Kurds (who make up 20 percent of Turkey's population), often referring to them as "Mountain Turks," and Kurdish language media in Turkey have long been banned or restricted. A book about the Kurds is appropriately entitled *No Friends but the Mountains.*[a]

[a] John Bullock, *No Friends but the Mountains* (London: Oxford University Press, 1993).

One of the great environmental and cultural tragedies of the late 20th century was Iraq's systematic draining after 1991 of the marshes adjoining the southern floodplains of the Tigris and Euphrates Rivers. To quell any opposition by the ancient Shi'ite culture of the **Ma'adan,** or **"Marsh Arabs,"** the Iraqi military built massive embankments and canals that drained and destroyed the marshes, forcing the Ma'adan to abandon their historic homeland. After the war of 2003, the embankments were destroyed and restoration of the marshlands began.

Throughout the 1990s, the regime of Saddam Hussein played a cat and mouse game with UN inspectors whose duty was to eliminate Iraq's ability to produce weapons of mass destruction (WMD), including chemical, biological, and nuclear weapons. United States, British, and other forces repeatedly struck Iraq as the country refused to allow full access by the UN inspectors and, in 1998, attacked scores of military targets when access was denied. The Baghdad regime flaunted the no-fly zones and restated its wartime position that Kuwait was an illegitimate country. International support for the sanctions against Iraq weakened greatly, particularly in the Arab world. President Saddam Hussein endured, much to the chagrin of his foes, and might have been able to hang onto power indefinitely if not for the 9/11 attacks.

Immediately after the Gulf War and during the Clinton administration of 1993–2001, the reasoning in Washington was that a unified Iraq under Saddam Hussein was better than the alternative: the possible fragmentation into the three entities of a Shi'ite south, a Sunni center, and a Kurdish north, some or all of which might be hostile to U.S. interests in the region. Although unsavory, Saddam Hussein could be contained. Under the new Bush administration, and especially after 9/11, the downfall of Saddam Hussein became an urgent priority, ranking alongside the war on terrorism. The Bush administration attempted to link Saddam Hussein with al-Qa'ida, particularly to garner public American support for a new war on Iraq, but such connections were not decisively established. It is possible that the administration saw a post-Saddam Iraq as a strong, central bridgehead against a long-term war on Islamist terrorism in the Middle East.

Publicly, Washington built the case that Iraq was harboring and continuing to develop weapons of mass destruction that might be used against the United States and its allies. The United States was able to prompt further United Nations pressure on Iraq, which finally allowed UN weapons inspectors to resume their work. When these inspections failed to turn up weapons, the United States insisted that Iraq was simply concealing them very effectively and that only the use of force could remove their threat. It was necessary to effect "regime change" in Baghdad. To take on Saddam Hussein, President Bush assembled what he called a "coalition of the willing" consisting mainly of Britain, Spain, and several Eastern European countries. It fell far short of the coalition assembled during the Gulf War. France, Germany, and a range of other traditional allies of the United States opposed the impending war against Iraq, and even within the United States there was considerable dissent against what was perceived as a war of choice rather than of necessity.

The U.S.-led ground and air assault on Iraq (dubbed **Operation Iraqi Freedom**) that began in March 2003 led quickly and with relatively few casualties to the downfall of the Iraqi

Figure 9.26 American troops in Iraq faced stiff resistance but also heard cheers from many Iraqis who had chafed against Saddam Hussein's rule.

regime. With about 130,000 troops, concentrated mainly in the Sunni-dominated center of the country around Baghdad, the United States settled in for a long and troubled occupation of the country (Figure 9.26). The United States hand-picked a **Governing Council** of Iraqis, broadly representative of the country's ethnic composition, and ran many of Iraq's security and other affairs with its own **Coalition Provisional Authority (CPA)** until late June 2004, when it turned over sovereignty to an interim Iraqi government. More than 100,000 American troops remained in country after this official transfer of authority. Although an undetermined percentage of Iraqis welcomed American soldiers as liberators, discontent with the American occupation grew, particularly among the Sunnis who had long backed Saddam Hussein. Using roadside bombs, car bombs, small arms, and rocket-propelled grenades, Sunni militants were able to inflict American casualties on an almost daily basis. Predictably, al-Qa'ida used the American occupation as a rallying cry for further resistance against the United States, and al-Qa'ida may have been behind some of the larger bombings in Iraq. British troops patrolled the country's south, where the country's Shi'ite majority initially cheered the war because it had driven out their longtime foe, Saddam Hussein. In the months that followed, the Shi'ites made it clear that in the new Iraq, they wanted political power proportionate to their majority numbers. Their cause was championed by a cleric named Ayatollah Ali al-Sistani.

Foreign intervention and costly conflict are not new to Iraq, and the modern problems there may be better appreciated through some understanding of the past. Iraq was a province of the Ottoman Empire (based in what is now Turkey) for almost 400 years. British forces defeated the Turks in Iraq in World War I, and Great Britain had a League of Nations mandate over the country until it gained its independence in 1932. Iraq's government was a pro-Western monarchy from 1923 to 1958, when an army coup made the country a republic. By the mid-1970s, Iraq had become an authoritarian one-party state ruled by President Saddam Hussein in the name of the Arab Baath Socialist Party. After 1973, increasing oil revenues from the country's huge reserves (about 11 percent of the world's total) underwrote new social programs, large construction projects, and the creation of new businesses. New high-rise buildings, busy expressways, and an active commercial and social life gave metropolitan luster to ancient Baghdad. Subsequent wars resulted in destruction in the city and in Iraq's economy. This country should have become rich from its oil, but in recent decades, it has used oil revenues only to make up some ground lost through hapless military adventures.

In 1980, perceiving internal weakness in the neighboring fledgling Islamic Republic of Iran and a means of securing land and mineral resources on the Iranian side of the Shatt al-Arab waterway (the river formed by the confluence of the Tigris and Euphrates Rivers), President Hussein launched an invasion of his more populous neighbor. Thus began an 8-year war that proved enormously costly to both sides, with an estimated 500,000 people killed or wounded. Iran mounted a far better defense than Iraq anticipated, expending its greater military personnel in bloody "human wave" assaults to counter Iraq's greater firepower from tanks, artillery, machine guns, and warplanes. A complex web of geopolitical relationships made the Iran–Iraq War a storm center in the political world, particularly as the superpowers and other countries sold sophisticated arms to both sides. The war ended in stalemate in 1988, with neither side gaining significant territory or other assets.

Oil and Upheaval in Iran

Iran, formerly known as Persia (see Figure 9.24), was the earliest Middle Eastern country to produce oil in large quantities. The main oil fields form a line along the foothills of the Zagros Mountains in southwestern Iran. British-built pipelines linked the large refinery at Abadan (on the Shatt-al Arab waterway) with the inland oil fields. Now nearly all exports of crude oil from Iran pass through a newer terminal at Kharg Island (see Figure 8.C), which allows supertankers to load in deep water.

Oil is the leading source of revenue for Iran, a country where a dry and rugged habitat makes it hard to provide a good living for its population of 67.4 million. Arid plateaus and basins, bordered by high, rugged mountains, are Iran's characteristic landforms (Figure 9.27). Encircling mountain ranges prevent moisture-bearing air from reaching the interior, creating a classic rain-shadow desert. Only an estimated 10 percent of Iran is cultivated. Enough rain falls in the northwest during the winter half-year to permit unirrigated cropping (dry farming), but rainfall sufficient for intensive agriculture is confined mainly to a densely populated lowland strip between the high, volcanic Elburz Mountains and the Caspian Sea. The western half of this lowland receives

Figure 9.27 Iran's rugged topography seen from space. This is a view of southeastern Iran, looking almost due south to the critical chokepoint of the Strait of Hormuz (just right of center near the top of the image).

more than 40 inches (c. 100 cm) annually and produces a variety of subtropical crops such as wet rice, cotton, oranges, tobacco, silk, and tea.

Agriculture elsewhere in Iran depends mainly on irrigation. Many irrigated districts lie on gently sloping alluvial fans (triangular deposits of stream-laid sediments) at the foot of the Elburz and Zagros Mountains; the capital, Tehran (population: 11.5 million), is in one of these areas south of the Elburz. Qanats (underground sloping tunnels) supply water to a large share of the country's irrigated acreage. However, they are expensive to build and maintain, and an increasing share of water comes from stream diversions and from deep wells. Wheat is the most important irrigated crop. The leading revenue earner, and Iran's third most important export after oil and its famous carpets, is the pistachio nut (Figure 9.28). Cattle are raised in areas where there is enough moisture to grow forage for them, but drier areas support sheep and goats. Many livestock belong to seminomadic mountain peoples—the Qashqai, Baktiari, Lurs, Kurds, and others—whose independent ways have long been a source of friction between these tribespeople and the central government. Before the Islamic revolution in Iran in 1979, it was government policy to sedentarize Iran's nomads so that they did not threaten the state.

Before World War I, Iran was a poor and undeveloped country, in marked contrast to its imperial grandeur when Persian kings ruled a great empire from Persepolis and to its architectural fluorescence under earlier Muslim leaders. In the 1920s, a military officer of peasant origins, Reza Khan,

Figure 9.28 Many of Iran's "Persian carpets" are handmade masterpieces of wool or silk that take years for an individual or family to complete. This carpet was made in the city of Isfahan in central Iran.

seized control of the government and began a program to modernize Iran and free it from foreign domination. He had himself crowned as Reza Shah Pahlavi, the founder of the new Pahlavi dynasty. Influenced by the modernizing efforts of Mustafa Kemal Ataturk in neighboring Turkey, the new shah (king) introduced social and economic reforms. In 1941, his young son took the throne as Mohammed Reza Shah Pahlavi and set out to expand the program of modernization and Westernization begun by his father. In the 1970s, Iran played a leading role in raising world oil prices. Mounting oil revenues underwrote explosive industrial, urban, and social development.

In the countryside, the shah's government attempted to upgrade agricultural productivity through land reform and better technologies. Iran made major efforts to increase the supply of irrigation water and electricity by building storage dams and hydropower stations. But agricultural reforms generally failed, and poverty-stricken families from the countryside poured into Tehran and other cities. From this devoutly Muslim group of new urbanites came much of the support for the revolution that ousted the shah in 1978–1979.

Oil money funded Iran's development under the Pahlavi dynasty, but the vast sums the regime spent on the military undercut the benefits to Iran's people. Arms came primarily from the United States. In 1978 and 1979, a revolution overthrew the shah and abruptly took Iran out of the American orbit. The central figure among the revolutionaries was the Ayatollah Ruhollah Khomeini, an elderly critic of the regime who had been forced into exile by the shah in 1964. From Paris, Khomeini sent repeated messages to Iran's Shi'ites that helped spark the revolution. A furious tide of Shi'ite religious sentiment rose against corruption, police heavy-handedness, modernization, Westernization, Western imperialism, the exploitation of underprivileged people, and the institution of monarchy, which Shi'ite beliefs regard as illegitimate. Growing numbers of people staged strikes and demonstrations in 1978, forcing the shah to flee the country and abdicate his throne. After hospitalization in the United States and life in exile in Panama and Egypt, the shah died in Cairo in 1981 and is buried there (Figure 9.29). The Ayatollah Khomeini returned in triumph to Iran in 1979. His followers held 52 U.S. diplomatic personnel as hostages in Tehran for 444 days. Due partly to his inability to resolve this crisis, U.S. President Jimmy Carter was defeated in his 1980 reelection bid.

Under Khomeini (who died in 1989), the country became an Islamic republic governed by Shi'ite clerics, who included a handful of revered and powerful **ayatollahs** ("signs of God") at the head of the religious establishment, and an estimated 180,000 priests called **mullahs.** Khomeini served as the Guardian Theologian, an infallible supreme leader enshrined by the principle of **divine rule by clerics** (*velayet-e-faqih*). These religious leaders continue to supervise all as-

Figure 9.29 The late 1970s and early 1980s were turbulent times in the Middle East. The exiled shah of Iran was welcomed only in the United States and in a single sympathetic country in the region: Egypt. Here a line of dignitaries, besieged by crowds, is following the casket of the shah to its resting place in a Cairo mosque in July 1981. At the far right center, wearing a military uniform, is Egypt's President Anwar Sadat, who was assassinated by Islamist militants 3 months later.

pects of Iranian life and perform many functions allotted to civil servants in most countries. They base their authority on Shi'ite interpretations of the Islamic faith (about 89 percent of Iran's population is Shi'ite).

In contrast to their status during the shah's reign, when they were encouraged to have stronger public roles, women in postrevolutionary Iran have lost many freedoms. Under the new regime, "polluting" influences of Western thought and media were purged. However, recent years have seen a general softening of restrictions on personal freedom of expression, and the Internet and other media have introduced broader worldviews to Iran's youth, a very important cohort, as two-thirds of the population are under the age of 30. Young people led an unprecedented series of prodemocracy protests in 2002.

Generally, there is a tug of war between the moderate, democratically elected, and reform-minded President Muhammad Khatami, who favors more openness, and the less tolerant supreme religious leader, Ayatollah Ali Khamanei (Khomeini's successor), who has lifetime control over the military and the judicial branch of government. Reformists won a majority of parliamentary seats in 2000 but lost to conservatives in 2004 after ruling clerics purged their ranks.

Western analysts alternatively perceive Iran as either developing a more accommodating attitude toward the international community or exporting ever more dangerous notions of revolutionary Islam to currently pro-Western

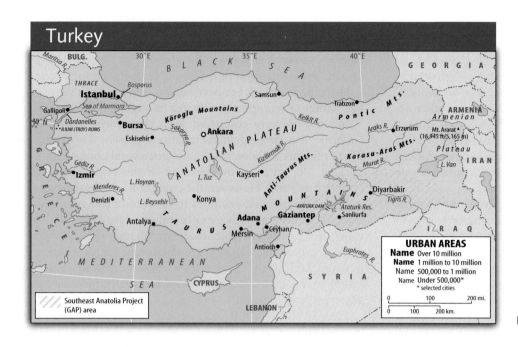

Figure 9.30 Principal features of Turkey

countries in the Middle East. Since 1998, relations have improved between Iran and the United States and between Iran and Great Britain. But there is a long way to go before the U.S. and Iran are friends again; Iran was one of the three countries (along with Iraq and North Korea) named by President Bush in 2002 as making up the **"axis of evil."** Washington is keeping an especially close eye on Iran's apparent nuclear weapons program.

9.7 Turkey: Where East Meets West

Like Iran, Turkey is a demographic heavyweight (population: 71.3 million people) and has long had an enormously important presence in the region (Figure 9.30). The Turks who organized the Ottoman Empire beginning in the 14th century originated as pastoral nomads from Central Asia, where Turkic cultures still prevail. From the 16th to the 19th century, their empire, based in what is now Turkey, was an important power. Modern Turkey was created from the wreckage of the old empire after World War I. Its founder, Mustafa Kemal Ataturk, was determined to Westernize the country, raise its standard of living, and make it a strong and respected national state. He inaugurated social and political reforms designed to break the hold of traditional Islam and open the way for modernization and Turkish nationalism.

Islam had been the state religion under the Ottoman Empire, but Ataturk separated church and state, and to this day, Turkey is the only Muslim Middle Eastern country to officially cleave them. Wearing the red cap called the *fez,* an important symbolic act under the Ottoman caliphs, was pro-

hibited, and state-supported secular schools replaced the religious schools that had monopolized education. To facilitate public education and remove further traces of Muslim culture, Latin characters replaced the Arabic script of the Qur'an. Slavery and polygamy were outlawed, and women were given full citizenship. Legal codes based on those of Western nations replaced Islamic law, and forms of democratic representative government were instituted, although Ataturk ruled in a dictatorial fashion.

Turkey has continued to have trouble establishing a fully democratic system, and there have been frequent periods of military rule (Figure 9.31). Turkey's constitution invests the

Figure 9.31 The military has a strong presence in secular Turkey. The soldiers are guarding an archeological site in eastern Turkey, a region of Kurdish rebellion against government rule.

military with the responsibility of protecting the country's secular government, and the army has used its powers. After decades of secularization, there was, for 2 years (1996–1997), a democratically elected government coalition comprised of avowedly secular politicians and Islamist leaders, including a prime minister who wanted to see Turkey reestablished on Islamic foundations. In 1998, however, Turkey banned government participation by the pro-Islamic Welfare (Refah) party that had put the prime minister in office. Backed by the army, the new government reaffirmed Ataturk's vow that Turkey remain secular. According to many observers, the regime's perception that Islamist tendencies threaten the state represents a setback for democracy and human rights in Turkey—there are even restrictions against women wearing Islamic-style head scarves—and raises the specter of violent dissent.

From a physical standpoint, Turkey is composed of two units: the Anatolian Plateau and associated mountains occupying the interior of the country and the coastal regions of hills, mountains, and small plains bordering the Black, Aegean, and Mediterranean Seas. Turkey in general has a physical presence quite unlike the popular image of the Middle East and North Africa as a desert region; it is in fact the only country in the region that has no desert, and cultivated land occupies about a third of Turkey's land area. The Anatolian Plateau, an area of wheat and barley fields and grazing lands, is bordered on the north by the lofty Pontic (Kuzey) Mountains and on the south by the Taurus Mountains. The plateau ranges from 2,000 to 6,000 feet (c. 600 to 1,800 m) and is highest in the east where it adjoins the mountains of the Armenian Knot. Its surface is rolling and windswept, hot and dry in summer and cold and snowy in winter, with a natural vegetation of short steppe grasses and shrubs. Production of cereals (especially barley, wheat, and corn) and livestock (especially sheep, goats, and cattle), employing both traditional and mechanized means, prevails in Anatolia (Figure 9.32). Coastal plains and valleys along and near the Aegean Sea, Sea of Marmara, and Black Sea are generally Turkey's most densely populated and productive areas. Here are grown cash crops such as hazelnuts, tobacco, grapes for sultana raisins, and figs. The slopes of the Pontic Mountains facing the Black Sea have copious rains in both summer and winter and support prolific tea harvests (see Figure 8.9).

Relative poverty compared with the nearby European countries is a striking characteristic of present-day Turkey. There remains a strong rural and agricultural component in Turkish life, with about one-third of the population still classified as rural. However, the country has embarked on a course of change that promises to modernize its agriculture, expand industry, and raise its general standard of living. The value of output from manufacturing (mainly textiles, agricultural processing, cement manufacturing, simple metal industries, and assembly of vehicles from imported components) is already greater than that from agriculture. Turkey does produce some oil.

Early in the 1990s, Turkey began an impressive $32 billion agricultural effort called the **Southeast Anatolia Project** (known by its Turkish acronym as **GAP**). Its aim is to convert the semiarid southeast quarter of the country into the "breadbasket of the Middle East." Surpluses of cotton, wheat, and vegetables are to be exported, along with value-added manufactures such as cotton garments and textiles. The centerpiece of this massive irrigation project is the Ataturk Dam on the Euphrates River (Figure 9.33). About 20 other dams on the Turkish Euphrates and Tigris are also part of the ongoing project to provide hydroelectricity as well as water. The GAP is controversial. It is inundating historic settlements and priceless archeological treasures and is compelling people to move. It is being developed in the part of the country where a restive Kurdish population is seeking recognition, autonomy, and among some factions, independence from Turkey. In addition, Turkey's Tigris and Euphrates waters flow downstream into neighboring states,

Figure 9.32 The wheat harvest on the Anatolian plateau in east central Turkey

Figure 9.33 Where there had been a bridge across the Euphrates River, there is now a ferry across the Ataturk Reservoir, upstream from the huge Ataturk Dam

raising serious questions about downstream water allocation and quality.

224 Turkey is an "in-between" country. Economically, it is well below the level of MDCs but above most of the world's LDCs. Culturally, it is between traditional Islamic and secular European ways of living. Part of Turkey—the section called Thrace—is actually in Europe, and Turkey as a whole aspires to become more European. It has been a member of the North Atlantic Treaty Organization (NATO) since 1952. It is an associate member of the European Union and is a candidate for full membership. The European Union, however, would first like to see Turkey become more European politically, for example, by diminishing the army's influence in government and by improving its human rights record. Turkey insists it has made huge strides toward satisfying the minimal democratic and human rights norms for EU member countries. It has abolished the death penalty and eased restrictions it imposed on the cultural expressions of its minority Kurdish population.

But at expansion talks in 2002, EU leaders rejected Turkey's bid and would not promise a date for membership negotiations. They argued that Turkey has still not gone far enough with democratic and human rights reforms. They probably feared the costs of absorbing Turkey's mainly low-income population into the union. Many Turks feel that their membership in the European Union was denied because they are a Muslim people. It is possible that the main reason Turkey was rejected was because EU leaders feared that Islamic terrorists would make their way from an EU Turkey into the heart of continental Europe. The EU's rejection of Turkey may fan the growing flames of sentiment in the Islamic world that the "crusader West" is waging a systematic campaign against Muslims.

Istanbul (population: 12.5 million), formerly Constantinople, is Turkey's main metropolis, industrial center, and port. One of the world's most historic and cosmopolitan cities, Istanbul was for many centuries the capital of the Eastern Roman (Byzantine) Empire. It became the capital of the Ottoman Empire when it fell to the Turks in 1453. However, the capital of the Turkish Republic was established in 1923 at the more centrally located and more purely Turkish city of Ankara (population: 3.6 million) on the Anatolian Plateau. Turkey's third largest city is the seaport of Izmir on the Aegean Sea.

Istanbul is located at the southern entrance to the Bosporus Strait, the northernmost of the Turkish Straits that connect the Mediterranean and Black Seas (Figure 9.34). The straits have long been a focus of contention between Turkey and Russia, who fought several wars in the past; their relations are much better now. Turkey's military posture has also been shaped by fragile relations with its NATO neighbor, Greece. The main questions recently at issue have been the status of Cyprus and conflicting claims to hundreds of islands in the Aegean Sea and that sea's oil and gas reserves.

Figure 9.34 Looking north along the Bosporus from the historic heart of Istanbul, with Europe on the left and Asia on the right

Already an EU member, Greece pressured Turkey to revise its policy toward Cyprus and to settle the Aegean Sea dispute as preconditions for Turkey joining the organization.

9.8 Rugged, Strategic, Devastated Afghanistan

High and rugged mountains dominate Afghanistan (population: 28.5 million), the only landlocked nation of the Middle East and North Africa (Figure 9.35). Historically, it has occupied an important strategic location between India and the Middle East. Major caravan routes crossed it, and a string of empire-builders sought control of its passes. Today, it has limited resources, poor internal transportation, little foreign trade, and much conflict; Afghanistan is one of the poorest of the world's LDCs.

Most Afghans live in irrigated valleys around the fringes of the high mountains that occupy a large part of the country. The country's second most populous area is the northern side of the central mountains, where people live in oases in foothills and steppes. Northern Afghanistan borders three of the five Central Asian countries, and millions of people on the Afghan side are related to peoples of those countries. The most heavily populated section is the northeast, particularly the fertile valley of the Kabul River, where the capital and largest city, Kabul (population: 2.3 million; elevation: 6,200 ft/1,890 m), is located. Most of the inhabitants of the southeast are Pashtuns (also known as Pushtuns or Pathans). Their language, Pashto, is related to Persian. The Pashtuns are the largest and most influential of the numerous ethnic groups that make up the Afghan state; the second and third largest are the Tajik and Hazara, respectively. The independent-minded tribal Pashtun people have always been loath to recognize the authority of central governments. In

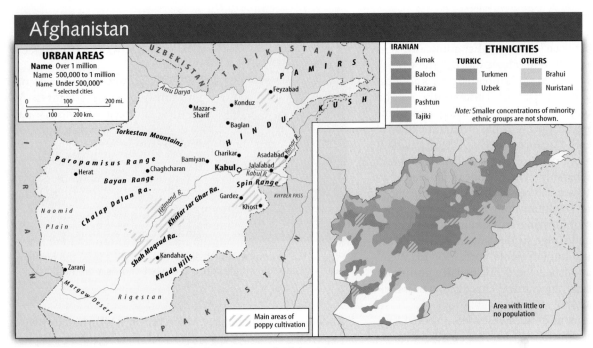

Figure 9.35 Principal features and ethnic groups of Afghanistan

the days of Great Britain's Indian Empire, the area saw warfare among tribes, tribal raids on British-controlled areas, and British punitive expeditions against the tribes. In those days, Peshawar—on the Indian (now Pakistani) side of the Khyber Pass into Afghanistan—became the hotbed for British, Russian, and other agents playing what came to be known as the **Great Game** of vying (and spying) for strategic influence in this part of the world.

Afghanistan is overwhelmingly a rural agricultural and pastoral country. It is so mountainous and arid that only about 12 percent is cultivated. Enough rain falls in the main populated areas during the winter for dry farming of grains. A variety of cultivated crops, fruits, and nuts are important locally. Raising livestock on a seminomadic and nomadic basis is widespread. In general, Afghanistan's agriculture bears many of the customary Middle Eastern earmarks: traditional methods, simple tools, limited fertilizer, and low yields. The most successful crop has been the opium poppy. A succession of rebel armies and then Afghanistan's ruling Taliban used revenues from opium and heroin to obtain their arms. Afghanistan was by far the world's largest opium producer when, in 2001, the Taliban imposed a ban on cultivation. There was virtually no opium harvest that year. After the United States ousted the Taliban, Afghanistan regained its former status, producing fully three-fourths of the world's opium in 2004.

Through most of the 20th century, this highland country was remote from the main currents of world affairs. After the Islamic revolution in Iran in 1979, however, Afghanistan's location next to that oil-rich country made it once again the target of foreign interests. The Soviet military intervention of 1979 and the ensuing devastation catapulted the country into world prominence. The USSR's motives for its invasion of Afghanistan may have included a desire to prevent by force the spread of Iranian-style Islamic fundamentalism into Afghanistan (which is 80 percent Sunni Muslim and 19 percent Shi'ite Muslim) or into the Central Asian Soviet republics deemed vital to the superpower's security. For its part, the United States warned the Soviet Union that it would not tolerate further Soviet expansionism.

In the ensuing war, there was widespread killing and maiming of civilians, destruction of villages, burning of crops, killing of livestock, destruction of irrigation systems, and sowing of land mines over vast areas (see Definitions and Insights, page 266). Soviet ground and air forces caused several million Afghan refugees to flee into neighboring Pakistan and Iran. Arms from various foreign sources filtered into the hands of the *mujahadiin,* the anti-Soviet rebel bands that kept up resistance in the face of heavy odds (Figure 9.36). The United States was one of the powers supporting the rebels and, in that sense, waged a proxy war against the Soviet Union in Afghanistan. In its waning years, the USSR recognized it could not win its "Vietnam War," and its troops withdrew from Afghanistan by 1989. It is estimated that, of the 15.5 million people who lived in Afghanistan when it was invaded in 1979, 1 million died, 2 million were displaced from their homes to other places within the country, and 6 million fled as refugees into Pakistan and Iran. By 2004, about 40 percent of those refugees had returned.

Support lent to the *mujahadiin* and sympathetic Arab

222

Joe Hobbs

Figure 9.36 When this photo was taken in 1987, these Afghan *mujahadiin* fighters were regularly involved in firefights with Soviet troops in Afghanistan. The *mujahadiin* movement and its success against the Soviets laid the groundwork for al-Qa'ida's war against the West.

fighters by the United States and moderate Arab states such as Egypt and Saudi Arabia soon came back to haunt those countries in a phenomenon that came to be known as **blowback**. Emboldened by their victory against the Soviets, the most militant Islamists among the fighters turned their attention to the United States and its Middle Eastern allies. Chief among them was Osama bin Laden, who developed "the base" (*al-Qa'ida* in Arabic) of thousands of Afghans, Arabs, and other anti-Soviet war veterans he could call on to wage a wider *jihad*. Bin Laden's organization trained an estimated 10,000 fighters in al-Qa'ida camps in Afghanistan, and from this unlikely, remote setting devised spectacular acts of terrorism: an assassination attempt on Egyptian President Hosni Mubarak, the bombing of U.S. military barracks in Saudi Arabia, the bombing of U.S. embassies in Kenya and Tanzania, the attack on the U.S.S. *Cole*, and the 9/11 attacks. The East African bombings in 1998 led within days to a cruise missile strike by the United States on six al-Qa'ida camps near Khost, Afghanistan. Many al-Qa'ida personnel were killed, but not its top leaders.

The *mujahadiin* succeeded in overthrowing the Communist government of Afghanistan in 1992, but after that, rival factions among the formerly united rebels engaged in civil warfare. During this period, most other countries utterly neglected Afghanistan, inadvertently promoting the growth of militant movements (which tend to thrive in remote, underdeveloped, and ungoverned regions). By 1996, one of the rebel factions, the **Taliban** (backed by Saudi Arabia and Pakistan), gained control of most of the country, including the capital of Kabul. Proclaiming itself the sole legitimate government of Afghanistan, the Taliban imposed a strict code of Islamic law in the regions under its control and gained international notoriety for its austere administration. The Taliban

removed almost all women from the country's workforce, forbade public education of girls, and outlawed "un-Islamic" practices such as dancing, flying kites, watching television, keeping birds, and trimming beards.

The Taliban continued to make advances against its opponents inside Afghanistan (particularly the **Northern Alliance**, whose leader, Ahmed Shah Massoud, was assassinated just days before and in apparent preparation for the 9/11 attacks) and, by 2001, controlled 95 percent of the country's territory. The neighboring Central Asian countries, Russia, and even Iran grew increasingly fearful of the spread of the Taliban's extreme interpretation of Islam into their nations. Russia ended up supporting some of the rebels it had fought because they were fighting the Taliban. Russia's once unlikely ally in supporting those rebels was the United States, which successfully lobbied the United Nations to levy economic sanctions against Afghanistan in 1999. The United States hoped that economic losses would pressure the Taliban into turning over bin Laden for prosecution and also announced a $5 million reward for information that would lead to bin Laden's capture or death.

The bin Laden bounty rose into the tens of millions of dollars after September 11, 2001. Named as the mastermind of the attacks against New York and Washington, bin Laden became the "world's most wanted" as U.S. President George W. Bush evoked a Wild West vow to have him captured dead or alive. In crafting its pronounced war against terrorism (which it promised to carry anywhere in the world necessary), the U.S. administration identified the Taliban as al-Qa'ida's mentor and targeted both organizations for elimination. Within a month of the attacks on the United States, American warplanes and special forces were operating against Taliban and al-Qa'ida facilities and personnel around Afghanistan. The British military assisted in the air campaign, and most of the ground fighting was carried on by the Northern Alliance, the Taliban's rival and nemesis. The United States successfully persuaded Pakistan to drop its support of the Taliban and join the effort, and much of the military campaign was based on U.S.–Pakistani cooperation. One after another of the Taliban's urban strongholds fell, including Kabul and the Taliban spiritual capital, Kandahar. Withering assaults from U.S. warplanes crushed the caves used by al-Qa'ida guerrillas, and many al-Qa'ida prisoners were taken to the U.S. naval base in Guantánamo Bay, Cuba. The elusive bin Laden, however, along with the Taliban leader Mullah Omar, slipped away, perhaps seeking safe harbor in the ethnic Pashtun region of western Pakistan.

The reconstruction of Afghanistan began almost as soon as the Taliban were driven from power. Its progress is slow because the country is so devastated and because of the tenuous security situation. American forces and the Afghan government headed by Hamid Karzai effectively took control of only a small area around Kabul, and much of the country remains in the hands of warlords whose allegiance is up for

Definitions + Insights

Land Mines

The scars of war will last a long time in Afghanistan. Now that the country's reconstruction has begun, one of the urgent tasks is to rid the landscape of millions of land mines sown during decades of war. These weapons have already killed or maimed more than 200,000 Afghans since 1979. They continue to kill or maim between 150 to 300 Afghans, the great majority of them civilians, every month.

Land mines are antitank and antipersonnel explosives, usually hidden in the ground, designed to kill or severely wound the enemy. There are many kinds: Soviet troops planted at least nine varieties in Afghanistan, including seismic mines, which are triggered by vibrations created by passing horses or people; mines with devices that pop up and explode when a person approaches; "butterfly" mines meant to maim rather than kill; and explosives disguised as toys, cigarette packs, and pens (Figure 9.B).

There may be as many as 10 million land mines in Afghanistan, or 40 per square mile (15/sq km). Reconstruction of this agricultural country cannot begin in earnest until the mines are cleared. The obstacles are enormous in rugged Afghanistan. Mines typically remain active for decades after they are planted and tend to last longer in arid places. The mountainous terrain prevents systematic clearance of the explosives. Many mines were scattered indiscriminately from the air and will be difficult to locate. Heavy rains move the mines downhill and away from known locations. And it is expensive to clear mines; Kuwait spent $800 million to rid its landscape of them after the Gulf War. So far, Western countries have donated more than $40 million toward the expense of clearing Afghanistan's land mines; Russia has contributed nothing. Land mines

are a global problem. Up to 70 million mines contaminate the land in 64 countries, and they kill or maim more than 25,000 civilians each year. The numbers and concentrations in some countries are staggering. The leader in numbers is Egypt, with 22 million mines; in density of mines, Bosnia and Herzegovina leads with 34 per square mile (13/sq km). The United Nations estimates that with current technology, it would take $33 billion and 1,100 years to clear the world's land mines.

Several developments may portend an eventual end to the land mine scourge. The world's most visible critic of land mines, Princess Diana of Wales, was killed in an automobile accident in Paris in 1997. In her wake, there were renewed calls for a moratorium on the manufacture and use of antipersonnel land mines. Two months after Diana's death, anti-land mine activist Jody Williams won the Nobel Peace Prize. Soon after, representatives from 125 nations met in Ottawa, Canada, to sign a treaty strictly banning the use, production, storage, and transfer of antipersonnel land mines. Many of the largest producers, users, and exporters of these explosives, including the United States, Russia, China, India, Pakistan, Iran, Iraq, and Israel, refused to sign. The United States argued that land mines were necessary to protect South Korea against an invasion from North Korea and failed to win an exemption on this count that would have allowed it to sign the treaty.

Too many countries regard these as legitimate weapons to make the prospect of a mine-free world likely. Costing as little as $2 each, they are very affordable weapons for cash-strapped warring LDCs (and so have been dubbed the "Saturday-night specials of warfare"). For now, these hidden horrors continue to be planted and to be cleared, as one demining specialist put it, "one arm and one leg at a time."

Joe Hobbs

Figure 9.B Inexpensive to manufacture and deploy, land mines are a scourge that outlast conflict and bring suffering to civilians.

grabs. Al-Qa'ida and Taliban fighters are still at large in the rugged countryside, despite ongoing efforts by U.S. troops to eliminate them.

Progress is being made in rebuilding the country's infrastructure. From the billions of dollars in aid pledged, a new highway has been built connecting Kabul with Kandahar. Schools and hospitals are being constructed to assist a population with almost no education and with infant and maternal mortality rates that are among the highest in the

world. There are plans to reforest landscapes that have been almost completely deforested. Afghanistan is not without resources of its own. Further prospecting may reveal that it has the world's largest copper deposit, and it is known to have large stores of high-grade iron ore, along with some gas, oil, coal, and precious stones. If peace can prevail, these might be exploited to help rebuild this shattered land and bring hope to a country that is all too symptomatic of the troubled Middle East and North Africa.

SUMMARY

- The United Nations Partition Plan of 1947 attempted a two-state solution to the dilemma Britain had created by promising land to both Arabs and Jews in Palestine. However, it envisioned geographically fragmented states, making each side feel vulnerable to the other. War prevented the plan's implementation. The Arab–Israeli conflict has continued from that time through numerous wars, crises, and terrorist actions, with Israel gaining more territory and the indigenous Palestinians failing to acquire a country of their own. Principal obstacles to peace between Israelis and Palestinians are the status of Jerusalem, the potential return of Palestinian refugees, and the future of Jewish settlements in the Occupied Territories.

- Among Israel's neighbors, Jordan has a majority Palestinian population. Lebanon is still recovering from a devastating civil war that was sparked when minority Christians failed to yield power to majority Muslims and was protracted by an Israeli invasion to crush the Palestine Liberation Organization. Syria is keeping its distance from the wider Arab–Israeli peace process.

- Egypt is a very populous country with a limited inhabitable area along the Nile. It has transformed the Nile Valley with thousands of years of habitation and recent large engineering works, but it remains a poor country dependent on outside assistance.

- Sudan is a transition zone between the Middle East and Africa south of the Sahara. Its Muslim Arab government has tried to impose its will forcefully on the non-Arab South, promoting civil war. The conflict shows signs of ending, but Sudanese Arabs have opened a new offensive on non-Arabs in the Darfur region of western Sudan.

- Most of Libya lies in the Sahara, and its primary natural resource is oil. The revenues generated from the exported oil have been used in part to develop a massive irrigation project using ancient underground waters. Colonel Qaddafi has given up support for international terrorism and weapons of mass destruction programs, leading to much improved relations with the West.

- The Maghreb ("the western land") is composed of Morocco, Algeria, Tunisia, and the disputed Western Sahara. Morocco is the region's poorest country, Algeria has the greatest mineral wealth, and Tunisia is the most progressive. Morocco controls Western Sahara, whose Saharawi people want independence.

- Two-thirds of the world's proven petroleum reserves are concentrated in a few countries that ring the Persian/Arabian Gulf. Saudi Arabia controls more than one-quarter of the world's oil.

The Gulf countries have earned enormous wealth from this resource, but their growing populations and the prospect of the oil running out present some critical challenges. Some of the countries, notably Saudi Arabia, are very traditional, whereas the United Arab Emirates and Bahrain have tolerated considerable social change. On the southern end of the Arabian Peninsula, Oman has enjoyed oil wealth, but Yemen remains poor.

- Under Saddam Hussein, Iraq squandered its oil wealth on military misadventures, including invasions of Iran and Kuwait. The Kuwait invasion led to an enormous U.S.-led counterattack that devastated the country's infrastructure and subjected it to years of economic sanctions. The U.S.-led invasion of Iraq in 2003 resulted in Saddam Hussein's downfall and an uncertain future for this ethnically diverse country (with large Sh'ite Arab, Sunni Arab, and Kurdish populations).

- Iran's oil revenues were used to modernize and Westernize the nation during the Pahlavi dynasty. Rapid social change and uneven economic benefits precipitated a revolutionary movement, which forced the shah to abdicate and flee. The new theocratic state, since the death of the Ayatollah Khomeini, is alternatively perceived to be either developing a more accommodating attitude toward the international community or exporting ever more dangerous notions of revolutionary Islam to the region. Iran's young population is agitating for change, but the ruling clerics are resisting.

- Turkey was formerly the seat of power for the Islamic Ottoman Empire. It is a secular nation with membership in NATO and hopes to join the European Union. It is blessed with fresh water resources that it has mustered with the hope of creating the "breadbasket of the Middle East."

- Afghanistan is one of the world's poorest nations. It is landlocked and has limited resources, poor internal transportation, and little foreign trade. Opium is its main export. Its historic location along the ancient caravan routes and adjacent to large oil reserves have made it the target of stronger powers. Backed by the United States, rebels succeeded in driving out a Soviet occupation force in the 1980s. That costly conflict was followed by a period of civil war during which outside countries neglected Afghanistan. The Islamist militant Taliban came to power and protected the presence and training of al-Qa'ida militants. Osama bin Laden and others planned the 9/11 and other attacks from Afghanistan and were driven from there in the wake of 9/11. Reconstruction of the country has begun.

KEY TERMS + CONCEPTS

Terms in blue are also defined in the glossary.

1948–1949 war (p. 234)
1967 war (Six Day War) (pp. 234, 235)
1973 war (p. 235)
1982 war (p. 243)

al-Aqsa Intifada (p. 239)
Alawite (p. 243)
Ashkenazi Jews (p. 236)
autonomy (p. 238)

"axis of evil" (p. 261)
ayatollah (p. 260)
barrage (p. 244)
basin irrigation (p. 244)

Battle of Beirut (p. 243)
blowback (p. 265)
Camp David Accords (p. 235)
Coalition Provisional Authority (CPA) (p. 258)
divine rule by clerics (p. 260)
Druze (p. 236)
ethnic cleansing (p. 247)
"facts on the ground" (p. 237)
Falashas (p. 236)
final status issues (p. 238)
foggara (qanat) (p. 249)
fossil waters (p. 248)
Gaza–Jericho Accord (p. 238)
ghurba (p. 237)
Great Game (p. 264)
Great Manmade River (p. 248)
internally displaced persons (IDPs) (p. 247)
Intifada (p. 237)
Iraqi Governing Council (p. 258)
Janjawiid (p. 247)
Jewish settlements (p. 237)

Judea and Samaria (p. 237)
kibbutz (pl. kibbutzim) (p. 240)
Kurdistan Workers Party (PKK) (p. 257)
Labor Party (p. 237)
land mines (p. 266)
Law of Return (p. 236)
Lebanese civil war (p. 242)
Likud Party (p. 237)
Ma'adan ("Marsh Arabs") (p. 257)
moshav (pl. moshavim) (p. 240)
mujahadiin (p. 264)
mullah (p. 260)
Northern Alliance (p. 265)
Occupied Territories (p. 236)
Operation Defensive Shield (p. 239)
Operation Iraqi Freedom (p. 257)
Oslo I and II Accords (p. 238)
Palestine Liberation Organization (PLO) (p. 238)
Palestinian Authority (PA) (p. 238)
Palestinians (p. 233)
perennial irrigation (p. 244)

Polisario Front (p. 250)
pre-1967 borders (p. 234)
remittances (p. 245)
"road map for peace" (p. 240)
right of return (p. 234)
Saharawis (p. 250)
salinization (p. 245)
"Saudization" (p. 251)
security fence (p. 240)
Sephardic Jews (p. 236)
Southeast Anatolia Project (GAP) (p. 262)
Sudanese People's Liberation Army (SPLA) (p. 246)
"Switzerland of the Middle East" (p. 242)
Taliban (p. 265)
"the gift of the Nile" (p. 244)
two-state solution (p. 234)
United Nations Resolutions 242 and 338 (p. 237)
Wahhabism (p. 252)
Wye Agreement (p. 238)

REVIEW QUESTIONS

WORLD REGIONAL Geography ⊛ Now™

Assess your understanding of this chapter's topics with additional quizzing and concept-based problems at http://earthscience.brookscole.com/wrg5e.

1. What precipitated the respective conflicts between Israel and its Arab neighbors? How did each of these conflicts rearrange the political map?

2. What lands did the peace process culminating in the Wye Agreement yield to Palestinian control? What is the current situation in those areas?

3. What are the major ethnic groups and territories of Lebanon?

4. How did Egypt transform the Nile Valley from a system of basin irrigation to one of perennial irrigation?

5. In what ways is Sudan's North different from its South?

6. What is the role of the Great Manmade River?

7. What are the main resources of the Maghreb countries? What issues do they have with countries outside the region?

8. Which Gulf countries have the most oil and natural gas? Have Oman and Yemen benefited from the oil boom?

9. What happened when Saddam Hussein invaded Kuwait? What happened when he invaded Iran?

10. What is the ethnic composition of Iraq? How did that consideration cause the United States to avoid invading the country in 1991?

11. Who and where are the Kurds?

12. What were some of the grievances against the Pahlavi dynasty that led to revolution in Iran?

13. In what ways is Turkey an "in-between" country?

14. What were Afghanistan's shifting fortunes between 1979 and the present?

DISCUSSION QUESTIONS

1. Discuss the claims of Palestinians and Israeli Jews to the lands of Israel and the Occupied Territories. How might both groups be peacefully accommodated? What faults, if any, existed in the failed Camp David plan of 2000?

2. Why did the Lebanese civil war happen? What caused the 1982 war in Lebanon? How is the country faring now?

3. What are some of the positive and negative outcomes of the construction of Egypt's Aswan High Dam?

4. How did differences between Sudan's North and South contribute to civil conflict?

5. Why did Libya's leader recently adopt a much more accommodating attitude toward the West?

6. Why did Algeria experience a costly civil war?

7. Who controls the Western Sahara? What is the fate of its Saharawi population?

8. Is the United States too dependent on Middle Eastern oil? What blessings and curses did oil bring the people who own it? What is the future of the oil-rich countries of the Gulf? How have the respective countries dealt with issues of women's rights, democracy, and relations with the wider world?

9. Recount the circumstances and events of the Gulf War of 1990–1991. How did this war differ from the 2003 U.S.-led war on Iraq? Why was the 2003 war fought? What have been its outcomes?

10. Why does Turkey want to become a member of the European Union? Why has it been rejected so far?

11. What are some of the current tensions within Iran and between Iran and the West?

12. What made Afghanistan an inviting base for al-Qa'ida? What is blowback?

13. Should the United States sign the treaty banning antipersonnel land mines?

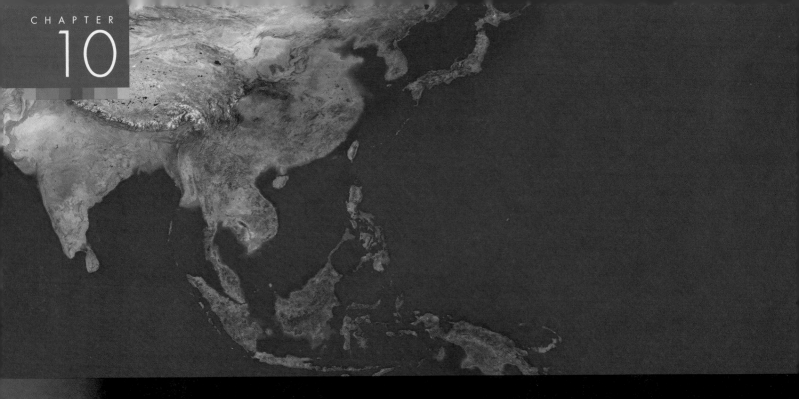

A Geographic Profile of Monsoon Asia

Sherpa farmers are able to coax a few crops out of high Himalayan fields near Mount Everest in Nepal.

chapter objectives

This chapter should enable you to:

- Recognize the role of seasonal monsoon winds and rains in livelihoods and perceptions

- Appreciate China and India as the demographic giants and surging economies of early 21st century Asia

- Learn how unique spatial and spiritual considerations have influenced the layout of Asian settlements

- Know the basic beliefs of Hinduism, Buddhism, Confucianism, and Daoism

- Become familiar with the pros and cons of the Green Revolution

- Understand the geopolitical dimensions of tension between India and Pakistan, North Korea and the West, and Islamists and governments in Pakistan and Indonesia

WORLD
REGIONAL
Geography ⊛ Now™

Look for this logo in the text and go to GeographyNow at http://earthscience.brookscole.com/wrg5e to explore interactive maps, view animations, sharpen your factual knowledge and geographic literacy, and test your critical thinking and analytical skills with unique interactive resources.

Monsoon Asia is a great triangle that runs from Pakistan in the west to Japan in the northeast and New Guinea in the southeast (Figure 10.1). In this land area of less than 8 million square miles (20,761,000 sq km)—about twice the size of the United States, or less than one-quarter of the earth's land mass—lives more than half of the world's population (Figure 10.2). The cultural makeup of these 3.5 billion people is as diverse and expressive as the physical environments they inhabit. The world's highest mountain peaks, some of the longest rivers, and Earth's most highly transformed and densely settled river plains are the settings for some of the oldest civilizations and the most modern economies. Just as this region has played a major role in human development over many millennia, it is a major force in shaping world affairs in the 21st century.

The Orient was the term used traditionally to refer to the countries of east Eurasia. The term *orient* comes from the Latin meaning "to rise" or "to face the east," while *occident* means "to set." The Europeans, who saw themselves in the center of the world, saw the sun rising from eastern Asia and called that eastern world "the Orient." As a term laden with Western stereotypes and misconceptions, the Orient has since fallen out of favor. This text uses the term *Monsoon Asia* for this region because it connotes a major physical attribute that affects most of this region and in gratitude to the achievements of geographer George Cressey who coined the term and wrote so well of the region.

10.1   Area and Population

Monsoon Asia encompasses the following regions: East Asia, which includes Japan, North and South Korea, China, Mongolia, Taiwan, and many nearshore islands; South Asia, which includes Pakistan, India, Sri Lanka, Bangladesh, the mountain nations of Bhutan and Nepal, and the island country of the Maldives; Southeast Asia, which includes both the peninsula jutting out from the southeast corner of the Asian continent with its countries of Myanmar (Burma), Thailand, Laos, Cambodia, Vietnam, Malaysia, and Singapore, and the island world that rings this peninsula, which includes the countries of Indonesia, the Philippines, Brunei, and East Timor (Table 10.1; see also Figure 10.1).

Perhaps the most critical fact of Monsoon Asia's geography is that this region is home to about 54 percent of the world's population (Figures 10.3a and b). Two countries alone, India and China, together have 2.4 billion people, or 37 percent of the world total. Their extraordinary demographic weight reflects thousands of years of human occupation of productive agricultural landscapes, combined with 20th-century advances in health technology that reduced death rates. The highest population densities correspond with the areas of most abundant precipitation or surface water, combined with the most productive soils; rivers and coastal plains are especially densely settled. Rugged mountains, almost waterless deserts, and until recently, tropical

Political Geography of Monsoon Asia

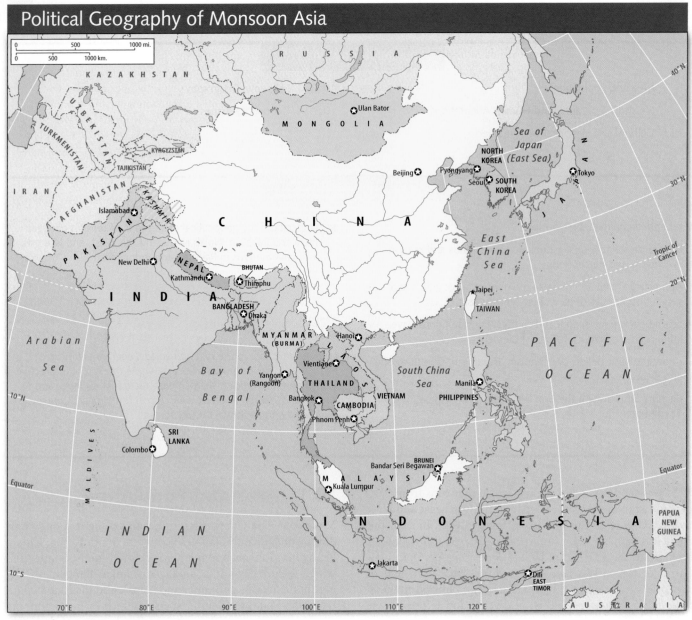

Figure 10.1 Monsoon Asia

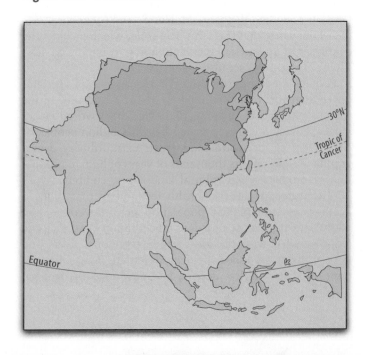

Figure 10.2 Monsoon Asia compared in latitude and area with the conterminous United States

Table 10.1   Monsoon Asia: Basic Data

Political Unit	Area (thousand/ sq mi)	Area (thousand/ sq km)	Estimated Population (millions)	Annual Rate of Natural Increase (%)	Estimated Population Density (sq mi)	Estimated Population Density (sq km)	Human Develop- ment Index	Urban Popula- tion (%)	Arable Land (%)	Per Capita GDP PPP ($U.S.)
South Asia										
Bangladesh	55.6	144.0	141.3	2.1	2541	981	0.509	23	62	1900
Bhutan	18.1	46.9	1	2.5	55	21	0.536	21	3	1300
India	1269.3	3287.5	1086.6	1.7	856	331	0.595	28	54	2900
Maldives	0.1	0.3	0.3	1.4	3000	1158	0.752	27	13	3900
Nepal	56.8	147.1	24.7	2.3	435	168	0.504	14	21	1400
Pakistan	307.4	796.2	159.2	2.4	518	200	0.497	34	28	2000
Sri Lanka	25.3	65.5	19.6	1.3	775	299	0.740	30	14	3700
Total	**1732.6**	**4487.4**	**1432.7**	**1.8**	**827**	**319**	**0.576**	**28**	**47**	**2685**
Southeast Asia										
Brunei	2.2	5.7	0.4	1.9	182	70	0.867	74	1	18,600
Cambodia	69.9	181.0	13.1	2.2	187	72	0.568	16	21	1900
East Timor	5.7	14.8	0.8	1.3	140	54	0.436	8	4	500
Indonesia	735.4	1904.7	218.7	1.6	297	115	0.692	42	11	3200
Laos	91.4	236.7	5.8	2.3	63	25	0.534	19	3	1700
Malaysia	127.3	329.7	25.6	2.1	201	78	0.793	62	5	9000
Myanmar	261.2	676.5	50.1	1.4	192	74	0.551	28	15	1800
Philippines	115.8	299.9	83.7	2.0	723	279	0.753	48	19	4600
Singapore	0.2	0.5	4.2	0.6	21,000	8108	0.902	100	1	23,700
Thailand	198.1	513.1	63.8	0.8	322	124	0.768	31	29	7400
Vietnam	128.1	331.8	81.5	1.2	636	246	0.691	25	20	2500
Total	**1735.3**	**4494.4**	**547.7**	**1.5**	**315**	**122**	**0.698**	**38**	**14**	**4059**
East Asia										
China	3696.1	9572.9	1300.1	0.6	352	136	0.745	41	15	5000
Japan	145.9	377.9	127.6	0.1	875	338	0.938	78	12	28,200
Korea, North	46.5	120.4	22.8	0.7	490	189	N/A	60	20	1300
Korea, South	38.3	99.2	48.2	0.5	1258	486	0.888	80	17	17,800
Mongolia	604.8	1566.4	2.5	1.2	4	2	0.668	57	1	1800
Taiwan	14	36.3	22.6	0.4	1614	623	N/A	78	24	23,400
Total	**4545.6**	**11773.1**	**1523.8**	**0.5**	**335**	**129**	**0.766**	**46**	**13**	**7560**
Summary Total	**8013.5**	**20755.0**	**3504.2**	**1.2**	**437**	**169**	**0.677**	**37**	**20**	**5019**

Sources: World Population Data Sheet, Population Reference Bureau, 2004; Human Development Report, United Nations, 2004; World Factbook, CIA, 2004.

rain forest vegetation have limited populations elsewhere. Some of the world's highest urban densities are found in the cities of Hong Kong, Macao, and Singapore, but in Mongolia and Nepal, population densities are extremely low. Singapore is one of the few nations to claim a 100 percent urban population (along with Kuwait and Monaco), while 92 percent of East Timor's population is rural. Only 3 percent of Bhutan is arable, but farmers in India and Bangladesh cultivate more than 50 percent of their lands.

It is not possible to generalize about population growth in Monsoon Asia. Growth rates range just 0.1 percent per year in Japan to a very high 2.5 percent per year in Bhutan. The

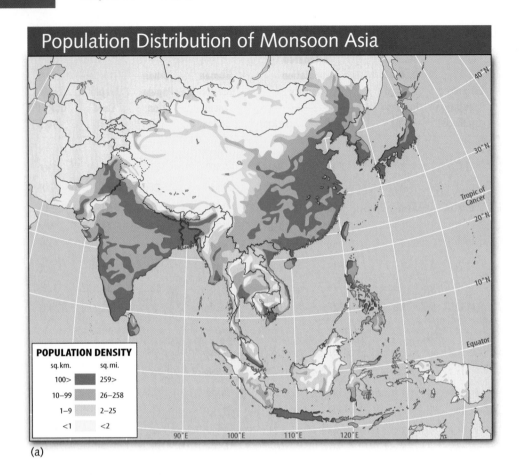

(a)

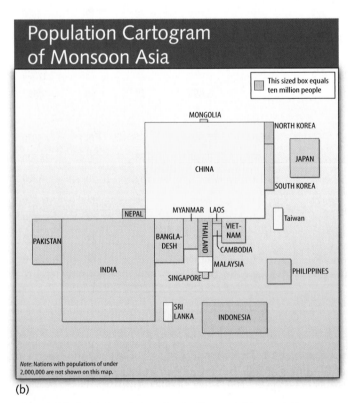

(b)

Figure 10.3 (a) Population distribution and (b) population cartogram of Monsoon Asia

region does reflect the general worldwide trends of poorer countries having higher growth rates and wealthier countries having lower growth rates. The region is demographically and otherwise recognizable as one of mainly less developed countries (LDCs), with a few very noticeable exceptions. Japan may be seen as a classic "postindustrial" country and, like Italy, must worry about a declining, aging population in which young people will have to shoulder an increasing economic burden to support their elders.

One of the outstanding anomalies in Asia's population equation is China. With a per capita GNP PPP of just $5,000, China might be expected to have a rather high population growth rate like India, which has a similar economic ranking. But in the 20th century, China's population was so alarmingly large and rapidly growing that its Communist leaders decided to reduce its growth drastically. China adopted a coercive "one child policy" that brought its growth rate to just 0.6 percent per year, representing a doubling time of 117 years (see Geographic Spotlight, pages 276–277). That is a remarkable turnabout from the situation in 1964, when China had a growth rate of 3.2 percent and a doubling time of 22 years.

To reduce its poverty, India has made great strides in reducing its population growth, but not in China's coercive fashion. With its growth rate of 1.7 percent per year, India is projected to overtake China as the world's most populous country in 2040, when it will have an estimated 1.5 billion people. A major wild card in India's population deck is HIV/AIDS. The scale of the problem there is not certain; 2004 estimates put the number of Indians infected with HIV at between 4 and 12 million, with the U.S. Central Intelligence Agency projecting a growth of up to 25 million cases by 2010. The potential for a pandemic in the world's second largest country is strong, and much will depend on how India's government chooses to fight the disease. About 80 percent of India's HIV infections are spread by heterosexual contact, and public health authorities have so far been shy to mount a strong campaign about sexual ethics and practices. China is thus far not in the peril that India is; 2004 estimates put the number of people infected at just below 1 million, and plans were announced to provide free HIV tests for all Chinese citizens.

10.2 Physical Geography and Human Adaptations

Monsoon Asia's physical geography may be described broadly as consisting of three concentric arcs, or crescents, of land: an inner western arc of high mountains, plateaus, and basins; a middle arc of lower mountains, hill lands, river plains, and basins; and an outer eastern arc of islands and seas (Figure 10.4).

The inner arc includes the world's highest mountain ranges, interspersed with plateaus and basins. In the south, the great wall of the Himalaya, Karakoram, and Hindu Kush Mountains overlooks the north of the Indian subcontinent. At the north, the Altai, Tien Shan, Pamir, and other towering

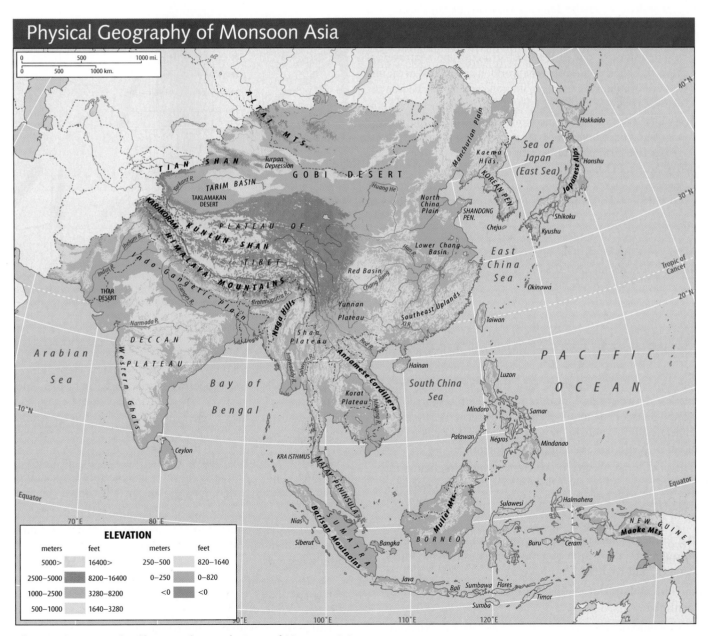

Figure 10.4 Major landforms and water features of Monsoon Asia

Geographic Spotlight

China's One Child Policy

China's population of 1.3 billion represents more than one-fifth of the world's people and is increasing by 7.8 million per year. Population is a very serious matter for a less developed country whose area of about 3.7 million square miles (9.4 million sq km) is only slightly larger than that of the United States, but whose inhabitants outnumber the U.S. population by about five to one. But a glimpse of China's age-structure diagram, with its wide middle ages but tapering youth, reveals that China has had one of the global success stories in population management (Figure 10.A).

The drive for smaller families was motivated by the fact that populations were surging while there was only modest growth in China's per capita food output. In 1976, China's Communist regime instituted one of the most stringent and controversial programs of birth control in the world. It culminated in the **one child campaign,** which aims to limit the number of children per married couple to one (Figure 10.B). The government tracks individual family birthing patterns—especially in the cities—and maintains surveillance through local authorities. It dispenses free birth control devices, free sterilization operations, free hospital care in delivery, free medical care for the child, free education for the child, an extra month's salary each year for parents who comply, and other favors and preferences to induce compliance with the one child per family norm. Those

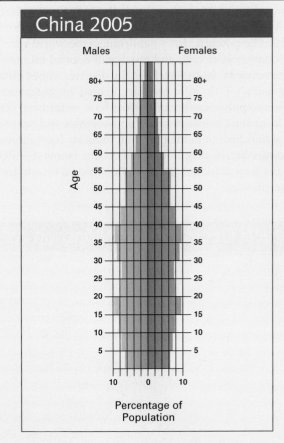

Active Figure 10.A The age-structure diagram of China reveals the country's success in slowing population growth. *See an animation based on this figure, and take a short quiz on the facts or concepts.*

---

ranges separate this region of Asia from the Central Asian countries. Between these mountain walls lie the sparsely inhabited Tibetan Plateau, at over 15,000 feet (c. 4,500 m) in average elevation, and the dry, thinly populated basins and plateaus of Xinjiang (Sinkiang) and Mongolia.

The middle arc, between the western inner highland and the sea, is occupied by river floodplains and deltas bordered and separated by hills and low mountains. The major features of this area are

the immense alluvial plain of northern India, built up through ages of meandering and deposition by the Indus, Ganges, and Brahmaputra Rivers;

the hilly uplands of peninsular India, geologically an ancient plateau;

the plains of the Irrawaddy, Chao Praya (Menam), Mekong, and Red Rivers in the Indochinese Peninsula

of Southeast Asia, together with bordering hills and mountains;

the uplands and densely settled small alluvial plains of southern China;

the broad alluvial plains along the middle and lower Chang Jiang (Yangtze River) in central China and the mountain-girded Red Basin on the upper Chang Jiang;

the large delta plain of the Huang He (Yellow River) and its tributaries in northern China, backed by loess-covered hilly uplands;

the broad central plain of northeastern China (Manchuria), almost enclosed by mountains.

The outer arc is an offshore fringe of thousands of islands, mostly grouped in great archipelagoes (clusters of islands)

Figure 10.B The Chinese government has been promoting "one couple, one child" for about three decades in an effort to slow down the annual population increase. This poster pleads to couples: "For the sake of a prosperous today and a beautiful tomorrow, limit your family to one child."

bearing so that population will stabilize before reaching projections of 1.4 billion by 2010 or 1.5 billion by 2025. The higher number may be more likely because government officials report widespread disregard of the "one couple, one child" policy. Reasons for noncompliance vary. In some cases, parents feel wealthy enough to pay the fines of 5,000–10,000 yuan ($600–$1,200) per extra child. Many rural-dwellers think their extra children will escape detection by authorities, who are most effective in the cities. Anticipating their years as elders, some couples want the "safety net" of additional children—especially boys. If a young urban couple has just one child and it is a girl, it is likely that she will marry and become, by Chinese tradition, such a part of the husband's family that she may not be available to care for her parents as they get older. Since China has traditionally used the son's family as the accommodation of choice for senior citizens, either a major shift in social services and associated taxation will be necessary or a full generation of seniors will be neglected.

Some parents want to produce a son if their first child was a girl. Female feticide is increasingly associated with desires to have a son. Couples often insist on an ultrasound test to determine the baby's gender, and about 90 percent of girls detected this way are aborted. In some provinces where the practice is widespread, three boys are born for every two girls. As in India, this practice is beginning to have serious repercussions as shortages of eligible brides have developed. There are many accounts of rural women abducted and trafficked as virtual slaves to other parts of the country where they become brides.

who violate the norm are subjected to constant social and political pressure, denial of the privileges accorded one child families, pay cuts, and fines. A woman who becomes pregnant after having had one child is pressured to have a free state-supplied abortion, with a paid vacation provided.

China's government hopes to continue to decrease child-

bordering the mainland. On these islands, high interior mountains (including volcanoes) are flanked by coastal plains where most of the people live. Three major archipelagoes include most of the islands: the East Indies, the Philippines, and Japan. Sri Lanka, Taiwan, and Hainan are large, densely populated islands outside these archipelagoes.

Between the archipelagoes and the mainland lie the China Seas and, to the north, the Sea of Japan (known in Korea as the East Sea). At the southwest, the Indian peninsula projects southward between two immense arms of the Indian Ocean: the Bay of Bengal and the Arabian Sea.

WORLD
REGIONAL
Geography ⊛ Now™

Click Geography Literacy to see an animation of plate boundary collisions.

Climate and Vegetation

Monsoon Asia is characterized generally by a warm, well-watered climate, but there are nine distinct types of climate in the region: tropical rain forest, tropical savanna, humid subtropical, humid continental (warm and cold types), desert, steppe, subarctic, and undifferentiated highland (Figure 10.5a). These generally are associated with predictable patterns of vegetation, which have been heavily modified by millennia of human use (Figure 10.5b).

The tropical rain forest climate zone and vegetation types are typically found within 5° to 10° of the equator and are characteristic of most parts of the East Indies, the Philippines, and the Malay Peninsula. Precipitation is spread throughout the year so that each month has considerable rain. The amount of precipitation is at least 30 to 40 inches (c. 75 to 100 cm) and often 100 inches (c. 250 cm). Average temperatures vary only slightly from month to month; Singapore, for

Climates of Monsoon Asia

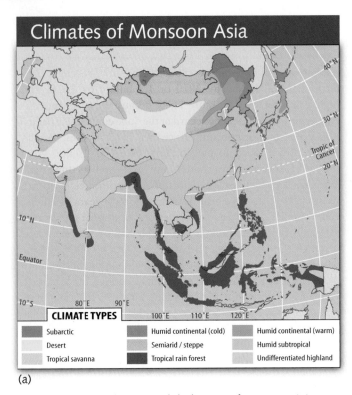

CLIMATE TYPES

Subarctic	Humid continental (cold)	Humid continental (warm)
Desert	Semiarid / steppe	Humid subtropical
Tropical savanna	Tropical rain forest	Undifferentiated highland

(a)

Biomes of Monsoon Asia

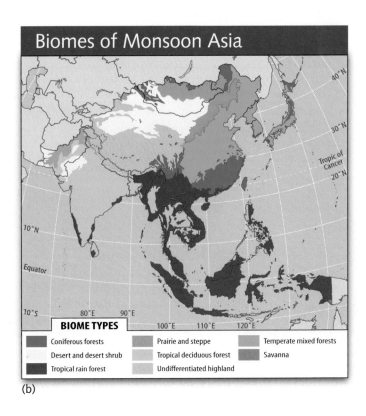

BIOME TYPES

Coniferous forests	Prairie and steppe	Temperate mixed forests
Desert and desert shrub	Tropical deciduous forest	Savanna
Tropical rain forest	Undifferentiated highland	

(b)

Figure 10.5 (a) Climates and (b) biomes of Monsoon Asia

example, has a difference of only 3°F between the warmest and coolest months. Heat prevails year-round, although some relief is afforded by a drop of 10° to 25°F (6° to 14°C) in the temperature at night, and sea breezes refresh coastal areas. In climates such as this, it is said that "nighttime is the winter."

The tropical savanna climate, like the tropical rain forest climate, is characterized by high temperatures year-round but is generally found in areas farther from the equator, and the average temperatures vary more from month to month. The savanna has a more well-defined dry season, lasting in some areas as long as 6 or 8 months each year, creating a problem for agriculture. Tropical savanna climates occur in southern and central India, in most of the Indochinese Peninsula, and in eastern Java and the smaller islands to the east. In Asia, the characteristic natural vegetation associated with this climate is a deciduous forest of trees smaller than those of the tropical rain forest. Tall, coarse grasses such as those found on African and Latin American savannas grow only in limited areas.

The humid subtropical climate zone prevails in southern China, the southern half of Japan, much of northern India, and a number of other countries. It is characterized by warm to hot summers, mild or cool winters with some frost, and a frost-free season lasting 200 days or longer. The annual precipitation of 30 to 50 inches (c. 75 to 125 cm) or more is well distributed throughout the year, although monsoonal tendencies produce a dry season in some areas. The natural

vegetation is a mixture of evergreen hardwoods, deciduous hardwoods, and conifers.

The humid continental climates characterize the northern part of eastern China, most of Korea, and northern Japan. They are marked by warm to hot summers, cold winters with snow, a frost-free period of 100 to 200 days, and less precipitation than the humid subtropical climate. Most areas experience a dry season in winter. The predominant natural vegetation is a mixture of broadleaf deciduous trees and conifers, although prairie grasses probably formed the original cover in parts of northern China. Extreme northeastern China and northern Mongolia have a subarctic climate associated with coniferous trees.

Steppe and desert climates are found in Xinjiang and Mongolia and in parts of western India and Pakistan. Due to limitations on agriculture, along with very high mountains, they are the region's least inhabited climate zones. The undifferentiated highland climate is most extensive in the Tibetan Highlands and adjoining mountain areas. It is made up of a broad range of montane microclimates that limit plant growth and vary from place to place according to elevation, aspect, and latitude.

The monsoons are the prevailing sea-to-land and land-to-sea winds that are the dominant climatic concern for people living in this world region. They play a significant role in both wet and dry environments and are especially influential in the coastal plains and lowlands of South Asia, the peninsula and islands of Southeast Asia, and the eastern third of China. The ways in which the sea and land absorb **insolation** (heat from the sun) are different, causing the instability in air masses that creates monsoonal winds. If coastal waters and the adjacent coastline receive approximately equal amounts of warmth from the sun, the land becomes warm or hot much more quickly than the seawater. As a result, air over the land begins to become unstable and to rise. This ascending air creates a low-pressure attraction for the air masses over the water, and marine winds begin to blow toward the land. In Asia, this is the **summer monsoon** characterized by high humidity, moist air, and generally predictable rains. Because of the moisture carried by such air masses, these wind shifts are the sources of major rainfall in the late spring, summer, and early fall seasons. Agriculture and patterns of human activity are keyed to these incoming rains. If there is a little bit of elevation in the landscape—hills or mountain flanks—the moist air is driven higher, where it cools and releases even more rain in the pattern of orographic precipitation.

In the **winter monsoon**, the land loses its relative warmth, while the sea and coastal waters maintain their warmth longer. As a result, the wind shifts and air masses begin to flow from the inland areas of Asia toward the sea. However, because there is little moisture in the source areas over land for the winter monsoon, there is little moisture picked up and much less rain. A monsoon climate, overall, is characterized by spring and summer precipitation and a long dry season in the low sun (winter) cycle. Japan, however, experiences an unusual pattern of winter precipitation related to the monsoon wind shifts. The winter winds blow out of China and East Asia, sweep across the Sea of Japan (East Sea) and pick up moisture, dropping heavy, wet snow on the west coast of the Japanese islands.

The wet summer monsoon is of critical importance to the livelihoods of a large portion of humanity, particularly the people of India. By early May in India, the land has typically been parched by months of drought, and the rising temperatures become almost insufferable. No thought can be given to farming until the rains come. Anticipation and discomfort rise steadily until the monsoon finally "breaks," bringing great relief in the form of much needed rain and lower temperatures (Figure 10.6). Sometimes the onset of the rain comes with particular ferocity, as related by this Indian writer:

> That year the monsoon broke early and with an evil intensity such as none could remember before. It rained so hard, so long and so incessantly that the thought of a period of no rain provoked a mild wonder. It was as if nothing had ever been but rain, and the water pitilessly found every hole in the thatched roof to come in, dripping onto the already damp floor. If we had not built on high ground the very walls would have melted in that moisture.[1]

[1] Kamala Markandaya, *Nectar in a Sieve* (New York: John Day, 1954), p. 57.

Figure 10.6 Torrential monsoon rains are regular and welcome features of land and life in much of the region. This is a flooded road in south central Sri Lanka.

Agricultural Adaptations

The wet and warm climates associated with the monsoons might be expected to offer outstanding opportunities for agriculture (Figure 10.7). However, the high temperatures and heavy rains promote rapid leaching of mineral nutrients and decomposition of organic matter, so many soils are infertile. The lush vegetation cover in these low-latitude areas is deceiving because most of the system's nutrient matter is in the vegetation itself rather than in the underlying soil. When Asian farmers clear the tree cover, they often find that the local soils will not support more than one or two poor harvests (see Definitions and Insights, page 281).

Agriculture in Monsoon Asia—with the principal exception of Japan—is characterized by the steady input of arduous manual labor. Known as **intensive subsistence agriculture**, it is built around the growing of cereals, especially rice, which is the premiere staple of Monsoon Asia and the crop of choice in areas with adequate rainfall or where irrigation waters are available (see "rice subsistence" on Figure 10.7). Farmers use organic fertilizers from both their animals and their own latrines (this is the key ingredient of "night soil")

Agricultural Land Use

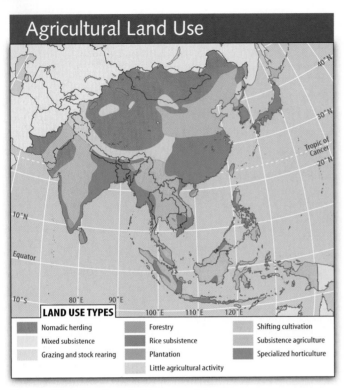

LAND USE TYPES

- Nomadic herding
- Mixed subsistence
- Grazing and stock rearing
- Forestry
- Rice subsistence
- Plantation
- Little agricultural activity
- Shifting cultivation
- Subsistence agriculture
- Specialized horticulture

Figure 10.7 Land use in Monsoon Asia

Figure 10.8 These meticulously crafted and maintained terraces produce a high yield of rice in Pakistan's northern Punjab.

329

and irrigate to the extent that they can or need to. Where natural conditions are not suitable for irrigated rice, grains such as wheat, barley, soybeans, millet, sorghums, or corn (maize) are raised. Where possible, farmers grow cash crops for local markets. Large-scale, commercial plantation agriculture is described in Chapter 12 on Southeast Asia, the subregion in which cash cropping has had the strongest role.

Shifting cultivation represents a precarious adaptation to local soil and climate conditions, capable of sustaining only small populations for brief periods at particular locales. At the other end of the spectrum of human adaptation to Asian lands is **wet rice cultivation,** capable of producing two to three crops per year and sustaining large populations over long periods of time. Wet rice agriculture takes place both in lowland floodplains and on carefully constructed upland **terraces** that are the results of enormous inputs of human labor (in both cases, the fields and the rice grown on them may be referred to as **paddy,** or **padi**). Terraces have been built almost everywhere in Asia where there is a combination of sloping land, available irrigation water, and productive soil (Figure 10.8). The dramatic stair-step terraces of the island of Luzon in the Philippines are the most famous, but equally dramatic cultural landscapes are in Pakistan's Punjab, in South China, and in many parts of Southeast Asia.

Terraces have the important function of preventing or dramatically slowing soil erosion, as the terrace walls allow heavy rainfall to run off without carrying topsoil away. Doing what the Chinese call "teaching water," people use canals and pipes to direct irrigation water from a source to

the terrace, which acts like a kind of dam to hold the water in place. Wet rice is actually immersed in standing water for extended periods, creating an environment unfavorable to many crop pests. The productivity of wet rice agriculture is sometimes exceptionally high, supporting as many as 500 people per square mile (200/sq km) on the volcanic soils of Java and Bali in Indonesia. This extraordinary output is a contributing factor to the high densities and overall populations of some of the Asian countries and regions, including Java. One anthropologist has observed that most of Monsoon Asia's rice is eaten within walking distance of where it is grown.[2]

Anthropologists have observed some unique cultural adaptations to opportunities for terraced wet rice. Group welfare takes precedence over individual interests, for example. Individual farmers cannot cultivate at will but must work with the community to ensure that crops are staggered through time. This allows water to be shared fairly and efficiently and gives some fields the chance to dry out while others are irrigated. This agricultural system also prevents the use of tractors and other mechanical technology. Over much of Asia today, just as in ancient times, wet rice cultivation is the productive of intensive work by many human hands.

In some cases, population growth has outpaced the ability of even the most productive wet rice farming to feed the

---

[2] John Reader, in the source for this discussion of wet rice irrigation: http://www.waterhistory.org/histories/terraces.

Definitions + Insights

Shifting Cultivation

Many Asian farmers, particularly in tropical rain forest regions, practice shifting cultivation (also known as swidden or **slash-and-burn** cultivation) or land rotation between crop and fallow years (Figure 10.C). The procedure begins with people clearing forest cover and then burning the dried vegetation during the dry season. Ash from the burn provides short-lived fertility to the soil in which farmers plant their sub-sistence crops. Most tropical soils lose their natural fertility quickly when they are cropped and, after 2 or 3 years, must be rested for several years (often 10 to 15 years and more) before they will again produce a crop. During this **fallow** period, the land reverts to wild vegetation.

The system of shifting agriculture is widespread in the world's tropics. Although it is not very productive, it does provide a bare living for people too poor to afford fertilizer or farm machinery. Traditionally, this system has minimized soil erosion because most of the land is not in cultivation at any given time. Unfortunately, there has been a recent widespread trend for farmers to reduce the length of the fallow period in the cycle of shifting cultivation or to abandon it altogether. This is a direct result of a surging growth in population: With more mouths to feed, in many places there is simply not enough land to provide the "luxury" of fallow.

The tragic consequence is that short-term overuse of the land is eroding the soil. On the **lateritic soils** characteristic of many tropical regions, excessive exposure of the land to the sun's ultraviolet radiation and to repeated wetting and drying turns the ground to a bricklike substance called laterite (plinthite), which is essentially impossible to cultivate. When the best available lands are exhausted, farmers are often "marginalized" to places unsuitable for farming, such as semiarid lands or slopes that are too steep to till without the consequence of disastrous erosion. Such dilemmas of land use are typical of the world's tropics from Indonesia to West Africa and Amazonia.

Joe Hobbs

Figure 10.C Slash-and-burn cultivation (swidden or shifting cultivation) is widespread in the world's tropical regions and can be sustained as long as the land is given sufficient time to recover.

36

39

people, creating the classic scenario of "people overpopulation." In Nepal, for example, high rates of population growth on limited areas of moderately sloped, terraceable land have been a push factor behind migration. After the 1950s, when malaria began to be eradicated in the lowland Terai region, millions of Nepalis migrated there, cleared the forests and planted crops. For some decades, the Terai served as a virtual "safety valve" for overpopulation in Nepal. Now that the Terai itself is mostly settled, there are no directions for landless Nepalis to go but up. They clear and cultivate increasingly steep lands on which it is impossible to build terraces. Without terraces in place to hold the topsoil, heavy monsoon rains cause relentless erosion of their farmlands. Locally, the consequences are that the eroded plots cannot be cultivated again, and landslides occur down slope, often causing loss of life.

Much farther away, in the lowlands of India and Bangladesh, the increased sediment load contributed by erosion in Nepal causes rivers to swell out of their banks and bring extensive flooding—again with crop losses and also loss of life. This sequence of events, beginning with overpopulation in Nepal and ending with flooding in Bangladesh, has

41

been described as the **theory of Himalayan environmental degradation.**[3]

Settlement Patterns

Monsoon Asia has some of the world's largest cities, including the greatest of all—Tokyo (metropolitan population: 31 million). About 1.3 billion people, or 37 percent of the region's total, live in urban settlements. But 2.2 billion, or 63 percent, do not. The main unit of Asian settlement is the village. Most of Asia's roughly 2 million villages are essentially groups of farmers' homes, although some villages are home to other occupational groups, such as miners or fisherfolk. Clusters of houses bunched tightly together are typical, and inexpensive and simple structures—often made of local clays and other building materials—are characteristic. Indoor plumbing continues to be somewhat exceptional in village homes, but electricity lines have been steadily expanding all through Asia. So have cell phones and satellite dishes; it is not unusual today to see what appear to be very "traditional" Asians enjoy-

[3] Jack D. Ives and Bruno Messerli, *The Himalayan Dilemma* (London: Routledge, 1989).

Geography of the Sacred

The Korean Village

In the West, people are accustomed to communities laid out on a perfect grid of streets intersecting at right angles, with their homes and businesses symmetrically fronting those streets. But in large parts of Asia, settlement and residence considerations are far more varied, especially because they are tied to spirituality, aesthetics, and topography. Villages and village homes in particular tend to be arranged for maximum harmony with the natural and spiritual environments—a practice known as **geomancy**—and constitute a special realm that may be considered sacred space. The traditional Korean village illustrates these unique Asian qualities.[a]

From ancient times, Koreans developed ways of determining the most auspicious village sites. The principles of *fengshui* and of Confucianism played strong roles in how they adapted settlements and homes to geographic conditions. *Fengshui* (pronounced "fuhng shway") is a traditional theory for selecting favorable sites for buildings, homes, and cities and for shaping many other urban and architectural planning decisions. It is essentially an ordering device for environmental planning and for daily living. Fengshui attempts to balance **yin** and **yang** (literally meaning "wind" and "water") to achieve harmony. Yin/yang stand for all dualistic, opposing forces that may be reconciled harmoniously, including female–male, night–day, moon–

sun, cold–hot, and right side–left side. Fengshui principles have begun to appear even in American suburban home architecture, as in houses built to conceal overhead beams and sharp protrusions (which are disharmonious) and provided with windows opening to the east (a harmonious situation).

The theory of fengshui and the Confucian view of nature have some common characteristics that affect site, settlement, and residence designs. Confucianism refers to the sociopolitical philosophy of the Chinese man named Confucius (see page 288). One of the main Confucian principles is that architecture and lived environment should be in harmony with nature. This may be seen in the Confucian institutes for higher learning, called *suwon,* which in Korea peaked in the Joseon Dynasty, around A.D. 1400. Each suwon had a lecture hall where Confucian scholars gathered to learn philosophy. The buildings were laid out in such a way that from the inside of the complex, the scholars could see distant landscapes. The scholars' minds and spirits were enhanced as they gazed at the mountains and contemplated nature's grandeur. They believed that the earth's power was an invisible force that would help them become men of dignity and eminence.

The principles of fengshui and Confucianism are also apparent in the design of the Korean villages. The village is laid out to achieve harmony with the mountains—an important consideration in a land that is 70 percent mountainous. Rising behind many villages is a mountain, called *jinsan,* which is a spiritual focus for the villagers. Jinsan serves as the village symbol and guardian. Typically a small stream flows in front of the village, and beyond it, there is a flat expanse of farmland. Beyond those fields is another mountain, called *ansan,* facing the village (Figure 10.D).

[a] This discussion is based largely on two unpublished manuscripts by, and the author's interview with, Korean geographer Sang-Hae Lee, of Sungkyunkwan University. The author is grateful for his contributions. The manuscripts are entitled "Living Spaces of Korean Architecture" and "In Tune with Nature: Rural Villages and Houses in Korea." A useful introduction to fengshui may be found on http://www.fengshuihelp.com/define.htm.

ing some of the latest innovations in global communications (Figure 10.9). In Monsoon Asia, as elsewhere in the developing world, many people are making the jump directly from 19th- to 21st-century technology. Wireless technology is especially appropriate where landlines and other infrastructure are difficult and expensive to install and where bureaucracies often slow down applications for access to utilities.

The original site selection for Asian villages was usually closely adapted to natural conditions and to perceived spiritual circumstances as well (see Geography of the Sacred, above). With consideration always given to the possibility of flooding monsoon rains, lowland villages tend to be situated on natural levees (raised riverbanks built up by deposition of sediments during floods), dikes, or raised mounds. Early villages in Indonesia were often built in defensible mountain sites. European colonists sometimes rejected traditional settlement patterns to serve their own interests. Dutch colonial

Joe Hobbs

Figure 10.9 Across Monsoon Asia, satellite dishes sprout incongruously from traditional dwellings.

Figure 10.D The Korean village of Yangdong is nestled between mountains that are separated by a river and lush rice fields.

the North. These guardians, named for figures in Chinese astronomy, could have varied manifestations. The Green Dragon, for example, could be a hill or a river to the east of the village, while the White Tiger could be a mountain or a road on the west side. One enters the village through a symbolic portal. For many villages, a pair of "spirit posts" stands at the foot of a nearby mountain. The posts mark the preliminary boundary to the outside, and they ward off evil spirits.

Not only is the village site-selection process different from that in the West, but the settlement pattern is uniquely Asian, too. Unlike in Western settlements, no more than two houses should be built side by side, they should not stand parallel to each other, and their gates should never face each other. The side street should curve naturally between the home lots. Overall, then, Koreans choose to emphasize the surrounding topography rather than geometrical order in the construction of their communities and homes.

The traditional Korean home is adapted to the natural environment, embracing its surroundings while offering protection, comfort, and order. As in the suwon, the layout brings outdoor scenery into the house itself. Homes are also adapted to the environment in their construction materials. A thatch roof and earthen walls effectively block heat in the summer and insulate in winter, a perfect combination for Korea's sweltering summers and raw winters (see Figure 10.9). The mulberry paper covering lattice windows diffuses sunlight, providing a pleasant natural light while absorbing sound and allowing some ventilation. The home's eaves are built so that sunlight penetrates deep into the house during the cold weather, but never during summer—a rather recent innovation in the West, where it is known as "passive solar" design.

These geographic conditions are economically advantageous to the community, providing it with productive fields and a good supply of water. But just as important, this situation is psychologically, spiritually, and symbolically beneficial to the villagers. The natural features are meant to embrace a village, as Koreans say a mother would hold a child. Korean historical records describe the traditional village as a place of comfort and serenity that safeguards against negative natural forces.

Symbolic guardians are associated with the topography. Ideally, there is the Green Dragon to the East, the White Tiger to the West, the Black Tortoise to the South, and the Red Bird to

administrators in Indonesia, for example, required that villages be built along main roads and trails in the lowlands, making it easier to exercise control, collect taxes, and draft soldiers or laborers for roadwork and other projects.

The majority of people in Monsoon Asia are rural, but the region's future, in terms of demographic weight, economic power, and cultural change, is unquestionably focused on the cities. Some of the countries are already highly urbanized, notably Japan (78 percent urban), Taiwan (78 percent), South Korea (80 percent), and entities like Singapore and Brunei that are effectively city-states. The layouts and architectures of the cities span a continuum from the ancient walled fortress city to the modern Western-style metropolis, and many, like Beijing in China and Lahore in Pakistan, accommodate both. A striking pattern across the region is the steady stream of **rural to urban migration,** consisting of people motivated by both push and pull factors to leave the relative poverty of the

countryside in search of employment and even prosperity in the city. Many Asian cities are hard-pressed to provide accommodation and services to their swelling populations, and where the political system allows (notably China), restrictions on internal migrations are imposed.

10.3 Cultural and Historical Geographies

Monsoon Asia has been home to some of the most important cultural developments of humankind in landscape transformation, settlement patterns, religion, art, and political systems. From Korea (not Gutenberg's Germany, as is widely thought) came the first movable printing type. From China came gunpowder, paper, silk, and china itself. From India came the great faiths of Hinduism and Buddhism. Handmade textile, artware, bone and leather and paper products, and precious metals and gems have flowed from Asia into global trade for millennia. Spices, foodstuffs, and exotic plants and products from the East Indies have been major commodities in Asian–European trade and commerce for more than five centuries. Some of the world's most important domesticated plants and animals originated in Monsoon Asia, including rice, cabbage, chickens, water buffalo, zebu cattle, and pigs.

As would be expected of such a vast region, in many places fragmented by islands and mountain barriers, the ethnic and linguistic composition is rich and complex (Figure 10.10). Central and South Asia gave birth to the **Indo-European language family,** which contains tongues as diverse as English and Hindi. **Hindi** and **Urdu,** which are major languages in India and Pakistan, respectively, belong in the **Indo-Iranian subfamily** that includes such other major South Asia languages as **Pashto** (spoken in western Pakistan and in Afghanistan), **Sanskrit** (which is extinct as a vernacular language but is still used in Hindu ritual), **Sinhalese** (the majority language of Sri Lanka), **Punjabi** (spoken on either side of the northern frontier between Pakistan and India), Nepal's national language of **Nepali, Bengali** (the language of Bangladesh), and **Romany,** the tongue of the Roma (Gypsies) of Europe.

When the Indo-European languages were carried from Central Asia into India—apparently by the ethnic group known as the **Aryans,** between 2000–1000 B.C., there was already an ancient and indigenous language family there: the **Dravidian,** spoken by peoples of much darker skin than the newcomers. The Dravidian languages, found almost exclusively in southern India and northern Sri Lanka (with the exception of a small pocket in Pakistan), include **Tamil, Malayalam,** and **Telegu.** Along the eastern side of the Himalayan mountain wall separating South Asia from China, there are speakers of the **Tibeto-Burman subfamily** languages of the **Sino-Tibetan family.** Farther north, in Tibet, and east, in Myanmar (Burma), the respective languages of **Tibetan** and **Burmese** are also in the Tibeto-Burman language subfamily.

The greatest numbers of people in China speak languages in the **Sinitic subfamily** of the Sino-Tibetan family, overwhelmingly **Chinese** and its subgroups such as **Mandarin, Xiang, Min,** and **Yue.** Almost all of these more than 1 billion Chinese speakers belong to the Han ethnic group, which is culturally, politically, and economically the dominant ethnic group of China. Northern and northwestern China, along with Mongolia, are home to speakers of languages in the **Altaic language family,** including **Uighar, Kazakh, Turkic, Mongolian,** and **Tungus.** Cut off from their Altaic-speaking kin are the **Korean** speakers of the Korean Peninsula and the **Japanese** speakers of Japan. Japan's indigenous **Ainu** language (also in the Altaic family) became extinct when its last speaker died in 1994.

The people of Taiwan include Chinese-speaking migrants from China and the indigenous speakers of an **Austronesian** language. In the context of Southeast Asia, many linguists view Austronesian as a subfamily of a greater **Austric language family.** Extreme southern China is linguistically complex, reflecting the presence of many minority ethnic groups. Several are kin to the peoples of peninsular Southeast Asia, including the **Hmong** of Laos, who speak a language in the **Hmong-Mien subfamily** of the Austric language family. Most people in Laos and neighboring Thailand speak **Lao** and **Thai** in the **Tai-Kadai subfamily** of the Austric family. The majority languages of Cambodia and Vietnam, respectively, are **Khmer** and **Vietnamese** in the **Mon-Khmer branch** of the **Austro-Asiatic subfamily** of the Austric language family.

Southward and eastward into Peninsular Malaysia, Indonesia, and the Philippines, the linguistic map is exceptionally rich and fragmented. The eight languages of the Philippines belong to the Austronesian group. Also in the Austronesian branch is a multitude of languages spoken in Indonesia, including **Malay, Dayak, Kayan-Kenyah, Barito, Javanese, Bugis, Tomini,** and **Timorese.** In many cases, a single language is spoken almost exclusively by an ethnic group inhabiting a single island. In other cases, as will be seen in Chapter 12, different ethnolinguistic groups sometimes live in conflicted proximity to one another on a single island. The large island of New Guinea and some of its smaller island neighbors are home to the 750 different **Papuan languages,** which are not related to Austronesian tongues.

Colonialism and regional economic enterprise led to the infusion of distant tongues into many parts of Monsoon Asia. British colonists brought English to India, where it still serves as a lingua franca in this linguistically complex land. The British also introduced Tamil-speaking Dravidians from south India into what is now Malaysia, where they worked the rubber plantations and tin mines and where their descendents remain as an important minority today. Another small but economically prosperous minority in several countries (and a majority in Singapore) is ethnic Han Chinese. Some people in Indonesia speak Dutch, reflecting that country's former colonial status, while the U.S. colonization of the Philippines is reflected in English being one of that country's two official languages.

Religions

Monsoon Asia is, along with the Middle East, one of the world's two great hearths of religion. Hinduism, Buddhism, and related traditions of Confucianism and Daoism—collectively practiced or observed by a whopping 1.3 billion, or 20 percent of the world's population—originated in this region. The basics of these belief systems are introduced here. The precepts of Islam and Christianity, also practiced in the region, are described in Chapter 8. The faiths with small numbers of adherents, like Zoroastrianism and Sikhism, are described in the contexts of particular countries in the following chapters. Religion is sometimes, unfortunately, at the root of domestic or international conflict, and these cases are also discussed in the appropriate chapters.

Languages of Monsoon Asia

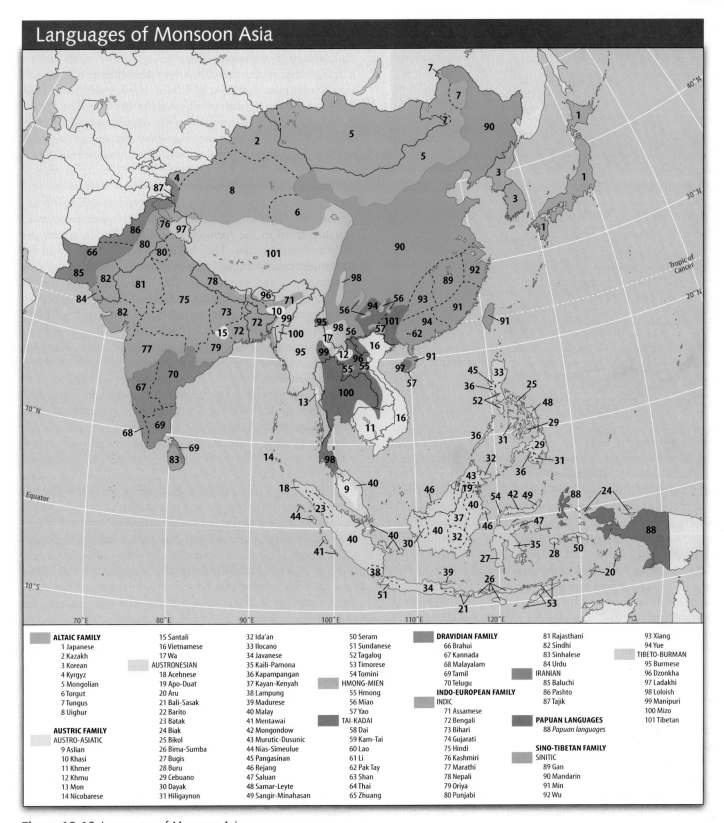

ALTAIC FAMILY	15 Santali	32 Ida'an	50 Seram	DRAVIDIAN FAMILY	81 Rajasthani	93 Xiang
1 Japanese	16 Vietnamese	33 Ilocano	51 Sundanese	66 Brahui	82 Sindhi	94 Yue
2 Kazakh	17 Wa	34 Javanese	52 Tagalog	67 Kannada	83 Sinhalese	TIBETO-BURMAN
3 Korean	AUSTRONESIAN	35 Kaili-Pamona	53 Timorese	68 Malayalam	84 Urdu	95 Burmese
4 Kyrgyz	18 Acehnese	36 Kapampangan	54 Tomini	69 Tamil	IRANIAN	96 Dzonkha
5 Mongolian	19 Apo-Duat	37 Kayan-Kenyah	HMONG-MIEN	70 Telugu	85 Baluchi	97 Ladakhi
6 Torgut	20 Aru	38 Lampung	55 Hmong	INDO-EUROPEAN FAMILY	86 Pashto	98 Loloish
7 Tungus	21 Bali-Sasak	39 Madurese	56 Miao	INDIC	87 Tajik	99 Manipuri
8 Uighur	22 Barito	40 Malay	57 Yao	71 Assamese		100 Mizo
	23 Batak	41 Mentawai	TAI-KADAI	72 Bengali	PAPUAN LANGUAGES	101 Tibetan
AUSTRIC FAMILY	24 Biak	42 Mongondow	58 Dai	73 Bihari	88 Papuan languages	
AUSTRO-ASIATIC	25 Bikol	43 Murutic-Dusunic	59 Kam-Tai	74 Gujarati		
9 Aslian	26 Bima-Sumba	44 Nias-Simeulue	60 Lao	75 Hindi	SINO-TIBETAN FAMILY	
10 Khasi	27 Bugis	45 Pangasinan	61 Li	76 Kashmiri	SINITIC	
11 Khmer	28 Buru	46 Rejang	62 Pak Tay	77 Marathi	89 Gan	
12 Khmu	29 Cebuano	47 Saluan	63 Shan	78 Nepali	90 Mandarin	
13 Mon	30 Dayak	48 Samar-Leyte	64 Thai	79 Oriya	91 Min	
14 Nicobarese	31 Hiligaynon	49 Sangir-Minahasan	65 Zhuang	80 Punjabi	92 Wu	

Figure 10.10 Languages of Monsoon Asia

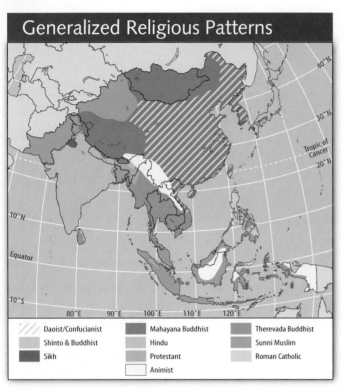

Figure 10.11 Religions of Monsoon Asia

The geographic distribution of the region's faiths is depicted in Figure 10.11. As this map reveals, Hinduism is dominant in India, although many other religions are practiced there as well. Islam prevails in Pakistan, Bangladesh, most parts of the East Indies, parts of the southern Philippines, parts of western China, and among the Malays of West Malaysia. Various forms of Buddhism are dominant in Burma, Thailand, Cambodia, Laos, Tibet, and Mongolia. Most people in Sri Lanka and Nepal are Buddhists or Hindus. The religious patterns of the Chinese, Vietnamese, and Koreans are more difficult to describe. Among them, Buddhism, Confucianism, and Daoism have been influential, often in the same household. In Japan, religious affiliations often overlap, with an estimated 84 percent of the Japanese population sharing Buddhism with the strongly nationalistic religion of **Shintoism,** itself a blend of nature reverence and Chinese ritual. The Philippines and East Timor, with large Roman Catholic majorities, are the only predominantly Christian nations in Asia, although Christian groups are found in virtually all other countries in this region.

Other customs include the **ancestor veneration,** which is especially prominent among Chinese, Japanese, Vietnamese, and Koreans. Many members of hill tribes in Southeast Asia are largely **animists,** believing that natural processes and objects possess souls. Quite often, indigenous animistic beliefs have been successfully incorporated into the mainstream religion, creating a vibrant and highly varied belief system. This blending is known as **syncretism** and has parallels elsewhere in the world, for example, among the Maya of Central America.

Hinduism is a regionally varied faith and a belief system so complex that it is extremely difficult to summarize. However, it has the following characteristic elements. It lacks a definite creed or theology. It is very absorptive, encompassing an unlimited pantheon of deities (all of which, most Hindus say, are simply different aspects of the one God, Brahman) and an infinite range of types of permissible worship (Figure 10.12). Most Hindus recognize the social hierarchy of the caste system, deferring authority to the highest (Brahmin) caste. They practice rituals to honor one or another of the principal deities (Brahman the creator, Vishnu the preserver, and Shiva the destroyer) and thousands of their manifestations (avatars) and lesser deities, attending temples to worship them in the form of sanctified icons in which the gods' presence resides. They believe in reincarnation and the transmigration of souls and revere many living things (see Geography of the Sacred, page 287). They are supposed to be tolerant of other religions and ideas. They participate in folk festivals to commemorate legendary heroes and gods. To earn religious merit and to struggle toward liberation from the bondage of repeated death and rebirth, they make pilgrimages to sacred mountains and rivers.

The Ganges is a particularly sacred river to Hindus, who believe it springs from the matted hair of the god Shiva. The pilgrimage to celebrate the festival of Purna Kumbh Mela at Allahabad, on India's Ganges River, drew 30 million pilgrims in January 2001. It was the largest pilgrimage in human history, and 30 million is the conservative estimate. Some sources put the number as high as 70 million! The most enduring Hindu pilgrimage destination is the Ganges city Varanasi (Benares) in the state of Uttar Pradesh. Many elderly people go to die in this city and be cremated where their ashes may be thrown into the holy waters. The Indian government is now attempting to clean up the Ganges, polluted in part by incompletely cremated corpses; many of the faithful poor cannot afford to buy the fuel needed for thorough immolation. Scavenging water turtles released into the river at Varanasi help dispose of the cadavers.

Figure 10.12 Hinduism has a complex and very colorful pantheon. This is Vishnu, preserver god, and his consort, Lakshmi, goddess of wealth and prosperity.

315

Geography of the Sacred

The Sacred Cow

Many attitudes, beliefs, and practices associated with cattle make up the world-famous but often poorly understood "sacred cow" concept of India. There are nearly 200 million cattle in India, representing about 15 percent of the world total and the largest concentration of domesticated animals anywhere on Earth. India's dominant religion of Hinduism forbids the slaughter of cows but allows male cattle and both male and female water buffalo to be killed. Reverence for the cow is well founded. Indians favor cow's milk, ghee (clarified butter), and yogurt over dairy products from water buffalo. They value cows as producers of male offspring, which serve as India's principal draft animal. Both cows and bullocks provide dung, an almost universal fuel and fertilizer in rural India.

Reverence for the cow in particular and cattle in general pervades Hindu religion and mythology. The bull Nandi is associated with the Hindu god Shiva. People allow cattle to freely roam the streets of Indian cities (Figure 10.E). As a symbol of fertility, cows are associated with (but not worshiped as) several deities. The mother of all cows, Surabhi, was one of the treasures churned from the cosmic ocean. Hindus honor cows at several special festivals. They use cow's milk in temple rituals. They believe the "five products of the cow"—milk, curds, ghee, urine, and dung—have unique magical and medicinal properties, particularly when combined. All over India, there are *goshalas,* or "old folks' homes," for aged and infirm cattle.

Remarkably, however, the subcontinent countries of Pakistan, Bangladesh, and India are major exporters of leather and leather goods. In India, the leather industry is mainly in the hands of Muslims, who do not share Hindus' restrictions on killing or eating cows. Muslims are forbidden to eat pork; therefore, pigs, a major food resource in many developing countries, are of little importance here. They are eaten mainly by Christians, very low-caste Hindus (some of whom are pig breeders), and tribal peoples.

Figure 10.E People and cows share street space across India.

Joe Hobbs

Buddhism is based on the life and teachings of Siddhartha Gautama, known as the Buddha or the Enlightened One (Figure 10.13). Buddha was born a prince in about 563 B.C. in northern India.[4] Although presumably born a Hindu, he came to reject most of the major precepts of Hinduism, including caste restrictions. When he was about 29 years old, he renounced his earthly possessions and became an ascetic in search of peace and enlightenment. Buddha eventually settled for a "middle path" between self-denial and indulgence, and he meditated through a series of higher states of consciousness until he attained enlightenment. From that point, he traveled, preached, and organized his disciples into monastic communities. In his sermons, he described the "Four Noble Truths" revealed in his enlightenment: life is suffering; all suffering is caused by ignorance of the nature of reality; suffering can be ended by overcoming ignorance and attachment; and the path to the suppression of suffering is the

Figure 10.13 The Buddha has many manifestations, especially in Mahayana Buddhism.

Joe Hobbs

---

[4]These descriptions of Buddhism and Daoism are based on the Buddhism and Daoism sections of the Microsoft Encarta Reference Library 2004.

Noble Eightfold Path, made up of right views, right intention, right speech, right action, right livelihood, right effort, right mindedness, and right contemplation.

An important Buddhist concept is the doctrine of *karma,* a person's acts and their consequences. A person's actions lead to rebirth, in which good deeds of the previous life are rewarded (for example, by which one could be reborn a human) and bad deeds punished (for example, by being reborn as a resident of hell). The goal of the practicing Buddhist is to attain *nirvana,* an enlightened state in which one is able to escape the cycle of birth and rebirth and all the suffering it brings.

Buddhism evolved into two separate branches: the **Theravada** ("Way of the Elders") and the **Mahayana** ("Great Vehicle"). The term Hinayana ("Lesser Vehicle") is used by the Mahayanists in a derogatory fashion to describe the Theravada. Theravada Buddhism is strongest in Sri Lanka, Myanmar (Burma), Laos, Cambodia, and Thailand. Theravada Buddhists claim to follow the true teachings and practices of the Buddha. Mahayana Buddhism originated in India and then diffused along the Silk Road to Central Asia, Tibet and China, and eventually into Vietnam, Korea, Japan, and Taiwan. Mahayana Buddhists accept a wider variety of practices than those espoused by the Theravada, have a more mythological view of the Buddha, and are interested in broader philosophical issues.

Confucianism is not as much a belief system as it is a sociopolitical philosophy that serves on its own or is blended with other religions, particularly among ethnic Han Chinese. It is based on the writings of Confucius (or Kung Futzu, or Master Kung, who lived from 551–479 B.C.), collected primarily in his *Analects.* Mencius (c. 371–288 B.C.) later became a major force in the widespread diffusion of this belief system throughout China. The major tenets of Confucianism are embodied in a system of ethical precepts for the proper management of society, emphasizing honor of elders and other authorities, hierarchy, and education. Confucius never proclaimed his beliefs to constitute a religion; he was simply attempting to create a social contract between different classes central to Chinese government and society. He viewed his philosophy as secular, and its diffusion to Korea, Japan, and Vietnam has been part of the cultural baggage taken abroad by a steady stream of Chinese emigrants to those places.

Another important Chinese school of thought, second only to Confucianism, and also widely blended with Buddhism among ethnic Han Chinese, is **Daoism** (or **Taoism**) (Figure 10.14). Its philosophy comes from a body of work known as the *Classic of the Way and Its Power,* or the *Daodejing (Tao-te Ching),* and is ascribed to the Chinese man Lao-tzu (Laozi), who lived in the sixth century B.C. Daoism encourages the individual to reject Confucian-style social conformity and seek to conform only to the underlying pattern of the universe, the "Way" (Dao, or Tao). To follow that way, the individual should "do nothing," meaning nothing unnatural or artificial. One can become rid of all

Figure 10.14 Daoist temples are often essentially Buddhist temples but are embellished with images of heroic figures.

doctrines and knowledge to achieve unity with Dao and thereby gain a mystical power. With that power, the individual can transcend everything ordinary, even life and death. Daoists prize the simple earthly life of the farmer.

European Colonization and Contemporary Asia

Like much of the developing world, Monsoon Asia is a region where Western colonialism reshaped many traditional geographic patterns (Figure 10.15). By the end of the 15th century, Portugal and Spain began to extend economic and political control over some islands and mainland coastal areas of South and Southeast Asia. In the 18th and 19th centuries, the pace of colonization and economic control quickened, and large areas came under European domination. By the end of the 19th century, Great Britain was supreme in India, Myanmar (Burma), Ceylon (Sri Lanka), Malaya, and northern Borneo; the Netherlands possessed most of the East Indies (Indonesia); France had acquired Indochina (Vietnam, Laos, and Cambodia) and small holdings on the coasts of India; and Portugal had Goa in India and Macao in China. Although retaining a semblance of territorial integrity, China was forced to yield possession of strategic Hong Kong to Britain in the 1842 Treaty of Nanking and to grant special trading concessions and extraterritorial rights to various European nations and the United States through the latter half of the 19th century. In 1898, the Philippines, held by Spain since the 1500s, came under control of the United States.

A few Asian countries escaped domination by the Western powers during the colonial age. Thailand (historic Siam) formed a buffer between British and French colonial spheres in peninsular Southeast Asia. Japan withdrew into almost complete seclusion in the mid-17th century but emerged in the later 19th century as the first modern, industrialized Asian nation and soon acquired a colonial empire of its own. Korea also followed a policy of isolation from foreign influences until 1876, when Japan began to colonize it. It is noteworthy that Japan, South Korea, and Thailand would emerge as very

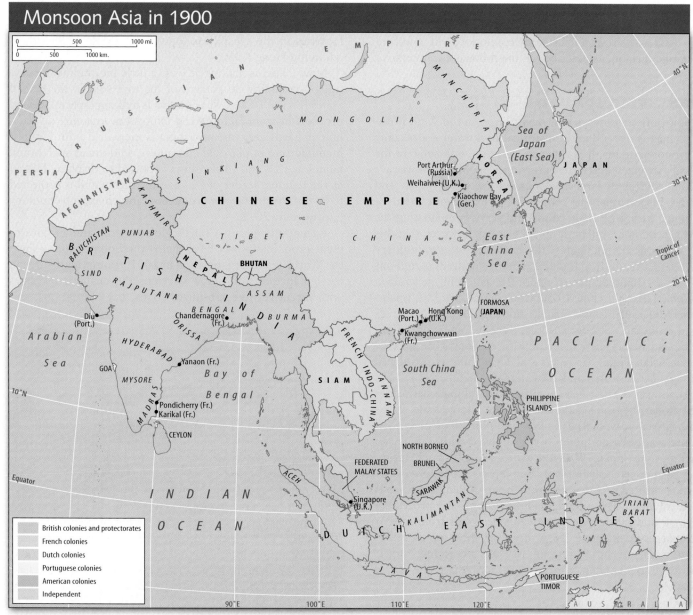

Monsoon Asia in 1900

Active Figure 10.15 Colonial realms and independent countries in Monsoon Asia at the beginning of the 20th century. *See an animation based on this figure, and take a short quiz on the facts or concepts.* Source: From "The World in 1900." in Atlas of World History, R. R. Palmer, ed. (Chicago: Rand McNally, 1957), pp. 169–171.

WORLD
REGIONAL
Geography⊕Now™

or relatively prosperous countries, whereas most of the former colonies lagged in economic development. Dependency theorists point to these discrepancies to affirm their argument that colonization hampered development.

Monsoon Asia was an extraordinarily profitable region for the colonizers. Western nations extracted vast quantities of tropical agricultural commodities and in turn found large markets for their manufactured goods. Westerners also invested heavily in plantations, factories, mines, transportation, communication, and electric power facilities. Some of the region's most important cities, including Shanghai in China,

Kolkata (Calcutta) in India, and Singapore, were developed mainly by Western capital as seaports serving Western colonial enterprises.

Western domination of Asia ended in the 20th century because the two world wars weakened the West's ability to conduct its colonial affairs in the region, because Japan rose to great power status and challenged the West early in World War II, and because effective anticolonial movements arose in nearly all areas subject to European control. After World War II, all colonial possessions in Asia gained independence. The last to revert were Hong Kong (returned by Britain to

China in 1997) and Macao (returned by Portugal to China in 1999). Until this era of independence, however, 20th century Asia was a region marked by revolution, war, and considerable turmoil, as described in the following chapters.

10.4 Economic Geography

Despite any stereotypes that might linger in the West, Monsoon Asia is not a quaint, historical backwater—especially in economic terms. It is a dynamic region with some of the world's strongest and fastest growing economies. The world's second and sixth largest economies are there: Japan and China, respectively. The economies of China and India are surging, with both their goods and services inundating the global economic system. It must be noted, however, that this growth is occurring against the backdrop of extraordinary poverty, particularly in India, and many hundreds of millions of people remain desperately poor. Despite being generally a region of less developed countries, Monsoon Asia is also home to the surprisingly strong, industrialized, or industrializing economies of South Korea, Taiwan, Singapore, Thailand, and Malaysia, which came to be known as the economic **Asian Tigers** (or Tigers and **Tiger Cubs** to distinguish stronger South Korea and Taiwan from the others).

Japan was the first Asian country to develop modern cities and modern types of manufacturing on a large scale. It has been a major industrial power for more than a century. China and India are much better supplied with mineral resources than Japan. With the world's largest populations, they also have cheaper labor. But Japan's labor force is, at least for the time being, generally more skilled, and it leads Asia in high **value-added manufacturing**—the process of refining and fabricating more valuable goods from raw or semiprocessed materials. Now, however, there are signs that Japan's preeminence is waning as China's economy surges. The biggest economic story in Monsoon Asia is the rise of China.

The Dragon Rises

In conventional terms of gross domestic product (total output of goods and services), China has the world's sixth largest economy. The more intriguing measure of purchasing power parity—which takes account of price differences between industrialized countries like the United States and emerging market countries like China—places China as the second largest economy, after the United States (India is fourth by this measure). However it is weighed, China's economic growth is soaring, with an average annual rate of about 8 percent between 1999 and 2004. China's exports (along with imports) doubled in those 5 years, growing as much as 15 percent in a single year. China's share of the world's goods and services also doubled between 1991 and 2004 to 13 percent (compared with 16 percent for the European Union and 21 percent for the United States). Just a few years ago, China was notable for making low value-added products—cheap toys, for example—that the United States and other richer

countries really had no interest in making. There were shirts and shoes, too, most of them manufactured in the 22,000 factories in the special economic zone (SEZ) of Shenzhen, adjoining Hong Kong.

Now China is making at least a little bit of almost everything—including 80 percent of the toys sold in the United States (Figure 10.16). The country is making especially rapid progress in manufactures and exports of information technology hardware. Already China is the world's third largest manufacturer of personal computers, sold under such brand names as Dell and IBM. China is poised to displace Taiwan as the regional leader of PC manufacturing and is projected to be the world's largest producer of information technology hardware by 2012.

Recent changes in China's global status and in government economic policy have fueled the boom. China joined the World Trade Organization in 2001, prompting a surge in foreign investment. China sweetens the deal for investors with incentives like subsidized loans, tax exemptions, and

Figure 10.16 All of these electronics goods in an American Radio Shack store—even those bearing the name of the Japanese firm Sony—were manufactured in China.

50 percent discounts on land prices. The addition of Chinese brawn and brains creates a nearly perfect investment climate. Chinese labor is much less expensive than is typical of most more-developed countries, and China has a rapidly growing pool of university trained engineers and other innovators.

China's economic boom is having enormous consequences for other countries, particularly its Asian neighbors. Japanese electronics companies are cutting costs by moving increasing amounts of production to China. Surging investment in China is linked to disinvestment elsewhere, especially in Southeast Asia. That region's Tigers or Tiger Cubs—Malaysia, Singapore, and Thailand—dominated the 1990s with manufactures and exports of medium- and high-tech industrial products. With its growing consumer appetite, China was buying such goods from these countries. But now the manufacturing is shifting to China from those countries, which are still smarting from an economic downturn in the late 1990s. South Korea's Samsung will soon make China the main base for production of its flat screens and computers. Japan's Toshiba makes televisions only in China. Dell recently moved some of its computer-making facilities from Malaysia to China. Between 1999 and 2004, Singapore lost more than 42,000 jobs, most of them to China. Motivated by lower costs in China, Japan's largest electronics manufacturer, Matsushita, has eliminated about 40 percent of its production and sales contractors in Southeast Asia and shifted new assets to China.

Formerly hot Southeast Asian economies may be increasingly relegated to supplying low-end products, especially food and raw materials, to China. In turn, they may end up buying cheap Chinese manufactured goods, harming their own fledgling industries. They may try to counter this trend by developing higher value niche products to meet demands in China. Thailand, for example, might boost manufactures of healthcare products, while Singapore focuses on biomedical products and financial services. Poorer Indonesia will probably be able to count on helping to satisfy China's growing appetite for natural gas to fuel its southern industrial cities. Meanwhile, the critical economies of Japan, Taiwan, and South Korea are increasingly vulnerable to China's ascendancy, as investors are lured from those countries by China's lower labor costs and its potential 1.3 billion consumers.

China's rise is also chipping away at U.S. economic dominance in the region. Most analysts feel it is just a matter of time before China eclipses the United States as the region's most essential trading partner. In the meantime, the U.S.–China relationship is growing, mainly in China's favor. Of China's exports, 35 percent go to the United States (its largest single trading partner), but its imports from the United States lag far behind. This U.S. trade deficit with China threatens to become a sore point in relations between the countries. Textile and other industries in the United States are pressuring the U.S. government to pass punitive tariffs on Chinese imports, a move that could precipitate a costly trade war.

Figure 10.17 Counterfeited goods made in China and sold throughout Asia bear the brand names of U.S. and other manufacturers but are sold to happy consumers at a fraction of the cost of the genuine product. Sellers like these can grab their DVDs and CDs and flee at the first sign of a police raid.

The United States is also furious at China's reluctance to crack down on counterfeiting or "piracy" of American products, particularly computer software and DVDs of Hollywood's latest films. China is the epicenter of prolific Asian trade in pirated products. Authorities in the region periodically crack down on sales of such goods, but the vendors are usually on a hair trigger, ready to disappear on a moment's notice with their goods in tow, only to reappear moments later when police leave (Figure 10.17). There is also growing concern in the United States over the perceived loss of jobs through outsourcing to India (see Definitions and Insights, page 292).

The Green Revolution

Most Asian countries do not want to improve their economies only by enhancing their outputs of services and manufactured goods. They also want to boost agricultural self-sufficiency and crop exports. Revolutionary changes in crops are at the heart of efforts to reshape Asia's agricultural economies. One of the most significant agricultural innovations in Asia during that past century is the development of new seed types and innovations in planting, cultivating, harvesting, and marketing of crops, efforts known collectively as the Green Revolution. These developments are not exclusive to Monsoon Asia, but this is where most of the growth and investment have taken place.

Ever since 1962, when the International Rice Research Institute (IRRI) was founded in the Philippines, there have been efforts to use science to increase food yields (particularly of rice) to stave off hunger and generate export income. Notable success has been achieved in breeding the new high-yielding varieties of seed stock, and there has been a large upsurge of production in certain areas where the new strains have been widely introduced. Many of the new varieties are bioengineered or **genetically modified (GM) crops**. This

Definitions + Insights

Outsourcing

Outsourcing (also known as **offshore outsourcing** and **offshoring**) is the flight of technology and other jobs from countries with high manufacturing and service costs to countries with low manufacturing and service costs. India is the world's largest recipient of these jobs, and the United States is the leading outsourcer.

In the United States, there is growing concern and even alarm over the trend of outsourced jobs to India. Many of these jobs are in computer programming and a wide variety of technical support, from computer problem phone calls to x-ray diagnosis. Many are mundane, like bill processing and order taking. India is an especially attractive venue for outsourcing for several reasons. It has a large population of well-educated people, most of whom speak excellent English. Those who work the telephone call centers typically receive training in vernacular American English and strive to perfect their American accents while shedding their Indian and British pronunciations. The bottom line is another major advantage: Indian skilled labor is cheap. The average computer programmer in the United States earns $80,000 per year, while his or her counterpart in India earns $20,000 per year.

Finally, the Internet and superb telephone communications, ironically combined with India's physical distance from the United States, provide some remarkable opportunities. While people in the United States sleep, the day shift is on in India. An American doctor can x-ray a patient in the afternoon and by early the next morning have the analysis—performed overnight in India—ready. On the down side, when the American consumer places a technical service call for a computer problem, he or she may be speaking with a very tired Indian employee working the graveyard shift. There are many accounts of fatigue and burnout among Indian call center employees.

What about the loss of American jobs due to outsourcing? An estimated 3.3 million U.S. service jobs will move offshore by 2015. While this sounds alarming, it represents somewhat over 200,000 jobs per year, which is a small fraction of the approximately 3.5 million jobs that are created in a typical year in the United States. Outsourcing companies generally argue that reducing the costs of their services reduces inflationary costs for their consumers and improves efficiency and productivity. This rationale is, of course, cold comfort to the American worker whose job has moved offshore.

---

means that through precise manipulation of the genetic components of a crop, it can not only produce a higher yield but also be more resistant to drought, flood, or pests or generate a higher amount of a desired nutritional component such as vitamin A. China, second only to the United States in the global biotechnology industry, has produced genetically engineered rice, corn, cotton, tobacco, sweet peppers, petunias, and poplar trees.

Because the underlying premise of the Green Revolution is economic—more crops will be produced at lower cost for higher revenue—most countries in the region are racing to increase their biotechnological output, especially in an effort to catch up with China and ensure they do not lose positions in the global marketplace. Malaysia is building "Biovalley," a biotechnology research park near Kuala Lumpur, and Indonesia is constructing a similar "Bioisland." One of Biovalley's goals is to create palm oil trees genetically modified to produce the raw material for specialized plastics used in medical devices (Figure 10.18). South Korea, Japan, and India have large and well-funded biotechnology research programs. Japan has so far resisted production of genetically modified food crops out of fears of possible health hazards, a concern that most Europeans and their governments have.

For Asian farmers to capitalize fully on the Green Revolution, they must overcome many obstacles. Success requires levels of capital that are often beyond means of peasant farm-

Figure 10.18 Palm oil trees in Malaysia are being genetically modified to produce specialized plastics. The palm oil comes from the cluster of nuts at the base of the fronds.

ers, landlords, and governments to provide. Such expenditures are needed for water-supply facilities (for example, the tubewells that have burgeoned by hundreds of thousands in the Indo-Gangetic Plain of the northern Indian subcontinent), chemical fertilizers, and chemicals to control weeds, pests, and diseases. And as agriculture becomes more mechanized, considerable increases in the costs of machinery, fertilizers, and fuel have to be borne by farmers who have, in many cases, had little experience with the cash economy. Governments must improve transportation so that the large quantities of fertilizer required by the new seed varieties can be delivered in a timely fashion. Not only must there be these associated infrastructural changes to support this "revolution," but the crop calendar of the farmer becomes much less forgiving because many of the newest seed grains demand more precise water, fertilizer, and cultivation requirements than traditional grains. Grain-storage facilities, now subject to plundering by rats, must be improved.

Overcoming the financial obstacles is rendered more difficult by the widespread system of share tenancy. Farmers who are share tenants generally have no security of tenure on the land and thus cannot be sure that money they invest in the Green Revolution will benefit them in the future. They may not wish to assume any additional risk, even if credit on reasonable terms is available. If they remain on their holdings, landlords may take up to half of the increased crop but bear little or none of the additional expense. Landlords in turn may be content to collect their customary rents without expending the additional capital necessary in this new mode of farming, or they may try to turn tenants off the land to create larger spatial units that they themselves can farm more profitably with machinery and hired labor. In fact, landowners with large holdings often become the chief beneficiaries

of the new technology, with many smaller farmers becoming a class of landless workers hired for low wages on a seasonal basis or migrants in the rural to urban migration stream.

There are other problems associated with the Green Revolution. Economic dislocations result when rice-importing countries become more self-sufficient, thus causing hardships for rice exporters. Large infusions of the agricultural chemicals associated with high-yield varieties have negative repercussions on natural ecosystems. On a much wider scope, there is concern that the development of a limited number of high-yield crop varieties grown in vast monocultures will dramatically reduce the genetic variability of crops, in essence interfering with nature's ability to adjust to environmental changes.

10.5 Geopolitical Issues

In the Middle East and North Africa, and in Russia and the Near Abroad, principal geopolitical concerns focus on the production and distribution of energy resources. In Monsoon Asia, by contrast, some of the most serious geopolitical issues are prospects for what may be done with weapons created from a particular energy source: nuclear energy. There are also concerns about traditional fossil fuels (see Geography of Energy), Islamist terrorism, and the security of shipping lanes.

Geography of Energy

The Spratly Islands

About 60 islands make up the Spratly Island chain, which lies in the South China Sea between Vietnam and the Philippines (see Figure 12.1, page 326). They are an idyllic tourist destination, where divers can hire luxury boats to explore the coral reefs and palm-lined beaches of remote atolls. However, the islands are much more significant for their strategic location between the Pacific and Indian Oceans. During World War II, Japan used the islands as a base for attacking the Philippines and Southeast Asia. Still more significant, as much as $1 trillion in oil and gas may lie beneath the seabed around the Spratlys.

Not surprisingly, many nations covet control of the Spratlys. Six nations claim some or all of the islands. China, Vietnam, and Taiwan claim sovereignty over all of them, while Malaysia, Brunei, and the Philippines claim some of them. During the Cold War, the competing claimants felt it was too hazardous to push their claims on the islands. As the Cold War drew to a close, however, the situation became more volatile. All the contenders except Brunei placed soldiers, airstrips, and ships on the islands. In 1988, the Chinese Navy invaded seven of the islands occupied by Vietnam, killing about 70 Vietnamese sol-

diers. In 1992, China again landed troops in the islands and began exploring for oil in a section of the seabed claimed by Vietnam. In 1995, China moved to expand its territorial claims on islands already claimed by the Philippines and, since then, has built what it calls "shelters," but which look like fortifications to Filipinos. Late in the 1990s, Malaysia occupied two disputed Spratly Islands and began building on one of them. In 2004, Vietnam began renovations of an airport on one of the islands, saying it was necessary to boost tourism there. China condemned the construction as a violation of its territorial sovereignty.

Indonesia, which has no claims on the Spratlys, sponsored unofficial workshops on joint efforts in oil exploration among the six claimants in an effort to defuse the emerging crisis. But so far, China, which claims not only the Spratlys but also all the South China Sea as its own, has refused to discuss the issue. There are fears that as the countries' petroleum needs grow, each will seek more aggressively to gain control of the Spratlys. The smaller powers fear that China could turn the islands into a kind of permanent aircraft carrier that could be used to dominate them militarily. An incident in these remote islands could trigger a much wider and more serious conflict in Asia.

Nationalism and Nukes on the Subcontinent

With its independence from Britain in 1947, India emerged as an avowedly secular democratic state. But after the 1998 victory of the Hindu nationalist party, the Bharatiya Janata Party (BJP), hopes for new privileges emerged among India's Hindu majority. Among BJP supporters, there was hope for a "Hindu bomb," a counterweight to a long-feared "Islamic bomb" in neighboring Pakistan.

On May 11, 1998, much to the surprise of U.S. and other Western intelligence agencies, India conducted three underground nuclear tests in the Thar Desert. With the blasts, India's government seemed to be trying to stake India's claim as a great world power, exhibit its military muscle to Pakistan and China, and garner enough political support for the BJP to form an outright parliamentary majority in the future. Initial reactions among India's vast populace were highly favorable; 91 percent of those polled within 3 days of the event supported the tests. Outside India, there was alarm. India had defied an informal worldwide moratorium on nuclear testing that went into effect in 1996, when 149 nations (not including India and Pakistan) signed the **Comprehensive Test Ban Treaty** (also known as the **Nuclear Non-Proliferation Treaty,** or **NPT**), which prohibits all nuclear tests.

The world's eyes quickly turned to India's neighbor to the west. Governments pleaded with Pakistan to refrain from answering India with nuclear tests of its own, arguing that Pakistan would have a public relations triumph it if exercised restraint: It would appear to be a mature and responsible power, whereas India revealed itself as a dangerous rogue state. But Pakistan's leaders felt obliged by their population's demands for a tit-for-tat response to India's blasts and soon followed with six nuclear tests. India and Pakistan thus joined only five other nations—the United States, Russia, China, the United Kingdom, and France—in acknowledging that they possess nuclear weapons.

There is disagreement about what this recent regional and global shift in the balance of power means. Some analysts fear an escalating nuclear arms race that, perhaps ignited by a border skirmish in Kashmir, could lead to the **mutually assured destruction (MAD)** of Pakistan and India. There are concurrent fears of nuclear war between China and India. Others, particularly within South Asia, argue that the weapons represent the best deterrent against conflict, as they did for decades between the countries of NATO and the Warsaw Pact. However, few people anywhere argue with the contention that the nuclear rivalry between India and Pakistan has taken an enormous economic toll in two nations that need to wage war on poverty.

To the United States, both India and Pakistan are pivotal countries, defined by influential historian Paul Kennedy as those whose collapse would cause international migration, war, pollution, disease epidemics, or other international security problems. The disposition of their nuclear arsenals is one of the main reasons for their inclusion on this list. If the government of Pakistan were to fall, Pakistan's nuclear weapons could come into the possession of a rogue element or a new government hostile to the West. India also has much to fear. Either a right-wing or an Islamist government would be far more likely than the current regime to take India on in the disputed Kashmir region. Such a confrontation could set the stage for a nuclear exchange between Pakistan and India, not only decimating those countries but sending economic shock waves around the world. India's recently surging economy has been a major force in India's recent diplomatic overtures to Pakistan; nothing discourages investment like war or the fear of it.

Pakistan became even more pivotal with the events of 9/11. Up to that point, the United States had slapped tough economic sanctions against Pakistan (and less severe ones against India) for the nuclear weapons tests. The United States pursued a diplomatic courtship with India, both as a counterweight against China (India's longtime foe in the region) and as a means of expressing displeasure with Pakistan (China's ally) for helping Afghanistan's Taliban to harbor Osama bin Laden. But the United States and Pakistan did an abrupt diplomatic about-face and embraced one another after 9/11. Pakistan instantly dropped support for the Taliban and quietly allowed the United States to use its territory to prepare for the assault on the Taliban and al-Qa'ida in Afghanistan. In return, the United States forgave much of Pakistan's debt to the United States and lifted its sanctions against Pakistan. To avoid isolating India, the United States also lifted its postnuclear test sanctions against that country. The United States has dramatically strengthened military ties with both countries, recognizing Pakistan in 2004 as a major non-NATO ally—one of just eleven countries to have that designation—and acknowledging India the same year as a "strategic partner." These designations open the doors of both countries to major new shipments of war matériel.

With the apparent defeat of the Taliban and al-Qa'ida in Afghanistan after 9/11, significant populations of both groups retreated to and regrouped in western Pakistan, particularly in the semiautonomous Federally Administered Tribal Areas (FATA; see Figure 11.17, page 316). Here the populace is mainly Pashtun and is sympathetic to the causes of their Taliban ethnic kin and their al-Qa'ida spiritual kin (Figure 10.19). This is an area where Pakistani government authority has long been kept at bay and where it was considered extremely difficult to insert Pakistani forces on a sustained, aggressive basis. In 2004, bowing to American pressure, Pakistan's government began to undertake military operations in the FATA.

On the other side of the border, U.S. military forces joined the hunt for al-Qa'ida and Taliban personnel. This cooperation carried enormous risks for Pakistani President Pervez Musharraf. His government's unprecedented intervention in the FATA, even without the direct insertion of U.S. troops, had the potential to lead to a popular uprising, a military coup, or both. Not only the Pashtuns, but the vast majority of all of Pakistan's ethnic groups, were strongly against U.S.

Joe Hobbs

Figure 10.19 The Pashtun of western Pakistan are a proudly independent people who are not receptive to American interests in the region.

interests in the region, and two factions within the army (one right-wing, another Islamist) stood by ready to challenge Musharraf's authority. If the government were to fall, there would be grave concerns about Pakistan's nuclear weapons.

East on the "Axis of Evil"

There are also major concerns about nuclear weapons in Northeast Asia. Ever since suffering Hiroshima and Nagasaki, Japan has had an official policy never to develop or use atomic weapons. But Japan worries about the potential nuclear threat from three adversaries, all of which possess nuclear weapons: Russia, China, and North Korea. The Japanese feel that the West dismissed potential threats from Russia too readily when the USSR dissolved. Japan still has territorial disputes with Russia, particularly involving the four Kuril Islands of Kunashiri, Etorofu, Shikotan, and Habomai, which the government of Josef Stalin seized at the end of World War II and which are just off the coast of northeast Hokkaido (see Figure 14.1, page 382). Japan and China also have a territorial dispute concerning the East China Sea, and if major oil reserves are discovered there, as anticipated, relations between the two countries could deteriorate. China continued to test nuclear weapons through 1996, adding to tensions in Japan. Finally, Japan fears reunification in Korea, which, as a former Japanese colony from 1910–1945, has a particular historic dislike of Japan. There is speculation that Japan may do the unthinkable and the officially disavowed—develop nuclear weapons—to counter the perceived threat of North Korea's nuclear weapons program.

The world looks nervously at the troubled relations between North and South Korea and between North Korea and the West. A crisis flared in 1994 when North Korea refused to permit full inspection of its nuclear facilities by the **International Atomic Energy Agency (IAEA)**. The country was suspected of separating plutonium that could be used in making nuclear bombs. Over time, the United States, South Korea, and Japan worked out an agreement in which North Korea would agree to freeze development of nuclear weapons in exchange for the others providing fuel oil and assistance in building nuclear power plants. Those nuclear reactors would be of the "light water" variety, much less likely than North Korea's plutonium-based reactors to be "dual use" for both military and civilian purposes. The Clinton administration trumpeted the agreement as a success, but it had several flaws. A major flaw was that North Korea did not have to dispose of its existing nuclear fuel; it simply had to stow it safely away—from where it could quickly be reactivated.

North Korea pressed ahead with a program to build missiles capable of carrying nuclear warheads and, in 1998, launched such a missile (minus the warhead) over Japan. This incident prompted the United States to renew its resolve to establish an antimissile defensive "shield" over the United States—the so-called **Star Wars initiative** dating to the Reagan administration of the 1980s. Some analysts believe the United States is now reluctant to see the two Koreas reunite because it would remove much of the justification on which the missile shield program is based. It might also put pressure on the United States to reduce its military presence in the western Pacific. They add that Japan, too, would not want to see Korean reunification because it would remove Japan's justification for building up its defenses, which are designed less for confrontation with North Korea than with China.

President George W. Bush came into office with a hard line against North Korea, effectively suspending dialogue and technical assistance for the still uncompleted nuclear power plants. In 2002, he proclaimed that North Korea was one of three countries making up an **"axis of evil,"** along with Iraq and Iran, signaling that the United States was more interested in confronting than accommodating North Korea. Late that year, North Korea dropped a virtual bombshell on the United States by admitting—after being confronted with evidence collected by U.S. intelligence agencies—that it did indeed have an active nuclear weapons program (developed, as it turns out, with Pakistani technical assistance). This admission followed earlier strident denials of such a program by North Korean officials. The American side, focused on developing the military campaign in Iraq, had been content to let the problem of North Korea's weapons program simmer quietly in the background. By most measures, North Korea should also have been content to keep quiet; the United States, South Korea, and Japan had for almost a decade been providing food and fuel to the economically beleaguered country, and any belligerence by North Korea could cut that aid.

What, then, was North Korea hoping to accomplish by revealing its nuclear weapons program? There are several

Joe Hobbs

Figure 10.20 A little humor eases the tension for U.S. troops serving along the world's tensest border separating North and South Korea. Even a minor incident here could touch off a conflagration that might take a huge toll in human lives.

possibilities. The least likely is that North Korea was instigating a military confrontation with the United States and its allies South Korea and Japan. In any war scenario, those allies would obliterate North Korea. The United States is obliged to defend South Korea in the event of a war with the North and would effectively wield its huge military advantage (Figure 10.20). However, that victory would come at an enormous price for South Korea and possibly Japan. North Korea has a standing army of nearly 1 million soldiers, an impressive military arsenal including short- and medium-range missiles, a stockpile of chemical and biological weapons that it might not be shy to use, and now, presumably, at least a few nuclear weapons that it could deploy. The casualties in South Korea would be enormous, whether from a conventional or unconventional assault from the North.

It is much more likely that by disclosing its nuclear weapons program, North Korea sought guarantees that it could avoid war and also gain even more assistance from the West. Since the early 1990s, its nuclear weapons program has been the only leverage that North Korea has had to coax desperately needed supplies from abroad. North Korea had only to mention the prospect of reactivating its nuclear weapons program to get more concessions from the United States and its allies—especially food aid during a succession of droughts and floods in the 1990s. In effect, North Korea was extorting money from the United States and its allies, which have been happy to ante up as a means of containing North Korea. By admitting to the nuclear weapons program, North Korea may feel it can get far more aid in technology, fuel, food, or other forms.

In addition, North Korea has been unable to get a promise it has sought from the United States since 1994: that the U.S. will not use nuclear weapons against it. North Korea is probably hoping that its nuclear program disclosure will lead to this pledge. It has a historical precedent for believing that a nuclear card can be played for protection; in 1962, during the Cuban Missile Crisis, U.S. President John Kennedy promised that the United States would not invade Cuba. In exchange for this and other assurances, the Soviet Union recalled the nuclear weapons it was dispatching to Cuba. In sum, North Korea feels it has little to lose, and may gain something, by spilling its nuclear secret.

In the long run, the United States will have to rely on China and Russia to shoulder some of the political and economic burdens of placating North Korea. China in particular is keen to play peacemaker with North Korea; this helps China gain more prominence on the world stage. The United States may even be able to "trade" this greater role for China in exchange for China agreeing to support American interests elsewhere—in Iraq, for example. North Korea, often dismissed as a runt rogue, has a seat at the table with the big players and knows how to play the game.

Islands, Sea Lanes, and Islamists

As related in Chapter 12, Indonesia—another pivotal country from the U.S. perspective—is a very ethnically complex nation whose integrity is threatened from a variety of secessionist movements in locales as far apart as Aceh in the west and Papua (formerly Irian Jaya) in the east (see Figure 12.F, page 348). There is a **domino effect** scenario for Indonesia that concerns foreign powers: If one province falls, many others will, and the largest country in the region will break up. Countries like the United States fear, for example, what would happen to vital international shipping lanes should several new and perhaps militant states emerge from Indonesia's fragmentation. There have already been scores of pirate

348

attacks on vessels plying the strategic Strait of Malacca, a critical chokepoint through which a quarter of the world's trade passes, including two-thirds of the world's shipment of liquefied natural gas and half of all sea shipments of oil (it ranks second only to the Strait of Hormuz as an oil shipping lane). Security in the Strait of Malacca is a source of great concern to Japan, which imports 80 percent of its oil on ships that pass through the strait. Aceh's vast natural gas deposits and Irian Jaya's copper and gold are also seen as critical in the global economy, so like Indonesia, many world powers are anxious to see that they stay in the "right" hands. The United States pumps a large amount of foreign aid into Indonesia, but the American presence is increasingly unwelcome because so many Indonesians perceive the post-9/11 United States as anti-Muslim.

U.S. intelligence agencies fear that Southeast Asia, and in particular, Indonesia, will emerge as an Islamist terrorist hearth to replace Afghanistan. Al-Qa'ida and its affiliate organizations are known to be active in Southeast Asia. Two of the nineteen men who carried out the 9/11 attacks planned part of the operation in Malaysia's capital, Kuala Lumpur. Malaysia is an unlikely refuge for groups like al-Qa'ida, mainly because it sees itself as a progressive engine of high-tech economic growth and would be loath to suffer the sanctions or other fallout associated with harboring terrorists. However, Malaysia is a convenient transit and staging point for terrorist interests, and some members of al-Qa'ida and affiliated groups are Malaysian nationals. Citizens of any predominantly Muslim country may enter Malaysia without obtaining a visa (and thus escape most opportunities for a background check).

Mainly Chinese, prosperous, aggressively policed Singapore is the region's most unlikely terrorist refuge of all. But with its 17,000 resident American population and its numerous Western commercial and military activities (including regular visits by U.S. naval vessels), Singapore is an attractive target. Evidence uncovered in an al-Qa'ida "safe house" in Afghanistan that was examined by U.S. intelligence experts revealed a plot that would have carried out ammonium nitrate truck bombings against U.S. and British embassies and other facilities in Singapore in 2002. An al-Qa'ida "sleeper cell" based in Singapore had been activated from Malaysia by Riduan Isamuddin (aka Hambali), an Indonesian citizen who hosted the two 9/11 hijackers, who was a suspect in the U.S.S. *Cole* bombing on their visits to Kuala Lumpur, and who was arrested by intelligence authorities in 2003. These sleepers, eight of whom had been trained by al-Qa'ida in Afghanistan, were Indonesians.

Indonesia is a secular rather than religious state, and a moderate form of Islam prevails in politics and in everyday life. What concerns Western intelligence authorities, and Indonesia's moderate leadership, are the more militant Islamic organizations that have emerged in the last decade and have apparently gathered strength since 9/11. One of these is **Laskar Jihad,** created in 2000 to mobilize Indonesian Muslims to fight Christians in Indonesia's Sulawesi and Muluku

Islands. Its leader, Jaffar Umar Tahlib, threatened to attack Americans throughout the country and to drive Indonesia's former president, Mrs. Megawati Sukarnoputri, from office (one of the refrains of the country's militant Islamic groups is that it is forbidden for a woman to govern a country).

Another organization with suspected terrorist links is **Jemaah Islamiah** (also known as **JI,** or the **Islamic Group**), led by Abu Bakar Baasyir (Abu Bakar Bashir), whose nominal role was preacher in an Islamic school in Solo on the island of Java. (Western authorities are much concerned about such religious schools, known as *pesantrens* in Indonesia. Rather like the *madrasas* of Pakistan and the Arab heartland of the Middle East, these are venues for both religious and secular learning, and some of the schools' curricula are imbued with virulent anti-Western content.) Baasyir admitted to having tutored 13 of the men arrested in connection with the plot to blow up Western targets in Singapore. These included Hambali, al-Qa'ida's point man in Southeast Asia, whose regional affiliation is with JI. Intelligence authorities believe that Baasyir was also the ringleader for a series of attacks that would have been carried out against American embassies and naval vessels in several Southeast Asian countries on the first anniversary of the 9/11 attacks. Finally, after the ferocious bombings of nightclubs on the Indonesian island of Bali in 2002 in which both al-Qa'ida and Baasyir were implicated, Indonesian authorities imprisoned Baasyir.

Indonesia has much going for it as a potential new hearth for al-Qa'ida affiliated activities. Much like Afghanistan was prior to 9/11, it has a predominantly Muslim population, including some extremist factions that would be willing to provide safe haven; it has a largely poor and otherwise disaffected population (in theory, providing recruits for a militant cause); it includes remote locales that make suitable weapons and tactics training grounds; and it has regions that the government is unwilling or unable to control. In 2003, the director of Indonesia's national intelligence agency acknowledged that al-Qa'ida had set up training camps in Indonesia. U.S. authorities now believe that Southeast Asia, with Indonesia as its centerpiece, has the world's highest concentration of al-Qa'ida operatives outside Pakistan.

Why, then, does the United States not have troops on the ground in Indonesia, or advisers there (as in the Philippines) to train Indonesian forces in antiterrorism tactics? The problem is that Indonesia is even more difficult than Pakistan for U.S. antiterrorism forces to operate in: It is more geographically complex and even less receptive to U.S. intervention. Perceived intervention by the United States could be enough to spark a now largely neutral populace into violent anti-Americanism and perhaps bring down Indonesia's government. Washington fears that many states, some of them anti-Western, could emerge in Indonesia's place and threaten American interests in the country—notably its oil, natural gas, copper, and location astride vital shipping lanes.

This concludes an introduction that has set the stage for further exploration of land and life in the subregions and countries of Monsoon Asia.

SUMMARY

- Monsoon Asia includes the countries of Japan, North and South Korea, China, Taiwan, Pakistan, India, Sri Lanka, Bangladesh, Bhutan, Nepal, Maldives, Myanmar (Burma), Laos, Thailand, Cambodia, Vietnam, Malaysia, Singapore, Indonesia, the Philippines, Brunei, East Timor, and numerous islands scattered along the edges of this major continental block. The term *monsoon* is used because of the central role this wind and precipitation pattern plays throughout this region.

- This is the most populous world region, with fully 54 percent of the world's people. It includes the world's most populous countries, China and India. Population growth rates in the region vary widely, from almost zero growth in Japan to almost 3 percent per year in Pakistan.

- Three concentric arcs make up the broad physiography of the region. The inner arc includes the highest mountain ranges of the Himalaya, Karakoram, and Hindu Kush. The middle arc is composed of major floodplains, deltas, and relatively low mountains. These include the Indus, Ganges, and Brahmaputra Rivers in South Asia; the Irrawaddy, Chao Praya (Menam), Mekong, and Red Rivers in Southeast Asia; and the Chiang Jiang (Yangtze) and Huang He (Yellow) in East Asia. The outer arc is an offshore fringe of thousands of islands, including the archipelagos of the East Indies, Philippines, and Japan.

- The region's climate types and biomes include tropical rain forest, savanna, humid subtropical, humid continental, steppe, desert, and undifferentiated highland. Varying types of agriculture are practiced depending on local conditions, including shifting cultivation and wet rice cultivation. Wet rice cultivation produces very high yields and is associated with dense human populations.

- Although Monsoon Asia has some of the world's largest cities, about 63 percent of the region's population is still rural. Many uniquely Asian traditions helped shape village settlement planning and home design.

- Monsoon Asia's ethnic and linguistic compositions are diverse. Major language families are Indo-European, Dravidian, Sino-Tibetan, Altaic, Austric, and Papuan.

- Major religions and sociopolitical philosophies of Monsoon Asia are Hinduism, Islam, Buddhism, Confucianism, Daoism, and Christianity.

- Great Britain, the Netherlands, France, and Portugal were the most important colonial powers in this region. Most of these domains were relinquished by the middle of the 20th century, and the later British return of Hong Kong and the Portuguese return of Macao (both to China) closed the colonial period.

- Japan has the region's strongest economy, and it is second only to that of the United States. Recent decades have seen strong economic growth among the Tigers and Tiger Cubs of the region, including South Korea, Taiwan, Singapore, Thailand, and Malaysia. The most important emerging economic power is China, whose large and inexpensive labor force has recently been attracting investment away from other parts of the region. India's well-educated and inexpensive labor force is contributing to its strong economic growth. There are fears in the United States about the outsourcing of American jobs to India.

- The Green Revolution is the name of a broad effort to increase agricultural productivity in dominant crops through expanded use of chemical fertilizers, new seed stock, a more exacting crop calendar, and better access to good market roads, markets, and credit. Biotechnology has produced crops that are more drought and pest resistant and capable of creating much higher yields, but genetic engineering of crops has perceived risks. Profits and other benefits from the Green Revolution are not uniformly spread. It has been difficult to overcome the entrenchment of strong traditional agricultural practices.

- There are several major geopolitical issues in Monsoon Asia. The traditional enemies Pakistan and India (both pivotal countries from the U.S. point of view) now possess nuclear weapons. There are fears that destabilization in Pakistan might allow the weapons to fall into the "wrong" hands. Pakistan's support of the United States in its war on terrorism is very risky because a popular backlash could bring down the government. North Korea has nuclear weapons and is apparently trying to use them as bargaining chips to get more food aid and other assistance from the West. Antigovernment and anti-Western Islamists are active in Indonesia, another country viewed as pivotal by the United States. Strife and fragmentation in Indonesia could threaten vital oceanic shipping lanes.

KEY TERMS + CONCEPTS

Terms in blue are also defined in the glossary.

Altaic language family (p. 284)
  Ainu (p. 284)
  Japanese (p. 284)
  Kazakh (p. 284)
  Korean (p. 284)
  Mongolian (p. 284)
  Tungus (p. 284)
  Turkic (p. 284)
  Uighar (p. 284)
ancestor veneration (p. 285)
animist, animism (p. 285)
archipelago (p. 276)

Aryan (p. 284)
Asian Tigers (p. 290)
Austric language family (p. 284)
  Austro-Asiatic subfamily (p. 284)
  Mon-Khmer (p. 284)
  Khmer (p. 284)
  Vietnamese (p. 284)
  Austronesian subfamily (p. 284)
  Barito (p. 284)
  Bugis (p. 284)
  Dayak (p. 284)
  Javanese (p. 284)

Kayan-Kenyah (p. 284)
Malay (p. 284)
Timorese (p. 284)
Tomini (p. 284)
Hmong-Mien subfamily (p. 284)
  Hmong (p. 284)
Tai-Kadai subfamily (p. 284)
  Lao (p. 284)
  Thai (p. 284)
axis of evil (p. 295)
Buddhism (p. 287)
chokepoint (p. 297)

REVIEW QUESTIONS

WORLD
REGIONAL
Geography ⊛ Now™

Assess your understanding of this chapter's topics with additional quizzing and concept-based problems at http://earthscience.brookscole.com/wrg5e.

1. What countries make up Monsoon Asia?

2. Use the concept of the three arcs to trace the outlines of Monsoon Asia's geography. What are the largest plains and river valleys of the region? What are the most important mountain ranges? What are the major island groups?

3. What are the region's most populous countries? Where are the highest and lowest population densities? Which countries have the highest and lowest population growth rates?

4. What are the monsoons? What roles do they play in climates and human activities?

5. What types of environments are shifting cultivation and wet rice cultivation practiced in? What are some of the basic methods and land-use considerations of both of these agricultures?

6. What roles have *fengshui* and Confucianism played in site selection and design of homes, schools, and villages in parts of Asia?

7. What are some of the innovations and other products that originated in Monsoon Asia and diffused from there?

8. What are the major linguistic and/or ethnic groups of Monsoon Asia? What are the principal religions and sociopolitical philosophies?

9. What were the major colonial powers in Monsoon Asia? What regions did they colonize?

10. What are the region's most significant economic trends?

11. Why are India, Pakistan, and Indonesia considered to be pivotal countries for U.S. interests? Has the United States been forced to take sides in the traditional enmity between India and Pakistan?

DISCUSSION QUESTIONS

1. Is China's one child campaign too coercive or is it justifiable intervention in view of the country's huge population?

2. Discuss how population growth leads to new and serious consequences in the contexts of both shifting and wet rice cultivation. What is the theory of Himalayan environmental degradation?

3. What practical and spiritual considerations went into the design of the traditional Korean village? How do these contrast with rural settlement patterns in your country?

4. Discuss the origins, spread, and major beliefs or ideas of Hinduism, Buddhism, Confucianism, and Daoism.

5. Discuss China's economy. What is the country producing?

What impacts does China's economy have on the region's other economies?

6. Discuss outsourcing. Is it good or bad for the countries that seem to be exporting jobs to India?

7. What are the major goals and results of the Green Revolution? What are the pros and cons of these innovations?

8. Why is the stability of Pakistan's government such an important consideration in regional and international security?

9. What is North Korea apparently hoping to gain by admitting that it has a nuclear weapons program?

10. What groups and events in Southeast Asia are tied to Islamist terrorism? Why might Indonesia become a haven for al-Qa'ida?

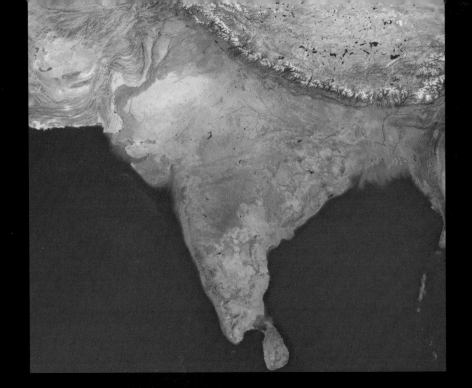

Complex and Populous South Asia

This Hindu *saddhu* is one face in the rich cultural fabric of South Asia. This ascetic is dedicated to prayer and abstention from worldly pursuits and depends on charity of the faithful.

chapter objectives

This chapter should enable you to:

- Recognize the tensions between Muslims and Hindus in South Asia that often give rise to violent conflict

- Appreciate how the partition of India created lasting problems in political relations, resource use and allocation, and industrial development

- See that food production has kept pace with enormous population growth in South Asia

- Consider the role of high technology in reshaping the economies and outlooks of South Asia's peoples

- Recognize the serious consequences of political insurgencies in Kashmir, Nepal, and Sri Lanka

WORLD REGIONAL
Geography ⊛ Now™

Look for this logo in the text and go to GeographyNow at http://earthscience.brookscole.com/wrg5e to explore interactive maps, view animations, sharpen your factual knowledge and geographic literacy, and test your critical thinking and analytical skills with unique interactive resources.

A triangular peninsula thrusts southward 1,000 miles (c. 1,600 km) from the Asian landmass, splitting the northern Indian Ocean into the Bay of Bengal and the Arabian Sea. The peninsula is bordered on the north by the alluvial plain of the Indus and Ganges (Ganga) Rivers, and north of these rise the highest mountains on Earth. The entire unit—peninsula, plain, and fringing mountains—is known as the Indian subcontinent. It contains the five countries of India, Bangladesh, Pakistan, Nepal, and Bhutan in an area about half the size of the 48 conterminous United States (Figure 11.1). Off the southern tip of India, across the narrow Palk Strait, lies the island nation of Sri Lanka, which shares many physical and cultural traits with the subcontinent. Farther to the southwest is the Maldives, a country of more than 1,000 small, low-lying islands. These seven countries make up the subregion of South Asia.

A world-class giant, India outranks the other South Asian nations in both population (1.09 billion) and area (1.1 million sq mi/3 million sq km; about the combined size of the U.S. states of Alaska, Texas, California, and Colorado). Bangladesh (141 million) and Pakistan (159 million) are much smaller than India in area and population, but both are among the world's 10 most populous countries. Nepal and Bhutan, on the southern flank of the Himalaya Mountains, are small, rugged, and remote, with fewer people (25 and 1 million, respectively). They serve as buffer states between India and China, which have engaged in sometimes violent border disputes since the early 1950s. Strife has not spared Sri Lanka, whose 20 million people have experienced a civil war for more than two decades. The 300,000 people of the Maldives enjoy relative peace and tranquillity on their isolated islands. Altogether, about 1.4 billion people, more than one of every five on Earth, lived in complex South Asia in 2004.

South Asia is a paradox. The true South Asia is found in the villages, where seven of every ten persons live. But the true South Asia is also found in its great cities: Mumbai, Kolkata, Karachi, and Dhaka are central to the identities and economies of this region's people. South Asia is ancient, modern, rural, and urban.

11.1   The Cultural Foundation

There is an enormous range of ethnic groups, social hierarchies, languages, and religions among regions of South Asia and even within single settlements. The subcontinent is the most culturally complex area of its size on Earth, and its civilizations have roots deep in antiquity. The Indian subcontinent is one of the world's culture hearths. Among the animals first domesticated in the region were zebu cattle (Indian humped cattle), which today are important livestock in many world regions. After 3000 B.C., some of the world's earliest cities (notably Mohenjo-Daro) developed on the banks of the

Overview of South Asia

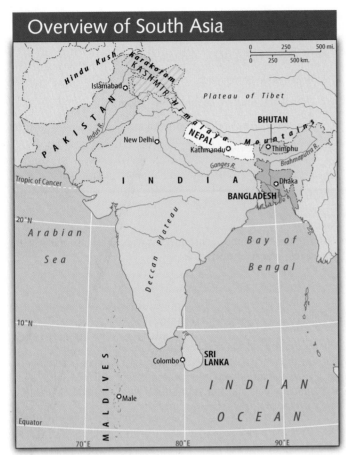

Figure 11.1 Political units of South Asia. The status of Kashmir, disputed between India and Pakistan, with some parts occupied by China, remains unresolved.

Figure 11.2 Mohenjo-Daro, on the banks of the Indus River in what is now central Pakistan, was one of the world's earliest cities, with an estimated 50,000 people in 2500 B.C. The written language of the Harappan civilization has not yet been deciphered.

Figure 11.3 For Indians and foreigners alike, the Taj Mahal in Agra symbolizes the cultural achievements of the Indian subcontinent.

Indus River in what is now Pakistan (Figure 11.2). Their peoples of the **Harappan civilization** were probably the ancestors of the Dravidians, the people represented today mainly by the Tamils of southern India.

Between 2000 and 1000 B.C. came the **Aryans** (or **Indo-Aryans**), herders of cattle and horses from central Asia who were the source of many ethnic and cultural attributes that are dominant in the subcontinent today. These include the lighter complexions than are typical of the Dravidians, the numerically dominant Indo-European languages such as Hindi, and the faith of Hinduism, which developed mainly from the beliefs and customs of the Aryans. In the eighth century, Arabs introduced the religion of Islam to what is now Pakistan. Waves of Muslim conquerors who came through mountain passes from Afghanistan and central Asia spread Islam farther afield in later centuries.

The last great Islamic invasion, by a Mongol-Turkish dynasty called the **Moguls** (**Mughals**), began in the early 16th century, soon after the Europeans first came to India by sea. Before it succumbed to rebellious local Hindus and Sikhs and to the British Empire, the Mogul empire ruled nearly the entire subcontinent from its capital at Delhi. Some of the region's most spectacular artistic and scientific achievements

date to the peak of Mogul power in the 16th and 17th centuries, including the Red Fort in Delhi and the tomb of Mogul Emperor Shah Jahan and Empress Mumtaz Mahal—the astonishing Taj Mahal in Agra (Figure 11.3).

There are major social and political divisions within the subcontinent that relate to religion. As will be seen later in the chapter, religious differences are often troublesome between the two largest religious groups: Hindus (whose religion is described in Chapter 10, page 286) and Muslims (whose religion is described in Chapter 8, pages 216–218). Hinduism was the dominant religion at the time Islam made its appearance in this region. It continues to have the largest number of adherents, including an estimated 81 percent of the population of India and 86 percent of Nepal. Muslims are majorities in Pakistan (97 percent) and Bangladesh (83 percent). Muslims make up about 12 percent of India's population, seemingly a small number except that it equals more than 130 million people, ranking India third among the world's countries with the largest Muslim populations, behind Indonesia and Pakistan.

Seldom in history have two large groups with such differing beliefs lived in such close association with each other as have Hindus and Muslims in the subcontinent. Islam holds to an uncompromising monotheism and prescribes uniformity in religious beliefs and practices. Hinduism is monotheistic for some believers but polytheistic for others and asserts that a variety of religious observances is consistent with the differing natures and social roles of humans. The exuberant and loud celebrations of the Hindu faith are a striking contrast to the austere ceremonies of Islam. Islam has a mission to convert others to the true religion, whereas most Hindus regard proselytizing as essentially useless and wrong; one is born or reborn as a Hindu, so conversion is not an issue. Islam's concept of the essential equality of all believers is in total contrast to the inequalities of the caste system endorsed in Hinduism (see page 315). Islam's use of the cow for food is anathema to Hinduism. And then there are religion-based politics, which often turn differences into discord and conflict, as will be seen later.

In addition to Muslims, there are other significant religious minorities in India. About 20 million **Sikhs** are concentrated mainly in the unusually prosperous state of Punjab. **Sikhism** is India's fourth largest religion, after Hinduism, Islam, and Christianity. Their faith emerged in the 15th century as a religious reform movement intent on narrowing the differences between Punjab's Hindus and Muslims. They regard men and women as equals and disavow the caste system recognized by Hindus. Sikh men are recognizable by their turbans, which contain their never-shorn hair, by a bracelet called a "bangle" worn on the right arm, and by their ubiquitous surname Singh, meaning "lion."

There are about 25 million Christians in India, most of whom live in the south of the peninsula; Buddhists, Jains, Parsis, and members of a variety of tribal religions make up the numerous smaller remaining religious minorities. Most of the estimated 7 million Buddhists are recent converts from among India's lowest castes. Numbering perhaps 75,000 and concentrated mainly in Mumbai, the famously entrepreneurial **Parsis** have attained wealth and economic power far out of proportion to their number. Their religion is the pre-Islamic Persian faith of **Zoroastrianism,** known mainly for its reverence of fire. Zoroastrians follow the teachings of the prophet Zarathustra, who was born around 600 B.C. in Persia. He preached that people should have good thoughts, speak good words, do good deeds, and believe in a single god, Ahura Mazda ("Wise Lord"). Persia's Zarathustras, as the Zoroastrians are also known, fled from Persia in two waves, first during the conquest by Alexander the Great around 300 B.C. and then again when Islam swept into Persia in the seventh century. They settled in China, Russia, and Europe, where they became assimilated with other cultures, and in India, where they maintained a separate existence.

India's 6 million **Jains** also have influence beyond what their numbers suggest, as they control a significant share of India's business. They live mainly in the western state of Gujarat. **Jainism,** founded on the teachings of Vardhamana Mahavira (a contemporary of Buddha, c. 540–468 B.C.), is renowned for its respect for geographic features and animal life. Jains believe that souls are in people, plants, animals, and nonliving natural entities such as rocks and rivers. Jainism has taken the principle of nonviolence (*ahimsa,* also present in Hinduism) to mean they should not even harm microbes. Therefore, Jain worshippers often wear masks to prevent inhalation of microscopic organisms, and Jains cannot be farmers because they would have to destroy plant life and living organisms in the soil (Figure 11.4).

In Sri Lanka there are two major ethnic groups, distinguished from each other by religion and language: the predominantly Buddhist **Sinhalese,** making up about 70 percent of the population, and the Hindu **Tamils,** about 15 percent (Figure 11.5). The light-skinned Sinhalese are an Indo-Aryan people who settled in Sri Lanka about 2,000 years ago. The dark-skinned Tamils, whose main area of settlement is the Jaffna Peninsula and adjoining areas in the north, are descendants of early invaders and more recent imported tea plantation workers from southern India. The migrations of Tamil tea workers from India to Sri Lanka, and their subsequent travails there, are part of the legacy of British colonial rule.

Figure 11.4 A priest in a Jain temple in Mumbai wears a mask so as not to breathe in and thereby harm microbes.

Figure 11.5 Buddhism is the religion of most Sri Lankans. These are the Colossi of Buddha and Ananda at Polonnaruwa.

11.2 Geographic Consequences of Colonialism and Partition

On a physical map, the Indian subcontinent looks like a discrete unit, marked off from the rest of the world by its mountain borders and seacoasts. But the social complexity of this area is so great that it has never been unified politically, except when Britain ruled. For more than a century, until 1947, India was the most important possession of the British Empire—the **"jewel in the crown."** India then consisted of the entire area of the present India, Pakistan, and Bangladesh. It had a classic colonial relationship with Britain, supplying the motherland with an almost endless pool of low-cost labor and valuable raw commodities like cotton and opium while providing Britain with a huge market for manufactured goods.

British colonialism left indelible marks on the subcontinent's cultural and economic landscapes, some favorable and others decidedly not. To serve the needs of their empire, the British developed the great port cities of Karachi, Bombay (now officially known as Mumbai), Madras (now Chennai), and Calcutta (now Kolkata) and India's current inland capital of New Delhi. The British created a lasting positive legacy in developing India's infrastructure, particularly the superior rail transportation networks that still assist industrial development in India, Pakistan, and Bangladesh. The vast rail network created by the British has been expanded and improved by the independent nations. In addition to transporting goods to internal markets and to ports for export, this system carries millions of passengers daily (Figure 11.6).

Around the world, there were numerous cases in which the colonizing power favored one ethnic or socioeconomic group over another, sowing the seeds for eventual discord. This was the case in British India, where the formerly subordinate Hindus came to dominate the civil service and most businesses. Many Muslims feared the results of being incor-

porated into a state with a Hindu majority, and their demands for political separation led with independence in 1947 to the creation of the two countries: the secular but predominantly Hindu nation of India and the Muslim nation of Pakistan. Pakistan had two parts, West Pakistan and East Pakistan, separated by Indian territory. In 1971, an Indian-supported revolt in East Pakistan led to the birth there of the new independent country of Bangladesh.

The partition of colonial India precipitated all kinds of problems for the new countries, some of them still prominent today. The largest problem is the disputed Kashmir region (see Problem Landscape, page 305). Immediately preceding and following partition in 1947, violence broke out between Hindus and Muslims, and hundreds of thousands of lives were lost in wholesale massacres. More than 15 million people migrated between the two countries. Pakistan today has a large population known as **Mohajirs,** or "migrants," the Muslim immigrants and their descendants who poured into the region from India when partition occurred. Particularly in Karachi and elsewhere in the southern province of Sind, violent conflict frequently occurs between rival factions among the Mohajirs and between the Mohajirs and other ethnic groups, including Pathans (Pushtuns) and Biharis.

As a consequence of the partition, India received almost all of the subcontinent's modern industries and most of its mineral wealth and energy supplies. Pakistan, including the present Bangladesh, was left with a small industrial infrastructure and fewer natural resources. Pakistan and Bangladesh are therefore still very far behind India in both high-tech and traditional industrial output. The burlap industry is a good example of how partition frustrated development in the region. The province of Bengal, in colonial India's eastern Ganges–Brahmaputra Delta region, is the world's greatest production area for jute, the raw material for burlap. With independence, most of the jute-growing areas were isolated on the Pakistani (now Bangladeshi) side, while the jute mills, which had developed in the Calcutta area, are in India.

Figure 11.6 The central rail station in New Delhi, India, is one of the hubs in a vast network dating to the British colonial period. Note New Delhi's modern skyline in the distance.

Problem Landscape

Kashmir

Since 1947, India and Pakistan have been in conflict over the status of Kashmir (called Jammu and Kashmir in India), a disputed province straddling their northern border (see Figure 11.11, page 312). Before independence, Kashmir was a princely state administered by a Hindu maharajah. About three-fourths of Kashmir's estimated population of 15 million is Muslim, which is the basis of Pakistan's claim to the territory. But under the partition arrangements, the ruler of each princely state was to have the right to join either India or Pakistan, as he chose. Kashmir's Hindu ruler chose India, which is the legal basis of India's claim.

After partition, fighting between India and Pakistan led to the demarcation of a cease-fire line—the "line of control" that still separates the forces—leaving eastern Kashmir, with most of the state's population, in India and the more rugged western Kashmir in Pakistan. Beginning in the mid-1950s, China pressed its own claims to remote northern mountain areas of Kashmir and occupied some of the border territories. Pakistan ceded part of the occupied territory to China and established friendly relations with its giant northern neighbor, but India rejected China's claims (these countries still have chilly relations). India's subsequent small-scale military actions failed to dislodge Chinese forces. India now holds about 55 percent of the old state of Kashmir, Pakistan 30 percent, and China 15 percent. India has another dispute with China over China's control of 34,750 square miles (90,000 sq km) in India's northeastern state of Arunachal Pradesh.

Three wars between India and Pakistan have effected little change in Kashmir. In 1965, conflict began in Kashmir, spread to the Punjab, and escalated to a brief but indecisive full-scale war involving tanks, airborne forces, and widespread air raids. Renewed hostilities in Kashmir in 1971, as part of the war in which India supported the revolt of Bangladesh against Pakistan, again did not alter the political landscape of Kashmir. During that conflict, the Siachen Glacier in the Karakoram Mountains, at about 20,000 feet (2,800 m), earned the title of "world's highest battlefield."

In 1989, Kashmir's Muslim majority escalated the campaign for secession from India, and the strife has continued ever since. India and Pakistan came perilously close to the brink of war over Kashmir in 2002. Late in 2001, armed insurgents attacked India's parliament in New Delhi in a failed attempt to assassinate Indian politicians. India's government identified the assailants as members of two Muslim rebel movements (Lashkar-e-Taiba and Jaish-e-Muhammad) based in Kashmir. India further accused Pakistan's government of supporting these rebel groups and therefore being ultimately responsible for the attack on India's parliament. Indian military forces began amassing on the line of control in Kashmir, and Pakistani troops responded in kind. Tremendous international diplomatic pressure was applied to both sides in an effort to prevent all-out war. The pressure was particularly intense for Pakistan. Pakistani President Musharraf responded by outlawing and beginning to crack down on five militant groups, including the two blamed for the December attack. It was a risky move, especially because many people within Pakistan believed their leader was forsaking a legitimate struggle in Kashmir and "selling out" to the United States and other Western powers in the wake of the 9/11 attacks. While Musharraf's actions did at least temporarily defuse the explosive situation in Kashmir, many observers remained concerned that Pakistan would prove unable to rein in Muslim attacks against Indian targets in Kashmir. There are also many questions about how much control Pakistan leaders have over their own intelligence service, the ISI (Inter-Service Intelligence Agency), which may contain rogue elements intent on continued support of the militants fighting India.

The Kashmir conflict has been very costly to both sides, with more than 65,000 people killed since 1965, about half of those since 1989. India's international reputation has suffered because of its unyielding position on the region. Pakistan spends about a third of its budget on defense, mostly focused on a potential engagement with India over Kashmir.

There would be economic benefits for both India and Pakistan if peace returned. "Kashmir is a Paradise on Earth," wrote Samsar Kour.[a] The stunning snow-crowned peaks and flower-laden valleys of this western Himalayan region once lured millions of Indian tourists and ranks of Western trekkers each year, and the tourism development potential is enormous (Figure 11.A). Fortunately, in 2003, Indian and Pakistani leaders met to speak about resolving the Kashmir problem and declared a formal cease-fire in the region for the first time in 14 years.

[a] Samsar Chand Kour, *Beautiful Valleys of Kashmir and Ladakh* (Mysore, India: Wesley Press, 1942), p. 2.

Figure 11.A Kashmir, "Paradise on Earth" but for the blood shed there. The mountain is Rakaposhi (25,550 ft/7,788 m) on the Pakistani side of the line of control.

The World's Great Rivers

Indus, Ganges, Brahmaputra

South Asia's Indus, Ganges, and Brahmaputra Rivers provided the waters on which some of Earth's most significant civilizations developed, and they are still vital to the livelihoods and spirituality of a large portion of humanity. The allocation of these precious and meaningful waters since India's independence from Britain has been problematic.

Since 1947, the region of the subcontinent called the Punjab, meaning "Five Waters" (referring to the five Indus River tributaries), has been split between two countries, with its rivers flowing through Indian territory before entering Pakistan (Figure 11.B; see also Figures 11.11 and 11.16). In geographic and geopolitical terms, India is an **upstream country**, and Pakistan is a **downstream country** where Punjab waters are con-

Figure 11.B The confluence of the Gilgit (left) and Indus (right) Rivers, in Pakistan's section of the disputed Kashmir region.

Joe Hobbs

cerned. There are no internationally accepted laws regarding water sharing between upstream and downstream states, and the upstream country often exercises its geographic advantage to siphon off what the downstream country views as too much water. Riparian (river-owning) countries often come close to, or actually go to, war over the problem. In many cases, formal treaties have averted what could have become major international conflicts.

After 1947, relations between India and newly independent Pakistan deteriorated in part because of the disputed waters of the Indus and the five Punjab rivers. For many years, the parties could not agree on how much of the water India could divert and how much should remain for Pakistan. The countries finally reached a satisfactory resolution in the 1960 **Indus Waters Treaty.** The agreement allocates the water of the three eastern rivers of the Punjab to India, which in return allows unrestricted and undiminished flow of the two others, and the Indus River itself, into Pakistan. The two countries have abided by the agreement ever since, despite their periodic wars and constant state of tension. Today, the biggest problem over Indus waters is within Pakistan, with downstream Sind Province complaining that upstream Punjab Province is committing "water robbery."

Another landmark water-sharing agreement was signed between India and Bangladesh in 1996. For two decades, these nations had a bitter disagreement about the allocation of water from the Ganges River. In the 1996 accord, which is meant to be in effect for 30 years, upstream India guarantees a larger share of water to downstream Bangladesh. This achievement may open the way to the integrated development of what is known as the Ganges–Brahmaputra–Barak (GBB) river basin, a watershed that is home to more than 500 million people in Bangladesh, Bhutan, China, India, and Nepal. Cooperation to settle the vital issue of water allocation among these countries could promote peace and reduce poverty in the region.

After independence, India struggled to promote jute cultivation and Bangladesh to promote jute processing. There were also problems in how to divide up fresh water among the newly independent countries (see The World's Great Rivers, above).

This book discusses many cases of actual or potential reunification between divided states—West and East Germany and North and South Korea, for example. As an indication of how bitter and final the partition of India was, there is never discussion of its reunification today. The greatest hopes are simply for reconciliation and peace between India and Pakistan.

11.3 Natural Regions and Resources

Physical conditions are very important in the lives of the rural majorities of South Asia. People's well-being depends largely on local agricultural resources, and industrialization is based mainly on other locally available natural resources. Fortunately, South Asia has some generous endowments of natural resources and a huge and talented labor force to work with them. It also has colossal natural hazards, including cyclones, floods, droughts, landslides, and earthquakes. The subcontinent may be roughly subdivided into three nat-

Tectonic History of India

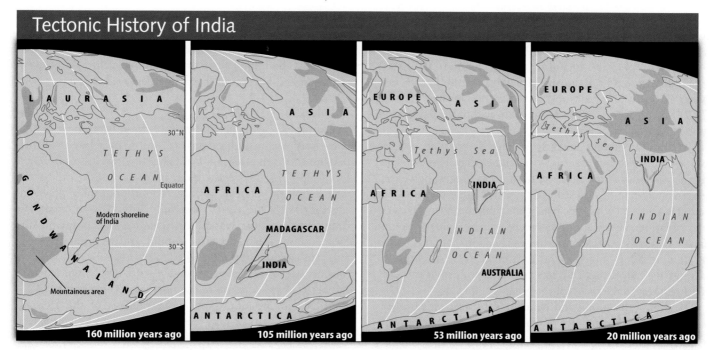

160 million years ago | 105 million years ago | 53 million years ago | 20 million years ago

Active Figure 11.7 Most of India was once an island. It was transported by tectonic movement across the Indian Ocean. When it collided with the Eurasian landmass, the world's highest mountains were formed. *See an animation based on this figure, and take a short quiz on the facts or concepts.*

ural areas: the outer mountain wall, the northern plain, and peninsular India (see Figure 10.4, page 275).

The Outer Mountain Wall

Between 40 and 50 million years ago, a triangular piece of the earth's crust completed its tectonic journey across the Indian Ocean and collided with southern Asia (Figure 11.7). The collision resulted in the thrusting up of a massive, almost solid, wall of mountains. That wall, made up mainly of the Himalaya Mountains and the Karakoram Range, forms an inverted U to define the northern boundary of the subcontinent. From each end of this massive wall, lower mountain ranges trend southward to the sea.

Pakistan's borders with Iran and Afghanistan traverse the rugged western section of the mountain wall. Desert and steppe climates reach upward to high elevations. These mountains provide only a few crossing places, including the storied Khyber Pass through the Hindu Kush Mountains on the border between Pakistan and Afghanistan. These passes have long been gateways into India for conquerors from the outer world, including the Aryans and the Arabs.

WORLD
REGIONAL
Geography⊛Now™

Click Geography Literacy to see an animation about plate boundary collisions.

The towering northern segment of the mountain wall has been much less passable, but a trickle of trade and cultural exchange has crossed these mountains for millennia. The Himalaya extend about 1,500 miles (c. 2,400 km) from Kashmir in the west, through the countries of Nepal and Bhutan, to the northeastern corner of India (Figure 11.8). Paralleling the Himalaya on the northwest, and separated from them by

Figure 11.8 The roof of the world: Looking across Nepal's boulder-strewn Gokyo Glacier toward Mt. Everest (hidden by clouds in the upper left).

the deep gorge of the upper Indus River in Kashmir, lies another lofty range called the Karakoram. The two ranges contain 92 of the approximately 100 world peaks greater than 24,000 feet (7,315 m), including the two highest: Mt. Everest (29,035 ft/8,859 m) in the Himalaya on the border between Nepal and China and Mt. K2, or Godwin Austen (28,268 ft/8,616 m), in the Karakoram within the Pakistani-controlled part of Kashmir.

Geographic surveys recently quieted a brewing debate over whether K2 might actually be higher than Everest: The Himalayan peak is the highest. Both great ranges, however, are young, dynamic, and still rising, so K2 could conceivably outgrow Everest. While the peaks are climbing, their glaciers are retreating. There is more ice here than anywhere else on Earth outside Greenland and the polar regions, but it is melting quickly. Some fingers point to human-made greenhouse gases as the culprit, and by one estimate, at current rates of melting, these glaciers would be gone by 2035.

The mountains along and near the subcontinent's border with Myanmar (Burma) are much lower than the Himalayan ranges but, being cloaked with dense vegetation and soaked with rain, are also nearly impenetrable. Cherrapunji in the Khasi Hills of India has an average annual rainfall of 432 inches (1,097 cm), more than nine-tenths of it falling in the summer. It ranks with a station on Kauai in the Hawaiian Islands as one of the two spots with the greatest annual rainfall ever recorded on Earth. Cherrapunji reportedly has the highest alcoholism rate in South Asia because so many people stay inside and drink on rainy days!

The Plains of India, Pakistan, and Bangladesh

Just inside the outer mountain wall lies the subcontinent's northern plain. This large alluvial expanse contains the core areas of the three major countries. In the west, the plain is split between Pakistan and India. The Indus River and its tributaries and distributaries cross the portion lying in Pakistan and water this country's two most populous provinces, Punjab and Sind, and its major cities of Karachi and Lahore. The climate of the western plains is desert and steppe; the dry region straddling the two countries is known as the Thar Desert, or the Great Indian Desert. Irrigation water from the rivers sustains highly productive agriculture and millions of people here.

East of the Punjab lies the part of the northern plain traversed by the Ganges (Ganga) River and its tributaries, especially the Jumna (Yamuna). The western portion of this region is known as the Doab, "the land between the two rivers." The northern plain is India's core region, containing more than two-fifths of the country's population as well as its capital New Delhi, many other major cities, and sacred places in the religions of Hinduism, Buddhism, and Jainism. This region of India is relatively well watered, especially toward the east, where agriculture based on irrigated rice supports high population densities. The southeastern part of the northern plain lies in the region of Bengal, which is essen-

tially the huge delta formed by the combined flow of the Ganges and Brahmaputra Rivers. Bengal is divided between India and Bangladesh and has an extremely dense population, with the major cities of Kolkata (Calcutta) on the Indian side and Dhaka on the Bangladeshi side.

Peninsular India

The southern peninsular portion of the subcontinent, which is entirely within India, consists mainly of a large volcanic plateau called the Deccan. It is generally low, less than 2,000 feet (c. 600 m) above sea level. Its rivers run in valleys cut well below the plateau surface. This topographic problem, together with the rivers' seasonal flow pattern, makes it difficult to use their waters for irrigation of the upland surface without large inputs of capital and technology. Wells and rain-catchment ponds called tanks are used, but water is in short supply and population density is low compared with the better watered parts of the northern plain (Figure 11.9).

Hills and mountains outline the edges of the Deccan. In the north, belts of hills (the Central Indian Hills separate the plateau from the northern plain. On both east and west, the plateau is edged by ranges of low mountains called

Figure 11.9 It is usually the task of women to draw water from communal supplies such as this one in a small settlement near Mumbai.

Ghats ("steps") overlooking alluvial plains along the coasts of the Bay of Bengal and the Arabian Sea. The low and discontinuous eastern mountains are known as the Eastern Ghats. The higher and almost continuous western mountains are the Western Ghats. The two ranges merge near the southern tip of the peninsula in clusters known locally as the Nilgiri, Palni, and Cardamon "hills." They are really rather high mountains, with some peaks rising above 8,000 feet (c. 2,400 m). The colonizing British founded several "hill stations" in them for recreation and administration during summer months, when the climate on the subcontinent's lowlands is stifling. They are now popular tourist destinations for Indians during that season. The coastal plains between the Ghats and the sea support very dense populations, especially in the far south along the Malabar Coast of western India and on the corresponding Coromandel Coast in the east. India's largest city, Mumbai, is on the peninsula's northwest coast, and the large city of Chennai is on its southeast coast. Inland are the major cities of Hyderabad and Bangalore.

Climate and Agricultural Conditions

Climatic conditions in South Asia range from Himalayan ice to the tropical heat of peninsular India and from some of the world's driest climates to some of the wettest. Climatic and biotic types in the subcontinent include undifferentiated highland climates in the northern mountains; desert and steppe in Pakistan and adjacent India; humid subtropical climate in the northern plains; tropical savanna in the peninsula; and rain forest along the seaward slopes of the Western Ghats and in parts of the Ganges–Brahmaputra Delta and the eastern mountains near Burma (see Figures 10.5a and b). Most of the region's original vegetation cover is gone, but there are numerous large national parks and protected areas, especially in India, designed to ensure that important remnants will remain. The region's wildlife still includes tigers, elephants, rhinoceroses, and in western India, the sole surviving population of the once widespread Asiatic lion.

Heat is nearly constant except in the mountains. The tropical peninsula is hot all year. The subtropical north has stifling heat before the break of the wet monsoon, and there are warm conditions even in the winter. Some of the warmest temperatures on the globe occur in the plains of Pakistan and northwestern India. Delhi averages 94°F (34°C) in both May and June. The highest temperatures over most of the subcontinent occur in May and June, and high humidities produced by the impending monsoon rains combine with the heat to create almost intolerable conditions. Then the monsoon rains break, bringing cooling relief—but sometimes catastrophic flooding—to the subcontinent's people.

India's monsoons include both the **wet southwest monsoon,** which brings the short, wet summer between June and September, and the **dry northeast monsoon,** which creates mostly arid conditions the rest of the year. There are two main arms of the wet, or summer, monsoon. One, approaching from the west off the Arabian Sea, strikes the Western Ghats and produces heavy rainfall on these mountains and the coastal plain. The amount of rain diminishes in the interior Deccan rain-shadow region to the east of the mountains. Here, there is barely enough rainfall for dry farming, and drought sometimes brings crop failures. The second arm approaches from the Bay of Bengal, bringing moderate rainfall to the eastern coastal areas of the peninsula and heavy rains to the Ganges–Brahmaputra Delta region and northeastern India. This is the monsoon that feeds cyclones (hurricanes). Wet monsoon winds pass up the Ganges Valley to drop moisture that diminishes in quantity from east to west.

Both arms of the monsoon bring some rain to Pakistan, but it is so little that semiarid or desert conditions prevail in most areas (Figure 11.10). Blowing mainly from land to sea, the dry monsoon brings dry and cooler weather to most parts of the subcontinent. An exception is in the far south of the peninsula, where the heaviest rainfall of the year falls along the eastern coast and in adjacent uplands from October through January.

With precipitation so concentrated within a short part of the year, it is essential to extend the crop season by irrigation. People in the subcontinent therefore expend an enormous amount of labor to get additional water onto the land. Modern technology does the job in places where huge dams impound rivers and where networks of canals distribute water from these reservoirs. Most dams are in the steppe and desert areas of the Punjab and the lower Indus Valley (Sind Province) in Pakistan.

A massive irrigation and hydroelectric power project is underway in the north central part of peninsular India, where the Sardar Sarovar Dam was recently constructed on the Narmada River. Largely because the dam project would displace 320,000 tribal and rural people (known as "oustees" in the project language) and inundate 28,000 acres of

Figure 11.10 Aridity and deforestation are among the urgent environmental problems faced by Pakistan. Foraging by goats and cutting trees for fuel in an already dry environment have created a lunar landscape in the country's Karakoram Mountain region.

279

cropland and 32,000 acres of forest, environmentalists and human rights advocates both in India and abroad opposed the dam. The World Bank withdrew its financial support in response to these protests, but asserting that its benefits will greatly outweigh its costs, India pressed ahead with the dam's construction. It is the centerpiece of the massive Narmada Valley Development Project, where 30 dams will ultimately displace about 1 million people by the completion date of 2025.

F10.3a, 274

Correlations between people and food supplies, both spatially and in raw numbers, are always critical in South Asia (see Regional Perspective, page 311). Staple food crops and population densities correlate spatially with the amount of water available in the subcontinent. Rice is the basis of life in the wetter areas and, with its high caloric yields per acre, is associated with the highest population densities. Rice dominates in the delta area of Bengal (in both India and Bangladesh), in the adjacent lower Ganges Valley, and in the coastal plains of the peninsula. India's Punjab (which produces 55 percent of India's food crops) and Pakistan's Sind and Punjab Provinces produce surplus rice, including the Basmati variety, which is an important export. Irrigated wheat is the staple crop and food in the drier upper Ganges Valley of India and in the dry Punjab of India and Pakistan. Here, the caloric yield and population densities are intermediate. Unirrigated sorghums and millets prevail over most of the Deccan plateau and in other areas where low rainfall cannot be supplemented much by irrigation. Caloric yields from these crops are low and so are population densities.

Maize (corn) is an important grain in the lower regions of the Himalaya. However, rice is the preferred food almost everywhere, and where enough water is available, there are patches of irrigated rice. Rice, cotton, jute, and tea are the main crops grown in the subcontinent for industrial use and export. Commercial agriculture is discussed further in the remainder of this chapter, which deals with economic and other important characteristics and problems of each of the South Asian countries.

11.4 India: Power, Courage, Confidence

Despite India's overwhelmingly rural population (72 percent, living in about 600,000 villages), the country boasts some of the world's **megacities** (a term used generally to describe the world's largest cities), notably Mumbai and Kolkata (Figure 11.11). These and many smaller cities are at the heart of India's current economic boom, which is based more on information technology than on old-fashioned heavy industry. It is difficult to make generalizations about India in economic or most other matters because there is such a vast continuum of Indian experience. Visitors to India may be astounded to see both the latest computer technologies and the labors of human hands performing, on a large scale, the tasks routinely done by machines in more developed countries—for

example, people making gravel by hammering boulders into pebbles. And always looming over India are questions of poverty, caste, and religious strife.

Urban and Economic Geography

India's recent high rate of economic growth (an average of 6 percent per year between 2001 and 2004) owes largely to important reforms and gives new meaning to its ancient three-lions symbol for power, courage, and confidence. Until the 1990s, India was intent on keeping out most foreign manufacturing in favor of industrial self-sufficiency. Most of the economy was socialized, with many state-owned industries. With these policies, India's industries expanded and diversified, but profitable exports were limited. After that, India opened up to more imports, privatized many state-run industries, and boosted exports. India's vast pool of well-educated people responded in creative ways to new opportunities. The result was remarkably strong economic growth.

India's diverse manufactures include old-economy products like cars, trucks, motorcycles, bicycles, and farm machinery, along with high-tech data-processing equipment, silicon chips for electronics, and computer software. India's leading factory industries are cotton textiles, jute products, food processing, and iron and steel. India is both a major user and exporter of cotton textiles and clothing. Favoring cotton manufacturing are a huge home market, inexpensive labor, and large domestic production of cotton.

The sector with the greatest current growth is information technology. India's software exports, two-thirds of which go to the United States, have been growing in recent years at 292 rates of up to 50 percent annually, and technology and service jobs have been increasingly outsourced to India from the United States. New technology venture firms are racing to establish satellite telephone service across the subcontinent. Improved phone service is closely linked to government and private efforts to wire India to the World Wide Web. Internet cafés are turning up even in remote villages, where their operators use diesel generators to keep computer terminals from failing during frequent power outages. Information technology is helping to bridge the digital divide separating the rich and poor in India.

Increasing access by the poor to the Internet may bring significant social and economic change in this poor country. For example, villagers might be able to use this technology to gain unbridled access to land records rather than having to bribe officials to get them. Those seeking work abroad could download passport applications instantly rather than spend time and money working through intermediaries. Farmers could get information on crop prices in different towns to improve their bargaining positions during harvests. This is already happening in parts of rural India, where a private company called ITC has set up Internet access so that farmers can check the price of soybean futures on the Chicago Board of Trade. Armed with this knowledge, the farmers can decide whether to sell their soybeans immediately or wait a while longer until prices rise. In this arrangement,

Regional Perspective

Keeping Malthus at Bay

Poverty and human health are serious problems in South Asia, and the prospect of continued high population growth raises the question of whether growth in food supplies can avert the proverbial Malthusian crisis (see Chapter 2, page 42).

India's population has surged since independence, from 352 million in 1947 to 1.09 billion in 2004. In recent decades, the overall trend has between toward lower population growth rates. Fifty years ago, the average Indian woman had six children; now she has three. That rate is still well above the natural replacement rate, however, that would keep the population steady. The population base is already so vast that even modest growth will add huge numbers. India is predicted to overtake China as the world's most populous country by 2040.

For many, the name "India" invokes an image of grinding poverty, and with good reason: An estimated 34 percent of the population is "abjectly poor," defined as living on less than $1 per day, and half of India's children are malnourished. But perhaps in defiance of Malthus, an increased population has so far succeeded in producing more economic resources overall. A few people in India are very wealthy, and there is an emerging middle class of as many as 300 million people—the world's largest middle class (Figure 11.C). India's economy has grown impressively in recent years. An apparent problem is that the benefits of the growing economy are unevenly distributed. Most of the economic growth benefits those who are already better off, widening the gap between rich and poor. In its bid for reelec-

tion in 2004, India's ruling BJP Party promised to make the country an economic superpower by 2010. In a stunning setback viewed as a referendum by the poor, who felt India's growth would continue to pass them by, the BJP lost the election to the opposition Congress Party. Population growth in the long run may further distort India's economic imbalances.

The issue of population growth is also critical for Pakistan, which has never mounted an effective family planning program, mainly because of opposition from Muslim religious authorities who regard birth control as an intervention against God's will (Pakistan's birth rate in 2004 was 34 per thousand compared with 25 per thousand in India). Muslim Bangladesh, along with Nepal and Bhutan, also have high birth rates, but they are much lower in Bangladesh than they once were. In the 1970s, the average woman had seven children while now she has three, and contraception use nationwide has increased from 4 percent then to 50 percent now. South Asia's success story in bringing population growth under control is Sri Lanka, with a birth rate of just 19 per thousand and annual population growth of 1.3 percent.

Agricultural output in South Asia has increased since independence. Most notably, despite its huge and growing population, India has achieved self-sufficiency in grain production, with a fourfold increase in output since 1950. Almost half a million "ration shops" sell subsidized food staples to the country's poorest people. The successes of agriculture in the subcontinent have been due mainly to the increased use of artificial fertilizers, the introduction of new high-yield varieties of wheat and rice associated with the Green Revolution, more labor provided by the growing rural population, better irrigation, the spread of education, and the development of government extension institutions to aid farmers.

One of the challenges confronting India in demographic and many other concerns is to improve the status of women. Despite a 1961 ban on dowries (the money and gifts given by a bride's parents to the groom), the practice is continuing. So is the rate of killing brides who do not provide enough dowry. The burden placed on the bride's family has prompted many parents to abort female fetuses, which can be detected with ultrasound technology. (India banned this use of ultrasound in 1996, but the practice continues.) Female feticide and infanticide have taken an estimated 25 million lives. In some of India's states, especially Haryana and Punjab in the west, one result is a serious shortage of local brides. The gap is being filled by marriages arranged with women from distant Indian states and Bangladesh.

Figure 11.C This Indian family visiting the Red Fort in Agra is among the growing ranks of the subcontinent's middle class.

Joe Hobbs

291

India

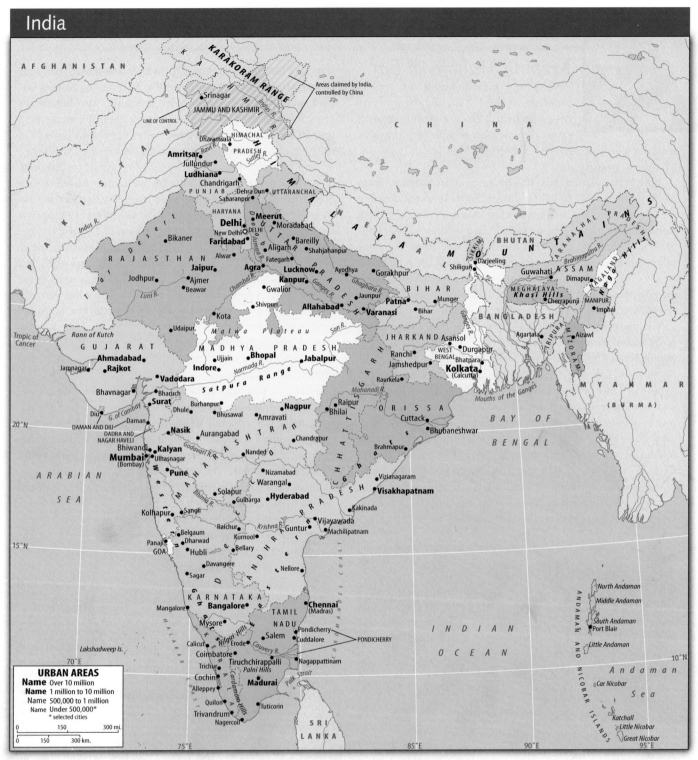

Figure 11.11 Principal features of India

known as "e-choupals," ITC is luring farmers to sell their better soybeans directly to the company rather than to the village traders or government markets from which ITC previously had to buy the crops.

India produces a wide range of chemical products, from vitamins to fertilizers. There is a flourishing pharmaceutical industry. Much to the chagrin of Western drug companies, their Indian counterparts routinely replicate foreign drug formulas and produce copycat medications for sale at a fraction of their cost in the West. Arguing that India is a poor country that needs low-cost medicines, the Indian government does not crack down on its drug "pirates." India's success in providing lifesaving medications to its population cannot be disputed; life expectancy has risen by 20 years since the country became independent. One of the big questions is if such improvements in the quality of life can be sustained for India's growing population.

Some of India's intellectual productivity is lost to other countries, perhaps slowing down potential economic growth. India suffers from a brain drain, the emigration of highly trained and highly educated individuals abroad. Chances are that the American reader of this text has been or will be treated in a hospital or clinic by a physician who emigrated from India, attracted by higher U.S. wages. The United States' gain, in the form of highly qualified medical personnel, is inevitably India's loss because better development requires better human resources. Indian computer scientists, engineers, nurses, and other professionals have in recent years followed the trend in "exports" of doctors to the United States and other countries. In American universities, there are more students from India (about 70,000) than from any other foreign country. Most return to stay in India, and others are eventually wooed into returning. Altogether, 20 million Indian citizens live abroad, with almost 2 million in the United States and about 3 million in the oil exporting countries of the Middle East. Another 2 million Bangladeshis and 1 million Pakistanis work in the Middle East, many in semi-skilled or unskilled jobs. The remittances they send home are vital to South Asia's economies.

Another challenge to India's growth has been to upgrade energy sources and distribution. India has some hydropower, coal deposits, and a modest oil output from fields in Assam, Gujarat, and offshore of Mumbai. Nuclear power produces about 2 percent of its electricity. Like the other South Asian countries, however, India is heavily dependent on imported energy.

India's industries are concentrated in its cities. On the peninsula's northwest coast is the second largest, Mumbai (Bombay; population: 17.3 million). The country's cotton mills are concentrated mainly in the metropolitan area of Mumbai and neighboring areas. Mumbai is India's Wall Street and also its Hollywood (it calls itself "Bollywood")—the center of a prolific and accomplished film industry (and the world's largest) that produces movies of international stature as well as the regular fare that a huge domestic market consumes passionately (Figure 11.12). In the south central peninsula is Bangalore, known as **India's Silicon Valley,** where bustling operations of Microsoft, Motorola, Intel, Texas Instruments, and a host of other high-tech firms present a startling contrast to the largely rural and unchanging image that many outsiders have of India. The largest metropolis of the far south, Chennai (formerly Madras) is a seaport on the Coromandel Coast.

At the western end of the Ganges–Jumna section of the subcontinent's northern plain, the old fortified capital of Delhi arose as a bastion against invaders from beyond the western mountains (Figure 11.13). The present adjoining capital of New Delhi was founded after 1912 as the capital of British-controlled India. Today, the population of Delhi–New Delhi is an estimated 17.4 million, making it India's

Figure 11.12 Mumbai produces the romance and action films that are the staples of Indian cinema. The advertising is misleading; Indian films tend to show little flesh.

largest metropolis. Its main industries are manufacturing, textiles, electronics, and chemicals. Far to the southeast is the English-founded port of Kolkata (Calcutta) in the state of West Bengal. A city of 14 million, Kolkata has a worldwide reputation for teeming slums and dire poverty—Rudyard Kipling called it "The City of Dreadful Night"—and it was this Indian city to which the extraordinary nun Mother Teresa dedicated her life's work of aiding the poor. Within India, Kolkata is recognized as a great center for culture and the arts. The surrounding Bengal region is an important industrial area, where the country's jute, iron, and steel industries are concentrated. The most valuable mineral resources in India, including iron ore, coal, manganese, chromium, and tungsten, are in this region. These resources have led to the rise of small- to medium-sized industrial and mining cities (notably Jamshedpur) in an area that otherwise is one of the least developed parts of the country.

Of the four major commercial crops in the subcontinent, only tea is exported on a sizable scale. Unlike the others (cotton, jute, and rice), tea is a plantation crop. The plantations

Figure 11.13 The Mogul Emperor Shah Jahan established Delhi as India's capital early in the 17th century. His monumental Red Fort (top left) still dominates this old city. Adjacent New Delhi is India's capital today.

Figure 11.14 A tea plantation in the Western Ghats of eastern Kerala state. These mountains have been widely cleared of native forest by logging and agricultural expansion.

were developed, and are still largely owned, by British interests. Production is greatest in the northeastern state of Assam, with another center in the mountainous far south of peninsular India (Figure 11.14). Cotton is grown mainly in the interior Deccan. India's large textile industry absorbs most of the production. Coffee, rubber, and coconuts are minor plantation crops, and India exports products made from coir (coconut fiber).

Social and Political Challenges

Partition from mainly Muslim Pakistan certainly did not solve India's problems. Very serious issues persist among the country's ethnic and religious groups. Some of the most problematic involve language, nationalism, and caste (see Definitions and Insights, page 315).

India has 17 major languages and more than 22,000 dialects. Many people speak two or even three languages, but overall literacy rates are low, and communication is often a problem. Since British colonial times, English has been the lingua franca and the de facto official language of the subcontinent. However, only a small percentage of the population, mainly the educated, is literate in English. This fact, along with nationalist sentiments, has prompted calls for indigenous languages to be adopted as official languages. After independence and partition, disputes arose in each country over which language should be chosen. India decided on Hindi, an Indo-European language spoken by the largest linguistic group, and made it the country's national language in 1965. This government decision resulted in strong protests, especially in the Dravidian south. The result has been a proliferation of official languages to satisfy huge populations; India now has 15, all of which appear on the country's paper currency. Rioting between speakers of different languages has occurred repeatedly, and demands by language groups have reshaped the boundaries of several Indian states.

Hindu nationalism is a serious issue in a country with such large religious minorities. The country is supposed to be secular, but in recent years, its politics have been heavily influenced by Hindu nationalists. The explicitly sectarian political party Bharatiya Janata Party (BJP) held a majority of parliamentary seats between 1998 and 2004. This Hindu nationalist party would like India's Hindu majority to have the strongest say in the country's affairs. Until recently, the BJP platform included the adoption of a code of civil law that would strip Muslims of separate laws in matters of marriage, divorce, and property rights.

Sectarian violence (known in the region as **communal violence**) between Hindus and Muslims is a constant threat to India's social and political fabric. Contention over places sacred to both faiths sometimes ignites widespread violence. The 16th century Babri Mosque in the city of Ayodhya in Uttar Pradesh was a place of prayer for Muslims on a spot revered by the Hindus as the birthplace of the god-king Ram, an incarnation of the god Vishnu. Backed by Hindu nationalists in the provincial government, a mob of about 250,000

Definitions + Insights

The Caste System

One ancient Hindu belief is that every individual is born into a particular caste, a socio-economic subgroup that determines the individual's rank and role in the society. Castes form a hierarchy, with the Brahmin caste at the top (comprising about 5 percent of all Hindus) and three others (Kshatriya, Vaisya, and Sudra) below. Caste membership is inherited and cannot be changed. Particular castes are associated with certain religious obligations, and their members are expected to follow traditional caste occupations. Marriage outside the caste is generally forbidden, and meals are usually taken only with fellow caste members.

At the bottom of the social ladder are the **Dalits,** or **untouchables** (about 20 percent of all Hindus), also known as the *scheduled castes*. They are not part of the caste system but are "outcaste," and according to Hindu belief, they are not twice-born. The untouchables are so called because they traditionally performed the worst jobs, such as handling of corpses and garbage; therefore, their touch would defile caste Hindus.

This ancient, rigid social system is changing. India's constitution outlawed the caste system in 1950, but a single decree could not undo thousands of years of tradition. Brahmin privileges are now being increasingly challenged and restricted by more laws. Indian law explicitly forbids recognition of untouchables as a separate social group. Confronting the fact that upper castes account for less than one-fifth of India's population but command more than half of the best government jobs, the Indian government has instituted an affirmative-action program to allocate more jobs to members of lower castes. Indian leaders of high caste have championed the Dalits' cause for both moral and practical reasons. Politicians use caste divisions to their advantage to bring out voters, promising if elected to act in the interests of their respective large social blocs. In 1997, an untouchable was elected president of India for the first time in the country's history. Since then, untouchables have won more regional and local offices.

The Dalits are voting in greater numbers than ever before with the conviction they can use India's democratic system to overcome ancient discrimination against them. They are making progress. Disintegration of the caste system has been especially rapid in the cities, where people of different castes must mingle in factories, public eating places, and public transportation. In India's vast rural areas, the caste system is more entrenched.

Hindu fundamentalists demolished the mosque in December 1992 with the intention of building a Hindu temple on the site. The ensuing communal violence between Hindus and Muslims throughout India, including even the cosmopolitan and usually tolerant urbanites of Mumbai, left thousands dead.

Ever since this incident, as a means of garnering votes from Hindu Indians, Hindu politicians have made veiled wishes to proceed with construction of the temple. In 1998, the BJP platform included a pledge to build a Hindu temple atop the ruins of the mosque at Ayodhya. His apparent support for the temple helped sweep the BJP leader Atal Vajpayee into power as prime minister that year. The conflict over this multireligious sacred site in India has parallels with the Temple Mount dispute between Israel and the Palestinians.

In India's Punjab, where Sikhs make up 60 percent of the population and Hindus 36 percent, the 1980s and early 1990s were violent years during which about 20,000 people died in armed clashes. Sikh factions were intent on establishing their own homeland of **Khalistan,** prompted to do so in part by New Delhi's plans to divert water from the Punjab. They challenged Indian authority and turned Amritsar's Golden Temple, the holiest site in the Sikh faith, into their military stronghold (Figure 11.15). In a controversial move to quell the revolt, Prime Minister Indira Gandhi ordered troops to storm the Golden Temple in June 1984. The resulting deaths and desecration led directly to Mrs. Gandhi's assassination

Paolo Koch/Photo Researchers, Inc.

Figure 11.15 The Golden Temple (established in 1577 and rebuilt in 1764) in Amritsar is the holiest shrine of the Sikh religion.

by her own Sikh bodyguards later that year. Indian authorities accused Pakistan of arming the Sikh militants.

Another trouble spot for India is the northeastern state of Assam, an extraordinarily diverse region with more than 200 different ethnic groups and followers of many different religions. Several rebel groups have been active in this region for years, agitating for autonomy and independence from India. Some based themselves in nearby Bhutan until a crackdown there in 2004.

11.5 Pakistan: Faith, Unity, and Discipline

Pakistan (Figure 11.16) is not the demographic giant that India is, but its 159 million people face many challenges, especially owing to a relatively small resource base relative to the population, a general shortage of industrial output, and local and geopolitical tensions that are destabilizing forces. Unlike India, Pakistan is not a secular country but an Islamic one, whose motto is "faith, unity, and discipline."

Urban and Economic Geography

As mentioned earlier, the 1947 partition left Pakistan with far fewer industries and less mineral wealth than India received. The country has yet to make up for that setback. Many of the country's bills are paid by foreign aid donors, with a large share of those contributions tied directly to Pakistan's perceived importance in the U.S.-led war on terrorism. Pakistan's future development may be enhanced by growing exports of its own significant natural gas reserves.

Pakistan has developed successful cotton-textile industries. Irrigated cotton in the Punjab and lower Indus Valley makes Pakistan the world's fourth largest cotton producer (after China, the United States, and India). Textiles, yarn, ready-made garments, and raw cotton constitute about two-thirds of Pakistan's exports. Rice is another important export.

Opium was recently a major illegal export crop of Pakistan, but stepped-up law enforcement has dramatically cut production. However, processing of the raw opium into

Figure 11.17 Lahore, in the Punjab region of Pakistan, is the country's cultural center. It boasts many fine Islamic monuments, including this 17th-century Badshahi Mosque, built by the Mogul emperor Aurangzeb.

heroin is still a significant industry in the rugged mountain region of the Northwest Frontier Province and nearby Federally Administered Tribal Areas, situated along the frontier with Afghanistan and inhabited mainly by Pashtun peoples. Domestic heroin addiction is a growing problem, with an estimated 3 million Pakistanis dependent on the drug.

Pakistan's major cities, which are also the main industrial centers, are on plains near the Indus River: the seaport of Karachi (population: 11 million) at the western edge of the Indus River Delta in Sind Province and the cultural and textile manufacturing center of Lahore (population: 6 million; Figure 11.17) in the Punjab. North of Lahore, at the foot of the Himalaya, is the capital city of Islamabad ("City of Islam"), a much smaller, very modern administrative metropolis.

Social and Political Challenges

Pakistan has internal problems centering on religious divisions, mainly between the majority Sunni Muslims (77 percent of the country's population) and Shi'ite Muslims (20 percent). Shi'ite Muslims opposed the government's efforts in the 1980s to establish an avowedly Sunni Islamic state, and there was violent conflict between members of the two sects. Pakistani factions who preferred a Western-style secular democracy also rioted in protest against this **Islamization** of the government. Democratic elections in the 1980s and 1990s twice installed Benazir Bhutto, a liberal, Western-educated woman, as the country's prime minister. Her regimes were so riddled with corruption that she was twice voted out of office to be replaced by Nawaz Sharif, who promised to cleanse the country of corruption and embrace the traditional Islamic system of law known as *shari'a*. Sharif in turn was ousted in a military coup in 1999.

Sharif's successor, General Pervez Musharraf, took the helm of a country torn between the secular wishes of a Western-oriented middle class and popular calls for greater Islamic influences and fewer ties with the West. The growing

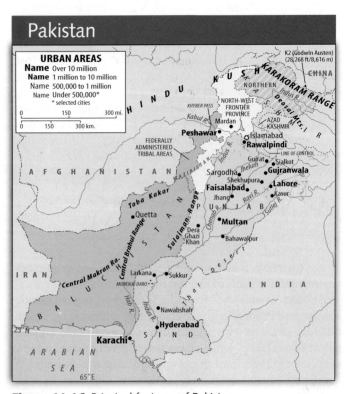

Figure 11.16 Principal features of Pakistan

Pakistan

URBAN AREAS
Name Over 10 million
Name 1 million to 10 million
Name 500,000 to 1 million
Name Under 500,000*
   * selected cities

Figure 11.19 Islamic education in Pakistan's *madrasas* begins at an early age. These boys are memorizing the Qur'an.

Figure 11.18 Pakistan's northwest frontier province city of Peshawar, has long been a haven for spies, smugglers, and insurgents.

Islamist sentiment is focused both on ousting India from disputed Kashmir and on resisting the U.S. and other Western intervention in the region in the wake of the 9/11 terrorist attacks. Much of the dissent is based in the Northwest Frontier Province, especially in and around its principal city, Peshawar (Figure 11.18). At the time of the attacks on the United States, there were already 3 million mainly ethnic Pashtun refugees from Afghanistan living in camps outside the city. Hundreds of thousands more poured in from Afghanistan in the weeks following 9/11, but a reverse flow has taken place since then. Within these impoverished, desolate encampments still simmers much anti-U.S. and generally anti-Western sentiment.

An even larger breeding ground for anti-Western (and anti-Indian) thought and action in Pakistan is the institution known as the **madrasa,** which may be translated as "Islamic school" or "seminary" (Figure 11.19). There are as many as 8,000 madrasas in Pakistan, where approximately 700,000 children (mainly boys) are educated. Their curriculum is based largely on study of the Qur'an and other Islamic texts, and some instructors promote an interpretation of the Islamic concept of **jihad,** or "struggle," in which Muslims are

urged to fight India, the United States, and other infidel entities. Many of the *jihadis,* or "holy warriors," who join groups like al-Qa'ida, Lashkar-e-Taiba, and Jaish-e-Muhammad have been educated in these schools. But madrasas also offer much legitimate instruction and help to compensate for Pakistan's poor public educational system. Pakistan's leaders are in a difficult position, at once called upon by the outside world to cleanse the madrasas of militant content and by their own people to uphold the long-cherished institution of the religious school.

11.6 Vulnerable Bangladesh

Formerly East Pakistan, Bangladesh is about the size of the U.S. state of Iowa or the European nation of Greece —not an important fact until one considers that it is home to more than 140 million people (Figure 11.20). With its limited resources, apparently "people overpopulated" Bangladesh is, after Bhutan and Nepal, South Asia's third poorest country on a per capita GDP PPP basis. Poverty is so deep that many Bangladeshi see neighboring India as a land of opportunity. About 20 million Bangladeshis are illegal immigrants living mainly in India's West Bengal Province, where authorities fear they will form permanent Muslim majority enclaves and perhaps threaten national security. Legal migrant labor is one of Bangladesh's major "exports"; more than 6 million Bangladeshis work abroad, sending home an estimated $2 billion in remittances each year.

The country's most notable success has been the development of cotton-textile industries. Chances are that the American reader of this text owns clothing made in Bangladesh. Textile, jute, leather, and seafood export industries are concentrated in Dhaka (Dacca; population: 10 million), the capital and largest city, and Chittagong, the main seaport.

The Bangladesh portion of the Ganges–Brahmaputra Delta ranks, along with the Indonesian island of Java, as one of the two most crowded agricultural areas on Earth.

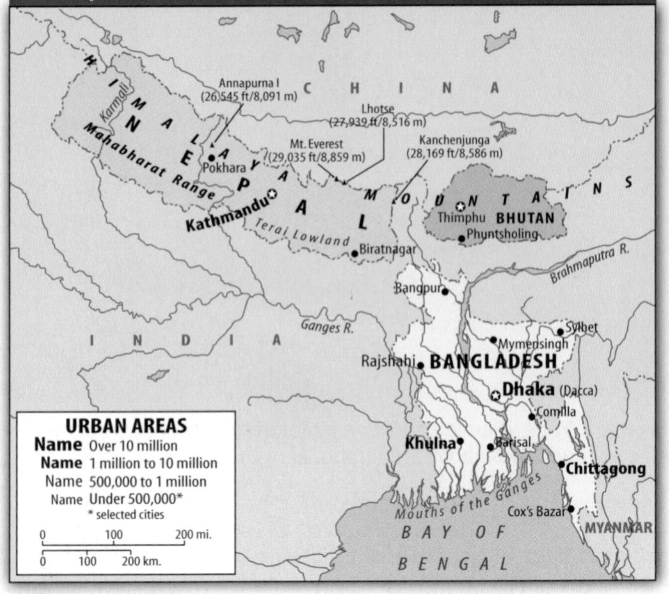

Figure 11.20 Principal features of Bangladesh, Nepal, and Bhutan

This area is subject to catastrophic flooding (Figure 11.21). Massive hurricanes, known locally as cyclones, ravage Bangladesh regularly in September and October. Deforestation upstream on the steep slopes of the Himalaya has increased water runoff and sediment load in the Ganges, also contributing to Bangladesh's flood problems. Bangladesh must also keep a wary eye on the potential for the sea level to rise if the world's temperatures increase, as predicted by many climate change models. The country's environment minister has insisted that 20 percent of Bangladesh will be under water by 2015 if nothing is done to control global warming.

11.7 Nepal and Bhutan: Mountain Kingdoms

The kingdoms of Nepal (capital, Kathmandu; population: 1.2 million; Figure 11.22) and Bhutan (capital, Thimphu; population: 62,000) are perched on South Asia's mountain wall (see Figure 11.20). Their landlocked locations and difficult topographies have contributed to their slow economic development. Both countries have productive agricultural regions in the "middle ranges," the bands of foothills between the vast lowland of the subcontinent and the towering Himalayan peaks. In Nepal, when the capacity of this productive area was saturated, many poor peasants moved to the lowland region known as the Terai. The shortage of quality farmland for Nepal's mainly rural (86 percent) population of 25 million is a critical problem. Nepal's mostly deforested landscape bears witness to relentless pressure to provide new land for agriculture and wood for fuel and construction (Figure 11.23).

Figure 11.21 Bangladesh is both a low-lying coastal land and a downstream country of several large watersheds. As a result of storms blowing in from the Bay of Bengal and of accelerated runoff from rainfall and snowmelt on deforested mountains upstream, it is subjected to devastating floods. This is an aerial view following the cyclone (hurricane) that struck Bangladesh on April 29, 1991, leaving more than 120,000 people dead and inundating most of the country's precious farmlands.

Nepal's Himalayan region is a popular destination for international tourists. About 300,000 visit each year, bound especially for treks in the Everest region in the east and the Annapurna area in the west. They bring in much-needed foreign currency to this poor country but also create problems by introducing social change into traditional cultures.

The conservative Buddhist kingdom of Bhutan has been more careful in opening its doors to tourism, placing restrictions on the numbers permitted in and on the routes they may follow. Tourism and hydroelectric power (mainly for Indian needs) are the most promising areas for future development of this poor country. This is the only country in the world whose politicians claim to disdain conventional eco-

Figure 11.22 Nepal's capital, Kathmandu, sprawls across a plain in the country's "middle" mountain range. In the foreground is the Royal Palace.

Figure 11.23 People are constantly collecting, hauling, storing, and using wood in Nepal, creating tremendous pressure on the country's remaining forests.

nomic growth; their mantra is **"gross national happiness"** for the people of Bhutan.

There are in fact religious and political troubles in the apparent "Shangri-la" of Bhutan. This is a monarchy ruled from the capital by a Buddhist king of the Drukpa tribe, to which most of the government ministers also belong. The government forbids political parties and is being challenged by a number of outlawed organizations, including the **Bhutan People's Party (BPP)**. The BPP represents ethnic Nepalis, who are Hindus and who for decades have been migrating out of their own overcrowded Himalayan kingdom to work the productive soils of Bhutan. Ostensibly as a means of promoting national unity, Bhutan's king outlawed the use of the Nepali language in Bhutan's schools and insisted that all inhabitants of Bhutan wear the national dress and hairstyle of the Drukpa tribe. Violent clashes ensued between ethnic Nepalis and Drukpas.

Nepal itself has suffered violent consequences of one-party rule and of palace intrigue. The country was an absolute monarchy until 1991, when prodemocracy riots and demonstrations forced King Birendra to accept a new role as constitutional monarch and permit elections for a national parliament. King Birendra and many other royal family members were shot to death at a dinner gathering in 2001, killed by Nepal's crown prince in a drunken rage over his mother's refusal to allow him to take the bride of his choice. Birendra's successor, King Gyanendra, soon declared a state of emergency and dismissed the democratically elected parliament. He claimed this was to quell a Maoist insurgency that began in 1996 with promises to alleviate rural poverty and rid the country of its caste system, which funnels wealth, power, and land ownership into relatively few hands. By 2004, conflict between government troops and the Maoist rebels of the **Communist Party of Nepal** had taken more than 9,000 lives, and the rebels claimed to control 80 percent of the country. The violence was exacting a toll on the country's vital tourism industry.

11.8 Sri Lanka: Resplendent and Troubled

Sri Lanka (population: 19.6 million), formerly Ceylon, is a tropical island country with affinities to the subcontinent. Its name conveys its physical beauty: Sri Lanka means "Resplendent Isle" in the native Sinhalese language, and medieval Arabs knew it as Serendip, the island of serendipity (Figure 11.24). The island consists of a coastal plain surrounding a knot of mountains and hill lands (Figure 11.25). Most

Figure 11.24 Verdant landscapes are typical of all but northernmost Sri Lanka. This view is near Kandy in the south central part of the country.

Figure 11.25 Principal features of Sri Lanka

people live in the wetter southwestern portion of the plain, in the south central hilly areas, and in the drier Jaffna Peninsula of the north. Coconuts and rice are the major crops of the low southwestern coast and Jaffna Peninsula, while tea and rubber plantations dominate the economy of the uplands. Colombo (population: 670,000), in the southwest, is the capital, chief port, and only large city.

Sri Lanka's economy is highly commercialized. Three cash crops—rubber, coconuts, and the world-famous Ceylon tea—supply about 20 percent of export earnings. Sri Lanka is the world's largest exporter of tea. Textiles and clothing manufactured in the Colombo area make up about two-thirds of the country's exports by value. Sri Lanka is nearly self-sufficient in rice production, and with declining rates of population growth, its development prospects are good, particularly if its civil conflict can be ended.

External influences have diversified the island's population and sown seeds of unrest. Centuries of recurrent invasion from India were followed by Portuguese domination in the early 16th century, Dutch in the 17th, and British from 1795 until the country was granted independence in 1948 as a member of the British Commonwealth. This eventful history, plus the longstanding commercial importance of the sea route around southern Asia, has given Sri Lanka a polyglot population that includes Burghers (descendants of Portuguese and Dutch settlers) and Arabs.

Sri Lanka's civil conflict reflects antagonisms between ethnic groups and discontent with economic and political conditions, especially among the minority Tamils (18 percent of the population). Since 1983, more than 63,000 people have been killed and many more made homeless as the Tamils of the country's north have fought for autonomy or independence from Sri Lanka's majority Sinhalese government. Tamils complain that the Sinhalese treat them as second-class citizens and deprive them of many basic rights. Tamil fighters want to establish their own homeland, **Tamil Eelam**, in the north and west of the country. In 1987, guerrilla attacks by the organization of Tamil separatists called the **Tamil Tigers**, or the **Liberation Tigers of Tamil Eelam (LTTE)**, led to military action by Indian troops invited to help restore order. India withdrew in 1990, but the troubles continued, and India worried that Tamil war refugees from Sri Lanka could flood southern India or that Sri Lankan infiltrators could inspire Tamil separatism within India itself.

The 1991 assassination in southern India of Indian Prime Minister Rajiv Gandhi, son of the slain Indira Gandhi, was linked to Indian sympathizers of the Tamil separatist movement in Sri Lanka. In 1995, Sri Lankan government troops succeeded in retaking the Tamils' geographic stronghold on the Jaffna Peninsula. The Tamil guerrillas retreated to the wild forests just south of the Jaffna and continued their insurgency with the sinking of Sri Lankan naval ships, a devastating car bombing of downtown Colombo, an attack on the Sinhalese Buddhists' holiest site (the Temple of the Tooth in Kandy), an assassination attempt on the Sri Lankan prime minister, and a spectacular assault on Colombo's airport that destroyed most of the national airline's planes. Many of these were suicide bombings, a favored tactic of the Tigers.

The country's leadership is being variously advised to crush the revolt militarily or to pursue a course of devolution that would turn over federal powers to outlying states—a possible track for granting autonomy in the north and east to the Tamils. Peace talks were underway in 2004, but the Tamil Tigers promised more violence if their basic demands for self-rule were not met. Until a formal agreement is reached, the rebels will govern their de facto state of Tamil Eelam; they have their own tax, legal, health, and educational systems and operate banks, newspapers, and radio stations.

11.9 Laid Back, Low-Lying Maldives

The roughly 1,100 islands that make up the Maldives (population: 300,000; capital, Male; population: 65,000), an independent country since 1965, appear to be the very essence of tropical paradise (Figure 11.26; for location, see Figure 11.1). So prized are its palm-blessed and coral-fringed beaches that almost half a million tourists, mainly Europeans, visit the country each year. More than 60 percent of the country's foreign currency earnings come from tourism, with fishing and clothing providing most of the rest.

But there may be trouble in paradise. More than 80 percent of this country's very limited land area consists of limestone atolls less than 3 feet (1 m) above sea level. If sea levels were to rise, as predicted by many common climate change scenarios, the entire country could be submerged. This prospect led to the Maldives' president's famous cry at a global warming conference: "We are an endangered nation!"

This concludes a detailed look at the geographies of South Asia. Despite linguistic, social, religious, and political difficulties, South Asia's accomplishments in recent decades have

Figure 11.26 The Maldives have everything the affluent, winter-weary sun-worshippers of Europe want: sun, sand, surf, and more sun.

been remarkable. India has lowered its birth rate, and its economy has grown. Despite its internal divisions and frictions, India has retained a representative democracy. But sectarian violence between Hindus and Muslims, a growing Hindu nationalist movement that threatens India's remarkable diversity, Sikh desire for autonomy, and the problem of Kashmir all pose long-term difficulties for this South Asian giant. Pakistan must walk the tightrope between support for its own majority Muslim interests and support for the secular West in its campaign against terrorism, while also avoiding war with India. Both Pakistan and Nepal must confront the problems of rapid population growth, deforestation, and land degradation where the balance between people and resources is already precarious. Bangladesh faces the huge challenge of dealing with the environmental consequences of upstream deforestation and repeated flooding assaults from the Bay of Bengal. The Maldivians hope and pray that sea levels remain constant but also lobby industrial powers to cut greenhouse gas emissions. Sri Lanka must accommodate its Tamil minority. All of these nations struggle with the chronic problem of poverty, which ultimately is the source of many of the region's most severe environmental and political dilemmas.

SUMMARY

- Most of South Asia is a triangular peninsula extending southward from the main mass of Asia and splitting the northern Indian Ocean into the Bay of Bengal and the Arabian Sea. The entire unit is often called the Indian subcontinent.

- The Indian subcontinent is one of the world's culture hearths. Some of the world's earliest cities developed on the banks of the Indus. The Aryans arrived between 2000 and 3000 B.C. and were the source of many of the dominant cultural attributes of the subcontinent today. The Tamils and other Dravidians predated the arrival of the Aryans.

- South Asia is a regional term used to describe the countries of India, Bangladesh, Pakistan, Nepal, Bhutan, Sri Lanka, and the Maldives in an area about half the size of the conterminous United States.

- The subcontinent is home to many different faiths, including Hinduism, Sikhism, Christianity, Islam, Buddhism, Zoroastrianism, and Jainism. Religious, ethnic, and other differences underlie serious and often violent conflicts. The principal ones have been the Hindus versus Muslims in India, Sikhs versus the government of India, and Tamils versus Sinhalese in Sri Lanka.

- Until 1947, what are now Pakistan, Bangladesh, and India formed the single country of India. For over a century, it was the most important unit in the British colonial empire—the "jewel in the crown."

- With independence, India was partitioned between avowedly Muslim Pakistan (which later divided into two countries, Pakistan and Bangladesh) and secular India. The partition had many consequences, including violence between Hindus and Muslims and their large-scale migrations, problems of how to share Indus River waters, and resources and industries stranded on either side of the new partition lines.

- The most severe product of the partition has been lasting conflict in Kashmir. Several wars have been sparked or fought in the still-volatile frontier province, which, despite its Muslim majority, was joined within India in 1947.

- The Himalaya and Karakoram, the world's highest mountains, enclose the subcontinent on its northern borders. They were produced by the tectonic collision of the former island of India with Asia.

- The subcontinent's northern plain contains the core areas of the three major countries of India, Bangladesh, and Pakistan. Three great rivers water this plain: the Indus River, the Ganges River, and the Brahmaputra River. Although mostly desert and steppe in Pakistan, the plain is relatively well watered in India where it supports unusually high population densities.

- The southern portion of the subcontinent is mainly a large volcanic plateau called the Deccan. Generally low in elevation, it is rimmed by highlands: the Western Ghats, the northern Central Indian Hills, and the Eastern Ghats.

- Climates vary, with undifferentiated highland in the northern mountains to desert and steppe in Pakistan and western India; humid subtropical occurs in the northern plain; tropical savanna is in the peninsula with the exception of the rain shadow in the lee of the Western Ghats; and there are rain forests in the Western Ghats' seaward slopes, the coastal plain to the south, and part of Bangladesh.

- The wet southwest monsoon is at its height from June to September, blowing in from the Arabian Sea and Bay of Bengal. The dry northeast monsoon occurs in the winter, with dry, cooler winds from the northeast coming from high-pressure systems over the continent.

- Although it is an overwhelmingly poor and rural country, India has many modern industries that, along with privatization of former state-run firms, have contributed to its substantial rate of economic growth in recent years. These include the software and computer industries of Bangalore and the film industry of Mumbai. Pakistan and Bangladesh are much less industrialized and rely heavily on textile exports. India and Bangladesh have been successful in lowering birth rates, but population growth is a serious issue for both countries. India has managed to feed its huge population.

- South Asia provides large numbers of workers to other countries, ranging from professional to semiskilled and unskilled. Remittances sent home by these workers are important to the South

Asian economies. Many of the professionals do not return, contributing to the region's brain drain.

- Some of India's large minorities have had serious difficulties with the majority Hindu population. Hindu nationalism is a problem in India, fanning the flames of communal violence between Hindus and Muslims. In the 1980s, some of the Punjab's Sikhs agitated for an independent country.

- Pakistan's government faces challenges from its Pashtun population in the country's west and northwest and from strong Islamist sentiment throughout the country.

- Bangladesh and the Maldives are threatened by serious natural hazards from the sea. Strong typhoons (hurricanes) regularly dev-

astate large parts of low-lying Bangladesh, and most of the Maldives would be inundated if sea levels were to rise by 3 feet (1 m).

- Nepal and Bhutan are picturesque but poor mountain kingdoms. Both have strong tourism potential, but Bhutan restricts tourists, and a violent Maoist insurgency in Nepal is scaring many tourists away.

- Sri Lanka, the world's largest tea exporter, has a Buddhist Sinhalese majority and a Hindu Tamil minority. For more than 20 years, Tamil factions led a violent struggle to establish an autonomous homeland in the island's north. The violence gave way to peace talks that have not yet arrived at a permanent solution.

KEY TERMS + CONCEPTS

Terms in blue are also defined in the glossary.

Aryans (Indo-Aryans) (p. 302)
Bhutan People's Party (BPP) (p. 319)
brain drain (p. 313)
caste (p. 315)
Communist Party of Nepal (p. 319)
culture hearth (p. 301)
Dalits (untouchables) (p. 315)
devolution (p. 320)
digital divide (p. 311)
distributaries (p. 308)
downstream country (p. 306)
dry northeast monsoon (p. 309)
e-choupals (p. 312)
"gross national happiness" (p. 319)

Harappan civilization (p. 302)
India's Silicon Valley (p. 313)
Indus Waters Treaty (p. 306)
Islamization (p. 316)
Jains (Jainism) (p. 303)
"jewel in the crown" (p. 304)
jihad (p. 317)
Khalistan (p. 315)
Liberation Tigers of Tamil Eelam (LTTE)
  (p. 320)
madrasa (p. 317)
megacity (p. 310)
Moguls (Mughals) (p. 302)
Mohajirs (p. 304)

Parsis (p. 303)
remittances (p. 313)
riparian (p. 306)
sectarian (communal) violence (p. 314)
shari'a (p. 316)
Sikhs (Sikhism) (p. 303)
Sinhalese (p. 303)
Tamils (p. 303)
Tamil Eelam (p. 320)
Tamil Tigers (p. 320)
upstream country (p. 306)
wet southwest monsoon (p. 309)
Zoroastrianism (p. 303)

REVIEW QUESTIONS

WORLD
REGIONAL
Geography ⊕ Now™

Assess your understanding of this chapter's topics with additional quizzing and concept-based problems at http://earthscience.brookscole.com/wrg5e.

1. What groups of peoples have been most significant in forming the cultural foundation of South Asia?

2. Using maps and the text, locate the outer mountain wall, the northern plain, peninsular India, and Sri Lanka. What are the major climates, resources, and other natural attributes of these areas? What and where are the three major rivers of the northern plain?

3. How were questions of water allocation between the region's countries resolved after independence?

4. Where are the main production areas for the export crops rice, cotton, jute, and tea?

5. What are the influences of the wet and dry monsoons in South Asia? What particular problems do Bangladesh and the Maldives have in terms of climate and climate change?

6. Why are there far fewer women than men in some Indian provinces? What problems are associated with this phenomenon?

7. What are the major nonagricultural industries of South Asia? Where is India's Silicon Valley? Its Hollywood? Its Wall Street? Where are the other major industries located? Which country is the most industrialized? Which lag behind and why?

8. What is the caste system? How does it affect employment and social relations in India? What evidence of change is there in this ancient system?

9. Where is Ayodhya? What site there is critical in relations between Hindus and Muslims in India?

10. What are Pakistan's *madrasas*? Why is their curriculum of so much concern to the government of Pakistan and to the West?

DISCUSSION QUESTIONS

1. How many people live in the region of South Asia and its largest countries? Where are populations especially dense or sparse? Why?

2. Why is the Indian subcontinent considered the most culturally complex area of its size on Earth?

3. Discuss the conflicts related to religious groups within India, between India and Pakistan, and within Kashmir. How might the Kashmir conflict be resolved peacefully?

4. What major civil problems are there in Sri Lanka, Nepal, and Bhutan?

5. What were some of the benefits and problems created by the British colonization of India? What problems resulted from the country's partition after independence?

6. How were the Himalaya and the Karakoram Mountains formed?

7. Why is India's Narmada Dam project so controversial?

8. Is the "Malthusian scenario" looming over India and its populous neighbors? What progress have the countries made in dealing with the equation between population and resources?

9. How might the Internet and other modern technologies improve the lives of rural peoples in South Asia?

10. What attracts workers from India and other South Asian countries to work abroad? What have your experiences been with expatriates from those countries? How might they illustrate the brain drain?

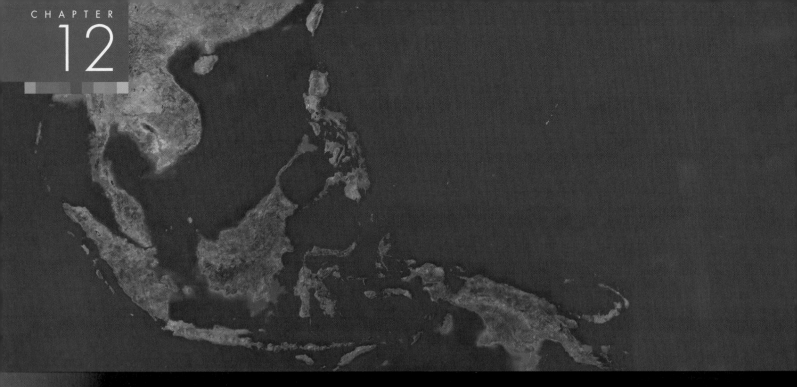

Southeast Asia: From Slash-and-Burn to Semiconductors

The peoples of Southeast Asia are modernizing rapidly but not sacrificing cherished traditions. These Sarawakian women are attending a jubilee festival in the city of Kuching on the island of Borneo.

chapter objectives

This chapter should enable you to:

- Recognize the correlation between environmental opportunity, environmental constraint, and population densities

- See how the interconnectedness of the global economic system led to "economic meltdown" in Southeast Asia in 1997

- Appreciate the economic strength of minority Chinese populations in many of the countries and how that prosperity provokes resentment in times of difficulty

- Understand some of the major political obstacles and environmental consequences associated with large dams, in this case on the Mekong River

- Consider how China's economic ascendancy has forced some of the countries to retool their industries

- Understand how demands for independence (rather than just autonomy) threaten the cohesion of Indonesia and raise alarm there and abroad

WORLD
REGIONAL
Geography ⊛ Now ™

Look for this logo in the text and go to GeographyNow at http://earthscience.brookscole.com/wrg5e to explore interactive maps, view animations, sharpen your factual knowledge and geographic literacy, and test your critical thinking and analytical skills with unique interactive resources.

Southeast Asia, from Myanmar (known until 1989 as Burma) in the west to the Philippines in the east, is a fragmented region of peninsulas, islands, and seas. East of India and south of China, between the Bay of Bengal and the South China Sea, the large Indochinese Peninsula projects southward from the continental mass of Asia (Figure 12.1). From it, the long, narrow Malay Peninsula extends another 900 miles (c. 1,450 km) toward the equator. To the south and east are thousands of islands, the largest of which are Sumatra, Java, Borneo, Sulawesi (Celebes), Mindanao, and Luzon. Another large island, New Guinea, is culturally a part of the Melanesian archipelagoes of the Pacific World, but its western half, held by Indonesia, is politically part of Southeast Asia.

Southeast Asia contributed domesticated plants (including rice), animals (including chickens), and the achievements of civilizations to a wider world. Innovations and accomplishments diffused from several centers, which themselves were influenced by foreign cultures. Indian and Chinese traits have been especially strong in shaping the region's cultural geography. Hinduism came from India, and Theravada

Buddhism came from India through Ceylon (Sri Lanka). The Chinese cultural influences of Confucianism, Daoism, and Mahayana Buddhism were particularly strong in Vietnam. China periodically demanded and received tribute from states in Burma, Sumatra, and Java. The Islamic way of life came with Arab traders from the west from the 13th through the 15th centuries and took permanent root. Southeast Asia is the eastern margin of the world of Islam, and Indonesia has more Muslims than any other country in the world today.

The historic and contemporary monumental architectures of Southeast Asia reflect these diverse origins. In the ninth century, the advanced Sailendra culture of the Indonesian island of Java built Borobudur, the world's largest Buddhist *stupa* (shrine) and still the largest monument in the Southern Hemisphere. In the 12th century, a Hindu king built the extraordinary complex of Angkor Wat as the capital of the Khmer empire based in Cambodia (Figure 12.2a). The first independent Thai kingdom emerged in the 13th century, and in the 14th century, Thai kings ruling from the great city of Ayuthaya (near modern Bangkok) extinguished

288

Southeast Asia

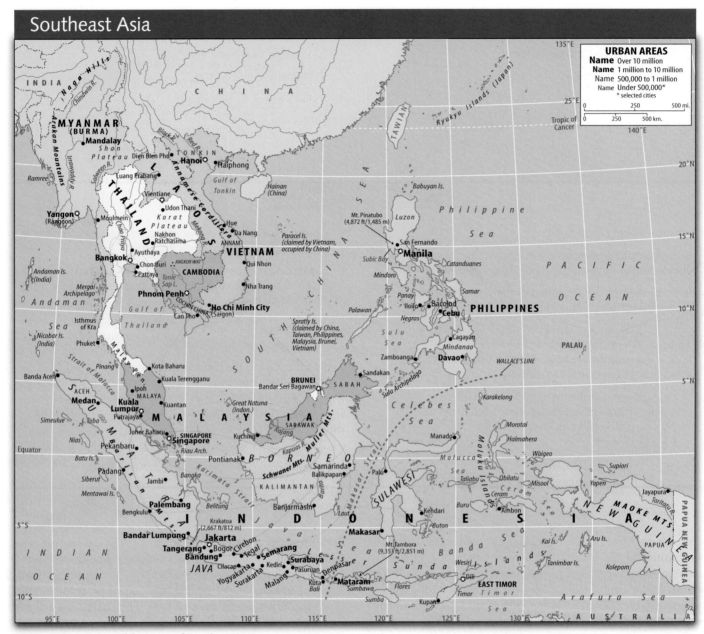

Figure 12.1 Principal features of Southeast Asia

the Khmer empire and extended their own control into the Malay Peninsula and Burma. The world's largest mosque is in modern Brunei, a tiny oil-rich country near the northern tip of the island of Borneo, and until being displaced by Taiwan's Taipei 101 building in 2004, Malaysia boasted the world's tallest buildings, the Petronas Towers in Kuala Lumpur (Figure 12.2b).

Southeast Asia today is composed politically of 11 countries: Myanmar, Thailand, Laos, Cambodia, Vietnam, Malaysia, Singapore, Indonesia, East Timor, Brunei, and the Philippines. With the exception of Thailand, which was never a colony, all of these states became independent from colonial powers after 1946. These 11 nations are fragmented both politically and topographically. Many of the countries are also culturally fragmented and have experienced strife between different ethnic groups. Outside intervention has sometimes complicated and worsened local discord, producing enormous suffering in the region and lasting trauma among the foreign soldiers who fought the determined inhabitants of Southeast Asia.

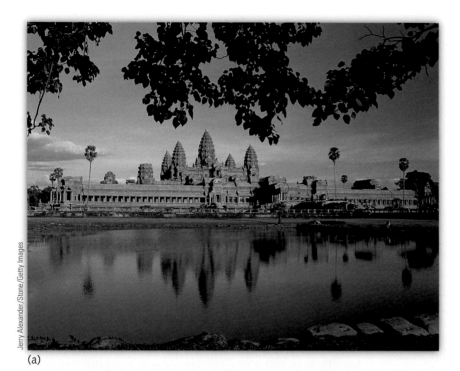

Jerry Alexander/Stone/Getty Images

(a)

Joe Hobbs

(b)

12.1 Area, Population, and Environment

It is about 4,000 miles (c. 6,400 km) from western Myanmar to central New Guinea and 2,500 miles (c. 4,000 km) from northern Myanmar to southern Indonesia. Despite these great distances, the total land area of Southeast Asia is only about 1.7 million square miles (4.4 million sq km), or about half the size of the United States (including Alaska) or China (see Figure 10.2). The estimated population of the region was 547 million in 2004, giving it an overall density of 315 people per square mile (122/sq km) (see Figure 10.3a and b). To put this in perspective, Southeast Asia has nearly four times the overall population density of the United States but well under half that of the Indian subcontinent or the densely populated humid eastern part of China, south of the Great Wall.

Southeast Asia has been less populous than India and China throughout history. Only about 10 million people lived in the region in the year 1800, and there were no widespread dense populations. Historically, geographic and environmental obstacles probably helped keep population growth low in Southeast Asia. By land, the region is rather isolated, cut off to some extent by the rugged mountains of the northern peninsula of Indochina. Over many centuries, the ancestors of most of the present inhabitants entered the area from the north as recurrent thin trickles of population, crossing this mountain barrier as refugees. Southeast Asia is tropical, with continuous heat in the lowlands, torrential rains, prolific vegetation difficult to clear and keep cleared, soils that are generally leached and poor, and a high incidence of disease. A 4- to 6-month dry season on the Indochinese Peninsula and on some of the islands presents further challenges.

More spectacular environmental difficulties result from the location of part of the region on the Pacific Ring of Fire, especially subjecting Indonesia and the Philippines to earthquakes, the tidal waves (tsunami) that earthquakes create, and volcanic eruptions (see Figure 2.9, page 26 and Natural Hazards, page 351). Violent wind-and-rain storms known locally as **typhoons** (known elsewhere as hurricanes

Figure 12.2 (a) Angkor Wat, the 12th-century capital of the Khmer empire, is a symbol of the rich cultural history of Southeast Asia. (b) Modernity and prosperity characterize some nations in Southeast Asia today. At 1,462 feet (446 m), the twin Petronas Towers in Kuala Lumpur are the world's second tallest buildings, after the 1,667-foot (508-m) Taipei 101 complex in Taiwan.

and cyclones) often strike, particularly in Vietnam and the Philippines.

Typical of less developed countries (LDCs), the introduction of modern medical, food-producing, and other technologies into the region softened the blows of environmental adversity and lowered death rates, producing a tremendous population increase in the last century and a half. This increase, which accelerated after World War II, is now slowing but still rapid. Between 1973 and 2004, population grew from an estimated 316 million to 547 million, a 31-year increase of 231 million people, or 73 percent. As will be seen later in the chapter, the combination of Dutch colonial practices and localized, very productive soils gave rise to an extraordinarily dense population on the island of Java. This Indonesian island's 130 million people represent almost one-fourth of the entire population of Southeast Asia. During recent times, the race between population growth and food supply has seen food supply winning in some countries but losing to population growth in others. Since the 1970s, increases in food production outpaced population growth in Malaysia and Indonesia. At the other end of the scale were Cambodia and Laos, in which war, repression, and inefficient communalization of agriculture slowed progress. These discrepancies are typical of Southeast Asia, home to a remarkable range of economies, from quite poor to very prosperous.

12.2 The Economic Pattern

Southeast Asia as a whole is among the world's poorer regions, yet some of its countries are rather prosperous. The countries' mean GDP PPP per capita of $4,059 in 2004 was about average for the world's less developed countries. In contrast, Singapore, Malaysia, the Philippines, Brunei, and Thailand were well above average for LDCs. Because of their high rates of industrial productivity and economic growth, Singapore, Malaysia, and Thailand are described either as "Asian Tigers" or "Tiger Cubs" (along with Hong Kong, South Korea, and Taiwan). The commercial-industrial city-state of Singapore, for example, has about the same GDP PPP per capita as Italy. High-tech industries brought spectacular economic growth to these countries through much of the 1990s. In 1997, a series of events precipitated an economic crisis that roared through the region and temporarily declawed these emerging economies, but conditions have since improved (see pages 332–333).

This region's economic activities span the entire continuum from Stone Age hunting and gathering (in Papua on the island of New Guinea) to slash-and-burn farming—one of humankind's earliest agricultural techniques—to the commercial agriculture typical of colonial times and from the export of raw materials like rubber to manufactured goods like clothing, shoes, and computer products. As Southeast Asians see it, their economic future will depend not so much on rubber, coffee, or even oil, which some countries have abundant supplies of, but on high-tech exports like semiconductors and software. But despite islands of technological innovation, rapid urbanization, and industrialization, the majority of the people in Southeast Asia are still farmers, with 62 percent of the people classified as rural.

Shifting and Sedentary Agriculture

Permanent agriculture has been difficult to establish in many parts of this largely mountainous, tropical realm. Much land is used only for shifting cultivation, in which a migratory farmer clears and uses fields for a few years and then allows them to revert to secondary growth vegetation while he moves on to clear a new patch. Since an abandoned field needs about 15 years to restore itself to forest that can again be cleared for farming, only a small portion of the land can be cultivated at any one time. This extensive form of land use can support only sparse populations. Unirrigated rice is the principal crop of shifting agriculture, with corn, beans, and root crops such as yams and cassava also grown. In some forested areas, hunting and gathering supplement agriculture. These migratory subsistence farmers are often of different ethnic groups from adjacent settled populations, having historically sought refuge in the backcountry from stronger invaders.

Most of Southeast Asia's people live in widely scattered, dense clusters of permanent agricultural settlements (see Figure 10.3a). These core regions stand in striking contrast to the relatively empty spaces of adjoining districts. Superior soil fertility is the main locational factor in these areas. However, there are a few districts of dense populations without high-quality soils to support them. Most of these rely on plantation agriculture of rubber and other less demanding cash crops.

The typical inhabitant of the areas of permanent sedentary agriculture is a subsistence farmer whose main crop is wet rice grown with the aid of natural flooding or irrigation (Figure 12.3). In some drier areas, unirrigated millet or corn replaces rice. Secondary crops that can be grown on land unsuitable for rice and other grains supplement the major crop. These secondary food crops include coconuts, yams, cassava, beans, and garden vegetables. Some farmers grow a secondary cash crop such as tobacco, coffee, or rubber. Fishing is important both along the coasts and along inland streams and lakes. Because of their importance in the food supply, many fish are harvested from artificial fishponds

Joe Hobbs

Figure 12.3 These terraced paddies in Borneo are ready to be planted in rice; a crop is already growing on the terrace between the two huts.

and flooded rice fields, in which they are raised as a supplementary "crop."

Subsistence agriculture in Southeast Asia exists alongside tropical plantation cash crops grown for export. The region is one of the world's major supply areas for such crops, which include oil palm, rubber, and cloves. This highly commercial type of production is a legacy of Western colonialism, which began in the 16th century. Until the 19th century, Westerners were interested mainly in the region's location on the route to China. They were content to control patches of land along the coasts and to trade for some goods produced by local people. During the 19th century, however, the Industrial Revolution in the West greatly enlarged the demand for tropical products. Europeans therefore extended their control over almost all of Southeast Asia. European colonists used their capital and knowledge, along with indigenous and imported labor, to increase export production rapidly. They focused on exporting some local produce such as copra (dried coconut meat from which oil is pressed) and spices. They also introduced some new commercial crops.

The usual method of introducing commercial production was to establish large estates, or plantations, managed by Europeans but worked by indigenous labor or labor imported from other parts of Monsoon Asia. There were of-

ten problems finding available local hands, so Europeans imported contract labor from India and China. Large migrations from these countries complicated an already complex ethnic mixture, setting the stage for today's unusual distribution of ethnic groups and their sometimes discordant relations.

Many export cash crops, some produced by small landowners, are produced in Southeast Asia. The major ones are rubber, palm products, and tea. Nearly three-fourths of the world's natural rubber is produced in the region, mainly in Malaysia, Indonesia, and Thailand (see Definitions and Insights, page 331). However, synthetic rubber, most of which is made from petroleum, now meets the greater part of the world's demand. Palm products include palm oil, coconut oil, and copra. Malaysia is the world leader in producing palm oil (a popular and inexpensive vegetable oil used in cooking throughout Asia), and Indonesia dominates the world output of coconut palm products. Indonesia is also the region's leading producer of tea and is among the world's top five producers. Other export crops in Southeast Asia include coffee from Indonesia, cane sugar from the Philippines and Indonesia, pineapples (of which the Philippines and Thailand have become the world's leading producers and exporters), and many others. However, the countries of Southeast Asia are generally coming to depend less on agricultural exports and more on mining and manufacturing.

One impact of Western colonialism on the economy of Southeast Asia was the stimulation of commercial rice farming in formerly unproductive areas. Three areas became prominent in commercial rice growing: the deltas of the Irrawaddy, Chao Praya, and Mekong Rivers, located in Myanmar, Thailand, and Vietnam and Cambodia, respectively. Almost impenetrable wetlands and uncontrolled floods had kept these deltas thinly settled, but incentives and methods for settlement have turned them into densely populated areas within the past century. Thailand, Vietnam, and Myanmar are the region's leading rice exporters. Other Southeast Asian nations import much of this produce.

Deforestation

Near areas of dense settlement in Southeast Asia, there are still some large areas covered in primary forest, but they are shrinking as population and development progress. Many environmentalists view the destruction of the region's tropical rain forest as an international environmental problem (Figure 12.4). In 2004, only 28 percent of Thailand's original forest cover remained. Neighboring, poorer Myanmar had a 51 percent forest cover. Vietnam lost more than one-third of its forest between 1985 and 2004, by which time only 30 percent of the original forest cover remained. Deforestation is advancing most rapidly in Malaysia, where 59 percent of the land area is still forested (official statistics often include palm oil plantations in forest estimates), and in Indonesia, where 55 percent is forested. Malaysia is now the

F10.18
292

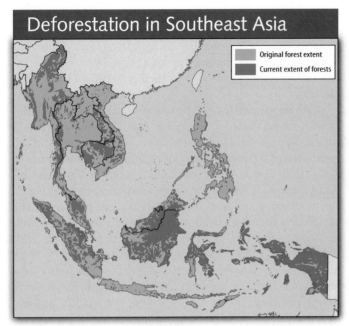

Deforestation in Southeast Asia

Original forest extent
Current extent of forests

WORLD
REGIONAL
Geography ⊛ Now™   **Active Figure 12.4** In Southeast Asia, natural forest cover has been rapidly reduced by many human uses, especially commercial logging and farming. *See an animation based on this figure, and take a short quiz on the facts or concepts.*

Figure 12.5 Commercial logging has removed all but small portions of the tropical rain forest in East Malaysia's states of Sarawak (seen here) and Sabah. The same process is underway in neighboring Kalimantan on Indonesia's side of the island of Borneo.

world's largest exporter of tropical hardwoods. Virtually all of the primary forests in Peninsular Malaysia have already been cleared, and it is projected that commercial logging will decimate those remaining outside protected areas in Malaysian Borneo within a decade (Figure 12.5).

Although Southeast Asia's tropical forests are smaller in total area than those of central Africa and the Amazon Basin, they are being destroyed at a much faster rate, most significantly by commercial logging for Japanese markets rather than by subsistence farmers. Environmentalists fear the irretrievable loss of plant and animal species and the potential contribution to global warming that deforestation in this region may cause. Malaysia and Indonesia have passed laws to slow the destruction, with Indonesia taking the aggressive step of banning the use of fire to clear forests. However, enforcing such legislation has thus far proven impossible. Hundreds of Indonesian and Malaysian companies, most of them large agricultural concerns (for example, palm oil plantations and pulp and paper companies) with close ties to the government, continue to use fire as a cheap and illegal method of clearing forests. The World Wildlife Fund estimates that 70 percent of Indonesia's exported trees are illegal, many of them removed by "timber barons" in the country's national parks while bribed park and police officials look the other way. There is a growing movement in the consuming countries to boycott the tainted timber by selling only certified "good wood." The U.S. company Home

Depot, for example, sells only imported woods bearing the Forest Stewardship Council (FSC) logo.

Deforestation in Southeast Asia has many actual and potential transboundary consequences. In 1997, a widespread drought attributed to the El Niño warming of Pacific Ocean waters turned the annual July–October burning season into a human-made holocaust. Deliberately set fires raced out of control over large areas of Sumatra and Borneo, resulting in a choking haze that shut down airports, closed schools, deterred tourists, and caused respiratory distress to millions of people in Indonesia, Malaysia, Singapore, the Philippines, Brunei, and Thailand. Less widespread but still severe smog crises struck Malaysia and Indonesia, particularly the island of Sumatra, in subsequent years, and Indonesian authorities fought back by revoking the licenses of plantation companies.

579

Definitions + Insights

Rubber

The earliest major cash crop introduced into Southeast Asia was rubber, first cultivated there in the 1870s. Englishman Sir Joseph Priestley gave the name "rubber" to the latex of the *Hevea brasiliensis* tree in 1770, when he discovered he could use it to rub errors off the written page. Extensive exploitation of this tree in its native Brazil gave rise to the "rubber boom" of the late 19th century, when Brazilian rubber traders were so wealthy that they used bank notes to light cigars and sent their shirts to be laundered in Europe.

In 1876, Englishman Henry Wickham smuggled 70,000 rubber seeds out of Brazil. Delighted botanists at London's Kew Gardens cultivated them and shipped them to Ceylon (Sri Lanka) and Southeast Asia for experimental commercial planting. Rubber trees proved exceptionally well adapted to the climate and soils of South, and later Southeast, Asia (Figure 12.A). Back in Brazil, however, a fungus known as leaf blight wiped out the rubber plantations, and by 1910, bats and lizards inhabited the mansions of Brazil's rubber barons. Fortunately, the Kew Gardens stock was free of the fungus; by 1940, 90 percent of the world's natural rubber came from Asian plantations.

Figure 12.A Rubber latex bleeding down a fresh incision and into the collection cup

Joe Hobbs

Many of the species that inhabit these forests are endemic (found nowhere else on Earth). Indonesia, which contains 10 percent of the world's tropical rain forests, is known, after Brazil, as the world's second most important **megadiversity country,** with about 11 percent of all the world's plant species, 12 percent of all mammal species, and 17 percent of all bird species within its borders. The Southeast Asian region is also particularly significant in biogeographic terms because of the so-called **Wallace's Line**—named for its discoverer, English naturalist Alfred Russel Wallace—that divides it (see Figure 12.1). Nowhere else on Earth is there such a striking local change in the composition of plant and animal species in such a small area on either side of this divide. East of the line (separating the Indonesian islands of Bali and Lombok, for example), marsupials are the predominant mammals, and placental mammals prevail west of the line. Similarly, bird populations are remarkably different on either side of the line.

WORLD
REGIONAL
Geography⊛Now™

Click Geography Literacy to take a virtual tour of Indonesia.

Mineral Production and Reserves

Petroleum is the most important mineral resource in Southeast Asia. Seven of the region's countries—Indonesia, Malaysia, Brunei, Myanmar, Thailand, Vietnam, and the Philippines—produce oil. Collectively, their proven oil reserves amount to about 1.1 percent of the world total. Their combined output is about 4 percent of world oil production, with Indonesia the leading producer. Indonesia's main oil fields are in Sumatra and Indonesian Borneo (known as Kalimantan).

Although it does not loom large on the world scene, Southeast Asian oil is very important to the countries that own and produce it. Oil and associated natural gas provide nearly all the exports of Brunei and make that tiny Muslim country one of the world's wealthiest nations as measured by GDP PPP per capita. That wealth is precarious, as Brunei has little to offer world markets except its finite fuels. Oil, gas, and refined products supply almost one-fourth of Indonesia's exports and about one-tenth of Malaysia's. Singapore profits as a processor of oil. The main customer for Southeast Asian oil and natural gas is Japan, the leading trade partner of several Southeast Asian countries. Natural gas is so important domestically for electricity production in several Southeast Asian countries that efforts are underway to link them in a network of pipelines, tapping mainly Indonesian and Malaysian gas reserves (Figure 12.6).

There is other mineral wealth in Southeast Asia. Tin is the most important ore mined, with about one-fifth of world output from Indonesia. The Philippines produces a variety of metals—copper, chromite, nickel, silver, and gold—in significant amounts. The most serious mineral deficiency in the region is the near absence of high-grade coal. However, more energy could come from the region's additional potential for hydroelectric power. Gemstones are important exports for Myanmar and Thailand (Figure 12.7).

Figure 12.7 The star sapphire from Myanmar is one of Southeast Asia's most prized gemstones.

The Meltdown and Its Aftermath

Following years of spectacular growth, in which economies across the region grew at rates up to 8 percent annually, 1997 saw a stunning reversal of fortunes in Southeast Asia. Up to 1997, the region's economic Tiger Cubs were praised for high rates of savings, balanced budgets, and low inflation. Export-driven growth rates were climbing, particularly with the increased manufacturing of computers and other electronics. Gradually, however, the appetite of the main consumer for the products, the United States, began to wane, and trade deficits developed in the producing countries. At the same time, the emerging economies of Southeast Asia pursued expensive public works projects, such as Malaysia's new capital of Putrajaya.

In some countries, corruption added to the potential for economic crisis as insolvent companies were propped up with infusions of cash from sympathetic politicians (in a phenomenon known as "**crony capitalism**"). Banks extended risky loans for both private and public businesses and construction projects. The loans were generally made in U.S. dollars. When local currencies rapidly lost their values, the borrowers were unable to repay these loans. Nervous foreign investors in the region's stock markets withdrew their investments; theirs was the "hot money" that can send stock markets soaring but can also be withdrawn quickly, with devastating results.

Economies melted down. Stock markets, which had seen explosive growth, imploded; Malaysia's market, for example, lost 80 percent of its value between September 1996 and September 1998. Among the region's stronger economies (excluding Vietnam, Cambodia, Laos, and Myanmar, which had far less to lose), Malaysia, Thailand, and Indone-

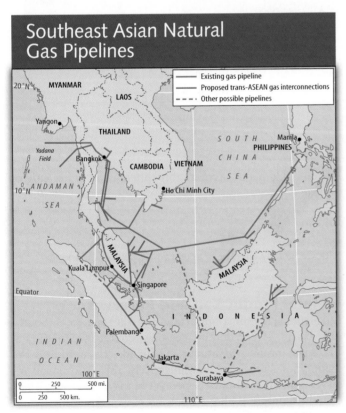

Southeast Asian Natural Gas Pipelines

——	Existing gas pipeline
——	Proposed trans-ASEAN gas interconnections
– – –	Other possible pipelines

Figure 12.6 Existing and proposed natural gas pipelines for Southeast Asia.

sia fell into **economic recession**, characterized by a relatively short fall in economic output and a moderate rise in unemployment. As the crisis dragged on, it threatened to tip into outright depression. Only the Philippines and Singapore escaped the bloodletting.

Economies gradually began to improve, but the economic meltdown had lasting impacts. Largely unchallenged in prosperous times, Southeast Asian ways of running governments and businesses have been under strong scrutiny since the hard times. Indications of this shift run the gamut from large-scale revolts against and/or impeachments of national leaders to parents fighting favoritism in school admissions, farmers protesting dam building projects, peasants protesting industrial pollution, and patients suing doctors for malpractice. Examples of these events are described in the discussions of the individual countries. Despite their internal problems and differences, the countries hoped that their common membership in the Association of Southeast Asian Nations (ASEAN) might unite them against future economic calamities (see the Regional Perspective below).

Although the countries of Southeast Asia have many similarities, each has its own distinctive qualities, with unique environmental features, indigenous and immigrant peoples, culture traits, and economic activities. Many of the characteristics and problems of these countries are the outcome of European imperialism, but in no two countries have the results of colonialism been the same. Here are portraits of the countries.

12.3 Myanmar (Burma)

Myanmar (population: 50 million) is centered in the basin of the Irrawaddy River and includes surrounding uplands and mountains. Within the basin are two areas of dense population: the Dry Zone, around and south of Mandalay (population: 1.2 million), and the Irrawaddy Delta, which includes the capital and major seaport of Yangon (known formerly as Rangoon; population: 4.5 million). The Dry Zone is the historical nucleus of the country. The annual rainfall of this area—34 inches (87 cm) at Mandalay—is very low for Southeast Asia. The people are supported by mixed subsistence and commercial farming, with millet, rice, and cotton the major crops. During the past century, the Dry Zone was surpassed in population by the delta, where a commercial rice-farming economy now provides one of the country's two leading exports (forest products is the other). The Irrawaddy forms a major artery of transportation uniting the two core areas.

The indigenous Burman people, most of whom live in the Irrawaddy Basin, number an estimated 68 percent of the country's population. About 90 percent of Myanmar's citizens are Theravada Buddhists. Distinct ethnic groups live in the Arakan Mountains of the west and the northern highlands. Ethnic Shans live in the northern part of the Shan Plateau of eastern Myanmar, and ethnic Karens live in the southern Shan Plateau and in the delta of the Salween River. These ethnic groups account, respectively, for about 9 and 7 percent of the country's population. The Shan Plateau in

Regional Perspective

The Association of Southeast Asian Nations (ASEAN)

The countries of Southeast Asia have sought to strengthen their economic and security interests by forming a regional organization. Known as the **Association of Southeast Asian Nations (ASEAN)**, the coalition was formed in 1967 to promote regional trade and minimize alignment with the two world superpowers: the United States and the Soviet Union. Singapore, Malaysia, Thailand, Indonesia, Brunei, and the Philippines were founding members. Vietnam joined in 1995 with the promise of strengthening regional ties and adopting the more open economic system that the Vietnamese call *doi moi*, or renovation. Myanmar and Laos joined in 1997, and finally Cambodia joined in 1999.

After U.S. troops withdrew from bases in the Philippines in 1991, ASEAN members grew concerned that they lacked a security "umbrella" and that China or possibly even Japan might become more aggressive in the region. Several ASEAN members, notably Indonesia, Malaysia, and Singapore, therefore strengthened ties with the U.S. military to deter other powers from attempts to dominate the region. The organization thus aspired to become both a large trading bloc and a substantial security counterweight to China.

By the end of the century, however, ASEAN members moved closer to China, increasingly seen as a benign and economically advantageous ally, and created an informal economic cooperation alliance with China, Japan, and South Korea known as ASEAN+3. The bloc's main goal is to adopt policies that would avert the type of economic meltdown that shook all the group's members in 1997 and 1998. Singapore and Thailand are calling for all of the ASEAN nations to have, by 2020, a European-style common market that would eliminate member country barriers to trade and investment. The countries are also trying to reduce dependence on the International Monetary Fund and other Western monetary agencies and countries in the event of future economic crises.

Regional Perspective

Sex, Drugs, and Health in Southeast Asia

Southeast Asia has long been one of the world's leading source areas for opium and its highly addictive semisynthetic derivative, heroin. For decades, global concern focused on combating the drug trade originating in this region and on dealing with the problem of heroin addiction in the consuming nations. Now, because of the HIV virus, the other strong Southeast Asian industries of prostitution and tourism have woven drugs into a leviathan problem of global concern.

The region's drug production is centered in the Golden Triangle, the remote region where Laos, Thailand, and Myanmar meet (see Figure 12.1). These countries have seldom exercised successful control over their territories in this area, although Thailand recently increased its presence. As in Afghanistan and Pakistan, the absence of a strong central government presence and ideal growing conditions have made the Golden Triangle one of the world's greatest centers of opium poppy cultivation (Figure 12.B). China White, or Asian-grown heroin, originates here and commands approximately 18 percent of the global heroin market. About 100 tons of heroin leave the Golden Triangle each year, bound for markets in the United States, Australia, and Western Europe. International authorities succeed in stopping only an estimated 10 percent of that flow.

In recent years, with new heroin producing areas cropping up in Mexico and elsewhere, the price of heroin for the consumer has dropped and its quality has risen. In the United States, these

Figure 12.B Raw opium oozes from slices in a poppy pod. Heroin, which is derived from opium, has a hearth in Southeast Asia's Golden Triangle region.

Joe Hobbs

trends are noticeable in the growing appeal of heroin among middle and upper income groups, by smoking and other previously rare forms of heroin consumption, and by the increasing numbers of deaths due to heroin overdose.

The U.S. Drug Enforcement Administration and other international antidrug authorities focus their concerns on Myanmar. Nearly all of the region's refining of heroin from opium is done in remote laboratories there. In northeastern Myanmar, armed ethnic minority groups specialize in the production of heroin and a potent methamphetamine known as *ya baa*. They smuggle raw opium into both China and India, and amphetamines into Thailand, providing large amounts of investment capital to contribute to the country's "development." Laundered drug money has paid for Myanmar's recent construction boom, and the government encourages such investment from the drug sector to make up for shortfalls in legitimate export revenue.

Not all of the associated problems are exported. A flood of cheap and potent heroin hit the market in Myanmar after the current regime came to power in 1988. Heroin addiction has soared ever since. The primary delivery is by needles shared in tea stalls. The numbers of users continue to grow; in 2004, an estimated 2 percent of adult men and 4 percent of adult women were addicts. Accompanying the heroin use is an AIDS epidemic, with an estimated 4 percent of Myanmar's adult population infected. The 2004 infection rate among heroin users in Myanmar stood at 60 percent.

The epidemic that began among heroin users spread quickly to Myanmar's sex industry. Year 2003 estimates put the HIV infection rate among prostitutes at 47 percent, twice the rate in

---

northeastern Myanmar is part of the **Golden Triangle,** which was long the world's leading source area for opium, until Afghanistan pushed it to number two in the 1990s (see the Regional Perspective above).

Great Britain conquered Burma in three wars between 1824 and 1885. It became an independent republic outside the British Commonwealth in 1948. There has been almost constant civil war since independence. Communism, targeted persecution, and ethnic separatism have motivated rebellions by ethnic Karens and Shans, a number of less numerous hill peoples, and Muslims of the Rohingya ethnic

group who fear the creation of a pure Buddhist state. Beginning around 1999, however, Myanmar's ruling military government reached cease-fire agreements with most of these ethnic groups (except the Karens) and allowed what it called **contingent sovereignty** to these groups, offering them more civil rights and economic opportunities than they had known previously. The government also encouraged profits made by some of these groups in the drug business to be "invested" in national development.

The economy has been badly damaged by fighting and for years was mismanaged under a form of rigid state control

neighboring Thailand. Many young men with AIDS are shunned by their families and go to monasteries to die. Tragically, their virus has spread among the populations of Buddhist monks (there are now 400,000 in Myanmar) who share razors to shave their heads. AIDS prevention is virtually unheard of. The government had outlawed condoms until 1993. They are now legal, but along with HIV tests, they are too expensive for most people. There are almost no anti-HIV drugs in Myanmar.

Myanmar's prosperous, cosmopolitan neighbor Thailand also has the scourge, with a 1.8 percent HIV infection rate among adults in 2004. Thailand's liberal social climate has given rise to the unique phenomenon of sex tourism, with package tours catering to an international clientele, particularly Japanese men (Figure 12.C). Thailand's overall annual income from prostitution is estimated to be between $2 and $4 billion, and prostitutes send as much as $300 million each year to relatives in the villages from which they migrated. However, the spread of HIV infection among intravenous drug users and among prostitutes in the red-light districts of Bangkok and the popular beach resorts of Pattaya and Phuket

threatens both the country's public health and its lucrative tourist trade. The Thai government pursued an all-out war on the drug trade in 2002 that resulted in the deaths of more than 2,000 people, most of them suspected drug dealers.

There are striking differences in how this region's countries are dealing with HIV/AIDS. Thailand's government has responded with an aggressive and increasingly successful anti-AIDS public awareness campaign, resulting in a decrease in the number of new HIV infections. Other Southeast Asian countries lag far behind. The government of the Philippines has been unsuccessful in mounting an anti-AIDS campaign because of opposition from the country's powerful and anticontraception Catholic clergy. Church officials have denounced the government's anti-AIDS program as "intrinsically evil" and have set boxes of condoms on fire at antigovernment demonstrations. Similarly, in the conservative and largely Islamic nations of Indonesia and Malaysia, Muslim clerics denounce their governments' anti-AIDS campaigns as efforts to encourage promiscuity.

Taimay Jones

Figure 12.C Bangkok's red light district was a breeding ground for HIV/AIDS until the government and nongovernmental agencies aggressively promoted a safe-sex campaign.

promoted as **"The Burmese Way to Socialism."** Myanmar became the poorest non-Communist country in Southeast Asia, a sad contrast with its reputation in early independence years for having the best healthcare, highest literacy rate, and most efficient civil service in Southeast Asia. In 2004, Myanmar ranked second from the bottom (just above Sierra Leone) in global quality of healthcare. In the 1990s, Myanmar began to abandon socialism in favor of a free-market economy and enjoyed considerable economic growth, with China its main trading partner and military ally. Most of the

fruits of this economic growth, however, lined the pockets of the ruling military.

The military government that seized power in Burma in 1988 (and changed the country's name to Myanmar) has yielded little to popular pressure for democratization. Myanmar remains one of the world's most repressive places to live. Access to the Internet was prohibited until 1999 and is now strictly regulated. Foreign journalists are banned, and citizens may not allow foreigners into their homes. It is illegal to gather outside in groups of more than five. Giant green billboards across the country read: "Crush all internal and external destructive elements as the common enemy."

There has been an ongoing and much publicized contest between the government and Nobel Peace Prize laureate Aung San Suu Kyi, who had been under house arrest after leading her antimilitary political party to overwhelming victory in the 1990 elections (the results of which the military government nullified). In 1995, the government released "The Lady," as Ms. Suu Kyi is popularly known. She then called for dialogue with the military junta and appealed for the parliament that was elected in 1990 to be convened. The government responded by again placing her under house arrest and then in prison, triggering strong economic sanctions by the United States.

Aung San Suu Kyi has called for would-be tourists to avoid Myanmar, which has abundant sites of interest but few visitors, arguing that tourism revenue would help strengthen the repressive government. She has also been successful in gaining support abroad for boycotts of U.S. and other Western companies doing business with Myanmar. American corporations including Pepsi and Phillips Petroleum have withdrawn from Myanmar, and the U.S. government has banned any new investments. The United States regards the country's government as illegitimate and therefore rejects use of the country name Myanmar. There is pressure on the American firm Unocal to suspend its contract to join a French firm in developing Myanmar's considerable offshore natural gas reserves in the Yadana Field of the Andaman Sea. The two companies have already built a pipeline from the field to a port west of Bangkok, Thailand, where the gas is used to generate electricity (see Figure 12.6). Human rights groups complain that slave labor was used to build the pipeline and that profits from gas sales help prop up Myanmar's repressive government. The pipeline has to be well defended in view of the government's many domestic opponents who might try to destroy it as a means of weakening the regime.

12.4 Thailand

Governed today as a democracy in which military influence is strong and in which a traditional monarchy has mainly symbolic functions, Thailand (formerly Siam; population: 64 million) was the only Southeast Asian country to preserve its independence throughout the colonial period. This unique status was due to Siam's position as a buffer between British and French colonial spheres.

Thailand is located in the delta of the Chao Praya River, known to Thais as the Menam ("The River"). Annual floods of the river irrigate the rice that is the most important crop in this Southeast Asian land of relative abundance and progress. Thailand is the world's largest exporter of rice.

On the lower Chao Praya is Bangkok (population: 9 million), the capital, main port, and only large city. It is a notoriously crowded and smoggy city; air pollution is responsible for as many as 400 deaths yearly. Environmental blight has prompted the Thai government's vision to build

"Heaven City," a satellite city of Bangkok intended to have little pollution or congestion. But Bangkok is also a city of stunning architectural marvels, particularly Buddhist temples. Bangkok's networks of canals (*klongs*), thronged with both commercial and residential activities, give it a distinctively amphibious character; it has been dubbed the **"Venice of the East."** Its "floating market" is especially renowned (Figure 12.8). Many of Bangkok's canals have been filled in with paved roads and buildings, thereby making the city vulnerable to flooding. Due to the pumping out of groundwater beneath the city, Bangkok's soils are settling downward, and the entire capital is sinking at a rate of about 2 inches (5 cm) yearly. Thai officials forecast that Bangkok will be below median sea level by 2050. If nearby sea levels rise, as predicted by many climate change models, the city would be imperiled.

Areas outside the river's delta are more sparsely populated and include the mountainous territories in the west and north, inhabited mainly by Karens and other ethnic groups, and the dry Korat Plateau to the east, populated by the Thai, related Laotian, and Cambodian peoples. To the south in the Kra Isthmus, a part of the Malay Peninsula, live more than 2 million ethnic Malays. Most Malays are Muslims, but an estimated 95 percent of the people of Thailand are Buddhists whose faith includes elements borrowed from Hinduism and local spiritualism (Figure 12.9). Spectacular Buddhist temples are a prime attraction for one of Southeast Asia's largest tourist industries, as are ceremonies celebrating Thailand's monarchy and other historical traditions.

The main ethnic minority of Thailand is Chinese (Figure 12.10). An estimated 8.9 million ethnic Chinese, or 14 percent of the population, reside in the country. They control much of the country's business, a situation that many Thais

Figure 12.8 The floating market of Damnern Saduak near Bangkok, Thailand, typifies the amphibious nature of life in Southeast Asia. Water transportation is vitally important in the river deltas and islands where most Southeast Asians live.

Figure 12.9 Piya Chat, a Buddhist nun, lives in a cave. Devout Buddhists having spiritual or family problems seek her counsel; they make appointments with her by cell phone and travel to her refuge for the consultations.

resent and that makes the loyalty of the Chinese an important issue.

Although Thailand has enjoyed more internal tranquillity than most of Southeast Asia, it has had its troubles. Since the late 1960s, guerrilla insurrections motivated by communism and ethnic separatism have persisted in the northeast and

Figure 12.10 Ethnic Chinese are a significant and generally prosperous minority in most countries of Southeast Asia. Pam and Zee Yoong, a lawyer and engineer, respectively, in the Malaysia city of Ipoh, are enjoying a favorite Chinese fruit, durian.

south. In the late 1970s, the country dealt with a mass migration of refugees from adjoining Cambodia. Recently, ethnic Malay guerrillas seeking independence for Thailand's mainly Muslim southernmost region have burned schools, killed Buddhist monks, and ambushed government troops in that area.

In spite of conflict and refugee problems, Thailand has had one of the strongest economies in Southeast Asia. From 1993 to 1996, the economy grew a robust 8 percent annually. The country shifted away from its traditional exports of garments, textiles, cut gems, and seafood in favor of medium-tech industries such as computer assembly. Tourism was the leading source of foreign exchange.

Then came 1997, when Thailand was the first "domino" to fall in the Asian economic crisis (see page 332). Thailand's currency, the baht, was pegged to the U.S. dollar, meaning that its value fluctuated in a narrow range relative to the dollar. In 1996 and 1997, the currencies of Thailand's main competitors, Japan and China, fell and put Thailand at an increasingly competitive disadvantage as the U.S. dollar rose in value. As the costs of Thailand's exports grew, its economy weakened and its banks, which had funneled billions of dollars into questionable investments, began to look vulnerable. Currency speculators, who profit by taking advantage of declining currencies, bet that the baht was due for a steep decline. The government exhausted its reserves of foreign currency and raised interest rates in a failed effort to strengthen the currency. The baht fell precipitously when its peg to the dollar was removed, and the increase in interest rates put a virtual halt to the country's economic development activities.

While the currency crisis and subsequent economic recession spread from Thailand throughout the region, Thailand's emerging consumer society suddenly found itself bereft. Economic growth had attracted tens of thousands of the rural poor to the factories of Bangkok, where they made garments for such major U.S. retailers as Gap, Nike, and London Fog. Now they were laid off and joined the swelling ranks of Thailand's poor. Even before the slump, Thailand had a huge gap between the wealthy and the poor, with more than 50 percent of the country's wealth in the hands of the richest 10 percent of its population. By the beginning of the new century, Thailand's economy was back on its feet. In recent years, it has recorded strong growth based largely on surging demand for electronic parts, petrochemicals, and rubber in China.

12.5 Vietnam, Cambodia, and Laos

Vietnam, Cambodia, and Laos are the three states that emerged from the former French colony of Indochina. Vietnam is the largest and most geographically complex, with three main regions. Tonkin, the northern region, consists of the densely populated Red River Delta and surrounding, sparsely inhabited mountains; Annam, the central region, includes the wild Annamite Cordillera and small, densely populated pockets of lowland along its seaward edge; and Cochin China, the southern region, lies mainly in the delta of the Mekong River and has a high population density.

Almost nine-tenths of Vietnam's people are ethnic Vietnamese (also known as Kinh), closely related to the Chinese in many cultural respects, including language and the shared religious elements of Confucianism, Buddhism, and ancestor veneration. A Catholic minority reflects the French occupation, and there are small indigenous religions. Fifty-two minorities enrich the country's ethnic makeup. Various groups known as the **hill tribes** or **Montagnards** live in the central highlands of Vietnam, where some have recently agitated for independence and religious freedoms. Lao people form a minority in the mountains of northern Vietnam, and Cambodians are a minority in southern Vietnam. A large Chinese minority lives in the major cities of the three countries.

Cambodia and Laos occupy interior areas except for Cambodia's short coast on the Gulf of Thailand. They are very different countries physically, and a particular culture dominates each. Cambodia is mostly plains along and to the west of the lower Mekong River, with mountain fringes to the northeast and southwest. Except for half a million ethnic Vietnamese, almost all the 13 million people of Cambodia are ethnic Cambodians, also known as Khmer people. Their language is Khmer, and their predominant religion is Buddhism.

Laos is mountainous, sparsely populated, and landlocked. It has 68 ethnic groups, the largest of which is the ethnic Lao, who are linguistically and culturally related to the Thais and whose dominant religion is Buddhism. Their largest numbers are in the lowlands, especially in and near the capital, Vientiane. Most of the ethnic minorities of Laos live in the mountain regions, including the country's second largest ethnic group, the Hmong.

The Vietnam War

The Vietnam War, which is such an important chapter in the American experience, had roots that preceded U.S. interests in the region. This conflict profoundly affected Southeast Asia and has a lingering legacy there today.

France conquered Indochina between 1858 and 1907. The French first extinguished the Vietnamese empire, which covered approximately the territory of today's Vietnam, and defeated the Chinese, whom the Vietnamese called upon for help. Later, the French took Laos (in 1893) and western Cambodia (in 1907) from Thailand. France's administration of Indochina, centered in Hanoi and Saigon, left a big cultural imprint. Its major economic success lay in opening the lower Mekong River area to commercial rice production, converting both Vietnam and Cambodia into important exporters of rice.

Japanese forces overran French Indochina in 1941, initiating five decades of warfare in the area. In World War II, a Communist-led resistance movement was formed to carry on guerrilla warfare against the Japanese. When French forces attempted to reoccupy the area after the end of World War II, these forces fought to expel them. After 8 years of warfare, the Vietnamese, with Chinese matériel support, destroyed a French army in 1954 at Dien Bien Phu in the mountains of Tonkin. France withdrew, ending a century of colonial occupation.

Four countries came into existence with France's departure. Laos and Cambodia became independent non-Communist states. North Vietnam, where resistance to France had centered, emerged as an independent Communist state. South Vietnam gained independence as a non-Communist state that received many anti-Communist refugees from the north. Vietnamese Catholics were prominent among those who fled southward. This partition of Vietnam into two countries was largely the work of U.S. power and diplomacy.

South Vietnam was from the outset a client state of the United States, whereas North Vietnam became allied first with Communist China and then with the Soviet Union. Warfare in Vietnam gained momentum. Between 1954 and 1965, an insurrectionist Communist force supported by North Vietnam and known as the **Viet Cong** achieved increasing successes in South Vietnam in its attempt to reunify the country. A similar situation existed in Laos. Increasing intervention by the United States in South Vietnam escalated from a small scale to a large scale in 1965 and eventually involved the full-scale commitment of half a million U.S. military personnel (Figure 12.11a). North Vietnam responded by committing regular army forces against American and South Vietnamese troops. The United States never invaded North Vietnam, although American air strikes there were devastating. American bombs, napalm, and defoliants such as Agent Orange caused enormous damage to the natural and agricultural systems of Vietnam, destroying an estimated 5.4 million acres (2.1 million hectares) of forest and farmland in an effort described in military parlance as "**denying the countryside to the enemy.**" Agent Orange also left a legacy of birth deformities among Vietnam's postwar generation.

The United States avoided leading an invasion of the North in part because of the risks posed by Chinese and Soviet support of the North. In limiting the theater of ground warfare, however, the United States found itself unable to expel from the South a North Vietnamese army that was determined, skillfully commanded, increasingly better equipped by its allies, accomplished in guerrilla tactics, and willing to bear heavy losses. More than 3 million Vietnamese soldiers and civilians died in the conflict, and about 58,000 U.S. soldiers and support staff perished. In 1973, almost all American forces were withdrawn from the costly war, which was extremely divisive at home. There are still 1,853 Americans listed as missing in action in Indochina.

North Vietnam completed its conquest of South Vietnam in 1975. Saigon was renamed Ho Chi Minh City (although many of its people still call it Saigon), after the original

Figure 12.11 (a) U.S. troops in combat in a Vietnamese village in 1972. The war inflicted enormous suffering on all sides. (b) Vietnam opened its doors to capitalism and Western goods in the mid-1990s.

leader of Vietnam's Communist Party. Vietnam entered a period of relative internal peace, but there were ongoing conflicts, including Vietnam's 1979 conquest of Cambodia, continued warfare against Cambodian and Laotian resistance groups, and a brief border war with China to resist Chinese aggression in 1979. Within its borders, a reunited Vietnam unleashed a repression strong enough to cause a massive outpouring of refugees, including the "boat people" (see Ethnic Geography, page 340). This situation intensified, especially for the country's ethnic Chinese, after China invaded Vietnam briefly in 1979. China and Vietnam gradually came to terms, however, finally normalizing relations in 1991 and solving longstanding disputes such as the demarcation of their territorial waters and corresponding exclusive economic zones in the Gulf of Tonkin. They have, however, been unable to resolve disputed claims to two island groups in the South China Sea: the Paracel Islands (with rich fishing grounds and potential for oil and gas), which China took from Vietnam by force in the early 1970s, and the Spratly Islands (see Chapter 10, page 293).

Vietnam Today

Vietnam (population: 81.5 million; capital, Hanoi; population: 1.4 million) is now restoring its war-torn landscape through large-scale reforestation, agricultural reclamation, and nature conservation programs, aided by international organizations such as the World Wildlife Fund. Animal species new to science (including two hoofed mammals) have been discovered in the Vu Quang Nature Reserve in the rugged Annamite Cordillera region. Such natural treasures, along with Vietnam's tigers, rhinoceroses, and elephants, are helping to lure international ecotourists who provide much-needed revenue. The protected areas that house these species are clustered mainly in the highlands near the Laotian border.

Since its embrace in 1986 of free-market reforms in the policy known as *doi moi* (renovation), Vietnam's economy has improved markedly (see Figure 12.11b). The poverty rate

fell by half in the 1990s, and in 2004, Vietnam had Asia's second fastest growing economy after China. The Communist government halted collectivized farming in 1988 and turned control of land over to small farmers under 20-year lease agreements. The results have been impressive; Vietnam since 1988 has become a major exporter of rice, second in the world only to Thailand. This is a vulnerable commodity, however, especially because production is concentrated along the Mekong River, which is prone to flooding in the May–October rainy season (see The World's Great Rivers, page 341). Vietnam is also pinning hopes on exports of its proven reserves of 600 million barrels of oil, especially to oil-hungry China.

In addition to joining ASEAN, Vietnam in 1995 restored full diplomatic relations with the United States, and U.S. businesses began scrambling for consumers in this large, promising market. After the two countries signed a trade agreement in 2000, Vietnam began exporting shoes, finished clothing, and toys to the United States. The dual personalities of North and South are being perpetuated in the new economic climate, with most investment and infrastructural improvement focused in the South. The North is a more austere economic landscape, and northerners dominate the civil service. Along with a growing gap between the rich and the poor, the differences between North and South could slow the country's overall development.

In an effort to better integrate the nation, Vietnam has begun a costly and ambitious effort to build a new road that will run the entire 1,000-mile (1,600-km) length of the country, along the historic Ho Chi Minh Trail. Environmentalists fear the highway's impacts on wildlife. Anthropologists are worried about its effects on the traditional peoples inhabiting remote areas along the route. And many in Vietnam's growing private business sector lament the road as a costly white elephant; the country already has a north–south highway and railway, they complain, and resources would be better spent on "new economy" infrastructure such as the Internet.

Ethnic Geography

The Boat People

The desperation of many people in Vietnam and its neighboring war-torn countries created the crisis of the boat people—refugees who fled onto the open ocean on large rafts, with chances of survival variously estimated at 40 to 70 percent. After 1974, almost 2 million Vietnamese fled Vietnam, and the United States provided permanent refuge for more than half of those who resettled. Canada, France, and Australia also received significant numbers (Figure 12.D). Large numbers of Vietnamese and Laotian refugees occupied camps in China near its borders with those countries. Boat people also filled refugee camps in Hong Kong, Malaysia, Indonesia, Thailand, and the Philippines.

The United States and other Western governments long ago reduced the size of the welcome mat for Southeast Asian refugees, but Vietnamese continue to immigrate to the United States at a rate of about 14,000 yearly. (Vietnam is in the top five of immigrant populations into the United States, behind Mexico, the Phillipines, India, and China.) Most of these immigrants are typical of an **expanding pyramid,** in which family members in the United States bring extended family members over from Vietnam.

Figure 12.D Many ethnic Vietnamese refugees settled in Canada, introducing their cuisine and culture to their new homeland.

About 2,000 are classified as political refugees who survived Vietnam's postwar "education camps," many of whom are **Amerasians,** the offspring of American soldiers who served in the Vietnam War. Vietnam is also among the leading foreign sources of children adopted by American couples. Altogether, the United States is now home to about 1.2 million ethnic Vietnamese. Approximately 100,000 visit Vietnam each year, bringing much welcome hard currency with them into Vietnam.

In 1996, Vietnamese refugee camps around Asia closed, and about 100,000 Vietnamese were compelled to return to their homeland. Just as those countries were closing their doors to Vietnam, Vietnam was undertaking political and economic reforms to provide incentives for the expatriates to return home and for resident Vietnamese to stay home. Since 1990, about 1 million overseas Vietnamese, mostly from the United States and France, have returned to Vietnam to stay. Their repatriation has also pumped economic capital into Vietnam's fledgling free-market economy. Many Vietnamese who once aspired to leave their home country have heard accounts of the difficulties refugees have faced in adjusting to their adopted countries and have chosen to remain in the hopeful nation of Vietnam.

Like that of many LDCs, Vietnam's progress has been threatened by the specter of uncontrolled population growth. Since the war with the United States ended in 1975, Vietnam's population has increased by more than 60 percent. In view of the country's limited size and resource base, the government has stepped in with strong economic incentives for couples to have no more than two children. Birth rates have begun to fall sharply.

Cambodia's Killing Fields

For a time, Cambodia escaped the kinds of ills known to Vietnam, but then it suffered even worse catastrophes. After 5 years of fighting in the country, a Communist insurgency overcame the American-backed military government in 1975. Known as the **Khmer Rouge** and led by a man with the nom de guerre of Pol Pot, this Communist organization ruled the country for 4 years with exceptional savagery. In 1973, the population had been estimated at 7.5 million, but by 1979, it was estimated at only about 5 million; a third of the population had been murdered. Some intended victims managed to escape as refugees from Cambodia's **killing fields** to adjoining Thailand. The Cambodian Communists' stated policy was to build a "new kind of socialism" by eradicating the educated and the rich, emptying the cities, and breaking up the family. The population of Phnom Penh, the capital and largest city, was expelled on a death march to the countryside. The city's population dropped from more than 1 mil-

The World's Great Rivers

The Mekong

The Mekong River, known as "Water of Stone" in Tibet, "Great Water" in Cambodia, "Nine Dragons" in Vietnam, and "Mother of Waters" (Mae Nam Khong, contracted to Mekong) in Laos and Thailand, is Southeast Asia's great river (Figure 12.E). From its source on the Tibetan Plateau in the Chinese province of Qinghai, the 2600 mile (4200 km) long Mekong snakes through China's Sichuan and Yunnan Provinces, forms boundaries between Laos and Myanmar and Laos and Thailand, winds across Cambodia, and fans out across Vietnam's densely populated Mekong Delta region before empting into the South China Sea.

Unlike the Chao Praya in Thailand, the Red River in Vietnam, and the Irrawaddy in Myanmar, the Mekong is not a great magnet for cities and civilizations. The biggest city on its banks is Cambodia's Phnom Penh, with just over a million people. The main factor limiting more settlement along the Mekong has been the difficulty in navigating the river above Phnom Penh. Historically, very little trade has been carried along much of its length; the oceans and overland routes have provided superior opportunities.

Figure 12.E The Mekong at Can Tho, Vietnam

Chris Lisle/Corbis

The Mekong is not economically unimportant, however. It is one of the world's most prolific fisheries, yielding about 2 million tons of fish each year, or twice the annual North Sea catch. There are more species of fish in the river (about 1,200) than in any other rivers but the Amazon and Congo. Unique creatures inhabit its waters, including about 70 Irrawaddy dolphins and the giant catfish that attain 10 feet (3 m) in length and 660 pounds (300 kg) in weight. Overfishing and the construction of dams have all but eliminated this superlative catfish throughout the Mekong Basin.

There is a big push on to capitalize on the Mekong's potential to produce hydroelectric power. Plans exist to build about 100 dams on the river and its tributaries. The first dam on the river itself (the Man Wan in China) was completed in 1993. China has since built another and has plans for four more. Vietnam is building five dams on the Mekong tributary called the Hoyt Dam, which forms part of its border with Cambodia. Thailand has dammed its main tributaries to the Mekong. Laos plans to build the most dams of all—20—and thereby become mainland Southeast Asia's biggest supplier of electricity.

These dams will have enormous impacts, some of them unforeseeable. It may be reliably predicted that the fish catch will fall dramatically because the main reason for the Mekong's productive fishery is its regular flooding, which will be dramatically reduced. Fishermen are already complaining of smaller catches since the dam building began. There are concerns that dams will ruin the ecosystem of Cambodia's Tonle Sap Lake, which is drained by a river that actually changes direction when downstream Mekong floodwaters inundate its basin. More than a million people make their living by fishing this lake.

As always with shared river systems, there are serious questions about how the Mekong waters should be divided. Vietnam, Cambodia, Laos, and Thailand have joined to form the **Mekong River Commission,** whose biggest responsibilities are to fix the minimum amount of water each country must discharge downstream on the Mekong and its tributaries and to agree on rules to ensure water quality. Both quality and quantity of water are threatened by upstream Myanmar and China, which have refused to join the Mekong River Commission.

223, 306

lion in 1975 to an estimated 40,000–100,000 in 1976 (its population in 2004 was 1.2 million).

The Communist/Nationalist imperialism of Vietnam came to dominate Cambodia's political landscape in 1979. There had been escalating border conflicts between the two countries, and in 1979, Vietnamese troops quickly conquered most of Cambodia and installed a puppet Communist government. Then the Khmer Rouge, supported by

China (Vietnam had by then become a close ally of China's rival, the Soviet Union), began guerrilla resistance against the Vietnamese. Two non-Communist Cambodian guerrilla forces also took the field. These three resistance forces often quarreled among themselves, but all maintained the conflict with the Cambodian/Vietnamese army of the Phnom Penh government, which periodically pursued them into Thailand and engaged Thai border military units. At first, the Vietnamese rejected international food aid for Cambodia and followed a policy of systematically starving the country, apparently to weaken resistance. Later, they attempted to promote recovery under their puppet government.

Peace negotiations among Cambodia's contending factions led, in 1991, to an agreement resulting in the restoration of democratic government under UN supervision. A new constitution with new leaders took effect in 1993, and Cambodia's former King Norodom Sihanouk returned to the throne. In 1997, Hun Sen, who had once helped lead but defected from and opposed the Khmer Rouge, ousted the popularly elected royalist leader Prince Norodom Ranariddh in a violent coup. Under much international pressure, Hun Sen permitted elections in 1998. Finally succeeding in forming a government legitimately and peacefully, he secured Cambodia's membership in ASEAN and regained Cambodia's seat at the United Nations.

Cambodia is a very poor country, but more stable conditions have helped the economy improve recently. Garment manufacturing is the largest industry, but there are fears that Cambodia will prove unable to compete with growing exports of cheaper clothing from China. Tourism is the fastest growing industry, with more than a million visitors arriving in 2004. Few visitors miss Angkor Wat, still coveted by some in Thailand because it once stood within that country's borders. Cambodians resent Thai claims to the site and also Thai ownership of many prominent tourism, telecommunications, and forestry businesses in Cambodia. That resentment led to riots in 2003—sparked by a Thai actress' claim that Angkor Wat belongs to Thailand—that did millions of dollars in damage to Thai businesses and the Thai embassy in Cambodia.

Even in this relatively peaceful time, the people of Cambodia must deal with a persistent scourge of war: land mines. Sown by hostile forces during years of warfare, the antipersonnel explosives remain active indefinitely. Mines have maimed more than 35,000 Cambodians, and about 50 more casualties occur each month. People sow new mines, imported mainly from China and Singapore, to protect their property in Cambodia.

The Lao Way

Laotians also suffer from ordnance dating to the Vietnam War. Between 1964 and 1973, U.S. warplanes dropped more than 2 million tons of bombs—more than the United States dropped on Germany in World War II—on the Laotian frontier with Vietnam in an effort to disrupt Communist supply lines on the nearby Ho Chi Minh Trail and to prevent Communist troops from entering Laotian cities. This Plain of Jars region retains the distinction of being the most heavily bombed place on Earth. An estimated 30 percent of the bombs failed to detonate, and today, millions of unexploded American cluster bombs remain. Smaller than a tennis ball,

the cluster bomb contains more than 100 steel ball bearings. Many curious Laotian children who play with them are killed or dismembered. About 200 people are killed or maimed yearly by these explosives. Of Laos's 17 provinces, 15 contain unexploded ordnance (UXO), according to UN studies, creating a serious deterrent to farming in a country with inadequate food supplies. Efforts are underway to clear the explosives, with only minimal economic assistance from the United States. The agency in charge of the cleanup hopes to de-mine enough land to produce 10,000 tons of rice yearly, which would feed 50,000 people.

The administrative capital of Laos is Vientiane (with only 200,000 people), located on the Mekong River where it borders Thailand. Vientiane replaced the royal capital of Luang Prabang ("Royal Holy Image"), located upriver on the Mekong when, in 1975, Laos came under Vietnamese occupation. That year, the Pathet Lao organization overthrew the Laotian monarchy and established a new People's Democratic Republic. The Communist dictatorship now rules Laos in a benign fashion, encouraging a market economy and foreign investment. The government calls its conservative approach to economic reform and social change "the Lao Way." To slow the potential flood of foreign worldviews and materials into the country, Laos has built only one bridge to Thailand across the Mekong River and has resisted appeals for the construction of a second bridge. However, there is an already important and growing gambling casino industry in Laos, patronized mainly by visitors from neighboring Thailand.

Despite gambling revenues, landlocked Laos is now one of Asia's least developed countries, ranking 135th of the 177 countries on the United Nations Human Development Index. About 80 percent of its people are peasant farmers. Exports are limited to timber, hydroelectric power (to Thailand), and garments and motorcycles assembled by inexpensive Laotian labor using imported materials. Gold and copper mining, which began in 2002, may bring much needed revenue to Laos.

Like those of Vietnam and Cambodia, the Laotian economy has suffered from the brain drain prompted by years of warfare, political turmoil, and underdevelopment—conditions that caused the best educated and most talented of the country's inhabitants to flee. After the 1975 takeover, 343,000 Laotians, mostly members of the Hmong tribe who had U.S. support to fight Laotian government troops during the war, fled to neighboring Thailand. Many Hmong refugees subsequently immigrated to the United States, where 250,000 ethnic Hmongs now live. A low-level Hmong insurgency continues against the Vientiane government.

Given the conditions of warfare, political turmoil, oppression, and inefficient economic systems, it is not surprising that the three Indochinese countries are, with Myanmar, the poorest in Southeast Asia. However, their outlook seems to be improving as tensions with the outer world relax, privatization of Communist economies increases, the supply of goods and services improves, and the number of returning refugees exceeds emigrants.

12.6 Malaysia and Singapore

The small island of Singapore (Figure 12.12) is just off the southern tip of the mountainous Malay Peninsula. It lies at the eastern end of the Strait of Malacca, the major passageway for sea traffic between the Indian Ocean and the South China Sea. For centuries, European sea powers competed to control this passageway. Britain gained control in 1824 and used the newly established settlement of Singapore to monitor the strait. With its central position among the islands and peninsulas of Southeast Asia, British Singapore developed into a large naval base and the region's major entrepôt. Singapore rivals Rotterdam as the world's largest port in tonnage of goods shipped.

From Singapore, the British extended their political hold over the adjacent southern end of the Malay Peninsula. This expansion gave Britain control over the part of the peninsula that is now included, along with northwestern Borneo (except Brunei), in the independent country of Malaysia. This southern end of the peninsula is known as Malaya, Peninsular Malaysia, or West Malaysia. The Borneo section includes the states of Sarawak and Sabah, together known as East Malaysia.

During the British period, Malaya developed a commercialized and prosperous economy. Tin and rubber were the major products, with oil palms and coconuts of secondary importance. Chinese and British companies pioneered the tin mining industry in the late 19th century. The building of railroads gave access to a line of tin mines and rubber plantations in the foothills between Malacca and the hinter-

land of Penang (George Town). Along this line, the inland city of Kuala Lumpur (population: 3.7 million) became Malaya's leading commercial center and then Malaysia's capital (see Figure 12.2b).

Although Kuala Lumpur will remain Malaysia's official capital, all government administration is to be transferred to Putrajaya, a city under construction about 25 miles (40 km) to the south (Figure 12.13). It is a high-tech capital stitched together by fiberoptic cable. Malaysian officials boast that the Internet and other forms of digital communication will replace paper in the country's new administrative center. Nearby is Cyberjaya, a new city that, as its name suggests, hopes to lure technology companies and become Asia's Silicon Valley. These cities are symbols of how Malaysia sees itself, a country leaving behind its colonial legacy as a "rubber republic" to become an engine of technical innovation and export. Putrajaya's $5 billion construction cost has drawn much criticism in the wake of Malaysia's financial crisis, however.

The development of a colonial economy gave Malaya an ethnically mixed population. The native Muslim Malays played only a minor role in this development. Chinese and Indian immigrants and their descendants became the principal farmers, wage laborers, and businesspeople of the tin and rubber belt. So heavy was the immigration that 24 percent of Malaysia's population of 26 million is Chinese and 8 percent is Indian. Today, the Chinese minority is quite prosperous, but the Indian population is generally disadvantaged. The growth of Singapore brought even heavier immigration so that now over three-fourths of Singapore's population of 4.2 million is Chinese.

Ethnic antagonisms between Malays and Chinese shaped the political geography of Malaysia and Singapore. Malaya accepted independence in 1957 only on the condition that

Figure 12.12 Singapore, located at a crossroads where the Pacific and Indian Oceans meet, is a very modern, prosperous city.

Figure 12.13 The Parliament Building and National Mosque in Putrajaya, Malaysia's brand new, controversial administrative capital.

Singapore not be included. This was to ensure that ethnic Malays would be a majority in the new state. In 1963, however, the Federation of Malaysia was formed, composed of Malaya, Singapore, and the former British possessions of Sarawak and Sabah in sparsely populated northern Borneo. The non-Chinese majorities in Borneo were counted on to counterbalance the admission of Singapore's Chinese. This experiment in union lasted only until 1965, when Singapore was expelled from the federation and left to go its own way as an independent state.

Within Malaysia, the government passed legislation to ensure that at least 60 percent of the university population was ethnic Malay, with similar measures to ensure business opportunities went primarily to Malays. Although ethnic Chinese and Indians still complain that such policies continue to shut doors of opportunity to them, the economic gap between these minorities and the majority Malays has narrowed considerably since the 1970s. But tensions between Muslims and non-Muslims in this country bear monitoring. Among Malaysia's majority Muslims, there are growing complaints that the government is too secular, although it is nominally Islamic (Figure 12.14). The leading champion of the adoption of **Islamic law** (*shari'a*) as the law of the land is

the **Parti Islam se-Malaysia (PAS)**, which previously administered state governments of Kelantan and Terengganu in Peninsular Malaysia.

Malaysia and Singapore are among the more economically successful of the world's formerly colonial states that received independence after World War II. Singapore is an outstanding center of manufacturing, finance, and trade along the Pacific Rim. Its industries include electronics, chemicals, oil refining, machine building, tourism, and many others. With its electronics sector now suffering from intense competition from China, Singapore is hoping to develop a strong new niche in biomedical research and other biotechnology industries; a huge biotechnology research complex called **"Biopolis"** has recently been built there. Singapore suffered less than Malaysia, Thailand, and Indonesia during the region's economic crisis of the late 1990s. This was due in part to a geographic advantage: its ability to conduct trade with a vast regional hinterland belonging to other countries without having to attend to the struggles of a large and poor rural population of its own. Singapore's prosperity is accompanied by strict senses of propriety and secularism. Chewing gum is banned, graffiti artists are flogged, and head scarves and other apparent religious symbols are banned in schools.

Malaysia still exports rubber and tin but has a diversified export portfolio that includes crude oil (from northern Borneo), timber, palm oil, automobiles, and electronic components. Malaysia is the world's largest exporter of semiconductors and air conditioners. Its annual GDP PPP per capita is only about one-third that of Singapore's, but within Southeast Asia, it falls behind only Singapore and Brunei and is about triple that of the average figure for the world's LDCs (see Table 10.1).

Malaysia has a national mantra of "Wawasan 2020" (Vision 2020) referring to the year by which national economic planners intend Malaysia to be a fully developed country. The vision became clouded after 1997, however, when following a decade in which the economy grew at a robust 8 percent annually, Malaysia fell into recession. The country quickly froze its ambitious and expensive development projects and adopted economic austerity measures, including even an appeal to individuals to cut down on the number of sugar lumps added to their tea. Like the other emerging markets, Malaysia had been benefiting from huge and often speculative foreign capital investments. And like the others, Malaysia suffered in 1997 and 1998 when worried foreign investors rapidly withdrew their capital.

Malaysia effectively pulled out of the orbit of globalization in 1998, when it took drastic actions to protect itself from the whims of foreign investment. It required foreign investors to wait a year before repatriating their funds. Rejecting the notion that free markets would continue to benefit his country, Malaysia's prime minister imposed strict currency controls, pegging the value of the Malaysian ringgit to that of the dollar. This made Malaysian money worthless overseas but ensured that Malaysian money abroad would return home. The fixed rate of the ringgit made it easier for busi-

Figure 12.14 Most ethnic Malays, like this woman, are Muslim.

nesses to plan for the future because they knew the value of the goods they exported and imported would not be subjected to wide fluctuations. Malaysian politicians blamed the International Monetary Fund and other institutions in the West for Asia's financial troubles, and in turn, these institutions condemned Malaysia for retreating from the principles of the global free market. Despite the ominous downturn, by the end of the decade Malaysia was once again registering economic growth—particularly as exports of computer chips and other electronics resumed—and was pulled out of the abyss in which neighboring Indonesia was still mired.

12.7   Indonesia and East Timor

Indonesia is by far the largest and most populous country of the region (population: 219 million) and is the fourth most populous country on Earth (after China, India, and the United States). Stretching across 3,000 miles (4,800 km), its 13,600 islands comprise an area about three times the size of Texas, representing over two-fifths of Southeast Asia's land area and about two-fifths of its population. The large population of Indonesia is related especially to the enormous concentration of people on the island of Java. Its 130 million people represent about 60 percent of Indonesia's population. The island's population density is about 2600 per square mile (1000/sq km). In contrast, the remainder of Indonesia, which is about 13 times larger than Java in land area, has an average density of only about 135 per square mile (52/sq km).

This extraordinary concentration of population on one island owes partly to the superior fertility of Java's volcanic soils (Figure 12.15). Other Indonesian islands (with the notable exception of Borneo) have areas of fertile volcanic soil

Figure 12.15 Molten lava illuminates the morning sky as it flows down the side of Mt. Merapi in central Java on February 19, 2001. Some have linked the mountain's growing rumblings to the rising political tension in the country. Mbah Maridjan, a sprightly 73-year-old Indonesian who was made the volcano's "gatekeeper" nearly 20 years ago by the Sultan of Yogyakarta—head of the royal family in Indonesia's nearby ancient capital—said mystics had been climbing the mountain to pray that it did not explode.

but lack the extremely high population densities found on Java. Their unfriendly coastlines, with their coral reefs, cliffs, and extensive wetlands, have hampered the economic development of some islands.

Java's extreme population density also has cultural and historical roots in the concentration of Dutch colonial activities there. The Dutch East India Company, after lengthy hostilities with Portuguese and English rivals and with indigenous states, secured control of most of Java in the 18th century. Large sections of the other islands, however, were not brought under colonial control until the 19th and early 20th centuries. The Netherlands exploited the natural wealth of Java with the introduction of the Culture System in 1830. Under this system, Dutch colonizers forced Javanese farmers to contribute land and labor for the production of export crops. From the Dutch point of view, this harsh system was successful. The Culture System was abolished in 1870, but commercial agriculture continued to expand, and its success supported increasing numbers of people. The introduction of the Culture System thus coincided with an enormous increase in Java's population, which is more than 20 times larger than it was a century and a half ago.

Java's historical leadership and the island of Sumatra's current supremacy in export production are seen in the distribution of Indonesia's larger cities. Of 23 cities having estimated metropolitan populations of more than 500,000, 12 are on Java. These include 5 "million cities," headed by Jakarta, the country's capital, main seaport, and the largest metropolis in Southeast Asia (population: 18 million). Java's second largest city is Bandung. Of the remaining Indonesian cities with more than half a million people, five are in Sumatra.

The Indonesian state has had a turbulent career. Increasing nationalism during the Japanese occupation of World War II led to a bitter struggle for independence from the Netherlands following the war. Indonesia finally won independence in 1949. Another confrontation with the Netherlands in 1962 took western New Guinea (West Irian) out of Dutch control. A United Nations plebiscite in 1965 ratified Indonesian control, and relatively friendly relations were established between Indonesia and the Netherlands.

In 1965 (known as The Year of Living Dangerously), an attempted Communist coup against the longstanding regime of Indonesian President Sukarno was unsuccessful. The ensuing retaliation resulted in the massacre of about 300,000 Communists and their supporters by the Indonesian army and by Islamic and nationalist groups.

From 1968 to 1998, Indonesia's president was an army general named Suharto, whose regime was regularly criticized for human rights abuses. Suharto ruled Indonesia undemocratically and often repressively but argued that he was a benevolent figure acting on behalf of his people. Indonesians generally tolerated his authority because the country's economy was advancing rapidly. The economic boom of the 1980s and early 1990s centered in the oil industry (in the hands of the state firm Pertamina) and in food production.

Oil fields located mainly in Sumatra and Kalimantan now produce about 2 percent of the world's oil output. Crude oil and liquefied natural gas account for about 20 percent of Indonesia's exports and have tied its economy very closely to that of Japan, which is the main market for these products. Indonesia is a member of OPEC, the Organization of Petroleum Exporting Countries. In addition to oil and gas, important contributions to the export trade include clothing, wood products, various tropical agricultural products, and increasingly, electronics.

Tourism has also brought welcome foreign exchange earnings to Indonesia. The country's principal attraction by far is the legendary island of Bali, home of Indonesia's only remnant of the India-born Hindu culture that permeated the islands from the 4th to 16th centuries (Figure 12.16). Bali is world-famous for its unique culture and forms of art and dance, which so far have endured and become even richer despite a massive influx of international tourism to this legendary island. Bali's popularity as a destination for Western tourists put it in the cross hairs of the al-Qa'ida-affiliated **Jemaah Islamiah,** which in 2002 carried out twin bombings of the nightclub district of Kuta. More than 200 people were killed, most of them young Australians; the Bali bombings have been called "**Australia's September 11.**"

Indonesia's economy collapsed under the force of the spreading Asian economic crisis of the late 1990s. The rupiah, Indonesia's currency, lost 80 percent of its value in less than a year. The Suharto government removed expensive subsidies on gasoline and kerosene. Resulting protests quickly grew into vast riots. The economic troubles were compounded by a prolonged drought attributed to El Niño's warming of Pacific waters, which caused significant declines in the country's rice harvest. Skyrocketing prices for rice and widespread rice shortages helped fuel widening popular discontent, which focused on two targets. First were the ethnic Chinese who represent 3 percent of Indonesia's population but control half the economy, and second was President Suharto. Rioters looted and destroyed hundreds of Chinese-owned shops in Indonesia's major cities, forcing thousands of ethnic Chinese to flee the country. Since the Chinese had dominated the food production business in Indonesia, food shortages intensified in the wake of the disturbances. Protestors and rioters called increasingly for the resignation of President Suharto, on whom they also blamed their economic troubles. Bowing to the relentless pressure, and averting a potential bloodbath, Suharto resigned from office in May 1998. He was succeeded by his vice president, B. J. Habibie, who in turn stepped down in 1999 following large protests.

Indonesia's first truly democratic elections in 40 years subsequently brought to power President Abdurrahman Wahid, a widely respected, nearly blind Islamic scholar. His "honeymoon" with the public was short-lived, and by 2000, an official investigation revealed he had embezzled millions of dollars of government funds. Wahid denied the allegations and insisted he would finish his term of office in 2004. He did not. Against a backdrop of massive anti-Wahid protests across Indonesia, the president was impeached by Indonesia's Supreme Court and fled the country in 2001. He was replaced in office by Megawati Sukarnoputri, daughter of former President Sukarno, who in turn was ousted by the election of secularist general Susilo Yudhoyono in 2004.

The Indonesian government faces some huge problems. There is a constant threat that this vast country might splinter along ethnic lines (see Problem Landscape, pages 348–349). Many observers believe that if Indonesia's secular leadership does not deliver economic improvement to the tens of millions of disadvantaged Indonesians, the country will turn to Islamic rule.

The scale and pace of Indonesia's economic decline were dizzying. Millions of people slipped below the poverty line, and unemployment surged (in 2004, one-third of the country's workforce was classified as unemployed or underemployed). The turmoil created a new generation of boat people, economic refugees who fled Indonesia for what they perceived as better lands, notably Malaysia. When Malaysia boomed, it had welcomed Indonesians to do much of the country's work, but with its own economic downturn, Malaysia tried to turn the unwelcome masses back.

Rapid population growth in Indonesia compounded the effects of the economic crisis. Indonesia's population grew by 91 million, or more than 70 percent, in the three decades between 1974 and 2004. The government lacks an aggressive family planning program and instead is exporting its "surplus" population by promoting emigration from Java, Bali, and other densely populated islands to some of the sparsely inhabited outer islands of the vast Indonesian archi-

Figure 12.16 The cultural arts of Bali's Hindu people are world-renowned, making this Indonesian island a popular tourist destination.

pelago. This **Transmigration Program,** one of the word's largest resettlement efforts, has been costly and controversial. With World Bank support, the Indonesian government has paid for more than 6.4 million people, most of them landless agricultural workers and their families, to relocate to new islands where they receive houses, farmland, and subsistence payments. The results have generally fallen short of expectations, however. The settlers' crop yields have been low, reflecting lack of experience in the new lands, shortages of family labor, and difficulties in marketing crops. The new farming has also contributed to Indonesia's deforestation.

12.8 The Philippines

The Philippine archipelago includes over 7,000 generally mountainous islands. The two largest islands, Luzon and Mindanao, make up two-thirds of the total area.

The Philippines were a Spanish colonial possession governed from Mexico from the late 16th century until 1898, when the United States achieved victory in the Spanish-American War and gained control of the Philippines. With the exception of the Japanese occupation of 1942 to 1944, the Philippines were controlled by the United States from 1898 until 1946, when the country became independent.

Spanish missionary activity in the Philippines created the only Christian nation in Asia. The country's Spanish legacy is still important. Ever since its founding in 1571, the Spanish-oriented capital of Manila (population: 14.8 million) has been the major metropolis of the islands. The society created by Spain was composed of a small upper class of Hispanicized Filipino landowners and a great mass of landless peasants. Problems created by this uneven distribution of agricultural land have persisted in the Philippines. Discontented peasants supported a Communist-led revolt after World War II. It was eventually suppressed, in large part through granting land (generally on sparsely populated Mindanao) to surrendered rebels. Revolts flared up again in the late 1960s, however, and have continued ever since. A Communist group known as the **New People's Army** leads an active insurgency in many parts of the country, frequently targeting military patrols and civilian infrastructure.

American economic, military, and other ties with the Philippines continued after independence. Preferential treatment in the U.S. market stimulated the growth of major Philippine export industries: coconut products, sugarcane, and abaca (a fiber used in making rope and fabrics). Education in English grew so that even now English is spoken by about 50 percent of the people and serves as a bridge between many of the diverse linguistic groups of the population. After independence, Tagalog, one of the most widely used indigenous Austronesian languages, became the principal base for the official national language of Filipino, spoken by 40 percent of the populace. English also remains an official

language. Despite the colonial history, very few people speak Spanish. This owes largely to the fact that Spanish missionary work in the Philippines was carried on in the local languages rather than in Spanish.

The Philippine economy has expanded considerably since independence. Until the 1970s, the most striking growth was in agricultural exports such as coconut products, sugar, bananas, and pineapples and in copper and other metal exports. More recently, there has been growth in manufacturing and food production. Labor-intensive manufacturing, such as that of electronic devices and clothing, now accounts for about 80 percent of all exports. Japanese and American capital, together with American agricultural science and technology, have been very important in Philippine economic development. The United States and Japan buy more than half of all Philippine exports. Inexpensive, skilled labor is another of the country's principal exports. About 7 million Filipinos work abroad, pumping $7 billion of **remittances** back into their country's economy.

Despite encouraging progress in recent decades, the Philippine republic remains poor and potentially explosive politically. It has not joined the ranks of Asian Tigers or Tiger Cubs. The country has taken a conservative approach to its development and has borrowed money carefully to invest in modest projects, avoiding the aircraft industry, the building of a new national capital, and other such showy and expensive enterprises undertaken in other Southeast Asian nations in the 1990s. Since the mid-1970s, the Philippines has also adhered closely to the terms of loans and financial reforms imposed by the International Monetary Fund. The country did not boom, but neither did it bust in the economic crisis that rolled across Asia in 1997 and 1998. During the regional economic recovery that followed, however, the Philippines' growth continued to lag. An estimated 36 percent of the population lived below the poverty line and unemployment stood at 13 percent in 2004.

Economic expansion in the Philippines has been hard-pressed to stay ahead of the more than 70 percent growth in population since 1975. The predominant Roman Catholic culture (Catholics make up 83 percent of the population, with 10 percent adhering to other Christian denominations) encourages large families. The country's leading cleric, Cardinal Sin, responded this way to a question about the population explosion: "The more the merrier." Church officials have labeled the government's current promotion of family planning as "demographic imperialism" masterminded by the United States. Condoms and other forms of birth control are too expensive for many to afford in a country where 40 percent of the population makes less than $1 per day. The collective result is that the Philippines has a rather high annual population growth rate of 2 percent. This is a marked contrast with Thailand. Both countries had about the same population size in 1975, but the Philippines now has about 20 million more people than Thailand, which has

Problem Landscape

The Balkanization of Indonesia?

The national credo of Indonesia, seen on banners and placards throughout the country, is: "One country. One people. One language." The government's constitution, in an effort to promote national unity, officially recognizes four faiths: Islam, Christianity, Hinduism, and Buddhism. But slogans and laws belie a grim reality: Indonesia may be in a process of balkanization, or fragmentation into ethnically based, contentious units like those found in the Balkans of southeastern Europe (Figure 12.F). The presence of some 300 different ethnic groups, combined with religious frictions, physical fragmentation, and economic problems, has made it difficult to attain peace, order, and unity in Indonesia. There is an official national language (Malay, known locally as Bahasa Indonesian), but more than 200 languages and dialects are in use. The largest ethnic group is the Javanese, who make up about 45 percent of Indonesia's population. Almost 90 percent of all Indonesians are Muslim, about 8 percent are Christian, and 2 percent are Hindu, living mainly in Bali.

Various groups in the outer islands have resented the dominance of the Javanese, who have traditionally asserted their power in colonial-like fashion over the larger, more resource-wealthy outer islands. Such animosities have escalated at times to armed insurrections. Under the Suharto regime, these militant expressions were countered by state ideology called *Pancasila.* Aimed mainly at suppressing militant Muslim aspirations, Pancasila was a pan-Indonesian nationalist ideology designed to neutralize all ethnic identities.

For a time, the Wahid government vowed to break with the repressive ways of Suharto and appeased would-be separatists with promises of **liberal autonomy,** meaning the provinces would have greater control over local administration and take more profits from the sales of local natural resources. With growing economic problems, pressure from nationalists, and accusations of corruption leveled against him, however, President Wahid shifted to using an iron fist against every Indonesian province that might aspire to follow East Timor's path.

Now independent East Timor (population: 800,000; capital, Dili) is a former Portuguese possession that Indonesia occupied after the collapse of the Portuguese colonial empire in the 1970s. East Timor may be regarded as a litmus test for what could happen to the political geography of Indonesia. The government carried on a long and bitter struggle against an independence movement led by the Catholics of East Timor. In 1998, Indonesia and Portugal finally reached an autonomy agreement that would give the Timorese the right to local self-government. The incoming Wahid administration was bolder, however, and in 1999 allowed the people of East Timor to vote on whether or not they wanted outright independence. The result was overwhelmingly in favor of independence.

Despite official Indonesian blessing of the result, progovernment militias reacted to the vote with a violent rampage that left a thousand Timorese dead, forced a quarter of the population to flee into neighboring West Timor, wrecked fishing boats and other assets, and destroyed 70 percent of East Timor's buildings, leaving tens of thousands homeless. The United Nations dispatched a peacekeeping force to oversee the complete transition to East Timor's full independence in 2002. The country's first president was José Alexandre Gusmao (who goes by his nom de guerre, Xanana), the Timorese rebel leader whom Indonesia had imprisoned in 1992 and released in 1999.

The economic future of this poor newborn country is actually rather bright. Between East Timor and Australia, in the portion of the Timor Sea known as the Timor Gap, lies a huge field of petroleum and natural gas. Because the distance between East Timor and Australia is less than 400 miles (640 km)—making it impossible to apply the international standard of a 200-mile (320-km) offshore territorial limit—the two countries must come to an agreement on how to divide these

Figure 12.F Troublesome provinces threaten to tear Indonesia apart.

Political Hot Spots in Indonesia

ACEH

NORTH MALUKU

PAPUA (IRIAN JAYA)

Jakarta

MALUKU

EAST TIMOR *(independent 2002)*

TIMOR GAP

125

undersea riches. Any deal is likely to yield significant wealth to East Timor.

Of outstanding concern for Indonesia's unity is the province of Aceh (population: 4.2 million) on the northernmost tip of Sumatra. Since 1976, the inhabitants of Aceh, who are a predominantly Muslim people of Malayan ethnicity, have sought independence from Indonesia. The central government had made them a promise of autonomy with Indonesia's 1949 independence from the Netherlands, but failed to keep it. The Acehnese insist that the Jakarta government is too secular for their Muslim tastes and that too little of the revenue from sales of Aceh's abundant natural gas reserves remains in the province (which also has a wealth of oil, gold, rubber, and timber).

The secessionists' main voice is the **Free Aceh Movement** (known by the acronym **GAM**), which would like to install as president a member of the indigenous royal family (Figure 12.G). While promising to allow a referendum by which the Acehnese could choose to adopt Islamic law, Indonesia refuses to even consider holding a referendum on independence. The Jakarta government and GAM reached a peace agreement in 2002 that would offer autonomy but not independence to Aceh and did not address the issue of greater profits from minerals for Aceh. The violence continued. As of 2004, more than 10,000 people, mostly civilians, had been killed in fighting between the military and the rebels. The government continues to see Aceh as vital to the nation's unity and economic viability, fearful that Aceh might inspire separatism elsewhere in the country and that GAM might act on its ambition to control the entire island of Sumatra.

Figure 12.G A supporter of the Free Aceh Movement (GAM) holds its flag above the Arun L.N.G. liquid natural gas plant, in December 1999 in Lhokseumawe, Indonesia. The Arun plant, part of a $3 billion production facility run jointly by Exxon Mobil and Indonesia's national oil company Pertamina, has scaled back operations in the province due to security threats presented by the separatist movement.

At the extreme eastern end of the Indonesian archipelago, Papua (formerly known as Irian Jaya) is home to 2.3 million people of 200 different tribes speaking 100 different languages. Most are Melanesian in ethnic origin, and most are at least nominally Christian. Collectively known as **Papuans,** they have little in common with the Javanese who control them from 2,500 miles (4,000 km) away in Jakarta. Their homeland on the western half of the island of New Guinea, which they call West Papua, is of great importance to the economies of Indonesia and the United States because it contains the world's largest copper and gold mines. The American company Freeport-McMoran, which provides more taxes to Indonesia's treasury than any other foreign source, operates these mines. There are also large assets of oil, natural gas, and timber. It is little wonder that Jakarta refuses to let its poorest, but its most resource-rich, province go. It has never recognized the province's 1961 unilateral declaration of independence. In 2000, when Papuan separatists demonstrated on the anniversary of this declaration, Indonesian troops responded with deadly force. The government banned the flying of the separatist "morning star" flag.

West of Papua lie North Maluku and Maluku, once known as the fabled Spice Islands of the Moluccas. Islamists there who are bent on establishing *shari'a* (Islamic law) in as many locales as possible in Indonesia have targeted the islands' majority Christian inhabitants. Over a period of 3 years, until a fragile peace was achieved in 2002, these Muslim fighters—often aided by regular Indonesian government troops—ethnically "cleansed" Christians from the North Maluku capital. Thousands of people became refugees. Humanitarian crises have become a familiar product of Indonesia's unrest. Scores of overcrowded refugee camps have sprung up in more than half of the country's 26 provinces.

Events in the Malukus reflect a broader wave of Islamist activism across Indonesia. Advocates of *shari'a* have succeeded in closing bars and discos across Java and have terrorized people they consider hostile to Islam. They have scapegoated Christians, who are generally wealthier than Indonesia's Muslims, for many of the country's economic and political problems.

Will East Timor's independence be followed by independence for Aceh, heralding the balkanization of Indonesia? Some argue East Timor is unique; it was the only province of Indonesia colonized by Portugal rather than Holland and is the only one forcibly brought into becoming part of Indonesia. Others point out that Indonesia itself is an artificial union inherited by Dutch colonialism and is bound for devolution, or the dispersal of political power and autonomy to smaller spatial units. The geopolitical concerns of larger powers may also influence the fate of some Indonesian provinces (see that discussion on page 296 in Chapter 10).

a population growth rate of 0.8 percent. Their contrasting age-structure diagrams are depicted in Figure 12.17.

Compounding the problem of population growth, land and wealth continue to be very unevenly distributed, as Philippine society continues to be dominated by only a few hundred wealthy families. The gap between haves and have-nots has inspired rebellion. On the island of Mindanao, rebels of the **Moro Islamic Liberation Front (MILF)** and a splinter organization, the **Abu Sayyaf,** have long fought the government in an effort to establish a separate Islamic state. In 2001, the MILF and the government reached a cease-fire, but Abu Sayyaf accelerated its campaign, employing terrorist tactics such as kidnapping and beheading civilians. In its declared worldwide war on terrorism, the United States in 2002 began providing military advisers and weaponry to Philippine troops fighting Abu Sayyaf.

This unrest has economic, political, and religious roots. Despite a wealth of oil, natural gas, fisheries, and forests, Mindanao is the Philippines' poorest region, but the predominantly Christian government invests little development aid in this mainly Muslim area. Hundreds of thousands of Mindanao's Muslims have fled as illegal immigrants to more prosperous Sabah, in the Malaysian portion of the island of Borneo.

The Philippines has seen some momentous political changes in recent years. In 1986, the country's dictator, Ferdinand Marcos, was ousted in a popular uprising and replaced by a democratically elected president, Corazón Aquino. Marcos and his wife, Imelda, had looted the country of billions of dollars over a period of 20 years and headed a corrupt and abusive regime virtually ignorant of the needs of the ordinary Filipino. The Marcos path seemed to be followed by Philippine President Joseph Estrada, who allegedly enriched himself in the 1990s through bribes, kickbacks, and embezzlement of public funds. Impeachment proceedings against Estrada crawled along, but he resisted their inevitable conclusion. Only massive street protests succeeded in forcing his resignation in 2001, when Gloria Macapagal-Arroyo replaced him as president.

The United States has maintained a concerned watch on the country's struggles. During the 1980s, many Filipino legislators resisted the prospect of renewing the leases under which the United States held the two military bases of Subic Bay and Clark Field on Luzon. Then, in 1991, the eruption of volcanic Mt. Pinatubo put an end to the debate over the fate of these two American outposts; volcanic ash forced their closure. The Philippines has 22 active volcanoes. With

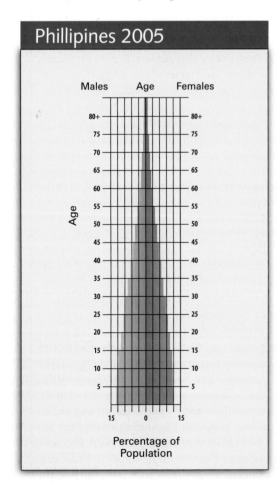

Phillipines 2005

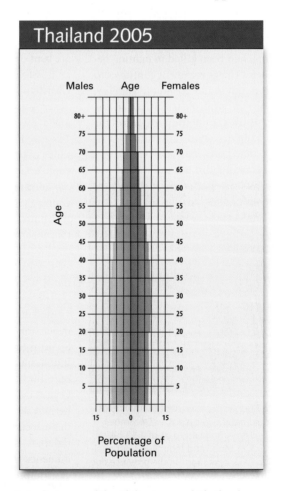

Thailand 2005

Active Figure 12.17 Age-structure diagrams of the Philippines and Thailand reveal the different population policies these countries have pursued since 1975, when both had about the same number of people. *See an animation based on this figure, and take a short quiz on the facts or concepts.*

Natural Hazards

The Great Tsunami of 2004

On December 26, 2004, the most cataclysmic natural event of modern times afflicted a vast portion of the Earth. Beneath the Indian Ocean, just off the northwest coast of the Indonesian island of Sumatra, a gargantuan "megathrust" earthquake measuring magnitude 9.0—with the force of 32,000 megatons (million tons) of TNT, or equivalent to as many as a million nuclear explosions—struck at 6:58 AM local time. In this tectonic event on the Java (Sunda) Trench, as much as 750 miles (1200 km) of faultline slipped 60 feet (20 m) along the subduction zone where the Indian Plate is sliding beneath the Burma Plate (part of the greater Eurasian Plate; see Figure 2.9, page 26).

As the seabed of the Burma Plate instantly rose several meters vertically above the Indian Plate, the massive displacement of seawater created a series of tsunami (**tsunami** is Japanese for "harbor wave" and is popularly but incorrectly known as a "tidal wave") that pulsed across the Andaman Sea and Indian Ocean at up to 500 miles (800 km) per hour. On the open ocean such waves are barely perceptible, but they break on coastlines with the force of giant storm surges. The tsunami roared ashore, at heights up to 45 feet (15 m), in 12 countries: Indonesia, Malaysia, Thailand, Myanmar, Bangladesh, India, Sri Lanka, the Maldives, Seychelles, Somalia, Kenya, and Tanzania (Figure 12.H).

The disaster unfolded just as this book was going to press, and its casualty and cost estimates as well as other implications were just beginning to emerge. Here are some of the early repercussions of the tragedy (as of the first week of January 2005).

These were the deadliest tsunami in history. The total number of dead exceeded 160,000. By far the greatest number of deaths came to Indonesia (101,000), Sri Lanka (46,000), India (9,700) and Thailand (5,100). Almost all the deaths occurred within a mile of the respective shorelines and were caused by blunt force from debris and by drowning. An estimated one third of the fatalities were children. There were many reports of children running out to investigate fish stranded by retreating seas—a characteristic precursor to a tsunami—only to be struck seconds later by a wall of water.

As many as 2 million persons were made homeless by the disaster. An unprecedented international relief effort, designed to feed and house these refugees and to contain the spread of epidemic diseases, began within hours.

Tsunami experts concur that countless thousands of lives would have been saved had the affected Indian Ocean region been equipped with the same kind of tsunami early-warning system now in place in the Pacific Ocean (see http://ioc.unesco.org/itsu). That investment had not been made because historically about 90 percent of the world's tsunami have occurred in the Pacific, and because many of the nations afflicted lacked the financial resources for the wave sensors and other infrastructure required by the system. Now, however, installing an early warning system in the Indian Ocean basin will be an urgent priority. In this event, Indonesia would have had just a few minutes' warning, but Thailand would have had two hours and Somalia seven hours to prepare for the disaster.

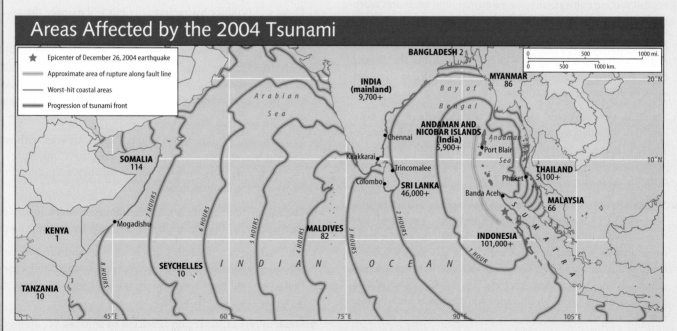

Areas Affected by the 2004 Tsunami

★ Epicenter of December 26, 2004 earthquake

Approximate area of rupture along fault line

Worst-hit coastal areas

Progression of tsunami front

Figure 12.H The epicenter of the earthquake, which registered magnitude 9.0, was located off the coast of Sumatra, Indonesia. It created a tsunami of tremendous force and scope.

volcanic eruptions, plus an average of 20 typhoon strikes yearly and regular monsoonal flooding, the Philippines earned the distinction of having more natural disasters in the 20th century than any other country. Mt. Pinatubo also ejected enough ash high into Earth's atmosphere to cool temperatures globally over the following several years, at least temporarily confounding efforts to track the trend in global warming. Meanwhile, the U.S. base at Subic Bay was converted into a thriving free-trade port that the Filipinos hope will become Asia's "New Hong Kong."

The Philippines is Southeast Asia in a nutshell: There are humanized landscapes never free of nature's whims, struggles in transforming a mercantile colonial economy to one based on high-tech, global trade, seemingly intractable odds to overcome in unifying diverse peoples under one flag, and wars on poverty that have yet to be fought.

SUMMARY

- Southeast Asia is composed politically of 11 countries: Myanmar, Thailand, Laos, Cambodia, Vietnam, Malaysia, Singapore, Indonesia, East Timor, Brunei, and the Philippines. With the exception of Thailand, all of these were formerly colonies of foreign powers.

- Southeast Asia contributed several domesticated plants and animals (including rice and chickens) and the achievements of civilization to a wider world.

- Relative to other parts of Asia, especially China and India, Southeast Asia has historically not had large populations or population densities. The most notable exception is on the Indonesian island of Java, where good soils and Dutch colonial polices produced a very dense population.

- This region poses numerous environmental challenges for its inhabitants. These include tropical climates, poor soils, disease, long dry seasons, deforestation, heavy erosion, tectonic processes, and monsoon climate patterns.

- There are more Southeast Asian villagers than urbanites. Farmers use shifting cultivation in the more marginal areas and sedentary cultivation in more productive regions.

- Plantation cash crops such as oil palm and rubber are grown for export and are the legacy of Western colonialism. The climate, cheap transportation, and abundance of land aided in the development of plantation agriculture.

- Southeast Asia's tropical forests are being destroyed at a faster rate than others of the world, mostly for commercial logging for Japanese markets. In biodiversity terms, Indonesia is a megadiversity country whose forests are under intense pressure, especially through illegal logging operations.

- Southeast Asia as a whole is among the world's poorer regions. The poorest countries—Myanmar, Cambodia, Laos, and Vietnam—have suffered from warfare and, in the case of Myanmar, from unwise political management. Peace has finally come to Cambodia, Laos, and Vietnam, and Vietnam is advancing economically. Singapore, Malaysia, and Thailand, because of their high rates of industrial productivity and economic growth, have often been described as Asian Tigers or Tiger Cubs.

- The region's mineral wealth includes oil, gas, and a wide variety of metal-bearing ores. Indonesia has the most oil. The most serious deficiency is the near absence of high-grade coal.

- Economic problems in Thailand in 1997 led rapidly to an economic meltdown in the region due to the interconnectedness of the regional and global economies. Most of the countries have since recovered from that setback, and several have experienced strong economic growth.

- Drug production, drug abuse, and HIV/AIDS are serious problems in Myanmar. Neighboring Thailand has some of these problems but has dealt more effectively with them than has Myanmar. Prodemocracy efforts in Myanmar have been thwarted by official repression.

- Thailand was never colonized and is more prosperous than its neighbors. French-colonized Cambodia, Laos, and Vietnam are much poorer. U.S. forces succeeded the French withdrawal from Vietnam in the 1950s. American efforts to win a war against communism in Vietnam failed. Normal relations now exist between the United States and Vietnam, and although Communist, Vietnam encourages private enterprise and has seen recent strong economic growth.

- Numerous dams are being constructed on the Mekong River. These will produce hydroelectricity but will damage the productive fishery of this basin.

- Cambodia has recovered politically from its devastating killing fields days and has joined or rejoined significant international organization, including the United Nations. Landlocked Laos is very poor but hopes to benefit from hydropower and mineral exports.

- Malaysia and Singapore are former British colonies with rather strong economies based on high-tech industries. Singapore has a very large and prosperous Chinese population and is developing a new niche in biotechnology.

- The region's demographic giant is Indonesia, which is very complex culturally and which has had a turbulent political career.

- The Philippines is a mainly Catholic country and a former colony of Spain. Population growth there is relatively high, and a gap between rich and poor threatens overall development. Large numbers of Filipinos work abroad.

- Much unrest in the region is related to tensions between ethnic groups. Following the regional economic crisis of 1997–1998, frustrations in Indonesia were directed against the prosperous Chinese minority. There are insurgents among the Muslim, ethnic Malay minority of southern Thailand. Indonesia is in danger of fragmentation. Catholic East Timor has gained impendence from Indonesia, and the provinces of Aceh and West Irian are seeking self-rule or independence. Muslim insurgents in the southern Philippines, and Communist rebels throughout the country, have stepped up their campaigns against the government.

KEY TERMS + CONCEPTS

Terms in blue are also defined in the glossary.

Abu Sayyaf (p. 350)
Amerasians (p. 340)
Asian Tigers and Tiger Cubs (p. 328)
Association of Southeast Asian Nations (ASEAN) (p. 333)
"Australia's September 11" (p. 346)
balkanization (p. 348)
Biopolis (p. 342)
boat people (p. 340)
brain drain (p. 342)
contingent sovereignty (p. 334)
crony capitalism (p. 332)
Culture System (p. 345)
"denying the countryside to the enemy" (p. 338)
devolution (p. 349)

doi moi (renovation) (p. 339)
economic recession (p. 333)
endemic species (p. 331)
expanding pyramid (p. 340)
Free Aceh Movement (GAM) (p. 349)
Golden Triangle (p. 334)
hill tribes (Montagnards) (p. 338)
Islamic law (*shari'a*) (p. 344)
Jemaah Islamiah (JI) (p. 346)
Khmer Rouge (p. 340)
killing fields (p. 340)
liberal autonomy (p. 348)
megadiversity country (p. 331)
Mekong River Commission (p. 341)
Moro Islamic Liberation Front (MILF) (p. 350)

New People's Army (p. 347)
Pancasila (p. 348)
Papuans (p. 349)
Parti Islam se-Malaysia (PAS) (p. 342)
plantations (p. 329)
remittances (p. 347)
Ring of Fire (p. 327)
shifting cultivation (p. 328)
"The Burmese Way to Socialism" (p. 335)
"the Lao Way" (p. 342)
Transmigration Program (p. 347)
tsunami (p. 351)
typhoon (p. 327)
"Venice of the East" (p. 336)
Viet Cong (p. 338)
Wallace's Line (p. 331)

REVIEW QUESTIONS

WORLD
REGIONAL
Geography Now™

Assess your understanding of this chapter's topics with additional quizzing and concept-based problems at http://earthscience.brookscole.com/wrg5e.

1. What are the countries of Southeast Asia? What are their major physiographic features?

2. What generalizations may be made about population numbers and densities in this region? What factors account for these characteristics?

3. What agricultural practices are typical of Southeast Asia? What are the main subsistence and commercial crops?

4. What countries are the largest producers and consumers of Southeast Asian tropical hardwoods?

5. What is the region's most important mineral resource? Which countries have most of it?

6. What are the major goals of ASEAN?

7. What major changes are coming to the Mekong River basin?

8. What political events have hampered development in Vietnam, Cambodia, and Laos?

9. Who are the boat people? Why did they leave their homelands? Where are they now?

10. How did Malaysia transform itself from a tin and rubber exporter to a high-tech economy?

11. What are the major political and resource-related issues faced by Indonesia and the Philippines?

DISCUSSION QUESTIONS

1. What were the main interests of colonial powers in Southeast Asia? Which powers were these? How did their activities transform landscapes and social and economic systems in the area? What country was never colonized?

2. What happened to Brazil's economy when rubber seeds were smuggled out to Southeast Asia? Can you think of other examples of dramatic economic effects from the introduction of exotic commercial plant species?

3. Why is deforestation in Indonesia of concern to that country's neighbors and to people living elsewhere on Earth? What can be done to stem the tide of illegal logging there?

4. Why did the region's economic meltdown occur? What were its impacts? Have the countries recovered from that crisis? What major industries typify the countries?

5. How are the problems of drug abuse and HIV/AIDS affecting

some of the countries? How well are governments and people dealing with these scourges?

6. What was the Vietnam War? Who participated in it, and what were its major outcomes?

7. What is the region's poorest country? What might account for that poverty?

8. How is China's growing economic strength casting a shadow over the region? What does Cambodian industry fear? What adjustments are being made in Singapore?

9. What are the prospects for and possible consequences of the balkanization of Indonesia? What provinces of Indonesia have sought autonomy or independence and why?

10. How do the governments of the Southeast Asian countries regard the issue of population growth? Compare the experiences of Thailand and the Philippines.

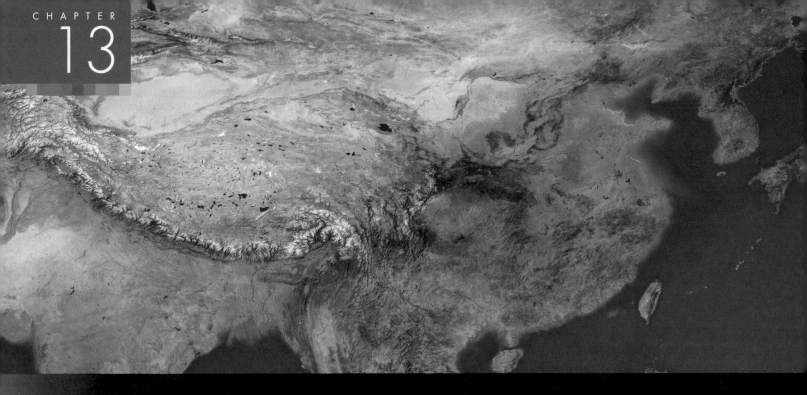

China: The Giant

Lee Sooi Fong, keeper of a Buddhist temple

Joe Hobbs

---

chapter objectives

This chapter should enable you to:

- Recognize China as a land empire in which a single ethnic group added vast peripheral areas to its eastern core

- Appreciate the constraints imposed by China's Communist system on personal, political, and economic freedoms and the recent substantial easing of those restrictions

- Balance the pros and cons of China's hugely ambitious plans to transform its natural environment through dams, canals, forest restoration, and other projects

- Consider the economic discrepancies between China's rural and urban populations and the forces behind large-scale rural to urban migration

- Understand the regional and geopolitical risks of Taiwan's aspirations for an identity independent of China

WORLD
REGIONAL
Geography⊛Now™

Look for this logo in the text and go to GeographyNow at http://earthscience.brookscole.com/wrg5e to explore interactive maps, view animations, sharpen your factual knowledge and geographic literacy, and test your critical thinking and analytical skills with unique interactive resources.

China has been home of the dominant culture of East Asia since very ancient times. Surrounding countries and regions show evidence of the longstanding influence of Chinese culture on their languages and writing, religion, agriculture, arts, crafts, institutions, ways of thinking, and histories. Chinese creativity is attested to by many early pioneering Chinese inventions and discoveries, such as gunpowder, paper, the wheelbarrow, the magnetic compass, and their system of writing. It is also witnessed by the organizational and engineering skills needed to build the Great Wall (the earliest version of which was completed in the third century B.C.) and the seventh-century construction of the Grand Canal to link north and south China.

From the earliest dynasties in the second millennium B.C. to the present, the story of China has been one of steady manipulation of a landscape that is not particularly fertile. There are ever-present challenges from nature, including floods and droughts. Today, while trying to feed its enormous population and raise fiber for its huge textile sector, China is pulling away from its past as a monumental agricultural nation and aggressively pursuing industrial development. With China's export-driven economy booming, its 2.5 million-soldier army backed by nuclear weapons, and its political influence represented with a permanent place on the UN Security Council, even the world's sole superpower,

the United States, must weigh most of its geopolitical decisions mindful of this world giant. Above all, China—officially the People's Republic of China (PRC)—must be recognized as the world's most populous country, home to 1.3 billion people, or 1 in every 5 on the planet. After a brief sketch of China's history, this chapter examines the physical resources for these people, and some of their culture and livelihood traits, by subregion. Agriculture, industries, and cities are covered. The chapter ends with discussions of Taiwan and Mongolia, countries whose histories and modern interests are closely linked to China.

13.1 Accomplishment, Subjugation, and Revolution

The "Middle Kingdom" is what China came to call itself, the middle realm between heaven and earth. Chinese culture is both very rich and very old. China was already a major political entity by the time of the Shang Dynasty (c. 1523–1027 B.C.). The Chinese had by then developed an agricultural system able to support large numbers of people on small areas of intensely worked land. During the Shang Dynasty, Chinese peasants occupied the great alluvial plain of the Huang

He (Yellow River) in North China. There they had a garden type of agriculture based on the hoe, irrigated crops, fertilizers, intensive use of flat alluvial lands, and the terracing of fertile hillsides. From its origins in the Huang He Basin, this system diffused by migration and conquest throughout the eastern part of the country and farther afield over a period of many centuries. The Huang He region is therefore recognized as one of the world's culture hearths.

In the third century B.C., the Qin (Ch'in) Dynasty accomplished the first unification of the Chinese people into one empire. Thus began China's development as a **land empire** that colonized its own vast hinterland. The conquerors were, as the dominant culture in China today is, ethnic Han Chinese. Under the Han Dynasty (206 B.C.–A.D. 220), China's influence spread as its leaders extended their authority through direct conquest and by accepting the fealty and tribute of less significant powers around them. Chinese armies pushed far westward across central Asia and almost made contact with the Romans in the vicinity of the Caspian Sea. Chinese ideas and materials traveled west along caravan routes to the Middle East and Europe. During times of internal disorder and weakness, China was conquered by nomadic invaders, notably the Mongols from the northwest and Manchus from the northeast. Under the Yuan (Mongol) Dynasty (1280–1367), the Chinese empire controlled Burma (modern Myanmar), Indochina, Korea, Manchuria, and Tibet. The leaders of China's last dynasty, the Manchu (Qing) of 1644–1911, were of the nomadic Jurchen tribes of northeastern China. They held large areas that eventually were added to the Asian part of the Russian empire.

China under the Manchus was increasingly subjected to unwelcome pressures from the West. European ships became frequent visitors to China in the 16th century, and Europeans began to press for more trade. The Chinese, who had high-quality cottons, silks, teas, and other products, were not much interested in what the West offered, and they also felt culturally superior to these intrusive "barbarians." The Chinese government strictly controlled trade, limiting it to the single port of Guangzhou (Canton) in southern China.

As European science, technology, and industrialization advanced, the Westerners grew in power and became more insistent. The weakening Manchu government was defeated by a British expedition in the Opium War of 1839–1842, forcing China to open its doors to opium from British-controlled India and to other British goods. Britain had carefully encouraged and fed the appetite for highly addictive opium in China, creating an enormous and very profitable market for itself.

Other European countries, Russia, and the United States joined the search for riches. China became a kind of joint colony of many competitive foreign exploiters, who got their way by occasional military actions and frequent threats. China was opened to Western products and then to Western-owned factories in its ports. Under the impact of Western machine-produced goods, China's handicrafts industries began to disintegrate. Foreigners in China were exempted from

the control of Chinese law in an arrangement known as **extraterritoriality**, and the Western powers gradually gained control of China's taxation system.

The foreigners also extorted territorial concessions, although only a small part of China had explicit colonial status. Britain secured the vital coastal port of Hong Kong. Portugal had earlier obtained nearby Macao. France and Germany won naval bases. Japan took Taiwan and the Ryukyu Islands, and Russia conquered parts of central Asia, Mongolia, and Siberia. **Treaty ports** were established, including Shanghai and Tianjin (Tientsin). In these ports, the foreign powers were granted special settlement areas under their own control. Chinese people could not normally reside in these areas and in some cases could not even enter them. European and American naval vessels patrolled China's coasts and her great rivers to enforce foreign privileges.

The Chinese did not accept this situation passively. Disorder mounted within the country, but the submissive Manchu sovereigns were kept on the throne by Western military power until they were finally overthrown by the **Chinese Revolution of 1911,** led by Sun Yat-sen. However, the new Republic of China had only nominal control over much of the country. Local warlords controlled many areas until, in the late 1920s, the army of the **Nationalist Party (Kuomintang,** or **KMT)** under Chiang Kai-shek fought northward from Guangzhou and managed to impose more control on most of the country.

The Nationalist regime, which established itself at Nanjing (Nanking), was able to abolish some foreign privileges, but its strength was sapped by almost continuing civil war against the **Chinese Communist Party (CCP)** led by Mao Zedong (Mao Tse-tung) and then by war against Japan. Chiang finally drove the Communists from their bases in the hills of southeastern China in 1934. Thus began the 2-year Long March—a more than 6,000-mile (9,660-km) flight on foot of nearly 100,000 Communists, pursued by Nationalist soldiers, from south central China north to Yenan, in Shaanxi Province, which became their wartime home base.

The Sino-Japanese War of 1937–1945 compelled Mao's Communists and the Nationalists to fight in common cause against Japan. But with the defeat of Japan in World War II, the CCP and the Nationalists returned to their civil war, in which Mao's forces triumphed. In 1949, 2 million Chinese Nationalists under Chiang Kai-shek fled to Taiwan, and Mao and the CCP claimed victory on October 1, 1949. Subsequent events in China are discussed throughout the chapter.

13.2 The Setting

China's area of about 3.7 million square miles (9.4 million sq km) is only slightly larger than that of the lower 48 United States (see Figure 10.2), but its inhabitants outnumber the U.S. population over 4 to 1. Two main environmental regions make up this populous country: an eastern **core** region, known as Humid China or China Proper, in which the country's population, developed resources, and productive capacity are heavily concentrated and a western **periphery**, or Arid China (Figure 13.1 inset). A rough boundary between the two extends from China's southern border with Myanmar to its northeastern border with Russia. This line corresponds generally to the stretch of land with average annual rainfall of about 20 inches (c. 50 cm), with Humid China to the east and Arid China to the west (see Figures 2.1 and 10.5a and b). Land uses of the regions are depicted in Figure 10.7.

Arid China

Much of China's unproductive and thinly settled land is in the western half of the country. Here are high mountains and plateaus, together with arid and semiarid plains, where rainfall is generally insufficient for agriculture. This arid region, sometimes called Frontier China, contrasts markedly with the better watered, more densely settled eastern half of the country and is home to only about 6 percent of China's population (see Figure 10.3a).

The principal regions of Arid China are the Tibetan Highlands, Xinjiang (Sinkiang), and Inner Mongolia (Nei Mongol). All three are identified historically with ethnic groups that are not Han Chinese (see Figure 10.10 and the discussion on pages 362–363). Now, after decades of politically motivated domestic migration from the east, Han Chinese comprise the great majority in Inner Mongolia, and recent estimates put them at a slight majority in Xinjiang.

The thinly inhabited Tibetan Highlands, including the province of Tibet (Xizang) itself and fringes of adjoining provinces, occupy about one-fourth of China's area. Most of this vast area is a very high, barren, and mountainous plateau averaging nearly 3 miles (c. 5,000 m) in elevation. To the northwest, there are many high basins with internal drainages, some containing large salt lakes. To the southeast, the plateau is cut into ridge and canyon country by the upper courses of great rivers such as the Tsangpo (the Brahmaputra of India), the Mekong, and China's own Chang Jiang (Yangtze Kiang, or Yangtze River) and Huang He (Hwang Ho, or Yellow River).

Lower elevations, warmer temperatures, and greater precipitation in parts of the southeastern plateau support some extensive grasslands and stands of conifers. Around the edges of the Tibetan Highlands are huge mountain ranges: the Himalaya on the south, the Karakoram and Kunlun Shan to the north and northwest, and the Qinling Shan and others to the east.

WORLD
REGIONAL
Geography⊛Now™

Click Geography Literacy to see an animation of tectonic plate boundaries.

Most of the 2.6 million people of the Tibetan Highlands are sedentary farmers and animal herders. A few hardy crops, especially barley and root crops, are basic to agriculture in this restrictive environment. In grasslands at higher elevations, some pastoral nomads graze their flocks of yaks, sheep, and goats. Yaks are particularly important not only as durable beasts of burden in these high elevations but also as providers of meat, milk and butter, leather, and hair traditionally woven into cloth (Figure 13.2).

Xinjiang (meaning "New Dominion") adjoins the Tibetan Highlands on the north. It has an area of roughly 635,000 square miles (c. 1.6 million sq km) and a population of about 19 million. It consists of two great basins: the Tarim Basin to the south and the Dzungarian Basin to the north, separated by the lofty Tien Shan Range. The Tarim Basin is rimmed to the south by the mountains bordering Tibet, and the Dzungarian Basin is enclosed on the north by the Altai Shan and other ranges along the southern borders of Russia and Mongolia. Both basins are arid and semiarid. The Tarim Basin is particularly dry because it is almost completely enclosed by high mountains that block rain-bearing winds.

The smaller adjoining Turpan Depression drops to 928 feet (283 m) below sea level and is the second lowest land spot on Earth, after the Dead Sea. These basins include the greater part of the desert of Taklamakan, whose name means "Once you get in, you'll never get out." From the Taklamakan and Mongolia's Gobi Desert, vast quantities of sand are airborne in spring sandstorms that darken skies and choke lungs downwind in Humid China and even Korea and Japan. In an effort to stem the desert's spread, since 1982 the Chinese have planted 42 billion jujube and other trees as "green walls" in Arid China. In 2002, the government announced the next decade's goal: to plant enough trees in this region to cover 170,000 square miles (440,000 sq km), about the size of California or Sweden.

The great majority of Xinjiang's population is concentrated near oases located mainly around the edges of the basins where streams from the mountains enter the basin floors. For many centuries, these oases were stations on caravan routes crossing central Asia from Humid China toward the Middle East and Europe on the historic Silk Road (see Regional Perspective, page 360). The ancient routes have been superseded by modern transportation. A railroad connects Lanzhou (Lanchow; population: 1.8 million), a major industrial center and supply base in northwestern China Proper, with Urumqi (Urumchi; population: 1.4 million), the capital of Xinjiang. Several major roads link the oasis cities, and several major airfields have been built in the region (see Figure 13.10).

These transport links are key elements in a drive to expand the economic significance of this remote part of China and bring it under firmer political control. Early in the Communist era of the middle 20th century, demobilized soldiers and urban youth from eastern China were organized into quasi-military production and construction divisions to expand irrigated land on state-operated farms in this arid landscape, promoting cotton as a major crop. Xinjiang is China's most mineral-rich region, with deposits of coal, petroleum, iron ore, and other minerals. Mining and manufacturing have expanded there. It continues to be a major focus in Chinese efforts to relocate significant numbers of ethnic Han

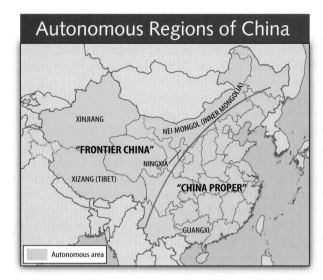

Autonomous Regions of China

XINJIANG

NEI MONGOL (INNER MONGOLIA)

"FRONTIER CHINA"

NINGXIA

XIZANG (TIBET)

"CHINA PROPER"

GUANGXI

Autonomous area

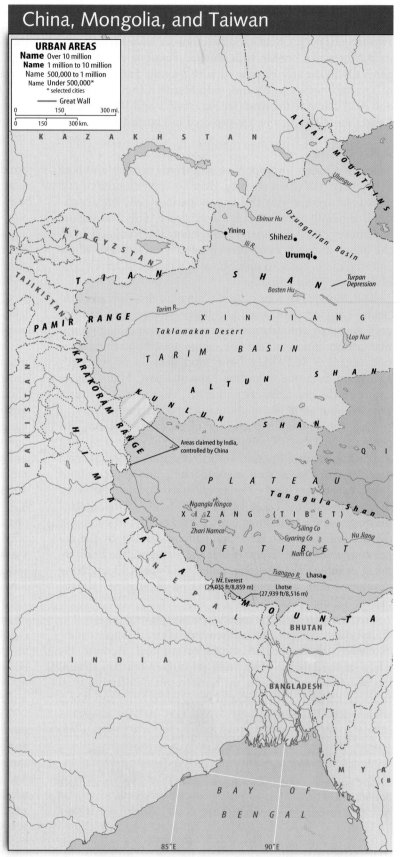

China, Mongolia, and Taiwan

URBAN AREAS
Name Over 10 million
Name 1 million to 10 million
Name 500,000 to 1 million
Name Under 500,000*
   * selected cities
——— Great Wall

0   150   300 mi.
0   150   300 km.

K A Z A K H S T A N

ALTAI MOUNTAINS

Ulungur

KYRGYZSTAN

Ebinur Hu
Dzungarian Basin

• Yining   Shihezi •
Ili R.

Urumqi •

TAJIKISTAN

T I A N   S H A N
Turpan Depression

Bosten Hu

PAMIR RANGE

Tarim R.

X I N J I A N G

Lop Nur

Taklamakan Desert

KARAKORAM RANGE

T A R I M   B A S I N

PAKISTAN

KUNLUN   A L T U N   S H A N

SHAN

Areas claimed by India, controlled by China

Q I

P L A T E A U

Tanggula Shan

Ngangla Ringco

X I Z A N G   (T I B E T)

Zhari Namco
Siling Co
Gyaring Co   Nu Jiang

O F   T I B E T

Nam Co

HIMALAYA

Tsangpo R.   Lhasa •

Mt. Everest
(29,035 ft/8,859 m)
Lhotse
(27,939 ft/8,516 m)

M O U N T A

BHUTAN

I N D I A

BANGLADESH

M Y A
(B

B A Y   O F

B E N G A L

85°E   90°E

Figure 13.1 Principal features of China, Mongolia, and Taiwan. In the inset map (top left), the line represents the division between Humid China, or China Proper, to the east and Arid China, or Frontier China, to the west. Humid China has the great majority of the Chinese population and economic activity, and Arid China is home to many minority peoples.

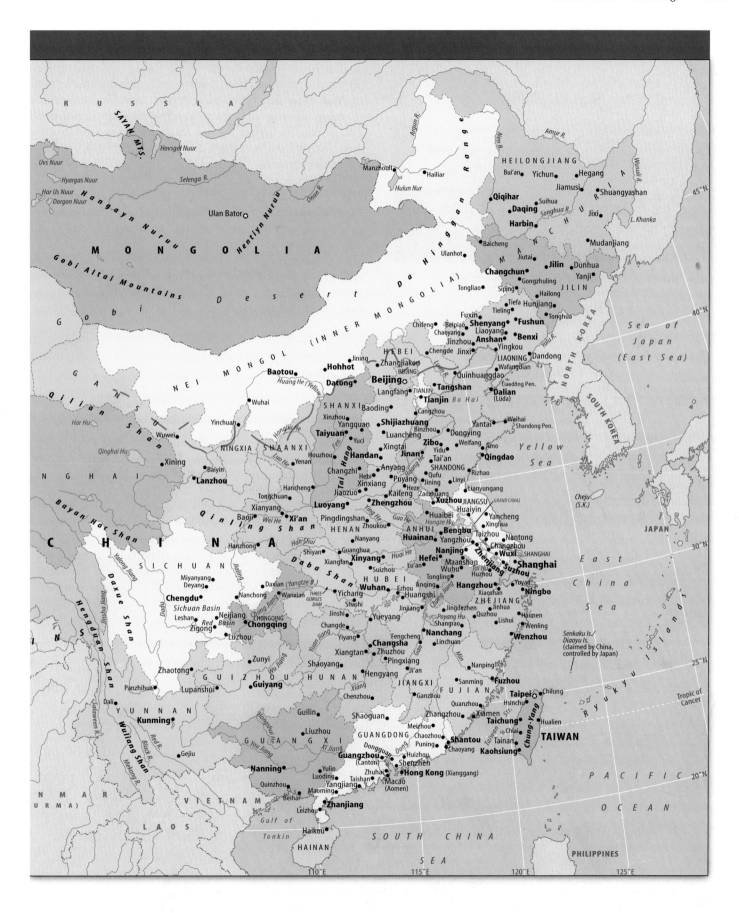

Figure 13.2 Yaks in a highland pasture of the Himalaya

Joe Hobbs

Chinese away from the coasts (see Ethnic Geography, pages 362–363).

Northwest of the Great Wall, rolling uplands, barren mountains, and parched basins stretch into the arid interior of Asia. Here lie slightly more than 1 million square miles (c. 2.6 million sq km) of dry, thinly grassed terrain divided almost equally between Inner Mongolia (Nei Mongol), the area nearest the Great Wall, and the country of Mongolia—known as Outer Mongolia in older literature. Most of the 24 million people of Inner Mongolia, which is an autonomous region of China, are ethnic Han Chinese concentrated in irrigated areas along and near the great bend of the Huang He (Yellow River). The greater amount of territory of Inner Mongolia is still home to a sparse population of ethnic Mongol herders. From these severe arid landscapes, the Mongols emerged to control China and, by the 13th century, an expanse of land that ranged from the Korean Peninsula in the east to the margins of Poland in the west. The desert land-

Regional Perspective

The Silk Road

In the early Han Dynasty (c. 206 B.C.– A.D. 8), the Chinese began a steady exchange of goods for items from the Mediterranean trading ports. For centuries to follow, China exported silk (not raised in the West until the sixth century), china, paper, and spices, and the west exported wool, gold, silver, glass, and metalware. The overland route for this exchange—called the Silk Road, or Silk Route— was approximately 4,000 miles (6,000 km) long, connecting the early Chinese capital of Chang'an (now Xi'an) to ports of

the eastern Mediterranean (Figure 13.A). The Road went alongside the Great Wall and the Taklamakan Desert, over the Pamir Mountains, and through Baghdad to the Mediterranean Sea. From there, goods were shipped to various European ports. This route also played a role in the diffusion of Buddhism into China in the first century A.D. and subsequently in the spread of languages, customs, and technologies. Until the Venetian explorer and trader Marco Polo did it in the 1270s, apparently no one ever followed the whole route. Travel was dangerous, and every caravan load had to pass numerous checkpoints and pay heavy tolls.

Figure 13.A The Silk Road

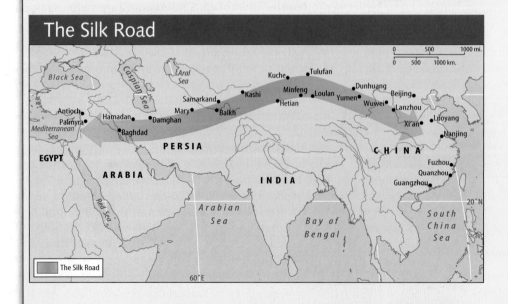

Karen Su/Corbis

Figure 13.3 The Great Wall of China. Although the wall's precursors date to the third century B.C., most of the wall standing today dates to the Ming Dynasty of the 15th century. The wall proved ineffective in keeping out the Ming enemies to the north, the Qing.

scape bears many traces of its rich past: caravan routes, temples hidden in desert caves along those trails, and remnants of the Great Wall (Figure 13.3). The Ming Dynasty began building this wall in the 15th century in an unsuccessful bid to stave off incursions from the north.

The traditional picture of nomadic Mongol tribespeople with their sheep, goats, camels, and horses is fast disappearing. The economic core of Inner Mongolia today is made up of agricultural areas near the Huang He. The irrigated areas produce mainly oats and spring wheat, while unirrigated fields are used mainly for drought-resistant millet and kaoliang, a sorghum grain used for animal feed and for liquor. The main city is Baotou (Paotow; population: 1.4 million), an expanding industrial center on the Huang. One of China's major iron and steel works was opened at Baotou with Soviet help in the 1950s to take advantage of nearby reserves of coal and iron ore.

Humid China

Humid China, sometimes referred to as Eastern China, Monsoon China, or China Proper—the core region—includes the most densely settled parts of the country (see Figure 10.3a). Humid China is commonly recognized in geographic literature as having two major subregions, North China and South China, which differ from each other in various physical, economic, and cultural respects. The Northeast, or Manchuria, is another subregion that straddles the line separating Humid China from Arid China, but it is important economically to the county and is therefore treated within the discussion of Humid China as the country's core region.

The Northeast

China's extreme Northeast region, better known to the West as Manchuria, lies north and east of the Great Wall. It is made up of three provinces: Liaoning, Jilin, and Heilongjiang. It

has an area of about 310,000 square miles (c. 800,000 sq km) and a population of approximately 110 million. To the west are the Mongolian steppes and deserts; to the east, the area is separated from the Sea of Japan/East Sea by the Korean Peninsula and by Russian territory. The Northeast consists of a broad, rolling central plain surrounded by a frame of mountains rising to about 6,000 feet (c. 1,800 m). The mountains on the east and north contain much valuable hardwood and softwood timber. The central plain, oriented northeast to southwest, is about 600 miles long (966 km) by 200 to 400 miles wide (322 to 644 km). The soils are very fertile, and there is usually enough rainfall for summer crops. Conditions are not favorable for irrigated rice, and the winters are too severe for winter grains. But soybeans, kaoliang, millet, corn, and spring wheat do well during the short frost-free season of 150 to 180 days.

This was once a sparsely populated frontier region, but political instability in the early 20th century prompted the ruling Manchu Dynasty to encourage migration to the relatively open lands of the Northeast. Millions of Chinese farmers moved to these provinces, pioneered its agricultural frontier, and gave it an ethnic Han majority by the outset of the Sino-Japanese War in 1937. As described on page 372, Japanese intervention began Manchuria's development as an important industrial region.

North China

Affected mightily by the monsoon wind patterns that characterize so much of Asia, North China has hot, humid, and rainy summers and cold, dry winters—traits typical of a continental climate. When the winter monsoon blows out of interior Asia, long, dark, cold winter days dominate North China. Many people cope with the cold with just the heat from the cooking stove or from the fire-heated brick bed called the *kang*.

Ethnic Geography

Han Colonization of China's "Wild West"

China's growth as a land empire has involved both subjugation of nonethnic Han peoples and the colonization of new territories by ethnic Han. There are at least 56 non-Han ethnic groups in China, in total about 100 million people living mainly in the west and southwest of the country. China's Communist government granted at least token recognition of the distinctive identity and rights of five large minorities by creating **autonomous regions**. These are the Guangxi Zhuang Autonomous Region, bordering Vietnam; the Nei Mongol and adjacent, small Ningxia Autonomous Regions in the north; the Xizang Autonomous Region of Tibet; and the Xinjiang Uighar Autonomous Region in the far west.

The Hui of Ningxia are Muslims of many different ethnicities who absorbed Chinese language and many Han customs. In contrast, the Uighars and Tibetans have defiantly resisted becoming Chinese—and so Han colonization focuses especially on their territories. The Chinese government aims to either assimilate these minority peoples into a broader, Han-based Chinese culture or to establish a strong enough Han demographic presence among them to discourage any aspirations of autonomy or independence.

Chinese authorities have also initiated what they call the **Western Big Development Project** with the stated objective of improving locals' livelihoods enough to diminish their ambitions for ethnic and political separatism. Focused on the autonomous regions and four other provinces in which there are large minority populations, this project seeks to improve infrastructure like airports and roads and to boost output from factories and farms. The Uighars, Tibetans, and other minorities generally fear that this "development" is just a cover for demographic, political, and economic domination by the Han.

Buddhism reached Tibet in the middle of the eighth century, and its arrival was followed by decades of conflict between Chinese and Indian Buddhists. The Chinese were defeated in this struggle and expelled from Tibet at the end of the eighth century. China sometimes reexerted loose control over Tibet, but Chinese authority vanished again with the overthrow of the Qing (Manchu) Dynasty in 1911. From 1912 until the Chinese Communists' violent conquest of 1951, Tibet existed as an independent state, with its capital and main religious center at Lhasa (Figure 13.B).

After the Chinese completed roads to Tibet in late 1954, an increase in restrictive measures by the Communist government contributed to the rise of Tibetan guerrilla warfare. This culminated in a large-scale Tibetan revolt in 1959 and the flight of the **Dalai Lama** (the spiritual and political leader of Lamaism, or Tibetan Buddhism) and many other refugees to India. Subsequently, the Chinese drove most of the monks from their monasteries, expropriated the large monastic landholdings, implemented socialist programs, and prohibited organized religion in the rest of China. Large-scale immigration by ethnic Han picked up so that Lhasa's population is now about 50 percent Chinese.

Beijing insists it is helping to develop Tibet, for example, with the construction of a railway between Lanzhou and Lhasa scheduled for completion in 2007 at a cost of $3 billion. But Tibetan resistance to Chinese rule continues. From his place of exile in Dharamsala in India, and on frequent international

---

North China receives between 15 and 21 inches (c. 35–55 cm) of precipitation annually but is plagued with high variability from year to year. Principal crops are adapted to this climate and include winter wheat, millet, and kaoliang, with some summer rice crops, corn, and a variety of vegetables. There is a history of droughts and floods. Floods in North China relate mainly to the problem of managing the Huang He (Yellow River; Figure 13.4). The Huang drops down from its source on the Tibetan Plateau and courses for more than 1,000 miles (1,600 km) through the loess soils of North China on its 2,500-mile (4,000-km) journey to the sea. These airborne soils blown from the Gobi Desert are not well consolidated, and they collapse easily into the river system. The river carries this yellow loess soil as silt and deposits it along the final 500–600 miles (800–960 km) of the river's flow to its delta.

This physical process is a mixed blessing. On the plus side, it has built up a broad and fertile arable area. On the nega-

Figure 13.4 The Huang He (Yellow River) passes through areas of intensively cultivated loess soils. Note the man working a small terrace just left of center.

Lowell Georgia/Corbis

Figure 13.B The Potala Palace in Lhasa, Tibet, is the ceremonial home of the 14th Dalai Lama, now in exile in India.

Hubertus Kanus/Superstock

192

157

However, China's government has so far refused to negotiate with the Dalai Lama.

China's government regards Xinjiang's Muslims, mainly from the 8 million-strong ethnic group known as Uighar, who speak a Turkic language and are kin to the ethnic Kazakhs to the west, as the most problematic minority groups. Many Uighars would like to pull at least a part of Xinjiang out of China's orbit to create an independent Chinese Muslim state. Two short-lived independent Uighar republics known as the Eastern Turkestan Islamic Republic existed in 1933 and again from 1944–1949, and some Uighar separatists are fighting for another. Beyond execution of those separatists deemed to be terrorists, China's response has been to press ahead with development projects and Han immigration.

At the beginning of the Communist takeover of China in 1949, the Han population in Xinjiang amounted to 6.3 percent of the province's population. Today, it accounts for between 50 and 55 percent, and some 250,000 more Han Chinese immigrate to the cities (especially Urumqi and Yining) annually, enticed in part by relatively high salaries and other incentives.

Human rights monitors report systematic Han abuses of the Uighar: Scores of mosques have been torn down, Uighar literature burned, Muslim clerics forced into "reeducation" training, Islamic schools closed, public prayer forbidden, head scarves banned, and prodemocracy demonstrations brutally quashed. China's struggle to maintain a grip on its ethnically distinct regions is reminiscent of Soviet efforts in non-Russian areas, and it remains to be seen whether the land empire of China might someday crumble as the Soviet empire did.

road trips, the Dalai Lama continues to appeal for Tibetan freedoms. He asks China to grant what he calls **genuine autonomy** to Tibet, meaning that Tibetans would have authority over most matters except foreign affairs and defense, which the Chinese government would control. His cause has many prominent backers, notably the American actor Richard Gere.

tive side, this sedimentation has elevated the actual stream channel of the river above the surrounding landscape, creating a perfect setup for flooding. A critical system of dikes follows the river all the way from the Tai Hang Mountains to the delta. But when flooding breaches these dikes, the river pours out of the channel and down to the lower, densely settled farmlands and cities of the North China Plain. Even after the ruptured dike is repaired and the river is contained once again, the great flow that issued forth from the elevated stream bed has nowhere to go but down into ponds, fields, and communities. This historic pattern on the North China Plain has earned the Huang He its nickname of **"China's Sorrow."** Despite these natural hazards, the Huang He and its Wei He tributary (east of Xi'an) were the historical focus for Chinese civilization.

Much trade between North and South China has been carried along the Grand Canal, a monumental civil engineering project of the Sui Dynasty (A.D. 581–618). This 1,050-mile (1,700-km) inland waterway—the longest ever built on Earth—links Hangzhou in the Chang Jiang Delta region to the area just southwest of Beijing. It was built principally to ensure a steady and safe supply of rice from the productive, subtropical southern lands to the north and to supply the south with wheat and coal. The canal was built at a great cost in tax and human labor but had the advantages of serving communities all along its banks and avoiding the hazards of coastal shipping. The Grand Canal is still an important waterway today and will be even more so when the eastern portion of the Jiang Chang Water Transfer project is complete (see Figure 13.D, page 365).

South China

The Qinling Mountains, reaching elevations of about 10,000 feet (3,048 m), mark the traditionally recognized dividing line between North and South China. These mountains run east to west and divide stream drainages between the Huang

Problem Landscape

The Three Gorges Dam

The Chang Jiang ("Long River") has been central to China's identity and welfare for thousands of years. The river delivers precious water and fertile soils, creating environments that have been intensively settled and farmed, especially for rice and wheat. But this great river has also delivered tragedy in the form of devastating floods. Flooding in August 1998 affected 300 million people along the river and did an estimated $24 billion in damage, for example. Inevitably, questions arose about the advantages to be gained—especially in flood control, drought relief, and hydroelectricity production—if the river were to be dammed. The concept of a giant dam on the river dates to 1919, when it was proposed by Sun Yat-sen (sometimes called the "George Washington of China"). His vision has been realized on a scale far grander than he could have imagined in the massive Three Gorges (Sanxia) Dam, begun in 1994 and scheduled for completion in 2009 (Figure 13.C).[a]

As with any large dam project, the Three Gorges may best be evaluated by considering both its pros and cons. On the plus side, this largest dam ever built—1.3 miles (2.1 km) and 610 feet (194 m) high—will create a 385 mile (620 km) long reservoir with a storage capacity that will dramatically reduce the threat of downstream flooding. Stored waters will also help to alleviate drought. As part of the larger **Chang Jiang Water Transfer Project,** the dam will also assist in the provision of water to more arid North China through three canals (Figure 13.D).

The Three Gorges Dam will improve navigation and therefore enhance trade, especially by connecting the burgeoning inland city of Chongqing to the world abroad. Historically, the Chang Jiang was able to accommodate freighters of 10-foot (3-m) draft all the way from the East China Sea to Wuhan. Above Wuhan, the shallower river was traditionally plied by smaller freighters to Yichang, situated just below the river's spectacular gorges. Yichang was a classic break-of-bulk point,

Figure 13.C The Three Gorges Dam in August 2003, viewed from space

with freight and passengers offloaded and put on flat-bottomed junks and sampans drawn upstream through the gorges to Wanxian by human trackers who pulled and rowed the vessels. With the completion of the Three Gorges Dam Project, vessels with much deeper draft will be able to sail 1,500 miles (2,400 km) upstream from Shanghai all the way to Chongqing in the Sichuan Basin. Small vessels will be able to pass over the dam in a ship elevator, a vast steel box that will lift the ship the height of the dam in 42 minutes. Most ship traffic will use a system of locks with five levels, taking 3 hours in transit. It is projected that shipping to Chongqing will increase fivefold. Some say the increased commerce will transform Chongqing into a new Hong Kong.

Also on the plus side, the world's largest hydropower plant that is built into the dam will provide about 85 billion kilowatt-hours of energy annually, enough to satisfy 15 percent of China's electricity needs. Some of the power may be sold to private users, helping to offset some of the dam's huge construction costs.

Most of the cons are related to the vast reservoir forming behind the dam. When filled in 2009, the 500 feet (150 m) deep lake will inundate 4,000 villages, 140 towns, 13 cities, numer-

[a] Much of the information in this feature is from a Discovery Channel film entitled *Three Gorges: The Biggest Dam in the World*, first broadcast on December 2, 2000.

He to the north and the Chang Jiang (Yangtze River) to the south. South China's agricultural dominance owes mainly to the 3,400-mile (c. 5,440-km) Chang Jiang that flows from its origin in the Tibetan Plateau to the coastal delta just north of Shanghai. This river has played a role in China quite similar to that of the historic Missouri River in the United States, serving as a major east–west transit corridor.

There are three large basins on the Chang Jiang. The two eastern ones—the Lower Chang Basin at the river's mouth and the Middle Chang Basin centered around Wuhan—are broad, densely settled areas with agricultural villages, small towns, and some expanding industrial cities (Wuhan being the most significant). Above the Middle Chang are the deep gorges now being transformed by the Three Gorges Dam (see

ous archeological sites, and 100,000 acres (40,000 hectares) of farmland. Because their homes have been or will be drowned, between 1 and 2 million people are being resettled. Many of the settlements to be vacated are quite ancient, and their inhabitants have deep cultural roots in them. The old villages are typically situated close to fertile farmlands that are also being inundated. In the new, higher communities into which people are being located, soil conditions are generally inferior.

The reservoir will have aesthetic consequences, too, replacing the world-class wonder of a wild river passing through scenic canyons. The natural marvels of the three gorges drew 17 million Chinese tourists and hundreds of thousands of foreign visitors every year. There are also practical concerns about

the silt that will build up behind the dam. It is possible that this increased soil load will have the ironic effect of causing flooding upstream from the dam, even as the dam prevents flooding below. Finally, there are fears of what is almost unthinkable: that the dam could burst, causing unimaginable suffering downstream.

Whatever the drawbacks, the government of China has pursued the mighty Three Gorges Dam Project without remorse and with no tolerance for dissent within China. The enormous cost of the project, nearly $30 billion, is born solely by China. Like megadams elsewhere, the Three Gorges Dam is a statement that its builder is a great power to be reckoned with on the world stage.

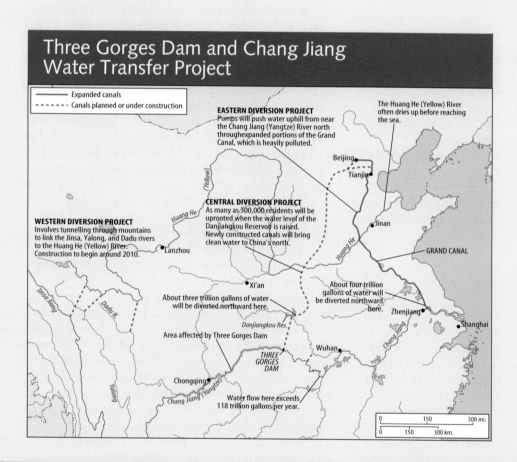

Figure 13.D The Three Gorges Dam and the Chang Jiang Water Transfer Project

Problem Landscape, above). Above the gorges, and also partially inundated by the Three Gorges Dam reservoir, is the Red Basin, home to more than 100 million people. Nature has provided little level land in the basin, but rice and other crops are grown on enormous numbers of small fields in narrow ribbons of valley land and on laboriously terraced hillsides. The Red Basin served as the center of the Chinese gov-

ernment during World War II because of its isolation and distance from Japanese bombers.

South of the Chang Jiang basins is the real cultural heartland of South China: the delta region at the mouth of the Pearl River Delta in the Guangdong lowlands. Its central features are the city of Guangzhou (Canton; population: 5.6 million) and the numerous small, densely farmed, and

Figure 13.5 Tower karst and paddy rice cultivation near Guilin

T. Waltham/Robert Harding/Getty Imager

settled lowlands on the Xi Jiang and its tributaries and distributaries. West of the Guangzhou lowlands, mostly in Yunnan and Guizhou Provinces, is a deeply dissected 6,000-foot-high (1,800 m) limestone plateau, with some of the world's most dramatic landscapes of karst, characterized by limestone pinnacles, caves, sinkholes, and "losing streams" that disappear from the surface into cavities below (Figure 13.5). The area is home to many of China's small minority populations who support themselves by hillside farming. The isolation of the region and its distance from the easy reach of national authority made it a perfect venue for opium poppy cultivation, until the Communist government cracked down in the 1950s.

13.3 Issues in Chinese Agriculture

China has had productive agriculture for thousands of years, helping both to create and to sustain the country's vast population. Today, more than 530 million Chinese (about 40 percent of China's population) live in rural settlements and are supported by agriculture (Figure 13.6). These peasants must wrest food for themselves and for the ever-expanding urban population from a landscape where 85 percent of the land is not in cultivation because of steep slopes, dryness, or short growing seasons. China's arable area, concentrated in the plains and river valleys of the east, is about 70 percent of the arable area of the United States, but through multiple-cropping practices (growing successive crops on a given field throughout the year), the sown acreage may actually be larger than the area farmed in the United States. China's territory provides, on the average, about two-thirds of an acre of arable land per person to the rural population and only one-fourth of an acre per person to the total population. For comparative purposes, arable land in the United States pro-

vides 7 acres of arable land per person to the rural population and 1.5 per person to the total population.

China's agricultural history has been characterized by efforts to raise yields per unit of land and to bring new land into cultivation. In the 1960s, China's Communist leader Mao Zedong proclaimed that "there are no unproductive regions; there are only unproductive people." During that decade, Mao launched a national campaign promoting self-reliance, especially through more productive agriculture. His policy of self-sufficiency in grain remains a central tenet of China's agricultural economy today, often forcing peasants to grow grain when they could make more money by selling fruits and vegetables.

Since Mao's early appeals, several measures have been taken to increase yields on already farmed land. Irrigation, water conservation, and flood-control measures have been part of this effort. The Chinese have struggled both to prevent water from running off in floods and to conserve it in reservoirs for when it is needed. To achieve these aims, local communities have built large numbers of small-scale dams, ponds, canals, and dikes. These are increasingly overshadowed by large-scale dams such as the Three Gorges and associated structures to control larger waterways. A major objective of water conservancy measures has been to increase the amount of land under irrigation and thereby increase output. China has more than doubled the amount of arable land under irrigation since Mao came to power in 1949. In arid areas, salinization is a challenge; too much irrigation deposits mineral salts from the water and reduces the land's productivity. Ongoing efforts at reforestation also relate closely to water control. Only 16 percent of China is forested. Under steady human pressure over thousands of years, most primary forests in humid China have been cut. Large areas of China are without significant vegetative cover, contributing to loss of water through rapid runoff.

Another approach to improving yields from existing lands lies in the application of scientific knowledge and technology to develop better seed, equipment, and farming techniques, together with the development of administrative means of promoting their adoption—China's version of the Green Revolution, described on pages 291–292. With more trained agricultural scientists and technicians, insecticides and herbicides are coming into wider use. These follow more rudimentary systems of pest control practiced during the 1950s and 1960s, during Mao's **Four No's Campaign** to eradicate rats, sparrows, flies, and mosquitoes. Observers were amazed at the sight of Chinese peasants fanatically hunting down the pests, motivated by government radio broadcasts over speakers placed in rural areas. A major objective was to reduce the annual losses of many millions of tons of grain. It was an unintended consequence that so many birds were killed that their natural prey, caterpillars, infested crops. Antipest campaigns against flies, mosquitoes, and rats had additional public health benefits. They were also politically important as they demonstrated the government's capacity to

Liu Liqun/Corbis

Figure 13.6 Rural China: The rice harvest in Guizhou province

marshal enormous numbers of people in a wide variety of national campaigns, many of which were focused on modifying the environment.

Increased fertilization has boosted farm productivity. The use of chemical fertilizers is growing alongside more traditional means. Chinese farmers have long fertilized their fields with pond mud, plant residues and ashes, oilseed cakes, nitrogen-rich crops that were raised to be plowed under for soil enrichment, and animal and human manures ("night soils"). Such traditional fertilizers, and the close contact between people and animals in rural China, play a role in the country's status as an "incubator" for disease (see Medical Geography, page 368).

Beginning in the 1950s, the Communists looked for ways to increase land cultivation. Small gains in the amount of cultivated land were made through reclamation of poorly drained land and extension of irrigation into deserts. However, there has actually been a decrease in the amount of land under cultivation, owing largely to the expansion of the urban base in China. Mechanization has developed slowly. The use of tractors is growing, especially in the northeast and in the west where more extensive farm fields are located, but human and animal power and simple hand tools are only slowly fading from the rural scene.

After 1949, when the Communist Party gained control of China, agriculture was not merely a matter for scientific and technological attention but for profound ideological change. China's leaders mandated a series of reorganizations of Chinese agriculture in successive attempts to solve the country's agricultural problems within the limits of Communist ideol-

ogy and control. Collectivization of agriculture began in 1955, with private farms reverting to state control in the form of agricultural cooperatives. The cooperatives were merged into even larger units in the agricultural **commune system** that began in 1958 as the cornerstone of an economic program called the **Great Leap Forward.** In the commune, 20 to 30 cooperatives consisting of over 20,000 people and 40 to 100 villages were merged into a single unit. Residents lived communally, sharing bathing, dining, and residence facilities. Any land and equipment of the former cooperative and any money or property still owned by peasants became the commune's property. An economic and administrative body controlled the labor and production and supervised all aspects of education and defense. Small-scale industries, most notably millions of "backyard steel furnaces," were also set up in the rural communes.

The legacy of this effort to bring industries to the countryside is still apparent in the Chinese economy today. There are about 25 million rural industrial enterprises with more than 125 million employees, representing about 40 percent of China's industrial labor force. Their rural-based, small-scale industries generate about half of China's annual industrial output, one-quarter of its total economic value, and more than one-third of China's exports. These successes owe largely to recent economic reforms to be discussed below, and not to the Commune Movement.

Collectivization and communes had disappointing results, related mainly to inefficiencies inherent in trying to manage such large units and to the lack of incentives for individual farmers. An estimated 30 million Chinese died from famine

Medical Geography

The Birth and Diffusion of SARS

Seemingly out of nowhere, **SARS (severe acute respiratory syndrome)** burst onto the world scene in spring 2003 and has cast a shadow of watchful anxiety ever since. But this extraordinarily virulent virus did come from somewhere: southeastern China.

As any recipient of a flu shot knows, the vaccine typically is targeted against a host of flu strains bearing Asian, usually Chinese, names such as "Shanghai" and "Beijing." SARS, like these influenza varieties, originated in a region where there is frequent close contact between people and animals. Farmers in China often cohabit with ducks, pigs, geese, and other domesticated animals and make extensive use of animal excrement as fuel and fertilizers. In these close quarters, the influenza strain often starts in a duck or other animal and jumps the species barrier to affect humans. Once the sickness manifests itself in a human host, it can spread rapidly to major cities in China and thence by airliner to distant parts of the globe.

The SARS virus had similar origins, although it probably arose from close contact between animals and animal merchants in southeastern China. In Guangdong Province, there are numerous large animal markets where poorly paid migrant workers carry out their sordid work of killing, skinning, plucking, gutting, and packing a wide variety of wild and domesticated creatures, including snakes, frogs, turtles, chickens, badgers, cats, rats, and civets, all bound for Chinese dinner tables.

Between November 2002 and January 2003, several of these animal merchants, along with food handlers and chefs further down the supply chain, began developing flulike and pneumonialike symptoms, including high fever. They had contracted SARS, an illness new to the world and a member of the coronavirus family that also includes the common cold. Some of them, perhaps 5 percent, died of the illness. Some sought medical attention in Guangdong hospitals. There they infected doctors, nurses, and other staff (30 percent of China's early SARS victims were hospital workers).

On February 21, a doctor from one of the hospitals managed, despite his high fever, to attend his nephew's wedding in Hong Kong. He stayed at the Metropole Hotel, where he passed the virus (probably by sneezing or coughing) on to two Canadians, an American businessman on his way to Vietnam, a man from Hong Kong, and three women from Singapore. In the days that followed, most of these sick people flew abroad. One of the Canadian women carried the illness home to Toronto, where she died after infecting her son and several health workers. At that point, SARS was poised to become a global epidemic. By June, 6,900 probable SARS cases had been reported in 29 countries, and 495 victims had died. In the United States, about 200 suspected SARS cases were reported in 38 states, with no fatalities.

Within less than 6 months, SARS had become an enormous problem that posed several threats. Of most immediate concern was its impact on human health. There is not yet a vaccine for the disease. It is especially difficult to develop a vaccine for an illness that affects both people and animals. The original animal host for SARS will have to be identified if an effective vaccine is to be made; an intensive search for that creature has been underway in southeastern China. Scrutiny of the civet "cat," actually a relative of the mongoose and a favorite wintertime dish for Chinese, was especially strong, and many mil-

---

attributable to the failures of the Great Leap Forward. Reforms began in the 1960s, with communes decentralized and some even divided into private farms. After the death of Mao and the 1978 ascension of Deng Xiaoping to the major leadership role in the Communist Party, free-market reforms began to take place in China. Individual households acquired long-term leases on the farms, which continued to belong to the state. They paid fixed amounts of their production to the state and were allowed to consume or sell the rest. Chinese farmers were finally allowed to sublet land, hire labor, own machinery, and make their own land-use decisions.

The results of this experiment with a freer semicapitalist system in agriculture were profound. Chinese agricultural output expanded by approximately 50 percent during the 1980s and 1990s, and rapidly enough—when coupled with the steady decrease in birth rate—generated an increase of 20 percent in the per capita food supply. Two statistics nicely reveal the overall outcome of China's agricultural efforts. In 1950, at the outset of the Maoist Revolution, the per capita amount of arable land was 2.7 *mu* (a *mu* is about one-sixth of an acre, or one-fifteenth of a hectare). By 1995, that figure had fallen to 1.2 mu, even with all of China's efforts to expand the country's arable base. Yet at the same time, the average per capita grain production had grown from 460 pounds (209 kg) per year to 837 pounds (380 kg) per year through increases in agricultural productivity. It is remarkable that the world's largest country has generally been able to keep agricultural growth apace with population growth and usually provide more than 90 percent of its own supplies of rice, corn, and wheat. The challenge will be for

Figure 13.E China's SARS epidemic prompted some new behaviors and fashions.

lions of the animals were destroyed in Chinese markets following the initial and subsequent SARS outbreaks.

There is no effective treatment; SARS does not respond to antibiotics, for example. In the absence of a prevention or cure, the disease follows its course. A high fever of 100°F or more is followed by chills, muscle aches and headaches, a dry cough, difficulty in breathing, and in some, severe pneumonia and even death. Death rates are as high as 15 to 20 percent among those infected. Although the highest mortality rates are among the elderly, one of the striking things about SARS is that it is quite capable of killing perfectly healthy, young adults.

Despite the very small number of cases to date, SARS has

proven to have strong impacts on human psyches and societies. The disease could appear anywhere and infect or kill anyone; thus, many people—especially those near the Asian epicenter—assume the worst during an outbreak. During the initial spread in spring 2003, untold numbers of people in China wore masks to reduce chances of contracting the disease (Figure 13.E). They no longer shook hands. Schools, theaters, dance halls, and Internet cafés closed in many communities. Shopping malls had few shoppers. Several Asian countries stopped granting visas to Chinese visitors.

The economic costs for such a limited outbreak were also extraordinary. In the globalized world of business, many firms halted interpersonal dealings with China. Toronto lost $30 million a day in cancelled conventions and other events during the brief period when the World Health Organization (WHO) maintained a warning that travel there should be avoided. In spring 2003, China and South Korea lost $2 billion in the tourism, retail sales, and manufacturing sectors, and Hong Kong, Taiwan, and Singapore perhaps lost $1 billion each. Sixty percent of U.S. corporations banned travel to Asia altogether. North American airline bookings to Hong Kong fell 85 percent, and international arrivals in Singapore were down by 60 percent. Anyone diagnosed as having a fever in Singapore airport's arrival hall was unapologetically led off to 2 weeks of quarantine.

In summer 2003, the disease peaked and began to decline in Canada, Singapore, and Hong Kong, and it had disappeared from Vietnam. But in its birthplace, China, it was still spreading. The number of new cases gradually subsided, but late in 2004, the World Health Organization, noting that the virus was still prevalent among civets in southern China, expressed fears of a new outbreak among humans.

China to continue this remarkable trend, especially as the Chinese develop new appetites for beef, beer, and other luxuries that require more inputs of energy and agriculture. Chinese agricultural planners are also alarmed by the accelerating loss of farmland to urban and other uses and by farmers' decisions to forsake grains in favor of more lucrative cash crops.

13.4 China's Industrial Geography

China today is one of the world's great industrial powers, ranking sixth in overall economic output. Remarkably, production of industrial goods did not start in China until the beginning of the 20th century. Prior to that time, Imperial China's handicraft industries produced high-quality goods that attracted Western traders in spite of China's official disinterest in such trade. To this day, handicraft and shop-scale industries continue to supply many of China's simpler needs.

China's early factory industries were mainly foreign owned and were attracted to China by the huge market for cheap goods, by inexpensive labor, and sometimes by certain natural resources. As in so many countries, the first large-scale factory industry to develop was the manufacture of cotton textiles. Japanese and other foreign firms, together with some native Chinese companies, developed a major cotton industry by the 1930s. This industry, and others, developed in coastal cities such as Shanghai and Tianjin. Meanwhile, the Japanese used the minerals of the Northeast (in their

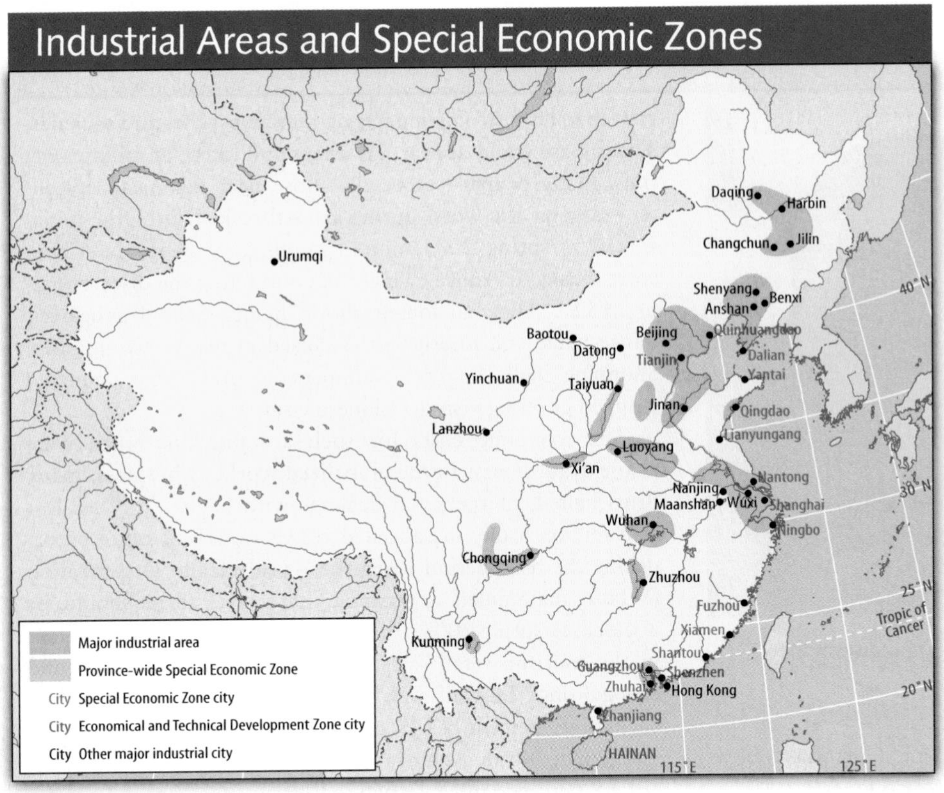

Industrial Areas and Special Economic Zones

Figure 13.7 Major industrial resources, industries, and industrial cities of China

puppet state of Manchukuo) to make that area the first center of heavy industry in China in the 1930s (Figure 13.7).

In their efforts to further industrialize after the 1949 victory over the Chinese Nationalists, Communist China directed economic development in four distinct national drives:

1. *Communization and repair of wartime damage, 1949–1952.* The industrial structure, which had been badly hurt by invasion and civil war from 1937 to 1949, was brought back to prewar production levels under the new system of state ownership. China meanwhile expended considerable capital and human effort in its involvement in the Korean War of 1950–1953.

2. *Heavy-industry development using the Soviet pattern, 1952–1960.* Aided by the Soviet Union, Chinese planners gave priority to coal, oil, iron and steel, electricity, cement, certain machine-building industries, and armaments. China's resources for these industries were considerable, and the country's centrally planned command economy—a system in which major economic activity is initiated by government planners—achieved major successes in developing them, as was the case in many Communist countries. There was also a strategic spatial component to this drive, with the Soviets advising the Chinese to diminish the coastal concentration of industrial plants. This strategy, meant to reduce the vulnerability of Chinese industry in the event of military conflict and to promote development outside the historic core, led to expansion into new industrial landscapes at Wuhan on the Chang Jiang, at Baotou in Inner Mongolia (Nei Mongol),

and even into the far west near the then Soviet border in Xinjiang. China rapidly accelerated its production of coal and steel.

Heavy industry has been built on a rich domestic resource base. China has about 12 percent of the world's coal reserves and is the world's largest producer and consumer of coal. China meets about 75 percent of its overall energy needs, and 75 percent of its electricity, with coal. This is the dirtiest of all the fossil fuels, producing far more particulate matter and carbon dioxide than oil or natural gas, and with generally lax environmental quality standards, China's skies are often badly polluted. The coal reserves are located mainly under the North China Plain and its bordering hills and in Liaoning Province in the Northeast. Iron ore is less abundant but is present in several locations, with the largest in Liaoning. Due to the resource concentration, the southern region of the Northeast has been the country's leading heavy-industrial region since the Japanese developed modern industries there. Coal and ore deposits facilitated Chinese Communist industrial development at other centers in China Proper, in Inner Mongolia, and along the Chang Jiang.

China's large production of many alloys and other metals has also facilitated development. China is today the world's largest producer of steel, aluminum, magnesium, and tungsten. Because of local deficiencies to meet the needs of its booming industries, China imports vast amounts of iron ore, lead, alumina, nickel, and copper. For a time, oil fields in the Northeast, Gansu, Sichuan, and in the Xinjiang Province met China's needs, and there were some surpluses for ex-

Definitions + Insights

Mao's Cultural Revolution

Between 1949 and his death in 1976, Mao Zedong (Mao Tse-tung; Figure 13.F) led the People's Republic of China through a demanding and sometimes counterproductive series of experiments in social reorganization. In 1957, he initiated the Great Leap Forward and the Commune Movement

Figure 13.F Chairman Mao greeting a crowd in 1971

Bettmann/Corbis

discussed earlier. The **Cultural Revolution** (1966–1972) was Mao's attempt to maintain an ongoing "permanent revolution" in China. Politically, the movement was meant to throw a newly established urban and intellectual elite off balance by promoting purges of virtually anyone who had offended Communist Party functionaries.

Local youth who donned the red armbands of the Red Guards became the agents of change in this campaign. The Red Guards were set loose on the populace to harass those who were accused of putting learning, skill, expertise, or personal matters above revolutionary enthusiasm or Marxist/Maoist goals and principles. Government oppression led to perhaps a million assassinations, executions, or suicides and to lifetime jobs for those deemed most loyal to the Maoist revolution, regardless of their job performance. Reforms initiated in the late 1970s finally began the process of removing unproductive people from their positions, but people whose lives were turned upside down by the Cultural Revolution continue to speak of themselves as a lost generation.

Despite some monumental setbacks and miscalculations, China under Mao regained a global prominence that had been denied for a century and a half under Western colonialism and interference in the nation's domestic affairs. Even though much of Mao's legacy has been dismantled, he continues to be a major icon of China's presence on the world stage.

ports. With the economy in overdrive, however, China became a major oil importer and by 2003 was second only to the United States.

3. *Costly political innovation under late Maoism, 1960–1978.* Communist China's industrial development began with the country's rift with the Soviet Union in 1958–1960 and continued until Mao's death in 1976. New economic policies were introduced in the late 1970s and early 1980s. The country's general industrial advance slowed in pace, technical development, quality of output, and modernization. These were consequences of China's near isolation from more advanced industrial countries, the Cultural Revolution from 1966 into the early-1970s (see Definitions and Insights, above), and the difficulties that nonmarket, centrally planned economies on the Soviet model typically had in creating consumer-oriented industries and advancing technology.

4. *Post-Mao liberalization, 1978–present.* The latest phase of China's industrialization began after Mao's death, when another veteran of the Long March, Deng Xiaoping, ascended to supreme power in China. Under his leadership, for the first time since 1949 China opened its doors to the West and began to devote more attention to personal incentives

for agricultural and industrial success. The liberalization of agriculture has already been described, and related measures were introduced into trade and industry, where emphasis shifted toward light industry, consumer products, and profit-oriented enterprises. Much more private enterprise was allowed, especially after the mid-1990s. State factories were required to operate at a profit, part of which they were allowed to retain and invest in expansion. The factories introduced pay incentives for superior performance and were no longer obligated to provide lifetime jobs regardless of output. Foreign investment capital was welcomed, and technical expertise from Japan and the West was sought. Great numbers of Chinese began to study abroad in graduate and professional programs. Turning its back on the isolationist days of Mao, China sought to become a global trading power. China has succeeded handsomely in this goal, as described on pages 290–291 in Chapter 10.

As for the future, there are risks for booming China, and it cannot be taken for granted that the country will become the economic dragon forecast by so many. China's extraordinary recent successes are tempered by a few concerns. One is freedom of expression. In 1989, the Chinese government cracked down brutally on prodemocracy demonstrations centered on Tiananmen Square in Beijing. Years later, there continue to be strictures on academic, media, artistic, and other freedoms that some observers believe are conducive to long-term economic well-being. China's government is one of a handful in the world (including Myanmar, Singapore, Saudi Arabia, and Iran) to censor the Internet. Another potential problem is that China's leaders are breathlessly trying to feed what they call the "socialist market economy" with gigantic projects, thereby hoping to create jobs and stimulate economic growth. They reckon the country needs to register at least 7 percent economic growth annually just to prevent massive unemployment and social unrest. And so one of Earth's most impressive building booms ever is now underway: new subway systems, railroads, highways, bridges, sewage systems, dams, water diversion projects, a natural gas pipeline, a national electricity project, urban improvements in Beijing for the 2008 Olympics, airports, industrial parks, and more. These efforts cost a lot of money, and the government is drawing down its national treasury to fund them. State banks are pouring billions of dollars into federal projects whose returns are uncertain.

A further hazard is that, with its galloping growth, China is developing a classic "economic bubble" that will be pricked, probably with costly results. Historically, few economies have been able to maintain torrid growth for long, especially when that growth is fueled by speculative cashflows and questionable bank loans, as has been the case recently in China.

Finally, there is concern that the benefits of China's economic growth have not been distributed evenly through the population. Prosperity and consumption have increased dramatically, but there is a remarkably unequal distribution of this wealth. Less than 20 percent of China's people controlled about 80 percent of the country's private wealth in 2004. Meanwhile, 18 percent of all Chinese lived on less than $1 a day—the World Bank's measure of abject poverty. These inequalities are comparable to those that existed when the revolution brought Mao's Communists to power on the promise of redistributing wealth. In 1980, China had one of the world's most even distributions of wealth, but that equation has changed dramatically. In sum, despite its phenomenal growth, China is still relatively poor (with a per capita GDP PPP of $5,000, comparable to Peru and El Salvador) and largely agricultural (see Table 10.1). The agricultural face of China is changing rapidly, though, as the cities boom.

13.5 China's Urban and Transportation Geography

When Mao came to power in 1949, about 85 percent of China's population lived in the countryside. Although there were massive cities such as Shanghai, Beijing, and Guangzhou (Canton), the great bulk of the Chinese population was still deeply rural and lived in small villages. Now China is about 59 percent rural and 41 percent urban (for comparative purposes, India is 72 percent and 28 percent and Japan is 22 percent and 78 percent).

As might be expected, despite the rural industries described earlier, China's industrial production is centered mainly in its cities. The country's 12 largest metropolitan cities are located in geographic clusters, revealing an important spatial component of China's economy. Four major clusters are especially notable: the distinctive group in the Northeast, the North China cities, the Chang Jiang cities, and the booming cities south of the Chang Jiang.

Strong neighbors have long coveted the cities and industries of China's Northeast. After a treaty signed in 1858, tsarist Russia landlocked Manchuria by extending Russia's coastal territory all the way to North Korea. Even today, China is eager to open a northeast access to the sea, even if only by leasing the Russian port of Zarubino. During the 20th century, the Northeast became a region of even more contention among China, Russia, and Japan. Russian interest focused on the Chinese Eastern Railway—a shortcut across the northern two provinces between Vladivostok and the Trans-Siberian Railroad east of Lake Baikal—and on the naval and port facilities of Dalian (formerly known as Darien–Port Arthur and as Lüda), which the Russians had used at various times.

Japan was long interested in developing the Northeast's impressive industrial resources and its potential for surplus food production. After the Russo-Japanese War of 1904–1905, fought mainly in Liaoning, Japan became increasingly active in the area and took over control in 1931 with the establishment of the puppet state of Manchukuo. In 1945, during the final days of World War II, Soviet forces took it from the Japanese. Eventually, the Chinese Communists incorporated the Northeast into the People's Republic, beginning this region's role as the center of broadly expanded industrial development. Significant factory plant and equipment were uprooted and shipped back to industrial centers in the former USSR by the Russians at the very end of World War II. However, the industrial and transportation infrastructure that remained, which had been created by the Japanese in the 1930s and 1940s during their occupation, was an important asset for subsequent Communist development of this region.

The cities of the Northeast include another three of China's twelve largest cities: the port of Dalian (population: 2.6 million); Shenyang (Mukden; population: 6.5 million), the main metropolis of a group of heavy-industry centers; and Harbin (population: 4.5 million), a rail center with connections to Russia and Korea. Since the Communists took over in 1949 and expanded China's industrial base, these cities of the Northeast have lost some of their industrial importance, but they continue to be centers of manufacturing, chemical production, and agricultural marketing and processing. The industries are centered around Shenyang. The largest center of iron and steel production is Anshan, south of Shenyang, where industry is based on deposits of iron ore in a belt crossing the southern part of the region. Substantial deposits of coal also exist, although the total reserves are far smaller than those of China south of the Great Wall. The Daqing oil fields, one of China's richest petroleum production areas, are located northwest of Shenyang in Heilongjiang Province.

The cities of North China include the historic national capital of Beijing (Peking; population: 10.3 million; Figure 13.8), the port city of Tianjin (Tientsin), and far to the southwest, the very ancient city of Xi'an (Sian). Each is located at an entryway to the North China Plain: Tianjin, by the sea; Beijing, near the end of a pass leading through the uplands to the north toward Mongolia and the Great Wall; and Xi'an, in the Wei River Valley, which leads upstream to interior China and downstream to the Huang Valley and the North China Plain.

The Chang Jiang Valley city group is a string of cities along and near the river. This group includes five of China's twelve largest cities, reflecting the importance of the Chang Jiang basins as a major axis of the national economy. The Chang Valley group includes a metropolis for each productive basin, plus an overseas global connecting point as well as a transfer and connecting point for interior China. At the western end of the axis in the Red Basin is Chengdu (population: 4.6 million), a provincial capital and node for transport lines connecting the axis with Arid China. It lies away from the river. On the river is another Red Basin metropolis, Chongqing (population: 7.3 million). Downstream lie Wuhan (population: 4.5 million) in the Middle Basin, an important manufacturing, transportation, and agricultural marketing center, and Nanjing (population: 3.6 million) in the Lower Basin. Nanjing (a temporary capital for China

Figure 13.9 Boomtown Shanghai

during the Sino-Japanese War in the mid-20th century) is a major agricultural marketing center.

On a navigable tributary (the Huang Po Creek) near the mouth of the Chang in the Lower Basin is the port of Shanghai (population: 14 million; Figure 13.9). With its location at the seaward end of this urban axis, Shanghai serves the whole Chang Valley and other areas. Built from a fishing village by foreign enterprise in the 19th century, Shanghai is the largest city and port in China.

The region south of the Chang Jiang has two of the country's twelve largest cities, including the port of Guangzhou (Canton; population: 5.6 million) in the Xi (Pearl) River Delta. Guangzhou and its vicinity now form a rising center of new industries, particularly around Shenzhen, a young city that in the 1980s was designated as one of five Special Economic Zones (SEZs) designated to attract investment and boost production through tax breaks and other incentives (see Figure 13.7). The other SEZs, all with access to the sea, are nearby Zhuhai, the entire island of Hainan, and further north Shantou and Xiamen on the Taiwan Strait, and finally Pudong, linked to Shanghai.

Close to the Xi river cluster of SEZs are Hong Kong (Xianggang; population: 8.2 million) and Macao (population: 453,000), which reverted, respectively, from British and Portuguese control in 1997 and 1999. Macao's economy is based on trade with inland China, fishing, tourism, and gambling. Prior to its reversion to Chinese control, Hong Kong was one of the world's busiest and most prosperous trading centers. China promised that Hong Kong would enjoy considerable autonomy and political liberalization under a unique **one country, two systems** formula after 1997, but since then, it has not been tolerant of democratic freedoms

Figure 13.8 Beijing

there. China has postponed elections for a chief executive and legislature until at least 2007 and 2008, respectively.

One of the remarkable patterns that emerges from this clustering of cities, and the economic power they represent, is that there is a serious problem of regional imbalance in China. Most of the country's industrial development and urban growth is strung along its east coast, near the traditional port and river cities that were strongly influenced by Western economic interests in China from the mid-19th century on. Even the rural areas of eastern China are experiencing much more growth relative to those in the west, especially because they are well linked to expanding markets and prospering cities. Meanwhile, western regions of China, mountain villages all through Humid China, and landscapes lacking in surface transportation in the southwest and parts of upland China are increasingly frustrated. They have seen relatively little evidence of China's much touted economic prosperity.

Predictably, large numbers of people have migrated from the less prosperous regions of the country to the promising SEZs and other urban centers of the east and south. Until recently, such movements were made difficult by a registration system called *hukou* that bound Chinese citizens to their home villages and cities. However, that practice has been largely abandoned, mainly in response to economists' and business owners' complaints that the policy hampered development by increasing labor costs.

There are also as many as 110 million migrant workers who leave their villages—some of these subsequently occupied mainly by children and the elderly—to work in construction or menial labors in cities, sending much of their meager monthly earnings of $50 to $100 home as remittances. People in the cities often discriminate against these migrants in education, housing, and jobs, treating them like illegal aliens. Many technically are illegal because they fail to sign work contracts that would both register them and require factories to pay fees to local cities. Police routinely "shake down" workers without papers, seeking bribes from the frightened migrants. Some of the factories are world-class, but others are **sweatshops** where working conditions border on the nightmarish, according to published reports. Management typically suppresses strikes, bans independent trade unions, and fails to enforce minimum wage laws for workers who toil through 18-hour workdays and catch brief rest in prisonlike dormitories.

In an effort to redress the regional imbalances that draw so many people to the east and southeast, and to integrate China's widespread resources and widely scattered producing and consuming centers into a functioning whole, the government has placed great importance on improving the transportation system (Figure 13.10). Although trucks play an important role, much commerce is still conducted by human porters, bicycles, pack animals, wheelbarrows, or carts

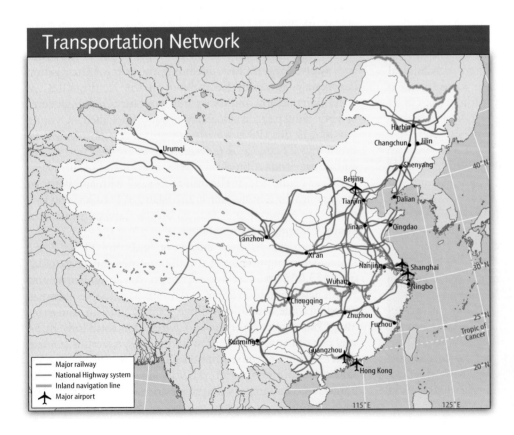

Figure 13.10 One of China's most ambitious efforts in expanding its infrastructural base for continuing economic development has been its transportation system.

pulled by animals or drawn and pushed by humans. To meet greater needs, the country has a growing highway system, much of it consisting of earth roads repeatedly lost to heavy seasonal rains. Strategic roads have been built in western China—for example, to Tibet, the Chinese-Indian border, and Nepal. China had extended the 1949 total of 50,000 miles (80,000 km) in the national road system more than 700,000 miles (1.2 million km) by the early 2000s. For comparison, the total road mileage in the United States is 3.9 million miles (6.3 million km).

There has been a major effort to improve and extend railways. Before 1949, the Northeast had a fairly extensive rail system developed by tsarist Russia and expanded by Japan. At the end of World War II, China's North and Northeast contained about 75 percent of China's rail mileage. Most of China south of the Chang Jiang floodplain and all of the western provinces were without rail lines. But the Communists made South China accessible by building the first three bridges ever to cross the Chang Jiang. They pushed rail lines into South China and the Red Basin and also into Xinjiang and across Mongolia to connect with the Trans-Siberian Railroad. Since 1949, the length of China's rail network has tripled, and the expansion continues. However, railroad development has lagged behind that of roads. Except for transport over short distances, railroads are much more efficient for hauling goods than trucks are. With China's economy booming, there are huge bottlenecks at ports and factories that suggest more railways will need to be built.

13.6 Taiwan

Known by Westerners for centuries as Formosa (Portuguese for "beautiful"), Taiwan is separated from South China by the 100-mile (160-km) wide Taiwan, or Formosa, Strait (see Figure 13.1). The island is nearly 14,000 square miles (c. 36,179 sq km) in area and is home to 22.6 million people, largely Chinese in origin. High mountains rise steeply from the sea on the eastern side of the island but slope on the west to a broad coastal plain (Figure 13.11). Taiwan lies on the northern margin of the tropics, and its temperatures are further moderated by the warm, northward-flowing Kuroshio, or Japanese, Current. Irrigated rice, the main food crop, is favored by the hot, humid summers in the lowlands and by abundant irrigation water from streams originating in the mountains. Sugar and pineapple are also important. The Japanese governed the island from 1895 to 1945, developing it mainly as a sugar-producing colony.

Driven from the mainland in 1949, the Chinese Nationalist government fled to Taiwan with remnants of its armed forces and many of its adherents. Nearly 2 million Chinese were in this migration. Here, protected by American sea power, the government reestablished itself as the Republic of China with its capital at Taipei (population: 7.8 million). The Nationalists, originally an authoritarian regime but with in-

Figure 13.11 Buddhist temple in the Taroko Gorge, Taroko National Park, Taiwan

creasing elements of political democracy, were very successful in fostering industrial development. They united inexpensive Taiwanese labor with foreign capital to build one of Asia's first urban-industrial countries and one of its Asian Tigers. Its major exports are machinery, electronics (especially computers), metals, textiles, plastics, and chemicals, especially to mainland China and the United States. Taiwan's per capita GDP PPP income is about $23,400, or over 4 times that of the People's Republic. The population is 78 percent urban, its infant mortality rate is very low, and its birth rate is one of the lowest in Asia (see Table 10.1). In almost all respects, Taiwan is clearly recognizable as a more developed country. A major hurdle against even stronger growth has been a lack of energy resources. Some coal and natural gas exist, and some hydropower has been harnessed, but the island depends heavily on oil imports.

Taiwan's Republic of China continues its claim as the legitimate government of China. Meanwhile, the People's Republic of China claims Taiwan as part of its own historic, and contemporary, territory. The United States backed the Nationalist claim for a time, but during the 1970s, the U.S. and the mainland People's Republic developed closer relations. After years of opposition, in 1971 the United States supported the revocation of Taiwan's seat in the United Nations and its replacement by the People's Republic. In 1979, bowing to the wishes of the People's Republic, the United States withdrew its official recognition of Taiwan, instead recognizing China's claim of sovereignty over Taiwan. This **One China Policy** prevents the United States from having formal diplomatic relations with Taiwan. However, the United States opposes the annexation of Taiwan to China by force, vowing to defend Taiwan from attack, just as it opposed Taiwan's call for a Nationalist effort to retake the Chinese mainland during periods of internal weakness there. In the meantime, the United States is a strong backer of Taiwan, supplying it with generous packages of weapons and economic aid.

WORLD
REGIONAL
Geography ⊗ Now™

Click Geography Literacy for links to the text of the China–US August 17 Communiqué and information about Taiwan–China relations.

Taiwan continues to resist mainland China's overtures for reunification, which include promises of broad autonomy, and is moving increasingly away from any type of reunion. That trend reflects Taiwan's ethnic makeup. Less than 15 percent of the island's people descend from the Nationalist refugees, who with their recent ties to the mainland are known as "mainlanders." Indigenous non-Chinese people make up about 2 percent of the population. The vast majority are descendents of Han peoples who emigrated from China as many as four centuries ago, notably the Fujianese who came from the southeastern coastal province of Fujian and the Hakka from Guangdong Province. This majority is much less mainland China-oriented in political outlook. Despite discrimination by the Nationalists, these "indigenous" Chinese of Taiwan have enjoyed increasing political power. In 2000 elections, their Democratic Progressive Party defeated the Nationalists for the first time in the country's history. With his reelection in 2004, President Chen Shui-bian of the Democratic Progressive Party came under pressure to fulfill his campaign vows to achieve formal independence for Taiwan within 4 years.

With Chen's election, the relationship between Taiwan and China became very tense. The two countries must confront serious choices about economic and political relations. Taiwan's economy has become very dependent on mainland China, where Taiwanese companies take advantage of low-cost mainland labor (about 80 percent cheaper than in Taiwan) for manufacture and assembly of products and where they have a huge market for their products. But there is strong pressure on President Chen to sever most ties with China and assert Taiwan's independence more forcefully. These steps could have large economic consequences and invoke military intervention from Beijing—and an American military response.

13.7 Mongolia

Mongolia (until recently, the Mongolian People's Republic and popularly referred to as Outer Mongolia) was once part of the Chinese empire, but it became a separate Communist country in 1924. Relations with the former Soviet Union were close, both politically and economically. Mongolia had the distinction of being the first Asian country to abandon

communism (in 1990). It has an area of about 600,000 square miles (960,000 sq km), with a population of about 2.5 million. Roughly 85 percent of the population is ethnic Mongol, and the largest minority (7 percent) is Turkic. Almost the entire population follows Tibetan (Lamaist) Buddhism. About 60 percent of Mongolia's population is urban. Ulan Bator (population: 820,000) is the capital and major urban center.

Mongolia was the home of the Genghis Khan and Kublai Khan who, in the second half of the 13th century, created the largest land kingdom history has ever known. From the Korean Peninsula in the east, all across the top of Asia to near the eastern border of the North European Plain, the Mongols—without benefit of a written language, cities, or most of the trappings associated with "civilization"—controlled an empire of major magnitude. And although they could not maintain that expansive empire, the Mongols were able to take over China in 1279 and control it until 1368, when the Ming Dynasty ascended.

The world's largest landlocked country, Mongolia contains large desert plains in the south and east, locally termed *gobis* (hence, the Gobi Desert), mountain ranges in the west, and wooded hills and mountains to the north. Most of the grassy valleys that are the archetypal steppes of Asia are also in the north (Figure 13.12). These steppes are favored lands for the vast herds of livestock, mainly sheep and goats, that Mongolians graze. Much of the population still derives its livelihood from animal husbandry. The previous Communist government made major efforts to promote farming (especially of wheat, barley, and potatoes), but the total amount of cultivated land amounts to no more than 1 percent of Mongolia's area. Mongolia's export income comes mainly from mineral products (especially copper), livestock and animal products, and textiles, including cashmere wool.

Figure 13.12 Many lives and much lore of the Mongolian steppe are focused on horses.

Political and economic changes are progressing rapidly. In 1990, the first multiparty elections ever held resulted in a coalition government committed to changing the Soviet-style "command economy" to a market economy. In the 2000 election, almost all seats in the Mongolian parliament were won by the Mongolian People's Revolutionary Party, characterized as a reformed Communist Party with democratic socialist inclinations. New private businesses are being established, and many state-owned enterprises and properties are being privatized. Commercial banks, business schools, and a stock market have opened, and most Russian technical and military personnel have left. Much economic hardship is accompanying the changeover. There is tension between the Mongolian majority and the Chinese minority that so far has prospered from capitalist reforms. In 2003, the government began the privatization of what was once the exclusively government-owned land, with every Mongolian now entitled to a free plot of land. There are fears that the purchase and parceling out of the grasslands will bring an end to Mongolia's ancient nomadic way of life, which is an extensive form of land use requiring unbridled access to pasture.

From Mongolia, the text turns to two lands inhabited mainly by distant kin of the Mongols: Japan and Korea. Both have had complex and often tense relations with China, which cast a large shadow on the western Pacific.

SUMMARY

- China is home to approximately 20 percent of the world's population on about 7 percent of the world's land. With 3.7 million square miles (9.6 million sq km) in area, it is the world's third largest nation and is the largest nation in East Asia.

- China's language, literature, government, agriculture, and religion have all been major cultural traits diffused from China to other parts of East Asia during the last several millennia.

- China was subjected to colonial interests in much of the 19th century, including Britain's successful bid to create large-scale and profitable opium addiction. Several European countries and Japan won territorial and trade concessions from China that did not end until after World War II with the Communist defeat of the Nationalists.

- China may be broadly divided into the arid west and the humid east physical regions. An arc drawn from Kunming in Yunnan north-northeast to Harbin is the rough dividing line between the arid west and the humid east. Population distribution is highly asymmetrical. The majority ethnic Han and non-Han realms are associated, respectively, with the more densely populated humid east and arid west. The source areas of the Huang He (Yellow River) and the Chang Jiang (Yangtze River) are in the Tibetan highlands.

- Humid China, also known as China Proper, is densely settled, focused on irrigated rice production, and highly dependent on the Chang Jiang and other river systems. This region is further subdivided into North China and South China. In subtropical South China, the agriculture is centered on rice. North China, which is drier than South China and has a more continental climate, is more dependent on winter wheat, spring wheat, corn, soybeans, and sorghum. The Huang He (Yellow River) deposits beneficial silt on the North China Plain but because of regular flooding is also called "China's Sorrow."

- The Northeast, historically called Manchuria by the West, was the early industrial heartland of China. It has coal, iron ore, petroleum, and well-developed transportation systems.

- The Three Gorges Dam, under construction on the Chang Jiang, is designed to control floods, provide hydropower, and allow transfer of water to the more arid north. The project has been both celebrated and lamented because of the scale of environmental change and social dislocation associated with it.

- Revolutionary campaigns in the first three decades of government after the 1949 defeat of the Chinese Nationalists led to experiments in communal agriculture, migration control, and backyard industrialization, all of which were less significant than China's opening to the Western free-market economy after the death of Mao in 1976. Since then, China has modeled much of its most dynamic economic growth after capitalist rather than communist economic and political models.

- Chinese agriculture in the past four decades has involved expansion of farmland, increase in irrigated land, opening new land in areas not previously farmed, and the use of more chemical fertilizers, new plant stock, and slowly increasing agricultural mechanization. There has also been much capital spent in increasing the railroad network and highway systems and in upgrading the continued use of canal and river systems.

- China's urban centers have continued to grow, with China now about 41 percent urban. China's aggressive migration management policies of the 1960s and 1970s have relaxed to allow inexpensive labor to help fuel the booming industries of the coastal cities. There is still considerable poverty in China, with the cities generally wealthier than rural areas and a minority that is becoming prosperous while the majority lags. The most prosperous areas are in the southeast, east coast, and farmlands near large cities, while regions in the west, southwest, uplands, and Arid China have dramatically less evidence of economic development.

- Taiwan (the Republic of China) continues to claim to represent the true government of China and has a tense relationship with the mainland's People's Republic of China. Taiwan's current president leans toward less political accommodation with the mainland, but his country is heavily invested in mainland China's development. The United States does not have diplomatic relations with Taiwan but supports it militarily and economically.

♻ One of the first countries to abandon communism in the final de-

cade of the 20th century, Mongolia has privatized its industries and lands, perhaps threatening the pastoral nomadic livelihood for which the country is famous.

KEY TERMS + CONCEPTS

Terms in blue are also defined in the glossary.

autonomous regions (p. 362)
break-of-bulk point (p. 364)
Chang Jiang Water Transfer Project
  (p. 364)
"China's Sorrow" (p. 363)
Chinese Communist Party (CCP) (p. 356)
Chinese Revolution of 1911 (p. 356)
command economy (p. 370)
commune system (p. 367)
core and periphery (p. 356, 357)
Cultural Revolution (p. 371)

Dalai Lama (p. 362)
extraterritoriality (p. 356)
Four No's Campaign (p. 366)
genuine autonomy (p. 363)
Great Leap Forward (p. 367)
land empire (p. 356)
loess (p. 362)
Long March (p. 356)
multiple cropping (p. 366)
Nationalist Party (Kuomintang, or KMT)
  (p. 356)

One China Policy (p. 375)
one country, two systems (p.373)
salinization (p. 366)
severe acute respiratory syndrome (SARS)
  (p. 368)
Special Economic Zones (SEZs) (p. 373)
sweatshops (p. 374)
treaty ports (p. 356)
Western Big Development Project (p. 362)

REVIEW QUESTIONS

WORLD
REGIONAL
Geography⊗Now™

Assess your understanding of this chapter's topics with additional quizzing and concept-based problems at http://earthscience.brookscole.com/wrg5e.

1. What are some of the innovations that began in China and spread through eastern Asia and even further afield?

2. What were some of China's experiences with foreign powers prior to 1949?

3. What were some of the major events and other milestones of China in the 20th century? Who were the major political figures associated with these events?

4. What are some of the major differences in physical and cultural geographies of Arid and Humid China? Of North and South China? What are the distinguishing characteristics of China's Northeast?

5. For what reasons were the Great Wall and the Grand Canal built?

6. What were the Great Leap Forward, the Commune Movement, and the Cultural Revolution?

7. How can China's economy be characterized? What is its "average" standard of living based on per capita GDP PPP? How effectively is the wealth distributed? Where is it concentrated geographically?

8. Where are the main clusters of China's cities? What geographic and historical factors help explain these concentrations?

9. Where are China's Special Economic Zones? What does their distribution suggest about China's economic geography?

10. What are the major environmental, economic, and political attributes of Taiwan and Mongolia?

DISCUSSION QUESTIONS

1. What were the major programs and other hallmarks of Communist Chinese reforms in agriculture and in industry?

2. Discuss China's efforts to "colonize" its western regions, particularly Tibet and Xinjiang. To what extent are these successful? How have the demographic and cultural characteristics of these areas been affected?

3. Discuss the pros and cons of the Three Gorges Dam Project and review the goals and routes of the Chang Jiang Water Transfer Project.

4. How did the SARS virus probably originate? How did it spread? What were some of its consequence on public health and on some economies?

5. What was the commune system? How effective was it? What happened to China's agriculture when the system was reformed?

6. What are some of the opportunities and costs posed by China's large coal reserves? (You may also want to consider the discussion of global warming in Chapter 2.)

7. China's economy has been booming, but what are the risks and possible repercussions of an economic collapse?

8. Why are people moving from rural areas of China to the cities of the east and southeast? What kind of social and living conditions do these migrants find in their new settings?

9. Discuss relations between Taiwan and mainland China. What position has the United States taken in this relationship? What are its main interests in the region?

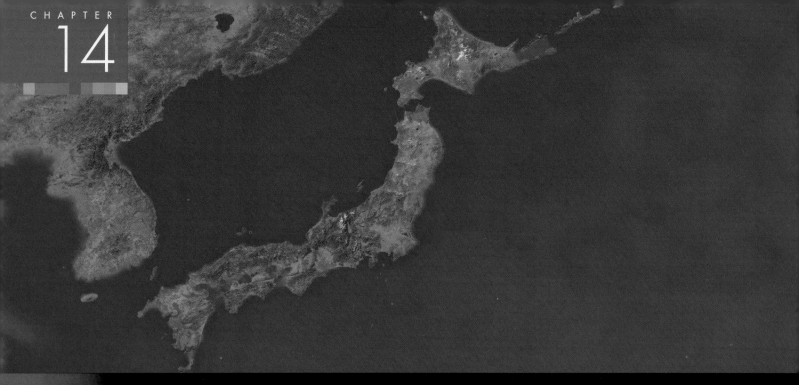

Japan and the Koreas: Adversity and Prosperity in the Western Pacific

Joe Hobbs

A Korean boy. Birth rates have been plummeting in both South Korea and Japan, where youngsters will grow up to bear the burden of looking after the elderly.

chapter objectives

This chapter should enable you to:

- Recognize the pattern of natural hazards associated with locations where plates of the earth's crust collide

- Appreciate how countries burdened by a lack of natural resources can become prosperous by marshaling their human resources

- Understand the unique challenges associated with a post-industrial society that has a prosperous and aging population

- See how a country's geographic location can make it a target of division and conquest

- Appreciate how different political and economic systems can produce dramatically different results for almost identical peoples

WORLD REGIONAL
Geography ⊕ Now™

Look for this logo in the text and go to GeographyNow at http://earthscience.brookscole.com/wrg5e to explore interactive maps, view animations, sharpen your factual knowledge and geographic literacy, and test your critical thinking and analytical skills with unique interactive resources.

On a cluster of islands off the eastern coast of Asia, the 128 million people of Japan operate an economy smaller only than that of the United States. This powerful economic engine is on mountainous, resource-poor, and crowded lands about the size of California, yet holding almost half as many people as the entire United States. Japan's dramatic rise from the ashes of World War II has drastically changed the world's economic geography. This chapter explains that phenomenal achievement and some of the challenges associated with it.

Although the Korean Peninsula has a seemingly marginal position on the Pacific Rim and the Eurasian landmass, it, too, has a central place in geopolitical affairs, mostly because of its strategic location between Japan, China, and Russia, the prosperity of South Korea, and the sometimes dangerous backwardness of North Korea. This chapter examines the problems of divided Korea and its prospects for reunification.

14.1 The Japanese Homeland

There are four main islands in the Japanese archipelago (Figure 14.1). On the west, the Sea of Japan—which Koreans know as the East Sea—and the Korea Strait separate the main islands from Russia and Korea. Japan faces the open Pacific on the east. Hokkaido, the northernmost island, is close to Russian-controlled islands to its north. The scenic and busy Inland Sea separates Honshu, the largest island,

from Shikoku and Kyushu. South of Kyushu, the smaller Ryukyu Islands run southwestward almost to Taiwan.

All of the main islands are mountainous, with the higher peaks between 5,000 and 9,000 feet (c. 1,500 to 2,750 m) above sea level. The islands are volcanic (Figure 14.2) and are subject to frequent and sometimes severe earthquakes (see Natural Hazards, pages 384–385). Located west of Tokyo, the volcano Mt. Fuji, whose name means "Fire Goddess," is Japan's highest mountain (12,388 ft/3,761 m; Figure 14.3). An **active volcano** (one that has had an eruption in recorded history), in the early 2000s Fuji began experiencing "swarms" of earthquakes that many geologists interpreted as signs of an approaching eruption; it would be the first since 1707. Fuji was sacred to the aboriginal Ainu people of Japan as long as 2,000 years ago. It is still a holy mountain in the national Shinto faith, and about 300,000 pilgrims and tourists visit its lofty summit each July and August, when it is free of snow. The journey to the peak of this most prominent symbol of Japan is a difficult one, as suggested in the Japanese proverb, "It is as foolish not to climb Fujisan as to climb it twice in a lifetime."

Climate and Resources

Japan's location east of the Eurasian landmass at latitudes comparable to those from South Carolina to Maine strongly influences its climates. A humid subtropical climate extends from southernmost Kyushu to north of Tokyo, a humid continental long-summer climate characterizes northern

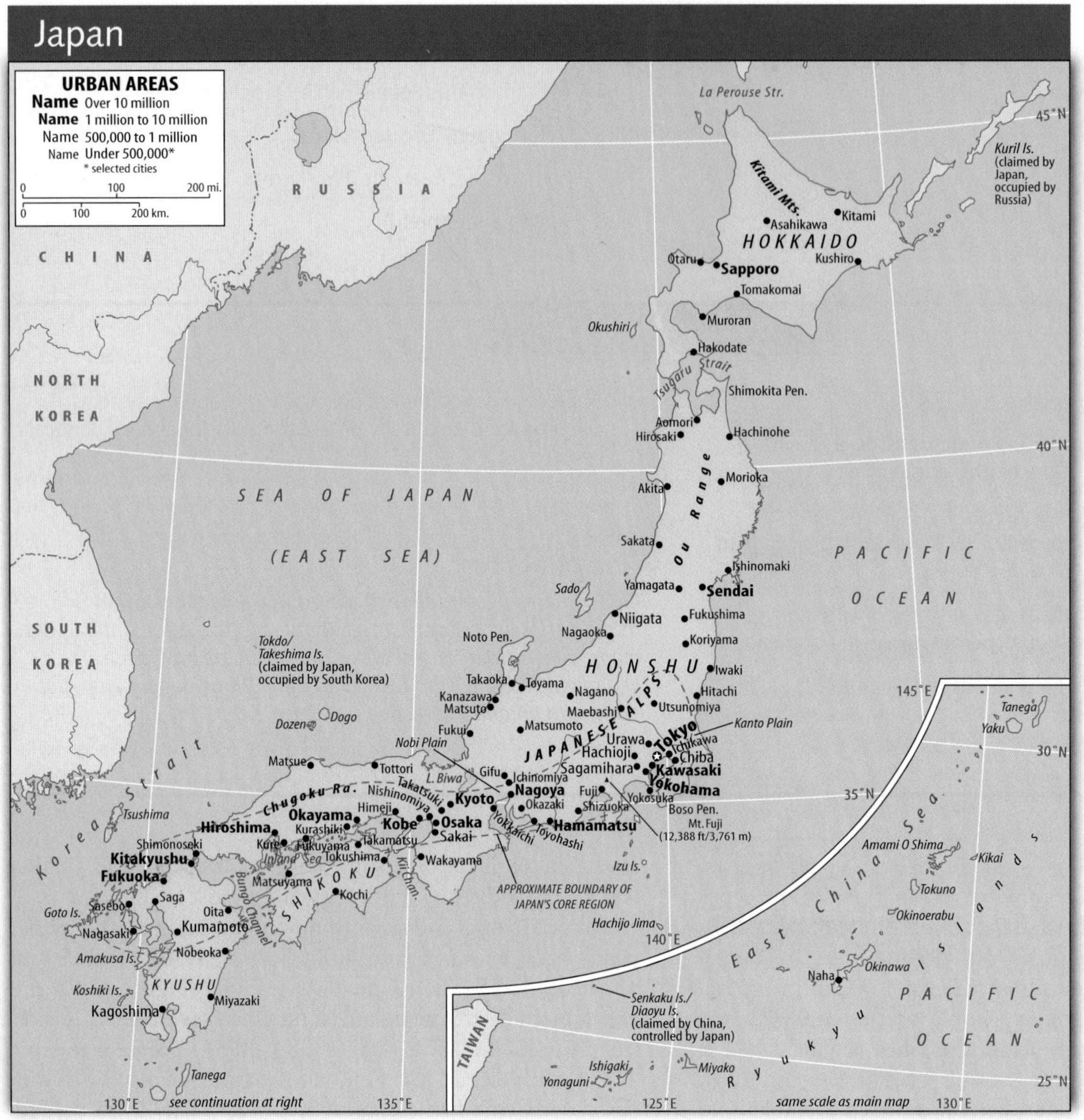

Japan

URBAN AREAS
Name Over 10 million
Name 1 million to 10 million
Name 500,000 to 1 million
Name Under 500,000*
* selected cities

Figure 14.1 Principal features of Japan

Honshu, and Hokkaido has a humid continental short-summer climate. In lowland locations, where most of the people live, the average temperatures for the coldest month range from about 20°F (−7°C) in northern Hokkaido to about 40°F (4°C) in the vicinity of Tokyo and about 44°F (7°C) in southern Kyushu. Averages for the hottest month are only about 64°F (18°C) in northern Hokkaido but rise southward to about 80°F (27°C) around Tokyo and above 80°F farther south. Conditions are cooler in the mountains, with snowfalls common in winter in Honshu and Hokkaido. Hokkaido has twice been the venue for the Winter Olympics.

Japan is a humid country. Annual precipitation of more than 60 inches (c. 150 cm) is common from central Honshu south, and 40 inches (c. 100 cm) or more fall in northern Honshu and Hokkaido. Their location off the east coast of Eurasia puts the islands in the path of the inward-blowing **summer monsoon** from the Pacific, and summer is the main rainy season in eastern Japan. But the islands are also exposed to the seaward-blowing **winter monsoon** after it has crossed the Sea of Japan, so much precipitation also occurs in the winter in western Japan. Japan is also subjected in summer and fall to powerful **typhoons,** the great low-pressure storm systems known as hurricanes in the western Atlantic region. These add to the woes of this hazard-prone country.

Japan's natural resource base is not large enough to meet the country's growing needs. Up to the middle of the 19th century, the country was nearly self-sufficient economically, although the standard of living was low. Japan was already crowded with about 30 million people, but disease, starvation, abortion, and infanticide slowed the rate of natural increase. Industrialization stimulated rapid population growth

Figure 14.2 Japan has numerous volcanoes—some active and others extinct. In 1991, Mt. Unzen, near Nagasaki in Kyushu, erupted for the first time since 1792. Thirty-five people were killed as a pyroclastic flow inundated the village of Shimabara below.

Figure 14.3 Volcanic Mt. Fuji loses its snow cover by late summer.

raw materials and energy supplies and about 60 percent of its food.

Forests, which now cover about 60 percent of the land surface, illustrate the limitations of Japan's natural resources. In a preindustrial and early industrial society, with a much smaller population, the nation's heavily forested mountainsides supplied the needed wood, but if today's demand for forest products were met with domestic reserves, the country would be rapidly deforested. There is a wide range of valuable trees, mainly broadleaf deciduous toward the south and needleleaf coniferous toward the north. Many mountainsides are quilted with rectangular plots of trees planted in rows. Most first-growth timber has long since gone, but some primary forests are still protected in preserves. Forests are important in Japanese culture, and this affluent country does not wish to lose them. With its enormous demand for wood products, Japan is therefore the world's largest importer of wood, mostly hardwoods from the tropical rain forests of Southeast Asia.

Kerosene stoves have largely replaced the charcoal used formerly for household cooking and heating, but the Japanese still use wood for many purposes (Figure 14.4). Japan's large publishing industry and other industries and handicrafts require a great deal of paper. Single-family dwellings continue to be built largely of wood, which provides more safety from earthquake shock waves than more rigid construction. However, wood makes the dwellings more vulnerable to the fires that typically sweep through Japanese cities after earthquakes rupture gas lines.

Small deposits of minerals, including coal, iron ore, sulfur, silver, zinc, copper, tungsten, and manganese, supported early

Figure 14.4 Wood is a vital resource in Japan. This wooden home is in Shirakawa, north of Tokyo on the island of Honshu.

in the second half of the 19th century and later. Population growth has recently almost ceased, however, as delayed marriages and the rising costs of child rearing have lowered birth rates. Meanwhile, affluence has increased dramatically. The material demands of so many people enjoying a high standard of living in a country with limited natural resources require Japan to import an astonishing 97 percent of its

Natural Hazards

Living on the Ring of Fire

The people of Japan must cope with one of the most hazard-prone regions on Earth. By far the greatest threats are related to Japan's location on the Pacific Ring of Fire, the name given to the boundaries of the Pacific plate of Earth's crust where earthquakes and volcanic eruptions are commonplace and often catastrophic (Figure 14.A). In the scheme of plate tectonics, Japan is in an area of *subduction* where one plate (the Pacific) is being pushed, as if by a conveyor belt, under another plate (the Eurasian), creating the country's characteristic volcanoes, earthquakes, and related features (see Figure 2.9, page 26).

Japan is most threatened by earthquakes on and near its islands but must also be on guard for the effects of distant earthquakes, even as far away as the shore of North America. A strong quake there can produce a **tsunami** (meaning "harbor wave" and commonly called a "tidal wave") that may travel thousands of miles to strike and flood coastal Japan.

Japan has a long history of seismic assaults. One of the most significant recent events took place on January 17, 1995, when a devastating earthquake released the tension that had built up on a transverse fault near historic Kobe, a port city of 1.5 million people. The Japanese soon labeled the 6.9-magnitude episode the Great Hanshin Earthquake, establishing its place in the annals of Japanese disasters alongside such events as the Great Kanto Earthquake (magnitude 7.9 on the Richter scale), which killed 143,000 people in and around Tokyo in 1923. Killing 6,400 people and injuring thousands more, damaging 190,000 buildings, displacing 300,000 people, and causing at least $100 billion worth of damage, the Kobe quake was the worst disaster to hit Japan since World War II. Kobe's remarkable lack of disaster preparedness and relief focused attention on a common Japanese complaint: that the government invests too much in commerce-oriented development and not enough in the welfare of its people.

Japanese authorities did not predict the Great Hanshin Earthquake. After the fact, there were reports that large schools of fish swam uncharacteristically close to the surface of the sea off Kobe in the days preceding the quake. Crows and pigeons reportedly were noisier than usual and flew erratically in the hours before the ground moved. In the weeks preceding the quake, levels of radon gas rose in a water aquifer below the city and then declined sharply after January 9. Animal behavior, water level and composition, and many other variables are often debated but still unproven barometers of an impending earthquake.

Since 1965, Japan has spent more than $1.3 billion in an effort to develop means of predicting earthquakes. During the 1990s, more than $100 million yearly went to this effort. But by 1998, no earthquake had been successfully predicted, and Japan suspended this expensive research program. Its critics argue that earthquakes are chaotic events and will always be impossible to predict. They also point out the hazards of issuing an earthquake warning that is not followed by a real quake. A false alarm in the Tokyo region would cost an estimated $7 billion per day in lost business and other economic disruptions. The bigger question is how such a great city could be evacuated safely, efficiently, and with minimal panic.

With the prediction program scrubbed, Japan has redoubled its efforts to prepare for future big ones. New structures must meet a tough safety code that aims at "earthquake-proof" construction. Much of the focus is on the Japanese capital, Tokyo, which is underlain by a complex system of active fault lines and which typically suffers a big quake every 70 years (in 2005, it had been more than 80 years since the last one). Current estimates are that a big one nearing 8.0 on the Richter scale would kill 6,717 people and destroy 300,000 structures (mostly by fire) in Tokyo. These numbers do not count the impacts of the estimated 27-foot-high (8-m) tsunami that would strike low-lying Tokyo within minutes of an offshore earthquake. Much of the devastation in Kobe resulted from the city's location on soft coastal soils and reclaimed lands, which also underlie great parts of Tokyo.

With such threats in mind, the government began serious discussions about moving most national administrative functions inland to a new capital, where tsunamis are not an issue and where mountain bedrock would reduce the devastating im-

industrialization in Japan. These minerals are in short supply today, and like most other raw materials, they must be imported. Japan's mountain streams provided much of the power for early industrialization, but they are so swift, shallow, and rocky that they are of little use for navigation. Most of the water used for irrigation in Japan comes from these streams.

Only about one-eighth of this small country's area is arable; Japan is not one of the world's major crop-growing nations. Most cultivable land is in mountain basins and in small plains along the coast. Terraced fields on mountain-sides supplement this level land. Irrigated rice, the country's basic food and most important crop, was once nearly everywhere on arable land (Figure 14.5). Production of this grain is heavily subsidized by the Japanese government and promoted by a powerful farm lobby, which aims at rice self-sufficiency. The country's farmers usually produce a net export of rice. However, in the 1990s, a scarcity of rice throughout East Asia and pressure on Japan to reduce its massive trade surplus with the United States prompted Japan to open its rice market to foreign imports for the first time.

pacts of seismic shock waves. Two sites on the island of Honshu have the most attention as possible new capitals: Hokuto, in the northeast near Sendai, and Tokai, west of Tokyo near Nagoya. Japan's struggle against the seismically inevitable is a national priority and a symbol of the nation's determination to maintain economic success despite environmental adversity.

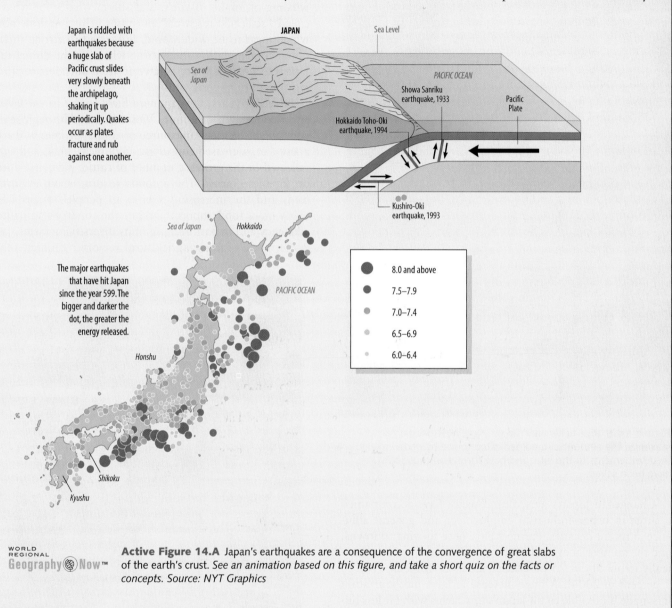

Japan is riddled with earthquakes because a huge slab of Pacific crust slides very slowly beneath the archipelago, shaking it up periodically. Quakes occur as plates fracture and rub against one another.

JAPAN

Sea Level

Sea of Japan

PACIFIC OCEAN

Showa Sanriku earthquake, 1933

Pacific Plate

Hokkaido Toho-Oki earthquake, 1994

Kushiro-Oki earthquake, 1993

Sea of Japan   Hokkaido

PACIFIC OCEAN

The major earthquakes that have hit Japan since the year 599. The bigger and darker the dot, the greater the energy released.

Honshu

8.0 and above

7.5–7.9

7.0–7.4

6.5–6.9

6.0–6.4

Shikoku

Kyushu

Active Figure 14.A Japan's earthquakes are a consequence of the convergence of great slabs of the earth's crust. *See an animation based on this figure, and take a short quiz on the facts or concepts. Source: NYT Graphics*

WORLD REGIONAL Geography Now™

Now, with silos overflowing with rice, and with the Japanese diet more dependent on wheat (in the form of bread and noodles) than on rice, the government is paying Japanese farmers to not grow rice. Japan's lavish subsidies of its declining, aging population of farmers are the subject of much condemnation by free-trade advocates.

Mild winters and ample moisture permit double cropping—the growing of two crops a year on the same field—on most irrigated land from central Honshu south. Rice is grown in summer, and wheat, barley, or another winter crop is planted after the rice harvest. Overall, about one-third of Japan's irrigated rice fields are sown to a second crop, and more than half of the unirrigated fields are double cropped. Intertillage—the growing of two or more crops simultaneously in alternate rows—is also common. In addition, dry farming produces a variety of crops: sugar beets in Hokkaido; all sorts of temperate-zone fruits and vegetables; potatoes, peas, and beans, including soybeans; and in the south, tea, citrus fruits, sugarcane, tobacco, peanuts, and mulberry trees to feed silkworms.

Figure 14.5 Rice fields near Tokyo. Note how people maximize the amount of land dedicated to rice by confining their settlements to the hills, and how forests on those hills have been saved.

Changes in Japan's agricultural sector have accompanied a surge in urban-industrial employment and general affluence in recent decades. Farmers have been moving to towns, and the number of agricultural workers has dropped dramatically (from about 25 percent of the labor force in 1962 to 7 percent in 2004). A huge increase in farm mechanization has made it possible to till Japan's tiny farms with far less human labor. With this technology and the heavy applications of fertilizers, plus improved crop varieties, Japanese agriculture has achieved the world's highest yields per unit of land.

The island people of Japan view the ocean as an important resource and promising frontier. The country has an ambitious "inner space" program of deep-sea exploration to mine the sea's living and mineral riches. Many food fish species are in the waters around Japan, and Japanese fishermen range widely through the world's major ocean fishing grounds. Japan takes more fish from the sea than any other nation—about 15 percent of the world's catch—and the Japanese have the world's highest per capita consumption

of fish and other marine foods, including *sashimi*, or raw seafood (Figure 14.6).

Historically, the Japanese have also enjoyed whale meat, but pressure from international environmentalists compelled Japan in 1985 to support a worldwide moratorium on commercial whaling. However, Japan has resumed small-scale whaling of minke, Bryde's, and sperm whales for what are called "scientific research" purposes. This loophole allows whale meat to be widely sold. Subjected to strong international condemnation for putting whale on the dinner table, Japanese whaling interests have accused outsiders of practicing "culinary imperialism."

In recent years, the Japanese have begun to eat less fish and rice in favor of a more Western diet. Although sea fish continue to be the main source of animal protein in the Japanese diet, increased consumption of meat, milk, and eggs is reflected in the growing number of cattle, pigs, and poultry on Japanese farms. The animals require heavy imports of feed, and the increasing variety in people's tastes means many more food imports. Despite the country's agricultural efficiency and in part because of its diversifying tastes, Japan is the world's largest agricultural importer country, relying

Figure 14.6 Seafood is a major element in the Japanese diet. This is the Tsukiji fish market in Tokyo.

on imports for 60 percent of its food needs. The United States is Japan's largest supplier of imported feedstuffs and food, and the United States exports more food to Japan than to any other country.

Japan's Core Area

Most of Japan's economic activities and people are packed into a corridor about 700 miles (c. 1,100 km) long. This **megalopolis** extends from Tokyo, Yokohama, and the surrounding Kanto Plain on the island of Honshu in the east, through northern Shikoku, to northern Kyushu in the west (see Figure 14.1). This highly urbanized **core area** contains more people than live in the Boston to Washington, D.C., megalopolis on the U.S. east coast. All of Japan's greatest cities and many less important ones have developed here, on and near harbors along the Pacific and the Inland Sea. They are sited typically on small, agriculturally productive alluvial plains between the mountains and the sea.

The largest urban complex in the world—Tokyo–Yokohama and surrounding suburbs (population: 31.2 million by one estimate; figures differ according to how the metropolitan area is defined)—has developed on the Kanto Plain, Japan's most extensive lowland. Tokyo grew as Japan's political capital and Yokohama as the area's main seaport. This urban complex functions today as Japan's national capital, one of its two main seaport areas (the other is Osaka–Kobe), its leading industrial center, and the commercial center for northern Japan (Figure 14.7).

Beyond Tokyo, in the entire northern half of Japan, the only major city is Sapporo (population: 2.2 million) on Hokkaido. About 200 miles (c. 320 km) west of Tokyo lies Japan's second largest urban cluster, in the Kinki District at the head of the Inland Sea. Here, three cities—Osaka, Kobe, and Kyoto—form a metropolis, together with their suburbs, of about 18 million people.

The inland city of Kyoto was the country's capital from the eighth century A.D. until 1869 (when it was moved to

Figure 14.8 Kyoto has a unique status in Japanese history and religious life. This is the 14th century Kinkakuji Temple, also known as the Golden Pavilion.

Tokyo) and is now preserved as a shrine city where modern industrial disfigurement is prohibited (Figure 14.8). Some U.S. military planners wanted to target Kyoto with a nuclear weapon in August 1945, but the decision that it should be avoided because of its historical importance spared the city. Kyoto continues to be the destination of millions of pilgrims and tourists annually.

Osaka developed as an industrial center, and Kobe as the district's deepwater port. Both cities now combine port and industrial functions. Another large metropolis, Nagoya, is situated directly between Tokyo and Osaka on the Nobi Plain at the head of a bay. It is both a major industrial center and a seaport.

West of the Kinki District, smaller metropolitan cities spot the Coreland. The largest are metropolises of more than 1.5 million people: the resurrected Hiroshima on the Honshu side of the Inland Sea, Kitakyushu (a collective name for several cities) on the Strait of Shimonoseki between Honshu and Kyushu, and Fukuoka on Kyushu. Hiroshima was destroyed on August 6, 1945, when the United States dropped the first atomic bomb ever used in warfare, detonating it directly over the center of the city. With no topographic barriers to deter the effects of the explosion, the city was obliterated, and an estimated 140,000 died by the end of 1945, when radiation sickness had taken its greatest toll (Figure 14.9).

Among the several smaller cities of northern Kyushu is

Figure 14.7 Tokyo, the capital of an industrious and affluent nation, is the world's largest city and one of its most congested. This is rush hour at the Shinjuko Station.

Joe Hobbs

Figure 14.9 On August 6, 1945, Hiroshima became the symbol of all the horror that nuclear war might bring the world. These are materials fused by the heat of the thermonuclear explosion.

Nagasaki, which, on August 9, 1945, was hit by a second U.S. nuclear bomb, the last used in warfare. Nagasaki was chosen during the flight mission after clouds obscured the primary target, Kokura. The bomb detonated over the suburbs rather than the city center, and hills helped to diminish the explosion's impact, but even so, an estimated 70,000 died by year's end. The bombing of Nagasaki brought Japan's surrender and the end of World War II. American decision makers insisted that the use of these weapons spared many thousands of lives, both American and Japanese, that would have been lost if the United States had instead undertaken an invasion of the Japanese homeland. Japan, which has always utterly condemned the bombing, has an official policy that it will never develop or use atomic weapons.

14.2 Historical Background

Japan has seen successive periods of isolationism and expansionism and of economic and military accomplishment and defeat. Today, it is near the top of most indicators of prosperity in the more developed countries (MDCs). The remarkable success story of Japan can best be appreciated by considering its turbulent history and the difficult home environment in which the Japanese have lived.

The Japanese, Japan's dominant ethnic group, are descended from peoples who reached Japan from other parts of eastern Asia in the distant past, especially after 7000 B.C. Most of these ancestors of the modern Japanese may have arrived at about the same times as the **Ainu.** Often described as the original indigenous group of Japan, the Ainu (also called **Utari**), numbering perhaps 20,000 today, are unrelated to the ethnic Japanese. Their ethnic and geographic origins are disputed by scholars today. The majority Japanese gradually drove Ainu into outlying areas, mainly in Hokkaido.

The oldest surviving Japanese written records date from the eighth century A.D. These depict a society strongly influenced by Chinese cultural and ethnic traits, often reaching

Japan by way of Korea. A distinct Japanese culture gradually evolved, and today, it is linguistically and in other ways different from its Chinese and Korean antecedents. Traditions about Japanese origins extend back to the reign of Jimmu, the first emperor, whose accession is ascribed to the year 660 B.C., but who may be a mythical rather than historic figure. Traditions of kingship are important in the Japanese worldview, which says that the islands and their emperor have a divine origin. Jimmu was said to have descended from the Sun Goddess, and the Japanese have known their homeland as the "Land of the Gods." The Japanese concept of Japan as unique and invincible developed early, shaped by legends such as that of the **kamikaze,** the "divine wind" that repelled a Mongol attack on Japan in the 13th century.

The early emperors gradually extended control over their island realm. A society organized into warring clans emerged. By the 12th century, powerful military leaders called **shoguns,** who actually controlled the country, diminished the emperor's role. Meanwhile, nobles, or **daimyo,** whose power rested on the military prowess of their retainers, the **samurai,** ruled the provinces. This structure resembled the European feudal system of medieval times.

Adventurous Europeans reached Japan early in the mid-16th century, beginning with the Portuguese in the 1540s. Most of these early arrivals were merchants and Roman Catholic missionaries. Japanese administrators allowed the merchants to trade and the missionaries to preach freely. There were an estimated 300,000 Japanese Christians by the year 1600.

Contacts with the wider world were curtailed as Japan entered a period when military leaders imposed central authority on the disorderly feudal structure. After winning the battle of Seikigahara in 1603 with arms purchased from the Portuguese, warlord Ieyasu Tokugawa proclaimed himself Nihon Koku Taikun ("Tycoon of All Japan") and became the first shogun to rule the entire country. During the era of the Tokugawa Shogunate, from 1600 to 1868, the Tokugawa family acquired absolute power and shaped Japan according to its will. To maintain power and stability, the early Tokugawa shoguns wanted to eliminate all disturbing social influences, including foreign traders and missionaries. They feared that the missionaries were the forerunners of attempted conquest by Europeans, especially the Spanish, who held the Philippines. The shoguns drove the traders out and nearly eliminated Christianity in persecutions during the early 17th century. After 1641, a few Dutch traders were the only Westerners allowed in Japan. Their Japanese hosts segregated them on a small island in the harbor of Nagasaki, even supplying them with a brothel so that they would not be tempted to venture out and pollute Japan's ethnic integrity. Under Tokugawa rule, Japan settled into two centuries of isolation, peace, and stagnation.

Foreigners succeeded in reopening the country to trade two centuries later, when Japan was ill prepared to resist Western firepower. Visits in 1853 and 1854 by American naval squadrons of black ships under Commodore Matthew Perry (who wanted to establish refueling stations in Japan for

American whaling ships) resulted in treaties opening Japan to trade with the United States; this was **gunboat diplomacy.** The major European powers were soon able to obtain similar privileges. Some feudal authorities in southwestern Japan opposed accommodation with the West. United States, British, French, and Dutch ships responded in 1863 and 1864 by bombarding coastal areas under the control of these authorities. These events so weakened the Tokugawa Shogunate that a rebellion overthrew the ruling shogun in 1868. The revolutionary leaders restored the sovereignty of the emperor, who took the name Meiji, or "Enlightened Rule." The revolution of 1868 is known as the Meiji Restoration.

The men who came to power in 1868 aimed at a complete transformation of Japan's society and economy. They felt that if Japan were to avoid falling under the control of Western nations, its military impotence would have to be remedied. This would require a reconstruction of the Japanese economy and social order. They approached these tasks energetically and abolished feudalism, but only after a bloody revolt in 1877. The Meiji leaders forged the model of a strong central government that would modernize the country while resisting foreign encroachment. Under this system, which lasted until 1945, democracy was strictly limited, with small groups of powerful men manipulating the government's machinery and the emperor's prestige. Military leaders were prominent in the power structure.

Following their fact-finding tours of Europe and the United States, Meiji leaders pressed their people to learn and apply the knowledge and techniques that Western countries had accumulated during centuries of Japan's isolation. Foreign scholars were brought to Japan, and Japanese students were sent abroad in large numbers. Japan adopted a constitution modeled after that of imperial Germany. The legal system was reformed to be more in line with Western systems. The government used its financial power, which it obtained from oppressive land taxes, to foster industry. New developments included railroads, telegraph lines, a merchant marine, light and heavy industries, and banks. Wherever private interests lacked capital for economic development, the government provided subsidies to companies or built and operated plants until private concerns could acquire them. Rapid urbanization accompanied a process of rapid industrialization.

So spectacular were the results of the Meiji Restoration that within 40 years Japan had become the first Asian nation in modern times to emerge as a world power. Outsiders often spoke of the Japanese in derogatory terms as mere imitators. The Japanese, however, knew which elements of Western technology they wanted, and they adapted them successfully to Japanese needs.

The shift from Tokugawa isolationism to Meiji openness put Japan on a pathway to empire and eventually to war. Japan was bent on expansion between the early 1870s and World War II, and by 1941 controlled one of the world's most imposing empires (Figure 14.10). Between 1875 and 1879, Japan absorbed the strategic outlying archipelagoes of the Kuril, Bonin, and Ryukyu Islands. Japan won its first acquisitions on the Asian mainland by defeating China in the Sino-Japanese War of 1894–1895, which broke out over disputes in Korea. In the humiliating 1895 Treaty of Shimonoseki, China ceded the island of Taiwan (Formosa) and

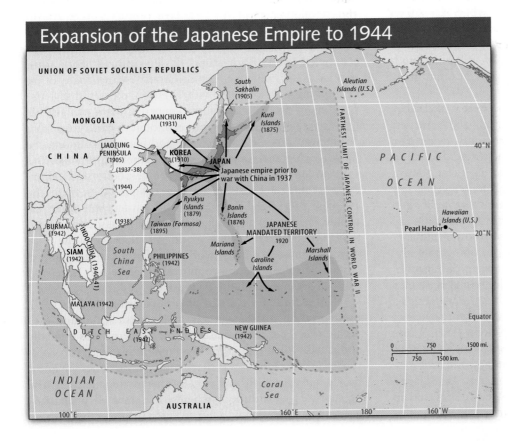

Expansion of the Japanese Empire to 1944

Active Figure 14.10 Overseas areas held by Japan prior to 1937 and the line of maximum Japanese advance in World War II. *See an animation based on this figure, and take a short quiz on the facts or concepts.*

Manchuria's Liaotung Peninsula to Japan and recognized the independence of Korea.

Feeling threatened by Japan's emerging strength so close to its vital port of Vladivostok, Russia demanded and won a Japanese withdrawal from the Liaotung Peninsula and leased the strategic peninsula from China. Japan went to war with Russia in 1904 by attacking the peninsula's main settlement of Port Arthur. Russia lost the ensuing Battle of Mukden, which involved more than a half-million soldiers, making it history's largest battle to that date. Russia's loss of the Russo-Japanese War of 1904 represented the first defeat of a European power by a non-European power. Victorious Japan regained the Liaotung Peninsula, secured southern Sakhalin Island, and established Korea as a Japanese protectorate; Korea was formally annexed to Japan in 1910.

At the end of World War I, the Caroline, Mariana, and Marshall Islands were transferred from defeated Germany to Japan as mandated territory. Japanese encroachments on China's region of Manchuria followed in the 1930s. In 1937, Japan began an all-out attack that overran the most populous parts of China. Japanese troops committed many atrocities in the assault on China, even employing biological warfare by airdropping plague-infested fleas on the city of Ningbo. The most notorious was the **Rape of Nanking** (now Nanjing) in 1937, when the invaders took control of China's temporary capital, killing an estimated 300,000 Chinese civilians and soldiers and raping about 20,000 Chinese women. All told, an estimated 2.7 million Chinese civilians were killed in Japan's effort to "pacify" China.

The encroachments continued. Japan seized French Indochina in 1940. After Japan's carrier-based air attack devastated the American Pacific Fleet at Pearl Harbor, Hawaii, on December 7, 1941 ("a day which will live in infamy," declared U.S. President Franklin Roosevelt), Japanese forces rapidly overran Southeast Asia as far west as Burma (modern Myanmar), as well as much of New Guinea and many smaller Pacific islands.

The motives for Japanese expansionism were mixed. They included a perception of national superiority and "manifest destiny," perceived security needs, a desire for recognition by the great powers, the wishes of military leaders to inflate their importance and gain control of the Japanese government, and ambitions of industrialists to gain sources of raw materials (Manchurian and Korean coal, for example) and markets for Japanese industries. The desire for materials and markets was based solidly on need. Expansion of industry and population on an inadequate base of domestic natural resources had made Japan dependent on sales of industrial products outside the homeland. Such sales provided, as they do now, the principal funds for buying necessary imported foods, fuels, and materials. As worldwide depression hurt the trade-dependent prewar Japanese economy, Japan chose to create a Japanese-controlled Asian realm that would insulate the Japanese economy from the volatile world economy and make the country a leading world power. In 1938, Japan therefore proclaimed its **New Order in East Asia,** and in 1940 and 1941, this widened into the **Greater East Asia Co-**

Prosperity Sphere, a euphemism for Japanese political and economic control over China and Southeast Asia.

Japan's militaristic and colonial enterprises proved disastrous at home and abroad. Japanese soldiers inflicted great suffering on civilians throughout the western Pacific. Even now, nations are demanding formal apologies from Japan. In August 1995, on the 50th anniversary of Japan's defeat in World War II, Japan's prime minister did issue a formal general apology for his nation's role in the war. Critics say Japan has not fully atoned for its war crimes and are trying, among other efforts, to block Japan's quest to gain a permanent seat on the United Nations Security Council.

August 1945 found Japan completely defeated. Its overseas territories, acquired during nearly 70 years of successful imperialism, were lost. Its great cities, including Tokyo—where 100,000 people died in napalm bombings—were in ruin. In addition to 1.8 million deaths among its armed services, Japan had suffered some 8 million civilian casualties from the American bombing of the home islands; most of its major cities were over half destroyed. Tokyo's population had fallen from nearly 7 million people to about 3 million, most of them living in shacks. In 1950, 5 years after the beginning of the U.S. military occupation of Japan, national production stood at only about one-third of its 1931 level, and annual per capita income was $32.

14.3 Japan's Postwar Miracle

Without colonies or empire, Japan has become an economic superpower since World War II. The nation's explosive economic growth after its defeat was one of the most remarkable developments of the late 20th century and is known widely as the Japanese "miracle."

Observers of Japan cite different reasons for the country's economic success. Proponents of dependency theory argue that Japan, never having been colonized, escaped many of the debilitating relationships with Western powers that hampered many potentially wealthy countries. Some analysts believe that the country's postwar economic miracle grew from an intense spirit of achievement and enterprise among the Japanese. Notably, many Japanese attribute this industrious spirit to Japan's geography as a resource-poor island nation. To overcome the constraints nature has placed on them, the Japanese people feel they must work harder. Still others attribute Japan's postwar achievements to its association with the United States following World War II. Some of the postwar U.S.-imposed reforms worked well economically, and relations between the two countries since the war have also generally stimulated Japan's industrial productivity.

The U.S. military occupation of Japan from 1945 to 1952 brought about major changes in Japanese social and economic life. The divine status of the emperor was officially abolished. When Emperor Hirohito announced Japan's surrender on the radio, it was the first time the stunned Japanese public had ever heard the voice of this mythic figure. A new U.S.-written constitution (which has remained unchanged ever since) made Japan a constitutional monarchy with an elective parliamentary government. With the rejection of divine monarchy, the strongly nationalistic Shinto faith lost its status as Japan's official religion, although worship at Shinto shrines was allowed to continue; most Japanese observe a faith that is a syncretism of Shinto and Buddhist beliefs and rituals. There was land reform in the countryside to do away with a feudal-style landlord system. Some major Japanese companies were broken up to reduce their monopolistic hold on the economy. Women were enfranchised. A democratic trade union movement was established.

Perhaps most significant for the economy, the United States forbade Japan to rearm, except for its small **Self-Defense Forces (SDFs)**. When the occupation ended in 1952, leaving U.S. bases in Japan but returning control of Japanese affairs to the Japanese, Japan was placed under American military protection. Without large military expenditures, much of Japan's capital was freed to invest in economic development. Only recently has Japan sought a wider role for its military—for example, by posting some forces to serve in postwar Iraq.

The U.S. military umbrella over Japan remains controversial today. Many Japanese regard it as an outmoded vestige of colonialism and the war. There have been several incidents of women being raped by U.S. soldiers stationed on the island of Okinawa, where three-quarters of the U.S. bases and more than half of the U.S. troops in Japan are stationed. The resulting protests and diplomatic appeals led the United States to agree to return control of a major airbase in Okinawa to Japan and to relocate some U.S. troops from Okinawa to mainland Japan. In a subsequent referendum, the people of Okinawa voted overwhelmingly in favor of further U.S. military withdrawal from the island.

Japan's special relationship with the United States boosted the economy long after the American occupation ended. The postwar economic recovery accelerated when the United States called on Japanese production of steel, textiles, and clothing to support American forces in the Korean War of 1950–1953. American economic aid continued to flow to Japan. The United States permitted many Japanese products to have free access to the American market and allowed Japan to protect its economy from imports. Japanese firms also had relatively free access to American technologies, which they often improved upon and marketed even before American firms did.

By the 1970s, Japan became an industrial giant with a gross national product far exceeding that of any countries except the United States and the Soviet Union. The United States and European nations found themselves at a disadvantage to Japanese competitors in many industries. They busily sought ways to match this competition or to protect themselves from it without doing too much damage to their exporting firms, consumers, and overall trade relations. By the 1980s, the United States and Japan were at odds over Japan's massive trade surplus with the United States.

Observers often point to several unique features of Japanese management and employment to explain the country's meteoric postwar gains. One is recruitment through an extremely challenging (some say brutal) educational system that emphasizes technical training. A rigorous and stressful testing system controls admission to higher education. Japanese management strategies emphasize benevolence toward employees, encouragement of employee loyalty, and participation of workers in decision making. About 20 percent of Japanese workers enjoy guarantees of lifetime employment in their firms, and many large Japanese companies help provide housing and recreational facilities for their employees.

One essential factor in Japan's postwar economic growth was a high level of investment in new and efficient industrial plants. Investment capital was freed by government policies that cut expenditures on amenities and services such as roads, antipollution measures, parks, housing, and even higher education. Japanese industrial cities grew explosively and became highly polluted areas of dense and inadequate housing, with few public amenities and snarled transportation. The money "saved" by not being spent on amenities and services was made available for investment in industrial growth, but the costs to Japanese society were high.

Some analysts cite elements of Japan's political culture to explain the country's economic successes. One political party, the **Liberal Democratic Party (LDP)**, has been repeatedly elected to power since Japan regained its sovereignty in 1952. It is a conservative and strongly business-oriented and business-connected organization. The party has promoted Japanese exports with policies to keep the yen (the Japanese unit of currency)—and thus, Japanese goods—inexpensive. The party has also cooperated closely with Japanese business in the development of new products and new industries. However, that very coziness between government and business—the "**crony capitalism**" seen in several other East and Southeast Asian countries—was partly to blame for the economic malaise that spread over Japan in the 1990s; thus, support for the LDP began to erode.

Despite all of these factors in Japan's economic favor, the Japanese miracle did not last. The peak of Japan's postwar success came in the 1980s. A powerful economic boom led to speculative rises in stock prices and land prices. At the end of the 1980s, the total value of all land in Japan was four times greater than that in the United States. The Tokyo Stock Exchange was the world's largest, based on the market value of Japanese shares. Then the "**bubble economy**" burst, and real estate and stock prices fell by more than 50 percent. Japan had near-zero economic growth through most of the 1990s and into the 2000s. Japanese industries whose growth had seemed unstoppable experienced an increasing loss of market share to U.S. and European producers. Part of the reason the economy could not regain its footing was official Japanese commitment to prop up faltering industries and farms with huge subsidies. Because of the disproportionate political influence of small towns and rural areas, the Japanese tried to modernize remote villages and islands with expensive and underutilized public works projects. Due to such projects, Japan is the world's most indebted country. Overall, the system succeeds in subsidizing its past at the expense of its future.

14.4 Japanese Industry

The gradual evolution of industry over more than three centuries that characterized Europe and the United States has been compressed into little more than a century in Japan. Early iron and steel development was concentrated in northeast Kyushu, on and near the country's main coal reserves. The products went mainly into a new railway system and into shipping (Figure 14.11). Japanese coal, supplemented increasingly by imports, powered the country's factories, railways, and ships. Hydroelectricity was developed on mountain streams and became important in the energy economy.

Japan's petrochemical and other oil-based industries reflect the country's great dependence on imports. Almost devoid of petroleum reserves, the country imports nearly all of its oil and gas, mostly from the Persian/Arabian Gulf region. Japan therefore downplays its alignment with the West in most Middle Eastern disputes.

Despite the unique stigma Japan associates with the power of the atom, the country has come to rely on nuclear power, with 53 plants supplying about 30 percent of the country's electricity needs. Hydropower supplies 8 percent, but with fossil fuels almost absent, the critical lack of energy has compelled the nation to develop the world's most energy-efficient economy and to persist with the nuclear technology to which many Japanese are opposed. Small fires, explosions, and other accidents in nuclear power and reprocessing plants in recent years have prompted new domestic concerns about the safety of nuclear energy, forcing a slowdown in Japan's nuclear industry.

There are also international fears. Japan imports large quantities of plutonium for use in its nuclear power industry, and the long-distance ocean shipment from Europe of this very hazardous material has contributed to Japan's poor reputation in environmental affairs. International protests against the plutonium shipments have caused Japan to postpone the construction of a series of nuclear breeder reactors, which use and create recyclable plutonium.

The phenomenal advance of industry since the early

Figure 14.11 The Port of Tokyo. The island nation of Japan thrives on seaborne trade.

Figure 14.12 Japanese-made consumer products are recognized throughout the world for their high quality and competitive prices.

1950s has been marked by a series of booms and declines in various sectors. In the 1960s, electronics, cameras and optical equipment, petrochemicals, synthetic fibers, and automobiles (notably those produced by Mazda, Honda, Toyota, Mitsubishi, and Nissan) became boom industries (Figure 14.12). In the 1970s and 1980s, Japan stepped up the manufacture of computers and robots. By the early 1980s, Japan led the world in the robotization of industry, but the United States took and continues to hold the lead in computer manufacturing. For the first time in the postwar era, Japanese industry in the early 2000s—when Japan began offshoring a growing share of its manufacturing to China—was not characterized by world supremacy in any single manufacturing sector. Japanese industry must now look for new directions in which to establish leadership. As always, to overcome the severe limitations imposed by small, resource-poor islands, the Japanese must live by their wits.

14.5 The Industrious People behind Japanese Industry

Japan has one of the world's most homogeneous populations; it is 99.5 percent ethnic Japanese. The largest non-Japanese group is the 600,000 strong community of ethnic Koreans, the people and their descendants whom Japan imported during World War II to do mainly hard manual labor. These ethnic Koreans have long suffered from discrimination in Japan. So have the ethnic Japanese social outcasts known as the **Burakim**, descendants of people who practiced "impure" trades like butchering and street sweeping as early as the eighth century. They still tend to be marginalized to low-paying jobs and live in ghettos. Other than Koreans, few non-Japanese have settled in the country. Only with the country's booming prosperity and growing labor shortages of the 1980s did Japan open its doors to only a few unskilled and low-skilled immigrants from countries such as Pakistan, Bangladesh, Thailand, Peru, and Brazil.

This shortage of racial and ethnic diversity has had mixed

results for Japan. Many observers believe that it has helped the country achieve a sense of unity of purpose, allowing the Japanese to persist through periods of adversity, especially the postwar years of reconstruction. However, the Japanese have also earned a reputation for intolerance of ethnic minorities. This attitude may be costly for Japan's future economic development. One way to increase productivity in the face of Japan's declining population would be to import skilled workers from abroad, but the country remains averse to immigration. Inward looking tendencies have also left Japan short of people capable of and interested in learning English, a language routinely sought by aspiring businesspeople in nearby South Korea and China.

Capitalist Japan has a remarkably egalitarian society, with about 80 percent of the population comprising its middle class. The richest third of the population has a total income just three times as great as that of the poorest third (compared with five times as much in the United States). The Japanese have historically embraced the concept of *wa*, or harmony, based in part on the principle of economic equality. The recessions of the 1990s and early 2000s, however, resulted in growing joblessness and homelessness in Japan, which tarnished the country's self-satisfied image. The country's official unemployment rate stood at 4.7 percent in 2004. There were loud international appeals for Japan to reform its economic system by allowing inefficient businesses to collapse (rather than strengthen them with expensive subsidies) and to resist the temptation to satisfy marginal rural populations with costly education and public works projects. The government balked at many of these suggestions because of the threat they posed to a socially harmonious Japan.

The legendary Japanese work ethic has had its advantages and drawbacks. It has helped the Japanese create a prosperous country, but it has also created a nation of workaholics beset with the same problems—stress, suicide, depression, and alcoholism—experienced by workers in the world's other MDCs. There is a growing incidence of what Japanese call *karoshi*, or death by overwork. The educational system encourages children to be highly successful but also to conform; critics say this stifles creativity and innovation. College graduates who are not hired during the annual recruiting season face difficult obstacles to entering the job market. And for those it affects, the practice of lifetime employment makes it difficult to change jobs. The country's welfare system, pensions, and social security are inadequate, compelling workers to work harder for savings. In their long hours at work, men are accustomed to spending little time with their spouses and children.

Women have not achieved parity with men in the workplace, and they complain increasingly of discrimination and sexism. In struggling to increase their footing in Japanese society, growing numbers of Japanese women are both working longer hours and marrying later, contributing to Japan's remarkably low birth rate (9 per 1,000 annually, one of the lowest in the world). Ironically, Japan's falling birth rates will only increase the burdens of its workers (Figure 14.13). The population is both shrinking (it is expected to go from 128

million in 2004 to 100 million in 2050 and 67 million in 2100) and aging (already one in five people were over age 65 in 2004, and 1 million people a year join those ranks). The Japanese of working age will face an increasing load of taxes and family obligations to meet the needs of older citizens.

Crowding and overdevelopment, accompanied by a surprising lack of amenities, are inevitable facts of life in Japan. Officially designated parks comprise 14 percent of Japan's land area, but many of these are developed with roads, houses, golf courses, and resorts. There has been enormous growth in the popularity of winter sports destinations in northern Honshu and Hokkaido and of hot springs and mineral bath resorts throughout the country. Reflecting dissatisfaction with too much growth and development, the conservation ethic is strong and growing in Japan. However, it is difficult for the Japanese to truly "get away from it all" at home.

Daily life for the Japanese is devoid of many of the benefits associated with the MDCs. Apartments and homes are generally very small. Only about half of the country's homes are connected to modern sewage systems. Smog from automobile exhaust and industrial smokestacks is a persistent health hazard. The ever-present traffic jams are known locally as traffic "wars." Japan's much-vaunted rail system, including its high-speed *Shinkansen*, or bullet trains, that carry traffic between northern Honshu and Kyushu at speeds averaging 106 miles (170 km) per hour, is chronically overburdened with passengers (Figure 14.14). Summarizing these living conditions, Japanese politician Ichiro Ozawa described Japan as having "an ostensibly high-income society with a meager lifestyle." Japan has long emphasized production of manufactured exports to compete with the United States and other large economies rather than sought to raise the material standard of living of its own citizens.

Japan had a troubled period of almost continuous economic recession from the early 1990s to the early 2000s. By 2004, there was growing consensus among analysts that the economy's worst troubles were over, thanks largely to painful cuts in payrolls and in inefficient businesses. Japan is poised to turn recent adversity into opportunity, as it has in the past. Japan's highly educated and skilled workforce is a major resource. Although natural assets are lacking, the country's favorable geographic location should also aid its recovery. As the western Pacific region experiences the dynamic economic growth that is propelled by China, Japan is well positioned to benefit from trade.

Japan will certainly continue to be an important member of the international community. Even during recession, it has maintained its status as the world's second largest donor of foreign aid (after the United States). Its long-term economic well-being is seen as critical to the overall stability of the global economy. The Japanese consume more goods than the roughly 2 billion people of the rest of East Asia combined. Any slowdown in Japanese spending decreases production in the factories of Indonesia, Malaysia, Thailand, and China that make goods for Japanese consumption and reexport; conversely, an economically stronger Japan raises all these economic boats. The United States supplies Japan with a wide

Japan's Population

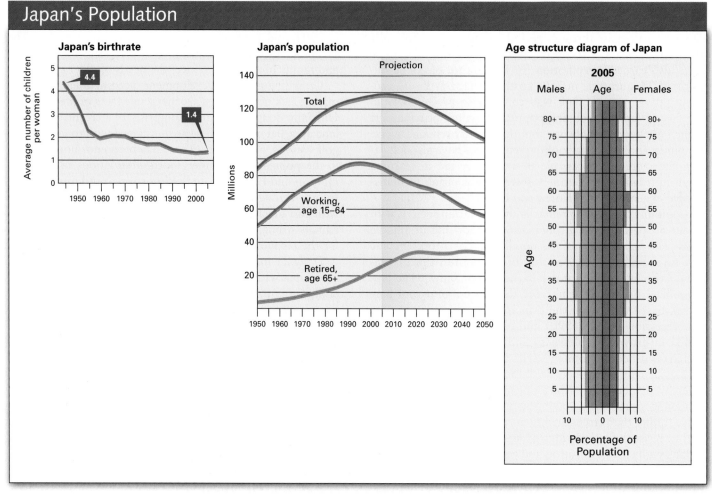

Japan's birthrate

Japan's population

Age structure diagram of Japan

Active Figure 14.13 Many countries face problems of overpopulation, but Japan is concerned about not having enough people to keep its economic engine running in the future. *See an animation based on this figure, and take a short quiz on the facts or concepts.*

Figure 14.14 A bullet train glides through the Ginza district of Tokyo. Japan has an efficient, high-speed rail network.

range of goods, including semiconductors, software, heavy machinery, movies, apparel, and agricultural goods, and it buys even more goods from Japan, so these two economies—the world's largest and second largest—are also extremely interdependent on one another's health.

14.6 Unfortunately Located Korea

Just 110 miles (177 km) from westernmost Japan is the Korean Peninsula, roughly the size of Minnesota or Portugal (Figure 14.15). Korea's political history during the 20th century, first as a colony of Japan and then as a divided land, has tended to obscure its distinctive culture and contributions to the world. Although the Koreans have been influenced by Chinese culture, and to a smaller extent by Japanese culture, they are ethnically and linguistically a separate people that have made important contributions to world civilization, including a printing press using moveable type.

The two Koreas have occupied an unfortunate location in historic geopolitical terms. North Korea adjoins China along

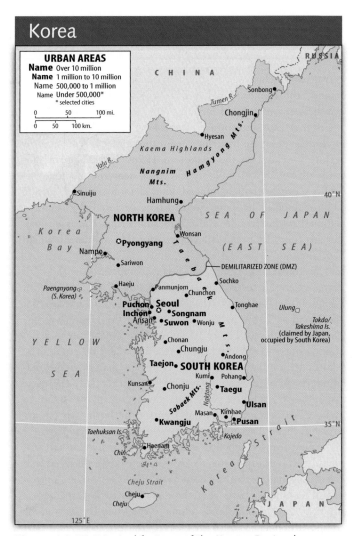

Figure 14.15 Principal features of the Korean Peninsula

a frontier that follows the Yalu and Tumen Rivers; it faces Japan across the Korea Strait; and in the extreme northeast, it borders Russia for a short distance. These small countries are thus located near larger and more powerful neighbors—China, Russia, and Japan—that have frequently been at odds with one another and with the Koreans. China and Russia traditionally feared Japanese rule in Korea because it might have served as a springboard for invasion of their home countries. Meanwhile, to Japan, Korea has been seen as a sharp dagger aiming at its heart to be used by China or Russia. Japan therefore wanted to occupy or block Korea. For many centuries, the Korean Peninsula has served as a bridge between Japan and the Asian mainland. From an early time, both China and Japan have been interested in controlling this bridge, and Korea was often a subject or vassal state of one or the other.

Despite external ambitions of conquest and division, from the late 7th century to the mid-20th century, Korea was a unified state, sometimes invaded and forced to pay tribute but never destroyed as a political entity. The decline of Chinese power in the 19th century was accompanied by the rise of modern Japan, whose influence grew in Korea, and from 1905 until 1945, Korea was firmly under Japanese control. In 1910,

it was formally annexed to the Japanese empire. A legacy of hostility and occupation continues to cast a shadow over relations between the Koreas and Japan, and disputes periodically emerge between them over issues such as control of islands and fishing rights (see Definitions and Insights, page 396).

Japan lost Korea at the end of World War II, setting the stage for the division between North and South Korea. To understand Korea's split, it is important to consider the Korean War, sometimes in the United States called "America's Forgotten War." The conflict cost the lives of an estimated 2 million North Koreans, perhaps 300,000 South Koreans, and 33,629 Americans.

In the closing days of World War II, the Soviet Union entered the Pacific war as an ally of the United States against Japan. The two sides drew up plans to accept Japan's surrender on the Korean Peninsula, arbitrarily drawing a line at the 38th parallel. The Soviet Union would accept Japan's surrender north of that line, and the United States south of that line. This was not meant to be a permanent boundary, but it did become one. On either side of the line, the Soviet Union and the United States moved to set up governments that would be friendly to them. By 1948, a Soviet-style, Soviet-backed, Soviet-armed Democratic People's Republic of Korea was established in the north, while in the south the Western-oriented Republic of South Korea was created under the auspices of the United Nations.

War came soon. On June 25, 1950, North Korean troops invaded the south and quickly took Seoul. In those Cold War days, the United States feared Soviet expansionism into and beyond South Korea. President Truman sent American forces in, and they soon fought under a United Nations flag with support from 16 other countries. Truman appointed General Douglas MacArthur as supreme commander of the effort. By September, MacArthur's forces pushed the North Korean advance all the way back to the Yalu River on the Chinese border. He proposed using nuclear weapons to press forward from there, and his dissention with President Truman over this issue cost MacArthur his job.

China, seeing enemy forces on its doorstep, reacted by sending huge numbers of forces across the border, driving the allies south across the 38th parallel, and taking Seoul on January 4, 1951. Seoul was recaptured on March 15, and the battlefront stabilized generally along the 38th parallel. An armistice was signed at the border site of Panmunjom on July 27, 1953, by the Chinese, North Koreans, and the United Nations command. It was only a cease-fire, and to this day, there has not been a full treaty and reconciliation between the Koreas.

The border between the two Koreas, often the tensest boundary on Earth, still follows this armistice line. The 150-mile-long (240-km), 2.5-mile-wide (4-km) **demilitarized zone** (**DMZ**) is a virtual no man's land of mines, barbed wire, tank traps, and underground tunnels. The world's largest concentration of hostile troops faces off on either side of it. Remarkably, with the near absence of human activity, rare birds like the Manchurian crane and other endangered animal species have taken refuge in this narrow strip (Figure 14.16).

Definitions + Insights

The Law of the Sea

Some 90 miles (145 km) between the shores of Japan and South Korea in the Sea of Japan lie two small, inhospitable islands known to outsiders as the Liancourt Rocks, to Japanese as the Takeshima Islands, and to Koreans as the Tokdo Islands (see Figure 14.15). South Korea has controlled the islands since 1956, but with only one Korean couple and some coast guard personnel living on the islands, Japan has periodically asserted its right to them. In the 1990s, they became the focus of a dispute between Japan and Korea, not because of any riches they contain, but because of a 1970s United Nations treaty known as the Convention on the Law of the Sea, which would permit their sovereign power to have greater access to surrounding marine resources.

The Law of the Sea was initiated in an effort to apportion ocean resources as equitably as possible and to avoid precisely the kind of conflict that developed between Japan and South Korea. The treaty gives a coastal nation mineral rights to its own continental shelf, a territorial water limit of 12 miles (19 km) offshore, and the right to establish an **exclusive economic zone (EEZ)** of up to 200 miles (320 km) offshore (in which, for example, only fishing boats of that country may fish). The power that controls offshore islands such as the Takeshima/Tokdo Islands can extend the area of its exclusive economic zone even further.

By early 1996, 85 nations had ratified the treaty. South Korea ratified the treaty late in 1995. Japan was preparing to ratify it in 1996 when news reached Tokyo that South Korea had plans to build a wharf on the islands. To avoid provoking Japan, North Korea, and China, South Korea avoided declaring an exclusive economic zone off its waters that would include the islands. Japan ratified the treaty. But fears that South Korea may build facilities and station more people on the islands as a step toward establishing an EEZ, and thereby excluding Japanese fishermen from the area, caused Japan to restate its claim to the islands in 1996. South Korean officials answered with military exercises near the islands, and South Korean civilians staged loud demonstrations outside Japan's embassy in Seoul. South Korea declared the islands a national monument in 2002. Still simmering, the issue may have to be settled in the same international legal arena in which it originated.

Even more important than Japan's competition with South Korea over territorial waters and islands is its contest with China. Both lay claims to the islands in the East China Sea that the Japanese know as the Senkakus and the Chinese as the Diaoyu Islands (see Figure 14.1). Again, both are interested in staking out an EEZ in which oil reserves might be found. The islands are uninhabited, but their control is championed by nationalists on both sides and is therefore a potentially explosive issue. In 2004, Chinese activists landed on one of the islands to stake China's claim there, but Japanese authorities promptly arrested them. This followed an earlier attempt that Japanese coast guard vessels repelled with water cannons. Japan's formal claim to the islands dates to 1895, when Japan declared that the islands had previously been *terra nullius,* or no one's property. But China insists that there is Chinese documentation about control of the islands dating to the Ming Dynasty in the 15th century. The islands are close to Taiwan, so any move by mainland China to secure access to them could also precipitate conflict between the two Chinas.

Figure 14.16 The DMZ is an unlikely nature refuge, but wherever a half-century passes without significant human manipulation of the landscape, plants and animals thrive.

14.7 Contrasts between the Two Koreas

The Korean armistice line divides one people, united by their ethnicity and language, into two very different countries (Figure 14.17; also see Table 10.1). North Korea has an approximate area of 46,500 square miles (120,400 sq km) inhabited by about 23 million people with a density of 490 per square mile (190/sq km). South Korea's corresponding figures are 48 million people in only 38,000 square miles (98,390 sq km), averaging 1,260 per square mile (490/sq km). South Korea is a republic that has fluctuated between attempts at democracy and a repressive military dictatorship. It has a capitalist economy heavily dependent on relationships with the United States and Japan. North Korea is a rigid and very tightly controlled Communist state that promotes what it calls *juche,* or self-reliance, allowing little economic or other exchanges with the outside world. Its principal Cold War ties shifted between China and the Soviet

(a)

(b)

(c)

Figure 14.17 The two Koreas are like night and day. (a) A satellite image taken at night shows the profound economic distinctions, with prosperous South Korea awash with urban and industrial lights while North Korea is nearly dark. (b) Affluence and freedom of movement are apparent in a shopping district of Seoul, South Korea (c) while military security preoccupies austere Stalinist North Korea.

Union. With the collapse of the USSR, China became North Korea's main ally and benefactor, and Russia is emerging as one of North Korea's few trading partners.

From World War II until his death in 1994, the dictator Kim Il Sung, whom his citizens called "The Great Leader," governed North Korea. His son and successor, Kim Jong Il (dubbed "The Dear Leader"), has perpetuated the country's militaristic character. North Korea spends a staggering 23 percent of its gross domestic product on its military, compared with South Korea's expenditure of only 2.7 percent, China's 4 percent, and Japan's 1 percent. North Korea's lavish expenditures on defense have been one cause of its severe economic decline. North Korea's economy is one-thirteenth the size of South Korea's. Perhaps the most telling contrast between the countries is the annual per capita GDP PPP of $1,300 in North Korea and $17,800 in South Korea.

There are marked physical contrasts between the two countries. Although both are mountainous or hilly, North Korea is more rugged. North Korea has a humid continental long-summer climate, with hot summers and cold winters. South Korea is mostly a humid subtropical area with shorter and milder winters. Both are monsoonal, with precipitation concentrated in the summer.

Rice is the staple food in both countries. Rice-growing conditions are best in South Korea and diminish northward. Unlike the North, South Korea is able to double crop irrigated rice fields by growing a dry-field winter crop such as barley after the rice has been harvested. The main supplementary crop in the North is corn, which must be grown in the summer and so cannot occupy the same fields as rice.

Beginning in 1995, successive waves of flood and drought brought famine to North Korea. The natural calamities only contributed to the already plummeting economic health of the country that came in the wake of the collapse of the Soviet Union, formerly North Korea's main benefactor and trading partner. North Korea's almost impenetrable veil of secrecy made it difficult to calculate the losses, but estimates of the number of people who died in the famine range from 900,000 to 2,400,000, or up to 10 percent of the country's prefamine population. As one crop after another failed, North Korea gradually and reluctantly sought food aid from abroad. It coaxed emergency supplies of wheat from the United States in part by threatening to withdraw from its 1994 agreement with the United States to halt its nuclear weapons program (see the discussion on pages 295–296 in Chapter 10). Although food aid from the United States, China, and South Korea helped alleviate the immediate crisis, the poor conditions of North Korea's health system, drinking water supplies, and electrical generation have perpetuated malnutrition and disease.

Another physical contrast between the two Koreas is that most of the nonagricultural natural resources are in North Korea. All of the peninsula's major resources—coal, iron ore, some less important metallic ores, hydropower potential, and forests—are more abundant in the North than in the South. North Korea was thus originally the more more industrialized state, featuring a typical Communist emphasis on mining and heavy industry, together with hydroelectric production and timber products. Viewed strictly in terms of resource potential for development, North Korea should have become the more prosperous power. Its decision to close its doors to the outside world and insist on strict state control of industry and agriculture provides a fascinating opportunity to appreciate how different political systems produce different economic results.

While the North focused on defense and strict socialism, the South took over industrial leadership in the 1970s and 1980s by developing a dynamic and diversified capitalist industrial economy.[1] South Korea's economy first began to boom between 1961 and 1980 under a state capitalism approach rather than free-market and free-trade capitalism. This was a time during which the government favored and strongly funded a few small companies that it nurtured into giant conglomerates called **chaebols**. Hyundai is a good example. Originally a small rice milk company, it diversified into trucks and buses. The government believed in its potential and supported Hyundai financially, supervising its growth into a world-class corporation renowned for the manufacture of automobiles and ships. A similar path was followed by Samsung, now one of the best electronics companies in the world (Figure 14.18). Altogether about 15 of these gigantic, interlocking, family-controlled conglomerates came to dominate the economy. Large investments from Japan and the United States, as well as access to markets in those countries, aided in the explosive development of *chaebol* industries. Also important was the availability of inexpensive and increasingly skilled Korean labor. Finally, the factor South Koreans most frequently point to in their success has been an emphasis on education. They sell valued possessions if need be to afford their children's educations at home or abroad, and their culture prizes teachers and professors. South Korea has the world's highest number of PhDs per capita.

Democratic and free-market challenges came to South Korea's economic, political, and social systems in the 1980s, with differences emerging between labor and management, military and civilian rule, and even men and women. In the late 1990s, an economic crisis occurred in part because of these challenges and also because South Korea's was hit by the shock waves of Asia's spreading economic crisis. The pain was almost universal because so much of the country's wealth was in the hands of the *chaebols,* which employed the majority of South Korea's working population and owned most of the banks. During the country's explosive period of growth, the banks had lent money back to their parent companies for risky investments all over the world, particularly throughout

Figure 14.18 An employee discusses Samsung's corporate work ethics and global ambitions.

Asia. As the Asian economic crisis spread, those projects were halted, and the borrowers defaulted on their loans. Foreign investors rushed to pull their capital out of South Korea, whose national currency and foreign reserves deflated.

With financial help from the International Monetary Fund (IMF), South Korea responded quickly to the crisis, especially by getting rid of bad loans and by restructuring the companies that borrowed too much. In the early 2000s, the country was once again experiencing economic growth, bringing sighs of relief around the world. South Korea is now among the top five countries worldwide in the production of automobiles, ships, steel, computers, and electronics. The development of its high-tech industries has concerned competitors in Japan and the United States; for example, Samsung has established an increasing market share in an industry dominated by Japanese firms. South Korea is the fifth largest trading partner of the United States, which has long held a **domino theory** that if South Korea's economy fell, so too would Japan's and perhaps the United States'.

South Korea's cities are visible evidence of the country's industrialization. The capital city of Seoul has a population of 20 million, ranking it among the top five most populous cities in the world (Figure 14.19). Other cities with more than 2 million are the port of Pusan and the inland industrial center of Taegu. By contrast, the only large city in North Korea is the capital, Pyongyang, with a population of 3.2 million. In an effort to begin to close the gap, North Korea opened the Rajin-Sonbong Free Economic and Trade Zone in the 1990s. Centered at the extreme northeast corner of North Korea around the twin cities of Rajin and Sonbong, and ringed by barbed wire, this trade zone was intended to attract foreign investment and industries that would capitalize on the site's favorable location near the borders of Russia

---

[1]For much of this discussion, Joe Hobbs is grateful to the Korea Society for sponsoring his field study of Korea in 2003. Professor Byong Man Ahn of the Hankuk University of Foreign Studies and Professor Taeho Bark of the Graduate School of International Studies at Seoul National University provided much of the information on Korean economies and politics.

Joe Hobbs

Figure 14.19 Seoul, South Korea's Capital

and China. When that effort foundered, a similar vision for the creation of a "new Hong Kong" was announced for the city of Sinuiju, in the extreme northwest, just across the Yalu River from China.

In sum, South Korea is one of Asia's success stories, an Asian Tiger that made an extraordinary postwar recovery, especially by using brain power to overcome its resource limitations, and that has enjoyed periods of rocketing economic growth. South Korea is, however, still something of an "in-between" country, not entirely an MDC or an LDC, not all urban but not all rural either, and both industrial and agricultural. The government still lavishes huge subsidies to protect its rice growers, for example, while also promoting the virtues of free trade in industry. The transition from LDC to MDC has brought some growing pains, seen especially in the social changes of more women in the workforce, higher divorce rates, and falling birth rates.

Meanwhile, North Korea is a poor country frozen in time, insistent on keeping alive the command economy model that failed in every other nation that employed it. Despite its motto of self-reliance, it depends on huge imports of food and energy, especially from Russia. Its principal exports are apparently missiles and other military hardware and even nuclear weapons fuel; heroin, amphetamines, and other synthetic and semisynthetic drugs; and counterfeit money. Its people, whose sentiments and hopes are unknown because they are not allowed contact with the outside world, hover perilously close to famine. The North Korean leadership has a dilemma: If it gives up on *juche* and opens its doors, the economy might improve, but North Koreans will discover the bounties just across the border and might revolt.

14.8 Sunshine for Korea?

States of cold war and periodic border incidents have persisted between North and South Korea. Relations between the two countries have periodically thawed, only to ice over again. A warming trend came to be known as the **Sunshine Policy** in 2000, when the leaders of North and South Korea held a historic, first-ever summit meeting between the countries. Their substantive focus was on reconciliation rather than reunification. The South proclaimed that it wanted to help protect North Korea's national security and assist in its economic development. In exchange for this help, South Korea expected North Korea to cease threatening the South and to provide reassurances of abandoning its nuclear weapons and long-range missile programs. The two countries agreed to restore a long-defunct rail link built during Japan's occupation of Korea. South Korea is hoping its resurrection will provide a large economic boost: Then it could ship its products overland to China and across the Trans-China Railway or Trans-Siberian Railway to Europe (Koreans are thus wistfully calling the link the "**Iron Silk Road**").

Koreans on both sides of the DMZ were delighted by the family reunions that occurred in the summit's wake. Seven million South Koreans who fled the north before and during the Korean War have relatives in the North. Since 2000, small numbers of North Koreans have been able to make short visits with relatives in South Korea, and vice versa, enjoying emotional reunions that have given hope on both sides of better times to come.

Any prospect for Korean reunification appeared to fade even as relations between the two countries improved. North Korea continues to argue that the two countries should unite peacefully and without foreign influence. Outside observers generally believe, however, that North Korea is not really interested in reunification because it would probably doom the country's regime. For its part, South Korea is fearful of the enormous economic cost it would have to pay for absorbing its much poorer neighbor, just as West Germany did in absorbing East Germany. Reunification would probably begin with a huge stream of poor northerners into the south. The nearby great powers are also quietly pleased with Korea's lingering division. The two Koreas provide two buffer zones between the historic adversaries China and Japan. And if the two Koreas united, finally taking advantage of North Korea's strong natural resource base, China and Japan could suddenly have a rival great power on their hands.

For now, there is little thought of Korean unification. There is much more concern with simply avoiding war. Any escalation of the tension between the United States and North Korea, or even an incident in the DMZ, could be a tripwire for a great conflict (Figure 14.20). U.S. troops would have to be involved, and Chinese troops might be. Some analysts fear that Korea may yet become a nuclear battlefield. Japan worries that it would also be drawn into any conflict between the Koreas. The economic and political stability of these three small countries on the western Pacific Rim is critical to the well-being of the global system.

Figure 14.20 Two South Korean soldiers keep a wary eye on their North Korean counterpart, just steps away in the Joint Security Area of the Panmunjom "truce village" in the DMZ. The low concrete slab running left–right in the central left portion of the photo marks the border between the two countries. From time to time, a North Korean defector runs across the border, sometimes pursued by North Korean soldiers. Firefights involving both U.S. and South Korean troops against North Korean soldiers have accompanied some of these incidents.

SUMMARY

- Japan's economy is the world's second largest, after the United States.

- There are four main islands in the Japanese archipelago: Hokkaido, Honshu, Shikoku, and Kyushu. The smaller Ryukyu Islands extend almost to Taiwan.

- All of the main islands of Japan consist largely of volcanic mountains and are subject to frequent and sometimes severe earthquakes. Other natural hazards also plague Japan.

- Japan went through periods of relative isolation. Then it opened up to the outside world and is today one of the most influential players in world affairs.

- Japan experienced phenomenal economic growth after its crushing defeat in World War II, but it has recently suffered economic recession.

- Due to its relative poverty in natural resources, Japan must import most of its raw materials, energy supplies, and a large share of its food.

- The people of Japan view the ocean as an important resource and promising frontier. The Japanese have the world's highest per capita consumption of fish and other marine foods.

- The quest for possession of offshore resources near islands is a source of tension between Japan and Korea and between Japan and China.

- Most of Japan's economic activities and a majority of its people are packed into a corridor about 700 miles (c. 1,100 km) long. This megalopolis extends from Tokyo, Yokohama, and the surrounding Kanto Plain on the island of Honshu, through northern Shikoku, to northern Kyushu.

- The hardworking Japanese are experiencing many symptoms of workaholism, and many complain about their poor standard of living relative to that of other prosperous countries.

- The Korean Peninsula has an unfortunate location in geopolitical terms, sandwiched between the greater powers of Russia, China, and Japan. Korea has had a turbulent recent political history, first as a colony of Japan and then as a land divided between North and South Korea.

- Although the Koreans have been influenced by Chinese culture, and to a smaller extent by Japanese culture, they are ethnically and linguistically a separate people.

- There are marked physical and economic contrasts between North and South Korea, with North Korea being much more rugged and having more natural resources for industry. North Korea has a humid continental long-summer climate, with hot summers and cold winters, whereas South Korea is mostly a humid subtropical area, with much shorter and milder winters. Capitalist South Korea is far more prosperous than Communist North Korea.

- North Korea has suffered famine recently but is reluctant to open up to trade and assistance because North Koreans might revolt.

- There are prospects for reconciliation and, more remotely, reunification, but also war between North and South Korea. Reunification is not favored by the South because it would be too costly. China and Japan would apparently be happier to have a weaker divided Korea than a stronger unified Korea.

KEY TERMS + CONCEPTS

Terms in blue are also defined in the glossary.

active volcano (p. 381)
Ainu (Utari) (p. 388)
bubble economy (p. 391)
Burakim (p. 392)
chaebol (p. 398)
Convention on the Law of the Sea
  (p. 396)
core area (p. 387)
crony capitalism (p. 391)
daimyo (p. 388)
demilitarized zone (DMZ) (p. 395)
domino theory (p. 398)
double cropping (p. 385)

exclusive economic zone (EEZ) (p. 396)
Great East Asia Co-Prosperity Sphere
  (p. 390)
gunboat diplomacy (p. 389)
intertillage (p. 385)
Iron Silk Road (p. 399)
juche (p. 396)
kamikaze (p. 388)
karoshi (p. 393)
Liberal Democratic Party (LDP)
  (p. 391)
megalopolis (p. 387)
Meiji Restoration (p. 389)

New Order in East Asia (p. 390)
plate tectonics (p. 384)
Rape of Nanking (p. 390)
Ring of Fire (p. 384)
samurai (p. 388)
Self-Defense Forces (SDFs) (p. 391)
shoguns (p. 388)
summer and winter monsoons (p. 382)
Sunshine Policy (p. 399)
Tokugawa Shogunate (p. 388)
tsunami (p. 384)
typhoon (p. 382)
wa (p. 393)

REVIEW QUESTIONS

WORLD
REGIONAL
Geography◉Now™

Assess your understanding of this chapter's topics with additional quizzing and concept-based problems at http://earthscience.brookscole.com/wrg5e.

1. What are Japan's four main islands? What are the outlying islands? Which are contested with other countries?

2. What elements of Japan's physical geography are related to the islands' location on the Pacific Ring of Fire?

3. What are Japan's principal climate types? In which parts of the country do they occur?

4. What are the principal features of Japan's agriculture? What position does rice have in the agricultural and economic systems?

5. To what extent does Japan draw on the resources of the sea?

6. What area is defined as Japan's core? What urban, industrial, and population characteristics definite it as a core region?

7. What is Japan's ethnic makeup? What relations have the Japanese majority had with internal minorities and foreign groups at various times?

8. How has Japan been able to overcome its poverty in natural resource assets by commercial, military, or other means?

9. What are some of the features of Japanese society today, especially those related to the country's economy?

10. What precipitated the division of the Korean Peninsula into North and South Korea?

11. Why are relations tense between Japan and South Korea?

12. What are the major differences between North and South Korea in population numbers, natural resources, political systems, and economic development?

DISCUSSION QUESTIONS

1. What are the characteristic natural and agricultural landscapes of Japan?

2. If a major earthquake were predicted in Tokyo—or San Francisco—should authorities alert the public to the danger? What are the potential advantages and drawbacks of issuing a warning? What other land-use and planning policies are related to earthquakes in Japan?

3. What is Japan's reputation among environmentalists? If Japan is so forested, why does the country import tropical hardwoods? Is Japan hunting whales? Why does it import plutonium if it is so averse to things nuclear?

4. Discuss the American decision to use nuclear weapons against Japan and the continuing legacy of that decision in Japan and in global affairs.

5. Discuss the milestones or general features of Japan's various periods of accommodation and isolation with the outside world, and its expansionist period.

6. Discuss the Japanese miracle by reviewing the various explanations for it.

7. What have been the principal features of Japan's relationship with the United States since World War II?

8. What were some of the causes and impacts of Japan's recession in the 1990s?

9. How many Japanese exports do you now have in your possession? How many do you have at home? Do you own any items made in Korea? What do these possessions reveal about today's global economy?

10. What are the major demographic characteristics of Japan and South Korea? What are the implications of these traits in the countries' future development?

11. How do some provisions in the Law of the Sea apparently provoke disputes over islands in this region?

12. Given similar cultures in North and South Korea, and better industrial resources in the North, how did South Korea's economy grow to dwarf that of the North?

13. What are the prospects for reunification of the Koreas? How do some outside powers and the Koreans themselves view reunification?

14. How has the Korean Peninsula been influenced by its proximity to China, Russia, and Japan?

A Geographic Profile of the Pacific World

Joe Hobbs

Australia is known as the "Red Continent." This is the Indian Ocean coast near the northwestern town of Broome.

chapter objectives

This chapter should enable you to:

- Appreciate the economic prominence of larger, Europeanized Australia and New Zealand in a vast sea of small, mainly indigenous political units

- Recognize the associations between the physical geographies of islands, their typical social and political organizations, and their economic characteristics

- Consider the impacts of the human agency, especially through deforestation and the introduction of exotic species, on island ecosystems

- Understand how minority control of most of the wealth generated by mineral and other resources has led to discontent and rebellion among majorities

- Hear the concern expressed by low-lying island countries about the production of greenhouse gases in faraway industrialized nations

- Recognize the peculiar dependence of some Pacific populations on the military interests of distant nations

WORLD
REGIONAL
Geography ⊕ Now™

Look for this logo in the text and go to GeographyNow at http://earthscience.brookscole.com/wrg5e to explore interactive maps, view animations, sharpen your factual knowledge and geographic literacy, and test your critical thinking and analytical skills with unique interactive resources.

Covering fully one-third of Earth's surface, the Pacific World is mostly water. The world's largest ocean, the Pacific, is bigger than all Earth's continents and islands combined. Before World War II, the Western world spawned legends about this ocean and its islands as a kind of utopia. However, there has long been trouble in this paradise. On many islands, foreign traders, whaling crews, labor recruiters, and other opportunists exploited the indigenous peoples, reducing their numbers and disrupting their cultures. The military battles of World War II further shattered the idyllic qualities of many islands. Yet today, the Pacific mystique, which has been perpetuated in books and films, forms part of the allure for booming tourism development across the region. This chapter presents the cultural and natural histories of the Pacific World, summarizes the impacts of colonial enterprises in the region, and discusses some of the obstacles to development confronting the countries of this watery realm.

15.1 Area and Population

The Pacific World region, often called Oceania, includes Australia, New Zealand, and the islands of the mid-Pacific lying mostly between the tropics (Table 15.1). The Pacific islands nearer the mainlands of East Asia, Russia, and the Americas are excluded here on the basis of their close ties with the nearby continents. Hawaii belongs to this region, but is dealt with mainly in Chapter 24's discussion of the United States.

Large areas of the eastern and northern Pacific that contain few islands are also discounted. Australia and New Zealand are included in the Pacific World region because of their strong political and economic interests in the tropical islands, their similar insular character, and the ethnic affiliations of their original inhabitants with the peoples of those islands. But they are also uniquely European in character and dwarf the other countries in economic and political clout. Therefore, they earn separate treatment in Chapter 16.

The Pacific islands are commonly divided into three principal regions: Melanesia, Micronesia, and Polynesia (Figure 15.1). The islands of Melanesia (Greek for "black islands"), bordering Australia on the northeast, are relatively large. New Guinea, the largest, is about 1,500 miles (c. 2,400 km) long and 400 miles (c. 650 km) across at the broadest point. It is divided between two countries. The western section, called Papua (formerly Irian Jaya), is part of Indonesia, and is discussed in Chapter 12. The eastern section is an independent country, Papua New Guinea. The other Melanesian countries are the Solomon Islands, Vanuatu, and Fiji. New Caledonia, also in Melanesia, is controlled by France.

Micronesia (Greek for "tiny islands") includes thousands of scattered small islands in the central and western Pacific, mostly north of the equator. The Micronesian countries are Palau, the Federated States of Micronesia, the Republic of the Marshall Islands, Nauru, and the western part of the

TABLE 15.1 Pacific World: Basic Data

Political Unit	Area (thousand/ sq mi)	Area (thousand/ sq km)	Estimated Population (millions)	Annual Rate of Natural Increase (%)	Estimated Population Density (sq mi)	Estimated Population Density (sq km)	Human Development Index	Urban Population (%)	Arable Land (%)	Per Capita GDP PPP ($US)
Australia and New Zealand										
Australia	2988.9	7741.3	20.1	0.6	7	3	0.946	91	6	29000
New Zealand	104.5	270.7	4.1	0.7	39	15	0.926	78	5	21600
Total	**3093.4**	**8011.9**	**24.2**	**0.6**	**8**	**3**	**0.942**	**88**	**6**	**27746**
Others										
American Samoa (U.S.)	0.2	0.5	0.05	0.0	250	97	N/A	90	10	8000
Cook Is. (N.Z.)	0.09	0.2	0.02	N/A	222	86	N/A	70	17	5000
Fiji	7.1	18.3	0.8	1.9	113	44	0.758	39	11	5800
French Polynesia (Fr.)	1.5	3.8	0.3	1.5	200	77	N/A	53	1	17500
Guam (U.S.)	0.2	0.5	0.2	1.6	1000	386	N/A	93	9	21000
Kiribati	0.3	0.7	0.1	1.8	333	129	N/A	43	2	800
Marshall Islands	0.1	0.2	0.1	3.7	1000	386	N/A	68	16	1600
Micronesia, Federated States of	0.3	0.7	0.1	2.1	333	129	N/A	22	5	2000
Nauru	0.009	0.02	0.01	1.8	1111	429	N/A	100	0	5000
New Caledonia (Fr.)	7.2	18.6	0.2	1.7	28	11	N/A	71	0	15000
Northern Mariana Islands (U.S.)	0.2	0.5	0.07	2.7	350	135	N/A	94	13	12500
Palau	0.2	0.5	0.02	0.8	100	39	N/A	70	8	9000
Papua New Guinea	178.7	462.8	5.7	2.2	32	12	0.542	15	0	2200
Solomon Islands	11.2	29	0.5	2.7	45	17	0.624	16	1	1700
Tonga	0.3	0.7	0.1	1.8	333	129	0.787	32	23	2200
Tuvalu	0.01	0.02	0.01	1.7	1000	386	N/A	47	0	1100
Vanuatu	4.7	12.1	0.2	2.2	43	16	0.570	21	2	2900
Total	**212.3**	**549.8**	**8.5**	**2.1**	**40**	**15**	**0.575**	**24**	**1**	**3921**
Summary Total	**3305.7**	**8561.7**	**32.7**	**1.0**	**10**	**4**	**0.846**	**71**	**5**	**21553**

Sources: *World Population Data Sheet,* Population Reference Bureau, 2004; *U.N. Human Development Report,* United Nations, 2004; *World Factbook,* CIA, 2004.

nation of Kiribati (pronounced KIHR-uh-bas). Micronesia also includes two possessions of the United States: Guam and the Mariana Islands.

Polynesia (Greek for "many islands") occupies a greater expanse of ocean than does either Melanesia or Micronesia. It is shaped like a rough triangle with corners at New Zealand, the Hawaiian Islands, and remote Easter Island. Excluding Hawaii and New Zealand, Polynesia's independent countries are the eastern part of Kiribati, Tuvalu, Samoa, and Tonga. Possessions of other countries in Polynesia include French Polynesia and France's Wallis Islands, American Samoa, New Zealand's Tokelau and Cook Islands, Britain's Pitcairn Island, and Chile's Easter Island.

The typical Pacific island country (excluding Australia and New Zealand) has about 100,000 to 150,000 people in an area of 250 to 1,000 square miles (c. 650 to 2,600 sq km),

consists of a number of islands, is poor economically, is an ex-colony of Britain, New Zealand, or Australia, and depends heavily on foreign economic aid. The total area of land coverage, including Australia and New Zealand, is 3.3 million square miles (8.5 million sq km), or about 90 percent of the size of the entire United States (Figure 15.2).

The region's total population is 32.7 million (Figure 15.3). Aside from Australia's approximately 20 million people, populations range from 5.7 million in Papua New Guinea, which is exceptionally large in population and area for this region but has a low population density, to 12,000 on tiny Nauru island. The highest population densities are in the smallest island groups, notably the Tuamotu Islands in French Polynesia and the Ellice Islands of Tuvalu. Population growth rates vary widely, from the predictably low 0.6 percent and 0.7 percent in Australia and New Zealand, respec-

Political Geography of the Pacific World

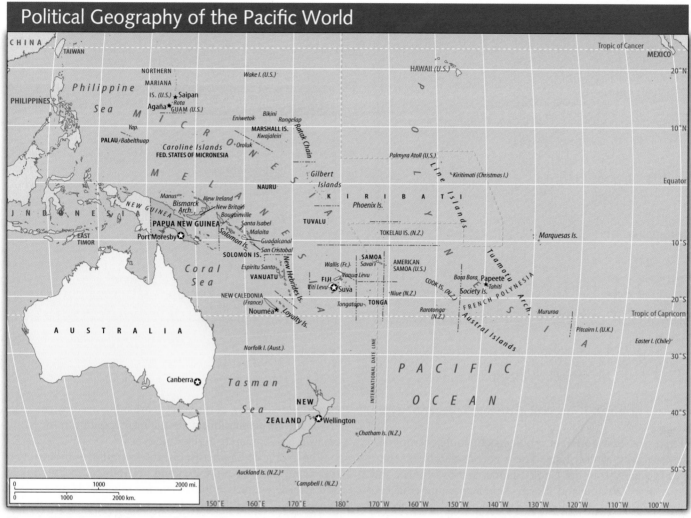

Figure 15.1 Principal features of the Pacific World. Note the peculiar jog in the International Date Line near Kiribati (along the equator).

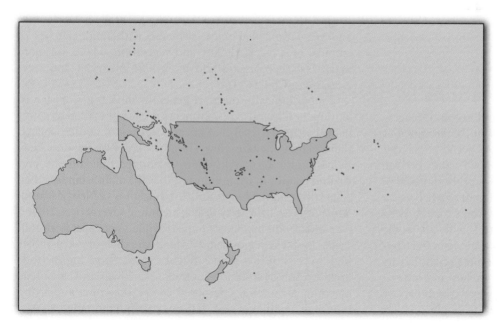

Figure 15.2 The Pacific World compared in area with the conterminous United States

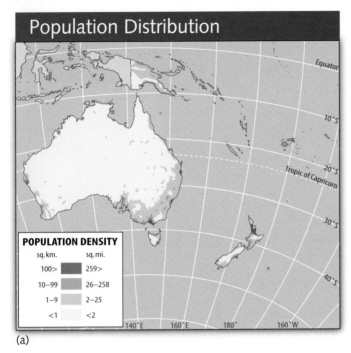

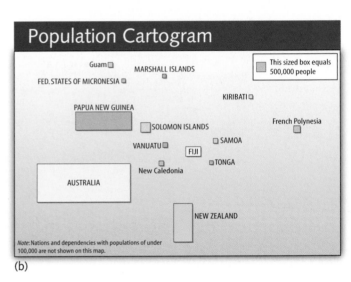

(a)

(b)

Figure 15.3 (a) Population distribution and (b) population cartogram of the Pacific World

tively, to a very high 3.7 percent in the much poorer Marshall Islands.

There are some apparent problems of people overpopulation, especially in Polynesia. There is considerable emigration, especially from the Cook Islands, Samoa, Tonga, and Fiji to New Zealand and North America. Many educated young Australians are also lured abroad, especially to the United States. With birth rates low in both Australia and New Zealand, there are the typical postindustrial concerns of not having enough people to support the countries' economies and aging populations. Both countries, however, are averse to liberal immigration policies that would boost their populations.

15.2 Physical Geography and Human Adaptations

The climates, vegetation, landscapes, and cultural geographies of the islands scattered across the vast Pacific vary, but a few notable patterns emerge (Figures 15.4 and 15.5a and b; note that only the islands of the southwest Pacific are large enough to have meaningful data depicted at this map scale). Climatically, most of the region is tropical; all of Micronesia and Melanesia, and most of Polynesia (except for northern Hawaii and New Zealand and its outlying islands), lie in the tropics. Most of New Guinea, which lies entirely in the tropics, has tropical rain forest climate and biome types, but its high mountains have undifferentiated climates and vegetation types, varying with elevation all the way to tundra above 11,000 feet (3,300 m). The great majority of the smaller

Pacific islands lying in the tropics are influenced by the easterly trade winds that bring abundant precipitation for tropical rain forest climate and biome types.

The Tropic of Capricorn bisects Australia. Southern Australia and most of New Zealand are influenced by cooling midlatitude westerly winds, bringing much of New Zealand and parts of coastal Australia a marine west coast climate with associated temperate mixed forests. Some coastal areas of southern Australia also have Mediterranean and semiarid/steppe climates, with associated Mediterranean scrub vegetation and prairie and steppe grasses. While much of the interior of Australia has a desert climate and vegetation, coastal regions of northern Australia generally have climates and vegetation of tropical savanna, with small patches of tropical rain forest climate and vegetation. There are more details on the climates and biomes of Australia and New Zealand in Chapter 16.

Topographically, there are generally three types of islands. Continental islands are continents or were attached to continents before sea level changes and tectonic activities isolated them. These include New Guinea, New Britain, and New Ireland in the Bismarcks; New Caledonia, Bougainville, and smaller islands in the Solomons; the two main islands of Fiji; Australia; and the North and South Islands of New Zealand. Some of these have lofty peaks, including several over 16,000 feet (4,877 m) in New Guinea, while Australia is overall rather low. The region's other islands are categorized as either high islands or low islands (Figure 15.6). Most high islands are the result of volcanic eruptions. The low islands are made of coral, a material composed of the skeletons and living bodies of small marine organisms that inhabit tropical seas.

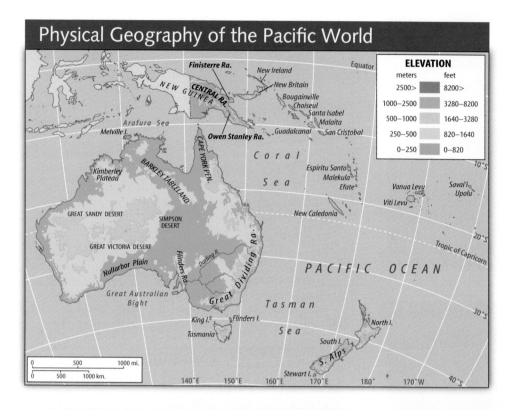

Physical Geography of the Pacific World

Figure 15.4 Major landforms of the Pacific World

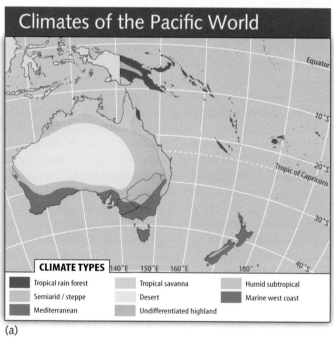

(a)

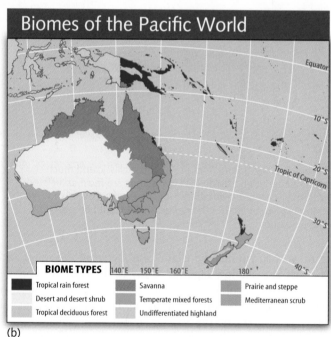

(b)

Figure 15.5 (a) Climates and (b) biomes of the Pacific World

These very different settings offer sharply different opportunities for human livelihoods (Figure 15.7). Overall, however, except in urban, industrialized Australia and New Zealand, the Pacific peoples are rural. They live in small farming or fishing villages where they have long depended on root crops (such as taro and yams), tree crops (such as breadfruit and coconuts), sea fish, and pigs. Outside Australia and New Zealand, there are only five cities with populations greater

than 50,000: Port Moresby and Lae in Papua New Guinea, Suva in Fiji, Nouméa in New Caledonia, and Papeete in Tahiti.

High Islands

Most of the region's high islands are not continental but volcanic in origin and have a familiar pattern. There is a steep central peak with ridges and valleys radiating outward to the coastline. Permanent streams run through the valleys. A

Roberto Arakaki /International Stock

(a)

Karl & Jill Wallin / FPG / Getty Images

(b)

Figure 15.6 (a) Bora Bora in Tahiti is a classic and much-romanticized high island. (b) Low islands are most associated with the "coconut civilizations" of the Pacific. Low islands are extremely vulnerable to the dangers that would be posed by global warming.

coral reef surrounds the island, and between the shore and the reef is a shallow lagoon. Many of the high volcanic islands are spectacularly scenic (see Figure 15.6a). Classic high islands in the region include the Polynesian groups of Hawaii, Samoa, and the Society Islands.

Some of the volcanic high islands of the Pacific comprise island chains. These are formed by the oceanic crust sliding over a stationary **geologic hot spot** in Earth's mantle where molten magma is relatively close to the crust (Figure 15.8). As the crust slides over the geologic hot spot, magma rises through the crust to form new volcanic islands. The Hawaiian Island chain is an excellent example. The big island of Hawaii is now situated over the hot spot and is still active; it is one of the chain's youngest members. The Pacific Plate of Earth's crust is moving northwestward here; thus, older volcanic islands that were born over the hot spot have moved northwestward, where they have become inactive. Still older

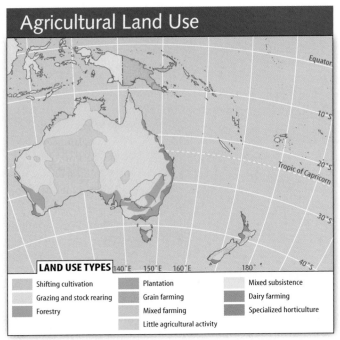

Agricultural Land Use

Figure 15.7 Land use in the Pacific World

islands farther northwest in the chain have submerged to become seamounts, or underwater volcanic mountains. Other seamounts have yet to break the surface, including one in the southeast that will break the surface in about 10,000 years to become Hawaii's newest island.

Geographer Tom McKnight explained that the high island's topography historically influenced its cultural and economic patterns. Each drainage basin formed a relatively distinct unit that was governed by a different chief. Within each unit, there were several subchiefs, each of them controlling a bit of each type of available habitat: coastal land, stream land, forested land, and gardening land. Since available resources were distributed relatively equitably among and between chiefdoms, there were few power struggles or monopolies over resources. It is important to note that these are general rather than universal patterns of association between physical geography and human adaptations in the Pacific. Geographers reject notions of environmental determinism that would insist on immutable connections between environment and society. *10*

Today, the main cities and seaports of the Pacific World, and the largest populations, are in the volcanic high islands and the continental islands. Valuable minerals are scarce, but the rich soils (typically derived from volcanic materials) support a diversity of tropical crops. Generally, these islands are more prosperous than the low islands.

Low Islands

The low islands are formed of coral. Most take the shape of an irregular ring surrounding a lagoon; such an island is called an atoll. Generally, the coral ring is broken into many pieces, separated by channels leading into the lagoon, but the whole circular group is commonly considered one island.

Creation of the Hawaiian Islands

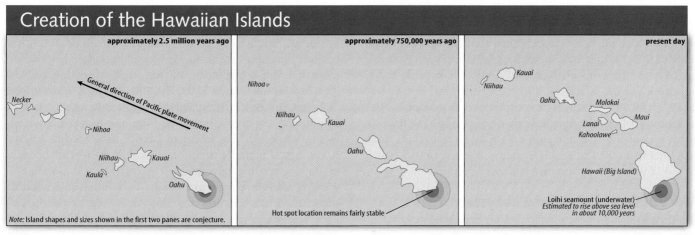

approximately 2.5 million years ago

General direction of Pacific plate movement

Necker

Nihoa

Niihau
Kaula
Kauai

Oahu

Note: Island shapes and sizes shown in the first two panes are conjecture.

approximately 750,000 years ago

Nihoa

Niihau
Kauai

Oahu

Hot spot location remains fairly stable

present day

Kauai
Niihau

Oahu
Molokai
Lanai   Maui
Kahoolawe

Hawaii (Big Island)

Loihi seamount (underwater)
*Estimated to rise above sea level
in about 10,000 years*

WORLD REGIONAL Geography Now™   **Active Figure 15.8** The Hawaiian Islands have been created as a piece of the earth's crust has slid over a geologic hot spot. *See an animation based on this figure, and take a short quiz on the facts or concepts.*

Charles Darwin devised a still widely accepted explanation for the three-stage formation of atolls (Figure 15.9). First, coral builds a fringing reef around a volcanic island. Then, as the island slowly sinks, the coral reef builds upward and forms a barrier reef separated from the shore by a lagoon. Finally, the volcanic island sinks out of sight, and a lagoon occupies the former land area whose outline is reflected in the roughly oval form of the atoll.

The low islands are generally smaller than the volcanic high islands and lack the resources to support dense populations. They are typically fringed by the waving coconut palms that are a mainstay of life and the trademark of the South Sea isles. The Gilbert Islands of the republic of Kiribati in Micronesia are typical of the picturesque low island atolls idealized by Hollywood and travel brochures (see Figure 15.6b). Other major atoll groups in Micronesia are the Caroline and Marshall Islands. The Tuamotu Archipelago is a Polynesian atoll group, and atolls are also scattered across Melanesia.

Despite their idyllic appearance, these islands pose many natural hazards to human habitation. The lime-rich soils are often so dry and infertile that trees will not grow, and shortages of drinking water limit permanent settlement. Their low elevation above sea level provides little defense against storm waves and **tsunamis,** the waves generated by earthquakes. Coastal regions of higher islands can also suffer from these events. In 1998, an underwater landslide triggered by an earthquake created a 30-foot-high (10-m) tsunami that swept over a 15-mile (24-km) stretch of northern Papua New Guinea, killing 2,100 people.

Before the intrusion of outsiders, the low island economies were based heavily on subsistence agriculture (with strong reliance on coconuts), gathering, and fishing. Geographer McKnight explained that sparse resources were distributed with relative uniformity among the low islands' populations. The political and social units were smaller and less structured than those of high islands. These resource-poor low islands have a long history of population limiting factors such as war, infanticide, and abortion.

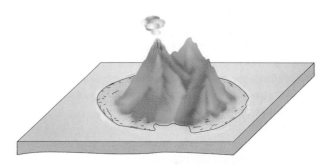

(a) Volcanic island with fringing reef

(b) Slight subsidence barrier reef

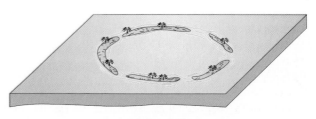

(c) An atoll

WORLD REGIONAL Geography Now™   **Active Figure 15.9** Charles Darwin's explanation of the development of an atoll. (a) First, a fringing coral reef is attached to the volcanic island's shore. (b) Then, as the island subsides, a barrier reef forms. (c) With continued subsidence, the coral builds upward, and the volcanic center of the island finally becomes completely submerged, forming an atoll. *Source: Gabler et al., 1987. Essentials of Physical Geography, p. 562. See an animation based on this figure, and take a short quiz on the facts or concepts.*

Many low island countries now rely heavily on modern commercialized versions of the coconut and fishing economy. The coconut is so important in their subsistence that many of these islands have developed what has been characterized as a **"coconut civilization."** Coconuts provide food and drink for the islanders, and the dried meat, known as copra, is the only significant export from many islands. The husks and shells of the nuts have many uses, as do the trunks and leaves of the coconut palms. People make baskets and thatching from the leaves, for example, and use timber from the trunk for construction and making furniture.

The Vulnerable Island Ecosystem

Island ecosystems in the Pacific region, like those across the globe, are typically inhabited by endemic species of plants and animals—those found nowhere else in the world. They result from a process in which ancestral species colonize the islands from distant continents and, over a long period of isolation and successful adaptation to the new environments, evolve to become new species. Gigantic and flightless animals, like the moa bird of New Zealand and the giant tortoises of the Galapagos (in the Pacific) and Seychelles (in the Indian Ocean), are typical of endemic island species. More than 70 percent of the native plant species of New Caledonia and Hawaii are endemics.

Generally developing in the absence of natural predators and inhabiting relatively small areas, island species have proved to be especially vulnerable to the activities of humankind. People purposely or accidentally introduce to the island alien plant and animal species, called exotic species and including rats, goats, sheep, pigs, cattle, and fast-growing plants, that often prey upon or overtake the endemic species. Habitat destruction and deliberate hunting also lead to ex-

tinction. For example, New Zealand's moas and the dodo birds of the Indian Ocean island of Mauritius are now "dead as dodos." Indigenous inhabitants of New Zealand set fires to hunt the giant ostrich-like moa and had already killed off most of the birds by the time the Maori people colonized New Zealand around A.D. 1350. By 1800, the moa was extinct.

From one end of the Pacific World to the other, there is heightened concern about such human-induced environmental changes. The Hawaiian Islands have become known to ecologists as the **"extinction capital of the world"** because of the irreversible impacts that exotic species, population growth, and development have had on indigenous wildlife there. Environmentalists are also concerned about the impact of commercial logging on the island of New Guinea (Figure 15.10). About 22,000 plant species grow there, fully 90 percent of them endemic. And ecologists consider nearby New Caledonia one of the world's biodiversity hot spots because of the ongoing human impact on its unique flora and fauna.

Human-induced extinctions complement a long list of natural hazards to island species. For example, island animals are particularly prone to natural catastrophes such as **typhoons** (hurricanes) and volcanic eruptions. Most animals find it difficult or impossible to evacuate in the event of a natural catastrophe, and the island's distance from other lands may hinder or prevent recolonization. So, although island habitats have a higher percentage of endemic species than would be found on mainlands, environmental difficulties sometimes cause the number of different kinds of species (that is, **species diversity**) to be lower. Far more than natural hazards, however, it is the human agency that has altered the Pacific island ecosystems, in some cases completely transforming the natural landscape (see Definitions and Insights, page 413).

Figure 15.10 Luxuriant forests of New Guinea contain an exceptional wealth of plant and animal species. This is early morning on the Karawari River.

Joe Hobbs

26

Definitions + Insights

Deforestation and the Decline of Easter Island

"Easter Island is Earth writ small," wrote naturalist Jared Diamond. Recent archeological and paleobotanical studies suggest that the civilization that built the island's famous monolithic stone statues destroyed itself through overpopulation and abuse of natural resources (Figure 15.A). Deforestation, or the removal of trees by people or their livestock, has serious repercussions wherever it occurs, but perhaps nowhere are these impacts so apparent as on Easter Island.

When the first colonists from eastern Polynesia reached remote Easter Island (known as Rapa Nui to locals) in about A.D. 400, they found the island cloaked in subtropical forest. Plant foods, especially from the Easter Island palm, and animal foods, notably porpoises and seabirds, were abundant. The hu-

man population grew rapidly in this prolific habitat. The complex, stratified society that emerged on the island grew to an estimated 7,000 to 20,000 people between 1200 and 1500, when most of the famous statues were built. Apparently in association with their religious beliefs, Easter Islanders erected more than 200 statues, some weighing up to 82 tons and reaching 33 feet (10 m) in height, on gigantic stone platforms. At least 700 more statues were abandoned in their quarry sites and along roads leading to their would-be destinations, "as if the carvers and moving crews had thrown down their tools and walked off the job," Diamond observed.[a]

"Its wasted appearance could give no other impression than of a singular poverty and barrenness,"[b] Dutch explorer Jacob Roggeveen wrote of the island on the day he discovered it—Easter Sunday 1722. Not a single tree stood on the island. The depauperate landscape bears testimony to the fate of the energetic culture that built the great statues. By A.D. 800, people were already exerting considerable pressure on the island's forests for fuel, construction, and ceremonial needs. By 1400, people and the rats they introduced to the island caused the local extinction of the valuable Easter Island palm. Continued deforestation to make room for garden plots and to supply wood to build canoes and to transport and erect the giant statues probably eliminated all of the island's forests by the 15th century. By then, people had hunted to extinction many terrestrial animal species and could no longer hunt porpoises because they lacked the wood needed to build seagoing canoes. Crop yields declined because deforestation led to widespread soil erosion. There is evidence that in the ensuing shortages, people turned on each other as a source of food. By about 1700, the population began a precipitous decline to only 10 to 25 percent of the number who once lived on this isolated Eden.

George Holton/Photo Researchers, Inc.

Figure 15.A The Polynesian people who erected these *moai* statues between A.D. 1200 and 1500 deforested most of Easter Island—in part to supply levers and rollers for the statues—and thus precipitated the demise of their civilization.

[a] Jared Diamond, "Easter's End." *Discover,* 16 (8), 1995, p. 64.
[b] Ibid.

15.3 Cultural and Historical Geographies

Their indigenous cultures having been greatly diminished in numbers and influence, Australia and New Zealand today are mainly European in culture and ethnicity. The region's other cultures, however, are overwhelmingly indigenous and quite diverse. Aside from Australia and New Zealand, with their majority European populations, and Fiji, New Caledonia, and Guam, each of which has populations about half indigenous and half foreign, approximately 80 percent of the Pacific's people are indigenous. The balance is 13 percent Asian and 7 percent European. Of the indigenous populations, Melanesians comprise about 80 percent, with 14 percent Polynesian and 6 percent Micronesian.

Indigenous Peoples

The Pacific region began to be settled about 60,000 years ago. Although the archeology is far from definitive on this point, there is a consensus that this is when the first people settled in Australia. They were the ancestors of today's **Aborigines.** Some ethnographers refer to the Aborigines as a distinct Australoid race. However, they share racial characteristics with other indigenous groups in Asia, including the Mundas of central India, the Veddahs of Sri Lanka, and even the Ainu of Japan. All of these groups, along with Australia's Aborigines, probably descended from a common ancestral race in Asia, and the Aborigines developed their distinctive features over tens of thousands of years of habitation of Australia. They probably arrived there in several waves either by sea from Timor or by land from what is now New

Guinea. During their migrations, sea levels were much lower, straits were narrower and easier to navigate, and there was a land bridge between New Guinea and Australia until about 10,000 B.C. Once settled in Australia, they developed a unique and enduring culture. Their **Aboriginal languages,** classified into the two major groups of **Pama-Nyungan** in the southern 90 percent of Australia and **non-Pama-Nyungan** in the north, are not clearly related to any languages outside the continent.

New waves of migrants known as Austronesians began a long process of diffusion across the western Pacific and eastern Asia after 5000 B.C. Their ancestral stock probably originated in Taiwan and southern China. Their descendants migrated to the mainland and islands (initially the Philippines and Indonesia) of Southeast Asia. Their livelihoods were based on fishing and simple farming, especially of taro, yams, sugarcane, breadfruit, coconuts, and perhaps rice.

Their domesticated animals included pigs and probably dogs and chickens, but they had no herd animals like cattle, sheep, or goats. From their Southeast Asian bases, over a period of several thousand years, the Austronesians embarked on voyages to Madagascar, New Zealand, Easter Island, and finally, Hawaii, "island-hopping" all the way and accomplishing extraordinary feats of navigation in their outrigger canoes. Their settlement of Polynesia came rather late in these adventures, within the last 1,500 years. Today's Micronesians and Polynesians are mainly their descendants, but there has been considerable mixing over the centuries.

These seafaring peoples spread their characteristic agriculture and a language called **Proto-Austronesian**—the ancestor of all modern Austronesian languages—across the Pacific. The Pacific region's linguistic diversity today is extraordinary (Figure 15.11). Most of the indigenous peoples of the region speak languages within the **Austronesian lan-**

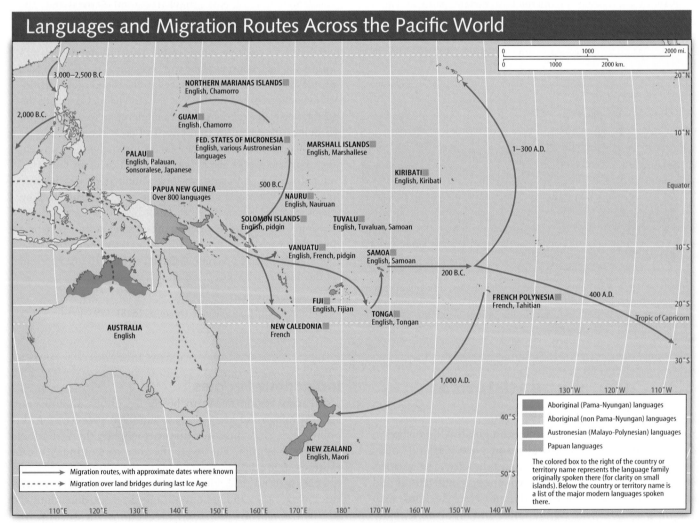

Languages and Migration Routes Across the Pacific World

Active Map 15.11 Languages and settlement routes of the Pacific World
See an animation based on this figure, and take a short quiz on the facts or concepts.
Sources: Bernard Comrie, et al., The Atlas of Languages, *Revised Edition (New York: Facts on File, 2003); settlement data from John Noble Wilford, "Seeking Polynesia's Beginnings in an Archipelago of Shards,"* The New York Times, *1-8-02, D1.*

guage family, which is divided between the **Formosan** and **Malayo-Polynesian language subfamilies.** The Melanesian peoples, however, speak **Papuan languages,** which are not a distinctive linguistic family but an amalgam of often unrelated and mutually unintelligible tongues. There are about 700 Papuan languages.

Papua New Guinea is home to 860 languages (mainly Papuan but also Austronesian), or 20 percent of the world's total number of languages, making it by far the world's most linguistically diverse country, with an average of roughly one language per 5,000 people. Vanuatu has fewer than 200,000 people and 105 identified languages, earning that country's population the distinction as the world's most linguistically diverse on a per capita basis, with roughly one language per 2,000 people. There are fewer of the mainly Austronesian languages in Micronesia and Polynesia, and more of them are mutually intelligible. Reflecting the colonial past, English and French are official languages in some of the islands and are spoken widely across the region. Another lingua franca is **pidgin,** comprised of English and other foreign words mixed with indigenous vocabulary and grammar. Pidgin is the official language of Papua New Guinea.

The Europeans

Europeans began to visit the Pacific islands during the 16th century, and their impact was especially profound after 1800. Spanish and Portuguese voyagers were followed by Dutch, English, French, American, and German explorers. Many famous names are connected with Pacific exploration, including Magellan, Tasman, Bougainville, La Perouse, and the most famous of all, British Captain James Cook, who undertook three great voyages in the 1760s and 1770s, only to be killed by Hawaiian Islanders in an unexpected dispute in 1779. Europeans were interested mainly in the high islands because of their relative resource wealth. They sought sandalwood, pearl shells, whales, and local people to indenture as whaling crews and as manual laborers. On island after island, European penetration decimated the islanders and disrupted their cultures. The intruders introduced venereal and other infectious diseases, alcohol, opium, forced labor, and firearms, which worsened the bloodshed in local wars.

They also came as Christian missionaries (see Perspectives from the Field, pages 414–416). Some ancient indigenous belief systems thrive in parts of the region, often nestled within Christianity and other recently introduced faiths (Table 15.2). There were even new faiths that arose based on the premise that foreigners were preventing locals from acquiring their just rewards. These were the **cargo cults** that emerged, particularly in New Guinea and elsewhere in Melanesia, and especially during World War II, as "Stone Age" people were exposed for the first time to bottles, metals, manufactured cloth, and other material trappings of Westerners. There were many variations among the cargo cults, but most were **millenarian movements** that maintained

TABLE 15.2 Religions of the Pacific World

Country or Territory	Major Religions
American Samoa (U.S.)	Christian Congregationalist, Roman Catholic, Protestant
Australia	Anglican, Roman Catholic, other Christian
Cook Is. (N.Z.)	Cook Islands Christian Church
Fiji	Hindu, Methodist, Roman Catholic, Muslim
French Polynesia (Fr.)	Protestant, Roman Catholic
Guam (U.S.)	Roman Catholic
Kiribati	Roman Catholic, Protestant
Marshall Islands	Protestant
Micronesia, Federated States of	Roman Catholic, Protestant
Nauru	Protestant, Roman Catholic
New Caledonia (Fr.)	Roman Catholic, Protestant
New Zealand	Anglican, Presbyterian, Roman Catholic, Methodist, other Christian
Northern Mariana Islands (U.S.)	Roman Catholic
Palau	Roman Catholic, other Christian, Modekngei religion
Papua New Guinea	Roman Catholic, Lutheran, Presbyterian, other Christian, indigenous beliefs
Solomon Islands	Anglican, Roman Catholic, United, Baptist, other Christian
Tonga	Christian
Tuvalu	Church of Tuvalu
Vanuatu	Presbyterian, Anglican, Roman Catholic, indigenous beliefs

Source: *World Factbook*, CIA, 2004.

there would be a cataclysmic set of events that would trigger a new and more prosperous age for their followers. Intervention by white foreigners would end, and the locals would be mystically delivered the massive "cargo" of material possessions long denied them. Some of the movements built jetties and storehouses in anticipation of those events. Overwhelmingly, however, the peoples of the Pacific adopted and continue to practice Christianity, with Methodists, Mormons, and Catholics the largest denominations. Where Asians have settled, there are pockets of Hinduism, Buddhism, and Islam.

Europeans created new settlement patterns, disrupted old political systems, and rearranged the demographic and natural landscapes. In Polynesia, for example, arriving Europeans typically sought shelter for their vessels on the lee side of an island. That safe place became the island's European port. Whatever chief happened to be in control of that spot typically, because of his association with the Europeans, became more powerful and wealthier than other island leaders.

Perspectives from the Field

With Aborigines and Missionaries in Northwestern Australia

In the austral winter of 1987, I served as leader and lecturer aboard a 30-passenger vessel that took Western tourists, Americans and British mostly, to the remote coastline of the Kimberley region in northwestern Australia. Except for the towns of Wyndham and Broome, respectively on the northeastern and southwestern edges of this region, this is a wilderness, virtually devoid of settlements, roads, and other infrastructure. The major population, quite small in number, is aboriginal, but there are a few European missionaries and cattle ranchers. It is a land of sandstone, spinifex grass, eucalyptus trees, and abundant wildlife, including the fearsome saltwater crocodile.

My journal entry from August 10 reflects on past and present cultural encounters in this tropical wilderness:

Last night we sailed from the Bonaparte Archipelago and made our way to the northern tip of the Kimberley. This part of the mainland is known as the Mitchell Plateau. It has long been an important aboriginal region, and today most of the lands in this coastal area are actually Aboriginal Reserve Lands. Traditionally very resistant to outsiders, these indigenous people succeeded in driving off Malays, Indo-nesians, and Europeans over a period of several hundred years. Within the past 100 years, various Christian groups successfully established several missions in the area.

Early in the afternoon we entered Napier Broome Bay and did a brief shore excursion. From the beach where our zodiacs [inflatable boats] landed, we climbed up to a rocky sandstone promontory. From there was an excellent view of the beach and the mangrove-lined inlets on either side. In the distance the smoke of bush fires lit by Aborigines could be seen. Returning to the beach, we saw a young green turtle poke its head out of the water for a look around. When he saw us he decided it wasn't safe, and submerged and paddled off rapidly into open water. Some of the crewmen strung out a long net from the beach in the hope of catching some mullet. Almost immediately, a small school blundered into the net. The catch became our gift to the aboriginal people we met later in the day.

Peter [Sartori, the vessel's captain] shifted the ship around Mission Bay, and we went ashore. We were received by a small but very enthusiastic party of aboriginal children, who enjoyed themselves immensely clambering in and out of the zodiacs. Their parents were our hosts for a very interesting couple of hours.

These people are members of the Wunambal tribe. They had come down from the Kalumburu Mission, about 18 miles inland on the King Edward River. One of them, a woman named Mary, was the last child born at the old Pago Mission on the shore of Mission Bay. The Pago Mission was founded by Benedictine (Spanish Catholic) monks in 1908. The missionaries' relationship with the Aborigines was not always peaceful; there were a few hostile encounters in which the missionaries came out the worse. However, the parties came to trust one another, and enjoyed good relations right up until the Pago Mission had to be abandoned in the 1930s, when its water supply ran out. The Kalumburu Mission was established to take its place.

Our hosts were very congenial and went out of their way to make us feel comfortable. We spoke freely with them, and found their English quite good; they had after all been educated at the Mission. The men, who had been wearing Western

Figure 15.B The Wunambal (with Basil at far left) performing a Corroboree by the shore of Mission Bay, Kimberley, Australia.

Joe Hobbs

clothes, disappeared for a while and then returned decked out with loincloths and their finest white ochre dance paint. Just before dusk, a signal was given and a traditional *corroboree*, or ceremonial dance, began (Figure 15.B). The women danced, too, but wore Western clothing, and had a minor role in the *corroboree*. The dance leader, Basil, was a real ham, boasting about his talents at every opportunity. But mainly he explained how each dance or skit told a story about some aspect of daily or ritual life of the Wunambal. It was quite clear that although they had been in a Christian mission environment for decades, these people held their ancestral beliefs about their origins and the world around them. Of the *Kalinda* dance, Basil said: "It has to do with cyclone Tracy. *Kalinda* is the song spirit. When we see people coming in a line to the *Pruru* or Dancing Place with the totems, we know it is the Song Spirits coming. When we know there are strong winds or cyclones coming, we take the totems to the hills for safety. When we are going to the hill where the Rolling Stone is, and we are halfway up the hill, the big stone rolls down and all the people stop until the stone falls to the ground. One man ventures out to see if it is safe; then all people come out to dance." Then there was the *Palgo;* Basil explained: "The spirits catch fish at night time. They light a bundle of sticks and take them to the waterhole. The fish are attracted by the light, and so we catch them." The last dance that told a story was the *Djarlarimirri* ("Witch Doctor"): "A man is killed by a spear, and the Witch Doctor is called. He performs his magic, and the man is restored to life." Finally, there was the *Djalurru,* a dance without spiritual significance, performed just for fun.

Some days later, I visited the ruins of the Pago Mission and the Kalumburu Mission that succeeded it, writing:

From the landing site I followed a dirt track through dry savanna to reach the old mission. Not much is left—some concrete floors, a few posts, and three wells that now have water in them—but its surroundings are attractive. A grove of old mangoes survives untended. The cemetery is overgrown. There is a billabong [pool] here used by water buffalo, and the locals say saltwater crocodiles lurk in it. The first monks who came to this place, in 1908, scrawled their names on a large boab tree still towering over the place. I learned more about their relations with the Aborigines. They spent 11 years here without having a single Aboriginal settle with them. The monks noted a shortage of males, and a population deficit overall, and attributed this to civil warfare. The monks proclaimed their peacekeeping role, hoping the mission would offer locals a refuge from strife and trouble. The Aborigines slowly began to drift in, attracted certainly by food. Most simply obtained food and supplies, visited with their settled kinsmen, and then disappeared back into the bush. Missionary records note that the Aborigines of the time practiced formalized warfare, at least once attacking the mission; in 1913, a missionary was speared.

Soon, government policy changed to insist that the Aborigines be protected and integrated into the larger society. The government began to give them yearly rations of cloth and flour so that the church became the purveyor of goods, prompting a kind of "cargo cult" among the Aborigines. They began to settle in earnest at Pago Mission. When they outgrew the water supply, the mission had to be abandoned in 1939 in favor of Kalumburu, whose construction had begun four years earlier up on the King Edward River. When the war broke out, the Australian armed forces asked the monks to build an airstrip at the mission, which they did with the help of Aboriginal labor. Bombers used it on their return runs from Timor. The airstrip brought tragedy to the mission. In 1943, Japanese bombers dropped their loads, apparently aiming for the runway but instead hitting the mission, killing one monk and five Aborigines.

Later in the day, I visited the Kalumburu Mission at the invitation of one of the nuns working there, the redoubtable Sister Scholastica, known affectionately as "Sister Scholly" (Figure 15.C), who sported about on an all-

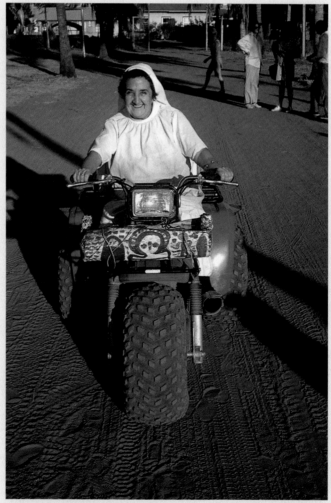

Figure 15.C Sister Scholly, Kalumburu Mission

(continued on page 416)

terrain vehicle. Of all things, rather than a cross as her motor vehicle's masthead, she had a bark painting of a *wandjina,* the aboriginal ancestral spirit of typhoons and monsoon rains. The Mission was celebrating the 55th anniversary of its founding. The entire community proceeded single file under a column of white smoke that symbolized purification. Everyone wore Western clothes—one young girl even had a Mickey Mouse T-shirt. And each wore distinctive Mission and aboriginal marks on this feast day: a blue headband symbolizing the Mission and traditional Wunambal white face paint. They all took their seats on a grassy area, and Father Saunders' outdoor mass got underway. The ceremony was quite interesting. It opened with a reading by Mary Pundilow, a 75-year-old woman who was the first child born at Pago Mission. She talked about how things use to be before the "White Men in Black Cloth" appeared and about the gradual acceptance of the missionaries and their peaceful efforts.

I walked around during part of the service, admiring the thriving gardens and grounds where peanuts, corn, tomatoes, mangoes, pineapples, papayas, and cashews grew. There were pens for pigs, geese, guinea fowl, and Brahman cattle. When mass ended, the Mission's population socialized in the tidy grounds around the clinic, church, and school.

Joe Hobbs

Eventually, he became high chief and often expanded his control to other islands; thus, the first Polynesian kings emerged.

Europeans, North Americans, and East Asians changed the natural history of the islands by introducing exotic crops and animals, including arrowroot and cassava; bananas; tropical fruits such as mangoes, pineapples, papayas, and citrus; coffee and cacao; sugarcane; and cattle, goats, and poultry. They established extensive sugar plantations on some islands, notably in the Hawaiian Islands, Fiji Islands (where sugar is still the major export), and Saipan in the Marianas. The valuable Honduran mahogany trees planted by British colonists in Fiji early in the 20th century are only now beginning to mature, and competition for export profits from this resource has become intense.

It is remarkable that as late as 1840 almost all of the Pacific islands had yet to be formally claimed by outside powers; only Spain had taken the Marianas. Then rapidly, Britain, France, Germany, and the United States vied for power. By 1900, the entire Pacific Basin was in European and American hands. Germany held Western Samoa (now Samoa), part of New Guinea, the Bismarck Archipelago, Nauru, and Micronesia's most important islands. Britain controlled part of New Guinea, the Solomon Islands, Fiji, and smaller island groups, and held Tonga as a protectorate. France controlled New Caledonia, French Polynesia, and with Britain, the condominium (jointly ruled territory) of the New Hebrides. The United States held Hawaii, Guam, and eastern Samoa.

When Germany lost its Pacific territories at the end of World War I, Japan took control of German Micronesia, New Zealand took Western Samoa, and Britain and Australia took the German section of New Guinea and the Bismarcks. Nauru came to be administered jointly by Britain, New Zealand, and Australia.

With the outbreak of World War II, Japan quickly overran much of Melanesia. The ensuing ferocious battles between Japan and the Allies for control of New Guinea, the Solomon Islands, and much of Micronesia brought enormous changes to the region. Many of the cultures were in the Stone Age as the war began, and by its end had seen the Atomic Age; Micronesia had become a nuclear proving ground. Everywhere, traditional peoples were drawn into cash economies and affected by other global forces originating far beyond their watery horizons.

Most of the colonists are gone, but the seeds of strife they planted are still germinating. Many conflicts in the region today reflect the demographic changes Europeans wrought to meet their economic interests. In many cases, the Europeans introduced laborers from outside the region because local peoples were too few in number or refused to work in the cane fields. Successive infusions of Chinese, Portuguese, Japanese, and Filipinos brought a polyglot character to Hawaii, with little apparent ethnic discord. But ethnic strife is ongoing in Fiji and the Solomon Islands, which became independent from Britain in 1970 and 1978, respectively.

The Solomon Islands' troubles date to 1942, when U.S. Marines fought to push Japanese troops off the island of Guadalcanal. The American forces enlisted the support of thousands of people from the neighboring island of Malaita. After the successful offensive, many of these Malaitans elected to stay on Guadalcanal, where under British colonial rule they came to dominate economic and political life. The indigenous people of Guadalcanal resented their dominance and in particular what they saw as the loss of their lands to the Malaitans (who purchased the lands). Conflict between the groups finally erupted in 1999, with the majority Guadalcanalans murdering scores of Malaitans and driving tens of thousands off the island; most returned to Malaita. Australia fielded a peacekeeping mission in 2003 in an effort to quell the unrest, but the **Guadalcanal Liberation Front** continued its resistance against the Malaitan-dominated government.

The violence in the Solomon Islands was apparently inspired by ethnic unrest in Fiji. In the early 1900s, colonial British plantation owners imported mainly Hindu Indians as indentured laborers to work Fiji's sugarcane plantations. Today, indigenous Fijians now only slightly outnumber ethnic Indians (**Indo-Fijians**). Here, too, the indigenous inhabitants resent the dominance of the foreign ethnic group; ethnic Indians control most of Fiji's economic life. A coup led by ethnic Fijians in 2000 succeeded in ousting Fiji's ethnic Indian prime minister and throwing out a 1987 constitution that granted wide-ranging rights to ethnic Indians. Accused of lawlessness, the coup leaders were threatened with economic sanctions by the European Union, which pays higher than market prices for most of Fiji's sugar crop, and by the United States, Australia, and New Zealand. The prospect of economic crisis led the army to take control of Fiji's affairs, imprison the coup leader, and promise a general election and return to civilian government in 2002. The unrest caused a crash in tourism, Fiji's most valuable industry. Prior to the coup, about 400,000 visitors came each year. Tourist numbers rebounded after a peaceful election that brought the indigenous Fijian political party back to power.

15.4 Economic Geography

A general lack of industrial development is characteristic of Pacific island economies (other than those of Australia and New Zealand), and the poverty typical of less developed countries (LDCs) prevails in the region. Contributing to this underdevelopment are what could be described as tyrannies of size and distance. Many of the countries consist of little more than small islands of limestone sprinkled with palm trees and situated hundreds and often thousands of miles from potential markets and trading partners. Most of the countries suffer large trade deficits and must import far more than they can export.

There are five major economic enterprises in the Pacific region, none of them involving value-added processing of raw materials: exports of plantation crops (especially palm products), exports of fish, exports of minerals, services for Western military interests, and tourism. Significant manufacturing outside Australia and New Zealand is limited to the Cook Islands and Fiji. Polynesia's Cook Islands, a self-governing territory in free association with New Zealand, has an industrial economy based on exports of labor-intensive clothing manufacture. Fiji also began exporting manufactured clothes after the mid-1990s, when Australia granted preferential tariff treatment to Fijian-made garments.

A few Pacific countries are beginning to participate, in rather unlikely and sometimes unscrupulous ways, in the information technology revolution that has generated so much global wealth since about 1990. One example is tiny Tuvalu in Polynesia. In 2000, an American entrepreneur agreed to pay the Tuvalu government $50 million to acquire Tuvalu's two-letter Internet suffix: tv. While the new owner of the dot-tv domain has been busily profiting by selling Internet addresses, Tuvalu is delighted with its windfall. It used part of the money to finance the process of joining the United Nations, built new schools, added electricity and other infrastructure to outlying islands, and expanded its airport runway. Vanuatu enjoys significant hard-currency revenue by permitting its country's phone code to be used for 1-900 sex lines. And as described below, Nauru and other island nations use telephone and digital technology to practice money laundering.

Mining takes place on only a few islands. The major reserves are on continental islands, notably petroleum, gold, and copper on the main island of Papua New Guinea, copper on Bougainville (an island belonging to Papua New Guinea), and nickel on the French-held island of New Caledonia. The 10,000 people of the tiny oceanic island of Nauru, which gained independence from Australia in 1968, once enjoyed a high average income from the mining of phosphate, the product of thousands of years of accumulation of seabird excrement, or guano. Until recently, Nauru's economy depended totally on annual exports of 2 million tons of phosphate and on revenues earned from overseas investments of profits from phosphate sales. (The tallest building in Melbourne, Australia, is a product of Nauruan investments and is known affectionately as Birdshit Tower.) This resource is approaching total depletion, however, and because the mining of phosphate has stripped the island of 80 percent of its soil and vegetation, turning it into a virtual lunar landscape, many people in Nauru are now looking for a new island home (Figure 15.12).

Others are content with what they see as Nauru's most recent and, they hope, future source of revenue: **money laundering**. With the phosphate stocks dwindling, Nauruan entrepreneurs have undertaken offshore banking, in which they can operate loosely controlled banks anywhere but in Nauru. In Nauru's case, these are "**shell banks,**" existing only on paper and specializing in turning money earned in illegal ways into money that looks legal. Nauruan law does not require any official oversight of offshore banking transactions, so no official records exist; therefore, any dirty funds moved through a bank registered in Nauru cannot be traced. Russian clients have registered half of Nauru's 400 shell banks, and the assumption is that much of the Russian "Mafia" money is laundered through them. Money laundering and tax haven schemes are also reported in Vanuatu, Palau, the Cook Islands, Marshall Islands, Niue, Tonga, and Samoa.

After 2001, hard-up Nauru began earning revenue by serving yet another unlikely function: detention center for refugees from the war in Afghanistan. By 2004, more than a thousand Afghan refugees who were apprehended at sea in transit to Australia, where they hoped to gain asylum, had been diverted to Nauru. Australia paid the government of

Nauru about $15 million yearly to hold the detainees while their cases were reviewed in the Australian courts.

Issues of control over mineral wealth have brought unrest to New Caledonia and Bougainville. Separatists demanding independence for nickel-rich New Caledonia clashed with French police in the 1980s. An accord signed after the confrontation guaranteed a 1998 referendum for New Caledonia's people (known as **Kanaks**), who would then choose to become independent or remain under French rule. In 1998, however, apparently to maintain its interests in the colony's nickel, France postponed the vote until 2014. In the meantime, France is allowing New Caledonia to have considerable autonomy; since 1999, a legitimate New Caledonian government has been permitted to look after civil affairs.

In the late 1980s, a crisis shook Bougainville, a former Australian colony that in 1975 became, at Australia's insistence, part of newly independent Papua New Guinea. Most Bougainvilleans were unhappy about their association with Papua New Guinea and wanted to be united again with the Solomon Islands, to whose people they are culturally and ethnically related. Angry landowners calling themselves the Black Rambos destroyed mining equipment and power lines to force the closure of the world's third largest copper mine, the Australian-owned Panguna mine. These generally younger landholders were receiving none of the royalties and compensation payments made by New Guinean and Australian mining interests to an older generation of the island's inhabitants. Residents of Bougainville who were concerned about the ruinous environmental effects of copper mining, disturbed that most mine laborers were imported from mainland Papua New Guinea, and angry that little mining revenue

remained on the island, joined the opposition. The growing crisis drove world copper prices to record high levels.

Opposition to the Australian-backed mining interests of Papua New Guinea in Bougainville grew after 1988 into a full-fledged independence movement led by the **Bougainville Revolutionary Army (BRA)**. War broke out in 1989 when the BRA leader declared independence from Papua New Guinea. Papua New Guinea responded with an economic blockade around Bougainville to prevent supplies from reaching the rebels, who salvaged World War II-era weapons and ammunition to fight superior troops. In a decade of fighting, Bougainville's 200,000 people suffered huge losses, with nearly 20,000 killed and 40,000 left homeless. A 1998 cease-fire finally brought to an end the longest conflict in the Pacific since World War II.

The most dynamic element of the Pacific economies is the rapid growth of tourism, with most visitors coming from the United States and Japan. Tourism is the largest employer in the Pacific region, providing substantial income and employment and stimulating overall economic growth by encouraging foreign investment and facilitating trade. The Pacific islands continue to exert an irresistible appeal to people awash in the conveniences and congestion of the industrialized world. The most popular destinations are Hawaii, Australia, New Zealand, and French Polynesia, but there has been strong growth in ecotourism, archeological tourism, and other tourism niches in destinations including Palau, Vanuatu, and Easter Island. Some of the countries have developed unique appeals for international tourists. In a bid to attract visitors by being the first country in the world to greet the new millennium, Kiribati unilaterally decided in 1997 to

Figure 15.12 The 10 square miles (25 sq km) of land that is Nauru have been devastated by the phosphate mining that once made Nauruans among the wealthiest people per capita on Earth.

shift the International Date Line eastward by more than 2,000 miles (3,200 km). The impact of this decision stands out on any map depicting the date line (see Figure 15.1).

Finally, the military interests of outside powers, particularly the United States and France, generate much revenue for some of the Pacific economies, notably Guam, American Samoa, and French Polynesia. This military presence is a mixed blessing, however, because of its environmental impacts and because it perpetuates dependence on foreign powers. These are among the geopolitical concerns discussed next.

15.5 Geopolitical Issues

Once entirely colonial, the Pacific World today is a mixture of units still affiliated politically with outside countries and others that have become fully independent. Since the end of World War II, the United States, Britain, Australia, and New Zealand have abandoned most of their colonies in the region. Only France has insisted on holding on to all of its colonies.

In some cases, colonial powers have delayed independence to would-be island nations. They explain that such territories are too small and isolated or are not economically viable enough for independence. Some islands remain dependent because they confer unique military or economic advantages on the governing power. For example, French Polynesia was until recently the venue for French atomic testing (see Regional Perspective, pages 420–421), and Guam and American Samoa are still useful to the United States for military purposes. Guam, which has been a U.S. possession since 1898 and a U.S. territory since 1950 (its residents are American citizens), is especially vital in U.S. strategic thinking. Pentagon planners classify it as a **power projection hub** that can be used as a forward base fully 5 days' sailing time closer to Asia than Hawaii is. It was the jumping-off point for American B-52 jets on bombing missions to Iraq and would be used in any U.S. military action in Korea or elsewhere in East Asia. In the event of such hostilities, the United States would need Japan's permission to operate from bases in Japan, but no such approval would be necessary to act from Guam. A generally unspoken but widely acknowledged U.S. military objective in building up readiness in Guam is that the island would be critical to any confrontation with China, should that giant country ever become a belligerent power.

Australia and New Zealand have Pacific-oriented defense and security agreements, somewhat precariously balanced with U.S. interests in the region. In 1951, they joined the United States to form the **Australia New Zealand United States (ANZUS) security alliance.** Australian troops supported American forces in the Vietnam War. Since 1987, however, when its liberal government declared that New Zealand would henceforth be a nuclear-free country, New Zealand's relations with the United States have been

strained. Legislation banned ships carrying nuclear weapons or powered by nuclear energy from New Zealand's ports. The United States refused to confirm or deny whether its ships violated either restriction, so New Zealand denied port access to them. This action removed New Zealand from ANZUS.

Despite its small army (only 24,000 strong), Australia has begun to provide a selective security blanket over its interests in the region and elsewhere. In 1999, Australian troops were deployed to head up the United Nations peacekeeping force to oversee the referendum that gave independence to East Timor and its subsequent transition to full independence. Australia also lent troops and other support in the U.S. buildup to and execution of the war against Iraq in 2003. Al-Qa'ida vowed revenge and apparently delivered it with the bombing in Bali, Indonesia, that killed so many young Australian tourists in 2002. Australia's apparent concern about a long-term threat from North Korea also prompted it to join the United States in the development of the SDI missile shield program. Australia is sometimes snubbed by its Southeast Asian neighbors in political and economic affairs because it is seen as something of a regional "police officer" or as a United States "deputy."

Another set of geopolitical concerns in the Pacific World is environmental. These remote islands, so far from the industrial Corelands of North America, Europe, and Japan, are actually the places most likely to suffer first and most from the most feared impacts of industrial carbon dioxide emissions. The islands would be the initial victims of any rise in sea level due to global warming. Sea levels have risen in recent years at a rate of 0.08 inch (2 mm) annually, and there are reports of unprecedented tidal surges on Pacific island shores. If the trend continues, the first Pacific islands to be totally submerged would be Kiribati, the Marshall Islands, and Tuvalu, while Tonga, Palau, Nauru, Niue, and the Federated States of Micronesia would lose much of their territories to the sea. These island nations are among the 39 countries comprising the **Alliance of Small Island States,** which argued forcefully but unsuccessfully at the 1997 Kyoto Climate Change Conference that, by 2005, global greenhouse gas emissions should be reduced to 20 percent below their 1990 levels. What they did get was a pledge by most industrialized nations (not including the United States), in the form of the Kyoto Protocol, to cut these emissions to at least 5 percent below their 1990 levels by 2012. Already claiming to be a victim of lost coastline, higher storm surges, and more storms as a result of global warming, Tuvalu lobbied other countries to join it in a lawsuit against the United States and Australia. Tuvalu wanted to make the case that these countries' failure to ratify the Kyoto Protocol is a principal cause of global warming. As so often in the past, the peoples of the Pacific today feel vulnerable to the actions and policies of more powerful nations far from their tranquil shores.

Regional Perspective

Foreign Militaries in the Pacific: A Mixed Blessing

The remote locales and nonexistent or small human populations of some Pacific islands have made them irresistible venues for weapons tests by the great powers controlling them (Figure 15.D). In the 1940s and 1950s, U.S. authorities displaced hundreds of islanders from Kwajalein atoll in the Marshall Islands to make way for **intercontinental ballistic missile (ICBM)** tests. During those years, the United States also used nearby Bikini atoll as one of its chief nuclear proving grounds (the island's fame at the time led French designers to name a new two-piece bathing suit after it). In 1946, U.S. government authorities relocated the indigenous inhabitants of Bikini to an-

other island, promising they could return when the tests were complete. Over the next 12 years on Bikini, there were 23 nuclear blasts out of a total of 67 conducted as **Operation Crossroads** in the Marshall Islands. A 15-megaton hydrogen bomb detonation on Bikini in 1954 had the power of 1,000 Hiroshima bombs, blowing a mile-wide crater in the island's reef and producing radioactive dust that fell downwind on the Rongelap atoll, where children played in the dust "as though it were snow," one observer wrote.[a] For decades, the U.S. Atomic Energy Commission claimed that a sudden shift in winds took the fallout to inhabited Rongelap, but recent records proved authorities knew of the wind shift 3 days prior to the test. The Marshallese today suffer a legacy of cancers and deformities linked to the weapons tests on Bikini and Eniwetok atolls. So far, the United States has provided more than $63 million to compensate Marshall Islanders made ill by the blasts.

The Bikinians have never accepted their exile. But returning to Bikini has been a hazardous prospect, given its legacy of radioactive soil. A repatriation of the Bikinians in the 1960s was found to be premature, as radiation persisted, and the nuclear nomads moved again. Recent tests have determined that Bikini has little ambient radioactivity and is essentially habitable, except for the cesium 137 that permeates the soil and becomes concentrated in fruits and coconuts. Bikinians do not want to go home to stay until they have assurances from the United States that the problem of poisoned produce can be resolved and that the island is completely safe. Along with the people of Eniwetok, they are seeking massive funding from the United States for complete environmental decontamination of their homeland and for further compensation for their travails.

Now that Bikini is at least safe to visit, some Bikinians are capitalizing on its formerly off-limits status. The island's lack of human inhabitants for more than 50 years has had the remarkable effect of rendering Bikini a true marine wilderness. In

US Navy/SPL/Photo Researchers, Inc.

Figure 15.D The nuclear explosion, code-named Seminole, at Eniwetok Atoll in the Pacific ocean on June 6, 1956. This atomic bomb was detonated at ground level and had the same explosive force as 13.7 thousand tons (kilotons) of TNT. It was detonated as part of Operation Redwing, an American program that tested systems for atomic bombs.

[a]Nicholas D. Kristof, "An Atomic Age Eden (but Don't Eat the Coconuts)." *The New York Times,* March 5, 1997, p. A4.

SUMMARY

⚬ The Pacific World region (often called Oceania) encompasses Australia, New Zealand, and the islands of the mid-Pacific lying mostly between the tropics. Tropical rain forest climates and biomes are most common, but Australia and New Zealand have several temperate climate and biome types.

⚬ The Pacific islands are commonly divided into three principal regions: Melanesia, Micronesia, and Polynesia.

⚬ The Pacific World is ethnically complex, having been settled by people with various Asian origins. Polynesia was the last to be populated. Papua New Guinea is the world's most linguistically diverse country. Christianity is the majority faith in this region.

⚬ Although countless islands are scattered across the Pacific Ocean, there are three generally recognized types: continental islands, high islands, and low islands. Continental islands are either con-the late 1990s, it opened as a tourist destination, with naval ships sunk by the atomic blasts a highlight for many divers. The dive masters and tour guides are Bikinians.

Like the United States, France has had a nuclear stake in the Pacific. In 1995, after a 3-year hiatus, France resumed underground testing of nuclear weapons on the Mururoa ("Place of Deep Secrets") atoll in French Polynesia. There was an immediate backlash from some national governments and from environmental and other organizations. Crew members of the *Rainbow Warrior 2,* the flagship of the environmental organization Greenpeace, confronted French Navy ships near Mururoa in July 1995. French commandos used tear gas to overwhelm the crew and seize the ship. Greenpeace managed to televise the event, inflicting enormous public relations damage to France. Strongly antinuclear New Zealand protested loudly against the nuclear tests. In Australia, customers boycotted imported French goods and local French restaurants, and postal employees refused to deliver mail to the French embassy and consulates. Antinuclear protestors firebombed the French consulate in Perth. Australia joined New Zealand in filing a case at the World Court in The Hague aimed at halting French nuclear tests. Worldwide, informal boycotts diminished sales of French wines.

In French Polynesia itself, activists for independence from France used the issue to highlight international attention to their demands. Over several days in September 1995, hundreds of anti-French demonstrators in Tahiti rampaged through French Polynesia's capital city of Papeete, looting and setting parts of the city on fire and causing millions of dollars of damage to the island's international airport, a vital tourist hub. Although French authorities continued to insist that the tests and the atoll itself were safe, France soon bowed to international pressure, ceasing its nuclear tests in the Pacific and signing the Comprehensive Test Ban Treaty in 1996.

Ironically, the greatest fear in the region now is that France will reduce its military—and therefore, economic—presence even more. Despite the nuclear tests, most inhabitants of French Polynesia have long favored continued French rule. Tourism and other local businesses contribute only about 25 percent of the territory's revenue; the remainder comes from French economic assistance, largely in the military sector. Many locals, however, whose parents and grandparents gave up subsistence fishing and farming for jobs in the military and its service establishments, are worried about their future.

Similar fears stalk Palau, the Marshall Islands, and the Federated States of Micronesia (FSM), three independent nations carved from the former U.S. Trust Territory of the Pacific. The United Nations awarded that territory, which includes the Bikini and Eniwetok atolls, to the United States in 1946 as the world's first and only political-military zone. In 1986, the United States granted independence to the Marshall Islands and the FSM in an agreement known as the **Compact of Free Association.** In this agreement, the countries agreed to rent their military sovereignty to the United States in exchange for tens of millions of dollars in annual payments and access to many federal programs. The United States uses its leases to conduct military tests—for example, on **Strategic Defense Initiative (SDI,** or **"Star Wars")** technology. The United States also provides large cash grants and development aid to Palau, which became independent in 1994.

In these cases, the American government handed over the monies with no strings attached. The funds have been embezzled and misspent, and there is little development to show for them. The United States has begun to reassess the payments with an eye to their eventual curtailment. Many islanders fear the bottoms of their economies will subsequently fall out. The people of the Marshall Islands and FSM are especially worried because more than half of the gross domestic product of each country is comprised of economic aid from the United States. The Marshall Islands' leaders, however, are taking a more defiant stand, saying the United States has no right to question what becomes of the aid money, and pointing out that the United States gets much in return in the form of extensive military rights. The United States is hard-pressed here because the Kwajalein Missile Range in the Marshalls is the only reasonable place in the world where, Pentagon officials say, they can test many of the Star Wars program components.

---

tinents themselves (such as Australia) or were connected to continents when sea levels were lower (such as New Guinea). Most high islands are volcanic. Low islands are typically made of coral, a material composed of the skeletons and living bodies of small marine organisms that inhabit tropical seas.

- ✪ The island ecosystems of the Pacific region are typically inhabited by endemic plant and animal species—species found nowhere else in the world. Island species are especially vulnerable to the activities of humankind such as habitat destruction, deliberate hunting, or the introduction of exotic plant and animal species.

- ✪ Europeans began to visit and colonize the Pacific islands early in their Age of Exploration and brought mainly negative impacts to island societies. However, a steady process of decolonization has accompanied a recent surge of Western interest and investment in the region.

◐ There are ethnic conflicts, related mainly to maldistribution of income, between Malaitans and indigenous Guadalcanalans on the Melanesian island of Guadalcanal, and between indigenous Fijians and Indo-Fijians on Fiji. Interest in securing more income from minerals has pitted the people of New Caledonia against the ruling power, France, and the people of Bougainville against Australian corporate interests in Papua New Guinea.

◐ Aside from a few notable exceptions, the poverty typical of less developed countries prevails throughout most of the Pacific region.

◐ In general, the Pacific islands' economic picture is one of nonindustrial economies. Typical economic activities include tourism, plantation agriculture, mining, and income derived from activities connected with the military needs of occupying powers. Several countries are profiting from offshore banking and 1-900 telemarketing.

◐ During the 1940s and 1950s, the United States used the Bikini atoll in the Marshall Islands as one of its chief testing grounds for nuclear weapons. Strong negative reaction arose throughout the region in the 1990s as the French resumed underground testing of nuclear weapons on the Mururoa atoll in French Polynesia. That testing has since ceased. The United States relies on the region for testing of its Star Wars missile technology. Many inhabitants of French and American military zones are fearful of the economic impacts that might accompany the withdrawal of military presence.

◐ Some of the low island countries are fearful that global warming might create higher sea levels that will inundate them, and Tuvalu considered legal recourse against the United States and Australia for failing to ratify the Kyoto Protocol.

KEY TERMS + CONCEPTS

Terms in blue are also defined in the glossary.

Aboriginal languages (p. 412)
  non-Pama-Nyungan (p. 412)
  Pama-Nyungan (p. 412)
Aborigines (p. 411)
Alliance of Small Island States (p. 419)
atoll (p. 408)
Australia New Zealand United States
  (ANZUS) security alliance (p. 419)
Austronesian language family (p. 413)
  Formosan language subfamily
  (p. 413)
  Malayo-Polynesian subfamily (p. 413)
  Proto-Austronesian (p. 412)
biodiversity hot spot (p. 410)
Bougainville Revolutionary Army (BRA)
  (p. 418)
cargo cults (p. 413)

"coconut civilization" (p. 410)
Compact of Free Association (p. 421)
continental islands (p. 406)
deforestation (p. 411)
endemic species (p. 410)
exotic species (p. 410)
"extinction capital of the world"
  (p. 410)
geologic hot spot (p. 408)
Guadalcanal Liberation Front (p. 416)
high islands (p. 406)
Indo-Fijians (p. 417)
information technology (p. 417)
intercontinental ballistic missile (ICBM)
  (p. 420)
Kanaks (p. 418)
low islands (p. 406)

Melanesia (p. 403)
Micronesia (p. 403)
millenarian movements (p. 413)
money laundering (p. 417)
Operation Crossroads (p. 420)
Papuan languages (p. 413)
pidgin (p. 413)
Polynesia (p. 404)
power projection hub (p. 419)
seamount (p. 408)
shell banks (p. 417)
species diversity (p. 410)
Strategic Defense Initiative (SDI,
  or "Star Wars") (p. 421)
tsunami (p. 409)
typhoon (p. 410)

REVIEW QUESTIONS

WORLD
REGIONAL
Geography ⊛ Now™

Assess your understanding of this chapter's topics with additional quizzing and concept-based problems at http://earthscience.brookscole.com/wrg5e.

1. What three principal subregions comprise the Pacific World? What are some of the countries and colonial possessions of each?

2. Describe the major differences between high and low islands. According to geographer Tom McKnight, how did the physical characteristics of the islands relate to social and political developments? What are the typical demographic and economic qualities of each?

3. What is a geologic hot spot and how does it relate to Pacific islands?

4. What unique ecological attributes do islands generally have? What are the major threats to them?

5. What is an atoll and how does it form?

6. What are the major ethnic groups, languages, and religions of the Pacific World?

7. What are two notable ethnic conflicts on mineral-rich islands in the Pacific World?

8. How was Nauru able for a time to avoid the poverty typical of most Pacific islands? How is Nauru's economy diversifying now that its principal asset is dwindling?

9. What are the major economic activities in the Pacific islands, excluding Australia and New Zealand? What explains the peculiar jog in the International Date Line?

10. What fears do some Pacific countries have about potential global warming?

DISCUSSION QUESTIONS

1. Describe the formation of islands due to geologic hot spots and to the sinking of volcanoes.

2. Discuss some of the principal natural and human-made hazards associated with the Pacific Islands.

3. How might the story of Easter Island serve as a parable for human activity elsewhere?

4. How and when did various peoples and languages spread across the Pacific region?

5. How did Christianity get to the Pacific region? Discuss the experiences of the Benedictine monks in Australia's remote Kimberley area. What are cargo cults?

6. How have some Pacific nations taken advantage of economic opportunities in the Information Age?

7. What is the future of the people of Nauru? Is Nauru like Easter Island?

8. Discuss in general terms the pros and cons of foreign military presence in the Pacific World. Why is it described in the text as a mixed blessing? Should the United States pay for the cleanup and repatriation of Bikini? Describe the worldwide reaction to French nuclear weapons testing in the Pacific World. How did various countries show their dissatisfaction with the tests?

9. How does the island of Guam serve American strategic interests?

10. What strategic roles has Australia designated for itself in the region and elsewhere? What are some of the apparent consequences of these for Australia? Why is New Zealand no longer a member of ANZUS?

11. Why did Tuvalu propose a lawsuit against the United States and Australia?

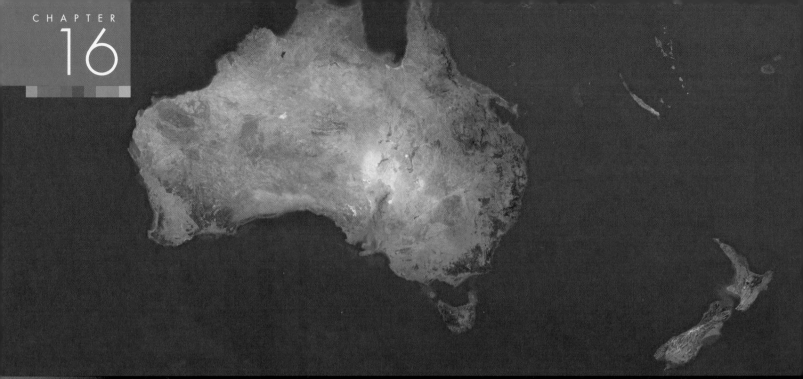

Prosperous, Remote Australia and New Zealand

Australia's Aborigines say that spirits "dreamed" the world and all its features into existence. This is a rock painting of one of those creator beings, the *wandjina*, a deity of wind and storms.

Joe Hobbs

chapter objectives

This chapter should enable you to:

- Recognize the parallels between the settlement of Australia and the United States

- Understand the obstacles faced by indigenous people in winning legal recognition of ancestral claims

- Evaluate the impacts of exotic species on island ecosystems

- Appreciate the process by which Australia and New Zealand are loosening ties with their ancestral European homeland and strengthening their regional orientation

- Consider the consequence of restrictions on immigration in these affluent and aging countries

WORLD
REGIONAL
GeographyNow™

Look for this logo in the text and go to GeographyNow at http://earthscience.brookscole.com/wrg5e to explore interactive maps, view animations, sharpen your factual knowledge and geographic literacy, and test your critical thinking and analytical skills with unique interactive resources.

Australia and New Zealand are unique. Far removed from Europe, their dominant cultures are nevertheless European. In the less developed Pacific World, they are prosperous. Distinctive plants and animals inhabit the often odd landscapes of these islands, stirring a sense of wonder among indigenous people, European settlers, and foreign tourists alike. Writer Alan Moorehead described the reactions of early European visitors: "Everything was the wrong way about. Midwinter fell in July, and in January, summer was at its height. In the bush there were giant birds that never flew and queer, antediluvian animals that hopped instead of walked or sat munching mutely in the trees. Even the constellations in the sky were upside down." [1]

16.1   Peoples and Populations

These two unusual countries are akin in population, cultural heritage, political problems and orientation, type of economy, and location. Australia and New Zealand are among the world's minority of prosperous countries. Australia's GDP PPP per capita of $29,000 is comparable to that of Japan and Canada, although well below that of the United States and the most prosperous European countries. Australia has been dubbed "The Lucky Country," and a recent World Bank study ranked Australia as the world's wealthiest country based on its natural resource assets divided by the total population (Figure 16.1). New Zealand is less affluent but is still prosperous compared with most nations of the world. In both countries, there are relatively few people to spread the wealth among; Australia's 20 million people plus New Zealand's 4 million amount to only about 70 percent of the people living in California.

Both countries owe their prosperity to the wholesale transplantation of culture and technology from the industrializing Great Britain to the remote Pacific beginning in the late 18th century. Australia (established originally as a penal colony for British convicts) and New Zealand are products of British colonization and strongly reflect the British heritage

Figure 16.1 For many at home and abroad, Australia epitomizes "the good life." This is Bondi Beach near Sydney.

[1] Alan Moorehead, *Cooper's Creek* (New York: St. Martin's Press, 1963), p. 7.

in the ethnic composition and culture of their majority populations. They speak English, live under British-style parliamentary forms of government, acknowledge the British sovereign as their own, and attend schools patterned after those of Britain. Despite their independence (for Australia in 1901 and New Zealand in 1907), both countries maintain loyalty to Great Britain. They still belong to the British Commonwealth of Nations. In both world wars, the two countries immediately came to the support of Great Britain and lost large numbers of troops on battlefields far from home. In World War I, Australia lost 60,000 of the 330,000 soldiers it sent to fight in Europe, by far the highest proportion of dead among any of the Allied countries. In World War II, U.S. forces helped to prevent a threatened Japanese assault on Australia. Since that time, Australia and New Zealand have sought closer relations with the United States, and British influence has waned (see Regional Perspective, pages 436–437).

Both Australia and New Zealand have minorities of indigenous inhabitants. The native Australian people are known as Aborigines. They have an extraordinarily detailed knowledge of the Australian landscape and its natural history and a complex belief system about how the world came about. According to Aborigines, Earth and all things on it were created by the "dreams" of humankind's ancestors during the period of **the Dreamtime.** These ancestral beings sang out the names of things, literally "singing the world into existence." The Aborigines call the paths that the beings followed the **Footprints of the Ancestors,** or the **Way of the Law;** whites know them as **Songlines.** In a ritual journey called **Walkabout,** the aboriginal boy on the verge of adulthood follows the pathway of his creator-ancestors. "The Aborigine clings to his native soil with every fiber of his being," wrote ethnographer Carl Strehlow. "Mountains and creeks and springs and water holes are to the Aborigine not merely interesting or beautiful scenic features. They are the handiwork of ancestors from whom he himself has descended. The whole country is his living, age-old family tree." [2]

The Aborigines numbered between an estimated 300,000 and 1 million when the first Europeans arrived in the 17th century. Colonizing whites slaughtered many of the Aborigines' descendants and drove the majority into marginal areas of the continent (see Ethnic Geography, pages 428–429). The Aborigines today number about 390,000, living mainly in the tropical northern region of the country. Some carry on a traditional way of life in the wilderness, but most live in generally squalid conditions on the fringes of majority white settlements.

When aboriginal populations began plunging precipitously, many white Australians assumed they would die out

entirely. When they did not, white Australia tried a program of **assimilation** of the Aborigines by settling them down, particularly in mission stations, and educating them in the service of whites. From 1910 to 1970, Australian authorities abducted about 100,000 aboriginal children (or between a tenth and a third of all aboriginal children) from their parents and, hoping to fully assimilate them, put them in the care of white foster parents. Today, members of the so-called **"stolen generations"** are seeking to identify their real kin and obtain apologies and compensation from the government. Thus far, the government has declined to make an official apology because that acceptance of responsibility might mean it would have to pay billions of dollars in reparation to the stolen children.

In New Zealand, the dominant indigenous culture is the **Maori,** a Polynesian people living mainly on North Island. Whites (known as *Paheka* to the Maori) broke the Maori hold on the land in a bloody war between 1860 and 1870. The Maori, who in 1900 seemed destined for extinction, have rebounded to make up almost 15 percent of New Zealand's population. Although their socioeconomic standing is poor, their situation is much better than that of the Australian Aborigines. The Maori are being increasingly integrated, both socially and racially, with New Zealand's white population.

At the same time, a Maori cultural and political movement is resisting the loss of traditional culture that this integration may cause. Like the Aborigines, the Maori are increasingly filing claims against the government to recover what they regard as traditional tribal lands. They assert that New Zealand has violated the spirit and letter of the **Treaty of Waitangi,** signed in 1840 by Maori tribes and the British government. In the agreement, the Maori gave up political autonomy in exchange for "full and undisturbed" rights to lands, forests, fisheries, and the "national treasures" of New Zealand. In one small but symbolically important concession to such rights, the government has made it legal for Maori to fish without a license.

16.2 The Australian Environment

Australia is truly a world apart. Located in Earth's Southern Hemisphere, where landmasses are few and far between, it is both a large island and a small continent (Figure 16.2). Its natural wonders and curiosities have provided endless inspiration, and sometimes consternation, to its inhabitants and to travelers and visiting writers. On arriving in Botany Bay—near Sydney Harbor—in 1788, British major Robert Ross declared, "I do not scruple to pronounce that in the whole world there is not a worse country than what we have yet seen of this. All that is contiguous to us is so very barren and forbidding that it may with truth be said here that nature is

---

[2] Quoted in Geoffrey Blainey, *Triumph of the Nomads* (Melbourne: Sun Books, 1987), p. 181.

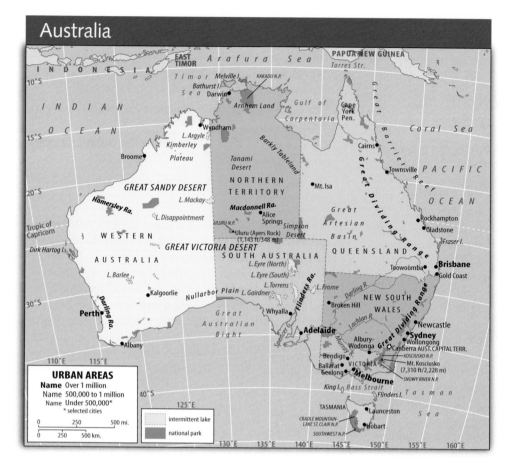

Australia

Figure 16.2 Principal features of Australia

reversed; and if not so, she is nearly worn out." [3] Mark Twain marveled at the Land Down Under in his book *Following the Equator:* "To my mind the exterior aspects and character of Australia are fascinating things to look at and think about. They are so strange, so weird, so new, so uncommonplace, such a startling and interesting contrast to the other sections of the planet." [4]

Including the offshore island of Tasmania, Australia has an area of nearly 3 million square miles (7.7 million sq km), about equal to the area of the contiguous United States (see Figure 15.2). On the basis of climate and relief, Australia has four major natural regions: the humid eastern highlands, the tropical savannas of northern Australia, the Mediterranean climate lands of southwestern and southern Australia, and the dry interior (see Figure 15.5a and b). Mainly because of extreme aridity, most of the continent is sparsely populated, and its people are concentrated where there is adequate fresh water and arable land. Most people live in the humid eastern highlands and coastal plains (especially in the cooler south),

in the areas of Mediterranean climate, and in the more humid grasslands adjoining the southern part of the eastern highlands and the Mediterranean areas. In only about three centuries of habitation, the mainly European peoples of Australia have transformed the island's ecosystems in profound ways, sometimes intentionally to suit their agricultural needs and sometimes accidentally—and disastrously (see Biogeography, page 431).

The Humid Eastern Highlands and Tasmania

Australia is the world's flattest continent. The only major highlands run down the east coast from just north of the Tropic of Capricorn to southern Tasmania in a belt 100 to 250 miles wide (161 to 410 km). These highlands are generally lower than 3,000 feet (914 m), but their highest summit and the highest point in Australia is Mt. Kosciusko, reaching 7,310 feet (2,228 m). The highlands and the coastal plains at their base are the country's only normally drought-free areas. South of Sydney and at higher elevations to the north, the climate is classified as marine west coast (despite the east coast location). North of Sydney, higher summer temperatures change the classification to humid subtropical, while

---

[3] From Howard Ensign Evans and Mary Alice Evans, *Australia: A Natural History* (Washington, D.C.: Smithsonian Institute Press, 1983), p. 12.

[4] Mark Twain, *Following the Equator: A Journey around the World* (Hartford, Conn.: American Publishing Company, 1897), p. 146.

Ethnic Geography

Australia's Indigenous People Reclaim Rights to the Land

As in the United States, newcomers to Australia forged a new nation by dispossessing the ancient inhabitants of the land. And as in the United States, in Australia there is a long history of white racism, discrimination, and abuse against people of color, who in Australia's case are the original inhabitants. Aborigines were not mentioned in the 1901 constitution, not allowed to vote until 1962, and not counted in the national census until 1967. The most disadvantaged group in Australian society, Aborigines suffer from high infant mortality (four times that of whites in Australia), low life expectancy (15 to 20 years lower than that of whites), and high unemployment (officially 20 percent, although aboriginal sources claim 50 percent, versus 7 percent for whites). Like Native Americans, a disproportionately large share of Aborigines fall prey to the economic and social costs of alcoholism. Statistics indicate they are 29 times more likely than non-Aborigines to end up in jail. Once in jail, they are more likely than non-Aborigines to commit suicide or be killed by guards. The Australian government has begun a program to improve the justice system for native Australians.

One of the largest issues of contention between Australia's indigenous people and the white majority is land rights. As in the United States, European newcomers pushed the native people off potentially productive ranching, farming, and mining lands into special reservations on inferior lands. The Europeans used a legal doctrine called *terra nullius,* meaning "the land was unoccupied," to lay claim to the continent.

Under the 1976 Land Rights Law, Aborigines were permitted to seek title to vacant state land ("Crown land") to which they could prove a historical relationship, but few succeeded. Following new legislation in 1992, the government began to return titles on a parcel-by-parcel basis—generally, in very marginal lands—to Aborigines who had argued their rights successfully. However, most Aborigines were unhappy with the terms of the returned titles, which allowed the native title to coexist with, but not supersede, the established Crown title. The government retained mineral rights to the land and could therefore lease native land to mining companies.

In 1993, aboriginal leaders expressed their objections to the legislation in a declaration known as the **Eva Valley Statement,** which made four demands:

1. Aborigines should have veto rights over mining and pastoral leases on native title land.

2. Mining and pastoral leases cannot extinguish native title to the land.

3. There should be no access by developers without aboriginal permission.

4. The federal government, not the states, should control native title issues.

Prime Minister Paul Keating rejected the Eva Valley Statement, claiming that the government had already made enough concessions. Aboriginal leaders took their case to the United Nations in Geneva, where they argued that existing legislation to validate mineral leases on native land was a breach of international human rights conventions, which Australia had signed.

In 1993, after the longest senate debate in the country's history, Foreign Minister Gareth Evans succeeded in pushing the **Native Title Bill** through Australia's parliament. The new legislation addressed the Aborigines' major objections and provided the following concessions:

1. Aborigines have the right to claim land leased to mining concerns once the lease expires.

2. Aborigines have the right to negotiate with mineral leaseholders over development of their land. However, they do not have the right of veto they requested.

3. Where their native title has been extinguished, Aborigines are entitled to compensation paid by the government.

4. With the help of a Land Acquisitions Fund, impoverished Aborigines can buy land to which they have proven native title.

The Native Title Bill is of great consequence in the states of South and Western Australia, where there is much vacant Crown land and many Aborigines are able to file claims on it. Those states are also heavily dependent on mining. Members of Australia's Mining Industry Council are unhappy with this legislation and worry that their mine leases will run out before the minerals do, which will require them to negotiate with the aboriginal titleholders.

Potential aboriginal claims to land expanded vastly in 1996 with another Australian High Court ruling on what is known as the **Wik Case** (named after the indigenous people of north Queensland who initiated it). The court concluded that Aborigines could claim title to public lands held by farmers and ranchers under long-term "pastoral leases" granted by state governments. These lands comprise about 42 percent of Australia's territory. The number of relevant aboriginal land claims awaiting judgment in the courts is 700 and growing.

There is bitter debate about what aboriginal claims to these lands would mean. Technically, in light of this court decision, Aborigines have a "right to negotiate," meaning they have a decision-making voice in how non-aboriginal leaseholders use the land. They also have a right to a share of the profits those leaseholders earn from their land uses. White farmers and ranchers protest that they would be ruined economically if they had to share profits. Generally, however, Aborigines have insisted their title rights would not mean running whites out of business but would simply confer nominal recognition of their title and access to the land, especially for visitation to sacred sites. Under pressure from white farmers, ranchers, and miners, Australia's conservative government under Prime Minister John Howard (elected in 1996) reacted to the Wik case with a 10-point plan. The plan would make it much more difficult for Aborigines to make native title claims and to challenge farming, ranching, and mining activities.

In the Northern Territory, which has its own land-rights legislation, vast tracts of land have already been given back to Aborigines. Aborigines there make up one-quarter of the population and now control more than a third of the land. Aborigines also hold title to Australia's two greatest national parks, Uluru (Ayers Rock) and Kakadu, both in the Northern Territory. In a **comanagement** arrangement that has become a model worldwide for management of national parks where indigenous people reside, the aboriginal owners of the reserves have leased them back to the Australian government park system, which comanages them with the Aborigines. Tourism programs in Uluru and Kakadu now highlight the cultural resources of the parks in addition to their natural wonders. At both parks, visitors can enjoy interpretive natural history walks that also emphasize aboriginal culture

Figure 16.A This aboriginal woman of the Anangu clan is a guide at Uluru (Ayers Rock) National Park.

and that are conducted by the land's oldest and most knowledgeable inhabitants (Figure 16.A). At Uluru, the Anangu Aborigines who own the rock hope that cultural awareness will cut down the number of tourists who climb the rock—estimated at 70 percent of the 400,000 annual total—because it is sacred to the Anangu.

Although there has been remarkable progress in recent years in this area and in others involving the Aborigines, the issue of rights to land has not been resolved to the satisfaction of all. "The handover of Ayers Rock is a turning point in Australia's race relations," declared Charles Perkins, then the country's top aboriginal civil servant. "It's a recognition that Aboriginal people were the original owners of this country." But aboriginal activist Galarrwuy Yunupingu declared, "I'd only call it a victory if the governor-general came and gave us title to the whole of Australia, and we leased that back to the Commonwealth." [a]

The Aborigines' struggle for legal recognition of traditional land tenure is representative of a growing trend among indigenous peoples in many parts of the world. These cultures are increasingly enlisting the aid of geographers, anthropologists, and other social scientists to document, measure, and analyze traditional land claims. The objective is to produce harder evidence of traditional land ownership that can be successfully presented in national and international court cases to win land rights. Such strategies are discussed at international meetings like the **International Forum on Indigenous Mapping**. Among the contributors at these meetings are geographers fresh from the field with Global Positioning System (GPS) data and maps testifying to traditional land claims. Many belong to the active Indigenous Peoples Specialty Group within the Association of American Geographers.

[a] Catherine Foster, "Australia's Ayers Rock: It's Not Just for Climbing Anymore." *Christian Science Monitor*, August 3, 1993, p. 10.

Figure 16.3 Large parts of Tasmania are rugged and wild. This is Wineglass Bay in Freycinet National Park.

still farther north, beyond approximately the parallel of 20°S, hotter temperatures and greater seasonality of rain create tropical savanna conditions.

Tasmania is a rugged, beautiful island off Australia's southeast coast (Figure 16.3). About 30 percent of its land is dedicated to national parks and other protected areas. The state's western region is heavily forested and dotted with lakes. Numerous rivers cut rapid, short courses to the sea, representing potential sources of hydroelectric power. At this latitude—roughly that of the U.S. state of Oregon—snow sometimes falls on the high peaks even in the height of summer (November–January). Tasmania has the unfortunate distinction of having lost its entire Aboriginal population to the afflictions of white colonialism. The island's economic prospects changed dramatically after 2001. Until then, it was a backwater losing population and employment. The 9/11 attacks in the United States, however, had the unlikely impact of making Australians feel less secure even in their vast continent. Many found what they perceived as safe harbor in Tasmania and snapped up homes and other real estate there.

The Tropical Savannas of Northern Australia

Northern Australia, from near Broome on the Indian Ocean to the coast of the Coral Sea, receives heavy rainfall during the season locals call "**the wet**"—the Southern Hemisphere summer season. The coastal zone is subjected in this season to hurricanes, known as cyclones in Australia. On Christmas Eve in 1974, a powerful cyclone flattened the city of Darwin, which was subsequently rebuilt. After such deluges, northern Australia experiences almost complete drought during the winter. This highly seasonal distribution of rainfall is the result of monsoonal winds that blow onshore during the summer and offshore during the winter. The seasonality of the

rainfall, combined with the tropical heat of the area, has produced a savanna vegetation of coarse grasses with scattered trees and patches of woodland (Figure 16.4). The long season of drought, the poverty of the soils, and the lack of highlands that would nourish perennial streams all combine to limit agricultural development. The alluvial and volcanic soils that support large populations in some tropical areas are almost absent in northern Australia.

The "Mediterranean" Lands of the South and Southwest

The southwestern corner of Australia and the lands around Spencer Gulf have a Mediterranean or dry-summer subtropical type of climate, with subtropical temperatures, winter rain, and summer drought. In winter, the Southern Hemisphere belt of the westerly winds shifts far enough north to bring precipitation to these districts, while in summer this belt lies offshore to the south and the land is dry. Crops introduced from the Mediterranean climate lands of Europe generally do well in these parts of Australia, but the shortage of highlands to catch moisture and supply irrigation water

Figure 16.4 Tropical savanna prevails in northern Australia. This is typical vegetation and terrain of the Kimberley region in the continent's extreme northwest.

Biogeography

Exotic Species on the Island Continent

Exotic species are nonnative plant and animal species introduced through natural or human-induced means into an ecosystem. Their impact tends to be pronounced and often catastrophic to native species, particularly on island ecosystems where resources and territories are limited. In Ecuador's Galapagos Island National Park, for example, cats, rats, pigs, goats, wasps, and other animals introduced by people have disrupted nests and food supplies and thereby reduced populations of endemic tortoises and other species.

The impacts of exotic species show how Australia truly is an island. Both inadvertently and purposely, people have introduced foreign plants and animals that have multiplied and affected the island, sometimes in biblical plague proportions. English settlers imported the rabbit for sport hunting, confident in the belief that, just as in England, there would be natural checks on the animal's population. They were mistaken (Figure 16.B). The rabbits bred like rabbits, eating everything green they could find, with ruinous effect on the vegetation. Observers described the unbelievable concentrations of the

Figure 16.B Rabbits were one of Australia's many human-caused scourges, and are still a problem, but not nearly as much as in the 1920s when people went on "rabbit drives" like these to chase, corral, and kill the declared vermin.

Hulton-Deutsch Collection/Corbis

lagomorphs as "seething carpets of brown fur." In the mid-20th century, the government mounted a huge eradication effort, employing fences, snares, dogs, guns, fires, and numerous other means, and exhorting combatants with the slogan "The Rabbit War Must Be Won!" An introduced virus called "white blindness" eventually helped reduce the vast numbers, but the animal is still prolific throughout large parts of the country.

Other exotic remedies have been applied, with varying success, to correct numerous other problem species, including mice, water buffalo, cane toads, and prickly pear cactuses. Many of the problem populations, notably cats and water buffalo, are feral animals—domesticated animals that have abandoned their dependence on people to resume life in the wild. An estimated 12 million feral cats infest the country and are held responsible for causing the extinction or endangerment of 39 native animal species. One Australian politician has been promoting legislation that would kill all cats on the continent by 2020.

Exotic livestock such as sheep and cattle are also a problem in the Australian environment. These hard-hoofed imports cut into the soil, promoting erosion and desertification. Some range-management scientists and conservationists believe that the future of Australian ranching lies with the country's 25 million kangaroos. These soft-hoofed marsupials are adapted to the Australian landscape and do not have such a damaging impact on the soil. Farmers and ranchers have traditionally eradicated about 2 million of them each year as pests. However, if more consumers at home and abroad could learn to appreciate kangaroo meat, there would be a strong incentive for ranchers to reduce their cattle and sheep herds and allow kangaroos to proliferate and be harvested. South Africa already imports more than 1,000 tons of kangaroo meat yearly from Australia, and nearly 4,000 more tons go each year to countries including Belgium, the Netherlands, Bulgaria, and the Czech Republic. Kangaroo skin is also used to make athletic shoes.

Australia is an unlikely exporter of another quadruped, the dromedary camel. The more than 200,000 camels in Australia are feral descendants of those brought from South Asia and Iran early in the 20th century to provide transport across the arid continent. Australian camels are exported even to Saudi Arabia, where their wild temperaments suit them to racing.

---

to the lowlands limits agricultural development. Australian wines produced in this region, particularly around Adelaide, are becoming increasingly popular abroad. Approximately 1,300 wineries in Australia now produce about 2 percent of the world's total wine output (Figure 16.5).

The Dry Interior

The huge interior of Australia is desert surrounded by a broad fringe of semiarid grassland (steppe) that is transitional to the more humid areas around the edges of the continent. This is the **outback** that has lent so much to Australian

Figure 16.5 Australian wine production is on the increase. This vineyard is in the Hunter Valley north of Sydney.

life and lore. Locals know it as **the bush,** the **back of beyond,** and the **never-never.** Altogether, the interior desert and steppe cover more than half of the continent and extend to the coast in the northwest and along the Great Australian Bight in the south. This huge area of arid and semiarid land is too far south to get much rain from the summer monsoon, too far north to benefit from rainfall brought by the westerlies in winter, and deprived of Pacific winds by the eastern highlands. The region is sparsely populated. The chief city is Alice Springs (population: 28,000; Figure 16.6), a famous frontier settlement that is the jumping-off point for visitors to Uluru (Ayers Rock).

WORLD
REGIONAL
Geography⊕Now™

Click Geography Literacy to take a virtual tour of Kimberly, "At the Top of Down Under." It includes many photos of the region.

Figure 16.6 Looking rather like a town in Nevada, this is Australia's frontier town of Alice Springs.

The western half of Australia is occupied by a vast plateau of ancient igneous, metamorphic, and hardened sedimentary rocks. Its general elevation is only 1,000 to 1,600 feet (c. 300 to 500 m), and its few isolated mountain ranges are too low to influence the climate or to supply perennial streams for irrigation. To the east, between the plateau and the eastern highlands, the land is even lower. The Lake Eyre Basin, the lowest part of this lowland—and of Australia—is the driest part of the continent. However, another part of the lowland, the Murray-Darling Basin, contains Australia's only major river system. It produces half of Australia's agricultural wealth and fully three-quarters of its irrigated crops. Some of the water comes from a massive engineering project, undertaken in the 1950s, to divert 99 percent of the waters of the Snowy River, which flowed into the Pacific, via underground tunnels to the basin. One unforeseen consequence is a dramatic rise in the water table, which has forced salts to the surface. This salinization is proving to be enormously destructive to the basin's cotton and rice. Consideration is now being given to growing more drought-tolerant olives and grapes and to returning more of the Snowy's water to its original drainage. Drought is a regular visitor to Australia, which weathered the worst drought episode in a century, with an accompanying plague of wildfires, in 2002–2003.

16.3 Australia's Natural Resource-Based Economy

Due to widespread aridity and the presence of highlands in its only humid area, Australia has little good agricultural land (see Figure 15.7). Although as much as 7 percent of the total land area is potentially cultivable, only about .05 percent of Australia is actually cultivated today. Fourteen percent of Australia's area is classified as forest and woodland, mostly of little economic value. More than 50 percent is classified as natural grazing land and about 20 percent as unproductive land. Despite its small proportion of arable area, however, Australia has one of the most favorable ratios of crop-producing land to population in the world.

The sectors of the Australian economy that regularly produce the country's leading exports are mining, ranching, and even with its small proportion of cultivable land, agriculture (which accounts for 3 percent of the country's gross domestic product and 17 percent of its total exports). Several factors support the country's emphasis on surplus production and export of primary commodities. There are large and dependable markets for Australian primary products in Japan, South Korea, China, the United States, Europe, and elsewhere. Growth has been particularly strong in recent years because of China's booming economy and the accompanying high demand for commodities and raw materials like copper, gold, and iron ore. These products can be taken to market with inexpensive and efficient transportation by sea. The Australian people can supply and import the necessary skills

290

and capital to exploit the continent extensively, utilizing a minimum of labor and a maximum of land and equipment. Australia's exports and its own service-oriented economy have given the country consistent, although not spectacular, growth since 1990. Australia has not experienced any period of recession since 1990, which is a remarkable achievement considering the country's strong economic relations with Asia and the United States, which did experience turbulent economic times.

Sheep and Cattle Ranching

In founding the first settlement around the excellent natural harbor of Sydney in 1788, the British government intended merely to establish a penal colony. There was to be enough agriculture on this "Fatal Shore" to make the colony self-sufficient in food and perhaps provide a small surplus of some products for export. Although self-sufficiency was elusive in the early days due to the poverty of the leached soils around Sydney, the colonists soon discovered that sheep did well, particularly the Merino breed that was first imported from Spain in 1796. The market for wool expanded rapidly with the mechanization of Britain's woolen textile industry. Australia thus found its major export staple in the early years of the 19th century, and until recently, sheep ranching was a strong and growing rural industry (Figure 16.7). Sheep grazers rapidly penetrated the interior of the continent, and by 1850, Australia was already the largest supplier of wool on the world market, a position it has never lost.

People have tried to raise sheep anyplace in Australia that seemed to offer any hope of success. Much of the interior has proven too dry, parts of the eastern mountain belt are too rugged and wet, and most of the north is too hot and wet in summer. The sheep ranching industry has thus become localized, concentrated mainly in a crescent-shaped belt of territory following the gentle western slope of the eastern highlands, from the northern border of New South Wales to the western border of Victoria (see Figure 15.7). Beyond this concentration, the sheep industry spreads northward on poorer pastures into Queensland and westward into South Australia. Still another area of production rims the west coast, especially near Perth. The sheep ranches, or "stations," in Australia vary in size from a few hundred to many thousands of acres.

The main product of the sheep stations is wool, which has been of great importance in Australia's development. Through most of the country's history, wool has been the principal export, amounting to as much as a quarter of all exports as late as the 1960s. Recently, however, it has declined rapidly in importance, partly because of the rise of other commodities in the Australian economy and partly because of wool's slowly losing battle with synthetic fibers. Today, Australian sheep are raised increasingly for meat. Each year, Australia exports large numbers of live animals to Saudi Arabia and other Muslim countries of the Middle East, where the devout slaughter them on the Feast of the Sacrifice commemorating the end of the pilgrimage to Mecca.

Lands in Australia unsuitable for sheep or for intensive agriculture are generally devoted to cattle ranching. Cattle stations are spread across the northern lands, with their hot, humid summers and coarse forage. The main belt of cattle ranching extends east–west across the northern part of the country, with the greatest concentration in Queensland (see Figure 15.7). On many of the remote ranches, especially in the Northern Territory, Aborigines are the principal ranch hands, known as "drovers" and "jackaroos."

218

Bill Bachman /Photo Researchers Inc.

Figure 16.7 Sheep are big business in both Australia and New Zealand. These motorized shepherds are mustering Merino sheep at the Narolah Station in southwestern Western Australia, on land that is productive for both sheep and wheat.

Beef is Australia's leading agricultural export, and Australia is the world's leading beef-exporting nation, a status achieved by the growing demand for meat that has accompanied increasing affluence in East Asia, Europe, and the United States. A single case of mad cow disease in the United States in 2003 also boosted Australian exports when several Asian countries banned American beef imports. The United States imports more Australian beef than any other nation. Australian cattle are pasture fed rather than grain fed and tend to be leaner than their U.S. counterparts. Americans prefer fattier meats, so much of the Australian beef is ground into hot dogs or mixed with American beef in hamburger.

Wheat Farming

Australia is also one of the world's leading wheat exporters (Figure 16.8). Wheat was the key crop in early attempts to make the settlement at Sydney self-sufficient. In the early years, the colony almost starved as a result of wheat failures, and the problem of grain supply was not solved until the development of the colony (now the state) of South Australia. There, settlers found fertile land and a favorable climate near the port of Adelaide, where wheat could be grown and shipped by sea to other coastal points.

Australian wheat production today is an extensive and highly mechanized form of agriculture, characterized by a small labor force, large acreages, and low yields per acre, but with very high yields per worker. The main belt of wheat production spreads from the eastern coastal districts of South Australia into Victoria and New South Wales, generally following the semiarid and subhumid lands that lie inland from the crests of the eastern highlands (see Figure 15.7). The main wheat belt has thus come to occupy nearly the same position as the main belt of sheep production. The two types of production are in fact often combined in this area.

Dairy and Sugarcane Farming

Although much less important than grazing and wheat farming for export revenues, dairy farming is Australia's leading type of agriculture in numbers of people employed. It has developed mainly in the humid coastal plains of Queensland, New South Wales, and Victoria (see Figure 15.7). Dairy farming has grown largely in response to the needs of the country's rapidly growing urban populations.

Australia is a sugar exporter with a unique history. The Queensland sugar-growing area is unusual in that it was a rare tropical area settled by Europeans doing hard manual work without the benefit of indigenous labor. Sugar production began on the coastal plain of Queensland in the middle of the 19th century to supply the Australian market. Laborers were initially imported from Melanesia to work as virtual slaves in the cane fields. However, when the Commonwealth of Australia was formed in 1901 by the union of the seven states (New South Wales, Victoria, Queensland, Northern Territory, South Australia, Western Australia, and Tasmania), Queensland was required to expel these imported "Kanaka" laborers in the interest of preserving a "white Australia." Anticipating that production costs would soar with the employment of exclusively white labor, the Commonwealth gave Queensland a high protective tariff on sugar. Operating behind this protective wall, the sugar industry has prospered there, though most of the original large estates have been divided into family farms worked by individual farmers.

Australia's Mineral Wealth

Unification of the separate Australian colonies into the independent federal Commonwealth of Australia encouraged the rise of industry, creating a unified internal market and tariff protection for manufacturing as well as for agricultural enterprises. Australia is rich in mineral resources,

Figure 16.8 Despite its limited arable land, Australia is a prodigious wheat producer. This is an aerial view of center pivot irrigation for wheat near Wyndham in northwest Australia.

and that wealth has facilitated industrial development. In addition to supplying most of its own needs for minerals, Australia has long been an important supplier to the rest of the world. In recent years, large new discoveries of iron ore, diamonds, and gold have been expanding this longtime role in international trade.

Mining has long had an impact on Australian life and landscapes. Gold played a large role in Australia's history, as it did in America's; some of the California forty-niners even continued their prospecting careers in Australia. Gold strikes in Victoria in the 1850s set off one of the world's major gold rushes and produced a threefold increase in the continent's population in just 10 years.

Ample supplies of coal also have aided Australian industrialization. Every Australian state has coal, but the major reserves are near the coast in New South Wales and Queensland. Most of Australia's steel-making capacity is located near the coal mines north of Sydney at Newcastle and at Wollongong to the south. Iron ore is brought from distant deposits along and near other sections of the Australian coast (Figure 16.9). Major new iron ore discoveries and developments in Western Australia have made Australia the world's leading exporter.

Figure 16.9 Australia is an important exporter of iron ore. This open-pit mine is in Western Australia.

Australia is by far the world's leading bauxite producer. Major reserves of bauxite are near the north Queensland coast, in the Darling Range inland from Perth, and on the coast of Arnhem Land. Aluminum plants supply the Australian market, and large quantities of the partially processed material called alumina are exported for final processing into aluminum metal at overseas plants.

Among Australia's other important mineral products are copper, lead, titanium, zinc, tin, silver, manganese, nickel, tungsten, uranium, and diamonds. Australia now produces 40 percent of the world's diamonds, most of them industrial cutting diamonds rather than gem diamonds. A diamond rush is under way, with prospecting and new finds in the country's northwest, both onshore and at sea. In the 1990s, Australia replaced Russia as the world's third largest gold-producing area, after North America and South Africa. Australia's most famous mining centers are three old settlements in the dry interior—Broken Hill in New South Wales, Kalgoorlie in Western Australia, and Mt. Isa in Queensland—all of which are still active. The Broken Hill Proprietary Company, which grew with the mines of that town, later branched into iron and steel production and other businesses to become Australia's largest corporation.

A shortage of petroleum has marred Australia's general picture of mineral abundance. There are only small oil fields. Most of these are on the mainland, but the main producing field is offshore Victoria. Although local supplies have aided the country's economy, Australia continues to import oil and to look for more domestic sources. Australia is blessed with a large endowment of natural gas, of which it is the world's fifth largest exporter, with China the main importer. A particularly large natural gas field called the Gorgon, off the northwest coast, is now being developed.

Urbanization and Industrialization

Although mining, grazing, and agriculture produce most of Australia's exports, most Australians—about 90 percent—are city-dwellers. This high degree of urbanization is one of the country's most notable and perhaps surprising characteristics. About 12 million people, or 60 percent of the population, live in the five largest cities: Sydney (the largest; population: 4.2 million; Figure 16.10), Melbourne, Brisbane, Perth, and Adelaide. All of these are seaports, and each is the capital of one of the five mainland states. The country's capital, Canberra (meaning "meeting place" in the aboriginal language), is one of its smaller cities (population: 320,000). Located inland about 190 miles (310 km) southwest of Sydney, Canberra is widely regarded as one of the world's most beautiful modern planned cities.

The Australian government has encouraged industrial development to support the country's growing population, to provide more adequate armaments for defense, and to increase stability through a more diversified and self-contained economy. A wealthy, urbanized, and resource-rich country

Regional Perspective

Australia and New Zealand: Less British, More Asian-Pacific

Australia and New Zealand have been remote satellites of Great Britain throughout most of their short histories. Even with independence, they pledged loyalty to the British Commonwealth of Nations. The United Kingdom was far more important than any other country in their trade relations, and as recently as the early 1970s, it accounted for about 30 percent of New Zealand's trade and 15 percent of Australia's. Their orientation is changing, however, as Australia and New Zealand seek stronger roles in the economic growth projected for the Pacific Basin. Great Britain is now a relatively insignificant trading partner of Australia and New Zealand. Already, seven of Australia's ten largest markets are in the Asia-Pacific region, with Japan, the United States, and China its leading trading partners (in that order). Australia, the United States, and Japan have become New Zealand's leading trade partners.

Perhaps the best symbol of Australia's new Pacific perspective is the debate over whether or not Australians should convert the country into a republic, ending more than 200 years of formal ties with Britain. Even now, according to Australia's constitution (which was written in 1901 by the British), Australia's head of state cannot be an Australian; it has to be England's king or queen (this continues to be the case in 16 of the 54 countries that make up the Commonwealth). Republic status would change that and allow Australians to have their own elected or appointed president. The monarch's portrait would be removed from Australian currency. The British Union Jack would be removed from its position in the upper left corner of the Australian flag.

In 1998, an Australian Constitutional Convention voted to adopt the republican model, and Australia's avowedly monarchist prime minister, John Howard, reluctantly agreed to put the issue to a referendum in 1999. In essence, the referendum asked Australians where they preferred their economic and political future to rest—either with the other nations of Asia and the Pacific or with Great Britain and the Commonwealth. A narrow majority (55 percent) voted to keep Australia's lot with Britain and the Commonwealth.

The debate has not gone away, however. Polls showed that at the time of the referendum, 60 percent of Australians favored a republic. Not all of these voted for republic status, however, because the referendum stated that the republic's president would be chosen by the parliament rather than elected by popular vote. If they were allowed to elect their president, the majority of Australians would probably vote for a republic at the next opportunity.

The monarchy/republic debate has led to something of an identity crisis for Australia and has focused more scrutiny on the sensitive issue of Asian immigration. Between 1945 and 1972, two million people, mainly British and continental Europeans, emigrated to Australia. They were allowed in by a **"white Australia" policy** that excluded Asian and black immigrants because of fears of invasion from Asia and a desire to increase the country's population with white, skilled, English-

Figure 16.10 One great attraction to international tourists is the beautiful port city of Sydney. The white structure is Sydney's famous Opera House.

Joe Hobbs

speaking immigrants. The rate of immigration has slowed since then, but it continues and has been opened considerably to skilled nonwhites. About half of the annual quota of 90,000 legal immigrants are Asian, who now make up about 7 percent of Australia's population. But Asians will form up to one-fourth of Australia's population by 2025, according to some projections. Australia has recently passed legislation requiring new immigrants to settle in cities other than Sydney, whose services threaten to be overwhelmed by growth.

Recent years have seen a marked surge in racism, targeted mainly at Asians. Pauline Hanson, a member of parliament and a candidate for prime minister in the 1998 election, said, "I believe we are in danger of being swamped by Asians," and vowed to halt Asian immigration if her One Nation Party took power. "A truly multicultural country can never be strong or united," she declared.[a] Ms. Hanson also insisted that Aborigines once practiced cannibalism and that therefore they are unworthy of political sympathy today. Her party was defeated soundly in the election. Since then, there has been a growing tide of illegal immigration from new sources, mainly Afghanistan and Iraq. These boat people are generally trafficked by Indonesian racketeers, who charge a refugee as much as $7,000 for the service. Prime Minister John Howard's vow to tighten restrictions against refugees and asylum-seekers helped win him a rare third term in office in 2001 (Howard won a fourth term in 2004).

[a] From Seth Mydans, "Sea Change Down Under: Drifting to the Orient." *The New York Times,* February 7, 1997, p. A4.

Ethnic tensions aside, Australia and New Zealand continue to forge important ties with their neighbors around the vast Pacific and with each other. Both countries adopted a free-trade pact in 1990 called the **Closer Economic Relations Agreement,** which eliminated almost all barriers between them to trade in farm and industrial goods and services. Both countries belong to the 18-member **Asia-Pacific Economic Cooperation (APEC)** group, which has agreed to establish "free and open" trade and investment between member states by the year 2020. Australia and the United States are the most aggressive member states in arguing for the abolition of trade quotas and the reduction of tariffs imposed on imported goods.

Australians refer to their Asian neighbors not as the "Far East" but as the "**Near North.**" Australia's Northern Territory city of Darwin is well positioned geographically to take advantage of new trading relations with the Near North. Darwin is closer to Jakarta, Indonesia's capital, than it is to Sydney, and Darwin is promoting itself as Australia's "Gateway to Asia." The city has established a Trade Development Zone to provide manufacturing facilities and incentives for overseas companies and for Australian companies wanting to do business overseas. A deep-water port was built in the 1990s, and a recently completed transcontinental Darwin–Adelaide railway called the Ghan is expected to help fuel trade between Australia and its neighbors to the northeast. Australia's Asia-oriented perspective has already changed the nature of one of the country's export staples—beef—as ranchers have been shifting to breeds of cattle better suited for live shipment by sea to Indonesia, the Philippines, and Thailand.

such as Australia should, like most more developed countries, be a major exporter of finished industrial products. However, approximately three-fourths of the country's exports come directly, or with only early-stage processing, from mining, agriculture, and grazing. Far from being an international industrial power of consequence, Australia does not meet its own demands for manufactured goods and relies heavily on overseas production. Australia is a big importer of manufactured goods from the United States, Japan, and China. The Labor Party government of Prime Minister Paul Keating was elected twice in part on its promise to correct this problem and to transform Australia from a protected agriculture- and mining-based economy to a lean, high-technology export economy. Its failure to achieve this goal contributed to the party's defeat in the 1996, 1998, and 2001 elections.

Tourism has become an important industry in Australia, particularly since the worldwide acclaim of the 1986 film *Crocodile Dundee.* (Crocodiles themselves occupy an im-

portant place in Australia's tourist business. In Kakadu National Park and other sections of northern and western Australia, many tours promise visitors a close-up look at "salties," the great saltwater crocodiles that are ancient and often deadly neighbors of people in the region.) Australia's attractions are sufficiently diverse to draw many different styles of international tourists, who number about 5 million yearly, to various destinations on the continent. The 1,200-mile-long (1,920-km) Great Barrier Reef, located off the northeast coast, is one of the world's most prized scuba-diving areas. Hotels and other services on nearby Queensland beaches of the Gold Coast cater to those looking for an idyllic, tropical, restful holiday. The geological features of Uluru (Ayers Rock) and the nearby Olgas in the country's center are on the itinerary of most international tourists. For many visitors, Sydney is as lovely a port city as San Francisco and Venice. Adventurous tourists raft the whitewaters and hike the mountains of Tasmania.

16.4 New Zealand: Pastoral and Urban

Located more than 1,000 miles (c. 1,600 km) southeast of Australia, New Zealand consists of two large islands, North Island and South Island (separated by Cook Strait), and a number of smaller islands (Figure 16.11). North Island is smaller than South Island but has the majority (about 77 percent) of the country's population of 4.1 million. With a total area of 104,000 square miles (269,000 sq km), the population density averages only 39 per square mile (15/sq km). New Zealand is thus a sparsely populated country, although not as sparsely populated as Australia.

Environment and Rural Livelihood

Much of New Zealand is rugged. The Southern Alps, often described as one of the world's most spectacular mountain ranges, dominate South Island (Figure 16.12). Film makers prize them; they featured prominently in *The Lord of the Rings* trilogy, for example. These mountains rise above 5,000 feet (c. 1,500 m), with many glaciated summits over 10,000 feet, and in the southwest, they are cut deeply by coastal fjords. The mountains of North Island are less imposing and extensive, but many peaks exceed 5,000 feet.

Figure 16.12 The Seaward Kaikoura Range of the Southern Alps

New Zealand's highlands lie in the Southern Hemisphere belt of westerly winds and receive abundant precipitation. While lowlands generally receive more than 30 inches (76 cm), distributed fairly evenly throughout the year, highlands often receive more than 130 inches (330 cm). Precipitation drops to less than 20 inches (50 cm) in small areas of rain shadow east of South Island's Southern Alps. Temperatures typical of the middle latitudes are moderated by a pervasive maritime influence. The result is a marine west coast climate, with warm-month temperatures generally averaging 60° to 70°F (16° to 21°C) and cool-month temperatures of 40° to 50°F (4° to 10°C). Highland temperatures are more severe, and a few glaciers flow on both islands.

Rugged terrain and heavy precipitation in the Southern Alps and in the mountainous core of North Island account for the almost total absence of people from one-third of New Zealand. About one-fifth of the country is completely unproductive, except for the appeal of the mountains to an expanding tourist industry. New Zealand's attractions for visitors include golfing, fishing, hunting, hiking, and camping in spectacular mountainous national parks, and geyser watching around the Rotorua thermal area of North Island.

Despite all of this wilderness, New Zealand has suffered the impacts that typically occur as a result of human activities in island ecosystems. Nearly 85 percent of the country's lowland forests and wetlands have been destroyed since the Maori arrived about 800 years ago. Of New Zealand's presettlement 93 bird species, 43 have become extinct and 37 are endangered (including the kiwi, the national bird and the informal namesake of all New Zealanders). As in Australia, exotic plant and animal species are a major threat to the indigenous varieties.

Well-populated areas are restricted to fringing lowlands around the periphery of North Island and along the drier east and south coasts of South Island. The climate of New Zealand's lowlands is ideal for growing grass and raising livestock (Figure 16.13). More than half of the country's total area is in pastures and meadows supporting major sheep and cattle industries. Pastoral industries contribute much to

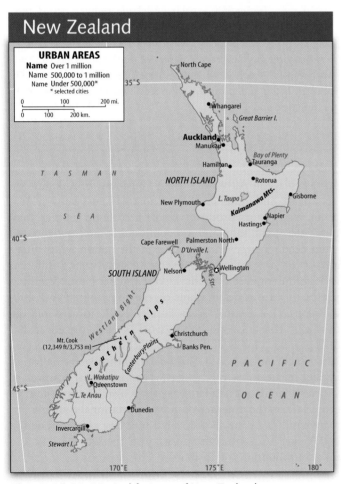

Figure 16.11 Principal features of New Zealand

Figure 16.13 A pastoral scene near Wellington

the country's export trade. Meat, wool, dairy products, and hides together account for about half of what New Zealand sells abroad. The country has developed a niche market in exports in very high-quality wines, allowing New Zealand to be competitive in an industry with a glut of cheaper vintages.

The earning power of its pastoral exports has given New Zealand a moderately high standard of living (per capita GDP PPP of $21,600, comparable to that of Spain). The United Kingdom was once the country's main market. When the United Kingdom joined the European Common Market in the 1970s, however, a new situation emerged because the Common Market (now the European Union) included countries with surpluses competitive with New Zealand's. In response, New Zealand tried to foster industrial growth and to expand trade with other partners, particularly in Asia. These efforts were not entirely successful. The economy suffered during the 1970s and 1980s, and the country slipped downward in the ranks of the world's affluent countries. A sizable emigration took place, mainly to Australia. These emigrants complained of not only the economic troubles but also boredom in an isolated, placid society in which tax policies made it difficult for almost anyone to make or keep a high income.

Today, there is a reverse flow of migrants, particularly of Asians immigrating to New Zealand's cities. Like Australia, New Zealand has begun to tighten up on immigration. Also like Australia, New Zealand is seeking to loosen or break its political and economic ties with the British Commonwealth of Nations and to reorient its economic orbit more toward Asia. New Zealand's economy felt the negative impact of the late 1990s Asian economic crisis as its exports to the region declined, but it has rebounded with renewed growth in Asia.

Industrial and Urban Development

In New Zealand, as in Australia, a high degree of urbanization and attempts to develop manufacturing supplement a basic dependence on pastoral industries (Figure 16.14). The two countries are similar in their conditions and purposes of urban and industrial development. New Zealand's resources for manufacturing do not equal those of Australia, but there

are modest reserves of coal, iron, and other minerals. There is much potential for hydropower development. New Zealand has magnificent natural forests, and production and export of forest products have become increasingly important as the country attempts to reduce its dependence on sheep and dairy exports.

New Zealand is not as urbanized as Australia. The country has only seven urban areas with estimated metropolitan populations of more than 100,000: Auckland (the largest; population: 1.1 million; Figure 16.15), the capital of Wellington (population: 345,000, briefly renamed "Middle Earth" to celebrate the premiere of *The Lord of the Rings* trilogy), at opposite ends of North Island; Napier-Hastings on the southeastern coast of North Island; Tauranga, on the Bay of Plenty; Hamilton, not far south of Auckland; and Christchurch and Dunedin on the drier east coast of South Island. Christchurch is the main urban center of the Canterbury Plains, which contain the largest concentration of cultivated land in New Zealand. About half of the country's population lives in these seven modest-sized cities (in comparison, Australia's Sydney metropolitan area has about the same population as all of New Zealand). Note that most of these cities have English names; as evidence of New Zealand's struggle to reidentify itself, there is a strong movement afoot to replace many of the country's Anglicized place names with their original Maori designations.

This concludes the survey of the Pacific World, with its far-flung islands and a single great island-continent. Although islands have an insular quality, this world region is, perhaps more than most, subject to influences from abroad because it depends so much on aid from and trade with greater continental powers. In the years to come, many islands will likely be forced to go it alone as foreign aid is withdrawn, and Australia may forge a new identity as a republic free of Great Britain. Rising sea levels as a result of global warming may also pose dilemmas for the region.

Figure 16.14 Queenstown, situated on the South Island, displays both the urbanity and wilderness that are New Zealand.

Joe Hobbs

Figure 16.15 Auckland, on North Island, is New Zealand's largest city.

SUMMARY

- Australia and New Zealand are products of British colonization. The British heritage is strongly reflected in the ethnic composition and culture of Australia's and New Zealand's majority populations.

- Both Australia and New Zealand have minorities and indigenous inhabitants. Native Australians are known as Aborigines, and the dominant indigenous group in New Zealand is the Maori.

- In Australia, one of the largest issues of contention between the Aborigines and the white majority is land rights. Aboriginal demands were declared in the Eva Valley Statement. In 1993, the Native Title Bill addressed the Aborigines' major concerns and made several land rights concessions. New Zealand's Maori have won some concessions in resource use.

- On the basis of climate and relief, Australia has four major natural regions: the humid eastern highlands, the tropical savannas of northern Australia, the "Mediterranean" lands of southwestern and southern Australia, and the dry interior.

- Exotic species are nonnative plants and animals introduced through natural or human-induced means into an ecosystem. Exotic species and human pressures have had catastrophic effects on native species and natural environments in Australia and New Zealand.

- The sectors of the Australian economy that regularly produce the country's leading exports are mining, ranching, and agriculture. Most exports go to Japan, the United States, and China.

- Most Australians—about 90 percent—live in cities. About 60 percent of the population lives in the five largest cities: Sydney, Melbourne, Brisbane, Perth, and Adelaide.

- New Zealand is located more than 1,000 miles (c. 1,600 km) southeast of Australia. The country consists of two large islands, North Island and South Island, and a number of smaller islands. South Island is very mountainous, and most people live on North Island.

- Pastoral industries, such as meat, wool, dairy products, and hides, contribute greatly to New Zealand's export trade.

- Australia and New Zealand have been reorienting their focus toward the Pacific Rim and away from the United Kingdom. An important expression of this change in focus is the ongoing debate about whether or not Australia should become a republic, thus ending more than 200 years of formal ties with Britain.

KEY TERMS + CONCEPTS

Terms in blue are also defined in the glossary.

Asia-Pacific Economic Cooperation group (APEC) (p. 437)
assimilation (p. 426)
Closer Economic Relations Agreement (p. 437)

comanagement (p. 429)
cyclone (p. 430)
Eva Valley Statement (p. 428)
exotic species (p. 431)
feral animals (p. 431)

Footprints of the Ancestors (Way of the Law, Songlines) (p. 426)
International Forum on Indigenous Mapping (p. 429)
Maori (p. 426)

Native Title Bill (p. 428)
"Near North" (p. 437)
salinization (p. 432)
stolen generations (p. 426)
terra nullius (p. 428)

the Dreamtime (p. 426)
the outback (the bush, the back of beyond,
  the never-never) (pp. 431, 432)
"the wet" (p. 430)

Treaty of Waitangi (p. 426)
Walkabout (p. 426)
"white Australia" policy (p. 436)
Wik Case (p. 428)

REVIEW QUESTIONS

WORLD
REGIONAL
Geography⊛Now™

Assess your understanding of this chapter's topics with additional quizzing and concept-based problems at http://earthscience.brookscole.com/wrg5e.

1. What features do Australia and New Zealand have in common?

2. Who were the first nonindigenous settlers of Australia? What brought them there?

3. What is the significance of the Dreamtime to Australia's Aborigines?

4. What are the four natural regions of Australia?

5. What advantages might kangaroo husbandry have over that of conventional livestock?

6. What are the major agricultural regions of Australia? What products come from them?

7. What factors support Australia's production and export of commodities and raw materials? What countries import these products?

8. What are Australia's largest cities? What characteristics do they share?

9. What are the notable landform features and natural resource assets of New Zealand?

DISCUSSION QUESTIONS

1. What are some popular nicknames for Australia and its regions? What do they tell us about the continent, the landscape, and the people? Describe some of the factors that make Australia unique as a continent.

2. Discuss the influence of Great Britain on the cultures of Australia and New Zealand.

3. Discuss some of the similarities between Australia's Aborigines and Native Americans. What other similarities are apparent between the settlement and experiences of Australia and the United States?

4. Discuss the struggles of the Aborigines and other indigenous peoples to obtain rights to traditional lands. What are some of the milestones and setbacks in their quest?

5. What is the history of Australia's sheep industry? For what purposes and to what destinations are sheep and sheep products exported?

6. What are some of the exotic plants and animals that have been introduced to Australia? What have been the effects of these exotics on Australia's natural environments?

7. What is notable about Darwin's geographic location? What does the city want to become?

8. What are some indications that Australia and New Zealand's orientation is beginning to focus on the Pacific Rim and is moving away from the United Kingdom?

9. What are the pros and cons, and the politics, of Australia's immigration policy? Compare Australia's immigration issues with those of the United States.

10. How do people in New Zealand make a living?

A Geographic Profile of Africa South of the Sahara

In the late morning, clouds begin to bank up around volcanic Mount Kirinyaga (Mount Kenya), Africa's second highest mountain.

Joe Hobbs

chapter outline

chapter objectives

This chapter should enable you to:

- Understand what caused Africa south of the Sahara to become and remain the world's poorest region

- Know how the region came to have the highest HIV/AIDS infection rates in the world and how the epidemic could be reversed there

- Appreciate the pressures on African wildlife and the unique approaches taken to protect the animals

- Know what is uniquely African about African cultures

- Recognize why, after more than a decade at the sidelines, Africa south of the Sahara is important again in geopolitical affairs

WORLD
REGIONAL
Geography⊛Now™

Look for this logo in the text and go to GeographyNow at http://earthscience.brookscole.com/wrg5e to explore interactive maps, view animations, sharpen your factual knowledge and geographic literacy, and test your critical thinking and analytical skills with unique interactive resources.

Geographers typically recognize the continent of Africa as having two major divisions: North Africa, which is the predominantly Arab and Berber realm of the continent, and ethnically diverse Africa south of the Sahara, often called sub-Saharan or Black Africa. In this text, North Africa including the Sudan is regarded as essentially Middle Eastern and is discussed in Chapters 8 and 9. This chapter sketches the broad outlines of the geography of Africa south of the Sahara, and Chapter 18 examines the major African subregions. The basic reference map for the region's 47 countries is Figure 17.1, and area and population data are in Table 17.1.

The region of Africa south of the Sahara is culturally complex, physically beautiful, and problem ridden. This introductory profile surveys the region's diverse environments, peoples and modes of life, large population concentrations, European colonial legacies, and major current problems. It includes a section dedicated to economic geography, but it is important to note at the outset that this is by far the world's poorest region. Africa ranks at the very bottom of every statistical indicator of global quality of life. Statistically, 76 percent of the region's people live in poverty, and 47 percent of them subsist on less than $1 per day. Africa is the only continent that has grown poorer in the last 25 years.

It is also a continent that has been plagued by strife, although in recent years conflict has abated significantly. Africa's plight may seem ironic because by some measures the region has the world's greatest variety of natural resources. But as these chapters will show, some of Africa's recent and current conflicts have not been about ideology or ethnicity but simply about control over resources such as diamonds, oil, and other minerals. A recent study of the world's civil wars since 1960 determined that there were three important risk factors for such conflicts: poverty, low economic growth, and high dependence on natural resources. Africa south of the Sahara has been richly endowed with these ingredients of war.

For many people, Africa south of the Sahara is the most poorly understood or misunderstood of the world's regions. It seems to have so many countries, so much violence and disease, and so much poverty that it is simply too difficult to think about. These chapters will show that getting to know Africa is manageable and can be both rewarding and interesting. There is much more to this region than suffering. Its diverse cultures and natural environments are sources of endless wonder, and conditions for many Africans are improving in a number of different ways.

17.1 Area and Population

Africa south of the Sahara (including Madagascar and other nearby Indian Ocean islands) has the second largest land area of all the major world regions described in this book. It

TABLE 17.1 Africa South of the Sahara: Basic Data

Political Unit	Area (thousand/ sq mi)	Area (thousand/ sq km)	Estimated Population (millions)	Annual Rate of Natural Increase (%)	Estimated Population Density (sq mi)	Estimated Population Density (sq km)	Human Develop- ment Index	Urban Popula- tion (%)	Arable Land (%)	GDP PPP Per Capita ($US)
The Sahel										
Burkina Faso	105.8	274.0	13.6	2.6	129	50	0.302	15	14	1100
Cape Verde	1.6	4.1	0.5	2.3	313	121	0.717	53	9	1400
Chad	495.8	1284.1	9.5	3.2	19	7	0.379	24	3	1200
Gambia	4.4	11.4	1.5	2.9	341	132	0.452	26	25	1700
Mali	478.8	1240.1	13.4	3.3	28	11	0.326	30	4	900
Mauritania	396.0	1025.6	3.0	2.7	8	3	0.465	40	0	1800
Niger	489.2	1267.0	12.4	3.5	25	10	0.292	21	3	800
Senegal	76.0	196.8	10.9	2.6	143	55	0.437	43	12	1600
Total	**2047.6**	**5303.2**	**64.8**	**3.0**	**31**	**12**	**0.353**	**27**	**3**	**1148**
West Africa										
Benin	43.5	112.7	7.3	2.7	168	65	0.421	40	18	1100
Ghana	92.1	238.5	21.4	2.2	232	90	0.568	44	16	2200
Guinea	94.9	245.8	9.2	2.7	97	37	0.425	33	3	2100
Guinea-Bissau	13.9	36.0	1.5	3.0	108	42	0.350	32	10	800
Ivory Coast	124.5	322.5	16.9	2.0	136	52	0.399	46	9	1400
Liberia	37.2	96.3	3.5	2.9	94	36	N/A	45	4	1000
Nigeria	356.7	923.9	137.3	2.9	385	149	0.466	36	31	900
Sierra Leone	27.7	71.7	5.2	2.1	188	72	0.273	37	7	500
Togo	21.9	56.7	5.6	2.7	256	99	0.495	33	46	1500
Total	**812.4**	**2104.1**	**207.9**	**2.7**	**255**	**98**	**0.460**	**37**	**20**	**1141**
West Central Africa										
Cameroon	183.6	475.5	16.1	2.2	88	34	0.501	48	13	1800
Central African Republic	240.5	622.9	3.7	1.7	15	6	0.361	39	3	1100
Congo Republic	132.0	341.9	3.8	2.9	29	11	0.494	52	1	700
Congo, Democratic Republic of	905.4	2345.0	58.3	3.1	64	25	0.365	30	3	700
Equatorial Guinea	10.8	28.0	0.5	2.6	46	18	0.703	45	4	2700
Gabon	103.3	267.5	1.4	2.1	14	5	0.648	73	1	5500
São Tomé and Principe	0.4	1.0	0.2	2.8	500	193	0.645	38	6	1200
Total	**1576**	**4081.8**	**84**	**2.8**	**53**	**20**	**0.404**	**35**	**4**	**1021**
East Africa										
Burundi	10.7	27.7	6.2	2.2	579	224	0.339	8	35	600
Kenya	224.1	580.4	32.4	2.3	145	56	0.488	36	8	1000

TABLE 17.1   (continued)

Political Unit	Area (thousand/ sq mi)	Area (thousand/ sq km)	Estimated Population (millions)	Annual Rate of Natural Increase (%)	Estimated Population Density (sq mi)	Estimated Population Density (sq km)	Human Development Index	Urban Population (%)	Arable Land (%)	GDP PPP Per Capita ($US)
Rwanda	10.2	26.4	8.4	1.9	824	318	0.431	17	40	1300
Tanzania	364.9	945.1	36.1	2.3	99	38	0.407	22	4	600
Uganda	93.1	241.1	26.1	3.0	280	108	0.493	12	26	1400
Total	**703.0**	**1820.7**	**109.2**	**2.4**	**155**	**60**	**0.449**	**22**	**9**	**963**
Horn of Africa										
Djibouti	9.0	23.3	0.7	2.3	78	30	0.454	82	0	1300
Eritrea	45.4	117.6	4.4	2.6	97	37	0.439	19	5	700
Ethiopia	426.4	1104.4	72.4	2.4	170	66	0.359	15	10	700
Somalia	246.2	637.7	8.3	2.9	34	13	N/A	33	1	500
Total	**727.0**	**1882.9**	**85.8**	**2.4**	**152**	**59**	**0.364**	**17**	**6**	**685**
Southern Africa										
Angola	481.4	1246.8	13.3	2.6	28	11	0.381	33	2	1900
Botswana	224.6	581.7	1.7	0.1	8	3	0.589	54	1	9000
Lesotho	11.7	30.3	1.8	1.1	154	59	0.493	17	11	3000
Malawi	45.7	118.4	11.9	3.1	260	101	0.388	14	23	600
Mozambique	309.5	801.6	19.2	1.7	62	24	0.354	29	5	1200
Namibia	318.3	824.4	1.9	1.6	6	2	0.607	33	1	7200
South Africa	471.4	1220.9	46.9	1.0	99	38	0.666	53	12	10,700
Swaziland	6.7	17.4	1.2	2.0	179	69	0.519	25	10	4900
Zambia	290.6	752.7	10.9	1.8	38	14	0.389	35	7	800
Zimbabwe	150.9	390.8	12.7	1.2	84	32	0.491	32	8	1900
Total	**2310.8**	**5984.9**	**121.5**	**1.6**	**52**	**20**	**0.509**	**38**	**5**	**5188**
Indian Ocean Islands										
British Indian Ocean Territory (U.K.)	0.02	0.05	0.003	N/A	150	58	N/A	100	0	N/A
Comoros	0.9	2.3	0.7	3.5	778	300	0.530	33	36	700
Madagascar	226.7	587.2	17.5	3.0	77	30	0.469	26	5	800
Mauritius	0.8	2.1	1.2	1.0	1500	579	0.785	42	49	11,400
Mayotte (Fr.)	0.1	0.3	0.2	3.2	2000	772	N/A	28	0	2600
Réunion (Fr.)	1.0	2.6	0.8	1.4	800	309	N/A	89	13	5800
Seychelles	0.2	0.5	0.1	1.0	500	193	0.853	50	2	7800
Total	**229.7**	**594.9**	**20.5**	**2.8**	**89**	**34**	**0.492**	**29**	**5**	**1663**
Summary Total	**8406.5**	**21,772.5**	**693.7**	**2.4**	**85**	**33**	**0.439**	**31**	**6**	**1767**

Sources: *World Population Data Sheet,* Population Reference Bureau, 2004; *U.N. Human Development Report,* United Nations, 2004; *World Factbook,* CIA, 2004.

Political Geography of Africa South of the Sahara

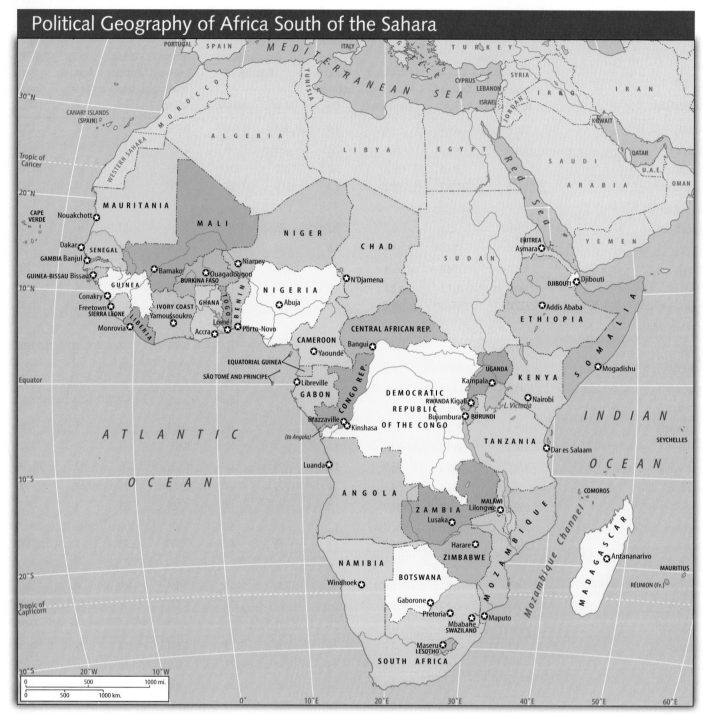

Figure 17.1 Africa south of the Sahara

covers 8.4 million square miles (21.8 million sq km) and so is more than twice the size of the United States (Figure 17.2). People overpopulation is apparent in some areas, and yet much of the region is sparsely populated. With a population of 694 million as of 2004, the region's average population density is slightly more than that of the United States. Even with the loss of population due to AIDS, the rate of natural population increase in Africa south of the Sahara is 2.4 percent per year, or about 4 times that of the United States. As in most less developed countries (LDCs), African parents

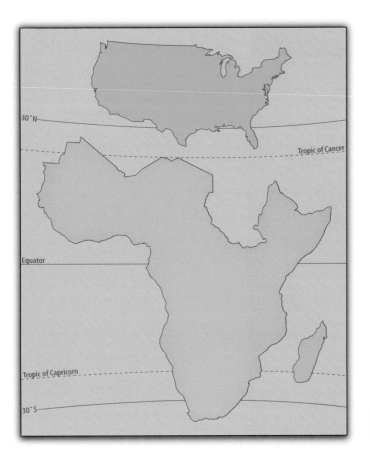

Figure 17.2 Africa south of the Sahara compared in area and latitude with the conterminous United States

generally want large families for several reasons: to have extra hands to perform work; to be looked after when they are old or sick; and in the case of girls, to receive the "bride wealth" a groom pays in a marriage settlement. Large families also convey status.

The majority of this region's people live in a few small, densely populated areas (Figure 17.3a and b). The main areas are the coastal belt bordering the Gulf of Guinea in West Africa from the southern part of Africa's most populous country, Nigeria, westward to southern Ghana; the savanna lands of northern Nigeria; the highlands of Ethiopia; the highland region surrounding Lake Victoria in Kenya, Tanzania, Uganda, Rwanda, and Burundi; and the eastern coast and parts of the high interior plateau of South Africa.

Africa is the world's most rural region (69 percent), with rural populations of most countries between 65 and 85 percent. The most rural are two East African nations with very fertile soils: Burundi (92 percent) and Uganda (88 percent). The most urbanized are Djibouti (82 percent), where there is no arable land and a small number of people are clustered in a port city, and Gabon (73 percent), where people are flocking to partake in an oil boom that is benefiting urbanites.

The region's major cities are discussed in Chapter 18. Life in villages is the rule, where a typical rural home is a small hut made of sticks and mud, with a dirt floor, thatched roof, and no electricity or plumbing (Figure 17.4).

This is the world's youngest population, where 44 percent of the region's people are younger than 15 years old (compared with about a third of the populations of Latin America and Asia). This is another indicator of how rapidly the region's population should grow, barring the vagaries of disease or famine. But the Malthusian scenario does seem to loom over Africa. Analysts fear the consequences of what they call the **"1 percent gap"**: Since the 1960s, the population of Africa south of the Sahara has grown at a rate of about 3 percent annually, while food production in the region has grown at only about 2 percent annually. The wild card in Africa's population deck is the **human immunodeficiency virus (HIV)**, and the disease it causes—**AIDS (acquired immunodeficiency syndrome)**—has inevitably been identified by some as the Malthusian "check" to the region's population growth (see Medical Geography, pages 449–451).

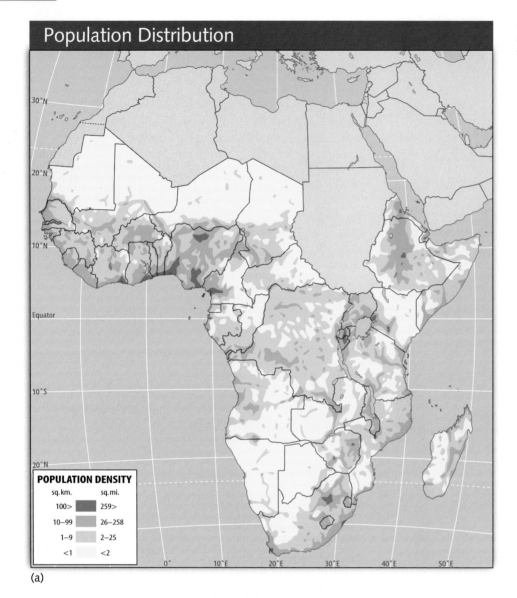

(a)

Figure 17.3 (a) Population distribution and (b) population cartogram of Africa south of the Sahara

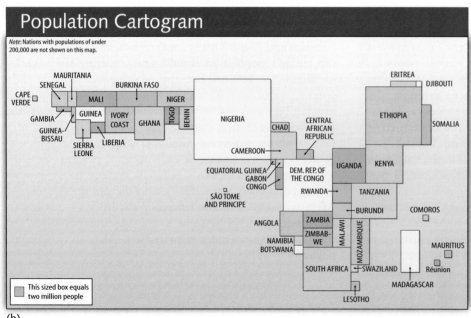

(b)

Figure 17.4 Homes in African villages are usually elevated above the ground to minimize the risk of flooding during the wet season, and their roofs are thatched. Streets and lanes are unpaved.

Joe Hobbs

Medical Geography

HIV and AIDS in Africa

AIDS is a global problem. It had killed 20 million people worldwide by 2004, and there are an estimated 5 million new infections around the world each year. Africa south of the Sahara is where the virus originated (apparently among chimpanzees, and the first human case was reported in the Congo in 1959) and is today the epicenter of this health crisis. The statistics are startling. In 2004, 64 percent of the world's estimated 39 million persons infected with HIV lived in Africa south of the Sahara. In this region, 7.5 percent of people aged 15 to 49 were HIV-positive in 2004. In 2003, 2.2 million died in the region. Public health authorities predict that 25 percent of the region's more than 700 million people will have died of AIDS by 2010, leaving behind 20 to 35 million orphans and slashing the regional economy by 25 percent.

The epidemic is most severe in southern Africa. In some countries, including Botswana and Zimbabwe, as much as one-fifth of the general population and more than a third of the adult population are HIV-positive. South Africa has more HIV cases (over 5 million in 2004) than any other country in the world. One in four adult South Africans, and one in nine of the general population, was HIV-positive in 2004. United Nations officials predict that if the epidemic continues apace, half of South Africa's 15-year-olds will die of AIDS in the coming years. Similar impacts are predicted for South Africa's neighbors. The effect would be to change the pyramid-shaped age-structure diagram typical of South Africa, Botswana, and Zimbabwe to a more tapered structure that demographers call "chimney"-shaped (Figure 17.A). In West Africa, more conservative, especially Islamic, mores about sexuality have slowed infection rates. However, AIDS is gaining ground rapidly in

that subregion. In Ivory Coast, for example, 7 percent of adults were HIV-positive in 2004.

Earth has not seen a **pandemic** (very widespread disease epidemic) like this since the bubonic plague devastated 14th-century Europe and smallpox struck the Aztecs of 16th-century Mexico. This scourge is causing sharp reductions in life expectancy in Africa and, unless contained, will dramatically alter recent projections of the region's population growth. Life expectancy in Botswana was 61 years in 1993, but because of the virus, it was 36 in 2004 and is projected to be 29 in 2010. Life expectancy in Zimbabwe and Namibia (41 and 47 years, respectively, in 2004) is projected to be 33 in 2010, and in South Africa (53 years in 2004), it is projected to be 35. In Zimbabwe, the population growth rate in 1998 was 1.5 percent rather than the projected 2.4 percent because of AIDS-related deaths. By 2004, Zimbabwe's growth rate was down to 1.2 percent. If the disease stays on course, the populations of Botswana, Zimbabwe, and South Africa will soon begin declining.

There is some potentially good news in the frightening picture of AIDS in Africa south of the Sahara. The number of new cases reported has declined slightly. Experts hope this will be a trend and an indication that the epidemic has reached what they call a "plateau." However, these authorities argue that this bit of hope is no reason to temper an all-out war on the problem.

This region is already the world's least developed, and HIV/AIDS is only worsening the prospects for development. Although it may seem logical that a lower population would mean more prosperity, the incidence of HIV infection is particularly high among the region's most educated, skilled, and ambitious young urban professionals, including teachers, white-collar workers, and government employees. These are the ones,

(continued on page 450)

HIV/AIDS in Africa

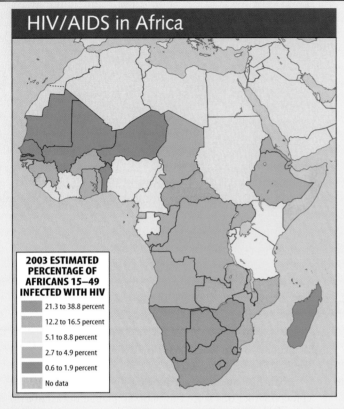

2003 ESTIMATED PERCENTAGE OF AFRICANS 15–49 INFECTED WITH HIV

- 21.3 to 38.8 percent
- 12.2 to 16.5 percent
- 5.1 to 8.8 percent
- 2.7 to 4.9 percent
- 0.6 to 1.9 percent
- No data

Botswana's Projected Age Structure Profile

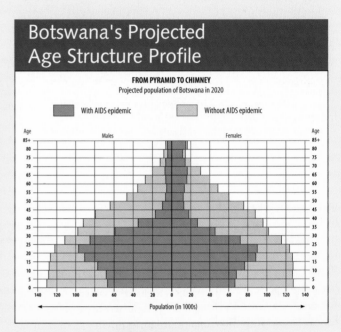

FROM PYRAMID TO CHIMNEY
Projected population of Botswana in 2020

- With AIDS epidemic
- Without AIDS epidemic

WORLD REGIONAL Geography⊗Now™

Active Figure 17.A HIV/AIDS has had a devastating impact in Africa south of the Sahara and threatens to redraw the demographic profiles of some countries in unprecedented ways. *Source: Data from UNAIDS. See an animation based on this figure, and take a short quiz on the facts or concepts.*

in addition to truckers and merchants, who travel the most, have higher incomes, and are therefore the most likely to have many sexual liaisons. Teachers are hit particularly hard. The typical pattern is that a young man trained in a city as a teacher takes his first post in a remote village. Because his salary is low, he often depends on the generosity of villagers for food and shelter. A village headman may assign a girl to cook and clean for him, and the teacher and his caretaker have sex. Girls in the village generally see the teacher as someone with elite status, and they vie for his attention, so he has multiple sexual partners. He will eventually come down with the virus and will die, but not before he has spread the virus further through the village population. In some villages, schools are closed because there are no teachers to replace the ones who have died from AIDS. These kinds of effects are incalculable but probably huge. They affect not only education but corporate profits, research, healthcare, tax revenue, and other indexes of progress.

There are other reasons, besides its unique way of causing a brain drain, that AIDS might contribute to underdevelopment in Africa. Sick people may be too weak to grow food. Sick workers are less productive. Family members must use precious funds to buy life-prolonging drugs rather than goods that would boost the economy. People anticipating short lives are

less inclined to save and invest money and less inclined to send their children to school. They may join gangs or engage in other reckless behavior, even joining the rebel armies that are a scourge in Africa (and that themselves are major transmitters of HIV).

Readers may wonder why the HIV infection rate is so high in this region, why people there do not practice safer sex, and why so little is being done to fight the epidemic. A number of factors come into play, and those listed here are generalizations that do not fit every African culture and country. First, attitudes about sexuality in some African societies are more relaxed than they are in American culture—for example, with less stigma associated with adultery or multiple partners. Second, both partners are often reluctant to use condoms. Even though she is most at risk of contracting the virus, the woman may insist that the man not use a condom because she does not want to be perceived as lacking virtue and be cast in the same high-risk category as prostitutes. Third, there is a widespread stigma about having AIDS. A person who contracts it is often shunned, even by family members, and often suffers and dies in neglect and isolation. The family or community will insist the person died of tuberculosis or some other ailment, so the real killer problem is never confronted. Fourth, despite the intensity of the problem, there is still a shortage of public awareness and education to fight the disease. Where leadership on this issue is needed most, it is sometimes absent. South Africa's President Thabo Mbeki, for example, was converted to the unusual belief that HIV does not cause AIDS, instead blaming poverty and malnutrition for the virus. Finally, there are inadequate health measures to fight the epidemic. Simple tests to determine the presence of HIV are seldom available, so those who have the virus pass it on unknowingly (the United Nations estimates that 90 percent of HIV-positive Africans do not know they have the virus). And mainly because of their high costs, anti-AIDS drugs are scarce. The drug nevirapine has a 50 percent success rate in blocking transmission of the virus from mother to child during childbirth. Its administration as a single pill just prior to delivery would thus have a substantial impact on reducing numbers of new HIV infections, but this drug has been prohibitively expensive.

What can be done about these difficult problems? Two countries that have had remarkable success in stemming the AIDS tide provide some answers. Uganda and Senegal have used relatively inexpensive public health tools, especially education, in their fights against the disease. To reduce the stigma associated with AIDS and to encourage counseling, testing, and condom use, the countries' political and religious leaders have been outspoken about AIDS. The results have been impressive. Uganda's adult infection rates dropped from 14 percent in the early 1990s to 8 percent by 2000 and 5 percent in 2004. Senegal's president spoke out forcefully about AIDS in the late 1980s, and infection rates have stayed below 2 percent ever since. Prostitution is legal in Senegal, but a vigorous education campaign about the virus has kept infection rates low among prostitutes and their clients.

The wealthier countries outside Africa could invest in combating the disease. Harvard University economist Jeffrey Sachs has suggested that if the rich countries combined to contribute just $10 to $20 billion per year to fight AIDS, malaria, and other disease in Africa, there could be huge progress in reducing death rates and thereby a great boost to the region's economic growth rate. This would not be a financial burden on the more developed countries, Sachs argues; in fact, they could afford such payments "without breaking a sweat."[a]

The United Nations in 2001 established its Global Fund to Fight AIDS, Malaria and Tuberculosis, setting an annual contribution goal of $7.5 billion, but contributions since then have totaled only one-fifth of that amount. American President George W. Bush pledged in 2003 that the United States would spend $15 billion over 5 years in its own effort to combat AIDS in twelve countries in Africa and two (Haiti and Guyana) in the Americas, with roughly half the monies spent on treatment, one-third on prevention, and the remaining one-sixth on care. Subsequent budget requests for actual funding, however, have fallen far short of that target. In sum, despite major goals and promises, wealthier countries and international organizations are not following through with major resources to fight HIV/AIDS.

Drug companies in the West have a unique opportunity to address Africa's AIDS problems. Some of them manufacture life-prolonging antiretroviral (ARV) drugs containing protease inhibitors that suppress replication of, but do not kill, the immunodeficiency virus. Importantly, these drugs do reduce the likelihood of the HIV-positive patient transmitting the virus to another person. The companies could help make such drugs more accessible; in 2004, only 4 percent of the 25 million HIV-positive Africans were receiving these medications. That would mean making the drugs inexpensive, which drug companies in the West have until recently been loath to do. An antiretroviral drug "cocktail" can cost $15,000 per patient per year in the United States.

However, other countries, notably Brazil and India, have taken advantage of a loophole in international patent rules to manufacture their own cheaper copies of the U.S.-made drugs. Called **compulsory licensing,** the loophole allows the countries to breach patents during national emergencies to manufacture generic versions of patented drugs. Citing AIDS as an emergency, Brazil produces virtually the same drug cocktail for $1,200 per patient per year, and India does it for $350 per year.

The Indian drug manufacturer Cipla has offered to sell the drugs to South Africa and other governments and to the organization Doctors without Borders for $140 per year (Doctors without Borders would then distribute the drugs free of charge). The pressure is now on U.S. companies to lower their prices on anti-AIDS drugs sold in Africa and elsewhere in the developing world. They hesitate to do so, not only because of the loss of revenue involved but out of fears that Americans would cry foul and demand lower costs on their own drugs.

The tide began to go Africa's way when a German company offered nevirapine free of charge for 5 years to developing countries. Soon some U.S. drug firms relented. Merck, for example, offered to make one of its key drugs available in Africa at $600 per patient per year, making no profit. The U.S. administration promised to keep the door open for even more, cheaper anti-AIDS drugs by declaring it would not seek sanctions against poor countries stricken with AIDS, even if American patent laws were being broken. Country-by-country decisions on how the multinational drug companies should distribute the drugs are now ongoing.

In the meantime, the challenge for MDC governments, aid agencies, and citizens is to help Africans pay for these drugs: Even $140 per year is out of reach. And the drugs will need to be correctly distributed and used; the manufacturers warn that if they are not, new strains of drug-resistant viruses could emerge.

---

[a] Jeffrey Sachs, "The Best Possible Investment in Africa." *The New York Times,* February 10, 2001, p. A27.

17.2 Physical Geography and Human Adaptations

Africa south of the Sahara is both rich in natural resources and beset with environmental challenges to economic development. It is home to some of the world's greatest concentrations of wildlife and to some of the most degraded habitats.

A Land of Plateaus and Rivers

Most of Africa is a vast plateau, actually a series of plateaus, with a typical elevation of more than 1,000 feet (305 m) (Figure 17.5). Near the Great Rift Valley in the Horn of Africa and in southern and eastern Africa, the general elevation rises to 2,000 to 3,000 feet (610 to 915 m), with many areas at 5,000 feet (1,520 m) and higher (see the accompanying

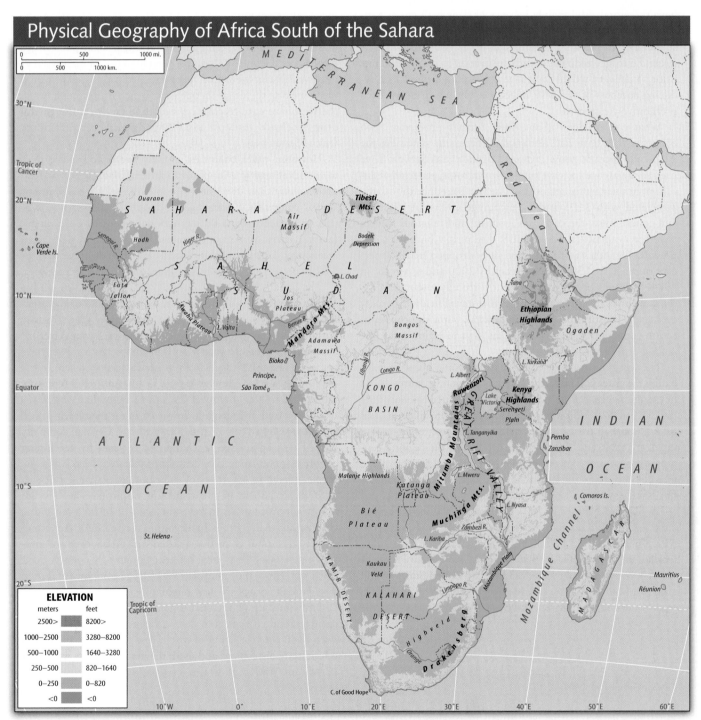

Physical Geography of Africa South of the Sahara

ELEVATION

meters	feet
2500>	8200>
1000–2500	3280–8200
500–1000	1640–3280
250–500	820–1640
0–250	0–820
<0	<0

Figure 17.5 Major topographic features, rivers, lakes, and seas of Africa south of the Sahara

Regional Perspective

The Great Rift Valley

One of the most spectacular features of Africa's physical geography is the Great Rift Valley, a broad, steep-walled trough extending from the Zambezi Valley (on the border between Zimbabwe and Zambia) northward to the Red Sea and the valley of the Jordan River in southwestern Asia (see Figure 17.5 and Figure 2.9 on global tectonics). Its relationship to the tectonic movement of crustal plates is still poorly understood. However, most earth scientists believe it marks the boundary of two crustal plates that are rifting, or tearing apart, causing a central block between two parallel fault lines to be displaced downward, creating a linear valley. This movement will eventually cut much of southern and eastern Africa away from the rest of the continent and allow seawater to fill the valley.

The Great Rift Valley has several branches. Lakes, rivers, seas, and gulfs already occupy much of it. It contains most of the larger lakes of Africa, although Lake Victoria, situated in a depression between two of its principal arms, is an exception. Some, like the 4,823-foot-deep (1,470-m) Lake Tanganyika (the world's second deepest lake after Russia's Lake Baikal), are extremely deep. Most have no surface outlet. Volcanic activity associated with the Great Rift Valley has created Mt. Kilimanjaro, Mt. Kirinyaga, and some of the other great African peaks, along with lava flows, hot springs, and other thermal features. Faulting along the Great Rift Valley in Ethiopia, Kenya, and Tanzania has also exposed remains of the earliest known ancestors of *Homo sapiens*.

WORLD
REGIONAL
Geography (✿) Now™

Click Geography Literacy for an animated figure of geological rifting.

Regional Perspective). The highest peaks and largest lakes of the continent are located in this belt. The loftiest summits lie within a 250-mile radius (c. 400 km) of Lake Victoria. They include Mt. Kilimanjaro (19,340 ft/5,895 m) and Mt. Kirinyaga (Mt. Kenya; 17,058 ft/5,200 m; Figure 17.6), which are volcanic cones, and the Ruwenzori Range (up to 16,763 ft/5,109 m), a nonvolcanic massif produced by fault-ing. Lake Victoria, the largest lake in Africa, is surpassed in area among inland waters of the world only by the Caspian Sea and Lake Superior. Other very large lakes in East Africa include Lake Tanganyika and Lake Malawi.

The physical structure of Africa has influenced the character of African rivers. The main rivers, including the Nile, Niger, Congo, Zambezi, and Orange, rise in interior uplands

Figure 17.6 Kenya's Mt. Kirinyaga (Mt. Kenya; 17,058 ft/5,200 m), an extinct volcano, is Africa's second highest mountain.

Joe Hobbs

Figure 17.7 The Kariba Dam straddles the Zambezi River between Zimbabwe and Zambia and provides hydroelectric power to both countries.

and descend by stages to the sea. At some points, they descend abruptly, particularly at plateau escarpments, so that rapids and waterfalls interrupt their courses. These often block navigation a short distance inland. Helping to offset this problem, Africa's discontinuous inland waterways are interconnected by rail and highways more than on any other continent. The Congo is used more for transportation than is any other river in the region. Unlike the Niger, Zambezi, Limpopo, and Orange, which have built deltas, the Congo has scoured a deep estuary 6 to 10 miles (10 to 16 km) wide that allows ocean vessels to navigate as far as Matadi in the Democratic Republic of Congo, about 85 miles (c. 135 km) inland.

The many waterfalls and rapids do have a positive side: They represent a great potential source of hydroelectric energy. There are major power stations on the Zambezi River at the Cabora Bassa Dam in Mozambique and at the Kariba Dam (Figure 17.7), which Zimbabwe and Zambia share; at the Inga Dam on the Congo River, just upstream from Matadi; at the Kainji Dam on the Niger River in Nigeria; and at the Akosombo Dam on the Volta River in Ghana. But only about 5 percent of Africa's hydropower potential has been realized (compared to about 60 percent in North America). Many of the best sites are remote from large markets for power. In some cases, geopolitical considerations pose obstacles to dam construction. Downstream Egypt, for example, has expressed concern and even hostile rhetoric about dams and water diversions of the Nile and its tributaries by upstream Sudan, Ethiopia, Uganda, Kenya, and Tanzania (see page 224 in Chapter 8).

Climate, Vegetation, and Water

The equator bisects Africa, so about two-thirds of the region lie within the low latitudes and have tropical climates and biomes; Africa is the most tropical of the world's continents (Figure 17.8a and b). One of the most striking characteristics of Africa's climatic pattern is its symmetry or regularity. This is due to the continent's position astride the equator and its generally level surface. Areas of tropical rain forest climate center around the great rain forest of the Congo Basin in central and western Africa. The forest merges gradually into a tropical savanna climate on the north, south, and east. This is the climatic and biotic zone supporting the famous large mammals of Africa. The savanna areas in turn trend into steppe and desert on the north and southwest. A broad belt of drought-prone tropical steppe and savanna bordering the Sahara Desert on the south is known as the Sahel. There is desert on the coasts of Eritrea, Djibouti, and Somalia in the Horn of Africa. In South Africa and Namibia, a coastal desert, the Namib, borders the Atlantic. The Kalahari Desert, which lies inland from the Namib, is better described as steppe or semidesert than as true desert. Along the northwestern and southwestern fringes of the continent are small but productive areas of Mediterranean climate, while eastern coastal sections and adjoining interior areas of South Africa have a humid subtropical climate. Bordering the subtropical climate region is an area of marine west coast climate.

Total precipitation in the region is high but unevenly distributed; some areas are typically saturated, whereas others are perennially bone dry. Even in many of the rainier parts of the continent, there is a long dry season, and wide fluctua-

Climates of Africa South of the Sahara

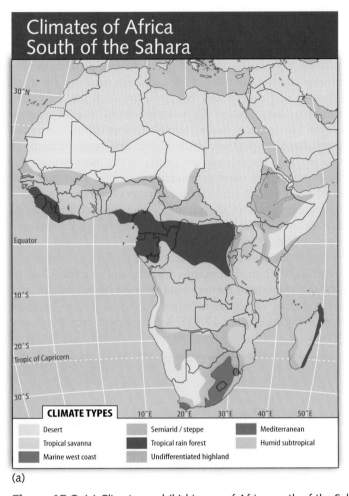

CLIMATE TYPES

Desert
Tropical savanna
Marine west coast
Semiarid / steppe
Tropical rain forest
Undifferentiated highland
Mediterranean
Humid subtropical

(a)

Biomes of Africa South of the Sahara

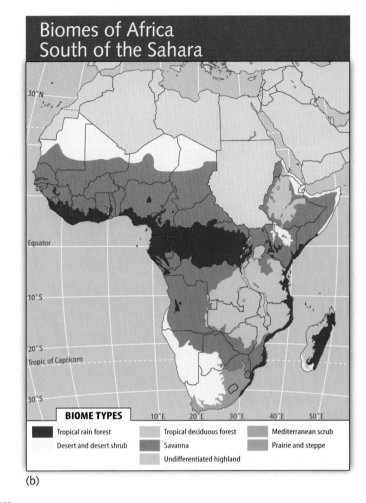

BIOME TYPES

Tropical rain forest
Desert and desert shrub
Tropical deciduous forest
Savanna
Undifferentiated highland
Mediterranean scrub
Prairie and steppe

(b)

Figure 17.8 (a) Climates and (b) biomes of Africa south of the Sahara

tions occur from year to year in the total amount of precipitation. One of the major needs in Africa is better control over water. In the typical village household, women carry water by hand from a stream or lake or a shallow (and often polluted) well. Use of more small dams would help provide water storage throughout the year.

Drought is a persistent problem in most of the countries. Although all droughts create problems, some last for years with devastating effects in this heavily agricultural region. Droughts have been particularly severe in recent decades in the Sahel and in the Horn of Africa, respectively in the northwest and northeast parts of the region.

Subsistence Agriculture and Pastoralism

The patterns of Africa's land use (Figure 17.9) reveal that the most productive lands are on river plains, in volcanic regions (especially the East African and Ethiopian highlands), and in some grassland areas of tropical steppes (notably the High Veld in South Africa). Soils of the deserts and regions of Mediterranean climate are often poor. In the tropical rain forests and savannas, there are reddish, lateritic tropical soils that are infertile once the natural vegetation is removed and can support only shifting cultivation.

Africa's soils favor **subsistence agriculture** (people farm-

ing their own food but producing little surplus for sale) and pastoralism, and over half of the region's people practice these livelihoods. Women do a large share of the farm work—they produce 80 to 90 percent of Africa's food—in addition to household chores and the bearing and nurturing of children (Figure 17.10). Mechanization is rare, fertilizers are expensive, and so crop yields are low. In the steppe of the northern Sahel, both rainfall and cultivation are scarce. The savanna of the southern Sahel, with its more dependable rainfall, is a major area of rain-fed cropping. Unirrigated millet, sorghum, corn (maize), and peanuts are major subsistence crops in the savanna. In the tropical savannas south of the equator, corn is a major subsistence crop in most areas, with manioc and millet also widely grown. Corn, manioc, bananas, and yams are the major food crops of the rain forest areas.

Many peoples, particularly in the vast tropical grasslands both north and south of the equator, are pastoral. Herding of sheep and hardy breeds of cattle is especially important in the Sahel. An increasing problem is that farmers often drive pastoralists from traditional grazing lands. Confined to smaller areas in which to browse and graze, the nomads' cattle, sheep, and goats often overgraze vegetation and compact the soil.

Agricultural Land Use

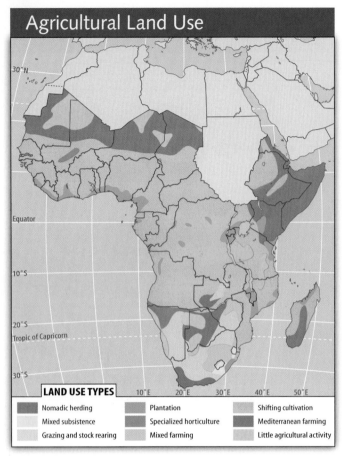

LAND USE TYPES

- Nomadic herding
- Mixed subsistence
- Grazing and stock rearing
- Plantation
- Specialized horticulture
- Mixed farming
- Shifting cultivation
- Mediterranean farming
- Little agricultural activity

Figure 17.9 Land use in Africa south of the Sahara

Although cattle raising is widespread through the savannas, cattle are largely ruled out over extensive sections both north and south of the equator by the disease called nagana, which is carried by the tsetse flies that also transmit sleeping sickness to humans (see Definitions and Insights, page 457). In tropical rain forests, tsetse flies are even more prevalent

Figure 17.10 Mother and child in Zimbabwe. Women plant and harvest most of Africa's food, while also rearing the children.

and few cattle are raised, but goats and poultry are common (as they are in tsetse-frequented savanna areas).

Most Africans who live by tilling the soil also keep some animals, even if only goats and poultry. Among African peoples such as the Maasai of Kenya and Tanzania and the Tutsi (Watusi) of Rwanda and Burundi, livestock not only contribute to daily diet but are an indispensable part of customary social, cultural, and economic arrangements. Cattle are particularly important, with sheep and goats playing a smaller role. The Maasai are probably the most famous African example of close dependence on cattle. They milk and carefully bleed the animals for each day's food and tend them with great care. The Maasai give a name to each animal, and herds play a central role in the main Maasai social and economic events through the year.

Madagascar is a good example of an African country in which cattle represent status, wealth, and cultural identity. Malagasy livestock owners tend to want higher numbers of the animals, rather than better quality stock, to enhance their standing (Figure 17.11). A Malagasy family practicing

Figure 17.11 Zebu cattle are extremely important in the cultures and household economies of rural Madagascar. There are about 10 million cattle in this country of 18 million people.

Definitions + Insights

Africa's Greatest Conservationist

Not much larger than the common housefly, the tsetse fly of sub-Saharan Africa packs a wallop. This insect carries two diseases, both known as **trypanosomiasis,** which are extremely debilitating to people and their domesticated animals. People contract **sleeping sickness** from the fly's bite, and cattle contract **nagana.** Since the 1950s, there have been widespread efforts to eradicate tsetse flies so that people can grow crops and herd animals in fly-infested wilderness areas (Figure 17.B). Where the efforts have been successful, people have cleared, cultivated, and put livestock on the land. The results are mixed. While people have been able to feed growing populations in the process of opening up these lands, they have also eliminated important wild resources and in many cases caused erosion, desertification, and salinization of the land. Where the tsetse fly has been eliminated, so has the wilderness. The diminutive tsetse fly thus may be characterized as a keystone species—one that affects many other organisms in an ecosystem. The loss of a keystone species—in this case, a fly that keeps out humans and cattle—can have a series of destructive impacts throughout the ecosystem. For its role in maintaining wilderness in Africa south of the Sahara, some wildlife experts know the tsetse fly as "Africa's greatest conservationist."

Figure 17.B Strenuous and successful efforts to eradicate tsetse flies have opened vast new areas of Africa to human use. This is a tsetse fly trap with two components: a jar containing a chemical liquid called "simulated cow's breath" that attracts the fly and a pesticide-soaked cloth that kills the fly.

the ritual commemoration of deceased ancestors sacrifices a large number of zebu cows for fellow villagers, and similar feasts accompany other important ritual dates. Due to such demands, the population of zebu cattle on the island is about 10 million. Their forage needs have grave consequences for Madagascar's rain forests and other wild habitats. People clear the forests and repeatedly set fire to the cleared lands to provide a flush of green pasture for their livestock, causing a rapid retreat of the island's natural vegetation.

Wildlife

Africa has the planet's most spectacular and numerous populations of large mammals. The tropical grasslands and open forests of Africa are the habitats of large herbivorous animals, such as the elephant, buffalo, antelope, zebra, and giraffe, as well as carnivorous and scavenging animals, such as the lion, leopard, and hyena. The tropical rain forests have fewer of these "game" animals (as Africans call them); the most abundant species here are insects, birds, and monkeys, with the hippopotamus, the crocodile, and a great variety of

fish present in the streams and rivers draining the forests and wetter savannas.

While film documentaries promote a perception outside Africa that the continent is a vast animal Eden, the reality is less positive. Human population growth, urbanization, and agricultural expansion are taking place in Africa, as elsewhere in the world, at the expense of wildlife. Hunting and competition with domesticated livestock also take their toll. Many species are protected by law, especially in the protected areas that make up about 14 percent of the region. Such laws are difficult to enforce, however, and poaching has devastated some species (see Regional Perspective, pages 458–459). Still, Africa remains home to some of the world's most extraordinary and successfully managed national parks, including South Africa's Kruger National Park, Tanzania's Ngorogoro Crater National Park, and Kenya's Amboseli National Park. International tourism to these parks is a major source of revenue for some countries, particularly South Africa, Tanzania, Kenya, Namibia, and Botswana.

Regional Perspective

Managing the Great Herbivores

Elephants and rhinoceroses are Africa's largest and most endangered herbivores. Their plight has accelerated in recent years and so have local and international efforts to maintain their populations in the wild. The problem has compelled countries with very different wildlife resources to work together toward solutions.

In 1970, there were about 2.5 million African elephants living on the continent. Poaching and habitat destruction reduced their numbers to 1.8 million by 1978. In 2005, there were an estimated 600,000. The main reason for the sharp decline is that elephants have something people prize: ivory. Poaching for ivory was reducing African elephants at a rate of 10 percent annually when delegates of the 112 signatory nations of the Convention on International Trade in Endangered Species (CITES) met in 1989. The organization succeeded in passing a worldwide ban on the ivory trade, and since then, the precipitous decline has halted.

CITES member states won the ivory ban over the strong objections of southern African states led by Zimbabwe. While the East African nations of Uganda, Kenya, and Tanzania were suffering crashing elephant populations, Zimbabwe was experiencing what it regarded as an elephant overpopulation problem (Figure 17.C). Zimbabwe had 5,000 elephants in 1900. In 2005, there were an estimated 88,000, and they were increasing at a rate of 4 percent annually. There are also healthy and growing elephant populations in Botswana, Malawi, Namibia, and South Africa. Before the worldwide ivory ban, these countries profited from the sustainable harvest and sale of elephant ivory, hides, and meat. At the 1989 CITES meeting, these countries argued that they should not be punished for their success in protecting the great mammals. They appealed for an exemption from the ivory ban so that they could earn foreign export revenue from a sustainable yield of their elephant populations. The majority of CITES members rejected this appeal, arguing that any loophole in a complete ban would subject elephants everywhere to illegal poaching, resulting in the loss of the African elephant. The elephant-rich countries reluctantly supported this position.

CITES reversed its policy in 1997 and elected to allow Zimbabwe, Namibia, and Botswana to sell ivory as a reward for their positive wildlife policies. These countries were permitted to export a combined total of 60 tons of ivory on a one-time basis only to Japan and only under the most tightly monitored circumstances. CITES granted a similar allowance to South Africa, Namibia, and Botswana in 2004. Critics argued that such loopholes in a complete ivory ban would promote increased elephant poaching across the continent and that CITES should resume the complete ban. Meanwhile, these elephant-rich nations continue to cull (kill) "excess" elephants. Zimbabwean officials argue that their country can support only 40,000 animals. Meat from the cull of about 5,000 to 7,000 elephants yearly goes to needy villagers and crocodile farms. This resource helped the country weather the drought and near famine of 1992. Licenses are also sold for male

Joe Hobbs

Figure 17.C Zimbabwe is blessed with elephants. These animals in Hwange National Park are feeding from a box of treats provided by the country's preeminent pachyderm ecologist, Alan Elliot (standing, wearing cap). Elliot is dead set against Zimbabwe's elephant cull program, which calls for 20,000 elephants to be killed in Hwange between 2005 and 2007.

elephants to be killed by foreign "trophy hunters," bringing a welcome source of revenue—$25,000 per animal—to this poor country.

If elephant ivory is like gold, rhinoceros horns are like diamonds. Men in the Arabian Peninsula nation of Yemen prize daggers with rhino horn handles (Figure 17.D). Although Western scientists deny the medicinal efficacy of powdered rhino horn, traditional medicine in East Asia (particularly in Taiwan, China, and Malaysia) makes wide use of it, including as an aphrodisiac. These demands, and the current black-market value of about $25,000 per horn, have led to a precipitous decline in population of black rhinoceroses in Africa. There were an estimated 65,000 black rhinos in Africa in 1982; poaching reduced their numbers to 2,300 by 1992. In 2004, thanks to strenuous antipoaching efforts, there were an estimated 3,600, which is still a critically low number.

Zimbabwe is on the front line in the hard-fought war to protect the black rhino. As recently as 1984, there were as many as 2,000 of the animals in the country. Having reduced the numbers in countries to the north, poachers turned to rhino-rich Zimbabwe. After 1984, there was a steady increase in the numbers of poachers crossing international boundaries to kill rhinos in Zimbabwe. Most of them came across the Zambezi River from Zambia. Zambia is a very poor country, and the prospect of making hundreds or thousands of dollars in a night's work was irresistible to many. Even the order to Zimbabwean wildlife rangers to shoot poachers on sight did not stop the slaughter. Armed with automatic weapons, poachers killed more than 1,500 rhinos between 1984 and 1991.

With fewer than 300 rhinos surviving in Zimbabwe in 1991, wildlife officials turned to more desperate measures. They began dehorning rhinos to make them unattractive to poachers. To do this, a sharpshooter tranquilizes the animal and two assistants use a chainsaw to remove the two horns. It is not a permanent solution because the horn grows back at a rate of 3 inches yearly, so each animal must be regularly re-dehorned. The program was not entirely successful. Poachers continued to kill the animals, perhaps out of spite, or because they could not tell in the dark if the prey has horns, or perhaps because they are content to have even a few inches of horn stump.

Nevertheless, the combination of efforts to combat the decline in rhino numbers has begun to take hold. By 2004, there were more than 450 rhinos in Zimbabwe. Ever vigilant to discourage poaching, Zimbabwean wildlife officials are now considering the possibility of opening a legal trade in rhino horns. They would raise rhino herds on state farms and regularly harvest their regrowing horns for sale. South Africa supports this idea of sustainable harvest. Like Zimbabwe, South Africa is sitting on a stockpile of tons of confiscated rhino horns and would profit greatly from their legalized trade. In addition, with about 1,100 living animals, South Africa is the last stronghold of the black rhino and does not want to become the next frontline state in the rhinoceros war.

Joe Hobbs

Figure 17.D Daggers are a nearly universal dress accessory for men in the Arabian Peninsula nation of Yemen. The most prized dagger handles are of rhino horn, a custom that has had a devastating impact on African rhinos thousands of miles distant.

17.3 Cultural and Historical Geographies

Many non-Africans are unaware of the achievements and contributions of the cultures of Africa south of the Sahara. The African continent was the original home of humankind. Recent DNA studies suggest that the first modern people (*Homo sapiens*) to inhabit Asia, Europe, and the Americas were descendants of a small group that left Africa via the Isthmus of Suez about 100,000 years ago. After about 5000 B.C., indigenous people were responsible for agricultural innovations in four culture hearths: the Ethiopian Plateau, the West African savanna, the West African forest, and the forest-savanna boundary of West Central Africa. Africans in these areas domesticated important crops such as millet, sorghum, yams, cowpeas, okra, watermelons, coffee, and cotton. From Africa, these diffused to populations in other world regions.

Civilizations and empires emerged in Ethiopia, West Africa, West Central Africa, and Southern Africa. In the first century A.D., the Christian empire based in the northern Ethiopian city of Axum controlled the ivory trade from Africa to Arabia. Ethiopian tradition holds that a shrine in Axum still contains the biblical Ark of the Covenant and the tablets of the Ten Commandments, which disappeared from the Temple in Jerusalem in 586 B.C. Several Islamic empires, including the Ghana, Mali, and Hausa states, emerged in West Africa between the 9th and 19th centuries. All of these agriculturally based civilizations controlled major trade routes across the Sahara. They profited from the exchange of slaves, gold, and ostrich feathers for weapons, coins, and cloth from North Africa. Three kingdoms arose between the 14th and 18th centuries in what are now the southern Democratic Republic of Congo and northern Angola. These included the Kongo kingdom, which had productive agriculture and was the hub of an interregional trade network for food, metals, and salt. In what is now Zimbabwe, the Karanga kingdom of the 13th to 15th centuries built its capital city at the site known as Great Zimbabwe. Its skilled metalworkers mined and crafted gold, copper, and iron, and merchants traded these metals with faraway India and China.

Some Common Culture Traits

In the 16th century, European colonialism began to overshadow and inhibit the evolution of indigenous African civilization. However, the artistic, technical, and entrepreneurial skills of the region's peoples continued to flourish. These traits are today part of a greater African culture that is poorly understood and often stereotyped in the wider world. In this diverse and often fragmented continent, there are many shared traits that together comprise what geographer Robert Stock, in the first edition of his geography of Africa, described as "**Africanity.**" He identified the following eight constituent elements of the African identity, based on observations by anthropologist Jacques Maquet:

1. A black skin color.
2. A unique conceptualization of the relationship between people and nature. Indigenous African religions emphasize that spiritual forces are manifest everywhere in the environment, in contrast with the introduced Christian and Muslim faiths, which tend to see nature as separate from God, and people as apart from and superior to nature (Figure 17.12).

Figure 17.12 The Central Mosque in Djenne, Mali. Spirituality is a strong component in each of Africa's many cultures and serves to influence many facets of everyday life.

3. An identity tied closely to the land, with many people dependent on hunting, herding, and farming. This dependence on the land reinforces the sense of closeness to nature. Africans tend to treat the land as communal rather than individual property.

4. Emphasis on the arts, including sculpture, music, dance, and storytelling, as essential to the expression of African identity.

5. A view of Africans as individuals making up links in a continuing "chain of life," in which reverence of ancestors and nurturing of children are virtues. Parents prefer to have many children to keep the chain growing and strive to educate them in the traditions of the ethnic group.

6. Extended rather than nuclear families, with parents and their children living and interacting with grandparents, cousins, nieces, nephews, and other relatives.

7. Respect for wise and fair authority, with village elders, "big men," and tribal chiefs endowed with powers they are expected to wield to benefit the group.

8. A shared history of colonial occupation that contributes to a unified sense that, in the past, Africans were humiliated and oppressed by outsiders.

Despite many apparently common cultural features, Africa south of the Sahara is culturally and ethnically diverse, with complex tribal and ethnic identities and modes of living.[1] In the second (2004) edition of his African geography, Stock warns that the notion of Africanity may be misleading because it focuses on rural cultures and does not appreciate the continent's cultural diversity. However, it would also be unfortunate to refuse to acknowledge that there is a cultural essence to Africa that can be appreciated both by Africans and outsiders. The eight elements of Africanity are therefore included here with the caution that they are only an introductory tool, are more relevant to rural than urban populations, and meant to open up rather than block doors to further exploration of the African cultures.

Languages

There is great linguistic diversity in Africa south of the Sahara (Figure 17.13). By one count, the peoples of this region speak more than 1,000 languages. Most of the peoples of this region belong to one of four broad language groupings:

1. The **Niger-Congo language family** (sometimes listed as a subfamily of the Niger-Khordofanian family) is the largest. It includes the many West African languages and the roughly 400 Bantu subfamily languages that belong in seven branches: **Benue-Congo, Kwa, Atlantic, Mandé, Gur, Adamawan,** and **Khordofanian.** Most of these are spoken south of the equator. The **Bantu** language of the Benue-Congo branch is the most widespread.

2. **The Afro-Asiatic language family,** including **Semitic** branch languages (such as the **Amharic** language of Ethiopia and **Arabic**) and tongues of the **Cushitic** (**Oromo** and **Somali** of the Horn of Africa, for example) and **Chadic** (especially the **Hausa** of northern Nigeria) branches. People living in the area adjoining the Sahara, from West Africa to the Horn of Africa, speak these languages. Even some of the Niger-Congo languages originating south of the Sahara, such as the **Swahili (Kiswahili)** tongue spoken widely in East Africa, have borrowed much from Arabic and other languages with roots elsewhere. The prominence of Arabic words in Swahili reflects a long history of Arab seafaring along the Indian Ocean coast of Africa; in fact, Swahili means "coastal" in Arabic.

3. The **Nilo-Saharan language family** of the central Sahel region, the northern region of West Central Africa, and parts of East Africa. This family includes about 100 languages in three branches: **Songhai** (a single language, spoken mainly in Mali) and the language groups **Saharan** and **Chari-Nile.**

4. The **Khoisan** languages of the San and related peoples in the western portion of southern Africa. Most languages in this family are extinct. The largest, **Khoekhoe (Damara),** is spoken by fewer than 200,000 people in Namibia.

The people of Madagascar speak a distinct language without African roots. This is **Malagasy,** an **Austronesian** tongue that originated in faraway Southeast Asia.

Africa's list of lingua francas—the languages most likely to be recognized on the continent and those that African can use to speak to the wider world—is provided by the six official languages of the African Union, the continent's supranational organization. They are English, French, Portuguese, Spanish, Swahili, and Arabic.

Religions

The religious landscape of Africa is complex and fluid. Spiritualism is extremely strong, but spiritual affiliations and practices are more interwoven and flexible than in most other world regions. It is not uncommon for parents and children to follow different faiths, for siblings to have different faiths, and even for an individual to change his or her religious beliefs and practices in the course of a lifetime.

Broadly, however, there are some dominant patterns of

---

[1] Politically and socially, most Africans have traditionally identified themselves by their tribe, recognizing members of the tribe as all those descended from kinship from a single tribal founder, or eponym. However, because *tribe* and *tribalism* have acquired connotations of primitive feuding between hostile rivals, many Africanists now prefer to use the terms *ethnic group* and *ethnicity* in their place. This blending of meanings can be misleading, however. Sometimes, as in Somalia, there is a single ethnic group made up of numerous tribes. Some countries, like Somalia, Lesotho, Swaziland, and Botswana, are very homogeneous ethnically, whereas Tanzania, Cameroon, and Nigeria have hundreds of ethnic groups.

412

Languages of Africa South of the Sahara

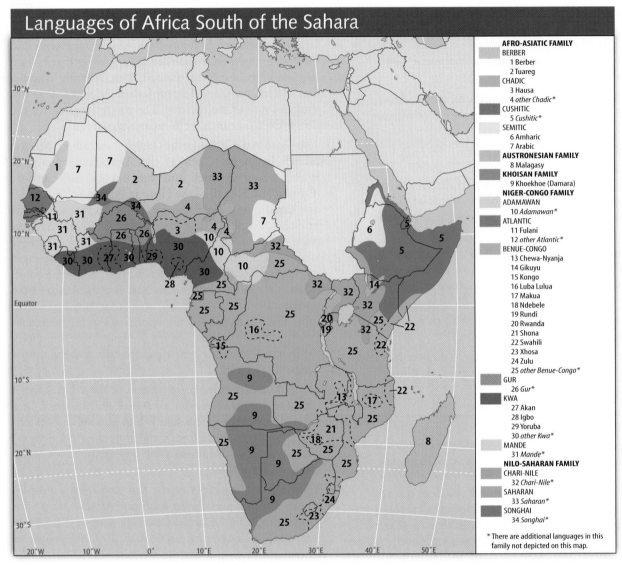

AFRO-ASIATIC FAMILY
BERBER
  1 Berber
  2 Tuareg
CHADIC
  3 Hausa
  4 *other Chadic**
CUSHITIC
  5 *Cushitic**
SEMITIC
  6 Amharic
  7 Arabic
AUSTRONESIAN FAMILY
  8 Malagasy
KHOISAN FAMILY
  9 Khoekhoe (Damara)
NIGER-CONGO FAMILY
ADAMAWAN
  10 *Adamawan**
ATLANTIC
  11 Fulani
  12 *other Atlantic**
BENUE-CONGO
  13 Chewa-Nyanja
  14 Gikuyu
  15 Kongo
  16 Luba Lulua
  17 Makua
  18 Ndebele
  19 Rundi
  20 Rwanda
  21 Shona
  22 Swahili
  23 Xhosa
  24 Zulu
  25 *other Benue-Congo**
GUR
  26 *Gur**
KWA
  27 Akan
  28 Igbo
  29 Yoruba
  30 *other Kwa**
MANDE
  31 *Mande**
NILO-SAHARAN FAMILY
CHARI-NILE
  32 *Chari-Nile**
SAHARAN
  33 *Saharan**
SONGHAI
  34 *Songhai**

* There are additional languages in this family not depicted on this map.

Figure 17.13 Languages of Africa south of the Sahara

religious geography (Figure 17.14). Islam is the dominant religion in North Africa and the countries of the Sahel on the southern fringe of the Sahara. The **Ethiopian Orthodox Church,** closely related to the Coptic faith of Egypt, makes Ethiopia an exception to the otherwise Islamic Horn of Africa region. Islam is also the prevailing religion of the East African coast, where Arab traders introduced the faith. Muslims are a majority or strong minority in rural northern Nigeria and Tanzania, and there are minority Muslim populations in cities and towns across the continent. Christians are a majority in southern Nigeria, Uganda, Lesotho, and parts of South Africa. Both Christianity and Islam are strongest in the cities, whereas traditional religions prevail or overlap with these monotheistic faiths in rural areas. There are Christian and Muslim efforts to dispose of "heretical" notions in African traditional belief systems, but these concepts are very resilient.

These broad strokes only begin to scratch the surface of African spirituality. Even where the large monotheistic faiths nominally prevail, and their followers are devout, there is a strong substrate of indigenous beliefs that has persisted and usually comfortably merged with the "official" faith—a syncretism also present in other cultures of the world. For example, as cited as one of the traits of Africanity, there is a strong reverence of ancestors, a spiritual belief that runs naturally with the strong social deference to the elderly and respect for preceding generations (see Perspectives from the Field, pages 464–465). Across the spectrum of African cultures, there is a widespread practice of using mediums to contact spirits, either of deceased ancestors, creator beings, or earth genies that can intercede on one's behalf. Many spirits are tied to particular places on the landscape, affirming the traits of Africanity that the identities of people themselves are tied closely to the land and that spiritual forces are

Generalized Religious Patterns

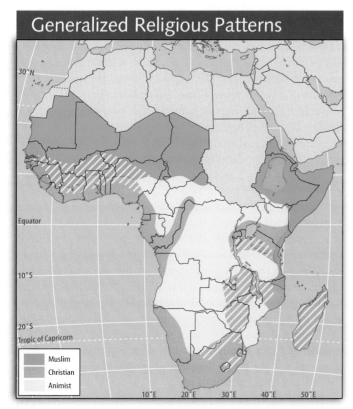

Figure 17.14 Religions of Africa south of the Sahara

manifest in the environment. Outsiders have tried, seldom successfully, to give accurate names to indigenous African belief systems, practices, and personnel: "ancestor worship," "animism," "*force vitale*" (living force), "living dead," and "witch doctor." While it would be gratifying to have accurate names and generalizations for African beliefs, they are unique and can best be understood by careful reading, observation, and discussion on a case-by-case basis.

The Geographic Impact of Slavery

Until about 1,000 years ago, the cultures of Africa south of the Saharan desert barrier remained largely unknown to the peoples north of the desert. Egyptians, Romans, and Arabs developed contacts with the northern fringes of this region and some trade filtered across the Sahara, but to most outsiders, the "Dark Continent" was a self-contained, tribalized land of mystery. Even at the opening of the 20th century, vast areas of interior tropical Africa were still little known to Westerners.

The tragic impetus for growing contact between Africa and the wider world was slavery (Figure 17.15). Over a period of 12 centuries, as many as 25 million people from Africa south of the Sahara were forced to become slaves, exported as merchandise from their homelands. The trade began in the seventh century, with Arab merchants using trans-Saharan camel caravan routes to exchange guns, books, textiles, and beads from North Africa for slaves, gold, and ivory from Africa south of the Sahara. As many as two-thirds of the estimated 9.5 million slaves exported between the

African Slave Trade

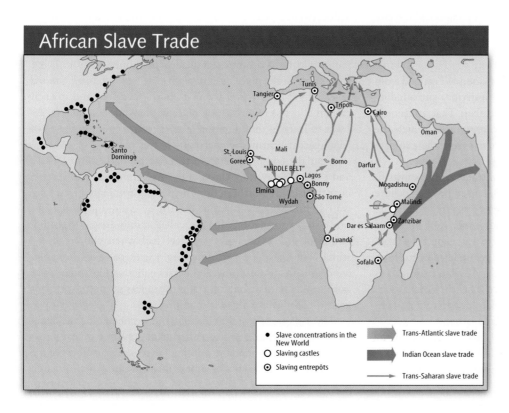

Figure 17.15 Slave export trade routes from Africa south of the Sahara *Source: Robert Stock, Africa South of the Sahara*

Under Madagascar

In the austral autumn of 2000, I went to Madagascar to investigate the relationship between people and caves on the island. This field mission, funded by the American Geographical Society, was part of a larger project I am working on, a book on human–cave interaction around the world since prehistoric times. Of all the countries I hope this project will take me to, Madagascar may have the most complex and fascinating history of human involvement with caves.

The air journey from the United States to Madagascar is one of the longest one may undertake on the planet. It was 42 hours from the time I left home in Columbia, Missouri, until I arrived in Antananarivo (Tana), Madagascar's capital. This included a 12-hour layover in Paris, where I visited some attractions of the legendary city. The fashionable shops and monumental opulence of Paris provided a huge contrast with the poverty and modest architecture of Madagascar's capital.

It is rare in my experience that such a long journey comes off without a hitch, so I was not surprised by the 3-hour delay of the Air Madagascar flight from Paris, during which excess weight, including my baggage, was unloaded from the aircraft. Twelve hours after takeoff, I arrived exhausted and without provisions (except for 30 pounds of camera and video gear in my hand luggage) but very excited about being in this exceptional country.

Fortunately, once in Tana, I was in the capable hands of an old friend and colleague, Steve Goodman. A zoologist based at Chicago's Field Museum, Steve is a pioneer in the natural history study of this extraordinary island, where nearly 90 percent of all plants and animals are endemic (occurring no where else on Earth). Steve has lived here since 1989 and has virtually become Malagasy. His wife, Asmina, is of the island's coastal Sakalava culture. Steve, Asmina, and their son Hisham reside with an extended family network, and as soon as I arrived from the airport, I was immersed in special festivities surrounding the impending wedding of a family member. That afternoon, these included the *fumba fumba,* a ceremony in which the bride's and groom's families seek approvals from their respective ancestors (*razana*) for the wedding.

It is impossible to overstate the significance of the spirits of one's ancestors in this culture. The Malagasy peoples believe them to influence or control virtually every aspect of daily life. Here, in Steve and Asmina's home, the *razana* were appeased with several long oratories and toasts of local rum. I managed an hour's nap before 10 P.M., when the group headed out for a raucous night of dancing held in, of all places, a prison mess hall (the inmates were tucked in their cells nearby). Here, your author danced to the jailhouse rock into the wee hours, concluding the first day of fieldwork in Madagascar.

But I had come to study caves and people, and when my baggage arrived 2 days later, Steve and I set out across the High Plateau, bound for the caves of northwestern Madagascar's region of Majunga. Not 3 hours into what should have been a 15-hour journey, Steve's vehicle developed a fuel line problem and began to hemorrhage gasoline. There was no choice but to return to Tana. At least this setback allowed me the opportunity to observe in more detail the environmental holocaust wrought by 2,000 years of human activity on the island. Relentless clearing of forest for rice cultivation and cattle pasture has created a virtual moonscape. The rivers run blood-red with what is left of the island's precious topsoil.

Steve was drawn away by other obligations, but I had the good fortune to be joined by his sister-in-law, Patty Vavizara, who would be my French and Malagasy interpreter and caving companion for the next 2 weeks. We flew to Majunga, hired a four-wheel-drive vehicle, and made our way to the vast and beautifully decorated caves of Anjohibe and Anjohikely. Rather than describe these, I will tell you of another cave, my favorite on this journey. We reached it late at night. We had not set out until dusk, and our two local guides had difficulty getting oriented in the darkness. There was no road to follow, and their views were obscured by grasses nearly twice the height of the vehicle. But they finally located Andoboara Cave, small and uninteresting but for its human and spiritual associations and therefore of huge interest to me.

The people of the nearby village of Ambalakedi consider this cave sacred because on three separate occasions, most recently in 1998, grief-stricken parents whose children had wandered into the forest had recovered them alive here. The children had been abducted by a sometimes malevolent spirit called *kalanoro* who sought to rebuke the parents for not taking proper care of their children. The parents sought guidance from another spirit called *tromba* who possessed a fellow villager and recited a laundry list of demands made by the *kalanoro* that would have to be met for the child to be found. Once the parents fulfilled their obligations by, for example, leaving honey and sacrificing zebu cattle in the place designated by the *tromba,* the *kalanoro* revealed via the *tromba* the lost child's location, and in each case, the joyous reunion took place in the Andoboara Cave. As I was to learn in the coming weeks, there are countless caves in Madagascar bearing rich accounts of this kind.

Near the end of my stay in this exceptional country, Patty and I did a 5-day walk through the Ankarana Special Reserve near the island's northern tip. The park is dominated by a long wall of Jurassic-era limestone eroded by wind and water into razor-sharp formations locally called *tsingy-tsingy* (meaning "ouch ouch," as in how your feet feel when you tread the stuff). The massif is perforated by scores of caves, some so enormous that even a powerful Petzl headlamp like mine could not illuminate their ceilings. Others are less grand in scale but graced with lovely calcite formations.

The Ankarana's sinkholes and the perimeter of the limestone massif contain protected remnants of dry tropical forest that are home to abundant wildlife. Each night, members of the primitive primate family of lemurs invaded our camp in search of food, calling so loudly that sleep came only through sheer exhaustion from the day's walking and caving. As a reptile lover, I was in paradise, indulging in the capture, photography, and release of boas and other snakes, riverine turtles, and my favorites, the chameleons.

But the caves and their human connotations were of greatest interest. At one night's campfire, a man from a village just outside the park told me how his ancestors, members of the royal family of the Antakarana tribe, sought refuge in the caves for nearly 3 years during the 1800s. They were escaping would-be slavery at the hands of the Merina tribe of highland Madagascar, whose goal was to unite (or pacify and enslave) the island's 17 ethnic groups. The Antakarana had taken advantage of the Merina's hopeless disorientation in the underground labyrinth of the Ankarana massif and survived in its caves while hunting, foraging, and growing some crops in the hidden karst canyons and sinkholes. The first four Antakarana kings were buried in caves here and are still visited on a ceremonial pilgrimage once every 5 years by the current king, Issa, and his subjects. I met with him to seek his permission to enter the royal cave tombs, but he said it was *fady* (taboo) even for him to do so, except on the specified occasion; he would welcome me to take part in the next event.

One particular day in the Ankarana was among the most extraordinary in my life. With an excellent guide named Aurelian Toly, we walked through the massive Andrafiabe Cave for several hundred meters before reaching a giant sunlit chasm containing what botanists call a "sunken forest," essentially a tropical paradise surrounded by karst desert wilderness. Lemurs howled and parrots cried in this lost world. I kept thinking, "Spielberg couldn't top this!" Crossing the forested ravine, we entered Cathedral Cave, whose entrance is littered with the hearths and pots used by the Antakarana in their time of refuge here. Well into the cave, we crossed a 600-foot-long (200-m) hill of guano built by a massive roost of insectivorous *Hipposideros* bats. Fresh deposits fell like raindrops on our helmets. Our noses were choked by the stench of ammonia and the dust of the guano trodden underfoot. Our headlights shined upon seething masses of cockroaches and centipedes feasting on the treasures dropped by the bats. We couldn't help but crunch this fauna as we walked, and with the shrieking of the annoyed bats, I thought again of a Spielberg movie set.

My last excursion into underground Madagascar was in Crocodile Cave. It is appropriately named, for running through it is a river in which crocodiles seek habitat while the world outside is parched in the long dry season of April–October. On the cave's sand banks, we saw abundant crocodile tracks, some fresh, and we crossed the cave stream with trepidation. Instead of finding the great reptiles (some up to 19.7 ft/6 m long have been seen here), about 3,300 feet (c. 1,000 m) into the cave we were astonished to come upon a dugout canoe beached on the bank. Next to the pirogue were three poles, each bearing on ei-

ther end a strand of eight to ten cormorants and other waterfowl tied by their feet. Some had already died and others struggled for life. In the next stretch of cave passage, our headlamps revealed the three poachers responsible, and a confrontation ensued. The two park guides in my party of seven cried out, "We are from the national park service. Come forward. Drop your weapons because the warden has a gun and will use it against you!" I was surprised to learn that I was the warden, and the GPS pouch on my belt bore the gun. The ruse was enough to convince the poachers, who now pleaded for leniency. They said they had known hunting was forbidden in the park, but the need to feed their families had driven them to take this long and strange canoe voyage into the cave.

I still did not understand. Why were there waterfowl in a cave? The park rangers demanded that the hunters return immediately to release the birds where they had taken them, and marching behind the poachers 1.25 miles (2 km) further into the cave, I had my answer (Figure 17.E). The cave opened up into another lost world, a great circular cavity in the Ankarana massif, this one filled not with forest but with marshes and lakes teeming with wildfowl. By this time, few of the birds were living, but the hunters did release them. The guides asked me (still feigning as warden) what to do with the criminals and their dead birds. "Would they do this again?" I asked. There was some discussion. Then the three poachers made what seemed a solemn vow that they would never again take wildlife from the park, lest the spirits of the ancestors of the Antakarana seek vengeance upon them. It was a particularly appropriate Malagasy way of dealing with a very understandable problem: the need of poor men to feed their families. For the geographer, the entire incident provided an especially fascinating insight into the relationship between people and caves in an extraordinary country.

Figure 17.E The poachers (with birds on poles) after being apprehended by the Ankarana park staff, here escorting them out of the Crocodile Cave

Joe Hobbs

years 650 and 1900 along this route were young women who became concubines and household servants in North Africa and Turkey. Male slaves usually became soldiers or court attendants (some of whom eventually assumed important political offices). From the 8th to 19th centuries, about 5 million more slaves were exported from East Africa to Arabia, Oman, Persia (modern Iran), India, and China. Again, most were women who became concubines and servants.

The notorious and lucrative traffic in slaves provided the main early motivation for European commerce along the African coasts, and it inaugurated the long era of European exploitation of Africa for profit and political advantage. The European-controlled slave trade was the largest by far. Between the 16th and 19th centuries, the capture, transport, and sale of slaves was the exclusive preoccupation of trade between the European world and West Africa. Portuguese and Spaniards began the trade in the 15th century, and a century later, English, Danish, Dutch, Swedish, and French slavers were active.

The business boomed with the development of plantations and mines in the New World. Populations of Native Americans in Anglo and Latin America were insufficient for these industries, so the Europeans turned to Africa as a source of labor. The peak of the trans-Atlantic slave trade was between 1700 and 1870, when about 80 percent of an estimated total of 10 million slaves made the crossing. In escape attempts, in transit, and in the famines and epidemics that followed slave raids in Africa, probably more than 10 million others died.

Slaves were a prized commodity in the triangular trade linking West Africa with Europe and the Americas (see Figure 2.17, page 32). European ships carried guns, ammunition, rum, and manufactured goods to West Africa and exchanged them there for slaves. They then transported the slaves to the Americas, exchanging them for gold, silver, tobacco, sugar, cotton, rum, and tropical hardwoods to be carried back to Europe. As "raw material" and as the labor working the mines and plantations of Latin America, the West Indies, and Anglo America, slaves generated much of the wealth that made Europe prosperous and helped spark the Industrial Revolution. The cycle of the triangular trade may be appreciated simplistically this way: New World cotton was transformed into textiles in Britain that were then traded in West Africa for slaves, who were bought by cotton profits in the New World, and who themselves grew more cotton.

While Europeans carried out the trade, their physical presence was limited to coastal shipping points. Africans were the intermediaries who actually raided inland communities to capture the slaves and assemble them at the coast for transit shipment. West African kingdoms initially acquired their own slaves in the course of waging local wars. As the demand for slaves grew, these kingdoms increasingly went to war for the sole purpose of capturing people for the trade. As the exports grew, so did the practice of Africans keeping African slaves. Even after Great Britain (in 1807) and the other European countries abolished slavery—finally bringing an end to the trans-Atlantic trade in 1870—slavery flourished within Africa. By the end of the century, slaves made up half the populations of many African states.

Slavery has not yet died out in the region (and indeed, a modern day form of slavery exists even in the United States; see page 602). In Mauritania, some Moors still enslave blacks, although the national government has outlawed this practice three times. Enslavement of children persists in West Africa. The typical pattern is that impoverished parents in one of the region's poorer countries, such as Benin, are approached by an intermediary who promises to take a child from their care and see that the boy or girl is properly educated and employed. This involves a fee (as little as $14) that, unbeknown to the parents, is a sale into slavery. The intermediary sells the child (on average, for $250 to $400) to a trafficker who sees that the child is transported overland or by sea to one of the region's richer countries, such as Gabon. There the child ends up working without wages and under the threat of violence as a domestic servant, plantation worker (especially on cocoa and cotton plantations), or prostitute. Benin is the leading slave supplier and trafficking center; Gabon, Ivory Coast, Cameroon, and Nigeria are the main buyers of slaves. Slavery also exists in the Sudan. In the 1990s, well-intentioned church groups and other organizations in the West began paying hard cash to buy freedom for slaves in Sudan. While many real slaves were freed, Sudanese profiteers also moved in with "counterfeit slaves," ordinary people hired to act as slaves until their "rescue" had been paid for.

Colonialism

Portugal was the earliest colonial power to build an African empire. The epic voyage of Vasco da Gama to India in 1497–1499 via the Cape route was the culmination of several decades of Portuguese exploration along Africa's western coasts. During the 16th century, Portugal controlled an extensive series of strong points and trading stations along both the Atlantic and Indian Ocean coasts of the continent. European penetration of the African interior began in 1850 with a series of journeys of exploration. Missionaries like David Livingstone, as well as traders, government officials, and now-famous adventurers and scientific explorers such as James Bruce, Richard Burton, and John Speke, undertook these expeditions. By 1881, when Africans still ruled about 90 percent of the region, their exploits had revealed the main outlines of inner African geography, and the European powers began to scramble for colonial territory in the interior. Much of the carving up of Africa took place at the Conference of Berlin in 1884 and 1885, when the French, British, Germans, Belgians, Portuguese, Italians, and Spanish established their respective spheres of influence in the region. By 1900, only Ethiopia and Liberia had not been colonized. Africa south of the Sahara had become a patchwork of European colonies—a status it retained for more than half a century (Figure 17.16)—and Europeans in these possessions were a privileged social and economic class.

Africa in 1914

Active Map 17.16 Colonial rule in 1914. Germany lost its colonies after World War I. *See an animation based on this figure, and take a short quiz on the facts or concepts.*

At the outbreak of World War II in 1939, only three countries—South Africa, Egypt, and Liberia—were independent. The United Kingdom, France, Belgium, Italy, Portugal, and Spain controlled the rest. But after the war, mainly in the 1960s and 1970s, there was a sustained drive for independence. Africa ceased to be a colonial region and came to hold more than one-fourth of the world's independent states. This was a peaceful process in most instances, but bloodshed accompanied or followed independence in several countries, including what are now Angola and the Democratic Republic of Congo.

Dependency theorists often point to Africa as a prime example of how colonialism created lasting disadvantages for the colonized. European colonization produced or perpetuated many negative attributes of underdevelopment. These included the marginalization of subsistence farmers, notably those who colonial authorities—intent on cash crop production—displaced from quality soils to inferior land. In addition, European use of indigenous labor to build railways and roads often took a high toll in human lives and disrupted countless families. Furthermore, the colonizers often corrupted traditional systems of political organization to suit their needs, sowing seeds of dissent and interethnic conflict.

The European colonial enterprise did have some positive impacts. The colonies, and the independent nations that succeeded them, were the beneficiaries of new cities and the transport links built with forced or cheap African labor; new medical and educational facilities (often developed through Christian missions); new crops and better agricultural techniques; employment and income provided by new mines and modern industries; new governmental institutions; and government-made maps useful for administration and planning. Such innovations were very helpful, but they were distributed unequally from one colony to another and were inadequate for the needs of modern societies when independence came.

One of the persistent problems in Africa's political geography is that many modern national boundaries do not correspond to indigenous political or ethnic boundaries. In most cases, this is another legacy of colonialism: British, French, and other occupying powers created artificial administrative units that were transformed into countries as the colonial powers withdrew. Nigeria is a good example of an ethnically complex, unnaturally assembled nation. On the political map, it appears as an integral unit, but its boundaries have no logical basis in physical or cultural geography. Some

Nigerians still refer to the "**mistake of 1914,**" when British colonial cartographers created the country despite its ethnic rifts. Simplistically, the greatest divide is between Muslim north and Christian south, but there are between 200 and 400 ethnic groups within the country. The British colonizers invested more in education and economic development in the south and built army ranks among northerners, whom they thought made better fighters; so the subsequent pattern is that the northern military leaders have tried to blunt the economic clout of southerners. Colonial administration and boundary-drawing thus sowed seeds of modern conflict in Nigeria. This problem and many like it are described in the subregion and country profiles in Chapter 18.

Although formal political colonialism has vanished, most countries still have important links with the colonial powers that formerly controlled them, and many foreign corporations that operated in colonial days still maintain an important presence. France has a long history of postindependence intervention in the political and military affairs of its former African colonies. France is the only ex-colonial power to keep troops in Africa (with the highest numbers in Djibouti, Ivory Coast, and Chad). In postcolonial Africa, France also has taken steps to ensure that most of its former colonies trade almost exclusively with France, and it in turn has supported national currencies with the French treasury. But France has found its paternalistic approach extremely expensive and has begun reducing its military presence and other costly assistance to its African clients.

17.4 Economic Geography

Great poverty is characteristic of Africa south of the Sahara; 15 of the world's 20 poorest countries are there (see Table 17.1). All of the economies except South Africa's are under-industrialized. Africa's place in the commercial world is mainly that of a producer of primary products, especially cash crops and raw materials (particularly minerals), for sale outside the region. In most nations, one or two products supply more than two-fifths of all exports—for example, oil in Angola and coffee and tea in Kenya (Figure 17.17). Such a country is vulnerable to international oversupply of an export on which it is vitally dependent. The value of imports far exceeds that of exports in Africa south of the Sahara, with imports consisting mainly of manufactured goods, oil products, and food.

Social and structural problems contribute to the region's underdevelopment. Most African societies lack a substantial middle class and the prospect of upward economic mobility. Instead, most are hierarchical, and any significant income tends to flow into the hands of a small elite controlling the lion's share of the nation's wealth. There are not enough schools to promote the economic welfare that can accompany literacy, and attendance in many is poor. Altogether only 60 percent of all children attend primary school, the lowest percentage of any world region. Bureaucratic obstacles and corruption can make starting up a new business a long, painful, and costly ordeal, discouraging investment both by Africans and foreigners.

Despite the overall grim picture of African economies, in recent decades there have been some changes for the better. Improved standards of health and literacy in many areas have resulted from the work of national and international governmental and nongovernmental organizations (NGOs). The extension of roads, airways, and other transportation facilities has promoted the marketing of farm products, including perishable items, from formerly inaccessible areas. Stores and markets in both rural and urban areas stock a variety of manufactured goods from overseas and African

Figure 17.17 Kenya's economy is highly dependent on the export of coffee. Cash crops and other raw materials are typical exports of Africa south of the Sahara.

sources. Modern factories have been established in many urban centers, and improved agricultural techniques have been introduced in many areas.

Such changes have affected some peoples and areas more than others, and their total impacts have only begun to lift the region from underdevelopment. There are outstanding problems and potential for development in agriculture, mining, and infrastructure for communication and transportation. Many international markets are closed to potential African manufactured goods and agricultural products. There is a pressing need in most countries to deal with a huge burden of debt to international lenders. And perhaps most critically to the resolution of these economic problems, better political leadership is in great demand.

Cash Crops

Per capita food output in most of the countries has declined or has not increased since independence. The average African eats 10 percent less than he or she did two decades ago. Rapid population growth and drought are partly responsible for the trend. Many regimes have also invested more in warfare than in getting food to their citizens. Food shortages also relate to government preference for cash crops over subsistence food crops. Coffee, cotton, and cloves, for example, provide a means of gaining foreign exchange with which to buy foreign technology, industrial equipment, arms, and consumption items for the elite. The proportion of crops grown for export to overseas destinations and for sale in African urban centers has therefore risen significantly in recent times.

Most export crops are grown on small farms rather than on plantations and estates. Large plantations have never become as established as they are in Latin America and Southeast Asia. Political and economic pressures forced many of

them out of business during the period of transition from European colonialism to independence. Governments of the freshly independent states nationalized and subdivided white-owned plantations, and many whites were obliged to sell their land to Africans (this process is still ongoing in Zimbabwe, which became independent in 1980). Now operated by Africans, many of these farms provide vital tax revenue and foreign exchange. One exception is South Africa, where the most important producers and exporters are the ethnic Europeans, who raise livestock, grains, fruit, and sugarcane. But throughout tropical Africa as a whole, the trend is toward export production from small, black-owned farms. The most valuable export crops are coffee, cacao, cotton, peanuts, and oil palm products (see Figure 17.9). Secondary cash crops include sisal (grown for its fibers; Figure 17.18), pyrethrum (used in insecticides), tea, tobacco, rubber, pineapples, bananas, cloves, vanilla, cane sugar, and cashew nuts.

Although cash crops are important sources of revenue in these poor countries, excessive dependence on them can be harmful to a country's economy. The income they bring often goes almost exclusively to already prosperous farmers and corporations. The prices they fetch are vulnerable to sudden losses amidst changing world market conditions. The plants are susceptible to drought and disease (see Biogeography, page 470). Finally, they are often grown instead of food crops, and cash crops do not feed hungry people.

Minerals

Mineral exports have had a strong impact on the physical and social geographies of Africa south of the Sahara. The three primary mineral source areas are South Africa and Namibia, the Democratic Republic of Congo–Zambia–Zimbabwe region, and West Africa, especially the areas near

Figure 17.18 A sisal plantation in eastern Kenya

Biogeography

Crop and Livestock Introductions as "The Curse of Africa"

Many important food and export crops of Africa are not native to the continent. For example, corn (maize), manioc, peanuts, cacao, tobacco, and sweet potatoes were introduced from the New World. Cattle, sheep, chickens, and other domesticated animals now characteristic of Africa south of the Sahara also originated outside the region. Such nonnative species are known as exotic species, or introduced species.

Some exotic species such as manioc and chickens have adapted well to the environments and human needs in Africa south of the Sahara. Others are more problematic. Known as "mealie-meal," milled corn has become the major staple food for southern Africans. However, the people of southern Africa came to regard corn as more of a scourge than a blessing during the severe regional drought of 1992. Introduced from relatively well-watered Central America, corn has flourished in African regions as long as rains or irrigation water has been sufficient. But the crop failed completely during the 1992 drought. The situation was particularly critical in Zimbabwe, where the government had just sold its entire stock of corn reserves in an effort to repay international debts. Zimbabwe suddenly became a large importer of corn. Many in the country proclaimed corn to be **"the curse of Africa,"** lamenting that, with its high water requirements, it should never have been allowed to become the principal crop of this drought-prone region. Agricultural specialists argued that native sorghum and cowpeas are much better adapted to African conditions and should be cultivated as the major foods of the future.

Corn was a major issue again in a severe drought in 2002. This time, the United States and other nations reacted early with major shipments of corn to especially hard-hit Zambia. Thousands of tons of American corn were refused by Zambian authorities, or locked up in silos, while Zambians went hungry. The reason: Zambia's concern that the U.S. corn was genetically modified (GM). Zambian authorities feared that if unmilled corn were to mix with Zambian seed stock, the anti–GM European Union might refuse future imports of corn from Zambia. They also worried of the potential health impacts of the milled corn.

Just as there are questions about corn, there are also growing concerns about whether the cow is an appropriate livestock species for Africa. Cattle frequently overgraze and erode their own rangelands. The traditional approach to improving cattle pastoralism in Africa south of the Sahara, and to boosting livestock exports, has focused on the need to develop new strains of grasses that can feed cattle and other nonindigenous strains of livestock because the native grasses of African savannas suit the wild herbivores but generally not the domesticated livestock. A fresh approach turns this problem around by asking whether it might be more appropriate to commercially develop the indigenous animals already suited to the native fodder and soil conditions. Kenya, Zimbabwe, South Africa, and other countries are now involved in **game ranching** in which antelopes—such as eland, impala, sable, waterbuck, and wildebeest—along with zebras, warthogs, ostriches, crocodiles, and even pythons are bred in semicaptive and captive conditions to be slaughtered for their meat, hides, and other useful products (Figure 17.F). Safari hunting and tourism provide supplemental revenue in some of these operations.

So far, these enterprises have demonstrated that large native herbivores produce as much or more protein per area than do domestic livestock under the same conditions. In commercial terms, the systems that have incorporated both wild and domestic stock have proven more profitable than systems employing only domestic animals. The World Wildlife Fund and other proponents of this **multispecies ranch system** argue that it is an economically and ecologically sustainable option of land use in Africa. They emphasize that ranchers who keep both wild and domesticated animals will be better able economically to weather periods of drought and will require a much lower capital input than they would expend on cattle or other domesticated livestock alone.

Having convinced many ranchers of the economic value of wildlife, the government of Zimbabwe also enlists peasant farmers in protecting, and thus benefiting from, game animals. In an innovative program known as **CAMPFIRE (Communal Area Management Program for Indigenous Resources)**, the government turned over management of wild animals to local councils, which are permitted to sell quotas of hunting licenses for elephant and other game. Profits from safari hunting return to the local communities to fund a wide range of development projects.

Figure 17.F Oryx antelopes maintained in a semidomesticated state on the Galana Game Ranch in Kenya. Husbandry of indigenous wildlife may prove to be a viable alternative to keeping cattle and goats, which have many harmful impacts on soil and vegetation in Africa south of the Sahara.

the Atlantic Ocean (Figure 17.19). Notable mineral exports from these regions include precious metals and precious stones (particularly diamonds; see Problem Landscape, page 472–473), ferroalloys, copper, phosphate, uranium, petroleum, and high-grade iron ore, all destined principally for Europe, the United States, and China.

Large multinational corporations, financed initially by investors in Europe or the United States, do most of the mining in Africa. Mining has attracted far more investment capital to Africa than any other economic activity. Money is invested directly in the mines, and many of the transportation lines, port facilities, power stations, housing and commercial areas, manufacturing plants, and other elements in the continent's infrastructure have been developed primarily to serve the needs of the mining industry.

Great numbers of workers in the mines are temporary migrants from rural areas, often hundreds of miles away. The recruitment of migrant workers has had important cultural and public health effects. Millions of Africans who work for mining companies have come into contact with Western ideas as well as those of other ethnic groups and have carried these back to their villages. Unfortunately, the phenomenon of migrant labor has also helped spread HIV, the virus that causes AIDS. Most miners are single or married men who spend extended periods away from their wives or girlfriends, and it is common that they contract the virus from prostitutes and then return home to their villages, where they spread it still further. The HIV infection rate among migrant miners working in South Africa is more than 40 percent.

Communications and Transportation Infrastructure

Poor transportation hinders development in Africa south of the Sahara (Figure 17.20). Few countries can afford to build extensive new road or railroad networks, and much of the colonial infrastructure has deteriorated. The region critically needs a good international transportation network, together with a lowering of trade barriers, to enlarge market opportunities. Poor transportation is also often a contributing fac-

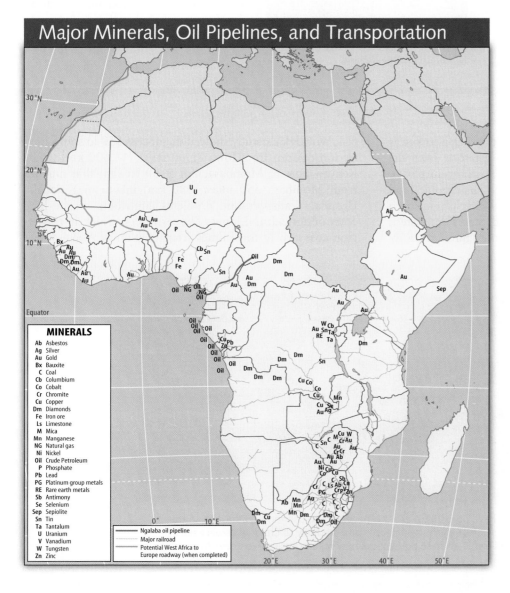

Figure 17.19 Minerals of Africa south of the Sahara

Problem Landscape

The Trade in Dirty Diamonds

One of the leading sources of revenue for governments and rebels fighting wars in Africa is a thing of great beauty, usually associated with eternity and conjugal bliss: the diamond (Figure 17.G). An estimated 4 to 15 percent of the world's annual total production of $8 billion worth of rough diamonds is made up of **dirty diamonds** (also called **conflict diamonds** or **blood diamonds**), defined by the United Nations as "rough diamonds used by rebel movements or their allies to finance conflict aimed at undermining legitimate governments." Gems of questionable origin have financed at least three African wars. In two of these, rebel movements controlled the mines: In Angola, the

Figure 17.G Does this symbol of love and beauty have sinister origins?

UNITA rebels held the diamond producing area until evicted by government troops in 1999. And for years in Sierra Leone, most of the mines were in the hands of the Revolutionary United Front (RUF) (whose hallmark was the amputation of civilians' limbs). Diamonds have also funded the government of the Democratic Republic of Congo (viewed widely as an illegitimate government) in its regional conflict.

Diplomatic, nongovernmental, and business efforts are underway to stem the tide of Africa's dirty diamonds. In 2002, following 4 years of negotiations, 45 countries endorsed a UN-backed certification plan, called the **Kimberley Process,** designed to ensure that only legally mined rough diamonds, untainted by violence, reach market. Rough diamonds must be sent in tamper-proof containers with a certificate guaranteeing their origin and contents. The importing country (the

509

489, 493

tor to famine. In Ethiopia, for example, the road network is so poor that it is extremely difficult to get food from the western part of the country, which often has crop surpluses, to the eastern part of the country, which has chronic food shortages. There have been instances in which it has proven cheaper to ship food from the United States to eastern Ethiopia than to truck it across the country. Similarly, a recent World Bank study showed that it cost $50 to ship a metric ton of corn from Iowa 8,500 miles (13,600 km) to the Kenyan port of Mombasa, but $100 to ship that amount from Mombasa 550 miles (880 km) inland to Kampala, Uganda. Transportation problems contribute to the high costs of agricultural inputs like fertilizers, which in this region cost two to three times what they do in Asia—where

Figure 17.20 Many African river crossings have no bridges, and it is necessary to use ferries, although some streams can be forded during the dry season. Here, vehicles and people board a ferry to cross an estuary north of Mombasa on the Kenyan coast. Backup of traffic at these bottlenecks may take hours or even days to clear.

biggest of which are Belgium and Israel) must certify that the shipments have arrived unopened and reject any shipments that do not meet the requirements. According to the plan, only countries that subscribe to the Kimberley Process will be allowed to trade in rough diamonds. Rule breakers would be suspended, making their diamond exports unlawful. Individual lawbreakers would be punishable by their own countries.

Enforcing these rules will be difficult. Millions of dollars in rough stones can be smuggled in a sock, and expert identification of their place of origin can be a challenge. Neighboring states are often complicit in the trade of dirty diamonds. The Republic of the Congo—with no diamond mines of its own—exports millions of dollars worth of diamonds each year to Belgium. Many or most of the gems are presumed to originate as dirty diamonds in adjacent Democratic Republic of Congo or in Angola. Rwanda, Uganda, and Zimbabwe also export probable dirty diamonds obtained from such sources. As long as the diamonds leave those countries in the certified, tamper-proof containers, the Kimberley Process has to accept them as clean and has no means of cracking down on the obvious breach of its goal. For the Kimberley Process to be effective, internationally acceptable tracking standards would be necessary. Special,

independent monitoring staff would have to keep their eyes on mines and on government diamond evaluation offices.

Diamond merchants are increasingly anxious to act on the Kimberley Process and ensure that blood diamonds never make it to the world's jewelry shops. The effort is being spearheaded by the world's leading diamond company (with about two-thirds of the market), a South African–based multinational named De Beers Consolidated Mines. De Beers was embarrassed by a report that it had bought $14 million worth of diamonds from Angolan rebels in a single year. In 2000, *The New York Post* ran a photo of a maimed child in Sierra Leone, indicating this was a result of buying diamonds. Perceiving that a public relations debacle could lead to a business disaster, such as happened with an organized boycott against fur products in the 1980s, De Beers seized the initiative. In the name of Africa's welfare, but certainly also a means of increasing demand and profit, De Beers introduced **branded diamonds,** or those marked as from nonconflict areas. The De Beers diamond is sold to the customer with a guarantee that it has not been used by an African ruler or rebel to finance civil war. Other diamond companies have followed suit, especially in an effort to please Americans—who buy half the world's diamonds.

---

fertilizer consumption is 10 times what it is in Africa south of the Sahara.

Poor communication also slows development. The growing numbers of radios and televisions appearing in homes across the continent, even in remote villages, are helping to convey an unprecedented wealth of information about the wider world. But more high-tech, interactive forms of communication are much in need. Telephone, fax, e-mail, and other technologies are critical to successful participation in the globalizing economy. Africa cannot take part because it is critically short of most of these assets. For example, while there are more than 500 telephone lines per 1,000 people in the more developed countries, in Africa there are 34 countries with fewer than 10 telephone lines per 1,000 people. Put another way, there are more phones in Tokyo than in all of Africa. Getting a phone line installed can be a costly and time-consuming procedure, and then the lines often do not work (a recent survey in 22 African countries showed a 60 percent faulty rate on phone lines). Calling rates, especially internationally, are very expensive, and few lines are reliable enough to handle Internet traffic.

Many of the line problems would be averted with mobile phones. So far, the main obstacle has been that many of the telecommunications companies are government owned and therefore profitable for governments, which are reluctant to

allow competition by private companies with cellular technologies. Some governments also like to monitor phone calls and e-mail, and more communication makes that a more challenging task. In some countries, the barriers are falling, however, and mobile phones especially are beginning to flourish in some of the world's most isolated places. About 50 million people in the region now use cell phones, up from 8 million in 1999, and there will be a projected 100 million users by 2009.

Internet cafés are also springing up, and the Internet may come to be Africa's answer to many of its communications problems. There are hopes that voice over Internet technology will make telephone calling more affordable for hundreds of millions of Africans. The West African nation of Ghana is jockeying to become an international service call center, comparable to Bangalore in India, using Internet telephony instead of conventional telephone carriers Access to the Internet remains limited. In 2004, the entire continent of Africa had only 13 million Internet users (of 813 million worldwide), but the number is growing quickly. Private companies and individuals cooperated to build a 20,000-mile (32,000-km) undersea fiberoptic cable system that, when completed in 2002, formed a ring of connections around Africa. There is much hope that such "digital bridges" will span the digital divide between Africa and the more developed world, helping to raise the standard of living and quality of life of many of Africa's peoples.

Trade, Debt, and Aid

One of the major obstacles to African economic development is that many countries outside the region have effectively closed their doors to African imports. These restrictions typically take the form of high tariffs or low quotas imposed on agricultural products (like cotton) or manufactured goods (like textiles). Potential importing countries, like the United States, impose these restrictions to protect their own industries.

An example of what reduced trade barriers might do for Africa is provided by the **African Growth and Opportunity Act (AGOA)**, passed by the U.S. Congress in 2000 and due to expire in 2008. AGOA reduced or ended tariffs and quotas on more than 1,800 manufactured, mineral, and food items that could be imported from Africa. African exports to the United States have surged, especially in the clothing industry. Apparel made in Lesotho, Kenya, Mauritius, South Africa, Madagascar, and elsewhere in Africa is now sold in Target and Wal-Mart. The United States is considering extending AGOA's benefits beyond 2008 with the group of countries comprising the **Southern African Customs Union (SACU)**: South Africa, Lesotho, Swaziland, Botswana, and Namibia.

Jobs and incomes for Africans have grown because of these new opportunities. Much of the profit, however, goes to the Asian firms that set up clothing factories in Africa because of limitations on clothing exports from their home countries. Those export quotas were set by the **World Trade Organization (WTO)**. After they expired in 2005, any country was able to export as much as it could from the home country. African countries have to compete with Cambodia and other poor countries, as the manufacturers look for the cheapest labor they can find. If costs are lower elsewhere, the up-and-coming African factories could fold.

Almost all of the countries of Africa south of the Sahara are heavily in debt to foreign lenders. Young nations undertook costly development projects with borrowed money, particularly after the mid-1970s. Western financiers, planners, and contractors gave optimistic assessments of the benefits to be expected from such projects, and African leaders were ready to accept loans as a way to reap quick benefits from newly won independence. Now, however, many countries are having great difficulty in meeting even the interest payments on their debts, and their efforts to do so often create further economic woes and have destructive environmental effects. The International Monetary Fund (IMF), World Bank, and other lenders have generally resisted requests of African debtor nations for rescheduled interest payments and new loans. As a condition of further support, the lenders often demand that African debtors put their finances in better order by such measures as revaluating national currencies downward to make exports more attractive, increasing the prices paid to farmers for their products, and reducing corruption, especially among government officials and urban elites. Although many African debtor nations have attempted to meet such demands, internal political factors slow progress. Loss of economic privileges angers urban elites on whom governments depend for support. Relaxation of price controls on food to stimulate production is often met by rioting among city-dwellers who have tight budgets already allocated for other costs of living.

Increased foreign aid in the form of gifts or grants to Africa is controversial. Many argue that foreign aid only increases dependence and deflects national attention away from problems underlying famine. International aid monies often free up funds from national treasuries to be spent on unfortunate uses. In the Horn of Africa, for example, foreign money was available for famine relief, so Ethiopia and Eritrea could more easily afford to purchase weapons used in their war. Aid monies are often simply stolen, and there has been inadequate monitoring by donors of how they are spent. Since about 1990, the donors have proceeded more cautiously, more often funneling money through private and religious groups working in healthcare, education, and small-business development, for example, through microcredit practices (see Definitions and Insights, page 475).

Costly Political Leadership

Africa scholars speak of the "**failed state syndrome**," a pernicious process of economic and political decay that has ruined countries like Somalia and Sudan and is eating away at Ivory Coast and Zimbabwe. Some African countries are little more than **shell states.** A shell state appears to have all the institutions of a country: a constitution, a parliament, ministries, and more. But most of the positions are occupied by friends and family of the country's ruler, who are happy to line their pockets rather than pursue development for their fellow citizens. Five of the world's 14 most corrupt countries are in this region, according to a leading watchdog group on corruption.

One of the worst cases was the government of Zaire (now the Democratic Republic of Congo) under Mobutu Sese Seko (1965–1997). Mobuto's minister of mines was regularly bribed by Zaireans and Europeans to obtain mining permits. He kept as much of the bribe money as he could, and the rest went to Mobuto. Little revenue from the lucrative mining industry ever made it into the state treasury or trickled down through the populace. That is how Zaire functioned, or failed to function, and how a resource-rich country succeeded in achieving one of the worst standards of living on the planet. In Liberia, former President Charles Taylor passed a law allowing him to personally sell any of the country's "strategic commodities," including mineral resources, forest products, art, archeological artifacts, fish, and agricultural products. The wealth taken by such countries' leaders generally does not find its way back into the economy but is either spent on grandiose personal accommodations or parked in foreign bank accounts.

Definitions + Insights

Microcredit

A recent promising alternative to the problem of national debt is microcredit, or the lending of small sums to poor people to set up or expand small businesses. Typically, poor people have no collateral and therefore cannot borrow from commercial banks. Local loan sharks may charge huge interest rates of 10 to 20 percent per day. But in the interest of fostering development, microlending organizations encourage poor borrowers to form a small group so as to cross-guarantee one another's loans. One member of the group may take out a loan of $25, for example, to help start up a small business such as a restaurant. Only when that individual pays it back may the

next person in the group receive a loan. Peer pressure minimizes the default rate.

Microlenders generally prefer to lend money to women because women are more likely to use the money in ways that will feed, clothe, and otherwise benefit their children, whereas men may be more likely to spend the money on alcohol or other frivolous pursuits. One fear in Africa is that microloans will not be repaid if the borrower or someone in the borrower's family contracts HIV or some other debilitating illness. Therefore, the loans are often extended only after the borrower receives some amount of public health training, for example, in the use of condoms and in general food preparation hygiene.

One of the assumptions of international lenders and aid agencies is that democracy helps to weed out such economic bloodletting. True democracy, they observe, is still short in the region. Between 1994 and 2004, 43 of the region's 47 countries held at least one multiparty election. However, few of these countries fully embrace political pluralism, and even they have problems with corruption, weak leadership, and declining education and health services. A 2004 study depicts only 11 "**free**" democracies in the region, with 20 "**partly free**" and 17 "**unfree**" (Figure 17.21). That is steady progress from 1990, when only 4 were considered free and 27 were unfree. But foreign lenders and donors want more, telling African leaders that if they do not institute democratic reforms, hold fair elections, or improve bad human rights records, they will not receive development aid or loans. Often, the leaders make just enough concessions to win the aid without instituting real reform—a phenomenon known as donor democracy. Some Africa observers say that positive development assistance must go where it is needed most, channeled directly into rebuilding destroyed institutions such as the civil service, courts, police, and the military.

Foreign powers might not be able to promote such reforms without becoming guilty of or being accused of neo-colonialism, but new African institutions are emerging to confront the problems themselves. In 2002, the 53-member **African Union (AU)** was formed to replace the old and often inconsequential **Organization of African Unity (OAU)**. One of the African Union's main articulated goals is to implement self-inspection or "peer review" to help promote democracy, good governance, human rights, gender equity, and development. It has formed an agency responsible for this monitoring, known as the **New Partnership for Africa's Development (NEPAD)**. NEPAD's premise is that improvements in such human affairs will improve the climate for international investment and assistance to Africa.

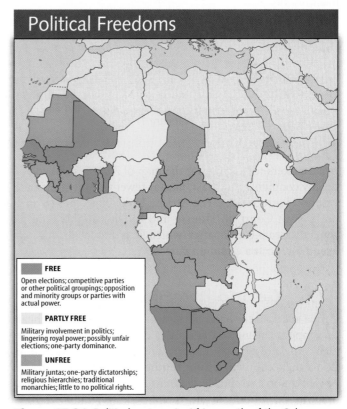

Figure 17.21 Political systems in Africa south of the Sahara, 2004. *Source: Abraham McLaughlin,* The Christian Science Monitor, *April 27, 2004.*

The African Union has also created a 265-member **Pan-African Parliament** that will begin making regional laws in 2007. To help ensure the stability needed for economic and political progress, the African Union is also establishing a continental military force, the **African Standby Force**, that takes orders from the organization's **Peace and Security Council**. The force will, in theory, field peacekeepers or

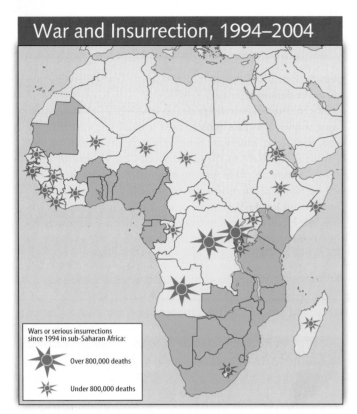

War and Insurrection, 1994–2004

Wars or serious insurrections since 1994 in sub-Saharan Africa:

Over 800,000 deaths

Under 800,000 deaths

Figure 17.22 Africa south of the Sahara was the world's most conflict-ridden region in the decade between 1994 and 2004. *Source:* Economist Magazine, *January 17, 2004*

peacemakers to ensure that the map of Africa, so filled with deadly conflicts in the past, has far fewer trouble spots in the future (Figure 17.22). The African Union has plans to create an African Court on Human and People's Rights, and like the European Union after which it is modeled, may also eventually seek a common currency.

17.5 Geopolitical Issues

Africa south of the Sahara has waxed and waned as a theater of geopolitical interest since the end of World War II. Initially, the great powers were interested in the region. To boost their competing aims during the Cold War years, the Soviet Union and the United States played African countries against one another, arming them with weapons with which to wage **proxy wars.** The superpowers also extended aid generously to many African nations.

Great power concerns about Africa changed with the end of the Cold War (c. 1990), and the nature of conflict in Africa changed. The great powers withdrew support, but their weapons remained to prolong smoldering conflicts. The United States helped build the arsenals of eight of the nine countries involved in the recent Democratic Republic of Congo conflict, for example. Cold War–era weapons like AK-47 assault rifles are still "dumped"—sold at low prices— in Africa, where they are not considered obsolete and where they fan the flames of conflict. Increasingly, fighting has be-

gun to spread across international borders (as in Congo and between Eritrea and Ethiopia). Previously, the United States and the Soviet Union maintained a kind of security balance that kept warfare from becoming internationalized, but that restraint no longer exists. The scenario of regional or even Africa-wide wars began to be feared and was realized with Africa's first "world war," centered on the Democratic Republic of Congo (see pages 492–493). In the absence of regional or international powers to keep the peace, other countries (for example, Sierra Leone) simply implode, fragmenting into fiefdoms run by factions that claim to be revolutionaries but are essentially profiteers. Meanwhile, some national governments (for example, Ethiopia) spend vast sums on expensive military aircraft while their people suffer from malnutrition.

The end of the Cold War constricted (and in the case of Russia, ruptured) aid pipelines. The United States cut its economic assistance to Africa south of the Sahara by 30 percent between 1985 and 1992 and reduced it another 10 percent between 1992 and 2001 (to $738 million yearly in 2001 for the entire region, about the same amount given as economic assistance to Egypt alone). Why the waning concern? American President George W. Bush was quoted as saying, "While Africa may be important, it doesn't fit into the national strategic interests, as far as I can see them."[2]

Then came 9/11. Al-Qa'ida's attack on the United States, and the U.S. counterstrike against al-Qa'ida in Afghanistan, raised fears that the Islamist organization might find new recruits and training grounds in Africa. Al-Qa'ida had already bombed U.S. embassies in Kenya and Tanzania in 1998 and in 2002 struck again in the Kenyan city of Mombasa. The United States maintains a focus on **terrorism hot spots** in the region, including Kenya, Somalia, Djibouti, Niger, Chad, and Mali. With a relatively low profile, U.S. forces have made the tiny Horn of African country of Djibouti a major military center for potential action against nearby Yemen or other countries where Islamist militants may be active. So far, U.S. interest in the hot spots has translated into military expenditures and partnerships with African governments but not wider development assistance. Some analysts point out that several of the hot spots receiving U.S. aid to combat terrorism are politically "unfree" countries and insist that terrorism can best be rooted out by encouraging democracy and development.

In addition to the terrorism hot spots, the United States has strong geopolitical interests in the region's oil producing countries. These, too, have become more critical since 9/11 and the U.S. war on Iraq, which only heightened the sense that Middle Eastern oil supplies are vulnerable. The United States is seeking more stable oil supplies and has pinned many hopes on Africa. Petroleum-rich Nigeria and Angola together supply more than 15 percent of America's oil, a figure projected by the U.S. Central Intelligence Agency to rise to 25 percent by 2015. Nigeria is also the largest U.S.

---

[2]Quoted in Ian Fisher, "Africans Ask if Washington's Sun Will Shine on Them." *The New York Times,* February 8, 2001, p. A3.

trading partner in Africa. American interest in West African security was apparent in the 2003 decision to field a small military contingent to separate rebel and government forces in Liberia. For almost a decade, following the loss of U.S. service personnel in Mogadishu, there was great reluctance to see American troops in African trouble spots. Now there is discussion of establishing U.S. military bases in several key countries other than Djibouti, beginning with Cameroon, Equatorial Guinea, and Gabon, and possibly securing air basing rights in Benin, Ivory Coast, and Nigeria. The U.S. Navy would like to establish bases in São Tomé and Príncipe and in Equatorial Guinea.

The United States views South Africa as the region's economic powerhouse and as the key to regional stability. South Africa's military strength, which included nuclear weapons until they were dismantled in 1990, contributes to its global significance. It is strategically important for its frontages on both the Atlantic and Indian Oceans, its possession of Africa's finest transport network (vitally important to many African countries that trade with and through South Africa), and its diversified mineral wealth.

Finally, HIV/AIDS in the region is also a geopolitical concern. President Bush's first secretary of state, Colin Powell, declared that Africa's AIDS epidemic is a national security issue for the United States. He was acknowledging some of the implications of globalization. Despite Africa's apparent remoteness from the United States, the two are linked by hundreds of air traffic routes traveled by thousands of people each day. Hundreds of HIV-positive Africans, perhaps most of them unaware of their virus, pass through U.S. airport gateways daily. Some African strains (subtypes) of the virus are more easily transmitted by heterosexual contact than the strains prevalent in the United States, so the risk to the population at large, not only homosexuals and intravenous drug users, is greater. This scenario seemed, to Powell, to represent a clear and present danger to the United States. There are other dimensions of the potential burden on the United States. As long as the epidemic rages in Africa, there may be large flights of "medical refugees" seeking asylum in the United States. There could also be AIDS-related political instability or civil wars that invite U.S. military intervention in the region.

Due to the HIV/AIDS epidemic and the region's other great problems, there will also be humanitarian crises that will tug on American and other Western heartstrings and purse strings. Since the mid-1980s, when television images of famine-wracked Ethiopia prompted American and Western European citizens to give generously, popular sympathy and support during African emergencies have waned. The bitter experience of the United States in Somalia in the early 1990s squelched its appetite to extend humanitarian assistance. This lack of apparent public and official interest in crises abroad is known as donor fatigue. How the rest of the world should respond to poverty and recurrent crises is one of the key questions facing the huge, resource-rich, problem-ridden region of Africa south of the Sahara.

SUMMARY

- The culturally complex region of Africa south of the Sahara is often called sub-Saharan or Black Africa.

- Africa is the cradle of humankind where hominids originated and from where they diffused. The region has seen many indigenous civilizations and empires and is ethnically and linguistically diverse.

- The majority of the region's people are rural. There are several clusters of dense population.

- HIV/AIDS is taking a huge toll, dramatically lowering life expectancy and projections for population growth, especially in southern Africa.

- A number of shared cultural traits may be referred to as comprising "Africanity." These do not apply to all African cultures, but they do provide a useful introduction.

- Africa south of the Sahara has a relatively low population density overall, but a majority of this region's people live in a small number of densely populated areas that occupy a small share of the region's 8.4 million square miles (21.8 million sq km).

- Most of Africa consists of a series of plateau surfaces dissected by prominent river systems such as the Nile, Niger, Congo, Zambezi, and Orange.

- One of the most spectacular features of Africa's physical geography is the Great Rift Valley, a broad, steep-walled trough extending from the Zambezi Valley northward to the Red Sea and the valley of the Jordan River in southwestern Asia. In terms of plate tectonics, the feature marks the boundary of two crustal plates that are rifting, or tearing apart.

- Although about two-thirds of the region lies within the low latitudes and has tropical climates and vegetation, Africa south of the Sahara contains a great diversity of climate patterns and biomes, some resulting from elevation rather than latitudinal position.

- Africa's diverse wildlife is often threatened by human population growth, urbanization, and agricultural expansion within the continent. Two of Africa's largest herbivores, elephants and rhinoceroses, are striking examples of Africa's endangered wildlife. Their management and protection require consideration of many issues, including international trade.

- Within the four culture hearths of Africa south of the Sahara—the Ethiopian Plateau, the West African savanna, the West African forest, and the forest-savanna boundary of West Central Africa—early indigenous people were responsible for several agricultural innovations. Within these culture hearths, Africans domesticated important crops such as millet, sorghum, yams, cowpeas, okra, watermelons, coffee, and cotton.

- The four major language families of the region are Niger-Congo, Afro-Asiatic, Nilo-Saharan, and Khoisan. Malagasy, spoken only on Madagascar, is an Austronesian language. Islam and Christianity are major faiths, but indigenous belief systems are often mixed with them or exist on their own in some locales.

- The tragic impetus for growing contact between Africa and the wider world was slavery. Over a period of 12 centuries, as many as 25 million people from Africa south of the Sahara were forced into slavery. Pockets of slavery still exist in the region.

- Most of Africa south of the Sahara fell under European colonialism after the Conference of Berlin in 1884 and 1885. During this conference, Africa was carved up, as the French, British, Germans, Belgians, Portuguese, Italians, and Spanish established their respective spheres of influence in the region.

- In general, the people of Africa south of the Sahara are poor, live in rural areas, and practice subsistence agriculture as their main occupation. In most countries, per capita food output has declined or has not increased since independence.

- Frequent droughts, lack of education, poor transportation, and serious public health issues have hindered development within Africa south of the Sahara. Most countries are underindustrialized and overly dependent on the export of a few primary products.

- Over half of the people of Africa south of the Sahara depend directly on agriculture or pastoralism for a livelihood. Export crops include coffee, cacao, cotton, peanuts, and oil palm products.

- The export of minerals has had a particularly strong impact on the physical and social geography of Africa south of the Sahara. Notable mineral exports include precious metals and stones, ferroalloys, copper, phosphate, uranium, petroleum, and iron ore.

Income from diamonds in many cases has funded and fueled conflict within and between countries.

- Exports of clothing from Africa to the United States have boomed since the U.S. Congress lowered trade barriers on numerous products, but the suspension of WTO-imposed export quotas in 2005 forced the region to compete with other low-cost producers.

- Most countries are heavily in debt to foreign lenders. Economic and humanitarian assistance to the region slowed considerably after the end of the Cold War, but it has picked up again since 9/11.

- Although many countries have been under authoritarian governments since independence, there has been some progress toward democracy. Serious political instability is characteristic of many African countries south of the Sahara. A diverse array of political, economic, and social ideas from the West, the Communist bloc, the Muslim world, and Africa south of the Sahara itself has influenced governments since independence. In addition, important links with the colonial powers that formerly controlled them exist in many countries of the region.

- The recently established African Union is a supranational organization dedicated to solving African problems without the intervention of outside powers.

- So often judged to be marginal in world affairs, Africa south of the Sahara deserves and is receiving increased international attention because of its humanitarian problems, the global implications of its public health and environmental situations, problems in the management of its natural resource wealth, its oil reserves, and concerns about terrorism.

KEY TERMS + CONCEPTS

Terms in blue are also defined in the glossary.

"1 percent gap" (p. 447)
African Growth and Opportunity Act (AGOA) (p. 474)
African Standby Force (p. 475)
African Union (AU) (p. 475)
Africanity (p. 460)
Afro-Asiatic language family (p. 461)
  Semitic subfamily (p. 461)
    Amharic (p. 461)
    Arabic (p. 461)
  Chadic subfamily (p. 461)
    Hausa (p. 461)
  Cushitic subfamily (p. 461)
    Oromo (p. 461)
    Somali (p. 461)
AIDS (acquired immunodeficiency syndrome) (p. 447)
antiretroviral (ARV) drugs (p. 451)
Austronesian language family (p. 461)
  Malagasy (p. 461)
blood (conflict, dirty) diamonds (p. 472)
branded diamonds (p. 473)
CAMPFIRE (Communal Area Management Program for Indigenous Resources) (p. 470)
compulsory licensing (p. 451)

digital divide (p. 473)
donor democracy (p. 475)
donor fatigue (p. 477)
Ethiopian Orthodox Church (p. 462)
exotic species (p. 470)
failed state syndrome (p. 475)
free, partly free, and unfree countries (p. 475)
game ranching (p. 470)
human immunodeficiency virus (HIV) (p. 451)
keystone species (p. 457)
Khoisan language family (p. 461)
  Khoekhoe (Damara) (p. 461)
Kimberley Process (p. 472)
microcredit (p. 475)
"mistake of 1914" (p. 468)
multispecies ranch system (p. 470)
nagana (p. 457)
New Partnership for Africa's Development (NEPAD) (p. 475)
Niger-Congo language family (p. 461)
  Adamawan subfamily (p. 461)
  Atlantic subfamily (p. 461)
  Benue-Congo subfamily (p. 461)
    Bantu (p. 461)

Swahili (Kiswahili) (p. 461)
Gur subfamily (p. 461)
Khordofanian subfamily (p. 461)
Kwa subfamily (p. 461)
Mandé subfamily (p. 461)
Nilo-Saharan language family (p. 461)
  Chari-Nile subfamily (p. 461)
  Saharan subfamily (p. 461)
  Songhai subfamily (p. 461)
Organization of African Unity (OAU) (p. 475)
Pan-African Parliament (p. 475)
Pandemic (p. 449)
Peace and Security Council (p. 475)
proxy war (p. 476)
shell state (p. 474)
shifting cultivation (p. 455)
sleeping sickness (p. 457)
Southern African Customs Union (SACU) (p. 474)
subsistence agriculture (p. 455)
terrorism hot spot (p. 476)
"the curse of Africa" (p. 470)
triangular trade (p. 466)
trypanosomiasis (p. 457)
World Trade Organization (WTO) (p. 474)

REVIEW QUESTIONS

WORLD
REGIONAL
Geography ⊛ Now™

Assess your understanding of this chapter's topics with additional quizzing and concept-based problems at http://earthscience.brookscole.com/wrg5e.

1. Where are the five principal areas of population concentration within Africa south of the Sahara?

2. What factors account for the high HIV infection rates in Africa south of the Sahara? What can be done to fight the epidemic? What is HIV/AIDS doing to life expectancy and the age-structure profiles in countries where it is epidemic?

3. What are the principal climatic zones and biomes of Africa south of the Sahara?

4. Where were the major slave trade routes in and from Africa south of the Sahara?

5. Where were the major colonial possessions of various European powers in Africa south of the Sahara?

6. What is the significance of cattle in many cultures of Africa south of the Sahara?

7. What are the region's major export crops? Why does dependence on them make a country vulnerable?

8. What is a keystone species? Why is the tsetse fly considered a keystone species in Africa south of the Sahara?

9. What is the 1 percent gap? What are its implications?

10. Why have clothing exports from Africa to the United States boomed in recent years?

11. What do the following terms mean and what do they reveal about Africa's political geography? shell state, failed state syndrome, donor democracy, donor fatigue

12. What impacts did the end of the Cold War have on many African countries?

DISCUSSION QUESTIONS

1. Describe the tectonic processes that have led to the development of the Great Rift Valley. What physical features are associated with the Great Rift Valley? What will be the final fate of areas adjacent to the rift?

2. What circumstances make teachers particularly likely to contract the HIV virus? What other segments of the society are hit especially hard and why? What steps could be taken to slow the advance of the epidemic and why haven't they been taken yet?

3. Describe the major impacts AIDS has had on Africa south of the Sahara. Use a Population Reference Bureau (PRB) world population data sheet or the PRB Web site to update HIV/AIDS infection rates, death rates, and population growth rates for countries south of the Sahara. Has the epidemic slowed, accelerated, or stalled since this book went into print? In your classroom, calculate the portion of occupied seats that would be HIV-positive if the class represented one of the afflicted countries.

4. Discuss the main reasons behind the sharp decline in elephants and rhinoceroses in Africa south of the Sahara. What efforts have been taken to preserve these animals? What are the pros and cons of culling elephants? Of the rhino dehorning program?

5. Describe the eight constituent elements of Africanity. Are these useful for becoming acquainted with African cultures? How do they resemble or differ from your culture traits?

6. Discuss the pattern of triangular trade. In addition to Africa south of the Sahara, what other regions of the world were involved in this pattern of trade? What goods were traditionally exchanged?

7. Discuss the modern slave trade in Africa and elsewhere. How could such practices continue in modern times? What factors explain slavery? Compare and contrast the slavery situation as described here with the discussion of the sex trade in Chapter 7.

8. Describe the positive and negative effects of European colonization of Africa south of the Sahara.

9. Why has corn been called "the curse of Africa" in Zimbabwe? What are some crops that might be better adapted to African conditions?

10. What are the apparent advantages of animals raised in game ranches over cattle and other domesticated livestock?

11. Discuss the positive and negative effects of mining on Africa south of the Sahara. What have been the cultural and public health effects of the recruitment of migrant mine workers?

12. What was the "mistake of 1914"? What is its legacy?

13. What is donor democracy? To what extent should the wealthy countries abroad participate in Africa's economic and political development? How can foreign assistance be put to the best uses possible?

14. What are the goals of the African Union? What bodies within the organization are responsible for meeting these goals?

15. Discuss the trade in dirty diamonds. What measures can realistically be taken to ensure that sales of these valuable gems do not finance conflict in the region?

16. Compare the maps of democratic freedoms and conflicts in Africa (see Figures 17.21 and 17.22). Are there any noticeable patterns? What might explain these?

17. Why and in what ways has the United States taken a renewed interest in Africa south of the Sahara?

The Assets and Afflictions of Countries South of the Sahara

Sarah Tsialiva, from the village of Mahamasina in Northern Madagascar. Almost half of the people in Africa south of the Sahara are of her generation, under the age of 15.

chapter objectives

This chapter should enable you to:

- Recognize how European colonial favoritism of some ethnic groups over others sowed seeds of modern strife and warfare

- Appreciate what corrupt leadership has done to impoverish people in countries with enormous oil and other natural resource wealth

- Recognize the political, economic, and demographic clout of a select number of countries—notably South Africa and Nigeria—in African affairs

- Understand how drought regularly triggers environmental degradation and famine and becomes a political as well as humanitarian issue in national and international affairs

- Consider regional and international efforts to prevent recurrence of the genocide and "ethnic cleansing" that marred Africa in the 1990s

- Evaluate the efforts to redistribute farmland from minority white to majority black control

WORLD
REGIONAL
Geography ⊕ Now™

Look for this logo in the text and go to GeographyNow at http://earthscience.brookscole.com/wrg5e to explore interactive maps, view animations, sharpen your factual knowledge and geographic literacy, and test your critical thinking and analytical skills with unique interactive resources.

In this chapter, sketches of the major subregions and their countries illustrate some of the characteristic and unique challenges of Africa south of the Sahara. The major subregions are the Sahel, West Africa, West Central Africa, East Africa, the Horn of Africa, southern Africa, and the Indian Ocean islands (Figure 18.1). The relevant area and population data are in Table 17.1.

18.1   The Sahel: Desertified and Regreened

The Sahel region extends eastward from the Cape Verde Islands to the Atlantic shore nations of Mauritania, Senegal, and The Gambia and inland to Mali, Burkina Faso ("Land of the Upright Men"; formerly Upper Volta), Niger, and Chad (Figure 18.2). In colonial times, The Gambia was a British dependency; Mauritania, Senegal, Mali, Burkina Faso, Niger, and Chad were French; and Cape Verde was Portuguese. Cape Verde's dry volcanic islands lie well out in the Atlantic off Senegal, but in population, culture, economy, and historical relationships, the islands are so akin to the adjacent mainland that they fit the Sahelian context.

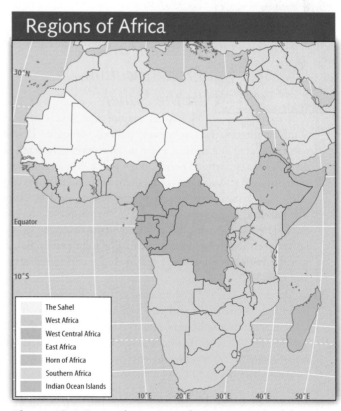

Regions of Africa

30°N

Equator

10°S

- The Sahel
- West Africa
- West Central Africa
- East Africa
- Horn of Africa
- Southern Africa
- Indian Ocean Islands

10°E   20°E   30°E   40°E   50°E

Figure 18.1 Regional groupings of countries south of the Sahara

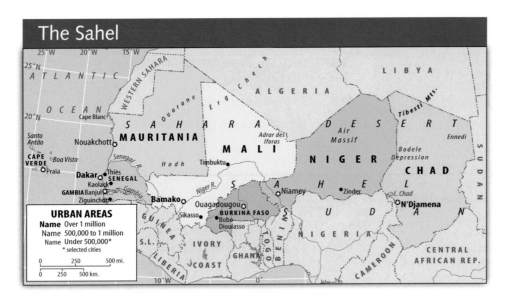

The Sahel

Figure 18.2 Principal features of the Sahel

Environments, Resources, and Settlements

Climatically and floristically, the region ranges from desert (in parts of the Sahara) in the north through belts of tropical steppe and dry savanna in the south (see Figure 17.8a and b). The name Sahel in Arabic means "coast" or "shore," referring to the region as a front on the great desert "sea" of the Sahara. Since the late 1960s, the area has been subjected to severe droughts, which, in combination with increased human pressure on resources, have prompted a process of desertification (see Regional Perspective, below).

This area has seen many dramatic changes in climate and vegetation. There is abundant evidence, particularly in the form of prehistoric rock drawings, that much of the now extremely arid northern Sahel and Sahara region was a grassy, well-watered savanna approximately 6,000 to 10,000 years ago. Many of the large mammals now associated with East

Regional Perspective

Drought and Desertification in the Sahel

The Sahel region, like much of Africa south of the Sahara, experiences periodic **drought.** This is a naturally occurring climatic event in which rain fails to fall over an area for an extended period, often years. When rain does return, the arid and semiarid ecosystems of the Sahel come to life with a profusion of flowering plants, insects, and herbivores like gazelles, whose populations climb when foods are abundant. These ecosystems have high **resilience,** meaning they are able to recover from the stress of drought and have mechanisms to cope with a natural cycle that includes periods of dryness and rain.

Desertification is the destruction of that resilience and the biological potential of arid and semiarid ecosystems. It is an unnatural, human-induced condition that has afflicted the Sahel periodically and severely since the late 1960s, and thus, it is different from the natural phenomenon of drought (Figure 18.A). However, the recent history of the Sahel suggests that drought can be the catalyst that initiates the process of desertification. During decades of good rains prior to the late 1960s, the Fulani (Fulbe), Tuareg, and other Sahelian pastoralists allowed their herds of cattle and goats to grow more numerous. The animals represent wealth, and people naturally wanted their numbers to increase beyond the immediate subsistence needs of their families. However, due to declining death rates, the numbers and sizes of families keeping livestock were also very large. Thus, the unprecedented numbers of livestock that built up during the rainy years made the Sahelian ecosystem vulnerable to the impact of drought on an unprecedented scale.

Drought struck the region in 1968, persisted through 1973, and returned over many years through 1985. Recent studies have determined that the cause of this prolonged drought was an abnormal warming of the Indian Ocean and Atlantic Ocean waters. The West African monsoon that brings rain to the Sahel is fueled by temperature differentials between land and water, and the warming waters disrupted those conditions.

With the drought, annual plants failed to grow, so the large herds of cattle and goats turned to acacia trees and other perennial sources of fodder. They ate all of the palatable vegeta-

Steve McCurry/Magnum

Figure 18.A Desertification in progress is shown graphically in this photo from Mali in the Sahel. As growing human populations intensify the impact of naturally occurring drought, the Sahara crept southward in the 1970s and early 1980s.

tion. This destruction was a critical problem because plants play an important role in maintaining soil integrity by helping to intercept moisture and funnel it downward through the root system. Plant litter and decomposer organisms working around the plant contribute to soil fertility and help stabilize the soil. With hungry livestock in the region eating the plants, the landscape changed. No longer anchored and replenished, good soils eroded. "Junk" plants like Sodom apple (*Calotropis*) replaced palatable plants, and an almost impermeable surface formed on the land. This degraded ecosystem lost its resilience so that even when rains returned, vegetation did not recover.

Meteorologists studying the Sahel drought discovered another important link between precipitation and the removal of vegetation. They noticed that when people and livestock reduced the vegetative cover, they increased the albedo, or the amount of the sun's energy reflected by the ground. Fewer plants meant more solar energy was deflected back into the atmosphere, lowering humidity and reducing local precipitation. This connection is known as the Charney Effect, named for the scientist who documented it. It was a tragic sequence: People responding to drought actually perpetuated further drought conditions.

The 1968–1973 drought devastated the great herds of the Sahel. An estimated 3.5 million cattle died. Two million pastoral nomads lost at least 50 percent of their herds, and many lost as much as 90 percent. Farmers dependent on unirrigated crops were also affected. Harvests were less than half of the usual crop for 15 million farmers, and for many, there was no harvest. The Niger and Senegal Rivers dried up completely, depriving irrigated croplands in Niger, Mali, Burkina Faso, Senegal, and Mauritania. The cost in human lives was estimated at 100,000 to 250,000. Environmental refugees poured into towns like Novakchott, the capital of Mauritania, which were unprepared to deal with such a large and sudden influx. Many pastoral nomads gave up herding and settled down to become

wage earners and, where possible, farmers, and have never returned to their former ways.

This crisis resulted in an international effort to combat desertification using recommendations developed by the United Nations in 1977. The UN report, however, soon became a case study in how *not* to deal with an environmental crisis in the developing world. Most of the recommendations were universal, high-technology solutions that proved impossible to implement in the villages and degraded pastures of the Sahel. For example, following the recommended guidelines, many international agencies attempted to relieve the suffering of Sahelian pastoralists using techniques such as digging deep wells to water livestock. But these agencies failed to anticipate that these wells would act like magnets for great numbers of people and animals; although there was plenty to drink, many animals starved to death when they decimated what vegetation remained around the new water supplies.

Since the disappointment of the 1977 UN plan, and with the lessons relief organizations learned as a result, there has been greater focus on sustainable development, with its local rather than universal solutions. An example is the technique of **rainwater harvesting,** in which villagers use local stones and tools to construct short rock barriers that follow elevation contours. Water striking these barriers backs up to saturate and conserve the soil, allowing more productive farming. Other low-tech water conservation measures, like excavating cisterns under drain spouts and subterranean pools to collect rainfall, are also effective. Such techniques, and a sustained period of good rains, have apparently combined to produce a remarkable regreening of much of the Sahel since 1985. Satellite imagery reveals significant regrowth of vegetation in southern Mauritania, northern Burkina Faso, northwestern Niger, central Chad, and southern Sudan. Annual millet and sorghum harvests in northern Burkina Faso have been about 70 percent above their pre-1985 levels. This change in the Sahel's fortunes is an important lesson that desertification is not inevitable.

46

Africa frequented the region. Now, even though under much less favorable conditions, the raising of sheep, goats, camels, and cattle, combined where possible with subsistence farming, is the major livelihood for peoples of the Sahel.

One of the most prominent features of the Sahel's landscape is Lake Chad, located north of landlocked Chad's capital, N'Djamena (population: 1.3 million). This large, shallow lake (between 7 and 23 ft/2–7 m deep) is situated in a vast open plain. It has no outlets and yet is not very saline because groundwater carries most of the salts away. Its waters are important economically for fish and irrigated agriculture. Characteristic of shallow lakes in arid lands, its surface area historically increased dramatically with rainy periods and shrunk in times of drought. However, the expansion of irrigation along the northward-flowing rivers that feed the lake has caused a huge and enduring shrinkage. Lake Chad's surface area in 2001 was 580 square miles (1,502 sq km) compared with 9,700 square miles (25,123 sq km) in 1963; that is a 95 percent loss. Four countries share the lake and its drainage area, so decisions to reverse the lake's shrinkage are difficult to reach, and the likelihood of future conflict brought about by water shortages is high. The situation is reminiscent of that of the Aral Sea in Central Asia.

Mining, manufacturing, and urbanization are generally insignificant in this subsistence-oriented region. The most remarkable recent improvement in the region's economy is the discovery of oil in southern Chad (see Figure 17.19). The reserves, with an estimated lifespan of 25 years, have recently begun to be tapped, and a controversial World Bank–financed pipeline completed in 2003 is carrying crude to Cameroon's Atlantic coast. Mauritania exports iron ore and Senegal exports phosphates. Niger exports uranium for France's nuclear power program. Salt is mined on Niger's sterile landscape, but its importance today pales in comparison with centuries ago, when a bar of Niger salt could be traded for a bar of gold.

The Sahel region in general was a crossroads for caravan traffic (especially northbound slaves, food, and gold and southbound salt, weapons, and cloth) between West Africa and North Africa. Ancient Mali was an especially important trading intermediary. Its legendary port of Timbuktu on the Niger River was an intellectual as well as a commercial center. In 1989, it was recognized by the United Nations Educational, Scientific and Cultural Organization (UNESCO) as a **world patrimony site.** Today, much of the traffic through this region is a northward migration of poor people from Nigeria, Niger, Guinea, Ghana, Cameroon, and the Democratic Republic of Congo seeking work in oil-rich Libya. Many subsequently use Libya as a jumping-off point for settlement in Europe. Some Europeans are fearful of a potential new stream of immigrants who might use a highway that, when complete, will for the first time allow vehicles to travel from Nigeria, Mali, and Mauritania all the way to the Strait of Gibraltar (see Figure 17.19).

The largest metropolises are Senegal's main port and capital of Dakar (population: 2.6 million) and Bamako (population: 1.3 million), the capital and main city of Mali. Dakar was once the capital of the immense group of eight colonial territories known as French West Africa. It is the commercial outlet for much of the peanut producing area, known as the peanut belt, in the Sahelian and West African savanna and is a considerable industrial center by African standards. Dakar is also a tourist and resort destination with fine beaches and an exotic blend of French and Senegalese cultures.

Ethnic Tensions

The Sahel is the border region between mainly Arab and Berber North Africa and mainly black Africa south of the Sahara. Relations between the ethnic groups are good in most countries, but there are problems in some. In Mauritania, for example, the government—led by the country's majority Arabic-speaking **Moors**—began expelling black Mauritanians across the border into Senegal in 1995. Analysts continue to fear that the Mauritanian government aims to eventually "cleanse" the country of its 40 percent black population.

Chad has north–south tensions that are reminiscent of those of Sudan and Nigeria. The northerners, who now dominate Chad's political life, are mostly Arabic-speaking Muslims whose historic mainstay has been cattle pastoralism. The southerners are non-Arab Christians (mainly of the **Sara** ethnic group) whose livelihood is farming. The French favored the Christian southerners and empowered them more than the northerners, and now that the tables are turned, the northern elite is discriminating against southerners. Northerners historically have enslaved southerners, and there are reports that they continue to do so. There are frequent land disputes as northern pastoralists overrun southern farms. There are fears that, as in Sudan and Nigeria, the dominant northern power will extract the most benefits from the newly developing oil industry and leave the oil-bearing region in poverty.

18.2 West Africa: Populous, Strife-Torn, with Oil

West Africa extends from Guinea-Bissau eastward to Nigeria (Figure 18.3). Its nine political units make up about 800,000 square miles (c. 2 million sq km), or nearly one-fourth of the area of the United States. These nine countries are Guinea-Bissau, Guinea, Sierra Leone, Liberia, Ivory Coast, Ghana, Togo, Benin, and Nigeria. The spatial, demographic, political, and economic giant of the region is Nigeria.

Environments and Resources

Climatic and biotic contrasts within this region are extreme, ranging from tropical steppe, dry savanna, and wetter savanna to areas of tropical rain forest along the southern and southeastern coasts (see Figure 17.8a and b). An estimated 90 percent of the region's tropical forest has been lost to commercial and subsistence deforestation. Most of the coun-

193

246, 488

West Africa

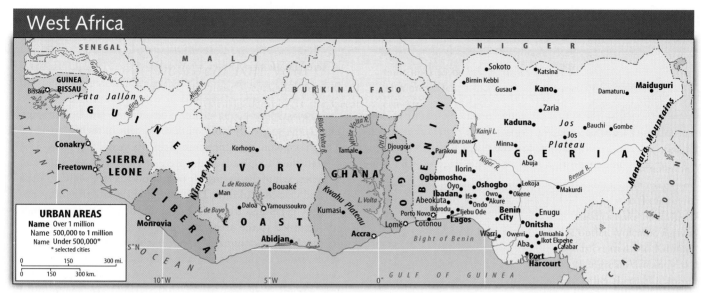

Figure 18.3 Principal features of West Africa

tries' more populous areas are in the zone of tropical savanna climate.

As is typical of much of Africa, the West African countries depend heavily on subsistence farming and livestock grazing, along with a few agricultural and/or mineral exports. In the wetter south, subsistence agriculture relies mainly on root crops such as manioc and yams and on maize, the oil palm, and in some areas, irrigated rice. In the drier, seasonally rainy grasslands of the north (in northern Nigeria and northern Ghana, for example), there is nomadic and seminomadic herding of cattle and goats, along with subsistence farming of millet and sorghum.

The major export specialties of West African agriculture are cacao (from which cocoa and chocolate are made), coffee, and oil-palm products in the wetter, forested south and peanuts and cotton in the drier north. Ivory Coast leads the world in cacao exports, producing almost half the global total in 2004. Nigeria was once the world leader in exports of palm oil and peanuts, but these exports almost vanished as Nigerian agriculture was neglected during the oil boom of recent decades.

The other main export industry of West Africa is mining (see Figure 17.19). The most important resource is petroleum, with OPEC member Nigeria leading the region in production and export. In 2004, when Nigeria stood as the world's seventh largest oil exporter (with 40 percent of those exports going to the United States), petroleum made up 90 percent of the country's foreign export earnings. Oil drilling takes place mainly in the mangrove-forested Niger River Delta and offshore in the adjacent Gulf of Guinea (see Problem Landscape, pages 486–487).

Other prominent mineral products of West Africa include iron ore, bauxite, diamonds, phosphate, and uranium. Iron ore exports are very important to Liberia. The main bauxite resources lie near the southwestern bulge of the West African

coast; Guinea is the main producer and exporter. Togo exports phosphate. Gold, bauxite, and manganese are important to Ghana, as are bauxite, titanium, and diamonds to Sierra Leone.

Population and Settlements

West Africa has an impressive total population of 208 million, or about 86 million fewer people than the United States. The most striking single aspect of West Africa's population distribution is that about two-thirds of the region's population reside within one country, Nigeria (whose 137 million also represent 20 percent of the population of the entire region of Africa south of the Sahara; see Figure 17.3a and b). This unusual concentration of people, which existed even before European contact, is related in part to the region's agricultural productivity and resource wealth.

Only about 37 percent of West Africa's people are city-dwellers, but the urban population is increasing rapidly. The world's large cities are associated mostly with trade, centralized administration, and large-scale manufacturing, but West Africa's big cities are different. They include national capitals and seaports, but as the region has only minor manufacturing, they have developed mainly in areas of productive commercial agriculture. In nearly every country in the region, the capital is a primate city, far larger than the nation's second largest city.

Nigeria has the most impressive urban development: Of the eleven largest metropolitan cities in West Africa, three are clustered in the ethnic Yoruba-dominated southwestern part of Nigeria, in and near the densely populated belt of commercial cacao production. The largest by far is Lagos (population: 10 million; Figure 18.4). It had long been Nigeria's political capital, but in the 1990s, most governmental functions were moved to a new capital at Abuja in the interior. The development of Lagos surged in the early 20th century

Problem Landscape

The Poor, Oil-Rich Delta of Nigeria

Oil is the economic lifeblood of populous Nigeria. Most of the production is concentrated in the Niger River Delta, home to about 7 million mostly Christian people (Figure 18.B). Some

Ethnicity in Nigeria

MAJOR ETHNICITIES

Chamba	Hausa & Fulani
Edo	Ibo
Efik-Ibibio	Ijaw
Gwari	Kanuri
	Nupe
	Tiv
	Yoruba
	mixed

Province names in RED are those practicing shari'a law.

F.C.T.: Federal Capital Territory
ANAM.: Anambra

Figure 18.B Nigeria's ethnic landscape

of that production comes from wells drilled in a 350-square-mile (910-sq-km) section of the Niger Delta region inhabited by 500,000 people of the **Ogoni** tribal group. According to their spokesperson, the Ogoni have derived few benefits and have suffered greatly from oil development in their homeland. They complain that for more than 30 years, oil spills have tainted their croplands and water, destroying their crops and fisheries, while the flaring off of natural gas has polluted their air and caused acid rain. They note that despite the enormous revenue generated from oil drilled on their land, little money has returned to the area; most goes to the oil companies and to the Muslim-dominated government in the north. Meanwhile, most Ogoni live in palm-roofed mud huts; half of the delta region lacks adequate roads, water supplies, and electricity. Schools have few books, and clinics have few medical supplies.

In an attempt to improve their plight, Ogoni activists founded the **Movement for the Survival of the Ogoni People (MOSOP)** in 1990. Ogoni chiefs and elders issued an **Ogoni Bill of Rights** in which they declared the right to a safe environment and more federal support of their people. One of the movement's leaders, a popular author, playwright, and television producer named Ken Saro-Wiwa, also called for self-determination for the Ogoni. However, after an antigovernment rally by hundreds of thousands of Ogoni people in 1993, government police razed 27 villages, killing about 2,000 Ogonis and displacing 80,000 more.

Figure 18.4 Lagos is Nigeria's vibrant and chaotic capital.

Daniel Laine/Corbis

In another violent incident, Ogoni activists killed four founding members of MOSOP in 1994. According to family members and supporters of the slain men, Saro-Wiwa incited their murders by stating that they "deserved to die" for not taking a more active position against the government and oil companies. Government authorities arrested Saro-Wiwa and 13 of his associates, charging them with responsibility for the murders. In late 1995, the government court found Saro-Wiwa and eight other defendants guilty and ordered their execution. Despite international protests that Saro-Wiwa and his codefendants were framed, the government had the men hanged. Many analysts regarded the executions as the government's effort to deter actions by other more serious rivals, especially within the army. As Nigeria is a thinly glued collection of ethnic groups, the government may also have seen the Ogoni struggle for minority rights as a Pandora's box that would lead to the breakup of the nation. Redistribution of oil revenue also would have diminished the income of the country's president and senior military officers.

Despite the return to civilian rule in Nigeria that came after Saro-Wiwa's death, dissent in the delta continued to grow. The Ogoni defiance and contempt of government and the oil companies have spread to the region's other major ethnic groups: the **Ibo (Igbo)**, **Yoruba**, and **Ijaw**. Reiterating the Ogoni protest that the region's oil generates $10 billion a year in revenue for the government, the Ijaw (who make up about 5 percent of Nigeria's population) in particular have pointed to the poverty and squalor of their settlements and demanded that some of the oil money be used to fund roads, electricity, running water, and medical clinics. Ijaw activists have spoken of secession and self-determination if their demands are not met. In 1998 and again in 2003, to register dissatisfaction with the delta economy, they seized and held onshore stations and offshore rigs belonging to foreign oil companies, temporarily disrupting Nigeria's oil exports on both occasions. The 2003 violence was in part directed by the Ijaw against the delta's ethnic Itsekiri people, whom the Ijaw say are collaborators with the oil companies. Questions of equity in a potentially rich country thus pose a threat to national, regional, and international stability.

Nigerians in the delta and elsewhere hope that their 1999 democratic election of President Olusegun Obasanjo was just the beginning of a long-lasting era of stability and economic reform in their land of promise. In 2004, Obasanjo delivered on a campaign promise to allot the oil producing states 13 percent of onshore oil revenues (compared with 5 percent previously) and some portion of offshore receipts. One of Nigeria's ironies is that in this oil-rich country, there are continuous shortages of gasoline and electricity. Neglect, mismanagement, corruption, and theft are all responsible. One of the cruelest and most repetitive Nigerian news stories is the immolation of scores and even hundreds of poor delta villagers who illegally puncture gasoline or oil pipelines and try to "scoop" and sell the hydrocarbons, only to ignite the fires that consume them.

when the British colonial administration chose Lagos as the ocean terminus for a trans-Nigerian railway linking the Gulf of Guinea with the north. During the oil boom of recent decades, migrants from all parts of Nigeria (and many from outside the country) flooded into the city. With its severe traffic congestion, rampant crime, inflated prices, and inadequate public facilities, Lagos gives the impression of urbanism out of control. It is also indicative of a disorderly national economy.

Close to the border with Niger is Kano, the most important commercial and administrative center of Nigeria's north and a major West African focus of Islamic culture. Immediately west of Nigeria on the Guinea Coast lies Cotonou, the main city and port of Benin. Further west, in southern Ghana—another agricultural area with a high degree of commercialization—is Accra (population: 2.9 million), Ghana's national capital. Until recently, it was the main seaport, despite the lack of deep-water harbor facilities. A modern deep-water port has been developed east of Accra at Tema, and that city is now Ghana's main port and industrial center. Well inland from Accra is the country's second largest city, Kumasi, in the heart of Ghana's cacao belt, which was once the world's greatest cacao producing area.

Today, the world's leading cacao producer and exporter is Ivory Coast (officially Côte d'Ivoire), with another of West Africa's largest cities, Abidjan (population: 4.2 million; Figure 18.5). Although the inland city of Yamoussoukro is the official capital, Abidjan is the de facto capital of a country that in the 1990s—prior to its descent into chaos—vowed to achieve economic prosperity comparable to that of the Asian Tiger countries. Also on the Guinea Coast are three more seaport capitals: Monrovia in Liberia; Freetown in Sierra Leone, on West Africa's finest natural harbor; and Conakry in Guinea.

Many of these port cities have poor connections with their countries' hinterlands. Rapids and lower water during the dry season limit the utility of rivers for transport. Some rivers do carry traffic, notably the Niger, its major tributary the Benue, the Senegal River, and the Gambia River. The Gambia River is the main transport artery for The Gambia—a re-

Figure 18.5 Only about one-third of the people in Africa south of the Sahara live in cities, but with considerable rural to urban migration, that number is growing. This is very modern Abidjan, the leading city of Ivory Coast, with St. Paul's Cathedral in the foreground.

markable country in that it is nearly surrounded by another (Senegal) and consists merely of a 20-mile-wide (32-km) strip of savanna grassland and woodland extending inland for 300 miles (c. 500 km) along either side of the river. Railways are more important than rivers in the region's transport structure, but they are few and widely spaced. Instead of forming a network, the rail pattern is a series of individual fingers that extend inland from different ports without connecting with other rail lines (see Figure 17.19).

Ethnic and Political Complexity and Conflicts

West Africa was an advanced part of Africa south of the Sahara when the European Age of Discovery began. A series of strong pre-European kingdoms and empires developed there, both in the forest belt of the south and in the grasslands of the north. In colonial times, it became a French and British realm, except for a small Portuguese dependency and independent Liberia. Of the present countries, The Gambia, Sierra Leone, Ghana, and Nigeria were British colonies; Guinea, Ivory Coast, Togo, and Benin (formerly Dahomey) were French; and Guinea-Bissau was Portuguese. Liberia is unique, having become Africa's first republic after 1822, when 5,000 freed American slaves sailed there and settled with support from the U.S. Treasury. The country's capital, Monrovia, was named for American President James Monroe. Its flag is a variant of the American stars and stripes, and its currency is the Liberian dollar. Many Liberians speak fondly of the United States as their "big brother."

In West Africa today, the English, French, and Portuguese languages continue in widespread use, are taught in the schools, and are official languages in the respective countries. European languages are a useful means of communication in

this region, where hundreds of indigenous languages and dialects, often unrelated, are spoken. Economic, political, and cultural relationships between West African countries and the respective European powers that formerly controlled them continue to be close. Britain maintains ties with **Anglophone** (English-speaking) **Africa** through meetings of the **Commonwealth of Nations,** to which all its former colonies in Africa south of the Sahara belong. France supplies economic and military aid to its former colonies in this and other parts of **Francophone Africa.** France perpetuates relations with its former dependencies by inviting their leaders to summit conferences to discuss matters of common interest. France also has some colonial-style economic relations with some of its former colonies; for example, it still controls the Central African Republic's "strategic resources," including its uranium.

As is often the case in Africa, the governments of the region must deal with serious problems related to ethnicity. European governments contributed initially to some present troubles by drawing arbitrary boundary lines around colonial units without proper concern for ethnicity. As a result, the typical West African country today, like most countries elsewhere in Africa south of the Sahara, is a collection of ethnic groups that have little sense of identification with the national unit. Nigeria, as described on page 467 in Chapter 17, is a good example.

The largest of Nigeria's ethnic groups (which number between 250 and 400) are the mostly Muslim Hausa and Fulani (together, 29 percent of the population) mainly in the north, the mostly Christian Yoruba (21 percent) in the southwest, the Ibo (Igbo) (18 percent) mainly in the southeast, and the Ijaw (10 percent) of the southern delta region. Recently, there has been escalating tension and violence between some of these groups, especially between the mainly Muslim Hausa and mainly Christian Yoruba, and each has formed militias. As Muslim northerners have traditionally dominated Nigerian politics, the ascendancy of the Christian Yoruba President Obasanjo may be seen as a concession by the northerners to prevent the country from fragmenting. However, mainly as a means of blunting Obasanjo's perceived Christian leanings, most of the Muslim northern states have since 1999 adopted *shari'a,* or Islamic law, and the Christian minorities of these regions fear persecution and intolerance (see Figure 18.B, page 486).

Christians have also attacked and killed Muslims elsewhere in the country, and violence between the groups has taken thousands of lives. A state of emergency was declared in 2004 in the central Plateau Province, epicenter of the worst violence. Some of the unrest here is related to livelihood as well as religion: Longtime residents are mainly Christian farmers, and they resent a recent influx of Muslim cattle herders. Cattle theft and crop burning are typically part of their skirmishes. The prospect of violent devolution hangs over Nigeria, reviving memories of deaths of a million people in 1967–1970 when the country's ethnic Ibo population struggled to establish a separate nation of Biafra.

A diamond purge and civil war plagued Sierra Leone between 1991 and 2002, taking more than 50,000 lives. Almost two-thirds of the country, including all of the diamond producing area, came under the control of the rebel **Revolutionary United Front (RUF)**. The RUF projected its power with a brutal combination of mutilation, rape, and murder. The country's prospects improved when its first democratically elected president, Ahmed Tejan Kabbah, took office in 1996. However, a military junta assisted by the RUF seized power a year later, forcing Kabbah into exile. Asserting its self-appointed identity as West Africa's regional superpower, and perhaps with an eye on Sierra Leone's mineral wealth, Nigeria responded in 1998 with a military attack on Sierra Leone that in turn forced the junta to flee and restored President Kabbah to office. RUF resistance continued. The 1999 **Lomé Agreement** gave the RUF some seats in the national cabinet and established RUF leader Foday Sankoh as the country's vice president. But the RUF broke the peace agreement and took hundreds of United Nations peacekeepers hostage, reigniting the war. As the fighting continued, British troops intervened, capturing Sankoh (who died while awaiting a war crimes trial) and helping Kabbah reassert government authority over this devastated country. In 2002, Kabbah was reelected to office, and the war was declared over.

In neighboring Liberia, the regional kingpin Charles Taylor long provided arms to Sierra Leone's RUF in exchange for diamonds. In 1989, American-educated Taylor ignited a civil war that persisted for 7 years. In 1997, Liberians elected him president because, analysts say, they feared that Taylor would restart the war if not elected. After that, he led the country on a path to ruin. Public services such as electricity were all but nonexistent. Schools and hospitals were downsized or closed. Taylor's family enriched itself by selling the county's assets, including its tropical hardwoods, and in exchange for Guinean diamonds, Taylor supplied guns to rebels in that country. By 2003, anti-Taylor rebels calling themselves the **Liberians United for Reconciliation and Democracy (LURD)** had mounted an effective insurgency, and Liberia was once again plunged into war. This time, the United States opposed its one-time ally Taylor, who fled the country, and briefly fielded a peacekeeping force to Liberia. An interim administration in which the government and the rebels shared power was established to control the country until elections in 2005.

Recently, there was violence in Ivory Coast, too. The conflict there was particularly unfortunate because it took place in a country long heralded as a beacon of hope and relative prosperity in troubled West Africa. For many years, it had the region's second strongest economy (behind Nigeria's), thanks mainly to exports of cacao. The commercial capital of Abidjan, dubbed the "Paris of Africa," prided itself for its cosmopolitan and prosperous urban ways. The country's leaders spoke of Ivory Coast's future as an "African Tiger," comparable to the emerging economic powers of East and Southeast Asia.

Ironically, Ivory Coast's economic promise helped sow the seeds of conflict and economic decline. The country grew prosperous from cacao and coffee exports in the 1990s. There were new roads, electricity and phone services, and schools. The middle class grew. Ivory Coast's fortunes made it a magnet for settlement by people from nearby, poorer nations. The government actively encouraged immigration because it needed extra hands and skills to assist in the country's development. Large numbers of mainly Muslim immigrants poured in from Burkina Faso and Mali; most of them settled in the predominantly Muslim northern part of Ivory Coast. As many as 15,000 were said to have been Malian children trafficked and sold into indentured servitude on the Ivoirian plantations. By the end of the decade, these newcomers made up an astonishing one-third to one-half of the country's population of 16 million.

Cacao prices took a nosedive after 2000, and so did Ivory Coast's economic good times. The once-welcomed Muslim immigrants in the north began to be resented and even despised by the mainly Christian southerners. The government, essentially a mirror of that southern population, began an active program of discrimination against the north. Southerners were portrayed as the only "true" Ivoirians—a concept they call *Ivoirité*—and laws were passed to exclude northerners from participation in many aspects of the country's economic and political life.

By 2002, three rebel groups in the north were sufficiently organized to pose a challenge to the government. After a failed coup, they began a military campaign that quickly earned them control over the northern half of the country, all the way to the northern suburbs of the political capital, Yamoussoukro. Leading the rebels was Guillaume Soro, the 30-year-old commander of the main rebel group, the **Patriotic Movement of Côte d'Ivoire (MPCI)**.

The rebel armies were then poised to take control of the rest of the county, and they may well have succeeded. But France, the country's former colonial power, intervened, sending 2,500 Legionnaires to block the rebel advance. The legendary fighting force quickly separated the warring sides and was joined by peacekeepers from the **Economic Community of West African States (ECOWAS)**. The French government managed to persuade leaders from both sides to attend peace talks in Paris as a means of averting all-out war and perhaps the country's partition. In a peace treaty signed in Paris in 2003, the southern-based government under President Gbagbo made large concessions to the rebels. It agreed to share power with its foes, giving nine cabinet seats to Soro and other rebel leaders. Additional seats were awarded to members of the RDR, the country's traditional (and nonviolent) opposition party. The new "national reconciliation" government was supposed to allow the rebels to achieve their main aim in a nonmilitary fashion: an end to the official discrimination against the mainly Muslim northerners.

The arrangement seemed promising. A degree of normalcy returned to Ivory Coast, and the economy began to recover. National road networks reopened, with cotton and sugar produced in the north again being trucked to Abidjan.

The European Union promised generous economic aid if the peace could hold. But the power-sharing arrangement unraveled, and rebels began speaking again of secession for the north. As of 2005, full-scale war had not resumed, but militia loyal to Gbagbo were said to be systematically killing immigrant workers in the cacao fields in the country's south. Refugees fleeing the fighting have joined earlier ranks of refugees from Ivory Coast, Sierra Leone, and Liberia living in camps in both Ghana and Guinea.

18.3 West Central Africa: Colonial "Heart of Darkness"

The subregion of West Central Africa is flanked by Cameroon and the Central African Republic to the north and the Democratic Republic of Congo to the south (Figure 18.6). It contains seven countries: Cameroon, Gabon, Central African Republic, Republic of Congo (often known by the shorthand Congo-Brazzaville), Democratic Republic of Congo (known by the shorthand Congo-Kinshasa, or DRC), São Tomé and Príncipe, and Equatorial Guinea.

Environments and Resources

Scattered parts of West Central Africa, primarily along the eastern and western margins, are mountainous. The most prominent mountain ranges lie in the eastern part of the

Democratic Republic of Congo near the Great Rift Valley and in the west of Cameroon along and near the border with Nigeria. Vulcanism has been important in mountain building in both instances (see Natural Hazards, page 491).

A narrow band along the immediate coast of West Central Africa is composed of plains below 500 feet (c. 150 m) in elevation. As in West Africa, the coast is fairly straight and has few natural harbors. Most of the region is composed of plateaus with surfaces that are undulating, rolling, or hilly. An exception to the generally uneven terrain is a broad area of flat land in the inner Congo Basin, once the bed of an immense lake. The greater part of West Central Africa lies within the drainage basin of the Congo River.

The heart of the region has tropical rain forest climate and vegetation grading into tropical savanna to the north and south. The biological resources are immense. The tropical forests that cover much of the two Congos, Gabon, Cameroon, Equatorial Guinea, and a small corner of the Central African Republic make up one-fifth of the world's remaining tropical rain forests (Figure 18.7).

One of the region's most problematic environmental and economic issues is the fate of these forests. In 2004, they were being cut, mainly for commercial logging, at a rate of 3,125 square miles (8,090 sq km) per year. At that pace, they will be gone by 2020. In an effort to slow the destruction, the World Wildlife Fund has reached an agreement with the World Bank whereby the World Bank will not extend loans in this region unless the borrowing governments can demonstrate they are using, or planning to use, their tropical forest resources in a sustainable manner. These forested countries do not like being, as they see it, held hostage by Western environmentalists, but they are making concessions. A good example is Gabon, which has an extensive area of almost untouched tropical rain forest. As its oil reserves dwindle, Gabon sees its future in selling tropical hardwoods. But in a recent pact, it has promised that part of its pristine rain forest, in the Lopé Reserve of central Gabon, will not be harvested by the Malaysian and French logging companies that have extensive concessions in the country.

Subsistence agriculture supports people in most areas of West Central Africa. There is small-scale production of export crops, mainly in rain forest areas, and some cotton is exported from savanna areas. Southern Cameroon is an important cash crop region, with coffee and cacao the main export crops, supplemented by bananas, oil-palm products, natural rubber, and tea. The farms occupy an area of volcanic soil at the foot of Mt. Cameroon (13,451 ft/4,100 m), a volcano that is still intermittently active. The mountain has extraordinarily heavy rainfall, averaging around 400 inches (1,000 cm) a year.

The island of Bioko has long been notable for its cacao exports and more recently for its oil boom. Bioko and the adjacent mainland territory of Río Muni, formerly held by the Spanish, received independence in 1968 as the Republic of Equatorial Guinea. Cacao production in Bioko developed on European-owned plantations, but the newly independent country's leaders expelled the plantation owners.

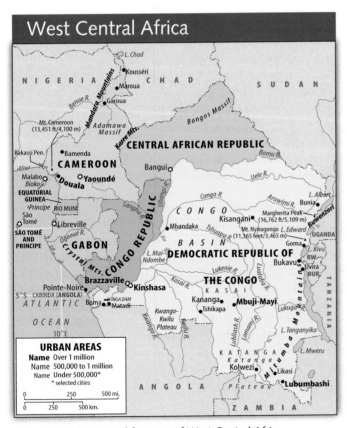

Figure 18.6 Principal features of West Central Africa

Natural Hazards

Deadly Lakes

The Cameroonian volcanoes region is also its **deadly lakes** region. Here, in the two volcanic-cone lakes Nyos and Monoun, carbon dioxide gas generated by volcanic activity deep in the earth builds up in the watery depths below a layer of gas-free fresh water (Figure 18.C). The boundary between these layers is called a chemocline. Sometimes an event such as a windstorm or landslide disturbs the top layer deeply enough to puncture the chemocline and release the carbon dioxide from its solution. It erupts explosively, rocketing gas and water upward and outward. On August 21, 1986, such an explosion in Lake Nyos killed 1,700 people who were simply suffocated by a roving cloud of carbon dioxide.

In recent years, U.S. and Cameroonian scientists and engineers have cooperated to install pipes that dissipate the carbon dioxide from these deadly lakes before it builds up to potentially catastrophic levels. The authorities are pleased with the progress but uncertain that enough venting pipes have been installed. Similar steps have yet to be taken in East Africa's Lake Kivu, where the same threat exists.

Vulcanism also poses more conventional hazards in this region. In 2002, the eruption of Mt. Nyiragongo, near the north-ern shore of Lake Kivu, sent rivers of molten lava into the Congolese city of Goma, incinerating the city and causing almost its entire population of 400,000 to flee as refugees.

Figure 18.C Cameroon's Lake Nyos after one of its eruptions of toxic carbon dioxide gas

Exports of minerals, principally oil, aid the financing of some countries, notably Gabon, Congo-Brazzaville, Cameroon, and most recently, Equatorial Guinea, where oil revenues since production began in 1996 have created economic growth rates of more than 60 percent per year and per capita GDP PPPs that are quite high for Africa south of the Sahara (see Table 17.1, page 444). Oil exploration began in the waters off São Tomé and Príncipe in 2004, and the 150,000 people of this cocoa producing country are hoping for a bonanza. Cameroon has been engaged in an oil-related dispute with Nigeria over control of the Bakassi Peninsula, which juts into the Gulf of Guinea near the border between the two

Figure 18.7 Tropical rain forest covers a large portion of West Central Africa. This bridge provides rare access to the forest canopy in Gabon.

countries. Control of the peninsula will give the dispute's victor ownership of promising offshore oil reserves of the Gulf of Guinea. Cameroon is also receiving royalties and lost-land compensation for the shipment of Chadian petroleum in a pipeline to the Atlantic through Cameroon's territory. Copper from the mineral-rich Katanga Province in the southeast part of Congo-Kinshasa was long the country's main export, but civil war disrupted the industry. There are also rich cobalt deposits there, and diamonds are in the adjoining Kasai region (see Figure 17.19). Another important resource in Congo is coltan, a rare mineral used in the manufacture of cell phones and fighter jets.

Population, Settlements, and Infrastructure

Within West Central Africa as a whole, population densities are low and populated areas are few and far between. Rough terrain, thick forests, wetlands, and areas with tsetse flies tend to isolate clusters of people from each other. The savanna lands and highlands are generally more densely populated than the rain forests.

Most of the countries have an overall population density well below that of Africa south of the Sahara as a whole. The Democratic Republic of Congo is Africa's second largest country in area (after Sudan); it is about the size of Western Europe, or Alaska, Texas, and Colorado combined, but its estimated population in 2004 was only 58.3 million, or well under one-half that of the continent's most populous country, Nigeria. Within the subregion, however, it is by far the most populous country. Nearly half of the Democratic Republic of Congo is an area of tropical rain forest in which the population is extremely sparse.

Cities are few in these countries. Most of the larger ones are political capitals and/or seaports, including the Republic of the Congo's capital city of Brazzaville (population: 1.2 million) and its seaport of Pointe Noire. Brazzaville, located directly across the Congo River from the much larger capital city of Kinshasa in the Democratic Republic of Congo, was once the administrative center of French Equatorial Africa. Kinshasa (population: 6.7 million) is the Democratic Republic of Congo's capital, largest city, and principal manufacturing center. Another important city is Lubumbashi (formerly Elisabethville) in the copper-mining region of the southeast.

Transportation infrastructure is generally poor due to political mismanagement, warfare, neglect, and poverty. Aside from a handful of rail fingers reaching inland from seaports, West Central Africa is largely devoid of railways, and roads only poorly serve most areas. Isolation and poor transport are major factors limiting economic development. There is not yet a paved road between Brazzaville and Pointe Noire, but the 320-mile (515-km) Congo-Océan Railway does connect the cities. One of the Democratic Republic of Congo's major arteries of transportation is a combination of railroad and river connecting Kinshasa with Lubumbashi. Another pair of cities connected by rail is Cameroon's seaport of

Douala (population: 1.5 million) and the country's inland capital, Yaoundé (population: 1.4 million).

Colonialism and Modern Struggles

In colonial days, France held three of the present units—Central African Republic, Republic of the Congo (Congo-Brazzaville), and Gabon—in the federation of colonies called French Equatorial Africa. Cameroon consisted of two trust territories administered by France and Britain. Belgium held what is now the Democratic Republic of Congo (Congo-Kinshasa), Spain held Equatorial Guinea, and Portugal held the islands of São Tomé and Príncipe. During the last quarter of the 19th century, the Congo Basin was virtually a personal possession of Belgium's King Léopold II, whose agents ransacked it ruthlessly for wild rubber, ivory, and other tropical products gathered by Africans. This lawless era inspired Joseph Conrad's famous novel *Heart of Darkness* (1902), in which the trader Kurtz, at the point of death, evokes the ravaged Congo region with the cry "The horror! The horror!" In 1908, the Belgian government formally annexed the greater part of the Congo Basin, creating the colony of Belgian Congo.

Pressure for independence built up rapidly in 1959, and Belgium yielded to the demands of Congolese leaders. In 1960, 3 months before adjacent French Equatorial Africa became the independent People's Republic of the Congo, the Congo colony became the independent Republic of the Congo. In 1971, it took the name Zaire, meaning "river." Following the overthrow of Zaire's government in 1997, the country was renamed Democratic Republic of Congo.

Hundreds of ethnic groups speaking many languages and dialects live in West Central Africa. The majority speak Bantu languages of the Niger-Congo family, while Nilo-Saharan language speakers predominate in the drier areas of grassland north of the rain forest. Small bands of Pygmies (Twa) live in the rain forest of the Congo Basin. The population of Congo-Kinshasa is especially complex, with some 250 ethnolinguistic groups speaking more than 75 languages.

In recent decades, the region's troubles have focused on its major power, the Democratic Republic of Congo (Congo-Kinshasa). During his tenure in office from 1965 to 1997, then-Zaire's autocratic ruler, Mobutu Sese Seko, had neglected the country's development but acquired a massive fortune. Political opposition and public discontent with unemployment and inflation led to unrest in the early 1990s. All-out war began after 1997, when Zairean army troops attempted to expel a force of indigenous Tutsis living in the country's far eastern region (for discussion of the region's Tutsis and Hutus, see page 497). The assault backfired when these Tutsis, supported by neighboring Tutsi-dominated governments in Rwanda and Burundi and by other sympathetic countries in the region, took up yet more arms and initiated fierce attacks on government forces.

The Tutsi counterattack quickly widened into an anti-government revolt comprised of many ethnic factions and led by a non-Tutsi guerrilla named Laurent Kabila. Kabila's forces pushed westward virtually unopposed through the huge country and, in May 1997, occupied Kinshasa with little loss of life. President Mobutu fled Zaire on the eve of Kabila's triumph. Kabila then declared himself president of the country he now called the Democratic Republic of Congo and promised that political reforms would live up to the country's new name. Foreign companies rushed to sign mining contracts with the fledgling government, and throughout Congo, there was hope that decades of neglect and poverty could be reversed in this resource-rich giant of West Central Africa.

But it was not to be. President Laurent Kabila's new regime proved to be a bitter disappointment to most of Congo-Kinshasa's citizens and neighbors and to the international community. Like his predecessor Mobutu, he funneled much of the country's wealth and power into the hands of family and friends. He also obstructed international human rights investigations into the deaths and disappearances of thousands of Hutu refugees during the heady days of rebellion against Mobutu. And he failed to bring stability to the far east of his huge country, where an environment of chaos had given birth to his own bid for power.

In 1998, Hutu forces based in the eastern Congo-Kinshasa launched successive devastating attacks on Tutsi interests in Rwanda. Although the Tutsis of Rwanda had helped bring Kabila to power in Congo-Kinshasa, they now turned against him, forming the backbone of a rebel alliance bent on bringing down Kabila's young regime. The unrest led to something rather rare in the region, a broader fight dubbed "**Africa's First World War**" that involved numerous countries, some of which were not even Congo-Kinshasa's neighbors. Each player became involved to protect its own national, separatist, or economic interests, largely at the expense of anything that would resemble national cohesion and stability in the Democratic Republic of Congo.

On one side was the so-called **Rebel Coalition** that sent forces to oppose the government in Kinshasa. Rwanda, to avenge Kabila's refusal to crack down on Hutus operating against this Tutsi-led country, and perhaps to establish a security buffer zone on its border with Congo-Kinshasa, supported rebels fighting Kabila. Rwanda also wanted protection for the Tutsi minority (known as Banyamulenge) living in extreme eastern Congo-Kinshasa. Like Rwanda, Burundi's Tutsi-dominated government fought Congo-Kinshasa to protect itself from Hutu attacks launched from inside that country. Uganda acted against Congo-Kinshasa in part to establish a security buffer zone that would help protect it from attacks by Ugandan rebels operating from the country. Angola's rebel UNITA forces also took on Kabila's forces in a bid to retain the bases inside Congo-Kinshasa that it used to attack the established government in Angola.

On the other side, the so-called **Congo Coalition** sent forces to support the government in Kinshasa. Zimbabwe supported Kabila, perhaps to divert attention from Zimbabwe's internal problems. By sending forces to aid Kabila, Zimbabwe also may have wanted to establish itself as a re-

gional superpower, akin to the position enjoyed by South Africa. And Zimbabwe acted to protect mining concessions and lucrative contracts that Congo-Kinshasa had awarded to Zimbabwean officials. The Angolan government wanted to crack down on UNITA rebel bases inside Congo-Kinshasa and so joined the fray on Kabila's side. Zambia's government perceived greater regional stability if Congo remained a single entity governed from Kinshasa, so it joined the coalition. Namibian firms had mining contracts with Congo-Kinshasa, so the government in Windhoek supported Kinshasa to protect these interests. The government of Sudan supported Kabila's Congo mainly because of Sudan's bitter enmity with Uganda, which supported southern Sudan's rebellion against the Khartoum government. Finally, Chadian troops were airlifted directly into the diamond-mining region of south central Congo-Kinshasa to protect Chad's investments there.

While the fighting was sometime ferocious, it was not warfare that directly claimed most of the estimated 3.3 million lives lost in this struggle between 1998 and 2001. Most of the casualties were civilians who died from starvation and disease or in widespread massacres directed against ethnic groups, usually Hutus killing Tutsis, but sometimes the reverse.

As the war progressed, Congo-Kinshasa effectively split into three entities: the rebel-held east, importing and selling goods through Uganda and Kenya; a south, trading through Zambia and South Africa; and a west, operating from Kinshasa and using the country's outlet to the sea. The country appeared to be on the verge of dissolution when Laurent Kabila was assassinated by a bodyguard in 2001. He was succeeded by his son Joseph, a Congolese leader seemingly, finally, committed to halting the country's decline. A formal peace treaty was reached in April 2003 when Joseph Kabila agreed to have several formal rebel groups and political parties join him in a government of national unity, meant to pave the way for national elections in 2005. Of his four vice presidents, two are former rebel leaders, and former rebel fighters were merged into the national army. In 2004, however, the main rebel group opposing Kabila, the **Rally for Congolese Democracy (RCD)**, was still fighting the government (despite its nominal role in it) and succeeded in taking the strategic eastern city of Bukavu despite a strong presence of United Nations peacekeepers. The Tutsi RCD is backed by Rwanda, and that country's interests may determine whether a wider war will reignite.

Evidence is emerging that conflict in Congo-Kinshasa has been enormously profitable for the warring factions, foreign governments, and some foreign companies. United Nations investigators have singled out Uganda and Rwanda for looting eastern Congo-Kinshasa of its gems, minerals, timber, agricultural produce, and wildlife, including elephant ivory from some of the country's national parks. African, European, and U.S. companies have been able to exploit the country's mineral wealth without paying corporate taxes. The quest for this country's spoils may continue to make peace elusive.

18.4 East Africa: Mauled but Healing

Five countries make up East Africa: Kenya and Tanzania, which front the Indian Ocean, and landlocked Uganda, Rwanda, and Burundi (see Figure 18.8). Their total area is 703,000 square miles (1.8 million sq km), roughly the size of the U.S. state of Texas, with a total population in 2004 of 109 million (more than one-third the population of the United States).

Environments and Resources

East Africa's terrain is mainly plateaus and mountains, with the plateaus generally at elevations of 3,000 to 6,000 feet (c. 900 to 1,800 m). All five countries include sections of the Great Rift Valley. International frontiers follow the floor of the Western Rift Valley for long distances and divide Lakes Tanganyika, Malawi, and Albert and smaller lakes among different countries. The more discontinuous Eastern Rift Valley crosses the heart of Tanzania and Kenya. Lake Victoria, lying in a shallow downwarp between the two major rifts, is divided between Uganda, Kenya, and Tanzania.

The two highest mountains in Africa, Mt. Kilimanjaro (19,340 ft/5,895 m) and Mt. Kirinyaga (formerly Mt. Kenya,

17,058 ft/5,200 m; see Figure 17.6), are extinct volcanoes. Both are majestic peaks crowned by snow and ice and visible for great distances across the surrounding plains; recent scientific observations, however, suggest that Kilimanjaro's legendary snows may be melted completely by 2015. The mountain has inevitably become a focus of the debate on global warming, but so far, conclusive proof on what is causing its ice to retreat has not been found. Most of East Africa's productive agricultural districts are in the fertile soils of these volcanic areas (Figure 18.9).

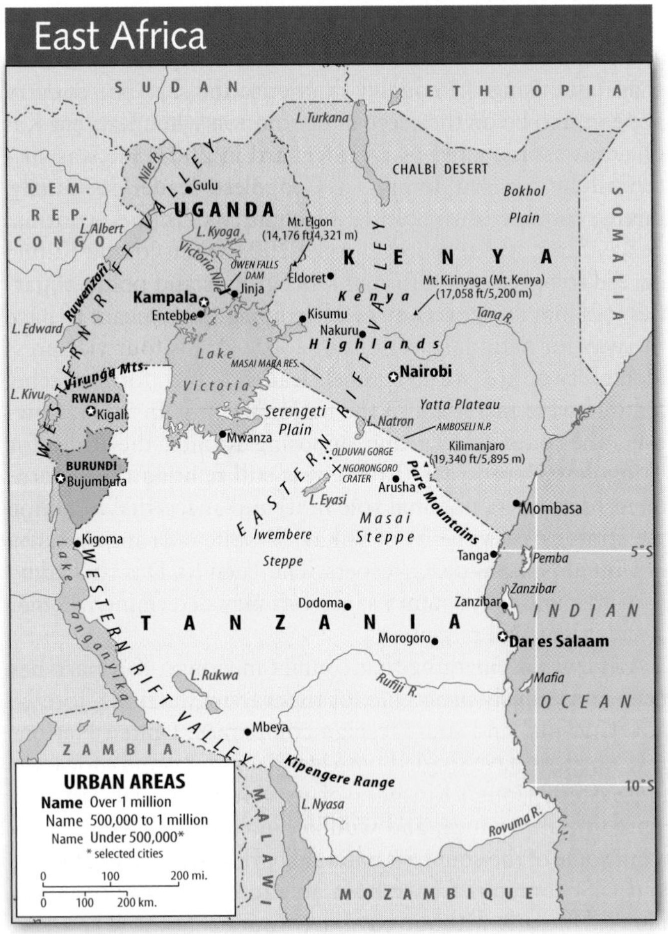

WORLD
REGIONAL
Geography⊚Now™

Click Geography Literacy for an animated figure of geological rifting.

Vulcanism was also important in building the fertile highlands of Rwanda and Burundi. The Virunga Mountains along the Rwanda–Uganda border include many active volcanoes. One of the region's most active is Mt. Elgon (14,176 ft/4,321 m) on the Uganda–Kenya border. On the Uganda–Democratic Republic of Congo border is Margherita Peak (16,762 ft/5,109 m), part of a nonvolcanic range known as the Ruwenzori. This range is often identified as the "Mountains of the Moon" depicted by early Greek geographers on their maps of interior Africa, but it should be noted that these geographers had little real knowledge of the region to work with.

The climate and biome types of most of East Africa are classified broadly as tropical savanna (see Figure 17.8a and b). In Uganda, Kenya, and Tanzania, park savanna—composed of grasses and scattered flat-topped acacia trees—stretches over broad areas and is the habitat for some of Africa's largest populations of large mammals. Great herds of herbivores and their predators and scavengers migrate seasonally across these countries to follow the changing availability of food and water. The countries manage many parks and wildlife reserves, and in Kenya and Tanzania,

Figure 18.8 Principal features of East Africa

Figure 18.9 East Africa is blessed with extensive areas of volcanic soil where commercial and subsistence crops thrive. This is a wheat field in the central highlands of Kenya.

Joe Hobbs

Figure 18.10 Savanna grasslands in east and southern Africa host the world's greatest concentrations of large mammals. This is Kenya's Maasai Mara National Park in part of the greater Serengeti ecosystem that also extends into northern Tanzania.

these are responsible for a huge share of the nations' hard currency revenues. Among the most famous are the Serengeti Plains and Ngorongoro Crater of Tanzania and the Maasai Mara and Amboseli National Parks of Kenya (Figure 18.10).

Over much of East Africa, moisture is too scarce for non-irrigated agriculture. There are great variations in the amount, effectiveness, and dependability of precipitation and in the length and time of occurrence of the dry season. Long and sometimes catastrophic droughts occur. From 1998 to 2001, Kenya suffered its worst drought in 40 years. More than 3 million people were in need of food aid as crops shriveled and livestock died. As the country's hydropower reserves were depleted, Kenyan cities suffered long periods without electricity. Tanzanian, Kenyan, and Ugandan efforts to dam and redistribute waters of the Nile Basin for irrigation, drinking water, and hydropower have aroused the ire of downstream Egypt (see discussion and map on pages 223 and 224 in Chapter 8).

The main subsistence crops are maize (corn), millets, sorghums, sweet potatoes, plantains, beans, and manioc. Elevation has such a strong effect on temperatures that in some areas midlatitude crops like wheat, apples, and strawberries do well. Kenyan tourist brochures boast that Nairobi (elevation: 5,500 ft/1,676 m) has a springlike climate year round. Some cultivators depend exclusively on crops for subsistence, and others also keep livestock. Cattle are the most important livestock animals.

The most valuable export crop for Kenya, the region's leading exporter, is tea, followed closely but cut flowers and fresh vegetables airfreighted to Europe. Coffee exports are also very important to Kenya. Kenya and Tanzania also produce sisal for its valuable fiber (see Figures 17.17 and 17.18). The islands of Zanzibar and Pemba support the millions of clove trees that provide much of world's supply. The islands' economies have been hurt drastically by a recent fall in world clove prices and by economic and political turmoil in Indonesia, which had long imported most of the clove crop for use in its aromatic *kretek* cigarettes. Seeking a new future, Zanzibar has announced plans to transform itself into a free

economic zone modeled after Singapore and Hong Kong. Tanzania, Kenya, and Uganda in 2001 revived the **East African Community,** which had been established in 1967 but fell apart 10 years later because of regional strife. The East African Community is meant to reduce trade barriers among its member states, establish a stock exchange, and adopt a common currency. Rwanda and Burundi have applied for membership, but the organization wants stronger assurances that conflict is over in these countries.

The East African nations generally lack mineral reserves and production. A project to mine titanium, very controversial because it will require the relocation of thousands of people, is underway near the Kenya coast. Manufacturing is limited mainly to agricultural processing, some textile milling, and the making of simple consumer items. There is almost no petroleum. There is a large cotton-textile mill at Jinja, Uganda, that uses hydroelectricity from a station at the nearby Owen Stanley Dam on the Nile. The Tana River project in Kenya produces hydropower and supplies irrigation waters.

Populations, Settlements, and Infrastructure

The highest population densities are in a belt along the northern, southern, and eastern shores of Lake Victoria; in south central Kenya around and north of Nairobi; and in Rwanda and Burundi, which have among the world's highest population densities, aside from those of city-states and some islands (see Figure 17.3a). As recently as the early 1990s, population growth rates in this region were so high that East Africa was a common academic setting for the Malthusian scenario. But after sharp reductions as a result of falling birth rates and increasing AIDS-related deaths, the population growth rates of the five East African countries were moderate, averaging 2.4 percent as of 2004. The region's populations are overwhelmingly rural. Only about 8 and 12 percent of the populations of Burundi and Uganda, respectively, are urban. Kenya and Tanzania are the most urbanized at only about 36 and 22 percent, respectively.

Joe Hobbs

Figure 18.11 Naval ships and grain silos in Kenya's Indian Ocean Port of Mombasa

The most agriculturally productive parts of East Africa are bound together by railways that form a connected system leading inland from the seaports of Mombasa in Kenya and Dar es Salaam and Tanga in Tanzania. Mombasa is the most important seaport in East Africa (Figure 18.11). From Mombasa, the main line of the Kenya–Uganda Railway leads inland to Kenya's capital, Nairobi (population: 3.4 million), the largest city and most important industrial center in East Africa. It is also the busiest crossroads of international air traffic and the main outfitting and departure point for safaris into East Africa's world-renowned national parks. It is a gritty, crime-ridden city that many locals call "Nairobbery." Nairobi was founded early in the 20th century as a construction camp on the railway, which continues westward to Kampala (population: 1.5 million), the capital and largest city of Uganda. Dar es Salaam ("House of Peace"; population: 2.5 million) is the capital, main city, main port, and main industrial center of Tanzania, with rail connections to Lake Tanganyika, Lake Victoria, and Zambia. Rwanda's capital is in the highlands at Kigali (population: 305,000), and Burundi's capital is on the northernmost point of Lake Tanganyika at Bujumbura (population: 340,000). All of the East African capitals are primate cities.

Political and Social Issues

The five East African countries face many social and political dilemmas. None has had much internal peace, order, or prosperity, and none has had easy relations with its neighbors. Ethnic rivalries and conflicts have beset all of them, but prospects are also now better for them than at any time in recent decades.

Aside from small minorities of Asians, Europeans, and Arabs, the population of East Africa is composed of a large number of African tribes; for example, Tanzania alone has about 120 groups. Among the large groups are the Kikuyu and Luo of Kenya, the Baganda of Uganda, the Sukuma of Tanzania, the pastoral Maasai of Kenya and Tanzania, and the Tutsi and Hutu of Rwanda and Burundi. Most East African peoples speak Bantu tongues of the Niger-Congo language family, but some in northern Uganda, southern and western Kenya, and northern Tanzania speak Nilo-Saharan languages. Swahili, a Bantu language drawing heavily on Arabic for vocabulary, is a widespread lingua franca (and is the national language of Tanzania), as is English. There are Christian, Muslim, and "animist" faiths in all the countries. The Muslim religion reflects a long history of Arab commercial enterprise in the region, including slave trading.

All of the countries are former European colonies. Rwanda, Burundi, and Tanzania were part of German East Africa until the end of World War I. With the end of the war, Tanzania joined Kenya and Uganda as British dependencies, and Rwanda and Burundi, then known as Ruanda–Urundi, were controlled by Belgium. They became two independent countries in 1962. Now a part of Tanzania, the islands of Zanzibar and Pemba, which lie north of the city of Dar es Salaam and some 25 to 40 miles (40 to 64 km) off the coast, were once a British-protected Arab sultanate. The city of Zanzibar was the major political and slave-trading center and entrepôt of an Arab empire in East Africa. While Arabs formerly controlled political and economic life, the majority of the population was, and is, African. Africans have been in control since 1964, when a revolution overthrew the ruling sultan. That same year, the political union of Tanganyika and Zanzibar was formed with the name United Republic of Tanzania. However, Zanzibar and Pemba have a parliament and president separate from Tanzania's.

Prior to independence, East Africa was home to about 100,000 Europeans, 60,000 Arabs, and 300,000 Asians. Arabs were primarily a mercantile class. Europeans were a professional, managerial, and administrative class. In the "White Highlands" of southwestern Kenya, about 4,000 European families owned and operated estates that were worked by African laborers to produce export crops and cattle. With independence in 1963, Kenya made a point of inviting the British and other Europeans to maintain their presence in the commercial crop sector and to train the black majority in the managerial and technical skills that had served white interests well. Many Europeans left, and

Africans acquired their farms, but those who stayed formed the core of the still-influential white Kenyans of today. Collectively known as the **Wahindi** in Swahili, the Asians included Indians, Pakistanis, and Goans (from the former Portuguese colony of Goa on the west coast of India). They dominated retail trade, owned factories, and produced export crops. Some were wealthy, and practically all were more prosperous than most Africans.

Since independence, the Asian minority has lost much of its commercially predominant position. Ugandan leader Idi Amin abruptly expelled most of the 80,000 ethnic Indians and Pakistanis from Uganda in 1972. Because they were so vital to the country's economy, it nearly collapsed. About 15,000 have returned in recent years, have successfully reclaimed lost land and other assets, and once again own and operate successful service and manufacturing businesses. The recent economic decline of Kenya has led to a slow but steady exodus of Asians, mainly to Australia, Britain, and the United States.

Episodes of bloodshed have punctuated East Africa's history. From 1905 to 1907, when mainland Tanzania was German controlled, German authorities put down a great revolt against colonial control, costing 75,000 African lives. Immediately prior to its independence, Kenya experienced a bloody guerrilla rebellion led by a secret society within the large Kikuyu tribe called the **Mau Mau.** Antagonism directed against British colonialism and European ways in Kenya cost the lives of about 11,000 Africans and 95 white settlers over the period 1951–1956.

From the 1960s through the 1980s, Kenya was regarded as a relative economic success story in Africa south of the Sahara. However, corruption, ethnic strife, crime, terrorism (including the 1998 bombing of the U.S. embassy in Nairobi, apparently by associates of al-Qa'ida), drought, and floods then combined to reduce Kenya's prospects for economic development. Another al-Qa'ida-linked bombing in Mombasa in 2002 took a big toll on the vital business of international tourism in Kenya, which normally contributes about half of the country's hard currency revenue. The country's growing HIV/AIDS problem also discouraged visitors. In 2003, half a million tourists visited Kenya, down half a million from the high in 1996.

Through much of the latter half of former President Daniel Arap Moi's 24 years in office, which ended in 2002, Kenya suffered from corruption and human rights abuses. Britain, the United States, and other international lenders cut off economic aid because Moi refused to implement democratic reforms. Mwai Kibaki, the candidate opposing Moi's political party, won the 2002 presidential election, and he has vowed a "zero tolerance" policy on corruption. Vital international aid to Kenya has resumed.

In postcolonial Uganda, a savage orgy of murder, rape, torture, and looting took place under the notorious regimes of Idi Amin and Milton Obote. The ravaged country had a respite after 1986 when Yoweri Museveni, the leader of a southern guerrilla force that had developed in opposition to Obote, took over the government. Despite some simmering conflicts among Uganda's 40 ethnic groups and despite the country's controversial interventions in regional conflicts, Uganda under Museveni has come to be regarded widely as a model for democracy, social progress, and economic development in Africa south of the Sahara. As of 2004, much international scrutiny was focused on Uganda's government to ensure that it would not promote ethnic cleansing of the Acholi people. Some Acholi support the rebel **Lord's Resistance Army** (**LRA**), which has been fighting government interests in the country.

Rwanda and Burundi have had tragic disputes between their majority and minority populations. About 85 percent of the population in Burundi and 90 percent in Rwanda are composed of the **Hutu** (**Bahutu**), and most of the remainder of the two populations is **Tutsi** (**Watusi**). These originally were not separate tribal or ethnic groups; the peoples speak the same language, share a common culture, and often intermarry. The distinction between Hutu and Tutsi rather was based on socioeconomic classification. Historically, the Tutsi were a ruling class who dominated the Hutu majority. Tutsi power and influence were measured especially by the vast numbers of cattle they owned. Tutsi who lost cattle and became poor came to be identified as Hutu, and Hutu who acquired cattle and wealth often became Tutsi.

Colonial rule in East Africa attempted to polarize these groups, thereby increasing antagonism between them. German and Belgian administrators differentiated between them exclusively on the basis of cattle ownership: Anyone with fewer than 10 animals was Hutu and anyone with more was Tutsi. Europeans mythologized the Tutsis as "black Caucasian" conquerors from Ethiopia who were a naturally superior, aristocratic race whose role was to rule the peasant Hutus. The colonists replaced all Hutu chiefs with Tutsi chiefs. These leaders carried out colonial policies that often imposed forced labor and heavy taxation on the Hutus. Education and other privileges were reserved almost exclusively for the Tutsis.

Ferocious violence between these two peoples has marred Rwanda and Burundi intermittently since the states became independent in the early 1960s. After independence, the majority Hutus came to dominate the governments of both Rwanda and Burundi. In 1994, the death of Rwanda's Hutu president in a plane crash (in which Burundi's president also died) sparked civil war in Rwanda between the Hutu-dominated government and Tutsi rebels of the **Rwandan Patriotic Front** (the **RPF**, based in Uganda). Hutu government paramilitary troops, still vengeful about decades of domination by Tutsis, systematically massacred Tutsi civilians with automatic weapons, grenades, machetes, and nail-studded clubs. Women, children, orphans, and hospital patients were not spared. An estimated 800,000 Tutsis and "moderate" Hutus (those who did not support the **genocide**) died. International media tracked these horrors, but outside nations did nothing to stop the fighting.

Well-organized and motivated Tutsi rebels seized control

Figure 18.12 Warfare and drought in postcolonial Africa have driven millions of Africans from their homes as refugees. This harrowing scene shows terrorized Rwandans fleeing into Tanzania in 1994 to escape the savage massacres of civilians in the country's civil war.

of most of the country and took power in the capital, Kigali, in 1994. That precipitated a flood of 2 million Hutu refugees mainly into neighboring Zaire, now the Democratic Republic of Congo (Figure 18.12). Although the Tutsi-dominated government promised to work with the Hutus to build a new multiethnic democracy, Hutu insurgents, based mainly in eastern Zaire, continued to carry out attacks against Tutsi targets in Rwanda. That violence and Rwandan interests in the Democratic Republic of Congo's civil war kept Rwandan forces active outside its borders until 2002.

Rwanda now has a national unity policy aimed at reconciling Hutus and Tutsis; both occupy important government posts, and although the Tutsi still hold the political upper hand, Tutsis have encouraged the resettlement of Hutu refugees who fled the country during its troubles. In 2004, marking the 10-year anniversary of the genocide with a new policy meant to prevent its recurrence, Rwanda's government outlawed ethnicity. Now any Rwandan who speaks or writes of Hutus and Tutsis may be fined or imprisoned for practicing "divisionism."

In Burundi, the Tutsi minority traditionally dominated politics, the army, and business. In 1993, there was a brief shift of power to the Hutu majority, as Burundians elected the country's first Hutu president. Tutsis murdered him in October 1993, and a large-scale Hutu retaliatory slaughter of Tutsis ensued. An estimated 300,000 died in a decade of violence. Burundi's Tutsis and Hutus now have a power-sharing agreement in which the presidency rotates between them. The tentative peace is being maintained by the first-ever deployment of the African Union's African Standby Force.

Collective shame and embarrassment in the international community about the failure to stop the killings in Rwanda

and Burundi led in subsequent years to numerous new antigenocide measures. These include a United Nations "early warning" system to prevent genocide and a new forum for crimes against humanity in the **International Criminal Court,** located in the Hague, Netherlands.

18.5 The Horn of Africa: Refuge for Judaism, Christianity, Islamist Militancy

In the extreme northeastern section of Africa south of the Sahara, a great volcanic plateau rises steeply from the desert. This highland and adjacent areas occupy the greater part of the Horn of Africa, named for its projection from the continent into the Indian Ocean. This subregion includes the countries of Ethiopia, Eritrea, Somalia, and Djibouti (Figure 18.13).

Environments and Resources

Much of the Horn of Africa lies at elevations above 10,000 feet (3,050 m), and one peak in Ethiopia reaches 15,158 feet (4,620 m). The highlands where the Blue Nile, Atbara, and other Nile tributaries rise receive their rainfall during the summer. Temperatures change from tropical to temperate as elevation increases. Bananas, coffee, dates, oranges, figs, temperate fruits, and cereals can be produced without irrigation. Highland pastures support sheep, cattle, and other livestock. The region's people are concentrated in the relatively well-endowed highlands; most of Ethiopia's 72.4 million people live there.

East of the mountain mass, in Eritrea, Djibouti, and So-

Horn of Africa

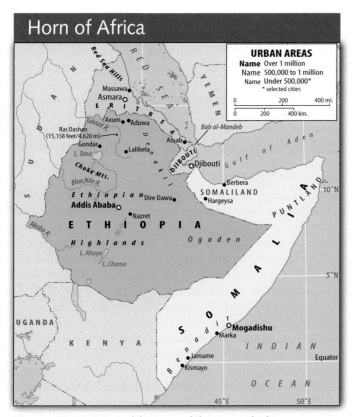

Figure 18.13 Principal features of the Horn of Africa

malia, lower plateaus and coastal plains descend to the Red Sea, the Gulf of Aden, and the Indian Ocean. Extreme heat and aridity prevail at these lower levels, and nomadic and seminomadic tribespeople make a living by herding camels, goats, and sheep. Dwellers of scattered oases carry on a precarious agriculture. The arid lowland sections have many characteristics more typical of the Middle East than other areas of Africa south of the Sahara. The populations of these countries are small compared to Ethiopia: Somalia with 8.3 million, Eritrea with 4.4 million, and Djibouti with only 700,000.

Ethiopia has substantial resources, perhaps including significant mineral wealth. However, its resource base remains largely undeveloped, and the country is impoverished. Coffee and animal hides are the main exports. With foreign assistance, a few textile and food-processing factories have been built, and with Chinese help, a massive dam is being built on a tributary of the Blue Nile to harness part of the country's large hydroelectric potential.

224 A poor transportation system—whose best feature is the 3,000 miles (c. 4,800 km) of good roads built by the Italians during their occupation in the 1930s—is a major factor keeping the country isolated and underdeveloped. The French-built railroad from Djibouti to Ethiopia's capital, Addis Ababa (population: 2.7 million), lacks feeder lines and has never been as successful as hoped. Many roads into the Ethiopian heartland are rough tracks, although better connections have gradually been developing. Air services—

especially Ethiopia's remarkable fleet of DC-3s, more than 50 years old but kept in top condition—help overcome the deficiencies in ground transportation. Many of Ethiopia's people live in villages on uplands cut off from each other and from the outside world by precipitous chasms (Figure 18.14). These formidable valleys were cut by streams carrying huge amounts of fertile volcanic silt to the Nile. Before the Aswan High Dam was built, they provided material for the nourishing floods in Egypt. *245*

Eritrea has gold, copper, and marble reserves but lacks oil, and little of its mineral potential has been developed. Its Red Sea coastline has superb coral reefs, and the country is hoping to attract international tourists to dive and snorkel among them. There is some cotton and oilseed agriculture, and fish are a mainstay in the diet.

Djibouti supports pastoralists but has almost no arable land. The country has little to offer economically except its strategic position near the mouth of the Red Sea, and this has been a windfall for Djibouti since 9/11. Its fine port, also named Djibouti, is a major staging ground in the U.S. war on terrorism. France has long maintained a significant military *225* presence and has pumped much-needed economic aid to the country.

Somalia is hyperarid, and only 2 percent of the country is cultivated. Nomadic and seminomadic pastoralism is the main economic activity. Until an outbreak of livestock disease in the 1990s, Somalia's main exports were live animals (goats, sheep, camels, cattle), shipped mainly to Saudi Arabia, Yemen, and the United Arab Emirates. Fish products and bananas are still shipped, but despite a ban in effect since 2002, charcoal is the main export to the Arabian Peninsula. The resulting environmental impacts on Somalia are severe because trees never were abundant, and there are fights between the tree-cutters and the pastoralists whose camels feed

Figure 18.14 This highland village in north central Ethiopia, like many in the country, is isolated by deep gorges (not visible in this view). The round structure to which all the village roads lead is the community's Christian church.

on tree fodder. An estimated 14 percent of Somalia was forested in 1992, but now less than 4 percent is.

Droughts have stricken this region, particularly Ethiopia, almost incessantly since the late 1960s. Ethiopian villagers are especially vulnerable to drought because of their heavy dependence on rain-fed agriculture. The volcanic soils are fertile, but the streams are entrenched so far below the upland fields that little irrigation is possible. In 1984 and 1985, drought in Ethiopia reached the point of widespread catastrophe. An estimated 1 million people perished. Only a massive and sustained international relief effort, some of it promoted by British and American rock musicians, helped reduce the suffering. Successive droughts have revisited the country, and chronic hunger reportedly affects between 4 and 5 million Ethiopians each year. Ethiopia's multimillion-dollar effort to resettle 2.2 million farmers from the apparently overcrowded highlands to the less populous lowlands by 2007 has run into major problems. The lowlands were less crowded to begin with because of malaria and sleeping sickness, which have taken a big toll on settlers.

Local Cultures, European Imperialism, and Regional Struggles

Ethiopia's people are ethnically and culturally diverse. About 45 percent, including the politically dominant **Amhara** peoples, practice **Ethiopian Orthodox Christianity**, an ancient branch of Coptic Christianity that came to Ethiopia in the fourth century from Egypt. The entire area has had important cultural and historical links with Egypt, the Fertile Crescent, and Arabia. The Ethiopian monarchy based its origins and legitimacy on the union of the biblical King Solomon and the Queen of Sheba, who, tradition holds, gave birth to the first Ethiopian emperor, Menelik. Until a Marxist coup brought an end to the emperorship in the 1970s, Ethiopia's rulers were always Christian. Ethiopia has many outstanding Christian artistic and architectural treasures, including the 11 churches of Lalibela, carved from solid rock in the 12th and 13th centuries (Figure 18.15). Most of the rest of Ethiopia's people are either Muslims (who make up about 40 percent of the population) or members of Protestant, Evangelical, and Roman Catholic churches. There are still small numbers of **Falashas,** or **Ethiopian Jews,** in Ethiopia, a remnant of a very ancient and isolated Jewish population. The majority, about 65,000, had fled to Israel in the 1980s and 1990s. Because this mountainous country has long served an isolated refuge for such unique groups, it has been nicknamed the "**Galapagos Islands of Religion.**"

Eritrea's population, like Ethiopia's, is about half Muslim and half Christian. Djibouti is about 95 percent Muslim and 5 percent Christian. In Somalia, about 99.8 percent of the people are Sunni Muslims and 85 percent are ethnic Somalis. They share the same language and the same nomadic culture. This is one of the most homogeneous populations in the world, a fact that would suggest peace and stability, but interclan rivalries tore the country apart in the 1990s, as related below.

Colonialism had a big impact on the region's peoples. European powers seized coastal strips of the Horn of Africa in the latter 19th century (see Figure 17.16). Britain was first with British Somaliland in 1882, and then France annexed French Somaliland in 1884. Finally, Italy took control of Italian Somaliland and Eritrea in 1889. These areas along the Suez–Red Sea route have never had much economic importance, but they have had great strategic significance.

Italy tried to extend its domain from Eritrea over more attractive and potentially valuable Ethiopia in 1896, but Ethiopians annihilated the Italian forces at Aduwa. Forty years later, in 1936, the Italians succeeded in taking Ethiopia. In World War II, however, British Commonwealth forces defeated Italian troops throughout the region. Ethiopia regained independence, and Eritrea was federated with it as an autonomous unit in 1952. Ethiopia incorporated Eritrea as a province in 1962.

Figure 18.15 The churches of Lalibela, carved from volcanic rock in the 12th and 13th centuries, are among Ethiopia's many Christian cultural treasures. Note man at left center for scale.

After World War II, Italian Somaliland was returned to Italian control to be administered as the trust territory of Somalia under the United Nations, pending independence in 1960 as a republic. The present country of Somalia officially includes both the former Italian territory and the former British Somaliland, which chose to unite with Somalia. The people of much smaller French Somaliland remained separate and eventually became independent as the Republic of Djibouti, with its capital at the seaport of the same name.

In 1974, Ethiopia's aging Emperor Haile Selassie (whom **Rastafarians,** mainly of Jamaica, regard as God) was deposed by a civilian and military revolt against the country's feudal order. A Marxist dictatorship, strongly oriented to the former Soviet Union, emerged. American influence, which had been strong under Haile Selassie, was eliminated. Munitions and advisers from the Soviet bloc poured in to equip and train one of Africa's largest armies, and the Soviets gained access to base facilities in Ethiopian ports. The United States, which had previously supported Ethiopia in its struggles against neighboring Somalia, now switched sides to back Somalia. Such foreign interventions and proxy conflicts contributed to the strife that has hindered development in the region.

Ethiopia had many internal troubles following independence. In what was then the province of Eritrea, both Muslims and non-Amhara Christians resented their political subjugation to Ethiopia's Christian Amhara majority. Terrorism and guerrilla actions escalated into civil war in the early 1960s. The Ethiopian army tried but failed to stamp out the revolts of two Eritrean "liberation fronts." At the same time, there was also a rebellion in Tigre Province adjoining Eritrea. In 1991, these combined rebel forces captured Addis Ababa and took over the Ethiopian government. The rebels in Eritrea established an autonomous regime, and Eritrea became independent in 1993. Eritrea's main city and manufacturing center, Asmara (population: 916,000), is located in highlands at an elevation of more than 7,000 feet (2,130 m) and has a rail link with the Red Sea port of Massawa.

Many geographers subsequently pointed to Eritrea as a model for sustainable development in Africa south of the Sahara. After independence, the country reduced its foreign debt, cut its population growth rate, and took other promising steps toward eliminating the burdens of underdevelopment. However, in 1998, Eritrea dimmed its bright prospects by launching a war against Ethiopia. The conflict emerged from economic rivalries and unresolved territorial issues between the nations. With Eritrea's independence, Ethiopia had agreed to become landlocked, as long as it could freely use Eritrea's Red Sea ports of Massawa and Assab. The concession was in large part Ethiopia's reward to Eritreans for helping with the overthrow of Ethiopia's Marxist regime. The countries' leaders in effect agreed on an economic union in which Eritrea would be the industrial power and Ethiopia the agricultural one. They promised to practice free movement of people and goods between the two states and to share a single currency, the Ethiopian birr.

Trouble began when Eritrea introduced its own currency and announced steep duties and fees on goods shipped to and from Ethiopia through Eritrean ports. Ethiopia then turned to Djibouti and Kenya to find new outlets to the sea. Deprived of an anticipated major source of revenue, Eritrea attacked Ethiopia. The war finally ended in 2000, when Ethiopia gained the upper hand by securing a border region disputed between the two countries. A demilitarized zone monitored by United Nations troops was established along the border. An estimated 100,000 Eritreans and Ethiopians had perished in the conflict. Eritrea's economy was hit particularly hard. With the Ethiopian market gone, Eritrea's exports declined 80 percent from their 1997 high. The port of Assab shut down.

Somalia has been in ruin since a rebellion overthrew the government in 1991 and ignited a civil war between regional Somali clans (the country has six major clans and numerous subclans). Contending forces inflicted great damage on the capital, Mogadishu (population: 1.2 million), from which large numbers of people fled to the countryside as refugees. Massive famine threatened the devastated country in mid-1992. Law and order disintegrated, and humanitarian assistance became difficult.

United States military forces then intervened with **Operation Restore Hope,** a mission to help protect relief workers and aid shipments. However, American policymakers decided to use military force to neutralize the power of one of the country's leading "warlords," Mohamed Farrah Aidid. In a gruesome sequence of events remembered today as the **Blackhawk Down incident,** 18 U.S. servicemen stranded in Aidid's territory in Mogadishu were killed in a single day in 1993. Osama bin Laden's al-Q'aida organization had equipped Somali forces with the weapons they used to assault the American troops. American public outcry at the tragedy brought an end to this U.S. military mission in 1994. Blackhawk Down had two lasting repercussions. It made the United States more reluctant to assist in humanitarian crises. It also emboldened al-Qa'ida, which became convinced that the United States could not stomach the loss of American lives. Osama bin Laden frequently cited the American withdrawal from Somalia as evidence of his group's successful *jihad* against the United States.

Since then, Somalia has fractured into three de facto states (see Figure 18.14). The southernmost, centered on Mogadishu (which is still much in ruin and without services), is Somalia, the only one with international recognition as a sovereign country. The country is barely functioning, as there is only a nominal government with little authority, there is no legal system, and several warlords (and one warlady) control their virtual territories. Internationally backed efforts to unify the country have picked up since 9/11 because Somalia's lawlessness and proximity to Yemen might make it an attractive base for Islamist activity. But unity is difficult to achieve in a country that has splintered. To the north is Puntland, which includes the tip of the Horn of Africa and considers itself autonomous (with a president of its own) but not independent of Somalia.

226

To the northwest, bordering Djibouti and Ethiopia, is So-maliland (the former British Somaliland). It proclaimed itself independent of Somalia in 1991 and has a president and broadly representative government based in the city of Hargeysa. Its port city of Berbera has served as Ethiopia's main trade outlet to the sea since Ethiopia's conflict with Eritrea began. Somaliland's economic mainstay was exports of cattle to the Muslim nations of the Arabian Peninsula, until an outbreak of Rift Valley Fever led to a complete cessation of the trade in 2001. Now Somaliland is pinning its hopes on exports of gemstones, frankincense, vegetable dyes made from henna, and products made from qasil plants, such as hair conditioners and body cleansers.

18.6 Southern Africa: Resource Rich, Finally Free

In the southern part of Africa south of the Sahara, five countries—Angola, Mozambique, Zimbabwe, Zambia, and Malawi—share the basin of the Zambezi River and have long had important relations with each other. The first two were former Portuguese colonies, and the latter three were formerly British. Still farther south are South Africa and four countries whose geographic problems are closely linked to South Africa: the three former British dependencies of Botswana (formerly Bechuanaland), Swaziland, and Lesotho (formerly Basutoland), and the former German colony of Namibia.

These 10 states are grouped here as the countries of southern Africa (Figure 18.16). Other than their geographic location, there is little in the way of economic or political integration that binds these diverse states together. This may change as a result of goals established by the **Southern African Development Community** (**SADC**), a regional bloc composed of these 10 countries plus Tanzania, Democratic Republic of Congo, and Mauritius. SADC's member nations have established an economic free-trade zone in which goods can cross borders without tariffs and taxes. The dominant power in that group, in the region, and on the continent is South Africa, and so it is treated in greatest detail here (with a dedicated map, Figure 18.17).

Environment and Resources

Water, Natural Regions, and Agriculture

Water is all important in this drought-prone region of a dry continent. South Africa is fairly well endowed with water, as two river systems—the Orange (including its tributary, the Vaal) and the Limpopo—drain much of the country's high (3,000 to 6,000 ft/c. 900 to 1,800 m) interior plateaus. Many dams have been constructed on these and other South African rivers to provide irrigation water and hydroelectricity.

The Zambezi is another of the region's great rivers. Rising in Zambia, it flows eastward to empty into the Indian Ocean

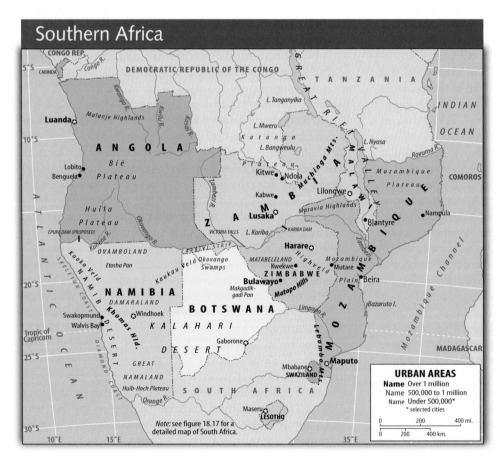

Figure 18.16 Principal features of Southern Africa

South Africa

Figure 18.17 Principal features of South Africa

on the Mozambican coast. On the middle course of the Zambezi River separating Zambia from Zimbabwe is the spectacular Victoria Falls, regarded traditionally as one of the seven Natural Wonders of the World. It is a significant international tourist destination today. Downstream from the falls, hydropower from the Kariba Dam (completed in 1962; see Figure 17.7) flows to both countries.

The driest countries are Namibia and Botswana. They can support pastoralism and, in some areas, limited subsistence agriculture, but commercial agriculture is lacking. The Namib Desert extends the full length of Namibia in the coastal areas (Figure 18.18). Inland is a broad belt of semi-arid plateau country, merging at the east with the Kalahari semidesert of Botswana. The dry zone extends southward into South Africa's Cape provinces and northward into western Angola. Chronic shortages of fresh water in Namibia have led the government to consider desalinating Atlantic Ocean waters (this option was rejected as too expensive); building the Epupa Dam on the Kunene River, which divides Namibia from Angola (this project is opposed by environmentalists and by human rights groups, who argue that the homeland of the Himba tribespeople would be inundated); and diverting water by pipeline southward from the Namibian portion of the Okavango River to Windhoek (widely re-

garded as Africa's cleanest and safest capital; population: 252,000). Arguing that the World Heritage Site of its Okavango Delta would be damaged or destroyed, Botswana is strongly opposed to this Namibian plan. With a wealth of wildlife attractions in the Okavango Delta and adjacent national parks, Botswana is a growing destination for international nature-loving tourists. Namibia has it own wetlands resource of international importance: Large wildlife populations responding to dramatic fluctuations in wet and dry conditions make Etosha National Park an attraction for the international tourists who represent an important source of hard currency to Namibia.

Not being landlocked, Namibia enjoys an additional resource that Botswana lacks. The arid western coast of southern Africa, swept by the cold, northward-moving Benguela Current, supports Africa's most important fisheries. **Upwelling** cool water brings up nutrients from the ocean floor and provides a fertile habitat for plankton, the basic food of fish. Namibia has established an exclusive economic zone (EEZ; see page 396) in the Atlantic stretching 200 miles (320 km) offshore, in which fishing and fish processing based at Walvis Bay have become important industries.

Tropical savanna prevails over most of southern Africa, including much of Zimbabwe, Zambia, Malawi, Mozam-

Figure 18.18 The Namib Desert starts right at the Atlantic shoreline of Namibia. This area has long been known as the "Skeleton Coast" for the fate of its shipwrecked sailors.

bique, eastern Angola, and northern South Africa (see Figure 17.8a and b). More than 90 percent of the annual rainfall comes in the cooler 6 months from November to April. The natural vegetation is primarily open woodland, although on the High Veld of Zimbabwe there is tall-grass savanna. The High Veld lies generally at elevations of 4,000 feet (c. 1,200 m) and higher and forms a band about 50 miles (80 km) broad by 400 miles (c. 650 km) long, trending southwest–northeast across the center of Zimbabwe. The High Veld was historically the main area of European settlement, and most of Zimbabwe's 50,000 whites still live on it. Until the year 2000, these included about 4,500 white farmers whose large holdings, worked by black laborers, produced most of the country's agricultural surplus to feed urban populations and to export. Tobacco from these farms was the country's most important export, followed by cotton, cane sugar, corn (maize), and roses. As discussed later, the future of these farms, their tenure, and Zimbabwe's commercial agriculture is uncertain.

South Africa's physical geography has played a major role in its unusual course of development. Lying almost entirely in the middle latitudes, it is a cooler land than most of Africa

south of the Sahara, with annual temperatures generally between 60° and 70°F (16°–21°C). In combination with the country's natural resource wealth, this climate made the area very attractive for European settlement; it was one of the "neo-Europes" of the colonial era and was historically Africa's main center of European settlement and advanced economic development.

Europeans favored the extreme east of the country's interior plateau, a region also known as the High Veld. The original vegetation here was a grassy steppe, similar to the midwestern prairies of the United States. This is South Africa's leading area of crop and livestock production. The principal farm animals are beef cattle, dairy cattle, and sheep, and the main crops are corn (maize) and wheat. Corn is the staple diet for South Africa's blacks. Surpluses are exported and used as livestock feed. In KwaZulu/Natal, there is hilly terrain between the Indian Ocean and the Drakensberg, a mountainous escarpment at the edge of the interior plateau. The leading commercial crop of that area is sugarcane.

South Africa's Western Cape Province has a Mediterranean climate. It is the only African area south of the Sahara with this climate type (Figure 18.19). Average temperatures are similar to those of the same climate area in southern California. The early settlers established vineyards in this region, and today, the area's grapes and wines sell both in South Africa and abroad, where they are acquiring a growing share of the world market. The province also produces oranges and other citrus fruits, deciduous fruits, and pineapples. Sheep ranching is a major occupation in the semiarid interior.

Mining and Manufacturing

Minerals, including petroleum, are by far the most valuable material assets of this region. South Africa has the greatest mineral endowment, and the country's relative prosperity owes in large part to that wealth. Immense British and other foreign investments pour into South Africa's mining industry. South Africa's diverse and abundant mineral wealth has

Figure 18.19 The Mediterranean climate of Western Cape Province is ideal for viticulture. This is a vineyard near Stellenbosch.

fueled its continental leadership in manufacturing. The country has about a third of Africa's manufactures and a third of its electric generation. Its diversified industries include iron and steel, machinery, metal manufacturing, weapons, automobiles, tractors, textiles, chemicals, processed foods, and others. Many thousands of jobs in Western countries are directly linked to mining and manufacturing in South Africa. The mines employ mainly black African workers recruited from South Africa and other African countries, particularly Lesotho, Swaziland, Mozambique, Botswana, and from as far away as Malawi. There is a large illegal immigration to South Africa, too; an estimated 2 to 8 million undocumented migrant workers are in this relatively prosperous country.

South Africa is the world's largest producer of gold (Figure 18.20), and gold is its most important export. The gold-mining industry is concentrated in the High Veld, especially the Witwatersrand (formerly Rand), an area in the province of Gauteng. Johannesburg, South Africa's largest city (population: 5.2 million), plus the adjacent impoverished and overcrowded black township of Soweto and nearby Pretoria, are located in Gauteng. Limpopo and Mpumalanga provinces and adjacent areas make up one of the world's most vital mineral-exporting regions; exports include copper, asbestos, chromium, platinum, vanadium, antimony, and phosphate.

Recovery of uranium from mine tailings and as a byproduct of current gold mining has made South Africa the largest uranium producer on the continent. And since 1867, the country has been an important source of diamonds from alluvial deposits along the Vaal and Orange River Valleys and from kimberlite "pipes" (circular rock formations of volcanic origin). Kimberley in Northern Cape Province is the administrative center for the diamond industry.

South Africa's mining industries require large amounts of electricity. Most of this is supplied by generating plants powered by bituminous coal, which is present in vast quantities near the surface on the interior plateau. These coal deposits, which are the largest in Africa, are immensely important to South Africa, as the country has only small petroleum and natural gas reserves and has only a modest hydroelectric potential. Coal is a major export, and it has facilitated development of Africa's leading iron and steel industry, concentrated at Pretoria and around Johannesburg. Iron ore comes by rail from large deposits in the northeast and central parts of the country. Some of South Africa's neighbors also have considerable mineral wealth (see Figure 17.19). Partly for this reason, South Africa is much engaged in their affairs, especially those of Lesotho and Swaziland, which South Africa surrounds. Botswana and Lesotho gained independence from Britain in 1966, and Swaziland, the last absolute monarchy in southern Africa, became independent in 1968. Since then, however, South Africa has occasionally exerted power on behalf of its interests in these weaker countries, particularly Lesotho. In 1998, for example, the prime minister of Lesotho perceived that a coup might be imminent in his nation and called on South Africa for assistance. South African troops immediately invaded Lesotho, but rather than quelling the unrest, this intervention led to widespread condemnation of South Africa for meddling in the internal affairs of a sovereign nation. Mining of coal, gold, and tin is an important source of revenue in Swaziland, as is tourism—which features gambling casinos—and sugar production. There is some diamond mining in Lesotho.

Diamonds make up about four-fifths of the total value of exports from Botswana, which after South Africa has the region's strongest economy. South African, British, and U.S. mining companies export Namibian diamonds and various metals. Zimbabwe has begun exporting titanium and is constructing a large aluminum smelting plant.

In Zimbabwe is an extraordinary geological feature called the Great Dike, an intrusion of younger, mineral-rich mate-

Jesse H. Wheeler, Jr.

Figure 18.20 Johannesburg, the "City of Gold," is surrounded by huge dumps of waste material from former gold-mining operations. One of these dumps occupies the center of the view. Note the automobile expressway at the bottom of the photo. In the distance lie residential areas occupied by whites.

NASA/SPL/Photo Researchers, Inc.

Figure 18.21 The stripe running top to bottom across this photo is Zimbabwe's Great Dike, seen from space. About 300 miles (500 km) long and 6 miles (10 km) wide, the mineral-rich Great Dike was formed about 2.5 billion years ago when molten lava filled and widened a fracture in Earth's crust. Note how faulting has offset the dike in two places.

rial into older rock. Ranging in width from 2 to 7 miles (3 to 11 km), the Great Dike crosses the entire nation and is readily visible from space as a giant stripe on the landscape (Figure 18.21). Along and near the Great Dike are the principal areas of commercial farming and of mines producing gold (Zimbabwe's second most important export, after tobacco), nickel, asbestos, copper, chromium, and other minerals, primarily for export.

The Great Dike thus forms the main axis of economic development in Zimbabwe. Along it lies a railway connecting the largest city and capital, Harare (formerly Salisbury; population: 2.4 million), in the northeast with the second largest city, Bulawayo in the southwest. Iron ore mined at Kwekwe, about midway between Harare and Bulawayo, is used there to produce steel. The Wankie coal field in western Zimbabwe supplies coal to smelt the ore. Until about 1998, Zimbabwe's manufacturing industries were more important, both in value of output and in diversity of production, than those of any other African nation south of the Sahara except Nigeria and South Africa. Manufacturing since then has plummeted because of the country's political and economic crises (described below), which have made it very difficult for factories to purchase fuel and spare parts.

Zambia's economy depends heavily on copper exported from several large mines developed with British and American capital. An urban area has developed at each mine, and the mining area, known as the Copperbelt, consists of a series of separate population nodes strung close to Zambia's frontier with the adjacent copper mining Shaba region of southeastern Democratic Republic of Congo. The Copperbelt, which contains the majority of Zambia's manufacturing plants and copper mines, is largely an urban region. Copper

composed 55 percent of Zambia's exports in 2004. However, exports are much diminished from their 1960s highs due to years of poor government control of the copper industry. The copper-mining industry is now entirely in private hands, and Zambia hopes to rise quickly from its position as the world's 12th largest copper producer. Other mineral production includes cobalt, a by-product of copper mining; zinc and lead mined along the railway between the Copperbelt and Zambia's capital and largest city, Lusaka (population: 1.8 million); and coal production from extreme southern Zambia.

The region's leader in petroleum production is Angola, which is increasingly important on the global energy stage. One of its vital oil producing areas is the coastal exclave of Cabinda, separated from the rest of the country by the Congo River mouth (which is controlled by the Democratic Republic of Congo). Other fields are scattered along the Angolan coast between the Congo mouth and the city of Lobito. Major new reserves have been discovered recently offshore Angola; in fact, more new petroleum reserves have been detected there in the past decade than in any other country in the world. U.S and other multinational oil companies are at work developing this resource. Crude oil and modest amounts of refined products made up nine-tenths of Angola's exports in 2004, and in that year, Angolan oil comprised 3 percent of U.S. oil imports. In addition to oil and its products, Angola exports small amounts of diamonds, coffee, and other items. Diamonds are concentrated in the northeast, where UNITA rebels mined them for years to finance their antigovernment operations.

Ethnicity, Colonialism, and Strife

South Africa: A Special Case

Visitors from Western Europe and North America will find in South Africa most of the institutions and facilities to which they are accustomed. However, the Europeanized cultural landscape does not reflect the majority culture of this unusual country. Whites represent only about 14 percent of the total population, and blacks make up 75 percent. There are two other large racial groups: the so-called coloreds, of mixed origin (8 percent), and the Asians (3 percent), most of whom are Indians. Many peoples make up the black African majority. The **Zulu** and **Xhosa** (pronounced Kнo-sa) are both concentrated in hilly sections of eastern South Africa near the Indian Ocean and are the most populous of the nine officially recognized tribal groups.

A huge economic gulf separates South Africa's impoverished black Africans from about 6.5 million whites of European descent who dominate the economy and enjoy a lifestyle comparable to that of people in Western Europe and North America. South Africa's overall per capita GDP PPP of $10,700 is typical of such low income MDCs as Poland and Croatia. For the unskilled and semiskilled labor needed to operate their mines, factories, farms, and services, the whites of South Africa have always relied mainly on low-paid black workers in most of the country and colored workers in the

Western Cape Province. The white-dominated economy could not operate without them, and the Africans in turn are extremely dependent on white payrolls. Although the races are interlocked economically, relations between them have historically been very poor and have been structured to benefit the whites. A country whose people were legally segregated by race, South Africa was one of the world's most controversial nations until its new beginning in 1994. Following is a summary of how the country's racial structures evolved.

South Africa's white population is split between the British South Africans and the **Afrikaners,** or **Boers** (Dutch for "farmers"). The Afrikaners speak Afrikaans, a derivative of Dutch, as their preferred language. They outnumber the British by approximately three to two. The Afrikaners are the descendants of Dutch, French Huguenot, and German settlers who began coming to South Africa more than three centuries ago. Their impact on indigenous Africans was catastrophic. White settlers killed, drove out, and through the unintentional introduction of smallpox decimated most of the original **Khoi** and **San** (**Bushmen**) inhabitants. They put the survivors to work as servants and slaves. When these proved to be insufficient in number, the colonists brought slaves from West Africa, Madagascar, East Africa, Malaya (now West Malaysia), India, and Ceylon (now Sri Lanka).

Most of the workers in the cane fields of Natal were indentured laborers brought from India beginning in the 1860s. The majority chose to remain in South Africa when their terms of indenture ended, and some Indians came as free immigrants. About 85 percent of South Africa's Indian population is still in KwaZulu/Natal, mainly in and near Durban, the province's largest city and main industrial center and port. Most Indians today are employed in commercial, industrial, and service occupations.

Like KwaZulu/Natal, the Cape provinces have a distinctive racial group: South Africa's coloreds (widely known as the "Cape Coloreds"). Nearly nine-tenths live in the provinces, primarily in and near Cape Town, the country's second largest city and one of its four main ports. This group originated in the early days of white settlement as a product of relationships between Europeans (Dutch East India Company employees, settlers, and sailors) and non-Europeans, including slaves and Khoi (**Hottentot**), Malagasy, West African, and various south Asian peoples. The resulting mixed-blood people vary in appearance from those with pronounced African features and others who are physically indistinguishable from Europeans. About nine-tenths speak Afrikaans as a customary language, and most of the others speak English.

Culturally, the coloreds are much closer to Europeans than to Africans, and many with light skins have "passed" into the European community. Coloreds have always had a higher social standing and greater political and economic rights than black South Africans, although they have ranked below Europeans in these respects. Most coloreds work as domestic servants, factory workers, farm laborers, and fishers and perform other types of unskilled and semiskilled labor.

There is a small but growing professional and white-collar class of coloreds. White owners of grape vineyards are donating portions of their fields to an increasing number of the coloreds who once worked as indentured servants on the white lands, leading to a sharp increase in the region's output and exports of wines. Such progress is a sign of how truly different the new South Africa is from the old.

Racial issues in South Africa have a long history of producing strife, even among the ruling white powers. Great Britain occupied South Africa—which it called the Cape Colony—in 1806, during the Napoleonic Wars. Friction developed between many Boers and the British authorities, who imposed tighter administrative and legal controls than the Boers were accustomed to. The British abolished slavery throughout their empire in 1833, contributing to the anti-British sentiments of the slaveholding Boers. Boer discontent resulted in the Great Trek, a series of northward migrations by which groups of Boers, primarily from the eastern part of the Cape Colony, sought to find new interior grazing lands and establish new political units beyond British reach. After some earlier exploratory expeditions, the main trek by horse and ox-wagon began in 1836. It resulted in the founding of the Orange Free State, the Transvaal, and Natal as Boer republics. Britain annexed Natal in 1845 but recognized Boer sovereignty in the Transvaal and the Orange Free State in 1852 and 1854, respectively.

While Boer disaffection with Britain was building, British settlers were coming to South Africa in increasing numbers. The area's natural resource wealth helped shape subsequent events. The British colonies and Boer republics might have developed peaceably side by side if diamonds had not been discovered in the Orange Free State in 1867 and gold in the Transvaal in 1886. The discoveries set off a rush to these republics of prospectors and other fortune hunters and entrepreneurs from outside the region, particularly from Britain. Ill feeling led to the Anglo-Boer War in 1899. British troops defeated the Boer forces decisively, ending the war in 1902 but leaving behind a reservoir of animosity that exists to this day.

The Union of South Africa was organized in 1910 from four British-controlled units as a self-governing constitutional monarchy under the British Crown. Dutch became an official language on par with English (this was later altered to specify Afrikaans rather than Dutch). Pretoria in the Transvaal became the administrative capital as a concession to Boer sentiment (however, the national parliament now meets at Cape Town and the supreme court sits at Bloemfontein in the Free State). In 1961, following a close majority vote by the ruling white population, South Africa became a republic outside the British Commonwealth.

Since 1910, and particularly since the **Nationalist** (now **National**) **party** took power in 1948, Afrikaners have dominated the political life of South Africa by virtue of their greater numbers and cohesion. Racial segregation characterized South African life from 1652 onward, but it was first systematized after 1948 under a comprehensive body of national laws supporting the official governmental policy of "separate development of the races," subsequently called "multinational development" and commonly known as apartheid. The new laws imposed racially based restrictions and prohibitions on the entire population, but they weighed most heavily on black Africans. They ruled out significant sharing of power by black people in a unified South Africa. They fragmented and displaced families and ethnic groups, rearranging the country's cultural landscape. Apartheid culminated in the scheme to establish tribal states, or homelands (also known as native reserves, Bantustans, and national states), which were nominally intended to achieve "independence." Whites ejected several million Africans not wanted in "white" South Africa from their homes and transferred them to homelands. They removed great numbers of blacks from urban areas and white farms and dumped them in poverty-stricken and crowded areas unequipped to handle new influxes of people.

The apartheid system drew furious criticism from many other nations. Most of the world community ostracized South Africa, and many countries, including the United States, imposed economic sanctions against it. Black unrest directed against apartheid and the general underdevelopment of the African majority became so widespread and violent between 1984 and 1986 that the government declared a state of emergency. Blacks killed many other blacks that they perceived to be collaborating with the whites. An intensive government crackdown resulted in the wholesale banning of political activity by antiapartheid groups, more intense restrictions on media, and the jailing of thousands of black Africans perceived as protest leaders. The government lifted the state of emergency in 1990, but widespread violence continued. Much of the fighting was between two large, tribally based rival factions: the Xhosa-dominated **African National Congress (ANC)**, led by Nelson Mandela, and the Zulu-dominated **Inkatha Freedom party (IFP)**, led by Chief Mangosuthu Buthelezi. Tension and violence between these groups have since largely abated.

A complete official turnabout on the issue of apartheid resolved South Africa's ongoing racial crisis. It began in 1989 after Frederik W. de Klerk, an insider in the Afrikaner political establishment, became president. Urged on by white South African business leaders, antiapartheid activists, the powerful force of South African and international political opinion, the impact of international economic boycotts against South Africa, and his own stated convictions concerning the injustices and unworkability of apartheid, de Klerk launched a broad-scale program to repeal the apartheid laws and put South Africa on the road to revolutionary governmental changes under a new constitution. Four years of intensive negotiations involving all interested parties resulted in an agreement to hold a countrywide, all-race election to create an interim government that would hold office for not more than 5 years. In 1994, voters turned out in massive numbers to elect a new national parliament, which then chose Nelson Mandela as president. (Only 4 years earlier, the white government had freed Mandela after a 27-year pe-

riod of political imprisonment.) This historic election ended white European political control in the last bastion of European colonialism on the African continent. The apartheid laws themselves became null and void.

Nelson Mandela, South Africa's "George Washington," retired in 1999 and was succeeded by a second black president, Thabo Mbeki. The country's political landscape is remarkably stable, but challenges remain. The races are continuing on their long path of reconciliation. Some of the healing process has been formalized in a national **Truth and Reconciliation Commission,** which allows those who were party to racial violence during the apartheid era to confess their misdeeds and, in a sense, be absolved of them. There is still a huge economic gulf between the haves and have-nots, but there are now more blacks—about 11 million—in the ranks of the middle and upper classes. Growing black wealth has boosted racial integration in neighborhoods. The black underclass continues to grow, however, and an overall unemployment rate of 40 percent fuels an ongoing epidemic of petty and violent crime, especially in the cities. Crime, the HIV/AIDS epidemic, a low level of education among workers, and relatively high labor costs have dampened foreign investment and economic growth in South Africa. No longer restrained by apartheid-era economic sanctions, South Africa has at the same time become a major investor in other African economies, buying banks, railways, cell phone networks, power plants, and breweries across the region.

As in Zimbabwe, there is real need for land reform in South Africa because the white minority owns about 70 percent of the productive land. Unlike Zimbabwe, however, South Africa has a plan to peacefully help reduce the imbalance: to redistribute 30 percent of the country's farmland from white to black hands by 2014, on a willing-seller–willing-buyer basis. Progress is slow, however, and some landless blacks vow a Zimbabwe-style takeover of white farms. A program is also underway to compensate the estimated 3.5 million blacks forcibly displaced by the government to the homelands between 1960 and 1982. More broadly, South Africa has an aggressive affirmative-action program to help redress the economic imbalance between the races. The **Employment Equity Act** does not impose quotas but requires employers to move toward "demographic proportionality" based on the national proportions of race and gender. In another effort to secure more income for the black majority, the South African government assumed control of all the country's mineral resources in 2002. Mining companies now can exploit these resources only under state license and, presumably, with more supervision to ensure that profits are not drained excessively abroad or to whites.

Troubles for Portugal's Ex-Colonies and Independent Namibia

Although Portugal was the first colonial power to establish interests in Africa south of the Sahara, stronger European powers came to prevail in the region. However, Portugal managed to retain footholds along both the Atlantic and Indian Ocean coasts. By 1964, Portugal's principal African possessions, Portuguese West Africa (now Angola) and Portuguese East Africa (now Mozambique), were the most populous European colonies still remaining in the world. Portugal fielded a large military force to combat guerrilla opposition in both areas. In 1974, Portugal's resistance broke, and an army coup in Portugal installed a regime in Lisbon that granted the two colonies independence.

When the Portuguese withdrew from Angola in 1975, a Marxist-Leninist liberation movement supported by the Soviet Union and Cuba took over the central government in the seaport capital and largest city, Luanda (population: 2.7 million; Figure 18.22). But this movement's authority was not recognized by other liberation movements, one of which—the **National Union for the Total Independence of Angola** (**UNITA**), rooted in the country's largest ethnic group, the Ovimbundu—came to control large parts of the country. South Africa and the United States supplied aid to this anti-Marxist movement headed by Jonas Savimbi, a member of Angola's largest ethnic group. Large quantities of Soviet arms, thousands of Cuban troops, and many Soviet and East German military advisers supported the government in Luanda. This cold war **proxy conflict** raged for years.

In 1988, representatives of Angola, Cuba, and South Africa signed a U.S.-mediated agreement under which Cuba and South Africa would withdraw militarily from Angola, and South Africa would grant independence to Namibia. But even after the withdrawals, hostilities continued between UNITA and the government until a peace agreement allow-

ing UNITA's participation in all levels of government was reached in 1994. This agreement soon faltered, however, and while some UNITA members took up their legitimate government posts, others, including UNITA leader Savimbi again took up arms. They fought especially hard to secure the oil fields of Cabinda and used revenue from diamonds to feed the war machine. Savimbi's death in 2002 led rapidly to a peace agreement in which UNITA agreed to demobilize and disarm, bringing an end to 27 years of civil war. Efforts have begun to clear the country of its estimated 9 million land mines—the most per person in the world.

Like Nigeria, Angola is a resource-rich country that so far has little to show for its oil wealth. The approximate $11 billion in annual revenue from the oil industry funded the war against UNITA and apparently have made the country's wealthy elite even wealthier but have done little for most Angolans; 70 percent live in poverty. Desperate for low-interest loans, Angola has promised the international community to triple its expenditures on health, education, and other social services.

Namibia became independent from South Africa only recently. South African forces overran this former German colony during World War I, and after the war, the League of Nations mandated it to South Africa under the name of South West Africa. Only the war in Angola released South Africa's grip. Backing UNITA, South African military forces had actively intervened in the Angolan struggle in the mid-1970s. Eventually, South African forces withdrew but continued to attack Angolan bases of the **South West Africa People's Organization** (**SWAPO**), a guerrilla group fighting

Figure 18.22 Luanda, the capital and largest city of Portugal's former colony of Angola, exhibits the common pattern of cities in Africa south of the Sahara: a modest cluster of high-rise buildings in the city center surrounded by a sea of small one-story houses, often on dirt streets and generally roofed with metal sheeting.

Sarah Errington/Hutchison Library

for Namibia's independence. In 1988, South Africa agreed to grant independence to Namibia, but only if the would-be Namibian government under SWAPO leadership could convince its main source of support, the government of Angola, to expel Cuban troops from Angola. The South African government granted independence to Namibia in 1990.

SWAPO is now the major political party, but its Marxist leanings have not prevented the country from promoting capitalist investment. White Namibians look anxiously at the confiscation of white farmlands in nearby Zimbabwe; just 4,000 of the country's 80,000 whites own 44 percent of Namibia's territory. The country has a willing-seller–willing-buyer policy designed to encourage sales of white farms to blacks, but little land has changed hands to date. Black farmers warn that Zimbabwe-style land invasions could ensue.

The anomalous sliver of territory in Namibia's extreme northeast known as the Caprivi Strip, which extends fully 280 miles (450 km) eastward to the waters of the Zambezi River, was important to the German colonists who ruled Namibia early in the 20th century. They insisted on this access to the Zambezi in the mistaken belief that it would allow navigation all the way to the Indian Ocean (the Victoria Falls stand in the way). Independent Namibia was permitted to retain the strip, despite protests from neighboring Botswana. The Caprivi is home to the **Barotse** (sometimes called **Lozi**) ethnic group, whose tribal lands also extend into Botswana, Angola, and Zambia. Barotse efforts both in the Caprivi Strip and in Zambia to establish an autonomous or independent homeland are strongly opposed by both Namibia and Zambia. With its scattered livestock forage, the Kalahari section of Namibia is peopled mainly by Bantu-speaking herders and by remnants of the Khoi and San populations. Some of the San are among the world's last remaining hunters and gatherers, but the majority have given up the nomadic way of life in recent decades and have settled as farmers and wage laborers.

Portugal's other ex-colony, Mozambique, is one of the world's poorest countries, but it has rather bright prospects. It suffered from years of guerrilla warfare, which first involved African insurgencies against the Portuguese and then warfare by African dissidents against the Marxist government that took over from the Portuguese in Maputo, the country's capital (population: 1.7 million). The highly destructive civil war costing 1 million lives between the Maputo government and the South African- and Zimbabwean-backed rebel organization called **RENAMO (Mozambique National Resistance)** dragged on until 1992, when a peace accord was signed and the United Nations sent a military peacekeeping force to Mozambique. Elections took place peacefully in 1994, and a large share of the 5 million people displaced by the war began returning to their villages. Only periodic disputes between the government and RENAMO have occurred since their reconciliation. Nature has dealt the strongest blows to Mozambique recently; devastating cyclones (hurricanes) in 2000 and again in 2001 brought flooding and suffering to much of the country.

Mozambique is being rebuilt. Transportation facilities in the Beira Corridor to Zimbabwe are being upgraded, and a new regional economic link with South Africa has been established. Known as the **Maputo Development Corridor,** this toll road joins Johannesburg's industrial and mining center with the deep-water port at Maputo. Low-tax zones were created to attract new businesses to the corridor. The country's estimated 2 million land mines are being cleared, and war refugees are returning to farm this fertile land. By 1998, the country was self-sufficient in food production. Major exports are Mozambique's famous huge prawns, cashew nuts, cotton, tea, sugar, and timber. Multinational efforts are underway to process titanium for export and to develop the country's pristine beaches into world-class tourist destinations. Impressed with Mozambique's progress, foreign lenders have agreed to cancel about 80 percent of the country's debt. Mozambique has acted aggressively on this windfall, with dramatic new investments in health, education, and other social services.

Zimbabwe's Struggles

Prior to the post–World War II movement for African independence, the countries of Zimbabwe, Zambia, and Malawi (then known as Southern Rhodesia, Northern Rhodesia, and Nyasaland, respectively) were British colonies. In 1953, the three became linked politically in the Federation of Rhodesia and Nyasaland, or the Central African Federation. It had its own parliament and prime minister but did not achieve full independence. The racial policies and attitudes of Southern Rhodesia's white-controlled government created serious strains within the federation, helping to bring about its dissolution in 1963. Northern Rhodesia then achieved independence as the Republic of Zambia, and Nyasaland became the independent Republic of Malawi.

WORLD
REGIONAL
Geography ⊛ Now™

Click Geography Literacy to take a virtual tour "Zimbabwe: How to Turn a Breadbasket into a Tin Cup."

But the government of Southern Rhodesia, the main area of European settlement in the federation, was unable to come to an agreement with Britain concerning Britain's demand for the political participation of the colony's African majority. In 1965, the government of Southern Rhodesia issued a **Unilateral Declaration of Independence (UDI)**, and in 1970, it took a further step in separation from Britain by declaring itself to be Rhodesia, a republic that would no longer recognize the symbolic sovereignty of the British Crown.

Britain and the Commonwealth responded with a trade embargo. Negotiators for Britain and the Rhodesian government unsuccessfully attempted to arrive at a formula to allow political rights and participation for the 95 percent of Rhodesia's population that was black and thus to achieve reconciliation and world recognition of Rhodesia's independence. Although the economic sanctions were ineffective, pressure on the white government increased in the mid-1970s with black guerrilla warfare against the government and the end of white (Portuguese) control in adjacent Mozambique. Independence for Rhodesia (renamed Zimbabwe) as a parliamentary state within the Commonwealth of Nations finally came in 1980.

This young country faces several ethnic dilemmas. First, the future course of relations between the country's two largest ethnic groups—the majority **Shona** (82 percent of the total population) and the Ndebele (14 percent)—remains uncertain. Many Ndebele people want autonomy for the Ndebele stronghold in the south known as Matabeleland. In addition, while most whites left Zimbabwe when it became independent, about 100,000 remained. Now about 50,000 strong, whites participated in Zimbabwe's political life, setting an early example to South Africa of an effective transition to majority black rule. But the apparent harmony has since shattered, particularly over the issue of land ownership. By the mid-1990s, millions of rural blacks were crowded onto agriculturally inferior **communal lands,** while whites owned about 70 percent of Zimbabwe's arable land. A poorly handled effort to address this imbalance has recently threatened to tear the country apart.

The problem of disproportionate white land ownership became explosive in 1997, when Zimbabwe's President Robert Mugabe (of the Shona ethnic group) used the issue to boost his party's prospects in impending parliamentary elections. He announced that the government would seize half of Zimbabwe's white-owned farms, "not pay a penny for the soil," and distribute the land to black peasants living in the overcrowded communal areas. Mugabe argued that the former colonial power, Britain, illegally occupied black lands in the first place and should now pay compensation to whites losing those lands. Insisting that the seizures violated its agreements with the government, and angry about Zimbabwe's support of the Kinshasa government in Congo's civil war, the International Monetary Fund (IMF) and most other foreign lenders cut off aid to Zimbabwe in 1999.

An economic crisis, so far unabated, followed; the value of the Zimbabwean dollar crashed, inflation soared, gold mines shut down, and tourism all but disappeared. Unemployment reached an estimated 70 percent by 2004, and a tide of illegal Zimbabwean immigrants flowed into South Africa. Mugabe continued to encourage blacks to challenge whites and their assets in advance of the 2002 presidential election, which Mugabe won, apparently fraudulently. Blacks, mostly either Mugabe loyalists or war veterans, moved onto the white-owned farms. All but 600 of the country's 4,500 white farmers had been uprooted by 2004, and the 300,000 black workers on the former white farms had lost their jobs.

Zimbabwe's Supreme Court ruled that the farm seizures were illegal, but Mugabe encouraged the squatters to stay. Few had any farming abilities or materials, and the country's agricultural output plummeted. In 2004, the same year that Zimbabwe outlawed private ownership of farmland, the United Nations estimated that two-thirds of Zimbabwe's population was "food insecure," meaning hungry and malnourished. Some white farmers fled to Zambia, where the government welcomed the new agricultural investors with open arms. Zambia hopes that their expertise with tobacco production in particular will boost exports and revenues.

18.7 The Indian Ocean Islands: Former Edens

The islands and island groups off the Indian Ocean coast of Africa are unique in their cultures and natural histories. Madagascar, Comoros, Réunion, Mauritius, and the Seychelles have had African, Asian, Arab, European, and even Polynesian ethnic and cultural influences (Figures 18.23 and 18.24). As island ecosystems, they are home to many endemic plant and animal species.

Madagascar and the smaller Comoro Islands are former colonies of France, and Réunion remains an overseas possession of France. Mauritius became independent of Britain in 1968 as a constitutional monarchy under the British Crown. The Seychelles, which gained independence in 1976, is a republic within the Commonwealth of Nations.

Madagascar (in French, *La Grande Ile*) is the fourth largest island in the world, nearly 1,000 miles (c. 1,600 km) long and about 350 miles (c. 560 km) wide. It lies off the southeast coast of Africa and has geological formations similar to those of the African mainland. Its distinctive flora and fauna include most of the world's lemur and chameleon spe-

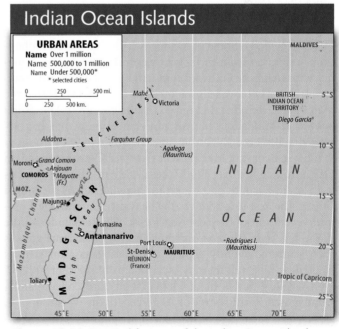

Figure 18.23 Principal features of the Indian Ocean islands

Figure 18.24 The 18 million people of Madagascar have a unique culture, with roots in Southeast Asia, the Pacific, and Africa.

cies (see Biogeography, page 513, and Perspectives from the Field, page 464). Some of Madagascar's early human inhabitants migrated from the southwestern Pacific region, bringing with them the cultivation of irrigated rice and other culture traits. There was also a later influx of Africans from the mainland. The official language of the country's 18 million people is Malagasy, an Austronesian tongue that has acquired many French, Arabic, and Swahili words.

The east coast of Madagascar rises steeply from the Indian Ocean to heights over 6,000 feet (1,829 m). Because the island lies in the path of trade winds blowing across the Indian Ocean, the east side receives the heaviest rain and has a natural vegetation of tropical rain forest. The remainder of the island has vegetation consisting primarily of savanna grasses with scattered woody growth. An eastern coastal strip is backed by hills and then by escarpments of the central highlands. Paddy rice is the principal food, and coffee, vanilla (Madagascar is the source of 70 percent of the world supply), cloves, and sugar represent most of the country's limited exports. There is a growing clothing export industry. The most

densely populated part of the island is the central High Plateau, where the economy is a combination of growing rice in valleys and raising cattle on higher lands. Antananarivo (formerly Tananarive; population: 1.7 million), the capital and largest city, is located in the central highlands. It is connected by rail with Toamasina (formerly Tamatave), the main seaport, located on the east coast.

Madagascar is not as economically developed as other large tropical islands such as Java, Taiwan, and Sri Lanka. Deep economic troubles and political dissension have marked its recent history under a socialist government. Unwise decision making made Madagascar fertile ground in the 1990s for international drug-money launderers, loan sharks, and con artists who purloined much of this poor nation's financial assets. In 2002, the contested results of a presidential election threatened to divide the country into two parts, one ruled from Antananarivo and the other from Toamasina. Until it was resolved successfully, the political crisis set back Madagascar's garment export industry that had boomed because of the American AGOA initiative (see page 474). Given Madagascar's extraordinary wildlife and scenic beauty, the ecotourism sector could grow considerably, but so far, the government has taken little interest in developing an infrastructure to encourage and sustain tourism.

The Comoro Islands, located about midway between northern Madagascar and northern Mozambique, are of volcanic origin. They have a complex mixture of African, Arab, Malaysian, and European influences. Aside from the island of Mayotte, which has a Christian majority and which, by its own choice, remains a dependency of France, the Comoro group forms the Union of Comoros (short form: Comoros), in which Islam is the official religion and both Arabic and French are official languages. The capital is Moroni (population: 63,000) on the island of Grande Comore. France has a defense treaty with the Comoros, which it invoked in 1995 to put down a coup led by a French mercenary. In 1997, two more Comoros islands (Mohéli and Anjouan) declared secession from the national government in Moroni and pleaded to become French dependencies in the hope they could enjoy the somewhat higher standard of living prevailing on nearby Mayotte. The country implemented a new constitution in 2002, giving each island its own president and self-governance on most issues. The federal presidency is rotated among the three island presidents.

Fishing, subsistence agriculture, and exports of vanilla provide livelihoods to the 700,000 inhabitants of these islands. The coelacanth, a primitive fish thought to have become extinct soon after the end of the Mesozoic Era (the Age of Dinosaurs), has been observed alive in Comoro Islands waters since 1938 (and quite recently in Indonesian waters).

Réunion (capital, Saint-Denis; population: 85,000), east of Madagascar in the Mascarene island group, is the tropical island home to 780,000 people mainly of French and African origin. The island was uninhabited until the French arrived in the 17th century. Elevations reach 10,000 feet (3,048 m). The volcanic soils are fertile, and tropical crops of great

Biogeography

Madagascar and the Theory of Island Biogeography

Harvard University entomologist Edward Wilson and Stanford University biologist Paul Ehrlich have asserted that if tropical rain forests continue to be cut down at the present rate, a quarter of all of the plant and animal species on Earth will become extinct by 2040.[a] They base their estimate on a model that correlates habitat area with the number of species living in the habitat. This theory of island biogeography emerged from observations of island ecosystems in the West Indies. It states that the number of species found on an individual island corresponds with the island's area, with a 10-fold increase in area normally resulting in a doubling of the number of species. If island A, for example, is 10 square miles in area and has 50 species, 100-square-mile island B may be expected to support 100 species.

What makes the theory useful in projecting species losses is the inverse of this equation: A 10-fold reduction in area will result in a halving of the number of species; therefore, 1 square-mile island C can be expected to hold only 25 species. In applying the model, ecologists treat habitat areas as if they were islands. Thus, if people cut down 90 percent of the tropical rain forest of the Amazon Basin, for example, the theory of island biogeography suggests they would eliminate half of the species of that ecosystem. Scientists caution that the theory is only a tool meant to help in making rough estimates; the actual number of species lost with habitat removal may be higher or lower.

As a rough guideline, the theory of island biogeography is useful in projecting and attempting to slow the rate of extinction in the world's biodiversity hot spots, such as Madagascar. More than 90 percent of Madagascar's plant and animal species are endemic, occurring nowhere else on Earth. Extinction of species was well underway soon after people arrived on the island; the giant, flightless elephant bird (*Aepyornis*) was among the early casualties. But human activities, particularly the clearing of forests to grow rice and provide pasture for zebu cattle, are eliminating habitat areas on the island at a more rapid rate than ever before. Meanwhile, scientists are anxious to learn whether some of Madagascar's remaining plants might be useful in fighting diseases such as AIDS and cancer. Already Madagascar's rosy periwinkle has yielded compounds effective against Hodgkin's disease and lymphocytic leukemia. Other species could become extinct before their useful properties ever become known.

How urgent is the task to study and attempt to protect plant and animal species in Madagascar? Scientists turn to the theory of island biogeography for an answer. Although people have lived on Madagascar for less than 2,000 years, they have succeeded in removing 90 percent of the island's forest, setting the stage for some of the most ruinous erosion seen anywhere on Earth (Figure 18.D). The theory of island biogeography suggests that in the process, they have caused the extinction of roughly half of the island's species. With Madagascar's human population projected to double in 23 years, and with pressure on the island's remaining wild habitats expected to increase accordingly, the task of conservation is extremely urgent.

Figure 18.D The subsistence needs of a growing human population have had ruinous effects on Madagascar's landscapes and wildlife. Before people came some 2,000 years ago, most of the island was forested, as in the Montagne d'Ambre National Park (top). Today, less than 10 percent of the island is in forest, and a characteristic landscape feature of its High Plateau is the erosional feature known locally as *lavaka* (bottom).

---

[a] Edward O. Wilson, "Threats to Biodiversity." *Scientific American*, 261 (3), 1989, pp. 60–66.

variety are grown. Cane sugar from plantations supplies 63 percent of all exports by value, with rum and molasses (derived from cane) supplying an additional 4 percent. Réunion also has a significant international tourist industry.

Sugar is the mainstay of the larger island of nearby Mauritius (capital, Port Louis; population: 145,000), also in the Mascarenes. Grown mainly on plantations, sugarcane occupies most of the cultivated fields on this volcanic landscape. Mauritius has a larger proportion of lowland area than Réunion. The island has developed a clothing industry that supplies more export value than sugar. International tourism also produces valuable revenues (about $250 million annually) for its 1.2 million people. The people of Mauritius are mainly descendants of Indians brought in by the British to work the cane fields, along with descendants of 18th century French planters as well as some blacks, Chinese, and racial mixtures. Mauritius is also home to about 5,000 Ilois, or "islanders," relocated from their homes in the Chagos Islands 1,250 miles (2,000 km) to the northwest. The Chagos comprise a British protectorate known as the British Indian Ocean Territory (BIOT). The British administered the islands from Mauritius until 1965 when, in the height of the Cold War, it leased Diego Garcia (the largest island) to the U.S. military and required all the Ilois to move out. Since then, U.S. forces have used Diego Garcia on many military forays, particularly against Iraq during the Gulf War and the 2003 war and against Afghanistan in the wake of the 9/11 attacks. Now many of the Ilois want to return to their home islands, a move opposed by the United States.

The unusual core of the Seychelles island group is composed of granites, which typically are found only on continental mainlands. Many foods have to be imported to the islands, which are generally steep and lack cultivable land. The country's largest exports consist of petroleum products, although the islands have no crude petroleum and no refining industry; the products are simply imported and then reex-

ported. Fish products are the only other important exports. The Aldabra atoll, far to the south of the main group of islands, is home to a population of giant tortoises and other endemic animals (Figure 18.25).

A tourist boom began in 1971 when an airport was built in the Seychelles (capital, Victoria; population: 23,000). The exceptional physical beauty of the islands and the renowned friendliness of its 100,000 people of French, Mauritian, and African descent have made the Seychelles a prime tourist destination among those wealthy enough to afford travel to this remote area. Outsiders are drawn to Seychelles, as they have long been to much of Africa south of the Sahara, as an imagined Eden. The realities of land and life in the region exceed the imagination and are most worthy of the world's attention.

Figure 18.25 A giant Aldabran tortoise on the beach of Aldabra atoll in the Seychelles. As in the Galapagos Islands, unique life forms have evolved through long periods of isolation from other animal populations.

SUMMARY

- Africa south of the Sahara can be divided into seven subregions: the Sahel, West Africa, West Central Africa, East Africa, the Horn of Africa, southern Africa, and the Indian Ocean islands.

- The Sahel region extends eastward from the Cape Verde Islands to the Atlantic shore nations of Mauritania, Senegal, and The Gambia and inland to Mali, Burkina Faso, Niger, and Chad. The name Sahel in Arabic means "coast" or "shore," referring to the region as a front on the great desert "sea" of the Sahara.

- Between the late 1960s and 1985, the Sahel was subjected to severe droughts, which, in combination with increased human pressure on resources, prompted a process of desertification.

- Oil production in Chad is the most significant recent development in the Sahel countries' economies.

- West Africa is the region extending from Guinea-Bissau eastward to Nigeria. Most West African countries maintain strong economic, political, and cultural relationships with the European powers that formerly controlled them.

- The West African countries' heavy dependence on subsistence crop farming, livestock grazing, and a few agricultural and/or export specialties is the customary economic pattern of the world's LDCs.

- Although a number of prominent mineral products are mined in West Africa, the most spectacular and significant mining development has been Nigeria's emergence as a producer and major exporter of oil. There are serious conflicts among ethnic, religious, and political groups in Nigeria, some resulting from the maldistribution of income from the country's oil wealth.

- West Africa's principal cities are also national capitals and seaports, but as the region has only minor manufacturing, its cities have developed primarily in its areas of major commercial agriculture.

- Major civil wars have recently taken place in Liberia, Sierra Leone, and Ivory Coast. The war in Ivory Coast had much to do with indigenous resentment of immigrants who came to work plantations during a boom in cocoa prices.

- The subregion of West Central Africa is flanked by Cameroon and the Central African Republic to the north and the Democratic Republic of Congo to the south.

- Subsistence agriculture supports people in most areas of West Central Africa. In addition, exports of cash crops and minerals aid the financing of some countries.

- Originally a colony of Belgium, the Democratic Republic of Congo (Congo-Kinshasa) has faced a number of challenges since its independence in 1960. Due to political mismanagement and war, the resource-rich country is in a state of extreme poverty. It was the epicenter of Africa's First World War during the late 1990s and early 2000s, and continued unrest in the east suggests that conflict may reignite.

- East Africa includes the countries of Kenya, Tanzania, Uganda, Rwanda, and Burundi.

- A wide array of crops is grown by the subsistence agriculturalists of East Africa. In addition, several export crops, such as coffee, tea, and cut flowers, provide a source of income for the region's inhabitants. In comparison with many other African countries, the East African nations lack mineral reserves production.

- Ethnic rivalries and conflicts have beset all of the East African countries. Warfare between the Hutus and Tutsis of Rwanda and Burundi in the 1990s was the most serious.

- The 2002 election that brought the opposition party to power in Kenya may have represented a turning point for this country, which had suffered years of corruption and declining international support.

- Comprised of a great volcanic plateau and adjacent areas, the Horn of Africa includes the countries of Ethiopia, Eritrea, Somalia, and Djibouti.

- After nearly 30 years of terrorism, guerrilla actions, and civil war, the former Ethiopian province of Eritrea became independent in 1993.

- In addition to the devastation wrought by war with Eritrea, droughts centering in its war-torn north have stricken Ethiopia since the late 1960s. In 1984 and 1985, an estimated 1 million people perished during a catastrophic drought that triggered a massive humanitarian relief effort.

- Southern Africa includes the countries of Angola, Mozambique, Zimbabwe, Zambia, Malawi, South Africa, Botswana, Swaziland, Lesotho, and Namibia.

- The dominant power in southern Africa is the country of South Africa. Its economic power, which is based on its diversified industries and mineral wealth, is comparable to that of the rest of Africa combined.

- Racial segregation characterized South African life from 1652 onward, but it was first systematized after 1948 under a comprehensive body of national laws known as apartheid.

- In 1994, Nelson Mandela was elected president of South Africa after an all-race election was held in the country. After years of conflict, this historic election ended white European political control in the last bastion of European colonialism on the African continent. The apartheid laws became null and void. Reconciliation between the races is continuing in postapartheid South Africa. The ranks of middle and upper class blacks are growing, but the gap between the haves and have-nots in South Africa is huge.

- After a long civil war, peace has returned to Angola. This country is becoming an important supplier of oil to the West, but oil exports have benefited few Angolans so far.

- Zimbabwe has experienced an economic crisis of its own making, triggered when President Mugabe redistributed commercial farms owned by the white minority to blacks who generally were not farmers. Agricultural production, factory output, and foreign investment plummeted.

- The Indian Ocean islands of Madagascar, the Comoro Islands, Réunion, Mauritius, and the Seychelles exhibit African, Asian, Arab, European, and even Polynesian ethnic and cultural influences. Madagascar has a wealth of endemic species under tremendous pressure by humans, who have cleared 90 percent of the original vegetation cover in the past 2,000 years.

KEY TERMS + CONCEPTS

Terms in blue are also defined in the glossary.

African National Congress (ANC) (p. 508)
Africa's First World War (p. 493)
Afrikaners (Boers) (p. 507)
albedo (p. 483)
Amhara (p. 500)
Anglophone Africa (p. 488)
apartheid (p. 508)

Barotse (Lozi) (p. 510)
biodiversity hot spots (p. 513)
Blackhawk Down incident (p. 501)
Charney Effect (p. 483)
chemocline (p. 491)
coloreds (p. 506)
Commonwealth of Nations (p. 488)

communal lands (p. 511)
Congo Coalition (p. 493)
cyclone (p. 510)
deadly lakes (p. 491)
desertification (p. 482)
devolution (p. 488)
drought (p. 482)

East African Community (p. 495)
Economic Community of West African
  States (ECOWAS) (p. 489)
Employment Equity Act (p. 508)
endemic species (p. 513)
Ethiopian Orthodox Christianity
  (p. 500)
Falashas (Ethiopian Jews) (p. 500)
Francophone Africa (p. 488)
"Galapagos Islands of Religion" (p. 500)
genocide (p. 497)
Great Trek (p. 507)
homelands (native reserves, Bantustans,
  national states) (p. 508)
Hottentot (p. 507)
Hutu (Bahutu) (p. 497)
Ibo (Igbo) (p. 487)
Ijaw (p. 487)
Inkatha Freedom Party (IFP) (p. 508)
International Criminal Court (p. 498)
Ivoirité (p. 489)
Khoi and San (Bushmen) (p. 507)
Liberians United for Reconciliation and
  Democracy (LURD) (p. 489)

Lomé Agreement (p. 489)
Lord's Resistance Army (LRA) (p. 497)
Maputo Development Corridor (p. 510)
Mau Mau Rebellion (p. 497)
Moors (p. 484)
Movement for the Survival of the Ogoni
  People (MOSOP) (p. 486)
National Union for the Total Independence
  of Angola (UNITA) (p. 509)
Nationalist (National) Party (p. 507)
Ogoni (p. 486)
Ogoni Bill of Rights (p. 486)
Operation Restore Hope (p. 501)
Patriotic Movement of Côte d'Ivoire
  (MPCI) (p. 489)
primate city (p. 485)
proxy conflict (p. 509)
rainwater harvesting (p. 483)
Rally for Congolese Democracy (RCD)
  (p. 493)
Rastafarians (p. 501)
Rebel Coalition (p. 493)
RENAMO (Mozambique National
  Resistance) (p. 510)

resilience (p. 482)
Revolutionary United Front (RUF)
  (p. 489)
Rwandan Patriotic Front (RPF)
  (p. 497)
Sara (p. 484)
shari'a (Islamic law) (p. 488)
Shona (p. 511)
South West Africa People's Organization
  (SWAPO) (p. 509)
Southern African Development
  Community (SADC) (p. 502)
theory of island biogeography (p. 513)
Truth and Reconciliation Commission
  (p. 508)
Tutsi (Watusi) (p. 497)
Unilateral Declaration of Independence
  (UDI) (p. 510)
upwelling (p. 503)
Wahindi (p. 497)
world patrimony site (p. 484)
Xhosa (p. 506)
Yoruba (p. 487)
Zulu (p. 506)

REVIEW QUESTIONS

**WORLD
REGIONAL
Geography ❀ Now™**

*Assess your understanding of this chapter's topics with additional quizzing
and concept-based problems at http://earthscience.brookscole.com/wrg5e.*

1. What are the seven subregions of Africa south of the Sahara?
   What are some of the most significant countries in each of these
   subregions?

2. In what ways is the Sahel both a physical and cultural boundary zone in Africa?

3. What country has the greatest demographic, political, and economic clout in West Africa? What unique difficulties does this
   country face? (Be prepared to discuss its recent turbulent political history.)

4. What are the notable natural resource assets of the Democratic
   Republic of Congo? Why isn't this country richer than it is?
   What are its relationships with its neighbors?

5. How did Idi Amin bring economic ruin to Uganda? What did
   Daniel Arap Moi do to reverse the bright prospects of Kenya?

6. What was the original distinction between the Hutus and the
   Tutsis? What issues brought them into conflict in the 1990s?

7. Why is Ethiopia known as the "Galapagos Islands of Religion"?

8. What are the main ethnic groups in South Africa? What were
   their traditional social and economic roles? How have these
   changed since the end of apartheid?

9. What were the main causes of the Anglo-Boer War of 1899?

10. What tensions emerged between blacks and whites over the is-
    sue of land tenure in Zimbabwe? What have been the impacts
    of government policy about land reform?

11. Who owns and who uses the island of Diego Garcia? What is
    its importance in military affairs?

DISCUSSION QUESTIONS

1. What is desertification? What are its major causes?

2. Summarize the main players and issues of current or recent
   conflict in Liberia, Sierra Leone, Ivory Coast, and Nigeria.

3. Has oil helped the people of Nigeria and Angola?

4. What African countries became involved in the Democratic Republic of Congo's recent civil war? What reasons motivated
   each of these countries to enter the conflict?

5. How has the Great Rift Valley affected the physical environment of East Africa? What natural hazards are associated with
   the rift system?

6. How did European colonialism sow the seeds for violence between Hutus and Tutsis?

7. What are Namibia's difficult choices in trying to increase its access to fresh water?

8. How important is tourism in Africa south of the Sahara? Identify some of the major tourist destinations and their attractions.

9. What geographic problems apparently led to war between Ethiopia and Eritrea?

10. What was the purpose of Operation Restore Hope? What was the outcome of U.S. intervention in Somalia? What have been the roles of Somalia and Djibouti in the war on terrorism?

11. What factors led to South Africa's advanced economic development compared to the rest of Africa?

12. Discuss the evolution and the annulment of the apartheid laws of South Africa. What are the major races in South Africa? How are they getting along today?

13. What is the theory of island biogeography? How exactly is the theory useful in understanding the relationship between habitat loss and species loss?

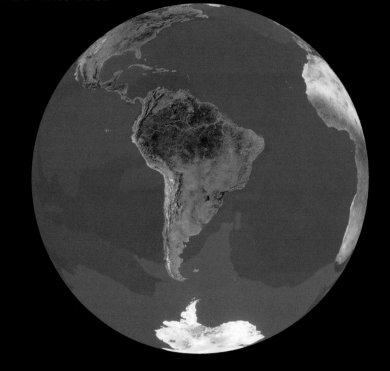

A Geographic Profile
of Latin America

Dawn on the Amazon River, near the borders of Colombia, Peru, and Brazil.

| chapter outline |

19.1   Area and Population

19.2   Physical Geography and Human Adaptations

19.3   Cultural and Historical Geographies

19.4   Economic Geography

19.5   Geopolitical Issues

| chapter objectives |

This chapter should enable you to:

- Appreciate how topographic variety creates a predictable range of environmental conditions and livelihood opportunities

- Know how diverse and accomplished the indigenous cultures of the region were and how European conquest and colonization decimated and changed these cultures

- Recognize the predominant ethnic patterns of the region and how ethnicity corresponds with livelihood, wealth, and political power

- Evaluate the region's efforts to shift from raw material

exports to manufacturing through participation in free trade agreements

- Understand how U.S. interests have shaped the region's political and economic systems

WORLD
REGIONAL
Geography⊛Now™

Look for this logo in the text and go to GeographyNow at http://earthscience.brookscole.com/wrg5e to explore interactive maps, view animations, sharpen your factual knowledge and geographic literacy, and test your critical thinking and analytical skills with unique interactive resources.

The land portion of the Western Hemisphere south and southeast of the United States is commonly known as Latin America. This name reflects the importance of culture traits inherited mainly from the Latin European colonizing nations of Spain and Portugal. Native American cultures were flourishing when Columbus initiated the conquest of the Americas in 1492. These cultures still persist to varying degrees and are particularly strong in Mexico, in the Central American country of Guatemala, and in several South American countries. But Latin influences are predominant today in the region as a whole.

In introducing important themes in land and life of Latin America, this chapter offers the reader an opportunity to become acquainted with the varied indigenous and European cultures that have shaped the region. The exceptional topographic and environmental variations within Latin America are introduced here, too. The economic prospects for this developing region are good compared with many others and are tied especially to relationships with the United States. These issues are described here, along with broader involvement of the United States in affairs south of the border. Exploration of the subregions and countries of Latin America continues through the following two chapters.

19.1 Area and Population

With its 38 countries, Latin America consists of a mainland region that extends from Mexico south to Argentina and Chile, together with the islands of the Caribbean Sea (Figure 19.1; Table 19.1). With a land area of slightly more than 7.9 million square miles (20.5 million sq km), Latin America is surpassed in size by the regions of Africa south of the Sahara, Russia and the Near Abroad, Monsoon Asia, and North America (including Greenland). However, its maximum latitudinal extent of more than 85°, or nearly 5,900 miles (c. 9,500 km), is greater than that of any other major world region. Its maximum east–west measurement, amounting to more than 82° of longitude, is also impressive.

Latin America is not as large as these statistics might suggest because its two main subregions are offset from each other. The northern part, known as Middle America (the subject of Chapter 20), includes Mexico, Central America (Guatemala, Belize, El Salvador, Honduras, Nicaragua, Costa Rica, and Panama), and the countries on Hispaniola, Cuba, and the smaller islands of the Caribbean. It trends sharply northwest to southeast from the north to south orientation of the continent of South America. South America

Political Geography of Latin America

Figure 19.1 Principal features of Latin America

TABLE 19.1 Latin America: Basic Data

Political Unit	Area (thousand/ sq mi)	(thousand/ sq km)	Estimated Population (millions)	Annual Rate of Increase (%)	Estimated Population Density (sq mi)	(sq km)	Human Develop- ment Index	Urban Popula- tion (%)	Arable Land (%)	GDP PPP Per Capita ($US)
Middle America										
Belize	8.9	23.1	0.3	2.3	34	13	0.737	49	3	4900
Costa Rica	19.7	51.0	4.2	1.4	213	82	0.834	59	4	9100
El Salvador	8.1	21.0	6.7	2.0	827	319	0.720	58	32	4800
Guatemala	42.0	108.8	12.7	2.8	302	117	0.649	39	12	4100
Honduras	43.3	112.1	7.0	2.8	162	62	0.672	47	9	2600
Mexico	756.1	1958.3	106.2	2.1	140	54	0.802	75	13	9000
Nicaragua	50.2	130.0	5.6	2.7	112	43	0.667	58	16	2300
Panama	29.2	75.6	3.2	1.8	110	42	0.791	62	7	6300
Total	**957.5**	**2479.9**	**145.9**	**2.1**	**152**	**59**	**0.774**	**68**	**12**	**7751**
Caribbean										
Anguilla (U.K.)	0.04	0.1	0.01	2.0	250	97	N/A	100	0	8600
Antigua and Barbuda	0.2	0.5	0.1	1.7	500	193	0.800	37	18	11,000
Aruba (Neth.)	0.07	0.2	0.07	0.5	1000	386	N/A	51	10	28,000
Bahamas	5.4	14.0	0.3	1.3	56	21	0.815	89	1	16,700
Barbados	0.2	0.5	0.3	0.6	1500	579	0.888	50	37	15,700
Cayman Islands (U.K.)	0.1	0.3	0.04	2.7	400	154	N/A	0	4	35,000
Cuba	42.8	110.9	11.3	0.5	264	102	0.809	75	33	2900
Dominica	0.3	0.8	0.1	1.0	333	129	0.743	71	6	5400
Dominican Republic	18.8	48.7	8.8	1.9	468	181	0.738	64	22	6000
Grenada	0.1	0.3	0.1	1.2	1000	386	0.745	39	6	5000
Guadeloupe (Fr.)	0.7	1.8	0.4	1.0	571	221	N/A	100	11	8000
Haiti	10.7	27.7	8.1	1.9	757	292	0.463	36	28	1600
Jamaica	4.2	10.9	2.6	1.4	619	239	0.764	52	16	3900
Martinique (Fr.)	0.4	1.0	0.4	0.7	1000	386	N/A	95	10	14,400
Netherlands Antilles (Neth.)	0.1	0.3	0.2	0.8	2000	772	N/A	69	10	11,400
Puerto Rico (U.S.)	3.5	9.1	3.9	0.7	1114	430	N/A	71	4	16,800
St. Kitts and Nevis	0.1	0.3	0.05	1.0	500	193	0.844	33	19	8800
St. Lucia	0.2	0.5	0.2	1.1	1000	386	0.777	30	6	5400
St. Vincent and the Grenadines	0.2	0.5	0.1	1.1	500	193	0.751	44	18	2900
Trinidad and Tobago	2.0	5.2	1.3	0.6	650	251	0.801	74	14	9500
Turks and Caicos Islands (U.K.)	0.1	0.3	0.02	3.0	200	77	N/A	46	2	9600
Virgin Islands (U.S.)	0.1	0.3	0.1	−0.1	1000	386	N/A	47	12	17,200
Total	**90.3**	**233.9**	**38.5**	**1.2**	**426**	**164**	**0.702**	**61**	**25**	**5632**
South America										
Argentina	1073.5	2780.4	37.9	1.1	35	14	0.853	89	12	11,200
Bolivia	424.2	1098.7	8.8	1.9	21	8	0.681	63	2	2400
Brazil	3300.2	8547.5	179.1	1.3	54	21	0.775	81	7	7600
Chile	292.1	756.5	16	1.2	55	21	0.839	87	2	10,100

(continued on page 522)

TABLE 19.1 (continued)

Political Unit	Area (thousand/ sq mi)	Area (thousand/ sq km)	Estimated Population (millions)	Annual Rate of Increase (%)	Estimated Population Density (sq mi)	Estimated Population Density (sq km)	Human Development Index	Urban Population (%)	Arable Land (%)	GDP PPP Per Capita ($US)
Colombia	439.7	1138.8	45.3	1.7	103	40	0.773	71	2	6300
Ecuador	109.5	283.6	13.4	2.1	122	47	0.735	61	6	3300
Falkland Islands (U.K.)	4.7	12.2	0.003	2.4	1	0	N/A	85	0	25,000
French Guiana (Fr.)	34.7	89.9	0.2	2.6	6	2	N/A	75	0	8300
Guyana	83.0	215.0	0.8	1.4	10	4	0.719	36	2	4000
Paraguay	157.0	406.6	6	2.5	38	15	0.751	54	7	4700
Peru	496.2	1285.2	27.5	1.7	55	21	0.752	72	3	5100
Suriname	63.0	163.2	0.4	1.5	6	2	0.780	69	0	4000
Uruguay	68.5	177.4	3.4	0.6	50	19	0.833	93	7	12,800
Venezuela	352.1	911.9	26.2	1.9	74	29	0.778	87	3	4800
Total	**6898.4**	**17,866.9**	**365**	**1.4**	**53**	**20**	**0.780**	**79**	**6**	**7239**
Summary Total	**7946.2**	**20,580.7**	**549.4**	**1.6**	**69**	**27**	**0.773**	**74**	**7**	**7262**

Sources: *World Population Data Sheet,* Population Reference Bureau, 2004; *U.N. Human Development Report,* United Nations, 2004; *World Factbook,* CIA, 2004.

protrudes much farther into the Atlantic Ocean than does the Caribbean realm or Latin America's northern neighbor, North America. In fact, the meridian of 80°W, which intersects the west coast of South America in Ecuador and Peru, passes through Pittsburgh, Pennsylvania. Brazil also lies less than 2,000 miles (3,200 km) west of Africa, to which it was joined in the supercontinent of Pangaea until some 110 million years ago. In area, Latin America totals approximately 7,941,000 square miles (20,568,000 sq km), with South America having 6,893,000 square miles (17,869,000 sq km) and Middle America 1,047,000 square miles (2,713,000 sq km). The total is about 2.5 times the size of the conterminous United States (Figure 19.2).

Latin America's 2004 population of 549 million represented 8.5 percent of the world total. The population has uneven distributions and densities (Figure 19.3a and b). Most of Latin America's people are packed into two major geographic alignments. The larger of these two areas, known as "the rim" or "the Rimland," is a discontinuous ring around the margins of South America. The second, generally a highland, extends along a volcanic belt from central Mexico southward into Central America.

The South American Rimland contains roughly two-thirds of Latin America's people. There are two major segments of the rim. The segment much larger in population and area extends along the eastern margin of the continent from the mouth of the Amazon River in Brazil southward to the humid **pampa** (subtropical grassland) around Buenos Aires, Argentina. The second segment is located partly on the coast and partly in the high valleys and plateaus of the adjacent

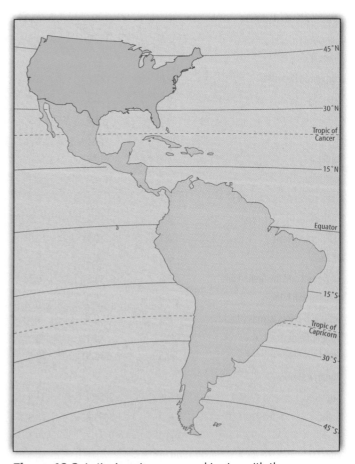

Figure 19.2 Latin America compared in size with the conterminous United States

Population Distribution of Latin America

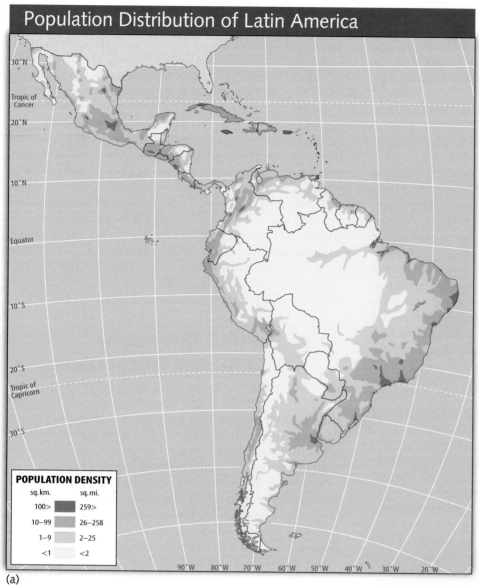

POPULATION DENSITY

sq. km.		sq. mi.
100>		259>
10–99		26–258
1–9		2–25
<1		<2

(a)

Andes Mountains. It stretches around the north end and down the west side of South America. This crescent begins in the vicinity of Caracas, Venezuela, on the Atlantic coast and arches all the way around to the vicinity of Santiago, Chile, on the Pacific side of South America.

This second segment is more fragmented, broken in many places by steep Andean slopes and coastal desert. A strip of hot, rainy, and thinly populated coast lies between the Amazon River mouth and Caracas. In the far south on the Atlantic side of the continent, the population rim is again broken in the rugged, relatively inaccessible, rain-swept southern Andes and the dry lands of Argentina's Patagonia in the Andean rain shadow. These territories have only a sparse human population, far exceeded by the millions of sheep that graze in Patagonia and adjacent Tierra del Fuego.

The second major alignment of populated areas in Latin America lies on the mainland of Middle America. Composed of the majority of Middle America's people, it extends along an axis of volcanic land that dominates central Mexico and

Population Cartogram

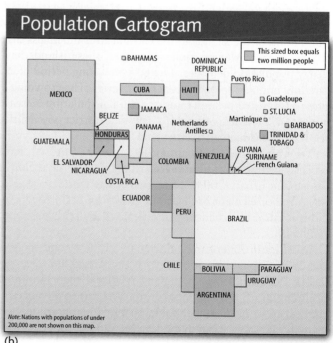

Note: Nations with populations of under 200,000 are not shown on this map.

(b)

from there reaches southeastward through southern Mexico and along the Pacific side of Central America to Costa Rica. This belt is characterized by good soils, adequate rainfall for crops, and enough elevation in most places to moderate the tropical heat but still permit the cultivation of tropical crops such as coffee. In Pacific Central America, the majority of people live in highland environments, but considerable numbers live in lowlands as well.

Average population densities conceal the fact that in nearly every Latin American mainland country there is a basic spatial configuration: a well-defined population **core** (or cores) with an outlying, sparsely populated hinterland. Densities within the cores are often greater than anything found within areas of comparable size in the United States. On the other hand, most of the mainland countries have a larger proportion of sparsely populated terrain than is true of the United States. Figures on population density for the greater part of Latin America are extraordinarily low, averaging fewer than two persons per square mile over approximately half of the entire region. This pulls down Latin America's average density—and that of most major Latin American countries—to levels somewhat below the average density for the United States (the U.S. figure is 79 people per sq mi/30 people per sq km).

The pattern of core and hinterland is particularly evident in Brazil and Argentina. Most Brazilians live along or near the eastern seaboard south of Belém, a port city in northeastern Brazil, whereas large areas in the interior are still thinly populated. In Argentina, approximately three-fourths of the population is clustered in Buenos Aires or the adjacent humid pampas—an area containing a little over one-fifth of Argentina's total land. Mexico and the Andean countries also reflect this demographic imbalance.

Not every country, however, displays the pattern of core versus hinterland. A very different picture of population density and distribution is evident in El Salvador and Costa Rica—both of which are composed of thickly settled volcanic land—and in most Caribbean islands. The combined population of the islands in the Caribbean is much less than that of central Mexico alone. However, these islands have heavy population densities, created over a long period of expanding population on restricted territories. The crowding is often aggravated by high rates of natural population increase and/or the prevalence of steep slopes that restrict possibilities for further farming and settlement.

Very high rates of urban and metropolitan growth have recently characterized Latin America. Overall, the region is 74 percent urban (2004), compared to the world average of 48. A major element in the rapid growth of Latin American cities is rural to urban migration caused by both push and pull factors.

Most Latin American countries are in the second stage of the demographic transition, clearly recognizable as less developed countries (LDCs) having relatively high (although declining) birth rates and low death rates due to advances in the spread of medical technologies. The highest population growth rates tend to be in the least developed countries, no-

tably those of Central America; Honduras and Guatemala have the highest in Latin America at 3 percent per year, and Nicaragua has the lowest GDP PPP of all the Central American countries. The lowest GDP PPP in all of Latin America belongs to the island nation of Haiti. It might be expected to have the highest population growth rate and probably would if not for the scourge of HIV/AIDS. Haiti has the highest infection rate in the Western Hemisphere, affecting some 6 percent of adults. Its annual population growth rate is tempered by HIV/AIDS to just 1.9 percent. The lowest annual rate of population change is 0.5 percent in Cuba, thanks more to steadfast Communist promotion of family planning rather than to overall prosperity. The overall rate of increase for the population of Latin America averages 1.6 percent, compared with an average of 1.5 percent for the world's less developed countries and 1.3 percent for the entire world. This regional rate of increase is down considerably from the high of 3 percent in 1960, thanks in large part to the fact that the region's economies are generally improving, if only at a slow rate.

19.2   Physical Geography and Human Adaptations

Dramatic differences in elevation, topography, biomes, and climates characterize Latin America (Figure 19.4). Low-lying plains drained by the Orinoco, Amazon, and Paraná-Paraguay river systems dominate the north and central part of South America and separate older, lower highlands in the east from the rugged Andes of the west. In Mexico, a high interior plateau broken into many basins lies between north to south trending arms of the Sierra Madre Mountains. High mountains within Latin America, largely contained within the Sierra Madre and the Andes, form a nearly continuous landscape feature from northern Mexico and the southern United States south to Tierra del Fuego at the southern tip of South America. Most of the smaller islands of the Caribbean Sea's West Indies are volcanic mountains, although some islands composed of limestone or coral are lower and flatter. The largest islands of the Caribbean have a more diverse topography, including low mountains.

WORLD
REGIONAL
Geography ⊛ Now™

Click Geography Literacy to see an animation of the formation of the Andes by plate tectonics.

Climates and Vegetation

The climatic and biotic diversity of Latin America is extraordinary (Figure 19.5a and b). Even within a single country, Ecuador, conditions range over a very short horizontal distance from sea level tropical rain forests to alpine tundra. The tropical rain forest climate and biome, with heavy year-round rainfall, continuous heat and humidity, and super-abundant vegetation dominated by large broadleaf evergreen trees, is generally a lowland type lying mainly along and near the equator, with segments extending to the tropical margins

Physical Geography of Latin America

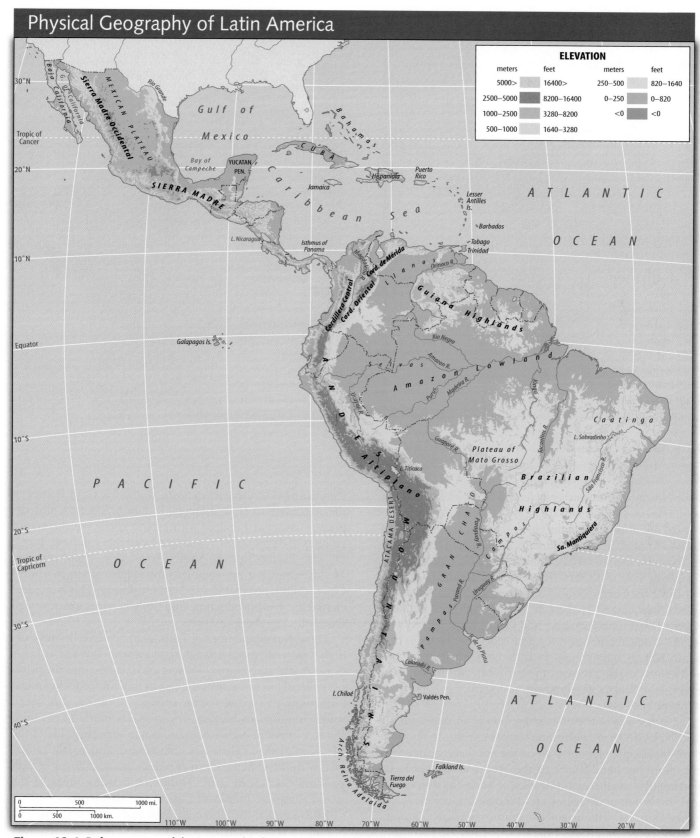

ELEVATION

meters	feet	meters	feet
5000>	16400>	250–500	820–1640
2500–5000	8200–16400	0–250	0–820
1000–2500	3280–8200	<0	<0
500–1000	1640–3280		

Figure 19.4 Reference map of the topographic features, rivers, and seas of Latin America

Climates of Latin America

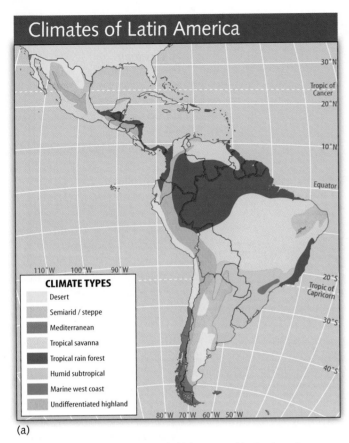

CLIMATE TYPES
- Desert
- Semiarid / steppe
- Mediterranean
- Tropical savanna
- Tropical rain forest
- Humid subtropical
- Marine west coast
- Undifferentiated highland

(a)

Biomes of Latin America

BIOME TYPES
- Tropical rain forest
- Tropical deciduous forest
- Mediterranean scrub
- Desert and desert shrub
- Savanna
- Prairie and steppe
- Coniferous forest
- Undifferentiated highland

(b)

Figure 19.5 (a) Climates and (b) biomes of Latin America

of the Northern and Southern Hemispheres. The largest segment, representing the world's largest continuous expanse of tropical rain forest, lies in the basin of the Amazon River system. Additional areas are found in southeastern Brazil, eastern Panama, the western coastal plain of Colombia, on the Caribbean side of Central America and southern Mexico, and along the eastern (windward) shores of some Caribbean islands, particularly Hispaniola (whose countries are Haiti and the Dominican Republic) and Puerto Rico. Much of the original tropical rain forest vegetation of the islands has been lost to human activity, and even the vast Amazon forest is under considerable pressure (this problem, which is largely a reflection of Brazil's development, is described in Chapter 21 on pages 584–586).

On either side of the principal region of tropical rain forest climate, the tropical savanna climate extends to the vicinity of the Tropic of Capricorn in the Southern Hemisphere and, more discontinuously, to the Tropic of Cancer in the Northern Hemisphere. In this climate zone, the average annual precipitation decreases and becomes more seasonal at higher latitudes. The mean temperature decreases, and the broadleaf evergreen trees of the rain forest grade into tall savanna grasses, woodlands, and deciduous forests that lose their leaves in the dry season.

Still farther poleward in the eastern portion of South America lies a large area of humid subtropical climate, with cool winters unknown in the tropical zones. Its Northern

Hemisphere counterpart is north of the Mexican border in the southeastern United States. In South America, this climate is associated mainly with prairie grasses in the humid pampas of Argentina, Uruguay, and extreme southern Brazil. On the Pacific side of South America, a small strip of Mediterranean or dry-summer subtropical climate in central Chile is similar to that in southern California. This is an ideal climate for wine production, and Chilean wines are now well established on world markets.

Those humid climates of Latin America have a fairly orderly and repetitive spatial arrangement. But the region's dry climates and biomes—desert and steppe—are the product of local circumstances such as mountains blocking moisture-bearing air masses or the presence of nearby cool ocean water. In northern Mexico, aridity is due partly to the rain-shadow effect of high mountain ranges on either side of the Mexican plateau. These dry climates are also partly associated with the global pattern of semipermanent zones of high pressure that create arid conditions in many areas of the world along the Tropics of Cancer and Capricorn.

In Argentina, the extremely high and continuous Andean mountain wall accounts for the aridity of large areas, particularly the southern region called Patagonia. Here, the Andes block the path of the prevailing westerly winds, creating heavy orographic precipitation on the Chilean side of the border but leaving Patagonia in rain shadow. However, in the west coast tropics and subtropics of South America, the Atacama

19,20

18

and Peruvian deserts cannot be explained so simply. Here, shifting winds that parallel the coast, cold offshore currents, and other complexities, as well as the Andes Mountains, are the conditions that combine to create the world's driest area. The mountains serve to restrict this area of desert to the coastal strip. Arid and semiarid conditions also prevail along the northernmost coastal regions of Colombia and Venezuela and in the region called the Sertão in northeastern Brazil.

Elevation and Land Use

One of the most significant features of Latin America's physical geography with respect to human adaptations is a series of highland climates arranged into zones by elevation. This zonation results from the fact that air temperature decreases with elevation at a normal rate of approximately 3.6°F (1.7°C) per 1,000 feet (304.8 m). At least four major zones are commonly recognized in Latin America (Figures 19.6 and 19.7): the *tierra caliente* (hot country), the *tierra templada* (cool country), the *tierra fría* (cold country), and the *tierra helada* (frost country). At the foot of the highlands, the tierra caliente is a zone embracing the tropical rain forest and tropical savanna climates. The zone reaches upward to approximately 3,000 feet (914.4 m) above sea level at or near the equator and to slightly lower elevations in parts of Mexico and other areas near the margins of the tropics. In this hot, wet environment, the favored crops are rice, sugarcane, bananas, and cacao. Latin America's blacks are concentrated in many of the tierra caliente zones, a legacy of the slave trade when they were forced to work the region's plantations.

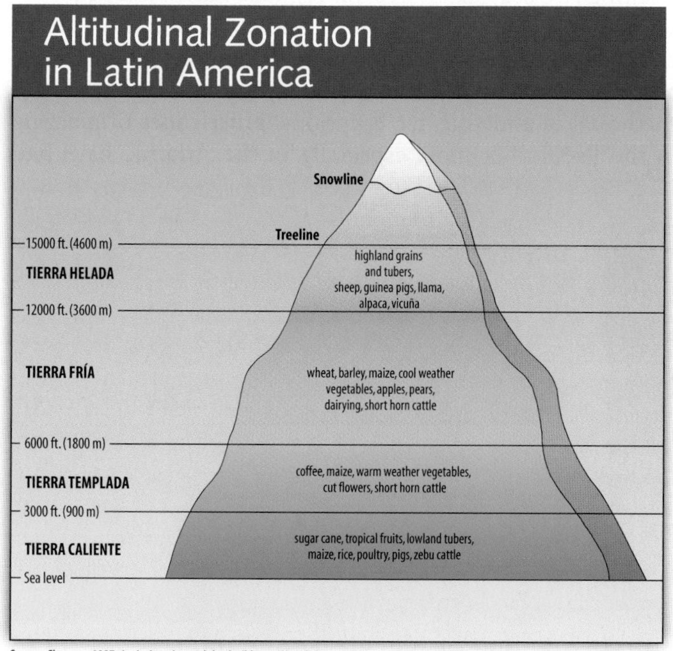

The tierra caliente merges almost imperceptibly into the tierra templada. Although sugarcane, cacao, bananas, oranges, and other lowland products reach their uppermost limits in the tierra templada, this zone is most notably the habitat of the coffee shrub. Here, coffee can be grown with relative ease, but at lower elevations, it often encounters difficulties of excessive heat and/or moisture. The upper limits of this zone—approximately 6,000 feet (1,800 m) above sea level—tend to be the upper limit of European-introduced plantation agriculture in Latin America. In its distribution, the tierra templada flanks the rugged western mountain ranges and is the uppermost climate in the lower uplands and highlands to the east. Densely inhabited sections occupy large areas in southeastern Brazil, Colombia, Central America, and Mexico. Although broadleaf evergreen trees characterize the moister, hotter parts of this zone, coniferous evergreens replace them to some degree toward the zone's poleward margins. In places such as the highlands of Brazil and Venezuela, where there is less moisture, scrub forest or savanna grasses appear, the latter generally requiring more water.

The tierra templada is a prominent zone of European-influenced settlement and of commercial agriculture. Seven metropolises exceeding 2 million in population—São Paulo, Belo Horizonte, Brasilia, Caracas, Medellín, Cali, and Guadalajara—are in this zone, and two others—Mexico City and Bogotá—lie slightly above it. Four smaller cities that are national capitals and the largest cities in their respective countries are located in the tierra templada: Guatemala City, San

Altitudinal Zonation in Latin America

TIERRA HELADA	
15000 ft. (4600 m) — Treeline	highland grains and tubers, sheep, guinea pigs, llama, alpaca, vicuña
12000 ft. (3600 m)	
TIERRA FRÍA	wheat, barley, maize, cool weather vegetables, apples, pears, dairying, short horn cattle
6000 ft. (1800 m)	
TIERRA TEMPLADA	coffee, maize, warm weather vegetables, cut flowers, short horn cattle
3000 ft. (900 m)	
TIERRA CALIENTE	sugar cane, tropical fruits, lowland tubers, maize, rice, poultry, pigs, zebu cattle
Sea level	

Source: Clawson, 1997. *Latin America and the Caribbean.* Wm. C. Brown. p 46

Figure 19.6 Altitudinal zonation in Latin America. Because of the profound differences in relief in the landscapes of Latin America, specific patterns of land use have evolved in relation to differing altitudinal zones. This map shows their elevations and general land-use patterns, which may also be compared with land uses depicted in Figure 19.7.

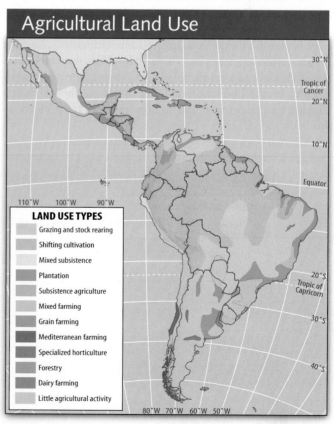

Agricultural Land Use

LAND USE TYPES

Grazing and stock rearing

Shifting cultivation

Mixed subsistence

Plantation

Subsistence agriculture

Mixed farming

Grain farming

Mediterranean farming

Specialized horticulture

Forestry

Dairy farming

Little agricultural activity

Figure 19.7 Land use in Latin America reflects a remarkable range of opportunities posed by varying latitudes and elevations.

Salvador, Tegucigalpa, and San José. Others, like Rio de Janeiro, which are situated at lower elevations, have close ties with predominantly residential or resort towns in these cooler temperature areas.

The tierra fría (at 6,000 to 12,000 ft/1,830 to 3,660 m) experiences frost and is often the habitat of a Native American economy with a strong subsistence component. In a classic process of marginalization, European colonization of Latin America drove some Native American settlements upslope and into the tierra fría zone, although some major populations (notably the Inca of Peru) had already selected upland locations for their settlement before the arrival of Columbus. These upland settlements are most extensive in Ecuador, Peru, and Bolivia and are also very evident in Colombia, Guatemala, and southern Mexico. The tierra fría is comprised of high plateaus, basins, valleys, and mountain slopes within the great mountain chain that extends from northern Mexico to Cape Horn. By far the largest areas are in the Andes, with significant areas also in Mexico (Figure 19.8). The upper limit of the zone is generally placed at about 12,000 feet (3,660 m) for locations near the equator and at lower elevations toward the poles. This line is usually drawn on the basis of two generalized phenomena: the **upper limit of agriculture,** as represented by such hardy crops as potatoes and barley, and the **tree line,** or upper limit of natural tree growth.

Another zone lies above the other three and consists of the alpine meadows, known as *páramos,* along with still higher barren rocks and permanent fields of snow and ice (Figure 19.9). This zone is known as the tierra helada. It generally occurs between 12,000 feet (3,660 m) and the lower edge of the snow line. It supports some grains and livestock (llama, alpaca, sheep) but is largely above the mountain flanks that are central to upland native settlement and agriculture. In the highlands of Latin America, the tierra fría tends to be a last retreat and a major home of the indigenous

Figure 19.9 The páramo at about 12,000 feet (3,700 m) in the Andes of Colombia

peoples, except in Guatemala and Mexico. Most of their settlements are rural villages based on subsistence agriculture. In some cases, notably in Bolivia and Peru, valuable minerals like tin and copper are located here, attracting large-scale mining enterprises into the tierra fría and the tierra helada.

Nature's Furies

Adjoining a large section of the Pacific Ring of Fire, and fronting two seasonal hurricane regions, Latin America is beset by natural hazards. A belt of land stretching from northern Mexico all the way to Tierra del Fuego, the "land of fire" at the southern tip of South America, has a violent history of earthquakes and volcanic eruptions. **Hurricanes** originating in the Pacific but most especially in the Atlantic have had

Figure 19.8 Characteristic tierra fría landscape and agriculture in Andean South America

sometimes devastating impacts. The monstrous Hurricane Mitch was an unusually late storm that made landfall in the Honduras on October 28, 1998, and roamed for 5 days over that country and Nicaragua, El Salvador, Guatemala, and Mexico. Its ferocious winds and rainfalls of up 2 feet (60 cm) brought floods, landslides, and storm damage—intensified by the region's widespread deforestation—that killed an es-

timated 15,000 people, inflicted enormous damage to property, crops, roads, power, and other infrastructure, and left disease epidemics in its wake. As if Mitch were not enough, the region had also suffered from the drought and fires visited on it in 1997 and 1998 by El Niño (see Natural Hazards, below).

Natural Hazards

El Niño

Every few years, things seem to go wrong with the winds and waters of the Pacific Ocean (Figure 19.A). Normally, the winds blow east to west, helping to promote the **upwelling** of nutrient rich, cold water from the depths to the surface of the tropical eastern Pacific. But sometimes, usually beginning in December, the winds reverse direction, blowing from west to east, suppressing the upwelling water, and thus raising the surface temperature of the water by as much as 20°F (11°C). Such an event is known in South America as *El Niño,* meaning "The Baby" in reference to the Christ child, because of its occurrence in December. Less prosaically, meteorologists know it as **El Niño Southern Oscillation** (**ENSO**).

El Niño conditions sometimes last a year or more, causing global climatic disruptions. These vary between events, but the typical El Niño pattern is illustrated in Figure 19.B. Closest to the source of the condition, Peru experiences torrential rains. Unusually high rainfall also occurs in southern Brazil, the southeastern United States, western Polynesia, and East Africa. Drought and/or unusually high temperatures settle in over northern Brazil, southeastern Canada and the northeastern United States, southern Alaska and western Canada, the Korean Peninsula and northeastern China, Indonesia and north-

ern Australia, and southeastern Africa. Warmer Pacific waters are far less nutrient rich, so typically, fish catches plummet and food chains are disrupted, leading to massive wildlife losses. The most severe El Niño events occurred in 1982–1983 and in 1997–1998, but it is not unusual for three or four moderate events to occur every decade.

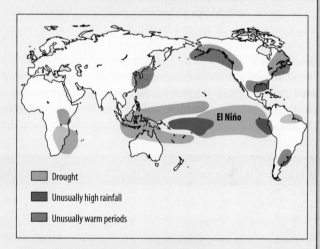

Figure 19.B Climatic impacts of El Niño

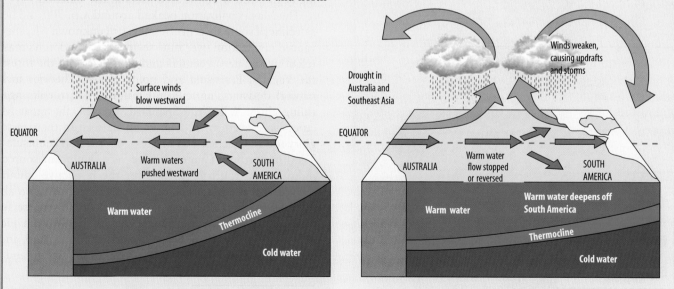

Figure 19.A Diagram of El Niño

See an animation based on this figure, and take a short quiz on the facts or concepts.

19.3 Cultural and Historical Geographies

It is perhaps unfortunate that this region came to be known as "Latin" America because there were no Latins among its inhabitants from about 10,000 B.C. to A.D. 1492. In 1492, when the Europeans arrived, the region was home to an estimated 50 to 100 million **Native Americans,** also known as **Amerindians,** or as Columbus (who thought he had landed in India) misidentified them, **Indians.** After their early migrations from the Old World, apparently across a land bridge now drowned by the Bering Strait, they developed many distinctive livelihoods and cultures. Some Native Americans practiced hunting and gathering, while others developed agriculture and took the same kind of pathway to urban life and civilization followed by several Old World cultures. Culture hearths associated with civilization emerged in the Andes region of South America and in southern Mexico and adjacent Central America (Figure 19.10).

In 1492, the **Maya** inhabited southern Mexico, Belize, and Guatemala. They practiced agriculture based on maize (corn), squash, beans, and chili peppers and had by A.D. 200 developed a highly complex civilization in both lowland tropical rain forests and highland volcanic regions. They created monumental religious and residential structures, including pyramids, temples, and astronomical observatories (Figure 19.11). They built stone roadways through dense

Figure 19.11 In the late classic period of the Maya, about A.D. 650, an estimated 70,000 people lived in Tikal.

forest areas but notably had no wheeled vehicles, nor did they have the sailing vessels, the plow, beasts of burden, and some other tools and trappings associated with civilizations elsewhere. The Maya had highly developed systems of mathematics, astronomy, and engineering, and an extremely precise calendar (which ends, some say ominously, on December 21, 2012).

They also had a hieroglyphic writing system that survives on stone monuments and in only a handful of books. Almost all of the written volumes were put to the torch by Spanish priests who regarded them as heretical. On July 12, 1562, Bishop Diego de Landa destroyed at least 30 Maya books in a bonfire outside a church in Mexico's Yucatán Peninsula. "They contained nothing in which there were not to be seen superstitions and falsehoods of the devil," de Landa later wrote. "We burned them all." Many Maya watched this event which, de Landa recalled, "they regretted to an amazing degree and which caused them much affliction."[1]

The Maya civilization peaked around A.D. 900 and began a decline of some six centuries. It is still unknown why this advanced civilization over time seemed to abandon its massive religious and urban centers and melt back into the forests of the Yucatán Peninsula and points south. Theories include natural disasters, agricultural failures, and revolts against ruling elites. When the Spaniards arrived, the great Maya cities were overgrown. Many can be seen in their restored glory today. Chichén-Itzá in Mexico's Yucatán Peninsula and Tikal in northern Guatemala are very significant sources of international tourism revenue for the respective countries.

North of the Maya realm, in the Valley of Mexico about 30 miles (48 km) from present-day Mexico City, arose Teotihuacán, the first true urban center in the Western Hemisphere. Its construction began about 2,000 years ago, and by A.D. 500, it covered about 8 square miles (21 sq km) and had

Figure 19.10 Major Native American groups and civilizations on the eve of the Spanish conquest (not including groups of the present-day United States)

---

[1] Quoted in Alfred M. Tozzer, "Landa's Relacion de las Cosas de Yucatan: A Translation." In *Papers of the Peabody Museum of American Archaeology and Ethnology* (vol. XVIII) (Cambridge, Mass.: Harvard University Press, 1941), p. 78.

200,000 residents, comparable to London in A.D. 1500. The city was laid out on a grid aligned with stars and constellations that were central to the ceremonial timekeeping of the **Teotihuacános.** There were some 2,000 apartment compounds, a 20-story Pyramid of the Sun, a Pyramid of the Moon, and a 39-acre (16 hectare) civic and religious complex called the Great Compound, or the Citadel (*Ciudadela*). The origins, language, and culture of Teotihuacános are still not fully understood, and ideally, future archeology will shed more light on these extraordinarily creative people.

The **Aztecs,** who called themselves the **Mexica** and came later than the Teotihuacános, honored their city as sacred space and a religious center when they took over the Valley of Mexico in the 1400s. They founded their capital, Tenochtitlán (now overlain by Mexico City), in A.D. 1325. With a population of about 250,000, Tenochtitlán had a sacred center surrounded by active market places, all linked by stone road surfaces to other parts of the urban settlement. There were some outlying districts that provided goods for trade, and craftspeople and merchants came to the city to work. In the middle of the 15th century, the Aztecs joined with rulers of two neighboring cities to form the **Triple Alliance.** From this base, the Aztecs and their allies developed an empire stretching across central Mexico. They tried but failed to conquer the kingdom of an adjacent civilization, the **Tarascan.** The Aztecs left a legacy of much merit in metallurgy (which diffused to Mexico from the Andean region), woodworking, weaving, pottery, and especially urban design.

In South America, the **Inca** built an empire about 2,000 miles (3,200 km) long from northern Ecuador to central Chile. They controlled it for only about a century before 1532 when the Spanish conquistadors, under Francisco Pizarro, began their subjugation. The Inca origins, dating to about A.D. 1200, are not fully known. From their governing center in Cuzco, Peru, the Inca engineered a system of roads, suspension bridges, and settlements that connected and supported their empire. The road system ran more than 2,500 miles (4,120 km) and was busy with movement of trade and tribute. The Inca also achieved great skill in stoneworking, terrace construction, and irrigation networks (Figure 19.12). Remarkably, they had neither paper nor a writing system but kept mathematical records by knotted ropes called *quipus.*

Still another group, the little known **Nazca** culture, left a unique legacy on the landscape of South America about 2,000 years ago. Their "Nazca Lines" in the desert of southern Peru is a complex of nearly 200 square miles (c. 520 sq km) of massive carvings in the sandy surface, depicting birds, insects, and fish—the smallest of which is an animal figure some 80 feet (c. 23 m) in length. These designs can only be appreciated from the air, giving rise to the improbable theory that they were built as landmarks for "ancient astronauts." No written records of the Nazca culture survive, so the purpose of their lines may never become known. Another of the less celebrated civilizations of South America is the **Chibcha** of what is now Colombia. The Chibcha lived in small agricultural villages rather than large cities, and their exquisite gold work is a major legacy of their culture. Speak-

Figure 19.12 Inca stone architecture was often exceptionally fine. It is frequently and correctly pointed out that even a knife blade cannot be inserted between some of their mortarless stone blocks, so precisely were they cut.

ers of Chibchan languages now live in the Andes of Colombia and Ecuador and as far north as Costa Rica.

Languages

With the exception of the Nascans (whose record ends about A.D. 600), the groups that have been discussed represent only the largest major civilizations at the time of the Spanish conquest. There were smaller civilizations and others that predated the conquest. Far more numerous were the groups that practiced hunting and gathering or agriculture but did not build cities and civilizations. Their legacy survives today in the cultures and languages of a rich variety of Native American ethnic groups. Linguistic geographers disagree sharply over Native American language distributions, so only very general patterns are depicted in Figure 19.13.

Six Native American language families are represented in Mexico and Central America. The **Hokan-Siouan** language family has speakers in the Baha Peninsula and in El Salvador and adjacent Honduras. The **Aztec-Tanoan** family includes the Nahuatl that the Aztecs spoke and that is still spoken in northern Mexico today. The **Oto-Manguean** language family includes seven languages spoken mainly in northern Mexico. There is a small **Totonac** language family region on the Gulf of Mexico. Southern Mexico is home to the **Penutian** and **Maya** language families. The indigenous cultures these languages represent form a significant minority overall in Mexico at 30 percent, but they comprise over 50 percent in some southern states of the country. The Maya language family has five major subfamilies represented by 61 million people living all across the realm of the ancient Maya, especially in Guatemala. Some Maya, particularly in highland Guatemala, never did mix with outsiders and are ethnically little changed from their forebears. The Maya of highland Guatemala tend to speak Mayan as their first language, but many are bilingual with Spanish. With its 23 Maya groups, Guatemala's population is more than 40 percent Native American overall.

South America—which had a Native American population as early as 12,500 years ago—has an even richer lan-

Figure 19.13 Indigenous and European languages of Latin America

guage palette, with about 600 languages belonging to 16 families. They include the **Quechu-Aymaran** languages, to which the language of the ancient Inca belonged. Quechuan speakers today are represented by five major subgroups and more than 7 million people, mainly in Peru but also in Colombia, Ecuador, Bolivia, and Argentina. Most still speak **Quechua** as a first language, but many are bilingual with Spanish. In general, indigenous languages are used far more widely in South America than elsewhere in Latin America, and there are more speakers that do not have a second language. Very high proportions of the populations in highland South America are Native Americans. Bolivia at 52 percent and Peru at 45 percent are estimated to be the highest. Native Americans are also a major population element in the basin of the Amazon River and in Panama, where scattered lowland people maintain their aboriginal cultures.

The outer world is pushing in on these indigenous peoples of South America despite some efforts on the part of governments to protect their ways of life. New constitutions expanded indigenous rights in Colombia in 1991, Peru in 1993, Bolivia in 1994, and Ecuador in 1997. In these countries, Native Americans now have more rights over land, and the governments recognize communal rights over some resources. Brazilian law forbids mineral extraction by outsiders on aboriginal lands, but illegal diamond and gold mining has led to many bloody skirmishes between Native Americans and prospectors in several Indian reserves of the country. On the negative side, as the Native Americans see it, many resources

of Brazilian and other national Indian reserves are placed under government control, and often aboriginal people are forbidden from setting up their own businesses.

European languages joined the linguistic map of Latin America beginning in 1492 (see Figure 19.13). At first, they were spoken only by the colonizers but became more widespread as colonial administration progressed. The distributions of the languages correspond with those of colonial rule. **Spanish,** now one of the world's **megalanguages** along with Chinese, English, Hindi, and Arabic, is the most widespread and is the prevalent European language throughout the region, with these major exceptions: **Portuguese** in Brazil; **French** in French Guiana, Haiti, Guadeloupe, and Martinique; **Dutch** in Suriname; and **English** in Guyana and Belize. These languages are also represented in the dependencies and former colonies of several other small Caribbean islands. The European languages generally became the sole official languages both of the colonial administrations and the independent countries that succeeded them. They serve as welcome lingua francas in countries like Mexico and Peru that have a bewildering diversity of indigenous languages. Bolivia, Peru, and Guatemala have official languages that are both European and indigenous.

Another linguistic product of European colonialism in Latin America was the evolution of **creole** languages. These are tongues that developed among the black slaves and indentured servants brought to work the plantations of the Caribbean islands and the Atlantic and Caribbean coasts of South and Central America. Their vocabularies come mainly from the respective colonizers' languages of Spanish, Portuguese, French, English, and Dutch. Sometimes a speaker of the parent language can recognize the creole; for example, in Costa Rican Creole English *"Mi did have a kozin im was a boxer, kom from Panama"* is easily enough understood as "I had a cousin from Panama who was a boxer."[2] But it is not always that easy. Creoles are distinct languages and have their own grammars.

None of the indigenous African languages that the slaves brought to the New World survived there to the present day. Some vocabulary and other elements of the ancestral African languages do survive in some of the creoles, such as in the Garifuna spoken on the coast of Belize.

The Conquest

The year 1492 brought more than a new language to the New World. That year marked the beginning of profound changes in almost every aspect of land and life in what would become Latin America (see Figure 21.12, page 583).

First came death, both deliberate and unintentional. Spanish conquistador Hernando (Hernán) Cortés landed on Mexico's coast near present-day Veracruz in 1519 and proceeded inland, building alliances with ethnic groups opposed to the Aztecs and killing those who would not join him. Unfortunately for the Aztecs, Cortés's arrival was identified

---

[2] Bernard Comrie, Stephen Matthews, and Maria Polinsky, *The Atlas of Languages* (New York: Facts on File, 2003), p. 151.

with that of Quetzalcoatl, a fair-skinned, bearded Aztec god that was prophesied to return one day from the east. In Tenochtitlán, the Aztec leader Moctezuma received Cortés and his army as emissaries of Quetzalcoatl. Cortés and just 30 men imprisoned Moctezuma and were unopposed, apparently because of the widespread belief about Quetzalcoatl's return. Using a divide-and-conquer strategy that enlisted local forces, Cortés achieved a military victory over the Aztecs in 1521, destroying Tenochtitlán and laying the foundation for Mexico City in its ashes.

Soon new expeditions were dispatched to conquer other peoples of Mexico and Central America. When Francisco Pizarro landed in Peru in 1531 with only 168 soldiers and 102 horses, he was warmly greeted by an Incan leader, Atahualpa. The Spanish conquistador imprisoned Atahualpa, freed him for a huge ransom, imprisoned him again, and then had him hanged (he was to have been burned at the stake, but Pizarro modified the execution because Atahualpa agreed to be baptized). Two years later, Pizarro's forces held the Incan capital Cuzco, and from that point, the Spanish and Portuguese colonization of all South America proceeded swiftly.

Disease brought death to the indigenous people on a vast scale. The Native Americans had never been exposed to, and thus had no natural immunities against, the host of diseases that the Spanish and then the Portuguese introduced: bubonic plague, measles, influenza, diphtheria, whooping cough, typhus, chicken pox, tuberculosis, and smallpox, joined later by malaria and yellow fever that came with African slaves. The worst was smallpox, which killed between a third and half of all Native Americans in affected areas. The disease, famine, and enslavement that came with Spanish and Portuguese colonization brought what was probably humankind's largest population decline ever. The numbers are elusive but were certainly huge. The Spanish chronicler de las Casas wrote that in Spanish-controlled Latin America, 40 million Native Americans died by 1560. Geographer David Clawson relates that, up to 1650, 90 to 95 percent of the Native Americans who had come into contact with Europeans died.

European settlement patterns emerged as the conquest proceeded. The development of ports as bases for subsequent penetration of the interior was the first major settlement task. Large coastal cities had not existed prior to European colonization. Many early ports eventually grew into sizable cities, and a few became major metropolitan centers. In some cases—notably Lima, Caracas, and Santiago—the main city developed a bit inland but retained a close connection with a smaller coastal city that was the actual port.

Around the ports, agricultural districts developed, spread, and shipped an increasing volume of trade products overseas. Plants and animals domesticated by Native Americans would revolutionize diets and habits in Europe and elsewhere in the Old World: tobacco, potatoes, corn, cacao, and turkeys were among them (Figure 19.14; see also Table 3.3, page 68). In time, the European-introduced horses, cattle, sheep, donkeys, wheat, sugarcane, coffee, and bananas would revolutionize New World agriculture. In the second half of the 1500s, slave ships began to bring the Africans who pro-

vided the principal labor on the plantations created and owned by Europeans, primarily in the tropical lowlands with the generally fertile alluvial soils of the coastal plains.

32, 463

The powerful lure of gold and silver stimulated deeper penetration of the Andes and the Brazilian Highlands in South America. After capturing the stores of precious metals that had been accumulated by the indigenous peoples for ceremonial use, tribute, and trade, the European newcomers opened mines or took over old mines and opened market centers to service them. Some highland settlements became centers of new ranching and plantation enterprises. A few highland cities, of which the largest is Bogotá in Colombia, grew into large metropolises. Separated from the seaports by difficult terrain, these cities were eventually connected to the coasts by feats of great engineering skill. Some of the main seaports and highland cities became important centers of colonial government as well as economic nodes, and several are national capitals today, including La Paz in Bolivia and Quito in Ecuador.

At the heart of each Spanish and Portuguese city established in Latin America, the newcomers placed their familiar combination of a plaza dominated by a massive Roman Catholic cathedral (Figure 19.15). **Roman Catholicism** was introduced, often forced, as the one and only acceptable faith of the New World. The map of faiths in Latin America reflects its dominance today (Figure 19.16). The British and Dutch brought their **Protestant faiths** to their possessions, and a growing movement of missionary **Evangelical religions** such as the **Seventh Day Adventist** is making inroads in traditionally Catholic communities throughout the region. Rastafarianism is a unique Jamaican faith dating to the early 20th century, with unlikely links to Ethiopia. They will not show up on a map because they are syncretized with Catholicism, but many indigenous beliefs dating to **pre-Columbian times** (before 1492) are alive and well among Native Americans in Latin America. The Catholic Maya of Guatemala, for example, still make offerings to the "Earth Lords" who dwell in caves (Figure 19.17).

501

Figure 19.14 A store of corn in a Maya home in Guatemala. Probably domesticated by the ancient Maya, corn is now a world staple.

Figure 19.15 Plaza and San Antonio Cathedral, Quito, Ecuador.

Generalized Religious Patterns

Roman Catholic

Other Christian

Muslim

Hindu

Animist

Figure 19.16 Religions of Latin America

Figure 19.17 Cave offering of the Q'eqchi' Maya, Alta Verapaz, Guatemala

Non-Native American Ethnic Patterns

Despite today's overall dominance in Latin America of culture traits derived from Europe, only 3 of this region's 38 nations—Argentina, Uruguay, and Costa Rica—have white **European** ethnic groups that have had little intermarriage with Native Americans or blacks (Figure 19.18a). These are mainly the descendents of relatively recent immigrants from Europe, especially during the late 19th and early 20th centuries. Scattered districts in other countries are also predominantly European.

Black Latin Americans of relatively unmixed African descent are found in the greatest numbers on the Caribbean islands and along the Atlantic coastal lowlands in Middle and South America (Figure 19.18b). These are the areas to which African slaves were brought during the colonial period, primarily as a source of labor for sugar plantations. An estimated 3 to 4 million were sold in Brazil alone. Slavery was gradually abolished during the 19th century, although not until the 1880s in Brazil and Cuba. By that time, slavery had generated large fortunes for many owners of plantations and slave ships. The African peoples it introduced to the region have made cultural contributions that today provide a varied, colorful, and important part of Latin American civilization.

Latin America is blessed in having escaped many of the racial tensions and violence that have battered much of the world. This may be because a majority of Latin Americans are of a mixed racial composition, with a primary mixture of Spanish and Native Americans that has resulted in a heterogeneous group known as **mestizos** (Figure 19.18c). **Mulatto** is the term meaning a person of mixed African (black) and European (white) ancestry, sometimes with Native American blood as well (Figure 19.18d); in some locales, the term used is **Creole**. It must be noted, however, that there are strong socioeconomic discrepancies that correspond with ethnicities in the region. Government and business sectors are often dominated by a thin stratum of people of mainly European origin, with mestizos and mulattos making up the bulk of the middle and lower classes. Native Americans almost always occupy the lowest standing, and in some countries such as Guatemala, there have been periods of systematic exclusion and violence directed against them.

Mestizos have not lost all traits of their Amerindian forebears. While it can be hazardous to generalize about ethnicities, Latin Americanist Simon Collier offers the following sketch of what the mestizos acquired from the Amerindians.[3] The mestizo inheritance from Native American roots includes the collective exploitation of land, the accumulation of life-long personal bonds, the devotion to a whole series of supernatural beings (who have Christian names but non-Western origins), the ritualized remembrance of deceased relatives, and reverence for political leaders. As will be seen

[3] Simon Collier, Harold Blakemore, and Thomas E. Skidmore, eds., *The Cambridge Encyclopedia of Latin America and the Caribbean* (Cambridge: Cambridge University Press, 1985), p. 159.

Figure 19.18 In addition to the Native Americans of Latin America, the leading racial types are (a) Europeans, (b) blacks, (c) mestizos, and (d) mulattos.

later in the chapter, reverence for political leaders cannot be taken for granted in this region today.

19.4 Economic Geography

Latin America is generally a region of less developed countries where people do not enjoy a high standard of living. It is not the worst off of the major world regions; in fact, its overall per capita GDP PPP is over four times that of Africa, south of the Sahara (see Table 1.2). But compared with North America or Europe, the Latin American region as a whole is quite poor. An estimated 44 percent of Latin Americans live in poverty, 20 percent live in extreme poverty, and the number of unemployed workers more than doubled between 1994 and 2004. The wealthiest countries on a per capita GDP PPP basis are the Bahamas and Barbados, while the poorest are Haiti and Nicaragua. The glitter of the great metropolises like Mexico City and Rio de Janeiro, with their forests of new skyscrapers, may mask some of this poverty, but portions of these cities are massive, squalid slums, called *favelas* in Brazil and *barrios* in Spanish-speaking countries (Figure 19.19). Such **shantytowns,** which are practically universal in the world's LDCs, are full of unemployed, underemployed, and ill-fed people. Many prefer their present plight to the conditions of the depressed rural landscapes from which they came. They aspire to the upward mobility that is at least within view in the urban areas. These squalid suburbs are thus sometimes called "slums of hope."

In an attempt to overcome their economic deficiencies and constrain popular discontent, many Latin American governments have borrowed heavily from the international banking community. In the early 2000s, unpaid loans had reached staggering proportions, and the Latin American debtor nations were having great difficulty in mustering even the annual interest payments, let alone generating surplus capital to pay toward the principal of these troubling loans—a dilemma all too characteristic of the world's less developed countries. In common with so many LDCs, Latin America also has a historic overreliance on non-value-added goods

TABLE 19.2 Primary Exports of Selected Latin American Countries

Country	Selected major exports
Argentina	Edible oils, fuels and energy
Bahamas	Fish, rum, salt
Brazil	Transport equipment, iron ore, soybeans, coffee
Chile	Copper, fish, fruits, paper and pulp
Colombia	Petroleum, coffee, coal
Costa Rica	Coffee, bananas, sugar
Cuba	Sugar, nickel, tobacco
Dominican Republic	Ferronickel, sugar
Guatemala	Coffee, sugar, bananas
Mexico	Manufactured goods, petroleum, silver
Nicaragua	Coffee, shrimp and lobster, cotton
Panama	Bananas, shrimp, sugar
Peru	Fish, gold, copper, zinc
Trinidad & Tobago	Petroleum, chemicals, steel products
Venezuela	Petroleum, bauxite, aluminum, steel

Source: *World Factbook,* CIA, 2004

like cash crops and minerals (Table 19.2). Free trade agreements form the foundations of a recent push to move away from raw materials and toward manufactured exports.

Commercial Agriculture

Although the total number of Latin Americans employed in agriculture has not declined much in recent decades, there has been a marked percentage decline in the value of agriculture in national economies. Dependence on agricultural exports has also dropped as national economies have become more diversified, but in many countries, more than half of all export revenue is still derived from products of agricultural origin, and some rely mainly on one or two agricultural commodities (see Table 19.2). Typical of less developed countries, overreliance on a narrow range of exports makes the Latin American countries economically vulnerable to changes in market conditions, competition from other sources, and changing consumer appetites. Reliance on coffee and bananas has whipsawed the economic fortunes of many countries, especially those in Central America that came to be known disparagingly as "banana republics" (see Definitions and Insights, page 538).

Types of Farms

Farms in Latin America may be divided into two major classes by size and system of production. Large estates with a strong commercial orientation are *latifundia* (sing. *latifundio*). These estates, whether called haciendas, planta-

Figure 19.19 Many of Latin America's poorest people live in ramshackle housing. This shantytown of Santa Marta overlooks scenic Rio de Janeiro.

David R. Frazier/Photo Researchers, Inc.

Figure 19.20 Joe Hobbs

Figure 19.20 Henequen (sisal) fiber, once produced extensively on plantations in the Yucatán

Joe Hobbs

Figure 19.21 Market scene in rural Ecuador

tions, or by some other name, are owned by families or corporations (Figure 19.20). Some have been in the hands of the same family for centuries. The desire to own land as a form of wealth and a symbol of prestige and power has always been a strong characteristic of Latin American societies. Huge tracts were granted by Spanish and Portuguese sovereigns to members of the military nobility who led the way in exploration and conquest. Some of this land has been reallocated to small farmers by government action from time to time. But in many countries, a very large share of the land is still in the hands of a small, wealthy, landowning class.

Minifundia (sing. *minifundio*) are smaller holdings with a strong subsistence component. The people who farm them generally lack the capital to purchase large and fertile properties; hence, they are relegated to marginal plots, often farmed on a sharecropping basis. Individuals who do own land are frequently burdened by indebtedness, the continued threat of foreclosure of farm loans, and the fragmented nature of farms, which are becoming smaller as they are subdivided through inheritance and the reversal of promised governmental programs of land distribution. Such farms produce food for family use and also for the local market (Figure 19.21). The crops most commonly raised, especially

in Middle America, are maize (corn), beans, and squash, although many other crops are locally important, especially in the different climatic zones of the highland regions.

Although these small farmers make up the bulk of Latin America's agricultural labor force, food has to be imported into many areas, and many of the people are poorly nourished. Productivity is low because of the marginal quality of the land and the generally rudimentary agricultural techniques in peasant farming. There is little capital with which to buy machinery, fertilizers, and improved strains of seeds. Soil erosion and soil depletion are serious problems in many areas. As elsewhere in the developing world, increasing numbers of young people are responding to difficult rural conditions and the perceived opportunities of the big city by abandoning the countryside and settling in urban shantytowns. The tide of rural to urban migration might be slowed or even reversed if there were effective **land reform,** but land reform has long been one the most contentious issues of

Definitions + Insights

Fair Trade Fruits

Bananas are the most popular fruit in the United States, with more than 8 billion sold in 2004—or about 84 bananas for each American. Now it is possible to be a discriminating banana buyer. At many North American and European grocery stores, one may find bananas, pineapples, other fruits, chocolates, and coffees bearing "Fair Trade Certified" stickers (Figure 19.C). These are higher in price than the standard products, but consumers who buy them are generally motivated by a sense of social responsibility. In standard trade transactions, only a tiny fraction—2 cents per pound of bananas, for example—of the final price paid in the supermarket actually goes to the grower. Intermediaries including importers, distributors, and the "ripener" (who sprays ethylene gas to turn the bananas yellow) all get a cut. Fair trade buyers deal directly with the farmer cooperatives they helped to organize, avoiding brokers and intermediaries and guaranteeing higher prices—typically 18 cents per pound for bananas—for the farmers. The fair trade organizations also help the farmers establish schools and health clinics.

The **Fair Trade movement** developed

Figure 19.C "Fair Trade" certification of Mexican coffee sold in a U.S. health food store

in Europe in recent years, when people began taking notice of the impacts of collapsing coffee prices. While farmers could earn $3.18 per pound for their coffee in 1999, a glut in the world coffee supply, caused largely by surging production in Vietnam, sent prices downward to 47 cents per pound by 2001—the lowest in a century, adjusted for inflation. Europeans learned that African and Latin American coffee farmers, never wealthy to begin with, were now destitute, so poor that their young children were compelled to work in the fields to help make ends meet. Green Party politicians and others stepped forward with the Fair Trade solution, seeking to guarantee so called "living wages" for the producers of these commodities.

Coffee is the world's second largest traded commodity (after oil), so Fair Trade could impact many people. Now more than a million coffee growers in Tanzania, Ecuador, Nicaragua, and other countries belong to the cooperatives that sell their products through Fair Trade channels instead of to commercial producers. Fair Trade products are becoming more mainstream in the United States; all Dunkin' Donuts espressos and some Starbucks coffees, for example, bear the label.

political and social change in Latin America (see Problem Landscape, page 539).

Minerals and Mining

Latin America is a large-scale producer of a small number of key minerals, notably petroleum, iron ore, bauxite, copper, tin, silver, lead, zinc, and sulfur. Only a handful of Latin American nations gain large revenues from exporting these minerals. Even in the countries that do have large mineral output—notably Mexico, Venezuela, Chile, Ecuador, and Brazil—much of the profit appears to be dissipated in the form of showy buildings, corruption, ill-advised development schemes, and enrichment of the upper classes and foreign investors. Even though benefits to the broader population from such mineral production is often minimal, that production has funded significant infrastructural development, including many new highways, power stations, water systems, schools, hospitals, and employment opportunities.

Most of the extracted ores from Latin America are shipped to overseas consumers in raw or concentrated form,

although the region has scattered iron and steel plants, as well as smelters of nonferrous ores, which process metals for use in Latin American industries or for export. Deposits of high-grade iron ore in Venezuela and central Brazil are the largest known in the Western Hemisphere and are among the largest in the world; the two countries are Latin America's main producers and exporters of ore. In Latin America, Brazil is the main producer of iron and steel by far, with Mexico second. Most of the region's production of bauxite—the major source for aluminum—comes from Jamaica, Brazil, Suriname, and Guyana. The deposits in these countries are located relatively near the sea and are of critical importance to the industrial economies of the United States and Canada. Huge unexploited bauxite deposits in Venezuela are a major resource for the future.

Chile is the largest copper producer in Latin America and also ranks number one in the world. Most of Latin America's known reserves of tin are in Peru, Bolivia, and Brazil, and these countries produce most of the region's output. The silver of Mexico, Peru, and Bolivia—sought from the very be-

Problem Landscape

The Troubled Fields of Land Reform

The most productive lands in Latin America, and those that have benefited most from technologically innovative and modern farming practices, were once owned by colonial powers and are now owned by large, and frequently absentee, landowners. The largest sector of the rural population, meanwhile, is made up of *campesinos*—the peasant farmers who sharecrop land owned by others or who have very small holdings and whose lives are dominated by poverty. Their prospects might improve, along with the nations' food security and agricultural export income, if land reform could effectively transfer more land from the wealthy to the poor.

The region's history is filled with attempts—sometimes very serious and bloody, sometimes less painful governmental innovations—to achieve a better balance between those who work the land and those who own the land. The main attention has been given to breaking up existing properties or bringing vacant land (whether owned by the public, by private individuals, or by the Catholic church) into cultivation, usually by small farmers. Some new farms have been structured as communal holdings, reflecting indigenous traditions or 20th century revolutionary plans of agrarian reform. In Cuba, for example, the Communist government that took control in 1959 placed the land from expropriated estates in large farms owned and oper-

ated by the state. Workers on these farms are paid wages. The Nicaraguan Sandinistas implemented land reform as part of their revolutionary efforts in the 1980s.

Land reform attempts and outcomes have varied sharply from one Latin American country to another. Such policies can help relieve poverty in the countryside, but they do little for the majority of Latin America's poor, who live in cities. In Mexico, there was great fanfare at the introduction of the *ejido* (communally farmed or grazed land in Native American villages) early in the 20th century. It has been so effective that ejidos currently account for 50 percent of the cultivated land in Mexico and produce nearly 70 percent of the beans, rice, and corn of this nation. The ejido, by making land available to farming communities, has been the most successful land reform program attempted in Latin America, including Cuba.

Puerto Rico has had public battles over *"pan, tierra, y libertad"* (bread, land, and liberty) because of the difficulty in effecting real distribution of farmland to the farming people. Virtually every nation in Latin America can point to events, martyrs, and programs birthed in protest or blood in an effort to achieve more equitable land ownership. Nevertheless, the great majority of the arable land in Latin America continues to be owned by wealthy farming families, agribusiness operations (both domestic and foreign owned), and the church. There is a world of small farmers in Latin America, but they seldom own as much as 20 percent of a given nation's arable land.

---

ginning of Spanish colonization—is found mainly in mountains or rough plateau country. Mexico and Peru are the largest Latin American producers of silver, lead, and zinc. All but a small proportion of Latin America's petroleum is extracted in the Caribbean Sea–Gulf of Mexico area, particularly the central and southern Gulf coast of Mexico and northern Venezuela. Other oil fields in Latin America are widely scattered, with the principal ones located along or near the Atlantic coast in Brazil, Argentina (in Patagonia), and Colombia or in sedimentary lowlands along the eastern flanks of the Andes in every country from Trinidad and Tobago to Chile. Natural gas is extracted in many areas that produce oil, but Latin American production is not yet of major world consequence. Currently, Mexico, Venezuela, Argentina, and Trinidad and Tobago are the largest producers in Latin America. Venezuela is also the only Latin American nation in the Organization of Petroleum Exporting Countries (OPEC) and was one of its founding members in 1961.

Latin America's heavy reliance on unprocessed minerals has periodically dealt crushing economic blows, especially when global recession reduced demand for such commodities and forced prices downward. But strong growth characterized Latin American economies after 2002, particularly

because of China's superheated economic growth. Brazilian iron ore, Chilean and Peruvian copper, and Venezuelan oil seem to be in ever-increasing demand to feed China's appetite for commodities. The flip side is that if China's boom were to go bust, Latin American economies would fall, too.

Free Trade Agreements

Many Latin American countries are trying to reduce dependence on raw materials and boost exports of value-added manufactured products, thus improving their financial situations. To do so, they have formed or joined free trade agreements (Figure 19.22).

The most economically significant is the **North American Free Trade Agreement** (**NAFTA**), of which Canada, the United States, and Mexico are members. Citizens of these countries make up about half of the 800 million people of the hemisphere. The essential goal in establishing this free trade agreement, like most, was to reduce duties, tariffs, and other barriers to trade between the member countries, theoretically helping to strengthen all the economies. The United States had a strong if perhaps understated NAFTA goal of reducing Mexican immigration into the United States by boosting Mexico's prosperity.

Economic Associations

NAFTA*
Central American Common Market
CARICOM
Andean Community
Mercosur
Mercosur associate
* *Note:* Canada is also a member of NAFTA.

Figure 19.22 Latin American countries belong to a variety of economic associations.

NAFTA was already very controversial before it went into effect in 1994, and it remains so today. Among U.S. citizens, there were fears that factories would close down and jobs would flow south to Mexico. American environmentalists feared that the U.S. environment would suffer because Mexican factories and other polluters, unable to afford the cleanup of their industries, would help set new and lower standards that U.S. companies could then match. Mexican officials feared that a surge in their consumption of American-made products would so increase their trade deficit with the United States that Mexico's monetary reserves would be exhausted.

More than a decade on, even as the agreement's provisions are still coming into force (they are being phased in gradually until 2009), some of NAFTA's long-term impacts are becoming clearer. One of the most profound early consequences was a virtual collapse of clothing industries in Jamaica and other Caribbean islands. Once duty fees were no longer imposed on Mexican clothing exports to the United States, Mexico immediately had a huge advantage over Jamaican and other Caribbean producers. Job growth in Jamaica stopped, and unemployment surged.

Some U.S. factories did close as production was cheaper south of the border—notably in the *maquiladora* factories (discussed on page 554 in Chapter 20)—but NAFTA was not the exclusive source of such closures. Lower cost manu-

facturing was an option for American companies elsewhere around the globe. NAFTA does require all its member countries to abide by **rules of origin** specifying that half or more of the components of any manufactured good must originate in Canada, the United States, or Mexico so that a fourth country cannot simply use Mexican labor in the assembly of its own component parts. Critics of NAFTA say the agreement should have demanded that an even larger proportion of the component parts originate in the member countries to boost their economies.

NAFTA did create more manufacturing jobs. However, farm jobs were lost to falling prices for corn, rice, beans, and pork because NAFTA virtually eliminated tariffs on agricultural imports from the United States. The heavily subsidized American farmer, receiving about $20,000 in government subsidies each year, can sell grain or livestock at prices far below production cost. A pig, for example, can be raised and sold in Iowa for one-fifth of what it would cost to raise and sell it in Mexico. Cheap American pork, corn, and other products have flooded the Mexican market, driving down corn prices, for example, by 70 percent since NAFTA was enacted. This makes it impossible for Mexican farmers to compete, driving many out of business; one-third of Mexico's 18,000 swine producers have gone out of business since NAFTA went into effect. On the positive side, Mexican companies have been able to turn the cheaper U.S. corn and pork into tortillas and sandwich meats sold back at a profit to American consumers. Overall, real wages for Mexicans have not increased since NAFTA was enacted. The hoped-for reduction of emigration to the United States has not materialized. Despite post-9/11 security concerns along the U.S.–Mexican border, about half a million illegal immigrants pass into the United States from Mexico each year. Overall, regarding the gains and benefits for Mexico from NAFTA, one international study called it a "wash," concluding that Mexico was neither much better nor much worse off for it.

A cluster of nations in South America—Brazil, Argentina, Uruguay, Paraguay, and Bolivia—belong to the **Southern Cone Common Market,** known as **Mercosur,** which was established in 1991. Mercosur reduced tariff and other long-standing barriers that had existed between the countries, promoting a huge increase in trade. Its members have worked to coordinate economic policies and have a better negotiating position with the United States and other powers. They have discussed but so far not adopted a common currency modeled after the European Union's euro. The more than 250 million people of this union can live and work in any of the member countries and have the same rights as citizens of those countries. This stipulation brought immediate relief to an estimated 3 million illegal workers.

Mercosur is not a lightweight: In value of products traded, this is the third largest trade group in the world, after the European Union and NAFTA. Mercosur and the European Union negotiated an important trade pact in 2004 in which

Europe offered the South American countries more generous import quotas for beef, dairy products, sugar, and coffee. In turn, Mercosur offered the European Union privileged access to investments in telecommunications, banking, and other services. The South American countries embraced the agreement in part at the urging of the regional heavyweight, Brazil, which did not want to be seen as bowing first and foremost to U.S. interests in the FTAA pact, discussed below.

Patterned after NAFTA, another U.S.-brokered trade organization was negotiated in 2004 and was due for ratification by the U.S. Senate in 2005 or 2006. This was the **Central American Free Trade Agreement (CAFTA)**, composed of the United States, Costa Rica, El Salvador, Guatemala, Honduras, Nicaragua, and the Dominican Republic. U.S. advocates of CAFTA promised that by eliminating duties on 80 percent of the U.S. goods that could be sold in the region, it would create the second-largest market in Latin America for U.S. exports, after Mexico. Politicians from textile-producing states like South Carolina argued that CAFTA will destroy American jobs by opening the U.S. market to cheap imports, and they promised to defeat the ratification of CAFTA in the U.S. Congress. Latin American critics of CAFTA argue that the trade agreement demands too many concessions by the poor Central American countries in protecting U.S. **intellectual property rights (IPR)**, meaning fundamentally that they would not be allowed to replicate American pharmaceutical drugs and sell them at lower prices. They also point out that CAFTA will still protect the U.S. sugar industry, hurting that important business in Central America, while flooding Central America with cheap U.S. corn and other grains that will drive Central American farmers out of business.

Using NAFTA as its foundation and CAFTA as a building block, the United States is taking the lead in establishing a hemisphere-wide trade organization called the **Free Trade Area of the Americas (FTAA)**. Like the political organization known as the **Organization of American States (OAS)**, it would include all of the countries of North, Middle, and South America, with the exception of Cuba. Originally envisioned for establishment in 2005, this agreement is proving very difficult to achieve because some of its loudest critics in Latin America—Brazil in particular—insist that the United States is including too many restrictions in its vision of "free" trade. Brazil wanted to see the United States drop the huge subsidies of many agricultural products and steel that keep American farmers and factory workers in business but that shut out Brazilian products. When the United States refused, Brazil retaliated by saying it would not endorse patent, copyright, and other IPR protection of U.S. industries. American negotiators of the agreement labeled Brazil the head of a league of "can't do" nations in free trade. For the time being, the United States chose to negotiate unilateral trade agreements with members of the proposed FTAA, beginning with Chile in 2004.

Figure 19.23 A Panama hat assembly plant in Ecuador. This small family-owned operation produces a modest but high-quality output of these famed sombreros for domestic and foreign consumption.

There are other regional trade associations, including the Central American Common Market, the Andean Community, and the Caribbean Community (CARICOM). The large maquiladoras of Mexico, churning out their automobiles, textiles, electronics, and computers, are the symbols of the new Latin American free trade-oriented economies. These have become an exception to the previous spatial pattern of Latin American manufacturing enterprises. Larger operations were almost always located in the larger cities, including the greatest producing metropolises of Mexico City, São Paulo, and Buenos Aires, representing the region's three greatest manufacturing countries of Mexico, Brazil, and Argentina. But maquiladoras are being situated where location serves them best: close to the border of the United States—the principal destination for most of their output—or deep within areas like the Yucatán where labor costs are lowest. It should be noted that the most numerous manufacturing establishments in Latin America are not like maquiladoras, but are household enterprises and small factories that employ fewer than a dozen workers and sell their products—such as textiles, ceramics, and wood products—mainly in home markets (Figure 19.23).

Sending Money Home

More and more Latin American families, particularly from Mexico, Central America, and the Caribbean, have come to rely on having at least one member work abroad to help out the family economy. Among the largest sources of income for Latin American economies are the **remittances**, or earned savings sent home by people working abroad, especially in the United States. In 2004, Latin Americans—typically working two jobs and earning less than $20,000 per year—sent home more than $30 billion, typically in periodic electronic transfers of $200–$300 each.

660

The greatest numbers of these estimated 6.6 million immigrant workers, of whom perhaps a third are **illegal** or **undocumented workers,** are in California, New York, Texas, Florida, and Illinois. An estimated 60 percent of them send monies home, in many cases to grandparents caring for the working parents' children; families often remain divided for many years while the father or both parents work abroad. Collectively, these remittances are so valuable—far surpassing the amount supplied to Latin America in foreign aid and foreign investment—that home countries are searching for ways to tax the income or otherwise find ways to funnel more of it into national development. Host and home countries both are searching for ways to make sending money home less expensive; prevailing systems often charge as much as $25 in transaction fees for a transfer of $200–$300.

The Tourism Business

During what many North Americans see as their too-long winters, they are bombarded by television and print advertisements of vacation in paradise: Jamaica and hosts of other destinations of the Caribbean basin and western Mexico. Tourism has become a major regional economic asset to Latin America, generating critical foreign exchange. A wide variety of vacation experiences are available, from the hedonistic resort complexes that tourists seldom leave (one resort on Jamaica actually is called "Hedonism"), to the insular sea voyage that offers day trips ashore in exotic destinations, to white water and rain forest adventures of the "ecotourist" variety (Figure 19.24). The region benefits from the proximity of its year-round warmth to the wealthy and mobile and seasonally chilled U.S. and Canadian populations. Tourism revenues reflect that **distance/decay** relationship: The highest tourism receipts in Latin America flow to Mexico, the nearest neighbor to the wealthy countries, but tend to fall off for more far-flung destinations. Countries in which tourist dollars amount to more than 50 percent of the nation's foreign exchange include Antigua and Barbuda, the Bahamas, and Barbados. Only the export of petroleum generates more foreign exchange overall for the region than does tourism.

19.5 Geopolitical Issues

The central geopolitical realities of Latin America are that it is in America's backyard and that, for better or worse, the United States has essentially staked its geostrategic claim to the region. Much of the United States is comprised of land wrested from Mexico in conflicts of the early 19th century. In 1823, U.S. intentions for Latin America were formulated as a policy known as the **Monroe Doctrine,** in which American President James Monroe said the United States would prevent European countries from undertaking any new colonizing activities in the hemisphere. Many subsequent interventions, wars, investments, and other American activities may be seen in light of this seminal policy. Even more consequential was the 1904 **Roosevelt Corollary** to the Monroe Doctrine, in which the United States declared it had the power to supervise the internal affairs of Latin American countries to ensure U.S. national security.

Ever since the Monroe Doctrine was established, the United States has had a particularly firm hand in Central America. Washington has dealt unapologetically and sometimes harshly with perceived threats to the United States from this region. Examples discussed in the following chapter include U.S. actions to support governments against nationalist insurgencies or insurgents against leftist governments: El Salvador and Nicaragua in the 1980s, Cuba in the 1960s, and Nicaragua in the 1950s were the most significant cases.

American hegemony in Central America was best symbolized by the Panama Canal. One of the world's most economically significant chokepoints, the canal is located centrally in this region. A 14-day transit via the canal cuts some 8,000 miles (12,800 km) and 20 days from the sea voyage between New York and San Francisco (Figure 19.25). That translates into enormous cost and time savings.

The Panama Canal is a product of U.S. commercial interest and investment, having been constructed between 1904 and 1914 by American contractors. For decades, it was sovereign U.S. territory, carefully monitored by an American military presence, and the United States kept more than half of the transit fees. In the late 1970s, however, the administration of U.S. President Jimmy Carter accepted the Panamanians' complaint that U.S. control of the canal was an outmoded vestige of colonialism. Carter negotiated the transfer of authority over the canal to Panama in a treaty that went into effect in 1999. Now, the revenues from ships transiting the canal are exclusively Panama's.

From a strategic and economic defense standpoint, however, the United States is still deeply involved in the canal's

219

Figure 19.24 Disney cruise ships call at the Bahamas' Castaway Cay, an island owned by this American company.

Joe Hobbs

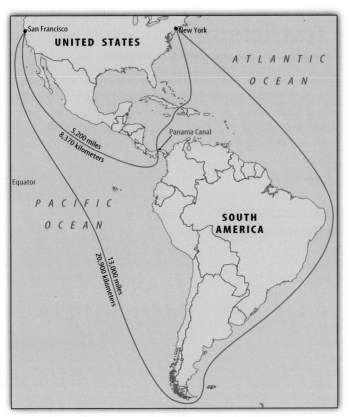

Figure 19.25 Distance, time, and money are saved by use of the Panama Canal.

fate. The Roosevelt Corollary was invoked even after Carter's treaty; in 1989, U.S. troops invaded Panama and ousted its ruler, Manuel Noriega, whom the U.S. accused of aiding the illegal trade of drugs into the United States. The U.S. "war on drugs" has been another central geopolitical issue of the region since the early 1970s and is discussed in the context of the South American countries most impacted by it in Geography of Drug Trafficking on pages 544–545.

The most consequential test of U.S. commitment to its interests in Latin America came with the **Cuban Missile Crisis** of 1962. Cuba is as close as 90 miles (144 km) from the United States. The frosty relations between the U.S. and Cuba's Marxist regime, led after 1959 by a firebrand revolutionary named Fidel Castro, were emblematic of the Cold War. Those tensions came dangerously close to hot war and even nuclear holocaust in 1962, when the Soviet Union began to build facilities in Cuba that would have accommodated nuclear missiles capable of striking Washington, D.C., New York, and other cities of the eastern United States. The presence of the Soviet missile silos and launchers was discovered through the use of aerial photo interpretation—a geographic research tool at that time growing in military and commercial utility. Although the missiles were ultimately withdrawn under intense pressure from the U.S. government, led at the time by President John F. Kennedy, Cuba has re-

mained a storm center of political affairs in the Western Hemisphere. After several attempts to assassinate Castro—even with the unlikely ploy of an exploding cigar—and a disastrous attempt by Kennedy's administration to overthrow him by military force (in an incident known as the **Bay of Pigs Invasion**), the U.S. administration settled on a course of economic and political isolation of Cuba.

Despite the resumption of diplomatic and economic relations with most of the remaining Communist countries in the world—China most notable among them—the United States has remained steadfast in imposing sanctions on Cuba. This anomaly to some extent reflects domestic political realities in the United States, where a large population of **Cuban Americans** (including 800,000 in Florida alone) who fled Castro's Cuba tend to vote for the politicians who keep up the most pressure against Castro. The Cuban Americans are against any relaxation on the effective ban on American tourism to Cuba, which is in place because of the **Trading with the Enemies Act** that forbids Americans from spending more than $300 in Cuba. The government of Cuba would like to see the country opened to American tourists. It would also like to see the United States withdraw from Guantánamo Bay, an enclave the U.S. secured in the 1903 treaty that addressed Cuba's fate following the Spanish-American war. The United States dutifully goes through the motion of paying Cuba a $4,000 annual fee for the lease of Guantánamo, and Cuba dutifully refuses to accept it. After 9/11, the U.S. established a prisoner camp for suspected al-Qa'ida and other terrorists at Guantánamo. Camp Delta's extraterritorial status allows the United States to suspend the usual legal rights awarded to criminal suspects, and is very controversial.

Several U.S. interests in Latin America converge on the nation of Colombia, now the fourth largest recipient of U.S. aid after Iraq, Israel, and Egypt. While the war on drugs was long the dominant concern, U.S. priorities focus increasingly on securing access to Colombia's considerable petroleum reserves. The United States already obtains more of its oil from Latin American than from the Middle East and views Colombia, Venezuela, and Mexico in particular as long-term counterweights to the volatile Middle East oil region.

Washington's thinking is that the best way both to obtain Colombia's oil and stem the drug tide is to take on Colombia's main rebel army, the Revolutionary Armed Forces of Colombia (FARC; see Chapter 21, page 576). FARC is the quintessential **narcoterrorist organization,** deriving most of its funding from the profitable cocaine industry based mainly in the areas it controls. But FARC and another rebel group, the ELN, also control areas containing much of Colombia's estimated 37 billion barrels of petroleum. Oil already accounts for about a quarter of the country's legal annual revenues, but Colombia dearly needs to address its economic problems by boosting oil production. FARC is making that difficult by attacking the petroleum infrastructure, especially

Geography of Drug Trafficking

The War on Drugs

From the Nixon administration of the early 1970s to the present, the United States has led a determined, expensive effort known as the **war on drugs**. The Nixon administration was the only one to focus significant efforts on combating demand for heroin, cocaine, and other drugs within the United States. All successive administrations have concentrated their efforts on the supply side, hoping to eradicate drugs at their sources—particularly in South America. Although it is far from over, the war on drugs has produced some definitive and sometimes predictable patterns (Figure 19.D).

Drug Production and Trafficking in Latin America

Legend:
- Main opium-producing areas
- Main coca-producing areas
- Major heroin trafficking route
- Other heroin trafficking route
- Major cocaine trafficking route
- Other cocaine trafficking route
- **0.3** Metric tons of heroin seized in 2002
- **5.1** Metric tons of cocaine seized in 2002

UNITED STATES 2.8 106
CANADA 0.2
MEXICO 0.3 12.6
CARIBBEAN 0.2 13
CENTRAL AMERICA 0.2 12.9
VENEZUELA 0.6 17.8
SURINAME 0.3
COLOMBIA 0.8 119
ECUADOR 5.1
BRAZIL 9.6
PERU 14.6
BOLIVIA 5.1
PARAGUAY 0.2
CHILE 2.3
ARGENTINA 1.6

to Europe
to Africa

WORLD REGIONAL
Geography Now™

Figure 19.D Production and trafficking of drugs in the Americas *See an animation based on this figure, and take a short quiz on the facts or concepts.*

the 500-mile long (800-km) pipeline between Canon Limon and Covenas. That pipeline is partially owned by Occidental Petroleum of Los Angeles. Occidental and other U.S. companies, including Exxon-Mobil, ChevronTexaco, and ConocoPhillips Petroleum, have spent tens of millions of dollars lobbying the U.S. government to help protect Colombia's oil from the rebels and therefore allow American firms to operate more freely and safely in the country.

Their lobbying has paid off. With U.S. funding and train-ing in an operation called **Plan Colombia,** which began in 2000, Colombian forces are undertaking conventional drug war tactics, such as stepped up herbicide spraying of coca crops, confiscation of traffickers' assets, and shooting down suspected drug-carrying planes, but are focusing much more on taking on FARC to both reduce drugs and secure oil. American military aid, previously targeted only against the drug trade, is now being used against the insurgents in an operation called **Plan Patriot.** It allows 800 U.S. military per-

One of the most consistent geographic patterns is the so-called "**balloon effect**," referring to how, when a balloon is pressed in one place, the air shifts elsewhere. In the case of drugs, concentrated eradication efforts in one place almost always shift production elsewhere, sometimes even from one continent to another. During the Nixon years, for example, U.S.-funded efforts succeeded in all but eliminating production of opium poppies (the plant from which heroin is derived) in Turkey. With demand still high in the United States, Europe, and elsewhere, however, poppy production surged in a region where it had been only modest: the "Golden Triangle" of Myanmar, Laos, and Thailand. Soon, increasing pressure on poppy producers in the Golden Triangle, the Bekaa Valley of Lebanon, and briefly in Afghanistan led to the eruption of the poppy industry in Mexico, Guatemala, and Colombia, oceans away from its Old World hearth.

In South America, increasingly successful eradication of coca (*Erythroxylon coca*, the shrub from which cocaine is processed) in the 1990s and early 2000s helped to diminish cocaine production in Bolivia and Colombia, but production grew in Peru, and within Colombia the eradication efforts led to some remarkable new developments in the industry. Colombia's coca crop had expanded threefold from 1995 to 2001, but there was a 21 percent decrease in coca growing there in 2003. The apparent successes in eradication have had two consequences. Cultivation dispersed into smaller and more widely scattered plots, often in new areas (including lowland Amazonia and in national protected areas, perhaps accounting together for 10 percent of total output). In addition, growers practiced selective breeding to produce plants that contain more active alkaloids, are more resistant to defoliants, and can grow with more shade cover, thereby escaping detection from the air.

Much of the force behind the balloon effect is economic: When eradication is successful in one area, supply falls and even more value is added to the already high value of cocaine and heroin. Growers and traffickers are easily tempted by the money to be made under such market conditions. Prices of

heroin and cocaine in the mid-2000s are low, and the purities of the drugs are high, testifying to a glut of both on world markets and suggesting that the war on drugs has so far not been won.

Alternative proposals to the long-fought, conventional war on drugs typically have two elements. First, much more attention needs to be paid to the demand side, using education, treatment, and other means to wean addicts from drugs and discourage others from using them. The second element focuses on the supply side. Peasant farmers who grow drugs are generally not trying to get rich (they do not, although many others up the supply chain do) but to secure a livelihood where jobs are scarce. Few crops offer the reliable market and certain returns that coca or poppies do, and **crop substitution** schemes to get farmers to switch to flowers or coffee almost always fail. Successful substitutions have occurred only in the context of so-called **alternative development** schemes, in which new crops are just one part of an overall effort to enhance the education, health, and livelihoods of rural communities.

A successful war on drugs, whether won though efforts on the supply side, demand side, or both, would have enormous environmental benefits, among others. Clearing of forest for new cultivation of coca has done enormous damage to South American forests, particularly on the steep eastern slopes of the Andes where probably no other cultivation would ever be attempted. The processing of coca leaves to coca paste and finally powdered cocaine involves the use of highly toxic chemicals that make their way into water supplies, harming ecosystems and human health. A successful war on drugs could also help to restructure in a positive way how the United States interacts with the countries of Latin America. At present, those countries deemed to be uncooperative in the war on drugs are subject to **decertification**, meaning that U.S. aid and investment in those countries are curtailed. This punishment is sometimes arbitrary; a country like Mexico often fails all the tests for certification but is too important to U.S. interests to be decertified.

sonnel and 600 civilian contractors to be on the ground in Colombia, working to subdue FARC's force of 18,000.

Between about 1980 and 2000, U.S. administrations exercised a very hands-on policy in Latin America. Through a combination of blunt U.S. military interventions and the countries' own efforts, Latin America witnessed an overall transformation of political systems from authoritarian to multiparty democratic. In the thinking of successive U.S. administrations, those greater political freedoms should have

ushered in more prosperity. This was one component of the so-called **Washington Consensus**. In this political-economic philosophy, the United States would press the democratic governments of Latin America to open markets to international trade, reduce tariffs, expand quotas, sell off state companies to private investors, allow more foreign investment, cut government spending, and reduce bloated bureaucracies. These reforms would bring the prosperity that would solve the region's problems. If serious problems emerged, the

United States could help; it did act, for example, to prevent a financial crisis that shook Mexico in 1994 from spreading.

After 2000 and especially after 9/11, aside from its boots on the ground in Colombia, the U.S. government appeared to adopt a more hands-off policy toward Latin America, perhaps distracted by regions that seemed more threatening in the war on terrorism. Since 9/11, one of the few connections between Islamist terrorism and Latin America has been the reported funding of Hizbullah and Hamas by drug monies and money laundering in the Paraguayan city of Ciudad el Este, where many immigrant Arabs live.

The Bush administration watched from the sidelines while the Argentine economy collapsed in 2001 and saw a succession of elections and polls that seemed to show a drift away from American wishes for the region. A 2004 poll conducted in 18 countries revealed that the majority of Latin Americans felt that prosperity was passing them by; 55 percent said they would support the replacement of a democratic government with an authoritarian one if it would produce economic benefits. Disillusionment with the economy seemed to be the overriding factor in Brazilians' choice in the 2002 presidential election: leftist labor union leader Luiz Inacio da Silva, known as "Lula," who vowed to fight the tide of globalization.

Similar changes in administration took place in Ecuador and Venezuela. In Ecuador, Lucio Gutierrez came to power in a 2000 coup and then, backed by a strong indigenous movement, was elected on promises of cutting poverty and scaling back market reforms. In Venezuela, President Hugo Chaves was elected in 1999, also with strong indigenous support, promising to use the country's oil wealth to fund social programs. All of these leaders have in common a stated intention to reverse the U.S.-backed push toward free trade and free enterprise, returning to state-owned companies and protected markets that would assist the poorest segments of their populations.

These leaders are supported by people who feel that the descendants of European colonizers of the New World, along with U.S. and other multinational firms, are robbing them of natural resources and any political clout. Such sentiments felt by the Native American Aymara majority of Bolivia actually succeeded in driving that country's president from office in 2003. President Gonzalo Sanchez de Lozada had been a strong advocate of a proposed development to export Bolivia's natural gas, mainly to the United States, through a pipeline that would cross Chile to a Pacific port. Native American Bolivian peasants, backed by labor unions, student groups, and opposition politicians, complained that royalties for Bolivia from the gas sales would be so low that the common people would never realize any benefit from them. They cited Bolivia's history as a colony stripped of its silver and tin to enrich the Spaniards and local aristocrats, and insisted that again only foreigners and Bolivia's elite would gain from the gas project. One unemployed miner summed up the protestors' views this way: "Globalization is just another name for submission and domination. We've had to live with that here for 500 years, and now we want to be our own masters."[4] These opponents wanted the gas to remain in Bolivia, where it could form the basis for new local industries. They also resented the pipeline passing through Chile, since one of Bolivia's historical resentments has been that Chile has cut it off from the sea. The self-proclaimed "ideology of fury" expressed by the Aymara in Bolivia is rattling nerves among the elite in other countries of Latin America with large indigenous populations such as Ecuador, Peru, Paraguay, Guatemala, and even Mexico. The following two chapters explore these and other issues among the peoples and lands of the region.

---

[4] Quoted in Larry Rohter, "Bolivia's Poor Proclaim Abiding Distrust of Globalization." *The New York Times,* October 17, 2003, p. A3.

SUMMARY

- Latin America lies south of the United States and includes Mexico, the Central American countries, the island nations of the Caribbean Sea, and all the countries of South America. It is called Latin America because of the post-1492 dominance of Latin-based languages and the Roman Catholic Church. The term *Middle America* refers to Mexico, Central America, and the Caribbean.

- A major tool for the analysis and description of Latin American landscapes and livelihoods is altitudinal zonation, which includes tierra caliente, tierra templada, tierra fría, and tierra helada. These zones go from sea level up through various altitudes, each with distinctive climate, soil, agriculture, and settlement patterns.

- Large areas of humid climates and biome types are in the lowlands of Latin America, with tropical rain forests occupying major river basins, especially in Brazil and Venezuela. Grasslands are found poleward of the tropics, and there is a zone of Mediterranean climate on the western coast of Chile.

- Dry climates and biomes, including desert and steppe, exist in some places because of orographic features that influence the movement of air masses. There are major desert regions on the west coasts of South America and northern Mexico.

- Latin America's population equals about 9 percent of the world's total, and its growth rate tends to be above the world average, especially in the Central American countries. The general demographic pattern in this region shows a majority of the people living on the rim of the land masses, an area often called the Rimland. There is also a major population band in the mountain valleys and, in some cases, on high plateaus or on the flanks of the Andes Mountains in western South America.

- Before the arrival of the European colonists and settlers in the late 1400s, Latin America had been home to some productive and highly advanced indigenous civilizations. The Aztecs, the Maya, and the Inca are the best known and most widely studied. There were also many other Native American groups that did not develop urban livelihoods. All of these cultures declined precipitously after the Europeans arrived in 1492. Although some of the losses are associated with war and with mining and other land-use demands, the biggest killer was European disease visited upon a population that had no immunity against Western diseases, especially smallpox.

- The region's major ethnic groups today are Native Americans, mestizos, mulattos, blacks, and Europeans. Political and economic systems still tend to favor people of European extraction, but non-Europeans have found increasing power through democratic political systems. Many reject the free trade and capitalist systems that they believe favor the ethnic Europeans.

- Latin America has been urbanizing rapidly, with now more than two-thirds of its population living in cities, compared to the world average of 48 percent. There continues to be steady rural–urban migration as well as a steady stream of both legal and illegal immigrants northward into the United States. The remittances of those workers are an important resource for Latin American economies.

- Latifundia and minifundia are two major agricultural systems of historical importance that continue to characterize major blocks of regional land use. There is also a strong plantation economy, especially in Central America, the Caribbean, and coastal Brazil,

and sugar and fruit play dominant agricultural roles. Farm landscapes range from peasant households that are largely subsistence farming, to haciendas, ejidos, and foreign-owned fruit and coffee plantations. A recent plunge in the price of coffee beans gave rise in Europe and later in the United States to a Fair Trade movement that raises prices for consumers of coffee, bananas, and other produce but helps ensure better wages for Latin American farmers. Ownership of farmland is concentrated in the hands of relatively few people, creating imbalances that land reform efforts have so far not redressed.

- Minerals have played a major role in the history, economic development, and trade networks of Latin America and include iron ore, bauxite, copper, tin, silver, lead, zinc, and sulfur. Gold has had local importance, and silver is at the center of centuries of European interest in the region. Petroleum is of growing importance, especially in Mexico and Venezuela.

- The Latin American countries have formed or joined several free trade agreements aimed at reducing trade barriers and promoting manufacturing. The largest in volume and value is the 1994 North American Free Trade Agreement (NAFTA). One of its goals is to promote economic growth in Latin America so that migration to the United States becomes less attractive to unemployed Latin Americans.

- Tropical Latin America's proximity to temperate and affluent North America has helped to make tourism one of the region's most important industries.

- U.S. interests in Latin America dominate the region's geopolitical themes and issues. The Monroe Doctrine and its Roosevelt Corollary were designed to justify U.S. activities and interventions in the region. Historic efforts have been directed at suppressing Communist or leftist interests in the region and at safeguarding passage through the Panama Canal. Modern interests focus on promoting trade, fighting drug trafficking, and guaranteeing secure access to oil in Venezuela and Colombia.

KEY TERMS + CONCEPTS

Terms in blue are also defined in the glossary.

alternative development (p. 545)
Aztec-Tanoan language family (p. 531)
Aztecs (Mexica) (p. 531)
balloon effect (p. 545)
Bay of Pigs Invasion (p. 543)
Central American Free Trade Agreement (CAFTA) (p. 541)
Chibcha (p. 531)
chokepoint (p. 542)
Cold War (p. 543)
core (p. 524)
creole languages (p. 532)
crop substitution (p. 545)
Cuban-Americans (p. 543)

Cuban Missile Crisis (p. 543)
culture hearth (p. 530)
decertification (p. 545)
distance/decay (p. 542)
Dutch (p. 532)
ejido (p. 539)
El Niño (El Niño Southern Oscillation, ENSO) (p. 529)
English (p. 532)
ethnic groups, non-indigenous (p. 535)
    black (p. 535)
    European (p. 535)
    mestizo (p. 535)
    mulatto (Creole) (p. 535)
Evangelical religions (p. 533)

Fair Trade movement (p. 538)
Free Trade Area of the Americas (FTAA) (p. 541)
French (p. 532)
hacienda (p. 536)
hinterland (p. 524)
Hokan-Siouan language family (p. 531)
hurricane (p. 528)
illegal (undocumented) workers (p. 542)
intellectual property rights (IPR) (p. 541)
Inca (p. 531)
land reform (p. 537)
latifundia (p. 536)
maquiladora (p. 540)
marginalization (p. 528)

Maya (p. 530)
Maya language family (p. 531)
megalanguage (p. 532)
minifundia (p. 537)
Monroe Doctrine (p. 542)
narcoterrorist organization (p. 543)
Native Americans (Amerindians, Indians) (p. 530)
Nazca (p. 531)
North American Free Trade Agreement (NAFTA) (p. 539)
Organization of American States (OAS) (p. 541)
Oto-Manguean language family (p. 531)
pampas (p. 522)
páramo (p. 528)

Penutian language family (p. 531)
Plan Colombia (p. 541)
Plan Patriot (p. 544)
Portuguese (p. 532)
pre-Columbian times (p. 533)
Protestantism (p. 533)
Quechu-Aymaran language family (p. 532)
    Quechua (p. 532)
remittances (p. 541)
Roman Catholicism (p. 533)
Roosevelt Corollary (p. 542)
rules of origin (p. 540)
Seventh Day Adventist (p. 533)
shantytown (*favela, barrio*) (p. 536)
Southern Cone Common Market (Mercosur) (p. 540)

Spanish (p. 532)
Tarascan (p. 531)
Teotihuacános (p. 531)
tierra caliente (hot country) (p. 527)
tierra fría (cold country) (p. 527)
tierra helada (frost country) (p. 527)
tierra templada (cool country) (p. 527)
Totonac language family (p. 531)
Trading with the Enemies Act (p. 543)
tree line (p. 528)
Triple Alliance (p. 531)
upper limit of agriculture (p. 528)
upwelling (p. 529)
war on drugs (p. 544)
Washington Consensus (p. 545)

REVIEW QUESTIONS

WORLD
REGIONAL
Geography Now™

Assess your understanding of this chapter's topics with additional quizzing and concept-based problems at http://earthscience.brookscole.com/wrg5e.

1. Where are the main areas of population concentrations in Latin America?

2. What and where are the principal climate and biome types of Latin America?

3. What are the four elevation zones? What human uses are generally associated with them?

4. What causes El Niño? What impacts are usually associated in particular regions with it?

5. What and where were the major Native American civilizations prior to 1492?

6. What and where are the major pre-Columbian and modern languages of Latin America? What are the principal faiths?

7. What are the principal cash crops and minerals exported from Latin America?

8. What are some of the main types of landholdings? What crops are associated with them?

9. What are the region's most important existing and planned free trade agreements?

10. What are remittances? What role do they play in Latin American economies?

11. Which of the region's countries benefit most from tourism revenue?

12. Tourism and mineral extraction both relate very closely to geographic conditions and characteristics. What traits are important in each and how are they expressed in Latin America?

13. List the countries included in, and explain the regional breakdown of, Latin America, Middle America, and South America.

14. What is the strategic and economic importance of the Panama Canal?

DISCUSSION QUESTIONS

1. How do the area and the extent of latitude and longitude of Latin America compare with those of the United States and Canada?

2. Demographically and economically, how does Latin America generally fit the profile of a region of less developed countries?

3. What were some of the major culture traits and legacies of the pre-Columbian civilizations in the region?

4. What were some of the major impacts of the Spanish conquest on Native American populations?

5. What are the four main racial types of Latin America? Where

do these groups tend to live? What, generally, is their economic and political status?

6. Why do Latin American countries want to move away from exporting raw materials and develop more manufacture-based economies?

7. What are Fair Trade products? What advantages do they have over conventionally traded products?

8. What kinds of land reform efforts have been undertaken in the region? Is there still an apparent need for land reform?

9. What are the main goals of free trade organizations? What are

some of the results of these groups in Latin America (and in the case of NAFTA, in the United States)?

10. What are the Monroe Doctrine and the Roosevelt Corollary? How have these policies been implemented in Latin America?

11. What are the principal interests of the United States in Colombia that might justify the huge U.S. aid investment there?

12. What are the principal efforts of the war on drugs? What have been some of its results? What might be the essential elements of a successful war on drugs?

13. What was the significance of the Cuban Missile Crisis?

14. What is the Washington Consensus? Has its sequence of accomplishments been realized in some Latin American countries? How do some people in the region apparently feel about its logic and about the roles of democracy and free trade?

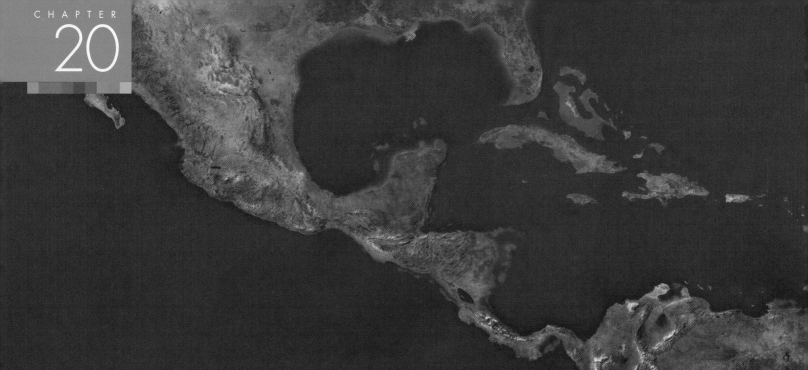

Middle America:
Land of the Shaking Earth

A girl named Lolita, of the Tz'utujil Maya, Santiago de Atitlan, Guatemala

Joe Hobbs

---

chapter objectives

This chapter should enable you to:

- Recognize how maldistribution of resources, particularly of quality farmland, has contributed to dissent and war

- Understand how the economic boom in China, where labor costs are very low, has taken a toll on Mexico's industries

- Recognize how fertile volcanic uplands came to be the core regions of most of the mainland countries

- Appreciate the vital role played by tourism in many of the nations' economies

- Understand how the limited resources of small island countries hampers their economies and leads them in search of Internet, banking, and other alternative ventures

WORLD
REGIONAL
Geography⊛Now™

Look for this logo in the text and go to GeographyNow at http://earthscience.brookscole.com/wrg5e to explore interactive maps, view animations, sharpen your factual knowledge and geographic literacy, and test your critical thinking and analytical skills with unique interactive resources.

The region most geographers prefer to call "Middle America," also known as Mesoamerica, includes Mexico (its largest country in area and population); the much smaller Central American countries of Guatemala, El Salvador, Belize, Honduras, Nicaragua, Costa Rica, and Panama; and the numerous island countries in and near the Caribbean Sea, the largest being Cuba, Haiti, the Dominican Republic, and Jamaica. The region is a vibrant patchwork of physical features, ethnic groups, cultures, political systems, population densities, and livelihoods. Most of the political units are independent, but a few are still dependencies of outside nations. This survey begins with the demographic and territorial leader, Mexico.

This is a region of spectacular natural landscapes, from highland mountain chains and towering, isolated volcanoes to lush, seaside tropical rain forests. Its location astride major zones of collision between plates of the earth's crust (see Figure 2.9) is responsible for much of the region's natural beauty, but has also brought suffering in the form of earthquakes and volcanic eruptions. Anthropologist Eric Wolf wrote a fine depiction of the Guatemalans and Mexicans living in the heart of the region entitled *Sons of the Shaking Earth*, which seems an appropriate name for all the peoples of Middle America.

20.1   Mexico: Higher and Further

The federal republic of Mexico—officially, *Los Estados Unidos Mexicanos* (United Mexican States)—is the largest, most complex, and most influential country in Middle Amer-

ica (Figure 20.1). The country's great population of 106 million in 2004 makes Mexico by far the world's largest nation in which Spanish is the main language. Compared to most of the world's countries, Mexico is a spatial giant, but its large size tends to go unrecognized because it lies in the shadow of the United States. Triangular in shape, Mexico's area of 756,000 square miles (c. 2 million sq km) is nearly eight times that of the United Kingdom, and its elongated territory would stretch from the state of Washington to Florida if superimposed on the United States. Its capital, Mexico City, stands now as the world's second largest city in population (after Tokyo), with a whopping 22 million inhabitants. Mexico is the most urbanized country of Middle America, with about 75 percent of its people living in cities. In these and many other attributes, Mexico in the Latin American context seems to live up to its motto: "Higher and Further."

Human Geographies

Mexico's modern cultural patterns are heirs to three major eras: the long Native American era prior to the Spanish conquest, the Spanish colonial era of the 16th to the early 19th centuries, and the Mexican era following Mexican independence in 1821. In the Native American era, the region was home to indigenous peoples speaking hundreds of languages and dialects. Many of these cultures were very advanced and among other things produced the remarkable architectural legacies of Teotihuacán and Chichén-Itzá. Richly varied indigenous cultures persist today (Figure 20.2). While some of the Native American groups are racially almost identical to their pre-Columbian ancestors, the greater part of the

530

Mexico

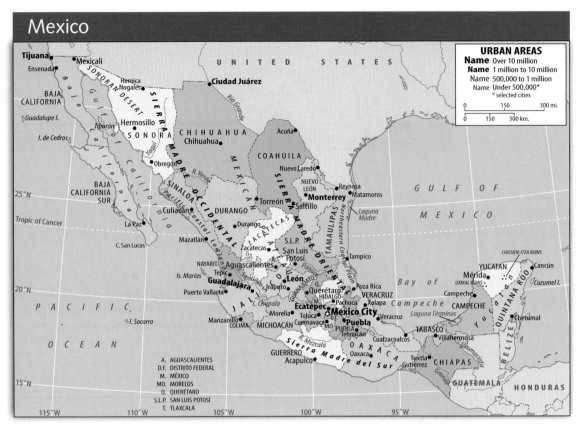

URBAN AREAS
Name Over 10 million
Name 1 million to 10 million
Name 500,000 to 1 million
Name Under 500,000*
* selected cities

Figure 20.1 Principal features of Mexico

Figure 20.2 This weaving was made by a Native American Huichol in north-central Mexico. This art reflects the artist's visions prompted by ritual use of peyote.

people of Mexico are *mestizos* of mixed Spanish and Amerindian ancestry. Of Mexico's population, Native Americans (locally called *indios* or *indigenas*) make up 30 percent, mestizos 60 percent, and ethnic Europeans (nearly all of Spanish ancestry) only 9 percent.

The Spanish colonial era lasted nearly four centuries. Following the overthrow of the Aztec empire by Hernando Cortés in 1521, the Spanish Crown established the **Viceroyalty of New Spain.** Ruled from Mexico City, the viceroyalty grew to encompass the area of modern Mexico, most of Central America, and about one-quarter of the territory of today's 48 conterminous United States. The **Hispanic** pattern of life still prominent today was established by Spanish administrators, fortune hunters, settlers, and Catholic priests. Spanish is spoken as a first language by 88 percent of the populace, and 89 percent practice the Roman Catholic faith that the Spaniards introduced into the region.

The present settlement pattern was strongly influenced by the distribution of Native Americans at the time the Spanish arrived. The conquerors sought out concentrations of Native Americans who would become laborers and Christian converts; thus, Spanish settlements tended to grow near existing population centers or were established as port cities with access to inland aboriginal settlements and resources. Innumerable towns and cities of today originated essentially as Catholic missions in the midst of indigenous populations. Spanish influences are found in architectural styles and urban layouts. Towns and cities characteristically have a Spanish-inspired rectangular grid of streets surrounding a plaza at the town center. The plaza functioned both as the major site for periodic markets and as a setting for the dominant Catholic church in the settlement (see Figure 19.15). The Spanish hunger for precious metals greatly expanded mining, particularly of silver, and many of Mexico's larger cities were founded as silver-mining camps or regional service centers for these mining areas.

When the Mexican era began in the early 19th century, Mexico was overwhelmingly poor, rural, illiterate, village centered, and church oriented. In the capital, an educated elite governed and exploited the country with little benefit to the Native American and mestizo farmers in the countryside. The bloody **Mexican Revolution of 1810–1821** ended Spanish control and instituted a short-lived Mexican empire that expired in 1823, to be succeeded by a federal republic in 1824. Then came a long period of political instability. From the 1830s through the 1840s, huge blocks of the nation's territory were lost to neighboring countries, including the Central American part of former New Spain, as well as Texas and areas now comprising California, New Mexico, Arizona, Nevada, Utah, and parts of other states.

During the 1850s and 1860s, a dynamic Native American reformist named Benito Juárez ruled, but then Mexico endured a long period (1876–1910) of despotic rule by President Porfirio Díaz. Under Díaz, democratic processes were suspended and the rural population sank deeper into poverty, even as heavy foreign investments were made in railroads, the oil industry, metal mining, coal mining, and manufacturing. Political discontent over national disunity and weakness, internal oppression and corruption, foreign exploitation, poverty, and cries for land reform then led to the **Mexican Revolution of 1910–1920,** led by Pancho Villa and Emiliano Zapata, which ousted Díaz and turned into a destructive struggle among personal armies of regional leaders. In 1917, a new constitution was enacted under which land reform was an urgent priority. After that, huge acreages were expropriated from *haciendas* by the government and redistributed or restored to landless farmers and farm workers, most particularly in a program that provided *ejidos,* or communal farmlands, to mainly Indian village farmers. Large haciendas still exist, especially in the dry north, but they contain only a very small proportion of Mexico's cultivated land.

Land-centered revolutions may not be over. In Mexico's poorest and southernmost state, Chiapas, dissidents mainly from several Maya ethnic groups formed a revolutionary movement called the **Zapatista Army for National Liberation (EZLN)** in 1994. Demanding indigenous autonomy and greater control over land and other resources, one of their chief complaints was the enactment that year of the North American Free Trade Agreement (NAFTA). They denounced globalization, calling Mexico's export industries a symbol of the country's slavery to rich nations. Despite arguments made that ecotourism development in Chiapas could help preserve their way of life and natural environments, the Zapatistas also denounced government tourism schemes,

saying they would turn the region into "an amusement park for foreigners." In a 2002 communiqué, the Zapatistas called ecotourism advocates "fools trying to change our lives so that we will cease being what we are: indigenous peasants with our own ideas and culture."[1] Zapatista seizures of mines and other assets led to clashes with government forces that caused numerous deaths and refugee problems. Negotiations led to the 1996 **San Andres Accords** that promised autonomy for many communities. The Mexican government has withdrawn its military from Zapatista-held areas and tensions have eased, but the Zapatistas complain that the government has not fulfilled its promises, so the situation remains volatile.

Agricultural Environments

A little over one-half of Mexico lies north of the Tropic of Cancer and is dominated by desert and steppe climates with hot summers and warm to cool winters, depending on elevation. South of the Tropic of Cancer, where about four-fifths of all Mexicans live, seasonal temperatures vary less, but elevation differences create conditions ranging from high temperatures in the *tierra caliente* of the lowlands along the Gulf of Mexico, through moderate heat in the *tierra templada* of the low highlands, to cool temperatures in the *tierra fría* of the highlands (for these elevation zones, see page 527 in Chapter 19). Nearly all the tropical areas are humid enough to support rain-fed crops, but irrigation is often used to increase crop diversity and productivity.

Agriculture is still a highly important source of livelihood and export revenue. Leading agricultural exports are sugar, coffee, corn, citrus fruit, and cattle. Corn (maize), first domesticated in Mexico about 9,000 years ago, is by far the most important food crop. Beans, squash, and chili peppers, also an inheritance from the Native American past, are universal Mexican foods, interpretations of which have become widespread in North American fast-food and kitchen cultures. Great agricultural variety exists despite the mountainous environment and the fact that only one-eighth of the country is cultivated. Good cropland exists mainly on the floors and lower slopes of mountain basins with volcanic soils and on scattered, small alluvial plains. Agricultural habitats vary according to elevation, soil types, physiographic history, and atmospheric influences. Mountains too steep to cultivate without the hazard of erosion dominate the terrain in most areas, but they are flanked by plains, plateaus, basins, and foothills that can be tilled productively where enough water is available. In many places, even the steep slopes are farmed as *milpas* (slash-and-burn plots) by peasant farmers desperate for access to land (Figure 20.3).

Mexico has a serious deforestation problem, losing about 1.3 million acres of trees annually, the fifth highest deforestation rate in the world. Much of this activity is illegal commercial logging operated by organized criminals. A con-

281

---

[1] Both quotes are from Tim Weiner, "Mexican Rebels Confront Tourism in Chiapas." *The New York Times,* March 9, 2003, p. TR3.

Joe Hobbs

Figure 20.3 Known in Mexico and Central America as *milpa,* swidden or slash-and-burn cultivation in the region requires the clearing and burning of forest before food crops like corn, beans, and squash are grown. When the soils are exhausted after 3 to 5 years, the farmer abandons the plot and moves on to clear another.

siderable amount of clearing is also done to open land for marijuana and opium cultivation. Environmentalists are particularly concerned with deforestation of the lowland tropics of Mexico's southernmost state, Chiapas.

The great biological diversity of these forests accounts largely for Mexico's ranking as the world's third most important **megadiversity country,** after Brazil and Indonesia. The jewel of these forests is the Montes Azules reserve of Chiapas, which the government of Mexico granted to a very traditional Maya group called the **Lacandon** in 1978. The Lacandon themselves were responsible for some of the destruction by granting logging leases to government-backed companies in the 1980s. More recently, the Lacandon have been blaming illegal squatters for the forest destruction in Montes Azules. Most of the squatters are indigenous peoples of the **Tzeltal Maya** and **Chol Maya** groups, and many of them are affiliated with the antigovernment Zapatista movement. Evicting them is therefore as much a political as an environmental problem.

Mining, Manufacturing, and Maquiladoras

Mexico's great environmental variety has produced a diversity of mineral resources. Mexico ranks highly among the world's producers of such commodities as metals, oil, gas, and sulfur, with oil the country's most important export. The Mexican oil industry is a government monopoly (carried on by a government corporation, Petróleos Mexicanos, or PEMEX), and the United States is its largest foreign customer, receiving fully 88 percent of the country's exports. Mexican oil represented 13 percent of oil imported to the United States in 2004, making Mexico the third largest foreign supplier.

The bonanza of oil strikes in the 1970s seemed so promising that the government borrowed heavily from foreign banks to finance both oil development and a wave of other

economic and welfare projects. Dependence on oil income grew to the point that crude oil represented two-thirds of all Mexican exports in value by 1983. Subsequently, however, a worldwide oil glut caused prices to skid, and by 1990, crude oil accounted for only one-third of exports. Economic reforms in Mexico during the 1980s reduced inflation and spurred growth, but by 1994, the country's external debt was nearly $128 billion, and a sizable share of export revenue was being spent on debt service. A collapse of the Mexican peso in 1994 was in large part due to the combination of massive foreign debt and diminishing oil revenues, a pattern that continued through the late 1990s. Nonetheless, recent restructuring of the economy and a steady increase in the value of oil exports have created a more promising outlook for the near future. As of 2004, oil exports comprised 13 percent of Mexico's exports by value. Oil production in Mexico is expected to peak in 2010 and effectively conclude by 2030.

Mexico has made substantial gains in its development of manufacturing and component assembly industries, especially since the initiation of the North American Free Trade Agreement. High-technology exports grew from 8 percent of total exports in 1990 to 22 percent in 2004. About one-third of all manufacturing employees in Mexico work in thousands of factories in metropolitan Mexico City. Most of those factories are small, and manufacturing is devoted largely to miscellaneous consumer items, although some plants carry on heavier manufacturing such as metallurgy, oil refining, and the making of heavy chemicals. The *maquiladoras* factories are typically located close to the U.S. border, but there is an important cluster in the Yucatán Peninsula, where labor is less expensive (Figure 20.4).

The maquiladora represents Mexico's effort to boost exports of manufactured goods by attracting industries from abroad with offers of tax exemptions, cheap labor, and other inducements. The maquiladora uses inexpensive local labor to manufacture finished goods, such as jeans, cars, and computers, mainly from components that can be imported duty free as long as they are assembled for export only. There has

Figure 20.4 Electronic assembly at a *maquiladora* in Tijuana

Annie Griffiths Belt/Corbis

been a surge in the growth and diversification of the maquiladoras since NAFTA was enacted; now more than 3,500 of these factories, employing more than 1 million Mexicans, are in the U.S. border region alone. They provide welcome jobs to Mexican workers. There has also been growth in industries that have grown up to support the maquiladoras, such as small-scale retail and transportation services. Among the most recent developments are the **energy maquiladoras**. These are power generation plants—built by American firms just inside Mexico, where pollution standards are much lower than in the United States—that provide electricity to the U.S. grid. Their power source, natural gas, comes from Texas, and Mexico supplies their coolant waters.

There are now signs that Mexico is beginning to lose the competitive edge represented by its inexpensive labor. More than 350 maquiladoras closed and more than 250,000 maquiladora jobs were lost between 2000 and 2004, mainly due to competition from China, especially in clothing, textile, and other low-tech industries. Many U.S. and other companies are finding much cheaper labor costs in China, and Mexico itself is being flooded with Chinese imports. India's English-speaking service sector is also more attractive to U.S. firms looking to **outsource** jobs to phone centers abroad. While conceding that they may have lost the low-tech, low-skill industrial advantage to China (and to some extent also nearby El Salvador, where labor is also cheaper), Mexican manufacturers are now studying ways to hang on to Mexico's edge in exports of flat screen televisions, computers, aircraft parts, and other more high-tech goods that are more expensive to ship from China. One of Mexico's outstanding development problems is that relatively few of its young people are gaining multilingual and multicultural educations abroad, while Chinese and South Koreans, for example, generally see study abroad as an important investment in their national economies.

Economic relations with the United States, so well symbolized by the border maquiladora, are crucial to Mexico. Trade between Mexico and the United States increased 240 percent between 1994 and 2004. Exports from Mexico to the United States have more than tripled. The trade is currently in Mexico's favor, with imports from Mexico exceeding exports to Mexico by 34 percent. The United States is Mexico's leading trade partner, with about 90 percent of its exports going to the United States. Mexico is the United States's second largest trading partner (after Canada), with 14 percent of U.S. exports bound for Mexico. Only Canada and China export more goods to the United States than does Mexico.

In spite of the NAFTA linkages, there are sensitive issues between Mexico and the United States. There are disparities between them in wealth, power, and cultural influence. Mexicans have not forgotten that Mexico lost more than one-half its national territory—including what became California, Texas, New Mexico, Arizona, Nevada, and Utah—to the *gringos* in the **Mexican War** (called the **American Invasion**

by Mexicans) in 1847. Mexicans have enormous national and cultural pride, and often feel misunderstood and underappreciated by their northern neighbors.

Regional Geography of Mexico

Mexico is strongly regionalized (for the country's physical geography, see Figure 19.4). Its major subregions are Central Mexico, the country's core; Northern Mexico—dry, mountainous, and intertwined with the United States through migration and maquiladoras; the Gulf Tropics, an area of hot, wet coastal lowlands along the Gulf of Mexico, together with an inland fringe of tierra templada in low highlands; and the Pacific Tropics, a poorly developed southern outlier of humid mountains and narrow coastal plains.

Central Mexico: Volcanic Highland Core Region

Many Latin American countries have a sharply defined **core region** where national life is centered, and Mexico is a classic example of this cultural and demographic pattern. About one-half of the population inhabits contiguous highland basins clustered along an axis from Guadalajara at the northwest to Puebla at the southeast. Toward the eastern end of the axis is massive Mexico City. The basins comprise Mexico's core region, often referred to as Central Mexico, the Central Plateau, or the Mexican Plateau. The basin floors vary in elevation from about 5,000 feet (1,524 m) to 9,000 feet (2,727 m), with each basin separated from its neighbors by a hilly or mountainous rim. Ash ejected from the region's many volcanoes has added fertility to the soils. The centers of the basins contain flat and sometimes marshy land. This swampiness has led populations to cluster on higher sloping land, and population pressure has induced cultivation of steep slopes, causing much soil erosion. Most areas are too high and cool for coffee or the other crops of the tierra templada. The prevailing tierra fría environment limits the growth of crops to those less susceptible to frost damage, such as corn, sorghums, wheat, and potatoes. Beef, dairy cattle, and poultry fare well in this environment.

In the mid-14th century, the Aztecs came to the Basin of Mexico from the north and built their capital, Tenochtitlán, on an island in the shallow, salty Lake Texcoco. The city was lined with canals and dike constructions to control flooding and protect water resources for the city, which was connected by causeways to the shore. This was the city Cortés conquered in 1521. Mexico City was later founded in this same general locale. Most of the lake was drained, and city structures have been built on the drained land over the past four and a half centuries. Gradual subsidence (settling of land and buildings) continues today as the lake sediments contract. In recent times, overpumping of ground water to support a metropolitan region of more than 20 million people has aggravated this problem. But despite the unstable subsurface, and the city's presence atop an active geologic fault zone, Mexico City sports imposing office towers. Fortunately, most have been built with foundations designed to withstand both subsidence and earthquake shocks. These generally withstood a strong test in the form of a magnitude 8.1 earthquake that rocked the city in 1985, killing 10,000 people and rendering 250,000 homeless.

The city has other planning achievements to its credit, including a modern subway system and a system to pump water to the city through the encircling mountain walls of the Basin of Mexico. Overall though, explosive urbanism has outrun the efforts of Mexico City's planners to cope with the capital city's growth. The city's inhabitants must cope with the all too familiar difficulties of developing world **megacities,** including dreadful air pollution, compounded in this case by its topographical setting. Being situated low in a valley with mountains surrounding, Mexico City (like Los An-

geles) often experiences a **thermal inversion.** This is a circumstance in which air that is cooled at night sinks down the mountainsides surrounding the city and settles underneath a layer of warmer air (Figure 20.5). The cooler air below is trapped by the warm air above, and without air circulation, the pollution from factories and millions of cars often rises to dangerous levels.

Rugged Northern Mexico and the Borderlands

North of the Tropic of Cancer, Mexico is characterized by ruggedness, aridity, and a ranching economy that supports a generally sparse population. There are some widely separated spots of more intense activity based on irrigated agri-

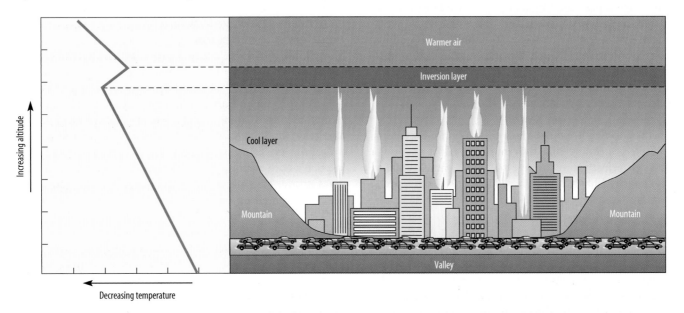

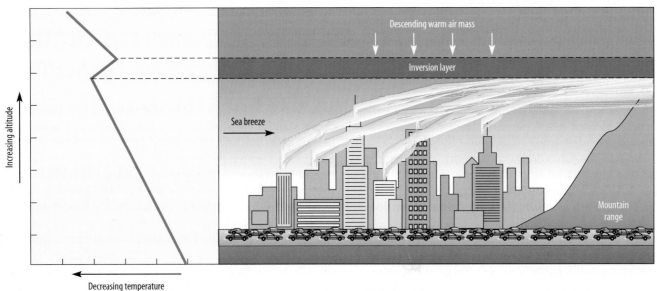

Active Figure 20.5 Model of the thermal inversion responsible for smog in Mexico City (above) and Los Angeles (below). *Source: From* Living in the Environment, *13th edition, by Miller. © 2004. Reprinted with permission of Brooks/Cole, a division of Thomson Learning, Inc. See an animation based on this figure, and take a short quiz on the facts or concepts.*

WORLD REGIONAL Geography⊛Now™

culture, mining, heavy industry, and diversified border industries adjacent to the United States. Most of Northern Mexico lies above 2,000 feet (610 m) in elevation, but there are lowland strips along the Gulf of Mexico, the Gulf of California, and the Pacific. The climate is desert or steppe, except where elevations are high enough to yield more rainfall and there are coniferous forests.

Two major north–south trending ranges of high mountains, the Sierra Madre Occidental in the west and the Sierra Madre Oriental in the east, tower above the surrounding areas. Surface transport across them is poor. Within Northern Mexico as a whole, ranching is a widespread source of livelihood, although the total number of people supported by it is small. Most of the forested land is in the Sierra Madre Occidental, which serves as the country's main source of timber.

Northern Mexico provides the country's principal output of metals, coal, and products of irrigated agriculture. Iron ore, zinc, lead, copper, manganese, silver, and gold are the main metals produced. The ores are smelted with coal mined in the region. Some coal is made into coke and used for the manufacture of iron and steel, particularly around the city of Monterrey, Mexico's leading center of heavy industry and Northern Mexico's most important business and transportation hub.

Irrigated agriculture in the dry north has expanded with the building of dams to store the water of rivers originating in the Sierra Madre and emptying into the Gulf of California and the Rio Grande. The main clusters of irrigated land are located in the northwest and along or near the Rio Grande. Cotton is a major product of Northern Mexico's irrigation, and varied fruits and vegetables (notably tomatoes and cucumbers) for shipment to U.S. markets also are grown. In some irrigated areas, the Mexican government has helped develop agricultural colonies of small landowning farmers who have migrated from more crowded parts of the country.

There has been a large upsurge of manufacturing, particularly since the enactment of NAFTA and the boom in maquiladoras, in the widely separated **border towns** adjacent to U.S. cities. The largest are Tijuana, adjacent to San Diego, California, and Ciudad Juárez, across the Rio Grande from El Paso, Texas. Much of the traffic in illegal drugs from South America and Mexico itself makes it into the United States through these border towns. The border towns are sin cities for U.S. citizens interested in indulging their pleasures in ways they would not dare at home: Prostitution is legal, the drinking age is 18, prescription drugs are sold under the counter, and illegal drugs are easy to obtain.

These cities are also collecting points for Mexican and Central American migrants who attempt to enter the United States illegally on the countries' common 2,000-mile (3,200-km) border, often succeeding despite the efforts of the U.S. Border Patrol. Periodic amnesties have pardoned illegal migrants who could prove long-term residence in the United States, but current laws try to halt the illicit flow. The draw of *El Norte*—the North—can be easily appreciated considering the ratio of 4 to 1 in the per capita GDP PPP figures for the United States and Mexico.

The Gulf Tropics: Oil and Tourists

On the Gulf of Mexico side, Mexico has tropical lowlands that have a much lower population density than Central Mexico but which are very important to the national economy. Their most valuable product is oil, discovered in huge quantities in the mid-1970s. Just as in coastal Texas and Louisiana, some fields are onshore and others lie underneath the gulf.

The Gulf Tropics are Mexico's main producer of tropical plantation crops. Cacao, sugarcane, and rubber are leading export crops in the lowlands, and coffee exports come from a strip of tierra templada along the border between the lowlands and Mexico's Central Plateau.

The greater part of Mexico's Gulf Tropics lies in the large Yucatán Peninsula, where the Maya civilization flourished in pre-Columbian times. The peninsula's east coast, fronting the Caribbean, is being developed and promoted as the "Maya Riviera." From Cancún southward, there is a swath of tourist resorts of varying degrees of harmony and disarray with the landscapes and seascapes (Figure 20.6). The sultry and exotic environment is an irresistible draw for U.S. tourists, who make up 80 percent of Mexico's foreign visitors and whose dollars provide Mexico with much needed revenue. Tourism overall is Mexico's third source of foreign currency earnings (after oil and remittances from Mexican workers abroad), with about 20 million visitors in 2004 generating an estimated $9 billion in income for the country.

The Pacific Tropics: Mountainous Southwestern Outlier

South of the Tropic of Cancer, rugged mountains with a humid climate lie between the densely populated highland basins of Central Mexico and the Pacific Ocean. This difficult terrain is far more thinly populated than Central Mexico. Native Americans carry on a traditional agricultural livelihood in this area of otherwise little economic activity. There are two exceptions to the comparative underdevelopment. One

Figure 20.6 Tourism developments are transforming the Caribbean coast of the Yucatán.

is the coastal resort of Acapulco, an international tourist destination located on a spectacular bay about 190 miles (c. 300 km) south of Mexico City. There is also a major iron and steel complex about 150 miles (240 km) up the coast from Acapulco near La Unión. Its operation is based on local deposits of iron ore.

20.2 Central America: Beyond Banana Republics

Between Mexico and South America, the North American continent tapers southward through an isthmus containing seven less developed countries known collectively as Central America. Five of these countries—Guatemala, Honduras, Nicaragua, Costa Rica, and Panama—have seacoasts on both the Pacific and the Atlantic Oceans (via the Caribbean Sea; Figure 20.7). The other two have coasts on one ocean only: El Salvador on the Pacific and Belize on the Caribbean. If the seven countries were one, they would have an area only about one-fourth that of Mexico (see Table 19.1), with a total population about 40 percent that of Mexico.

Fragmentation and Turbulence

Geographic fragmentation, manifest both in physical and political ways, is a major characteristic of Central America. Five countries—Guatemala, El Salvador, Honduras, Nicaragua, and Costa Rica—originally were governed together as a part of New Spain in a unit called the **Captaincy-General of Guatemala.** Composed of many small settlement nodes isolated from each other by mountains, empty backlands, and poor transportation, the unit never established strong geopolitical cohesion. Separate feelings of nationality became strong enough to fracture overall unity after the end of Spanish imperial rule in 1821. The area gained independence

from Mexico in 1825 as the **Central American Federation,** but in 1838–1839, this loose association segmented into the five republics of today. In recent decades, there were serious international tensions affecting the five countries, associated in part with antigovernment guerrilla warfare in various countries. However, by the mid-1990s, tensions had lessened, and the Central American area appeared to enter a new era of greater political stability.

In Guatemala, El Salvador, Honduras, and Nicaragua, the population is composed of groups that differ sharply in wealth, social standing, ethnicity, culture, and worldview. Wealthy owners of large estates—the latifundia—formed a social aristocracy that has lasted since the Spanish Crown made the first land grants in early colonial times. Members of this group tend to be ethnically European (Spanish). They have, in general, monopolized wealth, defended the status quo, and exercised great political power. Occupying a middle position in society are small landholding farmers, largely mestizos, who own and work their own land. At the bottom of the scale are millions of landless mestizo tenant farmers, hired workers, and the many Native Americans (especially in Guatemala) who live essentially outside of white and mestizo society. Also low in the economic ranks are blacks and mulattos descended from the black African slaves brought to the region beginning in the 16th century, and from the migrants who came later from Jamaica to work in the region's banana plantations. Poverty, unemployment, a young population, and weapons left over from the region's civil wars have provided ideal conditions for the emergence of an estimated 25,000 gang members that wreak havoc in the big cities of Honduras, El Salvador, and Guatemala.

WORLD
REGIONAL
Geography Now™

Click Geography Literacy to take a virtual tour "Study Abroad . . . in Guatemala . . . or Somewhere?"

There is popular sentiment in this region that the government has too much concern for governmental cabinets and high offices and too little for the needs of the common people. Central America historically has had revolutionary movements that embody these fears, and the recent histories of Nicaragua, El Salvador, and Guatemala have demonstrated a potential for popular uprising and often for U.S. intervention. In Nicaragua in 1979, the governing Somoza family—with a long history of U.S. support—was thrown out by the country's **Sandinista** revolutionaries. After a very contentious 11-year rule characterized by continual battles with the military-backed **Contras** (also supported by the United States), the Sandinistas were replaced by an elected president, and by the early 1990s, a reasonably stable government was established with some of the earlier Sandinista reforms institutionalized.

In El Salvador, a reformist-minded government attempted in 1980 to redistribute some of the country's better farmlands from the wealthy elite that controlled them to rural

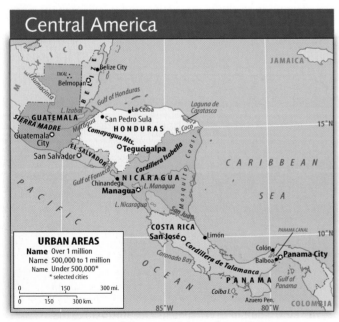

Figure 20.7 Principal features of Central America

peasant cooperatives. A violent civil war ensued between the right-wing landowners, supported by the military, and the left-wing peasants. An estimated 75,000 people were killed and 1 million made homeless in the 12-year conflict. Since the war ended, El Salvador has become a stable democracy with a reasonably strong economy based on clothing exports.

A 36-year civil war that finally ended in 1996 took more than 200,000 lives in Guatemala. The bloodletting began when the U.S. Central Intelligence Agency undertook a coup to oust the country's president. President Guzman had resisted the powerful influence of the **United Fruit Company (UFC)** in Guatemala, confiscating 400,000 acres of land from UFC with an aim to subdivide it among the landless poor. American President Eisenhower interpreted this action against a U.S. company as a sign that Guatemala might drift toward communism and ordered the coup. Most of the killings in the subsequent civil conflict were carried out by government forces in a scorched earth campaign against leftist rebels. The army leveled more than 400 villages and massacred tens of thousands of civilians, mainly from among the country's poor, disenfranchised, Native American Maya.

Costa Rica has had a different historical development from the other four countries of the old Central American Federation. Far from the center of the Central American Federation in Guatemala, it was left largely to its own devices in developing its economy and polity. Early Spanish settlers eliminated the smaller numbers of indigenous people there, and the ethnic pattern that developed was dominated overwhelmingly by whites of Spanish descent. More than 85 percent of the present Costa Rican population is white, much in contrast with the predominance of mestizos and/or Native Americans in El Salvador, Honduras, Nicaragua, and Guatemala. No precious metals of significance were found in Costa Rica, and no large landed aristocracy developed. The basic population of small landowning European immigrant farmers and their descendants managed to create one of Latin America's most democratic and economically productive societies, despite some episodes of dictatorial rule.

Costa Rica is also well known for its minimal military expenditures and its strong efforts to protect its natural heritage (and in the process, to lure ecotourists who spend welcome sums). National parks and other protected areas make up nearly one-quarter of the country, and tourism is the leading source of foreign currency. Costa Rica has a border dispute on its northwestern frontier with Nicaragua, where a separatist group would like to establish an independent country called Airrecu that would thrive on ecotourism income.

Panama and Belize were not part of the Central American Federation and achieved independence much later. Panama was governed from Bogotá as a part of Colombia until 1903, when it broke away as an independent republic. The United States was deeply involved in the political maneuvering that led to Panamanian independence. In 1904, the United States was granted control over the Panama Canal Zone and undertook the construction of the transoceanic Panama Canal, which opened in 1914 (see Regional Perspective, page 560).

Belize, on the Caribbean side of Central America, was held by Great Britain during the colonial era under the name of British Honduras. Remote, poor, underdeveloped, and sparsely inhabited, the colony was used mainly as a source of timber, with African slaves brought in to do the woodcutting. Guatemala also claimed Belize as part of its national territory (and the two countries still have a disputed border). Belize become independent in 1981 as a parliamentary state governed by a prime minister and bicameral legislature. The British monarch, represented by a governor-general, is the chief of state, and English is the official language. Ecotourism is a growing industry. Both the pristine tropical rain forests of the country and the laid-back culture of its islands, called *cayes* (pronounced KEYS), draw visitors from North America and Europe (Figure 20.8). There are concerns that the Chalillo Dam, a hydropower project under construction on the pristine Macal River, will diminish water quality, flood rain forests, and deter ecotourists. Dozens of similar dams

Figure 20.8 Large parts of Belize have not been cleared of their tropical rain forests, and the country is capitalizing on its potential to draw nature-loving tourists. This is the Macal River, downstream from where the controversial Chalillo Dam is under construction.

Regional Perspective

The Panama Canal

The Panama Canal, providing an inter-ocean passage through the narrow Isthmus of Panama, is of great importance to both Panama and the world (Figure 20.A). Some 1,000 ships a year, traveling 80 shipping routes and representing about 5 percent of the world's cargo volume, use the 50-mile (80-km) shortcut from the Atlantic to the Pacific Ocean. On an ocean voyage from New York City to San Francisco, nearly 8,000 miles are saved by using the canal. Currently, approximately one-seventh of all U.S. foreign trade is carried through it.

The idea of a shortcut between the oceans had been around for a long time before the site was chosen. For the first part of the 1800s, Nicaragua was the first choice because of natural landscape features that appeared better able to accommodate a

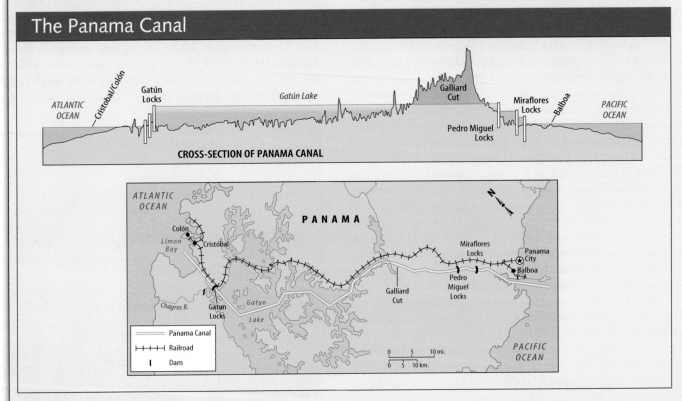

Figure 20.A Profile and plan of the Panama Canal

Environments, Settlements, and Economic Activities

The distribution of settlements and economic life in Central America relate strongly to three parallel physical belts: the volcanic highlands, the Caribbean lowlands, and the Pacific lowlands (see Figure 19.4).

All Central American countries except Belize share the volcanic highlands that reach south from Mexico. Here lie the core regions, including the political capitals, of Guatemala, El Salvador, Honduras, Nicaragua, and Costa Rica. The only cities with over a million people in their metropolitan areas are the capitals of those countries, respectively: Guatemala

are proposed or under construction across energy-starved Central America.

City, San Salvador, Tegucigalpa, Managua, and San José. Each of these is a primate city serving as the demographic, economic, and political heavyweight of the nation. Costa Rica, Panama, Nicaragua, and El Salvador are the most urbanized Central American populations, with roughly 60 percent of their peoples living in cities, while Guatemala is the least urban with about 40 percent.

The highlands are dotted with majestic volcanic cones and scenic lakes (Figure 20.9). Ongoing vulcanism periodically deposits ash into the highland basins, and this weathers into fertile soils that grow corn, beans, and squash for local food and the coffee that is Central America's largest export crop. Climatically, they are classed largely as tierra templada. Only a small fraction of the land is still forested. Between the highlands and the Pacific Ocean lies a coastal strip of seasonally

crossing without requiring massive construction; Nicaragua to this day is considering constructing a canal to connect its interior Lake Nicaragua with the Caribbean, thereby facilitating trade with the United States and Europe. In 1846, however, Colombia (which then controlled Panama) and the United States signed a treaty supporting a project to cut a canal across the much narrower Isthmus of Panama.

Interest in the project picked up substantially in 1849 with the discovery of gold in California. At that time, prospectors and speculators had to sail to the port of Colón on the east side of Panama, make the 50 miles of land crossing in any way possible, and then reboard a ship in Balboa and sail to San Francisco. By 1855, this pattern was modified by the construction—by a collection of New York business interests—of a railroad from port to port. This method of crossing, which required a classic break of bulk of any goods as well as passengers, would finally be superseded by the canal.

In 1878, Colombia granted permission to a French firm to build a canal across the isthmus near the railroad line. The engineering was designed by Ferdinand de Lesseps, who had seen the Suez Canal to completion in 1869. The French quickly bought control of the railroad line and began digging in 1882, with a plan to make their crossing a sea level transit that would require no **locks** (the devices that allow a ship to transit from one water level to another). This venture failed with bankruptcy in 1889 and was followed by another French attempt, which also failed. As difficult as the construction was in these efforts, failure was also prompted by the high death toll through yellow fever and other tropical diseases. By the beginning of the 20th century, the French had left. The United States had just participated in the 1898 Spanish-American War and took more geopolitical interest than ever before in this prospective waterway.

In 1903, Panama broke away from Colombia, and within weeks, the United States recognized the new government. Within months, an agreement had been struck between the new Panama and the U.S. governments for the construction of the trans-isthmus canal, which ultimately required 3 sets of locks. Construction began in 1904 and, at a cost of $380 million and many lives lost to yellow fever, bubonic plague, and malaria, was completed in 1914. The first crossing was made under the banner: "The Land Divided; The World United."

Work has been done continuously on the canal. A 20-year project, set for completion in 2014, is underway to widen the Gaillard Cut to make it a two-way passage. This is the portion of the canal on its highest point, on the **continental divide.** Since the turnover of the canal to Panama in 1999, there has been some concern in the United States that the two access points—Balboa in the Pacific and Cristóbal in the Atlantic—are now controlled by a company with strong Chinese backing. When and if the current locks are widened enough to accommodate massive oil supertankers, a new level of concern about security is bound to arise.

Many of the world's cargo ships are also too large to transit the Panama Canal, and there is growing consideration of constructing a cargo alternative to the canal. The normal transit time plus congestion bottlenecks and construction setbacks can make the Panama Canal unattractive to businesses interested in getting goods quickly to market. Therefore, the concept of a **"dry canal"** has emerged. This would not be a canal at all, but a rail system that, like the old Panama railway, would require break of bulk at its Atlantic and Pacific termini. The leading contenders so far are routes across Mexico's Isthmus of Tehuantepec and another across El Salvador and Honduras. Farther south, Colombia has its own alternative plan: a new wet canal across its narrow Darién region on the Isthmus of Panama.

Figure 20.9 The volcanic highlands of Central America have stunning landscapes and are home to most of the region's people. This is highland Guatemala.

dry lowlands where cattle and cotton are grown largely for an export market. By contrast, the Atlantic side of Central America is lower, hotter, rainier, and far less populous than the Pacific side. Here, on large corporate-owned plantations, are grown most of the bananas that are Central America's agricultural trademark and second most valuable export after coffee.

Commercial production of coffee, bananas, cotton, sugarcane, beef, and other commodities provides the majority of Central American exports (Figure 20.10). In the early 1900s, American-owned United Fruit Company (UFC) acquired a very profitable virtual monopoly of Central American banana production. The company established a network of railways to facilitate transport of the produce from growing areas to coastal ports. The UFC acquired an enormous

220

Joe Hobbs

Figure 20.10 Introduced by European commercial interests into Central America, bananas came to represent a strong portion of the region's economy and its dependence on outside powers.

influence over national political systems and came to be seen by Central Americans as an extension of official U.S. interests. An up-and-down history of banana growing saw many plantations eventually shift to the Pacific side of Central America because of devastating plant diseases on the eastern Caribbean side. Today, exports of bananas, grown in part on the Caribbean lowlands and in part on the Pacific lowlands, continue to be very important in the economies of Honduras, Panama, and Costa Rica. A trade dispute known as the "**banana war**" broke out in the late 1990s between the United States and the European Union over U.S. corporate banana production from this area, as the United States pressed to have Europe open its protected markets to Central American bananas.

All the countries except overcrowded El Salvador still have **pioneer zones**, especially on the Caribbean lowlands, where new agricultural settlement is taking place. Settlers generally come from overcrowded highlands, often aided by government colonization plans featuring land grants and the building of roads to get products to market and bring in supplies. Many peasants still practice the ancient *milpa* form of slash-and-burn agriculture.

There is some manufacturing in the few large cities, but the products are generally simple consumer goods for domestic markets. There are a few factories assembling foreign-made cars, but industries comparable to the maquiladoras of Mexico are still few. "Cut and sew" apparel exports are critical to El Salvador and Honduras, and baseballs handstitched from U.S. components are made in a free trade zone in Costa Rica. A large share of the profits from industrial enterprises makes its way into very few pockets.

20.3 The Caribbean Islands: From Bob Marley to Baseball and Communism

The islands of Caribbean America, also known as the West Indies, are very diverse (Figure 20.11). They encompass a wide range of sizes and physical types, ethnic variations, and political and economic systems. Altogether there are 13 independent countries and 12 possessions of other powers.

General Characteristics

Cuba, the largest island, is primarily lowland, with low mountains at the eastern and western ends. The next largest islands—Hispaniola (which includes Haiti and the Dominican Republic), Jamaica, and Puerto Rico—are steeply mountainous or hilly. These four large islands are known together as the Greater Antilles.

Most of the remaining islands are rather small and make up the group known as the Lesser Antilles. They fall into two broad physical types: low, flat limestone islands rimmed by coral reefs, including the Bahamas (northeast of Cuba) and a few others; and volcanic islands, such as the Virgin Islands, Leeward Islands (roughly the northern half of the Lesser Antilles), and Windward Islands (roughly the southern half of the Lesser Antilles), stretching along the eastern margin of the Caribbean from Puerto Rico toward Trinidad and Tobago. Each of the volcanic islands consists of one or more volcanic cones (most of which are extinct or inactive), with limited amounts of cultivable land on lower slopes and small plains (Figure 20.12). Barbados, east of the Windward Islands, is a limestone island with moderately elevated rolling surfaces. The country of Trinidad and Tobago, just off the northeast corner of South America, has a low mountain range in northern Trinidad, which is a continuation of the Andes, and level to hilly areas elsewhere.

The islands are warm all year, but extremely hot weather is uncommon. Precipitation varies, with windward northeastern slopes of mountains often receiving heavy rain, and

The West Indies

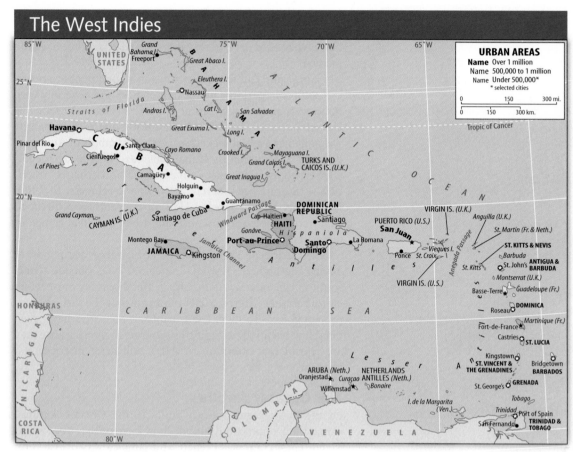

Figure 20.11 Principal features of the Caribbean Islands

leeward southwestern slopes and very low islands little rainfall. The natural vegetation varies from luxuriant tropical rain forests in wet areas to sparse woodland in dry areas. Most areas have two rainy seasons and two dry seasons a year. In late summer and autumn, hurricanes, approaching from the east and then curving northward, are a scourge throughout the region and northward all the way to the Middle Atlantic region of the eastern United States.

Plantation agriculture and tourism are the islands' economic mainstays. Sugarcane, the crop around which the island plantation economies were originally built, is still the main export crop in Cuba, Barbados, and a few other places. Other commercial crops include coffee, bananas, spices, citrus fruits, and coconuts. Production comes from large plantations or estates worked by tenant farmers or hired laborers, from small farms worked by their owners, and in Cuba, from state-owned farms. In pre-Columbian times, important crops in this region included tobacco, indigo, and cacao.

Mineral production is absent or unimportant on most islands, the most conspicuous exceptions being Jamaica (bauxite), Trinidad and Tobago (petroleum), and Cuba (metals). Manufacturing has increased and now provides the largest exports from some islands, including chemicals from

Barbados and clothing from St. Kitts and Nevis. Tourism, mainly from the United States, is a major source of revenue on many islands, most notably the Bahamas, Puerto Rico, the Dominican Republic, and Jamaica. Despite an official ban, increasing numbers of U.S. tourists are also visiting Cuba, already a favored destination for European tourists. Tourism has become Cuba's leading source of foreign exchange. Dominica and some of the other smaller island countries are also making money through Internet gambling and offshore banking, enterprises also undertaken by many small, resource-poor islands in the Pacific. Jamaica and Curacao (in the Netherlands Antilles) are developing their communication infrastructure with the aim of luring information technology (IT) services. Dominica and Grenada get an economic lift from medical schools established on the islands to serve medical students who failed to pass qualifying exams for American schools.

Overall, the region is quite poor, with a collective average GDP PPP of just $5,632. The wealthiest entity is the Cayman Islands, and the poorest is Haiti, whose indexes of quality of life are among the lowest in the world, comparable to many countries in Africa south of the Sahara. Caribbean poverty has long stimulated emigration. Millions of West Indians

417

Figure 20.12 A volcanic landscape typical of the Windward Islands

now live in the United States, the United Kingdom, Canada, France, and the Netherlands, mainly in cities. The **remittances** they send back to relatives are a major source of island income.

Haitian and Cuban immigrants reach, or attempt to reach, U.S. shores in periodic waves that correspond generally with political tumult in their home countries (Figure 20.13). The United States receives them differently. Cubans are generally welcomed as asylum seekers, serving as symbols of American defiance of Communist Cuba, whereas Haitians are generally turned away because—like so many other Latin Americans—they are seen merely as economic refugees. In 2004, U.S. policy further labeled Haitian migrants as a threat to U.S. national security because Haiti has provided refuge to terrorists.

The Caribbean region has the unfortunate distinction of having the highest HIV/AIDS infection rates outside Africa south of the Sahara. As in Africa, the disease is most widespread among heterosexuals who transmit the virus by unprotected sex. To date, the Caribbean countries have mustered few resources to fight this scourge, in part because they are fearful that attention to AIDS will deter international tourism.

The Greater Antilles

Of the five present political units in the Greater Antilles, Cuba, the Dominican Republic, and Puerto Rico were Spanish colonies, Haiti was French, and Jamaica was British (Figure 20.14). These units have developed and maintained distinct geographic personalities.

By any standard, Cuba is the most important Caribbean island. Centrality of location, large size, a favorable environment for growing sugarcane, some useful minerals, and proximity to the United States are important factors con-

Figure 20.13 Haitian refugees seem to bring to life the overcrowded lifeboat metaphor of Garrett Hardin (see page 45).

tributing to Cuba's leading role in the Caribbean. The island would stretch from New York to Chicago if superimposed on the United States and is nearly as large in area as the other Caribbean islands combined. Cuba lies only 90 miles (145 km) across the Strait of Florida from the United States.

At the western end of the island, Havana (population: 2.4 million) developed on a fine harbor as a major colonial Spanish stronghold, and it serves today as Cuba's main industrial center and its political capital, largest city, and main seaport. Gentle relief, adequate rainfall, warm temperatures, and fertile soils create very favorable growing conditions on the island. Agriculture has always been the core of the Cuban economy, and sugarcane has been the dominant crop. Cane sugar still accounts for the largest share of Cuban exports by value.

The 1959 Communist revolution, led by Fidel Castro, expropriated sugar and other estates from large private farms and shifted their ownership to the state. About 33 percent of all farmland is now under state control (down from almost 100 percent prior to 1993) and is worked by wage laborers. As happened in most Communist regimes, state-controlled agriculture led to significant decreases in crop production in

Cuba. The collapse of the Soviet Union in 1991 eliminated a pillar of foreign support and export market for Cuba's economy. The vital sugar industry lost its Soviet subsidies, forcing the closure of many sugar mills and the conversion of sugarcane fields to other crops.

A four decade old U.S. **trade embargo** against Communist Cuba has made it impossible for these countries to enjoy the trading relationship that could help make up for the loss of Cuba's partnership with the Soviet Union. Even with its economic troubles, to its credit Cuba's government has had marked success in raising the adequacy of medical care, housing, and education over the years, particularly in comparison to the levels of such services during the 25 years of the Batista regime that Castro replaced in 1959.

East of Cuba, the mountainous island of Hispaniola is divided by two different countries, cultures, and economies: the French-speaking Republic of Haiti and the Spanish-speaking Dominican Republic. In Haiti (population: 8.1 million; capital, Port au Prince, population: 1.8 million), the black population forms an overwhelming majority, and a small mulatto elite—perhaps 1 percent of the population—controls a major share of the wealth. Population density is

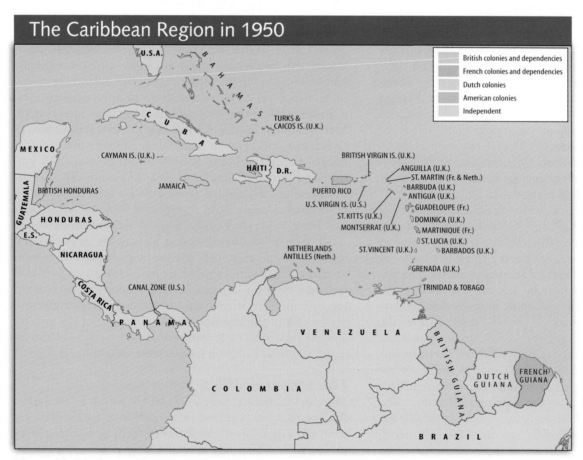

Figure 20.14 The Caribbean as a colonial Region

extreme and infrastructure is poor in this country where poverty rules. The overwhelmingly agricultural economy is based mainly on subsistence rather than export.

The French colonial regime was driven out of Haiti at the beginning of the 19th century. Independent since 1804, the republic of Haiti has a strong West African flavor in both culture and landscape. Unfortunately, deforestation, erosion, and poor transportation are major problems. Deforestation took place on a large scale as French colonists stripped the countryside of millions of its valuable mahogany trees. Commercial and subsistence exploitation continued after independence, and more than 90 percent of the country has been deforested. Aside from subsistence farming, Haiti's small economy concentrates on light manufacturing of clothing (aided by low-wage labor), and coffee growing.

The country has had a turbulent political history. Its first democratically elected president, Jean-Bertrand Aristide, was ousted by a military coup in 1991. U.S. military intervention restored him to power in 1994. Aristide was subsequently implicated in the trafficking of Colombian cocaine to the United States, and he fell out of favor with Washington. When rebels angry with the outcome of a flawed 2000 legislative election closed in on him in 2004, the United States did not intervene on his behalf and instead flew him out of the country. With Aristide's departure to Africa, U.S. troops and then a UN peacekeeping force were deployed to Haiti, and rebels continued to hold much of the countryside. There is evidence that the rebels are also involved in drug trafficking, and one goal of their revolt may have been to gain a bigger share of the estimated $1,500 per pound fee that Colombian traffickers of cocaine spend as payoffs to Haitian officials. An estimated 10 percent of the Colombian cocaine bound for the United States passes through Haiti, so the potential profits are enormous.

Haiti's neighbor to the east, Dominican Republic (population: 8.8 million) bears a mainly Spanish cultural heritage. It is more prosperous than Haiti. Santo Domingo (population: 2.5 million), its capital, is the oldest European city of continued habitation established in the New World (since 1496), with the first European church, school, university, and hospital. An independent country since 1844, the republic has a history of internal war, foreign intervention (including periods of occupation by U.S. military forces), and misrule, although recently there has been greater stability.

The increasingly diversified economy relies on tourism, mining, oil refining, and agricultural products such as sugar, meat, coffee, tobacco, and cacao. Maquiladora-style export production, patterned after Mexico's success with these assembly plants, is playing an increasing role in the republic's economy. Baseball is dear to the Dominican Republic. The country has sent some of the best major league baseball players to the United States—Pedro Martinez, Manny Ramirez, David Ortiz, and the Alou brothers, to mention a few—and produces baseballs for the U.S. market. Urbanization is more pronounced than in Haiti (64 versus 36 percent).

Puerto Rico (population: 3.9 million; capital, San Juan, population: 430,000) smallest and easternmost of the Greater Antilles, is a mountainous island with coastal strips of lowland. Spain ceded the island to the United States after its defeat in the Spanish-American War of 1898, and Congress administered Puerto Rico as a federal territory until 1952. It then became a self-governing commonwealth voluntarily associated with the United States. Puerto Ricans have U.S. citizenship but do not pay taxes, vote in U.S. elections, or receive federal U.S. benefits. The U.S. dollar is the official currency, and all of the trappings of American life prevail on the island, together with a strong population of tourists and expatriates from the U.S. mainland. Although a narrow majority of Puerto Ricans have chosen thus far to remain a commonwealth of the United States, there is continuing pressure within the island for statehood, and a small minority has expressed a desire for total independence.

Puerto Rico's pre–World War II economy was agriculturally based, with an emphasis on sugar. After the war, a determined effort by Puerto Ricans to raise living standards brought great changes. Hydroelectric resources were harnessed, a land-classification program provided a basis for sound agricultural planning, and development of manufacturing industries and tourism began, financed largely by the United States. A package of tax incentives and below market land options was organized to attract American businesses and industries. Manufacturing came to replace sugar as the island's main source of revenue, with petrochemicals, pharmaceuticals, electrical equipment, food products, clothing, and textiles among the major products. Sizable copper and nickel deposits were found and developed. But poverty is still a problem for Puerto Rico and has long motivated emigration to the U.S. mainland, especially to New York City.

Jamaica, independent from Great Britain since 1962, is located in the Caribbean Sea about 100 miles (c. 160 km) west of Haiti and 90 miles (c. 145 km) south of Cuba. Its population of 2.6 million (as of 2004) is largely descended from black African slaves transported there by the Spanish and (after 1655) the British. When slavery was abolished in 1838, most blacks left the sugar plantations on the coastal lowlands and migrated to the mountainous interior. Tourism, bauxite, and agriculture have constituted Jamaica's economic base for decades. Jamaican culture in the form of reggae music has become known worldwide, especially with the success of the late Bob Marley. He sang of the poverty and the humanity of Trenchtown, a poor section of the Jamaican capital, Kingston (population: 600,000).

The Lesser Antilles

More than 3 million people live on the islands of the Lesser Antilles that are scattered in a 1,400-mile (2,250-km) arc from the Virgin Islands (east of Puerto Rico) to the vicinity of Venezuela. This string of small islands defining the eastern edge of the Caribbean Sea contains a variety of political entities. The Netherlands Antilles and Aruba are Dutch dependencies evolving toward greater self-government. Guadeloupe and Martinique are French overseas departments, and Montserrat (an island overrun by volcanic eruptions in 1995 and 1997) and Anguilla are still British dependencies. Britain began its withdrawal from the islands with independence for Jamaica and Trinidad and Tobago in 1962. Since then, independence from British rule has been gained by Barbados (in 1966) and six smaller units. Britain also granted independence to the Bahamas in 1973, but retains its colonies of Bermuda, the British Virgin Islands, Cayman Islands, and Turks and Caicos Islands. All the former colonies are still members of the Commonwealth of Nations.

The economies of the islands suffer from many serious problems, including unstable agricultural exports, limited natural resources, a constantly unfavorable trade balance, embryonic industries, and chronic unemployment. A shared trait of the islands is their history of European colonialism and plantation slavery. Such generalizations contribute to some understanding of the islands but give little hint of the individuality these small units present when viewed in detail. Just one example—Trinidad and Tobago—is discussed here.

Trinidad and Tobago comprises the largest, richest in natural resources, and most ethnically varied of the island units in the eastern Caribbean. Trinidad, the larger of the two islands, has offshore oil deposits in the south. Yet Trinidad refines more oil than it pumps. Imported crude oil from various sources is refined and then shipped out, primarily to the United States. Some 40 percent of the population is black, most of whom live mainly in the urban areas and are often employed in the oil industry. Another 40 percent are the descendants of immigrants from the Indian subcontinent—a population stream that developed for indentured workers after the abolition of slavery. These Indians live mainly in the intensively cultivated countryside. Among their native sons is V. S. Naipaul, who has written fine portraits of Indian life on Trinidad and Tobago, including *The Mystic Masseur* and *The Suffrage of Elvira*. About 18 percent of the islanders are classified as "mixed," and there are small numbers of Chinese, Madeirans, Syrians, European Jews, Venezuelans, and others. This ethnic complexity of Trinidad is reflected in religion, architecture, language, diet, social class, dress, and politics.

From Trinidad and Tobago, the text takes a short jump with the next chapter to the nearby, great landmass of South America.

SUMMARY

- Middle America is made up of Mexico, the Central American countries, and countries in the Greater and the Lesser Antilles in the Caribbean Sea. The largest islands—Cuba, Hispaniola, Puerto Rico, and Jamaica – make up the Greater Antilles, and most of the smaller islands belong to the Lesser Antilles chain.

- Mexico is the largest, most populated, and economically most developed of the Middle American nations. Its area of 762,000 square miles (c. 2 million sq km) and its population of more than 100 million give Mexico a giant role in Middle America. It is also the largest nation in the world to have Spanish as its dominant language.

- The North American Free Trade Agreement (NAFTA) has played a major role in the pace of economic change in Mexico, particularly in the borderlands between Mexico and the United States. There has been steady demographic and economic growth in the cities lying just south of the U.S. border, and much of Mexico's industry occurs in the maquiladora factories of this region.

- Mexico's history is composed of three distinct eras: the Native American, the Spanish, and the Mexican. The Aztecs and Mayas were the most technologically advanced Native American populations. The Spanish or Hispanic presence is very strong, and the large population of mestizos represents the people who are a racial mix of the Spanish and indigenous groups in Mexico. The Spanish were a major force in establishing land-use patterns that continue today. The Mexican era began with the 1810–1821 period of the Mexican Revolution, and the country suffered

through many governments and the loss of much land to the United States in the following decades. The second period of revolution in Mexico was 1910 to 1920.

- Central Mexico is a volcanic highland area in which Mexico City, the world's second largest city (c. 22 million), is located. They city has problems with land subsidence and air pollution. It is Mexico's leading industrial center.

- The Gulf Tropics of Mexico are known for gas, oil, agricultural plantations and Maya archeological sites. Tourism is a vital industry. Mexican oil, a nationalized resource, has been important in economic development.

- Central America is made up of Guatemala, Belize, El Salvador, Honduras, Nicaragua, Costa Rica, and Panama. In Guatemala, Honduras, and El Salvador, there are sizable populations of Native American peoples, and there is a strong European presence in Costa Rica. Panama's history over the past century has been shaped powerfully by the construction of the Panama Canal in 1904–1914 and by the American presence there. In all of the Central American countries—although less in Costa Rica—there is a wide gap in economic prosperity between the very wealthy and the very poor.

- Insurgencies and civil wars were problems in most of the mainland Middle American countries in the latter part of the 20th century. El Salvador, Guatemala, and Nicaragua had the costliest conflicts. There are still concerns about Zapatista rebels in southern Mexico.

☙ Plantation agriculture continues to be important in the tierra caliente on both coasts of Middle America. Bananas are the major plantation crop, but sugar is also raised in Central and Middle America and the Caribbean. Deforestation is a major problem throughout the region, notably in southern Mexico.

☙ There are many newly independent countries in Caribbean America that have emerged from colonial connections to Europe in the past four decades. Some islands continue to have colonial relationships with Europe and maintain very strong economic and political affiliations with the countries that colonized them.

☙ Cuba is the most important Caribbean island because of its size, population, and geopolitical status. Sugar and tourism have long been central to its economy, but since Castro's clash with the United States in the mid-20th century, both tourism and the sugar trade have been constrained. Tourism is slowly gaining economic significance again, but Americans are dissuaded from visiting, and the United States continues to refuse to buy any sugar from Cuba.

☙ Puerto Rico is the smallest and easternmost of the Greater Antilles and is a self-governing commonwealth of the United States. There continue to be groups on the island seeking either U.S. statehood or complete independence. There is also a steady migration stream of Puerto Ricans from the island to the American Northeast, especially New York City. Agriculture has been steadily losing ground to industry as Puerto Rico has actively been pursuing stronger economic development.

☙ Other major islands of settlement and of political and economic significance in the Caribbean include Jamaica and, in the Lesser Antilles, the Netherlands Antilles, Aruba, Guadeloupe, Martinique, Barbados, the Bahamas, and the Virgin Islands. These represent a variety of political affiliations. Banking, tourism, and varied resources are important in the economic profiles of these locales.

KEY TERMS + CONCEPTS

Terms in blue are also defined in the glossary.

"banana war" (p. 562)
border towns (p. 557)
Captaincy-General of Guatemala (p. 558)
Central American Federation (p. 558)
Chol Maya (p. 554)
continental divide (p. 561)
Contras (p. 558)
core region (p. 555)
"dry canal" (p. 561)
energy maquiladoras (p. 555)

Hispanic (p. 552)
Lacandon Maya (p. 554)
lock (p. 561)
megacity (p. 556)
megadiversity country (p. 554)
Mexican Revolution of 1810-1821 (p. 553)
Mexican Revolution of 1910-1920 (p. 553)
Mexican War (American Invasion) (p. 555)
outsourcing (p. 555)
pioneer zone (p. 562)

remittances (p. 564)
San Andres Accords (p. 553)
Sandanistas (p. 558)
thermal inversion (p. 556)
trade embargo (p. 565)
Tzeltal Maya (p. 554)
United Fruit Company (UFC) (p. 559)
Viceroyalty of New Spain (p. 552)
Zapatista Army for National Liberation (EZLN) (p. 553)

REVIEW QUESTIONS

WORLD
REGIONAL
Geography❀Now™

Assess your understanding of this chapter's topics with additional quizzing and concept-based problems at http://earthscience.brookscole.com/wrg5e.

1. What are the major subregions of Mexico? What characteristics define the core regions of most of the Central American countries? What unique environmental problems must Mexico City deal with?

2. What are the particular attractions of the Mexican border towns for U.S. citizens? What drives the economies of these towns?

3. What are energy maquiladoras? What resources do they use, where is their product exported, and why are they located where they are?

4. What and where are the most important plantation crops of the region?

5. What role has oil played in Mexico's economy? What is its future?

6. Where and where are the capital cities of the Central American countries? What are the main economic activities of those countries?

7. What are the major geographic subdivisions of the Caribbean islands?

8. How does the United States deal differently with emigrants from Cuba and Haiti? Why?

9. Why does the text define Cuba as the most important Caribbean island?

10. What factors have helped to make Haiti the poorest country in the Western Hemisphere?

11. What is Puerto Rico's relationship with the United States? Does it appear likely to change?

DISCUSSION QUESTIONS

1. In what ways is Mexico the giant of this region?

2. What do the Zapatista rebels of southern Mexico want? How have issues of control over land been central to strife and war in many countries of Middle America?

3. Which countries benefit most from tourism? What styles of tourism are most popular?

4. What forces have been behind deforestation in the region? Which still-forested areas are of most concern?

5. What and where are the Mexican *maquiladoras*? How have China's booming economy and low labor costs impacted these industries?

6. What were the circumstances of the civil conflicts in Nicaragua, El Salvador, and Guatemala? In which of them, and how, did the United States intervene?

7. What roles have various ethnic groups played in the political and economic systems of Costa Rica, Guatemala, and Mexico?

8. What considerations went into the construction of the Panama Canal? What alternatives to the canal are now being considered?

9. Which countries benefit most from remittances earned abroad?

10. What roles have natural resources played in the development of some of the Caribbean islands? What are some of the more famous cultural products of the islands?

South America: Stirring Giant

A Native American Colorado of Ecuador. His red hair is dyed from the bean of the achiote. The Spaniards called this group Colorado ("red") after the hair.

Joe Hobbs

---

chapter objectives

This chapter should enable you to:

- Recognize the paradox of how a blessing of oil wealth can paralyze national economies and political systems and be rejected by the poor

- Appreciate the huge challenges that rebel forces have posed to national governments in the region

- Understand the liabilities of a landlocked location—in this region, Bolivia and Paraguay

- Appreciate how uneven distribution of national wealth promotes instability and hampers development

- Understand why "saving the rain forest" is not a simple problem and how various countries and groups within countries see the problem differently

- Appreciate how excessive dependence on raw materials can cause an uncontrollable cycle of economic boom and bust

WORLD
REGIONAL
Geography ⊛ Now™

Look for this logo in the text and go to GeographyNow at http://earthscience.brookscole.com/wrg5e to explore interactive maps, view animations, sharpen your factual knowledge and geographic literacy, and test your critical thinking and analytical skills with unique interactive resources.

South America has the lion's share of Latin America's total land area (see Figure 19.1, page 520). It is the fourth largest continent on Earth, amounting to 6,898,400 square miles (17,866,900 sq km). It is approximately 4,500 miles (7,200 km) from north to south and extends to a maximum width of 3,200 miles (5,120 km). It includes the world's largest Portuguese-speaking nation, Brazil, which is also home to the largest remaining tropical rain forest in the world. South America has a history of Native American civilization on the west side of the Andes Mountains, which stretch from Colombia in the north all the way to Tierra del Fuego at the southern tip of the continent. The continent also has a wide range of climates, landscapes, and ecological zones, ranging from the parched dry desert of the Atacama in Chile to the lush tropical rain forests of central Brazil and eastern Peru and Ecuador. In world affairs, South America is often depicted as a "sleeping giant," a land with huge potential for economic development and political leadership. This chapter explores its potential and the prospects for its awakening. There is also a brief discussion of the continent of Antarctica, which lies close to South America.

21.1 The Andean Countries: Lofty and Troubled

The South American countries of Venezuela, Colombia, Ecuador, Peru, and Bolivia—all within the tropics and traversed by the Andes—are grouped in this text as the Andean countries. Common interests among them were recognized in 1969, when five countries signed the Cartagena (Colombia) Agreement, which created a new free trade zone called the **Andean Group.** Initially, Chile was part of the group, but it later withdrew. Although the Andes Mountains run its entire length, Chile is not classified as an Andean country because its major traits are more akin to its southern neighbors. Venezuela came late to the Andean Group, which is now made up of Bolivia, Peru, Ecuador, Colombia, and Venezuela. These countries share the following traits:

1. *Environmental zonation.* This region has pronounced vertical and horizontal zonation, with extreme environmental contrasts and areas that are isolated by natural barriers.

2. *Fragmented settlement patterns.* Because of the broken topography of the Andean countries, patterns of settlement and economic development are fragmented. This is especially true in Peru.

3. *Populous highlands and sparsely settled frontiers.* Each country contains both well-populated highlands and an extensive fringe of lightly populated interior lowlands or low uplands now being settled by pioneers. Some of the interior areas have oil and other minerals.

4. *Productive coastal lowlands.* Each country except Bolivia has an economically productive coastal lowland region. (Bolivia originally had a Pacific coastal zone but lost it to Chile during the war of 1879–1883.)

5. *Significant indigenous populations.* All countries had large indigenous populations in the mountains, valleys, and basins at the time of the 16th century Spanish conquest. These old settlement areas are now major zones of dense population and strong political authority, especially in the highlands. The indigenous cultures have persisted despite the conquest. Today, Bolivia, Peru, and Ecuador have Native American majorities; the only other Latin American country with that distinction is Guatemala. In Colombia and Venezuela, Native Americans are just 1 or 2 percent of the total populations, with mestizos forming the largest ethnic element.

6. *Dominant Hispanic traditions.* Each of the Andean countries has strong Hispanic traditions. Spanish is the official language in all five countries (although Native American languages also are official in Peru and Bolivia), and over nine-tenths of each country's population is Catholic. The Spanish imposed these badges of Iberian culture on the advanced indigenous cultures of the Inca empire.

7. *Poverty.* All the countries in this region are relatively poor, despite oil production in most of them. Considerable capital flows into many of these countries through the cultivation, processing, and marketing of illegal drugs, especially cocaine, but most drug crop farmers remain poor.

8. *Primate cities.* In all countries except Ecuador and Venezuela, the national capital is a primate city, towering over all others in size and political and cultural importance.

This survey of South America's geographies continues with a closer examination of the Andean countries.

Venezuela

The population of Venezuela (Figure 21.1) is concentrated mainly in Andean valleys and basins not far from the Caribbean coast, as it was when the Spanish arrived in the 1500s. The highlands contain the main city and capital, Caracas (population: 3.7 million). At an elevation of 3,300 feet (c. 1,006 m), Caracas has a tierra templada climate. Around Caracas, the Spanish found a relatively dense indigenous population to draw on for forced labor, as well as some gold to mine and climatic conditions amenable to the commercial production of sugarcane and, later, coffee. Labor needs associated with sugarcane in this region led to the 16th century introduction of African slaves, but Venezuelan plantation development remained relatively small. About 10 percent of the population today is black. Until the middle of the 20th century, Caracas was a fairly small city, and its valley was still an important and productive farming area. Now the city overflows the valley, and Venezuelan agriculture is associated with other highland valleys and basins.

Oil resources in the coastal area around and under the large Caribbean inlet called Lake Maracaibo have been a

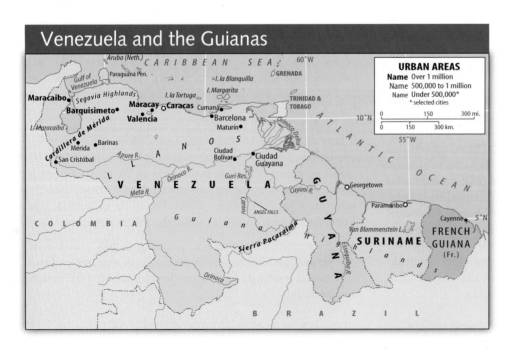

Figure 21.1 Principal features of Venezuela, Guyana, Suriname, and French Guiana

blessing for Venezuela. The country was one of the founding members of the **Organization of Petroleum Exporting Countries (OPEC)** in 1961 and is still a member. Oil revenues have fueled rapid urbanization, attracted foreign immigration, and financed considerable industrialization. However, in recent years, the country's income has suffered because of continuing fluctuations in world oil prices and because of unfortunate political events (see Geography of Energy). Venezuela is rather like Nigeria, an oil-rich country where the vast majority of people remain poor; Venezuela's per capita GDP PPP is little better or is worse than that of its oil-less neighbors.

Venezuela has some important mineral assets in the south. This sparsely inhabited section is mostly tropical savanna in the Guiana Highlands and in lowlands along the Orinoco River and its tributaries. For centuries, remote interior Venezuela has been Native American country (in the high-

lands) or poor and lightly settled ranching country (in the lowlands), but development is now quickening. Government-aided agricultural settlement projects are occupying parts of the lowlands, and major oil and gas reserves are being exploited. In the Guiana Highlands after World War II, enormous iron-ore reserves south of Ciudad Bolívar and Ciudad Guayana began to be mined for export by American steel companies, and huge bauxite reserves in the highlands continue to await exploitation. Venezuela has pushed the development of a major industrial center at Ciudad Guayana. Ocean freighters on the Orinoco River carry export products away and bring in foreign coal for industrial fuel. Major steel, aluminum, oil refining, and petrochemical industries are in operation, with further diversification planned.

Venezuela has dedicated an enormous amount of its area to protected parks and reserves, including a reserve for the

Geography of Energy

Venezuela's Petroleum Politics

The recent reversal of Venezuela's economic fortunes was engineered largely by the country's president, Hugo Chavez, who effectively handcuffed the country's oil production for some time, sending economic shock waves around the world. A former military officer, Chavez was twice elected to the country's highest office, most recently in 2000. After that second victory, the always left-leaning Chavez declared that he was leading a "**Bolivarian Revolution.**" His purported model was Simón Bolívar (1783–1830), the Venezuelan-born independence leader who led a regional quest for liberation from Spain. Chavez had his own ideas about liberation: He rewrote the constitution, put cronies on the country's supreme court, cracked down on the media, and jailed political foes. The Chavez agenda appealed to the poorest classes by promising a Cuba-like redistribution of wealth. Communism was not the objective, but rather what may be described as **military socialism,** with the armed forces and the president taking control of all but a small, and mainly foreign-owned, private sector.

Oil was the prize for Chavez. Venezuela's petroleum has long been controlled by the state-owned oil company Petroleos de Venezuela (PDVSA), and Chavez sought to bring it under his personal control. PDVSA, apparently backed by at least three-quarters of Venezuela's population of 25 million, resisted. In 2002, PDVSA joined business and labor interests across the nation in a general strike, pressuring Chavez to resign or call for a new election. Chavez resisted, and Venezuela's economy began a downward spiral that reversed only when world oil prices skyrocketed in 2004. A petition drive by the opposition led to a 2004 recall referendum that failed to unseat

Chavez. Venezuela's dependence on oil is illustrated well by the strike, which lasted only 2 months (December 2002 and January 2003). Prior to the strike, Venezuela was the world's fifth largest oil exporter. Petroleum provided 80 percent of the country's export income and 50 percent of its government's revenue. Within a month of the beginning of the strike, oil exports fell 90 percent, amounting to a loss of about $50 million each day. The economy contracted dramatically. Salaries of many workers fell more than 50 percent, and consumer spending dried up. Chavez's agents all but cut off supplies of gasoline to opposition stronghold regions and neighborhoods. Corn flour and cooking gas, so essential for Venezuela's poor, grew increasingly scarce and costly. Unemployment (officially 18 percent in 2004) and inflation rose. Chavez took steps to hold onto his faltering base among the poor, for example, by allowing landless peasants to occupy fallow farmlands.

For a world already rattled by instability in the Middle East, unrest in Venezuela came at a terrible time. Plummeting exports from Venezuela helped send oil prices to record highs. Venezuela's troubles produced instant economic and strategic headaches for the United States. Prior to the strike, Venezuela was its second largest oil supplier. The Bush administration cheered when a coup temporarily ousted Chavez in April 2002. After that, Washington was in a poor position to deal with Chavez. The United States did join a so-called "Group of Friends" that attempted to bring a diplomatic resolution to Venezuela's crisis. The other members were Spain, Chile, Mexico, Brazil, and Portugal. Chavez's only real friend among them was Brazil's president, Luiz Inacio "Lula" da Silva, who is in the vanguard of South American forces lined up against globalization and U.S. interests in the region.

Jay Dickman /Corbis

Figure 21.2 Angel Falls, the world's highest waterfall (3,212 feet /978 m), drains one of the dramatic *tepuis* of Venezuela. These mesas are sandstone remnants of the ancient Guyana Shield.

Native American Yanomamö that makes up fully 9 percent of the country's territory. National planners hope that investment in protected areas will have multiple returns, especially in revenue from ecotourism. Among the most spectacular of the country's natural features are the high sandstone outcrops called *tepuis* (Figure 21.2). Rising precipitously from the tropical rain forest, and often protruding from fog and cloud layers, they evoke prehistoric images that were said to have inspired Arthur Conan Doyle's *The Lost World* (which in turn inspired Spielberg's *Jurassic Park* films). From one of the tepuis spills Angel Falls (3,212 ft/c. 979 m), the world's highest waterfall.

The Guyana, Suriname, and French Guiana Trio

Three entities on the northeast edge of South America, although not classified as Andean, are best described in association with Venezuela: independent Guyana (formerly British Guiana, population: 800,000), independent Suriname (formerly Dutch Guiana, population: 400,000), and French Guiana (population: 200,000), which is an overseas department of France (see Figure 21.1). They have a long plantation history, with sugar most important in Guyana and French Guiana and tea and coffee in Suriname. The Dutch traded their settlement of New Amsterdam (later called New York City) for the rights to Suriname and its coastal richness for agricultural exploitation. French Guiana was also home to Devil's Island, a notorious colonial island prison used as recently as 1954. Today, the three countries economies are still mainly agricultural (especially with exports of sugar and timber) but have growing revenues from bauxite and tourism.

Colombia

Within its wide expanse of the Andes, Colombia (Figure 21.3) includes many populous valleys and basins, some in the tierra fría and others in the tierra templada. The great majority of Colombia's mestizos and whites are scattered among these upland settlement clusters, which tend to be separated by sparsely settled mountain country with extremely difficult terrain. Lower valleys and basins, along with some coastal districts, were developed by plantation agriculture that employed slaves. That legacy is apparent in Colombia's demographics, with 14 percent of the country's population mulatto and 4 percent black.

The Highland Metropolises: *Bogotá, Medellín, Cali*

Colombia's inland towns and cities have tended to grow as local centers of productive mountain basins. The three largest—Bogotá, Medellín, and Cali—are diverse in geographic character. The largest highland city in all of South America is Bogotá (population: 7.4 million; Figure 21.4), Colombia's capital. It lies in the tierra fría at an elevation of about 8,700 feet (c. 2,650 m—much higher than Denver, Colorado).

The Spanish seized the land in this region, put the indigenous people to work on estates, and improved agricultural productivity by introducing new crops and animals—notably, wheat, barley, cattle, sheep, and horses. However, exploitation and disease reduced Native American numbers so drastically that the population took a long time to recover. The Spaniards were not able to develop an export agriculture, as night coolness and frosts precluded such major trading crops as sugarcane and coffee, and early transport to the coast via the Magdalena River and its valley was extremely difficult. Mountain slopes were so challenging that a railroad connection with the Caribbean Sea was not completed until 1961. In recent decades, modern governmental functions and the provision of air, rail, and highway connections have led to explosive urban growth in upland Colombia. The Bogotá area has a large output of flowers grown for air shipment to the United States.

Colombia

Figure 21.3 Principal features of Colombia

Figure 21.4 The pilgrimage shrine of Montserrat, above Bogotá, is known as a place of miracles.

Medellín (population: 3.2 million) lies in the tierra templada at about 5,000 feet (1,524 m) in a tributary valley above the Cauca River. The area lacked a dense indigenous population to provide estate labor and was difficult to reach, so for centuries it remained a region of shifting subsistence cultivation of corn, beans, sugarcane, and bananas. The introduction of coffee and the provision of railroad connections changed this situation in the early 20th century. Excellent coffee could be produced under the cover of tall trees on the steep slopes, and the Medellín region became a major producer and exporter. Local enterprise and capital developed the city itself into a considerable textile-manufacturing center. In the 1980s and 1990s, Medellín and Cali became synonymous with the cocaine trafficking cartels that came to dominate the cities' economies until Colombian authorities cracked down on them.

Cali (population: 2.9 million) is located in the tierra templada some 3,000 feet (915 m) above sea level on a terrace overlooking the floodplain of the Cauca River. Cali had easier connections than did Medellín with the outside world via the Pacific port of Buenaventura. The Spanish developed sugarcane plantations in the Cali region and also produced tobacco, cacao, and beef cattle. Native Americans and later African slaves were put to work in these enterprises. From the late 19th century onward, dams on mountain streams provided hydropower for industrial development. This asset, along with cheap labor, attracted foreign manufacturing firms, and Cali grew at a rate comparable to that of Medellín.

Colombian Coasts and Interior Lowlands

There are sharp contrasts in environment and development between the Caribbean and Pacific coasts of Colombia. The Pacific coastal strip is the wettest area in all of Latin America, with some places receiving nearly 300 inches (762 cm) of precipitation annually. It has tropical rain forest climate and vegetation, and with adjacent sections of Panama and Ecuador, this Choco region is one of the world's biodiversity hot spots. In the north, mountains rise steeply from the sea, but there is a coastal plain farther south. The area is thinly populated, with a large proportion of the population of African ancestry. The main settlement is Buenaventura, an important port. Its principal advantage is that it can be reached from Cali via a low Andean pass.

The Caribbean coastal area is more populous. Three ports have long competed for the trade of the upland areas lying to the south on either side of the Magdalena and Cauca Rivers. Barranquilla and the historic Spanish fortress city of Cartagena are the larger and more important cities in terms of legitimate trade. But the smaller city of Santa Marta, known as a beach resort as well as a port, is probably the most important in overall trade. It is the control center for Colombia's huge illegal export trade in cocaine and marijuana.

Most of the Caribbean lowland section of Colombia is a large alluvial plain in which flooding has hindered agricultural development. Nevertheless, the region has a history of ranching and banana and sugarcane production. The Colombian government is now fostering settlement by small farmers who grow corn and rice.

Colombia's interior lowlands east of the Andes form a resource-rich but sparsely populated region contributing little to the national economy. It is a mixture of plains and hill

26

country, with savanna grasslands in the north and rain forest in the south. The grasslands are used for ranching and the forest for shifting cultivation. Considerable mineral wealth, especially oil, awaits exploitation. As in so many parts of Latin America, and especially in South America, there is intense isolation caused by difficult access to the region. Railway and road networks have not yet reached large portions of the countryside.

Colombia's Peril and Potential

Colombia has the resources for future economic development. Energy resources include a large hydropower capacity; coal fields in the Andes, the Caribbean coastal area, and the interior lowlands; oil in the Magdalena Valley and in the interior; and natural gas in the area near Venezuela's Lake Maracaibo. Iron ore from the Andes already supplies a steel mill near Bogotá, and large Andean reserves of nickel are under development. At least 40 percent of the country is forested, with most of the forest made up of mixed hardwood species in the tropical rain forest. Oil plays an increasing role in the country's export economy. In 2004, petroleum and its products represented approximately 25 percent of all legal Colombian exports by value, with coffee representing 6 percent. However, illegal cocaine and marijuana may have been the largest category of exports.

There should be more and wealthier Colombians, but 200,000 have died in four decades of conflict between leftist guerrillas and the government and its affiliated right-wing paramilitary forces. This civil conflict is Colombia's main concern and also of great concern to the United States, particularly because it is so closely linked to two resources of vital national and international interest: cocaine and petroleum. Here are the major contenders in that struggle.

The Colombian armed forces number almost 200,000, an apparently impressive number until one realizes that they control only about half of a country that is twice the size of Texas (see Figure 21.3). Substantial portions of the other half are under the de facto control of two leftist guerrilla groups: the FARC and the ELN. About 18,000 strong, **FARC (Revolutionary Armed Forces of Colombia)** is the government's principal enemy; the **ELN (National Liberation Army)** with about 6,000 fighters is a less significant but still dangerous force. The two forged a formal alliance against the government in 2003. Yet another nominally outlaw force is the **AUC (United Self-Defense Forces of Colombia)**, with 10,000 fighters. It is a paramilitary organization with some close (but illegal) ties with the Colombian army forces. Its self-proclaimed mission is to rid Colombia of the FARC and other terrorists. However, the paramilitary hardly emerges as a "good guy" in Colombia's bloody struggle. Like FARC, it obtains much of its funding by taxing coca growers—some of whom have been gunned down in revenge by FARC.

FARC controls most of three adjacent provinces in southern Colombia: Putumayo, Caqueta, and Huila. So tenacious was its presence in this contiguous zone that Colombia's previous president, Andres Pastrana, tolerated FARC autonomy there for about 3 years while conducting peace talks with the rebels. These negotiations broke down in 2002, and almost immediately, FARC brought its war to Colombia's cities. To demonstrate the government's vulnerability, FARC carried out a classic terrorist attack in the major southern city of Villavicencio, first exploding a dynamite stick in a crowded nightclub area to draw in curious spectators and then setting off a larger explosion that took numerous lives. (The parallel with the bombing of the Bali nightclub 6 months later was remarkable, but the Colombian attack bore the intellectual fingerprints of the Irish Republican Army; some IRA men were in fact arrested in Bogotá.) FARC forces also struck the capital, and in the northern city of Cali, several FARC fighters disguised as members of an army bomb disposal squad entered the government assembly office and kidnapped a dozen assembly members. These joined many other hostages, including a presidential candidate, a senator, and six congressmen that FARC held, hoping to exchange them for their own jailed commanders.

FARC has also undertaken a systematic campaign to recreate and enlarge its de facto enclave in the south. The typical FARC tactic is to leave a letter to a city's mayor wishing him good health and ordering him to leave the city within 24 hours or be declared a military target. One mayor in Caqueta has been killed and 22 others have fled; several have simply resigned. Courts in many small towns are closing. FARC has succeeded in cutting off electricity and newspaper service to some cities nominally under government control.

Most Colombians have had enough of this and are not interested in appeasing the rebels. In 2002, they elected Alvaro Uribe, an independent liberal who vowed as a candidate to make containing the guerrillas his top priority. He promised to double Colombia's ranks of police and professional soldiers, assemble a militia of 1 million civilians to help the armed forces, introduce tougher antiterrorism legislation, and negotiate with rebels only under the strictest conditions. Colombians voted overwhelmingly for Uribe, giving him a mandate to follow through. FARC welcomed him with an attack on the Presidential Palace during his inauguration, firing a dozen mortars that succeeded only in killing 19 people in a nearby slum (an attack that also bore an IRA signature).

Just barely in office, it was already time for Uribe to show his mettle. Five days after the inauguration/attack, he declared a state of "internal commotion," a constitutional provision allowing the government to carry out wide-ranging security measures to bypass the normal legislative process. His first use of the provision was to impose a wealth tax on the richest Colombians to help pay for the war against the rebels. The main audience for this decree was the U.S. Congress in Washington, which wanted to see some return for its huge investment in Colombia. Colombian forces seem to be making some headway in the struggle against the insurgents, but the outcome is far from certain.

Ecuador

The Andes form two roughly parallel north–south ranges in Ecuador (Figure 21.5). Between them lie basins filled with fertile volcanic ash. These basins are mostly in the tierra fría at elevations between 7,000 and 9,500 feet (c. 2,100–2,900 m). Quito (population: 1.8 million; Figure 21.6), Ecuador's capital and second largest city, is located almost on the equator, high up on the rim of one of the northern basins. Its

Figure 21.7 Cuenca is a colonial city, without skyscrapers.

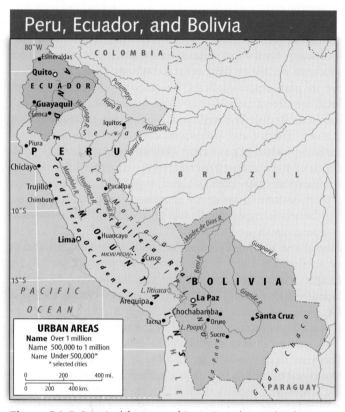

Figure 21.5 Principal features of Peru, Ecuador, and Bolivia

Figure 21.6 Quito lies in a high, fertile valley flanked by volcanoes.

elevation of 9,200 feet (2,800 m) yields daytime temperatures averaging 55°F (c. 13°C) every month of the year.

The first Spanish invaders came to Ecuador from Peru in the 1530s. They found dense Native American populations in Ecuador's Andean basins. The conquerors appropriated their land and labor and built Spanish colonial towns such as Cuenca (Figure 21.7), often on the sites of previous indigenous towns. But they found no great mineral wealth here, and both climate and isolation worked against the development of an export-oriented commercial agriculture. Even today, these Andean basins have an agriculture with a strong subsistence component and sales mainly in local markets. From this hearth, the potato diffused into Europe and into world trade.

Today, potatoes, grains, dairy cattle, and sheep are the major crops and livestock. The isolated Ecuadorian Andes historically attracted relatively few Spaniards, except to Quito, and did not require African slave labor. As a result, the mountain population has remained predominantly Native American, with only slight mestizo and white admixture. The isolation has recently decreased somewhat with the slight growth of industries such as food processing, textiles, and chemicals in Quito and smaller Andean cities.

A drastic shift of Ecuador's economy and population toward the coast took place in the 20th century. Through the 19th century, approximately nine-tenths of the people still lived in the highlands, but that proportion is now only about one-half. The growth zone near the coast, settled mainly by whites and mestizos, is a mixture of plains, hills, and low mountains, with a climate varying from rain forest in the north, through savanna along most of the lowland, to steppe in the south. The main focus of development has been the port city of Guayaquil, located at the mouth of the Guayas River, which drains a large alluvial plain where bananas are grown for export (Ecuador is the world's leading banana exporter) and rice is grown for domestic consumption. Coffee is cultivated on adjoining slopes. These agricultural developments have removed almost all of the area's original forest cover.

Along the Pacific, there is a fishing industry centered on anchovies. This industry regularly goes bust during the El Niño climatic events that warm the eastern Pacific waters. Guayaquil is the jumping off point for the Galapagos Islands, 600 miles (960 km) to the west in the Pacific Ocean. Claimed by Ecuador since 1832, with their extraordinary giant tortoises and other endemic species, the Galapagos are a significant resource for Ecuador because of the ecotourism revenues they generate (Figure 21.8).

Oil exploitation began in the 1960s in the little-populated, tropical rain forest of Ecuador's Amazon region. Ecuador became an oil exporting country as soon as a pipeline was built to cross the Andes to the small port of Esmeraldas on the Pacific. Oil and its products provide about 40 percent of the country's foreign-exchange earnings. Possible future mineral discoveries in the interior rain forest add urgency to a border dispute with Peru, which controls considerable territory claimed by Ecuador. In 1981, and again in

1994–1995, bouts of armed conflict between the two countries occurred along their undefined, mountainous eastern boundary. In 1998, the two nations signed a treaty resolving this issue for the present. At stake for both countries is the successful future growth of fossil fuel export industries, discussed in the context of the Amazon later in this chapter.

Peru

Peru is a poor nation with a population that is fragmented in distribution by great ruggedness (in the Andes), dryness (along the coast), and wetness (in Amazon lowland) (see Figure 21.5). More than one-half of the population, mainly unmixed Native Americans, lives in the Andean *altiplano,* the local name for the high intermontane plain that extends from southern Peru into west central Bolivia. These highlanders are in part descendants of the ancient civilization of the Inca, centered around Cuzco (population: 342,000) in Peru's southern Andes. The highlands, which are generally higher than in Ecuador and Colombia, are a poor environment for surplus-producing cash crop agriculture. Agricultural holdings tend to be in the tierra fría, where cool temperatures preclude export crops such as coffee, sugarcane, and cacao. The typical household economy is focused on potatoes, grains, and livestock, especially llamas and alpacas, raised for family consumption and local sale (Figure 21.9). Farmed areas in central and southern sections of the Peruvian Andes tend to be so dry that only irrigated areas are productive. Machu Picchu and other Inca sites in the highlands are important tourist magnets, generating welcome revenue for Peru. Growing, processing, and trafficking of coca and cocaine represent as much as 8 percent of the national economy.

A narrow, discontinuous coastal lowland lies between the Peruvian Andes and the sea. This is an extremely arid desert as a result of the cold ocean waters along the shore. Air moving landward becomes chilled over these waters. Onshore, a relatively cool and heavy layer of surface air underlies warmer air—a **thermal inversion** condition similar to the one that promotes smog in Los Angeles. Turbulence needed

Figure 21.8 Sea lion and marine iguanas in the Galapagos

Figure 21.9 Llama in a highland Andean pasture

to generate precipitation is absent, and only fog and mist yield moisture. Most rivers on the western slopes of the Andes carry so little water that they disappear before reaching the sea. Where they have water, irrigated agriculture has been practiced for millennia along these river valleys.

WORLD
REGIONAL
Geography⊛Now™

Click Geography Literacy to see an animation of a thermal inversion and take a short quiz on the facts or concepts.

In 1535, the Spanish founded Lima (population: 8.3 million) a few miles inland from a usable natural harbor. The city was well located for reaching the Native American communities and mineral deposits of the Andes, especially the Cerro de Pasco silver deposits that, for a time, led the world in silver production. Lima was designated the colonial capital of the **Viceroyalty of Peru** and became one of the main cities of Spanish America. In the late 19th century, foreign companies became interested in the commercial agricultural possibilities of the Peruvian coastal oases. As a result, commercially oriented oases now spot the coast, producing irrigated cotton, rice, sugarcane, grapes, and olives. Small ports associated with the oases are generally fishing ports and fish processing centers, as the typically cold **upwelling** waters along the coast are exceptionally rich in marine life. But this resource provides an uncertain livelihood. For a few years in the 1960s and early 1970s, Peru led the world in volume of fish caught (largely anchovies for processing into fishmeal and oil), and the port of Chimbote was the world's leading fishing port. But since then, periodic changes in water temperature associated with El Niño have led to a drastic decline and widespread unemployment in the industry. As of 2004, Peru was the world's third nation in total fish caught (after China and Japan).

Overall, Peru's agriculture is poor, and the country now imports a large part of its food. The country will depend increasingly on oil and other mineral exports to gain foreign exchange for these imports (see the discussion on page 587). At present, the overall export value of metallic ores—mainly copper, zinc, lead, and silver—exceeds that of oil, and Peru is the world's second largest silver producer. Metal mining sites are divided between the Andes and the coast.

More than one-half of Peru occupies the tropical rain forest of the Amazonian plains. Fewer than 2 million Peruvians inhabit this large area. The main settlement, Iquitos, originated in the natural-rubber boom of the Amazon Basin during the 19th century and has tended to be more oriented to outside markets via the Amazon than to Andean and Pacific Peru. The Amazon River is open to oceangoing vessels for 2,200 miles (c. 3,600 km), from its mouth all the way to Iquitos.

Economic development in Peru has been hampered by gyrations between civilian and military rule. The 1980s saw widespread guerrilla activity by the Maoist group **Shining Path (Sendaro Luminoso)**, and by the smaller **Tupac Amaru Revolutionary Movement (MRTA)**, which continued its assaults on government interests well into the 1990s. Altogether an estimated 69,000 people, mostly Native American Quechuan, died, perhaps half at the hands of the Shining Path and one-third killed by Peruvian armed forces. An ethnic Japanese president named Alberto Fujimori led a largely successful counterattack against the insurgents. Following his resignation in the wake of a fraudulent election, Fujimori was succeeded in 2001 by Alejandro Toledo, Peru's first Native American president. The new government promised peace and continued economic growth.

Landlocked Bolivia

Bolivia is a land of extreme alpine conditions (see Figure 21.5). It is the poorest Andean country and one of the poorest in all of Latin America, with an annual per capita GNP PPP of only $2,400, compared to $1,600 for Haiti and $7,262 for the Latin American average. Unlike that of several of its neighbors, Bolivia's economy has worsened over the past quarter century, with exports lower and unemployment higher.

Like the other Andean countries, Bolivia has a sparsely populated tropical lowland east of the mountains. But unlike the others, it has no coastal lowland to supplement the production of its mountain basins and valleys. Although limited by extreme tierra fría conditions combined with aridity, agriculture employs 47 percent of Bolivia's people. La Paz (population: 1.6 million) is the main city and the de facto capital (Sucre is the legal capital, but the supreme court is the only branch of the government located there). At an elevation of about 12,000 feet (c. 3,700 m), La Paz averages 45°F (8°C) in the daytime in its coolest month and only 53°F (12°C) in its warmest month.

A major rural cluster of people lives at a comparable elevation around Lake Titicaca on the Peruvian border. At 12,507 feet (3,812 m), this is the highest navigable lake in the world and boasts regular steamer service among lakeside ports high in the Andes. Smaller settled population clusters are scattered through Andean Bolivia, mainly along the easternmost of the two main Andean ranges and the altiplano. Export-oriented commercial agriculture is not possible at such high elevations. The climate is unfavorable, and products would have to reach the sea via 13,000-foot (c. 4,000-m) passes to the Pacific coast or perhaps by a long route to the Rio de la Plata to the east.

Bolivia is predominantly (more than 55 percent) Native American. Known as "Upper Peru" by the Spanish in the 16th and 17th centuries, Bolivia was settled by Europeans partly for the land and labor of its indigenous inhabitants and also for its minerals. In the late 16th century, the Bolivian mine of Potosí was producing about one-half of the world's silver output. Since the late 19th century, tin has been the mineral most sought in the Bolivian Andes, although exports of tin have now become far less valuable than exports of natural gas from Bolivia's eastern savanna lowlands. Since the 1920s, oil and gas discoveries have been made around the old lowland frontier town of Santa Cruz, and their future exports have become a very contentious issue for landlocked and mainly Native American Bolivia. Andean indigenous peoples, Europeans, and Japanese have settled around Santa Cruz in growing agricultural areas. Farther north, where the lowlands have a tropical rain forest climate and where intermediate Andean slopes lie in the tierra templada, other pioneer agricultural settlements have been attracting immigrants from the highlands who come to pan for gold, raise cattle, cut wood, and grow coffee, sugar, and coca.

Bolivia had a long history of benign use of the coca leaf by its indigenous population and became a major cocaine producer and exporter in the late 20th century. Most of the country's recently declining coca output comes from the Chapare of central Bolivia. With U.S. aid, the government has promoted the substitution of bananas and pineapples for coca. Coffee has also become a replacement crop in regions of the tierra templada and tierra fría.

21.2   Brazil: Populous Rain-Forested Giant

Brazil is the giant of Latin America (Figure 21.10). It has an area of about 3.3 million square miles (8.5 million sq km) and a population of 179 million as of 2004, second only to the United States in the Western Hemisphere. Brazil is 5 percent larger than the 48 conterminous states (see Figure 19.2). It has only about 60 percent as many people as the United States, but its annual rate of natural increase is so much higher 1.3 percent as opposed to 0.6 percent in the United States) that the population gap between the two countries is expected to narrow rapidly.

Brazil has an increasingly important role in hemispheric and world affairs. Its growth is especially notable in its overall population, in the explosive urbanism that has made São Paulo (population: 19.1 million) and Rio de Janeiro (population: 11.5 million; Figure 21.11) two of the world's largest cities, in the rise of manufacturing, and in the diversification of export agriculture. Brazil has the largest economy in South America. Its expanding domestic market and labor force have accelerated economic development in Brazil, but the burgeoning population also poses serious problems. Brazil is hard-pressed to keep economic output ahead of population increase and to maintain adequate education and other social services.

Brazil has its millionaires, a substantial and growing middle class, and large numbers of technically skilled workers, but the benefits from development have been distributed so unevenly that many millions still live in gross poverty. Brazil has the unfortunate distinction of having the world's largest maldistribution of national wealth, according to the World Bank, with the poorest 10 percent of the population receiving just 1 percent of the total income, and the richest 10 percent receiving almost half. (For comparison, in the United States, the poorest 10 percent receives 1.8 percent of the total income—making them the poorest population of all the MDCs—while the richest 10 percent gets almost one-third.) Brazil is also burdened with a huge foreign debt owed to the International Monetary Fund and other international lenders for its ambitious and costly development projects launched after the mid-1970s. However, Brazil has many things going for its future, including abundant and varied natural resources, many allies among the world's advanced economic powers, and a remarkable degree of internal harmony among the diverse racial, ethnic, and cultural elements that make up the country's population.

Brazil's Environmental Regions

Brazilian development must contend with many problems associated with humid tropical environments. The equator crosses Brazil near the country's northern border, and the Tropic of Capricorn lies just south of Rio de Janeiro and São Paulo. Only a small southern projection extends into the midlatitude subtropics. Within the Brazilian tropics, the climate is classed as tropical rain forest or tropical savanna nearly everywhere, with small areas of tierra templada in some highlands and a section of tropical steppe climate near the Atlantic in the northeast. The tropical rain forest and savanna ecosystems are characterized by lateritic soils that cannot tolerate cultivation for long once the native vegetation is removed. Overabundance of water sometimes creates general health and sanitation problems, while drought makes the provision of an adequate water supply difficult in the large savanna areas that have pronounced dry seasons.

Brazil does not have much high or rugged terrain but instead is made up mainly of low uplands and extensive lowlands, the latter primarily in the Amazon Basin and along the Atlantic coast. Brazil's three major landform divisions are the Brazilian Highlands, the Atlantic Coastal Lowlands, and the Amazon River Lowland. The country also includes a part of the Guiana Highlands that extends across the border from Venezuela, but this region of hills and low mountains is sparsely populated and not significant in the country's economy.

The Brazilian Highlands

The Brazilian Highlands is a triangular upland extending from about 200 miles (c. 325 km) south of the lower Amazon River in the north to the border with Uruguay in the south. On the east, it is separated from the Atlantic Ocean by a narrow coastal plain, although it reaches the sea in places. Westward, the highlands stretch into parts of Paraguay and Bolivia. The northern edge runs raggedly northeastward toward the Atlantic, steadily approaching the Amazon River but never reaching it. The topography is mostly one of hills and river valleys interspersed with tablelands and scattered ranges of low mountains. From Salvador to Porto Alegre, the highlands descend steeply to the coastal plain and the sea.

About half of Brazil's metropolitan cities with a million or more people are seaports spotted along this coastal lowland: In the north are Fortaleza, Natal, Recife, Maceió, and

Brazil

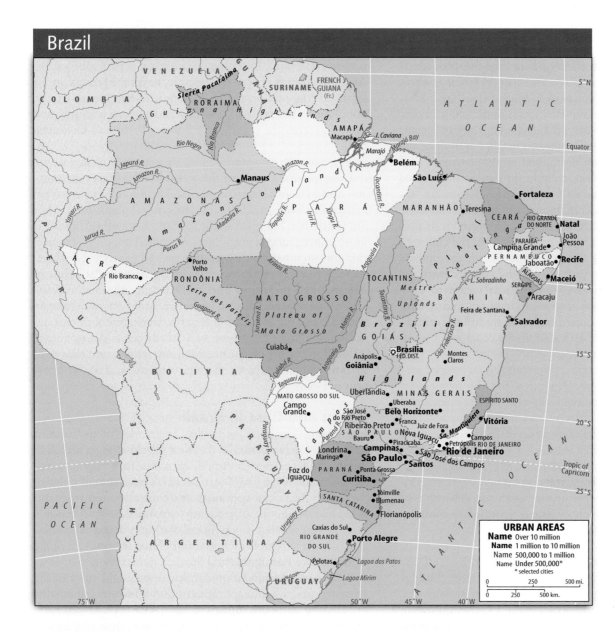

Figure 21.10
Principal features
of Brazil

Figure 21.11 The statue of Christ the Redeemer, Rio de Janeiro's most famous landmark, overlooks the city's spectacular harbor.

Salvador; in the south lie Rio de Janeiro, Vitória, Santos, and Porto Alegre. Two of Brazil's three historic capitals are among them: Salvador (formerly Bahia), the colonial capital from the 16th century to 1763, and Rio de Janeiro, which was made the colonial capital in 1763 and then, with independence in 1822, the national capital until 1960. In that year, the capital was moved to the new inland city of Brasília (population: 2.2 million). The shift to Brasília, a frontier city at the edge of the Amazon Basin, was symbolic of the country's hopes for future development of the Amazon. Designed to be home for an estimated 500,000 people, Brasília was also supposed to be highly integrated in socioeconomic classes and ethnic populations. It has far outgrown its projected population and has also become one of the country's most segregated cities. As elsewhere in Brazil, the opulent homes of the well-to-do are a distressing contrast with the dense *favelas* (shantytowns). Brasília was meant to be a growth pole attracting new settlement, industry, and governmental activity—attempting to break the traditional centrality of Rio de Janeiro and São Paulo on the coast—but its industrial growth has been negligible.

In the northeastern Brazilian Highlands is an interior area of semiarid steppe climate. This area and the tropical savanna around it are called the Caatinga, a word referring to its natural vegetation of sparse and stunted xerophytic forest. Wet enough to attract settlement at an early date, it has been a zone of recurrent disaster because of wide annual fluctuations in rainfall. Years of drought periodically witness agricultural failure, hunger, and extensive mass emigration. This region, also called the Sertão or the Northeast, is considered the least prosperous region of Brazil and always connotes both drought and poverty.

The Atlantic Coastal Lowlands

The narrow Atlantic Coastal Lowlands lie between the edge of the Brazilian Highlands and the Atlantic in most places, with the continuity broken occasionally by seaward extensions of the highlands. Climatically, the northern part of the discontinuous ribbon of plain is tropical savanna, the central part is tropical rain forest, and the southern part is humid subtropical. This was the first part of Brazil settled by Europeans, and it has been in agricultural use since the 1500s. These alluvial plains first attracted the Portuguese to introduce and raise sugarcane as the dominant Brazilian export crop for early trans-Atlantic trade. It was for this plantation economy and trade that African slaves were introduced to Brazil. Agricultural development has removed all but about 7 percent of the original forest cover. Ecologists lament this deforestation and identify Brazil's Atlantic forest as one of the critical biodiversity hot spots of the earth for the species richness of its remaining, threatened vestiges.

The Amazon River Lowland

Brazil's other major lowland area is the massive Amazon River Lowland, containing the world's largest continuous area of tropical rain forest, centered on what by many measures is the world's greatest river (see The World's Great Rivers, pages 584–587). This lowland extends and gradually widens westward from the Atlantic on either side of the Amazon's mouth. Far in the interior, beyond Manaus (at the strategic conjunction of the Amazon and Rio Negro Rivers; population: 1.6 million), the lowland widens abruptly and extends across the frontiers of Bolivia, Peru, and Colombia to the Andes.

The Amazon Lowland is made up mainly of rolling or undulating plains, except for the broad floodplains of the Amazon and its many large tributaries. The floodplains are often many miles wide and lie below the general plains level. They are nearly flat, but most of the lowland lies slightly higher, is more uneven, and is not subject to annual flooding. Tropical rain forest climate and vegetation prevail nearly everywhere, although southern fringes are classed as tropical savanna. Manaus and the ocean port of Belém are the major cities. Large parts of the Amazon region are still very undeveloped and sparsely populated, but Brazil's development in particular is populating the region at a steady pace. An estimated 23 million people live in Brazilian Amazonia, and that population is growing, partly through migration, at about 2 percent per year.

Boom and Bust in Brazil

In the 1494 **Treaty of Tordesillas,** Spain and Portugal agreed on a line of demarcation in the New World along a meridian intersecting the South American coast just south of the mouth of the Amazon (Figure 21.12). The eastward bulge of the South American continent was thus allocated to Portugal. The Portuguese subsequently expanded beyond this line into the rest of what is now Brazil, with international boundaries drawn in remote and little-populated regions east of the Andes where Portuguese and Spanish penetration met.

The colonial foundations of Brazil were laid early in the 16th century, when a sporadic trade with local Indians for brazilwood (a South American tree, Caesalpinia echinata, that produced a much-desired red dye in early trade) began. To combat French and Dutch incursions, Portugal promoted settlement beyond its first small trading posts in the 1530s. The coastal settlements became increasingly valuable as they rapidly developed exports of cane sugar to Europe and brought shiploads of slaves from Africa to work in the cane fields. Slave traders in search of Native American slaves penetrated the interior. Catholic missions were founded among the indigenous people, often bringing uncontrollable disease epidemics and slaving expeditions in their wake.

Expansion of coastal sugar plantations and increasing penetration of the interior dominated development in 17th century Brazil. Portugal's colony became the world's leading source of sugar. Although some sugarcane was grown as far south as the vicinity of the Tropic of Capricorn, the main plantation areas developed farther north in tropical areas that lay closer to coastal ports and to Europe. Meanwhile, missionaries penetrated up the Amazon, and a series of wide-ranging expeditions, originating largely from São Paulo,

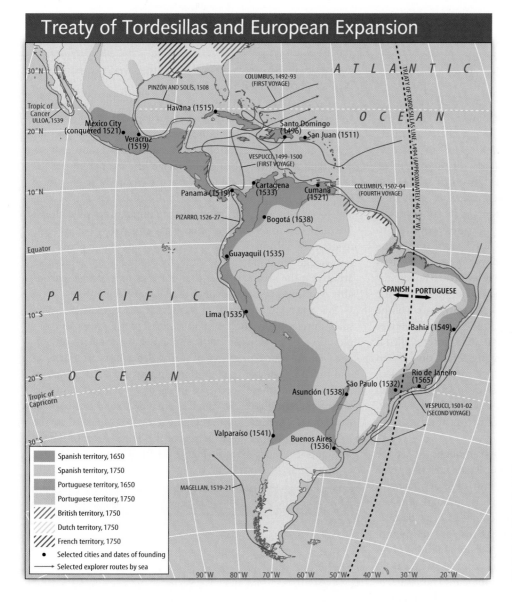

Treaty of Tordesillas and European Expansion

Figure 21.12 In 1494, Spain and Portugal met at Tordesillas, Spain, and made decisions about dividing up the non-Christian world. Spain acquired most of the lands of the New World, except for the eastern wedge of South America, which went to Portugal. Other European powers established outposts in subsequent centuries.

took place to enslave Native Americans and to locate new mineral deposits. In the late 1600s, sizable gold finds were made in the Brazilian Highlands hundreds of miles to the north and northwest of São Paulo in what is now called Minas Gerais (meaning "mines of many kinds") state.

During the 18th century, Brazil's center of economic gravity shifted southward in response to the expansion of gold and diamond mining in the highlands north and northwest of São Paulo and Rio de Janeiro. By the mid-1700s, Brazil was producing nearly one-half the world's gold. In the latter part of the century, this output declined, but by that time, much of the highland around the present city of Belo Horizonte (founded in 1896 as a state capital) had been settled, along with areas near São Paulo and spots in the far interior. Despite the difficulty of traversing the Great Escarpment between Rio de Janeiro's magnificent harbor and the developing interior, "Rio" became the port and economic focus for the newly settled territories. Gold and diamonds were ideal high-value commodities to move by pack animals over poor roads and trails to the port.

Besides the attraction of mineral wealth, the economic distress of the sugar coast to the north contributed to the rise of the new areas. The sugar decline resulted from increasing development of competing sugar production in other New World colonies (notably in the West Indies). Plantation owners and their slaves began moving to the newly developing areas in Brazil. A ranching economy, soon augmented by growing cotton, began in the Caatinga (Sertão) area inland from the sugar coast. However, 18th-century events set a still-unbroken pattern of economic hardship for Brazil's Northeast, characterized by hard-fought battles to produce sugar and cotton, regular visits by drought and floor, severe poverty, and the flight of the poor to other parts of Brazil.

During the 19th century, Brazil was changed by rubber and coffee booms, by much-intensified settlement of the southeastern section, by the beginnings of modern industry, and by extension of the ranching frontier far into the Brazilian Highlands. The country underwent these changes as a colony until 1822, as an independent empire with a strongly federal structure until 1889, as a military dictatorship until

The World's Great Rivers

The Amazon, Its Forest, and Its People

The Amazon is *the* world's great river (Figure 21.A). It is not the longest; its 3,915 miles (6,264 km) are surpassed in length by the Nile. But by all other standards, the Amazon rules. It has more volume than any other river. As much as one-fifth to one-fourth of all the world's available fresh water (excluding what is locked up in ice) is in the Amazon and its tributaries at any given moment. By some of the odd measures people use to evaluate such things, enough water is discharged out of the mouth of the Amazon each day to supply New York City's freshwater needs for 9 years. That is enough water to flood New York State to a depth of 6 inches! So much water flows from the mouth of the Amazon that it forces back the salty waters of the Atlantic for as much as 200 miles (320 km) offshore. The river is so impressively large that early Portuguese explorers called it **"the River Sea."** Many creatures normally found only at sea have adapted themselves to its freshwater vastness, including dolphins, sharks, and rays. The river is so deep so far upstream that oceangoing vessels can navigate it as far as Iquitos, Peru, nearly 2,200 miles (3,600 km) upstream from the Amazon's mouth.

The Amazon River is probably best thought of as the main drain for a vast basin, the Amazon Basin, covering some 2.7 million square miles (4.4 million sq km), or about ten times the size of Texas. About half of this basin is in Brazil, with the remainder sprawling into eight other countries. The basin is home to the world's largest remaining expanse of tropical rain forest and some of the world's most remote populations of indigenous peoples. It is also a region that promises, or seems to promise, prospects for economic development. The issues of pristine forest, Native American populations, and development are intertwined in many ways, some of them very contentiously. Here is a brief look at these problems.

The Amazon Basin rain forest is probably the world's largest storehouse of plant and animal species, the great majority of which have not yet been described by science. Ecologists argue that preservation of the rain forest's biodiversity is essential for ensuring the genetic variability nature requires for change and evolution and for helping to ensure valuable future supplies of food, medicines, and other resources for people. The forest also acts as a **carbon sink** to mitigate possible global warming due to excess greenhouse gas emissions. The governments of Brazil and other Amazon Basin countries generally acknowledge the importance of the area's natural state but also argue that the area is so vast that portions of it can be reasonably developed to help advance their economies.

Since the 1970s, Brazil in particular has aggressively pursued development in five major areas of the Amazon: exploitation of iron ore and other minerals; resettlement of "excess" populations from Brazil's crowded southeast; farming and ranching; timber exploitation; and hydroelectric development. Each of these serves as an agent of deforestation. All are ultimately linked to Brazil's expanded road network in Amazonia. To begin its development of the Amazon, in the 1970s, Brazil began construction of the **Trans-Amazon Highway** and its feeder system (Figure 21.B). The main line runs

30

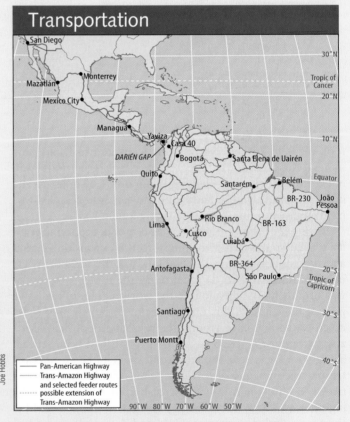

Figure 21.B The Trans-Amazon highway system and other select transport routes of Latin America

Figure 21.A Early Portuguese explorers called the Amazon "the River Sea" for its enormous width and volume. This view from midriver is about 1,900 miles upstream from the river's mouth.

Joe Hobbs

roughly east to west from the Atlantic toward its ultimate, long-awaited destination: the Peruvian coast. Planned construction for 2005–2009, part of an international project called the **Regional Initiative for the Infrastructure Integration of South America,** includes the final link that will cross the Andes and finally give Brazilian exporters their desired easier outlet to Asian markets. Principal feeder lines built so far include the controversial BR-364 through the Brazilian states of Acre and Rondonia and BR-163 running north–south through the heart of the region.

This road construction typically initiates the following sequence of events, illustrated well by Highway BR-364. First, the main road is cut as a swath through the forest (Figure 21.C). Landless peasants, lured by cheap and sometimes free land, cut and burn the rain forest along the main road and the smaller lanes linked to it. Typical of farmers' experiences with tropical slash-and-burn, or swidden, agriculture elsewhere, they may be able to wrest only 3 to 5 years of corn, rice, or other crops from the soil before it is exhausted (Figure 21.D). They then move on to clear and cultivate new lands—the classic pattern of shifting cultivation. If this used plot were left alone, it would return to mature tropical rain forest in a matter of decades. In Amazonia, however, the typical pattern is that large cattle operations move in to use the areas conveniently cleared for them by the farmers. As 57 million cattle

(one-third of Brazil's total cattle population) feed and trod over these Amazonian lands, the soils are exposed to excessive cycles of wetting and drying and to relentless bombardment by ultraviolet radiation. This turns the soil into the bricklike substance called laterite on which it is all but impossible for forest regrowth to occur. When ranchers have exhausted the land in this fashion, they move their cattle on to the next plots cleared by farmers. This is called **shifting ranching,** and following the shifting cultivation, it is part of a cycle that is exceptionally destructive of the rain forest.

This cycle and the other uses of the forest, especially a boom in commercial soybean cultivation, have resulted in the removal of about 16 percent of the original tropical rain forest cover of the Amazon Basin. Brazil lost about 4 percent of its Amazon forests between 1990 and 2003 (Figure 21.E). In recent years, the annual rate of deforestation in Brazil has been about 9,000 square miles (23,000 sq km), or about the size of New Hampshire. Amazon rain forest destruction has become a cause célèbre for environmentalists around the world, who denounce the development efforts of Brazil and the other countries. Most of the region's governments and people see it differently, denouncing foreigners as hypocrites who have cut down their own forests and who are intent on stunting development of the rain forest countries.

There are alternative development schemes, such as the

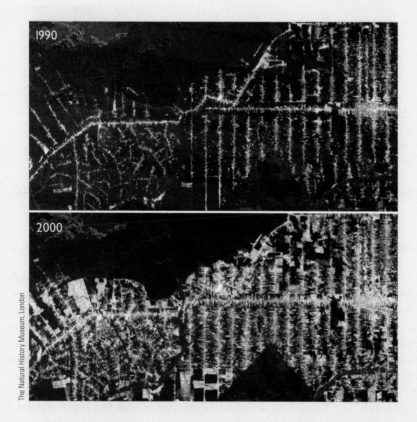

The Natural History Museum, London

Figure 21.C A "before" (1990) satellite image of the region of BR-364, Amazonia
Figure 21.D An "after" (2000) satellite image of the region of BR-364, Amazonia

(continued on page 586)

281

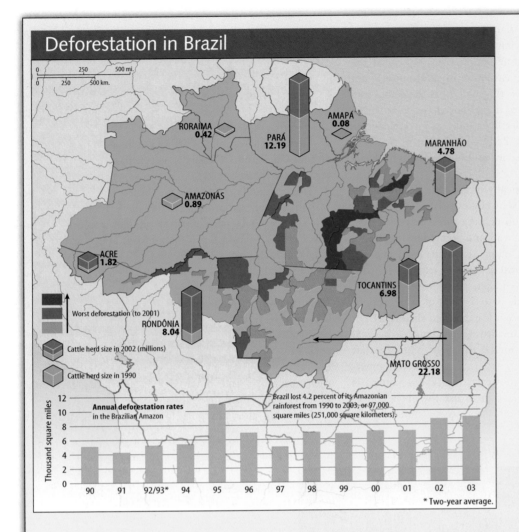

Deforestation in Brazil

0   250   500 mi.
0   250   500 km.

RORAIMA
0.42

AMAPÁ
0.08

PARÁ
12.19

MARANHÃO
4.78

AMAZONAS
0.89

ACRE
1.82

TOCANTINS
6.98

RONDÔNIA
8.04

Worst deforestation (to 2001)

Cattle herd size in 2002 (millions)

Cattle herd size in 1990

MATO GROSSO
22.18

Brazil lost 4.2 percent of its Amazonian rainforest from 1990 to 2003, or 97,000 square miles (251,000 square kilometers).

Annual deforestation rates
in the Brazilian Amazon

Thousand square miles

12
10
8
6
4
2
0

90   91   92/93*   94   95   96   97   98   99   00   01   02   03

* Two-year average.

WORLD REGIONAL
Geography⊕Now™

Active Figure 21.E Deforestation of the Brazilian Amazon   *See an animation based on this figure, and take a short quiz on the facts or concepts.*

commercial extraction of valuable exports like rubber and brazil nuts from intact tropical forests. These take place in areas like Brazil's **extraction reserves**, but they are very limited in economic impact relative to the conventional development enterprises in Amazonia, and their long-term prospects for growth are uncertain. In the meantime, Brazil and other Amazon Basin countries have passed legislation to slow or halt illegal logging and other unlawful uses that contribute to the rain forest's destruction. Authorities in Brazil have slowed the illegal logging of valuable mahogany trees, but the problem has

1894, and finally, as a civilian-ruled republic. Its political history was turbulent, and not until 1888 did the country emancipate its large slave population.

One of the major economic developments of the 19th century in Brazil was the boom in wild-rubber gathering in the Amazon Basin. This followed the discovery of vulcanization—the process that made rubber more flexible and durable and, consequently, made rubber a much more valuable material—by American Charles Goodyear. At that time, Amazonia was the only home of the rubber tree (*Hevea brasiliensis*). A "rush" up the rivers followed, and thousands of widely scattered settlers began to tap the rubber trees and send latex downstream. Manaus grew as the interior hub of this traffic and Belém as its port. But some rubber tree seeds

were smuggled to England, and seedlings transplanted to Southeast Asia became the basis of rubber plantation agriculture there. By 1910, Amazon rubber production, which in contrast depended on tapping scattered wild trees, was practically dead.

Coffee was the preeminent boom product of Brazil during the 19th century. It had long been produced in Brazil's Northeast in small quantities, and by the late 1700s, it was an important crop around and inland from Rio de Janeiro. During the 1800s, production spread in the area around São Paulo and then increased explosively in output during the last decades of the century. This growth was related to the broadening world market for coffee, the emergence of expanded and faster ocean transport, the nearly ideal climatic

331

boomed in the Peruvian Amazon. About 90 percent of the estimated 1.6 million cubic feet (45,000 cubic meters) of mahogany exported each year from Peru to the United States—where it becomes furniture, acoustic guitars, home decks, and coffins—is illegal, passing through a chain of corrupt officials and exporters until it reaches U.S. lumberyards.

Development in the Amazon has increasingly provoked conflicts with the region's indigenous populations. There is particular concern now about how fossil fuel production will impact the Kichwa, Achuar, Schuar, and other Native Americans. In Peru, there are plans to build 800 miles (1,280 km) of pipelines that would take natural gas from lowland rain forest areas, many inhabited by Native Americans, to coastal refineries and ports. Another big natural gas project is planned for a part of the Brazilian Amazon inhabited by indigenous people. In Ecuador, where oil already makes up half the country's exports by value, there are plans to double production, again mainly in areas where there are significant Native American populations.

In almost all cases, the Native Americans in the areas affected believe that little if any benefit will come to them from these developments. They are becoming increasingly activist and are using a combination of threats against the oil interests and pleas for more benefits from the projects to ensure that their concerns are heard. Some of the native groups have powerful activist allies abroad and have come to rely on the Internet to make their cause known; an example is the Kichwa people of Sarayaku (http://www.sarayacu.com), who live in one of the Peruvian oil blocks slated for development. Some Native American communities are developing their own ecotourist enterprises, hoping both to demonstrate to governments that there are alternative economic uses for the rain forest areas and to gain valuable support and publicity from international visitors (Figure 21.F).

Joe Hobbs

Figure 21.F Is ecotourism the way to protect indigenous people and natural environments in the Amazon?

and soil conditions in this part of the Brazilian Highlands, and the arrival of railway transport to move the product over land. A spectacular and critical transport development was the completion by British interests in 1867 of a railway linking the coffee-collecting city of São Paulo with the ocean port of Santos at the foot of the Great Escarpment. The coffee frontier crossed the Paraná River to the west and spread widely to the north and south of São Paulo.

Coffee production doubled and redoubled. In the late 1800s, over a million immigrants a year from overseas poured into Brazil, mostly into the region near São Paulo, to work in the coffee industry. São Paulo began its rapid growth as the business center and the collecting and forwarding point for the coffee industry. It also began its productive ca-

reer as an industrial center. These developments were aided by early hydroelectric production in the region.

But the coffee boom went bust. Brazil continued to lead the world in coffee production and export, as it does today, but increasing international competition depressed coffee prices. The coffee industry continues to struggle with wildly fluctuating demands and prices. Despite coffee's decline, industrialization continued to grow around Sao Paulo in the 20th century. The railway network expanded. Japanese immigrants began arriving in 1925 and made important contributions to agriculture. There was a major expansion of irrigated rice production in the Paraíba Valley between Rio de Janeiro and São Paulo. (Rice is a staple of the Brazilian diet, as are corn, beans, and manioc.)

Ready for Takeoff?

The pace of development in Brazil accelerated after 1945, with periodic fluctuations. Economic growth took place under elected governments between 1945 and 1964, but occurred most rapidly between 1964 and 1985 under a military dictatorship with technocratic leanings and great ambitions. Unfortunately, Brazil's hoped-for economic "takeoff" as an industrialized power coincided with the devastating impacts of a quadrupling of oil prices in the mid-1970s due to events in the volatile Middle East. In 1985, the dictatorship relinquished power to a new elected government, which then had to face the accumulated problems of that stalled takeoff. Among these were a crushing national debt that the military regime had contracted to finance the country's ambitious development projects and widespread economic misery caused by governmental attempts to keep up with payments on the debt.

Construction of an effective transportation system has been one of the country's priorities in the last few decades. This has involved building a network of long-distance paved highways, including the trans-Amazon system, and an emphasis on trucks as movers of goods. Major urban centers have been connected, and additional highways have been pushed into sparsely populated and undeveloped areas. The new roads have attracted ribbons of settlement, have facilitated the flow of pioneer settlers to remote areas, and have furthered the economic development of such areas by connecting them to markets (for rubber, minerals, timber, fish, some agricultural products, and tourism) and sources of supply. This entire move toward greater mobility has also stimulated a steady growth in the manufacturing of automobiles and trucks for both domestic and international markets. Brazil is Latin America's leading automobile producer.

Four major agricultural boosts in older settled territories have contributed to the country's recent economic growth. The government has succeeded in modernizing and reinvigorating the old sugar industry, and Brazil is the world's leading sugar exporter. In 1962, a severe freeze in Florida gave Brazil an entering wedge into the world market for orange concentrate, and by the 1980s, Brazil was overwhelmingly the world's largest exporter. It remains so today—producing half the world's total—despite a recent viral disease epidemic called "citrus sudden death." Brazil has the world's largest commercial cattle herd (160 million animals) and is third in global beef exports, behind the United States and Australia. In the 1960s, Brazil became a major soybean producer and exporter, with production centered in the subtropical southeast and in a new district in the tropical Brazilian Highlands (see Definitions and Insights, below). Soybean production is now pushing into the Amazon Basin and is a major source of deforestation there.

This expansion is in response to growing demand for soy products in China, the world's largest soy consumer. Brazil surpassed the United States as the world's leader in soybean exports in 2003 and is second only to the United States in soybean production. Soybeans account for 3 percent of the value of Brazil's exports. They are a major component of the overall 40 percent of Brazilian exports that are from the agribusiness sector. To gain an even larger market share, Brazil has recently given up on its longstanding opposition to **genetically modified (GM) foods** and is producing GM soybeans and other crops. Brazil is trying to maintain a delicate balance between its North American markets, which have no problem with GM foods, and its European markets, which shun them. Brazil claims it can produce GM and non-GM foods to satisfy both. American farmers look nervously at Brazil's surging agricultural exports. Having won the case it made to the World Trade Organization that U.S. farmers were too heavily subsidized, making Brazilian cotton and

Definitions + Insights

The Versatile Soybean

The soybean (Glycine max), originally domesticated in China, is the crop of the future for a number of world regions (Figure 21.G). The plant has a very high percentage of protein—approximately 30 to 35 percent—and no starch. It affixes nitrogen to the soil as it grows, making it a crop that is much less demanding on the soil than most. While nearly all of the U.S. production in soybeans is used for animal feed, the Brazilian crop is widely exported and used domestically for industrial production of chemicals, paints, ink, adhesives, insect sprays, and is now even used as a fuel blend (called soy diesel and akin to ethanol, a blend of corn and gasoline). The world market for soybeans is growing steadily not just because of the industrial uses of the bean but because it also can be blended with other products to create a high-protein, modestly priced food—such as tofu (bean curd)—that are increasingly important in global efforts to diminish world hunger.

Figure 21.G Soybeans

Joe Hobbs

other exports uncompetitive, Brazil would threaten to undercut American agricultural exports on the world market if the United States abides by the WTO ruling.

There has been rapid expansion in Brazil's manufacturing sector. The products involved today are not just the simple ones of early-stage industrialization, although textiles and shoes are still major items. Steel, machinery, automobiles and trucks, ships, chemicals, footwear and plastics are all important exports, as are weapons. Aircraft are also manufactured, mainly for the domestic market. São Paulo is the industrial core, producing nearly two-thirds percent of Brazil's manufactured goods. Brazil's metals make an important contribution to the country's industry. The Brazilian Highlands are very mineralized, with ores of iron, bauxite, manganese, tin, and tungsten. Brazil has about an eighth of the world's proven iron ore reserves and is the world's leading producer and exporter of iron ore. Because of these reserves, Brazil is now in the top 10 of world steel exporters, thanks to China's red-hot economy and its voracious appetite for this important product.

Brazil's greatest resource deficiency is in fossil fuels. Known supplies of coal are inadequate for the country's needs. Proven petroleum supplies are limited to the Northeast. Located mainly offshore, these fields were found during the intense exploration after the drastic rise in world oil prices in 1973. They supply 70 percent of Brazil's oil consumption, and crude oil has become a major Brazilian import. This is true despite a relatively successful program to develop and use **gasohol,** which contains 20 to 25 percent **ethanol** (alcohol derived from plants) currently supplied by sugarcane grown in the Northeast and in the area around São Paulo. Brazil is now the world's largest producer of ethanol. Unless the problem of oil supply can be solved, the country's continuing industrial development drive may stall.

The shortage of fossil fuel also translates into pressure on the country's tropical rain forests. In the central Amazon area of Grande Carajas, for example, there are large deposits of iron ore that elsewhere might be smelted for pig iron with the use of fossil fuel powered furnaces. Lacking the necessary coal or iron ore, Brazil uses rain forest hardwoods to power these furnaces. Numerous dams have been built and are scheduled for the Amazon to feed Brazil's appetite for electricity, almost 90 percent of which is produced by hydropower. Droughts like the one that afflicted Brazil in 2001 cause serious energy shortages and economic disruptions.

21.3 The Southern Midlatitude Countries: South America's "Down Under"

Four countries of southern South America—Argentina, Chile, Uruguay, and Paraguay—differ from all other Latin American countries in being essentially midlatitude rather than low latitude in location and environment (Figure 21.13; see also Figure 19.5a and b). Northern parts of all the countries except Uruguay do extend into tropical latitudes, but

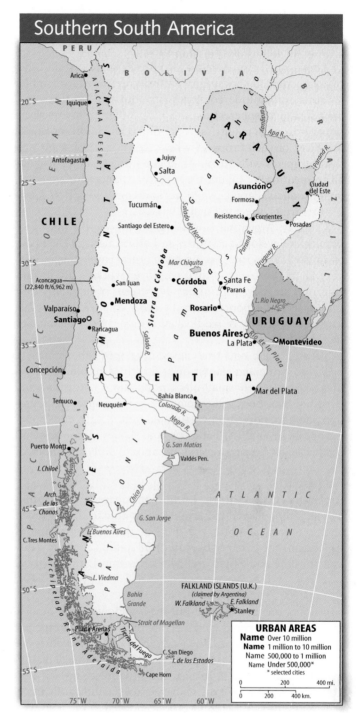

Figure 21.13 Principal features of Argentina, Chile, Uruguay, and Paraguay

the core areas lie south of the Tropic of Capricorn and are climatically subtropical. Argentina barely extends into the tropics in the extreme north, centers in subtropical plains around Buenos Aires, and reaches far southward into latitudes equivalent to those of northern Canada. Chile also extends south to these latitudes. Uruguay's comparatively small territory is entirely midlatitude and subtropical, and most Paraguayans live in the half of their country that lies poleward of the Tropic of Capricorn. Argentina, Paraguay, and Uruguay have the same humid subtropical climatic classification as the southeastern United States. Chile's midsection or core area has a Mediterranean (dry-summer subtropical) climate like that of southern California.

These four Latin American countries share certain other regional characteristics. Agriculturally, they are more similar to other midlatitude countries (in both the Northern and Southern Hemispheres) than to most other Latin American or other tropical countries. Argentina is an important export producer of such crops as wheat, corn, and soybeans and thus competes with North American farmers. Both Uruguay and Argentina are significant exporters of meat and other animal products. Paraguay exports soybeans and cotton. Chile exports fruits, vegetables, and wine.

Uruguay, Argentina, and Chile are among the most economically developed countries in Latin America. They have relatively low dependence on agricultural employment, with only 12 to 18 percent of their employed workers still in agriculture. Paraguay, with about 40 percent of its labor force still on farms, is poorer than the other countries, partly because of its history of landlocked isolation from the outside world.

Because these countries never developed labor-intensive and export-oriented plantation agricultures in colonial times, they have relatively small black populations. Argentina and Uruguay have been major magnets for European immigration and have populations estimated to be 85 percent "pure" European in descent. Paraguay and Chile, in contrast, attracted fewer Europeans and are about 90 percent mestizo.

Argentina

Argentina's area of nearly 1.1 million square miles (2.8 million sq km) is second in Latin America only to that of Brazil and is about one-third the area of the 48 conterminous American states. Its population of 38 million ranks it a distant third behind Brazil and Mexico within Latin America.

Argentine Agriculture

Argentina, one of the "**neo-Europes**" endowed with quality farming and ranching lands, is one of the world's principal sources of surplus food. The leading agricultural exports are wheat, corn, soybeans, and beef and are shipped mainly from the capital city of Buenos Aires (population: 13.2 million). Argentine wine exports have grown significantly in recent years, too. Agriculture is concentrated in the country's **core region**, known as the humid pampas (Figure 21.14). *Pampas* comes from a Native American word for "plains," and they have a prairie vegetation similar to that of the North American Great Plains (see Figure 2.7). The humid subtropical climate of the pampas is similar to that of the southeastern United States, featuring hot summers, cool to warm winters, and annual precipitation averaging 20 to 50 inches (c. 50 to 130 cm). The grasses of the original prairie provided an unusually high content of organic material (humus), just as they did in the North American prairie areas. The local beef industry is fed from the grasses and grain that

Figure 21.14 The Argentine pampas

flourish in this setting. Argentine per capita beef consumption is the second highest in the world (after Uruguay).

From the 16th century to the late 1800s, the humid pampas were sparsely settled ranching country, parts of which were still controlled by indigenous people. Then, in the late 19th and early 20th centuries, Argentina experienced a rapid shift to commercial agriculture and urbanization. There was heavy European immigration, and population grew rapidly. Commercial contact with Great Britain contributed much to Argentina's new dynamism. The British market for imported food was increasing rapidly. Refrigerated shipping came into use, allowing meat from distant sources such as Argentina to be imported into the United Kingdom. A British-financed rail network (the densest in Latin America) was built in Argentina to tap the countryside for trade. The network focused on Buenos Aires, which rapidly became a major port and large city. British cattle breeds, favored by the tastes of British consumers, were introduced to Argentine ranches, or *estancias,* along with alfalfa as a feed crop. European immigrant farmers assumed a major role in agriculture, generally as tenants on the large ranches that continued to dominate the pampas economically and socially. The immigrants came primarily from Italy and Spain but also from other countries, giving present-day Argentina a complex mixture of Spanish-speaking citizens from various European backgrounds. Soon an export agriculture developed in which wheat, corn, and other crops supplemented and then surpassed the original beef exports.

Argentine areas outside the humid pampas are less environmentally favored, less productive, and less populous. The northwestern lowland just inside the tropics, called *El Gran Chaco,* has a tropical savanna climate and supports cattle ranching and some cotton farming. Westward from the lower Paraná River and the southern part of the pampas, the humid plains give way to steppe and then desert, with production and population declining accordingly. Population in these western areas is concentrated in oases along the eastern foot of the Andes, where mountain streams emerging onto the eastward-sloping plains have long been impounded for irrigation. Vineyards and, in the hotter north, sugarcane are irrigated. Córdoba is the largest oasis service center and the second largest city in Argentina.

To the south of the pampas lie thinly populated deserts and steppes in the Patagonian plateau, a bleak region of cold winters, cool summers, and incessant strong winds. Native American country until the early 20th century, Patagonia now has millions of sheep on large ranches but little other agricultural production. Between 1999 and 2003, global wool prices tripled and mutton prices doubled, leading to a boom in the Patagonian sheep industry. Several factors were responsible: a decrease in a longstanding oversupply of sheep in Australia, rising oil prices (which made synthetic wools expensive and natural wool more attractive), and the mad cow disease that reduced European appetites for beef and increased them for mutton.

Urban and Industrial Argentina

Only about one-tenth of Argentina's labor force is directly employed in agriculture and ranching. The great majority of Argentineans are city-dwellers employed in services and manufacturing. Metropolitan Buenos Aires has about one-third of the national population. It was settled in the late 16th century as a port on the broad Paraná estuary called the Rio de la Plata. But until Argentina became reoriented toward British and then world markets, the city remained an isolated outpost of a country focused on the north and the western oases, with relatively little trade by sea. In the 19th century, Buenos Aires became the capital and grew explosively as a seaport, rail center, manufacturing city, and destination of mainly European immigrants.

Industrialization in Argentina followed a sequence found in many developing nations. The process began when the government imposed tariffs on overseas goods to help small-scale industries supply consumer items for sale within the country and to support meat packing and other plants processing agricultural exports. Some consumer industries such as textile manufacturing and shoemaking were expanded with government support when the country was cut off from imported supplies during World Wars I and II. After World War II, a state-led policy of industrialization was instituted, boosting manufacturing and diversification of industrial output to include chemicals, machinery, and automobiles. Imported coal to supply power was replaced by domestic oil (mainly from fields on the coast of Patagonia), together with natural gas from the archipelago of Tierra del Fuego and the Andean piedmont. Imported ore and coal supply steel mills. There is now considerable hydroelectric output and some nuclear power.

Argentina's industries have problems. Manufacturing has never become competitive on the world market, and the country's trade pattern is still dominated by agricultural exports and industrial imports. Manufacturing contributes less to national production than should be expected of a country of Argentina's size and potential. In 1913, Argentina was actually a prosperous country due to strong exports from the pampas, much foreign (mainly British) investment in railways and other infrastructure, and immigration of capital and talent from Spain and Italy.

Growth generally stalled after that, especially because of the country's disastrous politics, described below. Children and grandchildren of the hundreds of thousands of European immigrants who came to Argentina after World War II returned to their European "homes" by the thousands, especially during a recession in the late 1990s and an outright collapse of the economy in 2001, when more than half of the population fell below the poverty line. After that, Argentina's currency was devalued and the economy rebounded sharply, thanks especially to soaring prices for the country's exports of agricultural goods, natural gas and oil, and new output of honey, construction materials, cosmetics, clothing, and leather goods.

Disastrous Politics

Political turmoil and government mismanagement of the Argentine economy have taken a heavy toll on the country. After initial disturbances following independence, Argentina became a constitutional democracy controlled by wealthy landowners. This situation continued for many decades until 1930, when a military coup toppled the constitutional government. Between 1930 and 1983, the country was controlled either by military juntas or by Juan Perón or his party, with brief intervals of semidemocratic government. Although there were variations, the military generals and the Perónists tended to be ultranationalist and semifascist. They distrusted democracy, communism, and free-enterprise capitalism and believed in authoritarianism, violent repression of opposition (which itself was often violent), and state ownership and/or direction of major parts of the economy. In what was know as Argentina's **Dirty War**, their death squads were responsible for the disappearances and presumed deaths of an estimated 30,000 people—the so-called "**disappeared generation.**" They courted the support of a growing urban working class by providing it with jobs, pay, and benefits beyond what was justified by its productivity and output, and they paid for these things by inflating the currency and borrowing from abroad. Inflation rates eventually reached hundreds of percentage points a year, lowering savings and leading to the flight of investment capital and an economy heavily focused on currency manipulation and speculation rather than productive investment.

Chile

Like Argentina, Chile was set back by violent politics in the late 20th century. A U.S.-backed coup in 1973 resulted in the death of the country's socialist president, Salvador Allende, and swept in the brutal regime of General Augusto Pinochet. Pinochet began the free market reforms that were inherited and enlarged by a freely elected government in 1990. Chile after Pinochet has become the most prosperous countries of

South America on a per capita GDP PPP basis. It has enjoyed booming exports of copper, salmon, fruits, and wood products, particularly since concluding unilateral trade agreements with the United States in 2002, the European Union in 2003, and South Korea in 2004.

Chile's elongated shape conceals the country's size, which is more than double that of Germany or about that of the U.S. state of Texas. Chile stretches approximately 2,600 miles (c. 4,200 km) from its border with Peru to its southern tip at Cape Horn, but in most places, its is only about 100 to 130 miles (c. 160 to 210 km) wide (see Figure 21.13). The country has to struggle variously with rugged terrain, aridity, wetness, and cold, and due to such extremes, only one relatively small section in the center has any considerable population. The long interior boundary lies mostly near the high Andean crest, with the populous core areas of neighboring countries lying far away on the other side of the mountains. Although the Andean barrier has been passable historically, traffic across the mountains has been difficult and is still light.

The Middle Chilean Core Region

Chile's populous central region of Mediterranean (dry-summer subtropical) climate occupies lowlands between the Andes and the Pacific from about 31° to 37°S latitude. Mild wet winters and hot dry summers characterize this area, as they do southern California at similar latitudes on North America's west coast (see Figure 2.7). This strip of territory has always been the heart of Chile's core area. Topographically, it consists of lower slopes in the Andes, a hilly central valley, and low coastal mountains. The area was the southern outpost of the Inca empire and was occupied in the mid-1500s by a small Spanish army. The army made use of land grants from the king of Spain to create an isolated ranching economy. Grants varied in size by the military rank of the grantees, and this fact, together with the conquered status of the indigenous population, created a society characterized by great social and economic inequality. At one end of the scale was the aristocracy, composed of those holding enormous ranches. At the other end was a mass of landless cowboys and subsistence farmers. Men from all ranks of society married indigenous women freely, sometimes several at a time. This created the Chilean nation of today, which is over 90 percent mestizo.

Owners of the great haciendas dominated the country well into the 20th century, and ranching remained the primary pursuit in middle Chile. Irrigated and nonirrigated wheat, vineyards, and feed crops became agricultural specialties. During the 1960s and early 1970s, population pressure and resulting political demands led to large-scale land reform that reduced the role of many landowners. Growing population pressure in the core region has contributed to rapid urbanization in recent decades. The national capital of Santiago, founded at the foot of the Andes in the 16th century, has a metropolitan population of more than 5.4 million. Metropolitan Valparaíso, the core region's main port

and coastal resort, has 900,000, and the port of Concepción has about 930,000. Altogether, the area of Mediterranean climate is the home of about 80 percent of Chile's 16 million people. The country is now 87 percent urban, exceeded or equalled in South America only by Uruguay (93 percent), Argentina (89 percent), and Venezuela (87 percent).

North and south of the core area are transition zones with lower populations. On the northern edge of the core zone, the population declines across a narrow band of steppe to the arid and almost empty wastes of the Atacama Desert, the driest region on Earth. In the south, the transition zone is larger and more important economically. South of Concepción, rainfall increases, average temperatures decline, and the Mediterranean type of climate gives way to a wet version of the marine west coast climate, with natural vegetation of dense forests. The southernmost part of the Mediterranean climatic area and the adjacent northern fringe of the marine west coast area as far south as Puerto Montt have become part of Chile's core in the past century.

The indigenous Araucanian people of the southern forests fought off Inca expansion and resisted most Spanish expansion for three centuries. They were never conquered but eventually assimilated into Chilean society. Their full-blooded indigenous descendants are still numerous in this part of Chile. Also distinctive is a German element descended from a few thousand immigrants who settled on this wild frontier in the mid-19th century. Today, the German community is economically and culturally very important, even putting its stamp on architectural styles. Most people in the southern transition zone, however, are descendants of mestizo settlers who came from the core of Chile in the past century as population pressure increased. Agriculture in this green landscape of woods and pastures is devoted largely to beef cattle, dairy cattle, wheat, hay, deciduous fruits, and root crops.

Northern and Southern Chile

North of the core, Chile is desert, except for areas of greater precipitation high in the Andes. This part of Chile, the Atacama Desert, is one of the world's few nearly rainless areas. It is not a hot desert; mild temperatures and fog result from the almost constant winds from the cool ocean current offshore. Chile gained most of the Atacama by defeating Bolivia and Peru in the **War of the Pacific** (1879–1883). That is what deprived Bolivia of its outlet to the sea; Bolivia had previously included a corridor across the Atacama to the sea at the port of Antofagasta, and Peru owned the part of the desert farther north. The war was fought over control of mineral resources, primarily the Atacama's world monopoly on sodium nitrate, a material much in demand at that time for use in fertilizers and the manufacture of smokeless gunpowder. Since then, other mineral resources of the Atacama and the adjacent Andes, principally copper, have eclipsed sodium nitrate in importance. Chile has the world's largest copper mine, Escondida in the Atacama Desert, and has benefited from the recent surging demand for copper, prompted mainly by China's boom.

South of Puerto Montt, the marine west coast section of Chile continues to the country's southern tip in Tierra del Fuego, a cool island region of windswept sheep pastures and wooded mountains divided between Chile and Argentina. In this thinly populated strip, the central valley and coastal mountains of areas to the north continue southward, but here the valley is submerged, and the Pacific reaches the foot of the Andes along a rugged coastline penetrated by fjords. The spectacular pinnacles of the Paine Towers National Park are here. Extreme wetness, cool temperatures, violent storms, and dense mountain forests are characteristic. Little agriculture beyond sheep grazing is possible, and population is very sparse.

Chile has only limited coal, oil, and gas—mainly along and near the Strait of Magellan, which separates the South American mainland from Tierra del Fuego—and has come to rely on Argentine natural gas for about 40 percent of its electricity production. The countries have invested billions of dollars since the mid-1990s on cross-border pipelines and power plants. But in 2004, energy shortages in Argentina led to reductions in gas exports to Chile, leaving that country scrambling for new resources. Natural gas from Bolivia would be the obvious choice, but Bolivia's historical grudge against Chile makes that prospect unlikely. Meanwhile, there is talk about expanding highways to enable Argentine and Brazilian producers to transport soybeans, mineral resources, and manufactured goods to Chilean ports to promote trade across the Pacific Rim.

Uruguay: Buffer State

Small Uruguay, with an area of 68,500 square miles (177,400 sq km; about the size of Missouri or half the size of Germany) and bordered by Argentina and Brazil, is a part of the humid pampas (see Figure 21.13). Its independence from Argentina and Brazil resulted from its historic buffer position between them. In colonial times, there were repeated struggles between the Portuguese in Brazil and the Spanish to their south and west over possession of this territory flanking the mouth of the Rio de la Plata river system in southern South America. After Argentine independence from Spain, these struggles continued at the same time that a movement for political separation grew in the territory north of the Rio de la Plata. Great Britain eventually intervened, fearing Portuguese expansion south of the river, and the two contenders signed a treaty in 1828 that recognized Uruguay as an independent buffer state between the two larger countries.

Montevideo, Uruguay's national capital, is the only large city in the country, and its metropolitan population of 1.7 million includes about 50 percent of all Uruguayans. The city has dominated the country almost from Montevideo's inception in the 17th century, first as a Brazilian (Portuguese) and then as an Argentine (Spanish) fort. Its port facilities handle Uruguay's important waterborne trade. Montevideo directs an economy dominated by employment in government-owned services and industries. The city is the main center of Uruguay's manufacturing industries, focused on textiles,

woolens, and meat processing. Agriculture and ranching, with just 14 percent of the labor force, supply the country's meat-rich diet and the greater part of all Uruguayan exports.

Aside from hydropower, which supplies nearly all of the country's electricity, Uruguay has almost no natural resources to support industrialization and depends heavily on imported fuels, raw materials, and manufactured goods, which must be purchased with the proceeds from agricultural and textile exports. An extensive program is underway to privatize government-owned industries and diversify the economy beyond agricultural exports to Uruguay's neighbors.

Landlocked Paraguay

Landlocked and underdeveloped Paraguay has long been isolated from the main currents of world affairs (see Figure 21.13). The Spanish founded Paraguay's capital city of Asunción (population: 1.4 million) in the 1530s at a site far enough north on the Paraguay River to promise security against the Native Americans of the pampas to the south. In addition, high ground at the riverbank gave security from floods. Asunción became the base from which surrounding areas were occupied. But when independence came, Paraguay had no way to reach the sea except through Argentina or Brazil. The main route that developed reached down the Paraguay and Paraná Rivers to Buenos Aires. But this route is much longer than it appears on maps owing to meandering of the rivers, and navigation was hindered by difficulties such as shallow and shifting channels, fluctuations in water depth between seasons, snags, and sandbars. Transport costs were so high that Paraguay could not develop an export-oriented commercial economy, and the country remained locked into a predominantly self-contained economy based on the resources of its own territory.

West of the Paraguay River is the region known as the Chaco ("hunting ground"), which extends into Bolivia, Argentina, and Brazil. This is a very dry area. Precipitation averages 20 to 40 inches (c. 50 to 100 cm) annually, but high temperatures cause rapid **evapotranspiration,** and porous sandy soils absorb surface moisture. These conditions produce an open xerophytic forest in the east, which thins westward and then gives place to coarse grasses on a vast plain of flat alluvium where wide areas are flooded during the wet season. Among the small population of Paraguayans in the entire western region are the Mennonites who came here in the 1920s and 1930s to farm and practice their unique lifestyle.

Only 6 million people lived in all of Paraguay in 2004. Paraguay has been a country of emigration for most of its history and has never attracted any large immigration. The country has a few whites, and there is a high proportion of Native American Guaraní blood in the 95 percent of the population classed officially as mestizo. Urbanization and industrialization have not gone far. Asunción is the only urban area with more than 300,000 people, and about 40 percent of the country's labor force is still employed directly in agriculture. Per capita GDP PPP is far below that of Argentina, Chile, and Uruguay. There are modern road and air links with the outside world and between parts of Paraguay itself. With access to outside markets, exports of soybeans and cotton have grown rapidly and have come to dominate the country's trade.

The future may see larger and more dramatic changes due to three huge new hydroelectric installations along the Paraná River on the southeastern border of the country. Paraguay is now exporting electricity to pay its share of the costs of these projects, which it undertook jointly with Argentina and Brazil. The new hydroelectric output has given Paraguay a far more favorable energy base for the future than this fuel-deficient country has had in the past.

21.4 Antarctica: The White Continent

Antarctica is the world's fifth largest continent, with an area of 5.5 million square miles (14,245,000 sq km) lying south of the tip of South America and virtually filling the Antarctic Circle (Figure 21.15). It is the setting of enormous human dramas in exploration, bravery, and foolhardiness as people have crossed the continent with dogsleds, on skis, on foot, and in airplanes and helicopters. Some, including all members of the Scott Expedition of 1910–1912—which had hoped to beat the Norwegians to the South Pole, but arrived a few days too late—did not return alive.

The mystery of the place comes in part from its winter of darkness, with continual problems of "whiteouts" caused by light refraction on an extensive snow and ice surface covering some 95 percent of the continent. Average temperatures during the summer months barely reach 0°F (−18°C). Winter averages are the coldest in the world, with winter mean temperatures averaging −70°F (−57°C). It is also the world's windiest and driest continent, and a very sunny one during its half year of light; the South Pole on average receives more sunlight each year than the world's equatorial regions.

The whole continent is alive with ice formations created by the 2 to 10 inches (5 to 25 cm) of annual precipitation. The formations fed by this seasonal moisture then move toward the open seas of the South Atlantic. Glacial action of all sorts provides a continual breaking away of sea ice at the margins of the continent, forming icebergs. Sheets of sea ice adjacent to the continent have retreated in recent decades,

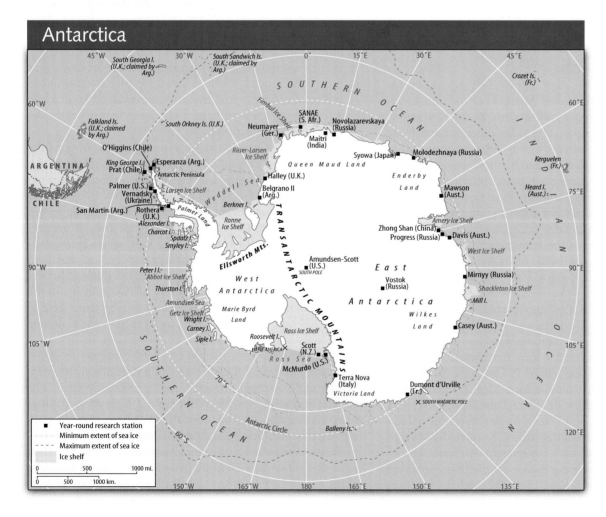

Figure 21.15
Principal features of Antarctica

possibly in connection with the worldwide trend toward warmer temperatures. Records show that the Antarctic Peninsula has warmed by 4.5°F (2.5°C) since 1944, helping to melt the peninsula's fringing ice sheet and in the process allowing the adjacent land glaciers to surge into the sea at an unexpectedly rapid rate. Such a sequence of events leading to rising sea levels is one of the long-anticipated outcomes of global warming. Antarctica has a central position in global concerns about thinning of the atmospheric ozone layer and the consequent increases in ultraviolet radiation, since the "ozone hole" in the earth's stratosphere is concentrated there. There is virtually no human settlement beyond the research teams that have constructed an array of semipermanent structures on the small areas of exposed land that lie near the outer edges, on the island of Little America in the Ross Sea, and at the South Pole itself. About 1,000 researchers are on the continent in the winter and about 4,000 in the summer.

There is growing interest in this distant world. Expensive trips on icebreakers allow international visitors to mingle with penguins and enjoy some of the wildest scenery on Earth. Seven countries have staked claims (some of which overlap) to portions of the continent as national territories: nearby Chile and Argentina, the Southern Hemisphere powers of Australia and New Zealand, the United Kingdom, France, and Norway. No other countries recognize these claims. Both the United States and Russia have avoided laying claim to parts of Antarctica but maintain that they have the right to do so. Such claims, which have been technically "frozen" since 1961, may be moot points. The **Antarctic Treaty,** signed by 45 countries including the 7 laying claims, forbids any exploitation of the continent's natural resources for several decades. There has long been speculation about the potential for considerable resource wealth lying beneath a cap of ice up to 10,000 feet (3,000 m) thick, but to date, no minerals deposits of significant economic value have been proven.

Antarctica is ringed by a number of small, largely volcanic islands. The largest are the Falklands, owned by the United Kingdom but also claimed by Argentina; these two countries fought a brief war over control of the islands in 1982. Fewer than 3,000 people live here, the majority employed in a lucrative fishing business. Other British possessions include South Georgia Island and the South Sandwich Islands, which were important whaling and sealing ports until 1965 but are now uninhabited. France controls Kerguelen and a host of smaller islands in the southern Indian Ocean, all uninhabited except for small teams of scientific researchers. These islands do contribute to France's enormous exclusive economic zone (EEZ), which is the world's largest.

The text now travels far to the north, to the last world region covered in the book, and the home to most of its readers.

SUMMARY

- The Andean countries include Colombia, Venezuela, Ecuador, Peru, and Bolivia. These countries have marked patterns of settlement both along the coasts in the tierra caliente and in the uplands in the tierra templada and tierra fría.

- Major characteristics of the Andean countries include varied environmental zonation, fragmented topographic patterns, populous highlands and sparsely settled frontiers, significant indigenous populations, dominant Hispanic traditions, poverty, and primate cities.

- In Venezuela, major settlement is also concentrated in the uplands, with the population in Caracas at more than 4 million. Discovery of oil in the coastal regions a century ago around Lake Maracaibo prompted rapid industrial growth. Iron resources in the Guiana Highlands support an active export flow, utilizing the Orinoco River in part.

- Venezuelan President Hugo Chavez led a self-proclaimed "Bolivarian Revolution" ostensibly aimed at using the country's oil wealth to enhance the prospects for the poor. It created a political backlash that included a boycott in the petroleum industry, temporarily hurting the country's economy.

- Colombia has three major highland metropolises: Bogotá, Medellín, and Cali. Coastal Colombia has less settlement than the uplands, with the Caribbean shores having the larger cities. Bananas and sugarcane have been the traditional crops, with corn and rice gaining new importance.

- Up to half of Colombia's territory is controlled by two rebel groups, the FARC and the ELN. With U.S. economic and military assistance, the government is fighting back and is also aided by the paramilitary AUC.

- Ecuador has major settlements in the Andean basins. There has been a steady increase in the proportion of Ecuadorian population locating in the lower slopes and the coastal areas. Oil in both Ecuador and Peru has begun to be extracted from the eastern side of Ecuador in the tropical lowlands that drain into the Amazon River. This important economic development is controversial because many indigenous peoples live on lands slated for oil development, and they are generally opposed to the industry.

- Peru's population is concentrated in the high plains region called the *altiplano.* Subsistence agriculture is important for the country's large peasant population, which includes a high proportion of Native Americans. Fish are an important export but are vulnerable to the natural hazard of El Niño.

- Bolivia is South America's poorest country, landlocked high in the Andes with its capital, La Paz, located at 12,000 feet (3,700 m). The Bolivian mine at Potosí produced nearly one-half of the world's silver in the 16th century as a result of the enormous investment of Native American labor and, later, African slaves in mining and refining precious metals for world trade. Natural gas is an important future resource. Bolivia has a majority Native American population. Coca production has declined recently in the country.

- Brazil is the giant of South America. It is larger in area than the 48 conterminous United States but has more than 100 million fewer people. Settlement tends to cluster more along the coast than in the thinly settled interior. In the late 1950s and early 1960s, Brazil undertook a campaign to promote greater settlement in the interior savanna of the country. A new capital, Brasília, was constructed, and in 1960, the seat of federal government was formally moved from Rio de Janeiro on the eastern coast to the new capital. This move has caused some demographic shift, but it is still along the coast that most of Brazil's population is concentrated. Brazil's economy is the largest in South America, and Brazil has the region's widest disparities between wealthy and poor people.

- Brazil has a large realm of tropical rain forest and tropical savanna climate patterns and does not have the highlands so characteristic of South American Andean countries. There is a broad coastal plain, and the Amazon River system drains almost 3 million square miles. Periods of economic change in Brazil have been marked by booms and busts in sugarcane, gold, diamonds, rubber, and coffee. Soybeans are increasingly important in Brazil's agricultural export economy. Beef and orange juice concentrate are also prominent exports. Iron ore and steel are vital nonagricultural exports.

- The Amazon has more volume of water than any other river in the world. The Amazon Basin extends into nine countries, and about half of it lies in Brazil. Development in the Amazon region is very controversial, in part because of the rich biodiversity in its rain forest. Brazilian development in the Amazon is focused on five major areas: exploitation of iron ore and other minerals; resettlement of "excess" populations from Brazil's crowded southeast; farming and ranching; timber exploitation; and hydroelectric development. The Trans-Amazon Highway and its feeder roads, which are scheduled to reach the Pacific soon, help to open Amazonian wilderness to development. Shifting cultivation is often followed by shifting ranching in a very destructive land-use cycle.

- Countries of the southern midlatitudes include Argentina, Uruguay, Paraguay, and Chile. The first three all possess a humid subtropical climate like that in the southeastern United States. Chile, however, is more similar to southern California since both regions are Mediterranean in climate.

- Argentina is the second largest nation in South America, with an area of more than 1 million square miles (2,776,884 sq km). It has major urban growth along the coast and in the alluvial plain of the Rio de la Plata, but it is the agricultural productivity of the *pampas*—the dominant grasslands—of the interior that largely defines Argentina. Beef and wheat are longstanding major Argentine export commodities, joined recently by sheep products. Argentina was relatively prosperous early in the 20th century but suffered setbacks, largely because of political troubles. The economy rebounded well after an economic collapse in 2001.

- Chile has worked to diminish its dependence on the export of copper for its foreign exchange, although it is still very important. The country is a strong advocate of globalization and has joined numerous trade pacts. Chile, which is resented by Bolivia for cutting off access to the sea, is finding it difficult to find adequate fuel sources in the wake of reduced supplies from Argentina. The world's driest desert, the Atacama, is in Chile.

- Uruguay was created as a buffer state between Argentina and Brazil. The country is resource poor, and its economy depends mainly on agricultural exports.

- Landlocked Paraguay has few mineral resources, and many of its people have emigrated. Hydroelectric development figures prominently in future development plans.

- Antarctica is the world's coldest, driest, and least populated continent. There is no permanent population, only scientific researchers. The region's melting sea and land ice may reflect global warming, and the "hole" in the earth's stratospheric ozone layer concentrates over Antarctica. Seven countries stake claims to the continent, but no other countries recognize these claims, and so far they have not had any economic significance.

KEY TERMS + CONCEPTS

Terms in blue are also defined in the glossary.

altiplano (p. 578)
Andean Group (p. 571)
Antarctic Treaty (p. 595)
biodiversity hot spot (p. 575)
Bolivarian Revolution (p. 573)
carbon sink (p. 584)
core region (p. 590)
"disappeared generation" (p. 591)
estancia (p. 590)
ethanol (p. 589)
evapotranspiration (p. 593)
extraction reserve (p. 585)
favela (shantytown) (p. 582)
gasohol (p. 589)
genetically modified (GM) foods (p. 588)

growth pole (p. 582)
internal commotion (p. 576)
laterite (p. 585)
military socialism (p. 573)
National Liberation Army (ELN) (p. 576)
neo-Europe (p. 590)
Organization of Petroleum Exporting Countries (OPEC) (p. 573)
ozone hole (p. 595)
Regional Initiative for the Infrastructure Integration of South America (p. 585)
Revolutionary Armed Forces of Colombia (FARC) (p. 576)
shifting cultivation (p. 585)
shifting ranching (p. 585)

Shining Path (*Sendaro Luminoso*) (p. 579)
tepui (p. 574)
the Dirty War (p. 591)
"the River Sea" (p. 584)
thermal inversion (p. 578)
Trans-Amazon Highway (p. 584)
Treaty of Tordesillas (p. 582)
Tupac Amaru Revolutionary Movement (MRTA) (p. 579)
United Self-Defense Forces of Colombia (AUC) (p. 576)
upwelling (p. 579)
Viceroyalty of Peru (p. 579)
War of the Pacific (p. 592)

REVIEW QUESTIONS

Assess your understanding of this chapter's topics with additional quizzing and concept-based problems at http://earthscience.brookscole.com/wrg5e.

1. What are the Andean countries of South America? What traits do they share?

2. What roles do coca/cocaine play in the national economies of Colombia, Peru, and Bolivia?

3. What factors have helped make Bolivia the poorest country in South America?

4. In what ways is Brazil the giant of South America?

5. What are the five major components of Brazil's development of the Amazon?

6. What cycle of land use is particularly destructive to the Amazon rain forest?

7. Why are Native Americans of the Amazon Basin generally opposed to fossil fuel development there?

8. What have been the principal booms and busts in Brazil's economic history? What exports are particularly strong today and why?

9. What rising and falling economic fortunes have been experienced in Argentina and why? What accounts for the current boom in sheep production?

10. Why has Chile's economy been particularly strong recently?

11. What claims exist to Antarctica's territory? Are these important for the countries or that continent?

DISCUSSION QUESTIONS

1. How and why did the oil-wealthy country of Venezuela plunge into an economic crisis? Who are the major contenders in Venezuela's struggles and what do they want?

2. Discuss these factions and their interests in Colombia: the government, the paramilitary, and the rebels. What are their contending interests? In what ways are some of the warring factions similar?

3. What are prospects for fossil fuel development in Colombia, Peru, and Brazil?

4. Why is it a problem in countries like Brazil that wealth is concentrated is so few hands?

5. What is the Trans-Amazon Highway system? What role does it play in the region's development and deforestation?

6. What are the main problematic issues in the development of the Amazon Basin? What policies and actions might be taken to de-

velop the region while also safeguarding its unique environments and cultures?

7. What are the risks to Brazil of using more genetically modified soybeans and other crops?

8. How did politics in Argentina contribute to the country's economic misfortunes?

9. What energy-related problems, worsened by geographic problems, exist between Chile and Argentina and between Chile and Bolivia?

10. Both Uruguay and Paraguay are small countries. Which is poorer and why?

11. What are Antarctica's important roles in global climatic issues?

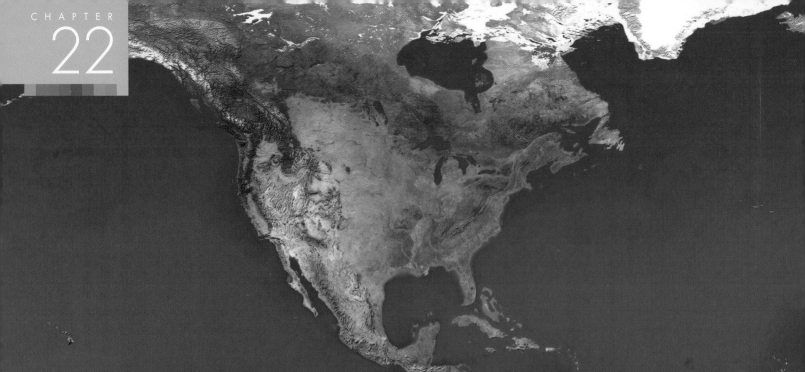

A Geographic Profile of the United States and Canada

Arizona's Monument Valley is one of North America's best recognized natural landscapes.

chapter outline

22.1   Area and Population

22.2   Physical Geography and Human Adaptations

22.3   Cultural and Historical Geographies

22.4   Economic Geography

22.5   Geopolitical Issues

chapter objectives

This chapter should enable you to:

⌖ Appreciate the wide range of Native American adaptations and ways of life in this region's diverse environments

⌖ Recognize Canada and the United States as countries shaped primarily by British influences but also by a wide range of foreign immigrant cultures

⌖ Understand how the United States acquired its vast land empire

⌖ View the prosperity of the United States and Canada in part as products of their vast natural resource wealth

⌖ Trace the rise of the United States to a position of global economic and political supremacy, and the emergence of its post 9/11 approach to the world

WORLD
REGIONAL
Geography ⊛ Now™

Look for this logo in the text and go to GeographyNow at http://earthscience.brookscole.com/wrg5e to explore interactive maps, view animations, sharpen your factual knowledge and geographic literacy, and test your critical thinking and analytical skills with unique interactive resources.

The United States and Canada, together with Mexico and Central America make up the continent of North America. Mexico and Central America are so linked with South America in culture, language, and tradition that they are best classified within the separate region of Latin America. The United States and Canada as a culture region are sometimes called Anglo America, emphasizing their British colonial origins and their contrasts with Latin America. Both Anglo America and Latin America were Native American regions well before the arrival of Columbus in 1492. Since then, both regions, but especially Anglo America, have been major target destinations for migrating peoples. The United States and Canada are today a complex cultural mosaic overlaying an essentially British political, cultural, and economic foundation. The island of Greenland belongs politically to Denmark but rises from the same continental crust on which Canada sits, so it is also described briefly in this region.

22.1   Area and Population

Canada is slightly larger in area than the United States (3.85 million sq mi/9.97 million sq km, compared with 3.72 million sq mi/9.63 million sq km) but is less wealthy and much less powerful on the global stage than is the United States (Figure 22.1). The region's map comparison

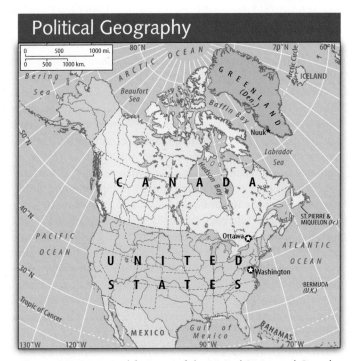

Figure 22.1 Principal features of the United States and Canada

TABLE 22.1 North America: Basic Data

Political Unit	Area (thousand/ sq mi)	Area (thousand/ sq km)	Estimated Population (millions)	Annual Rate of Increase (%)	Estimated Population Density (sq mi)	Estimated Population Density (sq km)	Human Development Index	Urban Population (%)	Arable Land (%)	GDP PPP Per Capita ($US)
Canada	3849.7	9970.7	31.9	0.3	8	3	0.943	79	5	29,800
Greenland (Den.)	836.3	2166.0	0.05	0.0	0.1	0.02	N/A	80	0	20,000
United States	3717.8	9629.1	293.6	0.6	79	30	0.939	79	19	37,800
Summary Total	**8403.8**	**21,765.8**	**325.6**	**0.5**	**38**	**15**	**0.939**	**79**	**10**	**37,013**

Sources: *World Population Data Sheet,* Population Reference Bureau, 2004; *U.N. Human Development Report,* United Nations, 2004; *World Factbook,* CIA, 2004.

with Europe is on page 59. Canada had a population of 31.9 million in 2004, compared with 293.6 million in the United States (Table 22.1 and Figures 22.2a and b).

The majority of this region's 325.6 million people live in the eastern half of the region, from the St. Lawrence Lowlands and Great Lakes south and east. The estimated average population density of the two countries in 2004 was only 79 per square mile (30/sq km) for the United States and 8 per square mile (3/sq km) for Canada, with its tremendous expanse of sparsely settled northern lands. With its vast icecap, Greenland is even more sparsely populated, with just 50,000 people living in an area of 836,000 square miles (2.17 million sq km), or 0.1 persons per square mile (.02/sq km).

Canada and the United States share the longest international border in the world—5,527 miles (8,895 km). About 90 percent of the Canadian population lives within 100 miles (161 km) of this border, inhabiting only 12 percent of Canada's territory. Conversely, only a small percentage of the U.S. population lives within 100 miles of the Canadian border.

The population distributions of Canada and the United States reveal the overwhelmingly urban character of these countries. City-dwellers account for 79 percent of Canadians and 79 percent of Americans. Canada's population clusters in the cities of four main regions: the Atlantic region of peninsulas and islands at Canada's eastern edge; the culturally divided **core region** of maximum population and development along the lower Great Lakes and St. Lawrence River in Ontario and Quebec provinces; the Prairie region in the interior plains between the Canadian Shield on the east and the Rocky Mountains on the west; and the Vancouver region on and near the Pacific coast at Canada's southwestern corner. Each region is marked by a distinctive physical environment and associated cultural landscape; each has a different ethnic mix and distinctive economic emphasis. Each plays a different role within Canada and has its own set of relationships with the United States and other countries. All the regions have important links with the Canadian North.

In the United States, populations are particularly concentrated in urban areas of the Northeast and portions of the West, with fully 53 percent of the country's population living in coastal areas. As of 2004, about 62 million people, or 21 percent of the national population, lived within the Northeast on 5 percent of the nation's area. The Northeast's population density is several times that of the South, Midwest, or West. The density is much more extreme within the narrow urban belt stretching about 500 miles (c. 800 km) along the Atlantic coast from metropolitan Boston (in Massachusetts) through metropolitan Washington, D.C. The belt is often called the Northeastern Seaboard, Northeast Corridor, Boston-to-Washington Axis, **Boswash,** or **megalopolis.** Here, seven main metropolitan areas, including the country's largest—New York—contain about 39 million people, or 1 in every 8 Americans. In the West are another 67 million people, a little more than one-fifth of the nation, more than half of whom live in California. Fully 36 percent of that state's population lives in the metropolitan area of Los Angeles.

Table 24.1, 650

From the beginning of European settlement until the 1970s, this world region had a rapidly growing population. By 1800, approximately two centuries after the first European settlements, there were more than 5 million people in the United States and several hundred thousand in Canada, and both immigration and natural increase continued to swell their ranks. Now, population growth rates of the United States and Canada are low (0.6 and 0.3 percent, respectively), testifying to the passing of these countries into the final phase of the demographic transition. The United States and Canada now account for about 5 percent of the world's population and about 13 percent of the world's land area (including Antarctica).

43

One of the striking features of the population geography of the United States is that despite a low rate of natural increase, this rate is rather high for a more developed country (MDC). This is due mainly to immigration rather than natu-

Population Distribution

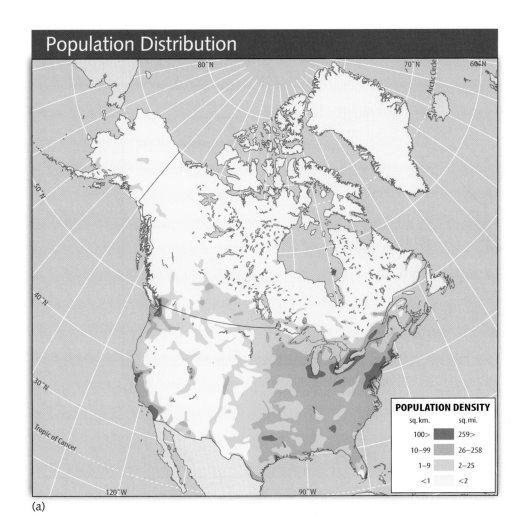

(a)

Population Cartogram

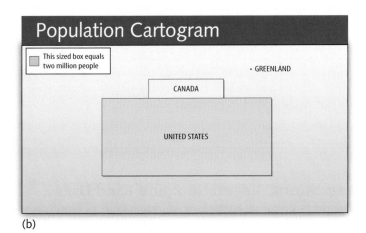

(b)

Figure 22.2 (a) Population distribution and (b) Population cartogram of the United States and Canada

ral increase, and both the United States and Canada are best appreciated as nations of immigrants and of continuing immigration. These themes are carried into the discussion of cultures later in this chapter and in the following two chapters, but modern migration flows are introduced briefly here.

Each year, approximately 1.1 million immigrants arrive in the United States and 183,000 in Canada (Figure 22.3; see also Figure 2.c). This number is well in excess of the legal immigration quota of 675,000 per year for the United States; Canada has no quota. The excess numbers are known as **illegal aliens** and include the **undocumented workers** that obtain temporary or other employment. Latin America is the largest source of both legal and illegal migration into the United States, with about 750,000 Latin Americans arriving each year. Other principal sources of migration are China, India, and Vietnam. The largest immigrant flows into Canada come from China, India, and the Phillipines. Unfortunately, not all immigrants come to the United States and Canada of their own will or to engage in the livelihoods they had hoped to; there is an all too large population of modern day slaves (see Geographic Spotlight, page 602).

Immigration to the United States and Canada

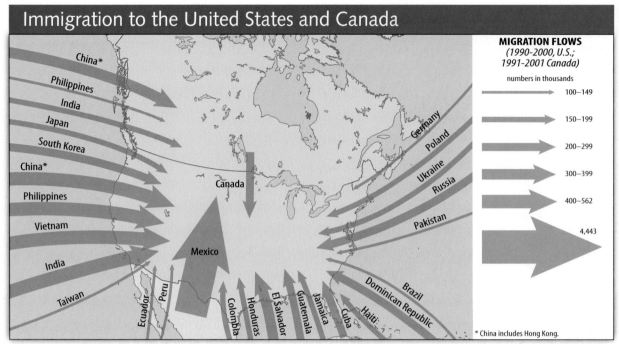

MIGRATION FLOWS
(1990–2000, U.S.;
1991–2001 Canada)

numbers in thousands

→	100–149
→	150–199
→	200–299
→	300–399
→	400–562
→	4,443

China*
Philippines
India
Japan
South Korea
China*
Philippines
Vietnam
India
Taiwan
Canada
Mexico
Ecuador
Peru
Colombia
Honduras
El Salvador
Guatemala
Jamaica
Cuba
Haiti
Dominican Republic
Brazil
Pakistan
Russia
Ukraine
Poland
Germany

* China includes Hong Kong.

Active Figure 22.3 Migration flows into the United States and Canada
See an animation based on this figure, and take a short quiz on the facts and concepts.

WORLD REGIONAL Geography Now™

Geographic Spotlight

The Modern-Day Slave Trade

Modern slavery is not confined to Africa. The London-based watchdog group Anti-Slavery International defines a **slave** as someone forced to work under physical or mental threat, who is completely controlled by an owner or employer, and who is bought or sold. By this definition, more than 30 million people worldwide are slaves.

Modern slaves may be found in the United States, where an estimated 50,000 people, mostly women, are trafficked each year to work as sweatshop laborers, strippers, and prostitutes. The victims are struggling to escape poverty in countries such as Thailand, India, and Russia. Promising legitimate work in a prosperous land, traffickers smuggle them into the country (particularly into Los Angeles), where they are whisked away to unfamiliar locations and put to their tawdry labors. The "master" holds the slave in submission by intimidation, such as reminding her that she is an illegal immigrant and insisting she might be raped by police if apprehended and threatening to have family members in the home country killed if she attempts to escape. Few of the slaves speak English, at least initially, making their escape from servitude more difficult. Their ordeals do not necessarily end when they do find freedom; they may be deported back into the hands of traffickers in their home countries and then be trafficked back to the United States again. New legislation in the United States will stiffen penalties for traffickers and make it easier for the victims of trafficking to stay freely in the United States.

179

22.2 Physical Geography and Human Adaptations

The natural environments of the United States and Canada are remarkably diverse, and include some of the most spectacular landscapes on the planet. This diversity reflects circumstances that have produced many microgeographies and landscapes with distinctive identities and use patterns.

Landforms, Vegetation, and Land Uses

The maps depicting the following discussions of landforms, climates and biomes, and land uses are Figures 22.4, 22.5, and 22.6. In the region's extreme northeast, the glacially scoured and thinly populated Canadian Shield is formed of ancient rocks that are rich in metal-bearing ores. The surface is generally rolling or hilly. Agriculture is limited and marginal due to poor soils and a harsh climate that is similar to

Physical Geography

ELEVATION

meters	feet
5000>	16400>
2500–5000	8200–16400
1000–2500	3280–8200
500–1000	1640–3280
250–500	820–1640
0–250	0–820
<0	<0

Maximum extent of winter sea ice

Typical minimum extent of summer ice

Permanent ice

Figure 22.4 Reference map of the topographic features, rivers, lakes, and seas of the United States, Canada, and Greenland

145 Russia's northeastern Siberia. Hydropower, wood, iron, nickel, and uranium are major resources. The United States section of the Canadian Shield borders Lake Superior and is called the Superior Upland.

The Arctic Coastal Plain occurs in two widely separated sections and is primarily unpopulated wilderness. The Alaska section along the Arctic Ocean contains rich oil deposits, some of which have been exploited and some of which are the source of heated debate about development (see page

682). The section along Canada's Hudson Bay is economically insignificant so far.

The Gulf–Atlantic Coastal Plain is low and mostly level, with generally sandy and infertile soils. There is agriculture in scattered areas of intensive development. Pine and mixed oak-pine forests cover large areas. Swamps, some with hardwoods, and marshes abound. The coast is indented by river mouths and bays on which many prominent seaports are located. Prairie grasslands were the original vegetation in some

Climates of the United States

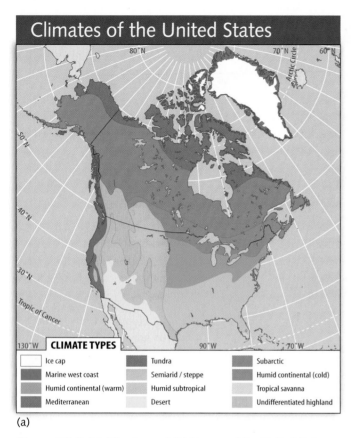

CLIMATE TYPES

☐ Ice cap	☐ Tundra	☐ Subarctic
☐ Marine west coast	☐ Semiarid / steppe	☐ Humid continental (cold)
☐ Humid continental (warm)	☐ Humid subtropical	☐ Tropical savanna
☐ Mediterranean	☐ Desert	☐ Undifferentiated highland

(a)

Biomes of the United States

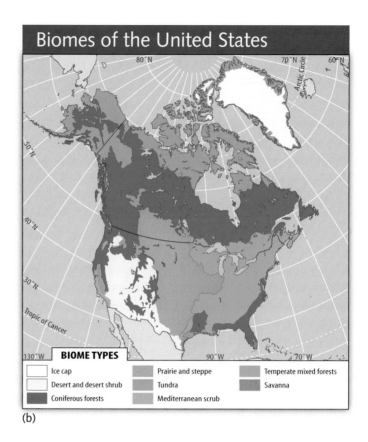

BIOME TYPES

☐ Ice cap	☐ Prairie and steppe	☐ Temperate mixed forests
☐ Desert and desert shrub	☐ Tundra	☐ Savanna
☐ Coniferous forests	☐ Mediterranean scrub	

(b)

Figure 22.5 (a) Climates and (b) biomes of the United States, Canada, and Greenland

Agricultural Land Use

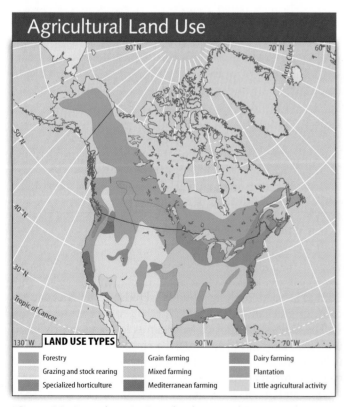

LAND USE TYPES

☐ Forestry	☐ Grain farming	☐ Dairy farming
☐ Grazing and stock rearing	☐ Mixed farming	☐ Plantation
☐ Specialized horticulture	☐ Mediterranean farming	☐ Little agricultural activity

Figure 22.6 Land use in Canada, the United States, and Greenland

coastal areas in Texas and southwestern Louisiana and in some inland areas underlain by limestone. The Lower Mississippi Valley is a subregion of productive soils. Oil, gas, sulfur, and salt are major resources in the Gulf Coast section, where they provide the basis for oil-refining and chemical industries.

The Piedmont, inland from the Gulf–Atlantic Coastal Plain, is formed of harder and more ancient rocks and is a bit higher and hillier. The soils are better, but in central and southern Piedmont areas, they were depleted as cotton, tobacco, and corn were grown on hilly areas with heavy rains. Many cities, such as Richmond, Virginia, Washington, D.C., and Baltimore, Maryland, are located along the fall line between the two physiographic regions, where falls and rapids mark the head of navigation on many rivers and have long supplied water power.

The Appalachian Highlands, lying west of the Piedmont, form a complex system in which mountain ranges, ridges, isolated peaks, and rugged dissected plateau areas are interspersed with narrow valleys, lowland pockets, and rolling uplands. The Appalachians are old mountains, so elevations are generally low, ranging from about 2,000 feet (600 m) to 6,684 feet (2,037 m). Most of the land is occupied by coniferous, deciduous, or mixed forests. Coal, hydropower, and wood are notable resources. The Appalachian region has a surprisingly large population for a highland area, and its importance in the regional economy is significant.

The Interior Highlands share many physical similarities with the Appalachian Highlands, but they are lower and have only minor coal deposits. The Interior Highlands are divided by the Arkansas River Valley into the dissected Ozark Plateau on the north and the east–west oriented ridges and valleys of the low Ouachita Mountains on the south.

The Interior Plains lie mostly within the drainage basins of the Mississippi, St. Lawrence, Mackenzie, and Saskatchewan Rivers. The western third of this region is called the Great Plains and is generally delineated from the eastern Interior Plains by the 20-inch (50-cm) **isohyet** (line of equal rainfall). Huge expanses of gently rolling or flat terrain have some of the world's finest soils and most productive farms here. The best soils have developed under prairie or steppe grasslands, parallel to similar settings in the region of Russia and the Near Abroad, but some very good soils in the east have developed under the humus-rich deciduous forest. Beef, pork, corn, soybeans, and wheat are the major agricultural products, and coal, oil, gas, and potash are the leading minerals. Agricultural and other developments have been responsible for elimination of about 90 percent of the original prairie habitats in the region. However, there is new interest in restoring prairie ecosystems, motivated by both nostalgia for historic landscapes and business interests in ecotourism.

The generally high and rugged Rocky Mountains consist of many linear ranges enclosing valleys and basins. Easy passes are few. Many major rivers rise in these mountains, including the Columbia, Colorado, Missouri, and Rio Grande. Ranching, mining (of both metals and fuels), and recreation are important activities. Much of the ranching is sustained by permits for ranchers to graze their cattle on public lands. The U.S. government has reevaluated this long-standing practice and its cost to taxpayers and is considering buying out the grazing leases (a rancher who grazes 600 cattle on federal land for 6 months would sell out that right for $262,000, for example). Environmentalists who have long denounced what they call **"welfare ranching"** are celebrating the plan as way of restoring ecosystems, but many ranchers argue that grazing promotes revegetation and helps prevent fires on public lands.

The Intermountain Basins and Plateaus consist generally of high plateaus trenched by canyons or, in the extensive Basin and Range Country of Nevada, Utah, and several adjacent states, by hundreds of linear ranges separated by basins of varying size. Shielded from moisture-bearing winds by high mountains on both the west and the east, the intermountain area is this continent's largest expanse of dry land. Mining of copper, oil, gas, coal, uranium, and other minerals is prominent in scattered spots, and numerous major dams impound irrigation water and generate much hydroelectricity along the Columbia and Colorado Rivers. The importance of water control at all scales is evident in this area.

The Pacific Mountains and Valleys lie parallel to the Pacific shore and have North America's highest mountains. California's flat, alluvial Central Valley (between the Sierra Nevada mountain system on the east and the Coast Ranges

on the west) has some of the world's most productive irrigated agriculture. Farther north, the Willamette–Puget Sound Lowland forms another major valley between the Coast Ranges and the Cascade Range of Oregon and Washington. Extreme western Canada and southern Alaska are dominated by high mountains, with many spectacular (but retreating) glaciers. The economy of the Pacific Mountains and Valleys has grown rapidly on the basis of major natural resources: Abundant forests, fisheries, rivers to produce hydroelectricity and supply irrigation water, oil and gas in California, gold that once drew settlers to California and British Columbia, and both climatic and scenic resources spur development in California.

Climatic Regions

The United States has more climatic types than any other country in the world, and even Canada is more varied than is commonly assumed. The wide range of economic opportunities afforded by climatic variety is the source of much of the economic strength of this world region.

Each climate region is depicted in Figure 22.5a, and a more detailed discussion of the climate types may be found on pages 21–25 in Chapter 2. Canada's and Alaska's tundra climate, characterized by long, cold winters and brief, cool summers, has an associated vegetation of mosses, lichens, sedges, hardy grasses, and low bushes. The subarctic climate zone, also in Canada and Alaska, has long, cold winters and short, mild summers, with a natural vegetation of coniferous (boreal) forest resembling the Russian taiga. In 2003, an unusual coalition of energy and forestry companies, environmental groups, and Native American communities joined an agreement to preserve at least 50 percent of Canada's subarctic forests and to develop the other half in sustainable ways. Population is extremely sparse in the subarctic and tundra climate areas. Scattered groups of people engage in trapping, hunting, fishing, mining, logging, and military activities; many live largely on welfare. The severe winter in vast sections of the tundra and subarctic zones requires unusual adaptations—like having to wait until the spring thaw to bury loved ones who died in the winter.

The humid continental climate with short summers ("humid continental cold" on the map) is characterized by long, cold winters and short, warm summers. Agriculturally, there is a heavy emphasis on dairy farming except in the extreme west, where spring wheat production is dominant. The humid continental long summer climate ("humid continental warm") has cold winters and long, hot summers; agriculturally, this belt, which includes the agricultural riches of the Midwest, is marked by an emphasis on dairy farming in the east and corn, soybeans, cattle, and hogs in the midwestern (interior) portion.

In the humid subtropical climate zone, winters are short and cool, although with cold snaps, and summers are long and hot. Agriculture varies by location, with some major specialized emphases on cattle, poultry, soybeans, tobacco, cotton, rice, peanuts, and a wide range of fruits and vegetables.

Within each of the three humid climatic regions just described, the pattern of natural vegetation is complex, with areas of coniferous evergreen softwoods, broadleaf deciduous hardwoods, mixed hardwoods and softwoods, and prairie grasses. Extreme southern Florida has a small area of tropical savanna climate, which is a major climate in adjacent Latin America. The state of Hawaii has a tropical rain forest climate, but most of the original associated vegetation has been cleared. Along the Pacific shore of the United States and Canada, the narrow strip of marine west coast climate is associated with the barrier effect of high mountains near the sea, ocean waters offshore that are warm in winter and cool in summer relative to the land, and winds prevailing from the west. The mild, moist conditions have produced a magnificent growth of giant conifers (most notably the redwood and the Douglas fir) that provides the basis for the lumber industry. Lush pastures support dairy farming, as they do in the corresponding climatic region in northwestern Europe.

The Mediterranean or dry-summer subtropical climate of central and southern California, in which nearly all precipitation comes in the winter season, is associated with irrigated production of cattle feeds, vineyards, vegetables, fruits, cotton, and a great range of other crops. These crops, together with associated livestock, dairy, and poultry produc-tion, make California the leading U.S. state in total agricultural output.

In the semiarid steppe climate, occupying an immense area between the Pacific littoral of the United States and the landward margins of the humid East and extending north into Canada, temperatures range from continental in the north to subtropical in the south. The natural vegetation of short grass, bunch grass, shrubs, and stunted trees supplies forage for cattle ranching, which is the predominant form of agriculture. Areas that are less dry are often used for wheat (both winter wheat and spring wheat are part of the agricultural scene), and other crops are grown in scattered irrigated areas, often associated with major rivers such as the Columbia, Snake, Arkansas, and Rio Grande.

The desert climate of the U.S. Southwest is associated with scattered irrigated districts and settlements emphasizing mining, recreation, and retirement. The high, rugged mountains of the Rockies and the Sierra Nevada have undifferentiated highland climates varying with latitude, elevation, and exposure to moisture-bearing winds and to sun. More detailed discussions of climatic regions and major landform divisions and associated landscapes, with many maps and photographs, are in Chapters 23 and 24. This wide range of environmental settings is associated with a wide variety of natural hazards (see Natural Hazards).

Natural Hazards

Nature's Wrath in the United States

The North American continent as a whole has more natural hazards than any other, and the United States has more natural hazards than any other country in the world (Figure 22.A). The western edge of the continent lies on a zone where two major plates of the earth's crust collide (see Figure 2.9, page 25), producing both earthquakes and volcanoes. The most consequential quake in U.S. history was the 1906 event of magnitude 8.0 that produced a displacement of up to 20 feet (6 m) on the San Andreas Fault near San Francisco, California. The earthquake itself, and especially the fires that followed rupturing gas lines in the city, destroyed roughly 30,000 of San Francisco's buildings and killed about 3,000 people. The San Andreas is just one of the largest and best known among the numerous fault lines that trend mainly north–south through California. The pressure built up by tectonic movement is enormous, and must eventually be released, so it is a certainty that many more earthquakes are in California's future. Some of them will be catastrophic, though the effects will be softened by building codes and emergency preparedness designed for those inevitable days. Anchorage and other cities in Alaska face threats from the network of faults at the edge of the continent. The coasts of Alaska, western Canada, and the western U.S.

must also be on guard for **tsunami,** the so-called "tidal waves" resulting when offshore earthquakes displace huge volumes of water.

351, 384

WORLD
REGIONAL
Geography ⊛ Now™

Click Geography Literacy to see an animation of the plate collision.

In the western U.S. states of Oregon and Washington lies the Cascade Range with its many volcanoes. Several of them, notably Rainier and Hood, still have the potential for eruption and, depending on wind flow and other variables, could do serious damage to Seattle and other populated areas. The destructive potential of these mountains was demonstrated on May 18, 1980, when Washington's Mount St. Helens blew up with an explosive force equivalent to 400 million tons of TNT. Few casualties occurred in this sparsely populated area.

Surprisingly, one of the most potentially destructive forces in the continental U.S. is a very ancient fault zone roughly paralleling the course of the southern Mississippi River. In 1811 and 1812, on the New Madrid fault in Missouri, three earthquakes of magnitude 8.0 or more on the Richter scale reportedly reversed the flow of the Mississippi River for some moments and rang church bells 1,000 miles (1,600 km) away in

Natural Hazards

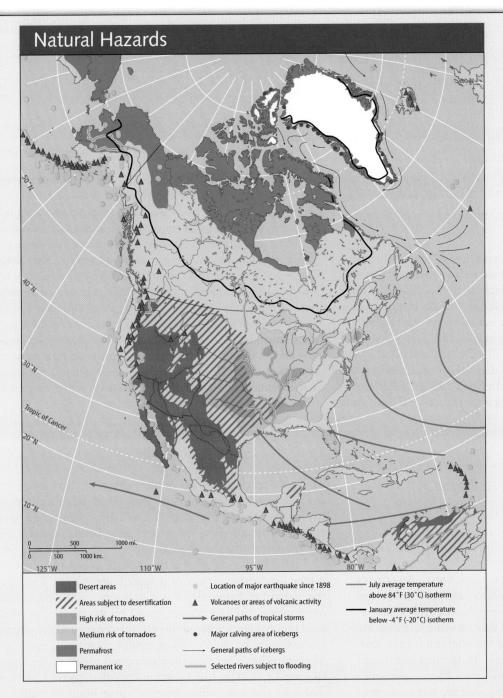

Figure 22.A Natural hazards of North America

Boston, Massachusetts. The mid-Mississippi region had a small population at the time and human losses were minimal, but a quake of that magnitude in the same region today would be very destructive to St. Louis, Memphis, and other cities.

Many Americans of the eastern United States think Californians are insane to live in an earthquake zone, but they regularly contend with their own menacing natural hazards. Storms often rule this part of the country. It is a generally humid region where the rainfall potential is high, and precipitation is triggered by a number of different agents. The convection associated with daytime heating and the collision of atmospheric fronts with air masses of different temperatures give rise to de-

structive thunderstorms, with heavy rain, lightning, hail, and one of the most powerful natural forces on Earth—tornadoes. Across a swath of the Midwest known as "**tornado alley**" (the "high risk of tornadoes" region of Figure 22.A), these intense vortexes of very low atmospheric pressure cut paths of destruction, sometimes causing large losses of life and property.

Occasionally, successive days or even weeks of thunderstorm activity produce localized or widespread flooding. In the summer of 1993, weeks of heavy rains over portions of the Missouri and Mississippi River watersheds produced a flood of historic proportions (Figure 22.B). Flooding and especially wind damage are typical results of the **hurricanes** that batter

(continued on page 608)

18

Joe Hobbs

Figure 22.B Floodwaters engorge the Missouri River in late July 1993.

the East Coast and Gulf of Mexico regions between June and September. The most catastrophic in U.S. history in terms of lives lost was a hurricane, which in 1900 killed more than 8,000 people, mainly in storm surges, in Galveston, Texas. In storm damages, the most costly was Hurricane Andrew, which came ashore south of Miami, Florida, in 1992, causing $27 billion in property losses. In the fall of 2004, a succession of less destructive hurricanes rampaged across Florida, inflicting cumulative losses that exceeded Andrew's toll. Yet another precipitation-related natural hazard is the **blizzard,** a combination

punch of heavy snowfall and high winds that strikes the United States—typically in the Midwest and Northeast—an average of 11 times each year.

WORLD
REGIONAL
Geography ⊗ Now™

Click Geography Literacy to see an animation of hurricane morphology.

A more pernicious natural hazard, a kind of disaster in slow motion, is the **drought** that sometimes rakes the country, especially in the West. The worst was the 8-year drought that produced the **Dust Bowl,** ruining crops and livelihoods and causing mass migrations of broken farming families to leave Oklahoma and adjacent areas, mainly to settle in the West. This story is told superbly in John Steinbeck's *The Grapes of Wrath.* As this text went to press, a drought that had already lasted 6 years was afflicting the western and especially southwestern United States, taking water supplies to record low levels. Climatological records suggest that such lower precipitation may be the norm for this region, which may have enjoyed decades of abnormally high precipitation. Human populations in the region boomed in those years, perhaps far beyond levels that can be sustained by local water supplies.

22.3 Cultural and Historical Geographies

As mentioned earlier, many geographers call the region of the United States and Canada "Anglo America," but the decision not to do so here is an acknowledgment of the rich ethnic roots and branches of this region. Native Americans began their migrations into the region as Asians, crossing what was then a land bridge between Alaska and Siberia at least 12,500 years ago—although there is some scientific evidence, still being scrutinized, that the migrations may have begun as early as 33,000 years ago. Those migrations persisted until about 11,000 years ago, and a diverse pattern of indigenous ethnic groups, languages, and life ways emerged.

Native American Civilizations

The indigenous cultures of Middle and South America have been described on pages 530–532 in Chapter 19. The indigenous cultures of what are now Canada and the United States, especially those of what is now the U.S. Southwest, were related to them in many ways. Some developed **civilizations,** the rather complex, agriculture-based ways of life associated with permanent or semipermanent settlements and stratified societies. In what are now the southwestern states of Arizona, Utah, Colorado, and New Mexico, three dominant civilizations emerged: the **Mogollon, Hohokam,** and **Anasazi.** Despite their arid environment, these peoples

developed productive agricultural systems that borrowed from and interacted with the Aztec and other cultures of Middle America. The same basic crop assemblages were found across all these cultures: Corn, beans, squash, and chili peppers were the staples.

The Hohokam of what is now southern New Mexico had a very productive system of irrigated agriculture. Their culture flourished between 100 B.C. and A.D. 1500. The Anasazi developed a dwelling pattern based on the **pueblo,** in which mud (adobe) brick interconnected residences and ceremonial centers, with beamed roofs of mud, sticks, and grass, were built into cliffsides or on the flat-topped mesas of the region (Figure 22.7). The Mogollon culture that thrived between 300 B.C. and A.D. 1400 inherited these building traditions and was effectively absorbed by the Anasazi, whose heartland was in the "Four Corners" region where Colorado, New Mexico, Arizona, and Utah meet. Around A.D. 1300, the Anasazi began to abandon their pueblos and the productive agriculture associated with them. The reasons for the Anasazi decline remain unknown, but drought, invasion by hostile neighbors, or simply political decisions to relocate may have been responsible. Some Native American groups of the region today, including the Hopi, Zuni, and Tiwa peoples, carry on the architectural legacy of the Anasazi by continuing to dwell in pueblos.

Farther to the east and north, in much of the eastern half of what is now the United States, the so-called **Mound**

Figure 22.7 Anasazi pueblo dwelling in Canyon de Chelly, Arizona.

Joe Hobbs

Builder civilizations developed. There were four main culture groups among them: the **Poverty Point** and the **Mississippian** cultures of the Southeast, the **Hopewell** in the Midwest, and the **Adena** in the Northeast. The earlier Poverty Point and Adena cultures (c. 2000 B.C. to A.D. 200) grew some crops but were mainly hunters and gatherers who thrived in the productive humid environments of the East. The Hopewell (c. 200 B.C. to A.D. 700) grew more crops, including corn, beans, and squash. The Mississippian (c. A.D. 700–1700), with an urban and suburban settlement pattern in many sites, was the greatest farming culture and had an extensive trading network. Of all the mound-building cultures, the Mississippian peoples were the greatest builders of earthen mounds. In all the cultures, these structures served as ceremonial sites, and in some, they were also tombs.

Indigenous Languages and Life Ways

The Native American groups not associated with complex societies, ceremonial centers, and civilizations were varied and numerous. As many as 18 different culture regions for these groups are recognized; this text uses 11 (Figure 22.8). The best way to categorize their essential features is to acknowledge some of the major groups within each region, and their original languages. Although the past tense is used below, because these cultures flourished in pre-Columbian times, none of the languages mentioned below is extinct, and speakers of all of them may still be found—although for most, English is a first or second language.

Seven Native American language families are represented in the United States, Canada, and Greenland by more than 250 languages. The **Aztec-Tanoan language family** that sprawled across most of northern Mexico was represented in the U.S. Southwest by languages including **Hopi, Comanche, Shoshone,** and **Papago.** Some of the groups that spoke these languages carried on the agricultural and pueblo-dwelling lifestyles pursued by the earlier Mogollon, Hohokam, and Anasazi civilizations, while others were nomadic hunter-gatherers who sometimes supplemented their diet by raiding the pueblo-dwelling peoples.

The larger part of what is now the contiguous United States was the hearth of the **Hokan-Siouan language family,** which includes such languages as **Iroquoian, Mohawk,** and **Lakhota.** Among its more famous culture groups were the **Sioux,** including the **Lakota** and **Dakota,** which developed their trademark subsistence pattern after the Europeans arrived in the New World. After the Spaniards and their successors introduced the horse, the Lakota and some other groups gave up what had been a village and farming way of life, migrating west to become horse-mounted hunters of the buffalo and other game on the Great Plains.

In what is now California and Oregon, the **Penutian language family** included **Miwok** and **Klamath-Modoc** languages, among others. Most of these peoples were not farmers but had extremely productive hunting and gathering economies that took advantage of the region's diverse and abundant resources. These natural bounties supported relatively dense human populations.

Farther to the north, but still on the Pacific coast, the **Mosan language family** included branches of **Chemakuan, Salish,** and **Wakashan.** The Northwest Coast peoples who spoke these languages also had rather dense populations that thrived on an abundance of sea mammals, fish, and land game. They generally developed complex societies in which the custom of **potlatch** played a prominent role. Potlatch was a system in which rank and status were determined by the quantity and quality of material goods one could give away—an interesting variant and inversion of the "keeping up with the Jones'" tradition of later Anglo North America.

The **Algic language family** members were spread across all 10 Canadian provinces and 20 U.S. states, mainly in the northern tier. Some of its better-known languages include the **Arapaho, Blackfoot, Cheyenne,** and **Cree.** The tribes that spoke the Algic languages relied on the region's vast forests for game animals, fuel, shelter, and tools, complementing their hunting and gathering with fishing and farming; some groups were seminomadic and lived in villages.

Indigenous Culture Groups

CULTURE AREAS

Arctic	Northwest Coast	California	Great Plains	Northeast
Subarctic	Plateau	Southwest	Great Basin	Southeast
	Uninhabited			

Figure 22.8 Native American culture areas and selected tribes of Canada, the United States, and Greenland

To the northwest of this linguistic region, in west central Canada and southeast Alaska, the **Na-Dene language family** was prominent. This group includes the **Athabascan** subfamily with its languages, including **Koyukon, Apache,** and in a remote outlier, the **Navajo** of the Southwest (the largest tribe in the United States, with more than 270,000 members). Most of these peoples lived either in the subarctic region of coniferous forests and wetlands or in the tundra, where agriculture was impossible. They hunted caribou and other game, and their populations were limited.

Across the rest of Alaska, throughout northern Canada, and in Greenland, the **Eskimo-Aleut language family** dominated. The peoples who carried these tongues into North America were among the last to migrate from Asia, probably between 2500 and 1000 B.C., and they did so in boats, as the former land bridge was inundated by then. The so-called **Eskimos**—an Algonquian term meaning "raw meat eaters"—prefer to call themselves **Inuit,** meaning "The People." The Inuit and the **Aleuts** certainly were characterized by a carnivorous diet, as their livelihood was based on hunting sea mammals, caribou (known as reindeer in the Old World), and fish.

What all these diverse groups had in common was an exceptionally well-developed set of adaptations for living in the local environment. Native American economies spanned a remarkably wide continuum from simple foraging and hunting to complex agricultural systems. These were not static adaptations, but changed and generally improved through time—for example, through greater mastery of finer tools. As the Lakota and others demonstrated by giving up village agriculture in favor of a nomadic life based on horses, these cultures were capable of dramatic adjustments to changing opportunities. It was once supposed that the Native American always lived within the limitations imposed on them by local ecosystems, but it is now known that they transformed landscapes in significant ways by fire and other means and may even have hunted populations of large animals to extinction; this is the Pleistocene overkill hypothesis (see page 31). One trait apparently shared by most if not all the Native American groups was their deep reverence for the natural world. Animals, forests, geological features, and the people themselves were tied together intimately in so-called **animistic belief systems** that emphasized the kinship between these diverse elements of life on Earth.

Settlement or Conquest

There are two very different versions of what took place in North America in the centuries following 1492. For European newcomers, these were times of settlement, development, taming the frontier, and "civilizing the savages." For the Native Americans, these were times of depopulation and cultural decimation. A popular T-shirt and bumper sticker on Indian reservations in the western United States these days reads: "Native Americans: Fighting Terrorism since 1492."

How many Native Americans lived in what is now Canada, the United States, and Greenland in A.D. 1492 is uncertain. The most common estimate is 1 million, with about three-quarters of that population in what is now the United States and almost all the rest in Canada. As in Latin America, the European contact initiated years of population losses among Native Americans through disease, famine, and warfare. In the United States, Native American populations crashed to an all time low of fewer than 250,000 between 1890 and 1910 but are now back up to 1.8 million. Their growth has been particularly strong in the Great Plains, where there are now more Native Americans (and bison) than there were in the late 1870s. In Canada, the current population of Native Americans is about 1 million. Collectively, these peoples refer to themselves as the **First Nations** in acknowledgment of their pre-Columbian cultures and claims to the land. In Canada, this definition excludes the Inuit, who were relative latecomers, but who are referred to as among the **First Peoples.**

In the United States, most of the Native American populations are in the west; in Canada, they are in the north and west (Figure 22.9). The two countries have dealt in different ways with their indigenous populations and claims to land and other rights, with government–Native American relations historically better by far in Canada. In the United States, most Native Americans today live on the **reservations** that the government established as remnants of their former tribal territories, or sometimes set up thousands of miles from native homelands, forcing the people to relocate there. These are among the poorest communities in the United States and are often plagued by high rates of incarceration, alcoholism, drug abuse, depression, unemployment, and suicide. The Lakota's Pine Ridge Reservation in South Dakota, for example, is characterized by an annual per capita income of just $6,500 (compared to the national U.S. average of $21,587) (Figure 22.10).

The Lakota and many other Native Americans are pinning future hopes on the economic development that could come from gambling revenues; the semiautonomous status of

Figure 22.9 Modern Indian reservations and other lands in the United States and Canada

Joe Hobbs

Figure 22.10 Lakota at the Wounded Knee Massacre Site, Pine Ridge, South Dakota

their territories allows casinos to circumvent many restrictions that prohibit casinos elsewhere (Figure 22.11). About one-third of the roughly 300 Native American tribes in the United States have developed casinos since Congress legalized gambling on their lands in 1998. This **gaming industry** has been a mixed blessing for the Native Americans, with a wide variety of successes and failures in different places. In some, gambling revenues have supported community education, healthcare, and cultural centers, but in others, they have lined a few pockets and failed to benefit many. Nonnative people often resent casinos because they view them as a windfall for individuals who have not earned special treatment. In all cases, casinos have introduced environmental and cultural changes.

The situation for indigenous peoples in Canada is unique, even on a worldwide scale. In 1999, Canada ceded nearly one-quarter of its total land area to the Inuit peoples, who called their land **Nunavut** ("our land" in Inuktitut; see Figure 22.9). These 733,600 square miles (1.9 million sq km) were carved from Canada's Northwest Territories and generally lie north of 60°N, surrounding the northwest part of

Joe Hobbs

Figure 22.11 Native American casino, Acoma Reservation, New Mexico

Hudson Bay and extending well within the Arctic Circle. The total population is approximately 30,000, scattered through about 30 communities. About nine-tenths of the capital required to keep the territory viable comes from Ottawa, the federal capital of Canada, making the per capita cost of maintaining Nunavut higher than for any other political unit in Canada. The Canadian government has allocated $1.2 billion to be distributed in the form of services, supplies, and subsidies until 2016 to the Inuit population, representing about $38,000 a year per person. Large subsidies also go to Native American populations in the Northwest Territories. All across Canada, settlements of indigenous claims to ancestral lands have led to federal government financial support of the communities, a measure of self-government for the indigenous peoples, and royalty income to them from mines and fossil fuel industries. Still, the average income of Canada's indigenous peoples is about a third lower than the population as a whole.

The New Lands, the Migrations, and the Immigrants

European settlement of what are now Canada and the United States took place in a series of waves propelled by religious persecution in Europe, colonization of new lands by European powers, and then the expansionist efforts of newly independent Canada and the United States. Canada's Coreland of Québec and Ontario developed with the French entry into the interior of North America. The French founded Québec City in 1608 at the point where the St. Lawrence estuary leading to the Atlantic narrows sharply. Fortifications on a bold eminence allowed control of the river. From Québec, fur traders, missionaries, and soldier-explorers soon discovered an extensive network of river and lake routes with connecting portages reaching as far as the Great Plains and, via the Mississippi River, to the Gulf of Mexico. Montréal, founded later on an island in the St. Lawrence River, became the fur trade's forward post toward the interior native-dominated wilderness. Between Québec City and Montréal, a thin line of settlement evolved along the St. Lawrence and formed the agricultural base for the colony. Population grew slowly in this northerly outpost where the winter was harsh and only French Catholics were welcome.

When finally conquered by Britain in 1759, French Canadians numbered only about 60,000. They were left in British hands when France was expelled from Canada in 1763. They did not join the English-speaking colonists of the Atlantic seaboard in the American Revolution and, by their refusal, laid the basis for the division of the continent of North America into two separate countries. Until the late 20th century, they increased rapidly in numbers. About 20 percent of Canada's people today are French in language and culture. These French Canadians are concentrated in the lowlands along the St. Lawrence River, where they form the majority population in the province of Québec. The French Canadians have clung tenaciously to their distinctive language and culture, attempting twice since 1980 to gain independence for the province of Québec.

A few British settlers came to Québec after the conquest, and this immigration increased rapidly during and after the American Revolution of 1775–1783, thus laying the foundations for the British segment of Canada's core. Many of the early English-speaking immigrants were Loyalist refugees from the newly independent United States, whose rebellion French Canada had refused to join. They were soon joined by more English settlers immigrating directly from Europe. These newcomers formed a sizable minority in French Québec, including a British commercial elite in Montréal that persists to this day, but they also settled west of the French on the Ontario Peninsula. By 1791, the British government decided to separate the two cultural areas into different political divisions, known at that time as Lower Canada (Québec) and Upper Canada (Ontario), taking the directional terms from their positions on the St. Lawrence River.

Farther west, the prairie region of Canada was settled much later, in the decades after 1890. Settlers found reasonably good land, with a climate of long and harsh winters, short and cool summers, and marginal precipitation. Only hardy crops could be grown. Early subsistence farming was succeeded by specialization in spring wheat for export, and the prairie region became the Canadian part of the North American Spring Wheat Belt. The settlers were predominantly English-speaking people from eastern Canada and the United States, but they included notable minorities of French Canadians, Germans, Ukrainians, and others. The ethnic composition of the region still reflects these elements. Canada's ethnic mix today includes about 7 million people of English origin, 2 million Irish, and 2 million Scottish (see Table 22.2).

Other ethnic groups, described below, would come to join Canadian society, which by law in the 1988 **Multiculturalism Act** was recognized as a multicultural society. These laws acknowledged that Canada's Native American and non-European groups would probably never become assimilated in Canadian society and that the country would probably be more stable—avoiding the kind of racial tensions that wracked the United States—if traditional identities were recognized and encouraged.

The first settlers of what is now the United States were Dutch, English, Swedish, and French, with the majority being English. Most of these early settlements were along the Eastern seaboard and were linked to Europe by the long but generally reliable sea passage across the Atlantic. Northeastern urban development began early. Boston was founded by English settlers and New Amsterdam (which became New York City) by the Dutch in the early 1600s, although New York was subsequently annexed and renamed by England in 1664. Philadelphia was founded by the English in the later 1600s and Baltimore in the early 1700s. All four cities were major urban centers by the time of the American Revolution (1775–1783).

Development of the interior began with exploration along waterways: the British up the Chesapeake Bay and its tributaries, the Dutch up the Hudson in the early 1600s, and the French up the St. Lawrence and Mississippi Rivers. Westward

F24.18f, 665

TABLE 22.2 Ethnic Groups of Canada and Their Origins

Major Ethnic Groups	Dates
Native peoples	pre-1600
French	1609-1755
Loyalists from the United States	1776-mid 1780s
English and Scots	mid-1600s on
Germans and Scandinavians	1830-50, c. 1900, 1950s
Irish	1840s
African Americans	1850s-70s
Mennonites and Hutterites	1870s-80s
Chinese	1855, 1880s, 1990s
Jews	1890-1914
Japanese	1890-1914
East Indians	1890-1914, 1970s on
Ukrainians	1890-1914, 1940s-50s
Italians	1890-1914, 1950s-60s
Poles	1945-50
Portuguese	1950s-70s
Greeks	1955-75
Hungarians	1956-57
West Indians	1950s, 1967 on
Latin Americans	1970s on
Vietnamese	1970s on
Filipinos	1990s

Source: From E. Herberg, *Ethnic Groups in Canada: Adaptations and Transitions.* Copyright 1989 Nelson Canada, Scarborough, Ontario; Census of Canada, 2001.

expansion proceeded very slowly; up to about 1800, there was little European settlement beyond the Appalachian region of what is now the eastern United States. The floodgates of expansion then opened, and wave after wave of immigration took place to fill the young U.S. ambition of its **manifest destiny** to settle the continent. The major territorial acquisitions that opened vast new frontiers for settlement were:

the Ohio and eastern Mississippi watersheds following the American Revolution in 1783;

the **Louisiana Purchase** of 1803, in which the United States bought from France for 3 cents per acre about 23 percent of what is now the lower 48 states, including much of the western watershed of the Mississippi and the lands of the Missouri River watershed;

the acquisition of Florida from Spain in 1819, followed by Texas, the Southwest, and California as spoils of war with Mexico in 1845 and 1848;

the **Gadsden Purchase** from Mexico of a small strip of present-day Arizona and New Mexico in 1853;

152

the establishment of the Oregon Territory in 1846;

and finally, Alaska, bought from Russia in 1867 (Figure 22.12).

The United States thus acquired a vast **land empire** not unlike that built earlier by Russia, although by different means.

The **Homestead Act** of 1862 made many of these new regions almost irresistible magnets for settlement, as it allowed a pioneer family to claim up to 160 acres of farming land for a fee of about $10. Two major trails, the ancestors of today's interstate highway system, were developed in the 1840s to feed the newcomers into the frontier zones (see Figure 22.12). From a trailhead in Independence (near Kansas City), Missouri, the **Oregon Trail** took a northerly track across what are now Nebraska, Wyoming, and Idaho to the fertile Willamette Valley of the Oregon Territory, with a southern branch (the **California Trail**) leading to what is now Sacramento, California. The southern route from Independence, known as the **Santa Fe Trail,** snaked across Kansas and New Mexico. Army posts were set up along the way to facilitate and protect the westbound immigrants. They also served as military garrisons in the forceful subju-

gation of Native American peoples, whose stories are told in broken treaties, massacres, and decimation of political leadership. Accessed in part by the Oregon and Santa Fe Trails, the U.S. Great Plains were settled rapidly between 1860 and 1890. A major impetus for settlement of the far west was the **California Gold Rush** of 1849. No Panama Canal existed at that point, so most of the **49ers** made their way either tortuously along the Oregon and California Trails or by sea for a 17,000-mile (27,200-km) and up to 7-month voyage around South America.

Most of the homesteaders, 49ers, and other settlers of the American frontier were ethnically of the English, German, Scandinavian, and other western and northern European ethnic groups. Ultimately, the United States developed into a nation in which about 99 percent of the people were either immigrants or of immigrant stock from a bewildering array of ethnic backgrounds. The larger groups represented today include more than 45 million of German origin, over 20 million of Irish origin, and over 20 million of English origin. Although some kept their own ethnic identities—still found in ethnic neighborhoods, especially in the East, or in small rural communities in the Midwest—most were blended into the melting pot that produced what might be called the

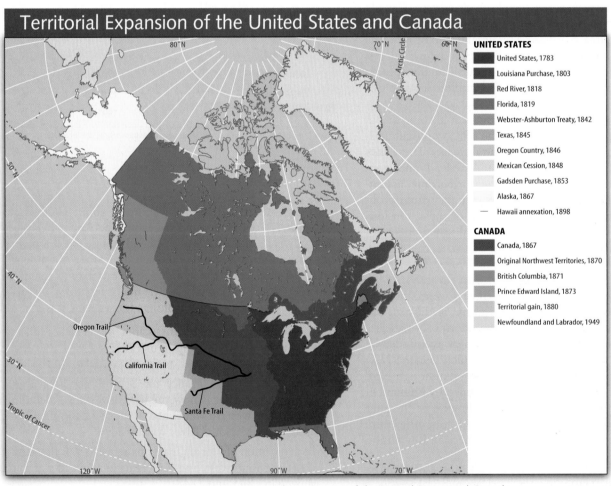

Territorial Expansion of the United States and Canada

UNITED STATES
- United States, 1783
- Louisiana Purchase, 1803
- Red River, 1818
- Florida, 1819
- Webster-Ashburton Treaty, 1842
- Texas, 1845
- Oregon Country, 1846
- Mexican Cession, 1848
- Gadsden Purchase, 1853
- Alaska, 1867
- — Hawaii annexation, 1898

CANADA
- Canada, 1867
- Original Northwest Territories, 1870
- British Columbia, 1871
- Prince Edward Island, 1873
- Territorial gain, 1880
- Newfoundland and Labrador, 1949

Active Figure 22.12 Territorial acquisitions of the United States and Canada
See an animation based on this figure, and take a short quiz on the facts or concepts.

WORLD REGIONAL Geography Now™

"**Mainstream European American Culture.**"[1] Ethnographer David Levinson describes that culture as one based on institutions such as the English language and British educational principles brought by the British colonists, and core American values that developed on U.S. soil. These include English as the national language, religious tolerance, individualism, majority rule, equality, a free-market economy, and the idealization of progress.

Added later to that mix, there were also immigrants from eastern and southern Europe, represented today by more than 11 million Americans of Italian origin, over 6 million of Polish origin, and over 5 million Jews. Many of these immigrants settled initially in the larger cities of the U.S. Northeast, where earlier immigrants of mainly northern and western European extraction often discriminated and directed violence against them. It took several generations for them to become part of the "mainstream" ethnic fabric of America. Other large groups, discussed below in some detail, have had more difficult careers with the white majority or are relative newcomers to the immigrant scene.

The plantation economy of the American South was built on the exploitation of more than 500,000 black slaves brought to America from Africa between 1619 and 1807. The U.S. Secretary of State Condoleezza Rice described slavery as "America's birth defect."[2] People of mainly African ancestry, known as **blacks** and **African Americans, Afro-Americans,** and formerly as **Negroes,** number about 40 million, or 14 percent of the U.S. population. The group also includes many people of mixed black and other ethnic extraction and blacks who came from regions other than Africa. Until the 2000 census, when the number of Hispanics slightly exceeded that of African Americans, blacks had been classified as the largest minority in the United States. Unlike the French Canadians, African Americans are not largely confined spatially to one part of the country. Some regions, especially in the eastern half of the United States, have a very strong black cultural presence (Figure 22.13), and African Americans are majorities in a number of large cities and particularly in the centers of those cities.

Race relations between whites and blacks in the United States were in general extremely poor until quite recently and have played an important role in the country's political and social history. Conflicting attitudes toward black slavery were among the issues that nearly split the United States into two countries in the American Civil War. The **Thirteenth Amendment** to the Constitution ended slavery in 1865, but the economically and politically subordinate position of blacks has been slow to change. As recently as the 1960s, there were legal struggles to end official **segregation** between the races in parts of the United States—for example, in the

[1] David Levinson, *Ethnic Groups Worldwide* (Phoenix, Ariz.: Oryx Press, 1998), p. 396.

[2] Quoted in Richard W. Stevenson, "New Threats and Opportunities Redefine U.S. Interests in Africa." *The New York Times,* July 7, 2003, pp. A1, 8.

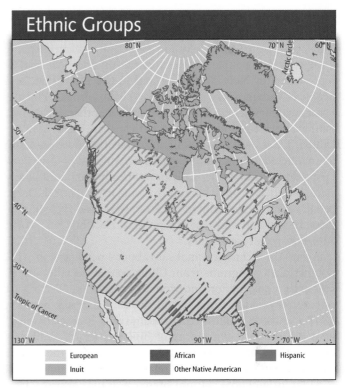

Ethnic Groups

European   **African**   **Hispanic**
Inuit   **Other Native American**

Figure 22.13 Major ethnic groups of the United States and Canada

state of Alabama, where there were separate sections for whites and blacks on buses and where the races were expected to live in different neighborhoods, attend different schools, and never intermarry. Today, while a segment of the African American population has moved into the middle class and above, a much larger portion resides in impoverished and troubled central cities, where there are problems in housing, jobs, and crime. About 44 percent of all persons incarcerated in U.S. prisons are black. All the legal battles to end segregation have ended, but "hearts and minds" have been slower to change, and African Americans continue to struggle for equity. Some of their efforts parallel earlier trials by American women, who finally obtained the right to vote in 1920 and whose wages are still generally lower than those of men in identical occupations. So far, a comprehensive **Equal Rights Amendment** to the U.S. Constitution, which would grant women official parity with men at all levels, has not moved through the legislative process.

In the 1840s, when the United States won from Mexico the great swath of the American Southwest, West, and Texas, it also acquired many Spanish-speakers of European or *mestizo* descent. The growth rate of people of Spanish or mixed Spanish background—the **Hispanic** (or **Latino**) population—has recently been more rapid (about 4 percent/year) than that of any other group, both domestically and through the major migration stream flowing actively from Latin America. Approximately half of the growth is from natural increase and another half from immigration, both legal and illegal. About 40 million people, or 14 percent of the U.S.

population, are Hispanic. The majority are of Mexican origin and are strongly concentrated in California, Texas, New Mexico, Arizona, and Colorado. Other important Hispanic groups include people of Puerto Rican origin, most heavily represented in New York City and in the Northeast, and people of Cuban origin, who live largely in southern Florida. Estimates vary as to how many million other Hispanics, mainly Mexicans and Central Americans, are resident in the United States as illegal immigrants, but the most common figure offered is 3 million. Like African Americans, American Hispanics are not numerically or politically dominant in any state, but their proportions are growing in many of the large urban centers, especially in the Southwest, and their importance in national and regional political elections has grown enormously in recent years.

Asian Americans now make up about 4 percent of the U.S. population. Immigrants from China were especially important in the construction and service sectors of the West Coast economies of both Canada and the United States between 1849 and 1882. Chinese laundries and the Chinese "coolies" who did the grueling labors of building western railways became the stuff of stereotype and legend, and there were sore points in Chinese–Anglo relations. Chinese immigrants brought an indelible cultural presence in the form of "Chinatown" landscapes in major cities, particularly in the 19th and early 20th centuries and especially in the western regions of both countries. Chinese food has become even more widespread than Mexican food across the continent.

Asian Americans have long been steadily moving out of the central city and into white and mixed neighborhoods; they include among their ranks not only the Chinese but large populations of ethnic Japanese, Korean, Vietnamese, and Filipinos, among others. The closest approach to Asian minority dominance of a state is in multiethnic Hawaii, where Asian Americans, especially of Japanese descent, make up more than 40 percent of the population. Race relations between the Anglo majority and Asian groups in both the United States and Canada have generally been good in recent decades. The last major episode of tension was during World War II, when large numbers of civilian Japanese Americans, suspected of being potential collaborators with the enemy, were rounded up and interned in prison camps in the West.

Asian Americans in the United States today have the distinction of being the most highly educated of all the ethnic/religious groups except for Mormons and Jews. Among their ranks, the ethnic Japanese, Koreans, Chinese, East Indians, and Filipinos have generally prospered economically, but the ethnic Vietnamese, Lao, and Cambodians are poorer. The largest Asian populations in the United States are the Chinese and Filipino, each over 1 million. The largest in Canada are ethnic Chinese (about 1.1 million, double their numbers in 1984), East Indian (mainly of the Sikh religion), and Filipino. Recent large Asian migrant flows into Canada include Koreans and Vietnamese.

These brief sketches only scratch the surface of the ethnic mosaics of the United States and Canada. Nearly 200 non-

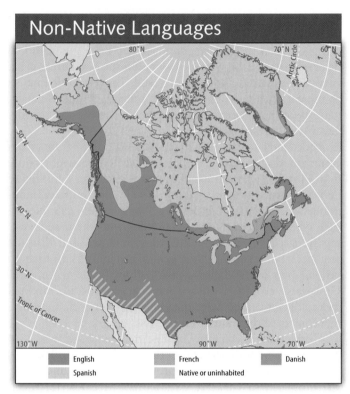

Figure 22.14 Non-Native American languages of the United States and Canada

Native American ethnolinguistic groups are represented in these countries, which are rather rare examples of nations that have generally welcomed immigrants as part of their social fabric. Discussions of ethnicity follow in the next two chapters.

Nonindigenous Languages and Faiths

The non-Native American religions and languages that prevail in Canada and the United States reflect both the early colonial influences and more recent waves of immigration. English and French are Canada's official languages, with French usage confined largely to the traditionally French southeast (Figure 22.14). The United States does not have an official language (despite repeated efforts to pass a constitutional amendment that would establish it as English), but 96 percent of U.S. residents speak English. The language map shows especially well the growth of Hispanic populations in the Southwest and Florida. "**Spanglish**," a hybrid tongue of Spanish and English, is moving from Hispanic neighborhoods of southern California into the mainstream media and culture of the United States. Spanglish uses "code switching," when words or phrases in one tongue are substituted for another ("vamos a la store para comprar milk," meaning "let's go to the store to buy some milk") and also coins new words, like "perro caliente" for "hot dog."[3] Within the larger cities of both the United States and Can-

---

[3] Discussion and quotes are from Daniel Hernandez, "A Hybrid Tongue or Slanguage?" *The Los Angeles Times*, December 27, 2003, pp. A1, 30.

ada, there is a dizzying array of languages associated with ethnic neighborhoods.

Religious freedoms are guaranteed by law in both the United States and Canada, and both countries have a rich fabric of faith (Figure 22.15). Christianity prevails in both countries, with numerous Christian churches represented. The largest single denomination in both countries is Roman Catholic (24 percent in the U.S. and 45 in Canada), brought to the United States by Irish, Italian, Polish, Filipino, Hispanic, and other immigrants and to Canada mainly by the French. The next largest group in the United States is Baptist (about 16 percent), then Methodist (about 7 percent), and Lutheran (about 5 percent). The other related monotheistic faiths are well represented by about 5 million Jews and 1 million Muslims, whose Islamic faith is the fastest growing religion in the United States.

U.S.–Canada Historic Relations

For many years after the American Revolution, which secured U.S. independence from Great Britain, the political division of the United States and Canada, between a group of British colonial possessions to the north and the independent United States to the south, was accompanied by serious friction between the peoples and governments. There were several sources of antagonism. These included the failure of the northern colonies to join the Revolution, and the use of those colonies as British bases during that war. In addition, a large proportion of the northern population had come from **Tory** stock driven from U.S. homes during the Revolution (Tories were "the friends of the king," the name for the colonists who wanted to maintain the political connections with the British government). Finally, there was uncertainty and rivalry over who had ultimate control of the central and western reaches of the continent.

The **War of 1812** was fought largely as a U.S. effort to conquer Canada, although this intent was not specified at the time. Even after its failure, a series of border disputes occurred, and the United States openly expressed ambitions to possess this remaining British territory in North America. Suggestions and threats of annexation were made in official American quarters throughout the 19th and even into the 20th century.

Canada's emergence as a unified nation was in good part a result of American pressure. After the American Civil War (1861–1865), the military power of the United States assumed a threatening aspect in Canadian and British eyes. Suggestions were made in some American quarters that Canadian territory would be a just recompense to the United States for British hostility to the Union during the Civil War. However, the **British–North America Act,** passed by the British Parliament in 1867, brought an independent Canada into existence by combining Nova Scotia, New Brunswick, and Canada into one dominion under the name of Canada. With that act, Great Britain sought to establish Canada as an independent nation capable of morally deterring U.S. conquest of the whole of North America.

Although hostility between the United States and its northern neighbor did not immediately cease with the establishment of an independent Canada, relations improved gradually. Canada and the United States today are strong allies with a connection only occasionally tested. Perhaps the strongest element in their relationship is the economic bond between them.

22.4   Economic Geography

The United States and Canada are very wealthy nations. Per capita GDP PPP figures for the countries are $37,800 and $29,800 respectively, compared with the world's highest, Luxembourg, at $55,100. The United States has the world's largest economy, is by far the world's largest producer and consumer of goods and services (representing overall about one-quarter of the total), and is the world's best example of consumption overpopulation. Comprising just 4.5 percent of the world's population, Americans each year consume, for example, more than 25 percent of the world's energy output, 50 percent of its diamonds, 60 percent of its illegal drugs, and 75 percent of its rubber.

This region, and particularly the United States, is blessed with very large endowments of some of the world's important natural assets. A number of geographic, political, and other circumstances have complemented this resource wealth in the development of these economically powerful countries. Both countries are large in area and are among the five largest in the world. A remarkable range of environmental settings has both allowed and stimulated full use of hu-

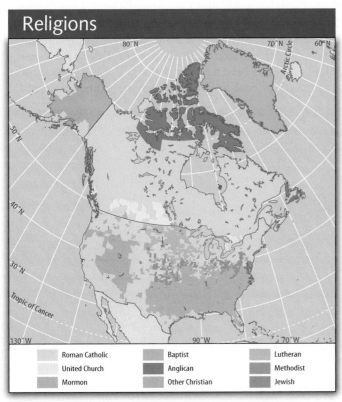

Figure 22.15 Religions of the United States and Canada

man ingenuity in economic development. The combined population of both countries is large, yet neither country is "people overpopulated." A large population represents both a large pool of prospective labor and talent to promote economic growth, and a vast market for the goods and services produced. Both countries developed mechanized economies rather early and under very favorable conditions. Innovations continue to replace human labor with machine labor, generally increasing efficiency and productivity (while also leading to unemployment in traditional manufacturing sectors). Peace and stability within and between these countries both prevail and have provided a good climate for investment and economic development; since the Civil War, the United States has managed to keep all of its major conflicts mainly off of its own shores. The generally cooperative relations between the United States and Canada also help. Both possess a comparatively effective internal unity despite episodes of poor race relations in the United States and the separatist movement in Québec. In both countries, there has been a history of reasonable continuity of political, economic, and cultural institutions. The combination of these influences creates a regional base that has a rich history of economic development and relative abundance for its populations.

The United States and Canada are not without significant problems of poverty and imbalance in the distribution of the products of this abundance. The United States in particular has large segments of the population that lack a "safety net" in the form of universal health insurance and other social services. Income distribution is skewed, with the wealthiest 10 percent of Americans taking in about one-third of the country's total income and the poorest 10 percent taking in less than 2 percent of the total; this represents the greatest rich–poor gap of all the more developed countries. Long dominated by liberal politics, Canada has a more socialistic approach to its population than does the United States and tends to be more in line with Europe than with the U.S. on most social and political issues. It has generally higher indexes of quality of life than the U.S. and a more equitable distribution of income than the U.S. Taken as a whole, however, Canada and the United States make up a region of continued strong potential for "the good life."

Agricultural, Forest, and Mineral Resources

The United States and Canada are the epitome of "**neo-Europes**," those lands that European immigrants chose to settle largely because of their resemblance to the European environments and their potential for production of wheat, cattle, and other vital products. The United States has more high-quality arable land than any other country in the world. A much smaller proportion of Canada is arable, but it has more arable acreage than all but a handful of countries. Such abundance has helped these two countries become the main food-exporting region in the world.

Canada and the United States also have abundant forest resources. Canada's forests are greater in area than those of the United States but are not as varied. Wood has been an essential material in the development of both countries, leading the less developed countries of the world today, such as Brazil and Malaysia, to cry "foul" when American environmentalists denounce tropical deforestation. Even now, the United States cuts more wood for lumber each year than any other country in the world, although Canada leads in the production of wood pulp and paper production. The United States both exports and imports large quantities, with the exports going largely to Japan (Figure 22.16) and the imports coming mainly from Canada. Canada's timber industry is considerably smaller, although it is still one of the greatest in the world, and Canada is the world's largest exporter of wood.

The United States and Canada are also outstandingly rich in mineral resources. These resources played a major role in the earlier development of the region's economy, and mineral output is still huge. Canada is a major exporter of mineral products, especially oil and natural gas, mainly to the United States, which in 2004 obtained 17 percent of its imported oil from Canada—more than from any other country. Although still a major mineral producer, the United States is also a major importer, especially of oil and industrial raw materials. This situation reflects the partial depletion of some American resources and the enormous demands and buying capacity of the U.S. economy.

Abundant energy resources have been essential to the region's economies. Coal was the largest source of energy in the 19th-century industrialization of the United States. Today, the country leads the world in estimated coal reserves (about 26 percent of the total). In coal production, China and the United States are first and second, with China producing 33 percent and the United States 22 percent. Due to high energy prices, there has been a recent boom in U.S. coal production, with scores of coal-fired electric power plants proposed for

Figure 22.16 Canadian timber en route to East Asia. Canada and the United States have strong trade with countries on the Pacific Rim of Asia.

construction. Environmental concerns about the impacts of this most polluting of fossil fuels may keep many or most of those plants from being built and raise long-term questions about what the United States can do with its generous coal endowment. As it stands, the United States does derive about 53 percent of its electricity from coal-fired plants, with natural gas supplying 15 percent and nuclear power 21 percent. Canada is also an important coal exporter, with reserves and production that are small on a world scale but large in terms of Canada's needs.

Petroleum fueled much of the development of the United States in the 20th century. For decades, the country was the world's greatest producer, having such large reserves that the question of long-term supplies received little thought. The United States is still exploiting its estimated 21 percent of the world's proven oil reserves, but its huge economy demands far more oil than the country can provide. The United States imports about 55 percent of its own domestic need, significantly more than the 36 percent it imported before the 1973 energy crisis. Canada's reserves, situated mainly in Alberta, are small in comparison, representing about 0.5 percent of the world total. Also in Alberta and neighboring Saskatchewan, Canada has the world's largest reserves of **tar sands.** These are mineral deposits that can yield a thick liquid petroleum known as bitumen through an energy-consuming process costing about $15 per barrel. When conventional oil prices are high (roughly, over $30 per barrel), tar sands become economically attractive to exploit. The price surge in oil beginning in 2003 led to strong growth in production from Canada's tar sands, with most of the output going to the United States.

The biggest obstacle to the tar sand development for domestic consumption is Canada's 2004 ratification of the Kyoto Protocol, which will compel the country to reduce its carbon dioxide emissions to 6 percent below their 1990 levels. This commitment will require a massive cutback in the use of fossil fuels. Alberta's provincial government was furious at the federal government's support of the Kyoto climate change treaty, arguing that it would cost the province billions of dollars in lost revenue and thousands of jobs. Alberta's politicians even threatened that the province might secede from Canada if the treaty were ratified. South of the border, American environmentalists pointed to Canada's ratification of the Kyoto Protocol as another example of how the United States, which refuses to ratify the treaty, is out of step with the rest of the industrialized world.

Natural gas has become an increasingly important fossil fuel to these two economies. The United States is the world's largest consumer of gas and is a major importer, especially from Canada. It ranks second in the world in output (behind Russia but far ahead of any other country), producing 21 percent of the world's gas from its 3 percent of proven reserves worldwide; that equation means that the present high output cannot be sustained for long. With the recent high energy prices, states like Montana and Colorado that still have reserves of natural gas are weighing whether or not to exploit them or to keep the lands above them relatively pristine for a "new economy" based on hunting, recreation, and tourism. Canada, with smaller but rapidly increasing proven reserves, produces about 7 percent of world output and is a major exporter.

The remaining potential for hydropower development in the United States and Canada is small on a world scale, but development has been so intense that the two countries rank with Russia as the world leaders in hydropower capacity and output (Figure 22.17). Canada generates about 56 percent of its electricity from hydropower, and the United States generates 7 percent. The dams are controversial because of the environmental changes they produce and, in the case of Canada, because of the inundation of indigenous lands, as described in the next chapter (page 640). Some dams in the United States, including two on Maine's Penobscot River, are actually in the process of **decommissioning** (major modification or removal) primarily for environmental reasons—for example, to allow salmon to reach historic upstream spawning grounds. Drawing on large uranium resources in each country, nuclear power is an important secondary source of electricity for both (see Geography of Energy, page 621).

Other mineral resources in North America have presented a similar picture of abundance (Table 22.3). Many of the ores easiest to mine and richest in content are now mostly depleted, and even large mineral outputs within the United States are insufficient to supply the huge demands of the domestic economy. In many such cases, Canada has been able to assume the role of principal foreign supplier. A major example is iron ore. The United States originally had huge deposits of high-grade ore, principally in Minnesota's Mesabi Range and in less significant ore formations near western Lake Superior, but by the end of World War II, this ore had been seriously depleted. New sources of high-grade ore then were developed in Québec and Labrador, principally for the American market. Meanwhile, the Lake Superior region

Figure 22.17 Hydroelectric power is important in the energy economies of the United States and Canada. These are turbines inside the Glen Canyon Dam on the Colorado River in Arizona.

TABLE 22.3 United States' and Canada's Shares of the World's Known Wealth and Production

Item	Approximate Percent of World Total			Item	Approximate Percent of World Total		
	U.S.	Canada	Total for Both		U.S.	Canada	Total for Both
Area	7	7	14	Crude oil production	10	3	13
Population	4.6	0.5	5.1	Crude oil refinery capacity	21	3	24
				Natural gas production	24	7	31
Agricultural and Fishery Resources and Production				Natural gas reserves	3	1	4
Arable land	13	3	16	Potential iron ore reserves	5	6	11
Meadow and permanent pasture	8	1	9	Iron ore mined	6	3	9
Production of:				Pig iron produced	8	2	10
Wheat	10	5	15	Steel produced	12	2	14
Corn (grain)	51	3	54	Aluminum produced	15	10	25
Rice	2	0	2	Copper reserves	17	5	22
Soybeans	47	2	49	Copper mined	11	5	16
Cotton	18	0	18	Lead reserves	22	13	35
Milk	16	2	18	Lead mined	15	5	20
Number of:				Zinc reserves	13	15	28
Cattle	7	1	8	Zinc mined	10	11	21
Hogs	7	1	8	Nickel mined	0	15	15
Fish (commercial catch)	5	1	6	Gold mined	14	6	20
				Phosphate rock mined	29	0	29
Forest Resources and Production				Potash produced	6	23	29
Area of forest/woodland	6	8	14	Native sulfur produced	21	0	21
Wood cut	14	5	19	Asbestos mined	1	18	19
Sawn wood production	28	17	45	Uranium reserves	9	12	21
Wood pulp production	34	16	50	Uranium produced	16	29	45
				Electricity produced	26	4	30
Minerals, Mining, and Manufacturing				Potential hydropower	4	3	7
Coal reserves	23	1	24	Production of hydropower	4	4	8
Coal production	20	2	22	Nuclear electricity produced	31	4	35
Coal exports	25	8	33	Sulfuric acid produced	26	3	29
Published proved oil reserves (tar sands and oil shale)	2	1	3				

Sources: United Nations Statistics Division, 2000; *World Population Data Sheet,* 2004.

revived its mining industry by extracting lower grade ore, which is present in vast quantities but must be processed in concentrating plants before it can be shipped at a profit to iron and steel plants. Major minerals in which both countries are deficient include chromite, tin, diamonds, high-grade bauxite, and high-grade manganese ore.

Mechanization, Services, and Information Technology

The high productivity and national incomes of this region's two countries are mainly the products of machines and mechanical energy, complemented recently by a boom in infor-

mation technology (IT). From an early reliance on waterwheels that powered simple machines, especially along the fall line in the American Northeast, the United States increasingly exploited the power generated from coal by steam engines. Later, the power used to drive machines increased enormously through the use of oil, internal-combustion engines, and electricity.

In achieving this unusually energy-abundant and mechanized economy, the United States was able to take advantage of a unique set of circumstances. For one thing, it had resources so abundant and varied that they attracted foreign capital and also made possible domestic accumulation of

Geography of Energy

Nuclear Power and Other Energy Alternatives in the United States

The United States is the world's largest producer of nuclear power and derives 21 percent of its electricity from that source. This energy source has an uncertain future in the United States, however. In 1979, an accident occurred in the Three Mile Island nuclear power plant in Pennsylvania, releasing only limited amounts of radiation to the environment but producing enormous public fear of nuclear power. The truly disastrous Chernobyl accident in 1986 in the former Soviet Union only heightened for many the perceived peril of this energy source. Due to such fears and to cost overruns and other problems, no nuclear power plants have gone online in the United States since 1978. The same concerns have troubled Canada, which gets 14 percent of its electricity from nuclear power. It has not built a new nuclear power plant since the 1970s but is considering building one on Lake Ontario.

More benign sources of alternative energy, such as windmills, photovoltaic panels, and hydrogen fuel cells, have received only modest federal funding and development since the energy crisis of the 1970s (Figure 22.C). Few tax breaks or other incentives exist for the development of these alternatives by private companies, and official policy is that alternatives should develop in the free market rather than through federal funding and legislation. The United States has also not endorsed the Kyoto Protocol or other international legislation that would compel it to embrace alternative energies.

Legislation has improved energy efficiency, however, especially in the vital transportation sector that accounts for about 27 percent of America's energy needs. In 1975, in the wake of the energy crisis, the U.S. Congress enacted the **Corporate Average Fuel Economy (CAFE) standard,** requiring that the fleets of passenger cars produced by all the auto companies achieve an average of 27.5 miles per gallon. The standard for light trucks was set lower, at 20.7 miles per gallon. Sport utility vehicles (SUVs) are classified as light trucks, allowing them to meet this lower standard—and the American consumers to pursue their love affair with these "gas guzzlers." Politicians sometimes speak of raising the CAFE standard to trim the nation's fuel bills and reduce its imports, but taking such a stand can be political suicide in a country where the car is king.

Joe Hobbs

Figure 22.C A wind farm near Palm Springs, California

capital through large-scale and often wasteful exploitation. There was also a labor shortage, which attracted millions of immigrants as temporary low-wage workers, but also promoted higher wages and labor-saving mechanization in the long run. A relatively free and fluid society encouraged striving for advancement. A strong work ethic, described as the **Protestant work ethic** but shared by people of many faiths and ethnicities, characterized the struggle for economic success and the **American Dream.** Only occasionally did national energies have to be diverted excessively into defense and war.

Most of the American conditions were duplicated in Canada. However, Canada lacked a large, unified, and growing internal market that allowed producers to specialize in the mechanized mass production of a few items, thus lowering the unit costs of production and further expanding the size both of the producing firms and of the market. In recent

decades, however, this U.S. advantage has been significantly eroded, along with some others, by the enormous expansion of the global market. Transportation and communication have become so relatively cheap and rapid, and so many artificial trade barriers have fallen, that the world market, rather than the domestic market, has become the essential one for major industries. The globalization of the economy has presented huge challenges to the United States, Canada, and the other more developed countries, which are trying to prevent their traditional industries from being undercut by cheaper output from abroad. The United States and its industrialized allies thus often find themselves in the awkward position of advocating free trade but erecting trade barriers (see Definitions and Insights, page 622).

The forces of mechanization and associated productivity have been increasingly replaced in recent decades with a new labor sector: services and information technology. The

36

Definitions + Insights

Trade Barriers and Subsidies

Many of the less developed countries of the world are convinced that the best way to develop is not to borrow money from the richer countries but to trade with them. After all, with their larger populations and lower labor costs, the LDCs should easily be able to churn out products at much more favorable prices than those sold by the MDCs. Buyers around the world should be snapping up those affordable products, pumping desperately needed money back into the developing countries. This exchange does not take place as freely as it could, however, because of two tools the MDCs wield to protect their own producers: subsidies and tariffs.

Subsidies are expenses paid by a government to an economic enterprise to keep that enterprise profitable, even if the market would not. **Tariffs** are essentially taxes that a government imposes on foreign goods, basically for the same reason, to ensure that domestic products make it to market, even though they are not produced at competitive costs. Alone or in combination, subsidies and tariffs can make it all but impossible for a less developed country to join the world economy as a trading partner.

There are often struggles between the MDCs and LDCs over the issues of subsidies and tariffs, especially in meetings of the **World Trade Organization (WTO)**. At these meetings, the LDCs have especially strong representation by the 21 countries that call themselves the **G-21 (Group of 21)** and whose most prominent members are Brazil, China, India, South Africa, Egypt, Indonesia, Thailand, the Philippines, Mexico, Argentina, Chile, and Pakistan. Representing most of the interests and economic strength of the MDCs are the so-called **G-7 (Group of 7)**: Germany, Japan, France, the United Kingdom, Italy, Canada, and the United States (also sometimes called the G-7 plus one—Russia—or the G-8, with Russia).

The bitter debate between these groups often focuses on the subsidies that the United States, the European Union, Japan, and other MDCs have used for several decades to prop up certain sectors of their economies, especially agriculture. The subsidies were created to help farmers and the countries get through especially difficult times: **the Depression** in the United States, when the livelihoods of so many farmers were threatened, and post–World War II Europe, when the specter of malnutrition hung over the Continent, for example. Over the years, the critical need for subsidies passed, but the sectors that received them were loath to part with them, and the federal suppliers of subsidies found political capital in maintaining them.

Now farming subsidies are something of a sacred cow in many of the MDCs, and any effort to reduce or remove them would have domestic consequences—especially at the ballot box.

Cotton is a fine example of how the subsidies and related tariffs work domestically and what their impact is in the LDCs. The United States spends $5.6 billion each year—more than $200 per acre—to subsidize its cotton farmers. The country produces far more cotton than it consumes, and most of the prodigious output is exported. The high level of production made possible by the subsidies drives the global price of cotton down to around 35 cents per pound. This is substantially less than the African production cost of around 50 cents per pound. The African cotton producer not only cannot compete in the world market but loses money doing so. Cotton is, or should be, big business for Africa. It makes up fully 10 percent of the gross domestic product, 40 percent of exports, and 70 percent of agricultural exports for the West African and Sahelian nations of Benin, Mali, Chad, and Burkina Faso. If the U.S. subsidies were reduced or removed and world cotton prices rose, the potential economic gains for those countries would be enormous. The United States, however, has been slow to reform cotton policies, except in ways that will clearly be in its own interest. In legislation signed by President George W. Bush, tariffs were reduced on some imported African textiles—but only those made from American cotton. The European Union, Japan, South Korea, and other countries are just as strident in protecting some of their products. Altogether the MDCs pay out a billion dollars a day in subsidies that support the farmers who make up less than 1 percent of their population.

The G-21 countries, which speak for 63 percent of the world's farmers, demand that the MDCs reduce or eliminate the subsidies and tariffs or be paid compensation for the revenue lost to subsidies and tariffs. They threaten to begin erecting tariffs of their own (a proposal the MDCs say is bad for the developing economies) if the MDCs do not yield. The MDCs typically give up little or no ground in such impasses, causing the G-21 delegates to walk away frustrated from trade talks.

Would removal of subsidies hurt the small American family farmer whose cause is championed by Willie Nelson and Neil Young? That would depend on how the subsidy cuts were targeted. Sixty-five percent of the subsidies go the top 10 percent of America's agricultural producers, the large corporate farming industries. It is possible that the unkindest cuts could be avoided and some sacrifices imposed on the bigger businesses, thereby perhaps helping some of the world's poorest countries to escape the poverty trap on their own.

469, 540

service sector, including insurance, finance, medical care, education, entertainment, retail sales, and other enterprises, has become a more significant employer than the manufacturing sector in the United States and Canada, as shoes, clothes, toys, cars, and other traditional manufactures have migrated outside the two countries. A widespread and economically dynamic sector has meanwhile grown up around the acquisition, manipulation, analysis, and distribution of information. As telecommunications technologies have interconnected North America with the rest of the world, the United States has acquired global leadership in information technology even while many of its old heavy industries have expired in the country's Northeast—its so called **Rust Belt.** The United States is the world leader in software and semiconductors. In the 1990s, there was such a speculative frenzy about IT that a classic **economic bubble** developed. Tens of millions of Americans invested for the first time, bidding up the values of companies to lofty heights—even companies that produced no tangible products. There was talk of a "new economy" in which the old paradigms of corporate investment and growth no longer applied. However, the **tech bubble** burst in 2000. Clever or lucky investors withdrew their investments in time, but others lost fortunes that had existed on paper. An economic recession set in. The economy gradually began to recover, with more attention paid to the old-fashioned brick-and-mortar industries, which actually produce tangible assets, and with more caution in the tech sector.

U.S.–Canada Economic Relations

A large volume of trade moves across the U.S.–Canada frontier and strengthens both countries economically and militarily. Each is a vital trading partner of the other, although Canada is much more dependent on the United States than is the United States on Canada. In 2004, for example, Canada supplied 17 percent of all U.S. imports by value and took in 23 percent of all U.S. exports, making Canada the leading country in total trade with the United States. In turn, in 2004, the United States supplied 60 percent of Canada's imports and took in more than 85 percent of its exports. The United States also supplies large amounts of capital to Canada, which has been an important factor in Canada's rapid economic development during recent decades. But it is also a source of Canadian disquiet because of the U.S. influence on the Canadian economy that the investment represents. However, Canadians in turn invest heavily in the United States. Except for Canadian exports of automobiles and auto parts to the United States, the main pattern of trade between the two countries is the exchange of Canadian raw and intermediate-state materials—primarily ores and metals, timber and newsprint, oil, and natural gas—for American manufactured goods.

Free trade agreements designed to open markets and increase revenues for both countries have facilitated this exchange. These followed periods when the countries tried to protect their industries from one another, often with poor re-

sults. A high-tariff policy to forestall American competition was adopted in Canada shortly after its confederation. Although it fell short of shutting out foreign industrial products, which Canada imported in large quantities, it did favor the growth of a large variety of Canadian industries, mainly in the eastern core region and often including branch plants of U.S. companies "jumping" the tariff wall. However, the policy aroused protest from other parts of Canada in which consumers preferred to buy cheaper American goods. It also fostered many high-cost Canadian industries that lacked international competitiveness. The exceptions were industries making products desired in the United States. For example, the Canadian pulp and paper and mining industries, with the U.S. market available for their products, tend to be large and efficient.

In the 1990s, there was a **"wheat war"** between the two countries as they struggled for greater market share in an environment of high yields and low prices. With agricultural prices for grains and pork dropping to the lowest levels in more than 50 years, many farmers both south and north of the U.S.–Canada border went bankrupt. U.S. producers claimed that Canada was **dumping** (selling the product for less than it cost to produce it) its durum wheat on the United States market.

There was also a **"salmon war"** between the two countries, marked by periodic skirmishes until a resolution was reached in 1999. Salmon are fish that hatch in Canadian and American rivers, spend most of their lives in the open seas, and return to their river birthplaces to spawn. The dispute centered on the issue of "interceptions," meaning that American fisherman often intercepted fish that hatched in Canadian upstream waters but migrated into U.S. waters.

In 2003, a new dispute emerged when, accusing Canada of dumping low-cost lumber in the United States, the U.S. imposed heavy new tariffs of 27 percent on softwood lumber imported from Canada into the United States. Until then, Canada had supplied about one-third of the lumber used in the construction of American homes. There are dozens of small water disputes across the border, including one over North Dakota's plan to divert excess water into the Red River, which flows north into—and might alter the environment of—Manitoba's Lake Winnipeg.

But despite such volleys occasionally fired on both sides of the border, the trend has been toward more cooperation and free trade. In 1965, the two countries negotiated a free trade agreement for the production, import, and export of automobiles. This sparked rapid expansion of the auto industry into southern Ontario, on the border with the U.S. state of Michigan. General Motors, Ford, and Chrysler developed plants there, some in Detroit's Canadian satellite of Windsor. However, metropolitan Toronto became the main center of the industry.

The two countries entered into the comprehensive **Canada–U.S. Free Trade Agreement** in 1988, and in 1994, Canada, the United States, and Mexico enacted the **North American Free Trade Agreement (NAFTA).** Its goals and impacts are described in detail, especially as they affect Mexico, on pages 540 and 554 in Chapters 19 and 20. NAFTA has brought a boom in trade between the three countries but also caused major industrial readjustments, generally characterized by the more labor-intensive firms (clothing especially) moving to Mexican locales from Canada's southeastern core region and from the industrial northeast of the United States.

Transportation Infrastructure

The governments of both the United States and Canada have sought to boost national unity and economic strength with internal transportation networks spanning their vast landscapes. In both countries, east–west transportation networks have been built over long distances, against the "grain" of the land, especially the great north–south trending ranges of the Rocky Mountains. The coasts of both countries were first tied together effectively by heavily subsidized transcontinental railroads completed during the latter half of the 19th century (Figure 22.18). Subsequently, national highway and air networks further enhanced transportation.

In the United States, a landmark development was the growth of the **interstate highway system** beginning in the late 1950s. Today, this system serves as the primary means for the shipment of cargo across the country, allowing large trucks to travel with few slowdowns for as long as the truckers themselves can stay at the wheel. In most other countries, railways generally meet these needs. However, the full potential of American railways is not being met, and few new lines are being built. Passenger rail traffic is particularly neglected, and once popular routes are no longer used. In many cases, the old rails have been torn up to become bicycle and walking paths (Figure 22.19). The stagnation or decline of U.S. passenger railways is a direct reflection of the American love affair with the car. Personal prosperity (or at least the ability to take out a loan) and the freedoms associated with being on the road in one's own automobile have diminished appetites to travel by rail and bus. Public transportation is only popular in the cities, where gridlock makes it an attractive alternative.

Figure 22.19 Missouri's Katy Trail State Park project has transformed an underutilized railway route into a popular recreation trail that runs the width of the state.

22.5 Geopolitical Issues

Just as the United States overshadows Latin America in geopolitical affairs, it overshadows Canada as well. Canadians, like Mexicans, often feel overlooked and underappreciated by the neighbors across the border. Canada maintains a rather low profile in international affairs, sometimes disagreeing with the United States on such critical issues as the Iraq invasion of 2003 and exerting a policy of relative independence and neutrality despite membership with the United States in the North Atlantic Treaty Organization (NATO) and other international agreements. Some American travelers in lands where Americans are often disliked have learned to exploit Canada's relative neutrality for their own protection by sewing badges of Canada's national symbol, the maple leaf, on backpacks and jackets.

The discussion here is mainly about the United States and, because America's clout is so large and persuasive, is limited to a few key points. This treatment is also brief because many of the geopolitical concerns of the United States are discussed in the sections on geopolitical issues introducing each world region in the book.

The United States is by any measure the most important economic, military, and diplomatic power in the world. Historically, it was not the same kind of empire-building power as many European countries were in establishing a network of overseas colonies. The United States did have its overseas possessions—the Philippines, Puerto Rico, and Guam, for example, and still holds a few—but its power has been and continues to be projected more through military action and trade.

The United States long enjoyed the geographic advantage of being something of an island situated far from the world's hot spots—especially Europe and the Middle East. In both world wars that ravaged Europe in the 20th century, the United States initially yielded to the **isolationism**—a view that those conflicts were Europe's, not Americas, and that the U.S. would do best to stay out of them. It was only late in

Figure 22.18 The Canadian Rail

World War I, on the widely unpopular insistence of President Woodrow Wilson, that the United States entered the war on the side of Britain and France. The United States lost 116,516 soldiers in that war but, in helping to win the war, also gained unprecedented influence and importance in world affairs.

Again in World War II, the United States managed to stay out of the conflict for more than 2 years after Hitler's troops invaded Poland and ignited conflagration across Europe. But on the infamous day of December 7, 1941, Japan forced the United States into the war with its surprise attack on Hawaii. And again, U.S. leadership and victory on both fronts of that war, which cost 405,399 American lives, helped the country achieve unprecedented strength in global affairs.

That war was succeeded by the Cold War, in which the two leading powers, the United States and the Soviet Union, faced off against one another, flanked by a host of often *72* strong but always less powerful allies. U.S. concerns about the spread of communism from the Soviet Union and China into the newly independent countries of the postwar world—and the so-called domino theory that one after another of these countries might fall to communism—led to numerous and sometimes very costly American military interventions. *338* The Vietnam War was the most significant of these.

The peaceful conclusion of the Cold War around 1990 was followed by a decade in which the United States sought a new role for itself on the world stage. There were military inventions to quell the conflict in the former Yugoslavia and to try to deliver lawless Somalia from famine, but there was a new sense of security—and, in hindsight, complacency. There was even an academic debate about "**the end of history**," meaning there was no longer any defining structure like east–west relations to international affairs. Terrorism was thought to be a problem that was far from home.

The attacks of 9/11/01 brought an end to the notion that the United States was protected by its geographic distance, and the events represented a great watershed in U.S. geopolitical history. Since that time, the United States has been developing a policy of **preemptive engagement**. This means that whenever and wherever the United States perceives a threat to its security, it will take military action if necessary to defuse that threat. The policy was founded mainly on the premise that such actions would prevent potentially devastating terrorist attacks on the U.S. homeland, perhaps with **chemical, biological, or nuclear (CBN) weapons of mass destruction (WMD)**. Such threats could emerge not only from transnational groups like al-Qa'ida but from "rogue states," *226* most of whom are on a list of **official state sponsors of terrorism**, including North Korea and Iran.

The U.S. administration of George W. Bush justified the invasion of Iraq in 2003 on the grounds that Iraq was a state sponsor of terrorism (and had links with al-Qa'ida) and had WMD it might use against the United States or its allies, especially Israel. The United States suffered a great setback in *257* the courts of world opinions when, in subsequent months, no conclusive evidence was found either of weapons of mass destruction or of links with al-Qa'ida. A June 2003 BBC poll found that 60 percent of Indonesians, 71 percent of Jordanians, and most disturbingly, 25 percent of Canadians viewed the United States as a greater threat than al-Qa'ida. Other detractors and analysts wrote of a "**new American imperialism**," proposing that U.S. post-9/11 actions marked the beginning of a new era of aggressive global involvement for the country; *Imperial Hubris* and *Superpower Syndrome* were among the books aspiring to document this new development. The Bush administration denied such intentions, insisting that its actions in Iraq and elsewhere had helped to protect Americans at home and strengthen democracy and the rule of law abroad.

The text now takes a closer look at Canada and the United States.

SUMMARY

- The United States and Canada comprise a region that is sometimes called Anglo America because of the strong British influences in both countries. But both countries had important pre-European indigenous cultures and have been shaped by many other cultures since 1492, particularly through immigration. Greenland is politically a part of Denmark but physically part of the region of the United States and Canada.

- Both the United States and Canada have large land areas of about the same size, but the United States is much more populous. The combined population of the two countries is about 326 million.

- The United States and Canada have a broad range of physical regions, including the Arctic Coastal Plains, Gulf–Atlantic Coastal Plain, Piedmont, Appalachian Highlands, Interior Highlands, Interior Plains, Rocky Mountains, Intermountain Basins and Plateaus, and Pacific Mountains and Valleys.

- Climatic patterns for the region are varied. Climate types include tundra, subarctic, humid continental with short summers, humid

continental with long summers, tropical savanna, and tropical rain forest. There are also undifferentiated highland and mountain zones. Patterns of vegetation and settlement are associated with these climate patterns.

- Particularly important resources for this region include agricultural land and forests. Both play a central role in exports of the two countries. Additional major resources include coal, petroleum and natural gas, water power, and minerals such as iron ore.

- Native American settlement of the region began at least 12,500 years ago. Some indigenous groups developed civilizations based on agriculture and permanent settlement, but most were at least partly nomadic and dependent on hunting and gathering. Great linguistic and cultural diversity existed among Native American groups. Almost all were extremely well adapted to local and changing environmental conditions, and all were changed profoundly by European contact. Native Americans are among the poorest populations in the United States but have fared somewhat better in Canada.

🌀 Demographic characteristics of the two countries show some similar patterns. Early European settlement displaced native populations, and subsequent waves of immigration continually changed the demographic, ethnic, and racial characteristics of the resident populations. In Canada, the French are the most significant minority. In the United States, African Americans and Hispanics are the largest minority groups. Asians are prominent minorities in both countries.

🌀 The United States built a land empire through battlefield victories and purchases. Significant expansions included the Louisiana Purchase (from France) of 1803; the acquisition of Florida from Spain in 1819, followed by Texas, the Southwest, and California as spoils of war with Mexico in 1845 and 1848; the Gadsden Purchase from Mexico of a small strip of present-day Arizona and New Mexico in 1853 ; the establishment of the Oregon Territory in 1846; and Alaska, bought from Russia in 1867.

🌀 Both countries became strongly industrialized in the 19th and 20th centuries. In the past several decades, the economic profiles of the countries changed as manufacturing became less important, while employment in service industries and information technology grew. Small towns have given way to middle-sized and large urban centers in both countries.

🌀 Canada and the United States are among the world's richest countries because of many factors. Both have large endowments of some of the world's most important natural assets. Both are large in area, with large—but not overwhelming—populations. Both became industrialized fairly early. Innovations in both have increased efficiency and productivity. Peace and stability at home have stimulated investment and economic development. Both have relatively strong internal unity and political stability.

🌀 The United States has the world's largest appetite for fossil fuel. Its energy efficiency has improved since 1973, but movement away from a fossil fuel economy has been slow. Nuclear energy production has stalled since the late 1970s, due mainly to environmental and health considerations.

🌀 The two countries have had generally good economic and political relations. The United States is the major customer for Canadian exports, and Canada plays a similar role for the United States. About 90 percent of the Canadian population lives within 100 miles of the international border. This proximity also causes some frustrations, as cultural influences from the United States sometimes seem to overwhelm Canada.

🌀 The United States is the world's strongest military and economic power. Its influence in world affairs was strengthened greatly in the wake of the two world wars of the 20th century. The terrorist attacks of September 2001 appear to have put U.S. foreign policy on a new course, based in part on the military doctrine of preemptive engagement.

KEY TERMS + CONCEPTS

Terms in blue are also defined in the glossary.

49ers (p. 614)
Aleuts (p. 610)
Algic language family (p. 609)
    Arapaho (p. 609)
    Blackfoot (p. 609)
    Cheyenne (p. 609)
    Cree (p. 609)
American Dream (p. 621)
animistic belief systems (p. 611)
Asian Americans (p. 616)
Aztec-Tanoan language family (p. 609)
    Comanche (p. 609)
    Hopi (p. 609)
    Papago (p. 609)
    Shoshone (p. 609)
blacks (African Americans,
    Afro-Americans, Negroes) (p. 615)
blizzard (p. 608)
"Boswash" (p. 600)
British–North America Act (p. 617)
California Gold Rush (p. 614)
California Trail (p. 614)
Canada–U.S. Free Trade Agreement
    (p. 623)
chemical, biological, nuclear (CBN)
    weapons (p. 625)
Cold War (p. 625)
consumption overpopulation (p. 617)
core region (p. 600)
Corporate Average Fuel Economy (CAFE)
    standard (p. 621)

Dakota (p. 609)
decommissioning (of dams) (p. 619)
drought (p. 608)
dumping (p. 623)
Dust Bowl (p. 608)
economic bubble (p. 623)
Equal Rights Amendment (p. 615)
Eskimo-Aleut language family (p. 610)
fall line (p. 604)
First Nations (p. 611)
First Peoples (p. 611)
G-7 (Group of 7) (p. 622)
G-21 (Group of 21) (p. 622)
Gadsden Purchase (p. 613)
gaming industry (p. 612)
globalization (p. 621)
Hispanic (Latino) (p. 615)
Hokan-Siouan language family
    (p. 609)
    Iroquoian (p. 609)
    Lakhota (p. 609)
    Mohawk (p. 609)
Homestead Act (p. 614)
hurricane (p. 607)
illegal aliens (p. 601)
information technology (IT) (p. 620)
interstate highway system (p. 624)
Inuit (Eskimos) (p. 610)
isohyet (p. 605)
isolationism (p. 624)
Kyoto Protocol (p. 619)

Lakota (p. 609)
land empire (p. 614)
Louisiana Purchase (p. 613)
Mainstream European American Culture
    (p. 615)
manifest destiny (p. 613)
megalopolis (p. 600)
Mosan language family (p. 609)
    Chemakuan (p. 609)
    Salish (p. 609)
    Wakashan (p. 609)
Multiculturalism Act (p. 613)
Na-Dene language family (p. 610)
    Athabascan subfamily (p. 610)
        Apache (p. 610)
        Koyukon (p. 610)
        Navajo (p. 610)
Native American civilizations
    (p. 608)
    Anasazi (p. 608)
    Hohokam (p. 608)
    Mogollon (p. 608)
    Mound Builders (p. 608)
        Adena (p. 609)
        Hopewell (p. 609)
        Mississippian (p. 609)
        Poverty Point (p. 609)
"neo-Europe" (p. 618)
"new American imperialism" (p. 625)
North American Free Trade Agreement
    (NAFTA) (p. 623)

North Atlantic Treaty Organization
(NATO) (p. 624)
Nunavut (p. 612)
"official state sponsors of terrorism"
(p. 625)
Oregon Trail (p. 614)
Penutian language family (p. 609)
    Klamath-Modoc (p. 609)
    Miwok (p. 609)
Pleistocene overkill hypothesis (p. 610)
potlatch (p. 609)
preemptive engagement (p. 625)
Protestant work ethic (p. 621)

pueblo (p. 608)
reservations (p. 611)
Rust Belt (p. 623)
"salmon war" (p. 623)
Santa Fe Trail (p. 614)
segregation (p. 615)
Sioux (p. 609)
slave (p. 602)
Spanglish (p. 616)
subsidies (p. 622)
tar sands (p. 619)
tariffs (p. 622)
tech bubble (p. 623)

The Depression (p. 622)
"the end of history" (p. 625)
Thirteenth Amendment (p. 615)
"tornado alley" (p. 607)
Tory (p. 617)
tsunami (p. 606)
undocumented workers (p. 601)
War of 1812 (p. 617)
weapons of mass destruction (WMD)
(p. 625)
welfare ranching (p. 605)
"wheat war" (p. 623)
World Trade Organization (WTO) (p. 622)

REVIEW QUESTIONS

WORLD
REGIONAL
Geography ⊛ Now™

Assess your understanding of this chapter's topics with additional quizzing and concept-based problems at http://earthscience.brookscole.com/wrg5e.

1. Where are populations of the United States and Canada concentrated?

2. What are the main topographic, climatic, and biotic zones and types of the region?

3. What are the principal natural hazards that threaten the United States?

4. What are some of the major Native American cultures of the United States and Canada? Historically, how did these people live in their diverse environments?

5. How did the French become such an important population in Canada?

6. What role has immigration had in shaping the populations of the United States and Canada?

7. What consequences did the American Revolution and the American Civil War have for Canadian development?

8. What are the major nonindigenous faiths and languages of the United States and Canada?

9. What have been the major trends and developments in the economies of the United States and Canada?

10. What are the major characteristics of economic and political relations between the United States and Canada historically and today?

DISCUSSION QUESTIONS

1. Is "Anglo America" an appropriate name for this world region?

2. What were the impacts of European settlement on Native American cultures and populations in the region? What struggles have other minorities had in this region?

3. May it fairly be said that a modern-day slave trade exists in the United States?

4. How did the United States acquire its land empire?

5. What and where were the major trails that opened the American west to settlement? What role did the California Gold Rush have in that settlement?

6. What are some of the reasons that Canada and the United States are among the world's wealthiest countries?

7. What major mineral and other resources are particularly important to the economies of the United States and Canada? How does the United States satisfy its huge energy appetite? What market price affects the exploitation of Canada's tar sands? Why did the government of Alberta oppose Canada's ratification of the Kyoto Protocol?

8. What changes have been made in energy use in the United States since the energy crisis of 1973? What are the prospects for future use of alternative energies? Which technologies are more and less appealing?

9. Much of America's economic progress is attributed to the Protestant work ethic and the American Dream. What are these, and if they exist, what impacts do they have on economy and society in the United States? Is there such a thing as a "mainstream European American culture"? If so, what does it consist of?

10. Discuss globalization and the economy. What impacts do subsidies and tariffs have on domestic and international economic conditions? What are the likely impacts of the removal of trade barriers? Why are the United States and other countries often reluctant to remove them?

11. How did the United States come to be the world's strongest power? What were the major mileposts in its evolution to that position? What are the hallmarks of the United States in world affairs today?

CHAPTER

23

Canada: From Sea to Sea

These schoolchildren on a museum visit reflect Canada's officially multicultural society.

Joe Hobbs

chapter outline

chapter objectives

This chapter should enable you to:

- ◔ Recognize how Canada's proximity to the powerful United States has shaped the country's economic geography

- ◔ Appreciate how resource depletion and distance from market and production centers contributed to the economic decline of the Atlantic region

- ◔ Understand why several sectors of Canada's population—the French, Inuit, and Prairie Province inhabitants—have pushed for more autonomy and even independence from Canada

- ◔ See how changing world energy and market conditions have altered indigenous ways of life in the North

- ◔ Understand the economic rationale for Greenlanders wanting to retain their ties with distant Denmark

WORLD
REGIONAL
Geography ⊛ Now™

Look for this logo in the text and go to GeographyNow at http://earthscience.brookscole.com/wrg5e to explore interactive maps, view animations, sharpen your factual knowledge and geographic literacy, and test your critical thinking and analytical skills with unique interactive resources.

Canada is a highly developed country: affluent, industrialized, technologically advanced, and urbanized. But in some ways, it is quite unlike most developed countries. These differences are related to Canada's internal geography and to its location adjacent to the United States. "From sea to sea" (this is the country's official motto), this chapter explores the geography of Canada, from its strongly French and English Coreland in the southeast to the "great white north," a land of Inuit peoples, tundra, and ice.

23.1 Canada's General Traits

Canada is very urbanized, with 79 percent of its people living in cities (the same percentage as the United States). About 36 percent of Canadians live in the country's three largest metropolitan areas: Toronto (population: 6 million), Montréal (population: 3.6 million), and Vancouver (population: 2.1 million) (Figure 23.1). These dynamic cities appropriately convey a positive image of an advanced and prosperous country. Impressively good national statistics on health, housing, and education provide further evidence of Canada's well-being.

Yet, Canada departs in some ways from the expected patterns of developed countries. It has an unusual pattern of trade. Canada's is among the world's top ten manufacturing nations. But despite fair-sized exports of automobiles, aircraft, industrial machinery, and some other fabricated items,

Canada is mainly an exporter of raw or semifinished materials, energy, and agricultural products: Wood pulp, timber, petroleum, natural gas, and hydroelectric power are among these. It is a major importer of manufactured goods, including machinery and equipment, motor vehicles, and consumer goods. By contrast, most developed nations generally export mainly manufactured goods and import a mixture of manufactured and primary products.

A second departure from other developed nations is Canada's overwhelming dependence on trade with a single partner, the United States. In 2004, the United States took in 86 percent of Canada's exports and supplied 60 percent of Canada's imports. No other developed country comes near these percentages of trade with just one other country.

Still another major divergence from other developed countries lies in the degree to which the Canadian economy is financed and controlled from outside the country; Canada again has an overwhelming dependence on the United States. More than three-quarters of all foreign investment in Canada is American. The extent of U.S. economic dominance in Canada has long been a sore point with many Canadians, who resent imputations that their country is an American economic "colony." These irritations tend to be heightened by the degree to which American mass-produced culture has permeated Canada and by how little Americans seem to know or care about Canada. Few U.S. schools offer courses on Canada, and a high percentage of American college stu-

Canada and Greenland

URBAN AREAS
Name Over 1 million
Name 500,000 to 1 million
Name Under 500,000*
* selected cities

0   200   400 mi.
0   200   400 km.

Figure 23.1 Principal features of Canada and Greenland

dents cannot even name Canada's capital (Ottawa). American news media give little coverage to Canada.

For some time, there has been a movement in Canada to achieve a more predominant position for Canadians within their own economy, to strengthen economic relations with industrial countries other than the United States, and to heighten cultural self-determination. But there seems to be no way for Canada to pull away from the United States without risking its own prosperity. The country's economy has

long been geared to a close interchange with the gigantic economy next door. This relationship became even more deeply rooted with the 1994 North American Free Trade Agreement (NAFTA).

Canada has a regional structure that presents some special problems. The country is divided between a thinly settled northern wilderness that occupies most of Canada's area and a narrow, discontinuous band of more populous regions stretching from sea to sea in the extreme south (see Figure

540
623

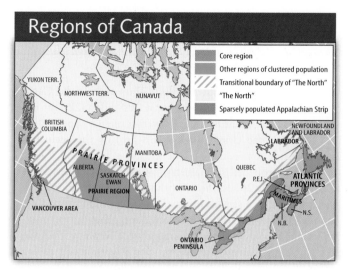

Figure 23.2 Canadian Regions

22.2A, page 601). These southern regions are sharply different from each other and sometimes conflict politically with one other and with the federal government in Ottawa. Following is a survey of Canada's regions (Figure 23.2).

23.2 Atlantic Canada: Hardscrabble Living

The easternmost of Canada's four main populated areas is characterized by strips or nodes of population, mainly along and near the seacoast in the provinces of Newfoundland and Labrador, New Brunswick, Nova Scotia, and Prince Edward Island. The four provinces are commonly called the Atlantic Provinces, and three, excluding Newfoundland and Labrador, have long been known as the Maritime Provinces. Their main populated areas are separated from the far more populous and more prosperous Canadian **core region** in Québec and Ontario by areas of mountainous wilderness. Prince Edward Island, Canada's smallest and most densely populated province, is a lowland with farms and small settlements. The main populated areas of New Brunswick, Nova Scotia, and Newfoundland occupy valleys and coastal strips of an upland. Politically, Newfoundland and Labrador includes the territory of Labrador on the mainland and the large island opposite the mouth of the St. Lawrence River. Formerly a self-governing dominion within the United Kingdom, in 1949 Newfoundland (whose name was changed to Newfoundland and Labrador in 2001) became the last province to join Canada. Bleak and nearly uninhabited Labrador, situated on the Canadian Shield, physically is part of Canada's North and is important for revenues from iron mining and hydroelectric production.

Isolation from markets afflicts most major enterprises in the Atlantic region. The market within the region is small and divided into many small clusters of people separated by sparsely populated land. The largest city in the Atlantic provinces—Halifax, Nova Scotia—has a metropolitan pop-

ulation of about 370,000. Only one other citiy—St. John's, Newfoundland, surpasses 125,000. The entire area has only 2.2 million people, or fewer than metropolitan Toronto or Montréal. The region is home to a large French population, especially in New Brunswick. It is one of the poorest of Canada's regions and, not surprisingly, is a region of out-migration. The poverty is somewhat surprising given the region's geographic advantage of being at the mouth of a major navigable river (the St. Lawrence Seaway) going into the heartland of a very prosperous country.

Conditions were not always like this. During the first two-thirds of the 19th century, the Maritime Provinces were a relatively prosperous area with a preindustrial economy. Local resources supported this development, including harbors along indented coastlines, fish for local consumption and export, timber for export and for use in building wooden sailing ships, and some agricultural land. The region carried on extensive commerce with Great Britain and the West Indies. Then followed an era of relative economic decline within an industrializing Canada. The ports were so far from the developing interior of the continent that most shipping bypassed them in favor of the St. Lawrence ports or the Atlantic ports of the United States. Halifax, Nova Scotia, and St. John, New Brunswick, are the main ports of the region today, and although each handles considerable traffic, neither is in a class with Montréal or the main U.S. ports. St. John's and Gander in Newfoundland played a significant role in serving as a refueling stop in the 1960s and 1970s for many trans-Atlantic air flights, but greater fuel efficiency and carrying capacity have now made that stopping point much less popular for commercial aviation.

The fishing industry was long the economic backbone of Canada's Atlantic region. Before the beginning of European settlement in the early 17th century, and probably before Columbus arrived in the Americas in 1492, European fishing fleets worked these waters, primarily on the "banks" along and near these shores. This area lies relatively close to Europe, and its waters were always exceptionally rich in fish.

But the bumper crop could not last forever, and fishing could not provide the basis for a prosperous modern economy. The United States imposed **tariff barriers** against Canadian fish to preserve its own domestic market. Competition from newly developed fishing areas grew in other parts of the world. Fishing fleets from other nations, often high-tech, began to drive Canadian fishermen from their own productive waters. The combination of expanding numbers of fishermen, more sophisticated and effective fishing technology, and steady overfishing led to a major and continuing decline in the number of fish in this traditionally rich area. The cod that were once the mainstay of this region's productivity are nearly depleted (see Problem Landscape, page 632). Populations of flounder, haddock, herring, and lobster are also in a calamitous decline.

Atlantic Canada is a difficult environment to farm. The four provinces lie at the northern end of the Appalachian Highlands and are mainly hilly. Under a cover of mixed forest, most soils are mediocre to poor. Agriculture has come

Problem Landscape

Overfished Waters

For many centuries, the western portion of the North Atlantic Ocean was a particularly rich resource base for fishing. Elevated portions of the sea bottom known as banks are located off the Atlantic coast from near Cape Cod to the Grand Banks, which lie just off southeastern Newfoundland. The shallowness of the ocean and the mixing of waters from the cold Labrador Current and the warm Gulf Stream foster rich development of the tiny organisms called plankton on which fish feed. These conditions helped put the Grand Banks at the heart of the fishing economy in Canada's northeastern region for nearly five centuries.

Inshore fisheries supplement the catch from the banks. Although many kinds of fish and shellfish are caught, the early fleets fished mainly for cod, which were cured on land before making the trip to European markets. Both the British and the French established early fishing settlements in Newfoundland, but the French—who had 150 vessels fishing these banks in 1577—were driven out by the British in 1763. The island has remained strongly British in culture, as its limited development has attracted very few other immigrants.

In 1977, worried about the effects of overfishing, the Canadian government began to enforce a 200-mile (322-km) offshore jurisdiction that prohibited overt foreign competition in the fishing of the Grand Banks. This held off the decline of available fish somewhat, but the greater efficiency of fishing technology propelled this industry toward difficult times. In the 1990s and early 2000s, the overfishing of cod and other fish in the banks led to a serious decline in fishing productivity, forcing the government to impose catch quotas that proved enormously unpopular with the fishing industry. Between 1992 and 2003, the government completely closed several cod fishing grounds in Newfoundland waters in an effort to stop the slide in fish populations. Tens of thousands of fishermen and fish processors lost their jobs. Provincial politicians introduced a bill in parliament in 2003 contending that the province should control its own waters. The fishing tradition that was for so long a major component of the economic base for Atlantic Canada appears to have suffered a great setback and may be restored only if fish stocks can themselves be replenished. Thus far, the moratoriums on fishing have failed to restore the fish populations for reasons that scientists have not yet determined.

---

to be concentrated on small patches of the best land (Figure 23.3). Climate also makes agriculture difficult. The Maritime Provinces have a humid continental short-summer climate, and Newfoundland has a subarctic climate. The entire region is humid and windy, with much cloud cover and fog. Strong gales are frequent in the winter. Summers are cool, and colder and less hospitable conditions occur as elevation increases; a good part of upland Newfoundland is tundra. Although most areas in the Maritimes and Newfoundland are agriculturally marginal, there are some favored lowlands with better soils, especially in Prince Edward Island and the Annapolis–Cornwallis Valley of Nova Scotia.

Agriculture in the Maritimes was also undermined by the building of railroads, which facilitated settlement of Canada's interior and thus brought cheaper farm products into the Atlantic region. It also provided easier access for local farmers considering a move west. Today, quite a few people in the Maritime Provinces still farm (especially potatoes, dairy products, and apples), but they do so on a small-scale, part-time basis while earning their living from other employment—a pursuit known as **hobby farming** in the United States. In this region, wood industries also suffered as overcutting depleted forests, ships began to be made of iron and steel instead of wood, and competition set in from new areas of forest exploitation farther west. Pulp and paper manufacturing is making a comeback because of the expansion of the international market for paper and wood products.

As these economic challenges have arisen, people of the Atlantic region have attempted to meet them by developing industry. Many small cotton-textile factories were built during the 19th century but failed when they were undercut by competition from mills in New England, central Canada, and Europe that were located closer to major markets. Coal reserves existed near the town of Sydney on northern Cape Breton Island in Nova Scotia, and iron ore was present on Newfoundland's Bell Island. On the basis of these resources, a coal mining industry and an iron and steel plant were developed at Sydney, which flourished for some time, largely by supplying steel for Canadian railway construction. But eventually, the plant closed and the coal mines declined, with consequent economic depression in the Sydney area. The plant at Sydney had been bought by the provincial government after World War II to forestall its closure, but government subsidies proved too costly to maintain. The local coal industry has been hurt by both the increasing cost of extracting the coal, which is deep lying, and the shrinking market for coal due to its widespread replacement by oil as a fuel. Many coal mines in this area have closed.

Currently, the basic economic activities that support Atlantic Canada's struggling economy are relatively small. In addition to the continued efforts to farm, fish, and develop wood products, there is some tourism. Canadian and American tourists alike are attracted by the region's stark beauty and picturesque villages. The export of electricity to adjacent

Francis Lepine/Animals Animals

Figure 23.3 This fertile valley in the Gaspé Peninsula is one of Atlantic Canada's rather few places for productive agriculture.

provinces and the United States, especially from Labrador's large Churchill Falls hydropower installation and from New Brunswick, has been a steady factor in the modest expansion of economic health in the region. The Hibernia Field off southeastern Newfoundland may possess enormous reserves of oil and gas, but extraction is particularly difficult and expensive because of local weather conditions and ocean floor characteristics. Oil and gas production in Canada's Atlantic waters has generally been disappointingly low to date.

Federal subsidies are extremely important to the economy of the Atlantic region. These come in the form of pension and welfare payments, the stationing of military forces in the region (Halifax is a major naval base), the presence of many federal administrative offices in the region, direct payments to provincial treasuries, and funding for economic development projects, with efforts to attract industry.

23.3 Canada's Core Region: Ontario and Québec

The core region of Canada, with about two-thirds of Canada's total population, has developed in a narrow strip along Lakes Erie and Ontario and seaward along the St. Lawrence River. In this area, the Interior Plains of North America extend northeastward to the Atlantic. They are increasingly constricted seaward on the north by the edge of the Canadian Shield and on the south by the Appalachian Highlands. The two provinces that divide the core region—Ontario and Québec—incorporate large and little-populated expanses of the shield; Québec has a strip of Appalachian country along its border with the United States. But most of the core region lies in lowlands bordering the Great Lakes and the St. Lawrence River from the vicinity of Windsor, Ontario, to Québec City, Québec. The entire lowland area is called the

St. Lawrence Lowlands, although the Ontario Peninsula between Lakes Huron, Erie, and Ontario is often recognized as a separate section.

Together, Ontario and Québec account for about two-fifths of the value of products marketed from Canadian farms. These two sections of the core region are markedly different in types of agriculture and cultural features. The Ontario Peninsula, which is strongly British in its roots, echoes the U.S. Midwest's cropbelts, with corn and livestock production, dairy farming, and the various specialty crops such as tobacco. Ontario, where 86 percent of the people speak English as a first language and only 5 percent speak French as a first language, is the largest in population of the nine predominantly English-speaking provinces.

In Québec, 82 percent of the people speak French as a preferred language and 8 percent speak English (Figure 23.4).

Joe Hobbs

Figure 23.4 French and English multilingualism is characteristic of Canada, especially in the east.

This is the only province in which French speakers predominate and control the provincial government, and Québec's potential devolution or even independence from Canada casts a long shadow over Canadian affairs (see Problem Landscape, page 636). The French heritage is clearly evident in the French "long-lot" pattern of strip-shaped agricultural landholdings (see Definitions and Insights, page 636) and in the region's large Catholic churches (Figure 23.5). The Québec Lowlands lie farther north than the Ontario Peninsula and have a harsher winter climate. Dairy farming is the predominant form of agriculture there.

Canada's core area contains more than 20 million people. Within this core, life focuses on Canada's two main cities: Toronto in Ontario and Montréal in Québec—the second largest culturally French city in the world after Paris (Figure 23.6). Other large cities of the core include Ottawa (population: 1.1 million), Québec City (population: 692,000), and Hamilton (population: 653,000). Ottawa, the federal capital, is located just inside Ontario on the Ottawa River boundary between the two provinces. Québec City, the original center of French administration in Canada, is the capital of Québec Province. Hamilton, a Lake Ontario port, is the main center of Canada's steel industry, supplying steel for the automobile and other metal-fabricating industries of Ontario. A good part of the urban development in the core is associated with manufacturing. Three-fourths of Canada's total manufacturing development is here, with somewhat more in Ontario than in Québec. But almost three times as many people are employed in the service industry, reflecting the region's status as the business and political center of the country.

Very important among the core region's advantages were its lowland and water connections between the Atlantic and the interior of North America. In eastern North America, the only other connection of this sort through the Appalachian

Figure 23.6 Montréal, on the St. Lawrence Seaway

barrier is the lowland route connecting New York City with Lake Erie via the Hudson River and Mohawk Valley. Rapids on the St. Lawrence River long barred ship traffic upstream from Montréal, but the lowland along the river facilitated railway construction, and the building of some 19th century canals bypassed the rapids and allowed small ocean vessels to enter the Great Lakes.

Finally, in the 1950s, the river itself was tamed by a series of dams and locks in the St. Lawrence Seaway Project (Figures 23.7 and 23. 8). Since then, good-sized ships have been able to reach the Great Lakes via the river. The Welland Canal, which bypasses Niagara Falls between Lakes Ontario and Erie by means of a series of stair-stepped locks, predated the seaway. The canal admits shipping vessels to the four Great Lakes above the falls. Farther up the lakes, the Soo Locks and Canals enable ships to pass between Lake Huron

Figure 23.5 Montréal's lovely Basilica of Notre Dame, built in the early 1800s and inspired by Paris's Sainte-Chapelle

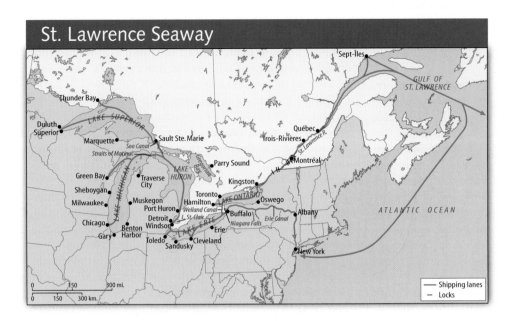

St. Lawrence Seaway

Figure 23.7 The St. Lawrence Seaway, more than 2,300 miles (c. 3,700 km) long, connects the Atlantic Ocean with the Great Lakes in the heart of the continent. This makes interior cities like Duluth, Minnesota, and Thunder Bay, Ontario, international ports.

Figure 23.8 Ice begins to lose its grip on the St. Lawrence Seaway in March.

and Lake Superior, and natural channels interconnect Lakes Erie, Huron, and Michigan. The total length of the St. Lawrence Seaway and associated waterways is 2,342 miles (3,769 km), and cities on the margins of the Great Lakes both in Canada and in the United States now serve as international ports. The seaway provides a 26-foot-deep (8-m) waterway in the heart of North America, and it moves wheat, iron ore, petroleum, and durable goods from mid-April until approximately mid-December with the assistance of icebreakers.

The St. Lawrence Seaway system has changed commercial geography in the heartland of North America in many ways. Through unwanted events, it has also changed the ecology of the region, particularly with the arrival of zebra mussels. The mussels, which originated in the Caspian Sea, diffused into these interior waters in the 1980s and have become a major blight on shipping and machine efficiency because of their

skyrocketing numbers and persistence. These shellfish attach themselves to the intake pipes of irrigation, drinking water, and power generation systems, impeding the flow of water and leading to significant machinery damage. These bivalves also deplete food resources for the native marine life.

The Rise of Commerce, Industry, and Urbanism in the Core

Until Canada was formed as an independent country in 1867, Québec and Ontario developed in a common preindustrial fashion, with economies based primarily on agriculture and forest exploitation. With its larger expanses of good land and a somewhat milder climate, the Ontario Peninsula became more agriculturally productive and prosperous than Québec. It specialized in the surplus production and export of wheat. On a superior natural harbor, the Lake Ontario port of Toronto developed and became the colonial (now the provincial) capital and Ontario's largest city.

But the bulk of Canada's exports at this time moved eastward to meet ocean shipping, and this movement favored Montréal. Until about 1850, most shipping from the Atlantic went no farther upstream than Québec City. Subsequent improvements along turbulent sections of the river enabled ocean ships to navigate safely as far inland as the major rapids (Lachine Rapids) at Montréal. Québec City was bypassed and its growth began to slow, although it has continued to be an active ocean port. In the later 19th century, Montréal decisively surpassed its French Canadian rival and became the leading port and largest city in Canada. Only recently has Montréal been exceeded in trade by the Pacific port of Vancouver and surpassed in population by Toronto.

As Montréal's port traffic grew, a rail network based in Montréal was developed. This network made use of lowland routes that radiated from the city in three directions: along the St. Lawrence Valley east to the Atlantic and west to the

Problem Landscape

The Québec Separatist Movement

Citizens of more developed countries are, on the whole, unaccustomed to the prospect that a major region of their own country might break off to form an independent country. This is a real prospect for Canadians, however, if the Québec separatist movement has its way. The modern separatist movement has its roots in early Canadian history: The French were early settlers in these lands, but after a major British military victory in 1763, the broad sweep of language, government, and authority became decisively British. A cultural dichotomy developed that gave the French dominant influence in Québec but established only modest influence for them beyond the St. Lawrence Lowlands region.

After the end of World War II, more widespread French Canadian dissatisfaction developed, in part because of increasing attention given worldwide to the plight of minorities. In Québec, this led to the formation of a separatist political party, the *Parti Québécois,* dedicated to the ultimate full independence of Québec from Canada. In 1976, the party achieved control of the Québec provincial government and made French the language of commerce, requiring that all immigrants settling in the province be educated in French. In 1980 and again in 1985, the separatists forced referenda on independence that lost both times. In 1995, the issue was taken to provincial voters again, and those who favored staying a part of Canada won by a very small margin of 51 percent.

Québec separatism is still very much alive and will likely stay on the national agenda. Separatists are frustrated by the steady influx of non-French-speaking immigrants seeking jobs and settlement in Québec City and other cities along the St. Lawrence Seaway, who tend to vote with the pro-English bloc, thereby thwarting separatist aspirations. Canadian law requires that there be no more than one separatist referendum every 5 years, but even that regulation may be overlooked as this process continues. It has the potential of turning Canada, seemingly a fortress of stability, over to political turmoil.

The effects of the separatist controversy have been considerable. The Canadian government has made French an official language of Canada—along with English—all across the nation. Laws now stipulate that the children of French Canadians born outside Québec may have their children educated in French. Canadian provinces have utilized the separatist strategies to gain more control of their own provincial governments. As Ottawa has given French Canadians more authority in Québec in the hope that they might find the idea of full independence less attractive, other provinces, especially in the Prairie region, have come forward with more demands for autonomy.

Definitions + Insights

Long Lots

The French **long lot** (Figure 23.A) is one of the most distinctive responses to living in a river environment. This land holding is a long and narrow lot that goes upslope from the river margin in the floodplain into the higher lands that generally parallel the riverbanks. The long-lot concept is said to have been designed to distribute the good and bad aspects of floodplain farming. To have frontage on the river was a benefit for trade, travel, possibly irrigation, and access to the water resource. However, such a location also meant the continual threat of flood. So, the French had the lot extend up to the higher land adjacent to the flood-prone areas. This meant that if there were a flood, then all the farmers living along the river's edge were flooded there, but they also had higher ground to fall back on. In Canada, the long-lot system is found mainly along the St. Lawrence River, especially in the province of Québec. It has become a distinctive landscape signature of French settlement not only in Canada but in almost all areas where the French were colonizers and settlers. In the United States, the long-lot pattern is part of the landscape along the lower Mississippi River, particularly where the French once lived in Louisiana.

Figure 23.A Long lots along the lower St. Lawrence River Valley, with the Laurentide Mountains in the distance

Great Lakes; along the Ottawa River Valley through which Canada's transcontinental railroads found a passage west and north into the shield; and along the Richelieu Valley–Champlain Lowland leading south toward New York City.

Britain created independent Canada in 1867 by confederating Québec, Ontario, New Brunswick, and Nova Scotia. By this time, the core area was already dominant in agriculture and population. In the following century, it was transformed into today's urban-industrial region. In its evolution to industrial dominance, the Coreland profited from a number of clear advantages over the rest of Canada: superior position with respect to transport and trade, accessibility to the largest markets, the most advantageous labor conditions, and the best access to significant resources.

There were many resources to support industrialization. Farmland good enough for large commercial agriculture, especially in Ontario, supplied materials for food processing industries and a market for farm equipment industries. Forests, mainly in the adjacent southern edge of the Canadian Shield, once supplied lumber for export and now supply wood for a large pulp and paper industry, mainly in Québec along the St. Lawrence and its north bank tributaries. Metallic minerals are varied and abundant in the shield to the north and west of the core area. These include large deposits of iron ore located near the western end of Lake Superior on both sides of the international border, and others, more recently developed, are in northeastern Québec and adjacent Labrador. A large development of metal processing and fabricating industries, especially in Ontario, uses these shield minerals, which are also exported.

There has also been abundant energy. Especially at the descent from the Canadian Shield to the plains, many high-volume, steeply falling rivers drive one of the world's larger concentrations of hydroelectric plants. Major hydroelectric plants are also present at Niagara Falls, along the St. Lawrence as part of the seaway project, and more recently, at far northern sites in the shield, including the James Bay plants discussed later (Figure 23.9). Coal requirements are met from nearby Appalachian fields in the United States, and oil comes from Atlantic or Middle Eastern ports by tanker up the St. Lawrence. Both oil and natural gas come by pipeline from western Canada.

Through the combination of these resources, their processing, and their distribution to both domestic and foreign markets, the Canadian core region has established a solid base for continuing economic development and relative prosperity. The region now accounts for more than three-fourths of the output of Canada's major manufactured goods. Within the core, there are marked differences between the industrial emphases of Ontario and Québec. During the formative decades of industrialization, Ontario had cheaper access to U.S. coal. This favored the development of iron and steel capacity and metal products industries there, and these types of industries are still more prominent in Ontario than in Québec. Montréal also has industries focused on making

Figure 23.9 Niagara Falls, from the Canadian side.

Joe Hobbs

transport equipment and other metal goods, but these are less typical of Québec than are more labor-intensive industries such as apparel manufacturing. Such industries were established primarily to deal with a traditional surplus of labor in Québec created by the rapid natural increase of Québec's Roman Catholic population. This labor surplus led to migration into other parts of Canada and the United States and also to the establishment of manufacturing industries that needed cheap labor. Textiles, shoes, furniture, and clothing are typical products. The differences between the two provinces still persist, and they have resulted in lower average wages and incomes in Québec, which has also experienced a steady decline in its rate of population growth.

23.4 The Prairie Region: Oil, Wheat, and Wilderness

Manitoba, Saskatchewan, and Alberta make up the Prairie Provinces. Isolation is one factor they have in common. Their populations are separated from other populous areas by hundreds of miles of thinly inhabited territory. To the east, the Canadian Shield separates them from the populous parts of Ontario, and on the west, the Rocky Mountains and other highlands separate them from the Vancouver region. To the north lies subarctic wilderness, and to the south are sparsely populated areas in Montana, North Dakota, and Minnesota.

F22.5B
604

The prairie environment that gives the provinces their regional name exists only in some of their southern sections, within the Interior Plains between the Canadian Shield and the Rockies. There is a triangle of natural grasslands, with its southern base resting on the U.S. boundary, extending to an apex about 300 miles (c. 500 km) north of the U.S. border. On all sides except the south, the grasslands are bordered by vast reaches of forest. Most of the southern part of the triangle is a northward continuation of steppe grasslands from the Great Plains of the United States. However, toward the forest edges, the soils are moister, and the original settlers found true prairies: taller grasses with scattered clumps and riverine strips of trees. Settlers were most attracted to the prairies, and these moister grasslands still form a more densely populated arc-shaped band near the forest edges.

Even though less than 5 percent of Canada's total area is suitable for farming and although only about 3 percent of the Canadian labor force is engaged in agriculture, Canada is a major world wheat exporter because of the Prairie region's soils and the spring wheat crop they support. To the south, the Great Plains region of the United States is a direct topographic continuation of this immense open plain. Agriculture is vital to the economy of the Prairie Provinces. The region produces about one-half the value of Canada's farm products. In addition to wheat, there is very productive agriculture of barley, rapeseed (for vegetable oil called canola, and fodder), and livestock, especially cattle.

Minerals are the foundation of the greatest recent advances in the Prairie Provinces. More than 60 percent of Canada's mineral production is in this region. Coal deposits underlie the Rocky Mountain foothills and the plains in Alberta as well as western Saskatchewan. Coal has recently begun to provide a major export market. Nickel, copper, zinc, and other metals come from the Canadian Shield of Manitoba. The Saskatchewan Shield produces uranium, and southern Saskatchewan produces major quantities of potash.

Petroleum and associated natural gas have had the most economic impact on the Prairie region. The region's oil industry began near Calgary, Alberta, in the 1940s and led to the explosive growth of both Calgary and Edmonton. The industry has remained predominantly in Alberta, with a minor portion in Saskatchewan and a tiny share in southern Manitoba. An important product extracted from natural gas in Alberta is sulfur, of which Canada is a dominant world exporter. It is estimated that **tar sands** in northern Alberta may contain two to three times the energy equivalent of all the oil now thought to exist in the Middle East. Extraction of this unconventional source of energy has surged as conventional oil prices have risen to record levels.

Urbanization and urban industries have become increasingly important in this region, especially in Alberta and Manitoba. The three main metropolises are Edmonton (population: 994,000) and Calgary (population: 1 million) in Alberta and Winnipeg (population: 674,000) in Manitoba. These cities have all grown near the corners of the Prairie triangle where connections to the outside world are focused. Winnipeg developed where the rail lines through the shield crossed the Red River and entered the Prairie region; Calgary grew near a pass over the Rocky Mountains that gave a route toward Vancouver; and Edmonton was established near another Rocky Mountain pass leading to the Pacific. Both Winnipeg and Edmonton developed additional functions as metropolitan bases for huge sections of Canada's North. None of the three Prairie Provinces is a major manufacturing center. The environmental setting has turned this region into an important tourist destination for Canadians and Americans alike. Calgary is the eastern gateway to the spectacular Rocky Mountain region, and its annual Calgary Stampede, which celebrates the region's colorful history of farming and ranching, is one of Canada's most famous and lucrative tourist attractions.

23.5 The Vancouver Region and British Columbia

Canada's transcontinental belt of clusters of population is anchored at the Pacific end by a small region centering on the seaport of Vancouver at the southwestern corner of British Columbia (Figure 23.10). The region contains no more than 5 percent of British Columbia's area but has over one-half of the province's population. Metropolitan Vancouver has 2.1 million people, and another 317,000 reside in the metropolitan area of the province's capital city, Victoria. Vancouver is located on a superb natural harbor at the mouth of the Fraser River. This valley provides a natural route for railways

Figure 23.10 The jewel of the Canadian west coast is Vancouver, British Columbia.

Jeff Greenberg/Visuals Unlimited

and highways coming from the rugged mountains and plateaus in Alberta and British Columbia. Victoria is located at the southern end of Vancouver Island, the southernmost and largest of a chain of islands along Canada's Pacific coast.

Victoria was the leading commercial and industrial center in British Columbia before the completion of Canada's first transcontinental railroad, the Canadian Pacific, in 1886. Vancouver then became the country's main Pacific port, and it has recently become the country's largest seaport. It has profited from continuing development and a buildup of foreign trade in British Columbia and the Prairie Provinces, particularly Canada's increasing exports of resource commodities to East Asia, primarily China and Japan. This city served as a major migration destination for Hong Kong Chinese in the years preceding the 1997 British handover of Hong Kong to the People's Republic of China. Vancouver's population is about 20 percent ethnic Chinese (Figure 23.11). Tourists from Asia (especially Japan, Korea, and China) and the United States come to Vancouver to enjoy its cosmopolitan and scenic attractions and to begin cruises up the spectacular coastlines of British Columbia and Alaska. The city's tolerance of cannabis consumption has earned it the nickname "Vansterdam."

Vancouver exports huge quantities of bulk commodities such as grain, wood in various forms, coal, sulfur, metals, potash, and asbestos. Sea trade is fundamental to Vancouver's economy, but the city is far more than a seaport. Despite its peripheral location within British Columbia, it is the province's main industrial, financial, and corporate administrative center. It has numerous links with production nodes scattered throughout the sparsely populated and mountain-

Figure 23.11 Ethnic Chinese are a strong and growing minority in Canada.

ous province of British Columbia, which is larger in area than California and Montana combined.

British Columbia, along with the Yukon Territory to its north and a narrow fringe of Alberta, is Canada's mountainous country (Figure 23.12). The fjorded Coast Ranges along the Pacific and the Rockies of the interior are high and spectacular, and the Intermountain Plateaus between them are rugged. Only the northeastern corner of the province is plains environment. This is Peace River Country, the extreme northern outpost of the Prairie region of grain and livestock agriculture. British Columbia is distinguished climatically within Canada by its coastal strip of marine west coast climate. Here, moderate temperatures and heavy precipitation

Figure 23.12 Spectacular British Columbia

support the dense coniferous forests that are one of the province's principal resources. The mild temperatures and spectacular scenery attract tourism and retirees. The climate of the interior is subarctic in the north and varies according to elevation and exposure in the south. As on the Coast Ranges, the interior is largely forested.

Scattered small communities exploit abundant natural resources. Forest products come from both the interior and the coast. The timber industry is not as valuable as it once was, thanks to competition from tropical countries like Brazil and Malaysia, which can export their hardwoods at lower cost. Salmon fishing continues along the coast as it has for more than a century. Salmon **fish farming** has become a very important industry to British Columbia and is posing a serious economic challenge to the "mom and pop" fishers of wild salmon in Canadian and Alaskan waters. With farmed salmon abundant, prices have plunged, so the traditional fisherfolk are promoting the better health and taste of their wild product in hopes of developing a market niche.

More important than forestry and fishing today is a boom in the mining of coal, natural gas, petroleum, and metals (copper, molybdenum, silver, lead, and zinc). Older metal production sites in the south are being supplemented by operations farther north. The driving factor is Chinese, Japanese, and U.S. market demand, with development being supported by foreign capital. Coal is shipped mainly to Japan, and electricity generated from the region's abundant waters is exported to the United States. Near the coastal town of Kitimat, hydroelectricity is used to produce aluminum from imported alumina (the second-stage material from bauxite ore). Wages tend to be high in the small and isolated production nodes scattered over British Columbia, but so is the cost of living, and cultural amenities are few. Most new residents of British Columbia settle in cosmopolitan Vancouver or in Victoria, known as the most English of Canadian cities.

23.6   The North: Lots of Land, Few People

About four-fifths of Canada has cold tundra and subarctic climates, with more than one-half in the glacially scoured and agriculturally sterile Canadian Shield, and a Pacific region dominated by rugged mountains. These conditions create the huge expanse of nearly empty land that Canadians call "the North." More than one-half of it lies in territories administered from the country's capital in Ottawa, Ontario, but large parts extend into seven of the ten provinces stretching across southern Canada that are political subdivisions of the country's federal system. In this text, the southern edge of the subarctic climate zone is used to define most of the North's southern boundary, but the island of Newfoundland is excluded because of its Atlantic associations and somewhat denser population.

F23.21
631

The Near North

The southern fringes of the North, often called the **Near North,** have widely scattered islands of development that are important to Canada's economy and, through trade connections, to the economy of the United States. Many of these are mining settlements. The Canadian Shield here has a wide variety of metallic ores, although geological processes have left it lacking in fossil fuels. The mining settlements range in size from small hamlets to small metropolitan areas. The largest is Sudbury (population: 148,000), located north of Lake Huron in Ontario. Noteworthy among the many metals produced in the region are nickel and copper around Sudbury, iron ore from Québec and Labrador, and uranium mined just north of Lake Huron and in northern Saskatchewan.

Some northern settlements manufacture pulp and paper and smelt metals. A scattering of towns producing wood pulp and paper is spread across the southern fringes of the North. Most are small, but in two places, clusters of mills combine with other functions to produce fairly sizable urban areas. A string of towns along the Saguenay River tributary of the St. Lawrence in Québec is supported by pulp and paper plants, by farming, and by the manufacture of aluminum from imported raw materials brought in by ship. The other urban concentration is at Thunder Bay, Ontario (population: 118,000), on the northwestern shore of Lake Superior. In addition to pulp and paper, Thunder Bay benefits from being at the Canadian head of Great Lakes navigation. Its port links the Prairie Provinces with the Ontario–Québec core area, the United States, and overseas ports.

Water resources are of major importance in the southern fringe of the North. Hydroelectricity supports local mines and mills and is sent to southern Canada and the United States. Rivers and lakes are so characteristic of the shield that from the air it appears to be an amphibious landscape (Figure 23.13). There is a cluster of hydroelectric plants where St. Lawrence tributaries from the north rush down through the Laurentide "Mountains"—the raised southern edge of the shield in Québec. But increasing demand, plus increasing ability to transmit electricity for long distances, has led to major developments much farther north. These are located near James Bay (between Ontario and Québec) and in Manitoba near Hudson Bay.

Hydropower development is especially controversial in the James Bay region because of its potential impacts on indigenous people and natural environments. In the 1990s, Québec's state-owned power utility Hydro-Québec pursued development of the **James Bay Project** to dam and divert a number of the area's wild rivers to generate electricity. The region's Native American **Cree** concluded that the diversions, reservoirs, and other changes would be detrimental to their caribou hunting, fishing, and other forms of livelihood and would inundate much of their tribal homeland. They waged an international public relations battle against Hydro-Québec. Their campaign in the U.S. media compelled New

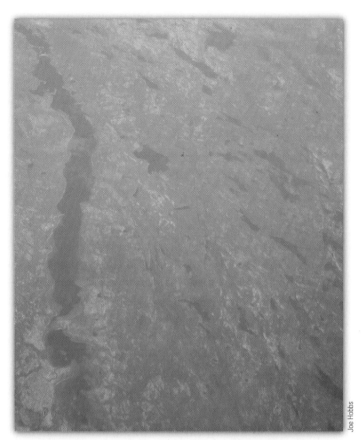

Joe Hobbs

Figure 23.13 The Canadian Shield is an often icy sheet of glacially scoured rock with little potential for agriculture.

York's governor to cancel a contract to buy some of the electricity that would have been produced by the new developments. Hydro-Québec eventually cancelled plans to build a dam on the Great Whale River, which would have had the largest impacts.

However, in 2002, the Cree reached a settlement with the Québec government in which they have agreed to allow the building of hydropower dams on two of the region's less significant rivers. The Cree will also drop billions of dollars in environmental lawsuits against the government. For the Cree, the deal includes cash payouts of $70 million a year for 48 years. The Cree will also obtain more control over their communities and economies, more power over decisions about logging, and more jobs with Hydro-Québec.

A remarkably small number of people live in this vast area. Agricultural settlements are rare, and most are declining. The port of Churchill on the western edge of Hudson Bay serves as the closest port for movement of the Prairie Provinces' wheat to the European market. It is also well known for a very specialized form of tourism, polar bear watching during the late fall season when the great predators linger on the Lake Hudson shoreline, awaiting the formation of the sea ice they will travel on through the winter.

Territorial Canada: Indigenous Landscapes

North of 60° latitude lies the part of Canada without provincial status. This wild area, known as Territorial Canada, is under the administration of the Canadian federal government and of indigenous people. There are three administrative territories (see Figure 23.1). In the northwest, the Yukon Territory is cut off from the Pacific by the panhandle of Alaska, once claimed by Canada but acquired by the United States in 1867. It is mostly rugged plateau and mountain country. The much larger Northwest Territories lie east of Yukon Territory. They include a fringe of the Rockies, a section of the Interior Plains along which the Mackenzie River flows northward to the Arctic Ocean, a vast area of plains and hills in the Canadian Shield, and the many arctic islands. Two huge lakes, Great Bear Lake and Great Slave Lake, both drained by the Mackenzie River, mark the western edge of the shield. The Yukon Territory and the western part of the Northwest Territories have mainly a subarctic climate and taiga vegetation, with tundra near the Arctic Ocean, but in the eastern part of Canada, tundra extends far to the south (see Figure 22.5a and b). The Nunavut Territory occupies eastern and northern areas withdrawn from the Northwest Territories in 1999 to form a semiautonomous region inhabited by **Inuit** peoples.

610
612

Europeans and Anglo Americans have been seeking resources from these far northern reaches since the 17th century, when England's Hudson's Bay Company first established fur-trading posts on the shore of Hudson Bay. The chief 20th-century quest has been for metallic and other minerals. There is active exploration for oil and gas in the Mackenzie Valley and in the Beaufort Sea section of the Arctic Ocean near the Mackenzie Delta. Future development of these resources is controversial because of potential environmental impacts. Among the issues under scrutiny is a plan to pipe natural gas from the Mackenzie Delta to Alberta, where it would provide energy for the extraction of oil from tar sands. In addition to income from the few mining enterprises, the territories also receive funding from federal government payrolls in the territorial capitals of Whitehorse (Yukon), Yellowknife (Northwest Territories), and Iqaluit (Nunavut; the town's Inuit name means "Place of Many Fish").

The population of Territorial Canada was only about 105,000 as of 2004. Less than 2 percent of the inhabitants in each of the present territories are the native **First Nation peoples,** which do not include the Inuit. In the Northwest Territories, about 9 percent—some 4,000—are Inuit, and in Nunavut, about 50 percent are Inuit. Most whites are government, corporate, or church employees, and they are often transient, working only until they can return south. First Nation peoples and Inuit still carry on some hunting, gathering, and fishing from homes in small fixed settlements. The trade in furs of seals, marten, beaver, fox, and other mammals

611

was once vital to their economies, but international protests about the methods by which the animals were caught and killed led to a decline in consumer appetite for fur products and to legislation by the European Union and other governments against fur imports. With the fur industry all but collapsed, many of the region's indigenous peoples have felt compelled to negotiate royalty arrangements with the natural gas and oil interests they once opposed on environmental grounds.

Practically all Inuit inhabit prefabricated government-built housing in widely separated villages along the Arctic Ocean or Hudson Bay (Figure 23.14). Unemployment among both First Nations folk and Inuit is very high, and a large proportion live on welfare. The nature of these people's lives has changed immeasurably with the diffusion of things both good and bad from Canada and the United States.

The North's isolation is somewhat mitigated by the airplane and by telecommunications bringing telephone service, radio, television, and the Internet to remote settlements. A network of schools and medical clinics blankets the area. In Iqaluit, the capital of Nunavut located on remote southern Baffin Island, the school system has undertaken an ambitious Internet linkage with schools all over the world in an effort to diminish the sense of isolation. Air transportation enables the seriously ill to be flown to the few hospitals in the region or to hospitals in cities such as Edmonton or Winnipeg, which serve as metropolitan bases for both the provincial and the territorial North. Meanwhile, the increasingly as-

Figure 23.14 Indigenous people still practice subsistence hunting in Canada's North. Here, hides of Arctic wolves, shot in the winter by this woman's snowmobile-mounted husband, cure in the summer sun in the village of Arviat on Hudson Bay. Canada's Inuit generally live in prefabricated houses like the one seen here. Fuel oil for household heating and cooking comes in oil drums brought by ship.

sertive First Nation and Inuit populations are pressing land claims to mineralized areas and pipeline routes and are demanding regulations to prevent further environmental disruptions by mining companies.

23.7 Greenland: A White Land

An unlikely cultural outlier of Europe, Greenland is geologically part of North America, and because it is closest to Canada, it is discussed briefly here (see Figure 23.1a). The world's largest island (840,000 sq mi/2,175,600 sq km), Greenland has been under Danish control since 1605. It became a Danish province in 1953, but in 1979, the Danes granted the large island self-government, while retaining control over its foreign affairs. It lies more than 2,000 miles (3,200 km) west of the European nation with which it is associated. In 1985, Greenland withdrew from the European Economic Community (EC; the precursor to the European Union) because of its frustration over EC fishing policies and the lack of any EC development aid. Today, however, the European Union gives Greenland almost $50 million each year in exchange for Greenland's granting of fishing quotas to EU member states in its territorial waters. Denmark subsidizes the massive island with an average support of more than $8,000 per person per year. There is increasing talk of independence among the Greenlanders, but severing the aid pipeline from Denmark and the European Union would make that a very costly proposition.

Greenland has nearly one-fourth the area of the United States, but about 85 percent of the island is covered by an icecap up to 10,000 feet (3,000 m) thick (Figure 23.15). Until recently, Greenland seemed to have bucked the worldwide trend toward melting of glacial ice, but it now appears to be experiencing significant melting of the outer edges of its ice sheet. Scientists are working to determine whether the melting is related to overall global warming and what its impacts on the "conveyer belt" of Atlantic oceanic waters and subsequent climates might be. Some models suggest that the ice sheet could shrivel completely over a period of about 1,000 years, which would raise the world's sea levels by 20 feet (6 m).

Ninety percent of Greenland's population is distributed along the southern coast. Greenland has a total population of just over 50,000, giving it a population density of less than one-tenth of a person per square mile. The capital and largest city is Nuuk (also called Godthab; population: 14,000) on the southwestern coast. The great majority of inhabitants are Inuit, with Danes making up about 15 percent of the population. The principal means of livelihood are fishing, hunting, trapping, sheep grazing, and the mining of zinc and lead. Under the auspices of NATO, joint Danish American air bases are maintained there, and the United States has a giant radar installation at Thule in the remote northwest.

The text turns finally to the United States.

Figure 23.15 Greenland's spectacular glacial landscape

Joe Hobbs

SUMMARY

- Canada is highly urbanized, with 79 percent of its population living in cities. It has a major global role in the production of pulp and paper, hydroelectricity, commercial motor vehicles, and aluminum. Nearly one-third of its total population lives in Toronto, Montréal, and Vancouver.

- Of all the world's more developed countries, Canada has the strongest economic relationship with a single country (the United States). About 86 percent of Canada's exports go to the United States.

- Ten provinces (excluding the territories and Nunavut) make up the political administration of Canada. More than one-half of Canada's area is in the North, with very low population densities. Nearly 90 percent of the Canadian population is settled within 100 miles (161 km) of the U.S. border, with the largest bloc of population settled in the southeast near the valley of the St. Lawrence River and Seaway.

- Atlantic Canada consists of the provinces of Newfoundland and Labrador, New Brunswick, Nova Scotia, and Prince Edward Island. The region is also called the Atlantic Provinces and, except for Newfoundland and Labrador, the Maritime Provinces. Cod fishing has been the mainstay of its economy ever since earliest settlement. The Grand Banks on the east of this region have long been a rich source of fish harvest because of the mixing of warm waters from the south and cold from the north. Recently, the Banks have become so overfished that government regulation has restricted the harvest, causing much unemployment. The Atlantic Provinces are seeking new revenues through specialized agriculture, tourism, forestry, energy, and minerals. Oil and gas production has fallen short of original expectations.

- A narrow strip across the southern parts of Ontario and Québec Provinces make up the core of Canada. Two-thirds of Canada's population live in this region. The margins of the St. Lawrence River and the Great Lakes are the primary areas of settlement and industrial activity. The most important economic resources for Québec and Ontario are agriculture, forestry, energy generation, and bulk commodity transport.

- Industrial development and trade moved upstream on the St. Lawrence steadily until the completion of the St. Lawrence Seaway in the 1950s. Québec City, the earliest urban settlement in Canada, was outgrown by Montréal, and then Montréal was surpassed by Toronto in size and economic importance (but Vancouver on the west coast has grown to be Canada's biggest port).

- The Ontario Peninsula is strongly British in origin and character, and Québec is strongly French, with 82 percent of the Québec population of French origin. Québec separatists have been a political presence for the past 40 years, and in 1995, the referendum for an independent Québec was defeated by less than 1 percent. In its push to receive additional accommodations from the federal government in Ottawa, Québec has gained more provincial autonomy. Other provinces, especially in the Prairie region, have sought the same benefits by threatening to change their relationship with Ottawa.

- The Prairie region is made up of Manitoba, Saskatchewan, and Alberta. Wheat, petroleum, and coal are the major economic resources. Major urban centers include Edmonton, Calgary, and Winnipeg. These centers provide linkage east and west with other major Canadian regions and also with the North.

◉ The Vancouver region is centered in Vancouver, British Columbia, at the mouth of the Fraser River. More than one-half of the province's population lives in the Vancouver area, which is the region's main industrial, administrative, financial, and cultural center. It has a strong ethnic Chinese population, and its trade is increasingly focused on East Asia. British Columbia has become a major tourist and retirement destination.

◉ The Canadian North is bounded at the south roughly by the southern limit of the region of subarctic climate. In this sparsely settled region, population densities tend to decline from south to north. Nickel, copper, and uranium are mined here and exported, and there are plans for expansion of the petroleum and natural gas industries. Forestry, pulp manufacture, and hydroelectricity are additional economic resources. Nunavut, the Inuit homeland, is the newest political unit. The fur trade was once vital to the economies of the North's indigenous peoples, but a backlash against the method by which furbearing animals were killed and new legislation have all but destroyed this industry.

◉ Greenland is geologically a part of North America but is a self-governing property of Denmark. Greenland's foreign affairs are controlled by Denmark which, along with the European Union, spends heavily on subsidies for the 50,000 inhabitants of the world's largest island. Scientists are watchful of the potential melting of the massive icecap that covers much of the island.

KEY TERMS + CONCEPTS

Terms in blue are also defined in the glossary.

core region (p. 631)
Cree (p. 640)
devolution (p. 634)
First Nation peoples (p. 641)
fish farming (p. 640)

(fishing) banks (p. 632)
hobby farming (p. 632)
Inuit (p. 641)
James Bay Project (p. 640)
long lot (p. 636)

Near North (p. 640)
Parti Québécois (p. 636)
St. Lawrence Seaway Project (p. 634)
tar sands (p. 638)
tariff barriers (p. 631)

REVIEW QUESTIONS

WORLD
REGIONAL
Geography◉Now™

Assess your understanding of this chapter's topics with additional quizzing and concept-based problems at http://earthscience.brookscole.com/wrg5e.

1. In what ways is Canada unique as a more developed country?

2. Why did fishing fail to provide the basis for a prosperous economy in Canada?

3. What are the major import–export patterns for Canada and the most important goods and partners in the trade?

4. What are the advantages of the long lot to a riverside farmer?

5. What two political units contain the core region of Canada? What are some of the major similarities and differences in their geographic characteristics?

6. What waterways make up the St. Lawrence Seaway system? What important roles do these waterways have in the economies of Canada and the United States?

7. What factors favored the development of Montréal?

8. What conditions have supported Canada's emergence as a major wheat producer?

9. What role has hydropower played in the development of Canada's Near North?

10. What nation is Greenland politically affiliated with? Why do most Greenlanders want to maintain these ties?

DISCUSSION QUESTIONS

1. What environmental factors limit agriculture in Canada? Which areas are most productive agriculturally and why? What are their major agricultural products?

2. What happened to the once strong economies of Canada's Atlantic Provinces? What has been the fate of its fishing industries and why?

3. What are the key economic and other advantages that the Coreland has over the rest of Canada?

4. What factors are behind Québec's push for independence? What are the prospects that this province might actually break away from Canada?

5. How is Vancouver's orientation toward Asia reflected in its population and in its trade?

6. What changing environmental and trade circumstances have altered the economic geography of British Columbia?

7. What plans did the government-owned Hydro-Québec company have for the James Bay region? How did the region's indigenous people react to these plans? What was the outcome of their dispute with the company and the provincial government?

8. How did information, changing fashions, and new laws all but destroy the fur trade that was once vital to the native peoples of Canada's North? What new industry has taken its place?

9. What is happening to Greenland's extensive ice cover and why? What would be the result of a complete melting of that ice sheet?

10. Several of the sparsely populated regions described in this chapter are supported by costly government subsidies. Why is such support deemed necessary? What apparent costs and benefits are associated with the subsidies?

The United States: Out of Many, One

One face of the diverse United States: A musician at a street fair in San Jose, California

Joe Hobbs

---

chapter objectives

This chapter should enable you to:

- Recognize how rivers, topographic boundaries, and other geographic circumstances promoted the development of many American cities

- Understand the environmental and political issues that complicate development of coal, oil, and natural gas in the United States

- Follow the decline of traditional heavy industries in the process of deindustrialization and their replacement by high-technology and service industries

- Evaluate the depopulation of the Great Plains, the decline of small towns, and the potential for communication technology and foreign immigration to reverse these trends

- Gain more insight into the country's ethnic geography and immigration issues

WORLD
REGIONAL
Geography 🌐 Now™

Look for this logo in the text and go to GeographyNow at http://earthscience.brookscole.com/wrg5e to explore interactive maps, view animations, sharpen your factual knowledge and geographic literacy, and test your critical thinking and analytical skills with unique interactive resources.

The book concludes with a country that is home to most of its readers (Figure 24.1). This chapter should give the American reader a great deal of new insight into the home country. There is a fair amount of emphasis here on the geographic circumstances that gave rise to the establishment and growth of the country's main cities. There is also much attention to the ways the country has been changing as it adjusts to new demographic realities at home and fresh economic and other challenges abroad. The chapter is organized around five vernacular regions: the Northeast, South, Midwest, West, and Alaska and Hawaii.

24.1 The Northeast: Center of Power

Of these five regions of the United States, the Northeast is the most intensively developed, densely populated, ethnically diverse, and culturally intricate. It incorporates the nation's main centers of political and financial activity, and it is the area in which relationships with foreign nations are the most elaborate (for example, the United Nations in New York City and the country's capital at Washington, D.C., are both located in this region; Figure 24.2). Strong traditions of intellectual, cultural, scientific, technological, political, and business leadership persist there. The country's political and economic systems had their main beginnings in the North-

east, and historically, it was the chief reception center for overseas immigrants.

The Northeast consists of six New England states (Maine, New Hampshire, Vermont, Massachusetts, Connecticut, and Rhode Island), plus five Middle Atlantic states (New York, New Jersey, Pennsylvania, Delaware, and Maryland), and the District of Columbia. Delaware and Maryland have important historical links with the states of the U.S. South, but their present character classifies them more appropriately as Middle Atlantic states, so they are included in this region.

About 62 million people, or 21 percent of the national population, lived within the Northeast, on 5 percent of the nation's area, in 2004 (Table 24.1). The regional population density is several times that of the South, Midwest, or West. But the density is much more extreme within a narrow urban belt stretching about 500 miles (c. 800 km) along the Atlantic coast from metropolitan Boston through metropolitan Washington, D.C. (see Figure 22.2a). In this **megalopolis**, seven main metropolitan areas contain about 39 million people: Boston (4.4 million), Providence (1.6 million), Hartford (1.1 million), New York (18.6 million), Philadelphia (5.7 million), Washington, D.C. (5.1 million), and Baltimore (2.6 million). Additional population within this belt amounts to roughly 10.6 million, giving it a total of about 50 million, or approximately 81 percent of the people in the Northeast. The leading belt cities are major centers of political decision making, corporations, finance, sales, and services.

600

United States

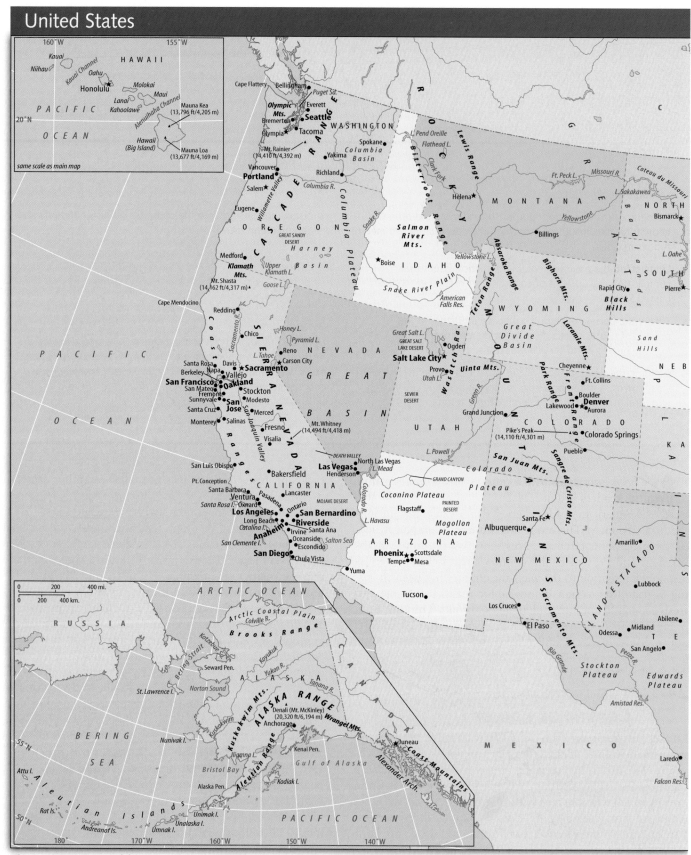

Figure 24.1 Principal features of the United States

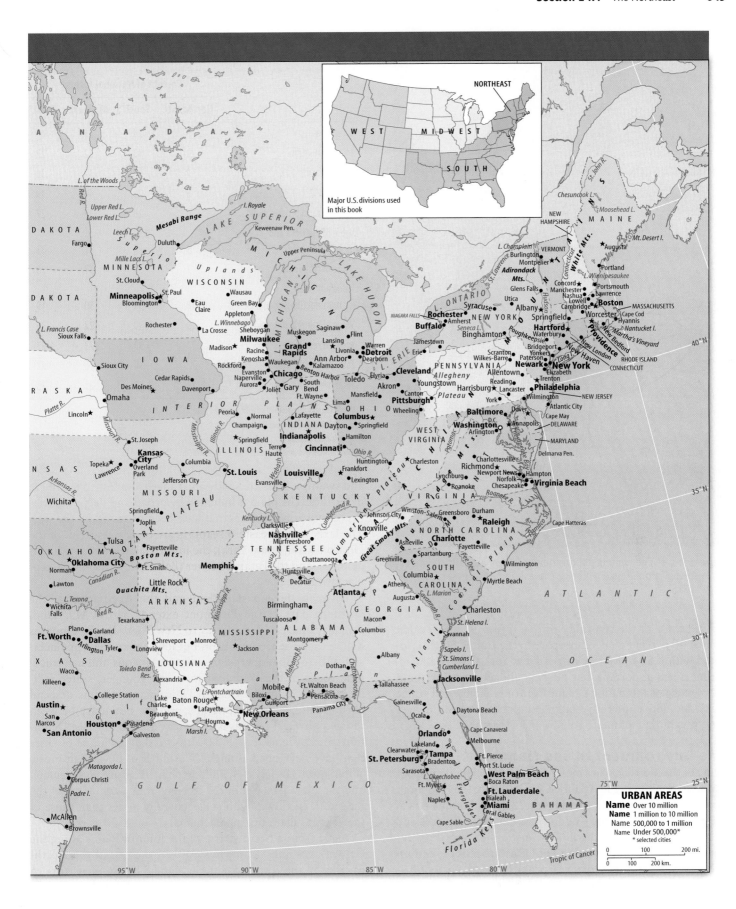

NORTHEAST

WEST   MIDWEST

SOUTH

Major U.S. divisions used in this book

URBAN AREAS

Name Over 10 million
Name 1 million to 10 million
Name 500,000 to 1 million
Name Under 500,000*
    * selected cities

0   100   200 mi.
0   100   200 km.

Figure 24.2 The United Nations headquarters are part of what gives special prominence to the northeast and its largest city, New York.

Joe Hobbs

Table 24.1 United States Regions: Comparative Data

	Percent of National Totals			
	Northeast	**South**	**Midwest**	**West***
Total area	5	24	21	49
Total population	21	33	22	22
Land in farms	2	29	37	30
Cropland**	5	29	46	19
Value of farm products sold	6	28	37	27
Value of livestock and products sold	5	23	37	33
Value of crops sold	7	34	37	20
Value added by manufacture shipments	18	31	30	19
Value of mineral output	9	30	22	38
Value of retail sales	21	32	23	23
Total personal income	22	30	20	23
Total metropolitan population	23	30	21	25

Note: Totals may not equal 100% due to rounding.

*The West includes Alaska and Hawaii.

**Totals exclude Alaska.

Source: U.S. Census Bureau, *Statistical Abstract of the United States,* 2004-2005.

and swamps. Soils are sandy and range from low to mediocre in fertility. The plain is indented by broad and deep estuaries (lower portions of rivers that have been drowned by the sea). The largest northeastern seaports are on the estuaries of the Hudson River (New York), the lower Delaware River (Philadelphia), and the Patapsco River (Baltimore) near the head of Chesapeake Bay.

Between the Coastal Plain and the Appalachians, the northern part of the Piedmont extends across Maryland, Pennsylvania, and New Jersey to New York. The Piedmont is higher than the Coastal Plain but lower than the Appalachians. Where the old, erosion-resistant igneous and metamorphic rocks of the Piedmont meet the younger, softer sedimentary rocks of the Coastal Plain, many streams descend abruptly in rapids or waterfalls, and the boundary between the two physical regions is known as the fall line (see Definitions and Insights, page 651). It is marked by a line of cities that have developed along it: Washington, D.C., Baltimore, Wilmington (Delaware), Philadelphia, Trenton (New Jersey), and the New Jersey suburbs of New York City. Most of the Piedmont is rolling, with soils generally better than those of the Coastal Plain or Appalachians.

Aside from a narrow strip of the Interior Plains along the Great Lakes and St. Lawrence River, the rest of the Northeast lies in the Appalachian Highlands. But in New York, a narrow lowland corridor—the Hudson–Mohawk Trough—breaks the Appalachians into two distinct subdivisions. This trough is composed of the Hudson Valley from New York

An indicator of the centrality of this region is the fact that almost 40 percent of all office space in the United States is located within 50 miles (80 km) of New York City. Manufacturing is less important, but the belt still accounts for about 18 percent of the nation's manufacturing. Congested traffic ways and active intercity commercial and cultural linkages bind this mosaic of metropolitan areas together.

The Environment of the Northeast

Most of the Northeast lies in the Appalachian Highlands, but small sections lie in the Atlantic Coastal Plain, the Piedmont, and the Interior Plains (see Figure 22.4). Atlantic Coastal Plain areas include Cape Cod in Massachusetts, Long Island in New York State, southern New Jersey, the Delmarva Peninsula, which includes nearly all of Delaware plus the eastern shore of Maryland (east of Chesapeake Bay) and parts of Virginia, and the western shore of Maryland inland to the boundary between Baltimore and Washington, D.C. The plain is low and flat to gently rolling, with many sand dunes, marshes,

Definitions + Insights

Fall Line and Break of Bulk

The fall line is a place where rivers flow from relatively resistant rock layers onto more easily eroded rocks and soils. Along the line, rivers cascade down across the hard rocks into the lowland that lies seaward of the falls. Before human settlement, such fall-line locations were of only modest significance. However, with the development of farms, village and town settlement, and increasing patterns of river traffic, these locations grew in importance. Where water flowed steeply and rapidly, it became evident that water power could be harnessed for machines and mills. Later, small industrial plants were located to take advantage of this waterhead (the potential power of moving water).

The fall-line location required goods transported up or down river to be reloaded on watercraft or other means of transport. This break-of-bulk point became a logical place to locate settlements. Slowly, cities grew up from these early fall-line settlements. The best-known fall-line cities along the east coast include Richmond, Virginia (on the James River), Washington, D.C. (on the Potomac River), Baltimore (on the Patapsco River), Philadelphia (at the confluence of the Delaware and Schuylkill Rivers), and Trenton (on the Delaware River). They all play major roles in the economic and demographic character in and at the margins of the Northeast megalopolis.

City northward to Albany and the Mohawk Valley westward from Albany to the Lake Ontario Plain.

Northeast and north of the trough, the old hard-rock mountains of New England and northern New York form several ranges characterized by rough terrain, cool summers, snowy winters, and poor soils. The Adirondack Mountains of northern New York rise as a roughly circular mass surrounded by lowlands. Heavily forested and pocked by numerous lakes, these glaciated mountains reach over 5,000 feet (1,524 m) in elevation. They are bounded on the east by the (Lake) Champlain Lowland, which reaches northward into Canada as a continuation of the Hudson Valley. East of it, the relatively low Green Mountains occupy most of Vermont and extend southward to become the Berkshire Hills of western Massachusetts and Connecticut.

Farther east across the narrow valley of the upper Connecticut River, the White Mountains occupy northern New Hampshire and extend into Maine. Here, Mt. Washington in New Hampshire rises to 6,288 feet (1,917 m), the highest elevation in the Northeast. It is in the Presidential Range of the White Mountains, which is characterized by rapid weather changes. Treks to the summit of this peak have fooled climbers, sometimes fatally, because of sudden drops in temperature and howling storm winds that come up very rapidly. Wind gusts of greater than 225 miles per hour (360 km/h) have been recorded on Mt. Washington. There are other attractions besides Mt. Washington; the Adirondacks and the mountains of New England are major recreation areas for the Northeast's urban populations, and their autumn colors are especially famous tourist magnets.

In New England, the hilly areas between the mountains and the sea, sometimes called the New England Upland, are considered part of the Appalachians. The original soils were stony (caused by the deposition of stones by glaciation), but the early settlers were able to use this area for subsistence agriculture once the stones and trees were laboriously cleared. Boulders from the fields were used to build New England's famous stone fences (these are like the hedgerows of Great Britain). Many of these have been torn down to make way for modern farming or urban development, but some derelict fences may still be seen keeping quiet guard where woodlands have overgrown abandoned farms.

The poor soils of the upland proved increasingly unable to support long-term commercial agriculture, and the late 18th and the 19th centuries saw a large migration of New Englanders from unproductive farms to northeastern cities and, especially with the completion of the Erie Canal in 1825, to more fertile agricultural areas to the west. This migration gave a strong New England cultural flavor to places scattered all the way to the Pacific coast.

South of the Hudson–Mohawk Trough, the Northeastern Appalachians include three distinct sections: the Blue Ridge in the east, the Ridge and Valley Section in the center, and the Appalachian Plateau in the north and west. They lie roughly parallel, with each trending northeast–southwest. The Blue Ridge—long, narrow, and characterized by old igneous and metamorphic rocks—has various local names. Often, it forms the single ridge its name implies, but it broadens into many ridges in the South. West of it, the Ridge and Valley Section, characterized by folded sedimentary rocks, consists of long, narrow, and roughly parallel ridges trending generally north and south and separated by narrow valleys. This valley-dominated eastern strip is essentially a single large valley known as the Great Appalachian Valley. Its limestone floor has decomposed into some of the better soils of the Appalachians.

The Appalachian Plateau lies west and north of the Ridge and Valley Section. The northern part is often called the Al-

83

legheny Plateau, whereas in eastern Kentucky and farther south (outside the Northeast region), the plateau is known as the Cumberland Plateau. Prominent east-facing escarpments—the Allegheny Front in the north and the Cumberland Front in the south—mark the eastern edge, from which elevations gradually decline toward the west. Except in New York, this plateau is underlain by enormous and easily worked deposits of high-quality bituminous coal. These have been extremely important in the economic development of the United States, but coal's contribution to the country's energy supplies is somewhat restrained by its impacts on the atmosphere and by the landscape impacts of coal mining. **Strip mining** of coal throughout much of Appalachia (the Appalachian region) involves the dynamiting away of mountaintops to expose seams of low-sulfur coal and then dumping the leftover waste into nearby valleys, creating sometimes very extensive (and unsightly) **valley fills.**

Early Development of the Seaboard Cities

Except for Washington, D.C., the largest metropolises of the Northeastern Seaboard gained their initial impetus as seaports, frequently located at river mouths. The federal capital was given its site as part of a political compromise in 1790. It was on the border between the agrarian southern states, with their large slave populations, and the more diversified northern states. The site also placed the capital on the fall line between seaboard and upcountry sectional interests. Many state capitals in the United States have been placed in response to similar locational compromises.

New York City was founded at the southern tip of Manhattan Island on the great natural harbor of the Upper Bay (Figure 24.3). It shipped farm produce from lands now occupied by built-up areas of metropolitan New York City and from estates farther up the Hudson River. Philadelphia, also on the fall line, was sited at the point where the Delaware River is joined by a western tributary, the Schuylkill. Philadelphia's local hinterland—the Pennsylvania Piedmont and Great Valley plus parts of southern New Jersey—was productive enough to make the city a major exporter of wheat and helped it become the largest city (population: c. 40,000 in 1776) in the 13 colonies. Baltimore was founded on a Chesapeake Bay harbor at the mouth of the Patapsco River and shipped tobacco and other farm products from its local hinterland in the Piedmont and the Coastal Plain.

Boston had a somewhat different early development because its New England hinterland produced little agricultural surplus for export. Instead, colonial New England placed its main emphasis on activities connected with the sea. Its forests yielded superior timber with which to build ships, and both timber and ships were exported, as were cod and other fish that were initially abundant along this coast. Whaling was important in the 18th and 19th centuries (Figure 24.4). New Englanders also developed a wide-ranging merchant fleet and trading firms with far-flung interests. Such activities characterized many ports, of which Boston was the largest.

Figure 24.3 New York's Manhattan Island. In this photo the twin towers of the World Trade Center, destroyed in the terrorist attacks of 9/11/01, still stand.

Transportation Geography and the Race for Midwestern Trade

As the American Midwest was settled, primarily between 1800 and 1860, its agricultural surpluses and need for manufactured goods expanded the trade of the cities of New York, Philadelphia, and Baltimore. The ports engaged in a race to establish transport connections into the interior. New York City surpassed its rivals, largely because of its access to the Hudson–Mohawk Trough, the only continuous lowland passageway in the United States through the Appalachians. The Erie Canal was completed along this corridor in 1825 (see Figure 23.7, page 635). Connecting Lake Erie at Buffalo to the navigable Hudson River near Albany, the canal was a key link in an all-water route from the Great Lakes to New York City. It helped reduce transport costs between New York and the Midwest to a small fraction of their previous level. This link stimulated growth in the Midwest, fed by a flood of trade through the port of New York. In 1853, using the same route, New York interests became the first to connect a growing Chicago with an east coast port by rail.

Joe Hobbs

Figure 24.4 Nantucket Island, off Cape Cod, was a major whaling center in the 19th century.

By the 1850s, New York had become by far the leading U.S. port, a position it lost only recently to the Los Angeles-Long Beach and Houston ports. The competing ports of Baltimore and Philadelphia also shared in trade with the Midwest. Major railways were eventually pushed through the Appalachians into the Midwest from both ports. Boston was never a real competitor in this race, although it has remained the largest seaport in New England.

Industrialization got its first big boost during the same period that the Northeast established connections to the interior. Favorable factors included accumulated capital from commercial profits, access to industrial techniques pioneered in Great Britain, skills from previous commercial operations and handicraft manufacturing, cheap labor supplied by immigrants, and the presence of energy resources and raw materials. Also favorable was a U.S. high **tariff** policy that restricted imports and protected much of the nation's market for its own emerging manufacturers. The first American factory, a water-powered textile plant using pirated British weaving technology, was established by Samuel Slater on the Blackstone River at Pawtucket (near Providence), Rhode Island, in 1790, and the subsequent growth of factory industry in the Northeast was rapid.

The circumstances and results of industrialization varied in different parts of the Northeast. New England had suitable water power near its ports, with many small and swift streams to turn early waterwheels, but this region lacked coal for metalworking. It specialized in textiles, drawing wool from New England farms, cotton from the South, and abundant labor from recent European immigrants, especially the Irish. However, one New England state, Connecticut, took a different path. It expanded a colonial specialty in metal goods, based on small local ore deposits, by becoming a major manufacturer of machinery, guns, and hardware (which it still is). Mill towns sprang up across southern New England and up the coast into southern Maine.

In the Middle Atlantic states, early industrialization also included textiles, but there was more emphasis on metals, clothing, and chemicals. The emphasis on metals reflected small local deposits of iron ore, especially in southeastern Pennsylvania. Then, in the second quarter of the 19th century, the largest deposits of anthracite coal in the United States became a major source of power. This coal lay deep underneath the Pennsylvania Ridge and Valley Section between Harrisburg and Scranton–Wilkes-Barre. It was made accessible despite the terrain by the determined building of canals and railroads and by large capital investment in the mines themselves. For many decades, anthracite was very significant industrially. Usable in blast furnaces without coking, it fueled large-scale iron and steelmaking in various eastern Pennsylvania cities. Iron and steel from the early plants contributed greatly to the first industrial surge in the Northeast by furnishing material for the manufacture of machinery, steam engines, railway equipment, and other metal goods.

The Dominance of New York City

The United States' largest city is centrally located within the Northeastern Seaboard's urban strip midway between Boston and Washington, D.C. An enormous harbor, an active business enterprise, and a central location within the economy of the colonies and the young United States had already made New York City the country's leading seaport before access to the Hudson–Mohawk route gave it an even more decisive advantage. Superior access to the developing Midwest then moved New York rapidly to unquestioned leadership among American cities in size, commerce, and economic impact.

New York City is centered on the less than 24 square miles (62 sq km) of Manhattan Island (Figure 24.5). The island is narrowly separated from the southern mainland of New York State on the north by the Harlem River, from Long Island on the east by the East River, and from New Jersey on the west by the Hudson River. Manhattan is one of five **boroughs** (administrative divisions) of the city proper. The others are the Bronx on the mainland to the north, Brooklyn and Queens on the western end of Long Island, and Staten Island to the south across Upper New York Bay. The broader New York metropolitan area also includes much of the population and industry of northern and central New Jersey, communities on Long Island east of Brooklyn and Queens, southernmost New York State just north and west of the Bronx, and southwestern Connecticut.

Expressways and mass-transit lines converge on bridges, tunnels, and ferries linking the boroughs to each other and to northern New Jersey. Port activity is heavily concentrated in New Jersey along the Upper Bay and Newark Bay, although some traffic continues to be handled in various New York sections of the waterfront. The whole area is held together in part by an amazing 722 miles (1,155 km) of subway system, parts of which are more than a century old. The city itself had a 2004 population of 8.1 million, and the metropolitan area claims 18.6 million people.

Size, financial and cultural complexity and variety, and national history give New York City a position of primacy that is uncontested in the United States. That primacy unfor-

New York: Three Perspectives

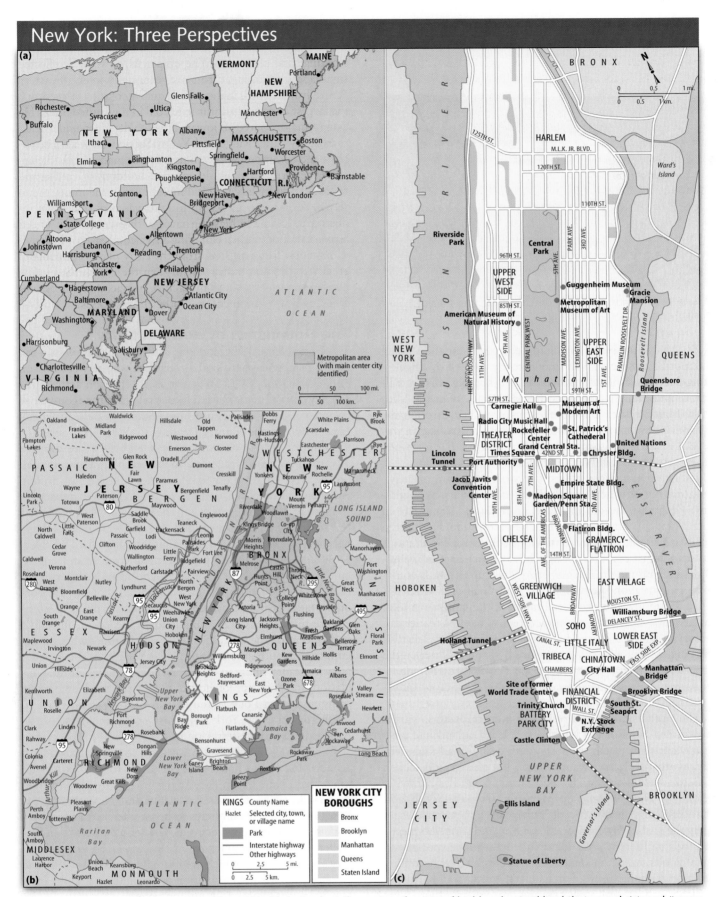

Figure 24.5 New York: three perspectives. New York stands at the center of a strip of highly urbanized land that was christened "megalopolis" by French geographer Jean Gottmann in 1961 (a). Many waterfronts and other areas shown on the two maps of New York City (b and c) are being redeveloped for new commercial, residential, and recreational uses. This is especially true of the piers along both sides of the Hudson River. In Manhattan (c), the Midtown hotel, shopping, theater, and office district includes most of the city's best-known

tunately put the city at the top of al-Qa'ida's hit list for the September 11 attacks and is always likely to make New York an inviting target for anti-U.S. insurgents.

Manufacturing in the Northeast

The Northeast rose to economic dominance through the production of ten specialty products. First has been *clothing design and manufacture,* carried on in many locations but concentrated in metropolitan New York City. The clothing industry in Manhattan was originally an outgrowth of New York's role as an importer of European-made cloth and clothing. It was favored by a large number of skilled immigrant workers in the later 19th century, among them Jewish tailors fleeing from persecution in tsarist Russia.

Then there were the powerful *iron and steel industries* that helped define American global industrialization from the late 19th century until after World War II. Pittsburgh, Pennsylvania, was the center, but factories and facilities expanded east to Baltimore and west to Chicago at the beginning of the 20th century. In recent years, employment and output have been practically eliminated because of competition from cheaper imported steel, large imports of products made from foreign rather than American steel, and the substitution of aluminum or plastics for steel in industrial processes, especially in automobile manufacturing.

Pittsburgh is a good example of what has happened in the old steel towns. After peaking in 1974, steel production plummeted there, and more than 75,000 workers in the industry lost their jobs. The city has struggled to reinvent itself as a center of "clean industries," such as bioengineering and robotics, and to expand its traditional role in providing education and healthcare services. The urban landscape of Pittsburgh is far cleaner and more appealing than in the days of steel, but not all the lost jobs have been replaced, and the city is heavily in debt. The U.S. steel industry itself may yet see a turnaround, especially if China's demand for steel remains strong and the cost of importing steel stays high because of freight charges and a weaker dollar.

Chemical manufacturing is very diversified and widespread in the region, especially in outlying areas of metropolitan New York City. *Photographic equipment manufacturing* is dominated by what is still (despite competition from Japan's Fuji) the world's largest photographic company, Eastman Kodak, which has both its headquarters and main plant in Rochester, New York. *Advanced electronics manufacturing,* centering on computers, is widespread in the Northeast. This industry is a major specialty in eastern Massachusetts, especially along Route 128 and other beltways that ring the Boston area. Some old textile towns

such as Lowell and Lawrence on the Merrimack River have largely converted to electronics or other high-tech industries. *Biotechnology* is booming in the Boston area, taking advantage of the region's top research universities, hospitals, and PhDs. Concentrated in southern New England, New York, and Pennsylvania, *electrical equipment manufacturing* is dominated by the General Electric Company, headquartered in Stamford, Connecticut, with major production facilities in Schenectady, New York.

The *manufacture of aircraft engines and helicopters* is heavily concentrated near Hartford, Connecticut, where it is carried on by United Technologies Corporation, maker of the famous Pratt and Whitney engines and Sikorsky helicopters. The *manufacture of Trident nuclear submarines* in the New London, Connecticut, area continues New England's long shipbuilding tradition as well as Connecticut's tradition of armaments manufacturing. *Publishing and printing* are industries in which the Northeast decisively overshadows the rest of the country. Most of the publishing industry is in the New York metropolis, although the Government Printing Office in Washington, D.C., is the world's largest publishing enterprise and printing plant. Many of the book and magazine printing phases of the industry have been steadily decentralized to other parts of the country, but editorial and management functions remain concentrated in the Northeast.

This assemblage of manufacturing, planning, design, decision-making, and management capabilities has also prompted the establishment of tens of thousands of smaller, complementary, and ancillary industries that make their living by supplying materials and information to these businesses.

From Rust Belt to Sun Belt

The vernacular term **Rust Belt** has come to describe parts of the Northeast and Midwest as a region of early urban growth and manufacturing of industries that became obsolete with time. The region contains the great majority of cities that were already prominent by the early 20th century. Most of the country's old **smokestack industries,** a term applied especially to the steel and other heavy metallurgy industries, emerged in the Middle Atlantic states and eastern Midwest. Today, these industries are characterized by outmoded buildings and equipment, depressed sales and profits, high unemployment, abandoned industrial buildings, landscapes of decline and abandonment, and complete factory shutdowns. These industries have been either replaced with "greener," more high-tech industries and services (such as described in Pittsburgh, Boston, and North Carolina in this chapter)—the classic process of **deindustrialization** that is

215

Figure 24.5 (continued) hotels, restaurants, department stores, specialty shops, theaters, concert halls, and museums. The Midtown area also includes numerous office buildings, most notably the cluster of skyscrapers at Rockefeller Center. The entertainment district around Times Square, and the historic Garment District, are at the southern end of this area. The lower Manhattan financial and office district includes the imposing cluster of office skyscrapers in the Wall Street area and was the home of the World Trade Center until 9/11. Note the complex of bridges, tunnels, and ferries that ties Manhattan Island to the rest of metropolitan New York.

Definitions + Insights

Adaptive Reuse and Gentrification

One of the characteristic landscape transformations in the Northeast is the adaptive reuse of older structures. This means taking abandoned commercial or factory buildings and converting them to some economic function that brings jobs, commercial activity, and people back to a once relatively productive area that has, in past decades, been largely abandoned. The process is also associated with gentrification—the refurbishment of old building stock by urban professionals for residential or office space.

In many downtowns that lost customers and businesses for some years, revitalization has occurred with the transformation of empty retail stores into specialty shops or trendy restaurants. Adaptive reuse takes place most often in old central business districts, but the process has also been occurring in old residential neighborhoods. Adaptive reuse and gentrification help give more value to the past, especially when historic buildings can be modified and used rather than razed. Although these processes are not unique to the Northeast, they have been particularly significant in New England.

70 also typical of Europe today—or have moved elsewhere in the United States and even abroad. The so-called Sun Belt of the South and West has been enjoying much faster growth in population and jobs, including those in manufacturing, in part due to migration from the Rust Belt. Sun Belt is an elastic term, but it connotes the region encompassing most of the South and the West at least as far north as Denver, Salt Lake City, and San Francisco.

Today, growth in the Northeast's economies focuses not on manufacturing but on services: banking and finance (notably Manhattan's financial district, including Wall Street), advertising (notably Manhattan's Madison Avenue), newspaper and television journalism, scientific research and education (especially in the "Ivy League" universities), and other "white collar" industries. But despite growth in these sectors, the region has been growing in population relatively slowly, with heavy out-migration from the Middle Atlantic region. The main exceptions to this slow growth are the peripheries of some metropolitan areas into which **suburban** and **exur-**ban development (with what are sometimes called "satellite developments" or "edge cities") is expanding. Notable examples include the expansion into New Hampshire from Boston, into Connecticut and New Jersey from New York City, and into Virginia from Washington, D.C. The state of Vermont, whose population growth has recently been high, is not suburban, but much of this quiet and scenic state is exurban, attracting "refugees" from metropolitan areas, owners of second homes and vacation homes, and some diligent commuters, including those traveling by air, and increasingly, professionals who are linked to urban centers through telecommunications networks. This expanding population can live in the country and still derive income from offices set right in the heart of the eastern cities. And despite the weight of economic and urban problems, the Northeast recently has been the scene of some strong new development in the great metropolises, both in old decaying cores and around the peripheries (see Definitions and Insights, above). Baltimore's Inner Harbor Development is one good example (Figure 24.6).

Figure 24.6 The Baltimore Inner Harbor Development, called Harborplace, is one of the most successful reclamations of an old warehouse and wharf area in the country. The current uses of this initial 18th-century harbor facility feature a two-story shopping mall with abundant outdoor restaurants and cafés, an aquarium, power plant buildings converted into elegant townhouses and condominiums, and access to the Camden Yards, home of baseball's Baltimore Orioles.

24.2 The South: Dixieland

As defined in this text, the South is a bloc of fourteen states: five along the Atlantic (Virginia, North Carolina, South Carolina, Georgia, and Florida); four along the Gulf of Mexico (Alabama, Mississippi, Louisiana, and Texas); and the five interior states of West Virginia, Kentucky, Tennessee, Arkansas, and Oklahoma (see Figure 24.1). Arguments could be made on various grounds for excluding some of these states. For example, West Virginia and Kentucky did not secede from the Union during the Civil War, and the western parts of Oklahoma and Texas lie in dry environments more typical of the West than the South. Although Missouri was a slave state and thus part of the South before the Civil War, today it is more typically midwestern and is not included here. In colloquial terms, the South has often since the Civil War been known as **Dixie** or **Dixieland**. Southerners also often refer to themselves as being below or south of the **Mason–Dixon Line**, which is an 18th-century delimitation across the southern border of Pennsylvania and the western border of Delaware that was drawn up by two surveyors (Mason and Dixon) to settle a land dispute.

The Physical Setting

Most of the South has a humid subtropical climate, with summers that are long, hot, and wet. January average temperatures range from the 30s (degrees Fahrenheit) (16°–22°C) along the northern fringe to the 50s (27°–33°C) near the Gulf Coast and the 60s (33°–39°C) in southern Florida and the southern tip of Texas. More than 40 inches (c. 100 cm) average annual precipitation is characteristic, rising to over 50 inches (c. 125 cm) in many Gulf Coast and Florida areas and more than 80 inches (c. 200 cm) in parts of the Great Smoky Mountains. High humidity is characteristic, intensifying heat discomfort in summer and chilliness in winter.

There are advantages and disadvantages to these climatic conditions. They foster rapid and abundant forest growth, and forest-based industries are of major importance. Agriculturally, they provide long growing seasons, heat, and moisture for a wide variety of crops. But insects and pests also flourish, and the heavy rainfall leaches out soil nutrients and can cause severe erosion and flooding. Florida's subtropical environment has provided haven to hundreds of exotic species of plants and animals that have crowded out and even eliminated native species. Brazilian pepper and Australian pine are among the more prolific exotic flora. Over 200 exotic animal species, many originating as escaped or released pets, slither, crawl, fly, and swim across Florida. These include Burmese pythons, boa constrictors, green iguanas, Cuban tree frogs, vervet monkeys, macaques, capuchin monkeys, parrots, and cockatiels.

Most of the South is classified topographically as plains, but the plains form sections of three major landform divisions: the Gulf–Atlantic Coastal Plain, the Piedmont, and the Interior Plains (see Figure 22.4). The Coastal Plain occu-

pies the seaward margin from Virginia to the southern tip of Texas, including all of Florida, Mississippi, and Louisiana and portions of all the other southern states except West Virginia. It is low in elevation, and large areas near the sea and along the Mississippi River are flat, but most inland sections have an irregular surface.

The Interior Plains section of the South occupies central and western Texas plus all of Oklahoma except the easternmost part. In Texas, the Balcones Escarpment forms an abrupt boundary between the "hill country" of the higher Interior Plains and the lower Coastal Plain (Figure 24.7). Most of Texas's larger cities, including Dallas, Austin, and San Antonio, lie on the Coastal Plain. Westward, the plains rise in elevation and become leveler. Irrigation, ranching of Angora goats and other livestock, and dry farming of wheat are important to the sparse population of this plains region.

The Piedmont extends across central Virginia, the western Carolinas, northern Georgia, and into east central Alabama. Its surface is mainly a rolling plain with some hilly areas. The natural cover is mixed forest in which broadleaved hardwoods, especially oak, tend to predominate over pines. Large areas have reverted to forest from former agricultural use. Soils tend to be poor, and many have suffered serious damage through past cultivation of clean-tilled row crops like cotton, corn, and tobacco on easily eroded slopes subject to frequent downpours. The region's fall line lies between the Piedmont and the Coastal Plain.

Large parts of the South are hilly and mountainous. In the central and eastern South, two large embayments of rugged country project into the South from the Northeast and Midwest. The larger of the two lies mainly in the Appalachian Highlands and also includes some rugged land west of the Appalachians known as the Unglaciated Southeastern Interior Plain. Farther west, the second large embayment is the southern part of the Interior Highlands. The three physical

Figure 24.7 Blessed with creeks and rivers, the Texas Hill Country is a refreshing getaway from the Lone Star State's big cities.

areas—Appalachians, Unglaciated Southeastern Interior Plain, and Interior Highlands—are often termed the Upland South.

Southern areas within the Appalachian Highlands extend from western Virginia to eastern Kentucky and southward to northern Alabama. The highlands include all the major physical subdivisions already distinguished in the Middle Atlantic states, with the total complex narrowing southward. The Blue Ridge extends from the Potomac River to northern Georgia. Its highest section is called the Great Smoky Mountains (Figure 24.8). Here, Mt. Mitchell in western North Carolina is the highest point in the eastern United States at 6,684 feet (2,037 m). The Ridge and Valley Section, including the Great Appalachian Valley, runs parallel to the Blue Ridge and inland from it, from western Virginia into Alabama. The southern section of the Appalachian Plateau, known as the Cumberland Plateau in parts of Kentucky and areas to the south, is wide in the north, where it occupies eastern Kentucky, most of West Virginia, and a small section of Virginia (Figure 24.9). It narrows across eastern Tennessee and then widens to occupy much of northern Alabama. As in Pennsylvania, it is bounded on the east by an abrupt escarpment, is mostly dissected into hills and low mountains, and is underlain by very large deposits of high-quality coal.

Central Kentucky, central Tennessee, and part of northern Alabama are in the Unglaciated Southeastern Interior Plain. This part of the Interior Plain, bounded by the Tennessee River on the west and south, was not subject to the glacial smoothing that occurred on the plains farther north. A good part of this "plains" area is actually hill country as a result of dissection by streams. True plains are present but not extensive. Two of them—the Kentucky Bluegrass region around Lexington and the hill-studded Nashville Basin around and south of Nashville, Tennessee—are underlain by limestone that has weathered into good soils. Both areas are famous for their agricultural quality in a region of poor to mediocre soils and difficult slopes.

In the South, the Interior Highlands occupy northwestern

Figure 24.9 The Cumberland Gap, a break in Kentucky's portion of the Appalachian Mountain chain

Arkansas and most of the eastern margin of Oklahoma. From Arkansas, they extend northward into the southern part of the midwestern state of Missouri. Composed of hill country, low mountains, and some areas of plains, these highlands are split by the broad east–west Arkansas Valley followed by the Arkansas River. North of the valley, the Boston Mountains form the southern edge of the extensive Ozark Plateau. South of the valley lie the Ouachita Mountains of west central Arkansas and adjacent Oklahoma. Most of the Ozark Plateau (known as "the Ozarks") has been dissected into gentle hill country, but some parts of the plateau, especially the Boston Mountains, have a mountainous character (Figure 24.10). The Ouachita Mountains consist of roughly parallel ridges and valleys running east and west.

The Southern Interior Highlands are not rich in resources. Most soils are poor, and many are stony. Most of the area is still forested, but it has been logged, and the present timber is generally rather poor. Mineral deposits are varied but often not economical to exploit. The main mineral wealth extracted today from the Interior Highlands, primarily lead, comes from southern Missouri. A number of large artificial lakes used for recreation are major resources in both Missouri and Arkansas.

Regional patterns of the South in population density, urbanism, ethnicity, and income are distinct from those of the other major regions. Until recently, this was a region of farms, villages, and small cities with low population density compared to that of the Northeast and Midwest. But for the past several decades, the South has been a region of rapid population growth, second only to the West, and its farm population has decreased dramatically. It has surpassed the Midwest in overall population density, although both are far behind the Northeast. Due to recent rapid growth of many small cities, the South has more small and medium-sized metropolitan areas than other regions of the country. However, it has fewer large cities than might be expected from its population. The largest urban complex, Dallas–Ft. Worth, has 5.5 million people. Houston (population: 5 million), Atlanta

Figure 24.8 Scene near Gatlinburg, Tennessee, in the Great Smoky Mountains National Park

Figure 24.10 The Missouri Ozarks

(4.6 million), Miami (5.3 million), and Tampa–St. Petersburg (2.5 million) are among the more than 20 American metropolitan areas with more than 2 million people. These southern cities continue to grow rapidly.

The population of the South has less ethnic variety than most other U.S. regions. The major groups are white Protestants of British ancestry, African American Protestants, and Mexican American Catholics. White Protestants of Anglo-Saxon origin are the largest contingent. The original European settlers along the Atlantic were British. Small groups of Spanish Catholics settled in Florida and Texas, and some French Catholics settled in Louisiana and elsewhere along the Gulf Coast, but their numbers were insignificant compared with the tide from Great Britain.

The South is distinguished ethnically by its high proportion of African Americans, due mainly to the importation of African slaves for plantation labor in the pre–Civil War South (see Figure 22.13). The five American states that rank highest in percentage of African Americans—from about one-fourth to more than one-third—are Mississippi, Louisiana, South Carolina, Georgia, and Maryland. Migration out of the South, mostly in the past 60 years, has given some northern states and California large African American minorities. New York and California both have larger African American populations than any southern state, but blacks make up only 17 percent of New York's people and 6 percent of California's. African American populations outside the South are concentrated in large metropolitan areas. Recently,

a reverse migration of African Americans from North to South has occurred as economic and social conditions in the South have improved while economic conditions of many northeastern cities have deteriorated.

Other distinctive ethnic elements in the South's population are more localized. Especially notable are **Mexican Americans, Cuban Americans,** and others of Hispanic origin, and **Cajuns** (of French descent). The five states leading in percentage of population of Hispanic origin are New Mexico (43 percent), California (34.3 percent), Texas (34.1 percent), Arizona (27 percent), and Nevada (21 percent) (see Figure 22.13). The country's Mexican American population tends to carry on its Hispanic identity with great vigor (Figure 24.11). In Texas, Hispanic people primarily of Mexican descent live mainly in a broad band of territory along the Gulf of Mexico and the Rio Grande, where they form a much higher proportion than the statewide 34 percent. Natural increase and immigration have rapidly expanded their numbers in Texas and elsewhere in recent decades (see Geographic Spotlight, page 660).

South Florida has received intermittent immigration of another Hispanic group, the Cuban Americans. Over one million refugees have fled from Communist Cuba since the revolution of 1959, and there are about 800,000 Cuban Americans and their descendents in Florida, most in metropolitan Miami. In 1994, tens of thousands of men, women, and children from Cuba (and thousands more from Haiti) launched themselves from the northern and western shores of their island, hoping to reach Florida or be picked up by the U.S. Coast Guard and ultimately be granted asylum in the United States. Although Cuban refugees had until then been generally granted political asylum in the United States, the Clinton administration stopped the flow in 1994, ostensibly to avoid further burdening Florida's resources. For his part, Cuban leader Fidel Castro argued that if the United States would relax its sanctions against his country, the Cuban economy would become strong enough that his citizens would not have to flee to the United States.

Figure 24.11 Alejandro and Julia Chavez-Lopez, Mexican American students

Geographic Spotlight

Illegal Immigration

Much of the ongoing Latino immigration into the United States is illegal, with the generally impoverished Mexican or other Latin American paying $2,000 to $10,000 to be smuggled by a so-called "coyote" or "chicken rancher" across weak points in the U.S. border defenses. The great majority of illegals caught within the United States are returned to the Mexican side of the border without prosecution, and an estimated half of them cross back into the U.S. Many are caught yet again, perpetuating a "revolving door" of illegal immigration that U.S. and Mexican authorities are now trying to address by returning the detainees deeper into Mexico. At the same time, the United States is weighing a plan to allow undocumented immigrants living in the U.S. to apply for a work permit for up to 3 years. After this plan was announced in 2004, the flow of illegal immigration into the United States spiked.

These apparently conflicting signals about immigration have led some observers to describe U.S. policy on immigration as "schizophrenic"; as an exasperated U.S. Border Patrol agent put it, "America loves illegal immigrants but hates illegal immigration."[a] Due to tighter security post-9/11 along the more populated California and Texas borders, Organ Pipe Cactus National Monument of southern Arizona, along with other wild sections of the Sonoran Desert frontier, is now a favored crossing point. Some immigrants traveling such desert routes in the extreme conditions of summer and winter succumb to the elements, and there are several "Good Samaritan" groups on the U.S. side that put out water and other provisions for them in anticipation of their travails.

[a] Quoted in Eilene Zimmerman, "Border Agents Feel Betrayed by Bush Guest-Worker Plan." *Christian Science Monitor*, February 24, 2004, p. 3.

Drawn largely from the middle and upper classes of pre-Communist Cuba, Cuban Americans have been upwardly mobile in American society. They have also been assimilating into the general population through marriage and adoption of Anglo American culture traits. Meanwhile, they have imprinted aspects of their own culture on south Florida. Cuban Americans have an important economic and political voice in the United States and may be largely credited (or blamed) for the continuing poor relations between the United States and the Communist Cuba they despise.

The largest non-Hispanic minority group in the South is the French-descended Cajun culture of southern Louisiana. Small French settlements, including New Orleans, grew along the Gulf Coast in the early 1700s. But it was the arrival of the refugee **Acadians,** expelled from Nova Scotia by the British in the mid-18th century, that implanted an enduring culture of French origin. That culture mixed with American elements to create the unique Cajun culture of southern Louisiana. The region's characteristic amphibious "bayou" landscape, a distinctive dialect with strong French features, the music of zydeco, and a culinary tradition of gumbo and hot peppers are trademarks of the Cajun region.

Agricultural Geography of the South

From tobacco to cotton to citrus, there is a strong agricultural tradition in the landscape and culture in the South. But change has overtaken tradition. In appearance and land use, the greater part of the South today is not agricultural. Over large areas, less than half the land is in farms, and the region's population is only about 1 percent agricultural. Existing farms tend to be small, low in production, and often part time. Animal products (especially beef and dairy cattle and broiler chickens) decisively surpass crops in total farm sales. Soybeans have become the most important southern crop. The leguminous soybean plant adds nitrogen to the soil and can be used for hay. But its greatest value lies in the beans that are pressed for soybean oil (for use in paints, soap, glycerin, printing ink, and many other chemical-based products and, now, as a fuel additive as well), with the residue (oilcake) fed to cattle and poultry. Grain sorghums, which have low moisture requirements, have become a mainstay for cattle feeding in drier parts of Texas and Oklahoma.

The greater part of Florida citrus fruits is now processed for frozen concentrate, and the residue is fed to cattle; the main market is the large ranch industry in southern Florida. Broiler chicken production has become a specialty of poorer agricultural areas, with major concentrations in the Interior Highlands of Arkansas and the Appalachians of Georgia and Alabama. In Arkansas, most of the workers on the chicken farms are Mexican Americans or new immigrants from Mexico, who are bringing a distinct new culture and language to a traditionally white bastion.

The historic staples of tobacco and cotton are still prominent in some parts of the South. Most of the nation's tobacco is grown in the eastern and northern South, especially in North and South Carolina, Kentucky, and Tennessee (Figure 24.12). Nearly all the South's remaining cotton is grown on large mechanized farms in the Mississippi Alluvial Plain or in west Texas. The region has other agricultural specialties, including citrus fruits (especially oranges) in the central part of the Florida Peninsula; **truck crops** (crops that are not processed before selling and that are directly used or sold fresh, such as lettuce, celery, and flowers) in the Florida Pen-

Joe Hobbs

Figure 24.12 Tobacco cultivation in Kentucky

insula; sugarcane in the Louisiana part of the Mississippi Lowland and near Lake Okeechobee in Florida; rice in the Mississippi Lowland and the prairies along the coasts of Louisiana and Texas; peanuts in southern Georgia and adjoining parts of Alabama and Florida; and race horses, bred particularly in the Kentucky Bluegrass region around Lexington, where the horse farms are showplaces of conspicuous wealth.

The construction of canals to favor commercial sugar production in southern Florida played a major role in the 80 percent shrinkage of the wetland Everglades region, from a wilderness that once featured a shallow, 40-mile-wide (64-km) river of "sheet flow," to a much tamer swampland encircled by cane fields. Corn, malted barley, and rye are principal ingredients of the whiskeys and bourbons that originate from Tennessee and Kentucky. Another key ingredient is the region's fresh spring water; all Jack Daniels whiskey, for example, is watered by a spring emerging from a small cave in Tennessee.

Industry in the South

Most industrial development in the South lies along the Piedmont, with extensions into the adjacent Coastal Plain and Appalachians. Within this concentration, which may be called the Eastern South Industrial Belt, textiles continue to be the most prominent branch of manufacturing, but clothing, chemical products, furniture, tobacco products, and machinery are also important. In the years since the 1994 signing of the North American Free Trade Agreement (NAFTA), however, textiles and clothing manufacture have experienced significant shifts to Mexico, Jamaica, and other "offshore" manufacturing sites. In an extraordinary turnabout of the **outsourcing** phenomenon, however, the Chinese company Haier has opened a large refrigerator factory in South Carolina. The Chinese managers argue that making the product here for the American market saves transportation costs and helps ensure speedier deliveries to retailers. These refrigerators are now on sale in Target, Wal-Mart, Costco, and Best Buy.

540

Much of this region's development, as well as that of adjacent areas in the South, was fostered by the **Tennessee Valley Authority (TVA)**. The TVA was founded in 1933 by the federal government during Franklin Roosevelt's New Deal era to promote economic rehabilitation of the economically depressed Tennessee River Basin. This public corporation built many dams, locks, and hydropower stations on the Tennessee River and its tributaries. These installations controlled floods, facilitated barge navigation, generated hydroelectricity, and provided recreation around reservoirs. The demand for electricity soon outran the capacity of the hydrostations, and many thermal plants (fired by coal and nuclear energy) were built. While northern industrial development focused on the metropolises that arose in the railway age, most development in the South came later and took advantage of the spread of long-distance power transmission and high-speed truck transportation.

The second main belt of southern industry lies along and near the Gulf Coast from Corpus Christi, Texas, to New Orleans and Baton Rouge, Louisiana. The chemical industry, including oil refining, is the dominant industrial sector within this West Gulf Coast Industrial Belt. It is based on deposits of oil, natural gas, sulfur, and salt. Including offshore oil fields in the Gulf of Mexico, which account for about 30 percent of domestic oil production in the United States, Texas and Louisiana produce over one-third of the oil, about three-fifths of the natural gas, and practically all the domestic sulfur output of the United States, with the Gulf Coast strip the leading area of production for all three. Machinery and equipment for oil and gas extraction, oil refining, and chemical manufacturing are also important products. Major refineries and petrochemical plants are clustered at several locations, most notably metropolitan Houston and Beaumont, Texas, and Baton Rouge, Louisiana.

Many other industries are present in the South, including food processing, machinery, furniture, and pulp and paper industries. Food industries are especially important in Florida. Pulp and paper plants abound near the Gulf Coast from eastern Texas to Florida and along the Atlantic Coast in Georgia. For raw material, these operations draw on the pine trees that grow rapidly in the wet subtropical climate. Aircraft and aerospace industries are prominent in the Atlanta and Dallas–Ft. Worth areas, and both auto assembly and the manufacture of auto components are expanding, especially in Tennessee, Kentucky, and Alabama. Just as U.S. firms often turn to outsourcing work abroad, Honda, Toyota, and some other foreign automakers take advantage of the region's skilled and relatively inexpensive labor force to create a product in the country where it is consumed.

Big Cities of the South

A recognizable Texas–Oklahoma metropolitan zone contains the two largest urban clusters in the South—Dallas–Ft. Worth and Houston—plus four other metropolises with populations of roughly 800,000 or more: San Antonio and Austin, Texas, and Oklahoma City and Tulsa, Oklahoma.

Houston (Figure 24.13) is a seaport by virtue of the Houston Ship Canal (or Channel), constructed inland to the city in the late 19th and early 20th centuries. The city has become a major port serving Texas and the southern Great Plains in addition to being the principal control, supply, and processing center for the oil and gas fields and the chemical industry of the Gulf Coast. Houston recently became the nation's number two port in terms of shipping volume. Dallas–Ft. Worth, San Antonio, and Austin developed at or near the western edge of the Coastal Plain in a north–south strip of territory known as the Black Land Prairie, where unusually fertile limestone-derived soils supported productive agriculture and towns and cities to service it.

Dallas–Ft. Worth and San Antonio also serve large trading hinterlands stretching far to the west. Dallas–Ft. Worth is the dominant business center for a huge region. San Antonio is a major military base and retirement center, and Austin is the site of the strongly funded University of Texas, a center for high-tech industries, and the seat of Texas state government. All these cities are involved in the oil and gas industry.

A second area of strong metropolitan development is the Florida Peninsula. Two Florida metropolitan areas have estimated populations of more than 2 million each—Miami and Tampa-St. Petersburg—and two have more than 1 million each: Orlando and Jacksonville. The distribution of Florida's main metropolitan areas reflects the importance of amenities for tourists and retirees. Proximity to the ocean is important, but winter warmth is even more crucial. However, Florida's cities are not entirely devoted to vacation and retirement functions. Miami is a major point of contact between the United States and Latin America in air traffic, tourist cruises, banking, and the illegal drug trade. Tampa Bay affords a harbor well placed for the trade of central Florida's citrus belt and phosphate mines. High-tech industries and regional corporate offices are attracted to Florida as the state's population grows, and as improved transportation and communication reduce the disadvantages of location in places remote from the old metropolitan belt of the Northeast.

The third major southern concentration of large metropolises lies along the Piedmont, the fall line, and the nearby Virginia Coastal Plain. These include Richmond, Virginia; Charlotte, Greensboro, and Raleigh, North Carolina; and Greenville, South Carolina. Raleigh, Durham, and Chapel Hill, North Carolina, are the figurative legs of the **Research Triangle Park** (**RTP**), a research park prominent in biotechnology and other high-tech research and development and a rival to Boston's thriving biotechnology business. Atlanta, Georgia (population: 4.6 million), ranks 4th in metropolitan population of the entire South (Figure 24.14). It has diverse industries but stands out as a major inland transport center. Its Hartsfield-Jackson Airport is the busiest in the United States as measured by volume of passenger traffic and operations size. The transport facilities and superior location have helped make Atlanta the largest southeastern business and governmental center. CNN, the world's largest 24-hour television news service, is headquartered in Atlanta, where its founder, tycoon Ted Turner, bases many of his business interests.

In the north on the Coastal Plain is the seaport of Norfolk, Virginia. Located on the spacious Hampton Roads natural harbor at the mouth of the James River, Norfolk ships Appalachian coal by sea and is the main Atlantic base for the U.S. Navy.

Figure 24.13 Suburban Houston, Texas: The American dream amidst the pines

Figure 24.14 Downtown Atlanta

The entire Upland South (Appalachians, Unglaciated Southeastern Interior Plain, and Southern Interior Highlands) contains only the following five metropolitan areas of more than 500,000 people. Louisville, Kentucky, began as a river port at the "Falls of the Ohio," where goods had to be transshipped around this break-of-bulk point. It was also the closest point of contact on the Ohio River for the earliest settled and most productive agricultural area in the state, the Bluegrass region. By 1830, when a canal was built to bypass the falls, Louisville was a leading presence on the river, and it has continued to grow as a diversified regional center.

Two comparable centers, Nashville and Knoxville, Tennessee, also began in locations associated with river transport and limestone soils. Nashville is located where the Cumberland River, a tributary of the Ohio, extends farthest south into the hilly but fertile Nashville Basin of central Tennessee. It is the undisputed capital of America's large and growing country music industry (but Austin, Texas, is an emerging contender for that title). Knoxville is situated far up the Tennessee River from the Ohio in an area of limestone valleys within the Ridge and Valley Section of the Appalachians.

The same themes of transport advantages, good soils, and growth as the major center for a large region are repeated in the case of Little Rock, Arkansas. The major exception to this general pattern of urban location and development in the Upland South is Birmingham, Alabama, which is the only large center of iron and steel industry in the South. Its development in the Ridge and Valley Section of the Appalachians was based on nearby large deposits of coal and iron ore.

Outside the upland, southern cities with metropolitan populations of more than 500,000 include New Orleans and Baton Rouge, Louisiana; Memphis, Tennessee; Charleston, South Carolina; and El Paso, Texas. The first three are Mississippi River ports. New Orleans was founded by the French just upstream from the head of the "bird's foot" delta where the Mississippi divides into separate channels flowing to the gulf. In the riverboat era before the Civil War, New Orleans was New York's main competitor as a port. It then declined as a port in the railway era, but it remains a major seaport, especially for shipping grains and soybeans from the Midwest and the Mississippi Alluvial Plain, as well as a major tourist center famous for its spring Mardi Gras. Upriver, Baton Rouge is accessible to medium-sized ocean tankers and is a major center of the Gulf Coast oil refining and petrochemical industries. Farther upriver, Memphis developed as a river port where the Mississippi flows adjacent to higher ground at the east side of its broad floodplain. The city's function as the main business center for the greater part of the Mississippi Alluvial Plain (and other nearby areas) dates back to antebellum days when a major sector of the so-called "Cotton Kingdom" developed in "The Delta"—a regional name then in use for the large part of the Mississippi Alluvial Plain adjacent to Memphis in northwestern Mississippi. One of the distinct industries of Memphis is tourism—verging on spiritual pilgrimage—to Graceland, the home and tomb of venerated musician Elvis Presley.

The other large metropolises lie at opposite ends of the South. Charleston was the port and commercial capital of the first major plantation area outside Virginia. It was the leading city of the South at the time of the American Revolution. It languished as the plantation economy moved westward but resumed rapid growth with the rise of the modern southern economy. El Paso serves as a major railway center and tourist and business entrance to Mexico. It lies across the Rio Grande from Ciudad Juarez, which also serves as a major transportation center for northern Mexico.

24.3 The Midwest: Big River Country

In this text, the Midwest region includes 12 states: in the east, Ohio, Indiana, Illinois, Michigan, and Wisconsin, and in the west, Minnesota, Iowa, North Dakota, South Dakota, Nebraska, Kansas, and Missouri (see Figure 24.1). The Mississippi River (whose Native American name means "**Father of Waters**") separates these eastern and western groups, except for part of the Wisconsin–Minnesota boundary. All states in the eastern Midwest border one or more of the Great Lakes, but only Minnesota does so in the western Midwest. In the eastern Midwest, Ohio, Indiana, and Illinois are bounded on the south by the Ohio River, which makes up 10 percent of the navigable waterways in the United States and carries 40 percent of its river traffic (Figure 24.15). In the western Midwest, all states except Minnesota front the Missouri River.

The term *Midwest* for these 12 states is somewhat arbitrary but can be justified by distinguishing spatial characteristics. These include its interior "heartland" location; a distinctive plains environment; important associations with major regional lakes and rivers; ethnic patterns; a productive combination of large-scale agriculture, industry, and transportation; and rural landscapes known for their symmetry and prosperous appearance.

Figure 24.15 The confluence of the Ohio and Mississippi rivers at Cairo, Illinois

Environment and Agriculture in the Midwest

These vast midwestern plains produce a greater output of foods and feeds than any other area of comparable size in the world. The Midwest typically accounts for more than two-fifths of all U.S. agricultural production by value. Iowa leads in output and is the third-ranking U.S state in value of farm products, exceeded only by much larger California and Texas. Ordinarily, five midwestern states—Iowa, Nebraska, Illinois, Kansas, and Minnesota—are among the nation's top seven in total agricultural output, and eight midwestern states are among the top fifteen. The lowest ranking midwestern states are Michigan, which has large areas of infertile sandy soil, and the Dakotas, which have dryness, rough land toward the west, and a shorter frost-free season.

Soils are exceptionally good in most of the Midwest. The best soils developed under grasslands, which supplied abundant humus, in the prairies stretching from northwestern Indiana in a broadening triangle to the eastern parts of the Dakotas, Nebraska, and Kansas and on into the South in Oklahoma and Texas (Figure 24.16). Such fertile soils normally accompany steppe climates in the middle latitudes, where precipitation is unfortunately so low as to be marginal for agriculture. But in the Midwest, these soils are located in a more humid continental climatic area. The Prairie region was probably enlarged by Native Americans using fire to improve hunting conditions and create grazing ranges for game. *31*

The excellence of many midwestern soils is also due to deposits of **loess.** These wind-borne soil materials accumulated in glacial times when strong winds transported the particles from dry surfaces where the ice had melted. In the eastern Midwest, the soils formed under natural broadleaf forests are neither exceptionally fertile nor poor. The main areas of poor soil in the Midwest are in the uplands of the Ozarks and the Appalachian Plateau; the sandy areas of coniferous forest in northern Michigan, Minnesota, and Wisconsin; and in the rougher western parts of the four westernmost states. After long use, the present quality of midwestern soils depends heavily on how well they have been handled. Aided by science and technology, the soils of the Midwest as a whole have higher yields today than ever before.

Corn (maize), the single leading product of midwestern farms, is grown in the legendarily productive Corn Belt stretching from Nebraska to Illinois (Figures 24.17 and 24.18a). Over four-fifths of American corn production and around two-fifths of world production come from the Midwest. Highly favorable conditions for corn are provided by fertile soils, hot and wet summers, and huge areas of land level enough to permit mechanized operations. **Biotechnology** has enhanced the quality and productivity of the grain. *71* Receipts from sales of beef cattle in the Midwest exceed those from corn, but here, too, corn is the key because the cattle are raised primarily on corn. Soybeans are another major element in Corn Belt agriculture. Surpluses of both corn and soybeans are used in the production of **ethanol,** an alcohol added to regular gasoline to produce what is marketed as "super unleaded" gasoline. Midwestern farmers hope that a good part of future alternative energy programs for the United States will rely on such fuels produced from grain and will lobby heavily for government support for that enterprise.

Figure 24.16 A prairie of the Flint Hills, Kansas

Figure 24.17 A cornfield in the Midwest

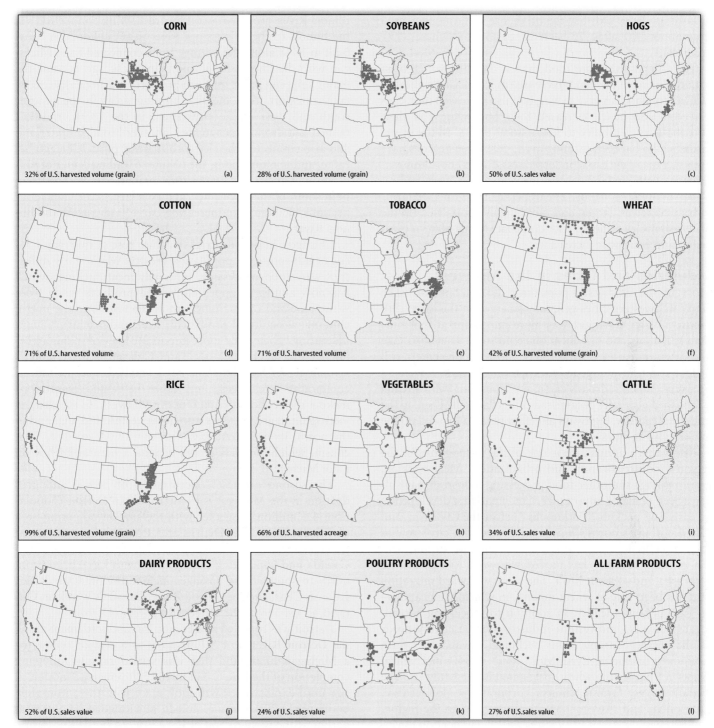

Figure 24.18 These 12 maps show the 100 leading U.S. counties for selected crops and farm products. Each map also indicates what percentage of the U.S. total production of this commodity comes from the aggregate of these counties.

Despite its typically impressive output, midwestern agriculture is a risky enterprise. Natural hazards such as drought and plant diseases increase the normal business risk. The main hazards are sharp fluctuations in world supply and demand (and therefore in market prices), and the prospect of the government cutting the generous subsidies that support many agricultural commodities. Such cuts could come from international pressure on the United States to open its markets to cheaper foreign farm imports. Waves of farmers have been driven out of business at various times. Between 1930 and 2004, the number of farms in the Midwest decreased from slightly over 2 million to about 700,000. Most of the land of defunct farms was bought or rented by surviving farmers who expanded their acreage. The amount of land in farming decreased, but the mean size of farms increased from about 180 acres (73 hectares) to about 487 acres (197 hectares).

The U.S. grows more food than it did 50 years ago on about 25 percent less land. Increased mechanization allowed farmers to work larger acreages, but also increased their capital costs and hence increased their borrowing and their household vulnerability to low market prices. Many of the farmers who have left farming in the past three decades have been victims of excessive indebtedness due to enormous machinery costs, and of course, mechanization itself has replaced many farm hands. The trend has been for U.S. farms to become fewer, larger, and more highly capitalized (in the Midwest and elsewhere), with corporate forms of farm ownership and enterprise increasing.

Industries and Cities in the Midwest

Most of the Midwest's population is metropolitan. The larger midwestern cities generally emerged in the early 19th century. Water transport was more important at that time than it is now, and city locations tend to be related to lake and/or river features. Even today, most major centers still benefit from advantageous locations for freight movement by water. When railroads came, the cities that were already established generally became rail centers as well. Then, in the mid-20th century, highways and airways also focused on the major existing cities.

Midwestern agriculture is one of the foundations of the region's industrialization and urbanization. Meat packing, grain milling, and other food processing industries have always been of major importance. Cincinnati first developed as an early hog-packing **"Porkopolis"** before the Civil War, and Carl Sandburg's poem *Chicago* celebrated that city's role as the "hog butcher for the world." The center of the midwestern packing industry has moved westward to east central Nebraska and western Iowa. The same westward migration also has been true of wheat milling, in which Minneapolis–St. Paul is now a leader. In addition to processing agricultural products, midwestern industry supplies many needs of the region's farms. Many technical advances in farm equipment (including Cyrus McCormick's 1840s mechanical reaper for harvesting wheat) originated in the Midwest. The world's largest farm-machinery corporations (John Deere, Caterpillar, and Navistar) are headquartered in the region, and many midwestern cities are heavily dependent on the industry. This activity helped foster the development of other types of machine manufacture. St. Louis, for example, has become second only to Detroit in the manufacture of the nation's automobiles.

The three largest midwestern metropolitan concentrations are around Chicago, Illinois, Detroit, Michigan, and Cleveland, Ohio. By far, the largest is Chicago, the country's third largest city (population: 9.3 million). The city began at the mouth of the small Chicago River, which afforded a harbor near the head of Great Lakes navigation from which the river projected farther into the prime farmlands of the Midwest. A few miles up this river, a portage to tributaries of the

Illinois River provided a southward-flowing connection via the Illinois to the Mississippi River near St. Louis. A canal built in 1848 along the portage route connected Lake Michigan with the Illinois River and helped Chicago become the major shipping point for farm products moving eastward by lake transport from the rich farming areas in the prairies south and west of the city. Then, in the 1850s, railways fanned out from Chicago into an ever-broader hinterland and also connected the city with Atlantic ports. Chicago became, and remains today, the country's leading center of railway routes and traffic. In the 20th century, new harbors were built south of the city center, and new large-scale canal connections facilitated an increasing barge traffic using the Illinois River.

Chicago's huge industrial structure rose at this major focus of transport and population growth. Major early industries were food processing, especially meat packing (which has now almost disappeared); wood products made from white pine and other timber cut in the forests to the north; clothing; agricultural equipment; steel; and machinery manufacturing. Today, the most important lines of manufacturing are machinery and steel, and Chicago is the country's leading center for both. The city's banks, corporate offices, commodity exchanges, and other economic institutions make it the economic capital of the Midwest. To the east, its metropolitan area reaches into northwestern Indiana, and to the north, its area coalesces with that of Milwaukee, Wisconsin, which also has grown around a natural harbor on Lake Michigan.

The second greatest concentration of people and production in the Midwest is in metropolitan Detroit (population: 4.4 million), together with the smaller adjoining area of Toledo, Ohio. The Detroit area extends along the water passages separating Michigan from the Ontario Peninsula in Canada and connecting Lake Huron with Lake Erie. From north to south, these passages include the St. Clair River, Lake St. Clair, and the Detroit River. Canada's Windsor, Ontario, is just across the Detroit River from Detroit.

Detroit was founded as a French fort and settlement along the Detroit River in 1701. It grew as a service center for a Michigan hinterland that was settled rapidly following completion of the Erie Canal in 1825. It remained modest in size until explosive growth took place with the automobile boom of the 20th century. Toledo was founded on a natural harbor at the western end of Lake Erie. In the 19th century, natural gas and oil from early fields in western Ohio furnished fuel for a glass industry there, and the metropolitan area now produces much of the glass for Detroit automobile production.

Several factors promoted the auto industry's growth in Detroit and nearby cities. Detroit labor and management had a background in building lake boats and horse-drawn carriages, and various inventors, including Henry Ford (a Detroit native) had conducted experiments in the area with internal combustion motors. The availability of steel and a central location relative to markets contributed to the industry's rise. When the industry began to develop during the early 20th century, large-scale production was necessary to achieve costs and prices low enough for mass sales. This required manufacturing to be centered in a few large plants. Steel was needed in huge quantities, as was a location from which distribution could reach the whole national market relatively cheaply. Detroit had a nearly optimal location. The national center of population then was in southern Indiana (it is now in central Missouri), and purchasing power was greater toward the north and east.

From southeastern Michigan, the auto industry spread over much of the eastern Midwest: north to Flint and Lansing, east to Cleveland, south to Dayton and Indianapolis, and west to the Milwaukee area. Many cities in this region became dependent on supplying auto parts. Assembly itself expanded to St. Louis and Kansas City and eventually to other major regions of the country. The American corporate giants of the industry—General Motors, Ford, and Chrysler (now Daimler-Chrysler)—still have headquarters and major plants in metropolitan Detroit. As with steelmaking, motor vehicle manufacturing provided the Midwest with a basic large industry that hired many people at high wages. However, like the steel industry, it became a focus of employment problems beginning in the 1980s because of decreasing plant productivity and overseas competition, especially in Japan. This industry has a history of boom and bust cycles.

The fourth-ranking metropolitan concentration in the Midwest is in northeastern Ohio, where the Lake Erie port of Cleveland has a metropolitan population (including nearby Akron) of 2.8 million. Cleveland developed in the 19th century as a center of heavy industry on a harbor at the mouth of the Cuyahoga River. Ore from the Lake Superior fields was offloaded there for shipment to Pittsburgh, and there was a return flow of Appalachian coal for Great Lakes destinations. Smaller Lake Erie ports also participated in this traffic. Movement of ore one way and coal the other made it logical to develop the iron and steel industry at Cleveland, at other Lake Erie ports, and at Youngstown, Ohio, on the route between Pittsburgh and the lake. Machinery and chemical plants developed, and eventually, the auto industry spread into the district. Akron rose as the leading center of tire manufacturing in the United States. Today, northeastern Ohio is afflicted by the slow growth, contraction, or closure of many older industrial enterprises. On the positive side, some of the urban blight has been relieved by adaptive reuse and gentrification. The environmental quality of Cleveland, a city once derided as the "mistake by the lake," and other centers of heavy industry is much improved. In 1969, the waters of the Cuyahoga River in Cleveland were so laden with fossil fuel waste that the river actually caught fire!

Except for Columbus, Ohio, and Indianapolis, Indiana, the remaining major metropolises of the Midwest originated at strategic places on the major rivers that were early arteries of midwestern traffic. These cities still benefit from riverside locations, although other forms of transportation have largely replaced river traffic.

St. Louis, Missouri, is the center of a metropolitan population of about 2.5 million in Missouri and Illinois (Figure

Figure 24.19 St. Louis, "Gateway to the West," on the Mississippi River

24.19). It was founded by the French in the 18th century and remained the largest city in the Midwest until the later 19th century. It is located at a natural crossroads of river traffic using the Mississippi to the north and south, the Missouri to the west, the Illinois toward Lake Michigan, and (via a southward connection on the Mississippi) the Ohio toward the east. Early traffic had to adjust to the fact that the Mississippi downstream from the Missouri mouth was notably deeper than the upstream channel or the Missouri. Hence, many riverboats plying the Mississippi below St. Louis drew too much water to use the channels above the city, and St. Louis became a break-of-bulk point where goods were transferred between different kinds of craft. Coal from southern Illinois was available to support industrialization, and the metropolitan area eventually developed varied industries.

In Minnesota, Minneapolis–St. Paul (population: 3.0 million) developed at the head of navigation on the Mississippi River (Figure 24.20). The Falls of St. Anthony, marking this point, supplied water power for early industries, especially sawmilling. Transcontinental rail lines were then built west from these Twin Cities to the Pacific. The spread of business westward along these lines made Minneapolis–St. Paul the

Figure 24.20 Downtown Minneapolis

economic capital of much of the northern interior United States. No competing center of comparable size emerged between Minneapolis–St. Paul and Seattle. The business hinterland of the Twin Cities thus stretches across the "spring wheat belt" (where the crop matures later than at points further south) to the foot of the Montana Rockies. Minneapolis–St. Paul has long been a major focus of the grain trade and flour milling and now has a diversified industrial structure. It is also home to the nation's largest retail mall—The Mall of America—which has a regular flow of tourists who come to the Twin Cities just to shop there.

A much older river city is Cincinnati (population: 1.9 million) on the Ohio River. Its metropolitan area includes the southwest corner of Ohio and adjacent parts of Indiana and Kentucky. Its situation once resembled Minneapolis–St. Paul in that it dominated a large hinterland to the west and north. The Ohio River turns sharply southward from the

city, and early 19th-century settlers moving down the river often disembarked at Cincinnati to proceed overland into Indiana. The city quickly developed a major specialty in slaughtering hogs from Ohio and Indiana farms and shipping salt pork downriver to markets in the South. This early meat packing function eventually was lost, but coal mining developed in the nearby Appalachian Plateau, and Cincinnati became the economic capital for much of the mining region. It also developed a diversified and relatively stable industrial base, built in part around tool and die machineworking for the region's industrial centers. New enterprises, including medical and scientific research and development, continued to attract professionals even as old industries died out (see Geographic Spotlight, below).

Kansas City (population: 1.9 million) straddles the Missouri–Kansas border and resembles Minneapolis–St. Paul in its situation and development. It began at what was

Geographic Spotlight

The Changing Geography of American Settlement

Across the United States, middle-sized cities of 750,000 to 2,000,000, like Cincinnati, are taking on a new importance as target destinations for manufacturing and service firms. The combination of established urban infrastructure and generally well-educated labor pools that are not as expensive as those of larger metropolitan areas has stimulated steady growth of these cities. The ever-expanding capacity of telecommunications has reduced the need for firms to locate in the central business district of a large metropolis. Computers, fax machines, the Internet, and airline links between middle-sized cities and large urban centers have changed the geography of business locations. Administrators of middle-sized cities often provide incentives to new businesses in the form of low-cost land, reduced taxes, and subsidized transportation, providing an even more attractive environment for business locations.

Smaller cities do not have some of the larger urban amenities in music, museums, shopping, and schools. But these middle-sized cities generally have less of the crowding, crime, environmental problems, and gridlock that large cities often do, and they attract companies for those reasons. Two critical populations are ready to work for firms in these locations. First are the semiskilled high school and college graduates willing to take starting salaries of $20,000–$40,000. There is also a population of well-educated young professionals who will accept a somewhat lower salary as a trade-off for a perceived better quality of life.

While the middle-sized American city has blossomed, the small town has withered. Between 1900 and 2000, while the American population more than tripled, eight of every ten U.S.

counties lost population. The decade between 1990 and 2000 saw a marked migration of people out of the rural Great Plains region in particular (Figure 24.A and B). A flight from small towns to larger cities was mainly responsible for this depopulation. Small towns had their heyday when they emerged as vital clusters of merchants and townsfolk located on some avenue of transportation—highway, railway, or river. These clusters were surrounded by the broad, open farmlands cleared and farmed in the late 18th and 19th centuries. Dirt roads made moving farm produce to local markets, to train stations, and to riverside wharfs difficult, so it made geographic sense for businesspeople and farmers to locate close together. The small towns that became county seats had a particular advantage because they were also the homes of local governments. Particularly fortunate towns became home to food processing firms, shoe or textile factories, and other small-scale industries that added a factory payroll to these minor urban centers.

But then, in the steady flow of landscape transformation, roads began to be surfaced so as to function in all seasons. Larger roads led into the ever-larger settlements that attracted most new businesses. By offering consistently lower prices and a wider selection of merchandise, Wal-Marts, and other large retailers drove out diverse businesses in small towns and in the middle-sized and larger towns in which they were opened; this so-called "**Wal-Mart Effect**" has had a huge impact around the country. Smaller towns lost their economic centrality and their residents.

Only in the last two decades has there been some reverse migration. Some small towns in rural states like North Dakota are actively "recruiting" international immigrants from countries including Vietnam, Sudan, and India to take up residence and help revive local economies. In 2003, the U.S. Senate introduced the **New Homestead Act**, designed to lure residents and

Depopulation of the Plains

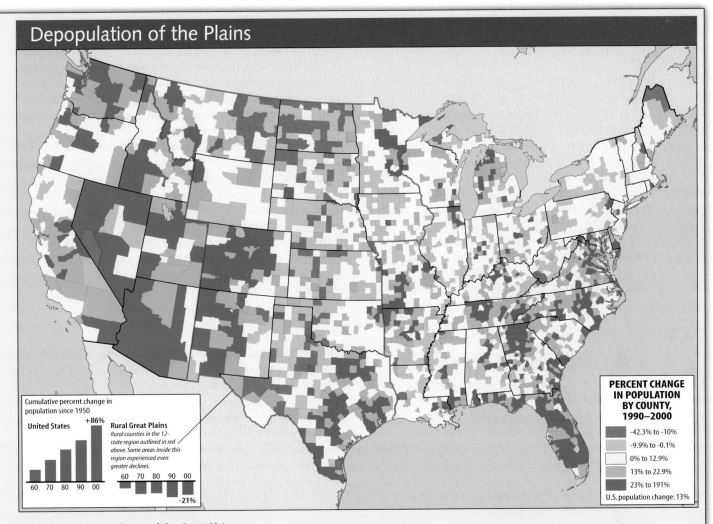

Cumulative percent change in population since 1950

United States

+86%

60 70 80 90 00

Rural Great Plains
Rural counties in the 12-state region outlined in red above. Some areas inside this region experienced even greater declines.

60 70 80 90 00

-21%

PERCENT CHANGE IN POPULATION BY COUNTY, 1990–2000

-42.3% to -10%

-9.9% to -0.1%

0% to 12.9%

13% to 22.9%

23% to 191%

U.S. population change: 13%

Figure 24.A Depopulation of the Great Plains

Joe Hobbs

Figure 24.B Many small towns across the Great Plains are dying—or even dead.

investment to rural areas by offering tax credits for businesses and home purchases, tax-exempt retirement accounts, and re-payment of up to half of college loans for those relocating to areas of high out-migration. Some urban professionals have been moving from large urban centers into small towns while still working for their big-city firms.

Another trend in the process of **exurbanization** has been for the disillusioned urban dweller to buy a rural "ranchette"—typically more than 35 acres, the lower limit of typical subdivision restrictions—well away from the big city. The ranchette boom in Colorado and other western states is having a profound impact on natural environments, and the trend's future is uncertain. Can networks of phone lines, fax machines, the Internet, and occasional trips back to the big city substitute for full-time residence in the major metropolitan area? Can the ranchette refugees from the big city coexist with traditional rural dwellers, while others fulfill their expectations of a picture-perfect Norman Rockwell life in the small town? Time will tell.

essentially the head of navigation westward on the Missouri River (whose nickname is the "**Big Muddy**"). Although river transport was possible upstream, it was subject to navigational hazards and led sharply northwest instead of directly westward. Consequently, two of the major overland trails to the West—the Santa Fe Trail and the Oregon Trail—had their main eastern terminals and connection with river transport in the Kansas City area. Later, cattle drives came to Kansas City from the Great Plains, and railways reached westward from the city. Today, despite a major competitor to the west in Denver, Colorado, Kansas City has a hinterland extending far to the west and is the economic capital of the "winter wheat belt," where crops mature earlier. Omaha, Nebraska (population: 792,000), is a smaller riverside city resembling Kansas City and Minneapolis–St. Paul in its functions and westward reach. Like them, Omaha is in the western edge of the intensively farmed part of the Midwest, with drier, less productive wheat and cattle country to the west.

24.4 The West: Booming and Thirsty

Eleven states make up the West (see Figure 24.1). The eight interior states (New Mexico, Arizona, Colorado, Utah, Nevada, Wyoming, Montana, and Idaho) are often called the Mountain states or Mountain West. Farther west are the three Pacific Coast states of California, Oregon, and Washington. The latter two are generally called the Pacific Northwest. Federal publications include Alaska and Hawaii in statistics for the Western states (as in Table 24.1), but these two states are so different that they are discussed in a separate section and are not included here in generalizations about the West.

Excluding Alaska and Hawaii, the West comprises 33 percent of the entire United States by area, or 38 percent of the conterminous United States. It contains large percentages of land owned by the federal government and supports major military facilities all across its mainly arid landscapes. The region is still lightly settled except for a huge cluster of people and development in central and southern California and a handful of smaller clusters in other widely dispersed locations. In 2004, the West had 67.4 million people (a little more than one-fifth of the nation), 35.9 million of whom lived in California. California is the West's economic colossus, producing almost 60 percent of the region's manufactured goods and one-half of its farm products (and 13 percent of the country's farm products). The eight Mountain states contrast sharply with California, with 19.8 million people in 2004 (6.7 percent of the national population). These states produced 8 percent of the country's agricultural output and 4 percent of the value of its manufactures in 2004.

The Physical Setting

The generally low intensity of development in the West results principally from dry climates, rough topography, and the absence of water transportation except along the Pacific Coast. Because settlement is clustered in places where water is adequate, the population pattern is oasislike, with oases separated from each other by little-populated steppe, desert, and/or mountains. The climates that affect Western settlement are strongly related to topography. Precipitation in this region is largely orographic, and dryness is largely due to rain shadow. Prevailing dry conditions often menace the region with wildfires and droughts, ironically punctuated by floods and landslides (Figure 24.21).

Figure 24.21 The west is a largely arid and semiarid region, frequently visited by droughts. The most severe drought of recent times settled in on much of the region in 1998 and was ongoing when this book went to press in 2005. Here, the normally high watermark of Arizona's Lake Powell is marked by the whitish line; at this point, in July 2004, the reservoir was at only about 40 percent of capacity.

Mountains border the Pacific shore (see Figure 22.4). The relatively low Coast Ranges provide an often spectacular coastline from the Olympic Mountains in Washington to just north of Los Angeles, where the mountains swing inland and then continue south to Mexico. Inland, a line of high mountains parallels the Coast Ranges. The northernmost of these inland ranges is the Cascade Mountains, extending southward from British Columbia across Washington and Oregon to northern California. Situated on the Pacific Ring of Fire, the Cascades include many volcanic cones, among which Mt. Hood (near Portland) reaches over 11,000 feet (3,330 m) and Mt. Rainier (near Seattle) over 14,000 feet (4,267 m). In southwest Washington is the Mt. Saint Helens volcano which, in its major explosion in 1980, killed 57 people and lost 1,300 feet (400 m) from its height. At its southern end, the Cascade Range joins the rugged Klamath Mountains, a deeply dissected plateau where the Cascades, Coast Ranges, and the Sierra Nevada Range almost meet. The Sierra Nevada Range increases in elevation southward to the "High Sierra," where Mt. Whitney, at 14,494 feet (4,418 m), is the highest peak in the 48 conterminous states (Figure 24.22).

Figure 24.22 The Sierra Nevada near Yosemite National Park, California

Joe Hobbs

WORLD REGIONAL
Geography Now™

Click Geography Literacy to see an animation of the subduction zone that causes these volcanoes.

Two large lowlands lie between the Coast Ranges and the inner line of Pacific mountains, and a smaller lowland lies outside the Coast Ranges in the extreme south. In Washington and Oregon, the Willamette–Puget Sound Lowland separates the Cascades from the Coast Ranges. The Oregon sector, extending from the lower Columbia River to the Klamath Mountains, is called the Willamette Valley. The Washington sector, north of the Columbia, is the Puget Sound Lowland, which is deeply penetrated by the deepwater inlet from the Pacific called the Puget Sound. The lowland surface in both states is generally hilly.

The second large lowland is the Central Valley, or Great Valley, of California, enclosed by the Klamath, Sierra Nevada, and Coast Ranges or by outliers that seem to weave the systems together. The Central Valley is a practically flat alluvial plain, some 500 miles (c. 800 km) long by 50 miles (80 km) wide (Figure 24.23). In California's far southwest, a smaller lowland lies between the coastal mountains and the shore. The wider northern part of this lowland is the Los Angeles Basin, and a narrower coastal strip extends southward to San Diego and Mexico. The Los Angeles Basin is a plain dotted by isolated high hills and overlooked by mountains.

Air masses moving eastward from the Pacific Ocean bring precipitation to this terrain, mainly in the winter. Summer conditions are desertlike in the south, although winter brings some rain and mountain snow. The mountains of the north receive the most precipitation, including the deepest snows in the United States in the higher areas. South of San Francisco, California, as precipitation declines, the magnificent Douglas fir and redwood forests of the north give way to the sparse forests and chaparral (drought-tolerant grasses and shrubs) that characterize the southern half of California.

The lowlands of this Pacific strip vary from wet to arid. The Coast Ranges of Washington and Oregon are mostly low enough, and the incoming moisture is great enough, that the Willamette–Puget Sound Lowland does not experience a rain-shadow effect and is instead wet and forested. Its temperatures are moderated throughout the year by incoming marine air, so this area has a marine west coast climate similar to that of Western Europe. The Central Valley of California is affected by a rain shadow cast by the Coast Ranges. Although generally classed as having a Mediterranean climate—with subtropical temperatures, wet winters, and dry summers—this valley is dry enough that its natural vegetation was steppe grass for the most part, grading into desert shrub in the extreme south. That has been utterly transformed by irrigation. The Los Angeles Basin also has a climate usually classified as Mediterranean subtropical, but the precipitation is so low that a steppe or even a desert classification could be argued.

Figure 24.23 California's fertile Central Valley, here adjoining the Coastal Range, is a breadbasket of the United States.

Figure 24.24 Arizona's Grand Canyon, carved deeply by the Colorado River, is but one of the characteristic canyons of the Colorado Plateau.

The rain-shadow effect becomes extreme to the east of the high mountains. A vast area reaching east to the Rockies and around their southern end in New Mexico is predominantly desert and steppe. Little moisture is received from air masses that already have provided precipitation to the mountains near the Pacific. This region is known physiographically as the Intermountain Basins and Plateaus. It includes large parts or all (in the case of Arizona) of nine Western states.

There are three major physical subdivisions of the Intermountain Basins and Plateaus. The Columbia–Snake River Plateau occupies most of eastern Washington and Oregon and southern Idaho. It is characterized by areas of relatively level terrain formed by past massive lava flows, together with scattered hills and mountains and deep-cut river canyons. The Colorado Plateau occupies northern Arizona, northwestern New Mexico, most of eastern Utah, and western Colorado. Here, rolling uplands lie at varying levels, cut into isolated sections by the deep river canyons of the Colorado and its tributaries (Figure 24.24). The rest of the intermountain area is the Basin and Range Country, which continues

beyond New Mexico into western Texas. The part between the Wasatch Range of the Rockies and the Sierra Nevada in California, called the Great Basin, has no river that reaches the sea. The whole Basin and Range Country consists of small basins separated from each other by blocklike mountain ranges. This is the driest part of the West.

East of the Intermountain Basins and Plateaus, the high and rugged Rocky Mountains extend southward out of Canada through parts of Idaho, Montana, Wyoming, Utah, and Colorado, and northern Mexico, where a number of peaks rise above 14,000 feet (4,267 m; Figure 24.25). The Rockies are comprised of many separate ranges, separated by valleys and basins. The continuity of the system is almost broken in the Wyoming Basin, which provides the easiest natural passageway through the Rockies and was the route through which the first transcontinental railroad, the Central Pacific, passed. The Rockies also are a barrier to moisture-bearing air masses from either west or east, creating rain shadows for both the Great Plains at their eastern foot and the intermountain area at their western foot. The mountains them-

Figure 24.25 Recreation area at the foot of Wyoming's spectacular Grand Tetons, which rise to 13,770 feet (4,197m).

selves receive much orographic precipitation and are forested. The highest elevations extend above the tree line to mountain tundra and snowfields.

The higher western section of the Great Plains (part of the Interior Plains) borders the Rockies in Montana, Wyoming, Colorado, and New Mexico. These plains are so high that Denver, at their western edge, is famed as the "**Mile High City.**" The most common landscape is that of an irregular surface interrupted by valleys. Variations include hilly areas, smooth plains, and isolated mountains, as in the Black Hills of Wyoming and South Dakota. Harney Peak (7,242 ft/2,207 m) in the Black Hills, sacred to the Lakota and other Native Americans of the area, is the highest point in the United States east of the Rocky Mountains. The semiarid plains are covered by steppe grassland and have temperatures that are continental in the north and subtropical in the south.

Water is a critical issue throughout the western United States. Historically, water needs were met by keeping settlements small. But with the explosive suburbanization following World War II, huge numbers of people moved west, particularly to California, in search of the American Dream. Populations soon exceeded local water supplies, and new sources needed to be found or redirected from elsewhere. Some needs are met by a 223-mile (357-km) gravity flow canal that brings water from the Sierra Nevada snowmelt through the Owens Valley to the Los Angeles metropolitan region. The largest source, however, is the flow of the Colorado River coming from the snowmelt of mountain peaks and flanks in north central Colorado and the Green River in Utah (Figure 24.26). Its waters are impounded behind two major dams: the Glen Canyon in Arizona (Figure 24.27) and the Hoover on the Arizona–Nevada border. Hoover Dam's

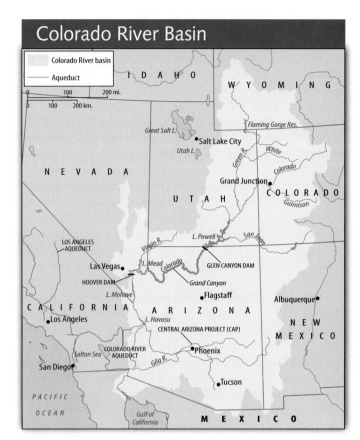

Figure 24.26 The Colorado River Basin

Lake Mead (which has the largest volume capacity of all American reservoirs) supplies a major portion of the water needs for about 25 million people in the three lower basin states of California, Arizona, and Nevada. Almost 20 million people in southern California drink these waters. Lake Mead

Figure 24.27 The Glen Canyon Dam, just upstream of the Grand Canyon, was constructed in the 1950s.

and Lake Powell (behind Glen Canyon Dam) also generate hydroelectricity for the region.

Colorado River water has been a source of litigation and federal and state disputes ever since Hoover Dam was completed in the mid-1930s. For decades, California has used more water from the river than planned, while the six upstream river basin states (Arizona, Nevada, New Mexico, Utah, Colorado, and Wyoming) have drawn less water than allocated. Meanwhile, the upstream states, especially Nevada and Arizona, have experienced exploding urban growth and demand for water. It took threats from the upstream states to actually cut off water to California to bring California to the negotiating table. In 2003, California agreed to reduce its use of Colorado River water over a period of 14 years to allow the six upstream basin states to use their share. There is unhappiness in California; water allocation is a classic **zero sum game** in which any benefit to one party equals a loss by another.

Most land east of the coastal states in the West is used for grazing, if at all, and the carrying capacity is so small that only low densities of animals and ranch population are possible. Unirrigated farming (dry farming) of some steppe areas supports slightly higher population densities, notably in the wheat-growing areas of the Great Plains and the Columbia Plateau (known also as the Palouse) of Washington. There, yields are relatively low, farms are large and highly mechanized, and population is sparse. Most Western agricultural output comes from a few large irrigated oases that grow both food crops and feeds, particularly alfalfa, for dairy and beef cattle. Some oases in California and Arizona grow cotton, fruits, vegetables, and sugar beets, and they depend on irrigation.

The West is strongly metropolitan in population. Most of these urban westerners are crowded into big clusters located in or near major oases. The Pacific Northwest cluster centering on Seattle and Portland might be considered an exception, but even this rain-fed area is a kind of oasis within the larger area of western drylands. These major urban clusters have diversified functions, among which the servicing of agriculture has long been important.

Southern California: Toward Megalopolis

The southern California population cluster is by far the West's largest. This area includes the Los Angeles Basin and the narrow lowlands north to Santa Barbara and south to San Diego, along with their interior mountains. The Chumash and other Native Americans who originally inhabited this area practiced no agriculture but subsisted on grass seeds, acorns, other wild foods, and some coastal fishing. In the 18th century, Spanish settlers from Mexico introduced small-scale irrigated agriculture and cattle ranching. By the 20th century, the Los Angeles area had become a large and important agricultural district. Today, sprawling urbanization has all but erased the orange groves and other agriculture that once characterized the Los Angeles Basin (Figure 24.28). About 16.4 million people lived in the metropolitan areas of Los Angeles, and another 2.9 million in the adjacent metro-

Figure 24.28 The Los Angeles basin presents a picture of urban sprawl, and often suffers under a blanket of smog (see Figure 20.5).

politan area of San Diego, in 2004. Packed into this small corner of the West are about one-half of California's people, one-third of the population of the entire West, and 1 of every 15 people in the nation.

San Diego and Los Angeles were founded as remote footholds on the northern frontier of Spain's empire in the Americas. Catholic priests founded missions in coastal locations as far north as San Francisco, including those in San Diego (founded in 1769) and San Gabriel (founded in 1781) in the Los Angeles area. The settlement at San Diego began on a magnificent natural harbor that is now the main continental base for the U.S. Navy's Pacific Fleet. The San Gabriel Mission and the Pueblo (civil settlement) of Los Angeles were sited near the small Los Angeles River. When Mexico gained independence from Spain in the early 19th century, the California settlements became part of the new country. They came into the possession of the United States at the end of the Mexican-American War in 1848. Until the late 1800s, San Diego and Los Angeles were little more than hamlets in one of the world's more isolated corners.

A nearly continuous southern California boom that has lasted to the present began with the arrival of railroad connections to the eastern United States in the 1870s and 1880s. The main lines (the Southern Pacific and the Santa Fe) crossed the area's encircling mountains via the Cajon Pass near San Bernardino. To promote settlement that would create a demand for rail traffic, the railroads offered extremely low fares westward. Oranges shipped in refrigerated cars were the main local product that could stand the transportation cost to far-off eastern markets, and irrigated orchards spread rapidly. Los Angeles was the main transport and business center. In the 1890s, a freeze in Florida put California in the lead in U.S. citrus production, and the state maintained this position until urban sprawl replaced many of the southern California groves in the mid-20th century (the Disneyland amusement park was constructed in a huge orange grove in 1955; see Geographic Spotlight, below). The climate that favored citrus also attracted a new population seeking health, comfort, and perhaps some of the glamour associated with Hollywood; the world's most famous filmmaking center is a small suburb of Los Angeles. In the early 1900s, pioneering moviemakers had moved from the Northeast to Los Angeles to take advantage of bright, warm weather and varied outdoor locales.

There were other resources to promote development in southern California. The "brea" of the La Brea Tar Pits was an oil-based resource used even by early settlers. In the 20th century, the region's petroleum grew in importance with the expanding automotive industry. The region is still a significant producer and refiner of petroleum, although environmental regulations prevent the offshore oil resources from being developed fully. Tuna and other fishing fleets plying Pacific waters were based in the area. In the 1920s and 1930s, aircraft manufacturing was attracted by the favorable climate, the amount of space still available for factories and runways, and the presence of a large labor force. This industry boomed during World War II and afterward joined with the electronics industry in aerospace development. Prior to the 1990s, the aerospace industry, heavily supported by federal military expenditures, had been southern California's largest manufacturer. Beneath its glittery Hollywood reputation, Los Angeles ("L.A.") had become a major industrial city—an aerospace Detroit with palm trees.

In the early 1990s, however, L.A.'s economy (and that of California generally) slumped, partly as a result of declining

Geographic Spotlight

Theme Park Landscapes

One of the most powerful forces in geography is the human capacity to remake the natural landscape into countless cultural forms, even some based on fantasy. In Disneyland, the first major theme park built (if the 1880s Coney Island on Long Island is not counted), developer Walt Disney created the worlds of Frontierland and Tomorrowland, approached on a mythical Main Street that reflected Disney's own small hometown of Marceline, Missouri. Disney's architects created Main Street buildings at seven-eighths scale so that people exploring this space would feel larger and would experience a city less threatening than Disney perceived the American city of the 1950s to be.

Theme parks have grown to become very important to economies and family experiences across the country. Magic Mountain, Six Flags, Busch Gardens, Sea World, and Universal Studios are standard fare of the American holiday. The largest complex is the Walt Disney World complex opened on 27,000 acres of former swampland near Orlando, Florida, in 1971 (Figure 24.C). In this massive complex, one of the major draws is the EPCOT Center (Experimental Prototype Community of Tomorrow), meant to be an educational as well as entertaining venue where people walk through the future. Orlando has become the biggest tourist draw in the United States, with more than 40 million visitors annually. Las Vegas, a city built on fantasy, has about 35 million. Even the Mall of America in Minneapolis, the nation's largest shopping complex, has adopted the theme park landscape by placing rides at its center.

Figure 24.C Walt Disney World brings Morocco to Central Florida.

federal defense outlays. For the first time in decades, California saw more people leave the state than enter it. By 1995, the patterns had reversed, with the economy and population growing again, only to give way in the first part of the following decade to a pronounced recession. A statewide crisis in energy services and the growth of an enormous budget deficit led in 2003 to the recall of California's governor and the election of a successor—appropriately for California, movie megastar Arnold Schwarzenegger (a.k.a. "The Governator"). In the long run, California can count on economic boosts from the tourist industry and from trade with the economically booming region of East Asia. Even with its economic troubles, California, if considered apart from the United States, would have had the world's fifth largest economy in 2004.

Non-Hispanic whites have recently become a minority in California for the first time in over a century, representing 16 million of the state's 35.9 million people. Active inmigration by Hispanics and Asians has caused most of the demographic shift. The changes have been particularly dramatic in Los Angeles, with a population that is 46 percent Hispanic, 30 percent white, 10 percent Asian, and 11 percent African American. The widely varied ethnic groups that add such a definitive flavor to California are not clustered only in homogeneous neighborhoods, such as the Koreatown and Chinatown of Los Angeles, but are scattered through many different and distinct sections of the metropolitan areas. Urban California is a veritable United Nations, with a multiculturalism that is perhaps unparalleled on the planet (Figure 24.29).

Not everybody is pleased with this vibrant diversity. With 1,400 people born in or moving to California each day, the state is showing some social strains. Now in the minority, many of the non-Hispanic whites are worried that growing immigration will overwhelm services and opportunities in the state. In 1994, California voters passed Proposition 187, described on the official ballot as "the first giant stride in ultimately ending the illegal alien invasion." The measure was designed to drive out undocumented aliens and to deter further immigration by cutting them off from medical and other public services and keeping their children out of schools. Governor Schwarzenegger was voted into office in part on his promise to repeal a controversial state law that allowed undocumented immigrants to obtain driver's licenses.

The San Francisco Metropolis

About 5.9 million Californians—31 percent of them Asian Americans—lived in the San Francisco metropolis, the 5th largest in the United States in 2004. San Francisco is located on a hilly peninsula between the Pacific Ocean and San Francisco Bay, but the metropolitan area includes counties on all sides of the bay (Figure 24.30). The climate is unusual. Marine conditions are apparent in San Francisco's moderate temperatures, ranging from an average of 48°F (9°C) in December to only 64°F (18°C) in September. Localized marine and topographic circumstances give rise to San Francisco's trademark fog, which can produce very chilly days even in the height of summer. Mark Twain is said to have quipped, "the coldest winter I ever spent was a summer in San Francisco." The area is in the border zone between the marine west coast and Mediterranean climates. Characteristic of the Mediterranean influence, about 80 percent of the area's rainfall comes in 5 months from November through March. Mean annual precipitation is only 20 inches (50 cm) in San Francisco, but the moderate temperatures and low rates of evaporation have enabled this amount of moisture to produce a naturally forested landscape. There are even groves of redwoods within a short distance both north and south of the Golden Gate Strait. East of San Francisco Bay, somewhat warmer temperatures and lower precipitation have produced a scrubby forest (chaparral or dispersed California oaks) and grassland.

San Francisco began as a Spanish fort (*presidio*) and a mission on the south side of the Golden Gate Strait. It developed first as a port for the gold mining industry and then for the expanding agricultural economy of central California. The **Gold Rush of 1849** centered on the gold-bearing alluvial gravels of Sierra Nevada streams entering the Central Valley north and south of the current city of Sacramento. San Francisco, with its spacious natural harbor, was the nearest port to receive immigrants and supplies for the gold camps, and the city boomed during the gold rush.

Today, the San Francisco Bay Area is a major domestic and international tourist destination, famed for its scenery, wines, arts, and other cultural attractions. Manufacturing in the metropolitan area is varied. In recent decades, computer-related and other electronics businesses have boomed from San Francisco south to San Jose, an area now known as **Silicon Valley,** named for the chips that carry microcircuits. The rise of Silicon Valley as a center of high technology is due in part to the attractiveness of the San Francisco area as a place to live and to the nearby presence of two outstanding universities: the University of California at Berkeley, near the in-

Figure 24.29 Multicultural California

Joe Hobbs

614

Figure 24.30 San Francisco, the city by the bay

dustrial center of Oakland on the east side of the bay, and Stanford University in nearby Palo Alto. Software firms have relocated to a number of other regions (notably Austin, Texas), but Silicon Valley continues to be the dominant U.S. center for this innovation, development, and venture capital. Much of the capital fled, and the region suffered a downturn, with the bursting of the "**dot com bubble**" (or, "tech bubble") at the turn of the century.

Seattle, Portland, and the Pacific Northwest

The Puget Sound Lowland of Washington and the Willamette Valley of Oregon to its south are not irrigated oases. They have abundant precipitation and luxuriant natural forests, as do the adjoining Coast Ranges and Cascades. But these humid conditions end abruptly near the crest of the Cascades. The area is thus oasislike, with a limited area of abundant water and clustered population within the generally dry West. This population is centered in two port cities, each located on a navigable channel extending inland through the Coast Ranges. Seattle (population: 3.1 million), Washington, located on the eastern shore of the Puget Sound, has access to the interior via passes over the Cascades. Seattle was home to the Boeing Aircraft Company, the world's largest aircraft producer until the headquarters moved to Chicago in 2001. The local early abundance of the Sitka spruce (used for early wooden plane construction) and the availability of cheap electricity led to the 1917 founding of the firm by William Boeing. Portland (population: 2 million), Oregon, is on the navigable lower portion of the Columbia River where the Willamette tributary enters it.

Fishing, logging, and trade built the cities of the lowland in the latter 19th and early 20th centuries. But agricultural

land in the Willamette Valley drew the first large contingents of settlers to Oregon Country in the 1840s. Agriculture is still of some importance, although hilly terrain and leached soils make it difficult. The leading products are milk and truck crops, and the Oregon wine industry is growing. Fishing, primarily for salmon, is still carried on, but the Pacific Northwest fish catch is now much smaller than any one of the three leading states in this industry—Alaska, Massachusetts, and Louisiana. Plummeting fish stocks off both U.S. coasts have led to a dramatic overall decline in the fishing industry. As federal fishing quotas have tightened, especially since 2000, the government has stepped in to buy out the faltering fishermen. On the West Coast, for example, boat owners have been paid about half a million dollars each to permanently remove their vessels from fishing.

Oregon and Washington are still the leading states in production of lumber, and they still have one-quarter of the saw-timber resources of the United States. The trees are mostly coniferous softwoods, and many, like the Douglas fir, are enormous. The timber industry is under constant attack by environmentalists but has many defenders because of its value to the economies of both states.

Another major industry in the Pacific Northwest is the Microsoft Corporation of Redmond, Washington. The firm was initially founded in Albuquerque, New Mexico, in 1975, but by 1979, the two founders—Bill Gates and Paul Allen—moved headquarters to the Seattle area. It has since grown to be the largest computer software company in the world and is now a source of major payroll and export revenues for the Northwest and the United States. Bill Gates, one of the world's richest men, has directed some of his enormous financial resources toward solving the global epidemic of HIV/AIDS.

Populations in the Mountain States

The eight inland states of the Mountain West are drier than those in the Pacific Mountains and Valleys. They lack the abundant irrigation water of California and the advantage of ocean transportation. These limiting factors have kept the population of the eight states to only 19.8 million people in 2004. About half of these are in a handful of metropolitan areas in irrigated districts. Outside these areas, the common pattern in a Mountain West state is that of a few hundred thousand people scattered across a huge area on ranches and farms, on Indian reservations, and in small service centers and mining towns.

The largest population cluster in the Mountain West lies in the Basin and Range Country of southern Arizona. Here live about three-quarters of Arizona's people in the metropolitan areas of Phoenix (population: 3.6 million) and Tucson (population: 892,000). Groundwater and waters from the Salt and Colorado Rivers have so far supported this development in a subtropical desert.

Phoenix was originally the business center for an irrigated area along the Salt River (Figure 24.31). The area expanded after 1912 when the first federal irrigation dam in the West (Roosevelt Dam) was completed on the river. Today, cotton is the area's main crop, and hay to support dairy and beef cattle is important. Growth in both Phoenix and Tucson accelerated after the introduction of air conditioning made them attractive retirement and business centers. Each is the site of a major state university, and Phoenix is the state capital. Tourism to both urban and nonurban destinations, the increasing income level of Sun Belt retirees, and copper mining are also basic to Arizona's economy.

Another important population cluster stretches along the Great Plains at the eastern foot of the Rocky Mountains. Within this area, called the Colorado Piedmont, ethnically diverse, metropolitan Denver (including Boulder and Fort Collins) was home to 2.8 million people in 2004, and more

Figure 24.31 In the desert city of Phoenix, water is lavished on golf courses and lakefront homes.

than half a million lived in metropolitan Colorado Springs and smaller cities. These cities developed on streams emerging from the Rockies. The stream valleys provided routes into the mountains, where a series of mining booms began with a gold rush in 1859. One mineral rush followed another through the later 19th century, and various minerals were exploited. The Colorado Piedmont towns became supply and service bases for the mining communities, which often were short-lived. Irrigable land along Piedmont rivers afforded a farming base to provide food for the miners, and agricultural servicing became an important function of the towns. In time, agriculture grew into a prosperous business that was stabler than mining.

The Piedmont centers have grown rapidly in the past century. Irrigated acreage has expanded, fostering a boom in irrigated feed to support some of the nation's largest cattle feedlots. The Colorado Rockies have also become an important producer of molybdenum, vanadium, and tungsten, which are rare alloys much in demand for high-tech industries. High energy prices in the 1970s led to expanded extraction of fuels from the state's huge coal reserves and scattered oil and gas deposits. So far, however, because of environmental concerns and shortages of water needed for processing, the region's **oil shales** have not been tapped significantly for the hydrocarbons they contain. Tourism in the Rockies has profited greatly from increasing American affluence, better transportation, and the rise in the popularity of skiing. Finally, Denver has also become a major center for the western regional offices of the federal government, with Colorado Springs the site of the U.S. Air Force Academy and the nearest large city to the headquarters of the **North American Aerospace Defense Command** (**NORAD**). Located in a massive complex hewn inside nearby Cheyenne Mountain, NORAD monitors potential airborne threats to the United States and Canada from abroad or within.

Third in size among the population clusters of the interior West is the Wasatch Front or Salt Lake Oasis of Utah, composed of metropolitan Salt Lake City–Ogden (population: 1 million) and Provo (population: 406,000) to its south. It contains about three-fourths of Utah's people, in a north–south strip between the Wasatch Range of the Rockies and the Great Salt Lake, with a southward extension along the foot of the Wasatch (Figure 24.32).

This oasis was developed as a haven for people of faith of Jesus Christ of the Latter Day Saints (popularly known as Mormons) when they were expelled from the eastern United States in the 1840s. After a difficult trek westward, they chose the Salt Lake Basin because they felt they would not be competed with or harassed in this difficult environment. More than 400 other, smaller Mormon settlements dot the West, but Utah has the most, and the state's population is about 70 percent Mormon. In Salt Lake City, the Mormons built the temple that is the symbolic center of their religion and culture. Using water from the Wasatch Range, they were able to develop a productive irrigated agricultural system.

Figure 24.32 Salt Lake City, at the foot of the Wasatch Range

With no large competing metropolis within hundreds of miles, the city today also is the business capital for the huge, sparsely populated area outside the Mormon core region. Salt Lake City is poised to become the epicenter of a fiberoptic network for high-speed Internet and other data communication with 17 other cities in Utah. Funded as a public works project, **UTOPIA (Utah Telecommunication Open Infrastructure Agency)** is designed to lure new high-tech firms and intellectual capital to the state.

The other five states of the interior West are even less populous than Utah. Their collective population of about 7 million, in states that are very large in area, is less than that of the Los Angeles metropolis. The people of New Mexico are concentrated along the north–south valley of the middle and upper Rio Grande, where Albuquerque (population: 765,000; Figure 24.33) is the main city, competing economically with El Paso, Texas, which is located on the Rio Grande just south of New Mexico. New Mexico's economy

is a typical Western mix of irrigated agriculture, ranching, mining, tourism, some forestry in the mountains, and federally funded defense contracting. As natural gas prices have risen, growing pressure to expand gas production has been resisted by Navajo concerned about defilement of sacred sites and by environmentalists. As in neighboring Arizona, many Navajo complain that energy companies have long cheated them by paying them low royalties for coal and other mineral leases (Figure 24.34).

Most of the population of Idaho is in oases along a river—in this case, the Snake River as it crosses the volcanic Snake River Plains in the southern part of the state. Agricultural output is more substantial in Idaho, nationally known for its potatoes. But tourism and federal money yield less income than they do in New Mexico. The capital and largest city of the state, Boise (which prides itself on quality of life), has only about 510,000 people.

Except for a small section that lies in the Sierra Nevada, the state of Nevada is desert, with very little irrigation and a history of booms and busts in gold and other minerals. Metropolitan Las Vegas (population: 1.5 million—the fastest growing major city in the United States with its growth of 112 percent between 1990 and 2004) and Reno (population: 374,000) have almost 90 percent of the state's population.

Figure 24.34 Many Navajo complain that "Navajo" power plants and the fuels they run on have not benefited them.

Figure 24.33 Albuquerque lies in the Rio Grande Valley, at the foot of the Sandia Mountains.

Figure 24.35 Las Vegas, the fastest growing city in the United States

Figure 24.36 Mammoth Hot Springs, Yellowstone National Park

Las Vegas (Figure 24.35) recently abandoned its effort to portray itself as a family-friendly destination, content instead to perpetuate its older reputation as "sin city"; a recent tourist promotion boasts "what happens in Las Vegas stays in Las Vegas." The state's laws have allowed gambling and prostitution to be Nevada's trademark industries, and gambling is an especially strong draw for tourists. Casinos and associated entertainment and convention business play a dominant role in the state's wealth. Military interests are another leading industry for Nevada, in which fully 83 percent of the land belongs to the U.S. government. Most U.S. nuclear weapons testing has been done in Nevada. Northwest of Las Vegas is Yucca Mountain, designated to become a repository for nuclear waste from around the country. As of 2005, however, technical and environmental concerns were still preventing the underground storage site from accepting the waste.

In Montana and Wyoming, the economies are based on farming, ranching, mining, and summer and some winter tourism. Wyoming is the country's largest producer of coal and its associated coal bed methane gas. Recent high energy prices have been a bonanza, as gas profits have filled Wyoming's treasuries in what have been lean times for most states. As in many other Western states, Wyoming has seen struggles during the presidency of George W. Bush between environmentalists and the Bush administration, which wants more energy production on public lands, even in national parks. Montana is known as the "**Big Sky Country**," famed for its wide-open spaces and the spectacular mountains for which the state was named. Neighboring Wyoming is also physically dramatic, particularly in the west, with its world-famous natural features of Yellowstone and Grand Teton National Parks (Figure 24.36). Both states have seen a surge in property investment by wealthy American entertainers and businesspeople; Ted Turner and actor Harrison Ford, among others, have ranches in this part of the country.

WORLD
REGIONAL
Geography ⊛ Now™

Click Geography Literacy to take a virtual tour "Earth-Shattering Events in Yellowstone."

24.5 Alaska and Hawaii: Final Frontiers

The United States began to acquire Pacific dependencies in the mid-19th century. Two of these, Alaska and Hawaii, in 1959 became the final states admitted to the Union. The United States acquired Alaska by purchase from Russia in 1867, and Hawaii was annexed in 1898. Acquisition of a Pacific empire was motivated partly by military considerations, and these have continued to play an important role in the settlement and development of both Alaska and Hawaii. Large military installations exist in both states, and defense expenditures are a major element in the economies of both.

In terms of physical geography, these states are about as different from each other as is possible and are quite distinct from the "**Lower 48,**" as Alaskans call the contiguous U.S. states (see Figures 22.4 and 24.1). Alaska (from *Alyeska,* an Aleut word meaning "the great land") makes up about one-sixth of the United States by area, but it is so rugged, cold, and remote that its total population was only 655,000 in 2004. The state has four major physical areas: the Arctic Coastal Plain, the Brooks Range, the Yukon River Basin, and the Pacific Mountains and Valleys.

Alaska's Arctic Coastal Plain is an area of tundra along the Arctic Ocean north of the Brooks Range. Known as "the North Slope," it contains the large deposits of oil and natural gas that make Alaska the biggest oil producing state in the United States and that pay every resident of Alaska about $1,500 a year just to live in the state (see The Geography of

Energy, page 682). The barren Brooks Range forms the northwestern end of the Rocky Mountains, with summits ranging in elevation from about 4,000 to 9,000 feet (c. 1,200 to 2,800 m; Figure 24.37). The Yukon River Basin lies between the Brooks Range on the north and the Alaska Range on the south. This is the Alaskan portion of the Intermountain Basins and Plateaus. The highest mountains are in the Alaska Range, where Denali (Mt. McKinley) rises to 20,320 feet (6,194 m) about 100 miles (160 km) north of Anchorage. This is also the highest peak in North America. The Yukon is a major river, navigable by riverboats for 1,700 miles (c. 2,700 km) from the Bering Sea to Whitehorse in Canada's Yukon Territory. In Alaska, the basin is generally rolling or hilly, with large areas of flat, often swampy alluvium. Its subarctic climate is associated with coniferous forest.

Fairbanks (population: 86,000) is interior Alaska's main town. The Alaska Highway connects it with the Lower 48 via western Canada, and both paved highways and the Alaska Railroad lead southward to Anchorage and Seward. Fairbanks is a major service center for the Trans-Alaska Pipeline from the North Slope oil area to the port of Valdez in the south. The city is located on the Tanana River tributary of the Yukon and is associated with a small agricultural area in the Tanana Valley.

The Pacific Mountains and Valleys occupy southern Alaska, including the narrow southeastern portion known as "the panhandle." The mainland ranges of the panhandle are an extension of the Cascade Range and the British Columbia Coastal Ranges, and the mountainous offshore islands are an extension of the Coast Ranges of the Pacific Northwest and the islands of British Columbia. The scenic Inside Passage from Seattle and Vancouver to the Alaskan panhandle lies between the islands and the mainland. It is a very popular route for cruise ships, as some of the world's most magnificent glaciers and mountains may be seen on its shores. On the mainland, some mountain peaks exceed 15,000 feet (4,572 m). Fjorded valleys extend long arms of the sea inland. A cool and wet marine west coast climate prevails, favoring coniferous forests. Northwestward, a majestic display of glaciers defines a long stretch of coast.

The state's capital, Juneau (population: 31,000), is located in this region. Juneau is the only U.S. state capital not reachable by road, but a proposed highway to Skagway may change that. This region is also home to Alaska's largest city, Anchorage (population: 339,000—about half of the state's total population). At the southwest, the Alaska Range merges into the lower mountains occupying the long and narrow Alaska Peninsula and the rough, almost treeless Aleutian Islands. The Aleutians mark the northern edge of the volcanically and seismically active Ring of Fire.

Immediately south of Anchorage, the Kenai Peninsula lies between Cook Inlet and the Gulf of Alaska, and still farther to the southwest is Alaska's largest island, Kodiak. These features mark the western end of the horseshoe of coastal mountains that curves across the north of the Gulf of Alaska from the southeastern panhandle. The subarctic climate of lower areas in this western coastal section is harsher than that of the panhandle. However, the general trend all over Alaska in recent decades has been toward warmer temperatures, with the average annual temperature increasing by 8°F (4.5°C) between 1974 and 2004. This trend has had significant impacts on the state's oil industry, cutting in half the number of days that industry vehicles are allowed to cross the tundra. There are other effects, too—for example, coniferous forests are sinking into areas where permafrost has melted, and river temperatures increased by 9°F (5°C) between 1984 and 2004, producing inhospitable habitat for salmon.

From Fur Trade to Oil Age

After about 200 years of occupation by Russia and the United States, the population of Alaska was still only 73,000 in 1940. Eighteenth-century Russian fur traders had been spread thinly along the southern and southeastern coasts. Their legacy is present today in the many Russian place names and in Russian Orthodox churches on the landscape. In 1812,

Figure 24.37 Wilderness of eastern Alaska

The Geography of Energy

The Arctic National Wildlife Refuge

The energy crisis of 1973 was a profound shock to U.S. and other industrialized economies. Among its many impacts was a push to develop more domestic sources of energy, particularly oil. The most significant development came on the North Slope of Alaska, where the Prudhoe Bay and nearby oil fields were established. Since 1977, their oil has been transported southward through the Trans-Alaska Pipeline, an extraordinary feat of engineering snaking 800 miles (1,300 km) across some of the most challenging environments on Earth, all the way to the port of Valdez on the south coast.

Prudhoe Bay (originally holding about 10 billion barrels of proven reserves) and nearby fields now produce about one-fourth of total U.S. domestic oil production. However, the fields' output peaked in 1988 and has declined steadily. Attention has inevitably turned east and southward to the estimated 3 to 8 billion barrels underneath the **Arctic National Wildlife Refuge** (**ANWR**, often spoken of as "Anwar"). When the 12,500 square miles (32,000 square km) of the refuge were set aside as wilderness in 1980, the fate of the oil beneath this tundra ecosystem was left for the U.S. Congress to decide at some later date.

There has been a bitter and polarizing debate over ANWR's oil since 1980. On the one side are the oil interests, Alaskan politicians, almost all Republicans in the U.S. Congress, and Republican presidents, who have argued that the United States must develop these reserves to give the country more independence from Middle Eastern oil. On the other is a coalition of environmentalists, Native Americans, and congressional Democrats who argue that oil production in ANWR will damage or destroy the coastal plain's unique environment and the indigenous people who depend on it.

The main Native American group involved is the **Gwich'in,** a 7,000-strong Athabascan tribe that derives much of its livelihood from exploiting caribou (the New World's counterpart of the reindeer). The Gwich'in live mainly south of ANWR, but the 180,000-strong population of caribou known as the Porcupine Herd spends the winter in Gwich'in territory. The Gwich'in insist that oil drilling and related activity in ANWR will disrupt caribou calving, decimate the herd, and thus endanger the Gwich'in way of life.

The indigenous stand on ANWR is not united, however; an Inuit group called the **Inupiat** supports oil production there, especially because of the jobs it would bring them. Environmentalists, citing the low end of the estimated oil reserves, insist that habitats and native cultures would be despoiled to provide just 6 months of U.S. oil needs, while the oil industry uses the higher figure to say that 20 years of U.S. supply is in ANWR. This difficult contest is sure to be fought in the court of public opinion and the halls of Congress for years to come.

the Russians established an agricultural colony on the California coast at Ft. Ross north of San Francisco Bay. This effort to better provision the Alaskan posts was abandoned in 1841, and then in 1867, remote and unprofitable Alaska was sold by Russia to the United States. Russia's selling price was 2 cents per acre. In the United States, the transaction was widely derided as "**Seward's Folly**," after William Henry Seward, the American secretary of state who negotiated it. The value of the oil pumped from Alaska each day now exceeds what the United States paid for its vast new territory.

By the time the United States acquired Alaska, hunting had nearly wiped out the sea otters, whose expensive pelts had provided the main incentive for Russia's Alaskan fur trade. Only recently, under American and Canadian governmental protection, has a substantial effort been made to replenish these mammals. There was a burst of American settlement connected with gold rushes in the 1890s. Then came only a slow increase connected mainly with fishing and military installations. There is commercial logging in Alaska, particularly in the Tongass National Forest of the panhandle region, but it is very unprofitable and is maintained only by

costly and controversial federal subsidies. The quest for opportunity and adventure in a remote land has attracted many more men than women (there are 114 single men for every 100 single women in Alaska, compared with the national U.S. figure of 86 single men for every 100 single women). One Alaskan cliché describes the single woman's prospects in these terms: "In Alaska, the odds are good . . . but the goods are odd."

Beginning with World War II, when Japanese troops invaded the Aleutian Islands, Alaska's growth became more rapid. Growth came again from increased military and governmental employment, the rise of the fishing industry (for salmon and king crab, especially), the boom in the oil industry after the 1973 energy crisis, the rise of air transport carrying both people and freight, and visits by increasing numbers of tourists, both domestic and foreign.

Today, metropolitan Anchorage at the head of Cook Inlet is the main population cluster. The rest of Alaska's people are unevenly distributed over the state but live mainly in or near Fairbanks, in the state capital of Juneau in the panhandle, or in other small and widely spaced towns and villages.

There is only a small rural population. Agriculture consists primarily of high-cost dairy farming in a few restricted spots, although the summer's midnight sun does promote some prolific growth of cabbages and other vegetables in the Matanuska Valley near Anchorage. Both food and other consumer goods come mainly from the lower 48 states. Of the total population, about 15 percent is composed of Alaska Natives (Inuit, or Eskimos; Native Americans; and Aleuts). Long the victims of exploitation and neglect by whites, these peoples received some redress in large land grants and cash payments under the Alaska Native Claims Settlement Act passed by Congress in 1971.

Hawaii: America in a South Seas Setting

The state of Hawaii consists of eight major tropical islands (see Figures 22.4 and 24.1) lying just south of the Tropic of Cancer about 2,000 miles (c. 3,200 km) west from California. Created by the movement of the Pacific plate of the earth's surface across a **geologic hot spot,** the islands are all volcanic and largely mountainous (see Figure 15.8). They have a combined land area of about 6,400 square miles (c. 17,000 sq km), a bit larger than the combined areas of Connecticut and Rhode Island or the European nation of Slovenia). The island of Hawaii is the largest and the only one with active volcanoes. The prevalent trade winds and mountains produce extreme precipitation contrasts within short distances. Windward slopes receive heavy precipitation throughout the year, rising to an annual average of 460 inches (1,168 cm) on Mt. Waialeale on the island of Kauai. One of the wettest places on the globe, lush Kauai stars as many tropical locations in Hollywood films. Nearby leeward and low-lying areas receive as little as 10 to 20 inches

(25 to 50 cm) annually. Some areas are so dry that growing crops requires irrigation. Temperatures are tropical except for colder spots on some mountaintops. The capital, Honolulu (on the island of Oahu), averages 81°F (27°C) in August and 73°F (23°C) in January (Figure 24.38). The city's average annual precipitation is 23 inches (58 cm), with a relatively dry summer and a wetter winter, bearing some similarity to Mediterranean climate patterns.

WORLD
REGIONAL
Geography ⊕ Now™

Click Geography Literacy to see an animation of the formation of the Hawaiian Islands.

Hawaii had a Polynesian monarchy when it was acquired by the United States (for more information on Polynesian geography and culture, see Chapter 15). Its population in 2004 was 1.2 million, with about 902,000 people concentrated in metropolitan Honolulu. Immigration from the Pacific Basin has given the state's population a unique mix by ethnic origin: 24 percent white, 17 percent Japanese, 6 percent native Hawaiian, 14 percent Filipino, 5 percent Chinese, and a small contingent of Koreans and Samoans. Although there has been ongoing tension between native islanders and the immigrants who come from the U.S. mainland and East Asia, much ethnic mixing has taken place. In general, the population is thoroughly Americanized even through there is a small Polynesian nationalist movement.

Hawaii's economy is based largely on military expenditures, tourism, and commercial agriculture. The main defense installations, including the Pearl Harbor naval base, are on Oahu. The main agricultural products and exports—

Figure 24.38 Honolulu

cane sugar and pineapples—have long been grown on large estates owned and administered by white families and corporate interests but worked mainly by Asians. Work on these estates was often the first employment of new immigrants. However, changing demographic patterns, land values, and the rapidly declining role of agriculture in Hawaii have all led to steadily changing landscapes in the islands. Agriculture now accounts for only about 18 percent of the annual gross state product. Sugar, once the mainstay of the agricultural economy, has lost importance.

Tourism has been the great growth industry of recent decades, propelled by the development of air transportation and the growing affluence of the population of mainland America and East Asia. The main attractions for tourists are the tropical warmth, beaches, spectacular scenery, exotic cultural patterns, and luxury resorts. However, some of the "South Seas" glamour of the islands has been eroded by the relentless spread of commercial and residential development. By some measures, Hawaii's natural environment has been altered more than that of any other U.S. state, with a higher proportion of native plant and animal species losing out to exotic competitors. It has even been described as the "extinction capital of the world." [26]

Hawaii thus has the last word in this book, which, it is hoped, has given you a strong geographic foundation for understanding our Earth.

SUMMARY

- The United States may be seen as having five regions: the Northeast, South, Midwest, West, and Alaska and Hawaii.

- The Northeast is comprised of Maine, New Hampshire, Vermont, Massachusetts, Connecticut, Rhode Island, New York, New Jersey, Pennsylvania, Delaware, Maryland, and the District of Columbia. In this region live about 21 percent of the nation's population, or some 62 million people. The primary center of this population is the belt of highly urbanized landscapes running from Boston at the north to Washington, D.C., at the south. There are seven major urban centers in this "Boswash" region: Boston, Providence, Hartford, New York, Philadelphia, Washington, D.C., and Baltimore.

- The Appalachian Highlands cover a large part of the Northeast, and the other three most significant physical regions are the Coastal Plain, the Piedmont, and the Interior Plains. The fall line represents the break in topography between the low-lying Coastal Plain and the inland rise in elevation in the Piedmont. The fall line has been significant as a break-of-bulk point for ships and small freighters coming inland that cannot negotiate the falls. Early manufacturing centers were also built around these drops in elevation, including Baltimore, Maryland, Washington, D.C., Philadelphia, Pennsylvania, and Trenton, New Jersey.

- The Erie Canal (opened in 1825), in conjunction with the Hudson–Mohawk Trough, stimulated a major migration flow to the northwestern territories, now the Midwest. By the 1850s, so much trade had flowed through that corridor that New York became the nation's leading harbor, and it still is of major importance, although Los Angeles-Long Beach has become the country's largest by volume. This passageway also led to significant migration from New England farming as whole farm households traveled to the states at the southern margin of Lake Erie in quest of better soils and farming situations.

- From the fall-line cities in the region, early industrialization expanded to western Pennsylvania as steel production began along the southern margins of Lake Erie and especially in Pittsburgh.

Additional development in flour milling, electrical goods, chemical and textile production, and photographic equipment took on regional and national importance. The greatest center of major industry, finance, textile manufacturing, publishing, and varied manufacturing was New York City. This city is still dominant in several of these industries. Biotechnology and other high-tech industries have grown strongly in Boston, and some of the other Northeast's former centers of heavy industry, such as Pittsburgh, are also striving to attract those businesses.

- The South includes Virginia, North Carolina, South Carolina, Georgia, Florida, Alabama, Mississippi, Louisiana, Texas, West Virginia, Kentucky, Tennessee, Arkansas, and Oklahoma.

- A humid subtropical climate with long hot summers is common in the South, but rainfall diminishes westward across the region. There is considerable regional variety in physiographic units from the Gulf–Atlantic Coastal Plain, the Piedmont, and the Interior Plains, with the fall line again lying between the Piedmont and the Coastal Plain. The Upland South has hills and low mountains, and the subunits of that area include the Appalachians, Unglaciated Southeastern Interior Plain, and the Interior Highlands.

- The South's main agricultural characteristics today include a diminishing role of agriculture overall in the region (only 1 percent of the population still lives on farms); the reduction of farming to a more part-time occupation, with wages being earned in nearby part-time city jobs; and the rise in beef cattle, broiler chickens, and dairy cattle, which have all become more important than crops. Tobacco has been the most recent crop to begin a strong decline. Citrus, peanuts, truck crops, and racehorses are some of the special farming activities that now characterize the region.

- Recent industrial development in the South has included textiles (although much of this industry has moved "offshore" to Latin America, Asia, and elsewhere), and the development of synthetic fibers from local resources has been important. In addition, energy from hydroelectric development, oil and natural gas, and to a smaller degree, coal has been developed. Texas and North Carolina are the region's two major industrial states.

- ↻ The industrial belts of the region lie along the Piedmont and in part are connected to the Tennessee Valley Authority of the 1930s. This regional industrial development was unusual because it focused as much on small towns and middle-sized cities as it did on the major metropolitan centers of the Northeast. The second major belt was along and near the Gulf Coast. The Texas–Oklahoma metropolitan zone has the two largest urban centers in the South: Dallas–Ft. Worth and Houston. The Houston Ship Canal has made that city a major shipping center. Florida is another major metropolitan zone, but with a focus more on service and tourism than on manufacturing. There are other metropolitan zones in the South, such as cities of the Upland South and the Mississippi Delta.

- ↻ The Midwest as a region encompasses 12 states: Ohio, Indiana, Illinois, Michigan, Wisconsin, Minnesota, Iowa, North Dakota, South Dakota, Nebraska, Kansas, and Missouri. It is especially prolific in the production of pork, farm machinery, and soybeans.

- ↻ Plains are the most distinctive topographic feature of the Midwest, with the areas of major relief lying at the margins of the region. Soils are generally very good, and toward the east, rainfall is generally adequate for productive farming. Rainfall diminishes westward, requiring more irrigation. Many soils benefit from the presence of loess—windborne dust particles blown from the west and north.

- ↻ The Corn Belt is a major part of the Midwest, and corn continues to be the single most important crop in midwestern agriculture, with soybeans gaining in importance. About 40 percent of the world's corn comes from the Midwest.

- ↻ Wheat is very important in the Midwest, with spring wheat more prevalent in the northern part of the region and winter wheat more so in the southern part. Associated with wheat are sorghums, barley, and livestock production. In the region overall, sales of animal products exceed those of crops. Cattle and hogs are the two primary animals raised in the Midwest. There are also specialty crops that include fruit, sunflowers, sugar beets, and potatoes.

- ↻ Detroit has long been the center of automobile manufacturing in the country, related in part to the development of a major steel concentration along the southern shore of Lake Erie, utilizing iron ore from Minnesota, coal from Pennsylvania, and the Great Lakes as the avenue of major movement into the St. Lawrence Seaway and world trade. Slowly, the steel industry has been moving away from its early 20th century locales, going farther south and west. St. Louis has now become second only to Detroit in the manufacture of automobiles.

- ↻ The region of the West includes New Mexico, Arizona, Colorado, Utah, Nevada, Wyoming, Montana, Idaho, California, Oregon, and Washington. The region's landscape is predominately arid and amounts to approximately 38 percent of the conterminous United States. About one-fifth of the nation's population lives in the West. Much of the land belongs to the federal government in the form of national parks, military reserves, and other designated areas.

- ↻ Aridity is the most pervasive geographic characteristic of the West, and settlement is often concentrated in relatively well-watered areas that are, in effect, oases. Topography plays a strong role in the climate patterns of the region. Major mountain ranges along the west coast (including the Cascades, Klamath Mountains, Coast Ranges, and Sierra Nevada) have the largest population centers on their windward sides, and population densities generally diminish quickly eastward. The Willamette–Puget Sound Lowland in the north and the Central Valley in California in the south are two very productive agricultural regions.

- ↻ Grazing is the most common use of the land in the West, except where irrigation is possible. Irrigated farmlands raise alfalfa for beef cattle and dairy herds, and support cotton and specialty crops, particularly in California. California's large population and productive agriculture place huge demands for water that cannot be met from local resources. California has long exceed its allowable allotment of Colorado River waters, and states upstream in the Colorado Basin have won court actions to increase their share and reduce California's.

- ↻ Southern California is the West's largest population cluster. It did not begin to grow significantly until water was brought from the distant Owens Valley to Los Angeles early in the 20th-century. As water from the Colorado River Project was also made available to cities in the southwest and in southern California, growth accelerated. Presently, about 7 percent of the United States population lives in southern California. The continual presence of natural hazards, including fire, flood, earthquakes, and intense drought, has done little to inhibit the growth of this area.

- ↻ San Francisco is also a major urban center, sharing many of the same hazards as southern California. It has grown from a different set of origins and has become well-known as the home area of Silicon Valley, the dominant software development center in the United States. Like Los Angeles and other Californian cities, the San Francisco Bay Area is very multicultural, with strong ethnic Asian and Hispanic populations. Some whites fear the growing numbers of nonwhites and the burdens they may place on state services and have tried and/or succeeded to pass legislation making life for immigrants more difficult.

- ↻ California's Central Valley is a very productive agricultural area, with grapes and cotton as the two leading crops. Irrigation, primarily by waters from the Sierra Nevada range, has long been central to such productivity. Milk and cattle are also major products from this region, and cities have developed in support of this productivity, Sacramento, Fresno, and Stockton among them.

- ↻ Seattle, Washington, its surrounding Puget Sound Lowland, and the Willamette Valley of Oregon have a marine west coast climate that provides abundant natural rainfall. Fishing, logging, and trade built these cities in the late 19th century, and aircraft manufacture (Boeing) and software industries of all sizes (Microsoft is the largest) have promoted development of this region.

⟲ In the Mountain West—the states that lie east of the Pacific coast states—water is more limited than on the coast, and population is clustered around centers that have a reliable water sources. Phoenix and Tucson, Arizona, Denver, Colorado, and Salt Lake City, Utah, are the four largest cities in this region

⟲ Alaska and Hawaii are the two most recent additions to the United States (they both became states in 1959), and they represent very distinctive landscapes and environmental locales. Large military installations exist in both states. Alaska stretches north of the Arctic Circle and has a total population of approximately 650,000. Alaska's varied landscapes are characterized by intense winter seasons, but recent decades have seen a pronounced warming that has already had noticeable environmental effects. Anchorage is the largest city and has nearly one-half of the state's total population. There is an ongoing dispute among the U.S. Congress, the oil industry, environmentalists, and indigenous peoples about whether or not to develop the oil reserves of northeastern Alaska's Arctic National Wildlife Refuge (ANWR).

⟲ Alaskan history chronicles fur trapping, gold rushes, oil exploration, and military installations. Sports fishing and hunting and general tourism play an increasingly significant role in the state's economy. Single men outnumber single women more in Alaska than in any other state.

⟲ Hawaii lies some 2,000 miles (c. 3,200 km) west of California and is comprised of eight major islands, all of volcanic origin. It has a great range of rainfall patterns, and agriculture played an important role in its early development. Today, agriculture contributes little to the state's economy, but sugar and pineapple exports to mainland markets continue. Military expenditures and tourism, especially from the U.S. mainland and Asia, have brought major capital to the islands.

KEY TERMS + CONCEPTS

Terms in blue are also defined in the glossary.

Acadians (p. 660)
adaptive reuse (p. 656)
Arctic National Wildlife Refuge (ANWR) (p. 682)
"Big Muddy" (p. 670)
"Big Sky Country" (p. 680)
biotechnology (p. 664)
borough (p. 653)
break-of-bulk point (p. 651)
Cajuns (p. 659)
carrying capacity (p. 674)
Church of Jesus Christ of Latter Day Saints (Mormons) (p. 678)
Cuban Americans (p. 659)
deindustrialization (p. 655)
Dixie, Dixieland (p. 657)
dot com bubble (p. 677)
estuary (p. 650)
ethanol (p. 664)
exotic species (p. 657)
exurban development (p. 656)

exurbanization (p. 669)
fall line (p. 651)
"Father of Waters" (p. 663)
gentrification (p. 656)
geologic hot spot (p. 683)
Gold Rush of 1849 (p. 676)
Gwich'in (p. 682)
Inupiat (p. 682)
loess (p. 664)
"Lower 48" (p. 680)
Mason–Dixon Line (p. 657)
megalopolis (p. 647)
Mexican Americans (p. 659)
"Mile High City" (p. 673)
New Homestead Act (p. 668)
North American Aerospace Defense Command (NORAD) (p. 678)
oil shale (p. 678)
outsourcing (p. 661)
permafrost (p. 681)
"Porkopolis" (p. 666)

Research Triangle Park (RTP) (p. 662)
Ring of Fire (p. 671)
Rust Belt (p. 655)
"Seward's Folly" (p. 682)
Silicon Valley (p. 676)
smokestack industry (p. 655)
strip mining (p. 652)
suburban development (p. 656)
Sun Belt (p. 656)
tariff (p. 653)
Tennessee Valley Authority (TVA) (p. 661)
truck crops (p. 660)
Utah Telecommunication Open Infrastructure Agency (UTOPIA) (p. 679)
valley fills (p. 652)
Wal-Mart Effect (p. 668)
waterhead (p. 651)
zero sum game (p. 674)

REVIEW QUESTIONS

WORLD
REGIONAL
Geography◍Now™

Assess your understanding of this chapter's topics with additional quizzing and concept-based problems at http://earthscience.brookscole.com/wrg5e.

1. What five regions make up the United States?

2. Approximately what percentage of the U.S. population lives in the Northeast? In the West?

3. What are the major topographic features that characterize the U.S. Northeast? What are the fall-line cities? What were their roles in early industrialization and settlement?

4. When did the Erie Canal open? What route did it follow? What demographic impact did it have?

5. What are the reasons for the primacy of New York City?

6. What are the topographic characteristics of the South? How do they relate to patterns of urban settlement and agriculture in the region? What are the region's most important cities?

7. Why are the soils so good in the Midwest? What and where are the most important agricultural products?

8. What are the major characteristics of industrial growth in the Midwest? What cities have been most influenced by these changes?

9. Why is the term *oasislike* used for urban settlement in the U.S. West?

10. What are the major topographic features of the West? What role do these features have in patterns of agriculture and urban settlement in the region?

DISCUSSION QUESTIONS

1. Cities generally develop at certain places to take advantage of geographic circumstances. What geographic conditions promoted the growth of Philadelphia, St. Louis, Minneapolis, Cincinnati, San Francisco, and other cities discussed in this chapter?

2. What roles have rivers played in the patterns of settlement, economic activity, and tourism in the United States?

3. How and why has America's steel industry changed? How does Pittsburgh reflect these changes? What alternatives are developing in steel's wake? What is the phenomenon of deindustrialization?

4. Why have the rural areas and small towns of the Great Plains been losing population? What hopes are there for the revitalization of America's small towns?

5. How have Colorado River waters been allocated and used? Why has there been contention over these waters? How has the major dispute over allocation been addressed?

6. Discuss the "Wal-Mart Effect." Has the world's largest retailer had impacts in your community? For better or worse?

7. Is the "revolving door" of illegal immigration into the United States a problem? How should communities, states, and the nation deal with illegal aliens?

8. What are the pros and cons of exploiting oil in Alaska's Arctic National Wildlife Refuge? Where else and in what ways has the development of fossil fuel resources been an issue of contention in the United States?

9. What makes Alaska a particularly challenging environment to live in? How does its demographic situation reflect some of those challenges?

10. What are the major industries and resources of remote Hawaii? What are the island's major environmental features?

Glossary

Absolute (mathematical) location Determined by the intersection of lines, such as latitude and longitude, providing an exact point expressed in degrees, minutes, and seconds.

Acid rain Precipitation that mixes with airborne industrial pollutants, causing the moisture to become highly acidic and therefore harmful to flora and bodies of water on which it falls. Sulfuric acid is the most common component of this acid precipitation.

Adaptive reuse Finding new uses for older buildings and stores, often accompanied by a shift from decline to steady renewal in an urban neighborhood.

Age of Discovery (Age of Exploration) The three to four centuries of European exploration, colonization, and global resource exploitation and trading led largely by European mercantile powers. It began with Columbus at the end of the 15th century and continued into the 19th century.

Age-structure profile (population pyramid) The graphic representation of a country's population by gender and 5-year age increments.

Agricultural (Neolithic or New Stone Age or Food-Producing Revolution) The domestication of plants and animals that began about 10,000 years ago.

Albedo The amount of the sun's energy reflected by the ground. Less vegetation cover correlates with high albedo, and vice versa.

Alfisol soils These productive soils found in steppe regions are among the world's more fertile. Also known as *chestnut soils*.

Altitudinal zonation In a highland area, the presence of distinctive climatic and associated biotic and economic zones at successively higher elevations. Ethiopia and Bolivia are examples of these kinds of zones. See also *Tierra caliente, Tierra fría, Tierra helada, Tierra templada.*

Anticyclone An atmospheric high-pressure cell. In the cell, the air is descending and becomes warmer. As it warms, its capacity to hold water vapor increases, and the result is minimal precipitation.

Anti-Semitism Anti-Jewish sentiments and activities.

Apartheid The Republic of South Africa's former official policy of "separate development of the races," designed to ensure the racial integrity and political supremacy of the white minority.

Arable Suitable for cultivation.

Archipelago A chain or group of islands.

Arid China In a climatic division of China approximately along the 20° isohyet, Arid China lies to the west of the line. It characterizes more than 50 percent of China's territory but accommodates less than 10 percent of the population.

Arms race Usually associated with the competition between the United States and the Soviet Union, it refers to rival and potential enemy powers increasing their military arsenals—each in an open-ended effort to stay ahead of the other.

Atmospheric pollution The modification of the blanket of gases that surround the earth largely through the airborne products of industrial production and the human consumption of fossil fuels.

Atoll Low islands made of coral and usually having an irregular ring shape around a lagoon.

Autonomy Self-rule, generally with reference to Palestinians' rights to run their own civil (and some security) affairs in portions of the West Bank and Gaza Strip allocated to them in the 1993–2000 peace agreements.

Balkanization The fragmentation of a political area into many smaller independent units, as in former Yugoslavia in the Balkan Peninsula.

Barrios Densely settled neighborhoods in city space that is characteristically inhabited by migrants of Latino or Hispanic origin.

Barter The exchange of goods or services in the absence of cash.

Basin irrigation The ancient Egyptian system of cultivation using fields saturated by seasonal impoundment of Nile floodwaters.

Bazaar (suq) The central market of the traditional Middle Eastern city, characterized by twisting, close-set lanes and merchant stalls.

Belief systems The set of customs that an individual or culture group has relating to religion, social contracts, and other aspects of cultural organization.

Benelux The name used to collectively refer to Belgium, the Netherlands, and Luxembourg.

Biodiversity hot spots A ranked list of places scientists believe deserve immediate attention for flora and fauna study and conservation.

Biological diversity (biodiversity) The number of plant and animal species and the variety of genetic materials these organisms contain.

Biomass The collective dried weight of organisms in an ecosystem.

Biome A terrestrial ecosystem type categorized by a dominant type of natural vegetation.

Birth rate The annual number of live births per 1,000 people in a population.

Black-earth belt An important area of crop and livestock production spanning parts of Russia, Ukraine, Moldova, and Kazakhstan. The main soils of this belt are mollisols.

Boat people Refugees who flee by sea because of conflict or political circumstances, including refugees from conflict in Indochina, Cuba, and Haiti.

Bourgeoisie In Marxist doctrine, the capitalist class.

Brain drain The exodus of educated or skilled persons from a poor to a rich country or from a poor to a rich region within a country.

Break-of-bulk point A classic geographic term describing a point in transit when bulk goods must be removed from one mode of transport and installed on another. A trainload of grain carried to a port for transshipment on cargo boats or barges is a common example.

Broadleaf deciduous forest Forests typical of middle-latitude areas with humid subtropical and humid continental climates. As cool fall temperatures set in, broadleaf trees shed their leaves and cease to grow, thus reducing water loss. They produce new foliage and grow vigorously during the hot, wet summer.

Buffer state A generally smaller political unit adjacent to a large, or between several large, political units. Such a role often enables the smaller state to maintain its independence because of its mutual use to the larger, proximate nations. Uruguay, between Brazil and Argentina, is an example.

Capital goods Goods used to produce other goods.

Carrying capacity The size of a population of any organism that an ecosystem can support.

Carter Doctrine President Jimmy Carter's declaration, following the Soviet Union's invasion of Afghanistan in 1979, that the United States would use any means necessary to defend its vital interests in the Persian Gulf Region. The "vital interests" were interpreted to mean oil, and "any means necessary" interpreted to mean that the United States would go to war with the Soviet Union if oil supplies were threatened.

Cartogram A special map in which an area's shape and size are defined by explicit characteristics of population, economy, or distribution of any stated product.

Cartography The craft of designing and making maps, the basic language of geography. In recent years, this traditional manual art has been changed profoundly through the use of computers and Geographic Information Systems (GIS) and through major improvements in machine capacity to produce detailed, colored, map products.

Cash (commercial) crops Crops produced generally for export.

Caste The hierarchy in the Hindu religion that determines a person's social rank. It is established by birth and cannot be changed.

Charney Effect Observed by an atmospheric scientist named Charney, this states that the less plant cover there is on the ground, the higher the albedo—solar energy deflected back into the atmosphere—and therefore the lower the humidity and precipitation.

Chemocline The boundary between lower, carbon dioxide-laden waters and higher, gas-free, fresh water in some African lakes. The puncture of this boundary can cause eruptions that are fatal to humans and other life around the lakes.

Chernobyl The site in the Ukraine where, in April 1986, the worst nuclear power plant accident in history occurred. It is thought that approximately 5,000 people died, and a zone with a 20-mile (32-km) radius is still virtually uninhabitable; 116,000 people were moved from the area, and cleanup continues to this day.

Chernozem A Russian term meaning "black earth." It is a grassland soil that is exceptionally thick, productive, and durable.

Chestnut soils Productive soils typical of the Russian steppe and North American Great Plains.

Chokepoint A strategic narrow passageway on land or sea that may be closed off by force or threat of force.

Choropleth maps Maps that are drawn to show the differing distribution of goods or geographic characteristics (including population) across a broad area. Such maps are good for generalizations but often mask significant local variations in the presence of the item being mapped.

Chunnel (Eurotunnel) The 16-mile (26-km) tunnel that links Britain with the European continent. It was completed in 1994 at a cost of more than $15 billion, making it the single most costly project in landscape transformation ever undertaken.

Civilization The complex culture of urban life.

Climate The average weather conditions, including temperature, precipitation, and winds, of an area over an extended period of time.

Climatology The scientific study of patterns and dynamics of climate.

Coal The residue from organic material compressed for a long period under overlying layers of the earth. Coals vary in hardness and heating value. Anthracite coal burns with a hot flame and almost no smoke. Bituminous coal is used in the largest quantities; some can be used to make coke by baking out the volatile elements. Coke is largely carbon and burns with intense heat when used in blast furnaces to smelt iron ore. Peat, which is coal in the earliest stages of formation, can be burned if dried. Concern over atmospheric pollution by the sulfur in coal smoke has recently increased the demand for low-sulfur coals.

Cold War The tense but generally peaceful political and military competition between the United States and the Soviet Union, and their respective allies, from the end of World War II until the collapse of the Soviet Union in 1991.

Collective farm A large-scale farm in the former Soviet Union that usually incorporated several villages. Workers received shares of the income after the obligations of the collective had been met.

Collectivization The process of forming collective farms in Communist countries.

Colonization The European pattern of establishing dependencies abroad to enhance economic development in the home country.

Colored A South African term for persons of mixed racial ancestry.

Command economy A centrally planned economy typical of the Soviet Union and its Communist allies, in which the government rather than free enterprise determines the production, distribution, and sale of economic goods and services.

Common Market An earlier name given to the (current) 25 countries that make up the European Union. In 1957, an initial six countries combined to form the European Economic Community (EEC), and this supranational community has grown to have considerable economic and political importance in Europe. See *European Union*.

Computer cartography Mapmaking using sophisticated software and computer hardware. It is a new, actively growing career field in geography.

Computer-controlled robots Microtechnology has allowed the creation of highly precise, robotic units that can be used in delicate manufacturing. Although expensive to create, they ultimately reduce per unit manufacturing costs because of predictable higher quality output and easier maintenance than human work organizations.

Concentric zone model A generalized model of a city. A city becomes articulated into contrasting zones arranged as concentric rings around its central business district.

Coniferous vegetation Needleleaf evergreen trees; most bear seed cones.

Consumer goods Goods that individuals acquire for short-term use.

Consumer organisms Animals that cannot produce their own food within a food chain.

Consumption overpopulation The concept that a few persons, each using a large quantity of natural resources from ecosystems across the world, add up to too many people for the environment to support.

Containerization The prepackaging of items into larger standardized containers for more efficient transport.

Continental islands Once attached to nearby continents, these are the islands north and northeast of Australia, including New Guinea.

Convectional precipitation The heavy precipitation that occurs when air is heated by intense surface radiation, then rises and cools rapidly.

Convention on the Law of the Sea A 1970s United Nations treaty permitting a sovereign power to have greater access to surrounding marine resources.

Coordinate systems A means of determining exact or absolute location. Latitude and longitude are most often used.

Cottage industry Handwork that is done in the evenings or in slack seasons in the rural world. Products created during such activity can include woven and sewn goods or piecework for local textile or manufacturing industries. This work brings cash into the rural economy, especially to those not completely involved in full-time farming activity.

Council for Mutual Economic Assistance (COMECON) An economic organization, now disbanded, consisting of the Soviet Union, Poland, East Germany, Czechoslovakia, Hungary, Romania, Bulgaria, Cuba, Mongolia, and Vietnam.

Crop calendar The dates by which farmers prepare, plant, and harvest their fields. Farmers who are involved in new agricultural patterns necessitated by greater use of chemical fertilizers and new plant strains are sometimes unable to adjust to the demands of a much more exact crop calendar than has been traditional.

Crop irrigation Bringing water to the land by artificial methods.

Crusades A series of European Christian military campaigns between the 11th and 14th centuries aimed at recapturing Jerusalem and the rest of the Holy Land from the Muslims.

Cultural diffusion One of the most important dynamics in geography, cultural diffusion is the engine of change as crops, languages, culture patterns, and ideas are transferred from one place to other places, often in the course of human migration.

Cultural geography The study of the ways in which humankind has adopted, adapted to, and modified the face of the earth, with particular attention given to cultural patterns and their associated landscapes. It also includes a culture's influence on environmental perception and assessment.

Cultural landscape The landscape modified by human transformation, thereby reflecting the cultural patterns of the resident culture.

Cultural mores The belief systems and customs of a culture group.

Culture The values, beliefs, aspirations, modes of behavior, social institutions, knowledge, and skills that are transmitted and learned within a group of people.

Culture hearth An area where innovations develop, with subsequent diffusion to other areas.

Culture System A system in which Dutch colonizers required farmers in Java to contribute land and labor for the production of export crops under Dutch supervision.

Cyclone A low-pressure cell that composes an extensive segment of the atmosphere into which different air masses are drawn.

Cyclonic (frontal) precipitation The precipitation generated in traveling low-pressure cells that bring different air masses into contact.

Death rate The annual number of deaths per 1,000 people in a population.

Debt-for-nature swap An arrangement in which a certain portion of international debt is forgiven in return for the borrower's pledge to invest that amount in nature conservation.

Deciduous trees Broadleaf trees that lose their leaves and cease to grow during the dry or the cold season and resume their foliage and grow vigorously during the hot, wet season.

Deflation A fall in prices, such as that caused by currency devaluations.

Deforestation The removal of trees by people or their livestock.

Delta A landform resulting from the deposition of great quantities of sediment when a stream empties into a larger body of water.

Demographic shift Major population redistributions as people move from the countryside to the city or from one region to another.

Demographic transition A model describing population change within a country. The country initially has a high birth rate, a high death rate, and a low rate of natural increase, then moves through a middle stage of high birth rate, low death rate, and high rate of natural increase, and ultimately reaches a third stage of low birth rate, low or medium death rate, and low or negative rate of population increase.

Dependency theory A theory arguing that the world's more developed countries continue to prosper by dominating their former colonies, the now independent less developed countries.

Desert An area too dry to support a continuous cover of trees or grass.

A desert generally receives less than 10 inches (25 cm) of precipitation per year.

Desert shrub vegetation Scant, bushy plant life occurring in deserts of the middle and low latitudes where there is not enough rain for trees or grasslands. The plants are generally xerophytic.

Desertification The expansion of a desert brought about by changing environmental conditions or unwise human use.

Development A process of improvement in the material conditions of people often linked to the diffusion of knowledge and technology.

Devolution The process by which a sovereign country releases or loses more political and economic control to its constituent elements, such as states and provinces.

Diaspora The scattering of the Jews outside Palestine beginning in the Roman Era.

Digital divide The divide between the handful of countries that are the technology innovators and users and the majority of nations that have little ability to create, purchase, or use new technologies.

Dissection The carving of a landscape into erosional forms by running water.

Distributary A stream that results when a river subdivides into branches in a delta.

Domestication The controlled breeding and cultivation of plants and animals.

Donor democracy Typical of countries in Africa south of the Sahara, this is a situation in which a government makes just enough concessions on voting rights or human rights to win foreign loans and aid, without instituting any serious democratic reform.

Donor fatigue Public or official weariness of extending aid to needy people.

Double cropping Growing two crops a year on the same field.

Drought avoidance Adaptations of desert plants and animals to evade dry conditions by migrating (animals) or being active only when wet conditions occur (plants and animals).

Drought endurance Adaptations of desert plants and animals to tolerate dry conditions through water storage and heat loss mechanisms.

Dry farming Planting and harvesting according to the seasonal rainfall cycle.

Ecologically dominant species A species that competes more successfully than others for nutrition and other essentials of life.

Ecology The study of the interrelationships of organisms to one another and to the environment.

Economic shock therapy Russia's economic transformation in the early 1990s from a command economy to a free-market economy. Overseen by Boris Yeltsin, this transformation was difficult for a country accustomed to government direction in all economic matters, thus the "shock."

Ecosphere (biosphere) The vast ecosystem composed of all of Earth's ecosystems.

Ecosystem A system composed of interactions between living organisms and nonliving components of the environment.

Ecotourism Travel by people who want to both see and save the natural habitats remaining on earth.

Ejido An agricultural unit in Mexico characterized by communally farmed land or common grazing land. It is of particular importance to indigenous villages.

Endemic species A species of plant or animal found exclusively in one area.

Energy crisis The petroleum shortages and price surges sparked by the 1973 oil embargo.

Environmental assessment The process of determining the condition and value of a particular environmental setting. Used both in aspects of landscape change and in evaluating environmental perception.

Environmental determinism The concept that the physical environment has played a sovereign role in the cultural development of a people or landscape. Also known as *environmentalism.*

Environmental perception The concept that how people view the world and its landscapes and resources influences their uses of the earth and therefore the condition of the earth.

Environmental possibilism A concept that rejects environmental determinism, arguing that although environmental conditions do influence human and cultural development, people choose from various possibilities in how to live within a given environment.

Equinox On or about September 23, and again on or about March 20, Earth reaches the equinox position. Its axis does not point toward or away from the sun, so days and nights are of equal length at all latitudes on Earth.

Escarpment A steep edge marking an abrupt transition from a plateau to an area of lower elevation.

Estuary A deepened ("drowned") river mouth into which the sea has flooded.

Ethnic cleansing The relocation or killing of members of one ethnic group by another to achieve some demographic, political, or military objective.

Ethnocentrism Regarding one's own group as superior and as setting proper standards for other groups.

Euro Part of the authority of the European Union has been the institution of a new currency that has become "coin of the realm" since early 2002. Not all EU nations accepted the euro.

European Community The name that was replaced in 1993 by the term European Union. The European Community continues to serve as a governing and administrative body for the EU.

European Economic Community An economic organization designed to secure the benefits of large-scale production by pooling resources and markets. The name has been changed to Economic Union. See also *Common Market.*

European Free Trade Association (EFTA) An organization that maintains free trade among its members but allows each member to set its own tariffs in trading with the outside world. Members are Iceland, Norway, Sweden, Switzerland, Austria, and Finland.

European Union The current organization begun in the 1950s as the Common Market. It now is made up of 25 nations (see Chapter 3).

Evaporation The loss of moisture from the earth's land surfaces and its water bodies to the air through the ongoing influence of solar radiation and transpiration by plants.

Exotic species A nonnative species introduced into a new area.

Extensive land use A livelihood, such as hunting and gathering, that requires the use of large land areas.

External costs (externalities) Consequences of goods and services that are not priced into the initial cost of those goods and services.

Fall line A zone of transition in the eastern United States where rivers flow from the harder rocks of the Piedmont to the softer rocks of the Atlantic and Gulf Coastal Plain. Falls and/or rapids are characteristic features.

Fault A break in a rock mass along which movement has occurred. A break due to rock masses being pulled apart is a tensional or "normal" fault, whereas a break due to rocks being pushed together until one mass rides over the other is a compressional fault. The processes of faulting create these breaks.

Feral animals Domesticated animals that have abandoned their dependence on people to resume life in the wild.

Fertile Crescent The arc-shaped area stretching from southern Iraq through northern Iraq, southern Turkey, Syria, Lebanon, Israel, and western Jordan, where plants and animals were domesticated beginning about 10,000 years ago.

Final status issues Issues deferred to the end of the Oslo Peace Process between Israel and the PLO. Finally dealt with at Camp David in 2000, they included the status of Jerusalem, the fate of Palestinian refugees, Palestinian statehood, and borders between Israel and a new Palestinian state. The negotiations broke down over these final status issues.

Fjord A long, narrow extension of the sea into the land usually edged by steep valley walls that have been deepened by glaciation.

Folding The process creating folds (landforms resulting from an intense bending of rock layers). Upfolds are known as anticlines, and downfolds are called synclines.

Food chain The sequence through which energy, in the form of food, passes through an ecosystem.

Formal region See *Region.*

Fossil waters Virtually nonrenewable freshwater supplies, the product of ancient rainfall stored in deep natural underground aquifers, especially in the Middle East and North Africa.

Four Modernizations The effort of the Chinese after the death of Mao Zedong in 1976 to focus on economic development in agriculture, industry, science and technology, and defense.

Front A contact zone between unlike air masses. A front is named according to the air mass that is advancing (cold front or warm front).

Frontal precipitation See *Cyclonic precipitation.*

Fuelwood crisis Deforestation in the less developed countries caused by subsistence needs.

Functional region See *Region.*

Gaia hypothesis A hypothesis stating that the ecosphere is capable of restoring its equilibrium following any disturbance that is not too drastic.

Gentrification The social and physical process of change in an urban neighborhood by the return of young, often professional, populations to the urban core. These people are often attracted by the substantial nature of the original building stock of the place and the proximity to the city center, which generally continues to have major professional opportunities. While this process brings an urban landscape back into a primary role as a tax base, it does dispossess a considerable number of minority peoples.

Geographic analysis By giving attention to the spatial aspects of a distribution, geographic analysis helps to explain distribution, density, and flow of a given phenomenon.

Geographic Information Systems (GIS) The growing field of computer-assisted geographic analysis and graphic representation of spatial data. It is based on superimposing various data layers that may include everything from soils to hydrology to transportation networks to elevation. Computer software and hardware are steadily improving, enabling GIS to produce ever more detailed and exact output.

Geography The study of the spatial order and associations of things. Also defined as the study of places, the study of relationships between people and environment, and the study of spatial organization.

Geologic hot spot A small area of Earth's mantle where molten magma is relatively close to the crust. Hot spots are associated with island chains and thermal features.

Geomorphology The scientific analysis of the landforms of the earth; sometimes called *physiography.*

Geopolitics The study of geographic factors in political matters, including borders, political unity, and warfare.

Glacial deposition In the process of continental and valley glaciation, the deposition of moraines that become lateral or terminal in the act of glacial retreat. This same process also leads to glacial scouring as moving ice picks up loose rock and reshapes the landscape as the glacier moves forward or retreats.

Glacial scouring See *Glacial deposition.*

Globalization The spread of free trade, free markets, investments, and ideas across borders and the political and cultural adjustments that accompany this diffusion.

Graben (rift valley) A landform created when a segment of Earth's crust is displaced downward between parallel tensional faults or when segments of the crust that border it ride upward along parallel compressional faults.

Great Rift Valley The result of tectonic processes, this is a broad, steep-walled trough extending from the Zambezi Valley in southern Africa northward to the Red Sea and the valley of the Jordan River in southwestern Asia.

Great Trek In what is now South Africa, a series of northward migrations in the 1830s by which groups of Boers, primarily from the eastern part of the Cape Colony, sought to find new interior grazing lands and establish new political units beyond British reach.

Green Revolution The introduction and transfer of high-yielding seeds, mechanization, irrigation, and massive application of chemical fertilizers to areas where traditional agriculture has been practiced.

Greenhouse effect The observation that increased concentrations of carbon dioxide and other gases in Earth's atmosphere cause a warmer atmosphere.

Gross domestic product (GDP) The value of goods and services produced in a country in a given year. It does not include net income earned outside the country. The value is normally given in current prices for the stated year.

Gross national product (GNP) The value of goods and services produced internally in a given country during a stated year plus the value resulting from transactions abroad. The value is normally expressed in current prices of the stated year. Such data must be used with caution in regard to developing countries because of the broad variance in patterns of data collection and the fact that many people consume a large share of what they produce.

Growth pole A new city, a resource, or some development in a heretofore undersettled area that begins to attract population growth and economic development.

Guano Seabird excrement that is also found in phosphate deposits.

Guest workers Migrants (mainly young and male) from less developed countries who are employed, sometimes illegally, in more developed countries.

Gulf Stream The strong ocean current originating in the tropical Atlantic Ocean that skirts the eastern shore of the United States, curves eastward, and reaches Europe as a part of the broader current called the North Atlantic Drift.

Gulf War The 1990–1991 confrontation between a large military coalition, spearheaded by the United States, and the Iraqi forces that had invaded Kuwait. The allies successfully drove Iraqi forces from Kuwait. Earlier, the Gulf War had referred to the 1980–1988 war between Iraq and Iran.

Hacienda A Spanish term for large rural estates owned by the aristocracy in Latin America.

Haj The pilgrimage to Mecca, the principal holy city in the Islamic religion. Every Muslim is required to make this pilgrimage at least once in a lifetime, if possible.

Han Chinese The original Chinese peoples who settled in North China on the margins of the Yellow River (Huang He) and who were central to the development of Chinese culture. Han Chinese make up more than 90 percent of China's population.

Hierarchical rule A system by which leadership for a culture group is defined by traditional and sometimes frustrating patterns of age–sex distinction. It is evident in most cultures and slowly modified by an increasing role of democratic elections and the establishment of a broader opportunity base for individual advancement.

High islands Generally the result of volcanic eruptions, these are the higher and more agriculturally productive and densely populated islands of the Pacific World.

High pressure cell (anticyclone) An air mass descending and warming because of increased pressure and weight of the air above. High pressure typically means low relative humidity and minimal precipitation.

Hinterland The region of a country that lies away from the capital and largest cities and is most often rural or even unsettled; it is often seen in the minds of economic planners as an area with development potential.

Historical geography A concern with the historical patterns of human settlement, migration, town building, and the human use of the earth. Often, this subdiscipline best blends geography and history as a perspective on human activity.

Holocaust Nazi Germany's attempted extermination of Jews, Roma (Gypsies), homosexuals, and other minorities during World War II.

Homelands Ten former territorial units in South Africa reserved for native Africans (blacks). They had elected African governments, and some were designated as "independent" republics, although they were not recognized outside South Africa. Formerly known as *Bantustans.* They were abolished in 1994. Also known as *native reserves* and *national states.*

Horizontal migration The movements of pastoral nomads over relatively flat areas to reach areas of pasture and water.

Hot money Short-term and often volatile flows of investment that can cause serious damage to the "emerging market" economies of less developed countries.

Human Development Index (HDI) A United Nations–devised ranked index of countries' development that evaluates quality of life issues (such as gender equality, literacy, and human rights) in addition to economic performance.

Humboldt (or Peru) Current The cold ocean current that flows northward along the west coast of South and North America. It plays a significant role in regional fish resources.

Humid China The eastern portion of China that is relatively well watered and where the great majority of the Chinese population has settled. An arc from Kunming in southwest China to Beijing in North China describes the approximate western margin of this zone.

Humid continental A climate with cold winters, warm to hot summers, and sufficient rainfall for agriculture, with the greater part of the precipitation in the summer half-year.

Humid pampa A level to gently rolling area of grassland centered in Argentina with a humid subtropical climate.

Humid subtropical A climate that generally occupies the southeastern margins of continents with hot summers, mild to cool winters, and ample precipitation for agriculture.

Humus Decomposed organic soil material. Grasslands characteristically provide more humus than forests do.

Hunting and gathering A mode of livelihood, based on collecting wild plants and hunting wild animals, generally practiced by preagricultural peoples. Also known as *foraging.*

Hydraulic control The ability to control water in irrigation systems, rivers, urban settlements, and for the generation of electricity. It has been central to the development of major civilizations and to culture groups throughout human history.

Ice cap A climate and biome type characterized by permanent ice cover on the ground, no vegetation (except where limited melting occurs), and a severely long, cold winter. Summers are short and cool.

Ideology A system of political and/or economic beliefs such as communism, capitalism, autocracy, or democracy.

Igneous (volcanic) rock Rock formed by the cooling and solidification of molten materials. Granite and basalt are common types. Such rocks tend to form uplands in areas where sedimentary rocks have weathered into lowlands. Certain types break down into extremely fertile soils. These rocks are often associated with metal-bearing ores.

Industrial Revolution A period beginning in mid-18th century Britain that saw rapid advances in technology and the use of inanimate power.

Inflation A rise in prices.

Information technology (IT) The Internet, wireless telephones, fiberoptics, and other technologies characteristic of more developed countries. Generally seen as beneficial for a country's economic prospects, IT is also spreading in the less developed countries.

Intensive land use A livelihood requiring use of small land areas, such as farming.

Intertillage Growing two or more crops simultaneously in alternate rows. Also known as *interplanting*.

Irredentism Movement by an ethnic group in one country to revive or reinforce kindred ethnicity in another country—often in an effort to promote succession there.

Irrigation The artificial placement of water to produce crops, generally in arid locations.

Island chain A series of islands formed by ocean crust sliding over a stationary hot spot in the earth's mantle.

Karst Landscape features associated with the dissolution of limestone or dolomite including caves, sinkholes, and towers.

Keystone species A species that affects many other organisms in an ecosystem.

Kyoto Protocol A treaty on climate change signed by 160 countries in Kyoto, Japan, in 1997, and put into force in 2005. It requires MDCs to reduce their greenhouse gas emissions by more than 5 percent below their 1990 levels by the year 2012.

Lacustrine plain The floor of a former lake where glacial meltwater accumulated and sediments washed in and settled. An extremely flat surface is characteristic.

Landscape A portion of the earth's land surface. Geographers are interested in the transformation of natural landscapes into cultural landscapes.

Landscape transformation The human process of making over the earth's surface. From initial human shelters to contemporary massive urban centers, humans have been driven to change their environmental settings.

Large-scale map A map constructed to show considerable detail in a small area.

Laterite A material found in tropical regions with highly leached soils; composed mostly of iron and aluminum oxides which harden when exposed and make cultivation difficult.

Latifundia (sing. latifundio) Large agricultural Latin American estates with strong commercial orientations.

Latitude A measurement that denotes position with respect to the equator and the poles. Latitude is measured in degrees, minutes, and seconds, which are described as parallels. Places near the equator are said to be in low latitudes; places near the poles are in high latitudes. The Tropic of Cancer and the Tropic of Capricorn, at 23.5°N and 23.5°S, respectively, and the Arctic Circle and Antarctic Circle, at 66.5°N and 66.5°S, respectively, form the most commonly recog-

nized boundaries of the low and high latitudes. Places occupying an intermediate position with respect to the poles and the equator are said to be in the middle latitudes. There are no universally accepted definitions for the boundaries of the high, middle, and low latitudes.

Law of Return An Israeli law permitting all Jews living in Israel to have Israeli citizenship.

Law of the Sea A United Nations treaty or convention permitting coastal nations to have greater access to marine resources.

Less developed countries (LDCs) The world's poorer countries.

Lifeboat ethics Ecologist Garrett Hardin's argument that, for ecological reasons, rich countries should not assist poor countries.

Location Central to all geographic analysis is the concept of location. Where something is relates to all manner of influences, from climate to migration routes. It is a crucial component in trying to understand patterns of historic and economic development.

Loess A fine-grained material that has been picked up, transported, and deposited in its present location by wind; it forms an unusually productive soil.

Long March The 1935–1936 flight of approximately 100,000 Communist troops under Mao Zedong from south central China to the mountains of North China. Mao was fleeing attack by General Chiang Kai-shek, and the success of the Long March led to the establishment of Mao's troops as a force with the capacity to overcome the much larger and better armed troops of Chiang.

Longitude A measurement that denotes a position east or west of the prime meridian (Greenwich, England). Longitude is measured in degrees, minutes, and seconds, and meridians of longitude extend from pole to pole and intersect parallels of latitude.

Low islands Made of coral, these are the generally flat, drier, less agriculturally productive, and less densely populated islands of the Pacific World.

Malthusian scenario The model forecasting that human population growth will outpace growth in food and other resources, with a resulting population die-off.

Map projection A way to minimize distortion in one or more properties of a map (direction, distance, shape, or area).

Map scale The actual distance on the earth that is represented by a given linear unit on a map.

Maquiladoras Operations dedicated to the assembly of manufactured goods, generally in Mexico and Central America, from components initially produced in the United States or other places. With the enactment of NAFTA in 1994, there was a massive expansion of maquiladora operations.

Marginalization A process by which poor subsistence farmers are pushed onto fragile, inferior, or marginal lands that cannot support crops for long and that are degraded by cultivation.

Marine west coast A climate occupying the western sides of continents in the higher middle latitudes; it is greatly moderated by the effects of ocean currents that are warm in winter and cool in summer relative to the land.

Marshall Plan The plan designed largely by the United States after the conclusion of World War II by which U.S. aid was focused on the rebuilding of the very Germany that had been its enemy in the war just concluded. Secretary of State George Marshall (who had been Chief of Staff of U.S. Army from 1939–1945) was central to the plan's design and implementation.

Material culture The items that can be seen and associated with cultural development, such as house architecture, musical instruments, and tools. In the study of a cultural landscape, items of material culture add considerable personality to cultural identity.

Medical geography The study of patterns of disease diffusion, environmental impact on public health, and the interplay of geographic factors, migration, and population. With the increasing ease of international movement, medical geography is becoming more important as the potential for disease diffusion increases.

Medina An urban pattern typical of the Middle Eastern city before the 20th century.

Mediterranean (dry-summer subtropical) A climate that occupies an intermediate location between a marine west coast climate on the poleward side and a steppe or desert climate on the equatorward side. During the high-sun period, it is rainless; in the low-sun period, it receives precipitation of cyclonic or orographic origin.

Mediterranean scrub forest The xeroyphytic vegetation typical of hot, dry summer, Mediterranean climate regions. Local names for this vegetation type include *maquis* and *chaparral*.

Meiji Restoration The Japanese revolution of 1868 that restored the legitimate sovereignty of the emperor. *Meiji* means "Enlightened Rule." This event led to a transformation of Japan's society and economy.

Melanesia (black islands) A group of relatively large islands in the Pacific Ocean bordering Australia.

Mental maps Every individual's mind has a series of locations, access routes, physical and cultural characteristics of places, and often a general sense of the good or bad of locales. The term *mental map* is used to define such geographies.

Mercantile colonialism The historical pattern by which Europeans extracted primary products from colonies abroad, particularly in the tropics.

Meridian See *Longitude*.

Mestizo In Latin America, a person of mixed European and Native American ancestry.

Metamorphic rock Rock formed from igneous or sedimentary rock through changes occurring in the rock structure as a result of heat, pressure, or the chemical action of infiltrating water. Marble, formed of pure limestone, is a common example.

Metropolitan area A city together with suburbs, satellites, and adjacent territory with which the city is functionally interlocked.

Microcredit The lending of small sums to poor people to set up or expand small businesses.

Micronesia (tiny islands) Thousands of small and scattered islands in the central and western Pacific Ocean, mainly north of the equator.

Microstate A political entity that is tiny in area and population and is independent or semi-independent.

Middle Eastern ecological trilogy The model of mostly symbiotic relations among villagers, pastoral nomads, and urbanites in the Middle East.

Migration A temporary, periodic, or permanent move to a new location.

Milpero A Latin American farmer who engages in swidden or shifting cultivation in forest lands. The plot is often called the *milpa*.

Minifundia (sing. minifundio) Small Latin American agricultural landholdings, usually with a strong subsistence component.

Mixed forest A transitional area where both needleleaf and broadleaf trees are present and compete with each other.

Mobility The pattern of human movement, typically changing because of continual improvement in means of transportation.

Mollisols Thick, productive, and durable soils, such as the *chernozem*, whose fertility comes from abundant humus in the top layer.

Monoculture The single-species cultivation of food or tree crops, usually very economical and productive but threatening to natural diversity and change.

Monsoon A current of air blowing fairly steadily from a given direction for several weeks or months at a time. Characteristics of a monsoonal climate are a seasonal reversal of wind direction, a strong summer maximum of rainfall, and a long dry season lasting for most or all of the winter months.

Montreal Protocol A 1989 international treaty to ban chlorofluorocarbons.

Moor A rainy, deforested upland, covered with grass or heather and often underlain by water-soaked peat. Also, Muslim inhabitants of Spain.

Moraine The unsorted material deposited by a glacier during its retreat. Terminal moraines are ridges formed by long-continued deposition at the front of a stationary ice sheet.

More developed countries (MDCs) The world's wealthier countries.

Mulatto A person of mixed European and black ancestry.

Multinational companies Companies that operate, at least in part, outside their home countries.

Multiple cropping Farming patterns in which several crops are raised on the same plot of land in the course of a year or even a season. Common examples are winter wheat and summer corn or soybeans in North America or several crops of rice in the same plot in Monsoon Asia.

Nation Commonly, a nation is a term describing the citizens of a state—or that state—but it also refers to an ethnic group existing with or without a separate political entity. See *Nation-state*.

Nation-state A political situation in which high cultural and ethnic homogeneity characterizes the political unit in which people live.

Nationalism The drive to expand the identity and strength of a political unit that serves as home to a population interested in greater cohesion and often expanded political power.

Natural levee A strip of land immediately alongside a stream that has been built up by deposition of sediments during flooding.

Natural replacement rate The highest rate at which a renewable resource can be used without decreasing its potential for renewal. Also known as *sustainable yield*.

Natural resource A product of the natural environment that can be used to benefit people. Resources are human appraisals.

Near Abroad Russia's name for the now-independent former republics, other than Russia, of the USSR.

Neocolonialism The perpetuation of a colonial economic pattern in which developing countries export raw materials to, and buy finished goods from, developed countries. This relationship is more profitable for the developed countries.

Neo-Malthusians Supporters of forecasts that resources will not be able to keep pace with the needs of growing human populations.

Newly industrializing countries (NICs) The more prosperous of the world's less developed countries.

Nonblack soil zone Areas of poor soil in cool, humid portions of the Slavic Coreland, suitable for cultivation of rye.

North Atlantic Drift A warm current, originating in tropical parts of the Atlantic Ocean, that drifts north and east, moderating temperatures of Western Europe. Also known as *Gulf Stream*.

North Atlantic Treaty Organization (NATO) A military alliance formed in 1949 that included the United States, Canada, many European nations, and Turkey.

North European Plain The level to rolling lowlands that extend from the low countries on the west through Germany to Poland. These are areas rich in agricultural development, dense human settlements, and a number of major industrial centers. The plain is broken by a number of rivers flowing from the Alps and other mountain systems in central Europe into the North and Baltic Seas.

Occupied Territories The territories captured by Israel in the 1967 War: the Gaza Strip and West Bank, captured respectively from Egypt and Jordan, and the Golan Heights, captured from Syria.

Oil embargo The 1973 embargo on oil exports imposed by Arab members of the Organization of Petroleum Exporting Countries (OPEC) against the United States and the Netherlands.

Open borders Political boundaries that have minimal political or structural impediment to easy crossing between two countries.

Organization for Economic Cooperation and Development (OECD) See *Organization for European Economic Cooperation (OEEC)*.

Organization for European Economic Cooperation (OEEC) An organization created in 1948 to organize and facilitate the European response to the 1947 Marshall Plan. In the early 1960s, the OEEC became the Organization for Economic Cooperation and Development (OECD), and this served as a multinational base for continued planning in economic and social development. The Common Market came to overshadow this organization, and finally, the European Union, in 1993, became the most powerful and influential multinational organization in Europe.

Orographic precipitation The precipitation that results when moving air strikes a topographic barrier, such as a mountain, and is forced upward.

Ozone hole Areas of depletion of ozone in Earth's stratosphere, caused by chlorofluorocarbons.

Palestine Liberation Organization The Palestinian military and civilian organization created in the 1960s to resist Israel and recognized in the 1990s as the sole legitimate organization representing official Palestinian interests internationally.

Parallel A latitude line running parallel to the equator.

Patrilineal descent system A kinship naming system based on descent through the male line.

People overpopulation The concept that many persons, each using a small quantity of natural resources to sustain life, add up to too many people for the environment to support.

Per capita For every person, or per person.

Per Capita Gross Domestic Product Purchasing Power Parity (GDP PPP) In this figure, annual per capita gross domestic product (GDP)—the total output of goods and services a country produces for home use in a year—is divided by the country's population. For comparative purposes, that figure is adjusted for purchasing power parity (PPP), which involves the use of standardized international dollar price weights that are applied to the quantities of final goods and services produced in a given economy. The resulting measure, per capita GDP PPP, provides the best available starting point for comparisons of economic strength and well-being between countries.

Perennial irrigation The year-round irrigated cultivation of crops, as in the Nile Valley following construction of barrages and dams.

Permafrost Permanently frozen subsoil.

Photosynthesis The process by which green plants use the energy of the sun to combine carbon dioxide with water to give off oxygen and produce their own food supply.

Physical geography The subdiscipline of geography most concerned with the climate, landforms, soils, and physiography of the earth's surface.

Piedmont A belt of country at an intermediate elevation along the base of a mountain range.

Pillars of Islam The five fundamental tenets of the faith of Islam.

Pivotal countries Those countries whose collapse would cause international refugee migration, war, pollution, disease epidemics, or other international security problems.

Place identity In geographic analysis of a given locale, the nature of place identity becomes a means of understanding people's response to that particular place. Determination of the environmental and cultural characteristics that are most frequently associated with a certain place helps to establish that "sense of place" for a given location.

Plain A flat to moderately sloping area, generally of slight elevation.

Plantation A large commercial farming enterprise, generally emphasizing one or two crops and utilizing hired labor. Plantations are commonly found in tropical regions.

Plate tectonics The dominant force in the creation of the continents, mountain systems, and ocean deeps. The steady, but slow, movement of these massive plates of the earth's mantle and crust has created the positions of the continents that we have and major patterns of volcanic and seismic activity. Areas where plates are being pulled under other plates are called subduction zones.

Plateau An elevated plain, usually lying 2,000 feet (610 m) above sea level. Some are known as tablelands. A dissected plateau is a hilly or mountainous area resulting from the erosion of an upraised surface.

Pleistocene overkill A hypothesis stating that hunters and gatherers of the Pleistocene Era hunted many species to extinction.

Podzol Soil with a grayish, bleached appearance when plowed, lacking in well-decomposed organic matter, poorly structured, and very low in natural fertility. Podzols are the dominant soils of the taiga.

Polder An area reclaimed from the sea and enclosed within dikes in the Netherlands and other countries. Polder soils tend to be very fertile.

Polynesia (many islands) A Pacific island region that roughly resembles a triangle with its corners at New Zealand, the Hawaiian Islands, and Easter Island.

Population change rate The birth rate minus the death rate in a population.

Population density The average number of people living in a square mile or square kilometer. It is a very handy statistic for generalized comparisons but often fails to provide a detailed sense of the real distribution of people.

Population explosion The surge in Earth's human population that has occurred since the beginning of the Industrial Revolution.

Postindustrial Description of an economy or society characterized by the transformation from manufacturing to information management, financial services, and the service sector.

Prairie An area of tall grass in the middle latitudes composed of rich soils that have been cleared for agriculture. The original lack of trees may have been due to repeated burnings or periodic drought conditions.

Primary consumers (herbivores) Consumers of green plants.

Primate city A city that dominated a country's urban scene, and usually defined as being larger than the country's second and third largest cities combined. Primate cities are generally found in developing countries, although some developed countries, such as France, have them.

Prime meridian See *Latitude*.

Producers (self-feeders) In a food chain, organisms that produce their own food (mostly, green plants).

Projection The distortion caused by the transfer of three-dimensional space on Earth's surface to the two dimensions of a flat map.

Push and pull forces of migration Emigration may be caused by so-called push factors, as when hunger or lack of land "pushes" peasants out of rural areas into cities, or by pull factors, as when an educated villager responds to a job opportunity in the city. People responding to push factors are often referred to as nonselective migrants, whereas those reacting to pull factors are called selective migrants. Both push and pull forces are behind the rural to urban migration that is characteristic of most countries.

Pyramid of biomass A diagrammatic representation of decreasing biomass in a food chain.

Pyramid of energy flow A diagrammatic representation of the loss of high quality, concentrated energy as it passes through the food chain.

Qanat A tunnel used to carry irrigation and drinking water from an underground source by gravity flow. Also known as *foggara* or *karez*.

Rain shadow A condition creating dryness in an area located on the lee side of a topographic barrier such as a mountain range.

Region A "human construct" that is often of considerable size, that has substantial internal unity or homogeneity, and that differs in significant respects from adjoining areas. Regions can be classed as formal (homogeneous), functional, or vernacular. The formal region, also known as a uniform region, has a unitary quality that derives from a homogeneous characteristic. The United States is an example of a formal region. The functional region, also called the nodal region, is a coherent structure of areal units organized into a functioning system by lines of movement or influence that converge on a central node or trunk. A major example would be the trading territory served by a large city and bound together by the flow of people, goods, and information over an organized network of transportation and communication lines. Vernacular or perceptual regions are areas that possess regional identity, such as the "Sun Belt," but share less objective criteria in the use of this regional name. General regions, such as the major world regions in this text, are recognized on the basis of overall distinctiveness.

Remote sensing Through the use of aerial and satellite imagery, geographers and other scientists have been able to get vast amounts of data describing places all over the face of the earth. Remote sensing is the science of acquiring and analyzing data without being in contact with the subject. It is used in the study of patterns of land use, seasonal change, agricultural activity, and even human movement along transport lines. This process relates closely to GIS. See also *Geographic Information Systems (GIS)*.

Renewable resource A resource, such as timber, that is grown or renewed so that a continual supply is available. A finite resource is one that, once consumed, cannot be easily used again. Petroleum products are a good example of such a resource, and because it takes too much time to go through the process of creation, they are not seen as renewable.

Resources Resources become valuable through human appraisal, and their utilization reflects levels of technology, location, and economic ambition. There are few resources that have been universally esteemed (water, land, defensible locales) throughout history, so patterns of resource utilization serve as indexes of other levels of cultural development. See also *Natural resource*.

Rift valley. See *Graben*.

Right of return, Palestinian The Palestinian Arab principle that refugees (and their descendants) displaced from Israel and the Occupied Territories in the 1948 and 1967 wars be allowed to return to the region.

Ring of Fire The long horseshoe-like chain of volcanoes that goes from the southern Andes Mountains in South America up the west coast of North America and arches over into the northeast Asian island chains of Japan, the Philippines, and Southeast Asia. This zone of seismic instability and erratic vulcanism is caused by the tension built up in plate tectonics. See also *Plate tectonics*.

Riparian A state containing or bordering a river.

Russification The effort, particularly under the Soviets, to implant Russian culture in non-Russian regions of the former Soviet Union and its Eastern European neighbors.

Sacred space (sacred place) Any locale that people hold in reverence, such as places of worship, cemeteries, and battlefields.

Salinization The deposition of salts on, and subsequent fertility loss in, soils experiencing a combination of overwatering and high evaporation.

Sand sea A virtual "ocean" of sand characteristic of parts of the Middle East, where people, plants, and animals are all but nonexistent.

Savanna A low-latitude grassland in an area with marked wet and dry seasons.

Scale The size ratio represented by a map; for example, a map with a scale of 1:12,500 is portrayed as 1/12,500 of the actual size.

Scorched earth The wartime practice of destroying one's own assets to prevent them from falling into enemy hands.

Scrub and thorn forest The low, sparse vegetation in tropical areas where rainfall is insufficient to support tropical deciduous forest.

Seamount An underwater volcanic mountain.

Second law of thermodynamics A natural law stating that high-quality, concentrated energy is increasingly degraded as it passes through the food chain.

Secondary consumers (carnivores) Consumers of primary consumers.

Sedentarization Voluntary or coerced settling down, particularly by pastoral nomads in the Middle East.

Sedimentary rock Rock formed from sediments deposited by running water, wind, or wave action either in bodies of water or on land and which have been consolidated into rock. The main classes are sandstone, shale (formed principally of clay), and limestone, including the pure limestone called chalk.

Segmentary kinship system The organization of kinship groups in concentric units of membership, as of lineage, clan, and tribe among Middle Eastern pastoral nomads.

Service sector The labor sector made up of employees in retail trade and personal services; it is the sector most likely to increase in employment significance in postindustrial society.

Settler colonization The historical pattern by which Europeans sought to create new or "neo-Europes" abroad.

Shatter belt A large, strategically located region composed of conflicting states caught between the conflicting interests of great powers.

Shi'a Islam The branch of Islam regarding male descendants of Ali, the cousin and son-in-law of the Prophet Muhammad, as the only rightful successors to the Prophet Muhammad.

Shifting cultivation A cycle of land use between crop and fallow years that is needed to work around the infertility of tropical soils. After clearing tropical forest, a farmer may get only a few years of crops before there is no fertility in the soil and must move on to clear more land. In the meantime, forest reclaims the previously farmed plots. Where new lands are not available, fallow periods on old fields are reduced or eliminated, resulting in soil deterioration. Also known as *swidden* or *slash-and-burn cultivation*.

Site and situation Site is the specific geographic location of a given place, while situation is the accessibility of that site and the nature of the economic and population characteristics of that locale.

Slavic Coreland (Fertile Triangle or Agricultural Triangle) The large area of the western former Soviet region containing most of the region's cities, industries, and cultivated lands.

Small-scale map A map constructed to give a highly generalized view of a large area.

Soil The earth mantle made of decomposed rock and decayed organic material.

Sovereign state A political unit that has achieved political independence and maintains itself as a separate unit.

Spatial Geographers recognize spatial distributions and patterns in Earth's physical and human characteristics. The term *spatial* comes from the noun "space," and it relates to the distribution of various phenomenon on Earth's surface. Geographers portray spatial data cartographically—that is, with maps.

Special Economic Zones (SEZs) In China, the urban areas designated after the 1970s to attract investment and boost production through tax breaks and other incentives.

Spodosols Acidic soils that have a grayish, bleached appearance when plowed, lack well-decomposed organic matter, and are low in natural fertility. Also known as *podzols*.

St. Lawrence Seaway Project This seaway has a total length of 2,342 miles (3,796 km) between the Atlantic Ocean and its western terminus in Lake Superior. It allows major oceanic vessels to reach the Great Lakes, thus enabling ports as far inland as the west side of Lake Superior to serve as international ports. The seaway was completed in 1959, although Canada and the United States began to anticipate such a waterway project in the last years of the 19th century.

State A political unit over which an established government maintains sovereign control.

State farm (sovkhoz) A type of collectivized state-owned agricultural unit in the former Soviet Union; workers receive cash wages in the same manner as industrial workers.

Steppe (temperate grassland) A biome composed mainly of short grasses. It occurs in areas of steppe climate, which is a transitional zone between very arid deserts and humid areas.

Subarctic A high-latitude climate characterized by short, mild summers and long, severe winters.

Subduction zone See *Plate tectonics*.

Subsidence The settling of land by the compression of lower layers of rock and soil, usually accelerated by the removal of water from below the surface. Of particular importance in urban centers, with Mexico City having one of the most persistent problems.

Summer solstice On or about June 22, the first day of summer in the Northern Hemisphere, the northern tip of Earth's axis is inclined toward the sun at an angle of 23.5° from a line perpendicular to the plane of the ecliptic. This is the summer solstice in the Northern Hemisphere. On or about December 22, the first day of winter in the Northern Hemisphere, the southern tip of Earth's axis is inclined toward the sun at an angle of 23.5° from a line perpendicular to the plane of the ecliptic. This is the summer solstice in the Southern Hemisphere.

Sun Belt A U.S. area of indefinite extent, encompassing most of the South, plus the West at least as far north as Denver, Salt Lake City, and San Francisco. This region has shown faster growth in population and jobs than the nation as a whole during the past several decades.

Sunni Islam The branch of Islam regarding successorship to the Prophet Muhammad as a matter of consensus among religious elders.

Sustainable development (ecodevelopment) Concepts and efforts to improve the quality of human life while living within the carrying capacity of supporting ecosystems.

Sustainable yield (natural replacement rate) The highest rate at which a renewable resource can be used without decreasing its potential for renewal.

Swidden See *Shifting cultivation*.

Symbolization The representation of distinct aspects of information shown on a map, such as stars for capital cities.

Taiga A northern coniferous (needleleaf) forest.

Technocentrists (cornucopians) Supporters of forecasts that resources will keep pace with or exceed the needs of growing human populations.

Tectonic processes Processes that derive their energy from within the earth's crust and serve to create landforms by elevating, disrupting, and roughening the earth's surface.

Tertiary consumers Consumers of secondary consumers.

Theory of island biogeography A theoretical calculation of the relationship between habitat loss and natural species loss, in which a 90 percent loss of natural forest cover results in a loss of half of the resident species.

Tierra caliente A Latin American climatic zone reaching from sea level upward to approximately 3,000 feet (914 m). Crops such as rice, sugarcane, and cacao grow in this hot, wet environment.

Tierra fría A Latin American climatic zone found between 6,000 feet (1,800 m) and 12,000 feet (3,600 m) above sea level. Frost occurs and the upper limit of agriculture and tree growth is reached.

Tierra helada A Latin American climatic zone above 10,000 feet (3,048 m) that has little vegetation and frequent snow cover.

Tierra templada A Latin American climatic zone extending from approximately 3,000 to 6,000 feet (914–1,829 m) above sea level. It is a prominent zone of European-induced settlement and commercial agriculture such as coffee growing.

Tokugawa Shogunate The period 1600 to 1868 in Japan, characterized by a feudal hierarchy and Japan's isolation from the outside world.

Tradable permits A proposed mechanism for reducing total global greenhouse gases in which each country would be assigned the right to emit a certain quantity of carbon dioxide, according to its population size, setting the total at an acceptable global standard. The assigned quotas could be traded between underproducers and overproducers of carbon dioxide.

Trade winds Streams of air that originate in semipermanent high-pressure cells on the margins of the tropics and are attracted equatorward by a semipermanent low-pressure cell.

Transhumance The movement of pastoral nomads and their herd animals between low-elevation winter pastures and high-elevation summer pastures. Also known as *vertical migration*.

Triangular trade The 16th- to 19th-century trading links between West Africa, Europe, and the Americas, involving guns, alcohol, and manufactured goods from Europe to West Africa exchanged for slaves. Slaves brought to the Americas were exchanged for the gold, silver, tobacco, sugar, and rum carried back to Europe.

Trophic (feeding) levels Stages in the food chain.

Tropical deciduous forest The vegetation typical of some tropical areas with a dry season. Here, broadleaf trees lose their leaves and are dormant during the dry season and then add foliage and resume their growth during the wet season.

Tropical rain forest A low-latitude broadleaf evergreen forest found where heat and moisture are continuously, or almost continuously, available.

Tropical savanna climate A relatively moist low-latitude climate that has a pronounced dry season.

Tundra A region with a long, cold winter, when moisture is unobtainable because it is frozen, and a very short, cool summer. Vegetation includes mosses, lichens, shrubs, dwarf trees, and some grass.

Underground economy The "black market" typical of the former Soviet region and many LDCs.

Undifferentiated highland A climate that varies with latitude, altitude, and exposure to the sun and moisture-bearing winds.

Vernacular region See *Region*.

Vertical migration Movements of pastoral nomads between low winter and high summer pastures.

Virgin and idle lands (new lands) Steppe areas of Kazakhstan and Siberia brought into grain production in the 1950s.

Warsaw Pact A military alliance, now dissolved, consisting of the Soviet Union and the European countries of Poland, East Germany, Czechoslovakia, Hungary, Romania, and Bulgaria.

Waterhead In attempting to control water for the generation of electricity or a gravity flow irrigation project, it is essential to pond up water in relative quantity so that there is an adequate force—waterhead—to achieve the desired goal. Dam structures are the most common engineering responses to this requirement.

Weather The atmospheric conditions prevailing at one time and place.

Weathering The natural process that disintegrates rocks by mechanical or physical means, making soil formation possible.

West The shorthand term for the countries of the world that have been most influenced by Western civilization and that link most closely with the United States and Western Europe.

Westerly winds An airstream located in the middle latitudes that blows from west to east. Also known as *westerlies*.

Western civilization The sum of values, practices, and achievements that had roots in ancient Mesopotamia, as well as Palestine, Greece, and Rome, and that subsequently flowered in western and southern Europe.

Westernization The process whereby non-Western societies acquire Western traits, which are adopted with varying degrees of completeness.

Winter solstice On or about December 22, the first day of winter in the Northern Hemisphere, the southern tip of Earth's axis is inclined toward the sun at an angle of 23.5° from a line perpendicular to the plane of the ecliptic. This is the winter solstice in the Northern Hemisphere. On or about June 22, the first day of summer in the Northern Hemisphere, the northern tip of Earth's axis is inclined toward the sun at an angle of 23.5° from a line perpendicular to the plane of the ecliptic. This is the winter solstice in the Southern Hemisphere.

Xerophytic Literally, "dry plant," referring to desert shrubs having small leaves, thick bark, large root systems, and other adaptations to absorb and retain moisture.

Zero population growth (ZPG) The condition of equal birth rates and death rates in a population.

Zionist movement The political effort, beginning in the late 19th century, to establish a Jewish homeland in Palestine.

Zone of transition An area where the characteristics of one region change gradually to those of another.

Place/Name Pronounciation Guide

The authors acknowledge with thanks the assistance of David Allen, Eastern Michigan University, in preparing an earlier pronunciation guide that the authors expanded and modified for this book. In general, names are those appearing in the text; pronunciations of other names on maps can be found in *Goode's World Atlas* (Rand McNally) or in standard dictionaries and encyclopedias. Vowel sounds are coded as: ă (uh), ā (ay), ĕ (eh), ē (ee), ī (eye), ĭ (ih), ōō (oo).

Afghanistan (af-*gann'*-ih-stann')
   Helmand R. (*hell'*-mund)
   Hindu Kush Mts. (*hinn'*-doo *koosh'*)
   Kabul (*kah'*-b'l)
   Khyber Pass (*keye'*-ber)
Albania (al-*bay'*-nee-a)
   Tirana (tih-*rahn'*-a)
Algeria (al-*jeer'*-ee-a)
   Ahaggar Mts. (uh-*hahg'*-er)
   Sahara Desert (suh-*hahr'*-a)
   Tanezrouft (*tahn'*-ez-rooft')
American Samoa
   Pago Pago (*pahng'*-oh *pahng'*-oh)
   Tutuila I. (too-too-*ee'*-lah)
Andorra (an-*dohr'*-a)
Angola (ang-*goh'*-luh)
   Benguela R.R. (ben-*gehl'*-a)
   Cabinda (exclave) (kah-*binn'*-duh)
   Luanda (loo-*an'*-duh)
   Lobito (luh-*bee'*-toh)
Anguilla (an-*gwill'*-a)
Antigua and Barbuda (an-*teeg'*-wah,
   bahr-*bood'*-a)
Argentina (ahr'-jen-*teen'*-a)
   Buenos Aires (*bwane'*-uhs *eye'*-ress)
   Córdoba (*kawrd'*-oh-bah)
   Patagonia (region) (patt'-hu-*goh'*-nee-a)
   Rio de la Plata (*ree'*-oh duh lah
      plah'-tah)
   Rosario (roh-*zahr'*-ee-oh)
   Tierra del Fuego (region)
      (tee-ehr'-uh dell *fway'*-goh)
   Tucumán (too'-koo-*mahn'*)
Armenia (ahr'-*meen'*-y-uh)
   Yerevan (yair'-uh-*vahn'*)
Aruba (ah-*roob'*-a)
Australia (aw-*strayl'*-yuh)
   Adelaide (*add'*-el-ade)
   Brisbane (*briz'*-bun)
   Canberra (*can'*-bur-a)
   Melbourne (*mell'*-bern)
   Sydney (*sid'*-nih)
   Tasmania (state) (tazz-*may'*-nih-uh)
Austria (*aw'*-stree-a)
   Graz (*grahts'*)

Linz (*lints'*)
Salzburg (*sawlz'*-burg)
Vienna (vee-*enn'*-a)
Azerbaijan (ah'-zer-*beye'*-jahn)
   Baku (bah-*koo'*)

Bahamas (buh-*hahm'*-uz)
   Nassau (*nass'*-aw)
Bahrain (bah-*rayn'*)
Bangladesh (bahng'-lah-*desh'*)
   Dacca (Dhaka) (*dahk'*-a)
Barbados (bar-*bay'*-dohss)
Belarus (*bell'*-uh-roos)
   Minsk (*mensk'*)
Belgium
   Antwerp (*ant'*-werp)
   Ardennes Upland (ahr-*den'*)
   Brussels (*brus'*-elz)
   Charleroi (*shahr*-luh-*rwah'*)
   Ghent (*gent'*)
   Liège (lee-*ezh'*)
   Meuse R. (*merz'*)
   Sambre R. (*sohm*-bruh)
   Scheldt R. (*skelt'*)
Belize (buh-*leez'*)
Benin (beh-*neen'*)
   Cotonou (koh-toh-*noo'*)
Bhutan (boo-*tahn'*)
   Thimphu (*thim'*-poo')
Bolivia (boh-*liv'*-ee-uh)
   Altiplano (region) (al-tih-*plahn'*-oh)
   La Paz (lah *pahz'*)
   Santa Cruz (sahn-ta *krooz'*)
   Sucre (*soo'*-kray)
   Lake Titicaca (tee'-tee-*kah'*-kah)
Bosnia-Hercegovina
   (*boz'*-nee-uh herts-uh-goh-*veen'*-uh)
   Sarajevo (sahr-uh-*yay'*-voh)
Botswana (baht-*swahn'*-a)
   Gaborone (gahb'-uh-*roh'*-nee)
Brazil (bra-*zill*)
   Amazonia (region) (am-uh-*zohn'*-ih-a)
   Belo Horizonte (*bay'*-loh haw-ruh-*zonn'*-
      tee)
   Brasília (bruh-*zeel'*-yuh)
   Caatinga (region) (kay-*teeng'*uh)
   Manaus (mah-*noose'*)
   Minas Gerais (*meen'*-us zhuh-*rice*)
   Natal (nah-*tal'*)
   Paraíba Valley (pah-rah-*ee'*-bah)
   Paraná R. (pah-rah-*nah'*)
   Pôrto Alegre (por-too' al-*egg'*-ruh)
   Recife (ruh-*see'*-fee)
   Rio de Janeiro (*ree'*-oh dih juh-*nair'*-oh)
   Salvador (sal'-vah-*dor'*)
   São Paulo (sau *pau'*-loh)
Brunei (*broo'*-neye)

Bulgaria
   Sofia (*soh'*-fee-a)
Burkina Faso (bur-*kee'*-na *fah'*-soh)
   Ouagadougou (wah'-guh-*doo'*-goo)
Burma (*bur'*-ma)
Burundi (buh-*roon'*-dee) ("oo" as in
   "wood")
      Bujumbura (boo-jem-*boor'*-a)
      (second "oo" as in "wood")

Cambodia
   Phnom Penh (*nom' pen'*)
Cameroon (kam'-uh-*roon'*)
   Douala (doo-*ahl'*-a)
   Yaoundé (yah-onn-*day'*)
Canada
   Montreal (mon'-tree-*awl'*)
   Nova Scotia (noh'-vuh *scoh'*-shuh)
   Ottawa (*aht'*-ah-wah)
   Quebec (kwih-*beck'*; Fr., kay-*beck'*)
   Saguenay R. (*sag'*-uh-nay)
Cape Verde (*verd'*)
Central African Republic
   Bangui (*bahng'*-ee)
Chile (*chill'*-ee)
   Atacama Desert (ah-tah-*kay'*-mah)
   Chuquicamata (choo-kee-*kah'*-mah-tah)
   Concepción (con-sep'-*syohn'*)
   Santiago (sahn'-tih-*ah'*-goh)
   Valparaíso (vahl'-pah-rah-*ee'*-soh)
China
   Altai Mts. (al-*teye*)
   Amur R. (ah-*moor'*) ("oo" as in
      "wood")
   Anshan (ahn-*shahn'*)
   Baotou (*bau'*-toh')
   Beijing (bay-*zhing'*)
   Canton (kan'-*tahn'*)
   Chang Jiang (chung jee-*ahng'*)
   Chengdu (chung-*doo'*)
   Chongqing (chong-*ching'*)
   Dzungarian Basin (zun-*gair'*-ree-uhn)
   Guangdong (gwahng-*dung'*)
   Guangzhou (gwahng-*joh'*)
   Hainan I. (*heye'*-nahn')
   Harbin (*hahr'*-ben)
   Huang He (*hwahng' huh'*)
   Lanzhou (lahn-*joh'*)
   Lhasa (*lah'*-suh)
   Li R. (*lee'*)
   Liao (lih-*ow'*)
   Lüda (*loo'*-dah)
   Nanjing (nahn-*zhing'*)
   Peking (pea-*king'*)
   Shandong (shahn-*dung'*)
   Shanghai (shang-*heye'*)
   Shantou (shahn-*toh'*)

Shenyang (shun-*yahng*′)
Shenzhen (shun-*zún*)
Sichuan (zehch-*wahn*′)
Taklamakan Desert (*tah*′-kla-mah-
    kahn′)
Tarim Basin (*tah*′-reem)
Tiananmen (*tyahn*′-an-men)
Tianjin (tyahn-*jeen*′)
Tien Shan (*tih*′-en *shahn*′)
Tientsin (*tinn*′-*tsinn*′)
Tsinling Shan (*chinn*′-leeng *shahn*′)
Ürümqi (oo-*room*′-chee)
Wuhan (*woo*′-*hahn*′)
Xian (shee-*ahn*′)
Xinjiang (shin-jee-*ahng*′)
Xizang (shee-*dzahng*′)
Yangtze R. (*yang*′-see)
Yunnan (yoo-*nahn*′) ("oo" as in
    "wood")
Colombia (koh-*lomm*′-bih-uh)
    Barranquilla (bah-rahn-*keel*′-yah)
    Bogotá (*boh*′-ga-*tah*′)
    Cali (*kah*′-lee)
    Cartagena (kahrt′-a-*hay*′-na)
    Cauca R. (*kow*′-kah)
    Magdalena R. (mahg′-thah-*lay*′-nah)
    Medellín (med-deh-*yeen*′)
Comoros (*kahm*′-o-rohs′)
Congo
    Brazzaville (*brazz*′-uh-vill′; Fr.: brah-
        zah-*veel*′)
    Pointe Noire (pwahnt *nwahr*′)
Costa Rica (*kohs*′-tuh *ree*′-kuh)
    San José (sahn hoh-*zay*′)
Croatia (kroh-*ay*′-shuh)
    Zagreb (*zah*′-grebb)
Cyprus (*seye*′-pruhs)
    Nicosia (nik′-o-*see*′-uh)
Czechoslovakia (check′-oh-sloh-*vah*′-kee-a)
    Bratislava (bratt′-ih-*slahv*′-a, braht-)
    Brno (*ber*′-noh)
    Moravia (muh-*rayve*′-ih-uh)
    Ostrava (*aw*′-strah-vah)
    Prague (*prahg*′)

Denmark
    Copenhagen (*koh*′-pen-hahg′-gen)
Djibouti (jih-*boot*′-ee)
Dominica (dahm′-i-*nee*′-ka)
Dominican Republic (duh-*min*′-i-kan)
    Santo Domingo (sant′-oh d-*min*′-goh)

Ecuador (*ec*′-wa-dawr)
    Guayaquil (gweye′-uh-*keel*′)
    Quito (*kee*′-toh)
Egypt
    Aswan (*ahs*′-wahn)
    Cairo (*keye*′-roh)
    Port Said (sah-*eed*′)
El Salvador (el *sal*′-va-dor)
Equatorial Guinea (*gin*′-ee)
Estonia (ess-*toh*′-nee-a)
    Tallin (*tahl*′-in)
Ethiopia (ee′-thee-*oh*′-pea-a)
    Addis Ababa (*ad*′-iss *ab*′-a-ba)
    Asmara (az-*mahr*′-a)

Fiji (*fee*′-jee)
    Suva (*soo*′-va)
Finland
    Helsinki (*hel*′-sing-kee)
    Karelian Isthmus (kuh-*ree*′-lih-uhn)
France
    Alsace (region) (al-*sass*′)
    Ardennes (region) (ahr-*den*′)
    Bordeaux (bawr-*doh*′)
    Boulogne (boo-*lohn*′-yuh)
    Calais (kah-*lay*′)
    Carcassonne (kahr-kuh-*suhn*′)
    Champagne (sham-*pahn*-yuh)
    Garonne R. (guh-*rahn*′)
    Jura Mts. (*joo*′-ruh)
    Le Havre (luh *ah*′-vruh)
    Lille (*leel*′)
    Loire R. (*lwahr*′)
    Lorraine (loh-*rane*′)
    Lyons (Fr., Lyon), both (*lee*-aw)
    Marseilles (mahr-*say*′)
    Massif Central (upland) (mah-*seef*′
        saw-*trahl*)
    Meuse R. (*merz*′)
    Moselle R. (moh-*zell*′)
    Nancy (naw-*see*′)
    Nantes (*nawnt*′)
    Nice (*neece*′)
    Oise R. (*wahz*′)
    Pyrenees (peer′-uh-*neez*′)
    Riviera (riv′-ih-*ehr*′-a)
    Rouen (roo-*aw*′)
    Saône R. (*sohn*′)
    Seine R. (*senn*′)
    Strasbourg (strahz-*boor*′)
    Toulon (too-*law*′)
    Toulouse (too-*looz*′)
    Vosges Mts. (*vohzh*′)
French Guiana (gee′-*ahn*-a)
    Cayenne (keye′-enn)

Gabon (gah-*baw*′)
Gambia (*gamm*′-bih-uh)
    Banjul (*bahn*-jool)
Georgia
    Abkhazia (ahb-kahz′-zee-uh)
    Adjaria (ah-*jahr*′-ree-uh)
    Caucasus Mts. (*kaw*′-kuh-suhs)
    South Ossetia (oh-see′-*shee*-uh)
    Tbilisi (tuh-*blee*′-see)
Germany
    Aachen (*ah*′-ken)
    Berlin (ber-*lin*′)
    Bremen (*bray*′-men)
    Chemnitz (*kemm*′-nits)
    Cologne (kuh-*lohn*′)
    Dresden (*drez*′-den)
    Düsseldorf (*doo*′-sel-dorf)
    Elbe R. (*ell*′-buh)
    Erzgebirge (*ehrts*′-guh-beer-guh)
    Frankfurt (*frahngk*′-foort′)
    Karlsruhe (*kahrls*′-roo-uh)
    Leipzig (*leyep*′-sig)
    Main R. (*mane*′; Ger., *mine*′)
    Mannheim-Ludwïgshafen
        (*man*′-hime *loot*′-vikhs-hah-fen)

Munich (*myoo*′-nik)
Neisse R. (*nice*′-uh)
Oder R. (*oh*′-der)
Ruhr (region) (*roor*′)
Stuttgart (*stuht*′-gahrt)
Weser R. (*vay*′-zer)
Wiesbaden-Mainz (*vees*′-bahd-en
    meyents)
Wuppertal (*voop*′-er-tahl) ("oo" as in
    "wood")
Ghana (*gahn*′-a)
    Accra (a-*krah*′)
    Kumasi (koo-*mahs*′-ee)
    Tema (*tay*′-muh)
    Volta R. (*vohl*′-tuh)
Greece
    Aegean Sea (uh-*jee*′-un)
    Piraeus (peye-*ree*′-us)
    Thessaloniki (thess′-uh-loh-*nee*′-kih)
Grenada (gruh-*nayd*′-a)
Guadeloupe (gwah′-duh-*loop*′)
Guatemala (gwah′-tuh-*mah*′-luh)
Guinea (*gin*′-ee)
    Conakry (*kahn*′-a-kree)
Guinea-Bissau (biw-*ow*′)
Guyana (geye-*an*′-a; geye-*ahn*-a′)

Haiti (*hayt*′-ee)
    Port-au-Prince (pohrt′-oh-*prants*′)
Honduras (hahn-*dur*′-as)
    Tegucigalpa (tuh-goo′-sih-*gahl*′-pah)
Hungary (*hun*′-guh-rih)
    Budapest (boo-duh-*pesht*′)

Iceland
    Akureyri (ah′-koor-*ray*′-ree)
    Reykjavik (*ray*′-kyuh-vik′)
India
    Agra (*ah*′-gruh)
    Ahmedabad (*ah*′-mud-uh-*bahd*′)
    Assam (state) (uh-*sahm*′)
    Bengal (region) (benn-*gahl*′)
    Bihar (state) (bih-*hahr*′)
    Brahmaputra R. (brah′-muh-*poo*′-truh)
    Deccan (region) (*deck*′-un)
    Delhi (*deh*′-lih)
    Ganges R. (*gan*′-jeez)
    Eastern, Western Ghats (mountains)
        (*gahts*′)
    Himalaya Mts. (himm-*ah*′-luh-yuh,
        -ah-*lay*-yuh)
    Jaipur (*jeye*′-poor)
    Jamshedpur (juhm-*shayd*′-poor)
    Kashmir (region) (*cash*′-meer)
    Kerala (state) (*kay*′-ruh-luh)
    Madras (muh-*drass*′, -*drahs*′)
    Punjab (region)
        (*pun*′-jahb, -jab; pun-*jahb*′, -*jab*′)
    Rajasthan (state) (*rah*′-juh-stahn)
    Uttar Pradesh (state) (*oo*′-tahr pruh-
        desh′)
Indonesia
    Bali (*bah*′-lih)
    Bandung (*bahn*′-doong)
    Celebes (Sulawesi) I.
        (*sell*′-uh-beez; sool-uh-*way*′-see)

Irian Jaya (ihr'-ee-ahn *jeye*'-uh)
Jakarta (yah-*kahr*'-tah)
Java (*jah*'-vuh)
Kalimantan (kal-uh-*mann*'-tann)
Medan (muh-*dahn*')
Palembang (pah-lemm-*bahng*')
Sumatra (suh-*mah*'-truh)
Surabaya (soo'-ruh-*bah*'-yah)
Ujung Pandang (oo'-jung
   pahn-*dahng*')
Iran (ih-*rahn*')
   Abadan (ah-buh-*dahn*')
   Elburz Mts. (ell-*boorz*')
   Isfahan (izz-fah-*hahn*')
   Meshed (muh-*shed*')
   Tabriz (tah-*breez*')
   Tehran (teh-huh-*rahn*')
   Zagros Mts. (*zah*'-gruhs)
Iraq (i-*rahk*')
   Baghdad (*bag*'-dad)
   Basra (*bahs*'-ra)
   Euphrates R. (yoo-frate'-eez)
   Kirkuk (Kihr-*kook*') ("oo" as in
     "wood")
   Mosul (moh-*sool*')
   Shatt al Arab (R.) (*shaht*' ahl ah-*rahb*')
   Tigris R. (*teye*'-griss)
Israel
   Gaza Strip (*gahz*'-uh)
   Eilat (*ay*'-laht)
   Haifa (*heye*'-fa)
   Tel Aviv (tel' a-*veev*')
Italy
   Adriatic Sea (ay-drih-*at*'-ik)
   Apennines (Mts.) (*app*'-uh-nines)
   Bologna (boh-*lohn*'-ya)
   Genoa (*jen*'-o-a)
   Milan (mih-*lahn*')
   Naples (*nay*'-pels)
   Palermo (pah-*ler*'-moh)
   Turin (*toor*'-in)
Ivory Coast (Côte d'Ivoire) (koht-d'-vwahr)
   Abidjan (abb'-ih-*jahn*')

Jamaica (ja-*may*'-ka)
Japan
   Fukuoka (foo'-koo-*oh*'-kah)
   Hiroshima (heer'-a-*shee*'-mah)
   Hokkaido (island) (hah-*keye*'-doh)
   Honshu (island) (*honn*'-shoo)
   Kitakyushu (kee-tah'-*kyoo*'-shoo)
   Kobe (*koh*'-bay)
   Kyoto (kee-*oht*'-oh)
   Kyushu (island) (kee-*yoo*'-shoo)
   Nagoya (nay-*goy*'-ya)
   Osaka (oh-*sahk*'-a)
   Sapporo (sahp-*pohr*'-oh)
   Shikoku (island) (shih-*koh*'-koo)
   Tokyo (*toh*'-kee-oh)
   Yokohama (yoh-ko-*hahm*'-a)
Jordan (*jor*'-dan)
   Amman (a-*mahn*')

Kazakstan (kuh-*zahk*'-stahn)
   Alma Ata (*al*'-mah ah-tah')
   Aral Sea (uh-*rahl*')

Karaganda (kahr'-rah-*gahn*'-dah)
Lake Balkhash (bul-*kahsh*')
Syr Darya (river) (*seer*' dahr-*yah*')
Kenya (*ken*'-ya)
   Mombasa (mahm-*bahs*-a)
   Nairobi (neye-*roh*'-bee)
Kiribati (keer'-ih-*bah*'-tih)
Korea, North
   Pyongyang (pea-ong-*yahng*')
   Yalu River (*yah*'-loo)
Korea, South
   Inchon (*inn*'-chonn)
   Pusan (*poo*'-sahn)
   Seoul (*sohl*')
   Taegu (teye-*goo*')
Kuwait (koo-*wayt*')
Kyrgyzstan (keer'-geez-*stahn*')

Laos (*lah*'-ohs)
   Vientiane (vyen-*tyawn*')
Latvia
   Riga (*reeg*'-a)
Lebanon (*leb*'-a-non)
   Beirut (bay-*root*')
Lesotho (luh-*soh*'-toh)
Libya (*lib*'-yah)
   Benghazi (benn-*gah*'-zee)
   Tripoli (*tripp*'-uh-lee)
Liechtenstein (*lick*'-tun-stine')
Lithuania (lith'-oo-*ane*'-ee-uh)
   Vilnyus (*vill*'-nee-us)
Luxembourg (*luk*'-sem-burg)

Madagascar
   Antananarivo (an-ta-nan'-a-*ree*'-voh)
Malawi (muh-*lah*'-wee)
   Lilongwe (lih-*lawng*'-way)
Malaysia
   Kuala Lumpur (*kwah*'-luh *loom*'-poor)
     ("oo" as in "wood")
   Sabah (*sah*'-bah)
   Sarawak (suh-*rah*'-wahk)
   Strait of Malacca (muh-*lahk*'-uh)
Maldives (*mawl*'-deevz)
Mali (*mah*'-lee)
   Bamako (*bam*'-ah-koh')
Malta (*mawl*'-tuh)
Martinique (mahr'-tih-*neek*')
Mauritania (mahr-ih-*tay*'-nee-a)
Mauritius (muh-*rish*'-us)
Mexico
   Acapulco (a-ca-*pool*-ko)
   Ciudad Juárez (see-yoo-*dahd*'
     wahr'-ez)
   Guadalajara (gwad'-a-la-*har*'-a)
   Monterrey (mahnt-uh-*ray*')
   Puebla (poo-*eb*'-la)
   Tijuana (tee-*hwah*'-nah)
Moldova (mohl-*doh*'-vah)
   Kishinev (*kish*'-in-*yeff*')
Monaco (*mon*'-a-koh')
Mongolia (mon-*gohl*'-yuh)
   Ulan Bator (oo-*lahn*' bah'tor)
Montserrat (mont'-suh-*rat*')
Morocco (moh-*rah*'-koh)
   Casablanca (kas-a-*blang*'-ka)

Rabat (rah-*baht*')
Mozambique (moh'-zamm-*beek*')
   Beira (*bay*'-rah)
   Cabora Bassa Dam (kah-*bore*'-ah
     bah'-sah)
   Maputo (mah-*poot*'-oh)

Namibia (nah-*mib*'-ee-a)
Nauru (nah-*oo*-roo)
Nepal (nuh-*pawl*')
   Katmandu (kat'-man-*doo*')
Netherlands
   Ijsselmeer (*eye*'-sul-mahr')
   The Hague (*hayg*')
   Utrecht (*yoo*'-trekt')
   Zuider Zee (*zeye*'-der zay')
Netherlands Antilles
   Curaçao (*koo*'-rahs-ow')
New Zealand
   Auckland (*aw*'-klund)
Nicaragua (nik'-a-*rahg*'-wah)
Niger (*neye*'-jer; Fr., nee'-*zhere*')
   Niamey (nee-*ah*'-may)
Nigeria (neye-*jeer*-ee-a)
   Abuja (a-*boo*'-ja)
   Ibadan (ee-*bahd*'-n)
   Kano (*kahn*'-oh)
   Lagos (*lahg*'-us)
Norway
   Bergen (*berg*'-n)
   Oslo (*ahz*'-loh)
   Stavanger (stah-*vahng*'-er)
   Trondheim (*trahn*'-haym')

Oman (oh-*mahn*')
   Muscat (*mus*'-kat')

Pakistan (pahk-i-*stahn*')
   Hyderabad (*heyed*'-er-a-bahd')
   Islamabad (is-*lahm*-a-bahd')
   Karachi (kah-*rahch*'-ee')
   Karakoram (Mts.) (kahr'-a-*kohr*-um)
   Lahore (lah-*hohr*')
   Peshawar (puh-*shah*'-wahr)
   Punjab (see India)
Papua New Guinea (*pap*'-yoo-uh new
gin'-ee)
   Port Moresby (*morz*'-bee)
Paraguay (*pair*'-uh-gwee')
   Asuncíon (a-soon'-see-*ohn*')
Peru (puh-*roo*')
   Callao (kah-*yow*')
   Cerro de Pasco (*sehr*'-roh day *pahs*'-
     koh)
   Chimbote (chimm-*boh*'-te)
   Cuzco (*koos*'-koh)
   Iquitos (ee-*kee*'-tohs)
   Lima (*lee*'-ma)
   Machu Picchu (ma-chu *pee*-chu)
Philippines (*fil*-i-peenz')
   Luzon (island) (loo'-*zahn*')
   Mindanao (island) (minn'-dah-*nah*'-oh)
   Mt. Pinatubo (pea'-nah-*too*'-buh)
   Visayan Islands (vih-*sah*'-y'n)
Poland
   Gdansk (ga-*dahntsk*')
   Katowice (kaht-uh-*veet*'-sih)

Krakow (*krahk'*-ow)
Lodz (*looj'*)
Posnan (*pohz'*-nan-ya)
Silesia (region) (sih-*lee'*-shuh)
Szczecin (*shchet'*-seen')
Wroclaw (*vrawt'*-slahf)
Portugal (*pohr'*-chi-gal)
Lisbon (*liz'*-bon)
Oporto (oh-*pohrt'*-oh)

Qatar (*kaht'*-ar)

Romania (roh-*main'*-yuh)
Bucharest (boo'-kuh-*rest'*)
Carpathian Mts. (kahr-*pay'*-thih-un)
Russia
Altai Mts. (*al'*-teye)
Amur R. (ah-*moor'*) ("oo" as in "wood")
Angara R. (ahn'-guh-*rah'*)
Arkhangelsk (ahr-*kann'*-jelsk)
Astrakhan (as-trah-*khahn'*)
Baikal, Lake (beye-*kahl'*)
Bashkiria (bahsh-*keer'*-ee-uh)
Bratsk (*brahtsk'*)
Buryatia (boor-*yaht'*-ee-uh)
Caucasus (*kaw'*-kah-suss)
Chechen-Ingushia
  (*chehch'*-yenn enn-*gooshia'*)
Chelyabinsk (*chell'*-yah-binnsk)
Cherepovets (*chehr'*-yuh-puh-vyetz')
Chuvashia (choo-*vahsh'*-i-a)
Dagestan Rep. (dag'-uh-*stahn'*)
Irkutsk (ihr-*kootsk'*)
Ivanovo (ih-*vahn'*-uh-voh)
Izhevsk (ih-*zhehvsk'*)
Kama R. (*kah'*-mah)
Karelia (kah-*reel'*-yuh)
Kazan (kuh-*zahn'*)
Khabarovsk (kah-*bah'*-rawfsk)
Kola Pen. (*koh'*-lah)
Kolyma R. (kuh-*lee'*-muh)
Krasnoyarsk (kras'-nuh-*yahrsk'*)
Kuril Is. (*koo'*-rill)
Kuznetsk Basin (kooz'-*netsk'*)
Ladoga, Lake (*lad'*-uh-guh)
Lena R. (*lee*-nah, *lay*-nah)
Lipetsk (*lyee'*-petsk')
Magnitogorsk (mag-*nee'*-toh-gorsk')
Moscow (*mahs'*-koh or *mahs'*-kow)
Murmansk (moor'-*mahntsk'*)
Nakhodka (nuh-*kawt*-kah)
Nizhniy Novgorod (*nizh'*-nee nahv-guh-rahd')
Nizhniy Tagil (tah-*geel*)
Novokuznetsk (naw'-voh-*kooznetsk'*)
Novaya Zemlya (*naw'*-vuh-yuh zem-lyah')
Novosibirsk (naw-vuh-suh-*beersk'*)
Oka R. (oh-*kah'*)
Okhotsk, Sea of (oh-*kahtsk'*)
Omsk (*ahmsk'*)
Pechora R. (peh-*chore'*-a)
Rostov (rahs-*tawf'*)
Sakhalin (sahk'-uh-*leen'*)
Samara (suh-*mahr'*-a)

Saratov (suh-*raht'*-uff)
Sayan Mts. (sah-*yahn'*)
Tatarstan (Tatar Rep.) (tah-*tahr'*-stahn')
Togliatti (tawl-*yah'*-tee)
Tuva Rep. (*too'*-va)
Ufa (oo-*fah'*)
Ussuri R. (oo-*soor'*-i)
Vladivostok (vlah'-dih-vahs-*tawk'*)
Volga R. (*vahl*-guh, *vohl'*-guh)
Volgograd (vahl-guh-*grahd'*, vohl'-)
Voronezh (vah-*raw'*-nyesh)
Yakutsk (yah-*kootsk'*)
Yekaterinburg (yeh-*kahta'*-rinn-berg)
Rwanda (roo-*ahn'*-da)

St. Kitts and Nevis (*nee'*-vis)
St. Lucia (*loo'*-sha)
St. Vincent-Grenadines (gren'-a-*deenz'*)
São Tomé and Príncipe
  (sau too-*meh'*, pren-*see'*-puh)
Saudi Arabia (*saud'*-ee)
Jiddah (*jid'*-a)
Riyadh (ree-*adh'*)
Senegal (sen'-i-*gahl'*)
Dakar (da-*kahr'*)
Seychelles (*say'*-shelz')
Sierra Leone (see-*ehr'*-a lee-*ohn'*)
Slovenia (sloh-*veen'*-i-a)
Ljubljana (*lyoo'*-b'l-yah-nah)
Somalia (so-*mahl'*-ee-a)
Mogadishu (mahg'-uh-*dish'*-oo)
South Africa
Bloemfontein (*bloom'*-fonn-*tayn'*)
Bophuthatswana (homeland)
  (boh-*poot'*-aht-*swahna'*)
Drakensberg (escarpment) (*drahk*-n's-burg')
Durban (*durr'*-bun)
Johannesburg (joh-*hann'*-iss-burg)
Kwazulu (homeland) (kwah-*zoo'*-loo)
Natal (province) (nuh-*tahl'*)
Pretoria (prih-*tohr'*-ee-uh)
Transkei (homeland) (trans-*keye'*)
Transvaal (province) (trans-*vahl'*)
Spain
Barcelona (bahr'-suh-*lohn*-uh)
Bilbao (bill-*bah'*-oh)
Catalonia (region) (katt'-uh-*lohn'*-ee-a)
Madrid (muh-*dridd'*)
Meseta (plateau) (muh-*say'*-tuh)
Seville (suh-*vill'*)
Valencia (vull-*enn'*-shee-uh)
Sri Lanka (sree *lahn'*-kah)
Sudan (*soo'*-dann)
Gezira (guh-*zeer'*-uh)
Juba (*joo'*-bah)
Khartoum (kahr'-*toom'*)
Omdurman (ahm-*durr'*-man)
Wadi Halfa (*wah*-di' *hawl'*-fah)
Suriname (*soor'*-i-nahm')
Swaziland (*swah'*-zih-land')
Sweden
Göteborg (yort'-e-*bor'*)
Skåne (*skohn'*)
Småland (*smoh'*-lund')
Switzerland

Basel (*bah'*-z'l)
Zürich (*zoor'*-ick)
Syria (*seer'*-i-uh)
Aleppo (a-*lep*-oh')
Damascus (da-*mas'*-kus)

Tajikistan (tah-*jick'*-ih-stahn')
Tanzania (tan'-za-*nee'*-a)
Dar es Salaam (dahr' es sa-*lahm'*)
Thailand (*teye'*-land')
Bangkok (*bang'*-kok')
Togo (*toh'*-goh)
Lomé (loh-*may'*)
Tonga (*tawng'*-a)
Trinidad and Tobago (to-*bay'*-goh)
Tunisia (too-*nee'*-zhee-a)
Tunis (*too'*-nis)
Turkey
Anatolian Peninsula (ann'-ah-*tohl'*-yun)
Ankara (*ang'*-kah-rah)
Bosporus (strait) (*bahs'*-puhr-us)
Dardanelles (strait) (dahr'-duh-*nelz'*)
Istanbul (iss'-tann-*bool'*)
Izmir (*izz'*-*meer'*)
Sea of Marmara (*mahr'*-muh-ruh)
Turkmenistan (turk'-menn-ih-*stahn'*)
Ashkhabad (*ash'*-kuh-bahd')

Ukraine (*yoo'*-krane')
Crimea (*cry'*-*mee'*-uh)
Dnepropetrovsk (d'nyepp-pruh-pay-trawfsk')
Dnieper R. (duh-*nyepp'*-er)
Donetsk (duhn-*yehtsk'*)
Kharkov (*kahr'*-kawf)
Kiev (*kee'*-yeff)
Krivoy Rog (*kree'*-voi *rohg'*)
Sevastopol (sih-*vass'*-tuh-*pohl'*)
Yalta (*yahl'*-tuh)
Zaporozhye (zah-puh-*rawzh'*-yuh)
United Arab Emirates
Abu Dhabi (*ah'*-boo dah'-bee)
Dubai (doo-*beye'*)
United Kingdom
Edinburgh (*ed'*-in-buh-ruh)
Glasgow (*glass'*-goh)
Thames R. (*timms'*)
United States
Albuquerque (*al'*-buh-ker'-kih)
Aleutian Is. (uh-*loo'*-sh'n)
Allegheny R. (*al'*-ih-gay'-nih)
Appalachian Highlands
  (ap'-uh-*lay'*-chih-un)
Coeur d'Alene (*kurr'* duh-*layn'*)
Des Moines (duh *moin'*)
Grand Coulee Dam (*koo'*-lee)
Juneau (*joo'*-noh)
Mesabi Range (muh-*sahb'*-ih)
Monongahela R.
  (muh-*nahng'*-guh-*hee'*-luh)
Omaha (*oh'*-mah-*hah'*)
Ouachita Mts. (*wosh'*-ih-taw')
Palouse (region) (pah-*loose'*)
Phoenix (*fee'*-nix)
Provo (*proh'*-voh)
Rio Grande (*ree'*-oh *grahn'*-day)

San Joaquin R. (sann wah-*keen'*)
Spokane (spoh*'-kann'*)
Tucson (*too'*-sonn)
Valdez (val-*deez'*)
Uruguay (*yoor'*-uh-gweye')
  Montevideo (mahnt'-e-vi-*day'*-oh)
Uzbekistan (ooz-*beck'*-ih-stahn') ("oo" as in
  "wood")
  Amu Darya (river) (ah-moo' *dahr'*-yuh,
    dahr-*yah'*)
  Bukhara (buh-*kah'*-ruh)
  Fergana Valley (fair-*gahn'*-uh)
  Samarkand
    (*samm'*-ur-*kand'*; Rus., suh-mur-
    kahnt')
  Tashkent (tash-*kent'*, tahsh-)

Vanuatu (van'-oo-*ay'*-too)
  Port Vila (*vee'*-la)
Venezuela (ven'-ez-*way'*-la)
  Caracas (ka-*rak'*-as)
  Maracaibo (mar'-a-*keye'*-boh)

Orinoco R. (oh-rih-*noh'*-koh)
Valencia (va-*len'*-see-a)
Vietnam (vee-et*'-nahm'*)
  Annam (region) (uh-*namm'*)
  Cochin China (*koh'*-chin)
  Hanoi (ha-*noi'*)
  Ho Chi Minh City (Saigon)
    (hoe chee *min'*; *seye'*-gahn)
  Mekong R. (may-*kahng'*)

Western Samoa (sa-*moh'*-a)
  Apia (ah-*pea'*-uh)

Yemen
  Aden (*ah*-d'n)
  S'ana (sah-*na'*)
Yugoslavia (yoo'-goh-*slahv'*-ee-a)
  Belgrade (*bell'*-grade, -grahd)
  Kosovo (*kaw'*-suh-voh')
  Montenegro (republic)
    (mahn'-tee-*nay'*-groh)

Zaire (*zeye'*-eer')
  Kananga (kay-*nahng'*-guh)
  Kasai R. (kah-*seye'*)
  Kinshasa (*keen'*-shah-suh)
  Kisangani (kee'-sahn-*gahn'*-ih)
  Lake Kivu (*kee'*-voo)
  Lualaba R. (loo'-ah-*lahb'*-a)
  Lubumbashi (loo-boom-*bah'*-shee)
  Matadi (muh-*tah'*-dee)
  Shaba (region) (*shah'*-bah)
Zambia (*zam'*-bee-a)
  Kariba Dam (kah-*ree'*-buh)
  Lusaka (loo-*sahk'*-a)
  Zambezi R. (zamm-*bee'*-zee)
Zimbabwe (zim-*bahb'*-way)
  Bulawayo (bool'-uh-*way'*-oh)
  Harare (hah-*rahr'*-ee)
  Que Que (*kway' kway'*)
  Wankie (*wahn'*-kee)

Index

Italicized page numbers indicate figures, illustrations, or tables.

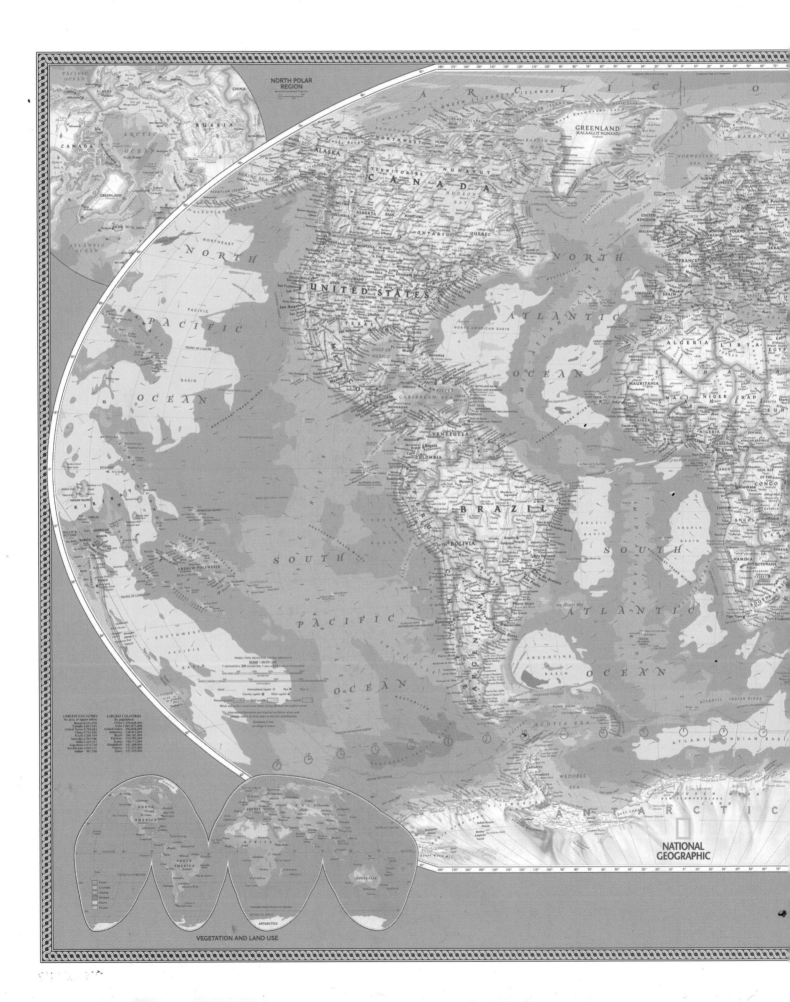

NORTH POLAR
REGION

VEGETATION AND LAND USE

NATIONAL
GEOGRAPHIC